Imperial College
London

the Library
www.imperial.ac.uk/library

THREE WEEK LOAN (STANDARD)

Please return or renew by the due date.
Fines may be charged on items returned late.

Central Library, South Kensington campus, London SW7 2AZ
Renewals: 020 7594 8810 / issuedesk@imperial.ac.uk

Progress in
Thermal Barrier Coatings

Progress in
Thermal Barrier Coatings

A John Wiley & Sons, Inc., Publication

Published by John Wiley & Sons, Inc., Hoboken, New Jersey.
Published simultaneously in Canada.

For general information on our other products and services or for technical support, please contact our Customer Care Department within the United States at (800) 762-2974, outside the United States at (317) 572-3993 or fax (317) 572-4002.

Wiley also publishes its books in a variety of electronic formats. Some content that appears in print may not be available in electronic format. For information about Wiley products, visit our web site at www.wiley.com.

Library of Congress Cataloging-in-Publication Data is available.

ISBN 978-0-470-40838-4

Printed in the United States of America.

10 9 8 7 6 5 4 3 2 1

Contents

DEVELOPMENTS IN PROCESSING

TESTING AND CHARACTERIZATION

THERMAL PROPERTIES

Introduction

Ceramics are used to coat other materials, usually metals, to protect them from high temperatures, moisture, oxygen, wear, corrosive fluids and body fluids. Thermal barrier coatings (TBCs) have their greatest application in protecting metal parts used in heat engines. The metal parts have the strength required for heat engine operation; however, they cannot withstand the high temperatures necessary for efficient and clean operation of the heat engine. TBCs provide this protection.

This edition of Progress in Ceramic Technology series is a compilation of articles published on TBCs by The American Ceramic Society (ACerS). These publications include the *American Ceramic Society Bulletin, Journal of the American Ceramic Society, International Journal of Applied Ceramic Technology, Ceramic Engineering and Science Proceedings* (CESP) and *Ceramic Transactions* (CT).

Papers in this edition are divided into five categories: Applications, Material Improvements and Novel Compositions, Developments in Processing, Testing and Characterization, Mechanical Properties, and Thermal Properties. The publication citations are included after each title in the table of contents.

Other articles on thermal barrier coatings can be located by searching the Society's website at www.ceramics.org.

Applications

CORROSION RESISTANT THERMAL BARRIER COATING MATERIALS FOR INDUSTRIAL GAS TURBINE APPLICATIONS

Michael D. Hill and Davin P. Phelps.
Trans-Tech Inc.
Adamstown, MD 21710 USA

Douglas E. Wolfe.
Assist Professor, Materials Science and Engineering Department
The Pennsylvania State University
University Park, Pa 16802 USA

ABSTRACT

Thermal Barrier Coatings are ceramic materials that are deposited on metal turbine blades in aircraft engines or industrial gas turbines which allow these engines to operate at higher temperatures. These coatings protect the underlying metal superalloy from creep, oxidation and/or localized melting by serving as an insulating barrier to protect the metal from the hot gases in the engine core. While for aircraft engines, pure refined fuels are used, it is desirable for industrial gas turbine applications that expensive refining operations be minimized. However, acidic impurities such as sulfur and vanadium are common in these "dirty" fuels and will attack the thermal barrier coating causing reduced coating lifetimes and in the worse case catastrophic failure due to spallation of the coating. The industry standard coating material is stabilized zirconia with seven weight percent yttria stabilized zirconia being the most common. When used in industrial gas turbines, the vanadium oxide impurities react with the tetragonal zirconia phase causing undesirable phase transformations. Among these transformations is that from tetragonal to monoclinic zirconia. This transformation is accompanied by a volume expansion which serves to tear apart the coating reducing the coating lifetime. Indium oxide is an alternative stabilizing agent which does not react readily with vanadium oxide. Unfortunately, indium oxide is very volatile and does not readily stabilize zirconia, making it difficult to incorporate the indium into the coating. However, by pre-reacting the indium oxide with samarium oxide or gadolinium oxide to form a stable perovskite ($GdInO_3$ or $SmInO_3$) the indium oxide volatilization is prevented allowing the indium oxide incorporation into the coating. Comparison of EDX data from evaporated coatings containing solely indium oxide and those containing $GdInO_3$ are presented and show that the indium is present in greater quantities in those coatings containing the additional stabilizer. Corrosion tests by reaction with vanadium pentoxide were performed to determine the reaction sequence and to optimize the chemical composition of the coating material. Lastly, select x-ray diffraction phase analysis will be presented.

INTRODUCTION

Thermal Barrier Coatings are ceramic materials that are deposited on metal turbine blades in aircraft engines or industrial gas turbines which allow these engines to operate at higher temperatures. These coatings protect the underlying metal superalloy from creep, oxidation and/or localized melting by serving as an insulating barrier to protect the metal from the hot gases in the engine core.

Several impurities common in fuels have been identified and associated with corrosion in EB-PVD coatings. These impurities include sodium, sulfur, phosphorus and especially vanadium. These impurities react with conventional YSZ turbine blade coatings, severely limiting the coating lifetime. Therefore, it is of great interest to develop alternative materials that react less readily with fuel contaminants and therefore increase the operating lifetime of the coating.

Standard 8YSZ EB-PVD coatings contain 8-weight percent yttria and crystallize in the metastable t' phase that is derived from a martensitic distortion of the "stabilized" cubic fluorite structure of zirconia. This rapidly cooled t' structure is the most desirable of all of the possible polymorphs in the yttria-zirconia system for TBC applications. Jones[1] described several mechanisms of chemical attack on 8YSZ coatings. These include chemical reaction, mineralization, bond coat corrosion and physical damage due to molten salt penetration. Of the four, only the first two mechanisms will be featured in this discussion.

Acidic species such as SO_3 and V_2O_5 have been shown to react with the yttria stabilizing the t' phase, destabilizing the Y_2O_3-ZrO_2 by extraction of the Y_2O_3. Of these, V_2O_5 has been determined to be the worst offender. Hamilton[2] and Susnitsky[3] have studied the reaction mechanism in detail. The reaction:

$$Zr_{1-x}Y_xO_{2-.5x}\ (t') + yV_2O_5 \rightarrow 2(1-y)\ ZrO_2\ (monoclinic) + 2y\ YVO_4$$

is especially deleterious to the TBC integrity. The vanadium has been shown to leach the yttria out of the zirconia leaving the yttria deficient monoclinic phase of zirconia remaining. The large volume expansion (7%) caused by this transformation leads to the TBC spalling therefore exposing the bond coat to further chemical attack.

Mineralization, on the other hand, describes a catalytic process by which a metastable phase (in this case, the t' phase) is broken into its stable phase assemblages by a catalyst or mineralizer. For example, ceria stabilized zirconia was investigated as a corrosion resistant coating due to the fact that ceria does not react with vanadium pentoxide.

$$Zr_{1-x}Ce_xO_{2-.5x}\ (t') + yV_2O_5 \rightarrow (1-x)ZrO_2\ (monoclinic) + xCeO_2 + yV_2O_5$$

However, vanadium does act as a mineralizer, destabilizing the t' phase without reacting to form the vanadate.

Alternate stabilizers for zirconia: A large number of cationic species act to stabilize the cubic and t' phases of zirconia. Therefore, one strategy toward finding corrosion resistant coatings was to find a stabilizer that is resistant to chemical attack by vanadium pentoxide. As mentioned above, ceria was investigated but found to be subject to a mineralization reaction[4]. Previous work at NRL[1] focused on studying acidic stabilizers to zirconia since basic stabilizers such as MgO and Y_2O_3 were especially susceptible to chemical attack by acidic vanadium pentoxide. Scandia (Sc_2O_3) and india (In_2O_3) in particular were examined in detail (Jones et. al.[5] Sheu et. al.[6]). Of these, india was found to be the most resistant to chemical attack by vanadium pentoxide.

India stabilized Zirconia as a TBC coating: Although india stabilized zirconia shows promise due to its relative inertness in vanadia containing atmospheres, there are still significant drawbacks in its use as a TBC material. First, india volatilizes at a lower temperature than zirconia. This resultin significant challenges for applyingplasma sprayed TBC's[1]. Although india stabilized zirconia coatings have been made in the t' phase (Sheu[6]), concerns about the volatility of indium oxide raise questions about the ability of india stabilized zirconia to form a homogenous coatings.

In_2O sublimes at 600 °C 10^{-4} torr at 650°C
In_2O_3 sublimes at 850 °C 10^{-4} torr at 850 °C

Jones, Reidy and Mess[5] were able to co-stabilize zirconia with yttrium oxide and indium oxide using a sol gel process. However, no attempt was made to provide ingot feedstock of this composition for EB-PVD testing. Furthermore, the high cost (> $300/kg) of In_2O_3 has also been a barrier for further research and development efforts.

Therefore, a logical approach was to incorporate the indium oxide into the ingot in a form that would make the indium oxide less volatile, therefore minimizing incidents of spitting, pressure fluctuations, and increase coating homogeneity while still providing enhanced corrosion resistant coating solely consisting of the t' phase. The strategy was to pre-react the indium oxide with a lanthanide oxide which forms either the $LnInO_3$ perovskite (La, Nd or Sm) or the hexagonal $LnInO_3$ (Gd or Dy). If the ingot contains zirconia and the $LnInO_3$ or just partially stabilized zirconia without free indium oxide, it was believed that a more homogeneous corrosion resistant coating could be deposited by electron beam physical vapor deposition (EB-PVD).

Advantages of Indate pre-cursor:

1) Perovskite indates ($LnInO_3$) are refractory compounds. The electropositive lanthanide ion (also stabilizers of the t' phase) stabilizes the In^{3+} state. It is the reduction to In^{1+} that leads to the volatilization of In.
2) Multiple stabilizing ions reduce thermal conductivity. The work of R. Miller[7] showed that TBC thermal conductivity decreases when numerous ions of different ionic sizes, valence and ionic weights are simultaneously incorporated into the zirconia as stabilizing agents. These are often referred to as oxide dopant clusters.

Lanthanide Selection: There are numerous factors that will determine the selection of the lanthanide ion accompanying the indium oxide.

1) Range of metastable t' phase field. Ideally one would like the largest range possible. Sasaki[8] found the t' phase between 15 and 20-mol % In_2O_3 when quenched from temperatures above 1500°C. Ideally this phase region would accompany the In mol% alone as well as the entire range up to the (Ln + In) mole percentage.
2) Melting temperature of $LnInO_3$ compound. The more refractory the compound, the better is the performance
3) Acidity/basicity of lanthanide ion. If La is used, this is likely to be strongly attacked by vanadium because of its basicity. As we progress through the heavier lanthanides (left to right on periodic table), the basicity decreases.
4) Ionic size and weight. Y is of the ideal atomic size for decreasing the monoclinic-tetragonal transformation temperature in ZrO2. (Sasaki[8]1993). As we move to smaller ions or larger ions this change in the transformation temperature is decreased. In addition, the greater the difference in ionic size and ionic weight between the In^{3+} and the Ln^{3+} ions, the lower the thermal conductivity (Miller[7]2004).

<u>Phase Diagram Information:</u> Only one ternary phase diagram exists containing any Ln_2O_3-In_2O_3-ZrO_2 ternary systems. That one is for Ln=Pr and it was produced by Bates [9] et.al in 1989. The compatibility relationships expressed in this diagram suggest that $PrInO_3$ perovskite would react with zirconia to form the $Pr_2Zr_2O_7$ pyrochlore and free indium oxide, the exact situation one should avoid. In addition, it has been shown [10] that the larger lanthanide ions (La-Gd) in zirconate pyrochlores react with the thermally grown oxide to form undersirable lanthanide aluminate phases. Therefore, the authors investigatedLn ions that formed stable binary oxides of the perovskite structure with In_2O_3 but did not form the pyrochlore structure or formed the pyrochlore structure sluggishly. Like the formation of the indate perovskites, the stability of the pyrochlore phase decreases as we proceed from the light to heavy lanthanides. The lanthanides of greatest interest are therefore Sm, Gd and Dy.

Sm_2O_3	Forms $Sm_2Zr_2O_7$ pyrochlore Stable to 1800°C (Yokakawa [11] 1992)	Forms $SmInO_3$ perovskite (Schneider, Roth and Waring [12] 1961)
Gd_2O_3	Forms $Gd_2Zr_2O_7$ pyrochlore Stable to 1575°C (Yokakawa [11] 1992)	Forms hexagonal $GdInO_3$ (Schneider, Roth and Waring [12] 1961)
Dy_2O_3	Does not form $Dy_2Zr_2O_7$ pyrochlore (Pascual and Duran [13] 1980)	Forms hexagonal $DyInO_3$ Stable to 1600 C (Schneider, Roth and Waring [12] 1961)

Lanthnides heavier than Dy do not form either the pyrochlore [13] or binary indate phases [12]. The samarium series is of interest because the indiate perovskite forms and since Sm is the most electropositive ion of the lanthanide series (to prevent In^{1+} formation and volatilization); however, Sm also forms the most stable pyrochlore which is undesirable. Conversely, the dysprosium series is of interest because it does not form the pyrochlore zirconate or the perovskite structure. The hexagonal compound that does form is unstable above 1600°C. Therefore the challenge is to find a compound indium oxide precursor that will prevent indium volatilization but will not react with zirconia to form a pyrochlore and thus liberate free (and volatile) In_2O_3.

In 2007, Mohan et. al. [14] reported that in addition to forming the zircon YVO_4 phase that YSZ will react with vanadate salts below 747°C to form the zirconium pyrovanadate (ZrV_2O_7) phase. The role this phase plays in the mechanical properties of YSZ coatings containing vanadium warrants further study.

EXPERIMENTAL

$LnInO_3$ materials were synthesized by blending yttrium, samarium, gadolinium or dysprosium oxides (loss on ignition determined at 1300°C for all starting oxides) with indium oxide in a ball-mill with yttria-stabilized zirconia (YSZ) media at 55% solids loading without dispersants for 4h. The slurry was pan dried and calcined at 1300°C for 8h. X-ray diffraction was used to evaluate the phase purity of the material by comparing with the appropriate JCPDS cards. If the reaction was incomplete, the milling and calcinations were repeated. The fully-reacted lanthanide indate compositions were then ball-milled with YSZ media until the median particle size was 2 microns or less.

Table I. - Physical and Chemical Properties of the Fired Ingot Material

Ingot Material	Fired Density	Phase Content	Evaporation Quality
6 mole% $SmInO_3$	4.81 g/cc	t-ZrO_2, m-ZrO_2 + $LnInO_3$	Poor - Spitting
6 mole% $GdInO_3$	4.85 g/cc	t-ZrO_2, m-ZrO_2 + $LnInO_3$	Poor - Spitting
6 mole% $DyInO_3$	4.80 g/cc	t-ZrO_2, m-ZrO_2 + $LnInO_3$	Extremely Poor
6 mole% $SmInO_3$ +3 mole% Y_2O_3	4.59 g/cc	t-ZrO_2, m-ZrO_2 + $LnInO_3$	Poor –Spitting
6 mole% $GdInO_3$ +3 mole% Y_2O_3	4.63 g/cc	t-ZrO_2, m-ZrO_2 + $LnInO_3$	Poor – Spitting

The indate precursors were then blended with zirconia to the desired composition and formed by cold isostatic pressing into the EB-PVD ingots. The materials were heat treated between 1430 °C and 1530°C for 10h to achieve a theoretical density between 60 and 70%. Table I shows the fired densities, the phase content and the evapoaration quality of the ingot material as a function of the chemical composition. XRD revealed the fluorite structure along with residual monoclinic zirconia and the indate perovskites as listed in Table I.

The ingots were evaporated onto platinum aluminide coated MAR-M-247 nickel based alloy one inch diameter buttons in an industrial prototype EB-PVD coating system at Penn State University. XRD and SEM microstructures were prepared for each coating, with selectEDX presented for semi-quantitative coating chemistry analysis.

Corrosion reactivity tests were performed by reacting the coated coupons with a thin coating of vanadium pentoxide and heated to temperatures between 400 – 650°C for 4 – 6 hours. X-ray diffraction was performed on the pre-reacted and as-reacted coating to identify any phases forming due to the reaction with vanadium pentoxide.

RESULTS

1) Evaporation: In general, the ingots evaporated poorly in the industrial scale EB-PVD coating unit. The material showed "spitting" and extensive cracking during evaporation. The spitting is most likely due to the difference in the vapor pressure between zirconium oxide and indium oxide containing phases in the ingot, but can also be the result of localized differences in ingot densities and degree of connected porosity. Cracking can also occur if the ingot density is too high or the ingot does not have sufficient thermal shock resistance. Despite the difficulties during ingot evaporation, coatings were obtained for each material studied. However, it should be noted that some "spits" or coating defects were observed on the surface of the coated coupons. Lastly, yttrium oxide was added into the composition as an evaporation aid during powder formulation and ingot fabrication, but it did not appear to substantially improve ingot evaporability.

2) Coating Properties: XRD revealed that all of the coatings were single phase with the desired t' structure. The coating microstructure as observed by scanning electron microscopy revealed a

columnar microstructure typical of those applied by the EB-PVD process. Figure 1 shows an SEM micrograph of the 6 mol% GdInO3 stabilized zirconia coating surface morphology. In addition, EDX was performed on the coating surface to determine semi quantitative compositional information regarding traces of rare earth and indium oxide compositions. These results are listed in Table II.

The first measure of success was to obtain a coating which contained the acidic stabilizer In_2O_3. Table II compares the ease of evaporability and the relative amount of india within the coating for the various compositions studied.   The two compositions containing samarium indate showed the highest amounts of residual indium followed by the sample containing both gadolinium and indium oxide. The ingot starting with 6 mole % indium oxide showed moderate amounts of indium remaining in the EDX trace although considerably less than either samarium containing composition despite starting with double the amount of indium oxide in the ingot.

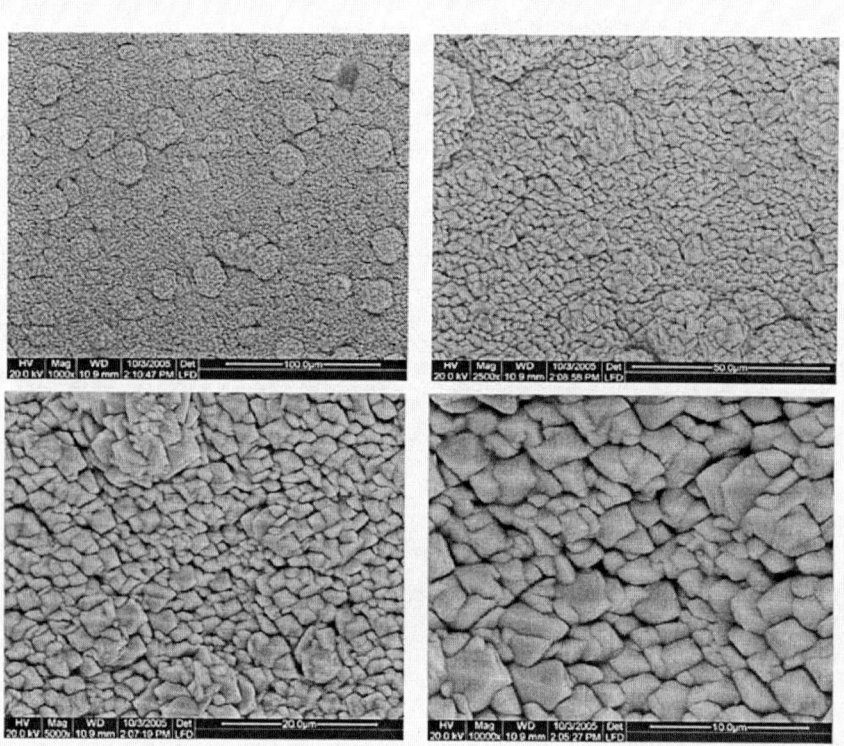

ESEM images showing the surface morphology of ZrO_2/Y/$GdInO_3$ deposited on a platinum aluminide bond coated MAR-M-247 button. Sample # S050923-1H
10/4/2005

Figure 1: SEM image of surface morphology of the EB-PVD coating obtained by evaporation of the 6 mol% $GdInO_3$-3 mol% Y_2O_3doped zirconia ingot composition. The coatings were applied on a platinum aluminide coated nickel base alloy. The top images show a lower magnification than the bottom images

3.) Reactivity Tests: Table III shows the results of the vanadium pentoxide reactivity tests. X-ray diffraction was performed on the various coatings before and after the reactivity tests in order to determine whether the coatings reactive with vanadium oxide. If any reactions occurred, the phases were identified. The sample containing samarium indate showed only the tetragonal prime phase until

500°C at 16h. Traces of the $LnVO_4$ phase with the zircon structure were observed in the samples containing the gadolinium and dysprosium indate at 400°C at 4 hours. Upon further testing at 500°C when exposed to vanadium oxide, traces of monoclinic zirconia and the ZrV_2O_7 phase appeared for $DyInO_3$ containing samples. With the exception of the coating that contained 6mol% $GdInO_3$, the t' phase completely disappeared at 650°C suggesting that these coatings reacted with the vanadium oxide to destabilize the yttria stabilized zirconia. The 6mol% $GdInO_3$ composition showed the most promising results with regards to resistance against vanadium oxide attack.

Table II: A comparison of properties for ingots of various compositions studied. Ease of evaporation, EDX In_2O_3 content, SEM microstructure and the phase content.

Composition	Ease of EB-PVD Evaporation (TD: theoretical density.)	Amount of In in coating	Microstructure	Coating Phase
6 mol% In_2O_3	Poor (62 % TD ingot)	some	TBD	t'
6 mol% $SmInO_3$	Poor (60 and 70% TD ingots)	most	Columnar	t'
6 mol% $SmInO_3$ 3 mol% Y_2O_3	Poor (60% TD ingot)	most	TBD	t'
6 mol% $GdInO_3$	Poor (60 and 70% TD ingots)	little	Columnar – not homogenous	t'
6 mol% $GdInO_3$ 3 mol% Y_2O_3	Poor (60% TD ingot)	most	TBD	t'
6 mol% $DyInO_3$	Poor (70% TD ingot)	some	Poorly formed columns	t'

DISCUSSION

All of the india containing compositions were difficult to evaporate as an ingot. This makes it unlikely that these materials would be useful for EB-PVD applications. EB-PVD is typically used for aircraft engine coatings. This application would typically use clean fuels devoid of acidic corrosive impurities. The material may be more useful as a plasma sprayed powder, which is a more typical TBC form for the industrial gas turbine industry with increased probability of being exposed to vanadium.

The samarium containing compounds showed the most residual india in the coating as determined by EDX. Sm is the most electropositive of the lanthanide co-stabilizers and is less likely to form pyrochlores than the lighter lanthanides like La or Nd. The Sm and Gd containing materials formed the typical columnar microstructure while the Dy containing sample showed poorly formed columns. It is not clear whether this was the result of processing difficulties caused by the ingot composition or phase stability.

The reactivity test showed that of all of the lanthanide co-stabilizers, the $SmInO_3$ containing composition showed the highest onset temperature before $LnInO_3$ formation. This is at least partly a result of the higher india content in the evaporated coating. Along with the formation of the $LnInO_3$ and the expected monoclinic zirconia, the ZrV_2O_7 phase appeared as well. There is no evidence of any influence of the stabilizing agent on the formation of this phase. It is uncertain whether this phase has a role on the mechanical durability or lifetime of the TBC. In the Gd and Dy containing coatings (which showed lessincorporation by EDX) the onset temperature for the appearance of $LnInO_3$ was the same as that for the formation of the ZrV_2O_7 phase. The appearance of the zircon structure vanadate either at lower temperatures than or concurrently with the monoclinic zirconia suggests that mineralization reactions are not taking place.

Table III. Table listing the reaction temperatures and phases observed when exposed to vanadium oxide at elevated temperatures.

Composition	Reaction with vanadia at 400C/6h	Reaction with vanadia at 400C/16h	Reaction with vanadia at 500 C/8h	Reaction with vanadia at 500 C/16h	Reaction with vanadia at 600 C/8h	Reaction with vanadia at 600 C/16h	Reaction with vanadia at 650 C/16h
6 mol% SmInO3	t'	t'	t'	t' + SmVO4 + trace ZrV2O7	t' + SmVO4 + mono (ZrO2) + ZrV2O7	t' + SmVO4 + mono (ZrO2) + ZrV2O7	SmVO4 + mono (ZrO2) + ZrV2O7
6 mol% GdInO3	t'	t' + trace GdVO4	t' + trace GdVO4	t' + trace GdVO4	t' + GdVO4 + mono (ZrO2) + ZrV2O7	t' + GdVO4 + mono (ZrO2) + ZrV2O7	t' + GdVO4 + mono (ZrO2) + ZrV2O7
6 mol% DyInO3	t' + trace DyVO4	t' + trace DyVO4	t' + trace DyVO4	t' + DyVO4 + trace mono (ZrO2)	t' + DyVO4 + mono (ZrO2) + ZrV2O7	t' + DyVO4 + mono (ZrO2) + ZrV2O7	DyVO4 + mono (ZrO2) + ZrV2O7

CONCLUSIONS

It has been shown that indium oxide is an alternative stabilizing agent to yttria which does not react readily with vanadium oxide. A processing technique has been developed to incorporate increased amounts of indium oxide by using rare earth oxides by pre-reacting the indium oxide with samarium oxide or gadolinium oxide to form a stable perovskite ($GdInO_3$ or $SmInO_3$). This resulted in reduced volatilization of the indium oxide and thus increased volume fractions of indium oxide being incorporated into the coating. Comparison of EDX data from evaporated coatings to the coatings produced after electron beam evaporation containing solely indium oxide and those containing $GdInO_3$ showed increased indium content present in greater quantities for those coatings containing the additional stabilizer. The primary findings of the presented work are summarized below:

1) That the addition of a lanthanide co-stabilizer (i.e.,Sm) will assist india incorporation into a EB-PVD thermal barrier coating. EDX revealed a greater india concentration in the 3 mol% coating as $SmInO_3$ than with 6 mol% In_2O_3.

2) The indate materials investigated in this effort do not appear to be ideal for EB-PVD coatings. This material combination is more likely to be better suited for plasma spraying.

3) Samples containing samarium indate showed the most resistance to reaction with vanadium pentoxide

4) The appearance of the $LnVO_4$ phase at temperatures below or concurrently with the monoclinic zirconia contra-indicates a mineralization reaction.

Continued efforts are suggested to further optimize the $LnInO_3$ content, to explore hot corrosion tests mimicking service conditions and to understand the role of the ZrV_2O_7 phase. In addition, additional efforts to prepare and field test plasma sprayed coatings of the india co-stabilized zirconia will be investigated. The materials described within are subject to a pending US patent.

REFERENCES

1) R.L. Jones **J Thermal Spray Technology** 6 [1] 1997 pp77-84
2) Hamilton and Nagelberg **JACerS** 67 [10] 1984 pp 686-690
3) Susnitzky, Hertl and Carter **JACerS** 71 [11] 1988 pp 992-1004
4) Jones and Mess **JACerS** 75 [7] 1992 pp1818-21
5) Jones, Reidy and Mess **JACerS** 76 [10] 1993 pp2660-2662
6) Sheu, Xu and Tien **JACerS** 76 [8] 1993 pp2027-2032
7) Miller, **Int. J. of Applied Ceramic Tech** 1 [2] 2004
8) Sasaki, Bohac and Gaukler **JACerS** 76[3] 1993 pp 689-698
9) Bates, Weber and Gatkin **Proc. – Electrochem. Soc. [Proc. Int. Symp. Solid Oxide Fuel Cells, 1st]** 1989 pp 141-156
10) C.G. Levi **Current Opinion in Solid State and Materials Science** 8 2004 pp77-91
11) Yokakawa, Sakai, Kaweda and Dokiya **Sci. Technol. Zirconia V [Int. Conf. 5th]** 1993 pp 59-68
12) Schneider, Roth and Waring **J. Res. Nat. Bur. Stds. Section A** 65 [4] 1961 pp345-374
13) Pascual and Duran **J. Mater.Sci.** 15 [7] 1980 pp1701-1708
14) Mohan, Yuan, Patterson, Desai and Sohn **JACerS** In Press

International Journal of
Applied Ceramic Technology
Ceramic Product Development and Commercialization

Industrial Sensor TBCs: Studies on Temperature Detection and Durability

X. Chen, Z. Mutasim, and J. Price

Solar Turbines Inc., San Diego, CA

J. P. Feist

Southside Thermal Sciences Ltd., London, U.K.

A. L. Heyes* and S. Seefeldt

Department of Mechanical Engineering, Imperial College, London, U.K.

This article describes recent developments of the thermal barrier sensor concept for non-destructive evaluation (NDE) of thermal barrier coatings (TBCs) and on-line condition monitoring in gas turbines. Increases in turbine entry temperature in pursuit of higher efficiency will make it necessary improve or upgrade current thermal protection systems in gas turbines. As these become critical to safe operation it will also be necessary to devise techniques for on-line conditions monitoring and NDE. Thermal barrier sensor coatings, which consist of a ceramic doped with rare-earth activator to provide luminescence, may be a possible solution. The thermo-luminescent response of such materials has been shown to be suitable for surface and sub-surface temperature measurement and possibly for material phase determination. Herein we describe a number of steps in the development of the sensor coating technology. For the first time sensor coatings have been successfully produced using a production standard air plasma spray (APS) process. Microscopic analysis of the coatings showed them to be similar to standard TBCs and thermal cycle testing of the coatings to destruction showed them to exhibit durability similar to that of standard TBCs suggesting that the addition of rare earth dopants to produce sensor coatings does not change the material structure or the longevity of coatings. Calibration of the coatings using the lifeteime decay response mode showed them to have a dynamic range for temperature measurement extending to just under 1000°C. However, it should be noted that newer compositions have been shown to respond up to 1300°C. Finally, a study of surface temperatures and film cooling has been conducted in a research combustor using APS sensor coatings and some preliminary results are presented.

Introduction

The efficiency of gas turbines is linked to the temperature at the entry to the turbine so that there is a motivation to find ways to increase this temperature.

*a.heyes@imperial.ac.uk

New materials would allow the turbine entry temperature to be increased but for the moment nickel-based super alloys are likely to remain the materials of choice and a better bet for short-term gains would seem to be offered by the development of improved cooling and insulation systems. Drawing air from the main gas path for cooling purposes reduces efficiency and can lead to an increase in noxious emissions. Hence, efficiency gains would seem to be best sought, in the short term at least, by improving and fully utilising the insulating properties of ceramic thermal barrier coatings (TBCs).

TBCs

Current TBCs consist of an intermetallic bond coat to provide oxidation resistance overlaid with a ceramic, which is typically yttria-stablized zirconia (YSZ). There are a number of features of the composition and structure that have made this the coating of choice but deficiencies, which may limit the maximum temperature at which it can be used. The durability of YSZ is temperature dependent and may be called into question if surface temperatures in excess of 1200°C are to be expected. If operated above this temperature on a long-term basis the material phase stability is undermined and a destabilizing phase transformation can occur. At these elevated temperatures sintering also becomes a problem. Furthermore, although the mechanism is not completely understood, failure of TBCs may occur via delamination at the TGO/bond coat interface leading to spallation.[2] Failure is correlated with the thickness of the TGO with the latter's growth rate a function of temperature.

In previous generations of gas turbines, failure of the TBC would result in increases in the temperature of the underlying metal that were still within design limits. However, in future designs, if the performance of every component is fully exploited, and gas stream temperature raised, then a similar failure would result metal temperatures beyond design limits. The inherent phase instability of YSZ above 1200°C, may make it necessary to look for new materials and indeed research is already underway Clarke and Levi.[3] Given the potentially severe consequences of coating failure, even if these become available and lifetime characteristics and failure modes become better understood, it will be necessary to monitor coatings to ensure temperature limits are not exceeded and to provide early detection of degradation and prediction of remaining life. Hence new non-destructive evaluation (NDE) technologies are required. These may be used in development testing or in service and ideally should enable temperature to be measured at critical regions such as the surface and bond coat/TGO interface. In addition, measurement of degradation on or off line, by erosion, changes in phase composition or hot corrosion, for example, to enable remaining life to be estimated would be of considerable value.

Thermal Barrier Sensor Coatings

The thermal barrier sensor coating (TBSC) concept proposed by Choy et al.[1] will, it is hoped, meet these requirements. Sensor coatings are made by modifying the composition of a TBC to include a small amount of a rare-earth element. These are the activators used in phosphors and hence imbue the TBC with phosphorescent properties. Phosphor thermometry is a well-known technique by which temperature may be deduced from the phosphorescent response of certain "thermographic phosphors." The method is thoroughly reviewed by Allison and Gillies[4] and in Heyes[5,6] and, for the sake of brevity will not be described herein.

Previous Work

The authors and colleagues have already demonstrated the viability of creating thermal barrier sensor coatings. The first step involved production of a thermographic phosphor based on YSZ as the host material.[7,8] Samples of 8YSZ (zirconia stabilized with approximately 8% by weight of yttria) doped with approximately 1.4% by weight of Eu_2O_3 were prepared as powders and coatings were laid down using a vapor deposition technique. With these samples temperatures could be measured up to around 830°C with both powder samples and coatings. Measurements were made at the substrate/TBC interface using a sample consisting of a layer of YSZ:Eu approximately 10 μm thick overlaid with about 50 μm of undoped YSZ. Hence the viability of sub-surface temperature measurements was demonstrated. Finally, differences in the emission spectra of samples differently processed were noted. These are thought to have been attributable to different phase compositions (and hence different phonon spectra) so that as suggested by Dexpert-Ghys, et al.[9] the lanthanide activator may act as a structural micro-probe to detect phase composition changes.

In the next phase, a YSZ:Dy coating was produced using electron beam physical vapor deposition (EBPVD) in an industry standard coating facility at Cranfield University.[10] The concentration of dysprosium in the coating was not exactly known due to the method of manufacture but was believed to be 5–10% by weight. The coating was shown to be suitable for temperature measurements up to at least 950 K. No changes to the emission spectra were noted and it was concluded that the phase stability of the coating remained intact at least to the temperatures tested.

In a subsequent investigation,[11] the effect of the concentration of dysprosium in yttria-stabilised zirconia on luminescent response and temperature dynamic range was studied. Concentration quenching was shown to be an important factor in the performance of YSZ:Dy with the brightest output obtained from a sample containing only 0.005 mol% of Dy. The temperature dynamic range at this concentration was shown to extend to 800°C. This is well below the target for gas turbine applications. However, new proprietary compositions devised by STS have also been tested and show a dynamic range extending to at least 1300°C.[12]

Current Objectives

The previous experiments described above have shown that thermal barrier sensor coatings can be manufactured based on the current standard material used for TBCs, i.e., YSZ and using the EBPVD industrial coating manufacturing method. Surface and sub-surface temperature measurements have been demonstrated as has the potential for phase change detection. On-going research is directed toward improving TBSCs to the point where they can be applied in an industrial setting on rotating or stationary components in a full-scale gas turbine. The technique is equally applicable in aero- and land-based gas turbines but the latter are regarded as the appropriate setting for a first application of the technology due to lower regulatory hurdles and the absence of weight restrictions. New TBC materials are under development but their composition is commercially sensitive and, at this stage, it is not clear what material if any will supersede YSZ. Our research is therefore currently concentrated on YSZ as the host material. Nevertheless, the basic technology remains the same and when new compositions become known the technique will be adapted to suite. The research objectives associated with bringing the technology to point of industrial application are outlined below.

Optimize Composition

Initial sensor materials and coatings based on YSZ showed a temperature dynamic range limited to around 800°C. More recently, compositions allowing temperatures of up to 1300°C to be measured have been established. However, further composition optimization is required with a dynamic range extending to at least 1400°C is the target. Material composition affects the dynamic range primarily as a result of the temperature dependent phonon spectra of the host lattice. However, other variables include activator concentration and the inclusion of sensitising agents. Concentration quenching is known to diminish luminescent output so that an optimum activator concentration should exist. Sensitizing agents may enhance performance by absorbing exciting radiation and pumping the activator by energy transfer.

Confirm Durability

TBC coating failure may occur as a result of phase instability, sintering or delamination at the bond coat/TGO interface. The durability of coatings has been shown to be a function of operating temperature, ceramic and bond coat composition, substrate composition and a number of variables related to the coating process. Hence any change in coating composition must be validated to ensure no degradation in durability. This involves testing by long-term exposure at typical operating temperatures and temperature gradients and thermal cycling between ambient and operating temperatures. In addition it may be necessary to re-define coating control parameters such as the substrate temperature to retain the desired morphology.

Instrumentation Development

On-line application of the TBSC technology will require a means to deliver exciting radiation to components inside the engine and to collect and return the luminescence emission. This represents a significant technical challenge since any probe must survive the harsh environment inside hot sections of an engine whilst continuing to perform for thousands of hours.

For commercial applications it would also be necessary to use a reliable and preferably low cost light source for excitation rather than high cost relatively temperamental pulsed laser sources used in laboratory experiments. To achieve this it will be necessary to characterize the excitation spectra of proposed TBSC compositions. This will enable the excitation wavelength to be optimized, for sub-surface measurements for example, and an appropriate light source selected.

Herein we review the manufacture and testing of the first sensor coatings to have been produced using a commercial plasma spraying process. A number of samples have been produced including coatings of various thicknesses and dual layer coatings with the doped layer underneath to enable depth selective temperature sensing. These coatings have been calibrated and a sample of the results is presented. A number of coatings have also been subjected to long term isothermal cyclic testing to establish their longevity relative to standard TBCs and some initial results are presented. The removable wall section of a development combustor at Imperial College has also been coated and a study of wall film cooling conducted. Some initial results of this study are also presented.

Coating Manufacture and Endurance Testing

The sensor TBC material was applied to 2.54 cm diameter by 0.635 cm thick Haynes 230 buttons for calibration and thermal cyclic testing and to two 23 cm by 11 cm nimonic plates for combustion rig testing. Both substrates were made from nickel-based superalloy. The coating system consisted of a NiCrAlY bond coating, 0.05 mm Dysprosium (Dy)-doped YSZ intermediate coating following by a 0.55 mm overcoat of commercial 6–8 wt% YSZ. All coatings were applied by the air plasma spray process. Thermal cyclic tests were conducted on the sensor TBC in a CM Rapid Temp Furnace (CM Furnaces, Inc., Bloomfield, NJ) at temperatures between 25°C and 1148°C. Holding time at the peak temperature (1148°C) was 10 h for each cycle. TBC failure was identified as having occurred when 20% of the area of the TBC or more had cracked and spalled from the substrate. Metallurgical evaluation was conducted using an optical microscope and a scanning electron microscope (SEM) with an energy-dispersive X-ray spectroscope (EDS). The coating

application, analysis and furnace test were conducted at Solar Turbines Incorporated.

Figure 1 shows a micrograph of the sensor TBC in as-coated condition and reveals that the microstructure was very similar to that of a conventional TBC. Figures 2 and 3 show the EDS spectra taken from Dy-doped YSZ intermediate layer and a conventional YSZ outer layer, respectively. The EDS analysis confirms the presence of Dy in the intermediate layer. The sensor TBC failed after an average of 122 cycles. Figure 4 shows a micrograph of the sensor TBC after failure. The sensor TBC failed at the interface between TBC and bond coat. The failure mode and the coating life were very similar to those of conventional TBCs.

Temperature Response Calibration

The experimental arrangement was as shown in Fig. 5. A pulsed YAG:Nd laser (Newport Corporation, Spectra-Physics, Lasers Division, Model GCR-201, Mountain View, CA) was used to excite samples housed in a furnace (Lenton Furnace Ltd., Hope Valley, England) capable of reaching temperatures of 1600°C and specially modified to provide optical access. The coated coupons were placed onto a solid ceramic stand providing good thermal connection with the furnace. The temperature controller displayed the temperature in 1° steps. The laser was operated either at 355 or 266 nm (Q-switch-mode), a repetition rate of 16 Hz and monitored continuously using the power meter shown in the

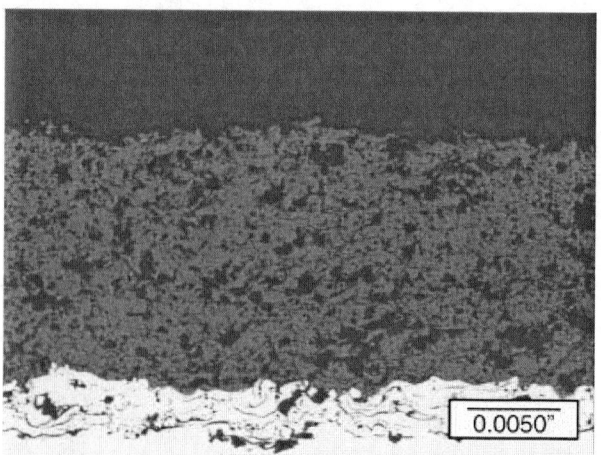

0.0050"

Fig. 1. Microstructure of the sensor thermal barrier coating in as-coated condition.

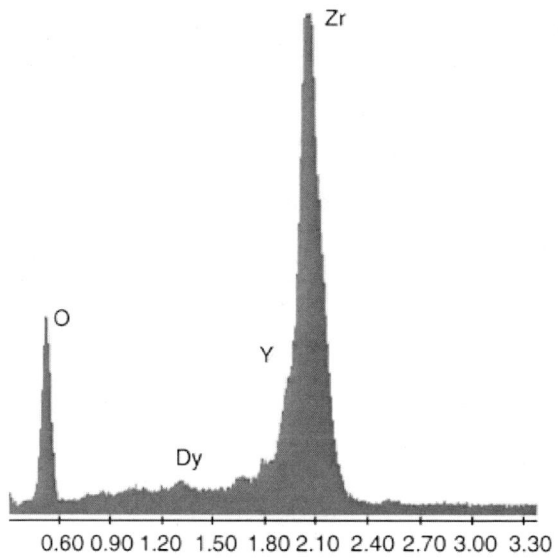

Fig. 2. *Energy-dispersive X-ray spectroscope spectrum taken from Dy-doped thermal barrier coating intermediate layer.*

figure. As shown in the figure, an external beam dump was incorporated to avoid accidental irradiation of the sample by "leaked" 532 nm emission from the laser. The beam was steered through a window into the furnace and the subsequent luminescence was observed through the same window using wavelength selective optics. The emission was collected using a 50 mm Ni-

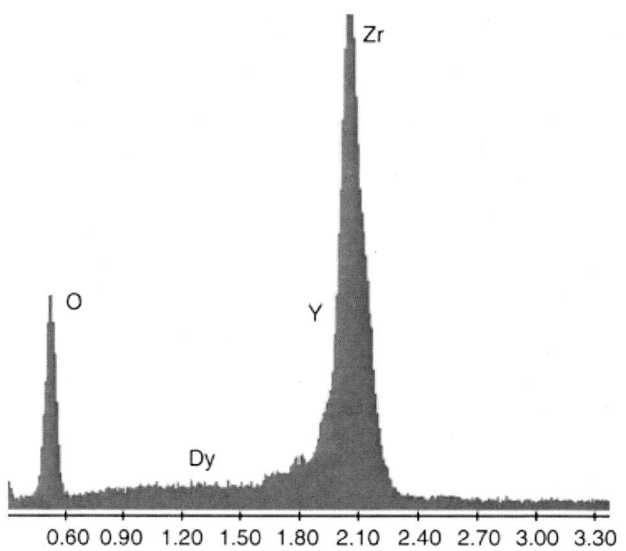

Fig. 3. *Energy-dispersive X-ray spectroscope spectrum taken from conventional yttria-stabilized zirconia thermal barrier coating outer layer.*

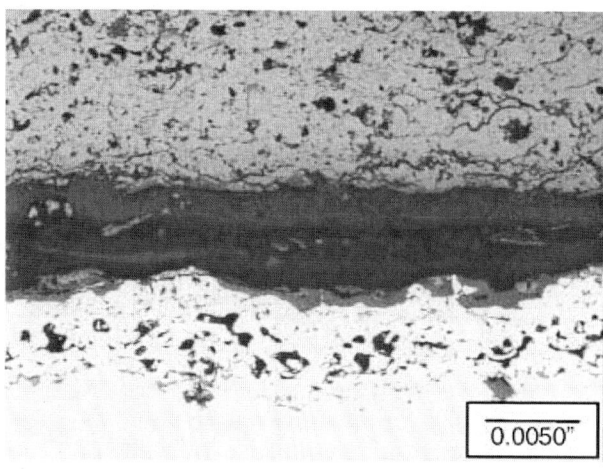

Fig. 4. *Sensor thermal barrier coating (TBC) failure along TBC/ bond coat interface.*

kkor lens, which focused an image of the sample on the entrance slit of a crossed Czerney–Turner spectrometer (TVC Jarrell–Ash MonoSpec 18, Grand Junction, CO; $f = 3.8$).

A photomultiplier tube (PMT; Hamamatsu, Photonics, Japan) was placed at the flexible exit slit of the spectrometer and used to measure the lifetime decay of the phosphorescent emission. An analog/digital converter (PICO Technology Ltd., Cambridge, England; ADC-200; 50 MHz) transferred data simultaneously from the PMT and the power meter to a PC and an exponential decay was fitted to each single exposure using commercial software. The powermeter data was used to monitor irradiation of the sample during testing and for triggering purposes. For an observation of the emission spectrum the photomultiplier system was replaced by a CCD linear array (Alton LS2000, Alton Instruments, Irvine, CA). In this configuration, the spectrometer could be used to compare the relative strengths of the various emission line for intensity ratio temperature measurements and to detect emission changes over a given wavelength range.

Figure 6 shows emission spectra from a powder sample of YSZ:Dy, an APS coating and from a powder sample of YAG:Dy all excited at 355 nm. From the figure, it can be seen that the powdered and sprayed samples show similar spectral features indicating that there was no change in structure as a result of the spraying process. The main emission bands of YAG:Dy and YSZ:Dy can be seen to be similar. This indicates that in both cases it is emission from dysprosium energy levels that is observed. YAG:Dy shows more distinct spectral features highlight-

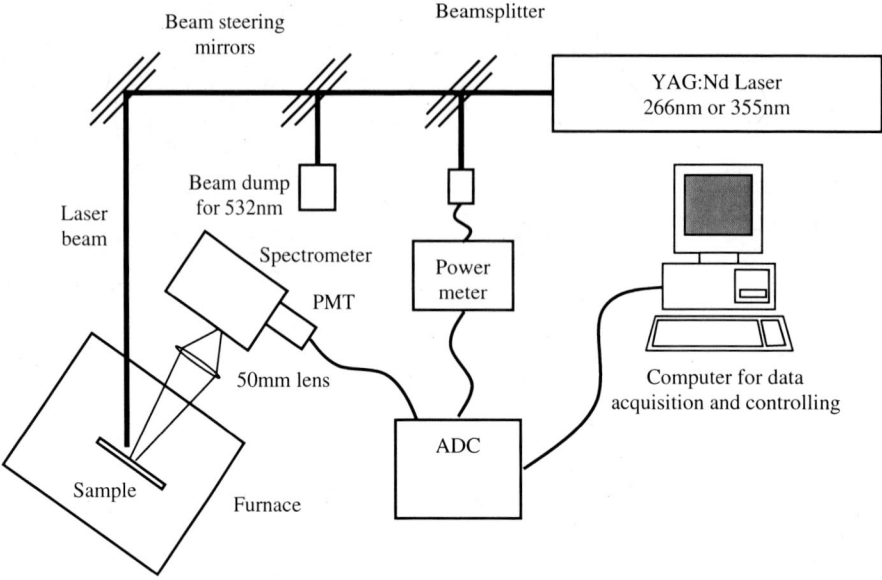

Fig. 5. Experimental setup for phosphor/sensor coating calibration.

ing the effect of the different host crystals lattice on the emission. The poor resolution of the lines in the YSZ:Dy spectra could be a result of the defect rich nature of this host. The emission spike at 532 nm is due to leakage from the second harmonic crystal of the YAG:Nd laser and is not a feature of either phosphor.

Figure 7 shows a lifetime decay calibration curve for one of the APS coatings derived from emissions at 584 nm. As can be seen, this particular composition provides a dynamic response up to almost 1000°C. At high temperatures, the signal becomes difficult to detect because it becomes progressively weaker and the re-

sponse becomes faster so that the frequency response limit of the detection system may be reached. The data shown is from a single layer coating, however, tests have also been conducted with dual layered coatings (the doped layer being beneath an undoped layer of YSZ). These have shown it to be possible to observe phosphorescence through a top layer up to 0.5 mm thick Feist *et al.*[10] It should also be restated that further optimized compositions than the ones discussed herein have shown a dynamic range extending to at least 1300°C.

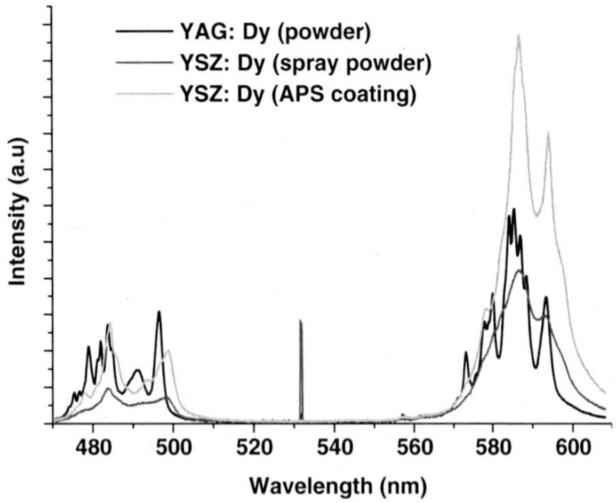

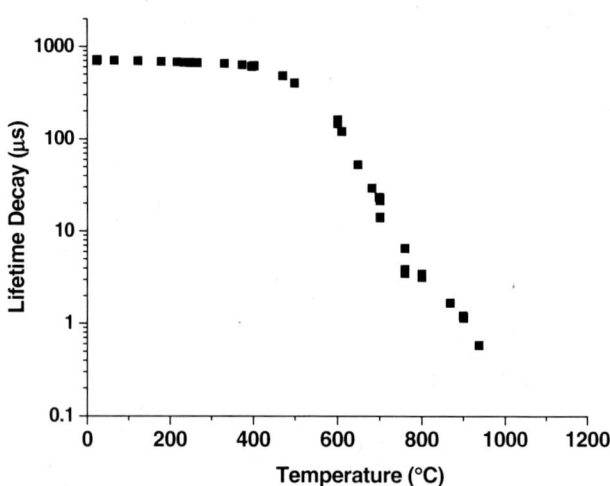

Fig. 7. Lifetime decay response of ytrria-stabilized zirconia (YSZ):Dy APS coating.

Fig. 6. Emission spectra of various Dy-based phosphors.

Combustor Wall Temperature Measurement

In addition to the button samples coated for the purposes of calibration and durability testing, a nimonic plate 23 cm × 11 cm was also coated. This plate forms the wall of a simple research combustor designed for the purpose of film cooling development studies. The plate is provided with an array of 108 cooling holes through which a metered supply of temperature controlled cooling air can be pumped.

A survey of the temperature distribution on the plate when exposed to a methane flame was conducted using the intensity ratio technique. With this approach, using a CCD camera to record the data a surface temperature distribution can be derived from a single or series of images.

The main components of the image acquisition system developed by the author's are an imaging stereoscope to enable two similar images of the subject to be recorded side by side on the same camera, a gated image intensifier and relay optics unit (LaVision IRO, Goettingen, Germany) and a CCD (LaVision Imager 3) camera. Triggering and acquisition was controlled via a PC with an integrated timing unit.

Excitation was achieved using the same pulsed Nd:YAG laser desribed above operating at 355 nm. The raw beam was approximately 10 mm in diameter and was expanded using a plano-concave lens to produce larger illumination areas as required. A ground fused silica diffuser (diffusion angle ± 10°) was also used to reduce spatial variations in intensity.

The coating was calibrated using the system shown in Fig. 5 over a temperature range from 300 to 900 K. Uncertainties were estimated and found to be temperature dependent due to the large discrepancy between emission intensities from the two lines at low temperatures and the overall low signal intensities at high temperatures. Over the range considered the standard error in temperatures measured varied from ± 1 to ± 6 K with lowest errors in the mid-range.

The imaging stereoscope has two apertures, which allow the images formed to be independently filtered so as to observe phosphorescence at different wavelengths. Bandpass filters centered at 458 nm (full-width at half-maximum (FWHM) ~ 10 nm, transmission ~ 50%) and 490 nm (FWHM ~ 10 nm transmission ~ 50%) were used in practice. This choice of wavelengths is not arbitrary and it is required that two emission lines derived from adjacent energy levels of the Dysprosium

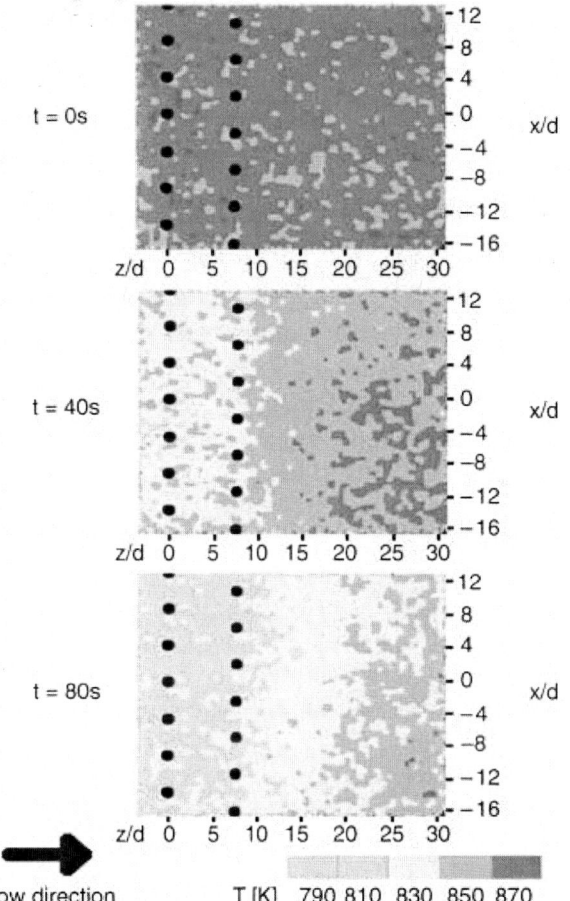

Fig. 8. Surface temperature distribution on the nimonic plate during transient cooling.

ion be selected. A more detailed discussion of this rationale and the emission spectrum of YSZ:Dy from 450 to 500 nm can be found in Feist et al.[10] A combination of two adjustable mirrors and a prism project the two images into the exit of the stereoscope, which is screwed onto the front of the standard 50 mm Nikon lens (Nikon UK Ltd., Kingston upon Thames, UK) (f/1.8) used to focus the image onto the CCD. The advantage of using a stereoscope is that two separate images can be taken with the use of a single camera and imaging optics whilst the main disadvantages is a loss of spatial resolution compared with a two camera setup. An additional 10% neutral density filter was used in the 490 nm light path. The luminescence for this emission line is considerably stronger especially at low temperatures and it was necessary to filter it to achieve comparable intensity levels.

The lens and stereoscope were attached to the image intensifier, which has a maximum gain of 300 camera counts per incident photoelectron. The relay optics produce a 2.17:1 demagnification of the image, which is projected onto the CCD camera. The minimum gating time is 100 ns. The camera used was a peltier cooled 12-bit CCD with a resolution of 1370 pixels by 1048 pixels.

Figure 8 contains a sample of the results obtained during this experiment. It shows a region of the plate that was allowed to acquire a steady-state temperature and subsequently cooled by the introduction of a cooling flow. The formation of a cooling film can clearly be observed as the temperature in the region of the cooling holes falls by almost 100°C over a period of 80 s.

Conclusions

The work described represents several steps in the development of the sensor coating technology. For the first time, sensor coatings have been made using a production standard APS process. The coatings were shown to exhibit similar response to that of the powder from which they were made so that it can be concluded that there were no structural changes as a result of the coating process.

Microscopic analysis of the coatings showed them to be similar to standard TBCs and thermal cycle testing of the coatings to destruction showed them to exhibit durability similar to that of standard TBCs. Hence it is suggested that the addition rare earth dopants to produce sensor coatings does not change the material structure or the longevity of coatings.

Calibration of the coatings using the lifeteime decay response mode showed them to have a dynamic range for temperature measurement extending to just under 1000°C. However, it was noted above that newer compositions have been shown to respond up to 1300°C.

A study of surface temperatures and film cooling has been conducted in a research combustor using the APS sensor coatings. Some preliminary results in the form of a transient temperature map have been presented.

Acknowledgments

The authors gratefully acknowledge Dr. S. Skinner and Ian Brooks of the Materials Department of Imperial College for production of YAG:Dy powders for comparison with the YSZ-based phosphors.

References

1. K. L. Choy, J. P. Feist, and A. L. Heyes, *Smart Thermal Barrier Coatings for Gas Turbines*, U.K., patent application No.: 9823749.
2. P. K. Wright and A. G. Evans, "Mechanisms Governing the Performance of Thermal Barrier Coatings," *Curr. Opin. Solid State Mater. Sci.*, 4 255–265 (1999).
3. D. R. Clarke and C. G. Levi, "Materials Design for the Next Generation Thermal Barrier Coatings," *Annu. Rev. Mater. Res.*, 33 383–417 (2003).
4. S. A. Allison and G. T. Gillies, "Remote Thermometry with Thermographic Phosphors: Instrumentation and Applications," *Rev. Sci. Instrum.*, 7 [68] 2615–2650 (1997).
5. A. L. Heyes, "Thermographic Phosphor Thermometry—Physical Principles and Measurement Capability," *VKI Lecture Series on Advanced Measurement Techniques for Aero and Stationary Gas Turbines*, Vol. Lecture Series 2004-04, eds. C. H. Sieverding and J.-F. Brouckaert. von Karman Institute for Fluid Dynamics, Brussels, 2004.
6. A. L. Heyes, "Thermographic Phosphor Thermometry—Applications in Engineering," *VKI Lecture Series on Advanced Measurement Techniques for Aero and Stationary Gas Turbines*. eds. C. H. Sieverding and J.-F. Brouckaert. von Karman Institute for Fluid Dynamics, Brussels, 2004.
7. J. P. Feist and A. L. Heyes, "Europium Doped YSZ for High Temperature Phosphor Thermometry, Proceedings of the Institution of Mechanical Engineers, Part L," *Proc. Inst. Mech. Eng. Part L, J. Mater.: Design Appl.*, 214 7–12 (2000).
8. K. L. Choy, J. P. Feist, A. L. Heyes, and J. Mei, "Microstructure and Thermoluminescent Properties of ESAVD Produced Eu Doped Y_2O_3–ZrO_2 Coatings," *Surf. Eng.*, 16 [6] 469–472 (2000).
9. J. Dexpert-Ghys, M. Faucher, and P. Caro, "Site Selective Spectroscopy and Structural Analysis of Yttria-Doped Zirconias," *J. Solid State Chem.*, 54 179–192 (1984).
10. J. P. Feist, A. L. Heyes, and J. R. Nicholls, "Phosphor Thermometry in an EBPVD Produced TBC Doped with Dysprosium," *Proc. Inst. Mech Eng. Part G, J. Aerospace Eng.*, 215 333–341 (2001).
11. J. P. Feist and A. L. Heyes, "Recent Developments in Thermal Barrier Sensor Coatings," *16th International Symposium on Airbreathing Engines*, Cleveland, OH: AIAA-2003–1049, 2003.
12. J. P. Feist, A. L. Heyes, C. Jaubertie, S. Seefeldt, and S. Skinner, "Latest developments in Thermal Barrier Sensor Coatings—Temperature, Degradation, Durability," *First International Conference on Gas Turbine Instrumentation*, Barcelona, Spain, 2004.

A. Kulkarni and H. Herman,
Center for Thermal Spray Research, State University of New York
Stony Brook, N.Y.

Industrial TBCs

Zirconia-based thermal barrier coatings (TBCs) are widely used for the thermal, oxidation and hot-corrosion protection of high-temperature components in gas turbines.[1,2] These coatings provide insulation to metallic structures in the hot section of turbine engines and offer three important benefits: increased operating temperature of the engine and, therefore, enhanced efficiency; enhanced durability and extended life of metallic components subjected to high temperatures and high stresses; and reduced cooling requirements to metallic components.[3,4]

TBCs are applied to turbine blades and combustor liners, transition pieces and nozzles. TBCs are comprised of a two-layer coating system on a superalloy turbine-blade substrate. The materials of interest are MCrAlY (where M is Ni, Co) alloys or Pt–Al-based oxidation-resistant bond coats followed by yttria-stabilized zirconia (YSZ) topcoats. The ceramic topcoats are deposited using atmospheric plasma spray (APS) or electron-beam physical vapor deposition (EB-PVD) processes, each producing distinctive microstructures.[5]

Plasma spray confers a splat-based layered structure with advantages of insulation and cost effectiveness.[6,7] EB-PVD offers coatings with superior strain tolerance and thermal-shock resistance, thus providing significant lifetime enhancements.[8,9] Each technique presents a myriad array of process-related defects: interlamellar pores, cracks and gas porosity in APS coatings; and intercolumnar pores, intracolumnar feathery cracks and fine pores in EB-PVD coatings. These imperfections can offer beneficial attributes, such as increased compliance to the EB-PVD coatings and reduction of thermal conduction due to phonon scattering in APS coatings.

> Microstructure–property studies were conducted on plasma-spray coatings having splat-based layered structures and on electron-beam physical vapor deposition coatings with columnar morphology.

The characteristics of these defects and their control is critical for the enhancement of system performance and reliability. The significant intrinsic anisotropy is dominant in the plane perpendicular to the deposition direction in the case of APS coatings because of brick-wall-like structure, where the splats are entwined in complex arrays. Because of the growth of individual EB-PVD, columns in the thickness direction during deposition, a high degree of intercolumnar porosity is developed, and the anisotropy is dominant in the plane parallel to the deposition direction.

COATING SYSTEMS

To explore the effects of feedstock particle size on YSZ coatings, three particle sizes (fine, medium and coarse) were sieved from the as-received plasma-densified powder and compared with the as-received/ensemble powder. Coatings were deposited using a plasma gun (Model 3MB, Sulzer Metco) at a 100 mm standoff distance. Details of the feedstock characteristics and spray parameters are presented elsewhere.[10]

The second set of coatings studied were industrial TBCs, each deposited under incrementally different spray conditions onto a NiCrAlY bond-coated superalloy (IN 718, General Electric, Schnectady, N.Y.). The coatings were labeled GE-1 to GE-4, going from conventional layered structure to dense vertically cracked (DVC) structure, respectively. The coating thickness was ~450 μm in each case. The EB-PVD TBC system under investigation was deposited at Chromalloy Gas Turbine Corp., Orangeburg, N.Y., and consisted of an 800 μm thick EB-PVD zirconia (stabilized with 7–8 wt% yttria) coating (topcoat) on a 50 μm thick NiCoCrAlY bond-coat, also deposited using EB-PVD, on a stainless-steel substrate.

Free-standing coatings were used for porosity and thermal conductivity measurements. Elastic modulus measurements were conducted on the coatings bonded to the substrate. Specimens were sliced and polished for modulus measurements. Morphological features were observed using optical microscopy.

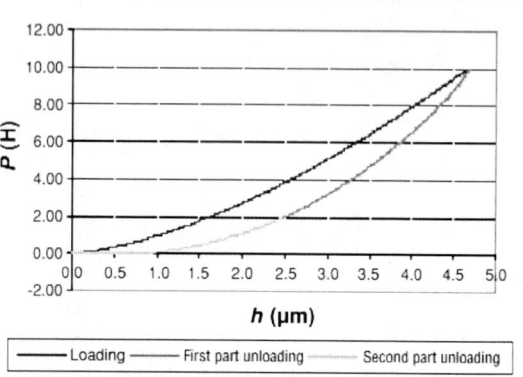

Fig. 1. Depth-sensitive indentation behavior of a PS TBC.

POROSITY, THERMAL CONDUCTIVITY, ELASTIC MODULUS

Surface-connected porosity was measured using mercury intrusion porosimetry (MIP; Autoscan 33, Quantachrome Corp., Bayton Beach, Fla.). The total porosity content was determined using the precision density (PD) method, where mass-over-volume ratios were obtained for a cut rectilinear specimen. The technique gave a fractional density (or porosity) uncertainty of standard deviation, ±1%, based on the average of 10 measured identical specimens and an assumed theoretical density of 6 g/cm^3.

Thermal conductivity measurements were conducted on a 12.5 mm diameter disk, coated with carbon, using a laser-flash thermal diffusivity instrument (Holometrix). Bulk density, thermal diffusivity and specific heat were used to calculate thermal conductivity.

A depth-sensitive indentation technique was used to extract the materials properties using the contact response of a small volume of material. A spherical indenter was used. Continuous measurements of load–displacement (Fig. 1) were performed on an instrument (Model Nanotest 600, Micro Materials Ltd., Wrexham, U.K.) with a 1/16 in. (1.6 mm) WC–Co spherical indenter with a maximum load of 10 N. The indentation procedure used consisted of 10–15 loading/unloading cycles. The elastic modulus was determined from the elastic recovery part of the load–displacement curve. Also, in-plane and out-of-plane elastic moduli measurements were conducted to examine the coating anisotropy.

POROSITY–THERMAL CONDUCTIVITY RELATIONSHIPS

Variations in porosity for the coatings studied have been determined using MIP and PD techniques; thermal diffusivity and thermal conductivity have been determined using the laser-flash technique (Table I). Although MIP measures only surface-connected porosity, the PD studies allow detection of open and closed porosity. The porosity increases with an increase in particle size. This is attributed to two factors: decreased melting efficiency of coarser particles in the plasma plume compared with fine particles; and increased fragmentation of splats that result in poor adhesion and porosity.

The microstructures of these coatings differ (Fig. 2). The thermal conductivity obeys the inverse relationship, decreasing with increasing particle size.

Table I. Porosity and Thermal Property Measurements

Material	PD porosity (%)	MIP porosity (%)	Thermal diffusivity (cm²/s)	Thermal conductivity (W/(m·K))
Fine	6 ± 0.4	6.4 ± 0.7	0.00403 ± 0.0006	0.77 ± 0.07
Medium	8.75 ± 0.8	9.3 ± 0.6	0.00378 ± 0.0005	0.68 ± 0.07
Coarse	15.7 ± 0.9	13.3 ± 0.9	0.00397 ± 0.0005	0.54 ± 0.06
Ensemble	12.3 ± 0.9	10.5 ± 0.9	0.00368 ± 0.0006	0.63 ± 0.07
GE-1	13.3 ± 0.7	14 ± 0.3	0.005 ± 0.0006	1 ± 0.1
GE-2	11 ± 0.7	10 ± 0.4	0.009 ± 0.0004	1.2 ± 0.04
GE-3	10.3 ± 0.6	7 ± 0.6	0.012 ± 0.0004	1.7 ± 0.03
GE-4	9 ± 0.5	9 ± 0.5	0.015± 0.0007	1.9 ± 0.06
EB-PVD	21.75 ± 1.45	20 ± 1	0.0079 ± 0.0008	1.9 ± 0.13

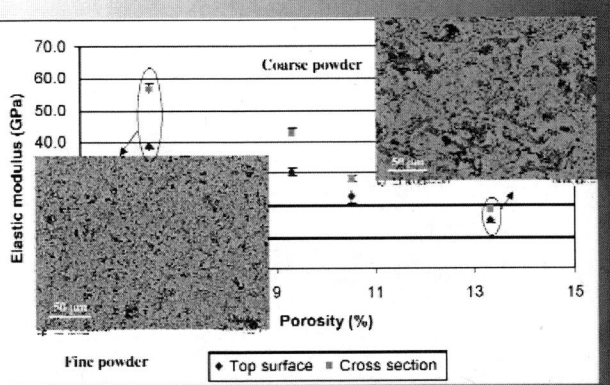

Fig. 2. *Microstructural variations explaining anisotropy in coatings with respect to elastic modulus.*

Cross-sectional micrographs have been made of the coatings sprayed under varied and controlled conditions (GE-1 and GE-4). A typical plasma-sprayed coating with a layered structure (GE-1) shows interlamellar porosity, which is a result of poor adhesion between splats (Fig. 3).

A significantly different structure with vertical macrocracks is observed for the GE-4 case. These cracks are considered to be beneficial for strain tolerance and component life during service. Although the density increases from GE-1 to GE-4, the MIP surface-connected porosity shows an opposite trend except for the DVC (GE-4). This is because macrocracks are accounted for as surface-connected porosity. The thermal diffusivity and conductivity values measured using the laser-flash technique show an inverse relationship with porosity.

For the EB-PVD coating, the observed morphological features (Fig. 4) show columnar grains 20–25 mm wide that grow perpendicular to the substrate plane. This unique morphology results in intercolumnar porosity (1–5 µm wide). The MIP surface-connected porosity is in close agreement with the precision density total porosity. Most of the porosity is open/connected porosity. The thermal conductivity values are higher than plasma-sprayed coatings (1.9 and 1 W/(m·K), respectively) because of the single-crystal columns growing through the thickness of the coating.

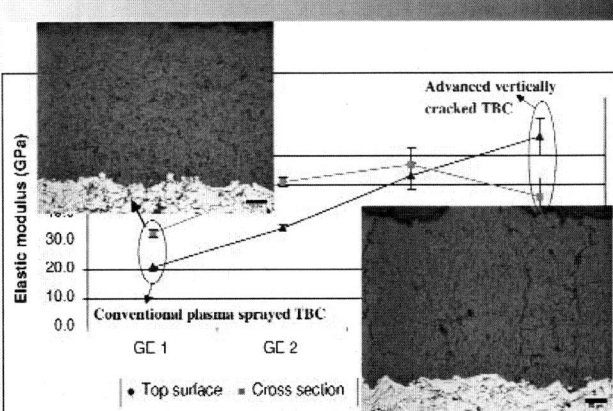

Fig. 3. *Comparison of in-plane and through-thickness elastic properties that show anisotropy in the coatings. Error bars indicate standard deviations for each measurement.*

ELASTIC MODULUS STUDIES

The modulus results have been plotted versus porosity, and microstructure photographs have been made to study particle-size effects (Fig. 2). Vast differences are observed; fragmented splats leading to poor splat–splat contact and formation of pores are evident. The coating prepared from fine powder shows well-adhered splats, whereas the unmelted and poorly adhered particles show a coating prepared from coarse powder.

The modulus, in through-thickness and in-plane direction, decreases with increasing particle size and obeys the same trend as the thermal conductivity.

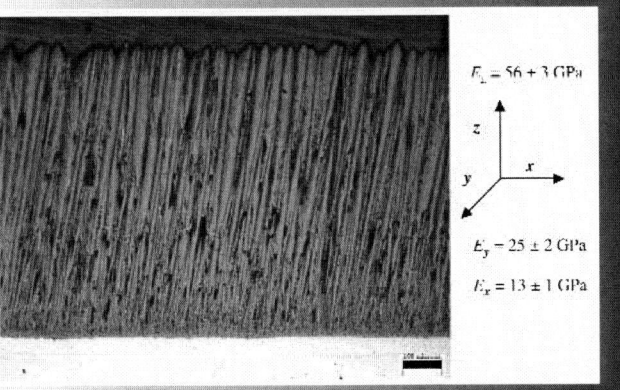

Fig. 4. *Morphological features observed in an EB-PVD TBC that show columnar grains with a large percentage of intercolumnar porosity.*

However, the well-adhered splats and the interlamellar pores and intrasplat cracks generate greater anisotropy in the case of coating made from fine powder particles. This anisotropy decreases with increasing particle size because of the microstructure developed from the unmelted particles.

The elastic modulus for the case of industrial TBCs shows that the out-of-plane (top surface) modulus increases consistently similar to the trend of thermal conductivity of the coatings, suggesting densification of coating microstructure (Fig. 3). The in-plane (cross-section) modulus, which is sensitive to the crack networks, increases except for the GE-4 (DVC) case. The lower in-plane modulus for the macrocracked GE-4 case is beneficial relative to strain tolerance, spallation resistance and component life during service.

The elastic modulus measurements for the EB-PVD case display significantly different anisotropy. Although the in-plane modulus is higher than the top-surface modulus for plasma-sprayed coating (except for GE-4 DVC), the reverse trend is observed for EB-PVD. Although the coating is stiff in the out-of-plane (through thickness) direction, the depth of penetration is larger in the in-plane (cross-sectional) direction; hence, the elastic modulus is significantly higher in the out-of-plane direction. However, because of the intrinsic nature of coating microstructural development (competitive growth among crystal nuclei by vapor condensation), a different scale of anisotropy is observed.

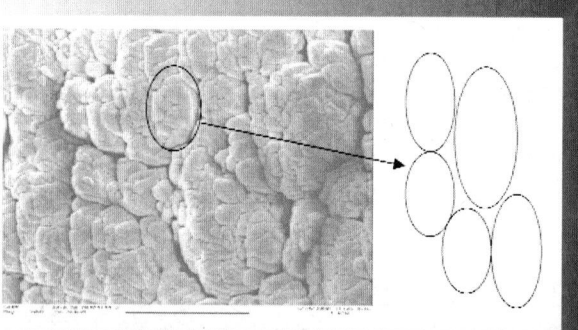

Fig. 5. Microstructural features observed on the top surface of an EB-PVD TBC.

During deposition, vapors are produced by heating the source material with an electron beam, and the evaporated atoms condense onto the substrate. Crystal nuclei are formed on favored sites, and they grow laterally and in thickness to form individual columns, between which develops a high degree of intercolumnar porosity. Thus, the coating microstructure displays property anisotropy in three directions (Fig. 4). This anisotropy in elastic modulus is explained by examining the top surface of the coating (Fig. 5): the columns grow in an elongated (elliptical) shape; thus, they introduce anisotropy in the two in-plane directions. ∎

Acknowledgments

This work was sponsored by the MRSEC program of NSF under Award No. DMR-0080021. The authors thank Dr. Curtis Johnson and Dr. James Ruud of GE Corporate Research and Dr. Stacy Fang and Dr. Paul Lawton of Chromalloy Gas Turbine for providing valuable coatings.

References

[1] R.A. Miller, *Surf. Coat. Technol.*, **30**, 1 (1987).
[2] W.J. Brindley and R.A. Miller, *Adv. Mater. Processes*, **8**, 29 (1989).
[3] S.M. Meier and D.K. Gupta, *J. Eng. Gas Turbines Power*, **116**, 250 (1994).
[4] J.T. DeMasi-Marcin, K.D. Sheffler and S. Bose, *J. Eng. Gas Turbines Power*, **112**, 521 (1990).
[5] R.L. Jones, "Thermal Barrier Coatings"; p. 194 in *Metallurgical and Protective Coatings*. Edited by K. H. Stern. Chapman and Hall, London, 1996.
[6] W. Mannsmann and H.W. Grunling, *J. Phys. IV*, **3**, 903–12 (1993).
[7] R. McPherson, *Thin Solid Films*, **83**, 297 (1981).
[8] T.E. Strangman, *Thin Solid Films*, **127**, 93–35 (1985).
[9] C.A. Johnson, J.A. Ruud, R. Bruce and D. Wortman, *Surf. Coat. Technol.*, **108–109**, 80–85 (1998).
[10] A. Kulkarni, A. Vaidya, A. Goland, S. Sampath and H. Herman, *Mater. Sci. Eng.* A, in review.

Progress in Thermal Barrier Coatings

Edgar Lara-Curzio ,Michael J Readey
Copyright © 2004 The American Ceramic Society

LOW THERMAL CONDUCTIVITY CERAMICS FOR TURBINE BLADE THERMAL BAR-RIER COATING APPLICATION

Uwe Schulz
DLR, Inst. Materials Research
51170 Cologne, Germany

B. Saint- Ramond
Snecma Moteurs
77550 Moissy-Cramayel
France

O. Lavigne
ONERA
F-92320 Châtillon
France

P. Moretto
EU-JRC IAM
1755 ZG Petten
The Netherlands

A. vanLieshout
Sulzer Metco coatings
5943 AD Lomm
The Netherlands

A. Börger
Montanuniversität, ISFK
8700 Leoben
Austria

INTRODUCTION

During the last 20 years, thermal barrier coating (TBC) developments have been mainly focused on the integrity of these complex coatings into the engine components and on their lifetime durability during service. There is no doubt that in the near future TBCs are needed with both increased temperature capability, i.e. long term operation above 1200°C, and reduced thermal conductivity. The challenge tackled by this study was to increase the thermal insulation performance of the ceramic topcoat for aero-engine turbine blade application through new chemistries. The other possibility to reduce thermal conductivity K, tailoring the microstructure of the TBC by introducing efficient porosity in the right amount, geometry and distribution combined with advanced image acquisition and image based calculation of K, is described for plasma sprayed TBCs [1] and for electron beam physical vapor deposition (EB-PVD) density related microstructure features [2] elsewhere. The current results have been jointly obtained as part of the European framework IV project HITS [3].

New zirconia based ceramic compositions have been under investigation for TBC applications since many years (e.g. alternative stabilizer for zirconia [48] or ternary oxide compositions [9, 10]. In recent years, non-zirconia TBCs became also attractive. New chemistries for TBC applications need to meet the following requirements:
- Thermal stability and chemical inertia at high temperature.
- Chemical and mechanical compatibility with a metallic bond coat layer.
- Manufacturability, best by use of existing coating technologies.

Reducing thermal conductivity of zirconia can be achieved by introducing a higher disordered crystalline lattice according to the following two ways:
- Introduction of oxygen vacancies (e.g. by increasing the total amount of stabilizer).
- Partial (or total) substitution of Y and/or Zr ions with ions of different mass and/or different ionic radius (mainly rare-earth metals).

Based on these considerations new zirconia based TBC compositions manufactured by EB-PVD were chosen as the prime candidate materials in this study.

EXPERIMENTAL

Since evaporation material of the new compositions in cylindrical form and suitable for EB-PVD processing was not available on the market, ingots were optimized and manufactured within the group. A powder based manufacturing route, consisting of the main steps mixing coarse and fine fractions, pressing, machining, and sintering was developed based on earlier results [11].

Flat metallic substrates of various geometries and with MCrAlY bond coats were coated with the new compositions under standard EB-PVD conditions. Reference systems with 7wt% yttria-zirconia (4P-YSZ) were manufactured from commercial evaporation material, too. Chemistry of the coatings was measured by X-ray fluorescence (XRF) and the microstructure of the TBCs was analyzed by field emission gun scanning electron microscopy (SEM).

Phase analyses was performed by quantitative X-ray diffraction (XRD) using the {111} and {400} peak areas to separate the zirconia polymorphs (for details see [12]). TBCs had been removed from the substrate in a hydrochloric acid solution which dissolves the bond coat. Some of the free-standing TBC material was carefully milled and than measured in a Siemens D-5000 diffractometer using copper radiation. The milling procedure was necessary to eliminate the strong texture of the EB-PVD coatings in order to get results that can be used for quantitative phase analyses. The free standing TBCs were heat treated for 100hrs at 1250°C in air, milled and measured again by XRD.

Thermal conductivity was determined after an initial pre-aging for 2hrs at 1080°C in vacuum from thermal diffusivity measurements by means of Laser flash analyses (LFA), geometrical density and literature heat capacity data. Measurements were performed under vacuum since initial studies revealed for EB-PVD TBCs only a minor effect of the measuring atmosphere on K. In order to evaluate the change in thermal conductivity during service, samples were aged at 1200°C for up to 100h in vacuum and re-measured by LFA. Vacuum annealing was necessary since the simple substrate - bond coat combinations used do not have high temperature capabilities under these severe annealing conditions. To increase accuracy of the measurement, two identical processed samples were measured and the average from these two samples was calculated. Additional ceramic bulk samples of nominal the same composition were sintered and measured by LFA.

RESULTS AND DISCUSSION

The composition of the three new EB-PVD TBCs, measured by XRF, was close to the measured ingot composition (the number in the version name gives the intended Mol%):

  4 DySZ: 11wt% Dy_2O_3
  12 DySZ: 29wt% Dy_2O_3
  12 YbSZ: 28wt% Yb_2O_3

Obviously, the vapor pressures of the alternative stabilizer do not differ too much from the values of zirconia, thus ceramic layers with homogeneous composition across the thickness were achieved (Fig. 1, after [13]). Only Gd-oxide that was not evaluated in the present study may cause some problems because the vapor pressure at higher temperatures is slightly different from the zirconia values.

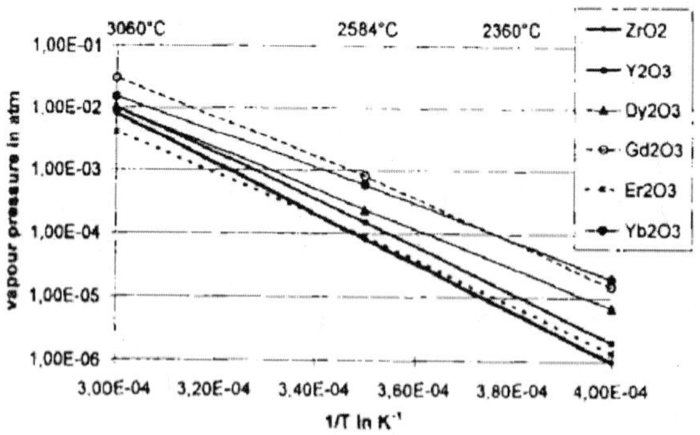

The microstructure of the TBCs in top view is shown in Fig. 2. In general, no major difference

Fig. 1: Vapor pressure of zirconia and rare earth oxides.

in the morphology of the TBCs can be found. All versions are columnar with a relatively regular shape and arrangement of the columns. Only the reference version has some variation in the column diameter on top, caused by the deposition conditions chosen.

In top view the fully stabilized versions (lower two pictures in Fig. 2) exhibit a fourfold shape which corresponds to the cubic phase structure while the partially stabilized versions show only one symmetry line that is parallel to the rotational axis. Cubic phases tend to easier form the ideal pyramidal shape of the columns tips that is caused by surrounding {111}-planes [14]. 4DySZ has the smallest column diameter compared to the other versions.

| 4 P-YSZ | 4DySZ |
| 12YbSZ | 12DySZ |

Fig. 2: Microstructure in top view of new EB-PVD TBCs.

XRD plots are shown in Fig. 3. In the as coated stage both partially stabilized versions (reference and 4DySZ) contain only the metastable tetragonal t' phase, indicated by the characteristic peak split in the {400}-region and by the presence of the {112}-peak which is forbidden for a cubic phase. The lattice parameters are similar (a=0.5105nm, c=0.517nm). On the other hand, the fully stabilized versions with 12 mol% stabilizer are both cubic. As a result of the larger ion radius, the 12DySZ version possesses the higher lattice constant compared to 12YbSZ indicated by the peak shift towards smaller angels. The broadening of the peaks of the latter one could be a consequence of either some small fluctuations of the stabilizer content or a very small grain size. A fluctuation of the composition was not detectable in EDS line scans in SEM but the loss of some stabilizer may support this assumption if one compares 31wt% Yb_2O_3 measured in the ingot to 27.6wt% found in the coating.

The phase content after 100hrs aging at 1250°C in air of both fully stabilized versions is unchanged. Peaks are getting sharper and the FWHM (full width at half maximum) decreases considerably. This can be associated with grain growth during the treatment. Obviously, the 12YbSZ version had a very fine grain size after deposition that increases during ageing towards standard level. For the partially stabilized versions, a decomposition into monoclinic (m), cubic (c), and

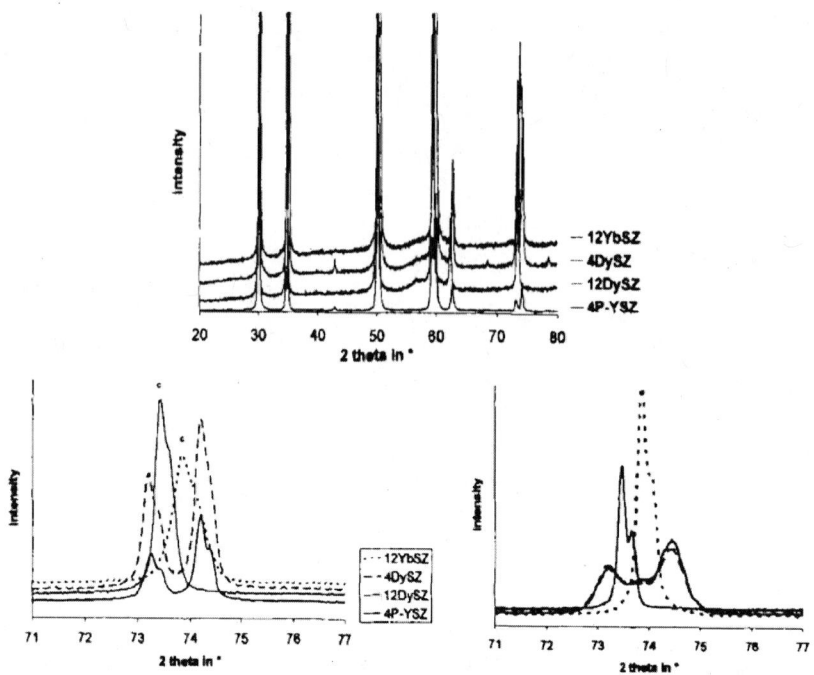

Fig. 3: XRD of powdered EB-PVD TBCs. Overview as coated (above), {400} region (below, left as coated, right after 100h@1250°C).

tetragonal (t) phases was observed. The phase contents after ageing are very similar for both versions:

4 P-YSZ	3% m + 78% t + 19% c
4 DySZ	2% m + 76% t + 22% c.

To summarize the phase transformation behavior, the cubic phases are stable under thermal exposure similar to Fully-YSZ. No difference between the thermal stability of partially stabilized reference PYSZ and DySZ was observed, only traces of monoclinic but remarkable amounts of cubic phase are formed. This is similar to previous investigation on the reference P-YSZ material [12].

Fig. 4 summarizes the thermal conductivity of the four TBCs in the as coated condition, measured under vacuum after a stabilizing heat treatment for 2hrs/1080°C. A drastically lowered conductivity of the 12DySZ version is notably [15]. This version had also the lowest density compared to the other versions. In table 1 thermal conductivity data are summarized. The fully stabilized versions show a slight increase of K with temperature while the partially stabilized versions have a higher K at room temperature which decreases with

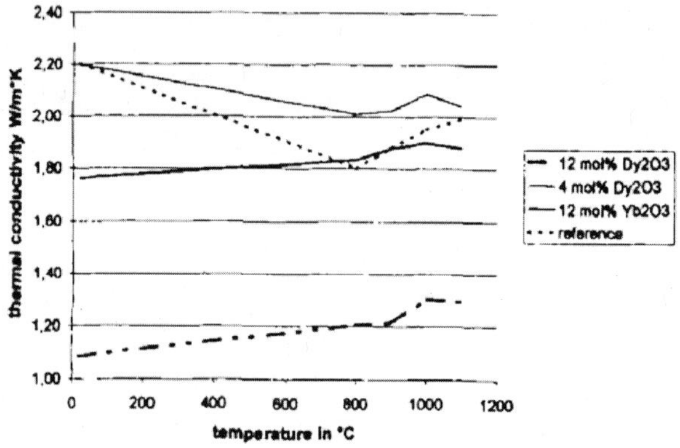

Fig. 4: Thermal conductivity of EB-PVD compositions after initial 2h@1080°C treatment.

increasing temperature which is in good agreement with literature data [16]. Since the starting values ("as coated") were obtained after an initial heat treatment that caused the major part of the rise in K, the increase in thermal conductivity after 50 and 100hrs at 1200°C was only moderate.

Only the 4DySZ showed some more increase in K after aging. With optimized process parameters on full scale industrial equipment and the additional incorporation of optimized morphologies an even lower thermal conductivity (number in brackets in table 1 for 12DySZ) was achieved. The results clearly show for DySZ that a larger content of the stabilizer lowers thermal conductivity. It remains unclear why the EB-PVD version 12YbSZ at high temperature did not show the expected lower thermal conductivity while on sintered samples this effect was found. A fine grained microstructure as observed for 12YbSZ is normal assumed to promote a low thermal conductivity, but this was not the case in this study. Overall, 12DySZ has a 39% lower thermal conductivity that remains low after annealing. This makes this version attractive for demanding TBC applications on turbine blades.

Table 1: Relative thermal conductivity K of new ceramic TBCs, measured at 1000-1100°C in vacuum, after initial 2h@1080°C and after 50h@1200°C annealing in vacuum. The reference composition was set 1. In addition, K at room temperature is given for dense sintered samples where the reference was set 1 as well.

Composition	Density [g/cm^3]	relative K	relative K heat treated	Relative K at RT for sintered samples
4P-YSZ	5.0	1	1.04	1
4DySZ	5.0	1.04	1.46	0.86
12YbSZ	5.3	0.96	1.08	0.64
12DySZ	4.8	0.66 (0.61)	0.76	0.61

The effect of lowering the thermal conductivity by adding trivalent ions is complex in nature. Not only ion radius, mass of the ion (see table 2 for numbers) and charging state must be considered, but also lattice distortion and stability of the created point defects in the lattice must be considered.

Table 2: Ion radius and atomic mass of rare earth oxides in comparison to Zr

Ion of:	Zr	Yb	Er	Y	Dy	Ce	Gd
ionic radius in Å	0,79	0,858	0,881	0,893	0,908	0,92	0,938
Atomic mass	91	173	167	89	162	140	157

To give an example, the combination of ionic radius and distortion of a- and c-axis causes different distorted lattices for the various stabilizer types although tetragonality in terms of c/a ratio is the same. For instance, GdSZ will have only a little shorter c-axis than pure tetragonal ZrO$_2$ but a much larger a-axis. The smallest ion, ytterbia, causes a much lower c-axis but only a little larger a-axis. It can be concluded that a combination of both high mass difference and large difference in ion radius is most effective to lower thermal conductivity while the degree of tetragonal distortion of the zirconia lattice is not of major importance.

CONCLUSIONS

The new EB-PVD ceramic composition TBCs 4DySZ, 12DySZ, 12YbSZ were investigated and compared against the reference 4P-YSZ material.

(1) EB-PVD manufacture of the new stabilized zirconia TBCs is easily possible by use of appropriate ingots and deposition parameters, leading to a columnar microstructure and homogeneous composition.

(2) Partially stabilized versions transform during annealing from tetragonal t' into monoclinic + tetragonal + cubic phases. Fully stabilized versions stay cubic.

(3) 12 DySZ has a 39% reduced thermal conductivity compared to 4P-YSZ. It is a promising candidate material for future TBC applications.

Acknowledgement

The authors acknowledge the contribution of many colleagues within the project, particularly M. Hangl (ISFK), E. Bullock (JRC), S. Alperine, V. Arnault and A. Bickard (Snecma), F. Ladru and A. Fischer (RWTH), M. Peters (DLR), R. Mevrel (ONERA), J. Wigren, A.-C. Leger and P. Bengtsson (Volvo Aero), F. Jansen and R. Damani (Sulzer Innotec), G. Marijnisen (Sulzer Coatings). Funding from the European commission within the 4th FWP under guidance of R. Simonini is highly acknowledged.

References

[1] O. Lavigne et al., "Microstructural characterisation of plasma sprayed thermal barrier coatings by quantitative image analysis", Quantitative Microscopy of High Temperature Materials, ed. A.S.a.J. Cawley, Sheffield, UK,: IOM Communications Ltd, (2001), 131-44.

[2] U. Schulz, J. Münzer, and U. Kaden, "Influence of deposition conditions on density and microstructure of EB-PVD TBCs," Ceramic Engineering and Science Proceedings, 23-4(2002), 353-60.

[3] B. Saint-Ramond, "HITS-High insulation thermal barrier coating systems," Air & Space Europe, 3(3/4)(2001), 174-77.

[4] S. Stecura, "New ZrO2-Yb2O3 plasma-sprayed coatings for thermal barrier applications", Thin Solid Films 150, (1987), 15-40.

[5] R. Hamacha, P. Fauchais, and F. Nardou, "Influence of dopant on the thermal properties of two plasma-sprayed zirconia coatings," J. Thermal spray technol., 5(4)(1996), 431-38.

[6] U. Schulz, K. Fritscher, and M. Peters, "EB-PVD Y_2O_3 and CeO_2/Y_2O_3 stabilized zirconia thermal barrier coatings - crystal habit and phase compositions," Surface and Coatings Technology, 82(1996), 259-69.

[7] B. Leclercq and R. Mevrel, "Thermal conductivity of zirconia-based ceramics for thermal barrier coatings", CIMTEC 2002, in press, (2002)

[8] F. Tcheliebou, M. Boulouz, and A. Boyer, "Preparation of fine-grained MgO and Gd2O3 stabilized ZrO2 thin films by electron beam physical vapor deposition co-evaporation," J. Mater. Res., 12(12)(1997), 3260-65.

[9] J.R. Nicholls, K.J. Lawson, A. Johnstone, and D. Rickerby, "Low Thermal Conductivity EB-PVD Thermal Barrier Coatings," Materials Science Forum, 369-372(2001), 595 - 606.

[10] D. Zhu, A. Nesbitt, T.R. McCue, A. Barrett, and R.A. Miller, "Furnace cyclic behavior of plasma-sprayed zirconia-yttria and multi-component rare earth oxide doped thermal barrier coatings." NASA TM-2002-211690 (2002)

[11] U. Leushake, W. Luxem, C.-K. Kröder, and W.-D. Zimmermann, "Process for coating with ceramic vaporizing materials," US 6,168,833, EP 0812930, DE 19623587 (1996)

[12] U. Schulz, "Phase transformation in EB-PVD yttria partially stabilized zirconia thermal barrier coatings during annealing," Journal American Ceramic Society, 83(4)(2000), 904-10.

[13] N. Jacobson, "Thermodynamic Properties of Some Metal Oxide-Zirconia Systems", NASA TM 102351, (1989)

[14] S.G. Terry and C.G. Levi, "The evolution of porosity in vapor-grown thermal barrier coatings," J. American Ceramic Society, accepted(2003)

[15] S. Alperine, V. Arnault, O. Lavigne, and R. Mevrel, "Heat barrier composition, a mechanical superalloy article provided with a ceramic coating having such a composition, and a method of making the ceramic coating," US 6,333,118, EP 1085109

[16] A. Azzopardi, R. Mévrel, B. Saint-Ramond, E. Olson, and K. Stiller, "Influence of aging on structure and thermal conductivity of Y-PSZ and Y-FSZ EBPVD coatings.," Surface Coatings and Technologies, (2003)

International Journal of
Applied Ceramic Technology
Ceramic Product Development and Commercialization

Thermal and Environmental Barrier Coatings for SiC/SiC CMCs in Aircraft Engine Applications*

Irene Spitsberg and Jim Steibel

GE Aircraft Engines, Cincinnati, OH

Accommodating relatively high engine component surface temperatures is a significant challenge that must be overcome in order to fully realize engine performance benefits associated with implementing CMC materials having relatively low density and higher temperature capability. The design surface temperature for selected components is expected to exceed the 1414°C (2577°F) melting point of silicon, a key CMC constituent, and thus protective coatings are required. This paper discusses challenges in developing the coating system, specifically addressing issues related to the selection of materials, designing the coating architecture, and the test methodology.

Introduction

Ceramic matrix composite (CMC) materials have recently received significant attention from both government and industry as a key material technology for meeting future propulsion needs, as CMCs provide a significant potential improvement in fuel consumption and thrust-to-weight ratio compared to metallic materials.[1] A ceramic composite material comprised of silicon carbide (SiC) fiber with a silicon carbide matrix is one such material. The SiC/SiC CMC material developed jointly under U.S. Government (NASA, DoD, and DoE) and GE programs utilizes the preferred Melt Infiltration (MI) process to achieve the desired thermomechanical properties. The MI SiC/SiC material has the capability of operating at an average temperature exceeding 1204°C (2200°F).

The feasibility of implementing SiC/SiC CMC material technology into turbine engines has been studied at GE Aircraft Engines over the last 10 years. As shown in Fig. 1, the candidate components include all major hardware throughout the turbine hot section. If the design requirements are met successfully for this wide range of components, engine performance benefits should be realized in a variety of applications including military and commercial.

The engine performance benefits associated with implementing CMCs are attributed to the material's relatively low density (approximately one-third of typical Ni-base superalloys) and higher temperature capability. Additional weight benefit may be realized when applying CMCs to rotating airfoils because of the impact on the entire rotor structure. In the future, the temperature capability of the current SiC/SiC CMC system should improve by an additional 93°C (200°F), or a potential benefit over superalloys exceeding 204°C (400°F). The higher temperature capability enables lower cooling air requirements, leading to a potential reduction in NO_x emissions for combustors and an increase in thrust-to-weight ratio

* Presented at the 27th Annual International Conference on Advanced Ceramics and Composites, Cocoa Beach, FL, January 27th, 2003, and at the Environmental Barrier Coatings for Microturbine and Industrial Gas Turbine Ceramics Workshop, Nashville, TN, November 19th, 2003.

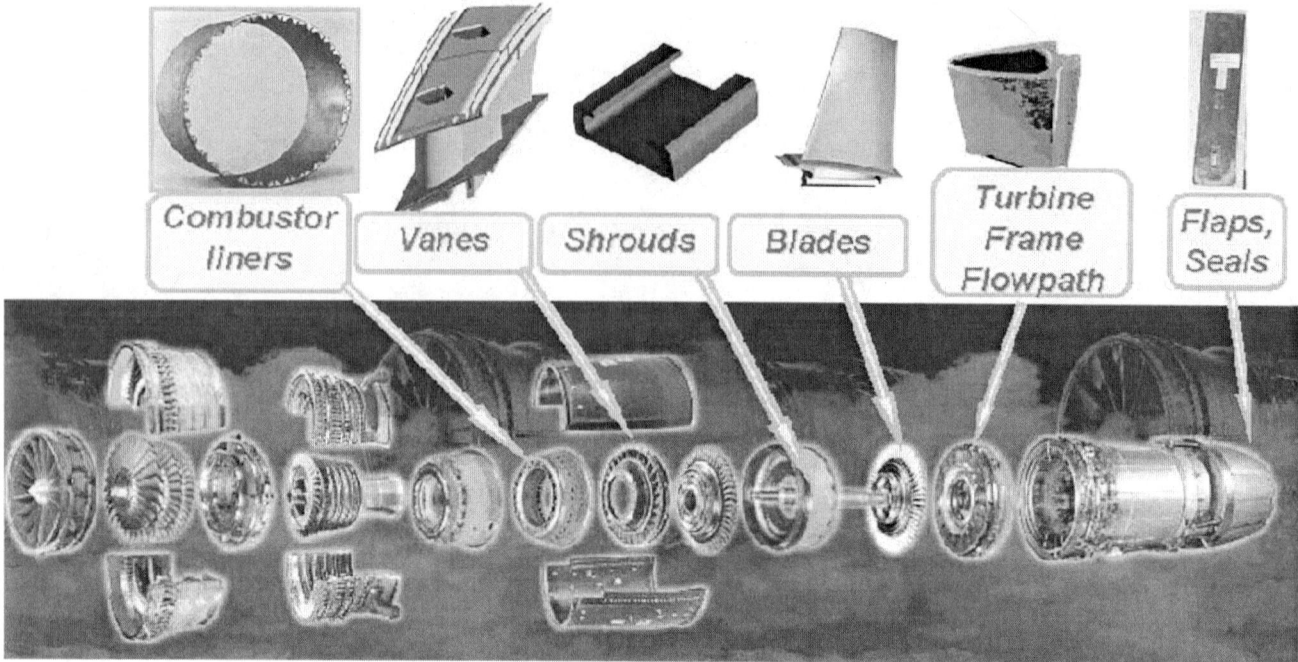

Fig. 1. CMC applications in the hot section of jet engine.

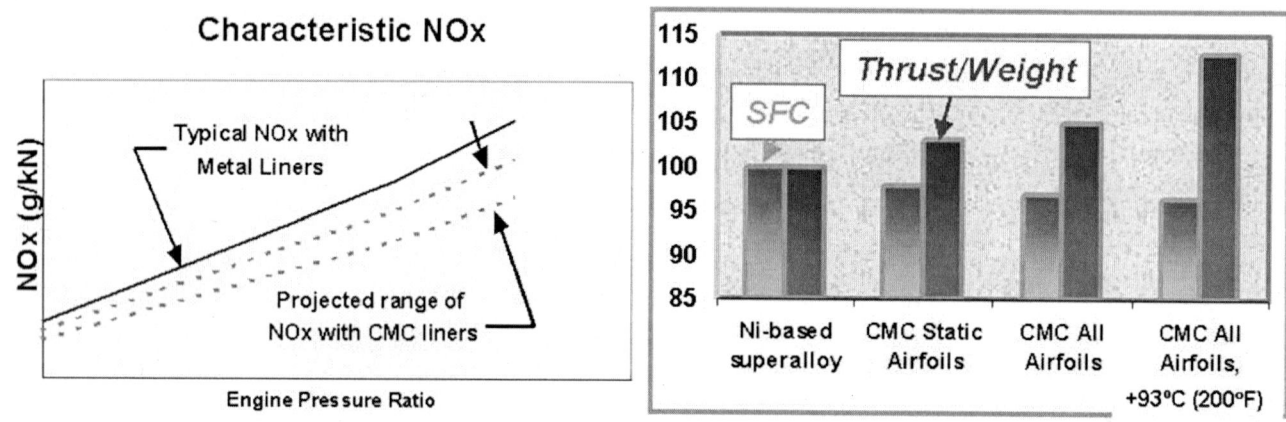

Fig. 2. CMC payoffs in NO_x emissions (left) and Specific Fuel Consumption (SFC) and Thrust-to-Weight ratio (right).

for turbines that incorporate CMC airfoils, as shown in Fig. 2. However, accommodating the relatively high component surface temperatures is a significant challenge that must be overcome in order to fully realize these engine performance benefits. The design surface temperatures are expected to exceed the 1414°C (2577°F) silicon melting point for selected components, thus, protective coatings are required. Even with the surface temperature below the melting point of key CMC constituents, environmental and thermal protection of the component surface most likely will be needed.

Current Status of Environmental and Thermal Barrier Coatings for CMCs

When exposed to an oxidative environment, SiC forms a protective slow-growing silicon oxide scale (SiO_2). However, it has been experimentally demonstrated[2-4] that SiC can undergo a significant weight loss in a water-vapor containing environment, as shown in Fig. 3. This occurs by the surface recession of the silica scale due to the reaction with the water vapor as follows:

$$SiO_2 + 2H_2O = Si(OH)_4 \text{ (gas)} \quad (1)$$

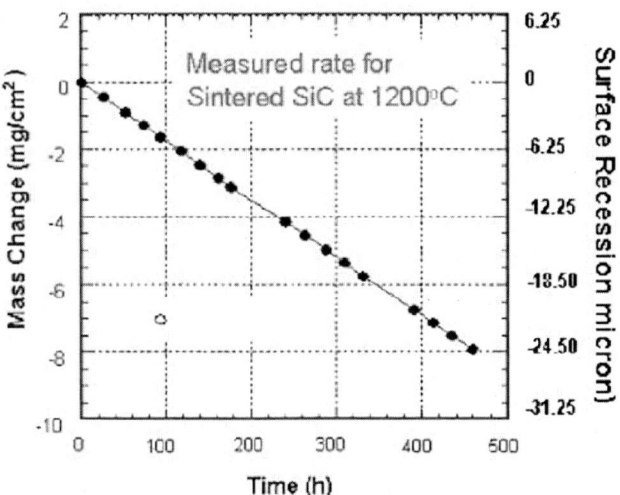

Fig. 3. *Mass change and surface recession for sintered SiC at 1200°C in 90% steam.*

The surface recession mechanism is very sensitive to the gas velocity and can be dramatically accelerated as the velocity increases.[5,6] For turbine engine conditions, the velocity factor is usually included in the gas Mach number and/or heat transfer coefficient. Surface recession was indeed observed during the first field engine tests of SiC/SiC CMC combustor liners in land-based power generation engines.[7,8]

Significant CMC surface recession can be expected in a typical combustion environment containing about 10% water vapor at a pressure of 10-30 atm, especially since these conditions must be withstood for lifetimes exceeding 1000 h. The recession may reduce the net-section thickness of the component wall or contribute to dimensional changes in the case of airfoils.

An environmental barrier coating (EBC) system capable of protecting the CMC component surface from recession over the life of the component was developed under the EPM program.[2,9] As shown in Fig. 4, the system consists of three layers: a silicon layer adjacent to the CMC, a transition layer consisting of a mixture of mullite ($2Al_2O_3 - 3SiO_2$) and barium strontium alumino-sili-

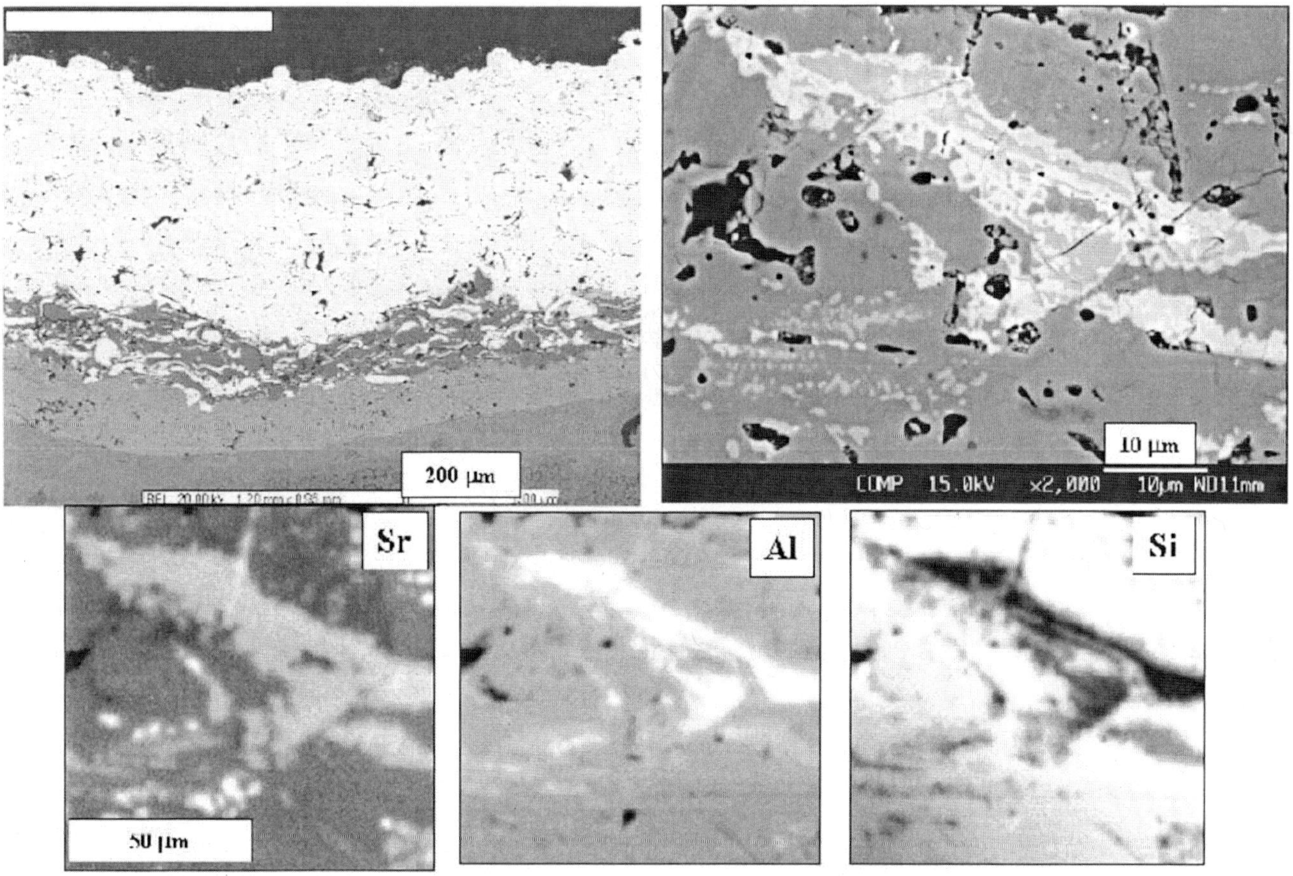

Fig. 4. *As-processed (by air plasma spray) and heat-treated 3-layer EBC system with BSAS environmental barrier layer; X-ray maps show higher Sr and Al content, as well as lower Si content in the secondary phase (brightly imaging phase within BASA layer).*

cate (BSAS), and a protective layer of stoichiometric BSAS (25 mol% BaO + SrO, 25 mol% Al_2O_3, 50 mol% SiO_2). As deposited using a conventional plasma-spray technique followed by a prolonged heat treatment,[2] the top layer predominantly exhibits a Celsian crystal structure, with a small percentage of a secondary phase with a strontium-rich composition (Fig. 4). This three-layer system is often referred to as a baseline or Gen 1 EBC coating system.

EBC development efforts have continued since the termination of the NASA EPM program in 1999. These activities have included microstructure and processing improvements, as well as scale-up and component level demonstration.[10,11] In addition, development efforts at General Electric have also been focused on developing coating systems with higher temperature capability for long-life applications in power-generation engines, as well as a thermal-environmental barrier coating (T/EBC) system with significantly higher surface temperature capability than the baseline BSAS system. The T/EBC coating system will potentially consist of a relatively low-conductivity ceramic top layer material selected from a material family other than silicates[12-14] due to the relatively high temperature capability requirement. Currently, the feasibility of establishing a T/EBC system that survives cyclic testing at a surface temperature approaching 1700°C (3100°F) for 50 h has been demonstrated. Further efforts to mature this coating system for the full life requirement are necessary, and the related challenges will be discussed in subsequent sections.

Coating Design Requirements

Prior to initiating development of a new coating system, the coating performance goals dictated by the component design must be defined and subsequently translated into specific requirements for the coating material system. Specific requirements for the coating performance in combustors and airfoils are summarized in Table I. The combination of high surface temperature, moisture content, and high gas velocities (Mach) generate harsh recession conditions in jet engines, and these conditions are typically more severe than power-generation engines. In addition, jet engine field experience on metallic components suggests that corrosion is also a concern. The corrosive environment may result from sea salt or sand-type deposits of calcium-magnesium-alumino-silicates (CMAS) that tend to melt at the high application temperatures. The coatings must also withstand erosion potentially induced by particles passing through the hot section of the engine. Larger particles (>500 μm) can potentially induce impact damage to the coatings, a phenomenon commonly referred to as foreign object damage (FOD).

The various engine operation conditions that the coatings must withstand are schematically summarized in Fig. 5. In addition to the specific requirements highlighted previously, Fig. 5a depicts the severe temperature gradients incurred through the coating thickness. The thermal shock conditions resulting from the cyclic nature of the engine operating cycle are shown schematically in

Table I. Typical Coating Design Requirements for Combustors and Airfoils

Engine Conditions	Combustor liners	Blades and Vanes
Surface Temperature	up to 1482°C (2700°F)	up to 1704°C (3100°F)
Temperature gradient	up to 315°C (600°F)	up to 537°C (1100°F)
Mach	up to 0.3	up to 1.25
Pressure	up to 30 atm	10 atm
Water vapor content	about 10%	about 10%
Others	Corrosion - sea salt - other	•Corrosion •CMAS •Particle erosion •Foreign object damage

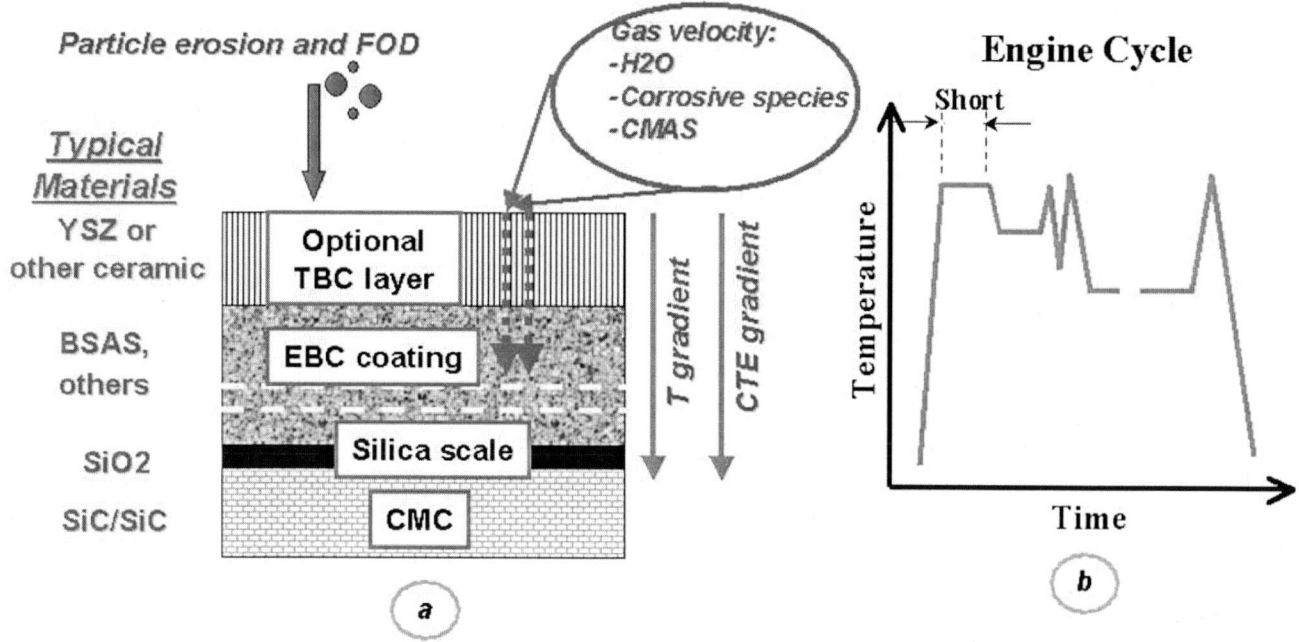

Fig. 5: a) Potential failure mechanisms for T/EBC coatings and b) schematics for typical jet engine temperature cycle.

Fig. 5b. Thus, maintaining the coating mechanical integrity is a significant challenge because a CTE differential between coating layers is inevitable.

Material System Development Challenges

Successful development of a new coating system for SiC-based CMC turbine components depends on the ability to address four major challenges: 1) Establishing the test methodology, or more specifically, demonstrating the performance goals on a laboratory-scale prior to introducing a coating system into the engine; 2) Identifying a material composition with adequate recession resistance; 3) Optimizing the coating deposition process to attain acceptable coating thickness variation; 4) Designing a coating system with the desired thermomechanical capability. These challenges are each discussed further in this section.

Testing Methodology

An effective laboratory testing approach must adequately assess the potential for the coating to survive in the engine. Besides investigating the capability of new coating materials, the testing methodology is complicated by the need to address the requirements for new components or different engines. These advanced engine requirements may exceed the capability of existing test equip-

ment. Consequently, existing test methods must be modified to support implementation of the coating system required for CMC turbine engine components. Table II summarizes currently employed laboratory tests and provides a comparison of the respective test capabilities relative to the engine conditions. Unfortunately, none of the existing tests simulates all of the required conditions. Therefore, a combination of the lab tests must be performed to assess the coating performance.

Selected test methods have been enhanced to enable adequate characterization of the CMC coatings. A state-of-the-art Laser Gradient Rig test has been developed at NASA[15] and adapted to enable the testing of EBCs and T/EBCs. Most recently, this test was upgraded to incorporate a steam environment. Also, a Flame Jet Gradient Test has been developed at GE Aircraft Engines. This test utilizes a mixture of propane and oxygen gases that impinge on the 25-mm disk-shaped specimen. The specimen is cooled from the back side with compressed air, and the temperature on the front and back sides is measured using a pyrometer. This system is capable of attaining surface temperatures up to 1760°C (3200°F) while imparting through-thickness temperature gradients approaching 537°C (1000°F), as well as various thermal shock conditions representative of the engine cycle. The gas velocities are relatively high compared to the Laser Gradient Rig, with the moisture content being about 7-10% at atmospheric pressure. Even though both of these

Table II. Comparison of Conditions of Laboratory Tests with the Engine Conditions.

Comparison of EBC Lab Tests with Engine Conditions						
		Cyclic Furnace		Gradient Rig		High gas velocity
		air	90% steam			
Test Variable:	Engine	FCT	HSCF	Laser (NASA)	Jet	Burner Rig
EBC interface adhesion	√	√	√	√	√	√
Water attack on interface	√	No	√	No	No	No
High gas velocity	√	No	No	No	√?	√
EBC thermal gradient/surf T	√	No	No	√	√	No
High heat/cool rate	√	No	No	No	√	√
Chemical effects (CMAS, salt)	√	No	No	No	No	√

Legend: FCT – furnace cycle test; HSCF – high steam cyclic furnace; Laser – Laser Gradient Test at NASA Glenn Research Center; Jet – Flame Jet Gradient Test at GE Aircraft Engines; Question mark - indicates that this condition is not precisely controlled.

90° set-up

15°set-up

Fig. 6. Burner Rig Test at GE Aircraft Engines.

conditions are not precisely controlled, the test can be sufficient for relative coating screening and ranking in terms of moisture and, in particular, gas velocity sensitivity. Comparative testing of the same T/EBC systems in both the Laser and Flame Rigs typically yields similar results in terms of number of hours "survived" in the test. The Flame test is somewhat more severe due to the radial temperature gradients induced by the localized flame jet as compared to the relatively uniform heating by the laser source in the Laser Rig.

The combined hot corrosion and oxidation test (CHCO), commonly referred to as a Burner Rig test and typically used for evaluation of metallic environmental coatings, has been adapted to accommodate testing of EBCs and T/EBCs. As illustrated in Fig. 6, the engine fuel is burned in the combustor at a specific fuel-to-air ratio and the exhaust gases are directed through the nozzle onto the coating specimen. The specimen orientation relative to the flame may be adjusted to alter the gas impingement angle. The gas temperature and Mach number are controlled in this test. The typical moisture content attained in this test from the jet fuel combustion is about 10%. In addition, a sea salt solution may be injected into the flame jet to enable testing under the hot corrosion conditions. Since the test is at atmospheric pressure, the salt concentration should be adjusted to account for the pressure difference between the test and application conditions. This general testing approach, described elsewhere,[16,17] is routinely used by GE Aircraft Engines for the CHCO testing of metallic environmental coatings for airfoils. Similarly, ceramic particles of selected size and quantities can be injected during the test cycle, in this case into the flame, to enable evaluating the coating response to erosion and impact conditions. This approach is analogous to the testing methodology employed on ceramic TBCs for airfoils.[18] Due to these advantages, the Burner Rig test is actively used as a part of the EBC and T/EBC evaluation by GE Aircraft Engines. The 1315°C (2400°F) temperature limitation should be noted, however.

Recession Issue for Combustors

CMC component surface recession can present an issue for both cooled and uncooled components. Thus, the EBC should be designed to minimize surface recession. Based on the SiC recession model developed by NASA,[8] the Gen 1 BSAS-type coating recession rate is about two orders of magnitude less than the SiC rate

(based on a generally agreed upon assumption for the estimated silica activity in BSAS of approximately 0.01). According to this model, which was developed for the SiC forming SiO_2 scale (with the unit silica activity) for laminar gas flow conditions, the surface recession K, in µm/h, due to silica volatilization by the reaction with water vapor (Eq. 1) is determined by the equation:

$$K \sim 10^{-5654/[T+273]} \times P^{1.5} \times V^{0.5} \times a_{SiO2} \qquad (2)$$

where P is total pressure at 10% H_2O, atm; V is the gas velocity, m/s; a is the activity of silica; and T is the temperature, °C.

The model was validated for gas velocities up to about 0.1 Mach, and the recession rate was shown to be independent of the gas impingement angle. An approximation of the field engine test results on BSAS-coated combustion liners in land-based engines[11] indicates that coating recession was reasonably close to the predicted behavior. An assessment of the BSAS recession rate for the aircraft turbine combustor liner applications is shown in Fig. 7. Total recession exceeding 200 µm is predicted at the higher temperature and pressure conditions. This approach likely underestimates the actual recession rates since these types of conditions are outside of the gas velocities for which the model was developed and validated. Indeed, BSAS surface recession was observed in the atmospheric Burner Rig test at temperatures above 1371°C (2500°F) and Mach 0.3 after only 50 h, even though water vapor content was only on the order of 10%, and the model would not predict a visible recession for these conditions. The initial stages of surface recession include pore formation and pitting, preceding the thickness loss at longer exposures, as opposed to the uniform reduction of the coating thickness. Surface phase changes somewhat similar to those observed in the engine tested liners after thousands of hours of exposure[11] were also observed in

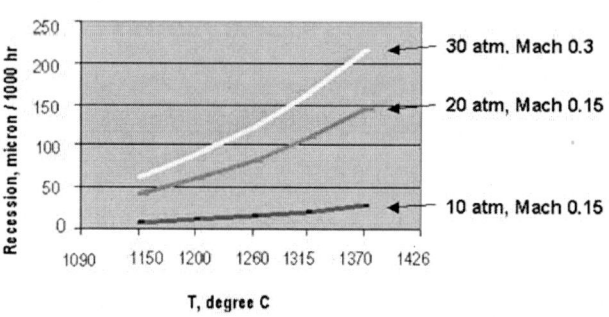

Fig. 7. Estimated recession rates at combustor conditions in accordance with the NASA model.[8]

the Burner Rig test. Additional work is needed to develop a mechanistic understanding of the BSAS recession mechanism and develop models adequately describing this mechanism at a given heat transfer condition. At the same time, it is clear that the coating surface recession for the BSAS-type coatings under the jet engine combustor application conditions can present a significant risk due to imposed variations in the coating thickness.

Coating Thickness Variation

Efforts focused on understanding the implication of coating thickness variation on the performance of a cooled combustor liner should consider the effects on the surface temperature and the CMC component thermomechanical stresses. If the EBC-coated CMC combustor wall is considered as one system under the temperature gradient, a significant portion of this gradient is "taken" by the coating, thus providing the potential to maintain the CMC component stress within the design limits of the CMC strength capability. As shown by schematic finite-element calculation results provided in the upper portion of Fig. 8, the minimum coating thickness requirement is driven by the CMC stress limit. Hence, one may be tempted to satisfy this minimal thickness requirement by increasing the nominal or mean coating thickness target for the as-deposited coating. However, as

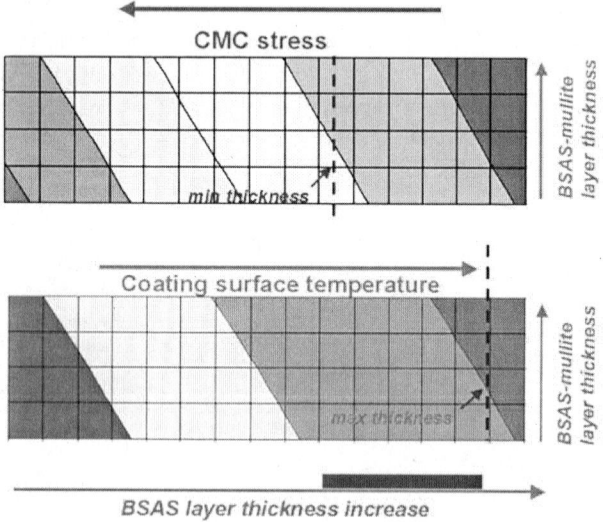

Fig. 8. Schematics of finite element calculations for the CMC stress and coating surface temperature as a function of the coating thickness in commercial combustor liner (blue bar in the bottom on the figure represents acceptable coating thickness range).

shown in the lower portion of Fig. 8, the coating surface temperature capability may be exceeded if the coating thickness is too high. Analysis shows that the acceptable thickness range, as indicated by the bar in the bottom of Fig. 8, may be as narrow as 100 μm. A 100-μm thickness tolerance imposes a strict requirement on the coating deposition process as well as tolerance for the surface recession. The upper thickness limit will be increased if the coating surface temperature capability is improved, and options include modifying the coating chemistry or adding a thermal barrier layer.

Approach for a High-Temperature T/EBC System

As was discussed in the previous sections, the design surface temperature requirements for the CMC components are expected to exceed the BSAS thermal stability limit. Thus, in addition to increasing the thickness range, high-temperature-capability materials with low thermal conductivity are required. At the same time, environmental protection of the CMC substrate from the recession and hot corrosion will still be needed. One approach to "building" a high-temperature T/EBC system would consist of adding a high-temperature ceramic layer, for example, zirconia-based, on top of the three-layer BSAS-type EBC system. A transition layer with an intermediate CTE will be needed to accommodate the CTE mismatch between the BSAS layer (about 5 ppm/°C) and the top layer (about 10 ppm/°C in the case of stabilized zirconia). This approach is schematically shown in Fig. 9. This approach will result in a five-layer system which presents challenges in terms of processing, meeting thickness requirements, and cost. Therefore, new types of materials that could replace the three-layer EBC system, provide the environmental protection, and serve as a bond coat for the top ceramic layer are currently sought.

For the multi-layer T/EBC system, the major material challenges arise in the areas of microstructure and phase stability as well as chemical compatibility of the individual layers, sintering of the ceramic top layer, and mechanical integrity of the system with the CTE gradient while operating under the temperature gradient cyclic thermal shock conditions.

While defining a five-layer T/EBC coating system architecture, the following factors are typically taken into account: temperature capability of the individual coating layers as well as interfaces between the layers, thermal conductivity of each layer, minimal thickness requirement for the environmental barrier, and the maximum total

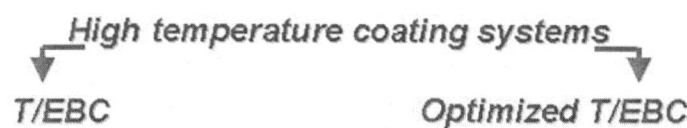

Gen-1 3-layer coating system

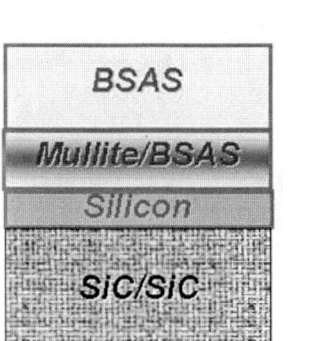

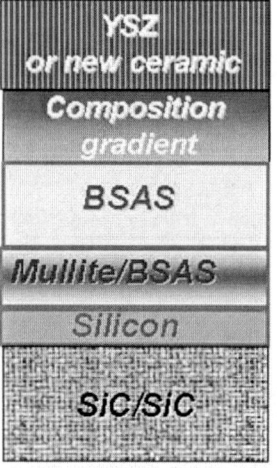

Fig. 9. Schematics of approach to designing high-temperature T/EBC system.

coating thickness requirement from the design. Cyclic exposures to the maximum temperature with the temperature gradient result in a complex stress state. An example of the stress distribution in a five-layer coating system with zirconia-based topcoat is shown in Fig. 10. The calculations are made for the cases of plasma-sprayed and EB-PVD deposited top layers, assuming a stress-free state at the top layer deposition temperature for each case.[19] Sintering and aging phenomena in the layers as well as the coating cracking will result in changes in the mechanical properties of the layers and the stress state. Further modeling efforts are required to adequately describe these changes and develop coating lifing models. Such models are critical for assessment of the coating performance as the coating architecture, processing methods, and the layer materials selection changes evolve as required by the design in various applications of the T/EBC CMC systems.

To date, the feasibility of a five-layer T/EBC coating system with EB-PVD top ceramic layer has been demonstrated for 50 1-h cycles at a surface temperature of 1675°C (3047°F) and a back-side temperature of 1106°C (2023°F) in the NASA Laser Gradient Rig. Coating spallation was not observed during this test, as shown in Fig. 11. Also, the performance has been validated in the Flame Jet Gradient test with similar temperature conditions but under thermal shock: 600 5-min cycles with 20 sec heat-up and cool-down cycle. This coating system has been success-

fully applied on CMC airfoil components for subsequent engine testing.

Upcoming engine tests planned at GE Aircraft Engines will be absolutely critical for correlating conditions of the lab testing to the real engine conditions. With the uncertainty in the recession mechanisms and predictions of the recession rates for the BSAS coatings, engine test results would provide the basis for refining the performance predictions. Engine testing on an EBC-coated

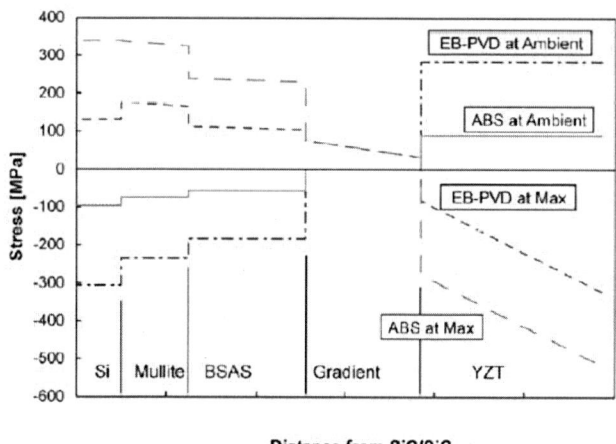

Fig. 10. Example of estimated stress state in a five-layer coating system with air plasma-sprayed and EB-PVD zirconia-based top layer.

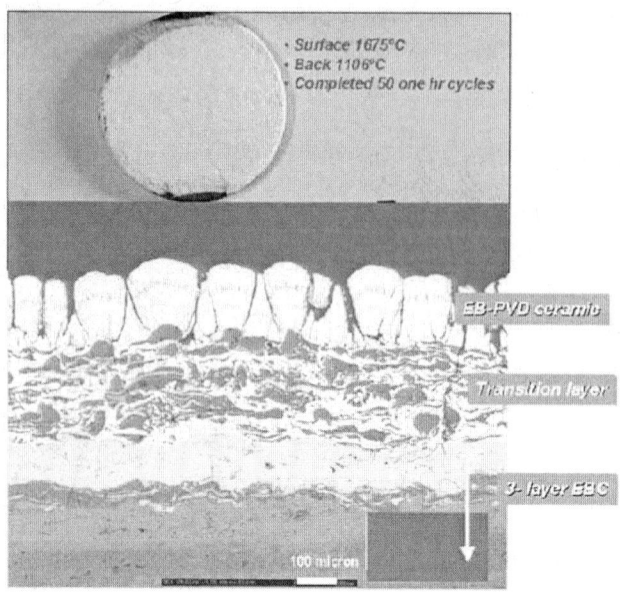

Fig. 11. Five-layer coating after 50 1-h cycles in the Laser Gradient Rig: coating sample view (top) and coating cross-section (bottom).

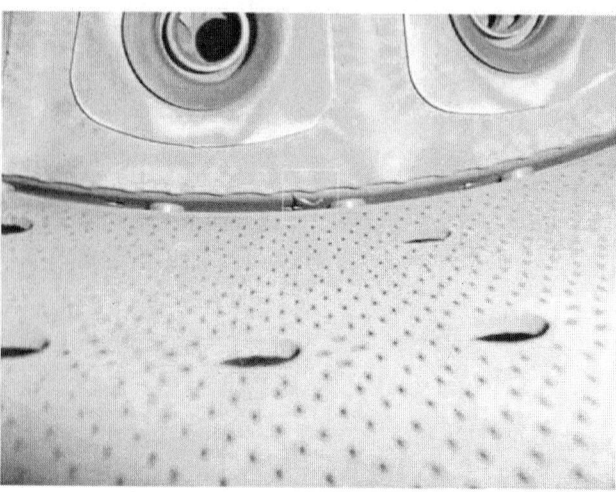

Fig. 12. Section of commercial combustor outer liner in the High-Pressure High-Temperature Combustor Rig after initial hours of testing.

CMC combustor liner is being currently planned at GE Aircraft Engines. The coating for this test was validated in the lab tests, described in the previous sections of the paper, and has undergone relatively short high-temperature, high-pressure Combustor Rig test. A section of the coated outer liner in the Combustor Rig, after initial hours of rig testing, is shown in Fig. 12. The upcoming engine test will provide a better assessment of the Gen 1 coating capability for the jet engine applications as well as of some other aspects of the coating technology.

Summary

Significant progress has been made in developing environmental and thermal barrier coatings for SiC/SiC CMCs over the last 10 years. The development efforts have resulted in three generations of the coatings, with their temperature capability increasing up to 1700°C (3100°F). EBCs for the lower-temperature applications (Gen 1) have been optimized, scaled-up, and demonstrated at the component level. Applications of the Gen 1 coatings are limited to temperatures below 1371°C (2500°F) due to the recession and material temperature stability issues. Efforts toward developing higher-temperature-capability coating systems (Gen 2 EBCs and Gen 3 T/EBCs) are continuing. The feasibility of utilizing these coatings at temperatures up to 1704°C (3100°F) has been demonstrated in the laboratory rig tests, and follow-on engine tests are planned at GE aircraft Engines. Success in maturation of these coatings will depend on improving the test methodology as well as achieving the higher temperature test capability. Establishing an understanding of the coating failure mechanisms and incorporating them into the coating life models is also critical. In addition, identifying new materials with the desired characteristics would enable a reduction in the number of layers for the T/EBC system and make these coatings more practical in terms of both manufacturing and prediction of performance.

Acknowledgments

The authors would like to acknowledge Mark Noe, Brian Hazel, Christine Govern, all of GE Aircraft Engines, and Y.-C. Lau of GE GRC for their contributions to the T/EBC development effort at GE Aircraft Engines; Karren More of Oak Ridge National Laboratory for collaborative work on characterization of the coatings; Dongming Zhu, Bob Draper, Dave Brewer of NASA Glenn Research Center, and Inna Talmy and Jim Zaykovski of Naval Warfare Research Center for their collaborative efforts; Lt. Jeremey Schroeder and Charles Stevens of Air Force Research Lab as well as Steve Fishman and Gill London of Office of Naval Research for support of T/EBC efforts at GE Aircraft Engines.

References

1. D.G. LaChapelle, M.E. Noe, W.G. Edmondson, H.J. Stegemiller, J.D. Steibel, D.R. Chang, "CMC Materials Applications to Gas Turbine Hot section Components", American Institute of Aeronautics and Astronautics, AIAA-98-3266.

2. H. Eaton et al, "Article having silicon-containing substrate and barrier layer and production thereof", US 6,387,456 (2001).

3. K. More, P. Tortorelli, M. Ferber, J. Keiser, "Observations of Accelerated Silicon Carbide Recession by Oxidation at High Water-Vapor Pressures", *J. Amer. Ceram. Soc*, 83 [1], 211-13 (2000).

4. K. More, P. Tortorelli, L. Walker, "Effects of High Water Vapor Pressures on Oxidation of SiC-Based Fiber-Reinforced Composites", *Materials Science Forum*, 362-372 (2001), 385-394.

5. R. C. Robinson, J.L. Smialek, "SiC recession Caused by SiO_2 Scale Volatility under Combustion Conditions:1, Experimental Results and Empirical Model", *J. Amer. Ceram. Soc*, 82 [7] 1817-25 (1999).

6. E. Opilia, R. Hann Jr., "Paralinear Oxidation of CVD SiC in Water Vapor", *J. Amer. Ceram. Soc*, 80 [1], 197-205 (1997).

7. H. Eaton, G. Linsey, K. More, J. Price, J. Kimmel, N. Miriyala, "EBC Protection of SiC/SiC Composites in the Gas Turbine Combustion Environment", presented at the International Gas Turbine & Aeroengine Congress & Exhibition, Munich, Germany – May 8-11, 2000, ASME 2000-GT-631.

8. N. Miriyala, J. Price, "The evaluation of CFCC Liners after Field Engine Testing in Gas Turbine-II", presented at the International Gas Turbine & Aeroengine Congress & Exhibition, Munich, Germany – May 8-11, 2000, ASME 2000-GT-648.

9. K. Lee, "Current Status of Environmental Barrier Coatings for Si-Based ceramics", *Surface and Coatings Technology*, 133-134 (2000) 1-7.

10. D.Mitchell, P. Meschter, K. Luthra, Y.-C.Lau, G. Corman, M. Schroder, K. Bruce, "EBC Development for Ceramic Matrix Composite Turbomachinery", presented at The 27th Annual International Conference on Advanced Ceramics and Composites, Cocoa Beach, Florida January 26-31, 2003.

11. K. More, P. Tortorelli, L. Walker, J. Kimmel, N. Miriyala, J. Price, H. Eaton, E. Sun, G. Linsey, "Evaluation of Environmental Barrier Coatings on Ceramic Matrix Composite Combustor Liners after Engine and Laboratory Exposures", presented at the International Gas Turbine&Aeroengine Congress & Exhibition, Amsterdam – June 3-6, 2002, ASME 2002-GT.

12. I. Spitsberg, H. Wang, "Thermal/Environmental Barrier Coating System for Silicon-Based Materials, US Patent 5,985,470 (1998)

13. D. Zhu, S. Choi, L. Ghosn, R. Miller, "Durability and Design Issues of Thermal/Environmental Barrier Coatings on SiC-based Ceramics under 1650ºC Test Conditions", presented at The 27th Annual International Conference on Advanced Ceramics and Composites, Cocoa Beach, Florida January 29, 2004.

14. D. Zhu, N. Bansal, R. Miller, "Thermal Conductivity and Stability of HfO_2-Y_2O_3 and $La_2Zr_2O_7$ Evaluated for 1650ºC Thermal/Environmental Barrier Coating Applications", Advances in Ceramic Matrix Composites IX, 331-343.

15. D. Zhu, R. Miller, B. Nagaraj, R. Bruce, "Thermal Conductivity of EB-PVD Thermal Barrier Coatings Evaluated by Steady-State Laser Heat Flux Technique", *Surface Science and Technology*, 138 [1] 1-8 (2001).

16. N. Jacobson, C. Sterns, J. Smialek, "Burner Rig Corrosion of SiC at 1000ºC", *Advanced Ceramic Materials*, 1 [2], 154-161 (1986).

17. D. Fox, N. Jacobson, J. Smialek, "Hot Corrosion of Silicon Carbide and Silicon Nitride at 1000ºC", *Ceramic Transactions*, 10, 227-249 (1990).

18. R. Bruce, "Development of 1232ºC (2250ºF)", "Erosion and Impact Tests for Thermal Barrier Coatings", *Tribology Transactions*, 41 [4], 399-410 (1998).

19. A. Karlsson, unpublished work.

International Journal of
Applied Ceramic Technology
Ceramic Product Development and Commercialization

Review on Advanced EB-PVD Ceramic Topcoats for TBC Applications

Uwe Schulz*, Bilge Saruhan*, Klaus Fritscher, Christoph Leyens

German Aerospace Center (DLR), Institute of Materials Research, D-51170 Cologne, Germany

Development of electron beam physical vapor deposited (EB-PVD) thermal barrier coatings (TBC) aims at low conductivity, increased temperature capability, and longer life. Considerable progress has been achieved by comprehensive understanding of the evolvement of the porous microstructure in columnar ceramic topcoats and its application to tailoring optimized microstructures. New ceramic compositions such as alternative stabilizers in zirconia, hafnia modified coatings, and pyrochlores are addressed. They have demonstrated their potential for future TBC applications. New results of both microstructure and chemistry are presented together with a summary of recent research results.

Introduction

TBCs (TBCs) are currently used in the hot gas path of aero-engines and land-based gas turbines. TBCs are utilized in the high-pressure turbine section for lifetime improvement of highly loaded turbine blades and vanes and to increase turbine efficiency. They typically comprise a ceramic topcoating that reduces the metal temperature and smoothens the temperature peaks during the transient stage of turbine operation, and a metallic bond coat for oxidation protection of the underlying superalloy part. There is ample literature available about TBCs, e.g., their manufacture, properties, behavior in service, lifetime prediction, and failure modes. Overview papers describe aspects of the whole coating system, the several interactions between the constituents, and new development trends.[1-4] The key features required for a material to become a TBC are a reasonably low thermal conductivity, sufficiently high temperature capability, and chemical and mechanical compatibility with the underlying layers. In the early days of thermal barrier application, the major research and development focused on feasibility and reliability issues, since coatings tended to spall and thus lose their protective properties. Today, TBCs are well-engineered systems that can survive tens of thousands of hours in service, although there is still tremendous potential for improvement of TBC lifetime, e.g., by bond coat optimization.

It is obvious that any improvement of TBCs aims at three major requirements: lower thermal conductivity, increased temperature capability, and improved lifetime. One more important issue concerns the overall cost of TBCs that are mainly caused by manufacturing but also by usage of TBCs (e.g., on blades for special handling and inspection requirements, quality control, new repair strategies, etc.). Other relevant TBC properties that should be improved are erosion resistance, tolerance to foreign object damage, and resistance to chemical reactions with gas impurities such as sulfur and vanadium, or with de-

* Member, the American Ceramic Society

posits like Calcium-Magnesium-Alumino-Silicate (CMAS).

During the last 5-8 years, many research activities worldwide aimed at improved ceramic topcoats. The current paper combines new developments with a comprehensive summary of major research results for ceramic topcoats only, exclusively manufactured by electron beam physical vapor deposition (EB-PVD). Other papers in the current issue deal with new TBCs made by plasma spraying.[5] Alternative techniques such as chemical vapor deposition or solution precursor plasma spraying, which are under development and have certainly the potential to generate advanced ceramic topcoats, are not included in this paper.

Guidelines for TBC Topcoat Improvement

Although thermal conductivity is the key physical property of a TBC, it had not been a major research and development topic until the early 1990s. Today, increasing TBC insulation capability clearly emerges as a technical and economic challenge for engine manufacturers.[6]

The benefits of low-conductivity coatings are quite obvious. They allow increased engine performance by improving the combustion efficiency (higher turbine entry temperature), reducing specific fuel consumption and internal cooling, lowering metallic component temperatures, and extending lifetimes. If coatings are made thinner, lower parasitic weight and, hence, less centrifugal loads on rotating blades result. Notably, the ceramic external surface temperature increases with decreasing thermal conductivity, partly explained by the fact that the heat flux is kept constant. Increased external surface temperatures may adversely affect the high-temperature stability of the ceramic by enhanced sintering which, in turn, results in thermal conductivity increase of the coating. Thermal conductivity of standard 6-8 wt% partially yttria-stabilized zirconia (P-YSZ) EB-PVD coatings in service is typically 1.8-2.0 W/mK. Reduction of this value to 0.6-1.2 W/mK, which is otherwise observed typically for standard plasma sprayed TBCs, could ensure further benefits from the use of EB-PVD coatings. Note that the low values for APS TBCs will further increase under high pressure and on sintering.

The thermal conductivity of a porous ceramic layer depends on the intrinsic thermal conductivity of the bulk ceramic, which is linked to its composition and structure. The architecture of the porous structure has a predominant effect, mainly due to volume fraction, geom-etry and distribution of the pores. Moreover, in the case of a multi-phase ceramic material (e.g., a multi-layer coating or a phase mixture), thermal boundary resistance and phase topology have to be taken into account.[7] Thus, lowering thermal conductivity of the ceramic layer can be achieved by engineering chemical composition and/or coating microstructure. Because there is valuable information about application of the theory of thermal conductivity to TBCs available that describes, e.g., the contribution of radiation (photon transfer) and lattice vibration (phonon transfer) to heat transfer,[4, 6, 8-13] this part is addressed here only briefly.

Changing the composition of zirconia should lead to a higher disordered crystalline lattice, achievable by:

- Introduction of oxygen vacancies (e.g., by increasing the total amount of stabilizer).
- Partial (or total) substitution of Y and/or Zr ions with ions of different ionic radius and/or different mass (mainly rare-earth metals).

Of course, alternative ceramics (other than zirconia-based) having intrinsically low thermal conductivity are also under development.

To optimize the microstructure of EB-PVD TBCs, one has to understand the origin of the different micro-structural features, their contribution to thermal insulation, and their changes during service. This is not only important for thermal conductivity but also for sintering. Fig. 1 shows the main porosity features of EB-PVD P-YSZ TBCs.

Columns and inter-columnar gaps (denoted type 1 in Fig. 1) originate from vapor phase condensation and operation of macroscopic shadowing caused by the curved columns tips, triggered by rotation of the parts during deposition. Since shadowing occurs primarily along the plane of vapor incidence, columns are significantly wider in the direction parallel to the rotation axis than in the direction perpendicular to it, leading to an anisotropy of the in-plane compliance with notable consequences to the strain tolerance of the TBC system.[14-18]

Globular and elongated spheroid type 2 pores are a consequence of rotation, too. They are arranged in layers inward from the edge to the center of the columns nearly parallel to the substrate surface, more precisely parallel to the individual column tip at that very location during growth. Each layer represents one revolution. Type 2 features are believed to consist mostly of closed porosity. Due to the sunset-sundown situation, bent sub-columns within each layer may be visible, depending on rotational speed, cutting direction with regard to the rotational axis, and

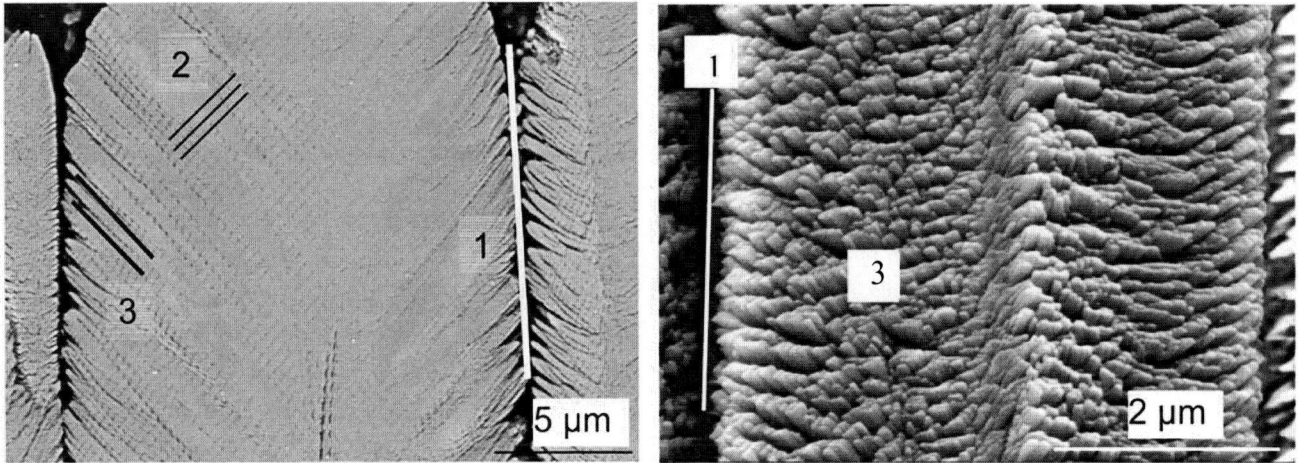

Fig. 1. Morphology of EB-PVD P-YSZ TBCs in polished (left) and fractured cross section (right) in as coated condition.

thickness of the TBC. The last microstructural feature, often referred to as "feather arms" (type 3 in Fig. 1), is a consequence of shadowing by growth steps on the column tips near the center of a column.[19] Since energetic reasons favor {111} planes to build the facets of a column tip while the column tip is curved for macroscopic shadowing reasons, a multitude of growth steps is formed that, during further growth, act as shadowing centers, leading finally to mostly opened porosity aligned under angles between 35-50° (often 45°) towards the main column axis. These feather arms might partly transform into lines of small nanometer-sized globular pores already during deposition, mainly driven by minimization of surface energy, overshadowing effects and gas entrapment. This is found more often in the center area of a column. Recent investigations by Ultra small angle X-Ray diffraction (USAX) and Small angle Neutron diffraction (SANS) indicate that the distribution of most type 2 and all type 3 pores is also highly anisotropic. Intra-columnar pores (type 2) are a combination of globular and elongated spheroids and range between 18-25 nm. Image analyses indicate opening dimensions of 200-250 nm at feathery features (type 3) with a typical aspect ratio of 1 to 10.[20, 21] All three porosity features together cause a large surface area of EB-PVD P-YSZ coatings, typically on the order of 4-6 m²/g.

To lower thermal conductivity, EB-PVD TBCs rely mainly on type 2 and 3 intra-columnar porosity while type 1 inter-columnar porosity is obviously less effective. Important factors regarding intra-columnar porosity are size, distribution, concentration, and morphology. It is well known that the microstructure of EB-PVD TBCs strongly depends on processing parameters such as deposition temperature, rotational speed, chamber pressure, pattern of vapor incidence, shadowing, etc. Hence, microstructure tailoring is feasible, within limits, which might be set by durability, processing cost issues, physical restrictions due to shadowing, etc.

During high-temperature loading of zirconia-based TBCs in service, significant changes to the microstructural features occur quickly. Inter-columnar sintering of type 1 pores increases the Young's modulus of the ceramic by formation of contact points between the columns and thus lead to stress increase in the ceramic topcoat during service.[17, 18, 22] The additional elastic energy stored in the ceramic provides further driving force for crack initiation and propagation and promotes spalling of the coating, thereby reducing the favorable strain tolerance of EB-PVD coatings. This will occur if application temperatures exceed 1200-1300°C, but changes in microstructure can be observed at temperatures as low as 900°C. These changes directly lead to an increase in thermal conductivity. For EB-PVD TBCs, this increase is temperature-dependent, but in the relevant temperature regime, only on the order of up to 20%,[6, 9, 23, 24] which is significantly lower than for plasma-sprayed TBCs. Nevertheless, the lowest possible thermal conductivity is an ultimate goal for TBCs. Sintering includes several aspects:[11, 15, 16, 18, 22, 25]

• Reduction of the internal free surface area mainly by transformation of the feather arms into lines of larger closed pores, creating a rounded smooth surface of the columns.

• Additional growth of both type 2 and type 3 pores in size accompanied by changes in pore fraction.

• Formation of sinter necks between contact points of columns in type 1 regions.

An example of the changes in microstructure is given in Fig. 2. Although the overall density is changed only slightly by these processes, pore surface area is reduced above 900°C with a high rate, indicating changes in pore size and pore distribution which are likely due to formation of large pores at the expense of smaller ones (e.g., <10 nm). Considering only pores larger than 1 μm that can be measured in Computer Micro-Tomography (CMT), porosity decreases drastically at temperatures below 1100°C at the root area of the coating due to the presence of many fine columns while the percentage of pores larger than 1 μm increases at the tip of the coating. If reduction of thermal conductivity is anticipated to primarily rely on microstructural optimization, excellent sintering resistance is required to maintain low conductivity during long-term application.

Phase stability of the ceramic topcoating becomes a concern for advanced TBC systems at surface temperatures of the ceramic above 1200°C. Although results in literature are not fully consistent, most studies show a rapid partitioning of the metastable t′ phase into low yttria portions (tetragonal t) and high yttria portions (tetragonal t″ or cubic).[11, 15, 26] Depending on cooling conditions and probably on the method of sample treatment (free standing or on substrate, metal, or sapphire substrate which rules the stress state in the TBC, textured or grinded after treatment), monoclinic phases may also be present at room temperature. Even no apparent phase transfor-

mation in Raman spectroscopy for P-YSZ TBCs on sapphire substrates after 350 h at 1400°C has been reported.[25] On real blades, especially on concave surfaces, already after 350 h at 1100°C, as well as after engine testing, some monoclinic phase was detected.[27, 28] Although phase stability is not a high-ranking goal for development of new compositions, non-transformable ceramic TBCs are needed. Formation of the monoclinic phase is of serious concern, since phase transformation from t′ to m is associated with a volume change on the order of 4-5 %, which places a significant load on the ceramic coatings, although recent results have shown that considerable amounts of monoclinic phase in an as-processed low-yttria EB-PVD coating can be tolerated.[29]

Since development of advanced TBC topcoats is not always clearly dedicated to one of the three major development goals (low conductivity, increased temperature capability, longer life), the following sections are sorted by chemical groups of materials rather than by development goal. In some parts only rudimentary information is available, mostly because development of new TBCs is a highly competitive area and, therefore, most of the improved TBCs are covered by patents. A comprehensive overview on this matter has recently been published.[4]

Development of Advanced EB-PVD Ceramic Topcoats

Improved Morphologies of EB-PVD P-YSZ TBCs

EB-PVD TBCs deposited at high chamber pressure and low substrate temperature possess a low density Fig. 3, with a microstructure of low-density TBCs characterized by larger gaps between the columns and an increasing column diameter with thickness.[30]

The effect of temperature on density is quite obvious: The lower the substrate temperature the lower is the ability for diffusion of the condensed particles into stable lattice positions, leading to a disturbed and imperfect microstructure having a low density. The correlation between density and thermal conductivity is nearly linear (Fig. 3b), offering an approximately 15% reduction potential by this way of microstructure tailoring. Larger reduction in thermal conductivity and improved cyclic lifetime have been found for TBCs deposited at high chamber pressure of 1.5×10^{-3} mbar by usage of mixtures of oxygen and an inert gas.[31]

Generating inclined columns is by nature an inherent possibility for PVD line-of-sight grown films. The

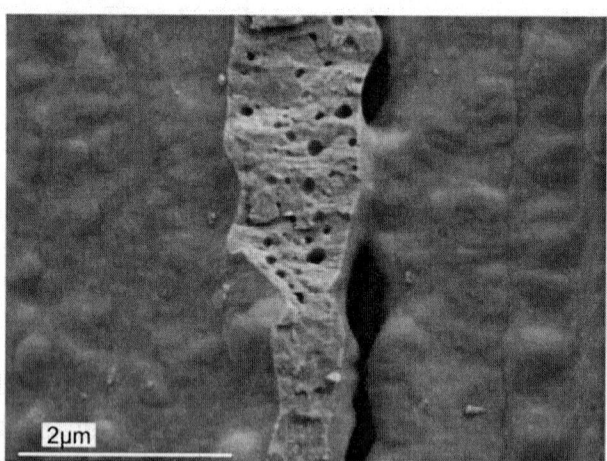

Fig. 2. Microstructure of EB-PVD TBCs after 9 x 100 h annealing at 1100°C.

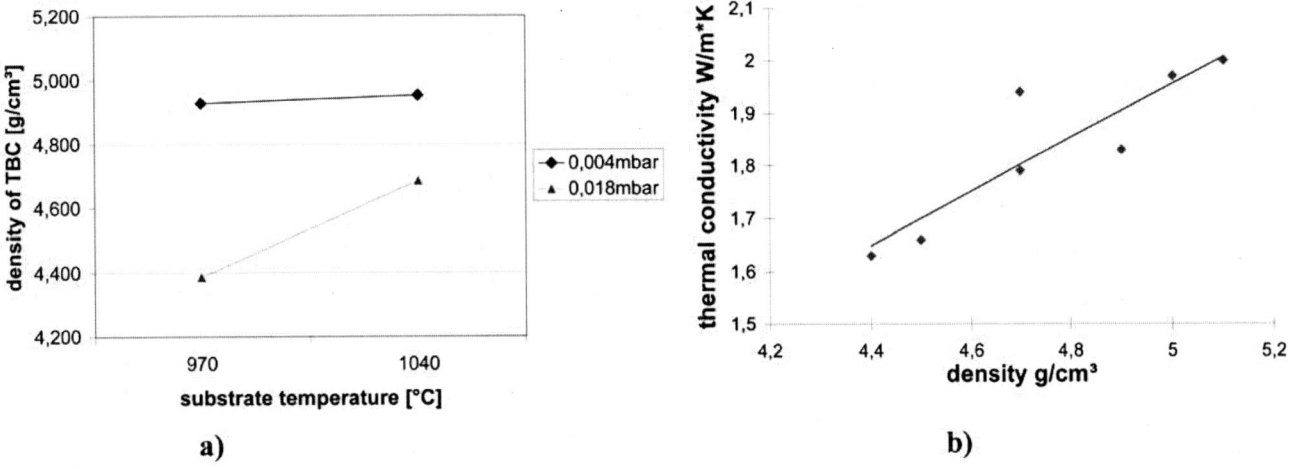

Fig. 3. Density as a function of process parameter (chamber pressure and substrate temperature (a), and thermal conductivity as a function of density (b).

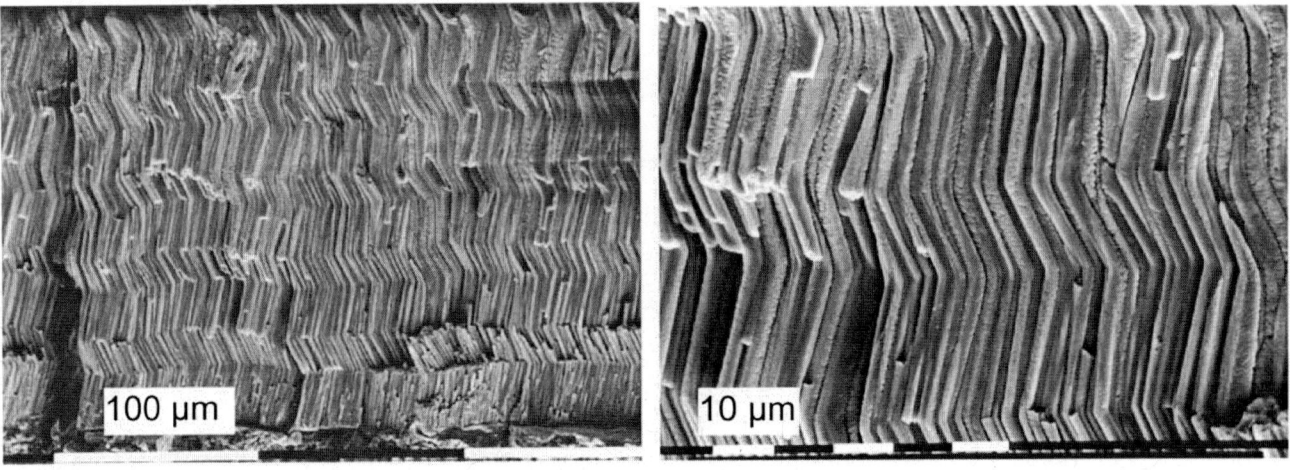

Fig. 4. Zig-zag structured EB-PVD TBC: overview (left) and detail of columns (right).

tilting angle of the columns is somewhat lower than the vapor incidence angle. In most cases, the empirical tan-rule describes the column tilt sufficiently well.[32] Any deviation from a perpendicular vapor incidence on a substrate, for whatever reason, will lead to inclined columns. The deviation may be caused by a tilted rotational axis,[33] a position of the part not directly above the evaporation pool, or in the case of stationary deposition by a tilted substrate.[34, 35] This knowledge was used to manufacture so called "Zig-Zag" or "Herringbone" structures that provide a substantial reduction in thermal conductivity up to 40%.[36] Due to increased porosity at the boundaries between the layers, the reduction is more than what can be calculated by simply increasing the path for heat transfer across the tilted columns, although the number of zig-

zag layers is not important. In Fig. 4, an example for such a microstructure manufactured in a similar way is shown.

The major problem with inclined columnar structures is their sensitivity for erosive destruction: The higher the column tilt, the lower the erosion resistance.[34, 36, 37] Burner rig and cyclic furnace life was shown to be equivalent to standard microstructures. Similar microstructures were obtained by changing both rotational speed and vapor incidence angle, the latter one in a continuous manner.[38] The achievable reduction in thermal conductivity was reported to be close to 20%. The same zig-zag microstructure was also successfully demonstrated for electron beam directed vapor deposition,[39, 40] although some different porosity types and dependence of conductivity from layer architecture was found compared to standard EB-PVD.

The microstructure of EB-PVD TBCs, especially type 2 porosity (see Fig. 1), can be easily changed by varying the rotational speed.[41, 42] Up to now no clear correlation has been presented between rotational speed and thermal conductivity,[34] although one could assume that a higher speed would increase the number of interlayer type 2 pores and should therefore lower conductivity. Generation of a club-like columnar structure by simultaneously modifying the rotational speed and the deposition temperature on EB-PVD processing resulted in a TBC showing exceptionally low Young's modulus and low tendency for sintering.[16] Another effective way is to take advantage of the anisotropy of the columns in the vertical direction. Close to the substrate a large number of small columns exist, most of which diminish with further growth and only a few favored columns become larger. This root area (usually less than 20-30 μm) represents a microstructure with a high number of boundaries in the heat path. Thus, thermal conductivity is significantly lower in this region.[8, 34, 43] Results obtained by laser flash measurements on EB-PVD TBCs having thicknesses of 50-370 μm are given in Fig. 5. The thinnest layer (52 μm) has a 35% lower thermal conductivity than a 350 μm TBC. This thickness dependence of the thermal conductivity of EB-PVD TBCs must be considered whenever thermal isolation properties of new topcoats are compared.

Maintaining a structure similar to the root area over the entire thickness of an EB-PVD coating requires application of interrupted-growth-mode deposition. This can be realized by sequentially inserting and withdrawing the sample carrier in and from the deposition chamber or by shutting the vapor cloud in intervals. Thermal conduc-

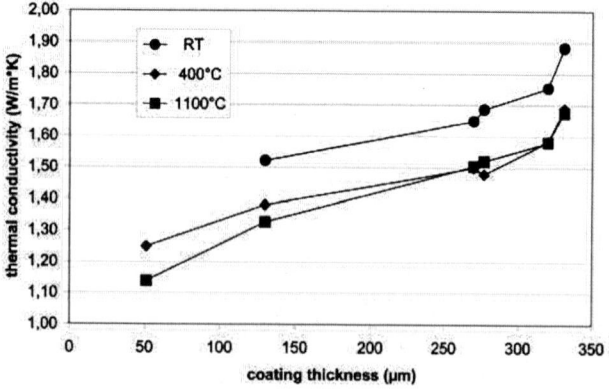

Fig. 5. Thermal conductivity values at different laser flash measurement temperatures of EB-PVD P-YSZ TBCs as a function of thickness. Measurements were done after an initial 2 h/1080°C heat treatment.[87]

tivity of coatings deposited with this mode is only slightly reduced, although a change in microstructure was observed mainly for the former mode which benefits from large temperature fluctuations.[44, 45] This led to a maximum 20% reduction in thermal conductivity. It has been noted that during high-temperature exposure up to 1316°C, the layered porous structure and the accompanied better thermal insulation has been maintained. Although its feasibility was demonstrated, industrial manufacturing might still be an obstacle to overcome for this technology, especially due to the prolonged deposition time, which may be an economic problem. Maintaining the microstructure of the root area in several layers was also achieved by periodically inserting a contaminant atmosphere into the deposition chamber that provides "regermination" of the ceramic layers, thereby providing a reduced thermal conductivity.[46]

A similar approach to provide a microstructural improvement by layering at a much finer scale (i.e., layers of about 1 μm thickness) is to periodically switch on and off a strong BIAS voltage within a plasma atmosphere. This procedure changes the density of the alternating layers and provides up to 37-45% reduction in thermal conductivity at room temperature.[8, 10, 43] Erosion properties are maintained while the long-term stability of any nanostructured coating might be a challenge.

Use of an outer layer covering the type 3 feather-arms introduced by infiltration has been suggested.[47-49] The advantage of the concept would be maintenance of low thermal conductivity and less increase in Young's modulus during thermal loading while the basic composition of the TBC remains unchanged. Feasibility studies carried out by sol-gel infiltration of TBCs with titania, demonstrated that the success of this concept depends on the availability of nanotechnological tools and further work is necessary to exploit the full potential of this method.

Alternative Compositions of EB-PVD TBCs

Basic Considerations for EB-PVD Technology

High-quality, reliable TBCs rely on robust processing. For EB-PVD issues such as melting and evaporation behavior of the source material, vapor pressures, deposition rate and efficiency, EB power, and oxygen addition are important factors.[50] During vapor phase processing most of the compounds decompose into their individual constituents. Therefore, a careful look into the vapor pressures and the evaporation behavior of these sub-oxides is necessary during the search for potential EB-PVD coat-

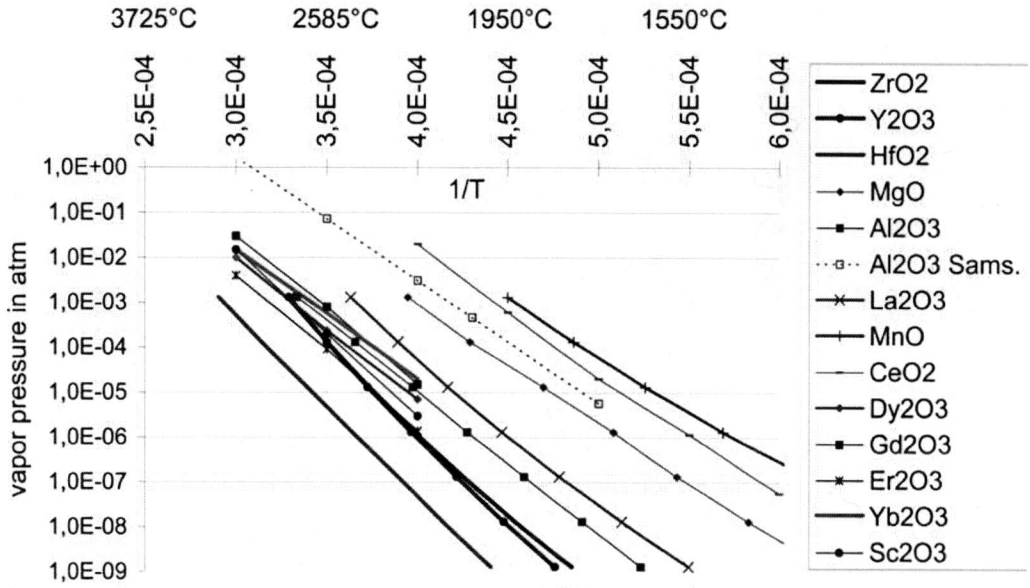

Fig. 6. Vapor pressures of various oxides as a function of inverse temperature.

ing candidates. Fig. 6 summarizes some constituent oxides that are under consideration for improved TBCs. Data were taken from Refs. [51-53].

The vapor pressure of most rare earth oxides is close enough to that of zirconia, hence no major composition problems are expected. Starting with La-oxide and all oxides with higher vapor pressure (oxides of Mg, Al depending on the reference, Ce, and Mn), there exists incompatibility with zirconia/yttria which most probably results in compositional fluctuation across the thickness of the TBC. This is true as long as mixtures of the constituents are evaporated from one single source. In addition, if highly different energy levels are needed for evaporation of two components in a mixture, the same problem is likely to occur.

The way to go is to separate the components with large differences in vapor pressure from each other, e.g., by evaporation of low and high vapor pressure constituents from two different sources. Hence, the desired coating chemistry is achieved by mixing of the species in the vapor phase as well as during coating growth. For industrial production, this fabrication technique brings about additional issues to overcome; one of them might be to obtain chemical uniformity over a large volume in the coater. Oxides that tend to sublime, such as ceria, magnesia, and silica, necessitate additional processing efforts. In any case, the quality of the ingot material in terms of homogeneity and spitting-free evaporation is one of the most critical issues for non-standard compositions.

Zirconia-Based Compositions

The easiest approach is to increase the amount of yttria within YSZ. The cubic phase in fully yttria-stabilized zirconia (F-YSZ) is in equilibrium and therefore stable against phase transformation. Thermal conductivity is approximately 10-30% lower than for P-YSZ, but sintering and loss of type 3 porosity (feather-arms) goes very fast. This may be partly caused by an initially much larger internal free surface area. Shorter lifetimes have been reported in both burner rig and cyclic furnace tests for F-YSZ EB-PVD TBCs and, more important, erosion resistance is very poor.[11, 15, 54-57] Surprisingly, reduced yttria contents below 7 wt% are recently reported to improve cyclic lifetime of EB-PVD TBCs.[29]

Ceria-stabilized zirconia (CeSZ) is considered as a potential EB-PVD candidate material, providing good corrosion resistance and superior phase stability at high temperature. Furthermore, the thermal conductivity is found to be lower than for P-YSZ, and benefits for lifetime and thermocyclic resistance are reported.[54, 56-61] As the vapor pressures of zirconia and ceria differ considerably, the evaporation from one source containing both zirconia and ceria turns out to be critical. Two-source evaporation was identified as offering a possibility to overcome the problem, but the sublimation of ceria in vacuum imparts new challenges for process control, especially if constant evaporation and condensation rates are mandatory. Fig. 7 summarizes improvements of CeSZ coatings,

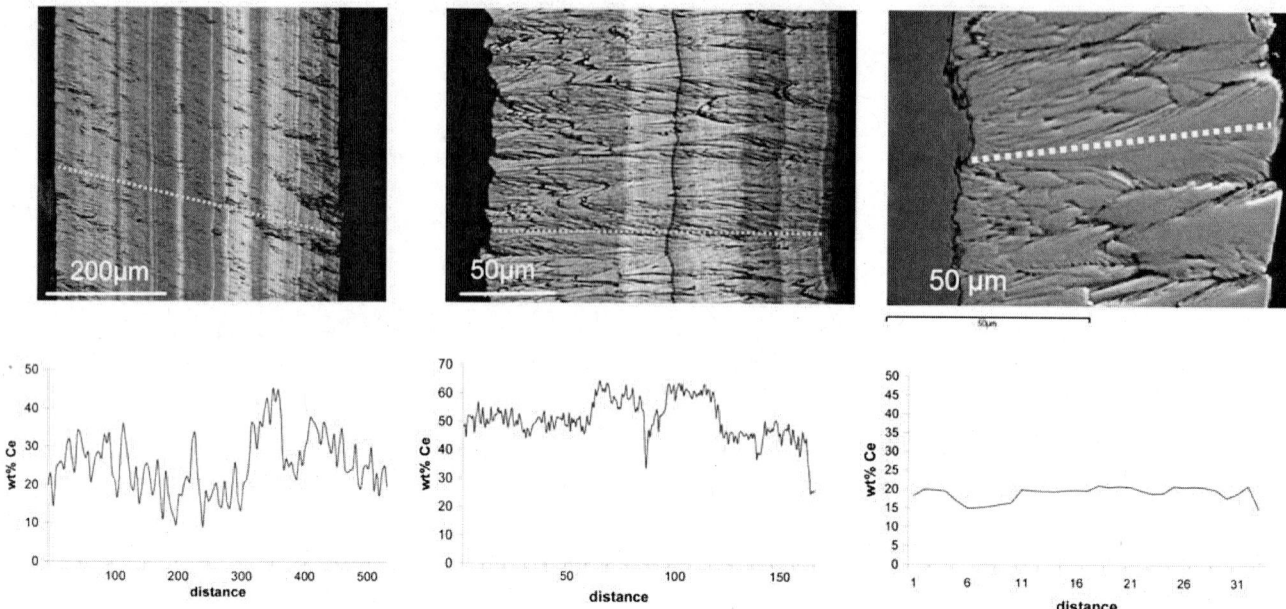

Fig. 7. *Ceria-stabilized EB-PVD TBCs in cross section with corresponding EDS line scans: single source (left), dual source (middle), improved dual source (right).*

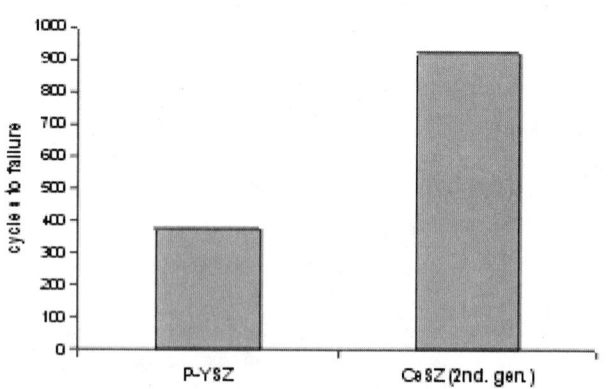

Fig. 8. *Cyclic furnace life at 1100°C (50 min heating/10 min cooling) for CMSX-4 bars with NiCoCrAlY bond coat.*

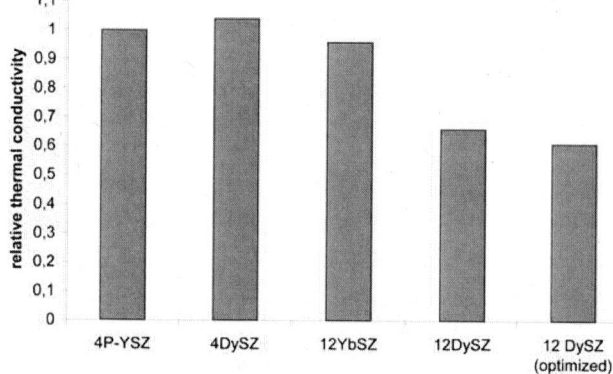

Fig. 9. *Relative thermal conductivity of alternative stabilized zirconia TBCs at 1000°C, measured after a 2 h/1080°C stabilizing treatment.*

mainly achieved by dual-source evaporation and enhanced pool control, beam pattern, and crucible design.

Reduced porosity of the thermally grown oxide in CeSZ TBC systems provides improved lifetime, as demonstrated for CMSX-4 substrates with NiCoCrAlY bond coats in cyclic furnace test (Fig. 8). By means of dual-source evaporation and rotation of samples within two overlapping vapor clouds, micro-layers with slightly changing Ce content are formed. These layers tend to equalize in composition during high-temperature exposure. Similar problems related to vapor pressure differences are re-

ported for ceria-based compositions (e.g., yttria-ceria). They were overcome by employing a special evaporation crucible design that separates the two oxide melt baths from each other,[62, 63] thereby providing real two-source evaporation conditions from sources that are close to each other and subsequent vapor mixing as described above. Thermal conductivity of yttria-ceria TBCs can be as low as 50% of the base line with only 15% increase in density.

Other rare earth oxide stabilizers such as dysprosia (DySZ) and ytterbia (YbSZ) behave similar to yttria: 4 mol% addition creates a metastable tetragonal phase while

12 mol% addition creates a stable cubic lattice. As shown in Fig. 9, a reduction of up to 40% in thermal conductivity was achieved with an optimized version of 12 mol% DySZ.[64, 65] The cyclic lifetime of these new EB-PVD TBCs was comparable to the standard P-YSZ version.

For scandia and scandia-yttria stabilized EB-PVD coatings, higher phase stability, excellent resistance to vanadate hot corrosion, longer oxidation lifetime, and easy manufacturing was demonstrated.[66, 67] This material has a special potential for applications with dirty fuels used in ship diesels or oil-fired turbines, but a drawback might be the high cost of the raw material. Feasibility of EB-PVD processing for zirconia stabilized with oxides of La,[57] Mg, and Gd[68] has been demonstrated as well. Zirconia, having 19 mol% Sm_2O_3, was reported to have a thermal conductivity of 1.26 W/mK at 760°C, equal to around 30% reduction compared to the base line.[69] Gadolinia-zirconia (which may contain small additions of yttria) with compositions that do not form the pyrochlore structure but instead a cubic fluorite lattice were found to yield a 50% reduction in thermal conductivity as well as reduced sintering rates. Exact compositions are not published.[70] Many more zirconia-based compositions are covered by patents but without giving real data on EB-PVD processed coatings or properties.

A promising concept is the use of binary or ternary oxides as addition to zirconia. La_2O_3 additions to 7 P-YSZ in the range of 1-5 mol% decreased the thermal conductivity at room temperature up to 63%, provided a low density, and lowered sintering.[71] Thus, the increase in thermal conductivity after annealing at 1200°C for 50 h was much lower than that of the standard material.

Multi-component, paired-cluster oxide dopants in zirconia[72] with tailored defect clusters are found to lower thermal conductivity in high-temperature laser rig testing up to 55%. Sintering was substantially reduced since the defects were thermodynamically stable. This concept of "a priori" stable defects that are not only frozen from high temperatures is very promising to retard any high-temperature activated processes such as diffusion, phase separation, or reduction of internal surface area. A minimum in thermal conductivity was reported in the range of 10 mol% total stabilizer concentration with a preference to have two additional oxides in the ratio 1/1 added to 7 P-YSZ. Binary oxide additions that have been successfully tested comprise additions of Nd/Yb, Gd/Yb, and Sm/Yb. Ternary additions of Gd/Yb/Sc and Nd/Yb/Sc did not further improve the thermal isolation properties. Cyclic furnace life was decreased with increasing stabilizer content, but some potential exists to optimize the multi-component oxide TBCs compared to only yttria-stabilized zirconia.[72]

A similar approach by adding oxides of Nd, Gd, Er, Yb, or Ni to YSZ demonstrated reductions in room-temperature thermal conductivity up to 46% when 8 mol% dopant was used.[10, 73] The effect was explained to be due to ions of differing ion radius (see introduction section) and by additional coloring of the TBCs, thereby reducing the radiation heat transport through the material.

Hafnia-Containing Compositions

Hafnia-containing TBCs are easy to manufacture by EB-PVD. The crystal lattice of zirconia and hafnia is isomorphous; a complete solubility exists. Although the vapor pressure of zirconia is slightly higher than that of hafnia (see Fig. 6), no major problems have been found in terms of composition up to now. Fig. 10 shows the effect of 5-10 mol% hafnia additions to the standard 7 P-

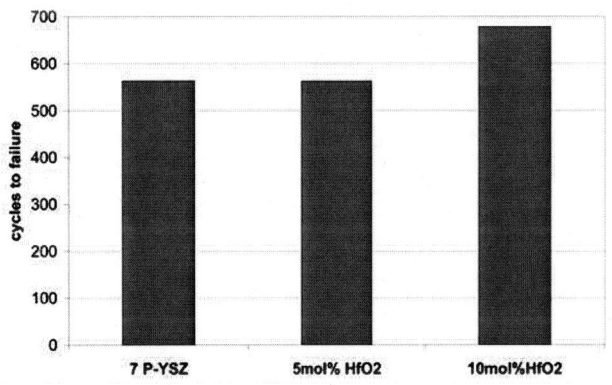

Fig. 10. Cyclic furnace life at 1100°C (50 min heating/10 min cooling) for IN100 substrates with NiCoCrAlY bond coats.

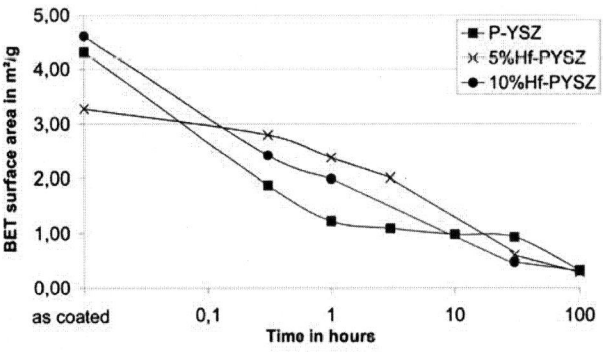

Fig. 11. BET values after 1100°C annealing for EB-PVD TBCs with varying hafnia content on alumina substrates.

YSZ material on cyclic lifetime at 1100°C. No significant influence of the hafnia content was found, although the average lifetime of the 10 mol% hafnia samples was slightly higher. The scatter in lifetime data was remarkably lower than for P-YSZ. In any case, the reference material always contained some 2 wt% hafnia as a natural impurity and never showed any detrimental effects.

Although gas adsorption measurements (BET) refer to some differences in the initial internal free surface area that were not visible in SEM pictures, and the surface recession rate is slower in the first stage for 5%Hf-PYSZ, the total reduction of surface area after annealing for 100 h at 1100°C is nearly the same for all three versions (Fig. 11). The final values are all around 0.3 m²/g. This clearly shows that diffusion-driven rearrangement of pores and sintering is not much influenced by small additions of hafnia. Thermal conductivity is 1.5 W/mK at 800°C for 5 mol% hafnia which is at the lower end of the typical range for the 7 P-YSZ standard TBC.

Larger additions of hafnia, e.g., 40 wt% zirconia + 40 wt% hafnia + 20 wt% yttria, reduce thermal conductivity further, but the largest effect of 30% reduction at high temperature was reported for zirconia-free 27 wt% yttria-stabilized hafnia.[45] The latter one showed a much denser and fine columnar microstructure and was less susceptible to sintering. Similar favorable lower shrinkage rates have been found for EB-PVD 7.5 wt% yttria-hafnia that were not rotated during deposition.[74] Our own experiments with 32 wt% yttria-hafnia (FYSHf) TBCs showed similar results. Evaporation from one source was possible without notable problems with nearly the same

composition in both ingot and TBC. The microstructure seemed to be slightly denser in terms of type 1 pores (Fig. 12), but the overall opened porosity is much higher than for the tetragonal standard 7 P-YSZ material (Fig. 13). Number and size of type 3 feather arm pores is obviously much higher, as can be seen in Fig. 12.

The feather-arms tend to be more frequent and reach farther down towards the column center. This is typical for most cubic phases we have investigated so far. A possible reason might be a larger number of initial growth steps on the surface of the column tips that would result in more type 3 pores.

The surface diffusion might be also different in the tetragonal and cubic phase during deposition, resulting in different internal free surface areas. While the fully yttria-stabilized zirconia that is included for comparison shows a dramatic reduction in internal free surface during annealing, and hence suffers from enhanced sintering, fully yttria-stabilized hafnia coatings have a lower rate of surface area decrease (Fig. 13). Thus, an increased sintering resistance can be attributed to this material. Phase stability is excellent, proved by XRD measurements after 100 h annealing at 1200-1400°C (Fig. 14). The only detected small change was the transformation of a notable amount of metastable tetragonal phase in the as-coated stage, indicated by the presence of the {112} peak into cubic phase during annealing. Grain growth can be deduced from peak sharpening with increasing annealing temperature.

The FYSHf possesses only 66% of the theoretical density for this composition while P-YSZ has around 80%

Fig. 12. Microstructure of EB-PVD yttria-hafnia (FYSHf) TBCs in fractured cross section (left) and column tips (right).

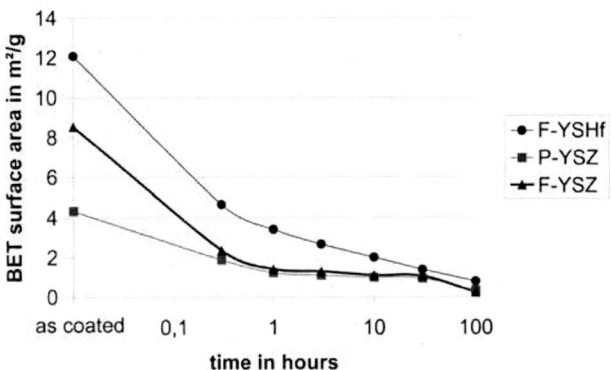

Fig. 13. *BET values after 1100°C annealing for EB-PVD cubic TBCs in comparison to standard 7 P-YSZ.*

of the theoretical density. Due to the higher porosity, the heavy element hafnium did increase the total density by only 15% compared to the standard material (5.5 g/cm³ compared to 4.8 g/cm³). This is a clear advantage for this TBC version if the low density is maintained during service.

Pyrochlores and Other Approaches

Rare-earth zirconates (e.g., $La_2Zr_2O_7$) have reasonable potential for TBC application. Reduced thermal con-

ductivity as well as improved sinter resistance have been found for EB-PVD pyrochlore TBCs, especially for $Gd_2Zr_2O_7$[70] and $Sm_2Zr_2O_7$.[75] Although not easily manufactured and some fluctuations in composition are encountered, a recent study has shown that doping $La_2Zr_2O_7$ with yttria in the range of 3-10 wt% reduces the compositional scatter during evaporation.[76, 77] Morphologically $La_2Zr_2O_7$-based coatings are similar to the cubic structured EB-PVD TBCs. The feather-arms are coarser and extend to the center of the columns which appear to grow tangled with each other. Although the microstructure changed significantly after 100 h of annealing at 1300°C, thermal conductivity remained low at about 1.4 W/mK.[78] Initial thermal conductivity was measured as low as 0.5 W/mK.

More exotic compositions such as $ZrSiO_4$,[59] doped La-hexa-aluminate with magnetoplumbite structure,[79] or $LaPO_4$,[80] are investigated for EB-PVD as well, but those are beyond the scope of this paper.

A combination of the morphology approach by micro-layering with new compositions yields multi-layers of different chemistry. Alternating layers of, e.g., zirconia and alumina,[81-83] P-YSZ and yttria-stabilized hafnia,[84] P-YSZ and yttria-ceria or P-YSZ and gadolinia-stabilized zirconia,[85, 86] P-YSZ and dysprosia-stabilized zirconia[59] have been investigated. With these systems, there are at

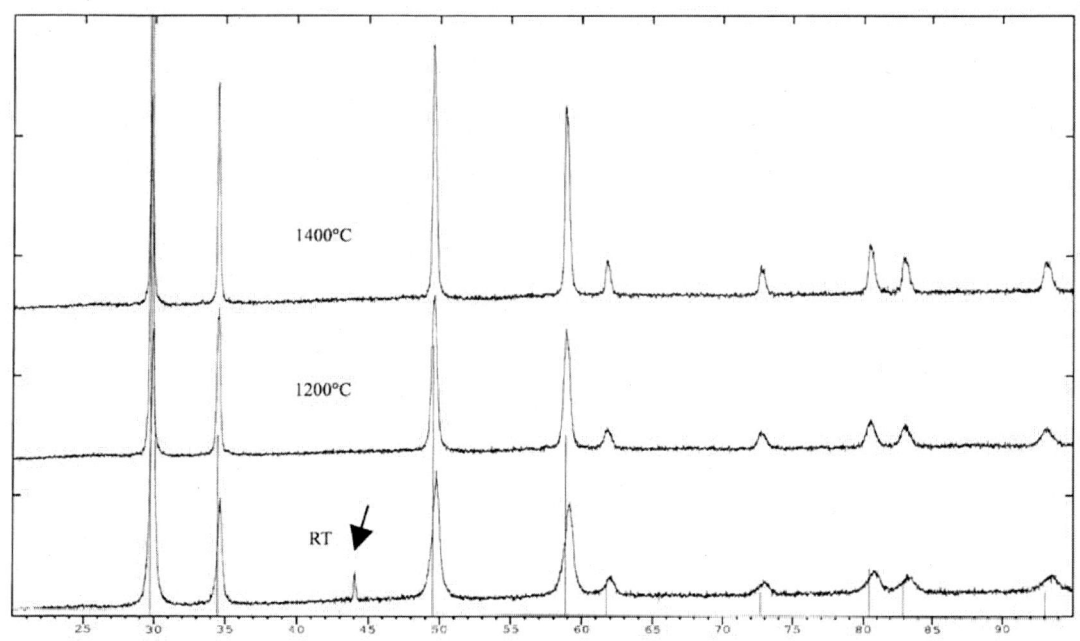

Fig. 14. *X-ray diffraction (Cu-K$_a$ radiation) of powdered EB-PVD FYSHf TBCs as coated and after 100 h annealing. The arrow indicates the {112} peak.*

least two factors contributing to a potential reduced thermal conductivity: First, thin films have been reported to exhibit lower thermal conductivities than the respective bulk materials, and second, additional boundaries are incorporated perpendicularly aligned to the heat flux. Attractive low thermal conductivities have been found in some cases; especially the long time stability of thin multi-layers at high temperatures remains to be proven. Even a combination of a thick EB-PVD gadolinia-zirconia layer as the strain tolerant "bond layer" with a plasma-sprayed low thermal conductivity layer on top has been investigated.[87]

Outlook

Many of the newly developed EB-PVD TBCs have been introduced into service or ground testing only with an underlying standard P-YSZ ceramic layer so far, partly because of chemical incompatibility with the alumina thermally grown oxide as found especially for pyrochlores,[4] and partly because the formation of the thermally grown oxide was altered and reduced lifetimes were observed for bare new compositions. Such double layers have been explicitly reported for topcoats of ceria-yttria, gadolinia-zirconia, dysprosia-stabilized zirconia, and Sm-zirconate. The accompanied increase in cost and complexity of the TBCs are a major disadvantage for those new materials. On the other hand, if the benefits of thermal conductivity or high-temperature capability outweigh the costs, those new TBCs may enter into production soon. However, most of these systems have not yet matured to commercial application due to the problems mentioned above.

Conclusions

Development of EB-PVD TBCs aims at low conductivity, increased temperature capability, and longer life. Research was guided into these directions by requirements of enhanced reliability, improved efficiency, and increased gas temperatures. Improvements have been achieved by modifying both microstructure and chemistry of the TBCs, resulting roughly in a 50% reduction in thermal conductivity and approximately 100-150 K temperature capability increase in the best scenarios. Customized multi-layers with multi-functionality seem to provide further progress, although a balance between all loads placed on the coatings and cost issues must be found.

Acknowledgment

The authors highly appreciate discussions with and contributions of their colleagues H.-J. Rätzer-Scheibe and M. Peters, as well as assistance given by J. Brien, C. Kröder, H. Mangers, D. Peters, A. Miksic, and A. Flores.

References

1. N.P. Padture, M. Gell, and E.H. Jordan, "TBCs for Gas-Turbine Engine Applications," *Science*, 296 [5566] 280-284 (2002).
2. C. Leyens, U. Schulz, and M. Peters, "Advanced TBC Systems: Research and Development Trends," pp 61-76 in *High Temperature Coatings-Science and Technology IV*, ed. N.B. Dahotre, J.M. Hampikian and J. Morral, TMS, Warrendale, PA, 2001.
3. C. Leyens, U. Schulz, K. Fritscher, *et al.*, "Contemporary Materials Issues for Advanced EB-PVD TBC Systems," *Z. für Metallkunde*, 92 762-772 (2001).
4. C.G. Levi, "Emerging Materials and Processes for Thermal Barrier Systems," *Current Opinion in Solid State and Materials Science*, 8 [1] 77-91 (2004).
5. R. Vassen, *Applied Ceramic Technology*, this issue, 2004.
6. S. Alperine, M. Derrien, Y. Yaslier, and R. Mevrel, "TBCs: The Thermal Conductivity Challenge," AGARD report 823, 1-1/1-10, 1998.
7. U. Schulz, C. Leyens, K. Fritscher, *et al.*, "Some Recent Trends in Research and Technology Of Advanced TBCs," *Aerospace Science and Technology*, 7 [1] 73-80 (2003).
8. J.R. Nicholls, K.L. Lawson, D.S. Rickerby, and P. Morell, "Advanced Processing of TBC's for Reduced Thermal Conductivity," AGARD report 823 "TBCs," Aalborg, Denmark: 6-1 to 6-9, 1998.
9. V. Arnault, R. Mevrel, S. Alperine, and Y. Jaslier, "TBCs for Aircraft Turbine Materials: Thermal Challenge and Materials," *La Revue de Metallurgie-CIT*, May 585-597, 1999.
10. J.R. Nicholls, K.J. Lawson, A. Johnstone, and D.S. Rickerby, "Methods to Reduce the Thermal Conductivity of EB-PVD TBCs," *Surface and Coatings Technology*, 151-152 383-391 (2002).
11. A. Azzopardi, R. Mevrel, B. Saint-Ramond, E. Olson, and K. Stiller, "Influence of Aging on Structure and Thermal Conductivity of Y-PSZ and Y-FSZ EB-PVD Coatings," *Surface and Coatings Technology*, 177-178 131-139 (2004).
12. B. Leclercq and R. Mevrel, "Thermal Conductivity of Zirconia-Based Ceramics for TBCs," CIMTEC 2002, Firenze, Italy, 365-372, 2002.
13. D.R. Clarke, "Materials Selection Guidelines for Low Thermal Conductivity TBCs," *Surface and Coatings Technology*, 163-164 67-74 (2003).
14. S.G. Terry and C.G. Levi, "The Evolution of Porosity in Vapor-Grown TBCs," *J. American Ceramic Society*, accepted 2003.
15. U. Schulz, K. Fritscher, C. Leyens, and M. Peters, "High-Temperature Aging of EB-PVD TBCs," *Ceramic Engineering and Science Proceedings*, 22 [4] 347-356 (2001).
16. K. Fritscher, F. Szücs, U. Schulz, *et al.*, "Impact of Thermal Exposure of EB-PVD TBCs on Young`s Modulus and Sintering," *Ceramic Engineering and Science Proceedings*, 23-4 341-352 (2002).
17. C.A. Johnson, J.A. Ruud, R. Bruce, and D. Wortman, "Relationships Between Residual Stress, Microstructure and Mechanical Properties of Electron Beam-Physical Vapor Deposition TBCs," *Surface and Coatings Technology*, 108-109 80-85 (1998).
18. U. Schulz, K. Fritscher, C. Leyens, and M. Peters, "Influence of Processing on Microstructure and Performance of Electron Beam Physical Vapor Deposition (EB-PVD) TBCs," *J. Engineering for Gas Turbines and Power*, 124 [2] 229-234 (2002).
19. S. Terry, "Evolution of Microstructure During the Growth of TBCs by Electron-Beam Physical Vapor Deposition", Ph.D. thesis, Materials Department, University of California, 2001.
20. A.A. Kulkarni, H. Herman, J. Almer, *et al.*, "Depth-Resolved Porosity Investigation of EB-PVD TBCs Using High-Energy X-rays," *J. Amer. Ceramic Soc.*, 87 [2] 268-274 (2004).

21. T.A. Dobbins, A.J. Allen, J. Ilavsky, *et al.*, "Recent Developments in the Characterization of Anisotropic Void Populations in TBCs Using Ultra-Small Angle X-Ray Scattering," *Ceram. Eng. and Sci. Proc.*, 24 [3] 517-521 (2003).
22. K. Fritscher, U. Schulz, C. Leyens, and M.Peters, "Aspects on Sintering of EB-PVD TBCs.," 7th International Symp. on Ceramic Materials and Components for Engines, ed. J.G. Heinrich and F. Aldinger, Goslar: Wiley-VCH, 517-5, 2000.
23. D. Zhu, R.A. Miller, B.A. Nagaraj, and R.W. Bruce, "Thermal Conductivity of EB-PVD TBCs Evaluated by a Steady-State Laser Heat Flux Technique," *Surface and Coatings Technology*, 138 [1] 1-8 (2001).
24. R.B. Dinwiddie, S.C. Beecher, and W.D. Porter, "The Effect of Thermal Aging on the Thermal Conductivity of Plasma Sprayed and EB-PVD TBCs," *ASME*, 96-GT-282 (1996).
25. V. Lughi, V.K. Tolpygo, and D.R. Clarke, "Microstructural Aspects of the Sintering of TBCs," *Materials Science and Engineering A*, 368 [1-2] 212-221 (2004).
26. U. Schulz, "Phase Transformation in EB-PVD Yttria Partially Stabilized Zirconia TBCs During Annealing," *Journal American Ceramic Society*, 83 [4] 904-910 (2000).
27. Y.H. Sohn, R.R. Biederman, and R.D. Sisson Jr, "Isothermal Oxidation of Physical Vapor Deposited Partially Stabilized Zirconia TBCs," *J. Materials Eng. and Performance*, 3 [1] 55-60 (1994).
28. Y.H. Sohn, K. Cho, E.Y. Lee, R.R. Biederman, and R.D. Sisson Jr, "Phase Analysis of Physical Vapor Deposited ZrO_2-8wt%Y_2O_3 TBCs," *Materials for Advanced Power Engineering PartII*, ed. K.A.P. D.Coutsouradis, *et al.*, 1345-1356, 1994.
29. R.W. Bruce, D.J. Wortman, R. Viguie, and D. Skelly, 1999, "TBC System," *US patent No. 5,981,088.*
30. U. Schulz, J. Münzer, and U. Kaden, "Influence of Deposition Conditions on Density and Microstructure of EB-PVD TBCs," *Ceramic Engineering and Science Proceedings*, 23-4 353-360 (2002).
31. D.V. Rigney, A.F. Maricocchi, D.J. Wortmann, R.W. Bruce, and J.D. Rigny, 2002, "Method of Forming a TBC System," *U.S. patent No. 6447854.*
32. J.M. Nieuwenhuizen and H.B. Haanstra, Philips Techn. Rev. 27, 87 (1966).
33. K. Fritscher and W. Bunk, "Density Graded TBCs Processed by EB-PVD," The First International Symposium on Functionally Gradient Materials, FGM 90, ed. M. Yamanouchi, *et al.*, Sendai, Japan: Functionally Gradient Materials Forum, 91-96, 1990.
34. K.J. Lawson, J.R. Nicholls, and D.S. Rickerby, "Thermal Conductivity and Ceramic Microstructure," *High Temperature Surface Engineering*, ed. D.R. J. Nicholls, D.Allen, Edinburgh: The Institute of Materials, London, UK, P8, 1997.
35. D.L. Youchison, M.A. Gallis, R.E. Nygren, J.M. McDonald, and T.J. Lutz, "Effects of Ion Beam Assisted Deposition, Beam Sharing and Pivoting in EB-PVD Processing of Graded TBCs," *Surface and Coatings Technology*, 177-178 158-164 (2004).
36. W. Beele, G. Marijnissen, E. Vergeldt, *et al.*, "Evaluation of Low Thermal Conductivity Ceramic TBCs for Aero-Engine Turbine Blades by the Herringbone Process," *Advanced Coatings for High Temperatures*, Nice, Forum of Technology, 1-7, 2002.
37. M. Kolloos and G. Marijnissen, "Burner Rig Testing of "Herringbone" EB-PVD TBCs," Turbomat, Bonn, Germany: DLR, German Aerospace Center, 18-21, 2002.
38. D.S. Rickerby, 2001, "Metallic Article Having a TBC and Method of Application Thereof," *U.S. patent No. 6183884.*
39. D.D. Hass, A.J. Slifka, and H.N.G. Wadley, "Low Thermal Conductivity Vapor Deposited Zirconia Microstructures," *Acta Materialia*, 49 [6] 973-983 (2001).
40. S. Gu, T.J. Lu, D.D. Hass, and H.N.G. Wadley, "Thermal Conductivity of Zirconia Coatings with Zig-Zag Pore Microstructures," *Acta Materialia*, 49 [13] 2539-2547 (2001).
41. D.V. Rigney, R. Viguie, D.J. Wortman, and D.W. Skelly, "PVD TBC Applications and Process Development for Aircraft Engines," NASA conference 3312, Cleveland: 135-149, 1995.
42. U. Schulz, K. Fritscher, H.J. Rätzer-Scheibe, W.A. Kaysser, and M. Peters, "Thermocyclic Behaviour of Microstructurally Modified EB-PVD TBCs," *Materials Science Forum*, 957-964 (1997).
43. J.R. Nicholls, K.J. Lawson, A. Johnstone, and D. Rickerby, "Low Thermal Conductivity EB-PVD TBCs," *Materials Science Forum*, 369-372 595-606 (2001).
44. J. Singh, D.E. Wolfe, and J. Singh, "Architecture of TBCs Produced by Electron Beam-Physical Vapor Deposition (EB-PVD)," *Journal of Materials Science*, 37 [15] 3261-326 (2002).
45. J. Singh, D.E. Wolfe, R.A. Miller, J.I. Eldridge, and D.-M. Zhu, "Tailored Microstructure of Zirconia and Hafnia-Based TBCs with Low Thermal Conductivity and High Hemispherical Reflectance by EB-PVD," *Journal of Materials Science*, 39 [6] 1975-1985 (2004).
46. Y.P. Jaslier, H.L. Malie, J.-P.J.C. Huchin, S.A. Alperine, and R. Portal, 2001, "Ceramic Heat Barrier Coating Having Low Thermal Conductivity, and Process for the Deposition of Said Coating," U.S. 6251504, *U.S. patent No. 6432478.*
47. R. Subramanian and S.M. Sabol, 2001, "TBC Resistant to Sintering," *U.S. patent No. 6,203,927 B1.*
48. B. Saruhan, A. Flores-Renteria, K. Fritscher, *et al.*, "Sintering Inhibition in Nanostructured EB-PVD-TBCs by Applying Liquid-Phase-Infiltration," 8th. Conf. European Ceramic Society, Trans Tech Publications Ltd, Switzerland, 2003.
49. T.E. Strangmann, 2002, "Durable TBC," *U.S. patent No. 6395343 (U.S. patent No. 5562998).*
50. R. Subramanian, S.M. Sabol, J. Goedjen, and M. Arana, "Advanced TBCs systems for the ATS Engine," ATS review meeting, 1999.
51. G.V. Samsonov, *The Oxide Handbook*, IFI/Plenum Press, New York, Washington, London, 1973.
52. N. Jacobson, "Thermodynamic Properties of Some Metal Oxide-Zirconia Systems," NASA TM 102351, 1989.
53. M.S. Chandrasekharaiah, "Volatilities of Refractory Inorganic Compounds," pp 495-507 in *The Characterization of High Temperature Vapors*, ed. J.H. Margrave, 1967.
54. M.F. Trubelja, D.M. Nissley, N.S. Bornstein, and J.T.D. Marcin, "Pratt&Whitney TBC Development," Advanced Turbine Systems, Proceedings of the Annual Program Review Meeting 1997, DOE/FETC-98/1057, DE98002004, CONF-971053, report 5.9, 1997.
55. O. Unal, T.E. Mitchell, and A.H. Heuer, "Microstructures of Y2O3-Stabilized ZrO2 Electron Beam-Physical Vapor Deposition Coatings on Ni-base Superalloys," *J. Am. Ceram. Soc.*, 77 [4] 984-992 (1994).
56. U. Schulz, K. Fritscher, and M. Peters, "EB-PVD Y_2O_3 and CeO_2/Y_2O_3 Stabilized Zirconia TBCs - Crystal Habit and Phase Compositions," *Surface and Coatings Technology*, 82 259-269 (1996).
57. U. Schulz, K. Fritscher, and M. Peters, "Thermocyclic Behavior of Variously Stabilized EB-PVD TBCs," *J. Engineering for Gas Turbines and Power*, 119 917-921 (1997).
58. B.A. Nagaraj and D.J. Wortmann, "Burner Rig Evaluation of Ceramic Coatings with Vanadium-Contaminated Fuels," *ASME J. Eng. Gas Turbine Power*, 112 536-542 (1990).
59. U. Schulz, K. Fritscher, and C. Leyens, "Two-Source Jumping Beam Evaporation for Advanced EB-PVD TBC Systems," *Surface and Coatings Technology*, 133-134 40-48 (2000).
60. U. Schulz, K. Fritscher, and W.A.Kaysser, "Cyclic Lifetime of PYSZ and CESZ EB-PVD TBC Systems on Various Ni-Superalloy Substrates," COST 2002, Liege: J. Lecomte-Beckers, M. Carton, F. Schubert, P.J. Ennis, 483-492, 2002.
61. C. Leyens, U. Schulz, and K. Fritscher, "Oxidation and Lifetime of PYSZ and CeSZ Coated Ni-Base Substrates with MCrAlY Bond Layers," *Materials at High Temperatures*, 20 [4] 475-480 (2003).
62. M. Maloney, H. Achter, and B. Barkalow, "Development of Low Thermal Conductivity TBCs," Proc. TBC Workshop, NASA, ed. N.L.R. Center, Cleveland, OH, 41, 1997.
63. M.J. Maloney, 2000, "Article Having a TBC Based on a Phase-Stable Solid Solution of Two Ceramics and Apparatus and Method for Making the Article," *EP patent No. 0972853, (US patent No. 6007880 and US patent No. 6187453).*
64. U. Schulz, B.S.-. Ramond, O. Lavigne, *et al.*, "Low Thermal Conductivity Ceramics for Turbine Blade TBC Application.," *Ceramic Engineering and Science Proceedings*, accepted 2004.
65. S. Alperine, V. Arnault, O. Lavigne, and R. Mevrel, 2001, "Heat Barrier Composition, A Mechanical Superalloy Article Provided with a Ceramic Coating Having Such a Composition, and a Method of Making the Ceramic Coating," *U.S. patent No. 6333118, EP patent No. 1085109.*
66. R.L. Jones, "Scandia, Yttria-Stabilized Zirconia (SYSZ): A Candidate Material for High Temperature TBCs," Proc. TBC Workshop, NASA, ed. N.L.R. Center, Cleveland, OH, 41, 1997.

67. R.L.Jones, "Experiences in Seeking Stabilizers for Zirconia Having Hot Corrosion-Resistance and High Temperature Tetragonal (t') Stability, " Naval Research Laboratory, NRL/MR/6170—96-7841 (1996).

68. F. Tcheliebou, M. Boulouz, and A. Boyer, "Preparation of Fine-Grained MgO and Gd$_2$O$_3$ Stabilized ZrO2 Thin Films by Electron Beam Physical Vapor Deposition Co-Evaporation," *J. Mater. Res.,* 12 [12] 3260-3265 (1997).

69. M.J. Maloney, M.F. Trubelja, S.G. Warrier, L.D. A., and N. Ulion, 2004, "TBCs with Low Thermal Conductivity Comprising Lanthanide Sesquioxides," *EP patent No. 1400610.*

70. M. Maloney, 2001, "TBC Systems and Materials," *U.S. patent No. 6177200, U.S. patent No. 617560.*

71. M. Matsumoto, N. Yamaguchi, and H. Matsubara, "Low Thermal Conductivity and High Temperature Stability of ZrO2-Y2O3-La2O3 Coatings Produced by Electron Beam PVD," *Scripta Materialia,* 50 [6] 867-871 (2004).

72. D. Zhu and R.A. Miller, "Thermal Conductivity and Sintering Behavior of Advanced TBCs," *Ceramic Engineering and Science Proceedings,* 23 [4] 457-468 (2002).

73. D. Rickerby, P. Morrel, and Y.A. Tamarin, 1998, "A Metallic Article Having a TBC and a Method of Application Thereof," *U.S. patent No. 6025078, EP patent No. 0825271.*

74. K. Matsumoto, Y. Itoh, and T. Kameda, "EB-PVD Process and Thermal Properties of Hafnia-Based TBC," *Science and Technology of Advanced Materials,* 4 [2] 153-158 (2003).

75. R. Subramanian, 2002, "TBC Having High Phase Stability," *U.S. patent No. 6258467, U.S. patent No. 387539.*

76. B. Saruhan, P. Francois, K. Fritscher, and U. Schulz, "EB-PVD Processing of Pyrochlore-Structured La$_2$Zr$_2$O$_7$-Based TBCs," *Surface and Coatings Technology,* 182 [2-3] 175-183 (2004).

77. B. Saruhan-Brings and K. Fritscher, "EB-PVD La$_2$Zr$_2$O$_7$-Based TBCs," Int. Symposium on Advanced TBCs and Titanium Aluminides for Gas Turbines TURBOMAT, German Aerospace Center (DLR), 180-182 (2002).

78. B. Saruhan, U. Schulz, R. Vassen, *et al.,* "Evaluation of Two New TBC Materials Produced by APS and EB-PVD," *Ceramic Engineering and Science Proceedings,* accepted 2004.

79. B. Saruhan-Brings, U. Schulz, and C.-J. Kröder, 2002, "Thermal Insulating Material with an Essentially Magnetoplumbitic Crystal Structure," *EP patent No. 1256636, (US patent No. 2002197503).*

80. O. Sudre, J. Cheung, D. Marshall, P. Morgan, and C. Levi, "Thermal Insulation Coatings of LaPO$_4$," *Ceramic Engineering and Science Proceedings,* 4 367-374 (2001).

81. D. Wortman, 1999, "Multi-Layer TBC," *U.S. patent No. 5,942,334, U.S. patent No. 5792521.*

82. K. An, K.S. Rvichandran, R.E. Dutton, and S.L. Semiatin, "Microstructure, Texture, and Thermal Conductivity of Single-Layer and Multi-Layer TBCs of Y$_2$O$_3$-Stabilized ZrO$_2$ and Al$_2$O$_3$ Made by Physical Vapor Deposition," *J. American Ceramic Society,* 82 [2] 399-406 (1999).

83. T. Krell, U. Schulz, M. Peters, and W.A. Kaysser, "Graded EB-PVD Alumina-Zirconia TBCs — An Experimental Approach," FGM 98, ed. W.A.Kaysser, Trans Tech Publications LTD, 396-401, 1999.

84. M. Peters, C. Leyens, U. Schulz, and W.A. Kaysser, "EB-PVD TBCs for Aeroengines and Gas Turbines," *Advanced Engineering Materials,* 3 [4] 193-204 (2001).

85. M. Maloney, 2002, "Method for Producing Ceramic Coatings Containing Layered Porosity," *U.S. patent 6365236.*

86. M. Maloney, 2000, "Ceramic Coatings Containing Layered Porosity," *U.S. patent No. 6057047.*

87. D. Gupta, D. Lambert, M.F. Trubelja, *et al.,* 2003, "Hybrid TBC and Method of Making the Same," *U.S. patent 20030152814 A1.*

88. H.J. Rätzer-Scheibe, U. Schulz, and T. Krell, "The Effect of Coating Thickness on the Thermal Conductivity of EB-PVD PYSZ TBCs," *Surface and Coatings Technology,* submitted 2004.

Material Improvements and Novel Compositions

CORROSION BEHAVIOUR OF NEW THERMAL BARRIER COATINGS

R. Vaßen, D. Sebold, D. Stöver
Institut für Energieforschung (IEF-1)
Forschungszentrum Jülich GmbH
Jülich, 52425, Germany

ABSTRACT

New thermal barrier coating (TBC) systems based on pyrochlores ($La_2Zr_2O_7$, $Gd_2Zr_2O_7$, $La_2Hf_2O_7$) have been produced by atmospheric plasma-spraying. Double layer systems consisting of a 200 μm yttria stabilized zirconia (YSZ) layer on the bondcoat and a 200 μm pyrochlore layer on top have been used. These coatings have been tested in a gas burner test rig. In the flame of the rig corrosive media have been injected. These media were Na_2SO_4 and $NaCl$ as a water solution and also kerosene. The lifetime of the coatings was largely reduced by the injection of the corrosive species. A detailed analysis of the failure mechanisms and a comparison to YSZ coatings will be made.

INTRODUCTION

Thermal barrier coatings systems typically consist of a metallic oxidation protection layer and an insulative ceramic topcoat. Electron beam physical vapour deposition (EB-PVD) and atmospheric plasma spraying (APS) are widely used processes to deposit the topcoat of these systems. The state of the art topcoat material for both processes is yttria partially stabilized zirconia (YSZ) [1, 2]. This material performs well up to about 1200 °C. At higher application temperatures, which are envisaged for a further improvement of the efficiency of the gas turbines, the YSZ undergoes two detrimental changes. Significant sintering leads to microstructural changes and hence a reduction of the strain tolerance in combination with an increase of the Young's modulus. Higher stresses will originate in the coating, which lead to a reduced life under thermal cyclic loading.

The second change is a phase change of the non-transformable t′-phase, which is present in the as-deposited YSZ coating. At elevated temperatures the t′-phase transforms into tetragonal and cubic phase. During cooling the tetragonal phase will further transform into the monoclinic phase, which is accompanied by a volume change and a high risk for a damage of the coating [3]. As a consequence, a considerable reduction of thermal cycling life is observed.

These disadvantageous properties of YSZ at high temperatures prompted an intense search for new TBC materials in the past. In [2, 4, 5] detailed overviews on the developments of new systems are given.

Among the interesting candidates for thermal barrier coatings, those materials with pyrochlore structures and high melting points show promising thermo-physical properties. Interesting candidates are especially $La_2Zr_2O_7$ $La_2Hf_2O_7$, $Gd_2Zr_2O_7$ or $Nd_2Zr_2O_7$. Previous investigations show excellent physical properties of theses materials, i.e. thermal conductivity lower than YSZ and high thermal stability [6]. However, relatively low thermal expansion coefficients and toughness values are observed in these materials [7]. As a result, the thermal cycling properties are worse than those of YSZ coatings. A way to overcome this shortcoming is the use of layered topcoats. The failure of TBC systems often occur within the TBC close to the bondcoat/topcoat interface. At this location YSZ is used as a TBC material with a relatively high thermal expansion coefficient and high toughness. The YSZ layer is then coated with the new TBC material (e.g. $La_2Zr_2O_7$) which is able to withstand the typically higher temperatures at this location. In the past years in several publications on $YSZ/La_2Zr_2O_7$ double layer systems we show that this concept really works [8, 9, 10, 11].

This study is now focused on the influence of corrosive media on the performance of new TBC systems. Corrosive media can be introduced from the environment e.g. in aviation engines from sand particles during landing and take-off. In addition, also the fuel contains corrosive constituents, the most harmful are vanadium, sulphur and sodium. The concentration can vary in wide ranges. While for aviation fuel the sulphur content is limited to 0.05 wt.%, it may reach 4 wt.% in heavy oil for stationary gas turbines [12].

A large number of investigations have been performed in the past to study the influence of the corrosive media in laboratory tests. An older review of the results has been made by Bürgel and Kvernes [13] and a more recent one by Jones [14], showing the most important reactions of the corrosive media with the TBCs. Under severe corrosive environments it is found that the stabilizing agent reacts with the corrosive media leading to a destabilizing of the coating. Also new stabilizing agents as Sc_2O_3 have been studied [15]. A distinct improvement of the corrosion behaviour of these materials compared to YSZ was not found.

Only few investigations on new TBC materials have been published. In [16] $La_2Zr_2O_7$ coatings have been compared to YSZ coatings under isothermal conditions. The new TBC material was relatively resistant against vanadia attack at 1000°C, while a fast decomposition in sulphur containing environment at 900°C was observed.

Corrosion testing under isothermal conditions in a furnace or in burner rigs testing [17, 18] show a major influence of the bond coat composition on the performance of the TBC systems. Typically higher Cr and Al contents improve the lifetime of the coatings. These results show that the corrosion of the bond coat and the attack of the formed thermally grown oxide (typically an alumina scale) play a major role. In the temperature range above 800°C so-called type I hot corrosion is expected. Reactions of the formed Al_2O_3 scale with SO_3 containing atmospheres can be as follows [19], with (1) being dominant at high temperatures (type I hot corrosion):

$$Al_2O_3 + O^{2-} \rightarrow 2\,AlO_2^- \quad \text{for low } SO_3 \text{ pressures} \tag{1}$$

$$Al_2O_3 + 3SO_3 \rightarrow Al_2(SO_4)_3 \quad \text{for high } SO_3 \text{ pressures} \tag{2}$$

In addition, under type I conditions, the sulphide formation of metals (Cr, Ni) at the slag/metal interface play an important role:

$$M + SO_2 \rightarrow \text{M-oxide} + \text{M-sulphide} \tag{3}$$

In the present investigation corrosive media are injected into the flame of a burner rig. With the used setup temperature profiles can be adjusted which are considered as relevant for modern gas turbines (i.e. bond coat temperatures in the range of 900 to 1000°C and surface temperature of about 1200°). Hence a rather realistic testing of TBC systems under corrosive conditions should be possible. In a previous paper already the results of the tests on YSZ coatings have been published [20].

EXPERIMENTAL

The investigated thermal barrier coating systems have been produced by plasma spraying with two Sulzer Metco plasma-spray units. Vacuum plasma spraying with a F4 gun was used to deposit a 150 μm $NiCo21Cr17Al13Y0.6$ bond coat (Ni 192-8 powder by Praxair Surface Technologies Inc., Indianapolis, IN) on disk shaped IN738 superalloy substrates. The diameters of the substrates used for thermal cycling tests were 30 mm, the thickness 3 mm. At the outer edge a radius of curvature of 1.5 mm was machined to avoid sharp edges.

The ceramic top coats with a thickness of about 400 μm were produced by atmospheric plasma spraying (APS) using a Triplex I gun. For the new TBCs a double layer structure was used with an about 200 μm YSZ layer on the bondcoat and an about 200 μm coating of the new TBC material on top. During the manufacture of the thermal cycling specimens also steel substrates were coated. These coatings were used to characterize the as-sprayed condition. The Argon and Helium plasma gas flow rates were 20 and 13 standard liter per minute (slpm), the plasma current was 300 A at a power of 20 kW.

Corrosion tests were performed in a gas burner rig setup, in which the disk shaped specimens were periodically heated up to the desired surface temperature of about 1200 °C in approximately 1 min by a natural gas/ oxygen burner. After 5 min heating the specimens were cooled for 2 min from both sides of the specimens by compressed air. The surface temperature was measured with an infrared pyrometer operating at a wavelength of 9.6 - 11.5 μm and a spot size of 5 mm. The substrate temperature was measured using a NiCr / Ni thermocouple inside a hole drilled to the middle of the substrate. All given values of temperatures are mean values taken over all cycles after completion of the heating phase. More details on the standard thermal cycling rigs can be found in [21]. In this investigation the central gas nozzle was used to introduce atomized liquids containing the corrosive media into the flame. Results on water based salt solutions and on kerosene will be presented.

The injection of water droplets into the flame led to a reduced stability of the flame and therefore larger variations in the temperature profile. In order to evaluate this effect also a pure water injection (without additional corrosive media) was examined for YSZ coatings. Here no significant influence of the water injection on the thermal cycling performance was found.

During the testing 1.4 g/min (salt solution) or 1.2 g/min (kerosene) of corrosive containing media has been injected into the flame. The corrosive media consisted of distilled water with 1 wt.-% Na_2SO_4, 1 or 5 wt.-% NaCl, or of kerosene. The total mean methane and oxygen gas flows during thermal cycling was about 550 l/min and 670 l/min giving about 0.22 wt.-% of corrosive salt species to natural gas for the 1 % solutions.

It turned out that the gas flows of the burner gases varied by about 25 % to adjust similar temperature profiles. This fact led to the rather large differences in the corrosive media concentration between 0.17 to 0.25 wt.-%. As the injection nozzle is located in the center of the flame the profile of the corrosive media concentration has a maximum in the centre.

The test was stopped when obvious degradation of the coatings (large delamination) occurred during the thermal cycling. This definition of failure results in an uncertainty in the lifetime data especially if partial delamination of the coatings occurs. However, as seen below the resulting error probably does not effect the interpretation of the results.

Sprayed specimens were vacuum impregnated with epoxy, and then sectioned, ground and polished. As for the preparation water was used as a media, water-soluble corrosion products might be removed by the preparation process. Cross-sections of the coatings were examined by optical microscopy and scanning electron microscopy (Ultra 55, Zeiss). The surface of some of the tested samples were analysed by X-ray diffraction using a Siemens D5000 facility at a wavelength of 1.5406 Å. For some samples also a special preparation technique was used. About 2 mm thick stripes were prepared from the center of the thermal cycling samples by laser cutting. In the center of these stripes an additional cut was made through the metal substrate. The ceramic coating was broken along this cut and the fracture surface investigated. This was made to check whether the water based preparation methods lead to a removal or redistribution of the corrosive species within the sample. It turned out that also on the fracture surface corrosive species were clearly found similar to the results of the polished samples. However, the quality of the images was purer than for the cross-sections. Therefore only results of the cross-sections will be presented here.

RESULTS AND DISCUSSION

The microstructure of the new TBCs and the YSZ coatings are shown in Fig. 1. The porosity level of the coatings measured by mercury porosimetry was about 12 vol.-%.

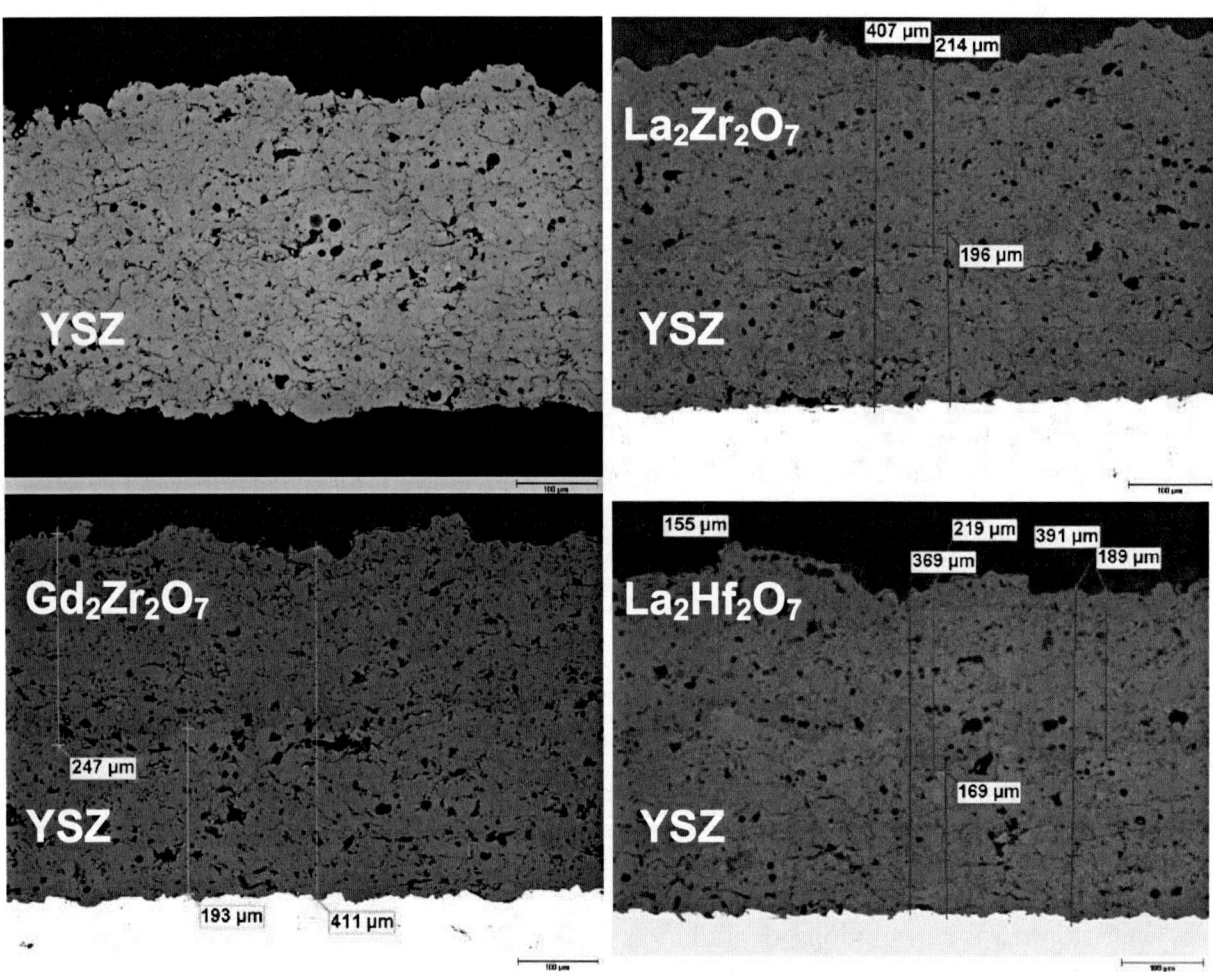

Fig. 1 Micrographs of the investigated TBCs.

Fig. 2 shows the performance of YSZ TBCs in a cyclic burner rig without the addition of corrosive species. The lifetime of the systems under the conditions used in this study ($T_{surface}$ between 1100 and 1250°C) is well above 1500 cycles. An investigation of the influence of the injection of pure water has been performed for YSZ coatings. A lifetime of more than 4000 cycles was found which is comparable to the lifetime of YSZ coatings without water injection for the given bondcoat and surface temperatures.

The results of the cyclic testing of the new double layers and YSZ coatings are also summarized in Fig. 2. It is obvious that the lifetime of single layer $La_2Zr_2O_7$ coatings is reduced compared to our standard YSZ coatings. The YSZ coatings show a strong reduction of lifetime at surface temperatures above about 1300 °C. Double layer systems made of YSZ and $La_2Zr_2O_7$ perform excellent even at much higher temperatures. Their temperature capability is under the given cyclic conditions in the range of 1450°C. Also the one tested YSZ/$Gd_2Zr_2O_7$ double layer system performed well. In contrast, the YSZ/$La_2Hf_2O_7$ coating showed an early failure with a spallation of individual spray splats from the surface. This result indicates a lower temperature capability of the hafnate compared to the zirconate materials.

At lower surface temperatures (< 1300°C) the double layer systems made of YSZ/$La_2Zr_2O_7$ perform similar as the YSZ coatings. Failure is then related to crack growth within the ceramic close to the bondcoat which is induced by the growth of the so-called thermally grown oxide (TGO) on top of the bondcoat. Here for both types of coatings a YSZ layer is present leading to similar performance data.

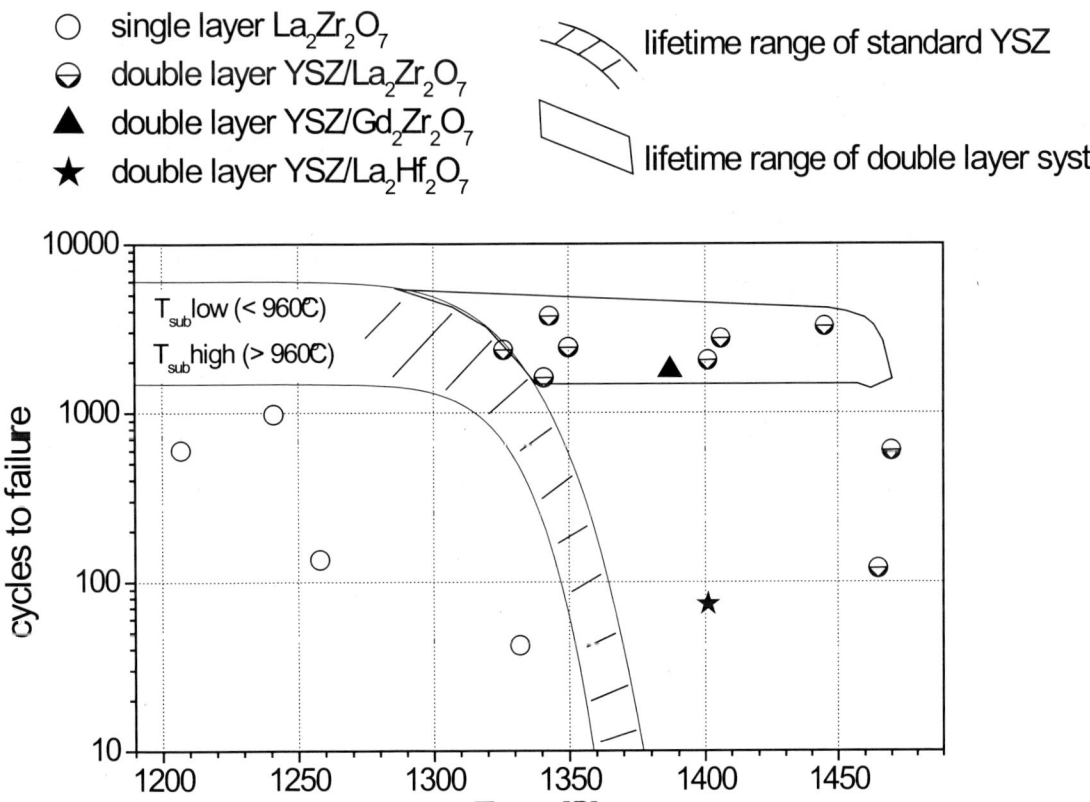

Fig. 2 Cycles to failure for single and double layer systems in a cyclic burner rig without addition of corrosive species as a function of the surface temperature.

In Fig. 3 the results of the cyclic testing with corrosive media are summarized. More details especially on the mean temperature during testing are given in Table I. While the substrate temperatures were in all cases between 890 and 950°C except for 5% NaCl, the surface temperature showed some larger variations. Typically values between 1150 and 1200°C could be established for the 1% salt and the kerosene addition. For the higher salt loading (5%) it was not possible to obtain sufficiently high surface temperatures. Also the substrate temperatures were reduced. The reason is possibly the melting of the NaCl on the surface of the samples. The heat of fusion of NaCl (melting temperature 801°C), leads to a heat consumption for the 5% media in an one cm² large area in the center of the sample of nearly 100kW/m², hence effectively cooling the surface.

A comparison with Fig. 1 shows that all samples failed earlier under the influence of the corrosive media than without. At the given substrate temperatures well below 950°C lifetimes of several thousand cycles are expected.

The highest reduction of more than a factor of 100 is observed for all samples for the media with the high concentration of NaCl (5 %). Compared to this result the reduced concentration (1 %) led to clearly increased lifetimes. Samples cycled with Na_2SO_4 media show lifetimes in between those of samples cycled with high and low NaCl concentration. The addition of kerosene had a reduced effect on the lifetime. Only for the YSZ/GZ coatings a pronounced reduction was observed.

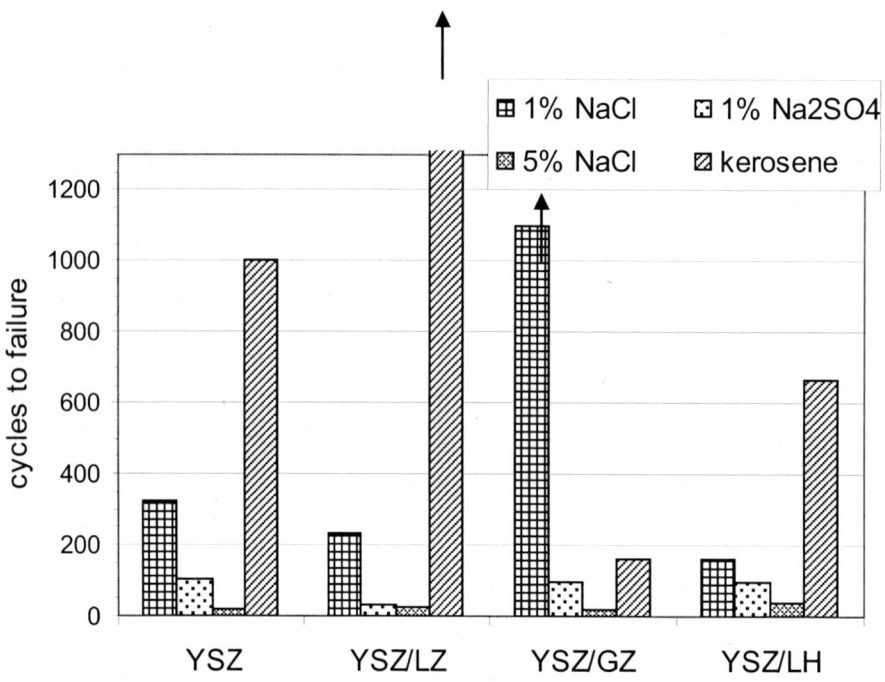

Fig. 3 Cycles to failure of YSZ and double layers of YSZ/$La_2Zr_2O_7$ (YSZ/LZ), YSZ/$La_2Hf_2O_7$ (YSZ/LH), and YSZ/$Gd_2Zr_2O_7$ (YSZ/GZ) in a burner rig test with addition of NaCl (1 % or 5 %, 1.4 g/min), Na_2SO_4 (1%, 1.4 g/min), and kerosene (1.2 g/min). The YSZ/GZ did not fail after 1100 cycles with 1 % NaCl addition. The lifetime of the YSZ/LZ sample with kerosene addition was 1750 cycles.

Photos of the thermally cycled samples are found in Fig. 4. The types of failure can be roughly divided into two groups. The first type is characterized by a spallation of the whole coating, correspondingly the bondcoat is visible on the photos in Fig. 2. In the second type of failure the ceramic coating itself is heavily attacked by the corrosive media leading to a partly delamination of the coating. With the exception of the YSZ/GZ system the 1% NaCl and the Na_2SO_4 addition led to failure mode one although some coloring of the coatings is observed, while 5% NaCl seem to promote the corrosive attack of the ceramic itself. The samples cycled with kerosene addition show some mixed kind of failure.

The failure mechanisms of the coatings cycled under the corrosive environment will be discussed in the following. It will also be investigated whether in failure mode 1 already an attack of the ceramic coating took place.

Table I. Summary of experimental details of the corrosion rig testing as mean substrate and surface temperature (T_{sub}, T_{surf}), cycles to failure, sample number (#), * indicates that the sample was not yet damaged.

	YSZ	YSZ/LZ	YSZ/GZ	YSZ/LH
corrosion media	1% NaCl			
#	884	860	872	879
Tsurface [°C]	1156	1182	1242	1215
Tsub[°C]	902	913	934	918
cycles to failure	325	232	1100*	159
corrosion media	1% Na2SO4			
#	885	861	873	880
Tsurface [°C]	1119	1163	1214	1142
Tsub[°C]	923	895	932	933
cycles to failure	102	32	98	97
corrosion media	5%NaCl			
#	887	863	875	872
Tsurface [°C]	1043	931	1004	1035
Tsub[°C]	843	858	804	913
cycles to failure	21	25	19	40
corrosion media	kerosene			
#	928	862	874	881
Tsurface [°C]	1167	1217	1184	1226
Tsub[°C]	947	931	940	933
cycles to failure	1004	1750	160	663

In Fig. 5 micrographs of some samples cycled with 1% NaCl and 1% Na_2SO_4 are shown. In most of the samples an increase of the porosity level is found in comparison to the as-sprayed condition (Fig. 1). This will be discussed later. In addition, also a dark layer of thermally grown oxide (TGO) is found on the bondcoat having a thickness between about 2 and 5 μm. This appears to be very thick for the low bondcoat temperature (< 980°C) and the short time at temperature (< 100 h). On the other hand the β-phase depleted zone in the top of the bondcoat is very thin. A depleted zone is typically seen in TBC bondcoats as the TGO is mainly Al_2O_3 which is formed from the Al of the β-phase (mainly NiAl). The reason for the thin depleted zone can be identified if the TGO is analyzed in more detail. Fig. 6 shows a SEM micrograph of a

YSZ/La$_2$Zr$_2$O$_7$ system and an EDX spectra of the indicated area after cycling with NaCl addition. Obviously, the oxide does not only consist of alumina, however especially on top more complex phases consisting of Ni, Co, Cr, Al and others are found. The corrosive species lead to a fluxing of the alumina scale forming a much thicker, complex oxide layer. This thick layer is expected to have low mechanical strength and introduces additional stresses into the coating system leading to an early failure of the TBC system. The failure of the investigated samples cycled with Na$_2$SO$_4$ and 1 NaCl addition is mainly the result of the TGO attack by the corrosive species. For the high NaCl concentration the time at high temperatures is too short to promote the corrosion of the TGO, for the kerosene addition the amount of corrosive species seems to be too low for a distinct attack.

1% NaCl

1% Na$_2$SO$_4$

5% NaCl

Kerosene

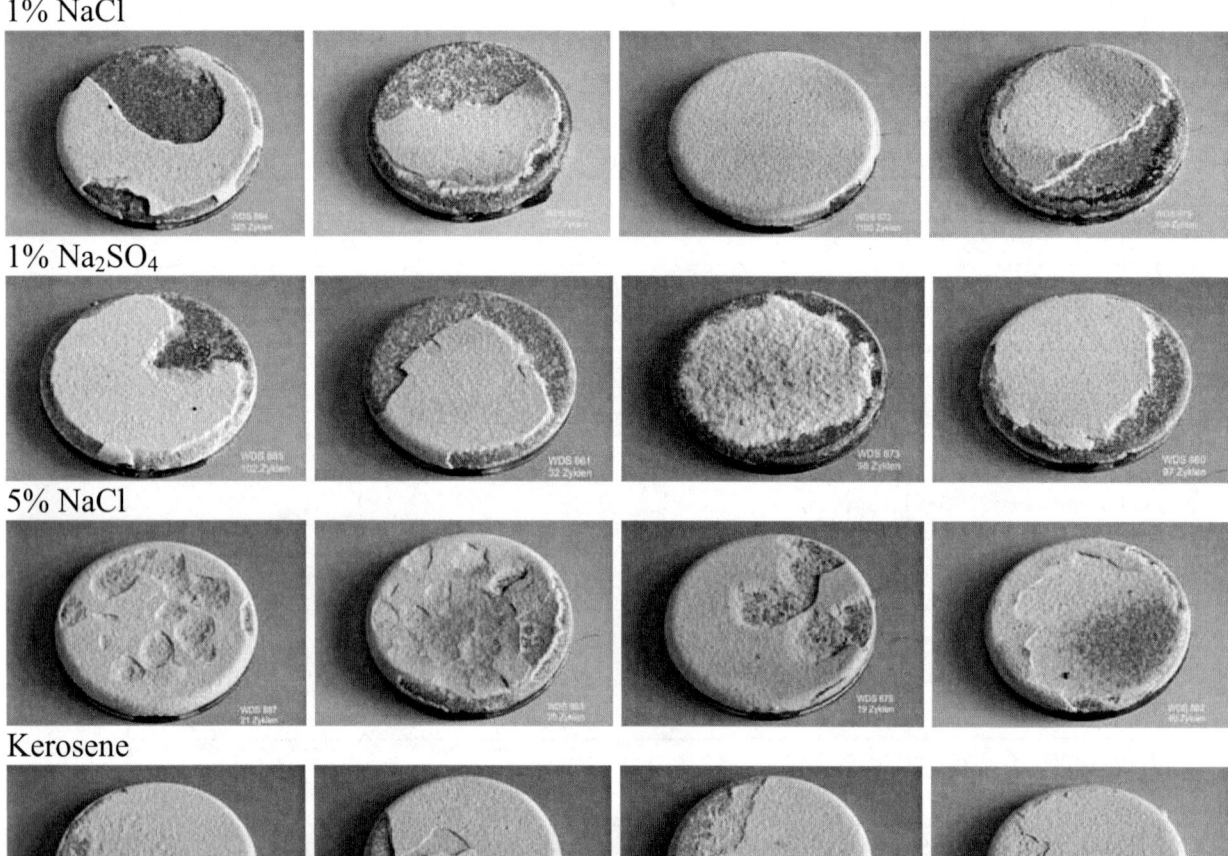

Fig. 4 Photos of the samples after corrosion tests, from left to right YSZ, YSZ/LZ, YSZ/GZ, YSZ/LH.

The attack of the ceramic layers will be discussed in the following. EDX analyis of the TBCs showed at many locations of the samples corrosive species containing sodium, sulfur, chloride but also chromium or nickel. These corrosive species could be especially detected in the fracture surfaces as there a removal of the water soluble species was avoided. These corrosive compounds lead to a stresses during thermal cycling and crack growth in the samples. Also a

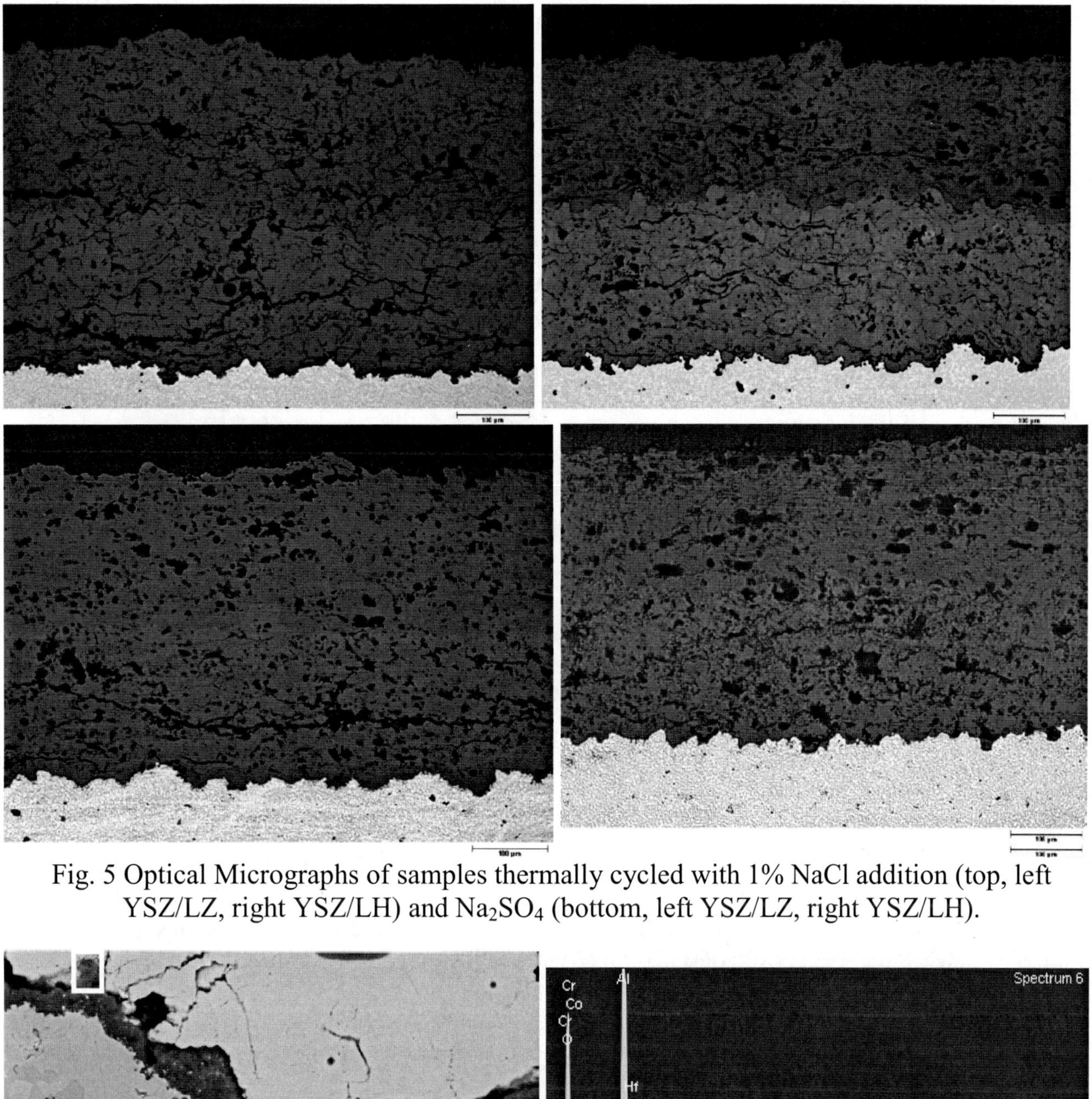

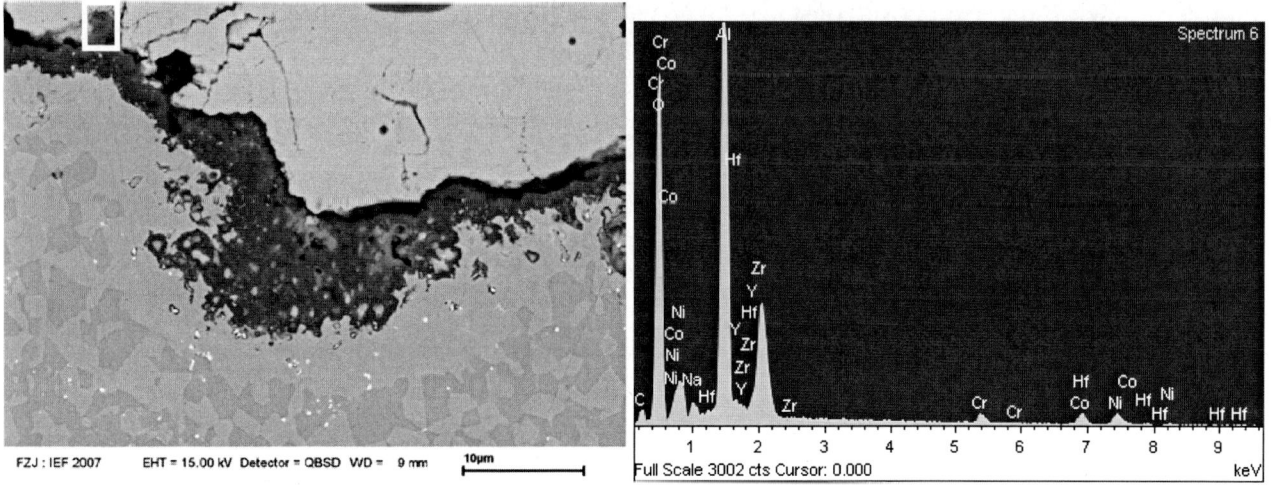

Fig. 5 Optical Micrographs of samples thermally cycled with 1% NaCl addition (top, left YSZ/LZ, right YSZ/LH) and Na_2SO_4 (bottom, left YSZ/LZ, right YSZ/LH).

Fig. 6 SEM micrographs of a $YSZ/La_2Zr_2O_7$ sample after thermal cycling with NaCl addition (top, left). The other graph shows the EDX analysis of the indicated area.

reaction with some of the ceramics was observed leading for example to the formation of non-stabilized zirconia and also to the formation of higher porosity levels. It should be mentioned here that part of the corrosive species within the ceramic could also come from the substrate holder made of a Ni base alloy.

In the case of the samples cycled with 5% NaCl addition the filling of the cracks with high amounts of NaCl salt result in a very early failure of the samples. It is assumed that the strain tolerance of the samples is lost due to the filled cracks and pores.

Due to the high number of investigated systems not all the results can be presented here. Therefore it was tried to summarize the results of the cyclic tests in Table II.

Table II. Summary of the thermal cycling behaviour under corrosive conditions. The numbers indicate the degree of corrosive attack: 0 hardly any attack, 1 some corrosion or crack formation, 2 distinct corrosion or crack formation, 3 massive attack. X: results of X-ray phase analysis.

Corrosive attack		YSZ	YSZ/LZ	YSZ/GZ	YSZ/LH
1% NaCl	BC	2	2	2	2
	TBC	2 (cracks, X: monoclinic ZrO_2)	1-2 (cracks, X: no phase changes)	1 (some cracks, no failure!, long test time)	3 (cracks, X: new phases)
1% Na_2SO_4	BC	3	3	3	3
	TBC	2-3	1-2 (short test time)	3 (new phases?)	2
5% NaCl	BC	0	0	0	0
	TBC	3 (X: no phase changes)	2 (cracks, X:monoclinic ZrO_2)	2 (cracks, X: no phase changes)	2-3 (top layer removed)
Kerosene	BC	1	1	1	1
	TBC	1-2	2 (long test time)	1 (short test time)	2-3

Under the given conditions $La_2Zr_2O_7$ and $Gd_2Zr_2O_7$ show a equal or slightly better performance than YSZ (with the exception of $Gd_2Zr_2O_7$ with Na_2SO_4 addition). $La_2Hf_2O_7$ has a reduced corrosive stability, however it should also be mentioned that the deposited coating showed some loss of lanthanum due to the thermal spray process which might influence the performance.

Table II also clearly indicates the more pronounced attack of the bondcoat by Na_2SO_4 compared to NaCl.

CONCLUSIONS

The results of a new burner rig allowing corrosion testing of various TBC systems under rather realistic conditions have been presented. The injection of NaCl, Na_2SO_4 and kerosene led to a distinct reduction of lifetime up to a factor of 100 for the highest concentration (> 1%). Failure of the coatings was for the moderate concentrations related to fluxing of the TGO oxide by the corrosive species. Clearly, the Na_2SO_4 addition attacked the TGO faster than the NaCl addition.

Also in the case of a TGO fluxing an additional attack of the ceramics was observed. Compared to YSZ $La_2Zr_2O_7$ and $Gd_2Zr_2O_7$ performed rather good, $La_2Hf_2O_7$ showed a lower stability under corrosive conditions. Failure was induced by the filling of the cracks with corrosive species and in some cases by a reaction of the ceramics with the corrosive species.

The addition of the solution with high NaCl concentration led to early failure which was explained by a significant loss of the strain tolerance by the filling of the crack network with salt species.

ACKNOWLEDGEMENT

The authors would like to thank Mr. K.H. Rauwald and Mr. R. Laufs (both IEF-1, FZ Jülich) for the manufacture of the plasma-sprayed coatings and Mrs. N. Adels and Mrs. A. Hilgers (both IEF-1) for the thermal cycling of the specimens. The authors also gratefully acknowledge the work of Mrs. H. Moitroux (IEF-1), Mr. P. Lersch (IEF-2), Mrs. S. Schwartz-Lückge (IEF-1), Dr. D. Sebold (IEF-1) and Mr. M. Kappertz (IEF-1) who supported the characterization of the samples by photography, XRD, optical and scanning electron microscopy, and sample preparation.

REFERENCES

1 W.A Nelson,. R.M. Orenstein, TBC Experience in Land-Based Gas Turbines, *Journal of Thermal Spray Technology* **6**, 2 , 176-180, (1997).
2 D.R. Clarke and C.G. Levi, Annu. Rev. Mater. Res. **33**, 383-417, (2003).
3 R. A. Miller J.L. Smialek, R.G. Garlick, Phase Stability in Plasma-Sprayed Partially Stabilized Zirconia-Yttria, in Science and Technology of Zirconia, Advances in Ceramics, Vol. 3, A.H. Heuer and L.W. Hobbs (eds.), The American Ceramic Society, Columbus, OH, USA, 241-251, (1981).
4 R. Vaßen, D. Stöver, Conventional and new materials for thermal barrier coatings, in Functional Gradient Materials and Surface Layers Prepared by Fine Particle Technlogy, NATO Science Series II: Mathematics , Physics and Chemistry - Vol. 16, Kluwer Acadmic Publishers, Dordrecht, The Netherlands 199-216, (2001).
5 J.R. Nicholls, Advances in Coating Design for High-Performance Gas Turbines, *MRS Bulletin*, Sept., 659-670 (2003).
6 H. Lehmann, D. Pitzer, G. Pracht, R. Vaßen, D. Stöver, Thermal Conductivity and Thermal Expansion Coefficients of the Lanthanum-Rare Earth Element-Zirconate System, *J. Amer. Ceram. Soc.*, **86,** 8 1338-44, (2003).
7 U. Bast, E. Schumann, "Development of Novel Oxide Materials for TBCs, *Ceramics Engineering & Science Proceedings*, 23, 4 (2002) 525-32.
8 R. Vaßen, G. Pracht, D. Stöver, New Thermal Barrier Coating Systems with a Graded Ceramic Coating, Proc. of the International Thermal Spray Conference 2002, Verlag für Schweißen und verwandte Verfahren DVS-Verlag GmbH, Düsseldorf, 2001, pp. 202-207.
9 R. Vaßen, G. Barbezat, D.Stöver, Comparison of Thermal Cycling Life of YSZ and $La_2Zr_2O_7$-Based Thermal Barrier Coatings, in Materials for Advanced Power Engineering 2002, eds. J. Lecomte-Becker, M. Carton, F. Schubert, P.J. Ennis, Schriften des Forschungszentrum Jülich, Reihe Energietechnik, **21,** 1, 511-521.
10 R. Vaßen, X.Q. Cao, D. Stöver, Improvement of New Thermal Barrier Coating Systems using a Layered or Graded Structure, *Ceramic Engineering & Science Proceedings*, **22,** 4, 435- 442 (2001).
11 R. Vaßen, M. Dietrich, H. Lehmann, X. Cao, g. Pracht, F. Tietz, D. Pitzer, D. Stöver, Development of Oxide Ceramics for an Application as TBC", *Materialwissenschaft und Werkstofftechnik* **8,** 673-677, (2001).

12 B.R. Marple, J. Voyer, C. Moreau, D.R. Navy, Corrosion of Thermal Barrier Coatings by Vanadium and Sulfur Components, Materials at High Temperatures, **17**, 3, 397-412 (2000).

13 R. Bürgel, I. Kvernes, Thermal Barrier Coatings, High Temperature Alloys for GasTurbines and Other Applications 1986, W. Betz et al. , ed. D. Reidel Publiching , 327-356(1986).

14 R.L. Jones, "Some Aspects of the Hot Corrosion of Thermal Barrier Coatings," J. of Thermal Spray Technology, **6**, 1, 77-84 (1997).

15 M. Yoshiba, K. Abe, T. Arami, Y. Harada, High-Temperature Oxidation and Hot Corrosion Behavior of Two Kinds of Thermal Barrier Coating Systems for Advanced Gas Turbines, J. of Thermal Spray Technology, **5**, 3, 259-68 (1996).

16 B. R. Marple, J. Voyer, M. Thibodeau, D. R. Nagy, R. Vaßen, Hot Corrosion of Lanthanum Zirconate and Partially Stabilized Zirconia Thermal Barrier Coatings, Transactions of the ASME. Journal of Engineering for Gas Turbines and Power, , **128**, 1, 144-52 (Jan. 2006), ASME, Journal Paper. (AN: 8664377).

17 P.E. Hodge, R.A. Miller, M.A. Gedwill, Evaluation of the Hot Corrosion Behaviour of Thermal Barrier Coatings Thin Solid Films, **73**, 447-453 (1980).

18 I. Zaplatynsky, Performance of Laser-Glazed Zirconia Thermal Barrier Coatings in Clyclic Oxidation and Corrosion Burner Rig Tests, Thin Solid Films, **95**, 275-284 (1982).

19 P. Kofstad, High Temperature Corrosion, Elsevier Appl. Sci., London, 1988.

20 R. Vaßen, D. Sebold, G. Pracht, D. Stöver, Corrosion rig testing of thermal barrier coating systems, on the 30th Int. Cocoa Beach Conf. & Exposition, 23th -27th January 2006, Cocoa Beach, Fl..

21 F. Traeger, R. Vaßen, K.-H. Rauwald, D. Stöver, A Thermal Cycling Setup for Thermal Barrier Coatings, Adv. Eng. Mats., **5**, 6, 429-32 (2003).

Thermal Conductivity of Plasma-Sprayed Aluminum Oxide—Multiwalled Carbon Nanotube Composites

Srinivas R. Bakshi, Kantesh Balani, and Arvind Agarwal[†]

Department of Mechanical and Materials Engineering, Florida International University, Miami, Florida

Aluminum oxide nanocomposites reinforced with multiwalled carbon nanotubes (MWNT) were prepared by atmospheric plasma spraying of blended and spray-dried powders. Thermal conductivity was measured using the laser flash technique for temperatures between 25° and 300°C. An aluminum oxide—4 wt% MWNT nanocomposite prepared from the blended powder showed the highest conductivity, followed by aluminum oxide without nanotubes, 8 and 4 wt% MWNT composite prepared from spray-dried powder in that order. The thermal conductivity values obtained are rationalized taking into account the crystallite size, porosity, MWNT content, microstructure, and the interfaces and metastable γ-Al_2O_3 content present in the nanocomposite.

I. Introduction

THERMAL conductivity is an important physical property, which is required in modeling heat transfer through solids and structures. It has also been used as a quality control parameter in the production and performance of nuclear fuels[1] and thermal barrier coatings.[2] Carbon nanotubes have shown excellent mechanical, thermal, and electrical properties due to which they have been proposed for a myriad number of applications.[3] Multiwalled carbon nanotubes (MWNT) have also shown[4] very high thermal conductivities in excess of 3000 $W \cdot (m \cdot K)^{-1}$. Hence, they serve as a first choice of materials as fillers for thermal conductivity enhancement in thermal management materials. The thermal conductivity of dense aluminum oxide has been reported[5] to be between 27 and 35 $W \cdot (m \cdot K)^{-1}$. In our previous work,[6] it has been shown that addition of carbon nanotubes to aluminum oxide resulted in a 43% increase in the fracture toughness. Addition of MWNT is also expected to increase the thermal conductivity of the composites, which is beneficial for many applications like electronic packaging. It is generally difficult to predict the thermal conductivity of plasma-sprayed coatings because of its complicated microstructure, which consists of splats, porosity, and interfaces. The goal of this paper is to study the thermal conductivity of plasma-sprayed aluminum oxide—MWNT nanocomposites and rationalize them by taking into account the crystallite size, porosity, MWNT content, matrix microstructure, interfaces, and metastable γ-Al_2O_3 content present in the coatings.

II. Experimental Procedure

(1) Plasma Spraying

The samples were fabricated using DC arc plasma spraying with a Praxair SG-100 gun (Praxair Inc., Danbury, CT). The powders were fed internally into the plasma using argon as a carrier gas. Four types of powders were sprayed. They are (a) a spray-dried nanoaluminum oxide (referred to as nano-Al_2O_3 hereafter), (b) a spray-dried nanoaluminum oxide blended with 4 wt% MWNT (referred to as Al_2O_3—4 wt% MWNT blended hereafter), (c) a spray-dried nanoaluminum oxide 4 wt% MWNT mixture (referred to as Al_2O_3—4 wt% MWNT spray dried hereafter), and (d) a spray-dried nanoaluminum oxide 8 wt% MWNT mixture (referred to as Al_2O_3—8 wt% MWNT spray dried hereafter). Coatings of thickness between 0.5 and 1 mm were sprayed onto a mild steel substrate. The details of the plasma spraying and the microstructure of the coatings are given elsewhere.[6] Carbon nanotubes were found to be well distributed in the partially molten regions and were also found in the intersplat and inter-particle regions.[6] It was also seen that the nanotubes were coated well with Al_2O_3, indicating wetting and better interfacial heat transfer between the two. Undamaged CNTs in the plasma-sprayed coatings were confirmed from scanning electron microscope (SEM, JEOL JSM-6330F, JEOL USA Inc., Peabody, MA) observation of fracture surfaces,[6] micro-Raman spectroscopy,[6] and high-resolution transmission electron microscopy (HRTEM, FEI Tecnai F30, FEI Company, Hillsboro, OR).

(2) Thermal Conductivity Measurement

To fabricate samples for thermal conductivity, the substrate with coating was cut into a 10 mm × 10 mm piece with a low-speed diamond saw (Buehler Isomet 11-1180, Buehler Ltd., Lake Bluff, IL). The diamond saw was used to cut through the substrate. The thin layer of mild steel attached to the coating was then removed by dissolving in nitric acid. Thus, free-standing samples of 10 mm × 10 mm area and 0.5–1-mm thickness were prepared. The bulk density of the samples was measured by the water displacement method using the Archimedes principle. Thermal diffusivity was measured using a Holometrix Micromet-300 Thermal Diffusivity Instrument (Metrisa Inc., Bedford, MA) by the pulse method for a number of temperatures between 25° and 300°C. The thermal diffusivity values were corrected for radiation heat losses using Cowan's method.[7] The nano-Al_2O_3 sample was coated with carbon using a carbon spray, to make it opaque to the laser radiation, and then the edges were ground to ensure that the carbon layer was only at the top and bottom surfaces. The MWNT-containing nanocomposites did not require any carbon coating as they were already opaque to the laser radiation. The error in the measured values of thermal diffusivity was within ±3%. The specific heat capacity of aluminum oxide was taken from the thermodynamic databank FactSage 5.0.[8] The specific heat of carbon nanotubes was taken to be the same as that of graphite and was also obtained from FactSage 5.0. Masarapu et al.[9] determined the specific heat capacity of aligned multi-walled carbon nanotubes and found it to be similar to that of graphite. The specific heat capacities of the 4 and 8 wt% MWNT composites were calculated using the Neumann–Kopp additive rule. Finally, the thermal conductivity was calculated from the following equation:

$$k = \alpha \cdot \rho \cdot C_p \tag{1}$$

Here, k is the thermal conductivity, α is the measured value of thermal diffusivity, ρ is the bulk density, and C_p is the specific heat capacity.

K. Watari—contributing editor

Manuscript No. 23313. Received June 6, 2007; approved August 19, 2007.
[†]Author to whom correspondence should be addressed. e-mail: agarwala@fiu.edu

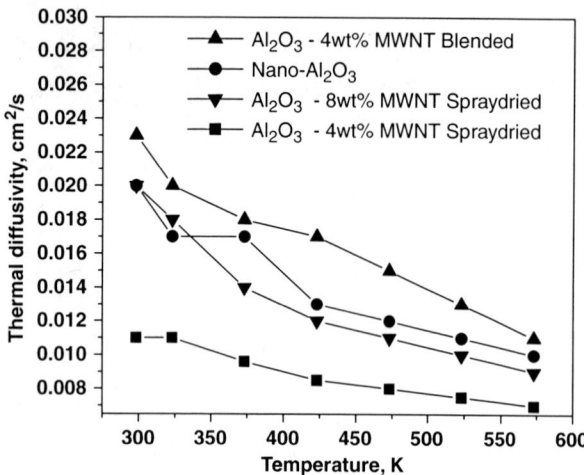

Fig. 1. Variation of thermal diffusivity with temperature.

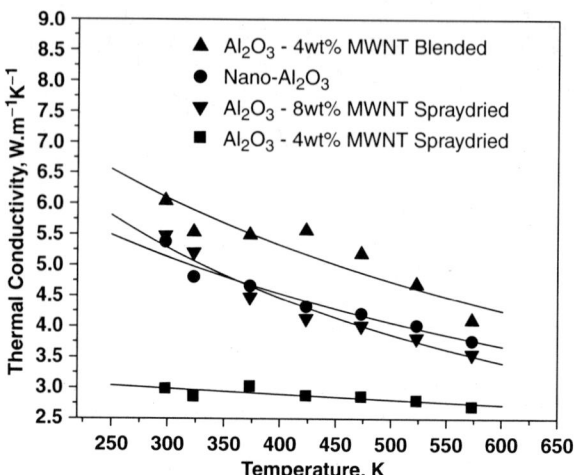

Fig. 2. Variation of thermal conductivity with temperature.

III. Results and Discussion

(1) Variation of Thermal Properties with Temperature

Figure 1 shows the variation of the thermal diffusivity of the nanocomposites with temperature. Similar to all dielectric materials, the thermal diffusivity decreases with an increase in the temperature. The Al_2O_3—4 wt% MWNT blended nanocomposite has the highest thermal diffusivity, while the Al_2O_3—4 wt% MWNT spray-dried nanocomposite has the lowest. The variation of the thermal conductivity has been plotted in Fig. 2. The solid curves in the figure represent the least square fit for the data with an equation of the type $(A+BT)^{-1}$ where A and B are constants and T is the absolute temperature. The thermal conductivity decreases with an increase in temperature endorsing the phonon mechanism of heat conduction.

It is observed that the Al_2O_3—4 wt% MWNT-blended nanocomposite coating has the highest thermal conductivity. For the Al_2O_3–MWNT nanocomposite synthesized from spray-dried powder, the thermal conductivity increases with the increase in the MWNT content.

The thermal conductivities of the nano-Al_2O_3 and the Al_2O_3—8 wt% MWNT spray-dried nanocomposite are almost similar. The thermal conductivity is affected by many factors like the crystallite boundaries, porosity, carbon nanotube content, interphase boundaries, and phase content. It is a complex function of the above features. In the following analysis, the thermal conductivity will be considered to be influenced by the crystallite size, porosity, and the MWNT content. This can be expressed by the following equation:

$$k_e = f_1(r) \times f_2(f_p) \times f_3(f_{MWNT}) \qquad (2)$$

where k_e is the overall thermal conductivity of the composite, r the crystallite size of α-Al_2O_3, f_p the volume fraction of porosity present in the coatings, f_{MWNT} is the volume fraction of MWNT present in the coatings, and f_1, f_2, and f_3 denote functions.

(2) Effect of Crystallite Size

Thermal resistance at the crystallite boundaries can lead to lowering of thermal conductivity. The thermal conductivity of a crystal with a known crystallite size can be expressed according to the relation[10]

$$\frac{1}{k_{mn}} = \frac{1}{k_m} + n \cdot R_{bound} \qquad (3)$$

where n is the number of crystallites per unit length, k_m the theoretical or the single-crystal thermal conductivity of Al_2O_3, k_{mn} the thermal conductivity of α-Al_2O_3 with nanocrystalline grains, and R is the boundary thermal resistance or the Kapitza resistance. The crystallite size and phase content in the coatings have been tabulated in Table I. The crystallite size of the phases in the initial powders and coatings was determined using the peak width of the XRD curves of the samples and sapphire reference. It is to be noted that the powder contains a 100% α-Al_2O_3 phase. Also, we see that the crystallite size of α-Al_2O_3 in the coatings is more than that in the powders.

This is due to the grain coalescence and growth during plasma spraying. In the Al_2O_3—4 wt% MWNT-blended sample, all the nanotubes reside on the surface of the spray-dried particle as compared with the Al_2O_3—4 wt% MWNT spray-dried powder. Hence, there is more absorption and retention of heat and consequently more coarsening in case of the Al_2O_3—4 wt% MWNT-blended coating. This also explains the increase in crystallite size with the increase in MWNT percent. Yang et al.[11] have measured the temperature dependence of the crystallite boundary resistance and have shown that they are efficient in reducing the thermal conductivity of YSZ. They found the Kapitza resistance to be between 0.4×10^{-8} and 2×10^{-8} $m^2 \cdot (W \cdot K)^{-1}$. Crystallite size and thermal conductivity values of α-Al_2O_3 were used for the computation because there are no data available for the thermal properties of γ-Al_2O_3. The value for the boundary resistance was taken to be equal to 1×10^{-8}

Table I. Phase Content and Crystallite Size of the Powders and the Coatings

Crystallite size and phase content	Powder α-Al_2O_3 crystallite size (nm)	Coating			
		α-Al_2O_3 crystallite size (nm)	α-Al_2O_3 content (%)	γ-Al_2O_3 crystallite size (nm)	γ-Al_2O_3 content (%)
Nano-Al_2O_3	44	55	77	22	23
Al_2O_3–4 wt% MWNT blended	46	95	68	13	32
Al_2O_3–4 wt% MWNT spray dried	45	73	82	15	18
Al_2O_3–8 wt% MWNT spray dried	43	76	75	21	25

Amount and crystallite size of α and γ phase of aluminum oxide present in the coatings as measured from the area under peaks and peak broadening in XRD patterns of the coating. Note the large crystallite size of α phase in case of Al_2O_3—4 wt% MWNT Blended coating.

Table II. Calculated Values of Thermal Conductivity Based on Crystallite Size of α-Phase

Sample	Crystallite size of α-phase (nm)	Number of crystallites per unit length (m^{-1})	Calculated thermal conductivity, k_{mn} (W·(m·K)$^{-1}$)
Nano-Al$_2$O$_3$	55	18 181 800	4.6
Al$_2$O$_3$–4 wt% MWNT blended	95	10 526 300	7.2
Al$_2$O$_3$–4 wt% MWNT spray dried	73	13 698 600	5.9
Al$_2$O$_3$–8 wt% MWNT spray dried	76	13 157 900	6.1

k_{mn}—Thermal conductivity of 100 percent dense matrix with nanocrystalline structure. Effect of crystallite size and hence amount crystallite boundary on the thermal conductivity. It is the main factor in reducing the thermal conductivity of the coatings and the main reason for the higher thermal conductivity of Al$_2$O$_3$–4wt% MWNT-blended coating.

m^2·(W·K)$^{-1}$, which is the average of that calculated by Yang et al.[11] and that has also been used by Poulier et al.[10] for alumina. The value of k_m was taken to be 30 W·(m·K)$^{-1}$. The calculated values of the thermal conductivity have been tabulated in Table II.

It can be seen that the crystallite size of the Al$_2$O$_3$–4 wt% MWNT-blended coating is the highest, and it is expected to have a higher conductivity due to the presence of lower number of interfaces. This is the main reason for the high thermal conductivity of the Al$_2$O$_3$—4 wt% MWNT-blended coating. The values tabulated in Table II will be used as the thermal conductivity values for the 100% dense samples (k_{mn}) in all calculations that follow.

(3) Effect of Porosity

The porosity content present in the plasma-sprayed nanocomposites has been tabulated in Table III. Porosities scatter phonons and hence the more the porosity, the lower the conductivity. Also, the porosities are filled with air, which has poor conductivity. One reason for the higher values of the thermal conductivity of the Al$_2$O$_3$—8 wt% MWNT spray-dried coating as compared with the Al$_2$O$_3$—4 wt% MWNT spray-dried coating is its lower porosity content. As mentioned earlier, the thermal conductivity of aluminum oxide has been reported[5] to be between 27 and 35 W·(m·K)$^{-1}$ at room temperature, which is very high compared with the values obtained in this work. The samples in the above-mentioned work were all hot-pressed or sintered and had an equiaxed microstructure with a comparatively coarser grain size than our coatings. Various relationships have been proposed for the porosity dependence of thermal conductivity based on the shape and distribution of the pores, especially in the context of nuclear fuels.[12–15]

The usual porosity relations that hold true for sintered or hot-pressed compacts do not hold true for plasma-sprayed coatings. The thermal conductivities of plasma-sprayed aluminum oxide and zirconium oxide coatings are much smaller than those predicted by these relationships.[16–17] This is due to the fact that the above relations are derived considering the pores to be spherical or prolate or oblate spheroids. Plasma-sprayed coatings have a lamellar microstructure with pores in between the lamellae. The pores are like very thin disks and because they are all aligned parallel to the thickness of the coating, they reduce the transverse thermal conductivity drastically. The formulas for the

thermal conductivity of the nanocrystalline matrix with pores according to various models are listed below:

Landauer[18]

$$k_{mnp} = \frac{1}{4}\left[k_p(3f_p - 1) + k_{mn}(2 - 3f_p)\right. $$
$$\left. + \left\{\left[k_p(3f_p - 1) + k_{mn}(2 - 3f_p)\right]^2 + 8k_{mn}k_p\right\}^{1/2}\right] \quad (4)$$

Meredith and Tobias[19]

$$\frac{k_{mnp}}{k_{mn}} = \left[\frac{2 - f_p}{2 + (W - 1)f_p}\right]\left[\frac{2(1 - f_p)}{2(1 - f_p) + Wf_p}\right] \quad (5)$$

where $W = \frac{1}{3}\left(\frac{1}{2F} + \frac{2}{(1-F)}\right)$, F is the shape factor, which is 0.1 for lamellar pores.

Shafiro and Kachanov[20]

$$\frac{k_{mnp}}{k_{mn}} = 1 - \left(\frac{2f_p}{\pi}\right)\left(\frac{d}{t}\right) \quad (6)$$

where f_p is the fractional porosity and d/t is the aspect ratio of the lamellar pores, which will be taken as 5 here. Here, k_{mnp} refers to the thermal conductivity of the nanocrystalline matrix containing the pores and k_p refers to the thermal conductivity of the pore. This value of the aspect ratio is taken in accordance to that observed by Wang et al.[21] The values of thermal conductivity of the porous coatings were calculated based on the porosity relations mentioned above by taking the value of k_{mn} from Table II and porosity values from Table III and $k_p = 0.4$ W·(m·K)$^{-1}$, and has been tabulated in Table IV below.

It can be observed that the values of the thermal conductivity calculated in this manner are close to that measured experimentally for the Al$_2$O$_3$–4 wt% MWNT blended and the Al$_2$O$_3$—8 wt% MWNT spray-dried coatings. The nano-Al$_2$O$_3$ coating exhibits a larger conductivity compared with the calculated one, which may be due to the smaller value of Kapitza resistance for the interfaces of α-Al$_2$O$_3$ crystallites. The effect of addition of carbon nanotubes has not been taken into account yet. The addition of carbon nanotubes will increase the computed values of the thermal conductivities of the composites. The thermal conductivity of plasma-sprayed zirconium oxide and aluminum oxide has been shown to be very low compared with the thermal conductivity of fully dense compacts produced by other methods.[16,17,21,22] Our values for aluminum oxide are in agreement with those reported by Kulkarni et al.[22] McPherson[23] has provided a model for thermal conductivity of plasma-sprayed coatings, considering that there are only a few regions of good contact between the lamella and existence of planar pores, which are essentially nonconducting at low temperatures. According to the model, the ratio of the thermal resistivity of the coating (R_c) and a fully dense bulk material (R_b) is given by

$$\frac{R_c}{R_b} = \frac{\pi a}{2f\delta} \quad (7)$$

where "$2a$" is the mean radius of the circular regions of contact between the splats, f is the mean fraction of the total area in true contact in any plane parallel to the coating surface, and δ is the

Table III. Density and Porosity Fraction Present in the Coatings

Coating	Density g/cm^3	Porosity %
Nano-Al$_2$O$_3$	3.47	13
Al$_2$O$_3$–4wt% MWNT blended	3.40	12.8
Al$_2$O$_3$–4wt% MWNT spray dried	3.52	9.8
Al$_2$O$_3$–8wt% MWNT spray dried	3.53	6

Note that the fractional porosity reduces of the coatings formed from the spray-dried powders reduces with increase in CNT content. Thermal conductivity of the Al$_2$O$_3$–4wt% MWNT blended coating is highest irrespective of its higher porosity.

Table IV. Calculated Thermal Conductivity of the Composites Based on the Porosity Relations at 298 K

Sample	Porosity fraction	Measured k at 298 K, $(W \cdot (m \cdot K)^{-1})$	Calculated k_{mnp} at 298 K $(W \cdot (m \cdot K)^{-1})$		
			Landauer Model	Meredith and Tobias Model	Kachanov Model
Nano-Al$_2$O$_3$ coating	0.13	5.4	3.8	3.3	2.7
Al$_2$O$_3$–4wt% MWNT blended	0.128	6.0	5.9	5.3	4.3
Al$_2$O$_3$–4wt% MWNT spray dried	0.098	3.0	5.1	4.6	4.0
Al$_2$O$_3$–8wt% MWNT spray dried	0.06	5.5	5.6	5.3	4.9

K_{mnp} is the computed value of the thermal conductivity of the nanocrystalline Al$_2$O$_3$ matric containing pores. The Meredith and Tobias model, which takes into account the pore shape, gives values close to the observed values. These values are used for further calculations.

thickness of the lamella or splats. From transmission electron microscope (TEM) images of the coatings, it was found that "$2a$" is comparable with "t" and the value of "f" can be taken to be around 0.2 from the fact that the values of elastic modulus of the coatings are around 0.2 times the theoretical values. Using these values, R_c/R_b is calculated to be around 4. This means that the thermal conductivity of plasma-sprayed coatings can be as low as 0.25 times the bulk value. Hence, the model predicts the observed values quite well, qualitatively.

(4) Effect of MWNT Addition

Improvements in thermal conductivity have been reported due to the addition of nanotubes.[24,25] Recently, Sivakumar et al.[26] have reported a 70% increase in thermal conductivity of SiO$_2$ due to the addition of 10 vol.% of MWNT. This increase is very low compared with the calculated values obtained from a simple rule of mixtures.

This indicates that the interfacial thermal resistance plays an important role in determining the effective thermal conductivity.

Conventional models of thermal conductivity of composites predict a large value for the thermal conductivity of MWNT composites and this has led to the development of new models based on the Effective Medium Approach (EMA) of Maxwell–Garnett.[27–29] These are summarized below in Table V.

Here, k_e is the effective thermal conductivity of the composite, k_{mnp} is the thermal conductivity of the matrix taken from Meredith and Tobias values of Table III, k_c is the MWNT thermal conductivity (3000 $W \cdot (m \cdot K)^{-1}$), f is the MWNT volume fraction, $\alpha = k_c/k_{mnp}$, and a_k is the so-called Kapitza radius, which is given as $a_k = R_k k_m$ where R_k is the Kapitza resistance or the thermal boundary resistance. The Kapitza radius is assumed to be 15 nm here. The MWNT used are 40–70 nm in diameter and 0.5–2 μm in length. The MWNT volume fraction f_{MWNT} is equal to 6.6% and 12.4% for the 4 and 8 wt%, composite, respectively. It can be seen that the value obtained from the rule of mixture formula is quite large. It is seen from Table V that the values obtained from these EMA models are larger than the observed values by a factor of 2–3. Huxtable et al.[30] have shown that an exceptionally small interfacial heat conductance

Table V. Calculated Thermal Conductivity Values at 298 K for Different MWNT Content

Reference	Model formulation	Calculated k_e at 298 K $(W \cdot (m \cdot K)^{-1})$		
		Al$_2$O$_3$—4wt% MWNT blended	Al$_2$O$_3$—4wt% MWNT spray dried	Al$_2$O$_3$μ—8wt% MWNT spray dried
Rule of mixtures[26]	$\dfrac{k_e}{k_{mnp}} = 1 + \dfrac{f_{MWNT} k_c}{3 k_{mnp}}$	77	71	129
Nan et al.[29]	$\dfrac{k_e}{k_{mnp}} = \dfrac{3 + f_{MWNT}(\beta_x + \beta_z)}{3 - 2\beta_x}$, $\quad \beta_x = \dfrac{2(k_{11}^c - k_{mnp})}{k_{11}^c + k_{mnp}}, \quad \beta_z = \dfrac{k_{33}^c}{k_{mnp}} - 1$ $k_{11}^c = \dfrac{k_c}{1 + \frac{2a_k}{d}\frac{k_c}{k_{mnp}}}, \quad k_{33}^c = \dfrac{k_c}{1 + \frac{2a_k}{L}\frac{k_c}{k_{mnp}}}$	11	10	15
Xue[28] with good dispersion	$\dfrac{k_e}{k_{mnp}} = \dfrac{1 - f_{MWNT} + (4f_{MWNT}/\pi)\sqrt{k_c/k_{mnp}} \times \tan^{-1}\left(\pi/4\sqrt{k_c/k_{mnp}}\right)}{1 - f_{MWNT} + (4f_{MWNT}/\pi)\sqrt{k_{mnp}/k_c} \times \tan^{-1}\left(\pi/4\sqrt{k_c/k_{mnp}}\right)}$	22	21	39
Xue[28] with poor dispersion	$\dfrac{k_e}{k_{mnp}} = \dfrac{1 - f_{MWNT} + 2f_{MWNT}\frac{k_c}{k_c - k_{mnp}} \ln\frac{k_c + k_{mnp}}{2k_{mnp}}}{1 - f_{MWNT} + 2f_{MWNT}\frac{k_{mnp}}{k_c - k_{mnp}} \ln\frac{k_c + k_{mnp}}{2k_{mnp}}}$	10	8	14

Thermal conductivities are (k_e: CNT composite, k_c: multiwalled CNT, k_{mnp}: nanocrystalline matrix with porosity). Models based on EMA approach are used to calculate thermal conductivity of CNT composites.

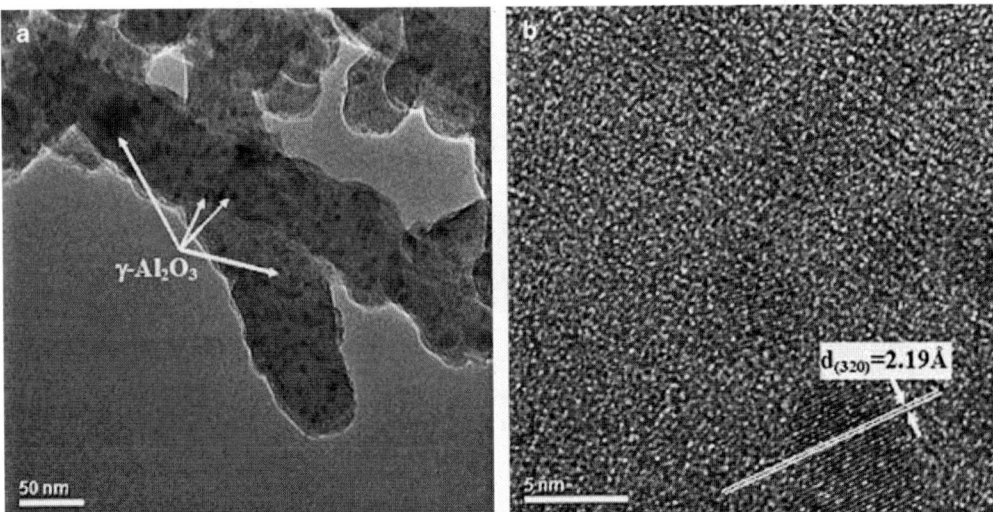

Fig. 3. Transmission electron microscopic image of the (a) Al$_2$O$_3$—8 wt% multiwalled carbon nanotubes (MWNT) spray dried coating showing the γ-Al$_2$O$_3$ nucleated on MWNT, and (b) HRTEM image of the nanotube surface showing a γ-Al$_2$O$_3$ precipitate.

of 12 MW·(m·K)$^{-1}$·K limits the enhancement of the thermal conductivity of CNT composites. Shenogina et al.[31] performed FEM calculations and found that as the distance between the two nanotubes approached zero, the rate of heat flow in the composite increased only marginally, indicating a possible lack of thermal percolation in carbon nanotube composites. Zhan et al.[32] have observed thermal conductivity of aluminum oxide—SWNT composites as compared with aluminum oxide prepared by spark plasma sintering. They also found that the thermal conductivity reduced with increasing vol.% of SWNT. They had attributed this to the lower thermal conductivity of SWNT bundles and possibly the high thermal resistivity of the SWNT–Al$_2$O$_3$ interface. However, there could be many reasons for the observed lower values of the thermal conductivity than tabulated in Table IV, which will be discussed below.

(A) Effect of Intersplat Thermal Resistance: It is known that during plasma spraying, the powder feedstock is introduced into the plasma where particles become heated. A molten/semi-molten particle accelerates and strikes the substrate at high velocities[6] of ∼200 m/s and forms a splat (as measured using in-flight particle diagnostic sensor Accuraspray G3, Tecnar Automation LTEE, St. Bruno, QC, Canada). Hence, the nano-composite has a layered morphology with interlamellar pores, intersplat interfaces, partially molten particles, and cracks due to thermal stress. Ravichandran et al.[17] have obtained values of thermal conductivity of plasma-sprayed alumina with 19% porosity as low as 1/10 (∼3 W·(m·K)$^{-1}$ at room temperature) of the theoretical value. According to their simple interfacial thermal resistance model,[21] the effective value of thermal conductivity of a layered splat-like microstructure is given by

$$k_{\text{eff}} = \frac{L}{\frac{n\delta}{k_m} + \frac{n-1}{h_i}} \qquad (8)$$

where L is the total thickness of the coating, n the number of splats making up the coating, δ the average splat thickness, k_m the thermal conductivity of the bulk material, and h_i is the interfacial heat transfer coefficient. Hence, for a coating of aluminum oxide made of 100 splats of thickness 1 μm each and assuming h_i to be 10^6 W·(m·K)$^{-1}$ and k_m to be 30 W·(m·K)$^{-1}$, one gets the k_{eff} value as low as 1 W·(m·K)$^{-1}$. So it is seen that interfacial resistance plays a significant role in reducing the thermal conductivity value. The value of the interfacial heat transfer coefficient will depend on many parameters including the area of contact, contact pressure due to residual stresses, impurity segregation, and gas pressure in the voids,

which are all dependent on many processing parameters. Hence it is not possible to assign a reasonable value or predict the value of the heat transfer coefficient at the interfaces. The MWNT distribution and morphology between the splats will also influence the interfacial conductance.

(B) Effect of γ-Al$_2$O$_3$ Formation: It is well known that during plasma spraying, metastable γ-Al$_2$O$_3$ forms during plasma spraying.[33,34] Guilemany et al.[34] have found that plasma-sprayed α-Al$_2$O$_3$ coatings contain between 17% and 23% of α-Al$_2$O$_3$ and the rest γ-Al$_2$O$_3$. The phase resulting from re-solidification of aluminum oxide is always the metastable phase γ-Al$_2$O$_3$. Figure 3 shows a TEM micrograph of the Al$_2$O$_3$—8 wt% MWNT spray-dried coating. It can be seen that CNT act as nucleation sites for γ-Al$_2$O$_3$. As stated, previously in the Al$_2$O$_3$—4 wt% MWNT-blended powder, the nanotubes reside on the surface and this leads to more absorption and retention of heat and consequently a higher degree of melting compared with the Al$_2$O$_3$—4 wt% MWNT spray dried and the Al$_2$O$_3$—8 wt% MWNT spray-dried coating. This, combined with the fact that γ-Al$_2$O$_3$ nucleates on MWNT, explains the higher amount of γ-Al$_2$O$_3$ in the Al$_2$O$_3$—4 wt% MWNT-blended coating. The more the amount of γ-Al$_2$O$_3$, the more the interfacial area between γ-Al$_2$O$_3$ and α-Al$_2$O$_3$, which will provide an additional interfacial resistance term and will reduce the thermal conductivity. The higher value of the thermal conductivity of the Al$_2$O$_3$—4 wt% MWNT-blended coating might be due to the overwhelming effect of an increase in the crystallite size of α-Al$_2$O$_3$.

The modeling of the effect of the γ-Al$_2$O$_3$ content on the thermal conductivity will require knowledge of the interface resistance and were involve the distribution and orientation of the γ and α interfaces and a value of thermal conductivity for the γ-Al$_2$O$_3$ phase that is not available in the literature because of its metastable nature. Also the γ-Al$_2$O$_3$ content and distribution is a function of many processing parameters and is difficult to quantify.

IV. Conclusions

The thermal conductivity of plasma-sprayed aluminum oxide blended with 4 wt% MWNT was higher than the coatings obtained from spray-dried powders containing 4 and 8 wt% MWNT. The thermal conductivity increases with an increase in the MWNT content. The crystallite size has a strong effect in reducing the thermal conductivity of the coatings. The addition of MWNT does not increase the values of the thermal conductivity significantly as predicted by the models based on the

Effective Medium Approach. The measured thermal conductivity values of the plasma-sprayed coatings are two to three times lower than that calculated by taking into consideration the boundary resistance. The lower thermal conductivity of the Al_2O_3–MWNT composite coatings can be ascribed to the inter-splat thermal resistance, which is difficult to quantify. MWNT acts as nucleation sites for γ-Al_2O_3 formation and the γ-Al_2O_3 serves as phonon-scattering centers, further reducing the thermal conductivity. It is challenging to quantify the effect of the crystallite size and content of γ-Al_2O_3 due to the unavailability of data on the physical properties and the complexity of the distribution of γ-Al_2O_3.

Acknowledgments

The authors would like to acknowledge the research funding from the Office of Naval Research (N00014-05-1-0398) for carrying out this research. One of the authors (S. R. Bakshi) would like to acknowledge Presidential Enhanced Assistantship from FIU and National Science Foundation (DMI-0547178) for funding. Kantesh Balani acknowledges Dissertation Year Fellowship from FIU.

References

[1]D. G. Martin, "A Re-appraisal of the Thermal Conductivity of UO_2 and Mixed (U, Pu) Oxide Fuels," *J. Nucl. Mater.*, **110**, 73–94 (1982).

[2]D. R. Clarke and S. R. Phillpot, "Materials," *Mater. Today*, **8**, 22–9 (2005).

[3]R. H. Baugman, A. A. Zakhidov, and W. A. de Heer, "Carbon Nanotube—The Route Towards Applications," *Science*, **297**, 787–92 (2002).

[4]P. Kim, L. Shi, A. Majumdar, and P. L. McEuen, "Thermal Transport Measurements of Individual Multiwalled Nanotubes," *Phys. Rev. Lett.*, **87**, 215502 (2001).

[5]N. P. Bansal and D. Zhu, "Thermal Conductivity of Zirconia Alumina Composites," *Ceram. Int.*, **31**, 911–6 (2005).

[6]K. Balani, S. R. Bakshi, Y. Chen, T. Laha, and A. Agarwal, "Role of Powder Treatment and CNT Dispersion in the Fracture Toughening of Plasma-Sprayed Aluminum Oxide–Carbon Nanotube Ceramic Nanocomposite," *J. NanoSci. Nanotech.*, (2007) doi: 10.1166/jnn.2007.851.

[7]R. D. Cowan, "Pulse Method of Measuring Thermal Diffusivity at High Temperatures," *J. Appl. Phys.*, **34**, 926–7 (1963).

[8]FactSage 5.2, http://www.factsage.com/ (GTT Technologies, Kaiserstr. 100, 52134 Herzogenrath, Germany, 2003)

[9]C. Masarapu, L. L. Henry, and B Wei, "Specific Heat of Aligned Multiwalled Carbon Nanotubes," *Nanotechnology*, **16**, 1490–4 (2005).

[10]C. Poulier, D. S. Smith, and J. Absi, "Thermal Conductivity of Pressed Powder Compacts: Tin Oxide and Alumina," *J. Eur. Ceram. Soc.*, **27**, 475–8 (2007).

[11]H.-S. Yang, G.-R. Bai, L. J. Thompson, and J. A. Eastman, "Interfacial Thermal Resistance in Nanocrystalline Yittria Stabilized Zirconia," *Acta Mater.*, **50**, 2309–17 (2002).

[12]S. K. Rhee, "Porosity-Thermal Conductivity Relations Correlations for Ceramic Materials," *Mater. Sci. Eng.*, **20**, 89–93 (1975).

[1,3]J. Francl and W. D. Kingery, "Thermal Conductivity: IX, Experimental Investigation of Effect of porosity on Thermal Conductivity," *J. Am. Ceram. Soc.*, **37**, 99–107 (1954).

[14]J.B. MacEwan, R.L. Stoute, and M.J.F. Notley, "Effect of porosity on the Thermal Conductivity of UO_2," *J. Nucl. Mater.*, **24**[1] 109–12 (1967).

[15]G. Ondracek and B. Schulz, "The Porosity Dependence of Thermal Conductivity of Nuclear Fuels," *J. Nucl. Mater.*, **46**, 253–8 (1973).

[16]T. A. Taylor, "Thermal Conductivity and Microstructure of Two Thermal Barrier Coatings," *Surf. Coatings Technol.*, **54/55**, 53–7 (1992).

[17]K. S. Ravichandran, K. An, R. E. Dutton, and S. L. Semiatin, "Thermal Conductivity of Plasma Sprayed Monolithic and Multilayer Coatings of Alumina and Yittria-Stabilized Zirconia," *J. Am. Ceram. Soc.*, **82**, 673–82 (1999).

[18]R. Landauer, "The Electrical Resistance of Binary Metallic Mixtures," *J. Appl. Phys.*, **21**, 779–84 (1952).

[19]F. Cernuschi, S. Ahmaniemi, P. Vuoristo, and T. Mantyla, "Modelling of Thermal Conductivity of Porous Materials: Application to Thick Thermal Barrier Coatings," *J. Eur. Ceram. Soc.*, **24**, 2657–67 (2004).

[20]B. Shafiro and M. Kachanov, "Anisotropic Effective Conductivity of Materials With Nonrandomly Oriented Inclusions of Diverse Ellipsoidal Shapes," *J. Appl. Phys.*, **87**, 8561–9 (2000).

[21]Z. Wang, A. Kulkarni, S. Deshpande, T. Nakamura, and H. Herman, "Effect of Pores and Interfaces on Effective Properties of Plasma Sprayed Zirconia Coatings," *Acta Mater.*, **51**, 5319–34 (2003).

[22]T. A. Kulkarni, S. Sampath, A. Goland, and H. Herman, "CMT Studies to Characterize Microstructure-property Correlations in Thermal Sprayed Alumina Deposits," *Script. Mater.*, **43**, 471–6 (2000).

[23]R. McPherson, "A Model for Thermal Conductivity of Plasma Sprayed Ceramic Coatings," *Thin Solid Films*, **112**, 89–95 (1984).

[24]S. U. S. Choi, Z. G. Zhang, W. Yu, F. E. Lockwood, and E. A. Grulke, "Anomalous Thermal Conductivity Enhancement in Nanotube Suspensions," *Appl. Phys. Lett.*, **79**, 2252–4 (2001).

[25]M. J. Biercuk, M. C. Llaguno, M. Radosavvljevic, J. K. Hyun, A. T. Johnson, and J. E. Fisher, "Carbon Nanotube Composites for Thermal Management," *Appl. Phys. Lett.*, **80**, 2767–9 (2002).

[26]R. Sivakumar, S. Guo, T. Nishimura, and Y. Kagawa, "Thermal Conductivity in Multi-Wall Carbon Nanotube/Silica-Based Nanocomposites," *Scripta Mater.*, **56**, 265–8 (2007).

[27]C.-W. Nan, Z. Shi, and Y. Lin, "A Simple Model of Thermal Conductivity of Nanotube-Based Composites," *Chem. Phys. Lett.*, **375**, 666–9 (2003).

[28]Q. Z. Xue, "Model for Thermal Conductivity of Carbon Nanotube Based Composites," *Physica B*, **368**, 302–7 (2005).

[29]C.-W. Nan, G. Liu, Y. Lin, and M. Li, "Interface Effect on Thermal Conductivity of Carbon Nanotube Composites," *Appl. Phys. Lett.*, **85**, 3549–51 (2004).

[30]S. T. Huxtable, D. G. Cahill, S. Shenogin, L. Xue, R. Ozisik, P. Barone, M. Usrey, M. S. Strano, G. Siddons, M. Shim, and P. Keblinski, "Interfacial Heat Flow in Carbon Nanotube Suspensions," *Nat. Mater.*, **2**, 731–4 (2003).

[31]N. Shenogina, S. Shenogin, L. Xue, and P. Keblinski, "On the Lack of Thermal Percolation in Carbon Nanotube Composites," *Appl. Phys. Lett.*, **87**, 133106 (2005).

[32]G.-D. Zhan, J. D. Kuntz, H. Wang, C.-M Wang, and A. K. Mukherjee, "Anisotropic Thermal Properties of Single-Wall-Carbon-Nanotube-Reinforced Nanoceramics," *Phil. Mag. Lett.*, **84**, 419–23 (2004).

[33]R. McPherson, "On the Formation of Thermally Sprayed Alumina Coatings," *J. Mater. Sci.*, **15**, 3141–9 (1980).

[34]J. M. Guilemany, J. Nutting, and M. J. Dougan, "A Transmission Electron Microscopy Study of the Microstructures Present in Alumina Coatings Produced by Plasma Spraying," *J. Thermal Spray Technol.*, **6**, 425–9 (1997). □

Infiltration-Inhibiting Reaction of Gadolinium Zirconate Thermal Barrier Coatings with CMAS Melts

Stephan Krämer,*,† James Yang, and Carlos G. Levi*

Department of Materials, University of California, Santa Barbara, California 93106-5050

The thermochemical interaction between a $Gd_2Zr_2O_7$ thermal barrier coating synthesized by electron-beam physical vapor deposition and a model $33CaO–9MgO–13AlO_{3/2}–45SiO_2$ (CMAS) melt with a melting point of $\sim 1240°C$ was investigated. A dense, fine-grained, ~ 6-μm thick reaction layer formed after 4 h of isothermal exposure to 1300°C. It consisted primarily of an apatite phase based on $Gd_8Ca_2(SiO_4)_6O_2$ and fluorite ZrO_2 with Gd and Ca in a solid solution. Remarkably, melt infiltration into the intercolumnar gaps was largely suppressed, with penetration rarely exceeding ~ 30 μm below the original surface. The microstructural evidence suggests a mechanism in which CMAS infiltration is arrested by rapid filling of the gaps with crystalline reaction products, followed by slow attack of the column tips.

I. Introduction

RARE-EARTH ZIRCONATES (REZ) have generated substantial interest as novel thermal barrier coatings (TBC) based primarily on their intrinsically lower thermal conductivity and higher resistance to sintering than the state-of-the-art ZrO_2—7-wt% Y_2O_3 (7YSZ).[1-4] In addition, the pyrochlore zirconates ($RE_2Zr_2O_7$) are stable as single phases at all relevant temperatures, from $\leq 1550°C$ for $Gd_2Zr_2O_7$ (GZO)[5] to $\leq 2300°C$ for $La_2Zr_2O_7$,[6] circumventing the phase stability problem that ultimately limits the temperature capability of 7YSZ.[7] In principle, this combination of attributes should enable higher gas path temperatures with a comparable coating thickness while maintaining the metal surface within allowable limits.

Increases in operating temperature associated with current TBC technology have clear benefits in engine efficiency[8] and also unintended consequences by introducing new modes of coating degradation that threaten further progress. Of particular interest to aircraft engines is the ingestion of siliceous particulates (dust, sand, volcanic ash, runway debris) with the intake air.[9] At lower temperatures, these particles may impact the airfoil surfaces and cause erosive wear or local spallation of the TBC.[10-12] As engine temperatures increase, the finer debris tend to adhere to the coating surface[13] and form calcium magnesium alumino-silicate (CMAS) melts that penetrate the open void spaces in the coating.[9] Upon cooling at the end of an operation cycle, the melt freezes and the infiltrated volume of the coating becomes rigid, losing its ability to accommodate strains arising from the thermal expansion mismatch with the underlying metal. The coating develops delamination cracks that lead to its progressive exfoliation with concomitant loss of insulation efficiency

and accelerated degradation of the metallic layers underneath.[14] Concurrent with the thermomechanical damage, 7YSZ coatings undergo significant chemical attack by the CMAS melt.[9,15]

The present investigation addresses the potential effect of CMAS on GZO as a candidate higher temperature TBC material. The focus is on coatings produced by electron beam-physical vapor deposition (EB-PVD), but similar phenomena are expected to occur in those deposited by an atmospheric plasma spray (APS). The results are discussed in the context of recent studies of a similar nature on 7YSZ.[15,16]

II. Experimental Procedure

$Gd_2Zr_2O_7$ was deposited on polycrystalline alumina substrates (25 mm $\times$ 25 mm $\times$ 0.6 mm, 99.5% purity, CoorsTek, Golden, CO) using an in-house dedicated EB-PVD facility[17] fed with prealloyed ingots (Trans-Tech, Adamston, MD). Ceramic substrates were selected to facilitate heating the system isothermally above the melting point of the CMAS ($> 1200°C$), which would cause excessive degradation of superalloy substrates. One important difference with 7YSZ is that GZO tends to react with Al_2O_3 to form a $GdAlO_3$ interphase, as reported elsewhere.[18] However, the times involved in the present experiments were relatively short and in practice did not considerably unduly compromise the coating adherence. Two hundred-micrometer thick coatings were deposited at ~ 2 μm/min on substrates held at 1000°C and mounted on a tubular ceramic holder rotating over the source at a rate of 8 rpm, as described elsewhere.[17]

A model CMAS composition of $33CaO–9MgO–13AlO_{1.5}–45SiO_2$ was selected for consistency with prior studies (henceforth, all compositions are given in mole percent of single cation oxide formula units). It was based on the average of melts found to have penetrated TBCs on turboshaft shrouds operated in a desert environment,[9] excluding the minor components believed to originate mainly from the engine (Fe and Ni). Earlier work revealed that this composition starts to melt at approximately 1235°C when made from the constituent oxides, is completely molten at 1240°C, and remains amorphous on cooling.[15] The model CMAS was prepared by mixing reagent-grade fine powders of the individual oxides and milling them in water to form a thick paste, which was subsequently applied to a localized portion of the TBC surface with an area density well in excess of the amount needed to infiltrate all the porosity in the coating (~ 8 mg/cm²). After drying, the specimens were heated to 1300°C for 4 h with ramp-up and -down rates of 6°C/min.

Cross sections of the specimens exposed to CMAS were cut along a plane perpendicular to the rotation axis. These were embedded in epoxy and subsequently polished for examination by scanning electron microscopy (SEM) in both secondary (SE) and back scattered electron (BSE) imaging modes. Transmission electron microscopy (TEM) specimens were cut from selected areas of the polished cross sections using a focused ion beam (FIB). This technique is particularly advantageous in the present specimens because it allows precise sampling of locations exhibiting specific microstructural features or reaction products. TEM analysis included bright-field (BF), dark-field (DF), high-angle annular dark-field (HAADF) imaging,

L. Pinckney—contributing editor

Manuscript No. 22963. Received March 21, 2007; approved October 8, 2007.
Research sponsored by the Office of Naval Research under contracts N00014-99-1-0471 and MURI/N00014-00-1-0438, monitored by Dr. David Shifler. The project benefited from the use of the UCSB-MRL Central Facilities supported by NSF under award No. DMR00-80034.
*Member, The American Ceramic Society.
†Author to whom correspondence should be addressed. e-mail: skraemer@engineering.ucsb.edu

as well as selected area diffraction and energy-dispersive X-ray spectroscopy (EDS).

III. Results

The first sign that gadolinium zirconate reacts significantly differently to the CMAS exposure than 7YSZ was immediately evident on the macroscopic scale. CMAS on 7YSZ heated to 1300°C penetrated the TBC quickly and locally and spread laterally within the coating, leaving behind a shallow glassy droplet completely contained within the limits of the original deposit (Fig. 1(a)). When applied to the GZO TBC under the same conditions, the CMAS melt spread over the entire sample, leaving a continuous layer with a flat glassy surface (Fig. 1(b)). Cross-sectional examination revealed the CMAS layer to be ~30-μm thick. At the boundary between CMAS and TBC, the originally pointed column tips had been replaced by an ~6–8-μm thick layer of reaction products with seemingly crystalline protrusions extending into the CMAS (Fig. 2). The intermediate contrast of the reaction product in BSE is consistent with the incorporation of lighter elements from the CMAS (darker) and heavier elements from the GZO (brighter).

The most striking observation was that the CMAS melt infiltrated the intercolumnar porosity in GZO only to a depth of ≤30 μm below the original TBC surface after a 4-h exposure (Fig. 2), whereas full penetration occurs within minutes in 7YSZ under the same conditions.[15] In some locations, CMAS penetrated no further than the lower boundary of the reaction layer in the neighboring columns. The distinction between open and infiltrated porosity is clearly made by combining SE and BSE imaging of the same area (Fig. 3). The inset in Fig. 3(a) also reveals traces of CMAS within the fine feathery porosity surrounding an otherwise empty intercolumnar pore beneath the main infiltration front, suggesting preferential spreading along the column walls. However, this effect is limited to a few micrometers, with no evidence of CMAS found in the lower three quarters of the TBC thickness.

The thermochemical interaction between GZO and CMAS is arguably linked to the arrest of the CMAS infiltration, and thus deserves further analysis. A closer examination of the reaction layer in Fig. 2 reveals a modulated structure with periodicity dictated by the intercolumnar spacing (Fig. 4). The microstructure can be conveniently divided into three types of regions. The first one is associated with the former intercolumnar gaps, which are now filled with reaction products and extend into the bulk reaction layer, leaving an intriguing "ghost" trace as denoted by the arrow in Fig. 4. These regions appear to show little or no residual CMAS relative to the rest of the reaction layer, and

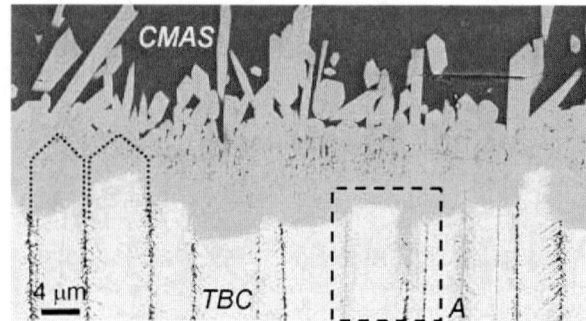

Fig. 2. Back-scattered electron image of the reaction zone between calcium magnesium alumino-silicate and a columnar Gd$_2$Zr$_2$O$_7$ thermal barrier coatings. Dotted lines on the left illustrate the approximate shape and height of the original column structure. The area marked *A* is shown in greater detail in Fig. 3.

retain features reminiscent of the feathery pattern within the partially penetrated gaps. The second type of microstructure within the reaction layer corresponds to the locations formerly occupied by column tips, which show a graded structure consisting of crystalline reaction products within an interpenetrating CMAS network. Both the relative amount of CMAS and the scale of the crystalline products decrease toward the interface between the reaction layer and the residual GZO, becoming irresolvable at the scale of the image in Fig. 4. The third and uppermost part of the reaction layer consists primarily of large acicular-faceted crystals protruding into the residual CMAS on top of the coating, interspersed with nonfaceted globular particles.

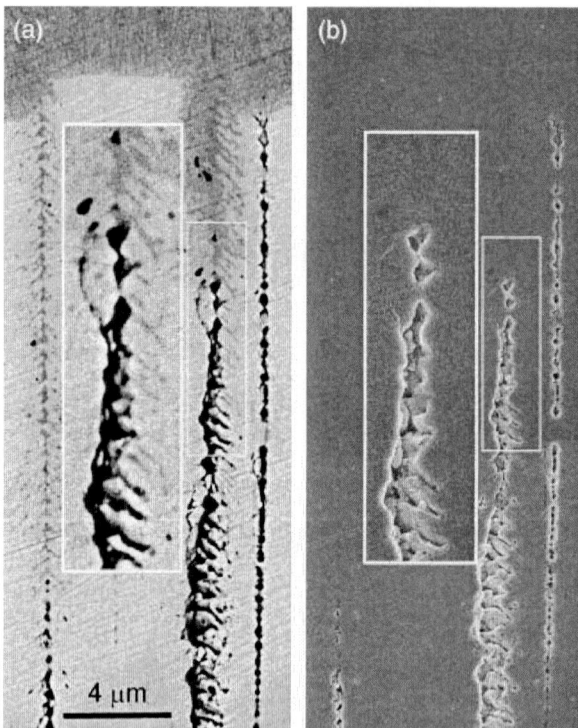

Fig. 3. Identification of open versus infiltrated porosity by combining secondary-scattered electron (SE) and back-scattered electron (BSE) imaging of the same area. (a) BSE contrast reveals the infiltrated spaces when filled by the reaction product whose Z-contrast is intermediate between calcium magnesium alumino-silicate (CMAS) and Gd$_2$Zr$_2$O$_7$, whereas (b) SE highlights empty holes due to charging at their edges that would not be evident if the pore were filled with unreacted CMAS. The insets show magnified (2×) views of the transition zone between filled and open porosity.

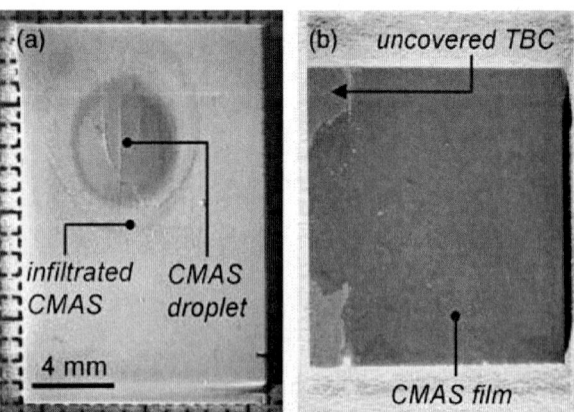

Fig. 1. Optical view of thermal barrier coatings surfaces after calcium magnesium alumino-silicate attack at 1300°C for 4 h: (a) ZrO$_2$–7-wt% Y$_2$O$_3$ and (b) Gadolinium zirconate. Note the distinctly different spreading behavior.

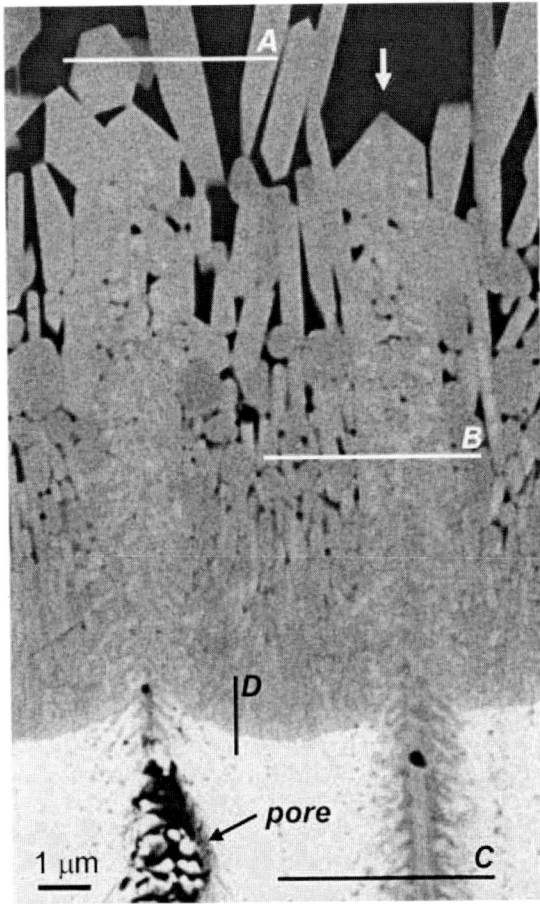

Fig. 4. Back-scattered electron image of the reaction zone. The imaging conditions were adjusted to maximize the contrast within the zone, which rendered the calcium magnesium alumino-silicate (above) indistinguishable from residual porosity (below). The lines mark positions where focused ion beam lamellae were extracted for transmission electron microscopy. "B" and "C" show exact locations for the specimens shown in Fig. 6. "A" and "D" show representative positions for the microstructures shown in Figs. 5 and 8.

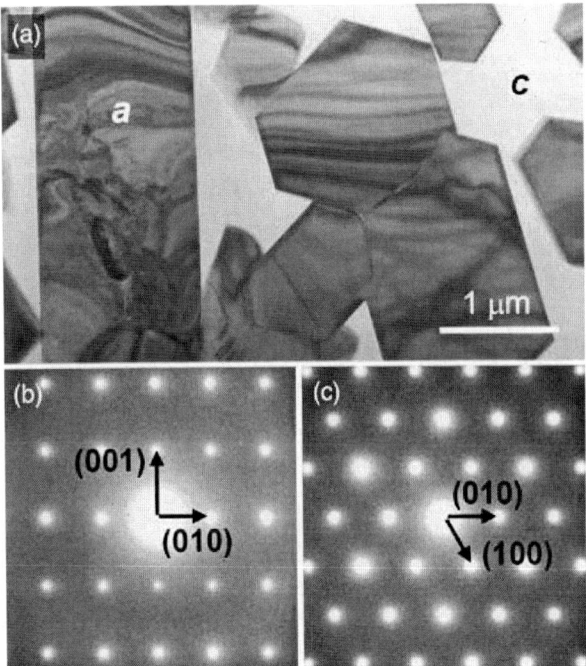

Fig. 5. Bright field transmission electron microscopy micrograph of an area similar to A in Fig. 4, containing crystalline needles extending in various directions into an amorphous calcium magnesium alumino-silicate matrix. The selected area diffraction patterns correspond to the $\langle 10\bar{1}0]$ and [0001] zone axes of the apatite grain marked with "a" in (a).

Microstructural and microchemical characterization of the reaction layer was undertaken by TEM analysis of FIB samples extracted from the locations schematically illustrated in Fig. 4. The issues of interest were (i) the identification of the reaction products, (ii) the structure of the infiltrated column gaps, both between the remaining columns and within the reaction zone, (iii) the constitution and morphology of the reaction zone above a column core, and (iv) the interface between the latter and the residual GZO.

The nature of the larger needles was investigated on an FIB lamella from location A in Fig. 4, whose BF image is shown in Fig. 5. Electron diffraction revealed a structure with hexagonal symmetry and lattice constants $a = 9.5$ Å and $c = 6.9$ Å, while TEM/EDS analysis suggested a silicate containing Gd, Ca, and a smaller amount of Zr (Table I). The evidence is consistent with an apatite-type phase similar to $La_8Sr_2Si_6O_{26}$,[19] which has been studied extensively owing to its potential as a solid oxide ion conductor.[20] The long axes of the needles are parallel to the [0001] direction of the apatite unit cell, while the facets bounding the transverse sections are of the $\{10\bar{1}0\}$ type. The amorphous matrix in Fig. 5(a) is CMAS modified with small amounts of Gd and Zr in a solid solution (Table I).

The structures along sections C and B in Fig. 4, corresponding to an infiltrated gap and its "ghost" extension into the bulk reaction zone, are depicted together with their surroundings in Fig. 6. The denser region in the middle of Fig. 6(a) indicates the former intercolumnar gap filled with reaction products, while the neighboring regions correspond to the microstructure within a former column tip. Both areas contain a mixture of apatite crystals of composition similar to the larger needles in the upper reaction zone and a fluorite phase containing Zr, Gd, and some Ca in a solid solution (Table I). However, while the region above the column core contains significant amounts of a residual glassy phase (darker pockets), much less is present in the former intercolumnar gap. Where surrounded by CMAS, the apatite and fluorite crystals can be readily distinguished by the faceted boundaries of the former and more globular appearance of the latter. This difference is less evident within the intercolumnar gap "ghost," where apatite crystals appear to be more abundant.

Transmission electron microscopy analysis of section C in Fig. 4, taken 2 μm below the lower boundary of the reaction zone at the same column gap as in section B, revealed a continuous chain of apatite grains between the two columns (Fig. 6(b)). The specimen in this figure was tilted so that both columns were close to the $\langle 100 \rangle$ zone axis, corresponding to the texture induced by the deposition conditions.[21] The strong diffraction condition darkened the columns and enhanced the visibility of the apatite phase (a) in the channel. Protrusions (z') extending from the column cores into the apatite phase were identified as fluorite ZrO_2 with a higher $GdO_{1.5}$ content (20%–30%) than in the upper reaction layer (~13%) but a similar amount of CaO (~3%). The diffraction contrast in Fig. 6(b) indicates that the fluorite protrusions grew epitaxially onto the parent column. Small pockets of other phases noted in Fig. 6(b) were identified as spinel- and residual-crystallized CMAS.

The variation in microstructure above the column core with distance from the GZO interface is shown in Fig. 7. The micrographs correspond to in-plane sections taken at distances of 5 μm (Fig. 7(a)), 2 μm (Fig. 7(b)), and 800 nm (Fig. 7(c)) from the interface with the remaining GZO column. The relative amount of CMAS clearly decreases toward the interface, forming a thin grain boundary phase with small pockets at grain corners in the lower parts of the reaction layer (Fig. 7(b)), which appears to be rather indistinct in Fig. 4. The scale of the reaction products also

Table I. Chemical Compositions of the Various Constituents in the Reaction Zone

Region (figure)	Constituent	CaO	MgO	$AlO_{1.5}$	SiO_2	ZrO_2	$GdO_{1.5}$
Upper RZ (5a, 6a)	CMAS	27±2	10±1	14±1	43±3	3	3
	Apatite	16±1	—	—	37±1	6±1	41±2
	Fluorite	4±1			—	83±1	13±1
Lower RZ (7c)	Apatite	12±1	—	—	37±1	4±1	48±2
	Fluorite	3±1	—	—	—	73±4	24±5
Infiltrated gap (6b)	Apatite	18±2	—	—	38±1	4±1	40±2
	Fluorite	3±1	—	—	—	76–66	20–30
	Spinel	—	32	63	1	2	2
	CMAS	30	9	26	26	5	4
TBC (9)	GZO	—	—	—	—	47	53

Values are based on cation ratios measured in the TEM EDS analysis. All compositions are given in mol% of cations. Standard deviations of average values are given.

decreases with distance from the interface, from ~400 nm at 5 μm (Fig. 7(a)) to ~100 at 800 nm (Fig. 7(c)). The faceted nature of the apatite grains becomes less distinct closer to the interface and the fluorite particles become less globular and meander around the apatite crystals, as illustrated in Fig. 7(c), similar to the protrusions in the intercolumnar gaps (Fig. 6(b)). The Gd content in the fluorite was found to increase from ~13% at the 5-μm level to ~24% 800 nm above the interface, whereas the Ca:Gd ratio in the apatite decreased concomitantly from 0.4 to 0.25 (Table I). Area analysis suggested approximately equal amounts of fluorite and apatite near the interface.

An FIB TEM lamella cut along location D in Fig. 4 revealed a structure of elongated grains within the reaction zone closest to the interface (Fig. 8). The grains, ~50-nm wide, comprise alternating apatite and fluorite phases, the latter with up to ~30% $GdO_{1.5}$ and ~2% CaO. The composition of GZO directly underneath the interface shows no deviation from bulk values. The thin dark lines in the Z-contrast image of Fig. 8 denote amorphous grain boundary films, which are distinguishable down to ~200 nm above the interface. Neither BF nor HAADF imaging could confirm an amorphous phase at the interface,

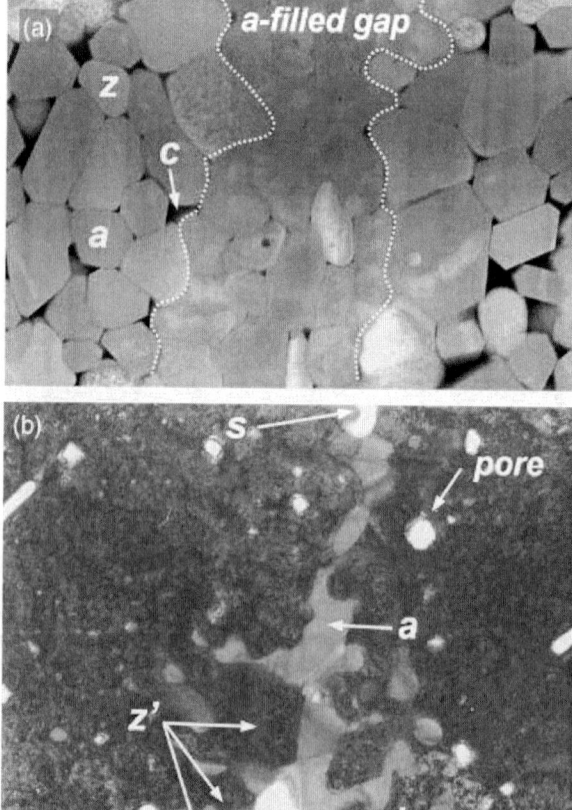

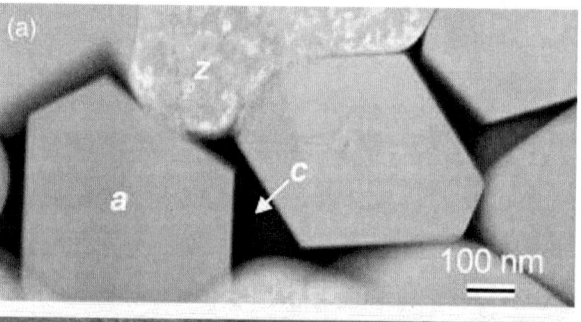

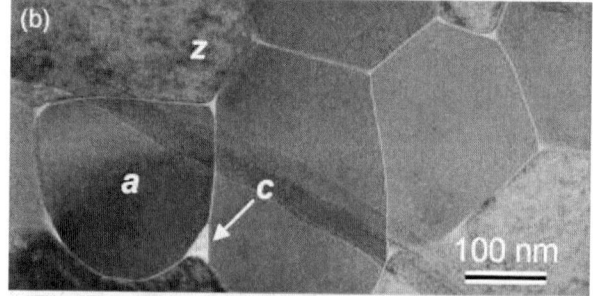

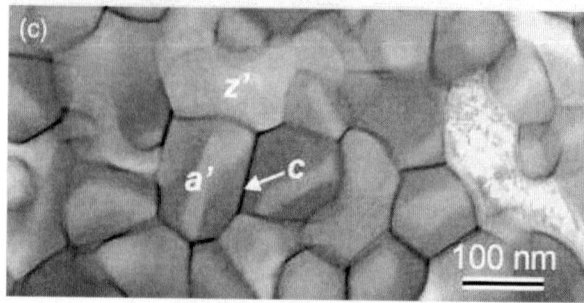

Fig. 6. (a) High-angle annular dark-field transmission electron microscopy (TEM) micrograph of the section marked "B" in Fig. 4. Z-contrast enables a reduction in diffraction contrast within the different grains and highlights the grain contours as well as the contrast between crystals and surrounding calcium magnesium alumino-silicate. (b) Bright field TEM micrograph of section marked "C" in Fig. 4. Phases: z, z': fluorite, a: apatite, s: spinel, p: pore, c, c': CMAS. See Table I for chemical compositions.

Fig. 7. Microstructures of in-plane sections within the reaction zone taken at distances of (a) 5 μm, (b) ~2 μm, and (c) 800 nm from the $Gd_2Zr_2O_7$ interface. (a, c) are recorded in the high-angle annular dark-field transmission electron microscopy mode, revealing the Z-contrast between crystallites and surrounding calcium magnesium alumino-silicate. The bright field micrograph in (b) highlights the glassy nature of the grain boundary films. See Table I for chemical compositions.

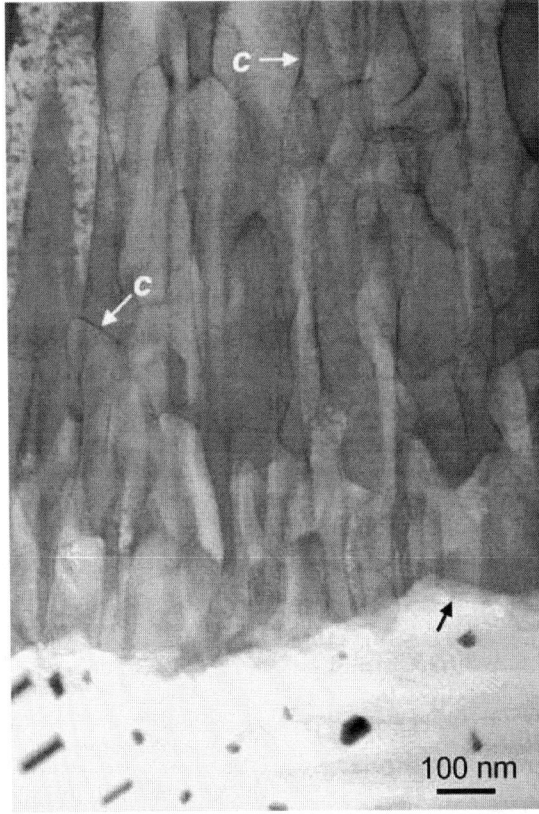

Fig. 8. High-angle annular dark-field transmission electron microscopy micrograph of the interface between $Gd_2Zr_2O_7$ and the lower part of the reaction zone, from a lamella taken at the position and orientation denoted by D in Fig. 4. White arrows indicate amorphous grain boundary films within the reaction zone. Ascertaining whether this film extends down to the interface is hindered by the roughness of the latter (black arrow).

Fig. 9. Bright field transmission electron microscopy micrograph of the interface between $Gd_2Zr_2O_7$ (GZO) and the lower part of the reaction zone with the electron beam parallel to the GZO $\langle 100 \rangle$ zone axis. The diffraction pattern (inset) at the interface reveals no special crystallographic relationship between GZO and the adjacent columnar grains of apatite and fluorite products.

possibly because of the associated roughness. However, DF analysis showed no evidence of epitaxial relationships between the parent GZO and the product fluorite (Fig. 9), such as those observed in the column gaps, suggesting the presence of a thin amorphous film at the interface. Figure 9 also shows that the growth of the apatite and fluorite phases near the interface is columnar, suggesting a cooperative mechanism. Diffraction analysis reveals that the apatite grains are textured with a preferred [0001] orientation that is tilted relative to the primary $\langle 100 \rangle$ axis of the GZO column but aligned approximately with the interface normal.

IV. Discussion

The crucial finding of this work is that CMAS infiltration into a columnar GZO TBC (and arguably any APS analog) can be largely suppressed even in the absence of a thermal gradient across the coating. To the limited extent that the intercolumnar channels were penetrated, typically ≤ 30 μm from the original coating surface, they were filled with a mixture of predominantly crystalline phases. This is in striking contrast with the results of identical experiments on 7YSZ coatings, where the channels were completely filled with amorphous CMAS, with only minor amounts of Zr and Y in a solid solution.[15]

The observations are consistent with a scenario wherein CMAS infiltration into GZO is suppressed by crystallization occurring concurrently with the inward flow of the silicate melt and at temperatures above the melting point of the original CMAS deposit ($\sim 1240°C$). This is obviously possible only if (i) the CMAS composition is dynamically modified by dissolution of the base GZO material as it penetrates into the coating, (ii)

the characteristic time scales of the penetration, dissolution, and crystallization kinetics are all comparable, and (iii) the process consumes a large fraction of the melt and yields sufficient volume of crystalline product to fill the gap close to the surface and preclude further percolation downward. (Possible effects of the modified composition on the viscosity and wetting ability of the melt are discussed later). It is therefore important to understand first the chemical interaction between CMAS and GZO and the nature of the reaction products.

(1) Reaction Products and Mass Balance

As a first approximation, the chemical reaction between the model CMAS and stoichiometric GZO may be written as

$$
\begin{aligned}
&C_{0.33}M_{0.09}A_{0.13}S_{0.45} + 1.6G_{0.5}Z_{0.5} \\
&\rightarrow 1.2C_{0.125}G_{0.5}S_{0.375}(\text{apatite}) + Z_{0.8}G_{0.20}(\text{fluorite}) \\
&+ 0.2M_{0.33}A_{0.67}(\text{spinel}) + 0.2C_{0.88}M_{0.12}(\text{residue})
\end{aligned} \quad (1)
$$

where the oxides are identified by the cation first letter for ease of notation and all formulae are given on the basis of 1 g atom of cations. There are three distinct crystalline products (apatite, fluorite, and spinel). Apatite and spinel are given here in stoichiometric form, the binary fluorite represents averaged experimental results (Table I), and the residue (to be elaborated later) is simply what is needed to balance the equation.

The key product is the apatite phase (hexagonal, $P\bar{3}$[19,20]), nominally based on $Gd_8Ca_2(SiO_4)_6O_2$, which melts congruently at $1930°C$.[22] The structure consists of isolated SiO_4 tetrahedra arranged around two types of channels running parallel to the c axis.[20] The larger of these channels, passing through the origin, contains a row of ionic O along its axis surrounded by cations that are coordinated by 7 anions (heretofore 7C sites). Analyses of related structures suggest these sites to be predominantly occupied by Gd^{3+}.[20,23] The smaller channels contain rows of alternating Ca^{2+} and Gd^{3+} at their center, each coordinated by 9 anions from the covalently bonded SiO_4 tetrahedra (9C sites). The apatite formula may thus be rewritten as $Gd_6[Gd_2Ca_2](SiO_4)_6O_2$, where cations in square brackets are 9C, those outside are 7C, and the ionic O is explicitly distinguished from that tied to Si tetrahedral units. The distinction between the sites is relevant to understanding the observed deviations from the apatite stoichiometry (Table I).

Apatite is the only intermediate compound in the quasibinary $CaSiO_3$–Gd_2SiO_5,[22] where it exhibits a finite homogeneity range $Gd_{2x}Ca_{(1-x)}(SiO_4)O_{2x-1}$, with $0.6 \le x \le 0.75$ at 1300°C. Studies in related structures suggest that excess Ca ($0.6 \le x \le 2/3$) is accommodated by substitution for Gd in the 9C sites, with the concurrent creation of vacancies in the ionic O lattice. Conversely, excess Gd is likely to substitute for Ca on 9C sites with the concomitant creation of vacancies in that same lattice. This view is supported by the site occupancies in the defect apatite $La_{9.33}(SiO_4)_6O_2$[19] also formed by Nd,[24] Gd,[25] and Dy.[23] Interestingly, these silicates show substantial solubility of Al,[24] but no significant Al was detected in the present crystals even though they grow from an Al-bearing melt. Instead, they incorporate Zr^{4+} (Table I), presumably in the 7C sites, given the smaller size of this ion relative to Gd^{3+} and Ca^{2+} and its preferred coordination in ZrO_2.[26] The measured Ca:Gd ratios vary from 0.45 to 0.25, the latter corresponding to the stoichiometric apatite. The excess Ca^{2+} helps compensate for the higher Zr^{4+} charge so that fewer anion or cation vacancies need to be created, with attendant benefits to the stability of the structure. Because the measured Zr^{4+} is generally in excess of the amount that can be compensated by the Ca^{2+}, the apatite of interest is likely to contain vacant 9C sites, whereupon its formulation becomes $Gd_{6-y}Zr_y[Gd_{2-(2z+y)/3}Ca_{2+z}](SiO_4)_6O_2$ ($y \ge z$). Taking the average concentrations of 15% CaO and 5% ZrO_2, the corresponding Si and Gd contents are $\sim 38\%$ and $\sim 42\%$, respectively, reasonably close to the measured compositions (Table I). The corresponding site occupancy would imply that $\sim 3\%$ of the 9C sites are vacant.

Because the apatite has a much higher Gd:Zr ratio than GZO, the observed ZrO_2-rich fluorite is expected from mass balance considerations. This phase is found to incorporate Gd over a wide composition range (0.12–$0.3 GdO_{1.5}$), but always as a minor component, as well as a small amount of Ca (0.03–$0.04 CaO$). In that respect, the reaction is similar to that with 7YSZ in which the TBC dissolves to precipitate a tetragonal phase with lower Y and some Ca incorporated from the CMAS melt.[15] This process, however, does not yield other crystalline products and the excess Y simply remains in the solid solution in the residual amorphous CMAS, whose composition does not change significantly. In contrast, the substantial depletion of Ca and Si from the melt when reacting with GZO to form apatite and the apparent inability of apatite and fluorite to incorporate significant amounts of Mg or Al lead to the supersaturation of these oxides in the melt and their precipitation as spinel, the third crystalline product of the reaction. This is fully consistent with the expected melt evolution path in the quaternary liquidus projection of the CaO–MgO–Al_2O_3–SiO_2 system (Fig. 2647 in Levin et al.[27]). The small amounts of Zr and Gd detected in the spinel presumably substitute for Al on octahedral sites.

Considering now the measured compositions of the different phases, on average, the balanced reaction may be rewritten as

$$C_{0.33}M_{0.09}A_{0.13}S_{0.45} + 1.20G_{0.53}Z_{0.47}$$
$$\rightarrow 1.19C_{0.15}G_{0.42}Z_{0.05}S_{0.38}(\text{apatite})$$
$$+ 0.65Z_{0.77}G_{0.20}C_{0.03}(\text{fluorite}) \qquad (2)$$
$$+ 0.21M_{0.33}A_{0.63}Z_{0.02}G_{0.02}(\text{spinel})$$
$$+ 0.15C_{0.86}M_{0.14}(\text{residue})$$

Note that the yield of crystalline products is 93% of the reactants on a molar basis. Given molar volumes (in cm^3 per gram-atom of cations) of 20.7 (CMAS), 20.3 (apatite), 22.1 (GZO), 20.8 (fluorite), and 13.3 (spinel),[‡] the volumetric yield of the reaction excluding the residue is $\sim 86\%$, just below the typical threshold for percolation. This is evidently an underestimate because the residue is diluted by additional reactants as indicated by the measured compositions of the fourth phase pockets in the reaction zone. Nevertheless, the reaction does lead in practice to the sealing of the intercolumnar gaps, probably because the "residual" CMAS observed in them is also crystalline. A comparison of reactions (1) and (2) indicates that the incorporation of Zr^{4+} into the apatite significantly reduces the amount of GZO that needs to be dissolved to produce the same volume of this crucial phase and, in principle, to seal the gaps. The broader aspects of the sealing mechanism and its interplay with the above reaction are now examined.

(2) Infiltration Inhibiting Mechanism

The microstructure of the reaction zone appears to have evolved in two major stages, as illustrated schematically in Fig. 10. The first one involves the sealing of the intercolumnar gaps soon after the deposit starts melting and the first liquid begins to flow into the coating. The mechanism is one of rapid dissolution of the GZO upon contact with the melt and nearly concurrent precipitation of multiple crystalline phases following approximately reaction (2). The very limited penetration suggests that the infiltration rate may be initially controlled by the availability of melt, and hence by the melting kinetics, rather than by the resistance to capillary-driven flow into a porous bed from a liquid reservoir at the surface, as assumed in the 7YSZ study.[15] (The details are not as critical in 7YSZ because the dissolution/reprecipitation process does not change significantly the volume of the melt or its flow/crystallization behaviors.) The envisaged process starts with the incipient melt trickling down the sides of the column gap, infiltrating the feathery porosity and thus the surface exposed to dissolution, and at some point downstream nucleating the apatite phase (Fig. 10(a)). Whether the fluorite phase might nucleate upstream at an earlier time is not critical to the apatite formation because its volume would be smaller than that of the dissolved GZO and would still lead to a net increase in the Gd content of the melt. The apatite crystals could nucleate within the melt and travel downstream until they become trapped at a narrow point in the channel, or heterogeneously on the column surfaces. The same process arguably occurs within the feathery pores, limiting the penetration of the melt into the column core, although factors such as pore breakdown upon thermal exposure could play an equally important role. As more melt trickles into the channel and additional apatite and fluorite form, both the area available for flow and the volume of melt decrease until the latter freezes into a mixture of spinel and residual aluminosilicate, hindering further penetration (Fig. 10(b)).

The melt-starved infiltration scenario has two important implications: (i) it makes the dissolution and crystallization processes more competitive with the infiltration kinetics, and (ii) because the incipient melt is unlikely to have the composition of the bulk CMAS, the relative proportions of crystalline products in reaction (2) could be different, presumably accounting for the smaller amounts of the spinel and residual CMAS observed in practice. (Sampling statistics, always an issue in TEM, could obviously play a role too.) Elucidation of these early infiltration stages is the subject of ongoing research and will be reported in a subsequent paper.

(3) Reaction with the Column Tips

The "ghost" traces of the intercolumnar gaps within the main reaction layer suggest that their filling with crystalline products occurred very early in the process, while the column tips had experienced only minimal attack. Once the channels are

---

[‡]Molar volumes were calculated from the CMAS density measured in Krämer et al.[15] and for the crystalline materials from the compositions in reaction (2) and lattice parameters for apatite from this work, and for the zirconate, fluorite, and spinel phases from the literature.[28,29]

Progress in Thermal Barrier Coatings

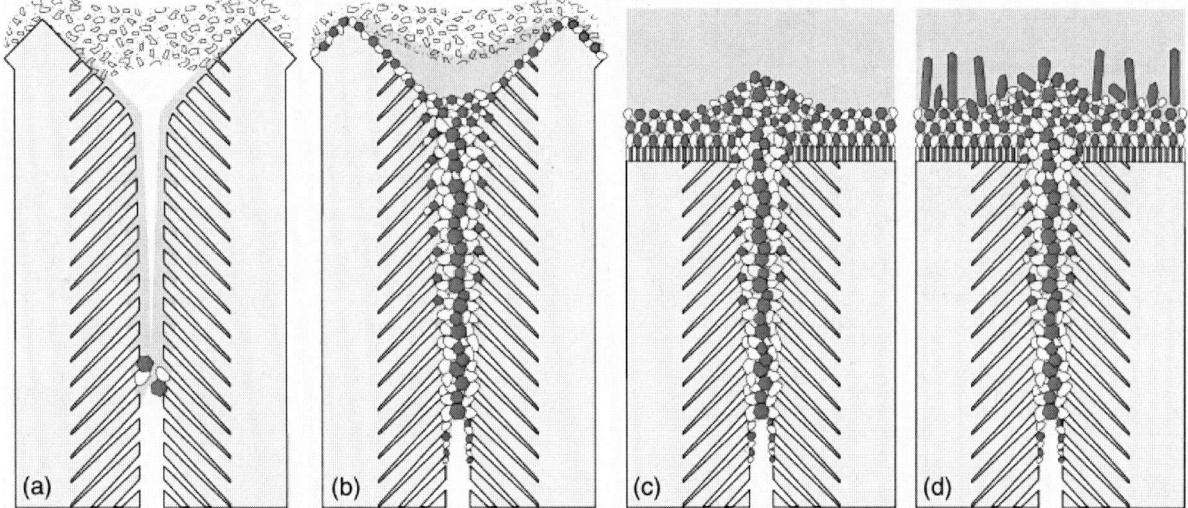

Fig. 10. Schematic representation of the envisaged mechanism leading to the microstructure of the reaction layer. (a) early stages wherein the incipient melt wicks down the column sides, leading to the nucleation of the reaction products within the channel (the granular material on top represents the unmelted deposit), (b) channel fills up with reaction products hindering further penetration of the melt, while column tips are only moderately attacked at this stage, (c) column tips are gradually attacked by the bulk calcium magnesium alumino-silicate melt forming a graded microstructure, and (d) additional growth of crystalline phases takes place during cooling.

blocked, the increasing volume of melt spreads laterally, covering the entire surface as noted in Fig. 1(a). The second major stage in the reaction mechanism thus involves the recession of the column tips as a result of further reaction with the melt (Fig. 10(c)). The larger volume of melt above the TBC, relative to that within the channels, is believed to hinder the formation of a dense "barrier" layer equivalent to the structure formed between columns, because it never becomes sufficiently saturated to crystallize completely. Indeed, no spinel or crystalline residual CMAS was found in the microstructure above the column cores, leaving a volume of melt through which the transport necessary for additional dissolution could take place.

The above scenario is broadly similar to the attack of the column tips on 7YSZ under similar conditions,[15] with some important differences. The upper reaction zone in 7YSZ comprises loose agglomerates of t-ZrO_2 globules embedded in a CMAS matrix sufficiently abundant to allow substantial contact with the underlying columns and the persistence of the dissolution–reprecipitation process. In contrast, the reaction with the GZO tips generates a layer of alternating apatite and fluorite elongated grains with an intergranular amorphous phase that becomes progressively thinner near the interface with the remaining GZO column core (Figs. 7–9). The transport of species needed to sustain the reaction, i.e., Ca and Si toward the interface, and excess Zr and Gd away from it, presumably takes place through this amorphous phase. Whether the latter extends to the interface, as suggested by the absence of a crystallographic relationship between GZO and the fluorite growing from it, and mediates the reaction is yet to be ascertained.

The details of the column core attack are still under investigation and are not essential to the main subject of this paper, i.e., the mechanism by which access of the CMAS melt to the majority of the pore space in the coating is impeded. The recession of the column cores arguably takes place over a much longer time scale than the sealing of the gaps. The structure of the near interface layer suggests that the rate decreases over time and becomes controlled by diffusion through the network of amorphous boundary films. The interfacial reaction apparently yields phases with compositions that are not fully compatible with the melt above, notably a higher Gd content in the fluorite phase. Away from the interface, the phases experience melt-mediated coarsening, adjusting their composition in the process. The larger acicular crystals at the top could evolve partially under these conditions but probably develop their elongated

shape as a result of additional growth upon cooling (Fig. 10(d)). Remarkably, the "ghost" microstructure produced by the earlier reaction within the intercolumnar gaps remains substantially unaltered, as can be seen in Fig. 4. This resilience is probably the result of phase compositions that are closer to equilibrium with the melt, and the absence of significant amounts of percolating amorphous phase as shown, for example, in Fig. 6(a). Whether some melt penetration and coarsening may occur near the surface, as suggested by Fig. 4, is not particularly critical as long as the gaps remain sealed further down. It is of some concern that the recession front might reach the bottom of the filled length within a gap and melt could leak down the coating, but this appears unlikely once the dense interfacial layer develops, as indicated by the open gaps right next to the bottom of the reaction layer in Figs. 2 and 3.

(4) Additional Considerations

Various other aspects of the CMAS–GZO interaction and its potential effect on the infiltration-governing parameters merit discussion. Modifications to the wetting characteristics of the CMAS melt by the dissolved Gd and Zr are undoubtedly possible, but experimental evidence suggests that the observed behavior is not due to diminished wetting ability. Indeed, the melt spreads over nearly the entire coating (Fig. 1(a)) and it readily penetrates the fine feathery pores on the column sides (Fig. 3(a)). It also wets the reaction products, as evident in Figs. 7 and 8. A more likely effect of the dissolved species could be on the viscosity of the melt and the ensuing inward flow. There are documented effects of rare-earth additions increasing the viscosity of aluminosilicate glasses,[30] but the latter contained no other modifiers like Ca or Mg, and much higher concentrations of rare earth in the melt. Conversely, the crystallization of apatite would increase the Ca:Si ratio in the melt, decreasing the connectivity of the network and, in principle, its viscosity. Arguably, the effects of the dissolution of GZO into the CMAS in promoting the crystallization of the melt are more important than those on the melt viscosity.

Finally, the potential of the infiltration-inhibiting reaction needs to be assessed in the context of more realistic conditions. The thermal gradient across the coating in actual engine operation should further limit the penetration, as it does in 7YSZ,[15] subject to the constraint that sufficient dissolution must take place for the melt to crystallize and fill the gaps. Conversely, engine operation involves thermal cycling, and hence the resistance of the reacted/infiltrated layer to delamination is a critical issue requiring evaluation. While the modified layer is substantially thinner than those observed in actual 7YSZ TBCs, the toughness of GZO is significantly lower[21] and the thermal expansion properties of the reaction products are not known. The CTE for the fluorite phase is probably similar to that of GZO (11.6 ppm/K[28]) but the literature for other silicate oxyapatites suggests that their CTEs tend to be significantly lower and anisotropic, e.g., 8.9 and 6.6 ppm/K along the a- and c-axes, respectively, for the closely related $La_8Ca_2(SiO_4)_6O_2$.[31] It is then possible that the thermal characteristics of the reaction products may partly counteract the benefits of the reduced penetration. These issues are under current investigation.

V. Conclusions

The current investigation has shown that the propensity of CMAS to react with most thermal barrier oxides of interest can be manipulated by careful selection of the coating material to mitigate the penetration of the compliance-inducing features of the microstructure and the detrimental effects on the strain tolerance of the coatings. Gadolinium zirconate has proven to be effective in this manner, and the details of the reaction suggest that other REZ may behave similarly. In essence, the mitigation mechanism relies on the dissolution of the zirconate into the melt and the ensuing conversion of the latter into a mixture of crystalline phases that fill the flow channels and prevent further penetration. The extent of penetration depends on the relative competitiveness of the infiltration, dissolution, and crystallization kinetics, all of which should be dependent on the temperature and the compositions of the melt and oxide material. Key to the effectiveness of the mechanism is the formation of a highly stable apatite phase incorporating Ca, Gd, Si, and some Zr. After the early sealing of the flow channels, the reaction continues slowly by interaction of the bulk CMAS with the column tips. The mechanism and product morphology are conducive to a progressive slowing of this attack, which is beneficial to the survivability of the reaction layer under thermal cycling. Further research on the mechanisms and assessment of the effectiveness of this mitigation approach is in progress.

Acknowledgments

The authors would like to thank Dr. D. Klenov (UCSB) for his guidance in Z-contrast imaging. Helpful discussions with Dr. M. Maloney (Pratt & Whitney) and Prof. A. G. Evans (UCSB) are gratefully acknowledged.

References

[1]M. J. Maloney, "Thermal Barrier Coating Systems and Materials"; U.S. Patent 6,117,560, 2000.

[2]R. Vassen, X. Cao, F. Tietz, D. Basu, and D. Stöver, "Zirconates as New Materials for Thermal Barrier Coatings," *J. Am. Ceram. Soc.*, **83** [8] 2023–8 (2000).

[3]M. J. Maloney, "Thermal Barrier Coating Systems and Materials"; U.S. Patent 6,177,200, 2001.

[4]R. Subramanian, "Thermal Barrier Coating Having High Phase Stability"; U.S. Patent 6,258,467, 2001.

[5]S. M. Lakiza, O. Fabrichnaya, C. Wang, M. Zinkevich, and F. Aldinger, "Phase Diagram of the $ZrO_2–Gd_2O_3–Al_2O_3$ System," *J. Eur. Ceram. Soc.*, **26**, 233–46 (2006).

[6]A. Rouanet, "Contribution a l'etude des systèmes zircone-oxydes des lanthanides au voisinage de la fusion," *Rev. Int. Hautes Tempér. Réfract.*, **8** [2] 161–80 (1971).

[7]C. G. Levi, "Emerging Materials and Processes for Thermal Barrier Systems," *Curr. Opin. Solid State Mater. Sci.*, **8** [1] 77–91 (2004).

[8]R. L. Jones, "Thermal Barrier Coatings"; pp. 194–235 in *Metallurgical and Ceramic Protective Coatings*, Edited by K. H. Stern. Chapman & Hall, London, 1996.

[9]M. P. Borom, C. A. Johnson, and L. A. Peluso, "Role of Environmental Deposits and Operating Surface Temperature in Spallation of Air Plasma Sprayed Thermal Barrier Coatings," *Surf. Coat. Technol.*, **86–87**, 116–26 (1996).

[10]J. R. Nicholls, R. G. Wellman, and M. J. Deakin, "Erosion of Thermal Barrier Coatings," *Mater. High Temp.*, **20** [2] 207–18 (2003).

[11]X. Chen, M. Y. He, I. T. Spitsberg, N. A. Fleck, J. W. Hutchinson, and A. G. Evans, "Mechanisms Governing the High Temperature Erosion of Thermal Barrier Coatings," *Wear*, **256** [7–8] 735–46 (2004).

[12]A. G. Evans, N. A. Fleck, S. Faulhaber, N. Vermaak, M. Maloney, and R. Darolia, "Scaling Laws Governing the Erosion and Impact Resistance of Thermal Barrier Coatings," *Wear*, **260**, 886–94 (2006).

[13]J. L. Smialek, F. A. Archer, and R. G. Garlick, "Turbine Airfoil Degradation in the Persian Gulf War," *J. Metals*, **46** [12] 39–41 (1994).

[14]C. Mercer, S. Faulhaber, A. G. Evans, and R. Darolia, "A Delamination Mechanism for Thermal Barrier Coatings Subject to Calcium-Magnesium-Alumino-Silicate (CMAS) Infiltration," *Acta Mater.*, **53** [4] 1029–39 (2005).

[15]S. Krämer, J. Y. Yang, C. A. Johnson, and C. G. Levi, "Thermochemical Interactions of Thermal Barrier Coatings with Molten $CaO–MgO–Al_2O_3–SiO_2$ (CMAS) Deposits," *J. Am. Ceram. Soc.*, **89** [10] 3167–75 (2006).

[16]S. Krämer, S. Faulhaber , M. Chambers, D. R. Clarke, C. G. Levi, J. W. Hutchinson, and A. G. Evans, "Mechanisms of Cracking and Delamination within Thick Thermal Barrier Systems in Aero-Engines Subject to Calcium–Magnesium–Almino–Silicate (CMAS) Penetration," *Mater. Sci. Eng.*, (2007), in press.

[17]S. G. Terry, "Evolution of Microstructure during the Growth of Thermal Barrier Coatings by Electron-beam Physical Vapor Deposition"; Doctoral Dissertation in Materials, University of California, Santa Barbara, CA, 2001.

[18]R. M. Leckie, S. Krämer, M. Rühle, and C. G. Levi, "Thermochemical Compatibility between Alumina and $ZrO_2–GdO_{3/2}$ Thermal Barrier Coatings," *Acta Mater.*, **53** [11] 3281–92 (2005).

[19]J. E. H. Sansom, D. Richings, and P. R. Slater, "A Powder Neutron Diffraction Study of the Oxide-Ion-Conducting Apatite-Type Phases $La_{9.33}Si_6O_{26}$ and $La_8Sr_2Si_6O_{26}$," *Solid State Ionics*, **139**, 205–10 (2001).

[20]J. R. Tolchard, M. S. Islam, and P. R. Slater, "Defect Chemistry and Oxygen Ion Migration in the Apatite-Type Materials $La_{9.33}Si_6O_{26}$ and $La_8Sr_2Si_6O_{26}$," *J. Mater. Chem.*, **13**, 1956–61 (2003).

[21]R. M. Leckie, "Fundamental Issues Regarding the Implementation of Gd Zirconate in Thermal Barrier Systems"; Doctoral Dissertation in Materials, University of California, Santa Barbara, CA, 2006.

[22]N. F. Fedorov, I. F. Andreev, and T. F. Korneeva, "System $CaSiO_3–Gd_2SiO_5$," *Inorg. Mater.*, **8** [12] 1919–21 (1972).

[23]S. T. Misture, S. P. Harvey, R. T. Francy, Y. Gao, S. DeCarr, and S. C. Bancheri, "Synthesis, Crystal Structure, and Anisotropic Thermal Expansion of $Dy_{4.67}(SiO_4)_3O$," *J. Mater. Res.*, **19** [8] 2330–5 (2004).

[24]U. Kolitsch, H. J. Seifert, and F. Aldinger, "Phase Relationships in the Systems $RE_2O_3–Al_2O_3–SiO_2$ (RE = Rare Earth Element, Y and Sc)," *J. Phase Equilibria*, **19** [5] 426–33 (1998).

[25]U. Kolitsch, H. J. Seifert, and F. Aldinger, "Phase Relationships in the System $Gd_2O_3–Al_2O_3–SiO_2$," *J. Alloys Compd.*, **257** [1–2] 104–14 (1997).

[26]S.-M. Ho, "On the Structural Chemistry of Zirconium Oxide," *Mater. Sci. Eng.*, **54**, 23–29 (1982).

[27]E. M. Levin, C. R. Robbins, and H. F. McMurdie (eds) *Phase Diagrams for Ceramists*, Vol. II. The American Ceramic Society, Columbus, OH, 1969.

[28]M. Zinkevich, C. Wang, F. M. Morales, M. Ruhle, and F. Aldinger, "Phase Equilibria in the $ZrO_2–GdO_{1.5}$ System at 1400–1700°C," *J. Alloys Compd.*, **398**, 261–8 (2005).

[29]N. W. Grimes and E. A. Al-Ajaj, "Low-Temperature Thermal Expansion of Spinel," *J. Physi.: Condensed Matter*, **4** [30] 6375–80 (1992).

[30]P. F. Becher and M. K. Ferber, "Temperature-Dependent Viscosity of SiReAl-Based Glasses as a Function of N:O and Re:Al Ratios (RE = La, Gd, Y and Lu)," *J. Am. Ceram. Soc.*, **87** [7] 1274–9 (2004).

[31]R. H. Hopkins, J. de Klerk, P. Piotrowski, M. S. Walker, and M. P. Mathur, "Thermal and Elastic Properties of Silicate Oxyapatite Crystals," *J. Appl. Phys.*, **44** [6] 2456–8 (1973).   □

SEGMENTATION CRACKS IN PLASMA SPRAYED THIN THERMAL BARRIER COATINGS

Hongbo Guo, Hideyuki Murakami and Seiji Kuroda

Materials Engineering Laboratory, National Institute for Materials Science (NIMS)

1-2-1 Sengen, Tsukuba

Ibaraki 305-0047, Japan

ABSTRACT

Thick thermal barrier coatings (TBCs) containing segmentation cracks have seen a successful application in combustion chamber parts. In this work, rather thin TBCs were produced by spraying yttria stabilized zirconia (YSZ) coatings onto two kinds of bond coats: CoNiCrAlY and Pt-Ir. The effects of process parameters including substrate temperature on the density of segmentation cracks were studied. The thermal cycling performance and failure mechanisms of the segmented thin TBCs were also investigated. The coating sprayed at 1073 K contains a segmentation crack density of approximately 5 mm^{-1}. The segmented coatings significantly improved the thermal cycling lifetime as compared with the non-segmented coatings. The segmented TBC with Pt-Ir bond coat attained a lifetime of more than 5000 cycles (each cycle includes 3 minute heating up to the maximum temperature and 7 minute holding at the temperature), revealing an excellent thermal cycling performance.

INTRODUCTION

Thermal barrier coatings (TBCs) perform the vital function of insulating gas turbine components, such as burner cans and turbine vanes or blades, which are subjected to excessive temperatures. The benefits of a TBC can be utilized in several ways: increased engine efficiency and combustion temperature, prolonged lifetime of parts, reduced transient stresses in parts, reduction of cooling air that results in higher cooling efficiency and lower emission.

Thermal barrier coatings are usually produced by electron beam physical vapor deposition (EB-PVD) or plasma spray. Compared to the EB-PVD TBC, the plasma sprayed coating has a better thermal insulation due to its multi-layered microstructure. Failure of plasma sprayed TBCs often occur at the interface between thermally grown oxide (TGO) layer and topcoat or within the topcoat due to the growth of TGO and thermal mismatch stresses [1-3]. Coating failures can decrease engine performance and accelerate substrate metal deterioration. Microstructure modifications of TBCs have been done in order to improve their thermal cycling performance. For traditional TBCs (usually<0.5 mm in thickness) sufficient durability is achieved by the use of porous and microcracked coatings [4]. For thick TBCs the thermal shock resistance can be significantly improved by inducing so-called segmentation cracks into the coatings as these cracks increase the compliance of coatings to substrates [5-7].

In TBC system, the bond coat is either MCrAlY (M=Ni, Co, etc) alloy or a Pt modified aluminide coating. To improve the protective performance of the PtAl coating, much attention has been focused on Pt-based binary alloy coatings, such as Pt-Ni and Pt-Rh [8,9]. Recently, a Pt-Ir coating was proposed [8,10,11], because Ir has the highest melting temperature (2716 K) among platinum group metals, excellent chemical stability, low oxygen diffusivity, and lower diffusivity into Ni-based alloys and lower cost than Pt.

In this paper, thin segmented TBCs are sprayed onto the MCrAlY and Pt-Ir bond coats, aiming at improving the thermal cycling performance of the thin TBCs. The effect of segmentation crack density on the thermal cycling lifetime of the thin TBCs will be studied. Also, the failure mechanisms of the thin TBCs will be investigated.

EXPERIMENTAL

A Ni-based single crystal superalloy, TMS-82+ with ⟨100⟩ orientation, with the compositions as given in Table 1, and an Inconel 718 Ni-based superalloy were used as substrate materials. Co-32Ni-21Cr-8Al-0.5Y powder (Sulzer Metco 9954) and ZrO_2-8mass% Y_2O_3 powder (Sulzer Metco 204 NS) were chosen for spraying MCrAlY bond coat and yttria stabilized zirconia (YSZ) topcoat, respectively. The MCrAlY bond coat of around 200 μm were sprayed onto the Inconel 718 substrates by low pressure plasma spray (Plasma Giken Corp., Japan). Pt-Ir coatings of approximately 7~10 μm were deposited on TMS-82+ substrates at 873 K by DC magnetron sputtering from Pt-Ir target. The deposition parameters have been reported elsewhere [11,12]. The Pt-Ir coated specimens were embedded in an aluminum retort containing a mixture of Al_2O_3, Al, Fe and NH_4Cl and aluminizing treatment was carried out at 1273 K for 5 h in flowing argon atmosphere.

Figs.1a and b show the micrograph of cross-section of Pt-Ir coated TMS 82+ specimen after aluminizing treatment and associated element distributions across the coating thickness, respectively. The top layer of around 25 μm basically consists of $PtAl_2$ and β-(Ni,Pt,Ir)Al phases and below this, an interdiffusion layer mainly comprising β-NiAl phase is formed on the substrate. YSZ coatings of around 400 μm were sprayed onto the CoNiCrAlY and Pt-Ir coated substrates in atmospheric plasma spraying using an SG 100 gun (Praxair, USA). Four sets of spray parameters, as shown in Table 2, were used for spraying YSZ coatings aiming at attaining different levels of segmentation crack densities. The choice of processing parameters for spraying segmented TBCs is based on the experimental details described in some literatures [13,14]. The main idea is that segmentation cracks were created by thermal tensile stresses during the deposition and hence, the heat input to the substrate should be enough high.

Progress in Thermal Barrier Coatings

Regarding to this, the substrate temperatures were varied by changing spray distance and plasma power. Condition 1 (C1), featured with a lower plasma power and a long spray distance as compared to other spray conditions, was used for spraying traditional non-segmented TBCs.

Table 1   Nominal chemical composition, wt%

Element	Ni	Co	Cr	Mo	W	Al	Ti	Ta	Hf	Re
	bal	7.8	4.9	1.9	8.7	5.3	0.5	6.0	0.1	2.4

Table 2   Spray parameters for YSZ topcoat (V: traverse speed of plasma gun; T_s: substrate temperature).

No.	Power (KW)	Ar (slpm)	He (slpm)	Distance (mm)	Feed rate (g/min)	V (mm/s)	T_s (K)
C1	23.1	50	27	120	20	150	593
C2	26.4	50	27	100	20	150	773
C3	34	50	27	80	20	150	923
C4	40.8	50	27	60	20	150	1073

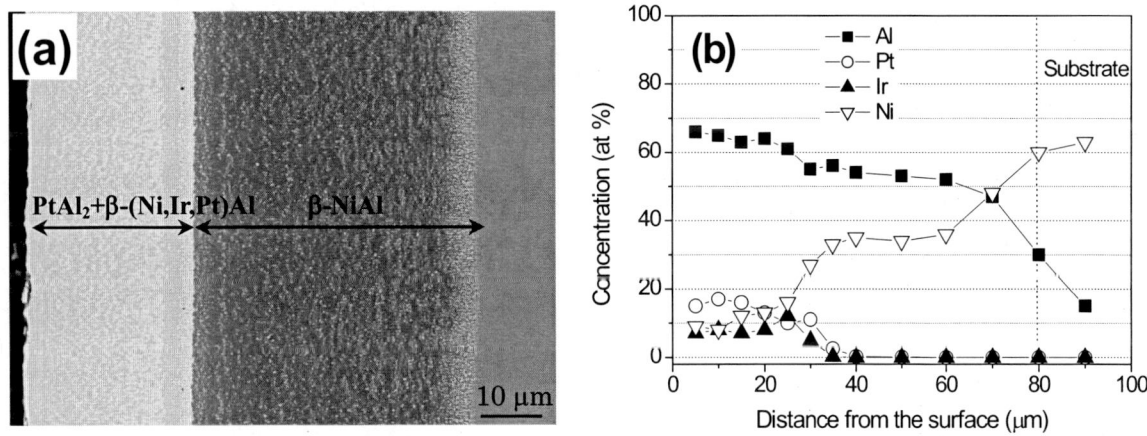

Fig. 1 (a) SEM micrograph of cross-section and (b) concentration profiles of the elements of the Pt-Ir coated TMS-82+ substrate after aluminizing treatment.

Thermal cycling testing of disk-shaped specimens (φ25mm×3mm) was performed in a high temperature furnace equipped with an automatic mechanical system to move 4 specimens in and out of the furnace simultaneously. The coated specimens were placed in Al_2O_3 holders that connected to a water tank so as to attain a certain temperature gradient across the specimen thickness during heating. The furnace temperature during the testing was set to be 1200 °C. The specimens were heated to the maximum temperature within 3 min and held at the temperature for 7 min and then cooled by air jets for 3 min from the coating side. The coating surface and substrate temperatures of the specimens were measured by thermocouples fixed to the specimens. The lifetime of a TBC specimen is defined as the number of cycles the specimen underwent before more than 1/3 surface area of the coating spalled off from the substrate.

The microstructures of the coatings before and after thermal cycling testing were examined by optical microscopy (OM), scanning electron microscopy (SEM) equipped with energy dispersive spectroscopy (EDS), and the phases in the coatings identified by X-ray diffractometry (XRD).

RESULTS AND DISCUSSION

Microstructures of segmented thin TBCs

Figs. 2a and b show the micrographs of the TBCs sprayed under C1 condition and C4 condition. The coating sprayed under C1 condition is highly porous since larger voids are abundant for the coating. Compared to the coating at C1 condition, the coating sprayed at C4 condition is much denser and contains some segmentation cracks running perpendicular to coating surface. In this study, the segmentation cracks are defined as the vertical cracks penetrating at least half the coating thickness.

The segmentation crack density of the as-sprayed coatings is shown in Fig. 3, in which the crack density is calculated by dividing the number of segmentation cracks found in a cross-section with the length of the cross-section. The crack density of the coating sprayed under C1 condition is nearly zero and hence the coating is considered as traditional non-segmented TBC. For those coatings deposited at higher substrate temperatures, the crack densities are much higher than that of the coating at C1 condition, especially for the coating at C4 condition, a crack density of approximately 5 mm^{-1} is achieved. Therefore, it can be concluded that the substrate temperature is an important factor in affecting the origin of segmentation cracks and a high substrate temperature gives rise to an increased segmentation crack density. The segmentation cracks initiate and propagate during the deposition phase, as a result of biaxial tensile stresses that arise from in-situ sintering. The mechanism for the development of segmentation cracks has been discussed in some literatures [15,16]. A good joining between lamellae is a key point for the propagation of segmentation cracks. Accordingly, high heat input to the substrate is one of most favorable factors in developing segmentation cracks because high substrate and particle temperatures promote the joining of adjacent lamellae.

Thermal cycling of the TBC with CoNiCrAlY bond coat

During thermal cycling tests, the maximum temperature at the coating surface is around 1393 K, while the substrate temperature is in a range of 1323 K to 1343 K. The lifetimes of the specimens are strongly dependent on the segmentation crack density, as shown in Fig. 4.

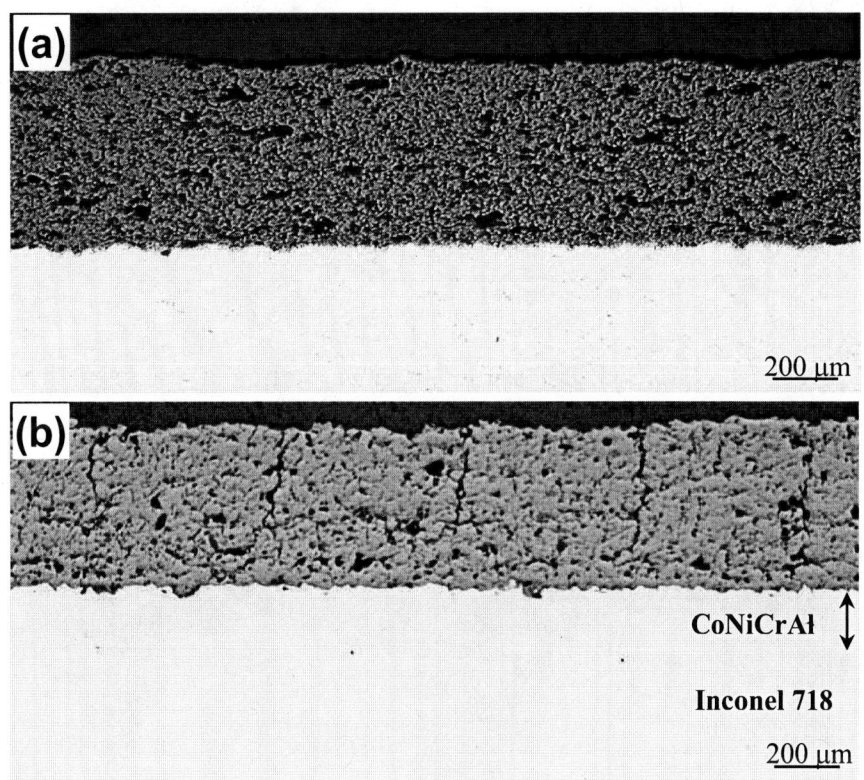

Fig. 2. Cross-sections of TBCs sprayed at C1 condition (a) and C4 condition (b).

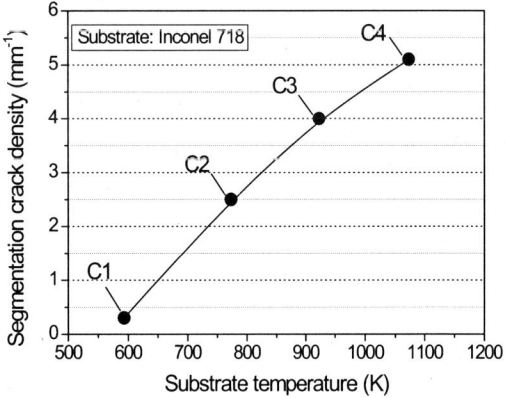

Fig. 3. Effect of substrate temperature on segmentation crack densities of TBCs with CoNiCrAlY bond coat.

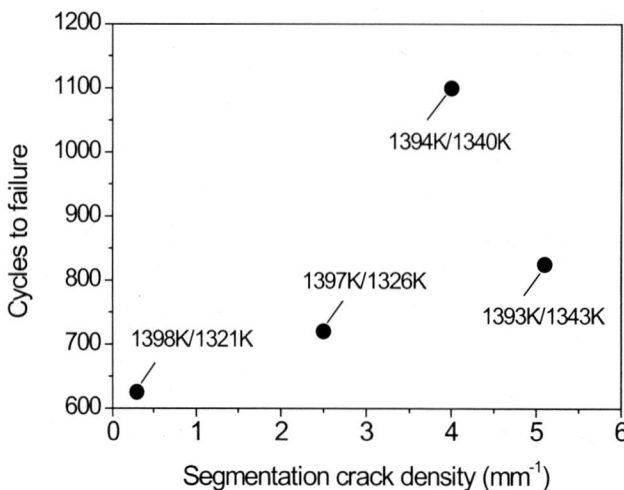

Fig. 4. Effect of segmentation crack density on thermal cycling lifetimes of the TBCs with CoNiCrAlY bond coat. The corresponding coating surface temperature (the right) and substrate temperature (the left) is also indicated.

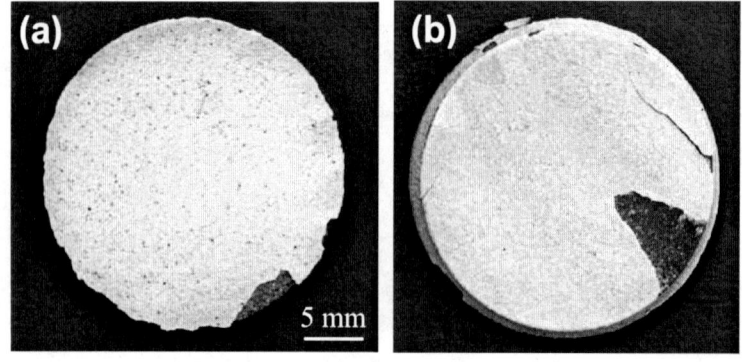

Fig. 5. Photographs of TBC sprayed under C1 condition cycled to 1398 K ($T_{surface}$) for 600 cycles (a) and TBC under C3 condition cycled to 1394 K ($T_{surface}$) for 1100 cycles (b).

The non-segmented TBC specimen has a lifetime of less than 650 cycles, while all of the segmented TBC specimens showed much longer lifetimes. A maximum lifetime of approximately 1100 cycles is achieved for the coating sprayed under C3 condition that has a crack density of around 4 mm^{-1}. It should be noted that the coating sprayed under C4 condition did not show the best thermal cycling performance, although the coating contains the highest crack density among the sprayed coatings. This can be explained by the processing parameters used for spraying the coating. For the coating sprayed under C4 condition, oxidation of the bond coat could occur due to quite high plasma power and short spray distance. On the other hand, too high a substrate temperature, up to 1073 K achieved during the spraying, could result in a high residual stress in the coating upon cooling down, which would degrade the bond strength between the bond coat and YSZ topcoat. Regarding to this, C3 condition is considered as the optimized condition for spraying TBC in terms of thermal cycling performance.

Both the non-segmented and the segmented TBC specimens show a spallation of large parts of ceramic coatings that occurred at the edge of the coatings, as shown in Fig. 5. This type of failure is primarily caused by the growth of TGO on the bond coat due to long time exposure at high temperature. The cross-sections of the above failed coatings were examined, as shown in Fig. 6. TGO layers were formed in the coatings and the TGO layer in the segmented coating (C3) is apparently thicker than that in the non-segmented coating (C4) due to relatively longer time exposure of the segmented coating. Some horizontal cracks propagated along the interface between bond coat and YSZ topcoat, indicating that the spallation observed in Fig. 5 close to the interface was induced by these cracks. On the other hand, the coatings after cycling is much thinner than the coatings before cycling and the coating surfaces even became rougher. This suggests that another kind of failure occurred in the YSZ coatings simultaneously during cycling by spallation or chipping close to the coating surface. Therefore, it can be concluded that the thin segmented TBCs reveal two kinds of failures, one of which occurred by spallation of large parts of coatings close to the YSZ/bond coat interface induced by the growth of TGO, and the other spallation or chipping close to the coating surface that was caused by thermal expansion mismatch stresses on cooling.

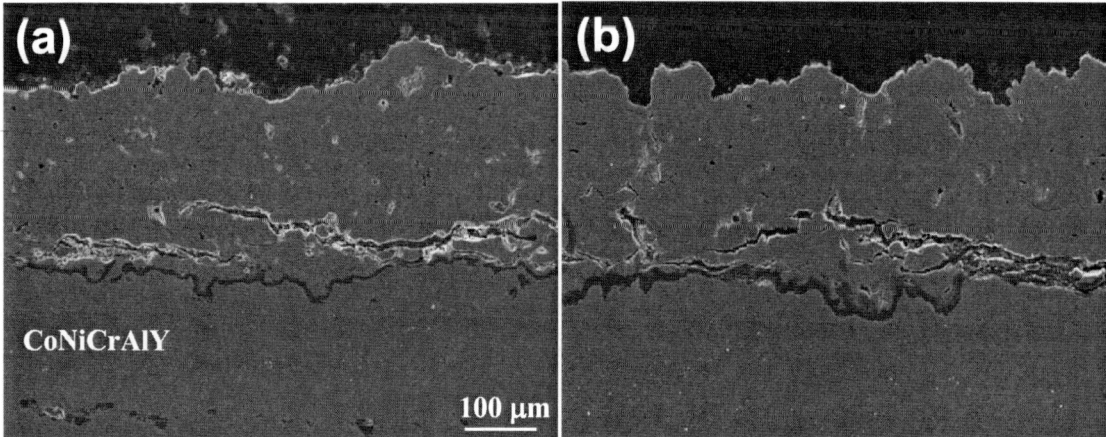

Fig. 6. SEM micrographs of cross-sections of TBCs sprayed under C1 condition after 600 cycles (a) and sprayed under C3 condition after 1100 cycles (b).

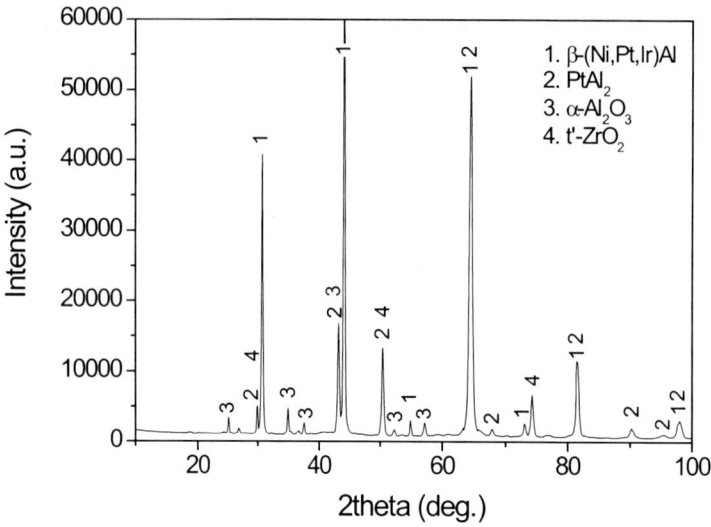

Fig. 9. XRD pattern of the Pt-Ir bond coated specimen after spallation of YSZ topcoat.

SUMMARY

Thin TBCs with different levels of segmentation crack density were sprayed onto the CoNiCrAlY and Pt-Ir bond coat. The substrate temperature played a crucial role in affect the origin of the cracks and the coating deposited at 1073 K attained a crack density of around 5 mm^{-1}. The segmented coatings showed longer thermal cycling lifetimes compared to the non-segmented coatings. Two kinds of failures occurred in the segmented coatings, one of which is chipping or spallation close to the coating surface, and the other spallation close to the interface between YSZ and bond coat primarily caused by the growth of TGO. The segmented TBC with Pt-Ir bond coat attained a lifetime of more than 5000 cycles, revealing an excellent thermal cycling performance.

ACKNOWLEDGMENTS

This work is supported by the Japan Society for Promotion of Science (JSPS) Fellowship program. Mr. M. Komatsu, Mr. M. Shibata and Mr. A. Yamaguchi from National Institute for Materials Science (NIMS) are gratefully acknowledged for their contributions in experiments.

REFERENCES

[1]D.R. Clarke, R.J. Christensen and V. Tolpygo, "The Evolution of Oxidation Stresses in Zirconia Thermal Barrier Coated Superalloy Leading to Spalling Failure", Surf. Coat. Technol., 94-5(1-3),

89-93 (1997).

[2]A. Rabiei and A.G. Evans, "Failure Mechanisms Associated with the Thermally Grown Oxide in Plasma-Sprayed Thermal Barrier Coatings", Acta Mater., 48(15), 3963-76 (2000).

[3]Nitin P. Padture, Maurice Gell and Eric H. Jordan, "Thermal Barrier Coatings for Gas-Turbine Engine Applications", Science, 296, 280-84 (2002).

[4]R. Vaßen, F. Traeger and D. Stöver, "Correlation between Spraying Conditions on Microcrack Density and Their Influence on Thermal Cycling Life of Thermal Barrier Coatings", J. Therm. Spray. Technol., 13(3), 396-404 (2004).

[5]T.A. Taylor, D.L. Appleby, A.E. Weatherill and J. Griffiths, "Plasma-Sprayed Yttria-Stabilized Zirconia Coatings: Structure-Property Relationships", Surf. Coat. Technol., 43/44, 470-80 (1990).

[6]P. Bengtsson, T. Ericsson and J. Wigren, "Thermal Shock Testing of Burner Cans Coated with a Thick Thermal Barrier Coating", J. Therm. Spray. Technol., 7(3), 340-48 (1998).

[7]H.B. Guo, R. Vaßen and D. Stöver, "Thermophysical Properties and Thermal Cycling Behavior of Plasma Sprayed Thick Thermal Barrier Coating", Surf. Coat. Technol., 192(1), 48-56 (2005).

[8]Y. Zhang, J.A. Haynes, W.Y. Lee, I.G. Wright, B.A. Pint, et al., "Effects of Pt Incorporation on the Isothermal Oxidation Behavior of Chemical Vapor Deposition Aluminide Coatings", Met. Mater. Trans. A, 32(7), 1727-41 (2001).

[9]G. Fisher, P.K. Datta, J.S. Burnell-Gray, "An Assessment of the Oxidation Resistance of an Iridium and an Iridium/Platinum Low-Activity Aluminide/MarM002 System at 1100 Degree C", Surf. Coat. Technol., 113(3), 259-67 (1999).

[10]F. Wu, H. Murakami, Y. Yamabe-Mitarai, H. Harada, H. Katayama and Y. Yamamoto, "Electrodeposition of Pt-It Alloys on Nickel-Based Single Crystal Superalloy TMS-75", Surf. Coat. Technol., 184(1), 24-30 (2004).

[11]P. Kuppusami and H. Murakami, "A Comparative Study of Cyclic Oxidized Ir Aluminide and Aluminized Nickel Base Single Crystal Superalloy", Surf. Coat. Technol., 186, 377-88 (2004).

[12]P. Kuppusami, H. Murakami and T. Ohmura, "Behavior of Ir-24at.-% Ta Films on Ni Based Single Crystal Superalloys", Surface Engineering, 21(1), 53-59 (2005).

[13]H.B. Guo, S. Kuroda and H. Murakami, "Microstructures and Properties of Plasma Sprayed Segmented Thermal Barrier Coatings", J. Am. Ceram. Soc., 89 (4), 1432-39 (2006).

[14]H.B. Guo, S. Kuroda and H. Murakami, "Segmented Thermal Barrier Coatings Produced by Atmospheric Plasma Spraying Hollow Powders", Thin Solid Films, 506-507, 136-39 (2006).

[15]P. Bengtsson and J. Wigren, "Segmentation Cracks in Plasma Sprayed Thick Thermal Barrier Coatings ", in: P.J. Maziasz, I.G. Wright et al (Eds.), Gas Turbine Materials Technology: Conference Proceeding from ASM Materials Solutions, Rosemont, IL, 1999, p. 92.

[16]H.B. Guo, R. Vaßen and D. Stöver, "Atmospheric Plasma Sprayed Thick Thermal Barrier

Coating with High Segmentation Crack Density", Surf. Coat. Technol., 186, 353-63 (2004).

[17]F. Wu, H. Murakami and H. Harada, "Cyclic Oxidation Behavior of Iridium-Modified Aluminide Coatings for Nickel-Based Single Crystal Superalloy TMS 75", Mater. Trans. JIM, 44(9), 1675-78 (2003).

DESIGN OF ALTERNATIVE MULTILAYER THICK THERMAL BARRIER COATINGS

H. Samadi and T. W. Coyle
Centre for Advanced Coating Technologies,
University of Toronto,
Toronto, Ontario, M5S 3E4 Canada

ABSTRACT

Increasing the combustion temperature in diesel engines is an idea which has been pursued for over 20 years. Increased combustion temperature can increase the power and efficiency of the engine, decrease the specific fuel consumption and CO emission rate. Ceramic thermal barrier coatings have been identified as the most promising approach to meeting these objectives. The most commonly used system is Yttria Partially Stabilized Zirconia (Y-PSZ). However, in contrast to the widespread use in aircraft and power generation turbine engines, Y-PSZ TBCs have not met with wide success in diesel engines. To reach the desirable temperature of 850-900°C in the combustion chamber, a coating with a thickness of at least 1mm is required. This introduces different considerations than in the case of turbine blade coatings, which are on the order of 100µm thick. The design of a multilayer coating employing relatively low cost materials with complementary thermal properties is described. Numerical models were used to optimize the thickness for the different layers to yield the minimum stress at the operating conditions while achieving the desired temperature gradient.

INTRODUCTION

Diesel engines are commonly used in buses, trucks and, outside of North America, in passenger cars. In a diesel engine, the fuel is compressed to a high pressure at which automatically ignites and burns. The idea of decreasing heat transfer from the combustion chamber is based on the knowledge that only 30-40% of the entering fuel energy is converted to useful work on the output shaft [1]. In the past few decades, the use of advanced ceramics to insulate the combustion chamber to reduce heat rejection has been widely investigated. The goal is to have an engine with around 50% efficiency rather than the typical 33%. [2].

In the 1980s there was an effort to use thermal barrier coatings (TBCs) in diesel engines in pursuit of advantages including higher power density, fuel efficiency, and multi-fuel capacity due to higher combustion chamber temperature (900°C vs. 650 °C) [3,4]. Preliminary studies showed that using TBCs in diesel engine could increase engine power by 8%, decrease the specific fuel consumption by 15-20% and increase the exhaust gas temperature 200K [5]. At the same time, TBCs should protect the metallic substrate against the corrosive attack of fuel contaminants (Na, V, and S). However the main problem was still unsolved: durability.

Ceramic thermal barrier coatings (TBCs) have been in use for some time as protective coatings in gas turbine hot sections. The super alloy turbine blades cannot tolerate the high temperature (more than 1000°C) and corrosive environment experienced in current turbine engines for extended periods without protection. A thin layer of Yttria Partially Stabilized ZrO_2 (YSZ) with a thickness of 100- 200 µm, fulfills the requirements.

However, the service environment of the coating in the turbine is markedly different from the diesel engine. In the former, the service temperature is high (more than 1100°C). The superalloy substrate's maximum service temperature is higher than 800°C. The thickness of coating is a few hundred microns and is applied to reduce the substrate temperature and protect it against oxidation, hot corrosion, thermo-mechanical fatigue and creep. Due to the high substrate temperature, oxidation of the bond coat plays a major role in coating failure. On the other hand, in the diesel engine the gas temperature, currently less than 650°C would ideally approach 900°C. The substrate temperature is limited to approximately 200°C, and therefore a thick coating (up to 1mm) is required which leads to a high thermal gradient . In a thick thermal barrier coating (TTBC) the bond coat temperature is too low for severe oxidation and creep [6].

In a thick thermal barrier coating, when the surface of the YSZ coating is heated, a compressive stress is developed in the surface. Stress relaxation through creep and sintering mechanisms may occur at the service temperature. Upon cooling, the stress at the surface may become tensile, initiating cracks (figures 1 and 2)[6]. Due to the mismatch in thermomechanical properties of the top coat, bond coat and substrate, these interfaces are sources for cracking and delamination [7].

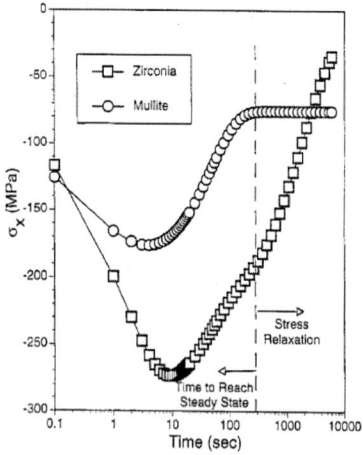

Figure 1. Surface Stress for Zirconia and Mullite during Transient Heating (q= 270 kWm^{-2}) (Reprinted from [6] with permission from Elsevier).

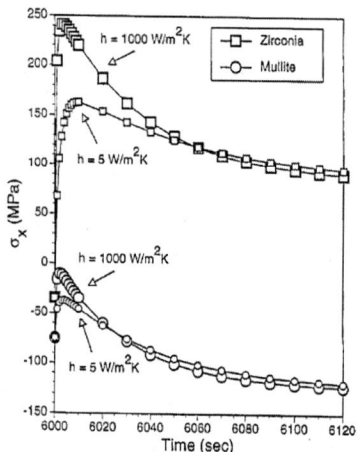

Figure 2. Surface Stress for Zirconia and Mullite during Transient Cooling (q=270 kWm^{-2}) (Reprinted from [6] with permission from Elsevier).

Another major difference between the two systems is that due to the on/off nature of diesel engine, TBC in a diesel engine is experiencing more thermal transients than a turbine blade. While being cooled the surface stress of the coating tends to become tensile leading to crack initiation and failure [6].

RESIDUAL STRESS IN THERMAL BARRIER COATINGS

One of the most important parameters controlling the durability of plasma spray coating is the residual stress [8]. There are two types of residual stress in plasma spray coatings: macro and micro. Macro residual stresses exist in the body of component over a range much larger than the grain size. Micro residual stresses vary at or below in the scale of the grain size [9].

Macro residual stresses may originate from non-uniform heating and cooling operations and coefficient of thermal expansion mismatch between phases. Chemical residual stresses related to chemical reactions, precipitation, or phase transformations may also produce long range stresses [9]. Residual stress is introduced in the TBC system because of coefficient of thermal expansion mismatch between the top coat and the bond coat (or substrate) and also oxidation of the bond coat [10]. In thermal spraying, the cooling time for a single splat is in the order of a few milliseconds. Thermal contraction due to cooling of an individual splat to the temperature of the substrate (or the previously deposited solid layer) results in a tensile stress in the splat which is known as the "quenching" stress [11]. Analytical and numerical approaches have been used to model stress in multilayer system with unequal coefficients of thermal expansion and elastic moduli.

In a thick TBC, a low CTE is desirable for the hot surface to minimize stresses due to temperature gradients and sensitivity to thermal shock. However, a relatively high CTE is desirable to avoid a large CTE mismatch with the metallic substrate which would limit coating adhesion. The MCrAlY bond coat fulfills this need to some extend in YSZ TBC systems. A multilayer system may permit these opposing requirements to be satisfied.

Most traditional, high temperature refractory ceramic materials are found in the $Al_2O_3 \cdot SiO_2 \cdot MgO$ system. Among these oxides, some have been considered as alternatives to YSZ in TBCs. A general advantage of materials in this system is their low price relative to YSZ (Figure 3) [12].

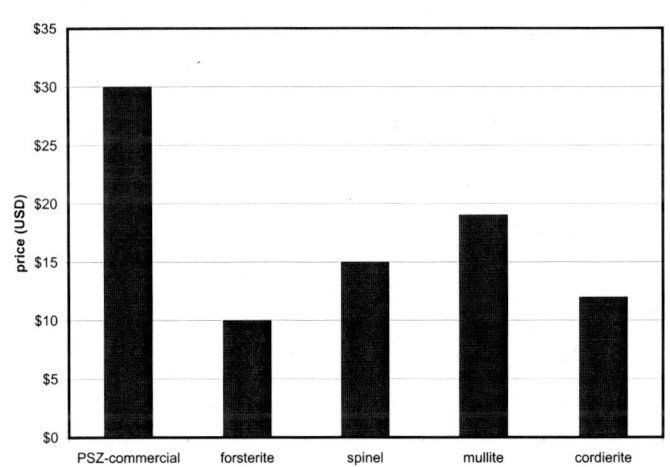

The multilayer system has been designed [12] consisting of forsterite, spinel and mullite. Table 1 contains the physical, mechanical and thermal properties of the substrate and coating materials. Forsterite has a CTE matching the steel substrate ($11\text{-}12.5 \times 10^{-6}$ K^{-1} [13]). Mullite has a low CTE (4.5×10^{-6} K^{-1} [6]), which may minimize thermal strain at the hot face. These two phases are not stable in contact with one another at elevated temperatures, therefore an intermediate layer of spinel is needed to prevent reaction. Spinel exhibits a CTE between that of forsterite and mullite.

Figure 3. A Comparison between Prices of YSZ vs. Some Oxides in $Al_2O_3.SiO_2.MgO$ Phase Diagram

Table 1. Physical, thermal and mechanical properties of the materials being used [13, 14].

Parameter	Stainless Steel	Forsterite	Spinel	Mullite
Thermal conductivity ($Wm^{-1}K^{-1}$)	44.5	3.5	8	2.5
Density (Kgm^{-3})	7850	2600	3100	2800
Heat capacity ($JKg^{-1}K^{-1}$)	475	800	750	920
Poisson's ratio	0.28	0.24	0.29	0.25
Young's modulus (GPa)	205	87.7	137	30
Thermal expansion coeff. (*e-6 K^{-1})	12.3	11	7.68	4.5

A finite-element based model of the multilayer system on a stainless steel substrate (1.5 mm thickness) has been developed. The thicknesses of the forsterite, spinel and mullite layers are 200, 50 and 640 μm, respectively. The stresses in the multilayer coating are compared to a conventional TBC with a MCrAlY bond coat (100 μm) and partially stabilized zirconia top coat (500 μm) on a 1.5 mm stainless steel which exhibits the same thermal insulating performance. To achieve this, the thickness of mullite in the multilayer system has been chosen such that both systems experience the same substrate-coating interface temperature when subjected to equal thermal flux on the surface of the TBC. Figure 4 illustrates the geometry and boundary conditions of both systems. The coatings were assumed to be isotropic and quenching stresses were not considered.

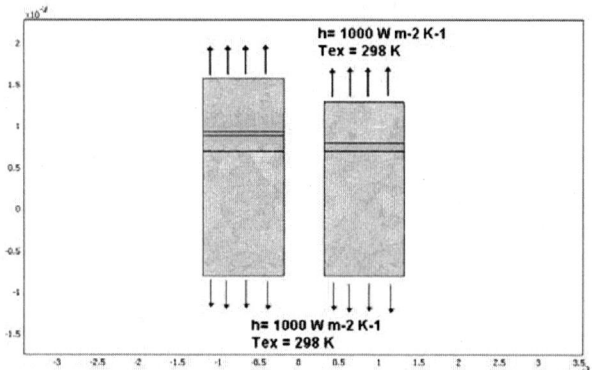 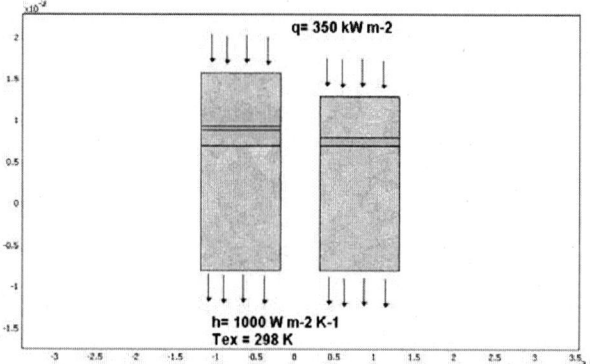

Figure 4. Geometry and boundary conditions of multilayer and duplex systems during a) cooling from the deposition temperature, and b) subsequent heating of ceramic surface.

Stresses were analyzed during cooling from the deposition temperature and during subsequent heating of the surface of the coating. In the first step, the whole system was considered to be initially at 650°C and stress free. It was then cooled from both sides assuming a heat transfer coefficient of 1000 $Wm^{-2}K^{-1}$ and an external temperature of 298 K. Subsequently a heat flux was imposed on the top coat (350 kW m^{-2}) while the back of the substrate was cooled with the same conditions as in the first step.

RESULTS AND DISCUSSION

Figure 5 illustrates the surface temperature of both systems. Due to its higher thermal diffusivity and heat capacity, the surface temperature of the mullite layer changes more slowly than that of the zirconia layer and the final surface temperature under the imposed heat flux is 150K lower than that of zirconia. Considering the lower creep rate of mullite [6], stress relaxation and densification in the mullite should be reduced compared to that occurring in zirconia top coats.

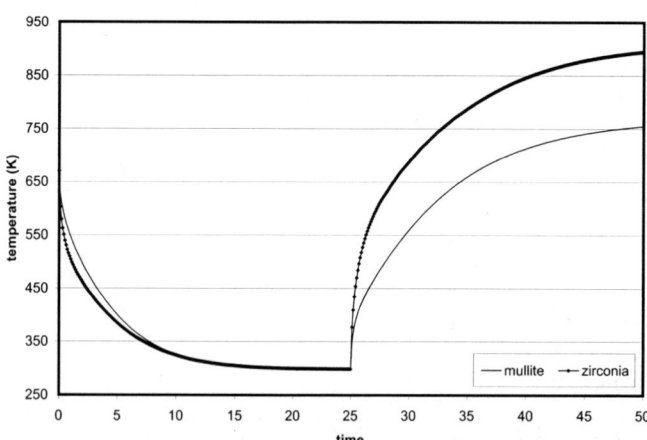

Figure 5. Surface temperature of both systems while cooling and heating.

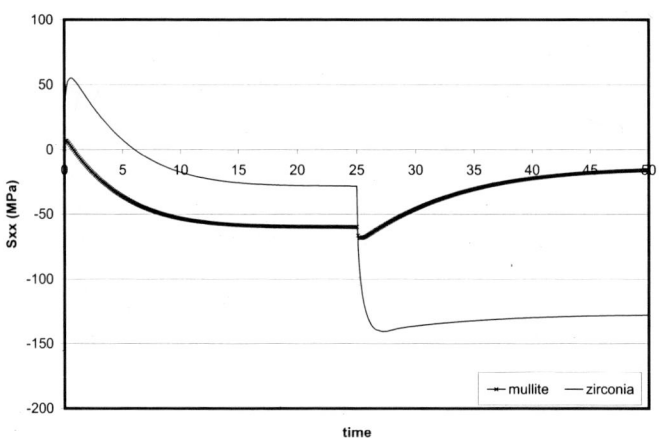

Figure 6. Surface stress (σ_x) of both systems.

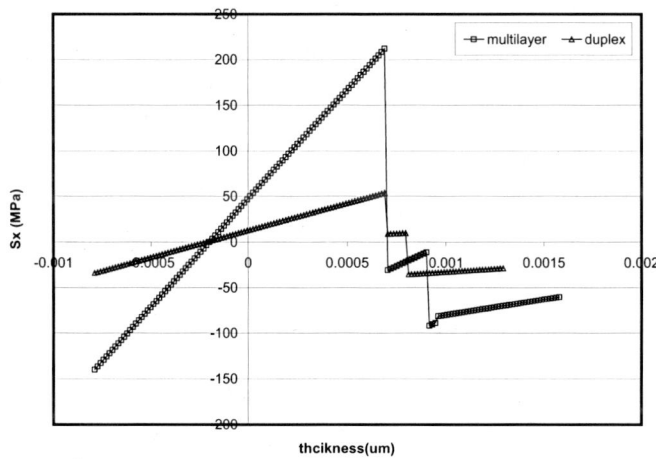

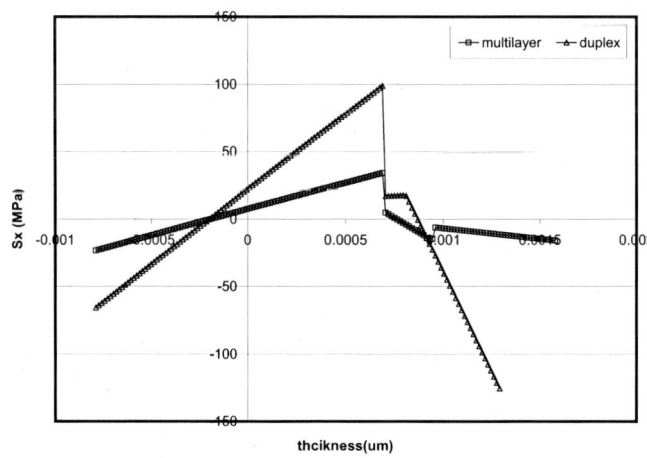

Figure 8. Stress distribution in both systems after the heating step

The in-plane surface stress, σ_x, of the systems is illustrated in figure 6. During the cooling stage, the surface of the zirconia first becomes tensile, and then as the underlying material begins to cool the stress gradually decreases, becoming compressive at room temperature. The surface of mullite reaches a much lower tensile stress at the beginning, and the residual stress at the end of the cooling stage is -60 MPa. After applying the heat flux, the stress in the surface of zirconia undergoes a sudden drop to about -140 MPa and the steady state stress of this coating is -130 MPa. Due to its lower CTE, no such a sudden drop is seen in mullite and the steady state stress is no greater than -15 MPa.

In figure 7, the stress distribution through the cross-section of both systems after cooling shows that in both cases, the ceramic coatings are in compression. During the heating step, a steady-state stress of 5 MPa was found for the forsterite layer at the interface of forsterite and the substrate, compared to 17 MPa for PSZ at the interface with the bond coat in the duplex coating (Figure 8).

Other than the small region of low tensile stress in the forsterite layer adjacent to the substrate, all three ceramic layers remain in compression through-out both the cooling and heating stages. In most thermally sprayed coatings, debonding and lack of cohesion is one of the main causes of failure; lower tensile stresses in the interfaces should make the coating more reliable. The multilayer system has this advantage.

CONCLUSION

A new multilayer thermal barrier coating was modeled using the finite element method. Comparing the internal stress of this system with a conventional duplex TBC, it was found that:

o During cooling, the surface of the multilayer coating experienced higher compressive stresses.

o During heating, the steady state compressive stress at the surface of the multilayer coating is lower.

o During the cooling/heating cycle, the difference between maximum and minimum surface stresses in the multilayer system is much less than in the duplex system.

These results suggest that the multilayer coating would be a promising system for the operating conditions considered here. More detailed analysis of the transient temperature and stress distributions are underway.

ACKNOWLEDGEMENTS

Support for this research from the Auto 21, Network of Centres of Excellence and Iran Ministry of Science, Research and Technology is gratefully acknowledged.

REFERENCES

[1] C. A. Amann, The low heat rejection diesel Advanced Diesel Engineering & Operation, Ellis Harwood Ltd, UK (1988) p. 173-180.

[2] R. A. Churchill, J. E. Smith, N. N. Clark, R.A. Turton, "Low-Heat Rejection Engines – A Concept Review", SAE Technical Paper Series 880014.

[3] P. Ramaswamy, S. Seetharamu, K. B. R. Varma, N. Raman and K. J. Rao, "Thermomechanical Fatigue Characterization of Zirconia (8% Y2O3-ZrO2) and Mullite Thermal Barrier Coatings on Diesel Engine Components: Effect of Coatings on Engine Performance", Proc. Instn. Mech Engrs., vol. 214, part.C (2000) 729-742.

[4] D. Zhu, R. A. Miller, "Thermal Barrier Coatings for Advanced Gas Turbine and Diesel Engines", NASA/TM—1999-209453.

[5] T. Hejwowski, A. Weronski, "The Effect of Thermal Barrier Coatings on Diesel Engine Performance", *Vacuum*, 65 (2002) 427-432.

[6] K. Kokini, Y. R. Takeuchi, B. D. Choules," Surface Thermal Cracking of Thermal Barrier Coatings Owing to Stress Relaxation: Zirconia vs. Mullite", *Surf. & Coat. Tech.*, 82(1996) 77-82.

[7] S. Rangaraji, K. Kokini, "Interface Thermal Fracture in Functionally Graded Zirconia-Mullite-Bond Coat Alloy Thermal Barrier Coating", *Acta Mater.*, 51(2003) 251-267.

[8] S. Kuroda, T. W. Clyne, "The Quenching Stress in Thermally Sprayed Coatings", *Thin solid films* **200** (1991) 49-66.

[9] F. A. Kandil, J. D. Lord, A. T. Fry, P. V. Grant, "A Review of Residual Stress Measurement Methods- A Guide to Technique Selection", NPL Report MATC (A)04, Project CPM 4.5 (2001).

[10] B. G. Nair, J. P. Singh, M. Grimsditch, "Stress Analysis in Thermal Barrier Coatings Subjected to Long-term Exposure in Simulated Turbine Conditions", *J. Mater. Sci.* **39** (2004) 2043-2051.

[11] Y. C. Tsui, T.W. Clyne, "An Analytical Model for Predicting Residual Stresses in Progressively Deposited Coatings, Part I: Planar Geometry", *Thin Solid Films*, **306** (1997) 23-33.

[12]R. Soltani, H. Samadi, E. Garcia, T.W. Coyle , "Developemt of Alternative Thermal Barrier Coatings for Diesel Engines", SAE World Congress, Detroit, MI, Paper No. 2005-01-0650 (2005).

[13]H. Wang, H. Herman, "Thermomechanical Properties of Plasma-Sprayed Oxides in the MgO-Al2O3-SiO2 System" *Surface & Coatings Tech.* **42** (1990) 203-216.

[14] Materials/ Coefficients Library, Comsol Multiphysics Software, Ver. 3.2 (2005).

LANTHANUM-LITHIUM HEXAALUMINATE – A NEW MATERIAL FOR THERMAL BARRIER COATINGS IN MAGNETOPLUMBITE STRUCTURE – MATERIAL AND PROCESS DEVELOPMENT

Gerhard Pracht, Robert Vaßen and Detlev Stöver
Forschungszentrum Jülich GmbH, Institute for Materials and Processes in Energy Systems 1, Jülich, Germany, D-52425 Jülich

ABSTRACT

The first study on a new material composition for thermal barrier coating applications is presented. The new material has the chemical composition $LaLiAl_{11}O_{18.5}$ and belongs to the group of hexaaluminate compounds. It is comparable to $LaMgAl_{11}O_{19}$, which has been discussed for TBC application for some years. The preparation and characterization of the composition as well as first plasma spraying experiments will be described. In particular, the thermal expansion mismatch of the ceramic coating on a steel substrate is discussed which leads to strong segmentation cracks of the plasma-sprayed and heat-treated coating. The substitution of lithium in the structure supported the segmentation process of the coating.

INTRODUCTION

Thermal barrier coatings are widely used in stationary and airborne gas turbines to thermally insulate air-cooled metallic components from the hot gases in the engine. Standard materials for TBC s are based on ceramic materials especially with yttria partially stabilized zirconia (YSZ). These coatings are typically applied by either plasma spraying or physical vapor deposition (EB-PVD).

In order to enhance the efficiency of gas turbines the temperature in the combustion chamber must be increased. However, this is not possible with YSZ coatings because the stability of this material is limited to a temperature of 1200 °C [1]. Additionally, sintering of the ceramic at operation temperature sets a limit to the lifetime of these coatings [2 - 4]. Besides the enhanced efficiency, a prolonged lifetime of the coating material is also desired.

To meet these requirements, new materials are under consideration [5, 6]. One very interesting candidate is $LaMgAl_{11}O_{19}$ – a hexaaluminate, which crystallizes in the magnetoplumbite structure [7]. Gadow et al. published first investigations on this material for TBC application as well as a patent [8 - 11].

On the basis of these investigations, compounds of the $ABAl_{11}O_{19}$ type were screened with respect to the influence of doping in order to study changes in their properties due to different A and B cations. The magnetoplumbite structural type was to be maintained and the compound should, if possible, be stable without secondary phases. Bi^{3+}, Gd^{3+}, und Y^{3+} were also tested as alternatives for the A site and Fe^{2+}, Ti^{4+}, Zr^{4+} und Li^{+} for the B site. The substitution of titanium, zirconium and yttrium did not lead to the desired structural type as the main phase, which is why these experiments were discontinued. Of the other substitutions, only the $LaLiAl_{11}O_{18.5}$ obtained will be discussed in detail here. The other substitutions will be described at a later time. For example, Saruhan et al. [12] also performed investigations on the substitution of manganese on the B site.

Although the magnetoplumbite structure is more complex than corundum or zirconia, it has a high hexagonal symmetry. The complex structure is important for a low thermal conductivity and high symmetry for good high-temperature stability. The hexaaluminates are based on an alumina structure and exhibit an additional plane in the composition filled up with lanthanum and magnesium atoms. In opposite to zirconia LaMgAl$_{11}$O$_{19}$ exhibits no ionic conductivity at high temperature. Furthermore, a high melting point is assumed.

Table I: Key properties of standard YSZ for application in gas turbines.

Property	Value / limits
Thermal expansion coefficient	$10 - 10.5\ 10^{-6}$ K^{-1} [13]
Thermal conductivity	< 2.3 Wm^{-1} K^{-1} (1000 °C) [13]
Sintering rate	$d(\Delta L/L_0)/dt \approx 0.1 - 0.35$ % (100 h)$^{-1}$
Temperature stability	Phase change above 1200°C
Young's modulus	$E = 50 - 200$ GPa [2, 13]
Porosity (APS coatings)	Standard $12 - 25$% [13]

New materials have to fulfill certain requirements such as higher temperature stability, lower thermal conductivity, adapted thermal expansion coefficient, lower sintering rate and a small Young´s modulus (s. table 1). From the industrial perspective, it is not necessarily a low thermal conductivity that is most important but rather a long lifetime. For a thermal barrier coating, apart from TGO formation and other properties, the lifetime depends in particular on the thermal mismatch.

In this paper, we present a new compound from the hexaaluminate family -LaLiAl$_{11}$O$_{18.5}$. This is not only a new material for thermal barrier applications, but it is also new with respect to the chemical composition itself. The properties obtained were therefore different as well as unexpected, when compared to standard YSZ, and similar to LaMgAl$_{11}$O$_{19}$.

EXPERIMENTAL

LaLiAl$_{11}$O$_{18.5}$ was prepared by a solid-state reaction of α-Al$_2$O$_3$, La$_2$O$_3$ and Li$_2$CO$_3$ powders. All substances used here had a purity ≥ 99% (Li$_2$CO$_3$ 99.0% Alfa Aesar, Karlsruhe; Al$_2$O$_3$ 99.8 % Martoxid MR70, Martinswerk, Bergheim, Germany; La$_2$O$_3$ 99,0% Projector GmbH, Duisburg, Germany). La$_2$O$_3$ was heat treated to eliminate any hydroxides. Than the powders were mixed and thoroughly milled in an ethyl-alcohol-based suspension to obtain a homogeneous reaction to a single phase. The powder was milled for longer than 12 hours to a particle size of less than $d_{50} < 1.5$ µm. The reaction temperature of the solid-state reaction was higher than 1450 °C [14] (different batches between 1450 and 1600°C). Lithium carbonate was decomposed during the reaction to lithium oxide.

All the prepared raw materials were analyzed by X-ray diffraction before being used for further investigations to guarantee high-quality powders. The phase analysis was conducted using an XRD (Model D5000, Siemens, Karlsruhe, Germany) in the 2θ range from 15° to 70° with CuK$_\alpha$ radiation (wavelength 1.5406 Å).

The physical properties of the new material were investigated on bulk samples in order to exclude the influence of the porosity and microstructure. The thermal expansion coefficient was obtained by dilatometric measurements and the thermal conductivity by the laser flash method.

A high-temperature dilatometer (Setaram, Caluire, France) was used to measure the thermal expansion coefficient. Prior to the dilatometer measurement, the required cylindrical specimens were cold-isostatic-pressed at a pressure of nearly 400 MPa and subsequently sintered at 1450 °C for 14 hours. Inside the dilatometer, the sample to be measured was positioned in an

Al$_2$O$_3$ cylinder and the thermal expansion was recorded by monitoring the change in height of an Al$_2$O$_3$ rod placed on the sample. In order to subtract the thermal expansion of the holder and rod, for every dilatometry measurement a blank measurement without sample was also carried out.

The thermal diffusivity of the specimens was measured using the laser flash method (Theta Industries, Port Washington, NY) with a neodymium-glass laser (wavelength 1065 nm). The laser-flash method requires dense specimens, because the thermal diffusivity of the bulk material is the point of interest. The specimens used here were disks with a diameter of 10 mm and a thickness of 2 mm. These specimens were fabricated by uniaxial hot pressing at a temperature of 1450°C and a pressure of nearly 87 MPa in an argon atmosphere. A detailed description is given by Lehmann et al. [16].

The thermal conductivity (λ) was calculated from the values for thermal diffusivity (α), specific heat capacity (C_p), and measured density (ρ), using the relation

$$\lambda = \alpha \cdot C_p \cdot \rho \qquad (1)$$

(α = thermal diffusivity and ρ = theoretical density)

The phase stability up to high temperature and values for the specific heat capacity were obtained on powder from DSC measurements (differential scanning calorimetry - DSC Model 404c, Netzsch, Germany) at up to 1300 °C in air. Data were recorded during the period of rising temperature at 20 K min^{-1} in platinum crucibles.

To use the material for plasma spraying, it was necessary to spray-dry the powder. Therefore a large batch of the initial powder was produced which contained only a very small amount of impurities (only perovskite phase). The chemical analysis is shown in table 2. A suspension for spray drying was prepared by mixing the milled powder (78 wt.%) with ethanol and 1.2 wt % of PEI ([CH$_2$CH$_2$NH]$_x$ m.w. 10,000, Polyscience).

The spray dryer was a Mobile Minor Ex Model H (Niro A/S, Denmark) with N$_2$ as the drying medium. A detailed description of the process is given by Cao et al. [15]. After spray drying, the powder was sintered at 1300 °C for 2 hours to remove the PEI binder. The tapped density of the spray-dried powder was 1.62 g cm^{-3}. For plasma spraying, a sieved fraction of between 36 and 125 μm of this spray-dried powder was used. The grain size distribution is characterized by D_{10} = 11.4 ; D_{50} = 23.4 ; D_{90} = 70.0 μm. The BET surface of this powder is 0.69 m^2 g^{-1}. An SEM micrograph of a spray-dried powder particle is presented in fig. 1.

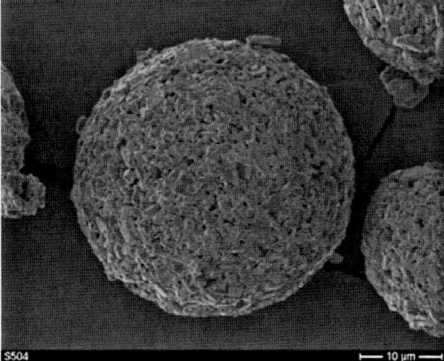

Fig. 1: SEM micrograph of spray-dried powder of LaLiAl$_{11}$O$_{18.5}$.

Plasma-sprayed coatings of LaLiAl$_{11}$O$_{18.5}$ were produced by atmospheric plasma spraying with a Sulzer Metco F4 gun on different substrates. Argon and hydrogen were used as plasma gases. Parameters and substrate pretreatment will be described below together with the results.

The coating porosity was determined by mercury porosimetry with a Pascal 140 and 440 supplied by CE instruments (Milan, Italy) operating in a pressure range between 0.008 and 400 MPa corresponding to pore diameters between 3.6 nm and 90 μm. These porosimetry measurements were performed on freestanding coatings and can only determine the open porosity.

RESULTS AND DISCUSSION

MATERIAL CHARACTERIZATION (POWDER AND BULK MATERIAL)

The X-ray pattern of the powder (s. fig 2) showed the same peaks as those of $LaMgAl_{11}O_{19}$. This new material therefore crystallized in the hexagonal magnetoplumbite structure similarly to $LaMgAl_{11}O_{19}$. The lattice constant of this hexagonal crystallizing composition was determined as a = 556.33 ± 0.55 pm and c = 2193.07 ± 2.86 pm. In fig. 2, an X-ray diffraction pattern of $LaLiAl_{11}O_{18.5}$ with a very small amount of the second phase of $LaAlO_3$ is shown. Additional phases often found during reaction to $LaLiAl_{11}O_{18.5}$ were $LaAlO_3$, α-Al_2O_3 and spinel phases.

Lithium oxide was used in the structure to fill up the vacancies in the lattice. Using X-ray diffraction it was not possible to measure whether all the vacancies in the lattice were completely filled, or whether some lithium oxide was lost during reaction because the magnetoplumbite structure is also formed without filling up the vacancies. In the X-ray diffraction pattern, the difference between $LaAl_{11}O_{18}$ (structure with empty vacancies) and $LaLiAl_{11}O_{18.5}$ or $LaMgAl_{11}O_{19}$ (structure with filled vacancies) therefore only leads to a slightly different pattern, which cannot be distinguished when typical secondary phases are formed hiding these different patterns.

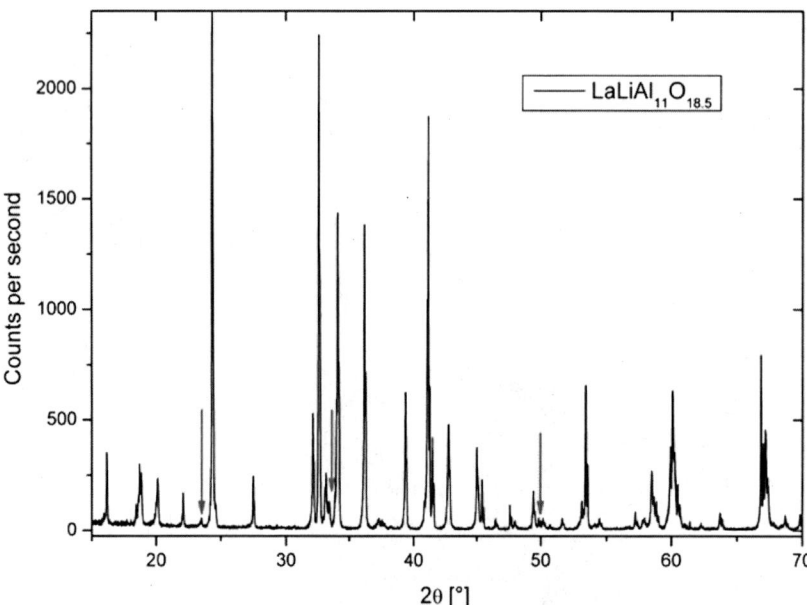

Fig. 2: X-ray diffraction pattern of $LaLiAl_{11}O_{18.5}$ fitting to magnetoplumbite structure. Gray arrows denote additional perovskite phases.

From chemical analyses (at ZCH, Forschungszentrum Jülich) it was confirmed that there was hardly any loss of lithium during powder preparation, spray drying and plasma spraying (tab. 2).

The thermal conductivity of this new material (investigations with bulk material) is 3.8 $Wm^{-1}K^{-1}$ at a temperature of 1000 °C. This is much higher than that of YSZ (2.2 $Wm^{-1}K^{-1}$ at 1000 °C [16]).

Table II: Chemical analyses by ICP-OES of two samples, one raw powder and the spray-dried powder, used for plasma-sprayed coatings. The amount of oxygen was analyzed by hot extraction in helium gas in combination with IR spectroscopy.

Wt. %	Powder for plasma spraying	Raw powder	Theoretical values
Al	40.1	40.1	40.2
La	18.2	18.3	18.8
Li	0.71*	0.77*	0.94
O	37.0	37.8	40.1
Si	0.044*	0.061*	-
Ca	0.026*	0.029*	-

* deviation of values ± 20%; all other values ± 3%

The measurements of the thermal expansion coefficient with a dilatometer led to unforeseen results. All the specimens showed shrinkage during heating although the samples were isostatically pressed and sintered for 14 hours at 1450 °C. The specimens were then heat treated again for 14 hours at 1450 °C, but the result did not change. The dilatometric analysis in figure 3 was obtained during heating, resulting in a thermal expansion coefficient between $6 \cdot 10^{-6}$ K^{-1} at 200 °C and $10.2 \cdot 10^{-6}$ K^{-1} at 1400 °C. Here it has to be mentioned that the sintering of the sample influenced the measurement significantly. The material could therefore display higher thermal expansion if sintering effects are excluded.

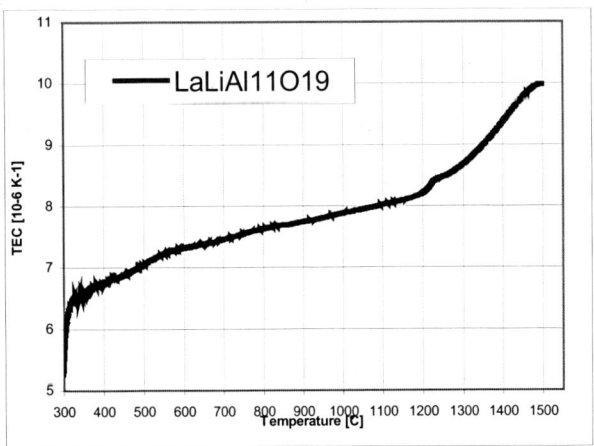

Fig. 3: Thermal expansion coefficient in a temperature range between 300 and 1500 °C.

The sintering was studied by keeping the sample at a high temperature (1500 °C) for 10 hours, as shown in figure 4. From this experiment it follows that a strong sintering process takes place during the whole period and was not completed even after 10 hours.

For plasma-sprayed coatings this sintering behavior is expected to be much stronger because of the porosity and cracks, which were dependent on deposition conditions. During deposition, the material is partly amorphous and these amorphous parts crystallize during heating. Shrinkage during crystallization is much stronger than pure sinter shrinkage. Further investigations to clarify the sintering mechanism are still in progress.

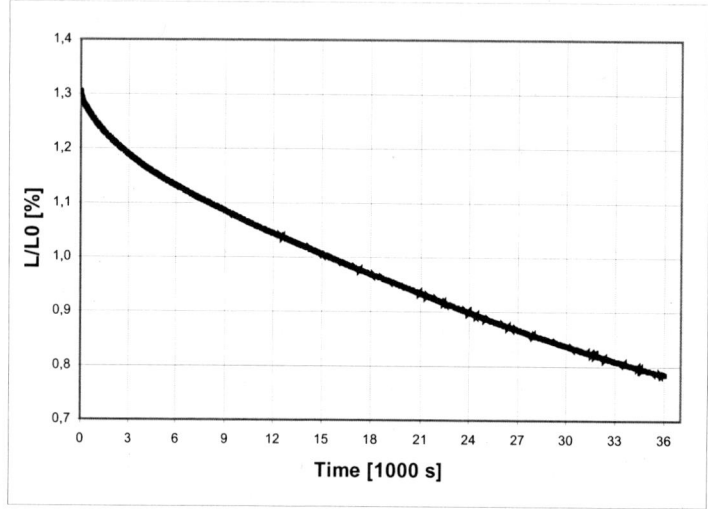

Fig. 4: Sintering of a dense sample of $LaLiAl_{11}O_{18.5}$ at 1500 °C. The sintering started at 1100 °C during heating and was not completed after 10 hours.

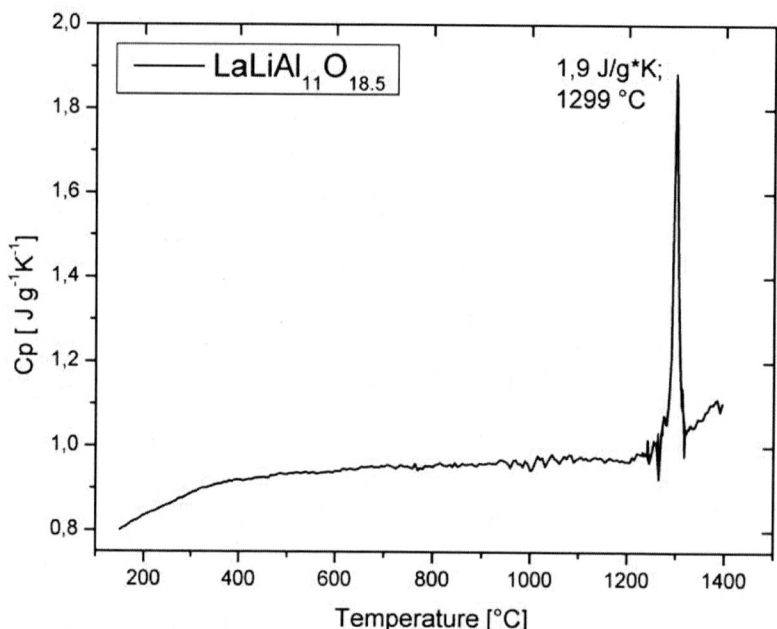

Fig. 5: DSC measurement of $LaLiAl_{11}O_{18.5}$ in the temperature range of 200 – 1400 °C. The material displays a significant endothermic peak at 1299 °C.

The DTA and DSC analyses of $LaLiAl_{11}O_{18.5}$ showed a reversible, endothermic peak at 1299 °C. In figure 5 a DSC diagram is presented. In the cooling curve the exothermic peak was at 1283 °C (not shown here). The powder showed no traces of melting. The reason for the peak at 1299 °C is unknown. In-depth investigations using high-temperature X-ray analysis on a platinum heating element in the area of 30 to 39 2θ at 1150, 1350 and 1400 °C found no phase transition. Only a small change in the content of main and secondary phases was observed. The amount of the perovskite phase decreased with higher temperature. During cooling this content increased again. So it is suggested from these high temperature X-ray results that the powder was not completely reacted to the composition of $LaLiAl_{11}O_{18,5}$.

COATING DEPOSITION AND CHARACTERIZATION

The spray-dried powder was used to deposit $LaLiAl_{11}O_{18.5}$ by plasma spraying using a F4 torch. Some typical parameters are shown in table 3.

Table III: Plasma-spraying parameters with an F4 torch for $LaLiAl_{11}O_{18.5}$. The plasma gas was usually argon (41 slpm) and hydrogen (10 slpm).

Coating no.	287	288	289
Distance [mm]	110	100	100
Power [kW]	48.4	40.9	48.1
Current [A]	601	500	601
Powder [g/ min]	13	4.3	4.3
Cycles	38	60	60
Translation speed [mm / s]	500	500	500
Porosity [%]*	9.4	14.7	7.9
Coat. thickness [μm]	1758	825	940

*Porosity determined by mercury porosimetry.

Fig.6: Micrograph of a cross section of a freestanding plasma-sprayed $LaLiAl_{11}O_{18.5}$ coating. No lamellar structure and no cracks were visible.

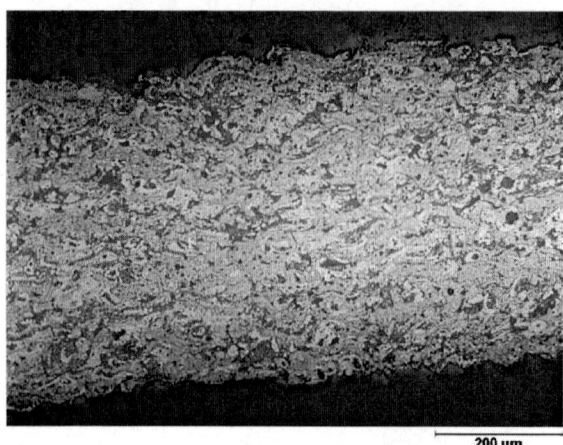

200 µm

Fig. 7: For comparison, a micrograph of a cross section of a plasma-sprayed LaMgAl$_{11}$O$_{19}$ coating.

The plasma-sprayed coatings of LaLiAl$_{11}$O$_{18.5}$ and LaMgAl$_{11}$O$_{19}$ shown in fig. 6, 7 and 8 were homogeneous and the microstructure looked very different compared to standard zirconia coatings because there is no lamellar structure visible as is typical of YSZ (fig. 7). In the cross-section micrographs in fig. 6 and 7, a great deal of open porosity is visible, specially for LaMgAl$_{11}$O$_{19}$. In the SEM pictures, LaLiAl$_{11}$O$_{18.5}$ appeared to be more dense than in the micrographs. This was typical of all coatings of LaLiAl$_{11}$O$_{18.5}$ in these investigations.

In fig. 8b some areas with small particles (marked by arrows) were obtained. These areas could have been formed by overspray during plasma spraying. However, from the evidence of a large number of evaluated pictures of LaMgAl$_{11}$O$_{19}$ and LaLiAl$_{11}$O$_{18.5}$ these particle areas appeared to be very typical of both these hexaaluminate materials. Friedrich et al.[17] gave a corresponding description of this microstructure.

During plasma spraying of hexaaluminate materials only a small amount of the material is deposited in a crystalline form. Especially for LaLiAl$_{11}$O$_{18.5}$ a large amount of amorphous phase was found in X-ray investigations (see fig. 9). This amorphous phase is partly soluble in acids, e. g. hydrochloric acid. Hence some coatings were deposited on a substrate which had been previously coated with a rock salt in order to remove the upper coating without using any acid. This is an easy way to obtain freestanding coatings by plasma-spraying.

FZJ - IWV 2004   EHT = 15.00 kV Detector = BSE   WD = 12 mm   100µm

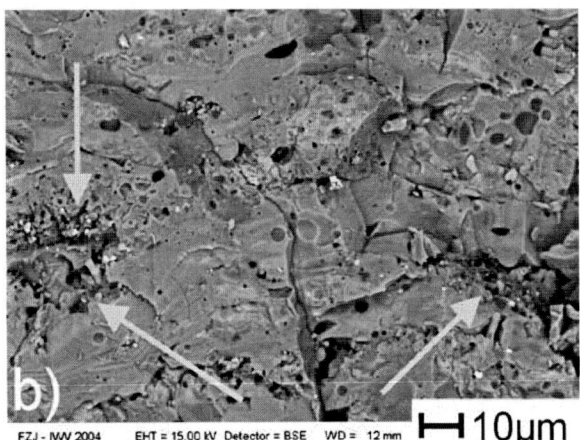

Fig 8 a+b: SEM pictures of fracture edges of plasma-sprayed coating (F4 torch; 287). The microstructure was significantly different to plasma-sprayed zirconia coatings. The arrows mark an area of small particles.

For a thermal barrier coating material, the properties at higher temperature are very important. For both hexaaluminates ($LaLiAl_{11}O_{18.5}$ and $LaMgAl_{11}O_{19}$) the partly amorphous phase in plasma-sprayed coatings have to crystallize during heat treatment. Additionally, for $LaLiAl_{11}O_{18.5}$ an intensive sintering was observed at higher temperature. Heat treatment of a plasma-sprayed coating of $LaLiAl_{11}O_{18.5}$ should therefore lead to a strong shrinkage due to sintering and recrystallization.

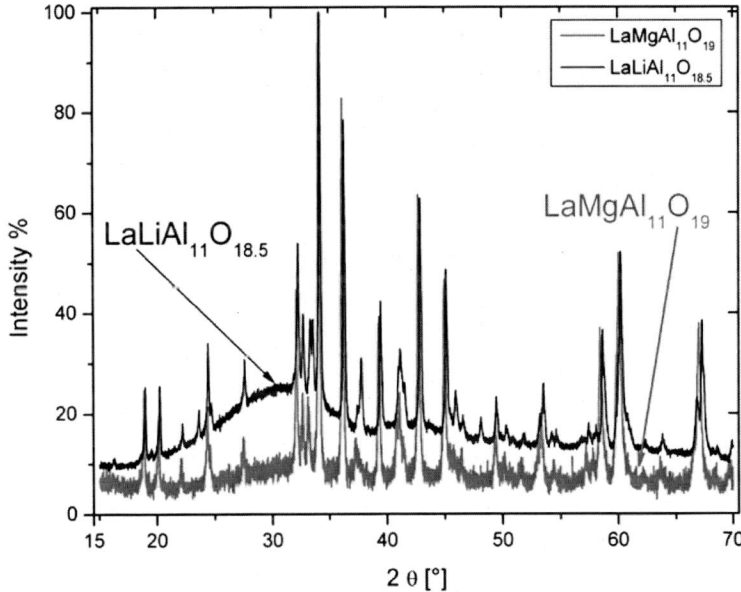

Fig.9: X-ray diffraction pattern of plasma-sprayed coatings of $LaMgAl_{11}O_{19}$ (gray) and $LaLiAl_{11}O_{18.5}$ (black). Both patterns display amorphous phase, but for $LaLiAl_{11}O_{18.5}$ the amount is significantly higher.

To check the shrinkage a simple test was performed on an steel substrate. Some test plates were coated by plasma spraying with YSZ, $LaMgAl_{11}O_{19}$, $LaFeAl_{11}O_{19}$ and $LaLiAl_{11}O_{18.5}$. The test substrates were stainless steel (type 1.4571) with a thermal expansion coefficient of $19.0 \cdot 10^{-6}$ K^{-1} (between 20 and 500°C). The coating thickness was between 300 and 1000 µm, always a single layer without a bond coat. These samples were heated together to

1200 °C in a vacuum oven for 5 hours. It was expected in the case of poor adhesion that the coating would spall off from the substrate because of the thermal mismatch of metal and ceramic and the reasons mentioned above. With another material, where the coating had good adhesion on the plate, it was observed that the metal plate became strongly bent after this heat treatment.

This test has an unforeseen result (s. fig. 10 - 12). The metal plates for thick coatings of YSZ, $LaMgAl_{11}O_{19}$ and $LaLiAl_{11}O_{18.5}$ were only weakly bent, but on the surface of all hexa-aluminates segmentation cracks were obtained. Both YSZ coatings had no segmentation cracks and did not spall off. Only for the thick YSZ coating was a weak bending obtained and some spallation in the marginal zone . The coating of the $LaFeAl_{11}O_{19}$ spalled off completely, even though segmentation cracks arose, too. The $LaLiAl_{11}O_{18.5}$ coating had very large cracks while all other hexaaluminates showed only small segmentation cracks (for details s. fig. 11).

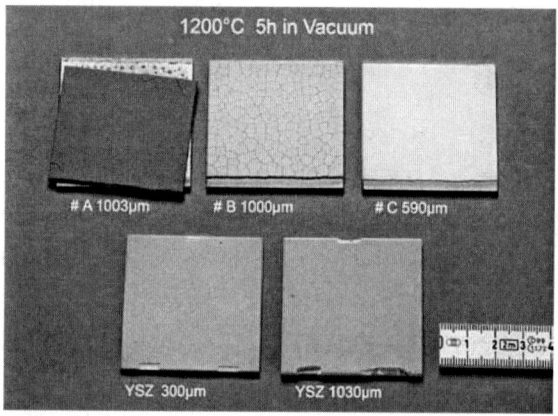

Fig. 10: Test plates with plasma-sprayed coatings after heat treatment (1200 °C 5 h in vacuum). A = $LaFeAl_{11}O_{19}$; B = $LaLiAl_{11}O_{18.5}$; C = $LaMgAl_{11}O_{19}$; YSZ = yttrium-stabilized zirconia. Different coating thicknesses between 300 and 1000 μm.

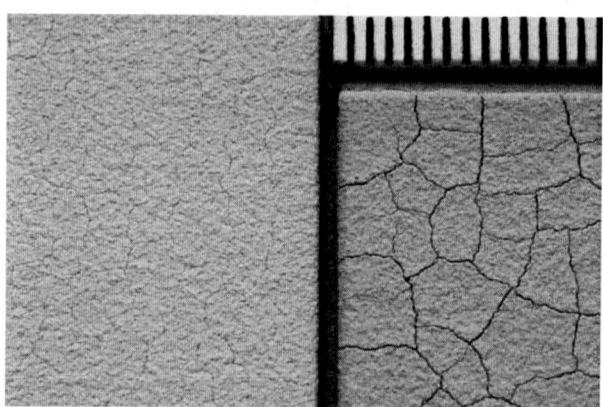

Fig. 11: Test plates with coatings of $LaMgAl_{11}O_{19}$ (Mg) and $LaLiAl_{11}O_{18.5}$ (Li) with higher magnification. The same plates as in fig. 10. On top right side, millimeter lines of a ruler.

The segmentation cracks had crack widths of 10 to 70 μm, which are visible in the micrograph in fig. 12.

The formation of segmentation cracks in the plasma-sprayed ceramic layers is rather unusual. However, in this test the high thermal expansion coefficient of the stainless steel substrate and the large layer thickness play an important part. A standard turbine alloy, such as IN738, has a thermal expansion coefficient of about $16 \cdot 10^{-6}$ K-1 and the usual layer thickness for an APS coating is only 300 μm. Nevertheless, the tendency to crack during high-temperature treatment has not been previously observed for materials such as zirconia or aluminum oxide in comparable coatings.

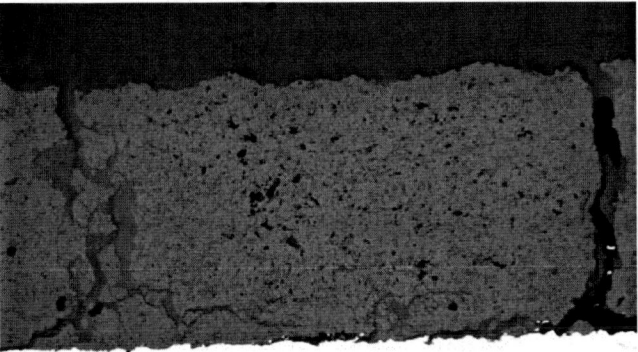

200 μm

Fig. 12: Micrograph of a cross section of the plasma sprayed and heat treated $LaLiAl_{11}O_{18.5}$ coating (different specimen than in fig.10; thinner coating). The segmentation cracks run from surface to substrate. Crack widths of 10 up to 50 μm can be observed. Coating thickness is 700 μm.

If one assumes that in the course of further heat action the cracks can no longer be closed by sintering then these cracks represent a special expansion tolerance for the hexaaluminates which is independent of thermal mismatch. This property should have a very positive effect on the lifetime of the thermal barrier coatings.

This property is particularly pronounced for $LaLiAl_{11}O_{18.5}$ especially since the adhesion to the substrate is sufficiently good, in contrast to $LaFeAl_{11}O_{19}$.

CONCLUSION

A new material was presented for application as a thermal barrier coating in gas turbines. This new material has a significantly higher thermal conductivity than comparable materials and demonstrates phase transition in the high-temperature range.

Nevertheless, the material has the unusual and probably very beneficial property that it can form segmentation cracks during heat treatment, which should lead to a considerable expansion tolerance. The extent to which the lifetime can be prolonged in applications as thermal barrier coating material still remains to be investigated.

ACKNOWLEDGEMENTS
The authors thank Dr. José Luis Marques for the dilatometric measurements and Dr. W. Fischer for the high-temperature X-ray investigations. Likewise thanks are due to P. Lersch for the X-ray measurements and K.-H. Rauwald and M. Kappertz for preparing the plasma-sprayed samples. The authors are also grateful to D. Pitzer (FZJ B-NZ) for laser flash and DSC investigations and to D. Sebold for the SEM measurements.

REFERENCES

[1] Jaeyun Moon, Hanshin Choi, Hyungjun Kim, Changhee Lee, "The effects of heat treatment on the phase transformation behavior of plasma-sprayed stabilized ZrO_2 coatings", *Surface and Coatings Technology* 155, 1 – 10 (2002).

[2] B. Siebert, C. Funke, R. Vaßen, D. Stöver, "Changes in porosity and Young's modulus due to sintering of plasma sprayed thermal barrier coatings", *J. Materials Processing Technology* **92 – 93**, 217 – 223 (1999).

[3] J. A. Thompson, W. Ji, T. Klocker, T. W. Clyne, "Sintering of the top coat in thermal spray TBC systems under service conditions", *Superalloys 2000*; T. M. Pollock et al. (eds) Seven Springs, USA TMS (The Minerals, Metals & Materials Society), 685 – 692 (2000).

[4] R. Vaßen, M. Ahrens, A. F. Waheed, D. Stöver, "The influence of the microstructure of thermal barrier coating systems on sintering and other properties", *International Thermal Spray Conference*, E. Lugscheider and C. C. Berndt (Eds.), Pub. DVS Deutscher Verband für Schweißen, Germany, 879 – 883 (2002).

[5] R. Vaßen, D. Stöver, "Conventional and New Materials for Thermal Barrier Coatings", *NATO Science Series II: Mathematics, Physics and Chemistry;* Kluwer Academic Publishers, Dordrecht, NL, 16, 199 – 216 (2001).

[6] D. Stöver, G. Pracht, H. Lehmann, M. Dietrich, J.-E. Döring und R. Vaßen, "New material concepts for the next generation of plasma-sprayed thermal barrier coatings"; *Thermal Spray 2003:* Advancing the Science and Applying the Technology; Proceedings of the 2003 International Thermal Spraying Conference, Vol. 2, 3rd International Conference on Thermal Spray Technology, Orlando Fl, USA, 1455-1462 (2003).

[7] Like Xie, A. N. Cormack, "Defect solid state chemistry of magnetoplumbite-structured ceramic oxides II: Defect energetics in $LaMgAl_{11}O_{19}$", *J. Solid State Chem.* **88**, 543 - 554 (1990).

[8] R. Gadow, G. W. Schäfer, "Thermal insulating material and method for producing same", *Patent* WO 99/42630, (1998).

[9] C. J. Friedrich, R. Gadow, T. Schirmer, "Lanthane Aluminate – A new material for atmospheric plasma spraying of advanced thermal barrier coatings", *Proceedings ITSC 2000*, Thermal Spray: Surface Engineering via Applied Research, C. C. Berndt (Ed.), Pub. ASM International, Materials Park, OH-USA, 1219 – 1226 (2000).

[10] S.R. Choi, N.P. Bansal and D. Zhu, Mechanical and Thermal Properties of Advanced Oxide Materials for High-Temperature Coatings Applications, Ceram. Eng. Sci. Proc., **26** [3] 11-19 (2005).

[11] R. Gadow, M. Lischka; "Lanthanum hexaaluminate – novel thermal barrier coatings for gas turbine applications – materials and process development"; *Surface and Coatings Technology* **151 – 152**, 392 – 399 (2002).

[12] B. Saruhan-Brings, U. Schulz, C.-J. Kröder, Wärmedämmmaterial mit im wesentlichen magnetoplumbitischer Kristallstruktur, "Thermal barrier coatings with mainly magnetoplumbite crystal structure", *European Patent EP* 1256 636 A2 (2002).

[13] Nitin P. Padture, Maurice Gell, Eric H. Jordan, "Thermal barrier coatings for gas-turbine engine applications", *Science*, 296, 280 – 284 (2002).

[14] R. C. Ropp and G. G. Libowitz, "The nature of the alumina-rich phase in the system La_2O_3-Al_2O_3", *J. Am. Ceram. Soc.* 61 [11 – 12], 473 – 475 (1978).

[15] X. Q. Cao, R. Vassen, S. Schwartz, W. Jungen, F. Tietz, D. Stöver, "Spray-drying of ceramics for plasma-spray coatings", *J. Europ. Ceram. Soc.* 20, 2433 – 2439 (2000).

[16] H. Lehmann, D. Pitzer, G. Pracht, R. Vaßen, D. Stöver, "Thermal conductivity and thermal expansion coefficients of the lanthanum-rare earth element-zirconate system", *J. Am. Ceram. Soc.*, **86** [8],. 1338 – 44 (2003).

[17] C. J. Friedrich, R. Gadow and M. H. Lischka, "Lanthanum Hexaaluminate Thermal Barrier Coatings", *Proceedings Cocoa Beach* 2001, USA, 375 – 382 (2001).

* Corresponding author: Gerhard Pracht e-mail address: gerdpracht@gmx.de

Applied Ceramic Technology

Ceramic Product Development and Commercialization

Thermal Barrier Coatings Design with Increased Reflectivity and Lower Thermal Conductivity for High-Temperature Turbine Applications

Matthew J. Kelly,* Douglas E. Wolfe, and Jogender Singh

Applied Research Laboratory, Penn State University, University Park, Pennsylvania 16801

Jeff Eldridge, Dong-Ming Zhu, and Robert Miller

NASA-GRC, Cleveland, Ohio 44135

High reflectance thermal barrier coatings consisting of 7% Yittria-Stabilized Zirconia (7YSZ) and Al_2O_3 were deposited by co-evaporation using electron beam physical vapor deposition (EB-PVD). Multilayer 7YSZ and Al_2O_3 coatings with fixed layer spacing showed a 73% infrared reflectance maxima at 1.85 μm wavelength. The variable 7YSZ and Al_2O_3 multilayer coatings showed an increase in reflection spectrum from 1 to 2.75 μm. Preliminary results suggest that coating reflectance can be tailored to achieve increased reflectance over a desired wavelength range by controlling the thickness of the individual layers. In addition, microstructural enhancements were also used to produce low thermal conductive and high hemispherical reflective thermal barrier coatings (TBCs) in which the coating flux was periodically interrupted creating modulated strain fields within the TBC. TBC showed no macrostructural differences in the grain size or faceted surface morphology at low magnification as compared with standard TBC. The residual stress state was determined to be compressive in all of the TBC samples, and was found to decrease with increasing number of modulations. The average thermal conductivity was shown to decrease approximately 30% from 1.8 to 1.2 W/m-K for the 20-layer monolithic TBC after 2 h of testing at 1316°C. Monolithic modulated TBC also resulted in a 28% increase in the hemispherical reflectance, and increased with increasing total number of modulations.

Introduction

A critical goal of both the Department of Defense and Department of Energy is doubling the thrust-to-weight ratio for high-operating-temperature turbine systems with the use of lightweight turbine structures. Therefore, there is a need to develop lightweight smart structures capable of operating at 3000°F. Coatings that are self-indicating of damage with tailored microstructure and chemistry may meet these challenges. The approach has to be simple, cost-effective, and compatible with current technology. Specific selection of materials will depend upon application and operating environment.

This research was sponsored by the United States Navy Manufacturing Technology (Man-Tech) Program, Office of Naval Research, under Navy Contract N00024-02-D-6604.

*mjk268@psu.edu

Thermal barrier coating (TBC) composed of partially stabilized zirconia ZrO_2–7wt% Y_2O_3, (7YSZ) has been used in the turbine industry for over 50 years and has played significant role in extending the life of components. Initially, TBC was applied by plasma spray process. However, for more than a decade, the preference is in applying TBC by electron beam-physical vapor deposition (EB-PVD) as it offers superior thermo-mechanical properties and lifetime performance over plasma spray process.[1] Recently, research efforts have been directed in reducing the thermal conductivity of TBC without sacrificing its high-temperature thermal stability and mechanical properties needed for turbine industry.

Thermal conductivity of ceramic materials is both a phonon and photon phenomena dependent on various factors including material structure, operating environment, and temperature. Scattering of anharmonic elastic waves, either by inelastic phonon–phonon or phonon–lattice interaction, and controls phonon conductivity. Phonon–phonon scattering or "Umklapp scattering" occurs when two or more anharmonic elastic waves interact, which give rise to a finite thermal conductivity contribution.[2] The number of phonon–phonon interactions are determined by the mean free path, and for next generation TBC operating at 3000°F the mean free path is approximated to be the lattice spacing, which leads to a fixed phonon thermal conductivity contribution for a given material system.[3] Lattice defects such as voids or dopants cause phonon–lattice interactions that reduce the mean free path and frequencies of oscillation permitted in the lattice thereby reducing phonon thermal transport.

Photon contributions to total heat transfer are often disregarded in calculations for thermal conductivity, because of the small effect at low temperatures. However, this effect can become quite large at elevated temperatures because of the fourth order dependency on temperature. Obeying Wien's Law, as the body increases in temperature, the emission spectrum becomes more pronounced and the peak of emission intensity shifts toward a smaller wavelength. For next generation turbines, coating material optical properties will contribute toward high temperature thermal conductivity. In particular, understanding the percentage of photons a material transmits, reflects, and absorbs for a given temperature is critical (Fig. 1). 7YSZ is 80% transparent (or transmits 80% of incoming photons) below 5 μm wavelength and nearly 100% opaque above 8 μm wavelength (or the material absorbs nearly all radiation),

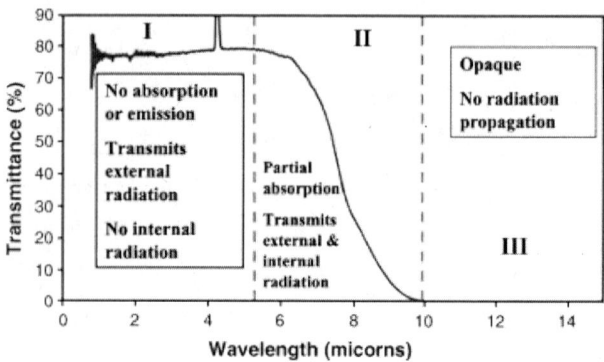

Fig. 1. Seven percent Yittria-Stabilized Zirconia Electromagnetic Transmission Spectrum.

and it is translucent in the range of 5–8 μm wavelength. Photon absorption occurs at interfaces either at the material boundaries or at atomic sites in materials that permit photon transport. Absorption causes a finite increase in energy, which then dissipates by atomic oscillations (phonon transport) and photon radiation at longer wavelength.

Partially stabilized zirconia has historically offered the best balance of mechanical and thermal properties needed for turbine application. Unfortunately, it is nearly transparent over a wide range of wavelengths. Because of this translucent property, in high-temperature environments significant heat can be transported through the 7YSZ to the base material in spite of its low thermal conductivity. The total amount of thermal energy transferred through TBC dictates oxidation rate and creep properties of the base material. By reducing the energy transportation rate, components may operate in higher temperature environments with the same lifetime or longer.

Significant progress has been made in developing low thermal conductive TBC with the same basic composition (ZrO_2–7wt% Y_2O_3) having rare earth oxide dopants. It has been reported that multi-component oxide dopants enhance the thermal stability and reduces thermal conductivity by 50%.[4,5] The reduction in thermal conductivity is because of presence of 5–100 nm defect clusters distributed through out the coating matrix, which limit the mean free path of both phonons and photons. New generation, low conductive thermal barrier coatings are based on:

$$ZrO_2 - Y_2O_3 - Nd_2O_3(Gd_2O_3, Sm_2O_3) - Yb_2O_3(Sc_2O_3)$$
Primary stabilizer   Oxide dopant cluster

Techniques for reducing the high temperature thermal conductivity focus on the rejection of incoming photons. Changing the optical properties of a material, often with a coating, will alter photon transport into the material. A common method of changing the transmission of electromagnetic waves is application of reflective coatings. Multiple interfaces throughout the thickness of the TBC cause incoming photons to be reflected. There are two approaches in creating multiple interfaces in the TBC. One approach is forming short-range alternate layers of high and low-density structures in monolithic materials (Fig. 2a), i.e., modulated microstructure with periodic density variation that will alter its refractive index properties. The second approach is creating a multilayer structure with two different materials (Fig. 2b) having different refractive indexes.

Single material coatings with periodic density variations have an advantage over inhomogeneous reflective coatings such that modulated density structures are not sensitive to the incidence angle of radiation. Modulated density variation in thermal barrier coatings was demonstrated in 7YSZ and Hf-26 wt% Y_2O_3 by periodically interrupting the incoming vapor flux during the deposition process. Hemispherical reflectance of 7YSZ and Hf-26 wt% Y_2O_3 was increased from ~30% to 45% (20 equal periodic vapor flux interruptions) and ~55% to 65% (with 40 equal periodic vapor flux interruption), respectively.[6] Significant benefit of this methodology is that there is no change in the coating chemistry, columnar morphology or crystallographic structure. The additional benefit is exhibiting lower thermal conductivity and better strain tolerance in comparison with standard 7YSZ coating. Monolithic (7YSZ) graded structures, as well as, multilayer coating materials with different refractive indexes and densities (7YSZ and Al_2O_3) will be discussed.

Experimental Procedure

All TBC systems were deposited in an industrial prototype Sciaky Inc. (Chicago, IL) EB-PVD unit consisting of six EB-guns (1–6) and a three-ingot continuous feeding system (A–C) as shown in Fig. 3. Before applying the TBC, 1 in. diameter platinum–nickel–aluminide plated Rene N5 discs were heat tinted for 30 min in air at 704°C and cooled to room temperature. The surface and the side of each test specimen were grit blasted in a Unihone grit blaster using high-purity 400 μm size aluminum oxide particles. The distance from the edge of the nozzle to the surface of the samples was approximately 15 cm with a pressure of 30 psi. The angle of the nozzle with respect to the sample surface was 45° in order to minimize the amount of embedded Al_2O_3 particles incorporated into the bond-coated surfaces. The grit blast time on each sample varied between 10–15 s and was performed until a uniform matte finish was obtained. The samples were then tack welded to strips of 304 stainless steel foil that were tack welded to a 5.08 cm diameter mandrel.

The sample holder with the mounted samples was then ultrasonically cleaned for 20 min in acetone, rinsed with de-ionized water, ultrasonically cleaned for 20 min in methanol, and then dried with nitrogen gas. The sample holder was then mounted in the evaporation unit with a source to substrate distance of 30.48 cm. The vacuum unit was evacuated to a base pressure of 7.5×10^{-6} Torr with the oxygen gas lines being evacuated. The samples were positioned directly over a 4.928 cm diameter 7YSZ ingot (TransTech Inc., Adamstown, MD) rotating at 7 rpm. Using two electron beams (#2 and #5 in Fig. 3a), the samples were indirectly heated to 1000°C under a graphite "A-frame" heater assembly. After a minimum of 20 min at temperature, the samples were allowed to soak at 1000°C for an additional 20 min while injecting approximately 150 sccm of oxygen ($P_{chamber} \sim 0.8 - 1 \times 10^{-3}$ Torr) into the chamber in order to form a thermally grown oxide (TGO). Immediately following the formation of the TGO, 7YSZ was evaporated (using electron beam #3 and TBC ingot position B in Fig. 3a) at an ingot feed rate of 0.8 mm/min. Again, ~150 sccm of oxygen was injected into the deposition chamber ($P_{chamber} = 1 \times 10^{-3}$ Torr) while simultaneously evaporating the 7YSZ ingot to maintain the oxygen stoichiometry of the condensing coating.

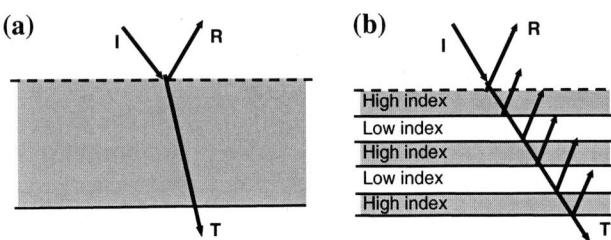

Fig. 2. *Two methods for reducing the photon transport through thermal Barrier Coatings. (a) single layer structure with modulated density, (b) multi-layer, multi-material structure.*

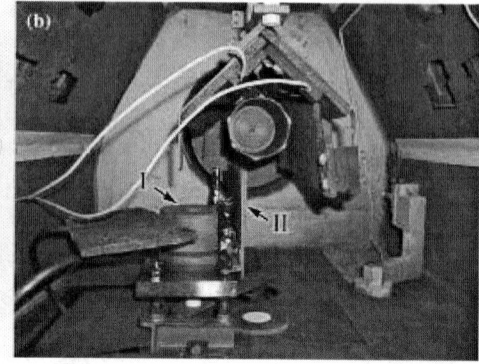

Fig. 3. (a) Sciaky Inc. electron beam physical vapor deposition (EB-PVD) unit with 6 EB-guns (1–6) and a 3-ingot continuous feeding system (A–C), (b) EB-PVD unit with multi-layer tooling configuration utilizing offset ingot (I) and vapor barrier (II).

Monolithic 7YSZ Coatings with Modulated Density

In order to introduce stable, microstructural features that would change the density, the condensing vapor flux was periodically interrupted using a graphite shutter. This periodic "shuttering" prevents the flux from depositing on the sample surface with little to no reduction (20–30°C) in substrate temperature. This method resulted in smaller strain fields consisting of one or more of the following: more diffuse interfaces, microporosity, vacancies, and smaller lattice strains, i.e., density/refractive index variation. In order to achieve the various number of diffuse layers within the TBC, the shutter was closed for 30 s for each layer (while continuously evaporating the 7YSZ ingot) to prevent the coating from condensing on the surface of the substrates which were maintained near 1000°C. Table I lists the primary deposition processing parameters for creating the microstructural changes for the 7YSZ layered coatings produced by the "shutter" methods. For both methods, at the end of the desired deposition time, the samples were retracted into the load lock chamber which was injected with oxygen gas to prevent

oxygen loss and allowed to cool for 10 min before venting to atmosphere.

In order to determine whether additional strain fields were incorporated into the TBC by the "shutter" method, the amount of deflection was measured for the various coatings deposited on PtAl-coated Haynes 188 alloys strips using a computer controlled machine (CMM). A special tooling holder was fabricated to ensure precise placement of the sample in the precoated and postcoated condition to ensure that the amount of deflection was measured accurately. Once the sample was positioned correctly, a probe traversed the surface of the strip using a six axes position scheme. The change in the vertical distance (z-position) was plotted as a function of the x–y coordinates and the difference from the original z-position was used for the deflection value.

7YSZ-A$_2$O$_3$ Highly Thermal Reflective TBC

Similarly, high reflective thermal barrier coatings were deposited in the industrial prototype EB-PVD unit with a modified experimental setup design as shown in

Table I. Deposition Parameters for Layered TBC Produced by "Shutter" Method

Sample #	Total # of layers	Evaporation time (min)	Additional evaporation time for layered structure (min)	Amount of time shutter remained open per layer (s)	Average coating thickness (μm)
A	1	73	0	4380	122.3
B	5	75	2.0	876	119.7
C	10	77.5	4.5	438	127.9
D	20	82.5	9.5	219	132.8

TBC, thermal barrier coatings.

Fig. 3b. Two sets of coatings were produced with (i) constant individual layer thickness of the 7YSZ (400 nm) and Al$_2$O$_3$ (100 nm) throughout the coating and (ii) with alternating individual layer thickness of 7YSZ (740 nm) and Al$_2$O$_3$ (765 nm) near the substrate coating interface, and decreasing in 5 nm increments to 7YSZ (90 nm) and Al$_2$O$_3$ (115 nm) near the surface of the coating. In order to achieve the desired individual layer thickness within the multilayer coating, modifications were made to the experimental setup as shown in Fig. 3b. A fourth crucible was added to the chamber containing aluminum oxide ingot and was offset by approximately 110 mm from the center of ingot B (7YSZ). By changing the relative source to substrate distances between the aluminum oxide and 7YSZ evaporant materials and material evaporating rates, the coatings were deposited to yield a 4:1 ratio of 7YSZ and Al$_2$O$_3$, respectively. In order to prevent intermixing of the vapor cloud, a vapor barrier was positioned between the evaporant materials (Fig. 3b). Similarly, the graded multilayer coatings of 7YSZ and Al$_2$O$_3$ with different individual layer thickness throughout the total coating thickness was achieved in this manner by varying the evaporation rate and sample rotational speed in order to get the desired individual layer thickness.

Fracture surfaces, cross-sections, and surface morphologies of the coated samples were examined by a Philips model PW6848/00 scanning electron microscope (SEM) to determine microstructural and morphology differences within the various coatings. Phase analysis, texture coefficients, and residual stresses of the coated samples were determined using Philips X'Pert model MPD and MRD X ray diffractometers. Thermal conductivity of the TBC was measured by the steady-state heat flux (CO$_2$ laser) technique at 1316°C,[7] and the IR hemispherical reflectance was measured in an FTIR spectrometer with an integrating sphere accessory at room temperature described elsewhere.[8]

Results and Discussion

Monolithic Modulated 7YSZ Microstructure

The typical microstructure of a TBC produced by EB-PVD can be divided into two zones. The inner zone (zone I) is the early part of multiple nucleation and subsequent growth of the columnar microstructure having large number of interfaces, grain boundaries, microporosity, and randomly oriented grains. The inner zone

ranges from 1 to 10 µm in thickness and exhibits lower thermal conductivity ($\sim$ 1.0 K/m-K).[9,10] With increasing thickness, the TBC microstructure is characterized by a high-aspect ratio columnar grain with dominant crystallographic texture. The thermal conductivity increases as the outer part of the coating becomes more crystallograhically perfect (zone II) with fewer grain boundaries (grain size increases with increasing coating thickness, i.e., larger grains). In this outer zone (II), the thermal conductivity approaches that of bulk zirconia (2.2 W/m-K). Thus, modifying TBC microstructures should offer the best properties available for commercial EB-PVD coatings: namely, low thermal conductivity, high hemispherical reflectance, high strain tolerance, and good erosion resistance. By altering the macrostructure on the micrometer and submicrometer levels through periodically introducing strain fields (i.e., density/refractive index changes by the incorporation of microporosity and surface restructuring), the thermal conductivity of TBC materials can be significantly reduced and hemispherical reflectivity increased. As previously discussed, layered periodicity in the coating will significantly reduce both the phonon scattering and photon transport resulting in lower thermal conductivity and higher hemispherical reflectance.

Through periodically interrupting the continuous flux of the vapor cloud by using a "shutter" mechanism, the temperature of the substrate remained almost constant ($\sim$ 30°C drop in substrate temperature) during the deposition process with no deposition occurring when the shutter is closed. During this interruption, it is believed that the surface mobility of the condensed species contributes to the surface relaxation of the deposited coating through restructuring. As the vapor flux is prevented from depositing on the surface, the surface atoms have enough time, energy and surface mobility to diffuse to regions of lower energy. As a result, the surface strains change resulting in more phonon scattering because of different strain energy fields (and potentially sub-micron grains, interfaces, and microporosity, thus resulting in lower thermal conductivity and higher reflectivity. When the shutter is opened, the new flux deposits on a slightly different strained surface, but which is not so strained as to result in significant lattice mismatch and act as nucleation sites for the new grains. This newly strained surface results in a very diffuse interface which may contain microporosity and intracolumnar morphology differences, as well as different strain fields. Throughout this atom coalescence, stable microporosity/strain fields and intracolumnar morphology differences develop with continued columnar growth without changing the surface morphology and columnar grain macrostructure as shown in Fig. 4. During the initial deposition, atoms form islands (Volmer–Weber growth mode) and grow until the flux rapidly increases resulting in grain coalescence and microporosity (resulting from grain coalescence) and tensile lattice strains. However, when viewed from the top surface, the grain size does not change, but the intracolumnar microstructure (i.e., microporosity, morphology, and strain field) is believed to be altered as shown in Fig. 4 for the 1-, 5-, 10-, and 20-layer 7YSZ coatings deposited by the "shutter" method. From the SEM micrographs (Fig. 4), it is difficult to determine whether microporosity is present. However, microporosity was confirmed in the coating through the thermal conductivity measurements as the reduction in thermal conductivity is directly related to the amount of microporosity.

As only a minimal temperature change occurs in addition to the disruption of the vapor flux, the long high-aspect ratio columnar grains continue to grow to the total length of the coating thickness similar to standard single layer 7YSZ. Comparison with a standard 7YSZ, the modified coating shows no distinct

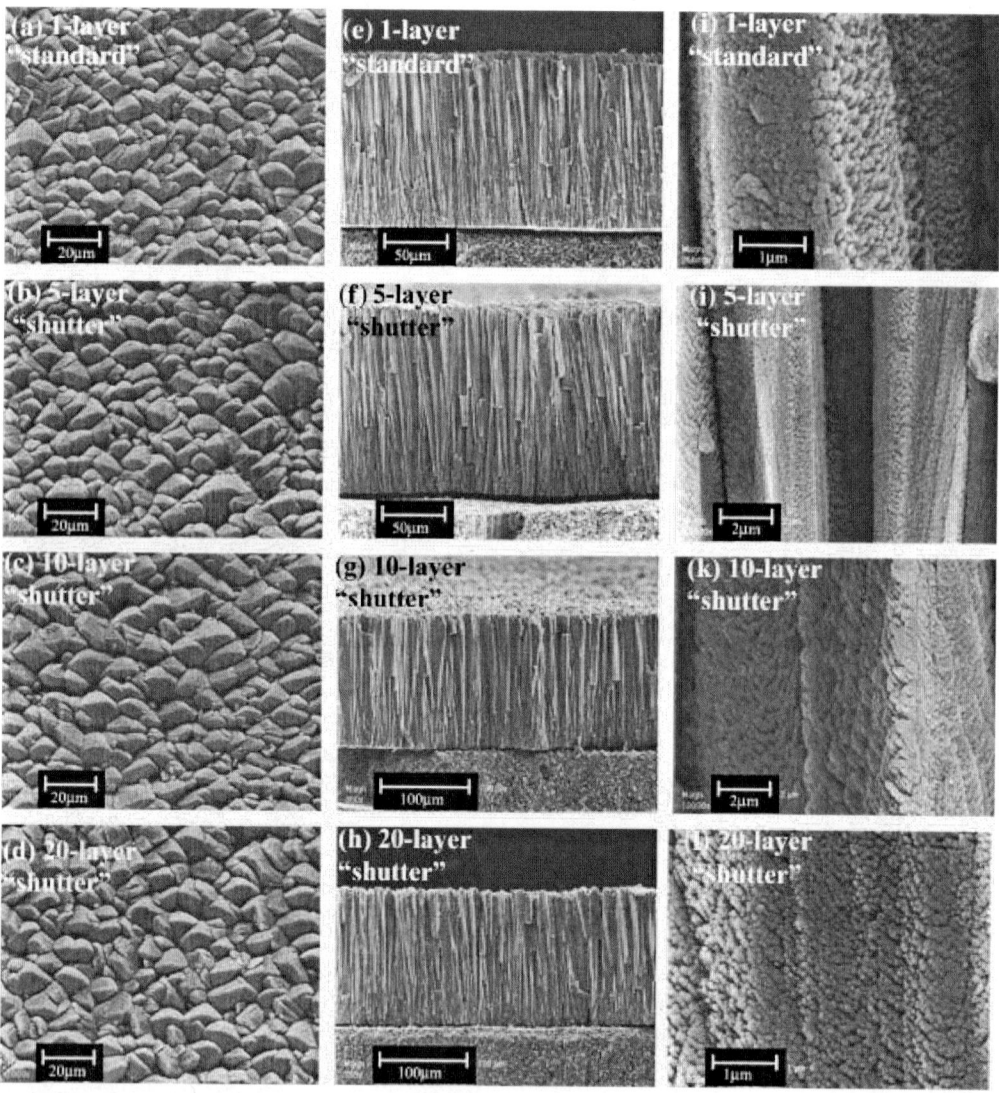

Fig. 4. Scanning electron microscope micrographs showing the surface morphology (a–d) and fracture surface (e–l) of EB-PVD 7YSZ deposited by the "shutter" method having 1-layer, 5-layer, 10-layer, and 20-layers, respectively, on CoNiCrAlY-coated MARM247 alloy.

differences in macrostructure. As the growth orientation of the new flux remains the same as the underlying grain, similar crystallographic texture also occurs, but varies slightly resulting from increased strain fields. The interface between the condensed flux and the newly arrived flux is diffuse (i.e., no sharp distinct interface) which results in different strain fields near the interface and within each columnar grain. Such microstructural modifications will have an impact on the thermal conductivity, as well as residual stress (i.e., strain tolerance).

Similar concept/phenomenon employed in the "shutter" approach can be correlated to the "feather-like" morphology often observed within each columnar grain of a standard TBC deposited by EB-PVD. The "feather-like" size and morphology are representative of the sample rotation speed as during each revolution, coating condenses and surface mobility changes. When the sample surface is opposite to the melt pool, little to no coating deposition occurs as well as a slight decrease in surface temperature as no radiant heat from the melt surface or heat of condensation occurs. As a result, micro strain fields develop on a much finer scale as compared with the "shuttering" method (30 s), because of restructuring or surface diffusion until the next high rate of condensation occurs when the sample surface is again facing the melt pool. In the present investigation, the sample rotation speed was 7 rpm, resulting in the surface of the samples being parallel to the melt pool approximately every 8.5 s. In comparison, using the 30 s delay with the "shuttering" method allows the samples ~3.5 times more time for micro "restructuring" to occur which results in increased tensile strains, and thus, lower overall compressive stresses as discussed later. The periodicity of the "shutter" results in regions of the TBC with periodic differences in the lattice size resulting in modulated strain fields. As previously discussed, the degree of surface restructuring was much greater for the "in and out" approach as compared with the "shuttering" method. However, it should be noted that there are many factors to consider as large changes in the substrate rotation speed can also affect the grain size and intercolumnar porosity. In addition, the size of the component, condensation rate, and other deposition parameters (i.e., substrate temperature, pressure, etc.) can greatly affect the coating microstructure.

Crystallographic Phase Analysis: X-ray diffraction (XRD) patterns were performed on the surface of the various layered 7YSZ coated samples. Normal Bragg-Brantano ($\theta/2\theta$) diffraction continuous scans were performed over the range of $2\theta = 15°$ to $90°$ at 2 s per step with a step size of $0.030°$. The diffraction patterns confirm that the coatings are polycrystalline and predominantly of the nontransformable tetragonal phase (t' phase) and have a strong (200) orientation which is highly desirable for strain tolerance during thermal cycling.[11,12] No evidence of the monoclinic or cubic phases was observed in any of the diffraction patterns.

Residual Stress: The preliminary investigation of the amount of strain within the various layered coatings supports the x-ray diffraction results (peak shifting) as the degree of deflection for the 7YSZ "shuttered" coatings decreased with increasing number of layers up to the 20-layer coating as shown in Fig. 5a. As the strip curvature was convex, the TBC is under a state of compression. From Fig. 5a, the amount of deflection decreased with increasing number of "shuttered" layers suggesting that the amount of tensile stress (strain fields) incorporated into the coating from restructuring phenomenon increased, resulting in a lower total compressive residual stress. To confirm the residual stress state of the deposited coatings, a four circle Philips X'Pert diffractometer (Almelo, The Netherlands) was used to determine the "as deposited" residual stresses using the $\sin^2$ technique discussed elsewhere.[13] Similar to the deflection measurement trend, the preliminary stress analysis (Fig. 5b) by X-ray diffraction shows the same trend of decreasing compressive stress with increasing total number of layers. This same phenomena has been observed with other multilayer materials deposited by EB-PVD with increasing number of layers.[14,15] The absolute values of the residual stress determined by the deflection of the strip are presently being correlated with the x-ray diffraction analysis of the coated buttons. Continued investigation of the amount of residual stresses within the various coatings is presently being performed to determine whether the strain fields/microporosity are stable as a function of time at elevated temperatures and will be presented later.

Thermal Conductivity: Several researchers have discussed the theory of thermal conductivity in ceramic materials and will only be discussed briefly here.[16,17] Electrons, phonons (lattice vibrations), and photons (radiation) are the three mechanisms in which heat is transported in crystalline solids. However, as 7YSZ is a

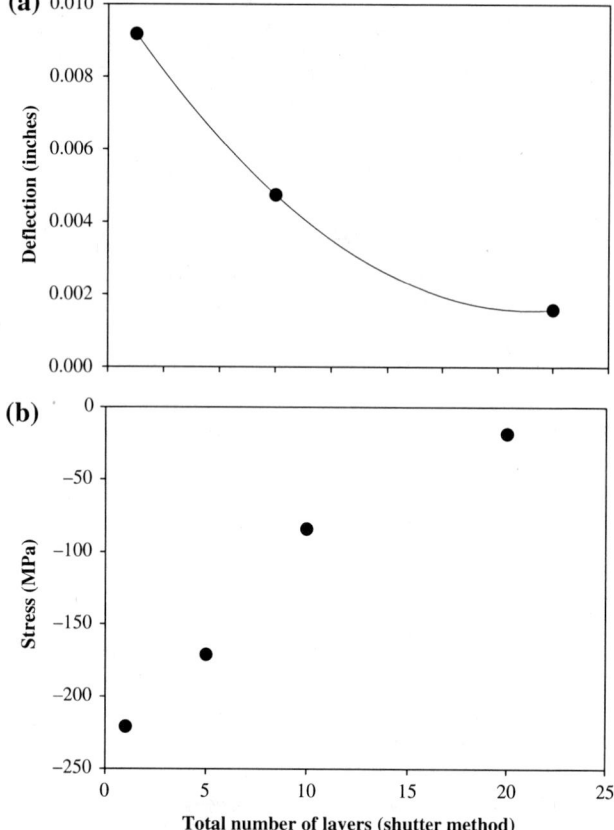

(a)

(b)

Fig. 5. (a) Deflection as a function of total number of layers for 7% Yittria-Stabilized Zirconia (7YSZ) (coating side) deposited by electron beam physical vapor deposition (EB-PVD) on PtAl-coated Haynes 188 alloy strips and (b) residual stress determined by x-ray diffraction of 7YSZ TBC deposited by EB-PVD as a function of total number of layers deposited by the "shutter" method.

ceramic insulator, electrons have little to no contribution to the thermal conductivity. Therefore, photons (k_r) and phonons (k_p) are the largest contributors to the thermal conductivity and the total thermal conductivity (k_t) can be written as their sum:

$$k_t = k_p + k_r$$

The thermal conductivity component resulting from phonons (k_p) can be further expressed by:

$$k_p = \frac{1}{3} \int C_v \rho v l_p$$

where C_v is the specific heat at constant volume, ρ is the density, v is the velocity, and l_p is the phonon mean free path. Whereas the radiation contribution (k_r) to thermal

conductivity can be written as:

$$k_r = \frac{16}{3} \sigma n^2 T^3 l_r$$

where σ is the Stephen–Boltzmann's constant, n is the refractive index, T the temperature, and l_r is the photon mean free path.

According to the above equations, in order to reduce the thermal conductivity, a reduction in the specific heat at constant volume, density, phonon velocity, refractive index, or mean free path is required. Of these, the specific heat at constant volume is constant above the Debye temperature. As a result, only changes to ρ, n, v, and mean free path will change the thermal conductivity. Phonon scattering generally occurs by interactions with lattice imperfections which include dislocations, grain boundaries, other phonons, vacancies, and atoms of different masses. The phonon mean free path (l_p) can be defined by

$$1/l_p = 1/l_v + 1/l_i + 1/l_{gb} + 1/l_s$$

where l_v, l_i, l_{gb}, and l_s represent the mean free path contributions because of vacancies, interstitials, grain boundaries, and strain, respectively.

Doped TBC materials with atoms or ions of different masses result in a distortion of the bond length which creates strain fields within the lattice and thus phonon scattering.[17] This is the similar concept used for the "shuttering" method in which period strain fields were developed through the TBC, but without changing the composition. Therefore, imperfections within the lattice can change the mean free path of phonons (scattering), and hence the thermal conductivity.

Figure 6a shows comparative thermal conductivity of a standard single layered 7YSZ, layered TBC with diffuse interface produced by "shutter" method. The thermal conductivity of the standard 7YSZ TBC produced by EB-PVD was found to be ~1.8 W/m-K. The 10-layer microstructurally "modified" TBC deposited using the "shutter" layering concept reduced the thermal conductivity to ~1.6 W/m-K.

In order to establish a relationship between the thermal conductivity as a function of total number of TBC layers, additional TBC coated samples were produced using the "shuttering" concept. It was established that the initial thermal conductivity (K_0) decreased near linearly as a function of increasing total number of TBC layers from 1 to 20 as shown in Fig. 6b. With the exception of the 5-layer coating, similar results were found

(a)

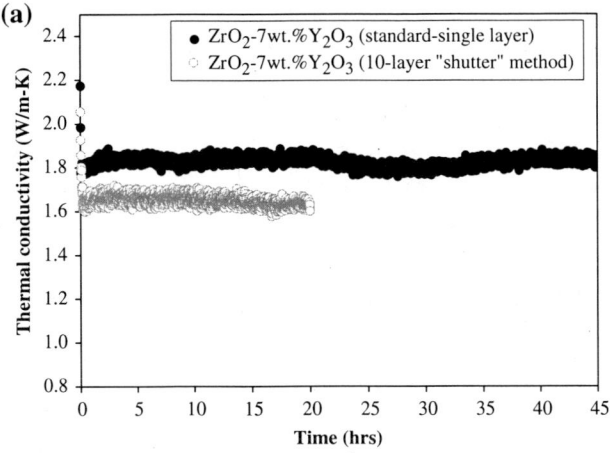

(b)

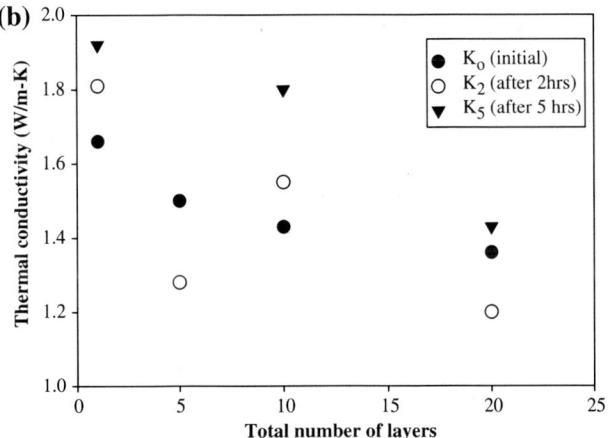

Fig. 6. Thermal conductivity of electron beam physical vapor deposition (EB-PVD) 7% Yittria Yittria-Stabilized Zirconia (7YSZ) thermal barrier coating (TBC) determined by a steady-state laser heat flux technique at 1316°C showing (a) thermal conductivity of EB-PVD 8YSZ produced by standard continuous evaporation and "shutter" method as a function of testing time, (b) thermal conductivity as a function of total number of layers produced by the "shutter" method measured at k_0 = as deposited, k_2 = after 2 h, and k_5 after 5 h of testing.

after 2 and 5 h of testing at 1316°C (Fig. 6b). After 2 h of testing, the thermal conductivity was found to decrease from 1.8 W/m-K for the standard TBC to ~1.2 W/m-K for the 20-layer "shuttered" TBC. This confirms that periodic interruption of the incoming flux by the "shutter" method results in lower thermal conductivity as compared with a standard single layer 7YSZ. The exact mechanism which contributes to the 20–30% reduction in thermal conductivity is still not fully understood, but is attributed to the localized density modulation. As previously discussed, the periodic interruption of the vapor flux results in surface relaxa-

tion or surface "restructuring" until the flux from the next layer deposits on the substrate surface. During this restructuring phase, it is believed that the surface atoms rearrange (surface mobility) their bonds and angles in order to minimize their free energy. When this occurs, the interatomic bond distance is believed to increase resulting in increased tensile strains. As the bond distances are now greater at the end of each layer, thermal transport between the material changes as there is a greater phonon distance resulting from the increase in bond length, which results in an overall lower thermal conductivity. In addition to these tensile strains, it is believed that microporosity forms at the start of the next growing layer resulting from island coalescence as previously discussed. Continued efforts are underway in order to study the heat conduction mechanisms resulting in the desired lower thermal conductivity values.

Hemispherical Reflectance: In addition to decreasing the thermal conductivity values, the TBC with modified microstructures was expected to also affect the hemispherical reflectance of the coating, and therefore, reduce radiative heat transport through the TBC. The hemispherical IR reflectivity of the layered TBC deposited by the "shutter" method resulted in an increase in reflectivity with increasing number of total layers as shown in Fig. 7a. The hemispherical reflectance increased from approximately 35% (1 layer) to 45% (20 layer) at the 1 μm wavelength which is approximately a 33% increase in the reflectivity as compared with the standard 7YSZ. This suggests that more heat will be reflected from the coating as the number of layers increase within the TBC, thus allowing higher engine operating temperatures and better fuel efficiency. Figure 7b shows the percent improvement in hemispherical reflectivity versus the total number of layers. The trend shows that the hemispherical reflectivity increases with increasing number of layers as expected. The slightly higher reflectivity values for the layered TBC produced by the "in and out" method as compared with the "shuttered" method is attributed to the sharp interfaces and larger coating thickness.

It is quite clear that the layering concept has opened an opportunity in engineering TBC with desired higher reflectance and lower thermal conductive properties through microstructural modifications.

Benefit of Interrupting Incoming Flux: As the novel method ("shuttering") of TBC deposition used in this

(a)

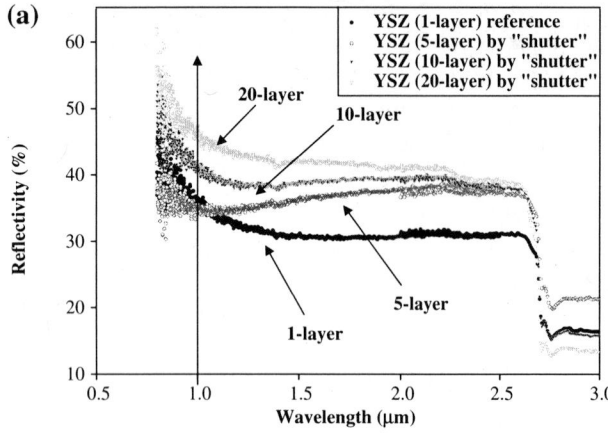

(b)

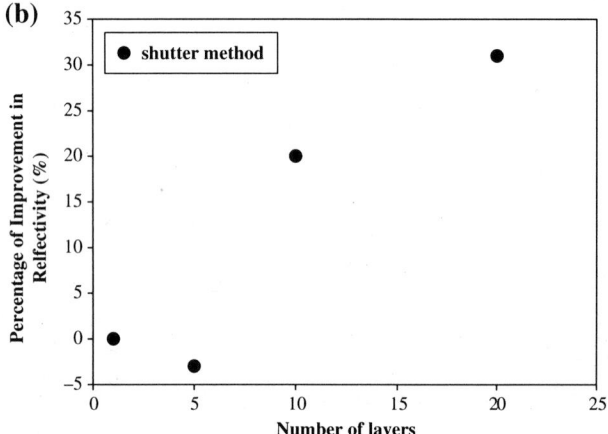

Fig. 7. (a) Hemi-spherical reflectance of layered (1, 5, 10, and 20 total layers) electron beam physical vapor deposition (EB-PVD) 7% Yittria Yittria-Stabilized Zirconia (7YSZ) thermal barrier coating (TBC) deposited by the "shutter" method after thermal exposure at 950°C for 20 h. (b) Percentage of improvement in hemispherical reflectivity versus the total number of layers within the EB-PVD 8YSZ TBC produced by "shuttering" method after thermal exposure at 950°C for 20 h.

investigation resulted in purely a microstructural effect (as there were no alloying additions, i.e., no compositional change), the concept can be applied to the next generation TBC materials including doped ZrO_2-based, HfO_2-based, etc. The shuttering concept was extended to HfO_2-based TBC with reproducible results and are discussed elsewhere.[18] In addition, depending on the total number of layers desired in the TBC, the additional time required to deposit the TBC is minimal. For example, in order to deposit a 10- or 20-layer "shutter" TBC, only an additional 5 or 10 min of deposition time is required, respectively. For the end user, this small

additional amount of time needed to deposit the 10- or 20-layer "shutter" coating is relatively insignificant when compared with the reduction in thermal conductivity (15–30%), increase in hemispherical reflectivity (28–56%), increase in thermal cyclic oxidation life (50–100%), better strain tolerance, and thus better performance of the component life with less down time of the engine. It should also be noted that deviations in the amount of time used for the "shuttering" concept could be further refined to yield better properties and coating performance.

7YSZ - Al₂O₃ Highly Thermal Reflective TBC

Multimaterial coatings with different refractive indices cause photon scattering obeying Snell's law. Each layer in the structure causes destructive interference to wavelengths that are odd integer multiples of the half-layer thickness and constructive interference to wavelengths that are even integer multiples of the half-layer thickness. Under this scenario, selection of materials is very important and it should be thermally stable at high temperature, minimum interdiffusion, and larger difference in refractive index. Typically, the refractive index (n) of ceramic TBC materials is: ZrO_2 (2.10), CeO_2 (2.35), HfO_2 (1.98), Al_2O_3 (1.60), SiO_2 (1.95), Y_2O_3 (1.82). Similarly, by controlling the thickness of each layer, reflectance of the coating can be controlled over a wide wavelength range.

It has been theorized that the thermal radiative properties of the ceramic coatings can be increased from ~35% (single layer ZrO_2 -7 wt% Y_2O_3) to nearly 100% by tailoring each coating layer thickness from 700 to 90 nm with alternating refractive index materials (ZrO_2 -7 wt% Y_2O_3 and Al_2O_3), Fig. 8. In addition, if the nanolayered structure and design is selected correctly, they will offer superior creep properties and impact resistance as compared with conventional single layered ceramic coatings. It is well documented that every 25°F (13°C) reduction in airfoil temperature will result in 2X component life improvement.[19] By increasing radiative reflectance using the nanolayered coatings, turbine blade surface temperature will be decreased up to ~Δ180°C (Fig. 9), resulting in longer component life by >10X. Combining highly thermal radiative barrier coating concept (i.e., nanolayered) with low thermal conductive material will offer a new class of high temperature coatings. The benefit of the tailored nanolayered coatings is enormous including no design constrain, low cost

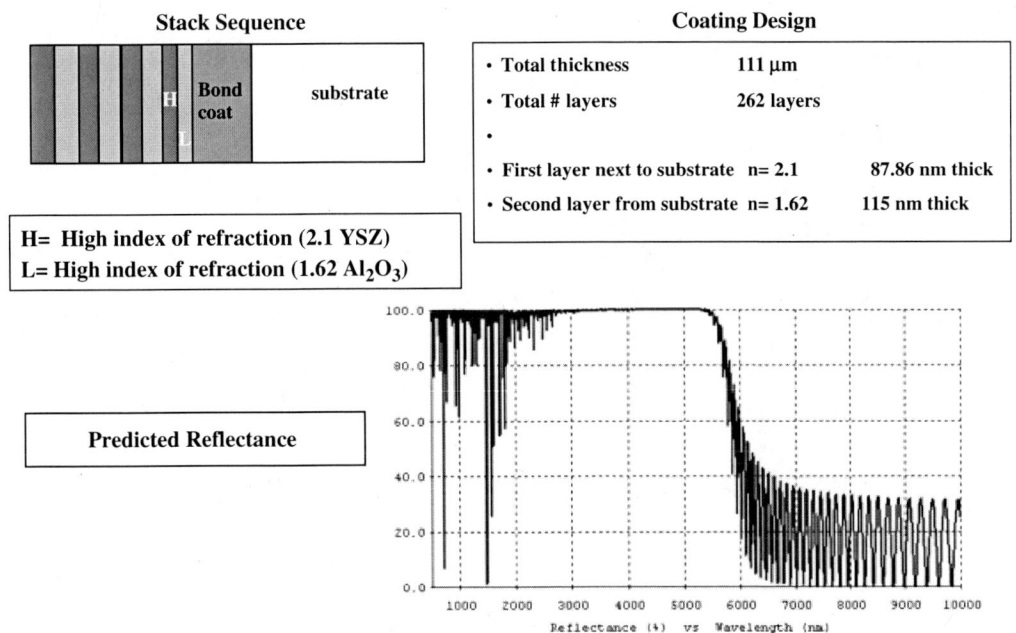

Stack Sequence

	Coating Design
• Total thickness	111 µm
• Total # layers	262 layers
•	
• First layer next to substrate n= 2.1	87.86 nm thick
• Second layer from substrate n= 1.62	115 nm thick

H= High index of refraction (2.1 YSZ)
L= High index of refraction (1.62 Al₂O₃)

Predicted Reflectance

Reflectance (%) vs Wavelength (nm)

Fig. 8. Predicted optical reflectance of multi-layer TBC with individual layers varying from 87.86 nm to 7.2 µm.

component use, reduction in engine weight and size, increasing power density (by 15%) or increasing Mach number (by 50%) and significant fuel savings. In addition, application of tailored nanolayered coatings will change the design concept for all propulsion engines including small single engines, aircraft, ship and land-based vehicle power plant.

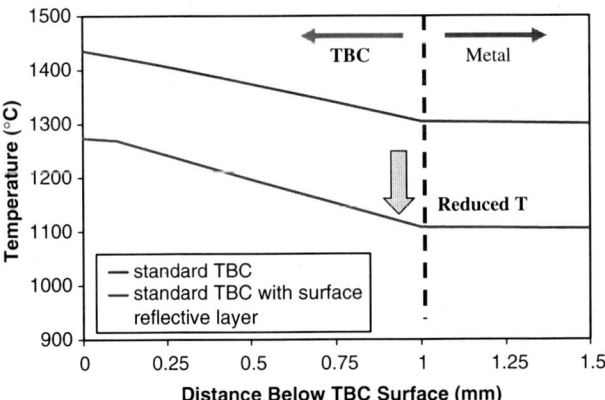

Fig. 9. Measured reduction in metal temperature because of modulated density (red line) and alternating material coating (green line).

To validate the nano-layered TBC concept and produce coatings composed of ZrO_2 -7 wt%Y_2O_3 and Al_2O_3 by EB-PVD, multilayer coatings of 7YSZ and Al_2O_3 were deposited. As shown in Fig. 8, the individual layer coating thickness near the coating surface was approximately 88 nm (Al_2O_3) and 115 nm (ZrO_2-7wt% Y_2O_3) and increased towards the substrate/coating interface.

Results and Discussion: For alternating layer coatings, the thickness of each Al_2O_3 layer was found to vary from ~75 to 100 nm (black color) while the thickness of 7YSZ remained constant at 400 nm (white color) when substrates were rotated at a constant rate shown in Fig. 10. Changing the rotation rate during deposition produced variable layer thickness coatings with a constant ratio of each material layer thickness.

The corresponding hemispherical reflectance of each coating is displayed in Fig. 11. Thermal conductivity of the TBC was measured by the steady-state heat flux (CO_2 laser) technique at 1316°C[8] and the IR hemispherical reflectance was measured in an FTIR spectrometer with an integrating sphere accessory at room temperature described elsewhere.[9] The reflectance of the fixed layer had a maxima of 73% reflection in the wavelength range of 1.85 µm , while variable layer coat-

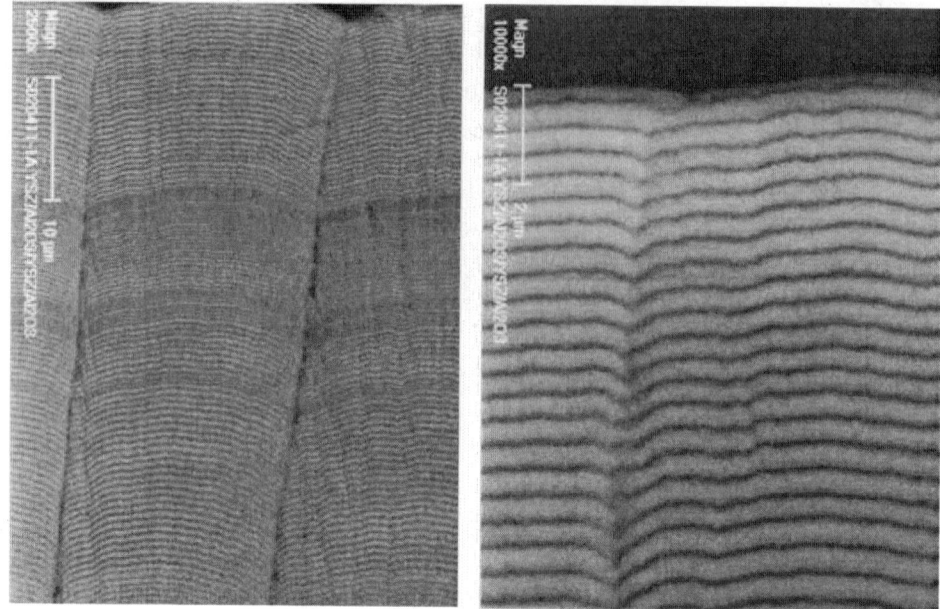

Fig. 10.   Scanning electron microscope micrographs of variable layer thickness coatings. Al₂O₃ layers dark, Yittria-Stabilized Zirconia layers light.

ings had an increased reflection spectrum over a 1 to 2.75 μm range. This preliminary experiment clearly showed that the reflectance of the coatings could be tailored to achieve increased reflectance over a desired wavelength range by controlling the thickness of individual Al₂O₃ and 7YSZ layers.

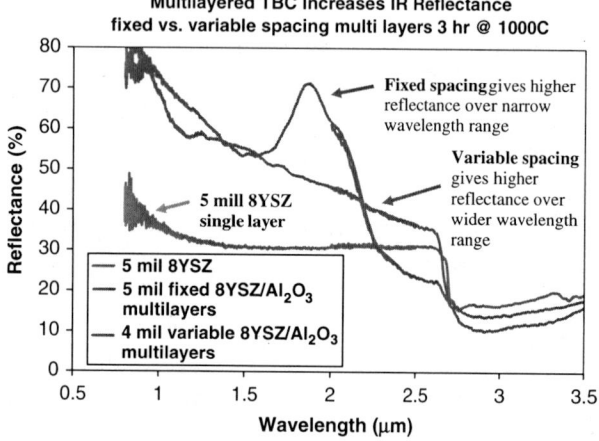

Fig. 11.   Hemispherical reflectance of modulated 7% Yittria-Stabilized Zirconia (7YSZ)/Al₂O₃ multilayer structures containing alternating 7YSZ/Al₂O₃ layer with uniform individual layer thickness and alternating layered structures through coating thickness.

Conclusions

Microstructural enhancements were used to produce low thermal conductive and high hemispherical reflective thermal barrier coatings deposited by EB-PVD in which the coating flux was periodically interrupted creating modulated strain fields within the TBC. These TBCs showed no macrostructural differences in the grain size or faceted surface morphology at low magnification as compared with standard TBC. The residual stress state was determined to be compressive in all of the TBC samples, and the amount of compressive stress was found to decrease with increasing number of modulations. The thermal conductivity value was shown to decrease primarily resulting from increased phonon scattering from the incorporation of strain fields caused by restructuring and microporosity. The average thermal conductivity was shown to decrease approximately 30% from 1.8 to 1.2 W/m-K for the 20-layer monolithic TBC after 2 h of testing at 1316°C. The monolithic layered TBC also resulted in a 28% increase in the hemispherical reflectance, and increased with increasing total number of modulations. Multilayer 7YSZ and Al₂O₃ coatings with fixed layer spacing showed a 73% IR reflectance maxima at 1.85 μm wavelength. The variable 7YSZ and Al₂O₃ multilayer coatings showed an increase in reflection spectrum from 1 to 2.75 μm.

Preliminary results suggest that coating reflectance can be tailored to achieve increased reflectance over a desired wavelength range by controlling the thickness of the individual layers.

Acknowledgments

Any opinions, findings, conclusions, or recommendations expressed in this material are those of the authors and do not necessarily reflect the views of the U.S. Navy.

References

1. S. M. Meier and D. K. Gupta, "The Evolution of Thermal Barrier Coatings in Gas Turbine Engine Applications," *Trans. ASME* 116 [1] 250–256 (1994).
2. W. D. Kingery, H. K. Bowen, and D. R. Uhlmann, (eds.), *Introduction to Ceramics*, 2nd Edition, Wiley, New York, NY, 1976.
3. D. R. Clarke and S. R. Phillpot, "Thermal Barrier Coating Materials," *Mater. Today*, 8 [6] 22–29 (2005).
4. D. Zhu and R. A. Miller, "Development of Advanced Low Conductivity Thermal Barrier Coatings," National Aeronautics and Space Administration, Cleveland, OH, E-14433, July 2004.
5. D. Zhu and R. A. Miller, "Development of Advanced Low Conductivity Thermal Barrier Coatings," *Int. J. Appl. Cerm. Tech*, 1 [1] 86–94 (2004).
6. D. E. Wolfe, J. Singh, R. A. Miller, J. I. Eldridge, and D. Zhu, "Tailored Microstructure of EB-PVD 8YSZ Thermal Barrier Coatings with Low Thermal Conductivity and High Thermal Reflectivity for Turbine Applications," *Surf. Coat. Technol.*, 190 132–149 (2005).
7. D. Zhu, R. A. Miller, B. A. Nagaraj, and R. W. Bruce, "Thermal Conductivity of EB-PVD Thermal Barrier Coatings Evaluated by a Steady-State Laser Heat Flux Technique," *Surf. Coat. Technol.*, 138 1 (2001).
8. J. I. Eldridge, C. M. Spuckler, K. W. Street, and J. R. Markham, "Infrared Radiative Properties of Yttria-Stabilized Zirconia Thermal Barrier Coatings," *Ceram. Eng. Sci. Proc.*, 23 417 (2002).
9. J. R. Nicholls, K. J. Lawson, D. S. Rickerby, and P. Morrell. Advisory Group for Aerospace Research and Development (AGARD)-R-823, Aalborg, Denmark, 6.1–6.9, 1997.
10. J. F. Bisson, D. Fournier, M. Poulain, O. Lavigne, and R. Mevrel, *J. Am. Ceram. Soc.*, 83 [8] 1993 (2000).
11. R. A. Miller, "Current Status of Thermal Barrier Coatings—An Overview," *Surf. Coat. Technol.*, 30 1 (1987).
12. F. C. Toriz, A. B. Thankker, and S. K. Gupta, "Flight Service Evaluation of Thermal Barrier Coatings by Physical Vapor Deposition at 5200 H," *Surf. Coat. Technol.*, 39/40 161–62 (1989).
13. P. S. Prevey, *ASM Handbook 10: Materials Characterization*. American Society of Metal, Materials Park, OH, 950 pp, 1990.
14. D. E. Wolfe, J. Singh, and K. Narasimhan, "Synthesis of Titanium Carbide/ Chromium Carbide Multilayers by the Co-Evaporation of Multiple Ingots by Electron Beam Physical Vapor Deposition," *Surf. Coat. Technol.*, 160 206 (2002).
15. D. E. Wolfe, J. Singh, and K. Narasimhan, "Synthesis and Characterization of Multilayered TiC/TiB2 Coatings Deposited by Ion Beam Assisted, Electron Beam-Physical Vapor Deposition (EB-PVD)," *Surf. Coat. Technol.*, 165 206 (2003).
16. J. R. Nicholls, K. J. Lawson, A. Hohnstone, and D. S. Rickerby, "Methods to Reduce the Thermal Conductivity of EB-PVD TBCs," *Surf. Coat. Technol.*, 151–152 383 (2002).
17. R. B. Peterson, "Direct Simulation of Phonon Mediated Heat Transfer in a Debye Crystal," *Trans. ASME J. Heat Transfer*, 116 815 (1994).
18. J. Singh, D. E. Wolfe, R. A. Miller, J. I. Eldridge, and D. Zhu, "Thermal Conductivity and Thermal Stability of Zirconia and Hafnia Based Thermal Barrier Coatings by EB-PVD for High Temperature Applications," *Mater Sci Forum*, 455–456 579–586 (2004).
19. O. Friedrich Soechting, NASA Conference on "Thermal Barrier Coatings Workshop" NASA's Publication 3312, 1995.

Delamination-Indicating Thermal Barrier Coatings Using YSZ:Eu Sublayers

Jeffrey I. Eldridge,[†] Timothy J. Bencic, and Charles M. Spuckler

NASA Glenn Research Center, Cleveland, Ohio 44135

Jogender Singh and Douglas E. Wolfe

Applied Research Laboratory, The Pennsylvania State University, University Park, Pennsylvania 16802

Nondestructive diagnostic tools that can reliably assess thermal barrier coating (TBC) delamination are needed to provide protection against premature TBC failure as well as to reduce the costs associated with unnecessary TBC replacement. A coating design for a TBC that is self-indicating for delamination has been successfully implemented by incorporating a europium-doped yttria-stabilized zirconia (YSZ) luminescent sublayer beneath the overlying undoped YSZ TBC. It was demonstrated that incorporation of the europium-doped YSZ layer could be achieved without disrupting TBC columnar growth by using multiple ingot electron beam physical vapor deposition. Both scanning luminescence mapping as well as luminescence imaging revealed greatly enhanced detected luminescence from scratch-induced delaminated regions. This enhanced detected luminescence arises due to high internal reflectivity of both excitation and emission wavelengths at the interface between the luminescent sublayer and the delamination crack. In particular, imaging of the enhanced luminescence associated with TBC delamination was fast and simple to implement, therefore showing great promise as a practical tool for inspecting for TBC delamination.

I. Introduction

ALTHOUGH thermal barrier coatings (TBCs), most commonly composed of yttria-stabilized zirconia (YSZ), provide beneficial thermal protection for turbine engine components, the risk of TBC spallation compromises their reliability. Consequently, only a fraction of the potential thermal protection provided by the TBC can be utilized, and TBC application is limited to temperatures at which the unprotected component can still survive. Nondestructive diagnostic tools that could reliably assess the damage state of TBCs would alleviate the risk of TBC premature failure by prompting TBC replacement before the level of TBC damage threatens performance or safety. Rather than following a timetable for TBC replacement based strictly on statistical patterns of TBC failure, TBC replacement could be guided by an informed assessment of TBC damage, thereby providing protection against premature TBC failure and also reducing costs associated with unnecessary TBC replacement. The nature of TBC failure progression demands that any successful TBC damage-assessment tool must probe through the full thickness of the TBC as TBC failure typically proceeds by the propagation of cracks along the bottom of the TBC, primarily above the thermally grown oxide (TGO) that forms on top of the bond coat.[1] As these buried cracks propagate, they

link together to produce TBC delamination and eventual TBC spallation. To address the requirement of assessing TBC damage below the coating surface, most previous efforts have utilized either electrical probes, such as eddy current[2] or electrochemical impedance,[3] or optical probes that take advantage of the TBC translucency, such as thermal wave imaging,[4] mid-infrared (MIR) reflectance,[5] and piezospectroscopy.[6–8] While significant progress has been made, interpretation of the data, especially in the presence of competing damage mechanisms, can sometimes be ambiguous and may require sophisticated modeling. As an alternative to developing sensing capabilities that can be applied to any TBC, advantages of simpler inspection with unambiguous interpretation can be gained by integrating a delamination-indicating functionality into the TBC itself. Other efforts have shown that temperature-sensing[9,10] and erosion-sensing functions[11,12] can be incorporated into TBCs by the incorporation of luminescent sublayers. Furthermore, Feist and Heyes[10] have demonstrated that YSZ doped with europium (YSZ:Eu) produces strong luminescence and can therefore be introduced into TBCs composed of YSZ by low level doping of the TBC itself, without needing to introduce a completely distinct phosphor layer that might sacrifice TBC performance. Eldridge et al.[9] have also demonstrated depth-penetrating TBC temperature measurements based on Eu^{3+} luminescence decay times by selecting an excitation wavelength of 532 nm, where YSZ is translucent, rather than the typical selection of an ultraviolet (UV) wavelength, where YSZ is opaque. Taking advantage of the demonstrated possibility of detecting luminescence from a buried luminescent sublayer, this paper will introduce a new approach utilizing luminescence-sensing of a buried YSZ:Eu sublayer to produce a TBC that is self-indicating for delamination.

II. Strategy for Delamination-Indicating TBCs

The strategy for producing delamination-indicating TBCs combines the contrast-producing mechanism of increased reflectance associated with subsurface cracks with the depth selectivity of sublayer luminescence. Eldridge et al.[5] showed that MIR reflectance can monitor the progression of delamination cracks in plasma-sprayed 8YSZ TBCs by detecting the increased overall reflectance associated with buried cracks. The propagation of buried cracks increases reflectance due to the significant fraction of radiation incident upon the crack with an angle of incidence greater than the critical angle for total internal reflection. The effect of the radiation beyond the critical angle for total internal reflection results in a very high diffuse reflectance (0.81 at a wavelength of 4 μm) at the TBC/crack interface. However, the overall increase in total reflectance due to crack progression is modest because of the large background reflectance from all the pre-existing scattering features (microcracks and pores) present throughout the TBC volume. The sensitivity to buried cracks therefore suffers due to a lack of depth discrimination that could extract changes in reflectance specifically from the depth where

D. Clarke—contributing editor

Manuscript No. 20871. Received August 12, 2005; approved April 6, 2006.
[†]Author to whom correspondence should be addressed. e-mail: jeffrey.i.eldridge@nasa.gov

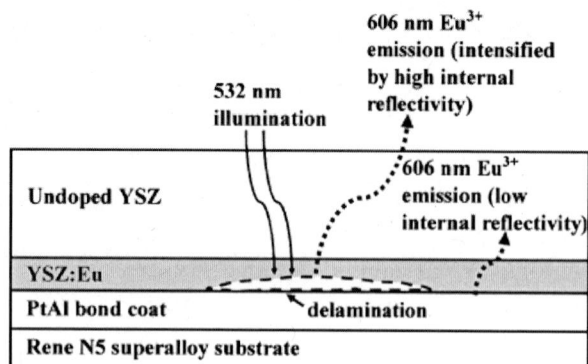

Fig. 1. Concept for thermal barrier coating (TBC) that is self-indicating for delamination. The TBC is translucent to the 532 nm excitation which can excite Eu^{3+} emission at 606 nm from buried yttria-stabilized zirconia (YSZ):Eu layer. 606 nm Eu^{3+} emission through the overlying undoped YSZ is enhanced by delamination-crack-induced high internal reflectance.

delamination is occurring. The strategy described here for delamination-indicating TBCs adds depth discrimination by sensing luminescence that originates exclusively from a luminescent sublayer that is placed along the bottom of the TBC, where delamination crack propagation typically occurs.

Figure 1 illustrates the coating design for a TBC that is self-indicating for delamination. The YSZ TBC incorporates an initial sublayer that is doped with europium (YSZ:Eu) below the overlying undoped YSZ. YSZ:Eu has several strong excitation peaks in the visible wavelength range as well as a strong emission peak at 606 nm (Fig. 2), that are similar, but not identical, to the excitation and emission spectra previously reported[9] for Y_2O_3:Eu (where the analogous strong emission peak is observed at 611 nm). Luminescence measurements from the buried YSZ:Eu layer will be restricted to excitation and emission wavelengths that can penetrate the overlying coating. Figure 3 displays hemispherical transmittance plots for freestanding electron-beam physical vapor-deposited (EB-PVD) and plasma-sprayed coatings of comparable thickness that had been prepared by burning off sacrificial carbon substrates at 800°C in air. These plots indicate significantly greater transmittance through the EB-PVD coatings than the plasma-sprayed coatings, especially at shorter wavelengths. This difference has been attributed to the more highly scattering microstructure typically associated with plasma-sprayed coatings.[13] These transmittance plots indicate that while any of the intense excitation peaks at 395, 465, 527, or 535 nm can penetrate through an EB-PVD TBC, the excitation wavelengths at 395 and 465 nm are not suitable for performing luminescence measurements from

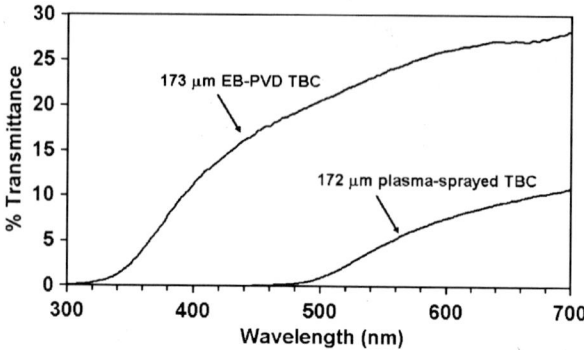

Fig. 3. Comparison of room-temperature hemispherical transmittance for freestanding electron-beam physical vapor deposition (EB-PVD) 7 wt% yttria-stabilized zirconia (YSZ) (173 μm thick) and plasma-sprayed 8 wt%YSZ (172 μm thick) thermal barrier coatings (TBCs).

beneath a plasma-sprayed TBC because of the near-zero transmittance at those wavelengths. While this paper will present results only for EB-PVD coatings, an excitation wavelength was selected that would also be applicable for plasma-sprayed coatings. Fortuitously, a second harmonic YAG:Nd laser produces light at 532 nm that sufficiently overlaps the YSZ:Eu 535 nm excitation peak to serve as an effective excitation source. Therefore, under 532 nm laser excitation, the YSZ:Eu layer produces a red luminescence emission peak at 606 nm (Fig. 2) that can be detected through the overlying undoped YSZ (Fig. 1). The enhancement of the 606 nm luminescence that indicates delamination is produced by the high internal reflectance of both the incoming 532 nm excitation and the 606 nm emission wavelengths by the delamination cracks beneath or within the YSZ:Eu layer (Fig. 1). High internal reflectance of the incoming excitation by delamination cracks prevents the excitation wavelength from being absorbed by the bond coat, essentially lengthening the potential optical path of the excitation light within the doped sublayer and therefore increasing the probability of producing luminescence. In addition, the same underlying cracks produce high internal reflectance of the luminescence emission, resulting in emission being reflected back out through the overlying undoped TBC instead of being absorbed by the underlying bond coat. Therefore, detecting delamination by sensing luminescence from a buried YSZ:Eu layer offers the advantages of contrast-producing enhancement from the high internal reflectance of both excitation and emission wavelengths as well as the intrinsic depth discrimination that arises because the detected luminescence can only originate from the doped luminescent sublayer, which can be strategically placed at the bottom of the TBC where delamination cracks propagate.

III. Experimental Procedure

The EB-PVD of the YSZ:Eu and undoped YSZ layers was performed without interruption so as not to disrupt the columnar growth that gives EB-PVD TBCs their desirable strain tolerance. The TBCs were deposited in an industrial prototype Sciaky EB-PVD chamber with six electron beam guns using two ingots positioned 16.2 cm apart. To produce the TBCs with buried YSZ:Eu sublayers by continuous deposition, the first ingot was composed of undoped 7 wt% yttria-stabilized zirconia (7YSZ) and the second ingot of Eu_2O_3 (both fabricated by Trans-Tech, Adamstown, MD). The PtAl bond-coated nickel-based superalloy Rene N5 substrates (25.4 mm diameter, 3.18 mm thick) were grit blasted with 400 μm-size aluminum oxide particles to remove the surface oxides. The Rene N5 substrates were mounted on a 50.8 mm diameter stainless steel mandrel positioned 30.5 mm above the ingots. Two electron beams were used to indirectly heat the substrates rotating at 7 rpm to 1000°C

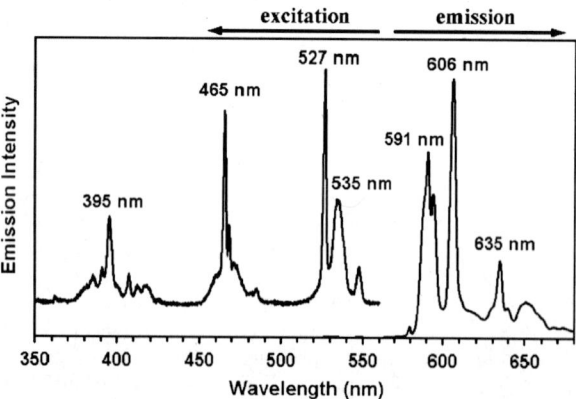

Fig. 2. Excitation spectrum (left) for emission at 606 nm and emission spectrum (right) for excitation at 532 nm for yttria-stabilized zirconia: Eu(3%).

under a graphite "A-frame" heating assembly. The TBC sublayer doping was achieved by co-evaporation of the 7YSZ ingot and the Eu_2O_3 ingot to produce the bottom YSZ:Eu layer, and single evaporation of the 7YSZ layer to produce the undoped 7YSZ top layer. The transition from Eu^{3+}-doped to undoped YSZ was accomplished by switching off the power for the electron beam gun that produced evaporation from the Eu_2O_3 ingot. YSZ:Eu sublayer deposition time was 6.2 min out of 68 min total TBC deposition time.

In order to obtain the desired sublayer dopant level for the buried YSZ:Eu layer, the substrates were positioned 51 mm from the centerline between the ingots toward the Eu_2O_3 ingot, and the electron beam power levels were adjusted to yield the desired evaporation rates. After the YSZ:Eu sublayer deposition, the specimens were repositioned directly centered over the 7YSZ ingot to continue deposition of the undoped YSZ layer without disrupting the columnar growth of the TBC. A set of 10 TBCs was deposited with a thin YSZ:Eu sublayer (7 μm thick YSZ:Eu layer below a 125 μm thick undoped YSZ layer). White light images and Eu^{3+} 606 nm luminescence images of the same area were obtained from metallographic cross-sections of TBC-coated specimens inspected with an optical microscope (Fig. 4). The luminescence images were acquired using UV illumination provided by a SuperBright 2000SW lamp (UV Systems, Renton, WA) that has a maximum intensity at 254 nm in conjunction with a bandpass filter centered at 606 nm (full-width at half-maximum = 10 nm) positioned in front of the camera. The luminescence image verified the deposition of a distinct YSZ:Eu luminescent sublayer (Fig. 4(b)), while the white light image (Fig. 4(a)) showed that this layering was achieved without significant interruption of the EB-PVD columnar growth. Eu^{3+} doping levels in the YSZ:Eu sublayers were verified by performing energy-dispersive X-ray spectroscopy (EDS) analysis of TBC cross-sections and by calibrating the EDS results against non-layered YSZ:Eu coatings for which doping levels were determined by X-ray fluorescence. Doping levels were determined in terms of atomic percent $EuO_{1.5}$. EDS analysis indicated an average doping level of 10.8 at.% in the 7 μm YSZ:Eu. However, melt-pool "spitting" during evaporation from the Eu_2O_3 ingot resulted in significant variation in Eu^{3+} doping layers

across the thickness of the YSZ:Eu sublayer, ranging from 8.7 to 13.8 at.%. These high rare-earth-dopant concentrations result in the presence of cubic phase YSZ:Eu, as confirmed by X-ray diffraction of specimens with coatings that consisted of a Eu^{3+}-doped layer without an undoped overlayer. Because the presence of cubic phase usually results in short TBC cyclic life, future depositions will aim for significantly lower Eu^{3+} concentrations. Because the high Eu^{3+} concentrations employed here produced well beyond the necessary luminescence intensity for delamination indication purposes, it is expected that significantly lower Eu^{3+} concentrations will still be sufficient to produce delamination-indicating TBCs. In addition, future Eu^{3+}-doped sublayer depositions will use a YSZ:Eu ingot for the source of Eu^{3+} doping rather than a Eu_2O_3 ingot and should result in a more stable melt pool during deposition and therefore more uniform Eu^{3+} doping levels.

To demonstrate their delamination-indicating capability, the TBC-coated specimens were first heated to 1000°C for 100 h in air (to insure the demonstration was not relying on an initial microstructure that might be transient). A scratch test (using a CSEM Revetest Automatic Scratch-Tester, Neuchatel, Switzerland) was then performed on the TBC-coated specimen in order to delaminate a local region of the TBC in the vicinity of the scratch. A single scratch was applied to the TBC-coated specimen using a hemispherical diamond indenter with a 109 μm radius-of-curvature. The load applied by the indenter was increased linearly from 0 to 80 N over the 8 mm length of the scratch. A backscatter electron image (Fig. 5) of a cross-section along a delaminated section revealed a clean separation between the YSZ:Eu sublayer and the underlying bond coat and therefore indicated that the delamination crack could be accurately modeled by an air gap between the TBC and its bond coat. The brighter intensity from the YSZ:Eu layer in the backscatter electron image (Fig. 5) is due to the higher average atomic number in the YSZ:Eu layer compared with the undoped YSZ. This contrast allows a higher magnification inspection of the boundary between the YSZ:Eu and undoped YSZ layers and shows that, even on a fine scale, columnar growth continues uninterrupted across the interlayer boundary.

Luminescence intensity mapping of the TBC-coated specimen with the scratch-induced delamination was performed with a Renishaw System 2000 (New Mills, UK) microscope equipped with a Prior ProScan II automated scanning stage. Excitation was provided by a Nanogreen (JDS Uniphase, San Diego, CA) pulsed frequency-doubled 532 nm YAG:Nd laser operating at a power of 4 μJ/pulse and a frequency of 5.2 kHz. The laser

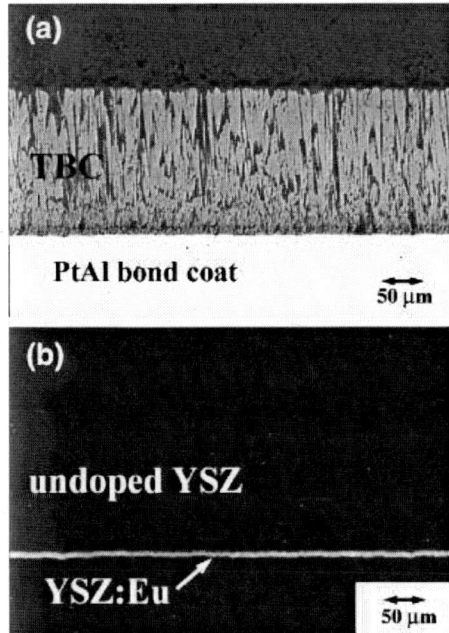

Fig. 4. Optical micrographs of cross-section of thermal barrier coating (TBC)-coated substrate. Thin (7 μm) initial yttria-stabilized zirconia (YSZ):Eu sublayer below 125 μm undoped YSZ overlayer. (a) White light image. (b) Eu^{3+} 606 nm luminescence image of same area.

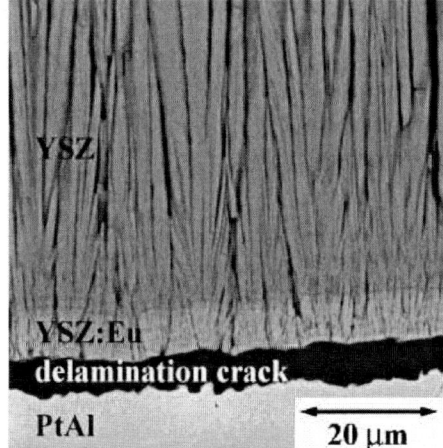

Fig. 5. Backscatter electron image of cross-section through delaminated section of thermal barrier coating consisting of 7 μm yttria-stabilized zirconia (YSZ):Eu sublayer (evident as bright intensity band) below 125 μm undoped YSZ overlayer. Clean separation is observed between the YSZ:Eu sublayer and the PtAl bond coat.

excitation was transmitted through a ×20 objective before striking the TBC-coated specimen placed on the automated scanning stage. The laser was defocused to a diameter of approximately 100 μm to average over local inhomogeneities in the coating. A scanning image was produced by rastering the specimen in a stepwise fashion (step size = 100 μm) under the stationary laser excitation to cover an 8 mm × 12 mm area that included the scratch-induced delaminated area. After each 100-μm step, a luminescence emission spectrum was collected and used to produce an 80 × 120 pixel image (total acquisition time ~8 h). The acquisition time for each emission spectrum was 1 s. The luminescence intensity at each pixel in the scanning image was determined by calculating the area under the 606 nm emission peak (Fig. 2), between 600.2 and 614.5 nm, after performing a baseline subtraction.

In addition to luminescence mapping, luminescence imaging was performed (without spectral acquisition) with image acquisition times (~4 s) that are much shorter than mapping acquisition times (~8 h). These much shorter acquisition times make detection of luminescence enhancement a much more practical inspection tool for TBC delamination. Luminescence images of TBC-coated specimens with scratch-induced delamination were collected using illumination from a green light emitting diode (LED; Luxeon # LXHL-NM98) with a peak output of 530 nm. The LED output was conditioned with a halographic diffuser and filtered with a 550 nm short pass interference filter to prevent longer wavelength light from being recorded by the camera. Images were acquired using a cooled slow-scan camera (Photometrics Series 300 with a SITe 502B CCD) coupled with a 105 mm lens equipped with a detection bandpass filter centered at 610 nm and a bandwidth (FWHM) of 10 nm. White light images were acquired using the same camera without the detection filter and using a light-guide-coupled white-light source equipped with an opal diffuser. In both lighting cases, the diffuser was used to produce a uniform illumination field using a single small source.

IV. Results

Figure 6 shows a white light image of a scratch-induced TBC delamination along with a scanning Eu^{3+} 606 nm luminescence emission map of the same area for a TBC with a YSZ:Eu sublayer (7 μm thick initial YSZ:Eu sublayer beneath a 125 μm thick undoped YSZ overlayer). In both the white light image and the luminescence map, there is a vertical line that corres-

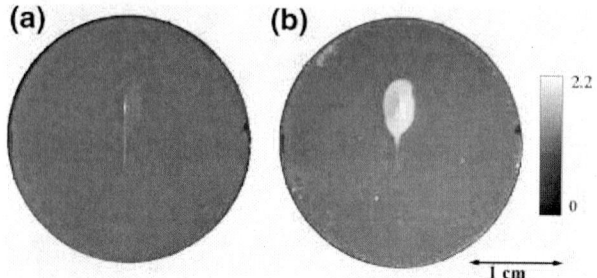

Fig. 7. Delamination indication by luminescence imaging using CCD camera and 610 nm bandpass filter (same delaminated region displayed in Fig. 6). Enhanced Eu^{3+} 606 nm luminescence detected from scratch-induced-delaminated region from 7 μm yttria-stabilized zirconia (YSZ):Eu sublayer beneath 125 μm undoped YSZ overlayer. (a) White light image and (b) Eu^{3+} luminescence images have been normalized to the area-averaged intensity level of the undelaminated thermal barrier coating and displayed using same gray scale.

ponds to the scratch that was applied with a load starting from 0 N at the bottom and increasing to 80 N at the top of the scratch. In both cases the scratch is surrounded by a bright area that corresponds to where the TBC has delaminated. The increased brightness observed under visible light for delaminated regions of EB-PVD TBCs has been previously reported.[14] The delaminated region is observed to broaden along the length of the scratch (as the applied load increased). In the white light image, the reflectance from the delaminated region is modestly higher than the reflectance from the surrounding adherent coating. In comparison, the contrast is tremendously enhanced in the luminescence map, making the delaminated area immediately evident. A comparison of emission spectra obtained inside and outside the delaminated area shows that the luminescence intensity, determined from the fit-derived area of the emission peak, is a factor of 4.1 times greater inside the delamination area (compared with a very modest reflectance enhancement factor of 1.09 in the delaminated region in the white light image).

Owing to much shorter collection times (4 s) and simpler image extraction, luminescence imaging provided a much more practical approach to delamination detection than scanning luminescence mapping. Figure 7 shows a white light image along with a Eu^{3+} 606 nm emission image under green LED illumination of the same TBC-coated specimen with the YSZ:Eu sublayer shown in Fig. 6. In both images, the original image intensity, $I_{raw\ image}$, has been normalized to the area-averaged intensity level of the TBC outside the delaminated region, $\overline{I}_{non-delaminated}$:

$$I_{normalized\ image}(x,\ y) = \frac{I_{raw\ image}(x,\ y)}{\overline{I}_{non-delaminated}} \tag{1}$$

where x and y are the pixel coordinates. This normalization produces an image where the pixel intensity is 1.0 outside the delaminated region and is equal to the luminescence enhancement factor inside the delaminated region. For direct comparison, both the white light and Eu^{3+} 606 nm emission images (Fig. 7) are displayed using the same intensity gray scale ranging from 0 to 2.2. As with the scanning luminescence maps, the delaminated region is immediately evident in the luminescence image (Fig. 7(b)) due to the tremendously enhanced luminescence from that region. Figure 8 displays line scans crossing the delaminated region from both the white light and luminescence images. These line scans indicate an intensity enhancement in the delaminated region of 2.2 in the luminescence image (compared with an enhancement of 1.09 in the white light image, ignoring the high intensity spike observed at the center of the scratch).

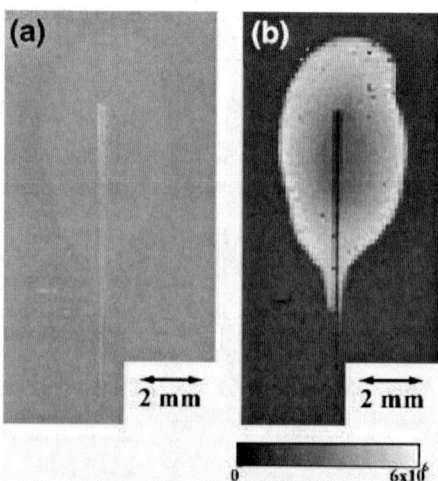

Fig. 6. Delamination indication by scanning luminescence mapping. Enhanced Eu^{3+} 606 nm luminescence detected from scratch-induced delaminated region from 7 μm yttria-stabilized zirconia (YSZ):Eu sublayer beneath 125 μm undoped YSZ overlayer. (a) White light image. (b) Scanning Eu^{3+} luminescence map.

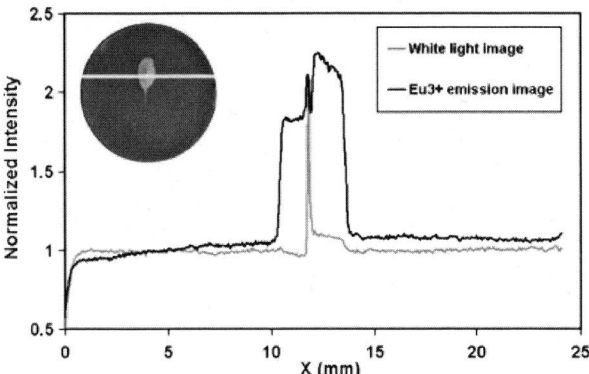

Fig. 8. Intensity line scans across delaminated region (location of line scan shown in inset) for white light and Eu^{3+} 606 nm luminescence images displayed in Fig. 7.

V. Discussion

The strong delamination-induced enhancement of luminescence originating from a YSZ:Eu sublayer residing underneath the standard undoped YSZ produces an immediate indication of TBC delamination (Figs. 6–8). This luminescence enhancement within the delaminated regions (up to a factor of 4.1 for the Eu^{3+} 606 nm luminescence maps) was much greater than the slight intensity enhancement factor of 1.09 observed in white light reflectance images that can be obtained without Eu^{3+} doping. In addition, the generation of luminescence at narrowly defined excitation and emission wavelengths (Fig. 2) will greatly reduce the effects of stray background light on the resultant images. Inspection of TBC cross-sections (Figs. 4 and 5) reveals that YSZ:Eu sublayers with well-delineated luminescence could be produced by multiple ingot EB-PVD without significant disruption of the desirable columnar growth associated with the EB-PVD process. Because the YSZ:Eu sublayer is incorporated without significant disruption in the EB-PVD growth process and with only the compositional disruption presented by the Eu^{3+}-doping, luminescence sensing using low Eu^{3+} doping levels is a promising approach for adding a simple and unambiguous delamination-indicating function to the TBC without sacrificing TBC performance.

The delamination indication by luminescence enhancement arises due to high internal reflectance of both the excitation and emission wavelengths at the TBC/crack interface. Luminescence intensity is altered by TBC delamination because of the change in the boundary condition for reflectance of radiative flux at the TBC/bond coat interface. Because delamination occurred cleanly between the YSZ:Eu sublayer and the underlying bond coat (Fig. 5), the delamination can be well modeled by an air gap between the YSZ:Eu sublayer and the bond coat. In this view, the crack essentially introduces two interfaces, a TBC/air-gap and an air-gap/bond-coat interface, to replace the TBC/bond coat interface. The reflectivity of the air-gap/bond-coat interface (hemispherical reflectance of bare bond coat measured to be about 0.40 near excitation/emission wavelengths) will be slightly greater than the reflectivity of the TBC/bond coat interface that it replaces. (The reflectivity increase predicted by Fresnel relations[15] for replacing the medium adjacent to the bond coat with an index of refraction of 2.18 (8YSZ)[5] with a medium with a lower index of refraction of 1.0 (air) is small because reflectivity is dominated by the bond-coat extinction coefficient.) However, the primary source for the overall increased reflectance is the insertion of the TBC/air-gap interface, which contributes a high internal reflectivity due to the difference in index of refraction across that interface (from 2.18 to 1.0). Because of the equiaxed grain structure at the base of the TBC, it is assumed radiation reaching the crack is diffuse due to scattering. By applying the integrated average of the Fresnel equation, the reflectivity for

diffuse external radiation at the air/coating interface, ρ_o, can be calculated[15,16]:

$$\rho_o(n) = \frac{1}{2} + \frac{(3n+1)(n-1)}{6(n+1)^2} + \frac{n^2(n^2-1)^2}{(n^2+1)^3} \ln\left(\frac{n-1}{n+1}\right)$$
$$- \frac{2n^3(n^2+2n-1)}{(n^2+1)(n^4-1)} + \frac{8n^4(n^4+1)}{(n^2+1)(n^4-1)^2} \ln(n) \qquad (2)$$

where n is the index of refraction of the coating ($n = 1$ for air is assumed). When considering internal reflectivity at the coating/air-gap interface, Richmond[17] showed that one must take into account the total internal reflection of the fraction of incident radiation with an angle of incidence above the critical angle so that the reflectivity for diffuse internal radiation, ρ_i, is

$$\rho_i(n) = 1 - \frac{1}{n^2}[1 - \rho_o(n)] \qquad (3)$$

The internal reflectivity for diffuse radiation is a single parameter function of the coating index of refraction, n. For $n = 2.18$ (index of refraction for YSZ at 500 nm wavelength), Eqs. (2) and (3) predict a high internal reflectivity at the TBC/crack interface of $\rho_i = 0.83$. This change in boundary condition by the introduction of a high reflectivity interface associated with the delamination crack is the source of the luminescence enhancement observed for delaminated TBC regions. While a two-wavelength (excitation and emission) radiative flux calculation using the new boundary condition for a delaminated TBC (Eq. (3)) should be able to quantitatively predict the luminescence enhancement, this will require determination of the nonisotropic scattering coefficients that vary across the thickness of the EB-PVD TBCs (owing to depth varying microstructure).

The collection of luminescence spectra at each point during scanning luminescence mapping allows emission peak intensity to be determined by post-acquisition spectral fitting and offers the flexibility of optimizing the wavelength intervals (effective bandpass width) over which the emission peak intensity is calculated and a choice of background subtraction methods. In addition, the luminescence spectra can reveal and therefore monitor phase changes associated with TBC degradation. While useful for fundamental research on TBC delamination progression, the time-consuming process of acquiring a large number of spectra while stepping a small-spot excitation beam across the region-of-interest prohibits adopting this method as a practical inspection method. By sacrificing advantages of spectral acquisition, luminescence imaging, using a bandpass filter to select the emission wavelength to be imaged, reduces acquisition times considerably (from hours to seconds) and is much simpler to implement, making this approach much more amenable to routine inspection. Both the scanning luminescence mapping and luminescence imaging reveal sharply defined regions of enhanced luminescence that can be used to identify the boundaries of the delaminated area. The higher luminescence enhancement factor (4.1 vs. 2.2) for the luminescence mapping may be due to the background subtraction inherent in the luminescence mapping intensity determination, whereas the luminescence imaging enhancement is a ratio of absolute intensities determined without background subtraction. In addition, a monochromatic pulsed laser source was used for scanning luminescence mapping while the more extended wavelength range continuous LED source was used for luminescence imaging; these two light sources may not produce the same luminescence-to-background ratios and may generate different relative contributions from specimen autofluorescence.

While the scratch-induced delamination that produces a clean separation between the TBC and the bond coat (Fig. 5) provided a simple demonstration of a TBC that is self-indicating for delamination, TBC failure due to thermal cycling is under the strong influence of the stresses produced by the growth of the TGO and can produce much more complex failure morphologies that will include cracks propagating within the TBC, within the TGO, as well as at the TBC/TGO and TGO/bond coat interfaces.[18] Despite the more complex nature of these failures, delamination indication should still occur as long as the crack propagates below at least a significant fraction of the YSZ:Eu layer and also does not propagate within the bond coat (which is opaque). However, the mixed failure modes will introduce greater variability in the degree of luminescence enhancement.

Refinements in the design presented here for delamination-indicating TBCs can be used to extend the applicability of this approach. For example, while cracks beneath a luminescent sublayer will produce enhanced luminescence as shown in this paper, cracks above a luminescent sublayer will produce diminished luminescence as high internal reflectance at the crack front will greatly reduce transmission of both the excitation and emission wavelengths across the crack. Because luminescence will be enhanced from sublayers above the crack and diminished from sublayers below the crack, multiple luminescent sublayers with different rare-earth dopants can be utilized to not only indicate the boundaries of delamination, but to also give an indication of crack depth. In addition, delamination indication by detected luminescence enhancement could be improved for plasma-sprayed coatings if dopants with longer excitation and emission wavelengths were utilized. Previous work[5] has shown that plasma-sprayed TBCs have much higher transmittance at infrared wavelengths than at visible wavelengths so that detection of emission from luminescent sublayers will be much more successful at infrared wavelengths. Therefore, a dopant such as Er^{3+}, which emits at 1550 nm and can be excited in the 800–1100 nm range, would be a better choice for delamination indication in plasma-sprayed TBCs.

VI. Conclusions

A coating design for delamination-indicating TBCs has been successfully implemented by incorporating a Eu^{3+}-doped YSZ luminescent sublayer at the base of a YSZ TBC produced by multiple ingot EB-PVD. It was demonstrated that incorporation of the Eu^{3+}-doped sublayer could be achieved without disruption of the EB-PVD columnar growth, suggesting that doped sublayer incorporation can be achieved without sacrificing the desirable columnar TBC microstructure. Owing to high internal reflectance of both excitation and emission wavelengths at the interface between the luminescent sublayer and the delamination crack, the detected luminescence from the sublayer is greatly enhanced in the presence of underlying delamination cracks. From this enhanced luminescence, regions of TBC delamination can be immediately identified that would otherwise be difficult to detect. Luminescence imaging, using only an LED illumination source and a bandpass filter in front of a camera, was very simple to implement and image collection times were only a few seconds, demonstrating that this approach can be adopted for routine inspection for TBC delamination. It has also been proposed that detection of enhanced luminescence for indicating TBC delamination could be extended to plasma-sprayed coatings by choosing dopants, such as Er^{3+}, that luminesce at infrared wavelengths that can be more successfully transmitted through the more highly scattering plasma-sprayed TBCs.

Acknowledgments

The authors thank Joy Buehler and Myles McQuater for expert metallographic preparation of challenging specimens and Dave Hull for EDS analysis used to determine rare-earth doping levels.

References

[1]J. T. DeMasi-Marcin, K. D. Sheffler, and S. Bose, "Mechanisms of Degradation and Failure in a Plasma-Deposited Thermal Barrier Coating," *ASME J. Eng. Gas Turbines & Power*, 112, 521–26 (1990).

[2]P. Crowther, "Non Destructive Evaluation of Coatings for Land Based Gas Turbines Using a Multi-Frequency Eddy Current Technique," *Insight Nondestruct. Test Cond. Monit.*, 46 [9] 547–49 (2004).

[3]N. W. Wu, K. Ogawa, M. Chyu, and S. X. Mao, "Failure Detection of Thermal Barrier Coatings Using Impedance Spectroscopy," *Thin Solid Films*, 457, 301–6 (2004).

[4]G. Newaz and X. Chen, "Progressive Damage Assessment in Thermal Barrier Coatings Using Thermal Wave Imaging Technique," *Surf. Coat. Technol.*, 190, 7–14 (2005).

[5]J. I. Eldridge, C. M. Spuckler, and R. E. Martin, "Monitoring Delamination Progression in Thermal Barrier Coatings by Mid-Infrared Reflectance Imaging," *Int. J. Appl. Ceram. Technol.*, 3 [2] 94–104 (2006).

[6]X. Peng and D. R. Clarke, "Piezospectroscopic Analysis of Interface Debonding in Thermal Barrier Coatings," *J. Am. Ceram. Soc.*, 83 [5] 1165–70 (2000).

[7]V. K. Tolpygo, D. R. Clarke, and K. S. Murphy, "Evaluation of Interface Degradation During Cyclic Oxidation of EB-PVD Thermal Barrier Coatings and Correlation with TGO Luminescence," *Surf. Coat. Technol.*, 188–189, 62–70 (2004).

[8]M. Gell, S. Sridharan, M. Wen, and E. H. Jordan, "Photoluminescence Piezospectroscopy: A Multi-Purpose Quality Control and NDI Technique for Thermal Barrier Coatings," *Int. J. Appl. Ceram. Technol.*, 1 [4] 316–29 (2004).

[9]J. I. Eldridge, T. J. Bencic, S. W. Allison, and D. L. Beshears, "Depth-Penetrating Temperature Measurements of Thermal Barrier Coatings Incorporating Thermographic Phosphors," *J. Thermal Spray Technol.*, 13 [1] 44–50 (2004).

[10]J. P. Feist and A. L. Heyes, "Development of the Phosphor Thermometry Technique for Application in Gas Turbines," 10th Int. Symp. on Applications of Laser Techniques to Fluid Mechanics, Lisbon, Portugal, 2000.

[11]J. I. Eldridge, J. Singh, and D. E. Wolfe, "Erosion-Indicating Thermal Barrier Coatings Using Luminescent Sublayers," *J. Am. Ceram. Soc.*, 89, 3252–4 (2006).

[12]M. M. Gentleman and D. R. Clarke, "Concepts for Luminescence Sensing of Thermal Barrier Coatings," *Surf. Coat. Technol.*, 188–189, 93–100 (2004).

[13]K. W. Schlichting, K. Vaidyanathan, Y. H. Sohn, E. H. Jordan, M. Gell, and N. P. Padture, "Application of Cr^{3+} Photoluminescence Piezo-Spectroscopy to Plasma-Sprayed Thermal Barrier Coatings for Residual Stress Measurement," *Mater. Sci. Eng. A*, 291, 68–77 (2000).

[14]A. Vasinonta and J. L. Beuth, "Measurement of Interfacial Toughness in Thermal Barrier Coating Systems by Indentation," *Eng. Fracture Mech.*, 68, 843–60 (2001).

[15]R. Siegel and J. R. Howell, *Thermal Radiation Heat Transfer*, 4th Edition, Taylor & Francis, New York, 2002.

[16]C. M. Spuckler and R. Siegel, "Refractive Index and Scattering Effects on Radiative Behavior of a Semitransparent Layer," *J. Thermophys. Heat Transfer*, 7 [1] 302–10 (1993).

[17]J. C. Richmond, "Relation of Emittance to Other Optical Properties," *J. Res. Nat. Bureau Stand.*, 67C [3] 217–26 (1963).

[18]I. T. Spitsberg, D. R. Mumm, and A. G. Evans, "On the Failure Mechanisms of Thermal Barrier Coatings With Diffusion Aluminide Bond Coats," *Mater. Sci. Eng. A*, 394, 176–91 (2005). □

Erosion-Indicating Thermal Barrier Coatings Using Luminescent Sublayers

Jeffrey I. Eldridge[†]

NASA Glenn Research Center, Cleveland, Ohio 44135

Jogender Singh and Douglas E. Wolfe

Applied Research Laboratory, The Pennsylvania State University, University Park, Pennsylvania 16802

A successful approach to producing thermal barrier coatings (TBCs) that are self-indicating for location and depth of erosion is presented. Erosion indication is demonstrated in electron-beam physical vapor-deposited (EB-PVD) TBCs consisting of 7 wt% yttria-stabilized zirconia (7YSZ) with europium-doped and terbium-doped sublayers. Multiple-ingot deposition was utilized to deposit doped layers with sharp boundaries in dopant concentration without disrupting the columnar growth that gives EB-PVD TBCs their desirable strain tolerance. TBC-coated specimens were subjected to alumina-particle-jet erosion, and the erosion depth was indicated under ultraviolet illumination by the luminescence associated with the sublayers exposed by erosion. Sufficiently distinct luminescent sublayer boundaries were retained to maintain an effective erosion-indicating capability even after annealing free-standing TBCs at 1400°C for 100 h.

I. Introduction

RELIABLE diagnostic methods are needed to evaluate the severity of erosion of thermal barrier coatings (TBCs) on either air-based or land-based turbine engine components. In particular, a useful diagnostic approach should provide adequate warning before the decreased thermal protection due to reductions in TBC thickness approaches performance- or safety-threatening thresholds, and the implementation should be amenable to simple, non-intrusive inspections during engine shutdowns. Several researchers have suggested embedding luminescent sublayers into TBCs to serve as erosion markers, so that when erosion exposes the luminescent "marker" layer, the luminescence characteristic of that layer will be produced when illuminated by the appropriate excitation wavelength.[1–3] Furthermore, Feist and Heyes[4] have demonstrated that yttria-stabilized zirconia (YSZ) doped with europia produces strong luminescence. Therefore, as TBCs are most commonly composed of YSZ, luminescent sublayers can be introduced into TBCs by low-level doping of the TBC itself, without needing to introduce a completely distinct phosphor layer that might detrimentally affect TBC performance.[5] However, for TBCs deposited by electron-beam physical vapor deposition (EB-PVD), the practical implementation of layered YSZ doping for erosion indication has not been addressed. In particular, it must be shown that layered YSZ doping can be achieved without disrupting the columnar growth of the TBC that gives EB-PVD TBCs their desirable compliance, and that dopant layering is

D. Clarke—contributing editor

Manuscript No. 21462. Received February 9, 2006; approved April 6, 2006.
[†]Author to whom correspondence should be addressed. e-mail: jeffrey.i.eldridge@nasa.gov

sufficiently retained after exposure to engine temperatures to maintain an effective erosion-indicating capability. Other researchers[6] have shown that multiple-ingot EB-PVD can be utilized to produce compositional changes in TBCs without disrupting columnar growth. As described in this communication, the multiple-ingot EB-PVD approach is used to produce erosion-indicating TBCs using rare-earth-doped YSZ luminescent layering without disrupting TBC columnar growth. These TBCs also exhibit sufficient retention of luminescent sublayer boundaries after heat treatments above expected engine temperatures to maintain effective erosion indication.

II. Experimental Procedure

The coating design for a TBC that is self-indicating for erosion is illustrated in Fig. 1. The YSZ TBC is divided into three equal-thickness sublayers: a top undoped YSZ layer, a middle Eu-doped YSZ (YSZ:Eu) layer, and a bottom Tb-doped YSZ (YSZ:Tb) layer. Under ultraviolet (UV) excitation, the YSZ:Eu layer produces an intense luminescence emission peak at 606 nm (red) and the YSZ:Tb layer produces an intense luminescence emission peak at 543 nm (green) (Fig. 1). By using short-wavelength UV excitation that does not penetrate YSZ, only layers exposed by erosion will produce luminescence, and this luminescence will indicate where the depth of erosion reaches each sublayer boundary.

The TBCs were deposited in an industrial prototype Sciaky EB-PVD chamber with six electron beam guns using a three-ingot (16 cm spacing) continuous feeding system. To produce the TBCs with luminescent sublayers by continuous deposition, the center ingot was composed of undoped 7 wt% YSZ (7YSZ), the first ingot of Eu_2O_3, and the third ingot of 90 wt% 7YSZ—10 wt% Tb_4O_7 (all fabricated by Trans-Tech, Adamstown, MD). The PtAl bond-coated nickel-based superalloy Rene N5 substrates (25.4 mm diameter, 3.18 mm thick) were grit blasted with 400 μm size aluminum oxide particles to remove the surface oxides. The Rene N5 substrates were mounted on a stainless-steel mandrel positioned 30.5 mm above the ingots. Several sacrificial carbon substrates (25.4 mm diameter, 1.59 mm thick) were also included with each deposition in order to produce free-standing coatings. Two electron beams were used to heat the substrates indirectly rotating at 7 rpm to 1000°C under a graphite "A-frame" heating assembly. The TBC sublayer doping was achieved by co-evaporation of the 7YSZ ingot and the 90 wt% 7YSZ–10 wt% Tb_4O_7 ingots to produce the bottom YSZ:Tb layer, co-evaporation of the 7YSZ ingot and the Eu_2O_3 ingot to produce the middle YSZ:Eu layer, and single evaporation of the 7YSZ layer to produce the undoped 7YSZ top layer. Sublayer transition was achieved by switching the power for the electron beam gun associated with each ingot between an "operating" power level that produces the desired ingot evaporation rate

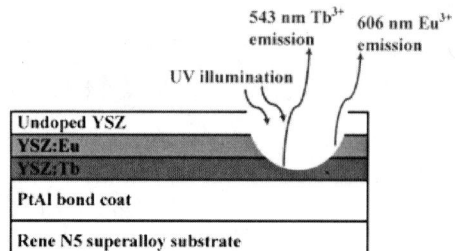

Fig. 1. Concept for erosion-indicating thermal barrier coatings. Ultraviolet illumination produces luminescence from sublayers exposed by erosion.

when evaporation of that ingot is desired and a "standby" low power that does not produce evaporation when evaporation of that ingot is not desired. In this manner, the full 165 μm thick coating with three equal-thickness sublayers is deposited continuously. Deposition times for the YSZ:Tb, YSZ:Eu, and undoped YSZ sublayers were 22.5, 27.0, and 27.0 min, respectively.

In order to obtain the desired sublayer sequence and dopant levels, substrate positions were changed at the transition between sublayers and the electron beam power levels were adjusted to yield the desired ingot feed rate. For the YSZ:Tb layer, the substrates were centered between the 90 wt% 7YSZ–10 wt% Tb_4O_7 and 7YSZ ingots, translated to the center line between the 7YSZ and Eu_2O_3 ingots for the YSZ:Eu layer, and directly centered over the central 7YSZ ingot for the undoped YSZ layer. Doping levels were verified by energy-dispersive X-ray spectroscopy (EDS) analysis of each sublayer and by calibrating the EDS results against non-layered doped YSZ coatings for which doping levels were determined by X-ray fluorescence. Doping levels were determined in terms of atomic percent $EuO_{1.5}$ or $TbO_{1.5}$. EDS analysis indicated an average Tb^{3+}-doping level of 1.3 at.% in the YSZ:Tb sublayer and an average Eu^{3+}-doping level of 1.3 at.% in the YSZ:Eu sublayer. X-ray diffraction analysis of both the top and the bottom of free-standing layer-doped TBCs indicated that, even with the additional rare-earth doping, the coatings consisted of only a tetragonal phase (primarily quenched non-transformable t' phase with a minor component of a metastable transformable t phase) with no evidence of cubic phase.

To demonstrate their erosion-indicating capability, the TBC-coated specimens were first heated to 1000°C for 100 h in air (to ensure this demonstration was not relying on an initial layering structure that might be transient), and then subjected to localized erosion using a Koehler (Bohemia, NY) K93700 Air Jet Erosion Tester with a 5 mm diameter nozzle through which a 50 μm diameter alumina powder impinges on the specimen at an angle of 30° off normal. To observe luminescence from the exposed sublayers, eroded specimens were viewed under UV illumination provided by a SuperBright 2000SW lamp (UV Systems, Renton, WA) that has maximum intensity at a wavelength of 254 nm. Metallographically prepared cross sections of the TBC-coated specimens were inspected by optical microscopy, both under standard white light illumination as well as under UV illumination by the same lamp used to inspect the eroded surface. In addition, free-standing coatings were prepared by heating TBC-coated carbon substrates in air at 800°C to burn off sacrificial carbon substrates. Free-standing TBCs were annealed in air at 1400°C for 100 h (which was too severe for TBC-coated Rene N5 substrates), and cross sections of untreated and annealed free-standing TBCs were then prepared and inspected using luminescence imaging.

III. Results and Discussion

The erosion-indicating capability of the layer-doped TBC is demonstrated in Fig. 2. Figure 2(a) shows an image obtained under standard white light illumination of the TBC-coated specimen after being subjected to 85 min of localized alumina-par-

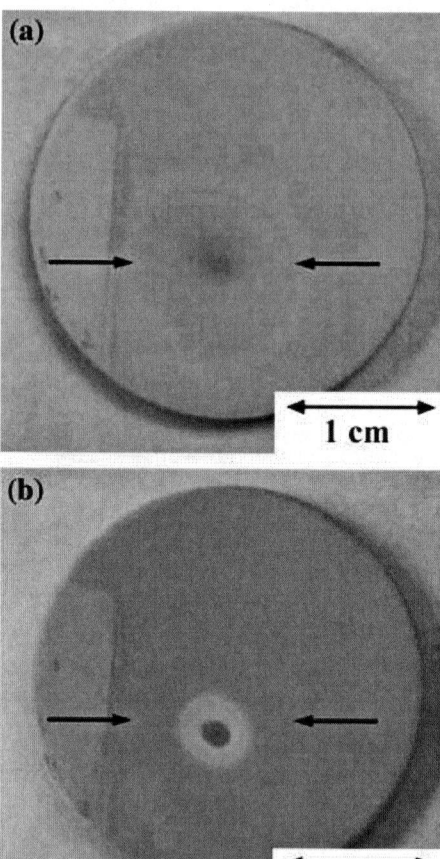

Fig. 2. Erosion-depth indication from YSZ:Tb and YSZ:Eu sublayers. Thermal barrier-coated specimen subjected to 85 min localized alumina-particle-jet erosion. The arrows indicate the edges of the erosion crater. (a) White light and (b) ultraviolet illumination. Green luminescence from exposed YSZ:Tb layer. Red luminescence from an exposed YSZ:Eu layer. YSZ:Eu, Eu-doped YSZ; YSZ:Tb, Tb-doped YSZ.

ticle-jet erosion. The thinner, eroded region of the TBC appears as a slightly grayish area. In contrast, UV illumination (Fig. 2(b)) excites strong luminescence from the exposed doped sublayers to immediately indicate the depth of erosion. Intense green luminescence is observed in the center of the eroded area, indicating that the bottom YSZ:Tb layer has been exposed, surrounded by an area of intense red luminescence indicating a region where erosion has reached the middle YSZ:Eu layer but has not yet reached the bottom YSZ:Tb layer.

Cross sections of free-standing sublayer-doped TBCs were inspected to establish whether the distinct luminescent sublayer boundaries that indicate erosion depth were achieved without disrupting the columnar growth of the TBC. Figure 3(a) displays an image obtained by an optical microscope with white light illumination of a cross section of a free-standing TBC before heat treatment and shows no microstructural demarcation of the sublayer boundaries, verifying that sublayer transition was achieved without disrupting columnar growth. Figure 3(b) displays a luminescence image under UV illumination of the same area and reveals that there are distinct boundaries between the contrasting luminescence produced by the doped sublayers. The variation in intensity in the red luminescence from the YSZ:Eu middle layer corresponds to a non-uniform evaporation of the Eu_2O_3 ingot, where melt-pool "spitting" produced fluctuations in the Eu-doping levels. Cross sections of free-standing TBCs subjected to 100 h at 1400°C were also inspected and showed that sufficient delineation between the contrasting luminescence of the sublayers is preserved even at the 1400°C heat treatment

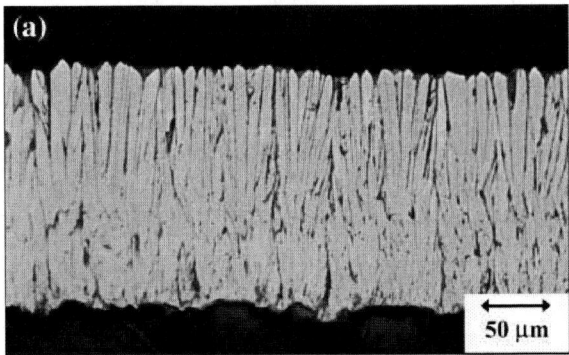

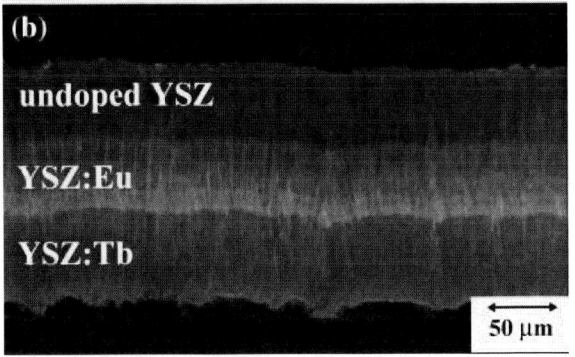

Fig. 3. Optical micrographs of cross section of free-standing sublayer-doped thermal barrier coatings (TBCs). (a) White light and (b) ultraviolet illumination of TBC.

to retain an effective erosion indication capability despite obvious coarsening of the columnar microstructure. Therefore, diffusion of the Eu^{3+} and Tb^{3+} dopants must not proceed over significant distances during coarsening of the columnar microstructure.

In view of the intense luminescence observed from the erosion-exposed sublayers (Fig. 2(b)), even much lower luminescent-rare-earth doping levels than the relatively low doping levels (1.3 at.%) used in this investigation should be sufficient

to produce easily visible luminescence for erosion indication. Reduced fluctuations in Eu-doping levels should be achieved by evaporating from a YSZ:Eu ingot, which is expected to have a more stable melt pool than the Eu_2O_3 ingot used in this work.

IV. Conclusions

A coating design for erosion-indicating TBCs has been successfully implemented by incorporating rare-earth-doped luminescent sublayers into TBCs by multiple-ingot EB-PVD without disrupting the columnar growth that gives EB-PVD TBCs their desirable strain tolerance. Under UV illumination, visible luminescence is excited from sublayers exposed by erosion, providing immediate indication of erosion location and depth. In addition, sufficiently distinct luminescent sublayer boundaries are preserved even after 100 h of heat treatment at 1400°C to maintain effective erosion indication. These positive results indicate that TBCs incorporating luminescent sublayers will enable simple, non-intrusive inspections during engine shutdowns to monitor TBC erosion.

Acknowledgments

The authors thank Joy Buehler and Myles McQuater for expert metallographic preparation of challenging specimens, and Dave Hull for EDS analysis used to determine rare-earth doping levels.

References

[1]K. Amano, H. Takeda, M. Tamatani, M. Itoh, and Y. Takahashi, "Thermal Barrier Coating"; U.S. Patent 4,774,150, September 27, 1988.

[2]K.-L. Choy, A. L. Heyes, and J. Feist, "Thermal Barrier Coating with Thermoluminescent Indicator Material Embedded Therein"; European Patent EP1105550B1, December 3, 2003.

[3]M. M. Gentleman and D. R. Clarke, "Concepts for Luminescence Sensing of Thermal Barrier Coatings," *Surf. Coat. Technol.*, **188–189**, 93–100 (2004).

[4]J. P. Feist and A. L. Heyes, "Development of the Phosphor Thermometry Technique for Application in Gas Turbines"; *10th International Symposium on Applications of Laser Techniques to Fluid Mechanics*, Lisbon, Portugal, 2000.

[5]J. I. Eldridge, T. J. Bencic, S. W. Allison, and D. L. Beshears, "Depth-Penetrating Temperature Measurements of Thermal Barrier Coatings Incorporating Thermographic Phosphors," *J. Thermal Spray Technol.*, **13** [1] 44–50 (2004).

[6]U. Schulz, K. Fritscher, and C. Leyens, "Two-Source Jumping Beam Evaporation for Advanced EB-PVD TBC Systems," *Surf. Coat. Technol.*, **133–134**, 40–8 (2000).   □

Rare-Earth Zirconate Ceramics with Fluorite Structure for Thermal Barrier Coatings

Qiang Xu,[†] Wei Pan, Jingdong Wang, Chunlei Wan, Longhao Qi, and Hezhuo Miao

State Key Laboratory of New Ceramics and Fine Processing, Department of Materials Science and Engineering,
Tsinghua University, Beijing 100084, China

Kazutaka Mori and Taiji Torigoe

Takasago Research & Development Center, Mitsubishi Heavy Industries, Ltd., Hyogo 676-8686, Japan

A series of rare-earth zirconate $Ln_2Zr_2O_7$ ceramics (Ln = Dy, Er, and Yb) with a fluorite structure (F-$Ln_2Zr_2O_7$) were prepared by pressureless sintering from zirconia and rare-earth oxide powders at 1600°C for 10 h in air. The microstructure experiments were performed by X-ray diffractometry (XRD) and scanning electron microscopy (SEM). The thermal conductivity and thermal expansion of these ceramics were evaluated using a steady-state laser heat-flux technique and high-temperature dilatometry, respectively. The XRD and SEM results demonstrate that $Ln_2Zr_2O_7$ ceramics with a single fluorite phase are synthesized and no other phases are found. The results of thermal conductivity show that their thermal conductivities (1.3–1.9 W/(m·K), 20°–800°C) are as low as those of the referenced $Ln_2Zr_2O_7$ ceramics (Ln = La, Nd, Sm, and Gd) with pyrochlore structure (P-$Ln_2Zr_2O_7$). It is concluded that rare-earth zirconate ceramics with a fluorite structure can be considered as candidate materials for future thermal barrier coatings.

I. Introduction

THE conventional thermal barrier coatings (TBCs) used in gas turbines have been yttria-stabilized zirconia (YSZ) coatings until now.[1,2] When the combustion-chamber temperature is increased further, the YSZ coatings will not fulfill the requirements for higher reliability and lower thermal conductivity at higher temperature. In the last few years, some novel ceramic materials with low thermal conductivity and high reliability at high temperature have been developed as candidate materials for future TBCs.[3–11] Significantly, the $A_2B_2O_7$-type rare-earth zirconate ceramics, such as $La_2Zr_2O_7$, $Nd_2Zr_2O_7$, $Sm_2Zr_2O_7$, and $Gd_2Zr_2O_7$, have received greater attention of researchers in the U.S.A. and EU.[11–19] For example, the $La_2Zr_2O_7$ ceramic had lower Young's modulus (175GPa) and lower thermal conductivity (1.6W/(m·K)) at 1000°C, when compared with YSZ.[14] Also, $La_2Zr_2O_7$ coating was reported to have higher phase stability against thermal treatment than conventional YSZ coating.[20] The fracture toughness of the pyrochlore material, which needed to be considerably improved, was reported to be about one magnitude lower than that of YSZ.[11] Afterwards, some other $Ln_2Zr_2O_7$ ceramics had been studied by Wu et al.[12] The results showed that all these ceramics had nearly the same thermal conductivities, all of which were ~30% lower than that of YSZ. Previous research on $A_2B_2O_7$-type ceramics for TBCs

mainly focused on the rare-earth zirconates with a pyrochlore structure (P-$Ln_2Zr_2O_7$). However, the rare-earth zirconates with fluorite structure (F-$Ln_2Zr_2O_7$) for TBCs have not been investigated in detail up to now.

In the present research, F-$Ln_2Zr_2O_7$ ceramics were prepared by pressureless sintering, and the thermal conductivities of these F-$Ln_2Zr_2O_7$ ceramics were investigated. The results showed that the thermal conductivities of F-$Ln_2Zr_2O_7$ ceramics were similar to those of the referenced P-$Ln_2Zr_2O_7$ ceramics. The possibility of these F-$Ln_2Zr_2O_7$ ceramics for TBCs will be discussed.

II. Experimental Procedure

Rare-earth oxide powders including dysprosium oxide, erbium oxide, and ytterbium oxide (Rare-Chem Hi-Tech Co., Ltd., Haizhou, Guangdong, China purity $\geq$ 99.99%) and zirconia (Farmeiya Advanced Materials Co., Ltd., Jiujiang, Jiangxi, China purity $\geq$ 99.9%, including 2.57% HfO_2) were chosen as the reactants. The oxides were weighed to the composition of $Ln_2Zr_2O_7$ stoichiometry. Materials were ball milled with ethanol, uniaxially cold pressed, and then sintered at 1600°C for 10 h.

Crystal-phase identification of the synthesized sample was determined by X-ray diffractometry (XRD, RIGAKU D/Max-rB, Tokyo, Japan) with Ni-filtered CuKα radiation (0.1542 nm) at a scan rate of 4°/min. The morphology of the sample was examined using scanning electron microscopy (SEM, JEOL JSM-6460LV, Tokyo, Japan). The bulk density of the sample was measured by the Archimedes method with an immersion medium of deionized water and compared with theoretical density to obtain the relative density and porosity of the product.

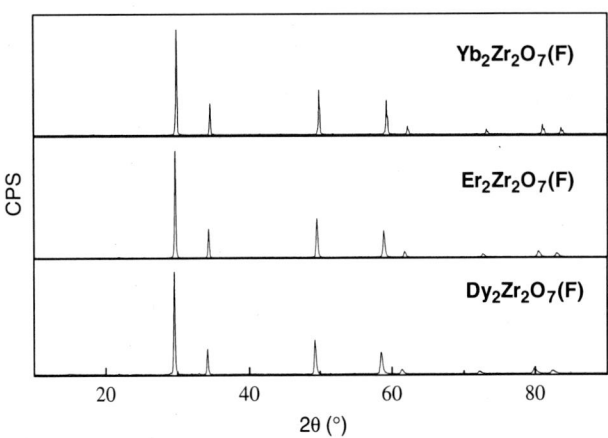

Fig. 1. X-ray diffractometry results of sintered F-$Ln_2Zr_2O_7$ (fluorite structure) samples.

J. Smialek—contributing editor

Manuscript No. 20484. Received April 29, 2005; approved June 28, 2005.
[†]Author to whom correspondence should be addressed. e-mail: xq@mail.tsinghua.edu.cn

Fig. 2. Microstructures of sintered F-Ln$_2$Zr$_2$O$_7$ samples (a) Dy$_2$Zr$_2$O$_7$ and (b) Er$_2$Zr$_2$O$_7$.

The thermal diffusivity (λ) for the sintered sample was measured, using the laser-flash method (Model NETZSCH LFA 427, Bayern, Germany), as a function of specimen temperature (in the range between ambient and 800°C) in an argon atmosphere. The present sample was about 10mm in diameter and about 1 mm in thickness. Before thermal-diffusivity measurements, both the front and the back faces of the specimen were coated with a thin layer of graphite. The specific heat capacity (C_p) was measured as a function of temperature (in the range between ambient and 600°C) using a differential scanning calorimeter (DSC) (Model NETZSCH DSC204, Bayern, Germany) at a heating rate of 10 K/min in air.

The thermal conductivity k is given by Eq. (1) with the heat capacity C_p, density ρ and thermal diffusivity λ:

$$k = C_p \lambda \rho \tag{1}$$

Table I. The Results of Densities of Sintered F-Ln$_2$Zr$_2$O$_7$ Samples

Material	Density (g/cm^3)	Theoretical density (g/cm^3)	Relative density (%)	Porosity (%)
Dy$_2$Zr$_2$O$_7$	6.89	7.273	94.8	5.2
Er$_2$Zr$_2$O$_7$	7.29	7.471	97.6	2.4
Yb$_2$Zr$_2$O$_7$	7.54	7.697	97.9	2.1

F, fluorite structure.

Table II. The Thermal Conductivities of Sintered F-Ln$_2$Zr$_2$O$_7$ Samples and the Referenced P-Ln$_2$Zr$_2$O$_7$ Samples (W/(m·K))

Material	Temperature (°C)					
	20	200	400	600	700	800
F-Ln$_2$Zr$_2$O$_7$						
Dy$_2$Zr$_2$O$_7$	1.46	1.45	1.45	1.40	–	1.34
Er$_2$Zr$_2$O$_7$	1.41	1.37	1.27	1.38	–	1.49
Yb$_2$Zr$_2$O$_7$	1.97	1.76	1.61	1.63	–	1.58
P-Ln$_2$Zr$_2$O$_7$						
Gd$_2$Zr$_2$O$_7$	–	–	–	–	~1.6[12]	–
Sm$_2$Zr$_2$O$_7$	–	–	–	–	~1.5[12]	–
Nd$_2$Zr$_2$O$_7$	~1.42[13]	–	~1.37[13]	–	~1.6[12]	~1.25[13]
La$_2$Zr$_2$O$_7$	–	~1.94[13]	~1.74[13]	~1.65[13]	~1.58[13]	~1.54[13]

F, fluorite structure; P, pyrochlore structure.

Because the sintered specimen was not fully (100%) dense, the measured thermal-conductivity value was modified for the actual data k_0 using Eq. (2):[11]

$$k/k_0 = 1 - 4/3\phi \tag{2}$$

where ϕ is the fractional porosity.

III. Results and Discussion

The XRD results of sintered F-Ln$_2$Zr$_2$O$_7$ samples at 1600°C are shown in Fig. 1. It can be seen that a single fluorite phase is formed in the Dy$_2$Zr$_2$O$_7$, Er$_2$Zr$_2$O$_7$, and Yb$_2$Zr$_2$O$_7$ samples. Increasing the atomic number from Dy to Yb leads to a shift of the X-ray spectrum to higher 2θ values.

Figure 2 shows the morphology of sintered F-Ln$_2$Zr$_2$O$_7$ samples. The grain of the product is inhomogeneous, and the average grain size is several micrometers. It can be also seen from Fig. 2 that the interface between the grains is clean and no other phases are found between the interfaces.

The densities of sintered F-Ln$_2$Zr$_2$O$_7$ samples presented in Table I increase with increasing atomic number. This may be due to decreasing rare-earth cationic radius.

The measured thermal conductivities of sintered F-Ln$_2$Zr$_2$O$_7$ samples and the P-Ln$_2$Zr$_2$O$_7$ reference samples are presented in Table II. Due to limitations of the measurement apparatus, the specific heat data are determined only from ambient temperature to 600°C. The specific heat value can be extrapolated to 800°C. It can be seen that the thermal conductivities of F-Ln$_2$Zr$_2$O$_7$ samples are comparable with those of P-Ln$_2$Zr$_2$O$_7$ samples.

IV. Summary

Dysprosium zirconate, erbium zirconate, and ytterbium zirconate were synthesized by pressureless sintering between rare-earth oxides and zirconia powders at 1600°C for 10 h. The results show that it is possible to produce stable, single phases with a fluorite structure. The thermal conductivities of F-Ln$_2$Zr$_2$O$_7$ samples are as low as the referenced P-Ln$_2$Zr$_2$O$_7$ samples.

From the investigation of F-Ln$_2$Zr$_2$O$_7$ samples reported here, they may be considered as novel prospective candidate materials for future TBCs.

References

[1]N. P. Padture, M. Gell, and E. H. Jordan, "Thermal Barrier Coatings for Gas-Turbine Engine Applications," *Science*, **296**, 280–4 (2002).
[2]C. G. Levi, "Emerging Materials and Processes for Thermal Barrier Systems," *Curr. Opin. Solid State Mater. Sci.*, **8**, 77–91 (2004).
[3]O. Sudre, J. Cheung, D. Marshall, P. Morgan, and C. G. Levi, "Thermal Insulation Coatings of LaPO$_4$," *Ceram. Eng. Sci. Proc.*, **22**, 367–74 (2001).
[4]C. J. Friedrich, R. Gadow, and M. H. Lischka, "Lanthanum Hexaaluminate Thermal Barrier Coatings," *Ceram. Eng. Sci. Proc.*, **22**, 375–82 (2001).

[5]R. Gadow and M. Lischka, "Lanthanum Hexaaluminate Novel Thermal Barrier Coatings for Gas Turbine Applications Materials and Process Development," *Surf. Coat. Technol.*, **151–152**, 392–9 (2002).

[6]N. P. Padture and P. G. Klemens, "Low Thermal Conductivity in Garnets," *J. Am. Ceram. Soc.*, **80**, 1018–20 (1997).

[7]M. Dietrich, R. Vaβen, and D. Stöver, "LaYbO₃, a Candidate for Thermal Barrier Coating Materials," *Ceram. Eng. Sci. Proc.*, **24**, 637–43 (2003).

[8]J. R. Nicholls, K. J. Lawson, A. Johnstone, and D. S. Rickerby, "Methods to Reduce the Thermal Conductivity of EB-PVD TBCs," *Surf. Coat. Technol.*, **151–152**, 383–91 (2002).

[9]D. M. Zhu and R. A. Miller, "Development of Advanced Low Conductivity Thermal Barrier Coatings," *Int. J. Appl Ceram. Technol.*, **1** [1] 86–94 (2004).

[10]D. M. Zhu and R. A. Miller, "Thermal Conductivity and Sintering Behavior of Advanced Thermal Barrier Coatings," *Ceram. Eng. Sci. Proc.*, **23**, 457–68 (2002).

[11]U. Bast and E. Schumann, "Development of Novel Oxide Materials for TBCs," *Ceram. Eng. Sci. Proc.*, **23**, 525–32 (2002).

[12]J. Wu, X. Zh. Wei, N. P. Padture, P. G. Klemens, M. Gell, E. García, P. Miranzo, and M. I. Osendi, "Low-Thermal-Conductivity Rare-Earth Zirconates for Potential Thermal-Barrier-Coating Applications," *J. Am. Ceram. Soc.*, **85**, 3031–5 (2002).

[13]H. Lehmann, D. Pitzer, G. Pracht, R. Vassen, and D. Stöver, "Thermal Conductivity and Thermal Expansion Coefficients of the Lanthanum Rare-Earth-Element Zirconate System," *J. Am. Ceram. Soc.*, **86**, 1338–44 (2003).

[14]R. Vassen, X. Q. Cao, F. Tietz, D. Basu, and D. Stöver, "Zirconates as New Materials for Thermal Barrier Coatings," *J. Am. Ceram. Soc.*, **83**, 2023–28 (2000).

[15]J. Wu, X. Zh. Wei, N. P. Padture, P. G. Klemens, M. Gell, E. García, P. Miranzo, and M. I. Osendi, "Thermal Conductivity of Ceramics in the ZrO₂–GdO₁.₅ System," *J. Mater. Res.*, **17**, 3193–200 (2002).

[16]G. Suresh, G. Seenivasan, M. V. Krishnaiah, and P. S. Murti, "Investigation of the Thermal Conductivity of Selected Compounds of Gadolinium and Lanthanum," *J. Nucl. Mater.*, **249**, 259–61 (1997).

[17]M. J. Malony "Thermal Barrier Coating Systems and Materials," U.S. Patent No. 6 284–323, 2001.

[18]R. Subramanian "Thermal Barrier Coating Having High Phase Stability," U.S. Patent No. 6 387–539, 2002.

[19]R. Vaβen, F. Traeger, and D. Stöver, "New Thermal Barrier Coatings Based on Pyrochlore/YSZ Double-Layer Systems," *Int. J. Appl Ceram. Technol.*, **1** [4] 356–61 (2004).

[20]X. Q. Cao, R. Vassen, W. Jungen, S. Schwartz, F. Tietz, and D. Stöver, "Thermal Stability of Lanthanum Zirconate Plasma-Sprayed Coating," *J. Am. Ceram. Soc.*, **84**, 2086–90 (2001).   □

Co-Doping of Air Plasma-Sprayed Yttria- and Ceria-Stabilized Zirconia for Thermal Barrier Applications

Zun Chen and Rodney Trice[†]

School of Materials Engineering, Purdue University, West Lafayette, Indiana 47907-2044

Hsin Wang, Wally Porter, and Jane Howe

Oak Ridge National Laboratory Oak Ridge, Tennessee 37831-6087

Matthew Besser and Daniel Sordelet

Ames Laboratory, Iowa State University, Ames, Iowa 50011

Co-dopants of either Yb^{3+} or Ca^{2+} were incorporated into 7.6 mol% $YO_{1.5}$–ZrO_2 (7.6YSZ) and 12 mol% CeO_2–ZrO_2 (12CeSZ) coatings by infiltrating porous spray-dried powders with salt solutions containing the appropriate co-dopant species prior to plasma spraying. Co-dopant concentration was varied from 2 to 5 mol%. Using a combination of transmission electron microscopy and energy-dispersive analysis, no secondary phase or Yb^{3+} segregation was detected at the grain boundary of either as-sprayed 2Yb/7.6YSZ or 2Yb/12CeSZ coatings. Dilatometer measurements showed that 2 mol% Yb^{3+} co-doped 7.6YSZ and 12CeSZ coatings shrank $\sim 0.6\%$ during a 5 h soak at 1400°C, approximately the same contraction as the baseline coatings (i.e. not co-doped). X-ray diffraction results show that the as-sprayed 7.6YSZ, 2Ca/7.6YSZ, and 2Yb/7.6YSZ coatings comprised of non-transformable, non-equilibrium composition tetragonal ZrO_2 (identified presently as t'-ZrO_2), while the 5Ca/7.6YSZ coating was a non-equilibrium composition of cubic ZrO_2. After a heat treatment of 100 h at 1200°C, the 2Yb/7.6YSZ coating was completely t'-ZrO_2, while the baseline and Ca^{2+} co-doped 7.6YSZ coatings showed evidence of partitioning. Therefore, it appears that co-doping of 7.6YSZ with 2 mol% Yb^{3+} increases the stability of t'-ZrO_2, whereas co-doping with 2 mol% Ca^{2+} decreases the stability of t'-ZrO_2. The volume fraction of m-ZrO_2 in the baseline 12CeSZ coatings was estimated to be 88% after a 100 h heat treatment at 1200°C. 2 mol% Yb^{3+} or Ca^{2+} co-doping limited the tetragonal to monoclinic phase transformation in 12CeSZ, with only 37% and 43% monoclinic phase observed, respectively, after a 100 h heat treatment at 1200°C; this was an improvement over the baseline 12CeSZ coating. As-sprayed 2Yb/7.6YSZ and 2Yb/12CeSZ coatings had slightly lower thermal conductivity than their baseline counterparts in the as-sprayed condition; after 100 h at 1200°C, their conductivity increased to that of the baseline coatings.

R. S. Hay—contributing editor

Manuscript No. 11277. Received August 18, 2004; approved December 7, 2004.
Part of this research effort was supported by a grant form the California Energy Commission, Grant # 51979A/00-22. The work performed at Ames Laboratory was supported by the U.S. Department of Energy through Iowa State University under Contract No. W-7405-ENG-82.
Research sponsored by the Assistant Secretary for Energy Efficiency and Renewable Energy, Office of FreedomCAR and Vehicle Technologies, as part of the High Temperature Materials Laboratory User Program, Oak Ridge National Laboratory, managed by UT-Battelle, LLC, for the U.S. Department of Energy under contract number DE-AC05-00OR22725.
Based in part on the thesis submitted by Z. Chen for the M.S. Degree in Materials Engineering, Purdue University, West Lafayette, Indiana, 2003.
[†]Author to whom correspondence should be addressed. e-mail: rtrice@purdue.edu

I. Introduction

VIRTUALLY every aircraft flying today uses thermal barrier coatings (TBCs) to protect metallic components within the gas turbine engine from temperature extremes and to allow them to operate at higher temperatures, thus increasing the overall operating efficiency. A TBC typically consists of two layers, e.g., a ~ 200 μm thick zirconia-based topcoat and a ~ 100 μm metallic bond coat (MCrAlY, where M can be Co, Ni, or Fe).[1] The zirconia layer is most often stabilized with Y_2O_3, but other oxides, like CeO_2, have been suggested.[2,3] These coatings can be applied via electron beam physical vapor deposition or plasma spray; the latter is the focus in the current work.

While TBCs are widely used to increase the durability of hot section components in land-based gas turbines used for power generation,[4,5] durability and reliability issues limit the benefits that can be derived from their use.[6] For example, the change in phase assemblage of the zirconia topcoat after multiple hours at temperatures at or above 1200°C can limit their durability. X-ray analysis of as-sprayed 7.6 mol% $YO_{1.5}$–ZrO_2 (7.6YSZ) coatings at room temperature reveals a single metastable tetragonal zirconia phase (referred to presently as t'-ZrO_2), with a composition of 7.6 mol% $YO_{1.5}$,[7] rather than m-ZrO_2 and c-ZrO_2,[8] as would be predicted by the phase diagram. However, this metastable t'-ZrO_2 phase is favored in TBC applications because of its high cyclic life.[9] The problem with this metastable phase is that it tends to partition into equilibrium composition t-ZrO_2 (~ 4 mol% $YO_{1.5}$) and c-ZrO_2 (~ 14 mol% $YO_{1.5}$) phases during long durations at high temperatures. Upon cooling, the now equilibrium composition of t-ZrO_2 will transform to m-ZrO_2, resulting in cracking of the coating. Ilavsky and Stalick[10] confirmed the partitioning of the t'-ZrO_2 phase after heat treatment at temperatures of 1100°–1400°C using neutron diffraction. Another study by Ilavsky et al.[11] showed that as the yttria content in the tetragonal phase approached a limiting concentration, the t-ZrO_2 phase would transform into m-ZrO_2 phase on cooling.

Lee et al.[12] studied the phase transformations of plasma-sprayed ZrO_2–CeO_2 (with 12–20.1 mol% CeO_2) TBCs. They found that as-sprayed CeSZ coatings with 12 and 13.6 mol% CeO_2 consisted of a single non-equilibrium t'-ZrO_2, while those with 15.2–20.1 mol% CeO_2 contained a mixture of t'-ZrO_2 and c'-ZrO_2 phases. During 45 min cyclic oxidation at 1135°C, the non-equilibrium t'-ZrO_2 and c'-ZrO_2 partitioned into the equilibrium t-ZrO_2 and c-ZrO_2 phases. Some of the tetragonal phases transformed to the monoclinic phase during cooling.

As the formation of m-ZrO_2 typically results in cracking of the coating, limiting the partitioning of the non-equilibrium phases that result during plasma spraying may lead to increased life in TBCs. Rebollo et al.[13] recently studied the phase stability of 7.6YSZ co-doped with rare earth elements (La, Nd, Sm, Gd, Y, and Yb). The authors found that the phase stability increased systematically as the size of the rare earth cation decreased for both single and co-doped compositions. In the research reported here, an approach similar to Levi and co-workers to limit the partitioning kinetics of the as-sprayed, metastable phase is investigated. However, in this study, we focus on preparing plasma-sprayed coatings made from co-doped 7.6YSZ or 12 mol% CeO_2–ZrO_2 (12CeSZ) powders to stabilize the t'-ZrO_2 phase. The co-dopants investigated included Ca^{2+} or Yb^{3+} ions.

II. Experimental Procedure

(1) Co-Dopant Infiltration and Coating Preparation

Two partially stabilized zirconia-based powders were investigated including one with 7.6 mol% $YO_{1.5}$ added as stabilizer (T14841/ZRO-236, Praxair Surface Technologies, Indianapolis, IN), and one containing 12 mol% CeO_2 (ZRO-248, Praxair Surface Technologies). These will be referred to as 7.6YSZ and 12CeSZ, respectively. Both 7.6YSZ and 12CeSZ powders were spray dried, with each powder comprising hundreds of particles <10 μm in diameter to form a porous structure. The diameters of 7.6YSZ and 12CeSZ powders were ~100 μm, with some powders finer than this being observed.

Co-dopants of either Ca^{2+} or Yb^{3+} were incorporated into the spray-dried powders by infiltrating them with solutions containing the desired impurity. These two co-dopants were chosen because of their relative ionic radius as compared with the Y^{3+} stabilizer: Ca^{2+} is larger and Yb^{3+} is smaller.[14] The infiltrant was prepared by dissolving hydrated salts of either $Ca(NO_3)_2$ (Stock # 30482, Calcium Nitrate, Alfa Aesar, Ward Hill, MA) or $Yb(NO_3)_3$ (Stock # 12901, Ytterbium Nitrate, Alfa Aesar) in ethanol. The molar amount of the dopants added was such that they would substitute 2% or 5% of the total cation sites in the powder to be infiltrated. The seven different compositions presently studied are listed in Table I. The co-dopant solution was poured into a beaker with the powders, and the resulting slurry was subjected to a vacuum of 3 Pa for 1 h. The slurry was then dried, stirring occasionally at 80°C to evaporate the ethanol. The resultant paste was moved into an alumina crucible and calcined at 1100°C for 2 h. The loosely agglomerated powders were broken up mechanically prior to spraying, and sieved through an aperture of 150 μm.

The 7.6YSZ, 12CeSZ, and co-doped powders were plasma sprayed (a Praxair SG-100 gun (Praxair, Danbury, CT) with a 730 anode, 729 cathode and a 112 gas injector were used) at Ames Laboratory to form coatings for investigation. The spray parameters used for all powders are listed in Table II; these parameters were not optimized for each powder type. Substrates of two geometries were used: a copper plate of 102 mm × 76 mm × 5 mm and a 203 mm long copper tube with an outer diameter of 9.5 mm and an inner diameter of 7.9 mm. Coatings removed from the cylindrical substrate were used for dilatometry studies; coatings removed from flat substrates were used in all other investigations. The substrates were grit blasted with 24-grit Al_2O_3 at 5.5×10^5 Pa prior to coating. Air jets were used to cool the substrates while plasma spraying, and ~30 spray cycles were used for each coating. The copper substrate was removed via a

1 h soak in HNO_3 to create either flat (i.e. from the plate) or cylindrical coatings (i.e. from the tube) for study. The amount of co-dopant incorporated into the coating was measured via mass spectroscopy techniques (NSL Analytical Services Inc., Cleveland, OH) on pulverized coatings.

(2) Coating Characterization

The bulk density and open porosity of the as-sprayed coatings were measured using the Archimedes method.[15] Approximately 1 g of specimen was used for each measurement. The accuracy of the scale was within 0.001 g.

Cross-sectional transmission electron microscopy (TEM) samples of as-sprayed 2Yb/7.6YSZ and 2Yb/12CeSZ were prepared using a tripod polisher (South Bay Technology, Santa Clara, CA). Both sides of each specimen were polished progressively using a sequence of diamond-coated polishing films until the thin end of the specimen began to recede. After the specimen was mounted on a nickel grid, an ion mill was used to further thin the specimen to electron transparency (model DMP 600, Gatan Ion Mill, Gatan Inc., Pleasanton, CA). TEM was carried out at Oak Ridge National Laboratory using a Hitachi HF2000 FEG-TEM (Hitachi High Technologies, Inc., Tokyo, Japan) equipped with a Thermo Electron Corporation electron-dispersive spectrometer (EDS) at 200 kV. EDS was used to measure the relative concentration of ytterbium content in 2Yb/7.6YSZ and 2Yb/12CeSZ within grains and across grain boundaries. The probe size was ~2 nm in diameter.

The phase stability of as-sprayed and heat-treated coatings was investigated using X-ray diffraction (XRD; Siemens D500 Kristalloflex, Karlsruhe, Germany) with CuKα radiation. Coatings of all compositions studied were heat treated at 1200°C for 10 and 100 h; the 2Yb/7.6YSZ coating was further heat treated at 1200°C for 400 h.

All seven coatings were analyzed for 2θ values of 20°–80° at a scan speed of 6°/min. The yttria-stabilized coatings were further investigated for 2θ values of 72°–76° at a scan speed of 0.5°/min to detect partitioning of the t'-ZrO_2 phase.[7,16] Deconvolution of the XRD peaks was performed using Rietveld analysis (TOPAS Software, Bruker AXS GmbH, Karlsruhe, Germany). For CeSZ coatings, 2θ values of 26°–34° were investigated at a scan speed of 1°/min. The integrated area intensity of {111} monoclinic and tetragonal/cubic phase was deconvoluted using Rietveld analysis. The volume ratio of the monoclinic phase to tetragonal/cubic phase in CeSZ was calculated using the following equation[12]:

$$\frac{V_m}{V_{t+c}} = 1.39 \left[\frac{I_m(11\bar{1}) + I_m(111)}{I_{t+c}(111)} \right] \qquad (1)$$

where V_m and V_{t+c} are the volume fraction of monoclinic phase and tetragonal+cubic phase; I_m and I_{t+c} are the integrated area intensity of the monoclinic phase and $t+c$ phase, respectively.

The shrinkage behavior of 12 mm tall stand-alone cylindrical coatings was characterized using a dilatometer (Orton 1600D push-rod type dilatometer, Orton Ceramic Foundation, Westerville, OH). Each sample was heated up at 10°C/min ramp to 1400°C and held for 5 h. A probe that can move in the axial direction was in contact with the sample to follow the expansion or contract upon heating. Data of temperature, probe displace-

Table I. Overview of the Coatings Investigated in the As-Sprayed Condition

Coating designation	Tested dopant concentration (mol%)[‡]	Average thickness (μm)	Bulk density (g/cm³)	Open porosity (%)
7.6YSZ	0.24 Ca[†]	~550	5.23±0.07	10.5±0.9
2Ca/7.6YSZ	1.75 Ca	~400	5.08±0.06	11.3±0.3
5Ca/7.6YSZ	4.10 Ca	~390	4.96±0.04	13.7±1.8
2Yb/7.6YSZ	2.69 Yb	~480	5.14±0.02	12.4±0.9
12CeSZ	0.11 Ca[†]	~570	5.59±0.03	8.0±0.6
2Ca/12CeSZ	1.43 Ca	~640	5.65±0.01	7.5±0.1
2Yb/12CeSZ	2.18 Yb	~490	5.57±0.02	9.1±0.9

[†]As an impurity in the starting powder. [‡]NSL Analytical Services Inc. 7.6YSZ, 7.6 mol% $YO_{1.5}$–ZrO_2; 12CeSZ, 12 mol% CeO_2–ZrO_2.

Progress in Thermal Barrier Coatings

Table II. Parameter of Plasma Spraying

Current (A)	900
Volts (at gun)	41.3
Arc gas (scfh)	54 (Ar)
Aux gas (scfh)	44 (He)
Carrier gas (scfh)	13 (Ar)
Powder feed rate (rpm)	1.5
Stand off distance (cm)	10.0

scfh, standard cubic feet per hour.

ment, and time were collected *in situ* by a computer. Shrinkage of the specimen was obtained by subtracting the thermal expansion of the apparatus and that of the sample being tested from probe displacement.

(3) Thermal Conductivity Measurements

The thermal conductivity, k_{th}, in units of W/m/K, as a function of temperature for each of the coatings was calculated using the following equation:

$$k_{th}(T) = \alpha(T) \cdot c_p(T) \cdot \rho \cdot 100$$

where $\alpha(T)$ is the thermal diffusivity as a function of temperature in units of cm^2/s, $c_p(T)$ is the specific heat as a function of temperature in units of J/g/K, and ρ is the as-sprayed density (g/cm^3) of the coating being investigated.

Thermal diffusivity (Flashline 5000 Thermal Diffusivity System, Anter Corporation, Pittsburgh, PA), α, was measured as a function of temperature using the laser flash method at the Oak Ridge National Laboratory. A detailed description of the system setup and measurement technique has been published elsewhere.[17] Disk-shaped samples, 12.7 mm in diameter, were used for all measurements. The disk samples were spray coated with a thin layer of colloidal graphite to ensure similar surface radiation characteristics in all samples prior to testing. Thermal diffusivity data were collected every 100°C from 100° to 1200°C. Specimens were tested in an aluminum furnace from 100° to 500°C and continued in a graphite furnace from 700° to 1200°C in a nitrogen atmosphere. An InSb detector was used in the low-temperature furnace and a silicon infrared detector was used in the high-temperature furnace. Three measurements were taken for each sample at each temperature and averaged. At least two samples of each coating were tested. The time–temperature curves were analyzed by the method of Clark and Taylor,[18] which takes into account radiation losses and uses the heating part of the curve to calculate thermal diffusivity. Specific heat as a function of temperature, $c_p(T)$, was measured from 25° to 1200°C at a heating rate of 20°C/min using a differential scanning calorimeter (Netzsch Instruments DSC 404C, Burlington, MA).

III. Results and Discussion

(1) Characterization of As-Sprayed 7.6YSZ and 12CeSZ Coatings

The measured co-dopant composition and physical properties of all as-sprayed coatings are summarized in Table I. The actual co-dopant concentrations were found to be within ± 35% of the expected concentrations for all samples; these differences may arise from the hydration/dehydration of the reactant chemicals. Coating depositions for all coatings were between 13 and 20 μm/pass. Densities were close to ~ 5 g/cm^3 for co-doped and baseline 7.6YSZ coatings and ~5.6 g/cm^3 for co-doped and baseline 12CeSZ coatings. Except for 2Ca/12CeSZ, the co-doped as-sprayed coatings displayed more open porosity than their baseline counterparts.

(2) TEM and EDS Investigations of 2Yb/7.6YSZ and 2Yb/12CeSZ Coatings

Investigation of both 2Yb/7.6YSZ and 2Yb/12CeSZ as-sprayed coatings in cross-section revealed typical plasma-sprayed micro-structures,[19,20] with lamellae structures composed of columnar grains observed. Intralamellar microcracks, located between grains, and lenticular-shaped interlamellar pores between two adjacent lamellae were also observed. Also, preliminary studies using high-resolution imaging of grain boundaries from either 2Yb/7.6YSZ or 2Yb/12CeSZ coatings showed no evidence of either an amorphous or crystalline grain boundary phase. In the current work, the location of Yb atoms in the grains and between their boundaries was investigated.

Figure 1 presents a TEM micrograph and corresponding EDS analysis from a region within a lamella of a 2Yb/7.6YSZ sample. As indicated with circles in the micrograph, analysis was performed at two regions inside each grain (points A and D), and at a grain boundary (point C). Note that the circles indicated on the image are to indicate probe location and that the actual probe size is much smaller than the diameter of the circle. The highest intensity peak corresponded to Zr-L$_\alpha$; an Yb-M$_\alpha$ peak was also evident in each location in the inset EDS spectra. It was found that the intensity ratio of the Yb-M$_\alpha$ and Zr-L$_\alpha$ peaks was approximately the same whether the location of probe was inside the grain or across the grain boundary. This would suggest that no segregation of Yb occurs at the grain boundary. Four other random regions were examined; despite slight variation of Yb content from region to region, no enrichment of Yb was detected at the grain boundary.

EDS and TEM analysis of 2Yb/12CeSZ revealed similar results. Figure 2 shows EDS spectra taken in a typical columnar grains area. EDS revealed that grain interior A, C and grain boundary B, D had the same small Yb peaks. Probing at other regions revealed no grain boundary segregation either.

Although grain boundary segregation is extensively observed in zirconia materials,[21,22] the conditions for this to occur are not well known. In our study, coatings were deposited on a substrate and cooled rapidly. McPherson and Shafer[23] has estimated the cooling rate to be as high as ~10^6 K/s. It is possible that the defect equilibrium does not have sufficient time to set up. As a result, no segregation was observed. Annealing at high temperature followed by a slow cooling may allow redistribution of cations and oxygen vacancies near grain boundaries driven by the space charge potential.

In summary, EDS results showed regions where Yb concentrations varied from a significant to a trace amount. The variation was probably because of the different diffusion distances of Yb. Since the spray-dried powder was composed of individual particles, and the co-dopant is added to the particles from the exterior, it was expected that particle exteriors would have a higher Yb concentration than the interior. While it was clear that a variation of Yb content did exist from region to region, in the same region, however, no evident variation was noted between grain interiors and grain boundaries within limits of the testing resolution.

(3) Linear Shrinkage Measurement

Four samples for each of 7.6YSZ, 12CeSZ, 2Yb/7.6YSZ, and 2Yb/12CeSZ were tested using a dilatometer. The final shrinkage at the end of the 5 h hold at 1400°C was recorded. The average and range of final shrinkage are listed in Table III. It was observed that the linear shrinkage difference between coatings is within the testing error. Therefore, no significant difference exists in the shrinkage behavior among the co-doped coatings and their baseline counterpart.

(4) Phase Stability of Co-Doped 7.6YSZ Coatings

Figure 3 shows the XRD peaks for both the as-sprayed co-doped and baseline 7.6YSZ coatings. For the 7.6YSZ, 2Ca/7.6YSZ, and 2Yb/7.6YSZ coatings, a non-equilibrium composition tetragonal phase was observed. This phase, identified presently as t'-ZrO$_2$, typically has ~7.6 mol% YO$_{1.5}$ and if applicable, the co-dopant ion, in solid solution with the ZrO$_2$. The 5Ca/7.6YSZ sample was completely cubic zirconia of a non-equilibrium composition (c'-ZrO$_2$).

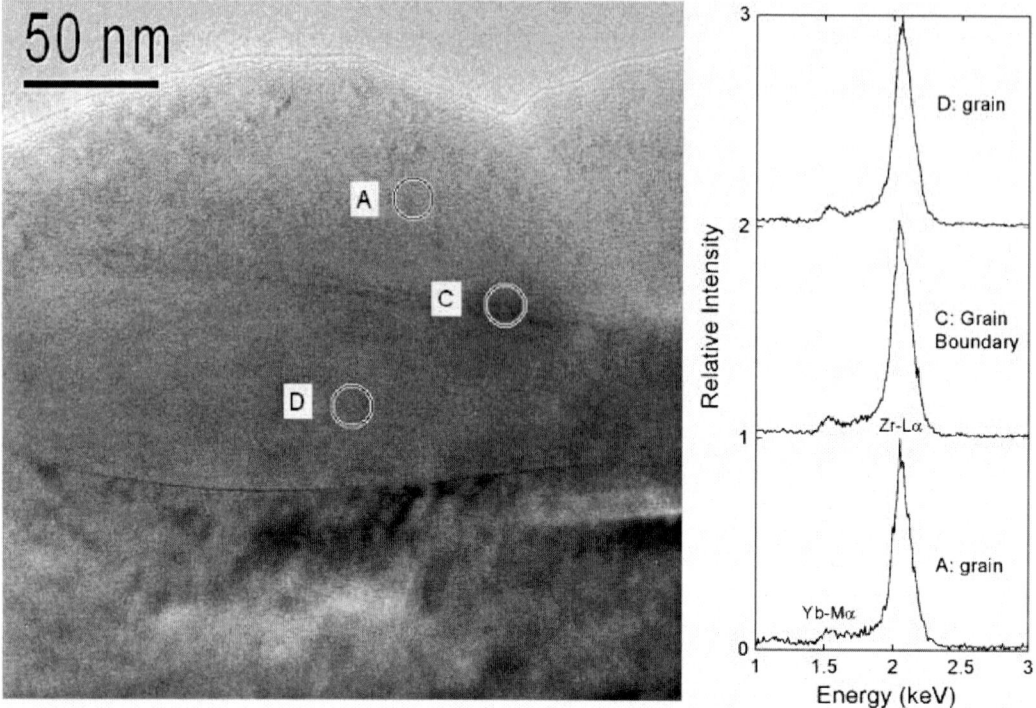

Fig. 1. Electron-dispersive spectrometer spectra and corresponding transmission electron microscopy micrograph showing probe location in a 2Yb/7.6 mol% $YO_{1.5}$–ZrO_2 coating. Note that because the Zr-L and Y-L energies are very similar, it is difficult to resolve these two peaks. However, inspection of the X-ray spectrum from 1–20 keV showed a Y-K^α line close to 14.9 keV; the ratio of yttrium peak height to zirconium peak height for their respective K^α lines did not vary substantially for points A, B, or C.

After a 10 h heat treatment at 1200°C, not shown, only the t'-ZrO_2 phase was observed in 7.6YSZ, 2Ca/7.6YSZ, and 2Yb/7.6YSZ; however, there was some shifting of the (004) peak of each coating composition to smaller 2Θ values. In 5Ca/7.6YSZ, the emergence of a tetragonal zirconia phase was observed. Thus, it appears that increasing the amount of Ca^{2+} co-dopant decreases the stability of 7.6YSZ.

Figure 4 presents the XRD peaks of the co-doped and baseline 7.6YSZ coatings after 100 h at 1200°C. Note that c-ZrO_2 has begun to form in the 7.6YSZ and 2Ca/7.6YSZ coatings. The excess

yttria in the t'-ZrO_2 phase forms the c-ZrO_2 phase, and its presence is a good indicator that destabilization of the t'-ZrO_2 phase is beginning. It should be noted that while the original metastable t'-ZrO_2 phase has begun to partition into an yttria-rich c-ZrO_2 phase, the remaining tetragonal phase, while it has less than 7.6 mol% $YO_{1.5}$, is still "non-transformable" in that its composition is greater than would be predicted by the equilibrium phase diagram. This was verified by XRD where no m-ZrO_2 was detected.

A c-ZrO_2 peak was also observed as in the 2Ca/7.6YSZ sample after 100 h at 1200°C. Thus, Ca^{2+} co-doping does not appear

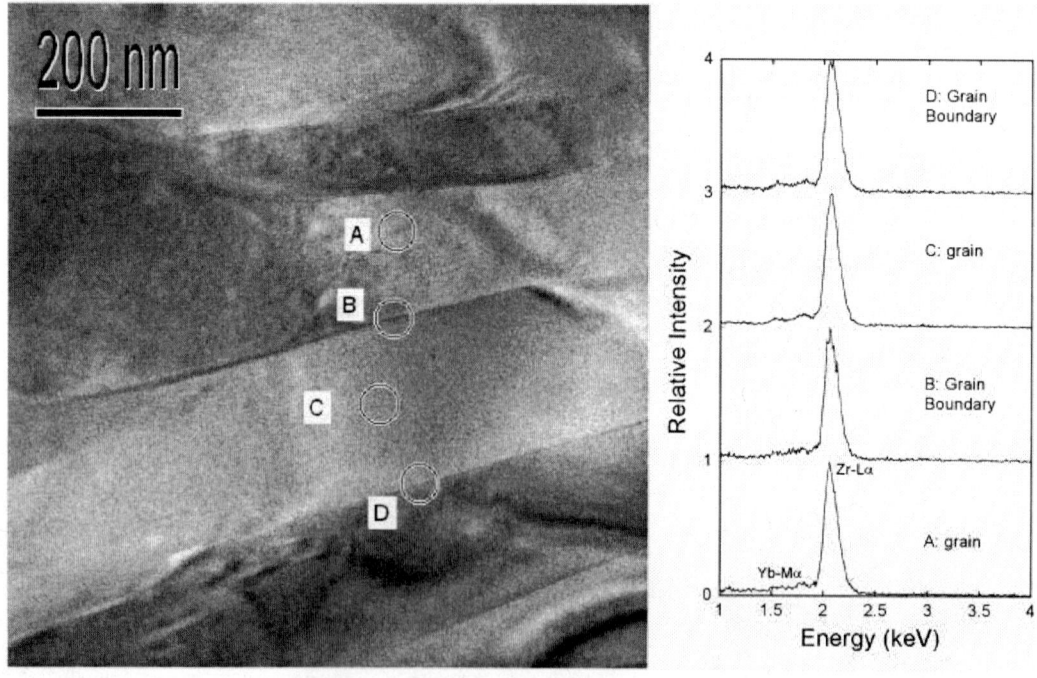

Fig. 2. Electron-dispersive spectrometer spectra and corresponding micrographs showing probe location in a 2Yb/12CeSZ coating.

Table III. Linear Shrinkage After 300 min at 1400°C

Coating	Linear shrinkage after 300 min at 1400°C (%)
7.6YSZ	0.56 ± 0.17
2Yb/7.6YSZ	0.68 ± 0.23
12CeSZ	0.63 ± 0.28
2Yb/12CeSZ	0.66 ± 0.24

7.6YSZ, 7.6 mol% $YO_{1.5}$–ZrO_2; 12CeSZ, 12 mol% CeO_2–ZrO_2.

to stabilize the t'-ZrO_2 phase. However, no change in the phase assemblage of 2Yb/7.6YSZ coatings was observed after 100 h at 1200°C. The separation between (004) and (400) tetragonal phase became more evident with longer heat-treatment time. Thus, the resistance to partitioning of the t'-ZrO_2 phase seemed to be enhanced by co-doping with Yb^{3+}. Furthermore, a heat treatment of 400 h at 1200°C on the 2Yb/7.6YSZ revealed that some c-ZrO_2 has formed, but no m-ZrO_2 was observed. The 5Ca/7.6YSZ coating was further characterized by destabilization of the c'-ZrO_2 phase.

According to the equilibrium Y_2O_3–ZrO_2 binary phase diagram[24] and the CaO–Y_2O_3–ZrO_2[25] (at 1250°C) and Yb_2O_3–Y_2O_3–ZrO_2[26] (at 1200°C) ternary phase diagrams, all four YSZ-based coatings, 7.6YSZ, 2Ca/7.6YSZ, 5Ca/7.6YSZ, and 2Yb/7.6YSZ, lie in the cubic and tetragonal two-phase field. The ternary compositions of 2Ca/7.6YSZ and 2Yb/7.6YSZ are both about in the middle of the tieline joining the equilibrium tetragonal and cubic boundary, and it may suggest a similar thermodynamic driving force for phase separation.[13]

In accordance with the phase diagrams, after 100 h heat treatment the phases in 7.6YSZ, 2Ca/7.6YSZ, and 5Ca/7.6YSZ were composed of tetragonal and cubic phases. Although it is difficult to determine whether the compositions of these phases are the equilibrium composition, phase separation of the as-sprayed coatings was evident. As previously stated, no phase separation was observed in the 100 h heat-treated 2Yb/7.6YSZ sample, indicating that phase separation in 2Yb/7.6YSZ is slower.

Phase separation in 7.6YSZ and 12CeSZ occurs via the diffusion of cations; the ionic radius of the relevant cations[22] for this study is as follows: $r_{Zr^{4+}} < r_{Yb^{3+}} < r_{Y^{3+}} < r_{Ca^{2+}}$. Since Ca^{2+} is larger in size than Yb^{3+}, it is expected to diffuse slower. However, 2Yb/7.6YSZ showed clear evidence of phase stabilization, while the 2Ca/7.6YSZ did not. Therefore, the slower phase separation in 2Yb/7.6YSZ cannot be explained by transport kinetics. And as suggested above, the thermodynamic driving force of 2Yb/7.6YSZ and 2Ca/7.6YSZ could be similar, and would not account for the difference either. Thus, it is most

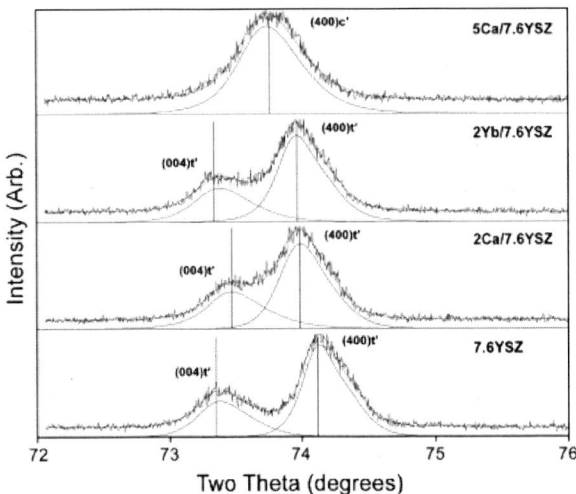

Fig. 3. Diffraction pattern of as-sprayed (a) 7.6 mol% $YO_{1.5}$–ZrO_2 (7.6YSZ), (b) 2Ca/7.6YSZ, (c) 5Ca/7.6YSZ, and (d) 2Yb/7.6YSZ for 2θ between 72° and 76°. All samples were pulverized prior to testing.

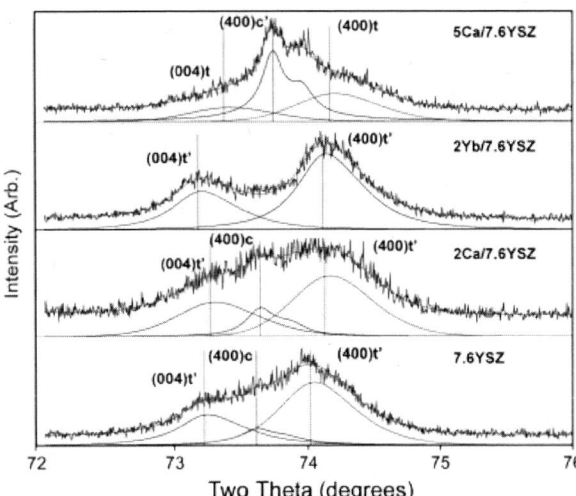

Fig. 4. Diffraction pattern of (a) 7.6 mol% $YO_{1.5}$–ZrO_2 (7.6YSZ), (b) 2Ca/7.6YSZ, (c) 5Ca/7.6YSZ, and (d) 2Yb/7.6YSZ after 100 h at 1200°C. All samples were pulverized prior to testing.

likely that the difference in phase stability between 2Ca/7.6YSZ and 2Yb/7.6YSZ arises from the coherency strains between the host and co-dopant ions in the structure. The larger co-dopant ions can increase the local distortion of the anion lattice and are believed to promote the nucleation of c-ZrO_2.[13] Therefore, it is expected that Yb^{3+} co-doping would accommodate distortion, while Ca^{2+} co-doping would induce more severe distortion because of the larger size mismatch with Zr^{4+} ions. A higher Ca^{2+} concentration induces more localized distortion and therefore more nucleation sites, which could explain why 5Ca/7.6YSZ began to partition after only 10 h at 1200°C.

(5) Phase Stability of Co-Doped 12CeSZ Coatings

XRD peaks of as-sprayed 2Ca/12CeSZ and 2Yb/12CeSZ (not shown) resembled those of baseline 12CeSZ, consisting of mostly a single non-equilibrium composition tetragonal phase, along with some m-ZrO_2. After a 10 h heat treatment at 1200°C, a significant increase of m-ZrO_2 phase was present in all three compositions of coatings; the amount of the m-ZrO_2 phase further increased after a 100 h heat treatment. The increase of m-ZrO_2 volume fraction with heat-treatment time at 1200°C is summarized in Table IV. These quantitative data were acquired from deconvolution of integrated intensity of $(111)_m$ and $(111)_{t+c}$ peaks in the $2\theta = 26°$–$34°$ region according to Eq. (1). As-sprayed coatings all started with about 7–8 vol% of the m-ZrO_2 phase. After 100 h heat treatment, the m-ZrO_2 phase amount was as much as 88 vol% in 12CeSZ, and only 43 and 37 vol% in Ca^{2+} and Yb^{3+} co-doped coatings, respectively. This result suggests that co-doping may stabilize the as-sprayed tetragonal phase in ceria-stabilized zirconia, and therefore limit the tetragonal-to-monoclinic transformation.

(6) Thermal Conductivity Measurements

Figure 5 presents the thermal conductivity of as-sprayed baseline and co-doped 7.6YSZ coatings as a function of measure-

Table IV. Volume Fraction of Monoclinic Phase in As-Sprayed and 10 and 100 h Heat-Treated Samples of 12CeSZ, 2Ca/12CeSZ, and 2Yb/12CeSZ

	Vol% of m-ZrO_2 phase		
	As-sprayed	10 h at 1200°C	100 h at 1200°C
12CeSZ	8	84	88
2Ca/12CeSZ	8	33	43
2Yb/12CeSZ	7	36	37

Error in each measurement is ± 2%. 12CeSZ, 12 mol% CeO_2–ZrO_2.

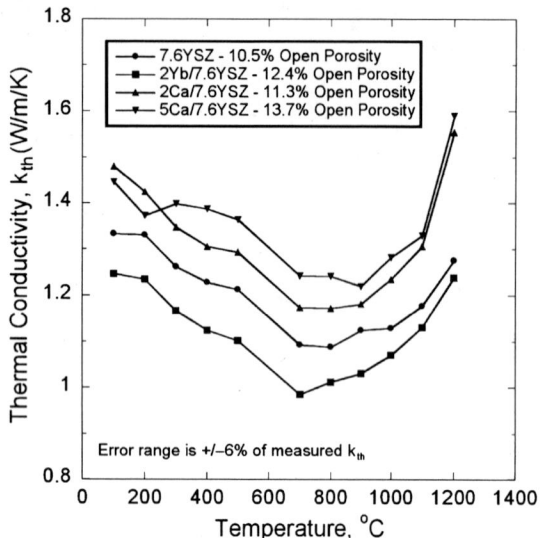

Fig. 5. Thermal conductivity of as-sprayed plasma-sprayed Yb^{3+} and Ca^{2+} co-doped 7.6 mol% $YO_{1.5}$–ZrO_2 (7.6YSZ) coatings at different measuring temperatures. The amount of open porosity in each coating is also indicated.

ment temperature. For all samples, it was observed that the conductivity decreases with temperature until it reaches ~700°C, and then increases through 1200°C. In the temperature range investigated, 2Yb/7.6YSZ exhibited a lower value of thermal conductivity than 7.6YSZ. However, co-doping with 2 and 5 mol% Ca (2Ca/7.6YSZ and 5Ca/7.6YSZ) resulted in coatings with higher k_{th} than the baseline 7.6YSZ coating. Ca^{2+} co-doped coatings also exhibited a more dramatic increase in k_{th} as the temperature increased beyond 1000°C.

Figure 6 shows the thermal conductivity of as-sprayed baseline and co-doped 12CeSZ coatings. Again, the Yb^{3+} co-doped coating demonstrated the lowest thermal conductivity among the three tested, and 2Ca/12CeSZ exhibited thermal conductivity comparable with the baseline 12CeSZ coating.

Defects play a critical role in reducing the intrinsic thermal conductivity of zirconia. In yttria-stabilized zirconia, the substi-

Table V. Calculated Oxygen Vacancy Concentration and Measured Open Porosity in Baseline and Co-Doped Coatings

Coating	Oxygen vacancy concentration (mol%)	Open porosity (%)
7.6YSZ	1.9	10.5
2Ca/7.6YSZ	2.9	11.3
5Ca/7.6YSZ	4.4	13.7
2Yb/7.6YSZ	2.4	12.4
12CeSZ	0	8.0
2Ca/12CeSZ	1	7.5
2Yb/12CeSZ	0.5	9.1

7.6YSZ, 7.6 mol% $YO_{1.5}$–ZrO_2; 12CeSZ, 12 mol% CeO_2–ZrO_2.

tution of Zr^{4+} with Y^{3+} is accompanied by the creation of oxygen vacancies to maintain the electroneutrality of the lattice. One expects oxygen vacancies to be much stronger phonon scatterers than Y^{3+} cations, which have only a slightly different mass and size than Zr^{4+} cations. However, the concentration of oxygen vacancies is dependent on the concentration of Y^{3+} in zirconia. A theoretical calculation performed by Bisson et al.[27] predicts that the thermal conductivity of YSZ should decrease monotonically with increasing oxygen vacancy concentration; however, their k_{th} measurements on a YSZ single crystal showed that there was a minimum value for thermal conductivity at ap-

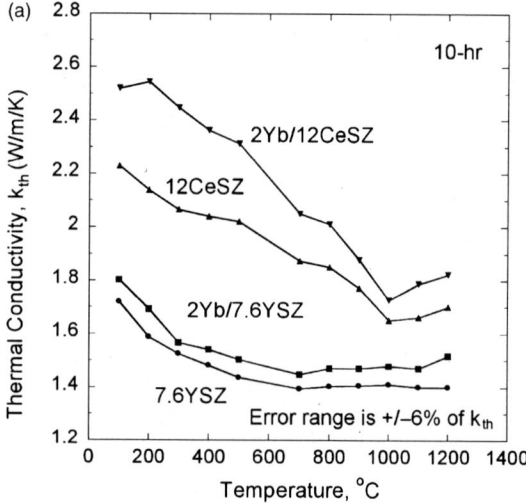

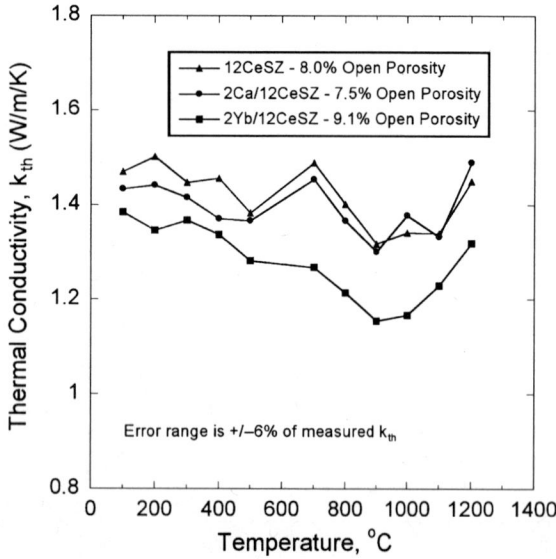

Fig. 6. Thermal conductivity of as-sprayed Yb^{3+} and Ca^{2+} co-doped 12 mol% CeO_2–ZrO_2 (12CeSZ) coatings at different measuring temperatures. The amount of open porosity in each coating is also indicated.

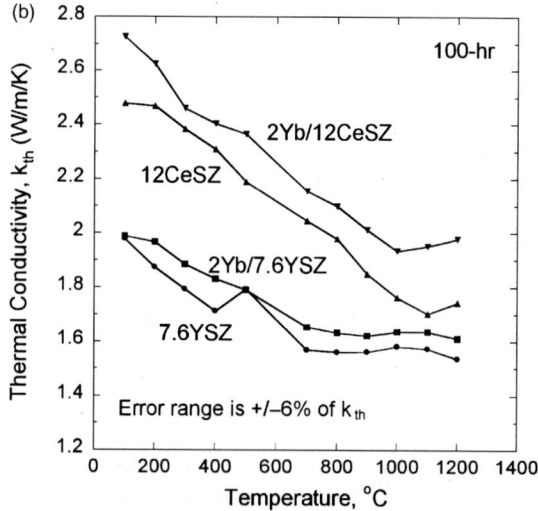

Fig. 7. The thermal conductivity of baseline and Yb^{3+} co-doped 7.6 mol% $YO_{1.5}$–ZrO_2 (7.6YSZ) and 12 mol% CeO_2–ZrO_2 (12CeSZ) coatings after (a) 10 h and (b) 100 h heat treatments at 1200°C.

proximately a 5 mol% oxygen vacancy concentration. They argued that the increase in k_{th} beyond 5 mol% oxygen vacancies was because of the local ordering of the point defects. Recent research on defect structure of cubic YSZ supported this hypothesis.[28]

The calculated oxygen vacancy concentrations for the baseline and co-doped coatings are listed in Table V, along with the measured open porosity values. We observed that the as-sprayed 7.6YSZ and 2Yb/7.6YSZ fit the above-mentioned trend, i.e. they demonstrated a decrease in thermal conductivity with increased oxygen vacancy concentration. It should also be noted that the decrease in thermal conductivity observed in Fig. 5 could also be attributed to the higher porosity observed in the 2Yb/7.6YSZ coating. A previous study by Trice et al.[29] showed that a 2% increase in porosity resulted in an ~ 0.1 J/m/K decrease in thermal conductivity. However, while the Ca^{2+}-doped samples have a larger concentration of oxygen vacancies and greater open porosity, they exhibited greater thermal conductivity as compared with the 7.6YSZ coating. Therefore, some factors other than oxygen vacancy concentration or porosity should account for this behavior. One of these factors could be point defect clustering caused by the coherency effect of different dopant species. Some literature suggested such a link between thermal conductivity and co-doping, but the fundamental aspect is not yet clear.[30]

Figure 7(a) and (b) shows a comparison of the thermal conductivity of 7.6YSZ and 12CeSZ baseline and Yb^{3+} co-doped coatings after 10 and 100 h heat treatments at 1200°C. It was found that unlike the as-sprayed coatings, the heat-treated Yb^{3+} co-doped coatings show no advantage over either baseline 7.6YSZ or 12CeSZ coatings. In general, the CeSZ-based coatings have higher thermal conductivity than YSZ-based coatings, probably because the latter contains more oxygen vacancies that tend to scatter the lattice waves.

IV. Conclusions

Co-dopants of either Yb^{3+} or Ca^{2+} were successfully incorporated in 7.6YSZ and 12CeSZ coatings by infiltrating porous spray-dried powders with salt solutions containing the appropriate co-dopant species prior to plasma spraying. No Yb^{3+} segregation was detected at the grain boundary of either as-sprayed 2Yb/7.6YSZ or 2Yb/12CeSZ coatings. Based on the dilatometer measurements, Yb^{3+} co-doped coatings did not demonstrate any difference in shrinkage behavior as compared with baseline 7.6YSZ and 12CeSZ coatings.

XRD results show that the as-sprayed 7.6YSZ, 2Ca/7.6YSZ, and 2Yb/7.6YSZ coatings comprised of non-transformable, non-equilibrium composition t'-ZrO_2. Both 7.6YSZ and 2Ca/7.6YSZ partitioned into c-ZrO_2 and t-ZrO_2 after a 100-h heat treatment at 1200°C; however, no m-ZrO_2 was observed. The 2Yb/7.6YSZ coating retained a single t'-ZrO_2 phase after a 100 h heat treatment. Thus, it can be concluded that 2 mol% Yb^{3+} co-dopant increases the stability of the t'-ZrO_2 phase as compared with the baseline coating made from 7.6YSZ. The as-sprayed 5Ca/7.6YSZ was a single c'-ZrO_2 phase, but began to partition after a 10 h heat treatment at 1200°C. The volume fraction of m-ZrO_2 in 12CeSZ was estimated to be 88% after a 100 h heat treatment. Yb^{3+} and Ca^{2+} co-doping limited the $t \rightarrow m$ phase transformation in 12CeSZ, with only 37% and 43% monoclinic phase observed, respectively, after a 100 h heat treatment at 1200°C.

The thermal conductivity of plasma-sprayed 7.6YSZ is lower than that of 12CeSZ, probably because the former contains more oxygen vacancies that tend to scatter the lattice waves. As-sprayed 2Yb/7.6YSZ and 2Yb/12CeSZ coatings had lower thermal conductivity than their undoped counterparts. The decrease in thermal conductivity of Yb^{3+} co-doped coatings is attributed to both an increase in oxygen vacancies or more porous structure. However, after heat treating for 10 and 100 h at 1200°C, Yb^{3+} co-doped, as-sprayed 2Ca/7.6YSZ, 5Ca/7.6YSZ, and 2Ca/

12CeSZ showed little or no effect in decreasing the thermal conductivity from the baseline coatings.

References

[1]R. A. Miller, "Current Status of Thermal Barrier Coatings—An Overview," Surf. Coat. Technol., 30, 1–11 (1987).

[2]S. Sodeoka, M. Suzuki, S. K. Ueno, H. Sakuramoto, T. Shibata, and M. Ando, "Thermal and Mechanical Properties of ZrO_2-CeO_2 Plasma-Spray Coatings," J. Therm. Spray Technol., 6 [3] 361–7 (1997).

[3]J. Wilden, M. Wank, H. D. Steffens, and M. Brune, "New Thermal Barrier Coating System for High Temperature Applications," Proc. Int. Thermal Conf., 2, 1669–73 (1998).

[4]R. Hillery, Coatings for High Temperature Structural Materials. National Academy Press, Washington, DC, 1996.

[5]A. Rabiei and A. Evans, "Failure Mechanisms Associated with the Thermally Grown Oxide in Plasma-Sprayed Thermal Barrier Coatings," Acta Mater., 48 [15] 3963–76 (2000).

[6]A. Freborg, B. Ferguson, W. Brindley, and G. Petrus, "Modeling Oxidation Induced Stresses in Thermal Barrier Coatings," Mater. Sci. Eng. A, A245 [2] 182–90 (1998).

[7]K. Muraleedharan, J. Subrahmanyam, and S. B. Bhaduri, "Identification of t' Phase in ZrO_2-7.5 wt.% Y_2O_3 Thermal Barrier Coatings," J. Am. Ceram. Soc., 71 [5] C226–7 (1998).

[8]H. G. Scott, "Phase Relationships in the $ZrO_2-Y_2O_3$ Systems," J. Mater. Sci., 10 [9] 1527–35 (1975).

[9]N. R. Rebollo, O. Fabrichnaya, and C. G. Levi, "Phase Stability of Y+Gd Co-Doped Zirconia," Z. Metallkd., 94 [3] 163–70 (2003).

[10]J. Ilavsky and J. K. Stalick, "Phase Composition and its Changes During Annealing of Plasma-Sprayed YSZ," Surf. Coat. Technol., 127, 120–9 (2000).

[11]J. Ilavsky, J. K. Stalick, and J. Wallace, "Thermal-Spray Yttria-Stabilized Zirconia Phase Changes During Annealing"; pp. 1185–9 in Proceedings of International Thermal Spray Conference, Montreal, 2000.

[12]E. Y. Lee, Y. Sohn, S. K. Jha, J. W. Holmes, and R. D. Sisson Jr., "Phase Transformation of Plasma-Sprayed Zirconia-Ceria Thermal Barrier Coatings," J. Am. Ceram. Soc., 85 [8] 2065–71 (2002).

[13]N. R. Rebollo, A. S. Gandhi, and C. G. Levi, "Phase Stability Issues in Emerging TBC Systems"; pp. 431–42 in High Temperature Corrosion and Materials Chemistry IV. Electrochemical Society Proceedings, Vol. PV-2003-16, Edited by E. Opila, P. Hou, T. Maruyama, B. Pieraggi, M. McNallan, D. Shifler, and E. Wuchina. Electro-Chemical Society, Paris, 2003.

[14]Y. M. Chiang, D. Birnie III, and W. D. Kingery, Physical Ceramics: Principles for Ceramic Science and Engineering. Wiley Publishers, New York, 1997.

[15]ASTM C373-88, Standard Test Method for Water Absorption, Bulk Density, Apparent Porosity and Apparent Specific Gravity of Fired Whiteware Products (Reapproved 1999). ASTM, Philadelphia, 1999.

[16]P. A. Langjahr, R. Oberacker, and M. J. Hoffman, "Long-Term Behavior and Application Limits of Plasma-Sprayed Zirconia Thermal Barrier Coatings," J. Am. Ceram. Soc., 84 [6] 1301–8 (2001).

[17]H. Wang, R. B. Dinwiddie, and P. S. Gaal, "Multiple Station Thermal Diffusivity Instrument"; pp. 119–27 in Thermal Conductivity, Vol. 23, Edited by K. E. Wilkes, R. B. Dinwiddie, and R. S. Graves. Technomic Publishing, Basel, 1996.

[18]L. M. Clark III and R. E. Taylor, "Radiation Loss in the Flash Method for Thermal Diffusivity," J. Appl. Phys., 46 [2] 714–9 (1975).

[19]R. McPherson, "A Review of Microstructure and Properties of Plasma-Sprayed Ceramic Coatings," Surf. Coat. Technol., 39/40, 173–81 (1989).

[20]R. W. Trice, Y. J. Su, J. R. Mawdsley, K. T. Faber, A. R. De Arellano-López, H. Wang, and W. D. Porter, "Effect of Heat Treatment on Phase Stability, Microstructure, and Thermal Conductivity of Plasma-Sprayed YSZ," J. Mater. Sci., 37, 2359–65 (2002).

[21]J. A. Hines, Y. Ikuhara, A. H. Chokshi, and T. Sakuma, "The Influence of Trace Impurities on the Mechanical Characteristics of a Superplastic 2 mol% Yttria Stabilized Zirconia," Acta Mater., 46 [15] 5557–68 (1998).

[22]S. Hwang and I. Chen, "Grain Size Control of Tetragonal Zirconia Polycrystals Using the Space Charge Concept," J. Am. Ceram. Soc., 73 [11] 3269–77 (1990).

[23]R. McPherson and B. Shafer, "Interlamellar Contact in Plasma-Sprayed Coatings," Thin Solid Films, 97, 201–4 (1982).

[24]R. Stevens, Zirconia and Zirconia Ceramics. Magnesium Elektron Ltd, Manchester, 1986.

[25]E. R. Andrievskaya, I. E. Kir'yakova, and L. M. Lopato, "Phase-Equilibria in the Systems $HfO_2-Y_2O_3$-CaO and $ZrO_2-Y_2O_3$-CaO at 1250°C," Inorg. Mater., 27 [10] 1839–44 (1991).

[26]G. S. Corman and V. S. Stubican, "Phase-Equilibria and Ionic-Conductivity in the System $ZrO_2-Yb_2O_3-Y_2O_3$," J. Am. Ceram. Soc., 68 [4] 174–81 (1985).

[27]J. Bisson, D. Fournier, M. Poulain, O. Lavigne, and R. Mevrel, "Thermal Conductivity of Yttria–Zirconia Single Crystals, Determined with Spatially Resolved Infrared Thermography," J. Am. Ceram. Soc., 83 [8] 1993–8 (2000).

[28]J. P. Goff, W. Hayes, S. Hull, M. T. Hutchings, and K. N. Clausen, "Defect Structure of Yttria-Stabilized Zirconia and its Influence on the Ionic Conductivity at Elevated Temperatures," Phys. Rev. B, 59 [22] 14202–19 (1999).

[29]R. W. Trice, Y. J. Su, K. T. Faber, H. Wang, and W. Porter, "The Role of NZP Addition in Plasma-Sprayed YSZ: Microstructure, Thermal Conductivity and Phase Stability Effects," Mater. Sci. Eng. A, A272 [2] 284–91 (1999).

[30]D. Zhu and R. A. Miller, "Thermal Conductivity and Sintering Behavior of Advanced Thermal Barrier Coatings," Ceram. Eng. Sci. Proc., 23 [4] 457–68 (2002).

Ta_2O_5/Nb_2O_5 and Y_2O_3 Co-doped Zirconias for Thermal Barrier Coatings

Srinivasan Raghavan,[†] Hsin Wang,[*,‡] Ralph B. Dinwiddie,[‡] Wallace D. Porter,[‡] Robert Vaßen,[§] Detlev Stöver,[§] and Merrilea J. Mayo[†,¶]

Department of Materials Science and Engineering, The Pennsylvania State University, University Park, Pennsylvania 16802

High Temperature Materials Laboratory, Oak Ridge National Laboratory, Oak Ridge, Tennessee 37831

Institute for Materials and Processing in Energy Systems, Forschungszentrum Jülich GmbH, 52425 Jülich, Germany

Zirconia doped with 3.2–4.2 mol% (6–8 wt%) yttria (3–4YSZ) is currently the material of choice for thermal barrier coating topcoats. The present study examines the ZrO_2–Y_2O_3–Ta_2O_5/Nb_2O_5 systems for potential alternative chemistries that would overcome the limitations of the 3–4YSZ. A rationale for choosing specific compositions based on the effect of defect chemistry on the thermal conductivity and phase stability in zirconia-based systems is presented. The results show that it is possible to produce stable (for up to 200 h at 1000°–1500°C), single (tetragonal) or dual (tetragonal + cubic) phase chemistries that have thermal conductivity that is as low (1.8–2.8W/m K) as the 3–4YSZ, a wide range of elastic moduli (150–232 GPa), and a similar mean coefficient of thermal expansion at 1000°C. The chemistries can be plasma sprayed without change in composition or deleterious effects to phase stability. Preliminary burner rig testing results on one of the compositions are also presented.

I. Introduction

ZIRCONIA stabilized with 3.2–4.2 mol% (6–8 wt%) yttria (3–4YSZ) is the current material of choice for thermal barrier coating (TBC) topcoats.[1] It is, however, limited to temperatures <1200°C,[2] is susceptible to hot corrosion,[3] and is still not considered reliable enough to increase the design temperature of the turbine.

Improving on current TBC topcoats requires changing its microstructure,[1,4] chemistry, or both (as in layered systems[5]). Alternative chemistries,[6–17] the route with which this paper is concerned, have been evaluated for topcoats with limited success. The research reported in this paper involved zirconia-rich chemistries (Table I) from the ZrO_2–Y_2O_3–Ta_2O_5 and ZrO_2–Y_2O_3–Nb_2O_5 systems. Previously, the thermal conductivities and hot corrosion behavior of these ternary chemistries have been discussed in detail.[18,19] This paper, apart from dealing in depth with

phase stability and plasma spray processing, is a summary—from material design to preliminary burner rig evaluation—of the properties of these materials from a perspective of TBC application.

II. Rationale

The rationale behind choosing the chemistries listed in Table I was based on the possible effect of their point defect chemistries, given by Eq. (1), on their thermal conductivity and phase stability.[20–24] It is based on the assumption that all the elements would be present in a single phase.

$$xZ_2O_5 + yY_2O_3 = 2xZ'_{Zr} + 2yY'_{Zr} + (5x + 3y)O_O + (y - x)V_O^{\cdot\cdot} \quad (1)$$

The chemistries in Table I can be divided into two types, as outlined in the following sections.

(1) Chemistries Containing Excess Yttria, y > x in Eq. (1) (Except N5 and T13)

To a first approximation, a tetragonal phase with a stability similar to the conventional 3.2–4.2 mol% or 6–8 wt% yttria-stabilized zirconia can be obtained by keeping the population density of oxygen vacancies equal to 3.2–4.2 mol%; i.e., setting $(y - x)$ in Eq. (2) to 3.2–4.2.[25,26] Then, by increasing both y and x simultaneously, the substitutional cationic defect concentration can be increased to provide additional (to the oxygen vacancies) defects for a further reduction in thermal conductivity without changing the phase stability. Thus, the excess yttria chemistries were designed to mimic the 3–4YSZ in phase stability but with lower thermal conductivity. The slight difference in excess yttria (in design), e.g., 3.6 vs. 3.2, is necessary to keep the tetragonality (c/a) ratio and hence the stability of the tetragonal phase in the co-doped samples the same as in the 6–8YSZ.[26]

Again, the above arguments are valid only if all the elements are present in a single phase. Available phase diagrams[25] and previous studies[18] on pressed and sintered samples show that this is not possible. Thus, the current paper explores the compositional limits that might be suitable for TBC applications despite phase splitting, the effects of aging, and the possibility of retaining all the elements in a single unstable or metastable phase by plasma spraying, as is done in 3–4YSZ.

(2) Chemistries with y = x (N5 and T13)

These $YTaO_4$- and $YNbO_4$-stabilized zirconias[22] do not have any extrinsic oxygen vacancies, and the only defects introduced by doping are substitutional cationic defects. Results in literature[25,27] and our previous studies indicated that as much as 20 mol% of these compounds can be dissolved in a single tetragonal phase, resulting in thermal conductivities as low as the 3–4YSZ.[18]

N. P. Padture—contributing editor

Manuscript No. 187195. Received January 22, 2002; approved August 27, 2003.
Presented in part at the 102nd Annual Meeting of The American Ceramic Society, St. Louis, Missouri, May 3, 2000 (Basic Science and Engineering Ceramics Divisions, Paper No. B3–019-00).
Based in part on the thesis submitted by Srinivasan Raghavan for Ph.D. degree in Materials Science and Engineering, Pennsylvania State University.
Supported by the U. S. Department of Energy under contract number DE-F602-98ER45700.
*Member, American Ceramic Society.
†Department of Materials Science and Engineering.
‡High Temperature Materials Laboratory.
§Institute for Materials and Processing in Energy Systems.
¶Currently Director, GUIRR, National Research Council, Washington, DC 20418.

Table I. Phase Compositions of Samples

ID	As-designed composition	ICP measured composition	1400/4/C	1400/4/T	1500/5/M	1500/5/C	1500/5/T
N2	4.6Y1.1NZ	4.3Y1.1NZ	25.8	74.2	10.0	30.3	59.8
N3	6Y1.5NZ	5.9Y1.6NZ	34.7	65.3	7.5	38.8	53.7
N4	9Y4.4NZ	8.4Y4.3NZ	37.6	62.4	0.0	37.1	62.9
N5	10Y10NZ	9.6Y9.6NZ^A 9.6Y9.4NZ^B	0.0	100.0	0.0	0.0	100.0
N6	12Y7.5NZ	11Y7.2NZ	35.6	64.4	0.0	42.0	58.0
T4	4.8Y1.3TZ	4.3Y1.1TZ	21.4	78.6	32.4	23.6	44.0
T5	6.1Y1.5TZ	5.7Y1.3TZ	37.5	62.5	31.4	68.6	0.0
T8	6.1Y2.5TZ	5.7Y2.2TZ	25.1	74.9	42.3	57.7	0.0
T9	9Y4.4TZ	8.5Y4.0TZ	35.2	64.8	0.0	43.3	56.7
T10	9Y5.4TZ	9.2Y5.1TZ	30.1	69.9	0.0	32.7	67.3
T13	10Y10TZ	9.9Y8.5TZ^A 9.9Y9.8TZ^B	0.0	100	0.0	0.0	100
T15	12Y7.5TZ	11.6Y6.5TZ	37.5	62.5	0.0	37.2	62.8

Phase compositions (wt%) of aged (1500°C, 200 h) samples

ID	Monoclinic	Tetragonal	Cubic
T9	35	65	
T10		59	41
T13		100	
N4		56	44
N5		100	

Column headings give sintering schedule and wt% of phase present as temperature (°C)/time(hours)/phase(C-cubic, T-Tetragonal, M-Monoclinic) present. Superscripts A and B in column 2 refer to results from two different labs.[19] In columns 2 and 3, numbers refer to mol% yttria (Y), niobia (N), tantala (T) in zirconia (Z).

Available ternary diagrams of the two systems indicate that the tetragonal phases are equilibrium phases at 1500°C.[25,27] If this were true at the lower temperatures of TBC operation as well, they would be immune to phase-splitting problems (like cubic 8YSZ)[1,5], in contrast to the 3–4YSZ, which is not. Such immunity, however, would be useless, as in 8YSZ, if cyclic life were to be poor. Furthermore, the 20 mol% $YTaO_4$-stabilized zirconia (T13 in Table I) is known to have a Young's modulus of 150 GPa as compared with the 200 GPa modulus of 3–4YSZ.[25] The lower modulus and the possibility of a higher coefficient of thermal expansion (CTE)[28] were expected to reduce stresses owing to thermal expansion mismatches and, therefore, be beneficial to TBC lifetimes as well.

III. Experimental Details

This section deals only with the details not described in Ref. 18.

(1) Spray-Dried Powders for Plasma Spraying

Powders synthesized by the co-precipitation technique[18] were heat-treated at 1200°C for 2 h to reduce their specific surface area. A slurry containing 80% of the heat-treated powders, and 20% of ethanol + polyethyleneimine (PEI) solution (88% ethanol +12% PEI by weight) by weight was prepared and homogenized with an ultrasonic horn. This slurry was then spray dried at 55 mL/min in a closed-circuit spray dryer (Mobile Minor™, Model H, Niro A/S, Soaborg, Denmark) with nitrogen at 1 bar pressure and 175°C temperature as the drying medium. The spray-dried powders were sieved to obtain the −75 to +36 μm fraction. Then the powders were heated at 1°C/min to 500°C followed by cooling down at the same rate to drive off the binder and yield granules for plasma spraying.

(2) Sample Preparation

Air plasma spraying was done using 20 slpm (standard liters per minute) argon and 13 slpm helium as plasma gases, 1.5 slpm nitrogen as carrier gas, 110 mm spray distance and 21 KW plasma power at 300 Amp. For thermal diffusivity samples, an 800-μm-thick coating, plasma sprayed onto a grit blasted mild steel substrate, was stripped from it by soaking in 32% HCl solution at room temperature for ~4 h. Following ultrasonic cleaning in distilled water and ethanol, disks, 600 μm in thickness and 12 mm in diameter, were machined (core drilled and ground) from the stripped coatings. For thermal expansion measurements parallelepiped-shaped samples 25 mm in length, were prepared from both the spray-dried powders (by pressing with 2% PEI binder and sintering to 80% at 1500°C) and plasma-sprayed coating (by abrasive wheel sectioning and acid stripping). For X-ray analysis pellets (98% dense) were made by green pressing and sintering in air. Pellets were heated at 2°C/min, held at temperature (1000°–1500°C) and then cooled at 10°C/min. Furnace temperatures were measured using two S-type thermocouples. For modulus measurements and aging studies, disks, 6 mm in diameter and 1 mm in thickness, were fabricated by pressing and sintering the co-precipitated powders. These were aged at 1500°C for 200 h to 99%+ densities. The aged samples were polished down to a 0.25 μm diamond finish to obtain flat and parallel surfaces for modulus measurements. For thermal cycling disk-shaped substrates, 30 mm in diameter and 3 mm thick with a beveled edge,[29] were machined out of IN738, a nickel-based superalloy. The substrates were coated with a 150-μm-thick layer of Ni-21Co-17Cr-12Al-0.6Y (Ni-192–8, Praxair, Seattle, WA) bond coat by vacuum plasma spraying (VPS). The bond-coated substrates were subjected to a diffusion bonding heat treatment (−25°–20°C/min–1120°C–120 min; 20°C/min–25°C– 20°C/min– 845°C–24 h– 20°C/min–25°C) in argon at 1100 mbar pressure. A 300 μm layer of T13 was then applied as the topcoat by APS.

(3) Testing and Characterization

Elastic modulus was calculated using standard equations[30] from sample densities, longitudinal and shear wave velocities. Sound-wave velocities were measured using a 100 MHz ultrasonic transducer (Ultran Labs, Boalsburg, PA). CTE measurements were done in a dual push rod differential dilatometer (Theta Instruments, Port Washington, NY) using sapphire as a reference. Thermal cycling was performed in a gas burner facility as explained in Ref. 16. Surface temperature-bondcoat temperature combinations used were 1250°–980°C and 1340°–1010°C, respectively. X-ray diffraction (Model 85–1852 05, Scintag, Cupertino, CA) for phase analysis and composition calculations using relations from literature[21,31] are described in detail in Ref. 18. To calculate the compositions of the individual phases, it was assumed that the cubic phase does not contain any pentavalent dopant. Data in the literature[25] shows that the cubic phase can contain Ta_2O_5.

However, as will be seen later, calculations done using this approximation explain the trends observed adequately.

IV. Results and Discussions

(1) Plasma Spraying

As an experiment, two of the chemistries, T10 and T13, and 3.8YSZ were deposited as coatings by plasma spraying. No loss of tantalum oxide during plasma spraying could be detected by inductively coupled plasma atomic emission spectrometry (ICPAES) chemical analysis of the T10 and T13 coatings. Given the similarities between Nb_2O_5 and Ta_2O_5, we do not expect any problems with plasma spraying the niobia-doped compositions, though it has not been attempted in the present work. The relative densities of all three coatings were determined by the Archimedes technique to be ~88%, and their microstructures exhibited the typical splat morphology of plasma-sprayed coatings. To a first approximation, the similarity in microstructure—given the similar bulk densities and thermal conductivities in the dense state[18]—is best characterized by the similarity of their thermal conductivity-temperature curves plotted in Fig. 1.

(2) Phase Stability

(A) Pressed and Sintered Samples: For excess yttria compositions, N2-N4, N6, T4-T10, and T15, it is obvious from data in Table I and Figs. 2(a) and (b) that, contrary to initial expectations, the samples cannot be kinetically limited to a single (unstable or metastable) phase on sintering even for periods as short as 4–5 h at 1400°–1500°C. Samples sintered at 1400°C, other than N5 and T13, are a mixture of the tetragonal and cubic phases, whereas some of those sintered at 1500°C also contain the monoclinic phase. The diffraction patterns of sample T8 in Fig. 2(b) serves as an example. Perusing Table I, it can be seen that the

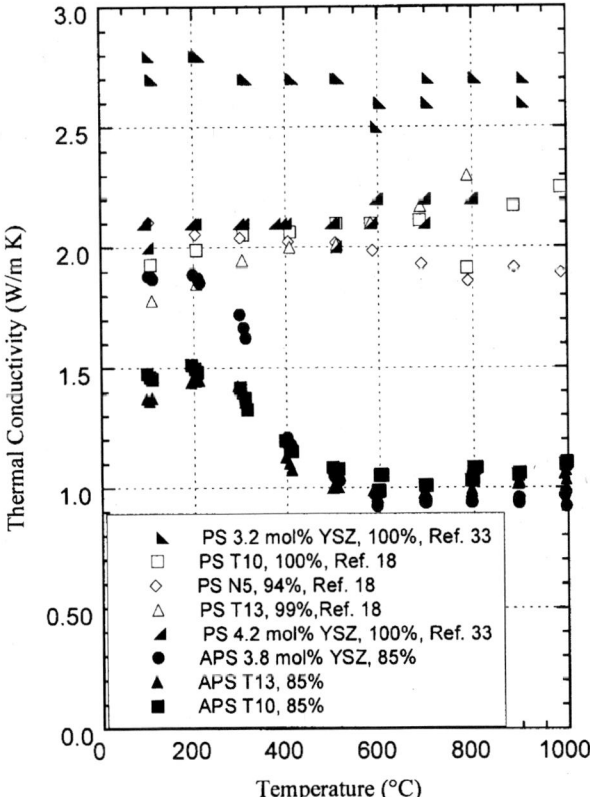

Fig. 1. Thermal conductivities of various pressed and sintered (PS) and air plasma sprayed (APS) samples. Bulk relative densities as percentage of theoretical have been indicated.

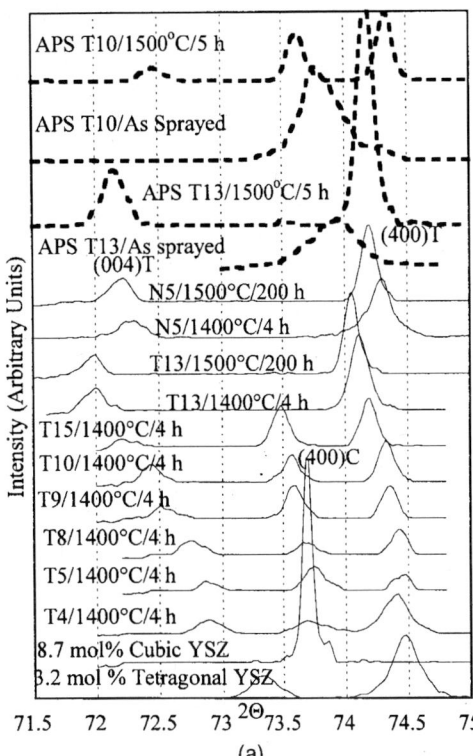

(a)

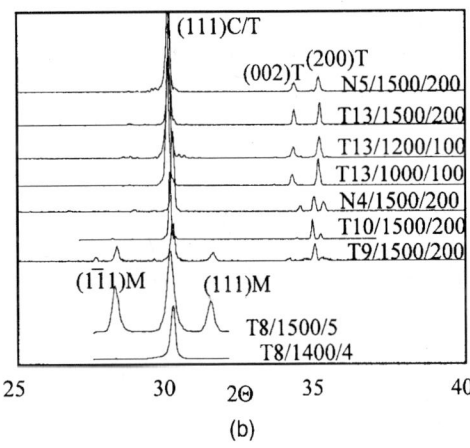

(b)

Fig. 2. (a) Representative X-ray patterns of the various pressed and sintered and APS samples. Numbers represent heat treatment. (T = Tetragonal, C = Cubic). (b) Representative X-ray patterns of the various compositions. Pattern captions five ID/temperature (°C)/time (hours). In peak captions, M = Monoclinic, T = Tertagonal, C = Cubic.

monoclinic phase is present only in samples containing low dopant concentrations.

Calculated chemical compositions of the phases (Table II) indicate that the tetragonal phase in the samples contains the two dopants in a ratio close to 1:1, just as in the intentional no-vacancy compositions, N5 and T13. This is in agreement with the available ternary section.[27] The excess Y_2O_3 along with some Ta_2O_5 is then sequestered in the cubic phase. Difficulties in deconvoluting the peaks in the (400) region for the 1500°C monoclinic-containing samples prevented calculations of phase chemical compositions for those samples.

The c/a ratios of the tetragonal phase in the 1400°C samples, as calculated from the 2θ positions of the (400) peaks (without any assumptions) are listed in Table II. The rise in c/a ratios (compare

Table II. Chemical Composition (mol%) of the Phases and c/a Ratio of the 1400°C Tetragonal Phase

ID	Y_2O_3 in C phase		Y_2O_3 in T phase		Ta_2O_5/Nb_2O_5 in T phase		c/a ratio
	1400°C/4 h	1500°C/5h	1400°C/4h	1500°C/5h	1400°C/4h	1500°C/5h	
N2	7.02		3.33		1.43		1.0179
N3	7.92		4.82		2.38		1.0182
N4	10.37	9.85	7.17	7.50	6.81	6.76	1.0224
N5			9.58	9.58	9.64	9.64	1.0238
N6	11.37	11.08	10.77	10.92	11.15	12.37	1.0226
T4	7.82		3.38		1.40		1.0178
T5	7.43		4.72		2.14		1.0182
T8	8.60		4.68		2.97		1.0197
T9	10.18	9.49	7.52	7.67	6.17	7.05	1.0214
T10	10.56	10.50	8.62	8.58	7.24	7.51	1.0221
T13			9.93	9.93	8.52	8.52	1.0252
T15	12.10	11.31	11.25	11.72	10.41	10.34	1.0244

Chemical compositions (mol%) of phases after 1500°C/200h

	Y_2O_3 in T phase	Ta_2O_5/Nb_2O_5 in T phase
N4	7.0	7.9
T10	8.0	9.3

separation of (004) and (400) peaks) of the tetragonal phases is consistent with the enrichment of the tetragonal phase with the compounds $YTaO_4$ and $YNbO_4$ and subsequent stabilization of the tetragonal phase owing to ordering of the dopant cations within the zirconium sublattice[22] and thus confirms the analysis.

To determine the effect of aging, from both grain coarsening and compositional changes, the compositions N4 and T9, the two lowest dopant level compositions that did not contain the monoclinic phase after the 1500°C/5 h sintering schedule, were aged at 1500°C for 200 h. The monoclinic phase was detected by X-ray diffraction (Fig. 2(b)) in the aged sample T9 but not in N4. Also, the chemical compositions (Table II) of the phases in N4 aged for 200 h remain approximately the same as those determined from the samples sintered at 1500°C for 5 h. However, no monoclinic phase was detected in sample T10, with the next higher level of dopant concentration, following aging. Despite the transformation having occurred in T9, large-scale cracking of samples was not observed. Thus, when the chemistries are such that the tetragonal phase following phase splitting contains at least 12–14 mol% $YTaO_4$ or $YNbO_4$, they are much more stable (at least for 200 h at 1500°C) than the 3–4YSZ, which undergoes destabilization on cooling following phase splitting at temperatures >1200°C.[2]

Because the chemical composition calculations involve assumptions, the c/a ratio can be used as a criterion to decide between stable and unstable compositions. From Table II, it can be noted that a room temperature c/a ratio of ~1.023 (12–14 mol% $YTaO_4$) separates the stable and unstable compositions in the 1400°C sintered samples. As will be discussed in the next section, this is similar to the c/a ratio of pure tetragonal, 1.0234, at its transformation temperature of 1150°C.[22]

Finally, in a manner very similar to the 3–4YSZ system, there appear to be compositional boundaries in the current system below which the tetragonal phase is unstable ($YTaO_4$ <12 mol%) or above which it is totally stable ($YTaO_4$ >16 mol%) against the martensitic transformation. Thus, analogous to the 3–4YSZ determination of the optimum composition for best TBC performance within this range, requires a systematic variation and evaluation of compositions as was done for the 3–4YSZ.[4]

For the compositions without oxygen vacancies, the XRD patterns of T13 and N5 aged for up to 200 h from 1000°–1500°C are shown in Figs. 2(a) and (b), respectively. Samples were found to be made of a single tetragonal phase on cooling to room temperature. This tetragonal phase is stabilized by the association of substitutional defects in the cationic lattice.[23] Whereas the aging result for composition T13 is in agreement with the literature,[25] that for composition N5 is not. Composition N5 is supposed to be a two-phase mixture of a tetragonal phase and $YNbO_4$.[32]

The most striking feature of the X-ray patterns of compositions T13 and N5 is the large separation in their (400) peaks as compared with that of the 3.2 mol% YSZ, seen from Fig. 2(a). The c/a ratio corresponding to this separation of ~1.025 at room temperature is similar to literature data[25] and similar to the c/a ratio of pure tetragonal zirconia of 1.0234 above 1150°C.[21] Because the coefficient of thermal expansion along the c-axis is greater than that along the a-axis[21] for pure zirconia, it is expected that the c/a ratio of pure zirconia at room temperature would be <1.025, i.e., lower than the c/a ratio in T13 and N5 compositions at room temperature. The larger c/a ratio of the T13 and N5 composition is the result of defect association in the cationic lattice between the large Y^{+3} and the smaller Ta^{5+} or Nb^{5+} ion. This association and the larger c/a ratio thereof is what is responsible for stabilizing the tetragonal phase in compositions N5 and T13 down to temperatures <1150°C, at which pure zirconia transforms.

Figure 2 shows that compositions N5 and T13 can be used for extended periods of time above 1200°C, unlike the 3–4YSZ as far as phase stability is concerned. It is emphasized again that the improved phase stability would not be of any use if the cyclic lives of these compositions were as poor as that of the totally stable 8YSZs. For these types of chemistries, data in the literature[25,27] show that only compositions >16 mol% $YTaO_4$ and 10 mol% $YNbO_4$ can be used for TBCs, as those containing lower amounts of the dopants destabilize into a monoclinic phase above room temperature. One particular chemistry that might be of interest is zirconia doped with 16 mol% $YTaO_4$. This chemistry is known to have phase stability as good as the 20 mol% $YTaO_4$-doped zirconia (T13),[25] is expected to have a thermal conductivity of 2 W/m·K,[18] and in addition has a fracture toughness of 4 MPa√m,[25] which is twice that of the 3.2–4.2 mol% yttria-stabilized zirconia in the as-sprayed form.

(B) Phase Stability of Plasma-Sprayed Coatings: Representative XRD patterns of the APS samples discussed below are shown in Fig. 2(a). The T10 composition in the as-sprayed state is either a single cubic phase or a mixture of a cubic phase and a tetragonal phase with a c/a ratio close to 1. However, within ~10 h of aging at 1200°C and 5 h at 1500°C, the as-sprayed T10 reverts to a two-phase mixture as in the pressed and sintered samples examined in prior work.[18] This result implies that any of the advantages of the single phase in the as-sprayed state cannot be exploited for TBC applications.

The as-sprayed T13 coating is made up of a single phase that seems almost cubic but is most probably a tetragonal phase with a very small c/a ratio, as can be seen from the asymmetry of the peak in Fig. 2(a). Following aging of the APS samples for 5 h at 1500°C, the peaks are better resolved. The diffraction pattern of aged coatings thus resembles that of the same material in the pressed and sintered coating, which exhibits a c/a ratio large

Table III. Results of Thermal Cycling of 20YTaO₄SZ Coatings

ID	$T_{Surface}$	$T_{Substrate}$	$T_{Bondcoat}$[†]	Number cycles[‡]	d_{TGO} (μm)	$d_{depleted\ zone}$ (μm)
58	1236	969	1013	559/?/717	4.2/4.4/3.4	16.6/17.6/11.8
55	1254	986	1030	616/650/710	6.4/5.4/5.4	17.8/18.6/14.2
57	1329	1009	1062	427/?/600	5.2/3.6/<1	18.4/8.8/<5
56	1350	1050	1099	4/13/300	4/4/<1	7/11/<5

[†]The bondcoat temperature was calculated from surface and substrate temperature assuming a thermal conductivity of 1 W/m/K for the TBC. [‡]The three numbers indicate the number of cycles at the last inspection of the intact sample, the number of cycles at failure (not always given), and the total number of cycles. In addition, the thickness of the thermally grown oxide (TGO) and of the depleted zone is given at the left, the middle, and the right location of the sectioned and polished sample.

enough to give two well separated (400) peaks. This is because the substitutional defects in the cationic lattice are quenched in at random in the as-sprayed state. On aging, they rearrange, resulting in defect association and larger c/a ratios as discussed earlier.

(3) Thermal Conductivity

Thermal conductivities of the APS samples and PS samples of two representative compositions T10 ($y > x$ category) and T13 ($y = x$ category) that were plasma sprayed are plotted in Fig. 1 and compared with 3–4YSZ. It is observed that all the plasma-sprayed samples whose bulk porosities were similar, have similarly low thermal conductivities as that of the 3–4YSZ.[33]

(A) T10, an Excess Yttria Chemistry: If in the as-sprayed state the defects are indeed quenched in randomly, in a single phase, in T10, then this results implies that even the random presence of heavy substitutional tantalum ions in the zirconium lattice in addition to the vacancies in the oxygen lattice does not contribute to a further lowering of thermal conductivity below 2 W/m·K. The latter scenario would mean that the phonon mean free paths, as a result of defect scattering, have reached their minimum possible values and a thermal conductivity of ~2 W/m·K is, therefore, the minimum attainable in stabilized zirconia systems, by point defect scattering. However, the presence of porosity of complicated morphology prevents a totally valid comparison. A strict evaluation of point defect scattering would require processing a dense sintered sample containing all the elements in a single phase, which is not possible in these systems due to the phase splitting.

(B) T13, a Chemistry without Extrinsic Oxygen Vacancies: As discussed in detail elsewhere,[18] the most interesting feature is that the data from the pressed and sintered samples indicate that niobia seems to have as much influence on reducing thermal conductivity as tantala. Consideration of mass difference effects alone, as has been done to model the thermal conductivity of doped zirconias,[20] would predict a much lower thermal conductivity for T13 than N5, as well as a different trend in temperature. Defect association between the larger Y^{+3} ion and the smaller Ta^{5+}/Nb^{5+} ions, which is driven by the need to reduce lattice strain energy, would nullify the effect of ionic size difference. Thus, the similarly low thermal conductivities in both the niobia- and tantala-doped cases, despite the huge difference in the cationic masses of these oxides, is attributed to the overriding effect of a change in the force constants of the bonds in these co-doped oxides. That this could be so can be seen from the fact that the measured room temperature moduli change on doping zirconia with as much as 10 mol% Y_2O_3 or 14 mol% Ta_2O_5 is small; 220 and 217 GPa, respectively, compared with the 232 GPa modulus of pure zirconia. In contrast, the moduli of the co-doped compositions T13 and N5 are 146 and 162 GPa, respectively.

A final observation concerning T13 is that it represents the maximum amount of $YTaO_4$ that can be dissolved in a single phase.[25] The conductivity of T13 and N5 is thus expected to be the minimum conductivity possible in the pseudobinary system, ZrO_2–Y(Ta/Nb)O_4, just as the near solvus compositions in the MgO–NiO system represent the minimum thermal conductivity for that system.[34]

(4) Burner Rig Testing

T13 was considered to have the most promising combination of properties among all the samples examined in the present study

because of its low thermal conductivity (2 W/m·K at 1000°C), excellent phase stability (1500°C/200 h), low elastic modulus (150 GPa for T10 and 210 GPa for 3–4YSZ) and mean CTE (1.05×10^{-5} at 1000°C vs. 1.06×10^{-5} for 3–4YSZ). Thus, it was chosen for preliminary evaluation in a burner rig. Four samples of this composition were tested. The results are listed in Table III. Spallation along the beveled edge of the sample was considered as failure. The number of cycles to failure are plotted in Fig. 3 as a function of the surface temperature and compared with results found on YSZ coatings. As is obvious, the T13 coatings have lower cyclic lives than the 3–4YSZ coatings. A comparison of the thermally grown oxide (TGO) and depleted zone thickness of the T13 coatings given in Table III reveals that the values are similar except for the sample tested at the highest temperature. This indicates that, at least for temperatures <1350°C, failure seems to be correlated with the growth of the TGO as is also observed in the YSZ-based TBC system. However, the typical critical TGO thickness at which failure occurs for 3–4YSZ is typically higher at 6–8 μm. On the other hand, the fact that we observe a considerable TGO growth at all (Fig. 4) is certainly promising. Most of the new TBC materials, such as pyrochlores or perovskites investigated as-single layer coating, showed a failure at the curved edges of the sample without TGO growth.[5,35] In addition, long cracks in the TBC close to the bondcoat are found, see Fig. 4, which are also observed in YSZ coatings after thermal cycling.

For sample 56, failure at the outer rim was observed after 13 cycles. Although the bondcoat temperature was very high for this sample, the TGO thickness after this number of cycles is expected to be of the order of 0.1 μm. Obviously, this low TGO thickness in combination with high bondcoat and surface temperatures is sufficient to promote failure at approximately one order of magnitude earlier than that observed for YSZ coatings at the edge. This shortcoming can probably be overcome by using a double-layer structure consisting of YSZ at the bondcoat and T13 coatings on

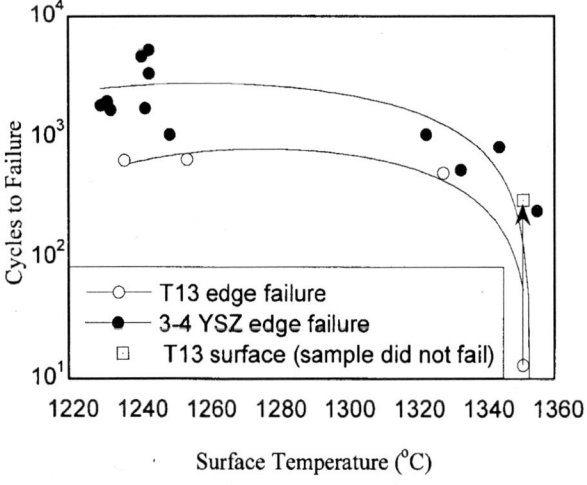

Fig. 3. Results of thermal cycling of 20YTaO4SZ as a function of surface temperature. Comparison results of YSZ coatings are also shown (Ref. 35).

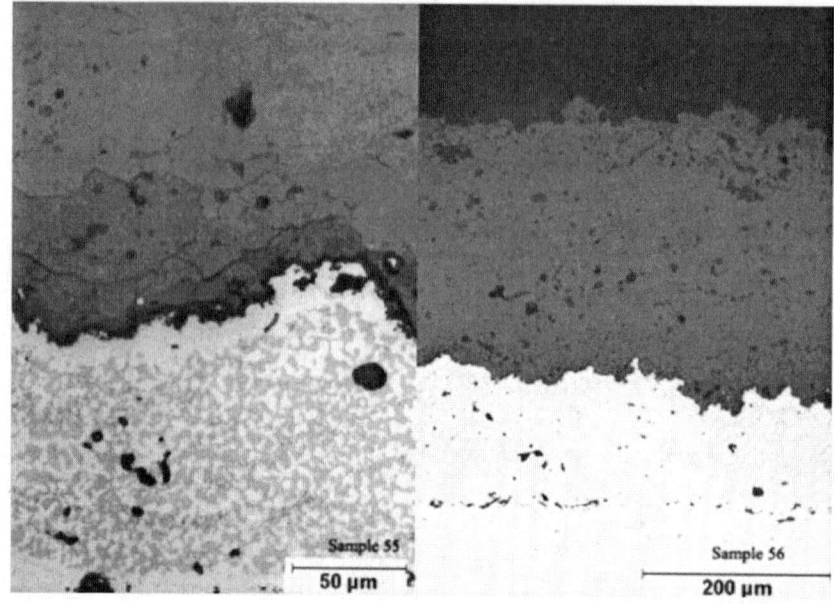

Fig. 4. Micrographs of sample 55 at an unspalled region after 710 cycles. The TGO appearing as a black line between bondcoat and TBC, the depleted zone, and the presence of long cracks at the interface are clearly visible. Micrograph of sample 56 after 300 cycles shows the coating to be intact away from the edge (see Table III for details).

top. Results on pyrochlore materials clearly show the improved performance at high surface temperatures in comparison with conventional YSZ coatings.[35] However, this approach makes sense only if the temperature capability of the T13 coatings is sufficiently high. Whether this is true can be evaluated from the microstructure of the samples tested at high temperatures. For sample 56, the cycling was continued on purpose after the failure at the outer rim to investigate the influence of the cycling on the central part of the coating. Sample 56 tested at 1350°C for 300 cycles shows no considerable change in the appearance of the surface in comparison with the other samples and the as-sprayed condition. In addition, micrographs of these samples also reveal no significant changes (Fig. 4). In contrast to these results, the YSZ samples cycled at temperatures >1330°C show massive spallation of sprayed splats from the surface in combination with large cracks developing close to the surface of the TBC.[35] The result shows that composition T13 could be capable of withstanding higher temperatures than the 3–4YSZ.

V. Conclusions

The properties of zirconias doped with Ta_2O_5 and Nb_2O_5 were investigated for potential thermal barrier applications. Single- and dual-phase compositions that are stable despite aging at 1500°C for 200 h and having a thermal conductivity as low as the conventionally used 3–4YSZ were synthesized. In addition the $YTaO_4$- and $YNbO_4$-doped zirconias have a modulus 50 GPa less than the 3–4YSZ. Zirconia doped with 20 mol% $YTaO_4$ can be deposited as a coating by plasma spraying without a change in chemistry, has a CTE that is similar to the 3–4YSZ at 1000°C, and showed promise in preliminary burner rig tests.

Acknowledgments

The authors gratefully acknowledge the financial support of the U. S. Department of Energy and the research sponsored by the Assistant Secretary for Energy Efficiency and Renewable Energy, Office of Transportation Technologies, as part of the High Temperature Materials Laboratory User Program, Oak Ridge National Laboratory, managed by Lockheed Martin Energy Research Corp. for the U. S. Department of Energy. One of the authors would like to acknowledge the help of the staff of the Institute for Materials and Processing in Energy Systems, Forschungszentrum, Jülich, Germany, in preparing and characterizing plasma-sprayed samples, and Dr. Mel Jackson, GE CRD, of Schnectady, New York, for providing the IN 738 Ni-based superalloy.

References

[1]R. A. Miller, "Thermal Barrier Coatings for Aircraft Engines: History and Directions," *J. Therm. Spray Technol.*, **6** [1] 35–42 (1997).

[2]U. Schulz, "Phase Transformations in EB-PVD Yttria Partially Stabilized Zirconia Thermal Barrier Coatings during Annealing," *J. Am. Ceram. Soc.*, **83** [4] 904–10 (2000).

[3]R. L. Jones, "Some Aspects of the Hot Corrosion of Thermal Barrier Coatings," *J. Therm. Spray Technol.*, **6** [1] 77–84 (1997).

[4]S. Stecura, "Optimization of the Ni-Cr-Al-Y/ZrO_2-Y_2O_3 Thermal Barrier System," *Adv. Ceram. Mater.*, **1** [1] 68–76 (1986).

[5]R. Vaßen, X. Cao, and D. Stöver, "Improvement of New Thermal Barrier Coating Systems Using a Layered or Graded Structure," *Ceram. Eng. Sci. Proc.*, **22** [4] 435–42 (2001).

[6]S. Stecura, "Two Layer Thermal Barrier Coatings for High-Temperature Components," *Am. Ceram. Soc. Bull.*, **56** [12] 1082–85 (1977).

[7]F. C. Toriz, A. B. Thakker, and S. K. Gupta, "Thermal Barrier Coatings for Jet Engines," *Am. Soc. Mech. Eng.*, No. 88-GT-279, 1988.

[8]J. W. Holmes and B. H. Pilsner, "Cerium Oxide Stabilized Thermal Barrier Coatings"; pp. 259–70 in *Proceedings of the National Thermal Spray Conference, Thermal Spray: Advances in Coating Technology* (Orlando, FL, 1987). Edited by D. L. Houck. ASM, Metals Park, OH, 1988.

[9]J. R. Brandon and R. T. Taylor, "Thermal Properties of Ceria and Yttria Partially Stabilized Zirconia Thermal Barrier Coatings," *Surf. Coat. Technol.*, **39/40**, 143–51 (1989).

[10]S. Appiano and P. Vincenzini, "Durability of Zirconia Thermal Barrier Coatings," *J. Mater. Synth. Process.*, **1** [1] 17–24 (1993).

[11]S. Sodeoka, M. Suzuki, T. Inoue, and K. Ueno, "Thermal and Mechanical Properties of Plasma Sprayed ZrO_2–CeO_2–Y_2O_3 Coatings"; pp. 295–302 in *Thermal Spray: Practical Solutions for Engineering Problems*. Edited by C. C. Berndt. ASM, Metals Park, OH, 1996.

[12]N. Mifune, Y. Harada, H. Taira, and S. Mishima, "Field Evaluation of $2CaO$–SiO_2–CaO–ZrO_2"; pp. 299–303 in *Thermal Spray: A United Forum for Scientific and Technological Advances*. Edited by C. C. Berndt. ASM, Metals Park, OH, 1997.

[13]M. Leoni, R. L. Jones, and P. Scardi, "Phase Stability of Scandia-Yttria-Stabilized Zirconia TBCs," *Surf. Coat. Technol.*, **108–109**, 107–13 (1998).

[14]P. Ramaswamy, S. Seetharamu, K. B. R. Verma, and K. J. Rao, "Evaluation of CaO–CeO_2 Partially Stabilized Zirconia Thermal Barrier Coatings," *Ceram. Int.*, **25**, 317–24 (1999).

[15]C. J. Friedrich, R. Gadow, and T. Schirmer, "Lanthanum Aluminate: A New Material for Atmospheric Plasma Spraying of Advanced Thermal Barrier Coatings"; pp. 1219–26 in *Thermal Spray: Surface Engineering via Applied Research, Proceedings of the 1st International Thermal Spray Conference* (Montreal, Canada). Edited by C. C. Berndt. ASM, Metals Park, OH, 2000.

[16]R. Vassen, X. Cao, F. Teitz, D. Basu, and D. Stöver, "Zirconates as New Materials for Thermal Barrier Coatings," *J. Am. Ceram. Soc.*, **83** [8] 2023–28 (2000).

[17]R. Vaβen and D. Stöver, "Conventional and New Materials for Thermal Barrier Coatings"; pp. 199–216 in NATO Science Series II: Mathematics, Physics and Chemistry, Vol. 16, *Functional Gradient Materials and Surface Layers Prepared by Fine Particle Technology*. Edited by Kluwer Academic Publishers, Dordrecht, The Netherlands, 2001.

[18]S. Raghavan, H. Wang, W. D. Porter, R. B. Dinwiddie, and M. J. Mayo, "Thermal Properties of Zirconia Co-Doped With Trivalent and Pentavalent Oxides," *Acta Mater.*, **49** [1] 169–79 (2001).

[19]S. Raghavan and M. J. Mayo, "Hot Corrosion Resistance of 20 mol% $YTaO_4$ Stabilized Tetragonal Zirconia and 14 mol% Ta_2O_5 Stabilized Orthorhombic Zirconia for Thermal Barrier Coating Applications," *Surf. Coat. Technol.*, **160**, 187–96 (2002).

[20]P. G. Klemens, "Thermal Conductivity of Zirconia"; pp. 209–20 in *Thermal Conductivity*, Vol. 23. Edited by K. E. Wilkes and R. Dinwiddie. Technomics, Lancaster, PA, 1996.

[21]D. J. Green, R. H. J. Hannik, and M. V. Swain, *Transformation Toughening of Ceramics*. CRC Press, Boca Raton, FL, 1989.

[22]P. Li, I.-W. Chen, and J. E. Penner-Hahn, "Effect of Dopants on Zirconia Stabilization—An X-ray Absorption Study: II, Tetravalent Dopants," *J. Am. Ceram. Soc.*, **77** [5] 1281–88 (1994).

[23]P. Li, I.-W. Chen, and J. E. Penner-Hahn, "Effect of Dopants on Zirconia Stabilization—An X-ray Absorption Study: III, Charge-Compensating Dopants," *J. Am. Ceram. Soc.*, **77** [5] 1289–95 (1994).

[24]P. Li, I.-W. Chen, and J. E. Penner-Hahn, "Effect of Dopants on Zirconia Stabilization—An X-ray Absorption Study: I, Trivalent Dopants," *J. Am. Ceram. Soc.*, **77** [1] 118–28 (1994).

[25]D. J. Kim and T. Y. Tien, "Phase Stability and Physical Properties of Cubic and Tetragonal ZrO_2 in the System ZrO_2–Y_2O_3–Ta_2O_5," *J. Am. Ceram. Soc.*, **74** [12] 3061–65 (1991).

[26]D. J. Kim, "Effect of Ta_2O_5, Nb_2O_5, and HfO_2 Alloying on the Transformability of Y_2O_3-Stabilized Tetragonal ZrO_2," *J. Am. Ceram. Soc.*, **73** [1] 115–20 (1990).

[27]D. Y. Lee, D. J. Kim, and D. H. Cho, "Low Temperature Phase Stability and Mechanical Properties of Y_2O_3 and Nb_2O_5 Co-Doped Tetragonal Zirconia Polycrystal Ceramics," *J. Mater. Sci. Lett.*, **17** [3] 185–87 (1998).

[28]M. F. Ashby, "On the Engineering Properties of Materials," *Acta Metall.*, **37** [5] 1273–93 (1989).

[29]C. Funke, J. C. Mailand, B. Siebert, R. Vaβen, and D. Stöver, "Characterization of ZrO_2–7 wt% Y_2O_3 Thermal Barrier Coatings with Different Porosities and FEM Analysis of Stress Distribution during Thermal Cycling of TBCs," *Surf. Coat. Technol.*, **94–95**, 106–11 (1997).

[30]M. C. Bhardwaj and K. A. Tripett, "Nondestructive Characterization of Green and Sintered Ceramics"; pp. 1–7 in *Proceedings of the 1st International Symposium on Science of Engineering Ceramics Japan*. Edited by S. Kimura and K. Niihara. The Ceramic Society of Japan, Tokyo, Japan, 1991.

[31]R. A. Miller, J. L. Smialek, and R. G. Garlick, "Phase Stability in Plasma Sprayed Partially Stabilized Zirconia Yttria"; pp. 241–53 in Advances in Ceramics, Vol. 3, *Science and Technology of Zirconia*. Edited by A. H. Heuer and L. W. Hobbs. The American Ceramic Society, Columbus, Ohio, 1981.

[32]D. J. Kim, H. J. Jung, J. W. Jang, and H. L. Lee, "Fracture Toughness, Ionic Conductivity, and Low-Temperature Phase Stability of Tetragonal Zirconia Co-doped with Yttria and Niobium Oxide," *J. Am. Ceram. Soc.*, **81** [9] 2309–14 (1994).

[33]S. Raghavan, H. Wang, R. B. Dinwiddie, W. D. Porter, and M. J. Mayo, "The Effect of Grain Size, Porosity and Yttria Content on the Thermal Conductivity of Nanocrystalline Zirconia," *Scr. Metall. Mater.*, **39** [8] 1119–25 (1998).

[34]W. D. Kingery, H. K. Bowen, and D. R. Uhlmann, *Introduction to Ceramics*. John Wiley and Sons, Singapore, 1991.

[35]R. Vaβen, G. Barbezat, and D. Stöver, "Comparison of Thermal Cycling Life of YSZ and $La_2Zr_2O_7$ Based Thermal Barrier Coatings"; pp. 511–21 in *Materials for Advanced Power Engineering*, Vol. 21. Edited by J. Lecomte-Becker, M. Carton, F. Schubert, and P. J. Ennis. Forschungszentrum, Jülich GmBH, Jülich, Germany, 2002. ☐

International Journal of

Applied Ceramic Technology

Ceramic Product Development and Commercialization

New Thermal Barrier Coatings Based on Pyrochlore/YSZ Double-Layer Systems

R. Vaßen, F. Traeger, D. Stöver

Institut für Werkstoffe und Verfahren der Energietechnik (IWV1), Forschungszentrum Jülich GmbH, Germany

Pyrochlore materials $La_2Zr_2O_7$ and $Gd_2Zr_2O_7$ have been used to produce thermal barrier coating systems by atmospheric plasma spraying. The materials have been applied as single-layer coatings with only a topcoat made of pyrochlore material. In addition, double-layer systems with a first layer of yttria-stabilized zirconia (YSZ) and a top layer made of pyrochlore material were produced. These systems have been tested in thermal cycling test rigs at surface temperatures between 1200-1450°C and the results were compared to the behavior of YSZ coatings. Single-layer coatings had a rather poor thermal cycling performance. On the other hand, double-layer systems showed similar results to YSZ coatings at temperatures below about 1300°C. At higher temperatures the double-layer coatings produced from our own powders revealed excellent thermal cycling behavior. At the highest test conditions, lifetime was thereby orders of magnitude better than that of YSZ coatings. Results indicate that an increase of the maximum surface temperature in gas turbines by at least 100 K becomes possible with the new coatings. Coatings produced from commercial powders showed a somewhat reduced performance.

Introduction

Thermal barrier coating (TBC) systems typically consist of a metallic oxidation protection layer and an insulative ceramic topcoat. Currently, two processes are widely used to produce the topcoat of these systems: electron beam physical vapor deposition (EB-PVD) and atmospheric plasma spraying (APS). The state-of-the-art topcoat material for both processes is yttria partially stabilized zirconia (YSZ).[1,2] This material performs quite well up to about 1200°C. At higher application temperatures, which are envisaged for further improvement of the efficiency of the gas turbines, the YSZ undergoes two detrimental changes. Significant sintering leads to microstructural changes and, hence, a reduction of the strain tolerance in combination with an increase of the Young's modulus.[3] Higher stresses will originate in the coating, which lead to a reduced life under thermal cyclic loading.

The second change is a phase change of the nontransformable t′-phase, which is present in the as-sprayed YSZ coating. At elevated temperatures the t′-phase transforms into tetragonal and cubic phase. During cooling the tetragonal phase will further transform into the monoclinic phase, which is accompanied by a volume change and a high risk for damage to the coating.[4] As a consequence, a considerable reduction of thermal cycling life is observed.

These disadvantageous properties of YSZ at high temperatures prompted an intense search for new TBC materials in the past.[5-10] Detailed overviews on the development of new systems are also available given.[11-13]

Among the interesting candidates for TBCs, those

materials with pyrochlore structures and high melting points show promising thermophysical properties. Especially interesting candidates are $La_2Zr_2O_7$ $La_2Hf_2O_7$, $Gd_2Zr_2O_7$, or $Nd_2Zr_2O_7$. Previous investigations show excellent physical properties of theses materials, i.e., thermal conductivity lower than YSZ and high thermal stability.[7,14] However, the thermal expansion coefficient is typically lower ($9\text{-}10 \times 10^{-6}$/K) than that of YSZ ($10\text{-}11 \times 10^{-6}$/K), which leads to higher thermal stresses in the TBC system as both substrate and bondcoat have higher thermal expansion coefficients (about 15×10^{-6}/K). In addition, relatively low toughness values are observed in these materials,[15] as no toughening effects are expected as observed in YSZ.[16]

As a result, the thermal cycling properties are worse than YSZ coatings. We believe that this problem is relevant for most of the new TBC materials, as the need for thermal stability seems to contradict the ability of transformation toughening. A way to overcome this shortcoming is the use of layered topcoats. The failure of TBC systems often occurs within the TBC close to the bondcoat/topcoat interface (white failure[17]). At this location YSZ is used as a TBC material with a relatively high thermal expansion coefficient and high toughness. The YSZ layer is then coated with the new TBC material (e.g., $La_2Zr_2O_7$) which is able to withstand the typically higher temperatures at this location. We could show in the past years in several publications on $La_2Zr_2O_7$/YSZ double-layer systems that this concept really works.[18-21] Recently, other groups have also adopted these ideas and promising results on EB-PVD double layers have been presented.[22]

In the present paper, new results on $La_2Zr_2O_7$- and $Gd_2Zr_2O_7$-based TBC systems cycled at extremely high temperatures are presented together with the old results. Certainly, the success of the concept was proven once again. However, it also became clear that the starting powder seems to play a key role in gaining excellent performance.

Experimental

The investigated TBC systems were all produced by plasma spraying with Sulzer Metco plasma-spray units. About 150-μm-thick vacuum plasma-sprayed NiCoCrAlY bondcoat was used. In most cases the bondcoat was applied with an F4 gun using Ni 192-8 powder (Praxair Surface Technologies Inc., Indianapolis, IN). In some cases,[19] indicated by "Amdry 997 BC," Sulzer Metco produced the bondcoat using an Amdry 997 NiCoCrAlY powder with about 4 wt% Ta. Disk-shaped nickel-base

superalloy IN738 substrates with our standard thermal cycling sample geometry (see below) were used. The ceramic topcoats with a thickness between 300-500 μm were produced by atmospheric plasma spraying (APS) using a Triplex I gun. During the manufacture of the thermal cycling specimens, steel substrates were also coated. These coatings were used to characterize the as-sprayed condition. The porosity level was measured for free-standing coatings with mercury porosimetry. The standard conditions for the deposition of pyrochlore and YSZ coatings were Argon and Helium plasma gas flow rates of 20 and 13 standard liter per minute (slpm), a plasma current of 300 A at a power of 20 kW, and a spray distance of 90 mm. Lower currents (245 A and 270 A) and one larger spray distance (130 mm) were also used in some cases. Several coatings were produced using a Y-shaped powder injection system, in which the two different carrier gases containing YSZ and $La_2Zr_2O_7$ powders were mixed.[18] The standard injection system uses only one carrier gas. The double-layer coatings consisted of a YSZ coating directly deposited on the bondcoat and a pyrochlore topcoat. The thickness of each coating was about half the total coating thickness.

For comparison, the results of single-layer YSZ coatings are also given which have been reported earlier.[23] Results of coatings supplied by Sulzer Metco are also included.[19] The YSZ powder used was a 7.8 wt% yttria-stabilized zirconia powder (Metco 204 NS) delivered by Sulzer Metco GmbH, Hattersheim, Germany.

The $La_2Zr_2O_7$ powder was produced by a solid-state method using La_2O_3 (99.9%, Aldrich) and ZrO_2 (99%, Aldrich). From the starting powders, a suspension was made which was then ground and mixed for about 24 h in a planetary mill using zirconia containers and zirconia milling balls. After complete drying of the suspension, the solid-state reaction took place in zirconia crucibles for 24 h at 1400°C in air.

The synthesized powders were then spray-dried as described in detail in a previous paper.[24] The obtained powder had good flowability and high density, and a particle size of between 30-80 μm was obtained by sieving. The particle size distributions were measured by laser scattering in a Fritsch analysette.

In addition, $La_2Zr_2O_7$ and $Gd_2Zr_2O_7$ (supplied by Praxair Services GmbH & Co. KG, Wiggensbach, Germany) were used in the investigations. Information on the used powders is given in Table I and scanning electron microscope (SEM) pictures are shown in Fig. 1. XRD patterns of the three powders are shown in Fig. 2.

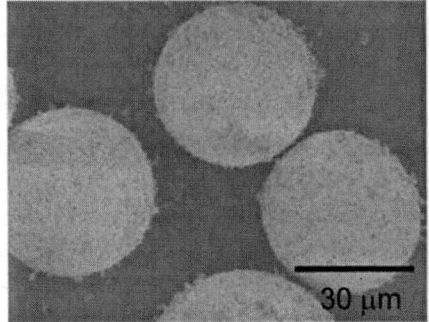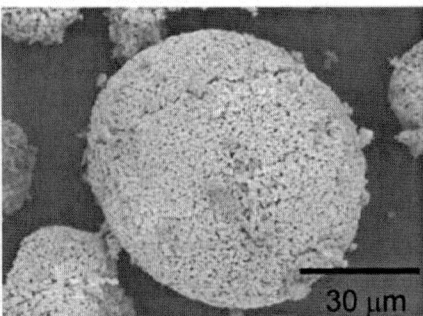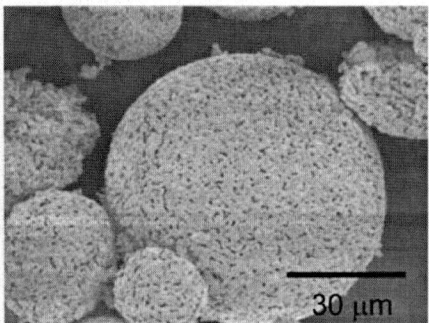

Fig. 1. Micrographs of the used powders (LZ-FZJ, LZ-Prax, GZ-Prax, from left to right).

Table I. Information on the Used Pyrochlore Powders

short name	LZ-FZJ	LZ-Prax	GZ-Prax
producer	FZJ	Praxair	Praxair
powder	$La_2Zr_2O_7$	$La_2Zr_2O_7$	$Gd_2Zr_2O_7$
mean particle size [μm]	46	45	46

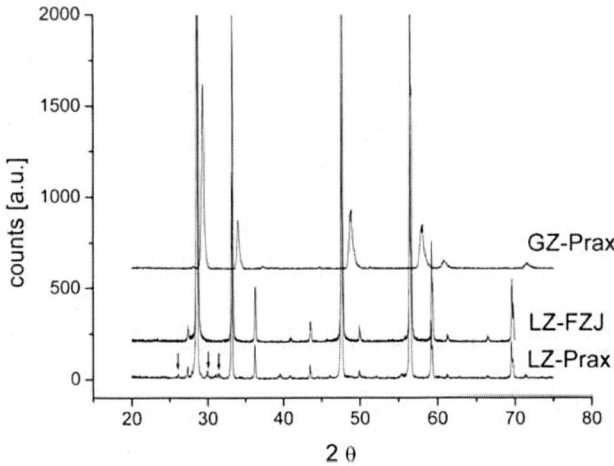

Fig. 2. XRD patterns of the used powders. Arrows indicate the presence of monoclinic ZrO_2 and La_2O_3.

Gas burner test facilities operating with natural gas and oxygen were used to evaluate the thermal cycling behavior of the new systems.[25] The substrates of the thermal cycling samples had a diameter of 30 mm and a thickness of 3 mm. In these rigs the substrates were cooled by compressed air from the back. The surface temperature was measured with a pyrometer operating at a wavelength of 8-13 μm and a spot size of 12 mm. For YSZ, the emissivity for this wavelength range was determined to be close to 1. The same value was taken for the different pyrochlore materials. More exact calculations taking into account the

emissivity of the coatings in the relevant wavelength range show that the underestimation of the surface temperatures might be rather high (about 70-100 K).

Additionally, the substrate temperature was measured by a thermocouple. This thermocouple was located in the center of the substrate. The surface temperature was varied between 1200-1450°C, the substrate temperature was adjusted between 930-1070°C. Using the thermal conductivities of the coatings and the substrate, one can estimate that the bondcoat temperature is about 30-40 K higher than the substrate temperature. In the test facility, a gas burner with a broad flame was used giving a homogeneous temperature distribution in the center of the sample. After heating for about 20 s the maximum temperature is reached. After 5 min the burner is automatically removed for 2 min from the surface and the surface is cooled at an initial rate of more than 100 K/s using compressed air. After about 1 min of cooling, both surface and substrate temperatures are below 100°C.

Cycling was stopped when a clearly visible spallation (about 5 × 5 mm² at least) of the coating occurred at the central part. At present, no automatic system is installed to detect spallation. As a result, a certain number of cycles will be performed even after spallation, as the specimens are not inspected after each cycle, given a certain error of the results.

Metallographic cuts have been prepared from all samples to investigate the microstructure.

Results and Discussion

Microstructural Characterization

All coatings showed a microcracked microstructure as often found in plasma-sprayed coatings. Micrographs of double-layer coatings are shown in Fig. 3. For the LZ-FZJ and the GZ-Prax powder, two different porosity val-

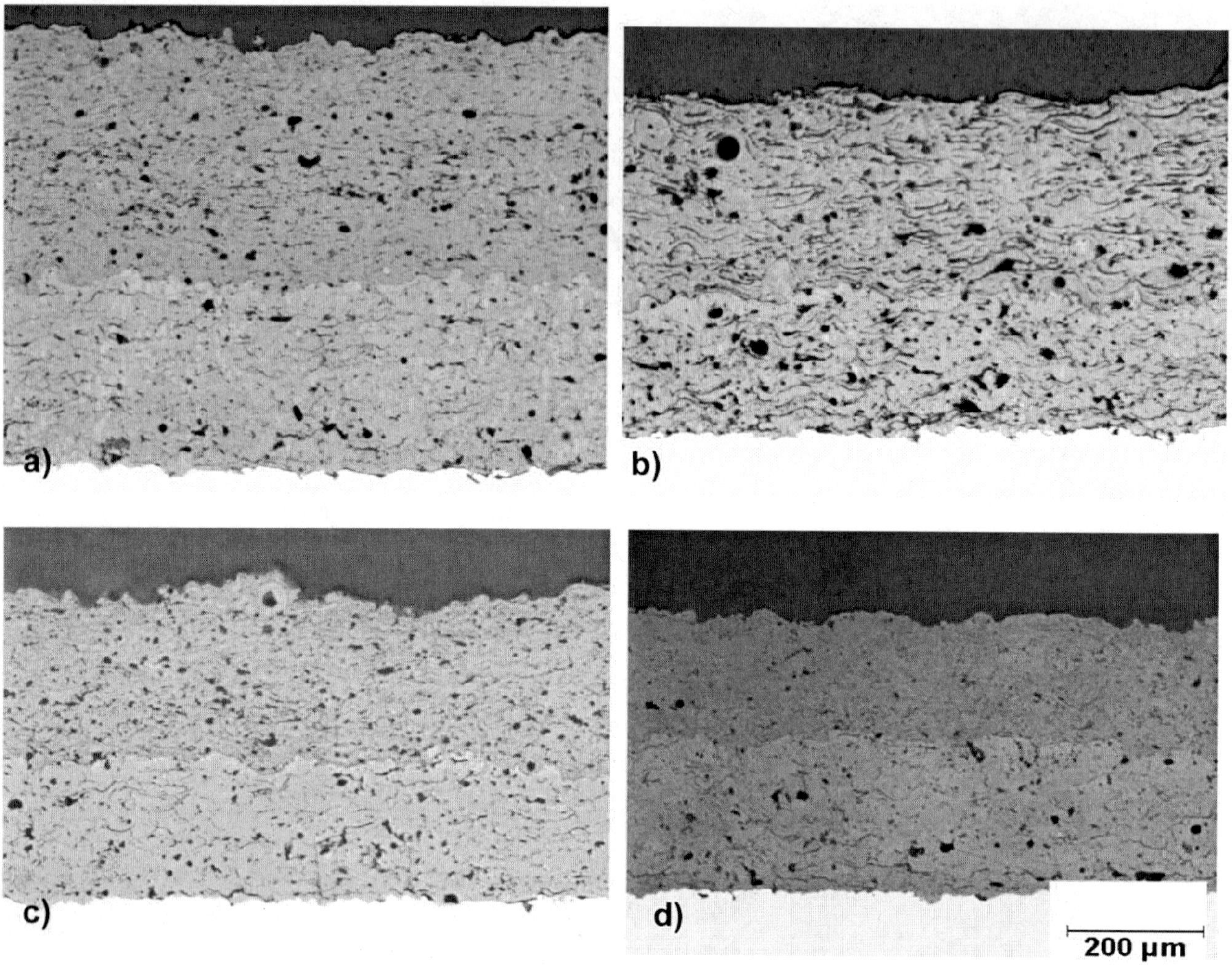

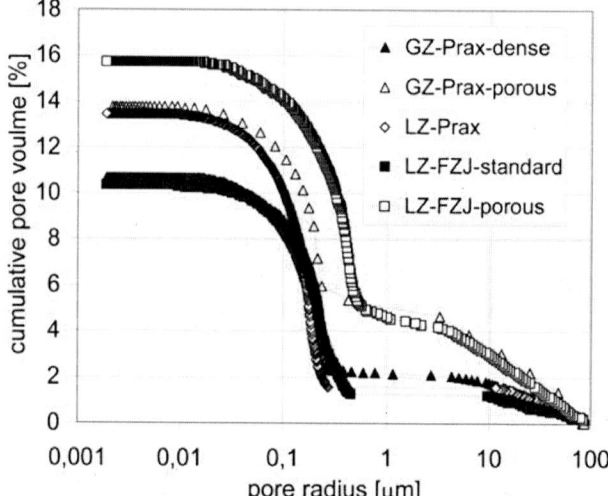

Fig. 3. Micrographs of double-layer coatings used for thermal cycling produced from the powders: a) LZ-FZJ (dense), b) LZ-FZJ (porous), c) LZ-Prax, d) GZ-Prax (dense). Dense and porous refer to the density of the coatings as explained in the text.

ues were used, referred to as dense and porous. The porosity levels were determined on about 0.5-mm-thick freestanding coatings deposited with the same spraying conditions as those used for the cycling samples.

The results of the corresponding mercury porosity measurements are shown in Fig. 4. The porosity levels of double-layer coatings from Metco 204 NS and LZ-FZJ powders with the standard injection system were about 10.4%. The coatings produced with the Y-shaped injection system had a significantly higher porosity level of about 15.4% due to the fact that the higher feeding gas rate led to non-optimal particle injection into the plasma.

For the Praxair powders, several spraying conditions were investigated before deposition of the thermal cycling specimens. For the LZ-Prax powder, the spray conditions of coatings selected for testing had a porosity of 13.5%. For the Gd-Prax powder, two porosity levels were used (10.6% and 13.7%).

Fig. 4. Mercury porosity measurements of single-layer (Praxair powders) and double-layer coatings (LZ-FZJ) corresponding to those used in thermal cycling experiments.

Progress in Thermal Barrier Coatings

The YSZ coatings used for comparison and as the first layer in the double-layer systems had porosity levels of 11-13%. One exception was the YSZ layer used in the porous double layers with $La_2Zr_2O_7$ made from LZ-FZJ. Due to the use of the Y-shaped injection system the porosity is higher (see Fig. 4).

Thermal Cycling Results

YSZ Coatings

The results of the thermal cycling tests are shown in Fig. 5. For comparison, the lifetimes of standard YSZ coatings are also indicated by the given range.[23] Typically it is found that above 1300°C, a drastic decrease in lifetime is observed due to the limited temperature stability of YSZ. At lower temperatures, a rather large variation of the data is observed as the surface temperature is no longer the most significant parameter to determine lifetime. In this range a more appropriate plot is shown in Fig. 6, where the cycles to failure are plotted above the bondcoat temperature. In this plot the data for the YSZ coatings split into two sets: one corresponding to the standard bondcoat and one belonging to the Amdry 997 bondcoat. The last bondcoat shows a higher lifetime at lower temperatures, which can be correlated to the lower growth rate of the thermally grown oxide (TGO) at these temperatures. At temperatures of about 1100°C the difference between the two bondcoats seems to vanish. Samples

tested in this temperature range also typically show quite high surface temperatures (>1300°C), as the total temperature drop across the TBC is limited by the available heat flux. The maximum heat flux is about 1 MW/m² for the used gas burners, which gives a typical temperature drop of 300 K for a 300-μm-thick coating with a thermal conductivity of 1 W/m/K. For these high surface temperatures the temperature capability of YSZ is reached and failure typically takes place within the ceramic close to the free surface of the coatings. The samples showing this kind of failure are marked by "( )" in Fig. 6. In Fig. 5 these samples lie in the sharp drop of the lifetime range of YSZ above about 1320°C. An example of a YSZ coating failed in the high temperature range is shown in Fig. 7. In the central part of the micrograph the original coating thickness of about 310 μm is already reduced by about 60 μm. This thinning is due to a subsequent spallation of individual spray splats[26] indicated by the long crack in the TBC close to the surface on the left side in Fig. 7.

Single Layers of $La_2Zr_2O_7$ and $Gd_2Zr_2O_7$

In the following, the single-layer coatings made of new ceramics will be discussed. Fig. 5 clearly shows that the lifetime of these coatings is well below those of standard YSZ coatings. The type of failure is similar for all these samples. An example of a specimen made of GZ-Prax powder with the higher porosity level is shown after thermal cycling in Fig. 8. Long cracks within the TBC

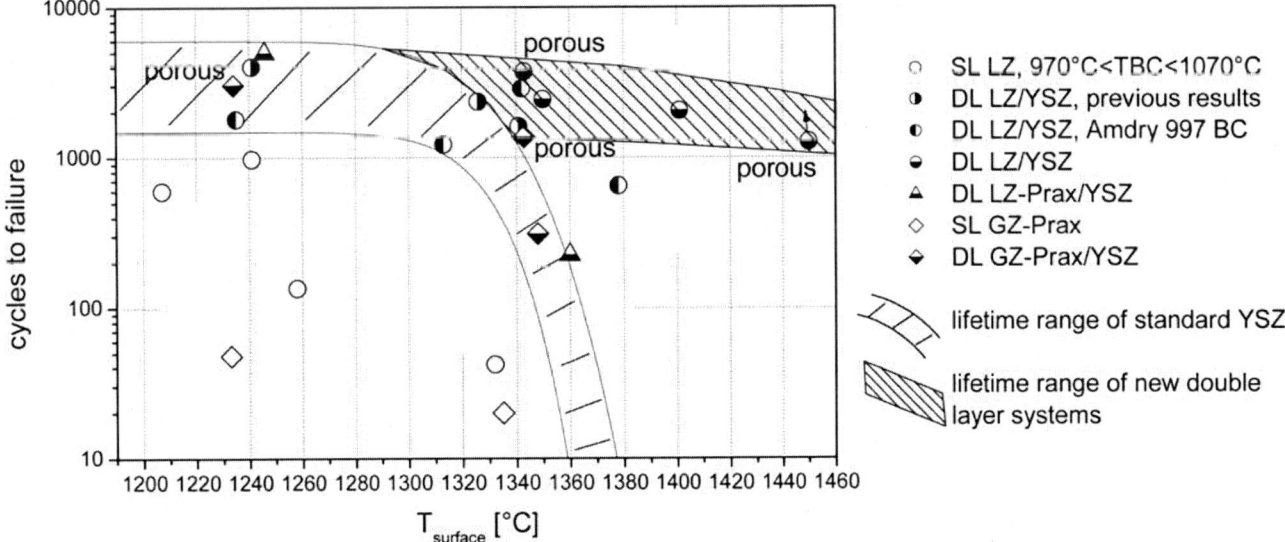

Fig. 5. Cycles to failure for the investigated single-layer (SL) and double-layer (DL) systems produced from powders given in Table I. For comparison the range of the lifetime of single-layer YSZ coatings is also given. In coating systems in which two different porosity values were used, the porous ones are marked correspondingly.

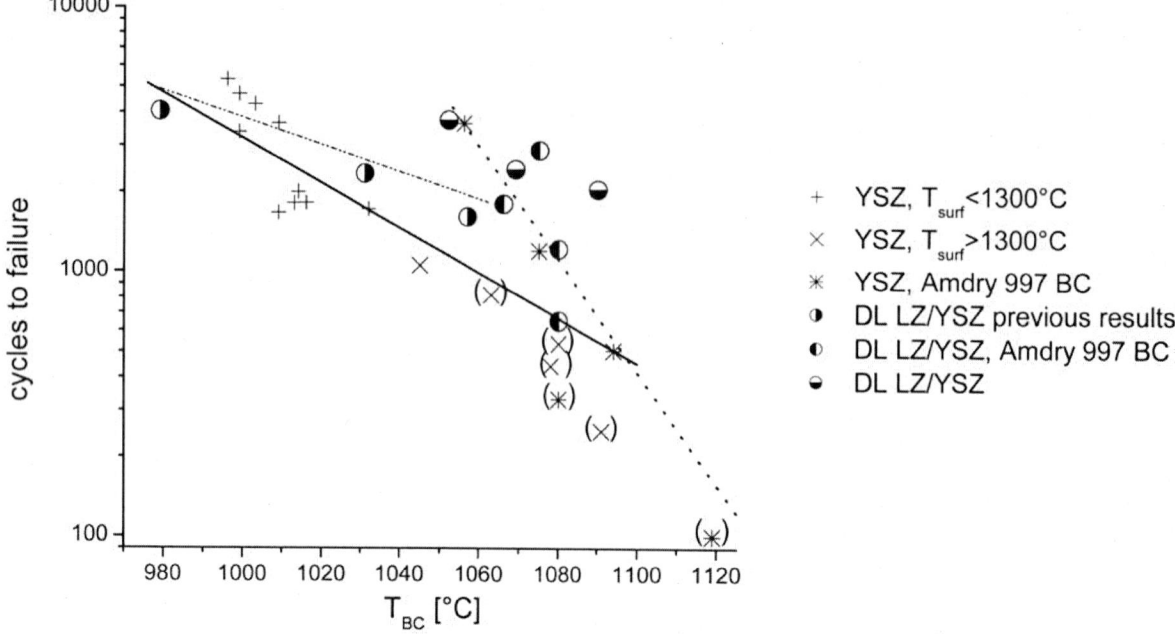

Fig. 6. *Cycles to failure for the La₂Zr₂O₇-based double-layer (DL) systems produced from LZ-FZJ powders as a function of the bondcoat temperature. For comparison the lifetimes of single-layer YSZ coatings with different bondcoats are also given (same samples as in Fig. 5). The different lines are guides for the eye for the YSZ coatings on standard (——) and Amdry 997 bondcoat (······) as well as the double-layer systems on standard bondcoat (-·-·). Marked with ( ) are those samples that fail within the ceramic close to the surface.*

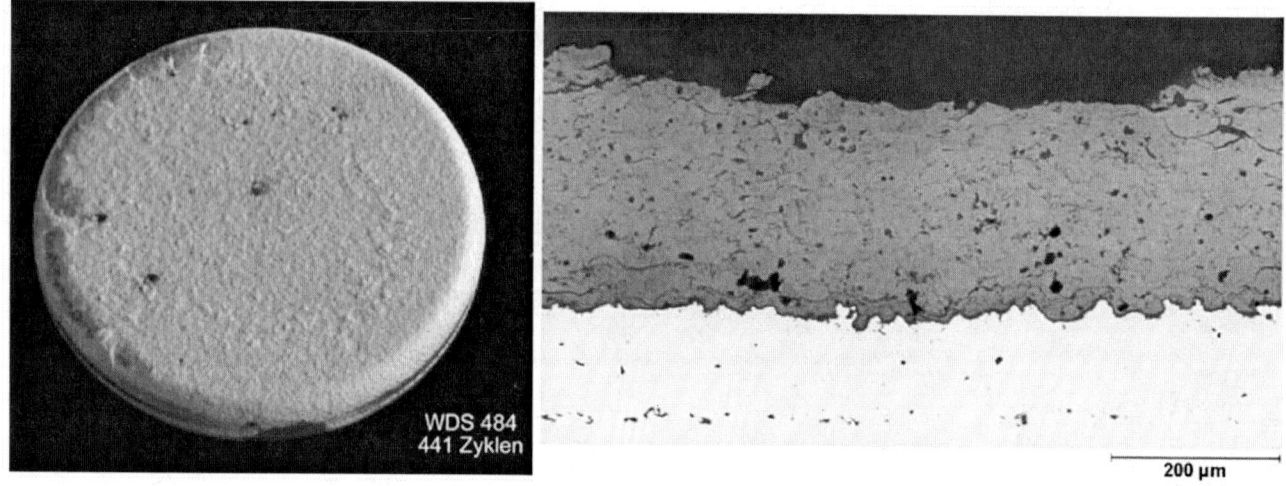

Fig. 7. *Photograph and micrograph of YSZ coating cycled for 441 cycles at 1341°C surface and 1078°C bondcoat temperature.*

close to the bondcoat/TBC interface are formed even without significant thermally grown oxide (TGO). In standard YSZ coatings, the additional stress levels associated with the growth of the TGO are necessary to drive the crack growth.[17] The low toughness of most of the new ceramics[15] seems to allow a crack growth even with lower stress levels. These findings were the basis for the development of the double-layer approach as outlined in the introduction.

Double Layers, Effect of Outer Rim

For all double layer coatings, significantly higher lifetimes have been found. However, several of the systems tested in the present investigation showed a spallation of the outer layer made of new ceramics after only a limited number of cycles (about 100 cycles) at the outer rim of the thermal cycling specimens. After this type of failure, the coatings typically remain unchanged for a large num-

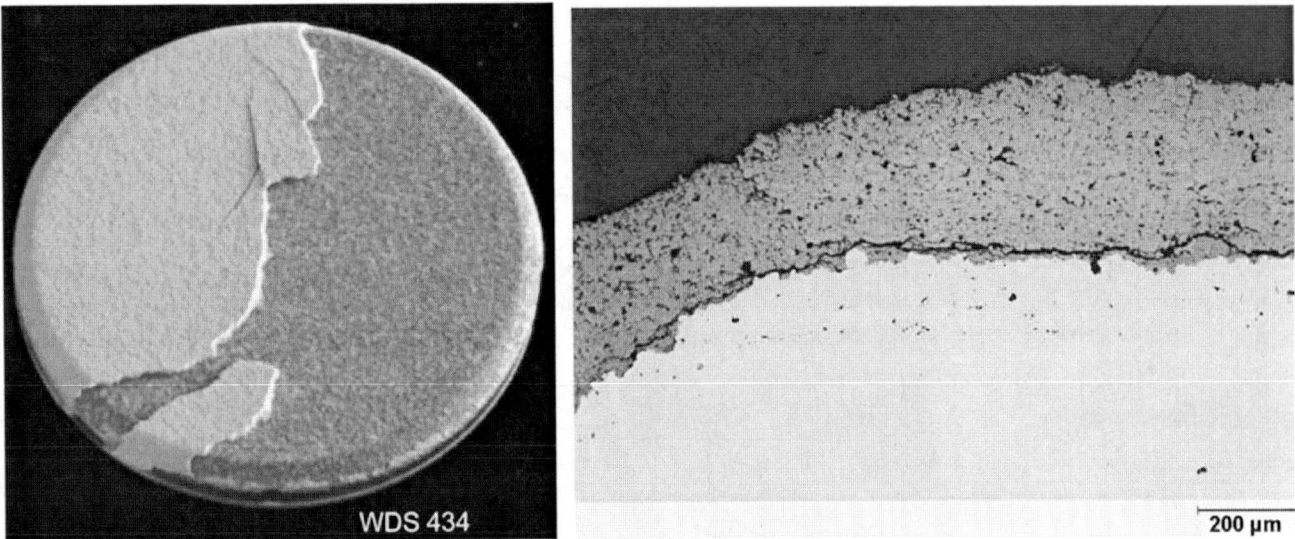

Fig. 8. Photograph (left) and micrograph of a porous single-layer coating made of GZ-Prax powder after 48 cycles at 1233°C and 1038°C surface and bondcoat temperature, respectively.

ber of additional cycles. In fact, the number of cycles given in Figs. 5 and 6 do not take into account this early failure. The reason for this failure is probably related to the small radius of curvature at the outer rim (1.5 mm). After stress relaxation at high temperatures, the cooling down of the TBC system will lead to tensile stress levels in the ceramic topcoat perpendicular to the interface. These stress levels are inversely proportional to the radius and reduce approximately linearly from a maximum value at the interface to zero at the free surface of the TBC.[17] For a double-layer system with a coating made half of YSZ and half of new ceramic, this results in a stress level at the interface YSZ/new ceramic of about half the level at the interface. This stress is probably high enough to promote crack growth in the new ceramics for the used sample geometry. As the radius of curvature and, hence, the stress is smaller by a factor of at least 3 than radii found in conventional TBC-coated components of gas turbine (e.g., small blades of aero engines), this type of failure is expected to be not critical for an application of the double-layer systems. This statement is further justified by the fact that this type of failure had not been observed in earlier investigations. Obviously, the stress levels are at the limit to be sufficient to promote crack growth.

Double Layers Prepared from GZ-Prax Powder

For this powder, three thermal cycling samples have been tested. Certainly, this is not sufficient to finally evaluate the potential of this powder; however, it gives a first impression. At moderate surface temperatures the system meets the expectations, as the performance is similar to those of YSZ- or $La_2Zr_2O_7$-based double-layer systems (see Fig. 5). At higher surface temperatures (1342°C), the more porous coating behaves slightly better than YSZ; however, the denser sample tested at slightly higher surface temperatures (1348°C) seems to perform equally to YSZ coatings. The photographs of both samples reveal at least partial failure within the ceramic, indicating limited temperature capability (Fig. 9). This is further underlined by the micrographs showing a long crack running in the ceramic and parts of the new ceramic already spalled (Fig. 9). In earlier investigations on $La_2Zr_2O_7$ it was found that free La_2O_3 is extremely critical for the performance, as it leads to the formation of $La(OH)_3$ in water vapor. However, in XRD patterns (see Fig. 2) no indication of free Gd_2O_3, which might be similarly critical, was found. XRD studies revealed that the defect fluorite structure found in the as-sprayed coatings transforms more slowly into the ordered pyrochlore structure than $La_2Zr_2O_7$ coatings. This indicates lower diffusion rates in the lattice and, hence, a better temperature capability. The reason why an early failure was observed is not clear; it might be related to impurities not detectable in the XRD patterns.

Double Layers Prepared from LZ-Prax Powder

For the LZ-Prax powder, two coatings with a double-layer structure were tested (Fig. 5). One cycled at relatively low temperature showed good performance com-

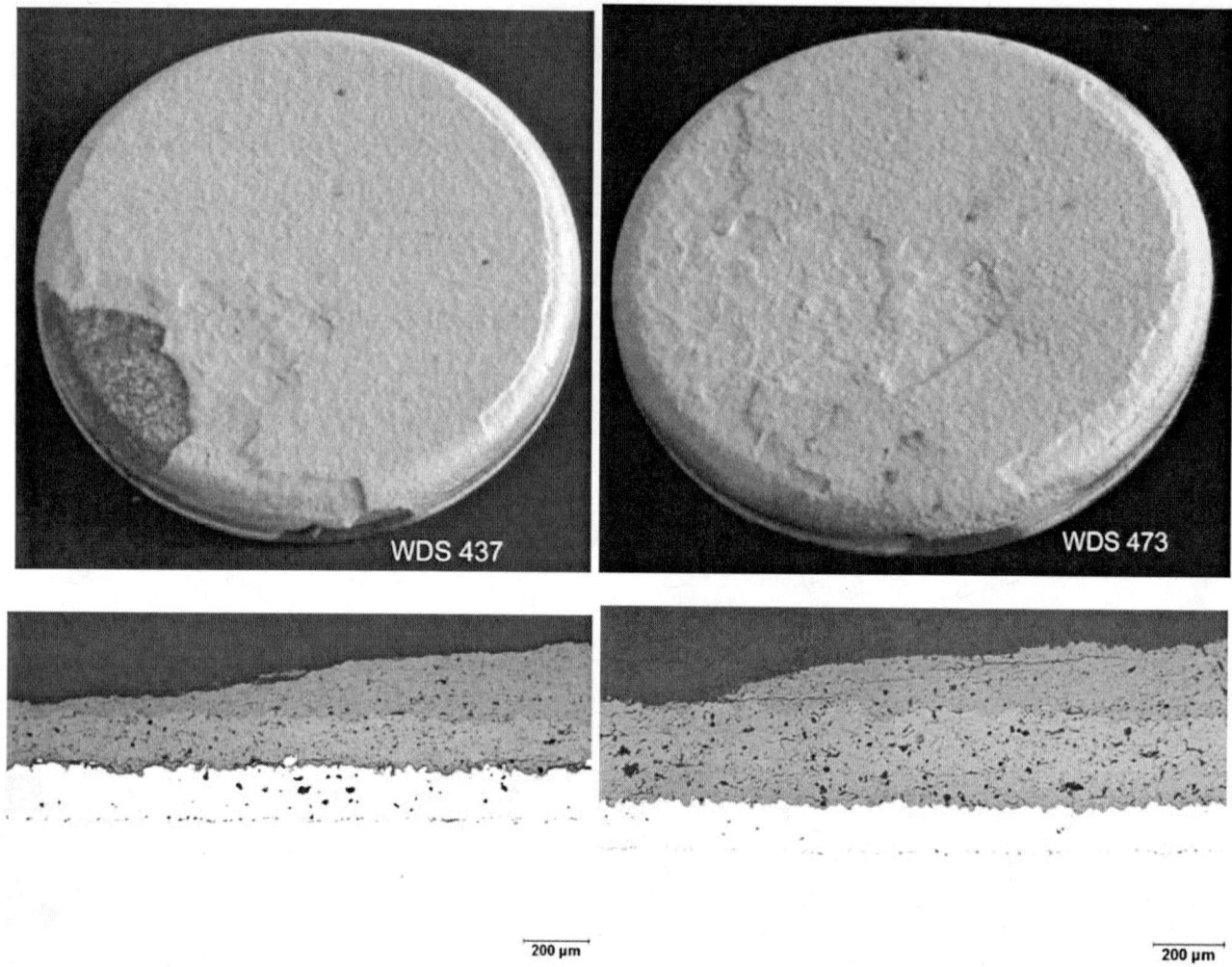

Fig. 9. *Photographs (top) and micrographs (bottom) of double-layer coatings prepared from GZ-Prax powder cycled for 1368 (left) and 310 cycles (right) at 1343/1348°C and 1070/1081°C surface and bondcoat temperature, respectively.*

parable to YSZ coatings. The other sample was tested at a surface temperature of 1360°C. After 245 cycles it showed a rather massive failure within the central part of the coating (Fig. 10). These findings are similar to the ones discussed above for the GZ-Prax powder. They indicate a limited temperature capability of the coatings prepared from these powders for temperatures above about 1350°C. In contrast to the GZ-Prax powder, a rather clear indication of free La_2O_3 and also ZrO_2 is found in the LZ-Prax powder (Fig. 2). Both impurities are certainly critical due to the reaction to hydroxide in water vapor or due to phase transitions during cycling, respectively. Obviously, the formation of $La_2Zr_2O_7$ is not complete in this powder.

On the other hand, the numbers of cycles to failure at the lower temperature were high. If impurities play a major role it would be likely that these would also affect the performance at lower temperatures.

So, in conclusion, a clear explanation for the results of the commercial powders at the highest test conditions is not yet available. Further investigations are necessary.

Double Layers Prepared from LZ-FZJ Powder

In previous investigations it was already shown that double-layer coatings prepared from our powders showed, at relatively low surface temperatures (<1300°C), a thermal cycling behavior similar to that of YSZ coatings. In this investigation the focus was put on the determination of the temperature limit of these coatings and, hence, the testing was performed at extremely high temperatures. All double-layer coatings tested at temperatures above

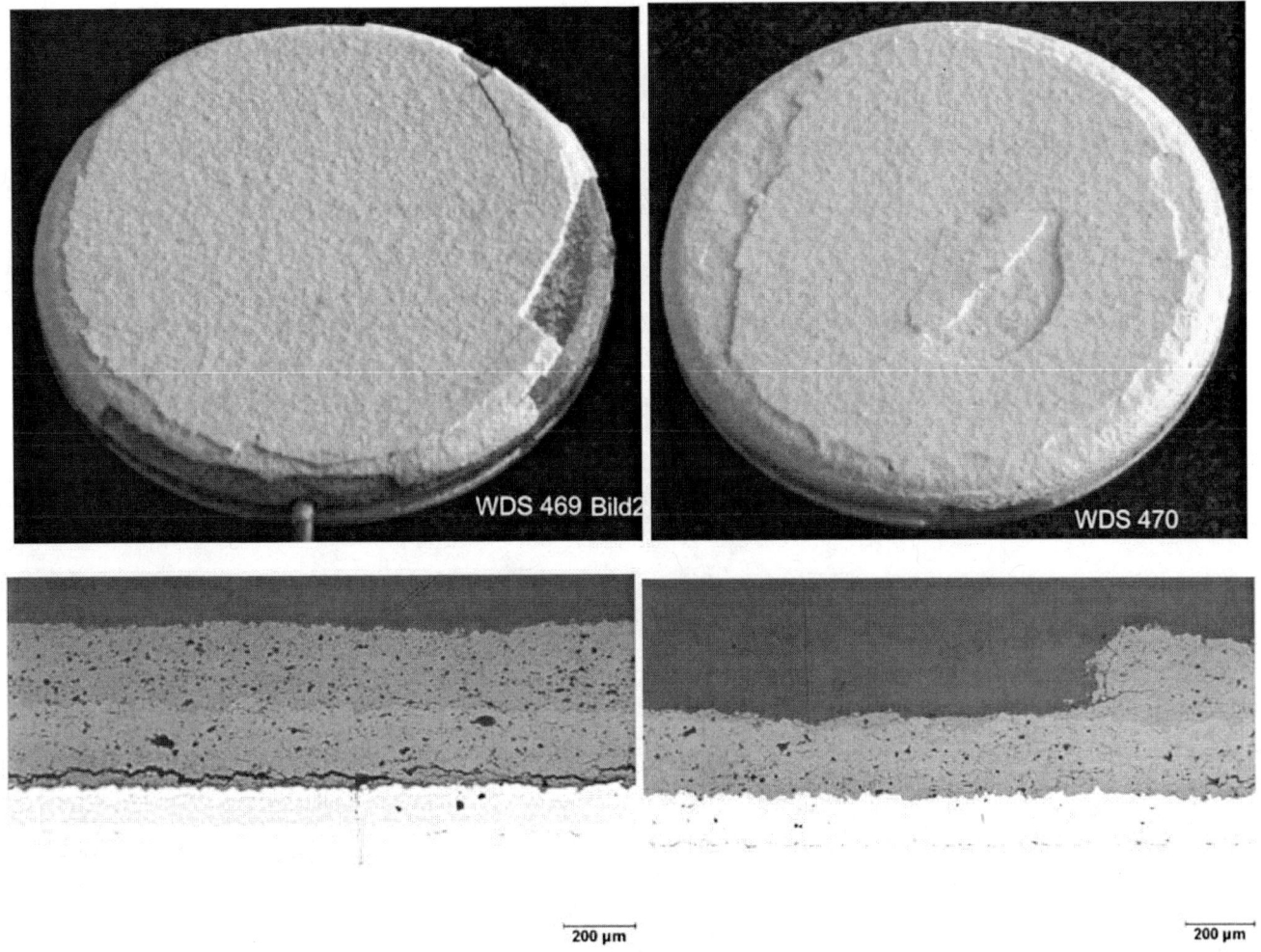

Fig. 10. Photographs (top) and micrographs (bottom) of double-layer coatings prepared from LZ-Prax powder cycled for 4987 (left) and 225 cycles (right) at 1246/1360°C and 1041/1083°C surface and bondcoat temperature, respectively.

about 1340°C showed a better performance than standard YSZ coatings (Fig. 5). Two samples tested at about 1350° and 1400°C are shown in Fig. 11. For both coatings the major failure (outer rim failure was discussed above) took place close to the bondcoat, indicating that failure was driven by the bondcoat oxidation and not by the limited temperature capability of the new ceramic. This statement is also proved by the micrographs in which no long cracks with a wide crack opening are found in the new ceramic (Fig. 11). Only for the sample tested at 1400°C, and somewhat more pronounced for the one tested at 1450°C, a slight roughening of the surface is observed which might indicate the beginning of cracking of the new ceramic. On the other hand, the coating withstood more than 1200 cycles at 1450°C without massive failure and this shows the excellent temperature capabil-

ity of the new double-layer systems. Fig. 5 indicates that the maximum temperature is increased by at least 100 K by this system compared to standard YSZ coatings. The reduced lifetime of the coating cycled at 1380° can be explained by the high bondcoat temperature (see Fig. 6). It should also be mentioned here that no significant difference between the porous and the more dense coatings has been found. This is certainly an advantage with respect to processing, as the system will forgive certain deviations in the processing conditions without massive degradation of performance.

Plotting the lifetime data of the double-layer systems as a function of bondcoat temperature (Fig. 6) shows that these data lie in the intermediate- and high-temperature range well above the data of standard YSZ coatings. For the highest bondcoat temperature this is quite clear, as

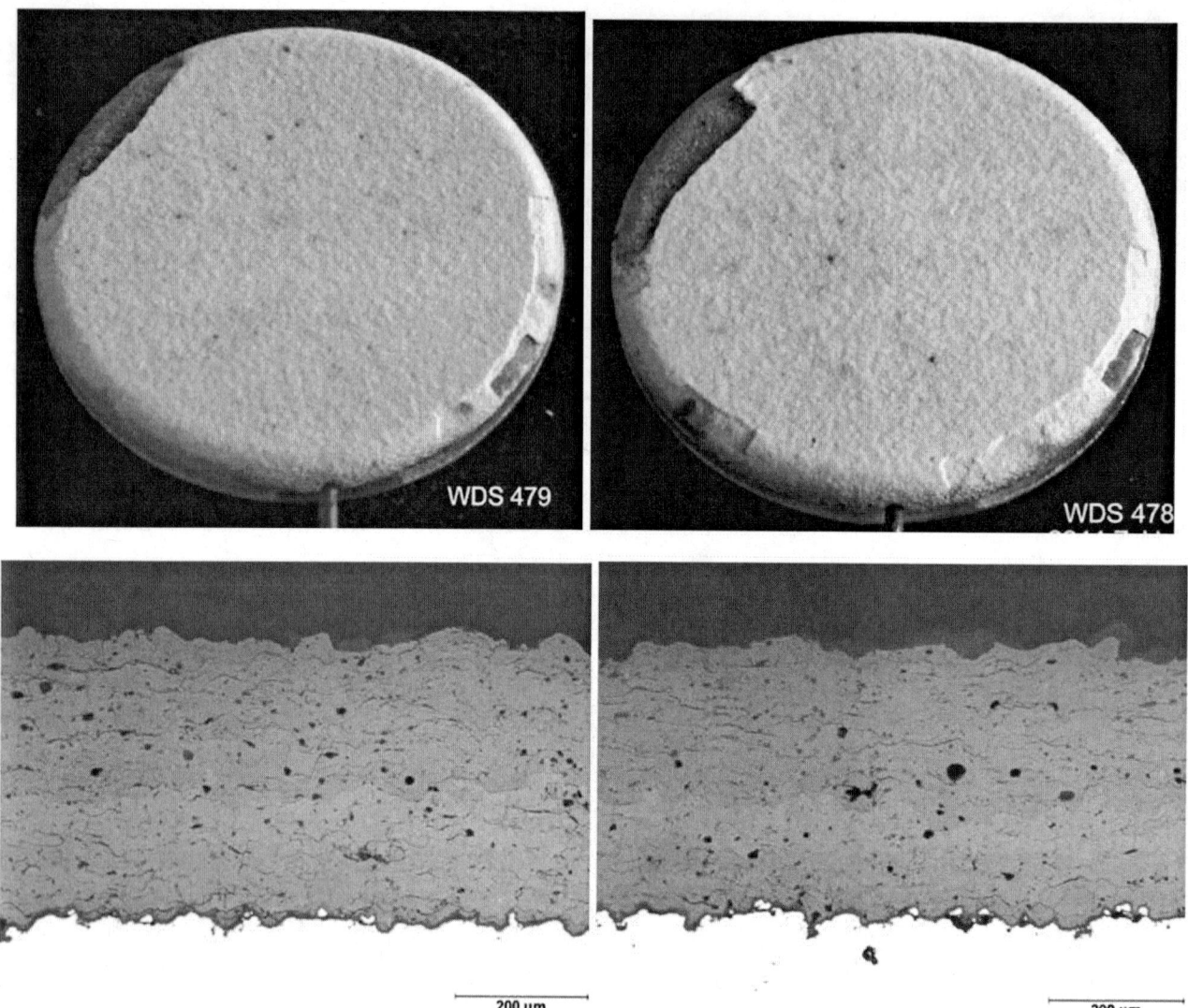

Fig. 11. Photographs (top) and micrographs (bottom) of double-layer coatings prepared from LZ-FZJ powder cycled for 2432 (left) and 2044 cycles (right) at 1350/1401°C and 1069/1090°C surface and bondcoat temperature, respectively.

lifetime is reduced in the YSZ coatings due to failure within the ceramic. However, at intermediate temperatures, at which failure is typically related to bondcoat oxidation, an increased lifetime is found for the double-layer coatings. The reason for this behavior might be the better thermal stability of the double-layer coatings, which leads to a slower relaxation at high temperatures and, consequently, a slower buildup of compressive stress levels at room temperature. These stresses are considered most critical for the failure correlated with TGO growth.[17]

Summary

TBC systems made of $La_2Zr_2O_7$ and $Gd_2Zr_2O_7$ have been produced by plasma spraying and the performance of these coatings has been tested in gas burner thermal cycling facilities. Single-layer coatings made of pyrochlore materials had shorter lifetimes than standard YSZ coatings. In contrast, double-layer coatings consisting of a YSZ-layer at the bondcoat and a subsequent coating made of pyrochlore materials showed a much better performance. The tested double-layer coatings in the low surface temperature range (<1300°C) showed performance similar to, or, for medium bondcoat and surface temperatures, slightly better than the YSZ coatings.

The TBCs prepared from commercial pyrochlore powders at higher temperatures showed a limited temperature capability as compared to YSZ.

In contrast, double-layer coatings prepared from our own $La_2Zr_2O_7$ powder revealed excellent high-temperature capability by orders of magnitude better than YSZ. Surface temperatures up to 1450°C were applied without fast degradation of the coating. The results indicate that a surface temperature increase of at least 100 K compared to standard YSZ seems possible with the new double-layer approach.

Conclusions

TBC systems having a 100 K higher temperature capability than YSZ allow greater efficiency of gas turbines by either increasing the combustion temperature or by increasing the surface temperature of the structural components. The second measure will reduce the heat transfer from the combustion gases and, hence, the amount of cooling gas. For pyrochlore materials with their low thermal conductivity, this is possible even without changing the coating thickness. As a result, these new coatings will provide significantly greater efficiency and they can be quite easily adapted to existing gas turbine systems.

Acknowledgment

The authors would like to thank Mr. K.H. Rauwald and Mr. R. Laufs (both IWV1, FZ Jülich) for the manufacture of the plasma-sprayed coatings and Mrs. A. Lemmens for the thermal cycling and the porosity measurements of the specimens. The authors also gratefully acknowledge the work of Mrs. H. Moitroux (IWV1), Dr. D. Sebold (IWV1), Mr. P. Lersch (IWV2), Mr. D. Weigand (IWV1), and Mr. M. Kappertz (IWV1), who performed the characterization of the samples by photography, SEM, XRD, and optical microscopy.

Special thanks to Mr. J. Munroe from Praxair, Indianapolis, for his cooperation.

References

1. W.A Nelson,. R.M. Orenstein, "TBC Experience in Land-Based Gas Turbines," *Journal of Thermal Spray Technology*, 6 [2] 176-180 (1997).
2. D. Stöver, C. Funke, "Directions of Developments of Thermal Barrier Coatings in Energy Applications," *Materials Processing Technology*, 92-93 195-202 (1999).
3. C. Funke, B. Siebert, R. Vaßen, D. Stöver, "Properties of $ZrO_2 – 7$ wt% Y_2O_3 Thermal Barrier Coatings in Relation to Plasma Spraying Conditions," 277-284 in the *Proceedings of the United Thermal Spray Conference* (September 15-19, 1997, Indianapolis, Indiana), ed. C.C. Berndt, ASM International, Materials Park, OH, 1998.
4. R. A. Miller J. L. Smialek, R.G. Garlick, "Phase Stability in Plasma-Sprayed Partially Stabilized Zirconia-Yttria," 241-251 in *Science and Technology of Zirconia, Advances in Ceramics*, Vol. 3, eds. A.H. Heuer and L.W. Hobbs, The American Ceramic Society, Columbus, OH, USA, 1981.
5. R.L. Jones, R.F. Reidy, D. Mess, "Scandia, Yttria Stabilized Zirconia for Thermal Barrier Coatings," *Surface Coating and Technology*, 82 70-76 (1996).
6. R. Vaßen, F. Tietz, G. Kerkhoff, R. Wilkenhöner, D. Stöver, "New Materials for Advanced Thermal Barrier Coatings," 1627-35 in *Proceedings of the 6th Liège Conference, Part III, Materials for Advanced Power Engineering*, eds. J. Lecomte-Beckers, F. Schubert, P.J. Ennis, Forschungszentrum Jülich GmbH, Jülich, Germany, 1998.
7. R. Vassen, X. Cao, F. Tietz; Basu, D. Stöver, "Zirconates as New Materials for Thermal Barrier Coatings," *J. Am. Ceram. Soc.*, 83 [8] 2023-28 (1999).
8. Schäfer, G. W.; Gadow, R., "Lanthane Aluminate Thermal Barrier Coating," *Ceram. Eng. Sci. Proc.*, 20 [4] 291-297 (1999).
9. M. Dietrich, V. Verlotski, R. Vaßen, D. Stöver, "Metal-Glass Based Composites for Novel TBC Systems," *Materialwissenschaft und Werkstofftechnik*, 8 669-672 (2001).
10. X.Q. Cao, R. Vassen, S. Schwartz, W. Jungen, F. Tietz, D. Stöver, "Chemical and Thermal Stabilities of Lanthanum Zirconate Plasma-Sprayed Coatings," *J. Am. Ceram. Soc.*, 84 [9] (2001).
11. R. Vaßen, D. Stöver, "Conventional and New Materials for Thermal Barrier Coatings," 199-216 in *Functional Gradient Materials and Surface Layers Prepared by Fine Particle Technlogy*, NATO Science Series II: Mathematics, Physics and Chemistry - Vol. 16, Kluwer Academic Publishers, Dordrecht, The Netherlands, 2001.
12. D. R. Clarke, C. G. Levi, "Materials Design for the Next Generation Thermal Barrier Coatings," *Annu. Rev. Mater. Res.*, 33 383-417 (2003).
13. J. R. Nicholls, "Advances in Coating Design for High-Performance Gas Turbines," *MRS Bulletin*, Sept. 2003, 659-670.
14. H. Lehmann, D. Pitzer, G. Pracht, R. Vaßen, D. Stöver, "Thermal Conductivity and Thermal Expansion Coefficients of the Lanthanum-Rare Earth Element-Zirconate System," *J. Amer. Ceram. Soc.*, 86 [8] 1338-44 (2003).
15. U. Bast, E. Schumann, "Development of Novel Oxide Materials for TBCs," *Ceramics Engineering & Science Proceedings*, 23 [4] 525-32 (2002).
16. P.D. Harmsworth, R. Stevens, "Phase Composition and Properties of Plasma-Sprayed Zirconia Thermal Barrier Coatings," *J. Mat. Sci.*, 24 611-615 (1992).
17. R. Vaßen, G. Kerkhoff, D. Stöver, "Development of a Micromechanical Life Prediction Model for Plasma Sprayed Thermal Barrier Coatings," *Materials Science and Engineering A*, A303 100-9 (2001).
18. R. Vaßen, G. Pracht, D. Stöver, "New Thermal Barrier Coating Systems with a Graded Ceramic Coating," 202-207 in *Proc. of the International Thermal Spray Conference 2002*, Verlag für Schweißen und verwandte Verfahren DVS-Verlag GmbH, Düsseldorf, 2001.
19. R. Vaßen, G. Barbezat, D.Stöver, "Comparison of Thermal Cycling Life of YSZ and $La_2Zr_2O_7$-Based Thermal Barrier Coatings," in *Materials for Advanced Power Engineering*, eds. J. Lecomte-Becker, M. Carton, F. Schubert, P.J. Ennis, Schriften des Forschungszentrum Jülich, Reihe Energietechnik, 21 [1] 511-521 (2002).
20. Robert Vaßen, Xueqiang Cao, Detlev Stöver, "Improvement of New Thermal Barrier Coating Systems Using a Layered or Graded Structure," *Ceramic Engineering & Science Proceedings*, 22 [1] 435- 442 (2001).
21. R. Vaßen, M. Dietrich, H. Lehmann, X. Cao, G. Pracht, F. Tietz, D. Pitzer, D. Stöver, "Development of Oxide Ceramics for an Application as TBC," *Materialwissenschaft und Werkstofftechnik*, 8 673-677 (2001).
22. U. Schulz, B. Saint-Ramond, O. Lavigne, P. Moretto, A. van Lieshout, A. Borger, J. Wigren, "Low Thermal Conductivity Ceramics for Turbine Blade Thermal Barrier Coating Application," to be published in the *Ceramics Engineering & Science Proceedings*.
23. F. Traeger, M. Ahrens, R. Vaßen, D. Stöver, "A Life Time Model for Ceramic Thermal Barrier Coatings," *Materials Science and Engineering*, A 358 255-65 (2003).
24. X. Cao, R. Vassen, S. Schwartz, W. Jungen and D. Stöver, "Spray-Drying of Ceramics for Plasma-Spray Coating," *J. Eur. Ceram. Soc.*, 20 2433-2439 (2000).
25. F. Traeger, R. Vaßen, K.-H. Rauwald, D. Stöver, "A Thermal Cycling Setup for Thermal Barrier Coatings," *Adv. Eng. Mats.*, 5 [6] 429-32 (2003).
26. Robert Vaßen, Franziska Träger, Detlev Stöver, "Correlation Between Spraying Conditions and Micro Crack Density and their Influence on Thermal Cycling Life of Thermal Barrier Coatings," 1573-1582 in *Proceedings of the 2003 International Thermal Spray Conference*, Orlando, FL, USA, May 5-8, 2003, eds. B.R. Marple, C. Moreau, ASM International, Materials Park, OH.

International Journal of
Applied Ceramic Technology

Ceramic Product Development and Commercialization

Development of Advanced Low Conductivity Thermal Barrier Coatings

Dongming Zhu* and Robert A. Miller

NASA John H. Glenn Research Center at Lewis Field, 21000 Brookpark Road, Cleveland, OH 44135

Advanced multi-component, low-conductivity oxide thermal barrier coatings have been developed using an approach that emphasizes real-time monitoring of thermal conductivity under conditions that are engine-like in terms of temperatures and heat fluxes. This is in contrast to the traditional approach where coatings are initially optimized in terms of furnace and burner rig durability with subsequent measurement in the as-processed or furnace-sintered condition. The present work establishes a laser high-heat-flux test as the basis for evaluating advanced plasma-sprayed and electron beam-physical vapor deposited (EB-PVD) thermal barrier coatings under the NASA Ultra-Efficient Engine Technology (UEET) Program. The candidate coating materials for this program are novel thermal barrier coatings that are found to have significantly reduced thermal conductivities and improved thermal stability due to an oxide defect-cluster design. Critical issues for designing advanced low-conductivity coatings with improved coating durability are also discussed.

Introduction

Ceramic thermal barrier coatings (TBCs) are receiving increased attention for advanced gas turbine engine applications. The TBCs are considered technologically important because of their ability to further increase engine operating temperatures and reduce cooling requirements, thus achieving higher engine efficiency, lower emission, and increased performance goals. In order to fully take advantage of the TBC capability, an aggressive design approach—allowing greater temperature reductions through the coating systems and less cooling air to the components—is required whenever possible. Advanced TBCs that have significantly lower thermal conductivity, better thermal stability, and higher toughness than current coatings using advanced design approaches must be developed for future ultra-efficient and low-emission engine systems[1].

Higher surface temperatures and larger thermal gradients are expected in advanced thermal barrier coating systems as compared to conventional coating systems. As illustrated in Fig. 1, TBCs with lower thermal conductivity can be used in thin coating configurations while still achieving sufficient temperature reductions at higher engine operating temperatures. The low conductivity coatings will have a significant advantage over the conventional ones, particularly for rotating engine components (such as turbine blades), where a reduced weight is highly desirable. Considerable efforts have been made in order to develop advanced TBCs with low conductivity and high thermal stability by modifying current ZrO_2-(7-8)$wt\%Y_2O_3$ coating microstructures and porosity[2-6], using alternative oxide ceramic compounds[7-13], and doping the ZrO_2- or HfO_2-based solid solution alloy systems[1,5,14-18]. The multi-component doped, oxide alloy defect-clustered TBCs have been shown to offer the low conductivity and

* Member, the American Ceramic Society.

high stability required for future high-temperature engine applications[15-18].

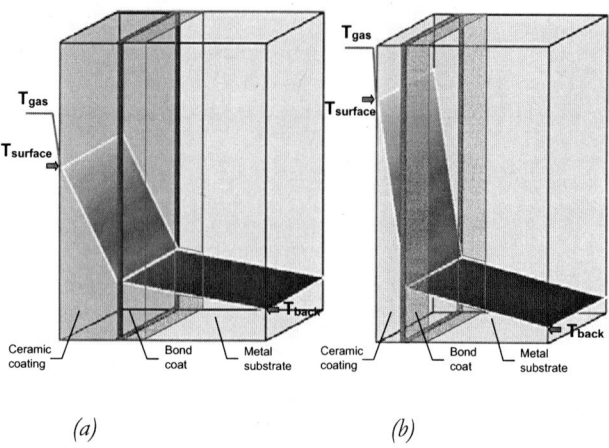

Fig. 1. Advanced TBCs with lower thermal conductivity and better temperature stability will allow the use of a thinner coating system to achieve a larger temperature reduction at higher engine operating temperatures. The substrate temperature can be maintained at a lower level while the cooling can be significantly reduced. A thin coating system is highly desirable for engine rotating components such as airfoils where a reduced weight is critical. (a) Current TBCs; (b) Advanced TBCs.

The development of low-conductivity and high-temperature-stability TBCs requires test techniques that can accurately and effectively evaluate coating thermal conductivity at high surface temperatures, typically in the range of 1300-1400°C. It is known that the coating thermal conductivity can increase significantly due to coating sintering and/or phase structure changes after a long-term thermal exposure. Therefore, evaluation of the initial and post-exposure thermal conductivities, and the rate of conductivity increase is crucial in characterizing the coating's performance. In this study, a laser high-heat-flux test technique has been developed for evaluating advanced plasma-sprayed and electron beam-physical vapor deposited (EB-PVD) thermal barrier coatings under the NASA Ultra-Efficient Engine Technology (UEET) program. The test approach emphasizes real-time monitoring of thermal conductivity and, therefore, the conductivity increases at high temperature under simulated engine thermal gradients to determine the optimum coating compositions. Novel candidate TBC materials are developed using an oxide defect-clustering concept that incorporates paired rare-earth oxide cluster dopants into zirconia-

yttria or hafnia-yttria systems[16,18], thereby achieving low thermal conductivity and sintering-resistant coating systems. The coating durability issues and the dopant effect on coating furnace cyclic behavior are also discussed.

Experimental Materials and Methods

Advanced defect-clustering oxide TBCs

Advanced low-conductivity and high-stability TBCs were developed using a multi-component oxide defect-clustering approach[1, 15-16]. The advanced oxide coatings were designed by incorporating multi-component, paired-cluster rare-earth oxide dopants into conventional zirconia- and hafnia-yttria oxide systems. The dopant oxides were selected by considering their interatomic and chemical potentials, lattice elastic strain energy (ionic size effect), polarization, as well as electro-neutrality within the oxides. The added dopant oxides were intended to effectively promote the creation of thermodynamically stable, highly defective lattice structures with essentially immobile defect clusters and/or nanoscale ordered phases, thus reducing oxide coating thermal conductivity and improving coating sintering resistance[15].

In the present study, selected oxide cluster TBC systems including ZrO_2-Y_2O_3-Nd_2O_3(Gd_2O_3,Sm_2O_3)-Yb_2O_3(Sc_2O_3) were synthesized, and their conductivity and sintering behavior were investigated. Emphasis was placed on the effect of total dopant concentrations on the coating thermal conductivity, sintering resistance, and durability. The advanced TBC systems, typically consisting of a 180-250-μm ceramic top coat and a 75-120-μm NiCrAlY or PtAl intermediate bond coat, were either plasma-sprayed or electron-beam physical vapor deposited onto the 25.4-mm-diameter and 3.2-mm-thick René N5 disk substrates. The plasma-sprayed coatings were processed using pre-alloyed powders. The ceramic powders with designed compositions were first spray-dried, then plasma-reacted and spheroidized, and finally plasma-sprayed into the coating form. The advanced EB-PVD coatings were deposited using prefabricated evaporation ingots that were made of the carefully designed compositions. The EB-PVD coatings were processed into test coating specimens by two different vendors (General Electric Aircraft Engines, Cincinnati, Ohio, and Howmet Coatings Corporation, Whitehall, Michigan).

Laser test approach for evaluating advanced TBCs

A 3.0-kW CO_2 laser (wavelength 10.6 μm) high-heat-flux thermal conductivity rig was established for evaluating advanced TBCs. The general approaches used for coating conductivity measurement under the high-temperature and high-thermal-gradient conditions have been described in detail elsewhere[19-22]. During the testing, a large thermal gradient in the ceramic coating can be established by the laser surface heating and backside air-cooling. A given constant laser-delivered heat flux was applied to the coating surface throughout a 20-hr steady-state test period. Thermal conductivity of each candidate ceramic coating was determined in real-time during the laser test, based on the applied laser heat flux and the measured temperature gradient across the coating. The surface temperature for all test specimens was at 1316°C at the beginning of the tests. The coating/metal interface temperature was approximately in the range of 950-1100°C, depending on the coating thermal conductivity and applied laser heat flux. Since the coating conductivity increases with time due to ceramic sintering, the coating surface temperature will continuously drop under the fixed laser heat-flux condition. The measured initial coating conductivity (k_0), the conductivity at 20 hr (k_{20}), and the conductivity rate of increase were used for evaluating the candidate coating performance. It should be mentioned that for some of the EB-PVD oxide coating systems, the coating conductivity after 5 hr testing (k_5) was used for characterizing the coating behavior. This is a viable approach for effectively reducing the testing time, because the EB-PVD coatings usually reached a steady-state conductivity increase stage after 5 hr of testing, with a relatively low subsequent rate of conductivity increase.

Experimental Results

Thermal conductivity of advanced TBCs

Fig. 2 illustrates the high-temperature thermal conductivity of plasma-sprayed oxide cluster TBCs as a function of test time. The advanced oxide coatings investigated in this study consisted primarily of ZrO_2-Y_2O_3, but were also co-doped with additional paired rare earth oxides Nd_2O_3-Yb_2O_3 or Gd_2O_3-Yb_2O_3 (i.e., ZrO_2-$(Y,Nd,Yb)_2O_3$ and ZrO_2-$(Y,Gd,Yb)_2O_3$ oxide systems). As a comparison, the thermal conductivity of a baseline coating, ZrO_2-$4.55mol\%Y_2O_3$ (i.e., ZrO_2-$8wt\%Y_2O_3$, or 8YSZ), is also plotted in Fig. 2. It can be seen that the coating conductivity generally increased with time. The advanced oxide cluster coating systems exhibited much lower thermal conductivity and conductivity increases than the conventional baseline coating. As shown in Fig. 2(a), approximately one-third of the 20-hr baseline coating conductivity value was achieved for some of the best coating systems after the 20-hr laser high-temperature tests. Fig. 2(b) shows that the plasma-sprayed ZrO_2-$13.5\%(Y,Nd,Yb)_2O_3$ coating achieved almost one-order-of-magnitude-lower conductivity rate of increase as compared to the ZrO_2-$4.55mol\%Y_2O_3$ baseline coating.

Fig. 3 shows thermal conductivity and the rate of conductivity increase of various plasma-sprayed oxide cluster TBCs as a function of total dopant concentration. Fig. 3(a) illustrates the initial and 20-hr conductivity values of the coatings. It can be seen that the baseline ZrO_2-$4.55mol\%Y_2O_3$ coating had an initial conductivity of about 1.0 W/m-K. The conductivity of the baseline coating increased to about 1.4 W/m-K after 20 hr of high-heat-flux testing. In contrast, the oxide cluster coatings, including ZrO_2-$(Y,Nd,Yb)_2O_3$, ZrO_2-$(Y,Gd,Yb)_2O_3$, and ZrO_2-$(Y,Sm,Yb)_2O_3$ systems, exhibited lower initial and 20-hr thermal conductivities than the baseline coating. Thermal conductivity of the oxide cluster coatings generally decreased with increasing total dopant concentration. However, a very low-conductivity region was observed in the concentration range that contains 6-13 mol% of the total dopants. Similar behavior was observed for the rate of conductivity increase data, shown in Fig. 3(b). A minimum region for the rate of increase was also observed in the dopant concentration range of 6-13 mol%, corresponding well with the low-conductivity valley region for the conductivity of the coating systems.

In order to investigate the effect of the cluster dopant concentration ratio on conductivity, plasma-sprayed ZrO_2-Y_2O_3-Nd_2O_3-Yb_2O_3 oxide coatings with decoupled cluster dopant concentrations were

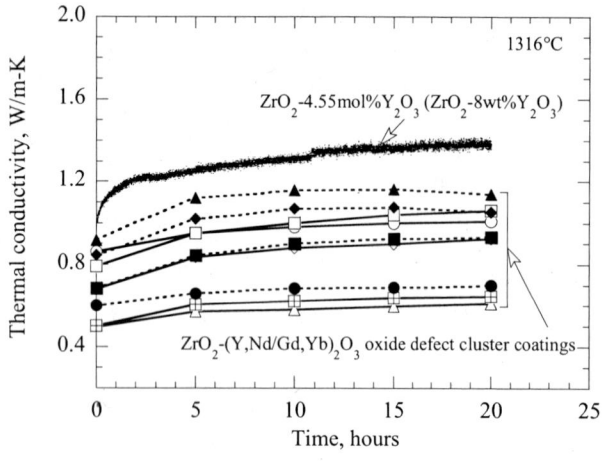

(a)

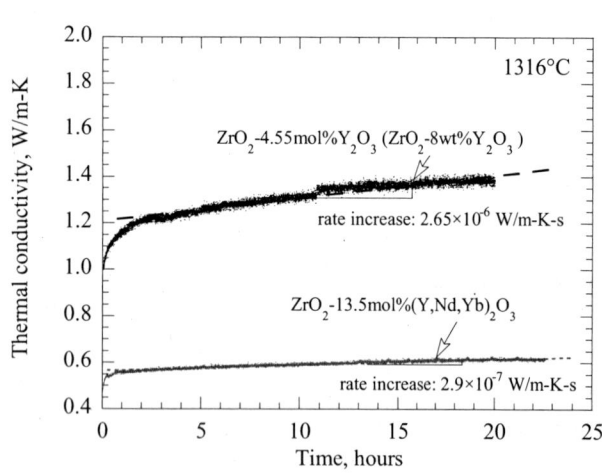

(b)

Fig. 2. *Thermal conductivity of plasma-sprayed multi-component ZrO₂-(Y,Nd/Gd,Yb)₂O₃ and ZrO₂-4.55mol%Y₂O₃ coatings as a function of time, determined using a steady-state laser heat-flux technique at the surface temperature of 1316°C. (a) Thermal conductivity of various composition ZrO₂-(Y,Nd,Yb)₂O₃ and ZrO₂-(Y,Gd,Yb)₂O₃ oxide coating systems. (b) Thermal conductivity rate of increase for the ZrO₂-13.5mol%(Y,Nd,Yb)₂O₃ and ZrO₂-4.55mol%Y₂O₃ coatings.*

designed and prepared near the optimum low-conductivity region. This set of oxide coatings had compositions ranging from ZrO₂-Y₂O₃ (i.e., YSZ) only, YSZ plus a single Nd₂O₃ or Yb₂O₃ dopant, YSZ plus both the Nd₂O₃ or Yb₂O₃, but in varying relative concentrations (with either equal or non-equal cluster

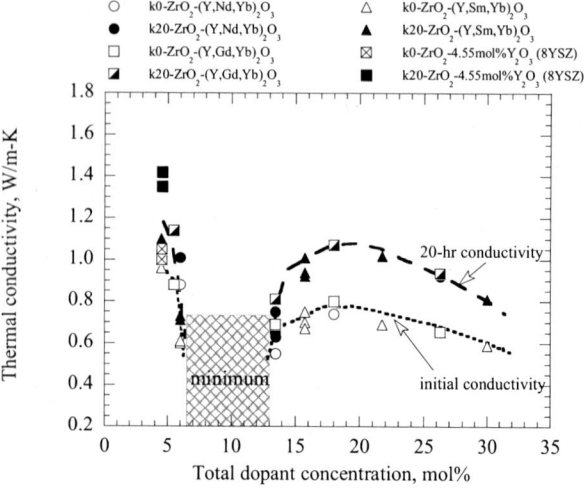

(a)

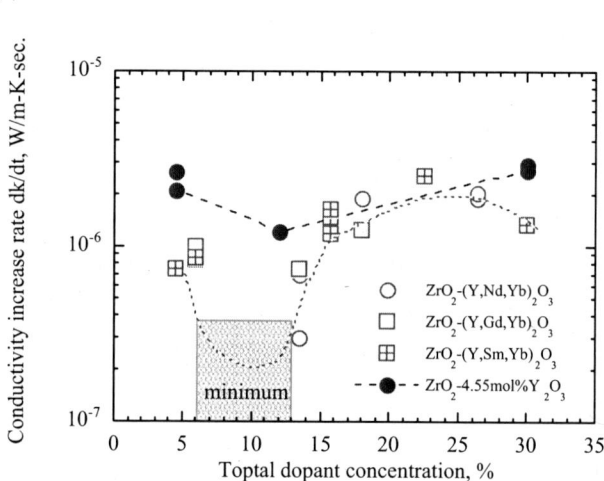

(b)

Fig. 3. *Thermal conductivity and the conductivity rate of increase of ZrO₂-(Y,Nd,Yb)₂O₃, ZrO₂-(Y,Gd,Yb)₂O₃, and ZrO₂-(Y,Sm,Yb)₂O₃ plasma-sprayed oxide defect-cluster coatings as a function of total dopant concentration, determined using a steady-state laser heat-flux technique at the surface temperature of 1316°C. A low conductivity and low-rate-increase regime is observed in the range of 6-13 mol% total dopant concentration. (a) Initial and 20-hr thermal conductivity; (b) Corresponding conductivity rate of increase.*

dopant concentrations). Fig. 4 shows the thermal conductivity results of the ZrO₂-(Y,Nd,Yb)₂O₃ oxide TBCs as a function of total dopant concentration and cluster dopant concentration ratio (ratio of Yb₂O₃ to Nd₂O₃ in mol%). It can be seen that TBCs of ZrO₂-

Y₂O₃, and ZrO₂-Y₂O₃ with a single cluster dopant, Nd or Yb, showed typically higher thermal conductivities than the coatings of ZrO₂-Y₂O₃ with paired dopant additions (Nd₂O₃+Yb₂O₃). The oxide cluster coatings with equal amount of cluster dopants added (Yb₂O₃/Nd₂O₃ = 1) often showed the lowest conductivity at a given total dopant concentration. The paired dopants (with equal cluster dopant concentrations) especially showed significant beneficial effects in reducing the coating conductivity at about 10 mol% dopant concentrations.

the baseline coating value of 1.3×10^{-6} W/m-K-s by the addition of the cluster dopants.

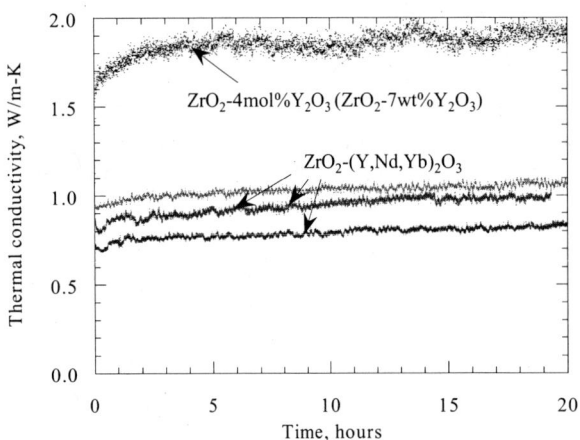

Fig. 5. Thermal conductivity of EB-PVD-processed ZrO₂-(4~6)mol%(Y,Nd,Yb)₂O₃ oxide defect cluster coatings as a function of time. The cluster coatings exhibit significantly lower thermal conductivity and conductivity rate of increase than the ZrO₂-4%molY₂O₃ (ZrO₂-7wt%Y₂O₃) coating.

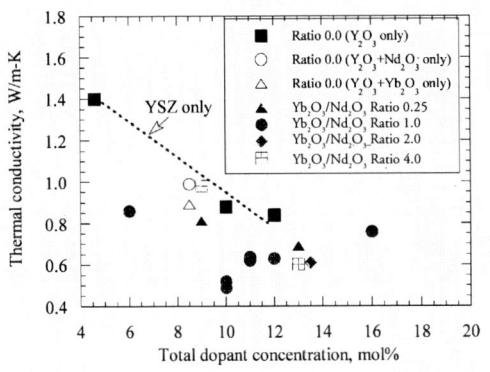

Fig. 4. Thermal conductivity of plasma-sprayed ZrO₂-(Y,Nd,Yb)₂O₃ oxide defect-cluster coatings as a function of total dopant concentration and cluster dopant concentration ratio of Yb₂O₃ to Nd₂O₃ (in mol%) near the optimum low-conductivity region. The cluster oxide coatings with the equal cluster oxide dopants (Yb₂O₃/Nd₂O₃ = 1 in mol%) showed the lowest conductivity at 10 mol% total dopant concentration.

Thermal conductivity of EB-PVD oxide cluster TBCs was also investigated using the laser heat-flux technique. Fig. 5 shows typical conductivity changes as a function of time for EB-PVD-processed ZrO₂-(4~6mol%)(Y,Nd,Yb)₂O₃ oxide cluster coatings. It can be seen that the oxide cluster coatings exhibited lower thermal conductivities and rates of conductivity increase compared to the baseline ZrO₂-4.55mol%Y₂O₃ coating. The conductivity for the oxide cluster coatings can be as low as 0.85 W/m-K after the 20-hr high-temperature testing, as compared to the conductivity of 1.85-1.90 W/m-K for the baseline coating. The rate of conductivity increase was also reduced to $0.8 \times 10^{-6} - 1.0 \times 10^{-6}$ W/m-K-s from

Fig. 6 illustrates thermal conductivity of various oxide cluster TBCs as a function of total dopant concentration after 5-hr or 20-hr laser high-heat-flux tests at 1316°C. The conductivity data were plotted for selected NASA composition oxide cluster coatings that were prepared at General Electric Aircraft Engines Company and Howmet Coatings Corporation. Note that for some coating systems, the 5-hr conductivity data k_5 (instead of 20-hr conductivity data k_{20}) were used. This is still an acceptable approach based on the consideration that there are only small differences between k_5 and k_{20} for the oxide cluster EB-PVD coatings, simply because the coatings have reached a steady-state conductivity increase stage and also have relatively low rates of conductivity increase. It can be seen that the EB-PVD coating systems generally showed lower thermal conductivity than the YSZ coatings at any given total dopant concentration. In addition, similar to the plasma-sprayed coatings, the EB-PVD coating systems also exhibited a low-conductivity region that is centered around 10 mol% total dopant concentration.

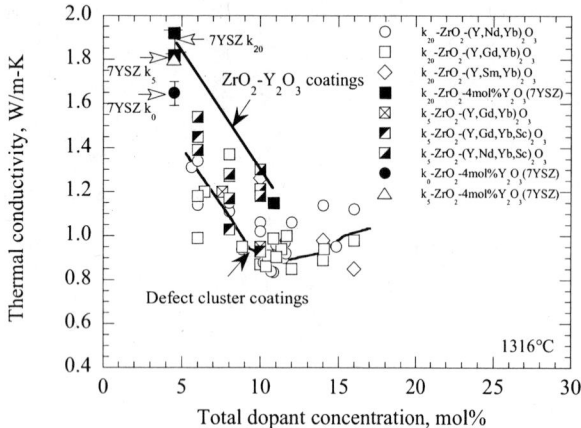

Fig. 6. *Thermal conductivity of various composition EB-PVD oxide defect-cluster coatings as a function of total dopant concentration. Similar to the plasma-sprayed coatings, the EB-PVD coating systems also possess a low-conductivity region at about 10 mol% total dopant concentration.*

Furnace Cyclic Behavior of the Advanced Thermal Barrier Coatings

Furnace cyclic oxidation tests have been carried out to evaluate durability of the advanced oxide coating systems. The coating specimens were thermal cyclic tested at 1163°C using a tubular or a box furnace with 45-min hot time cycles[23]. Fig. 7 summarizes the test results for various coating compositions which were processed from different batches. It can be seen that, regardless of the relatively large scatter, the coating cyclic life generally decreased with increasing total dopant concentration. The oxide cluster coatings followed a similar trend as compared to the yttria-zirconia (YSZ only) coatings in the furnace cyclic behavior. However, the present results suggest some beneficial effect in improving coating cyclic durability by the addition of cluster oxide dopants. The multi-component cluster oxide coatings typically showed better cyclic durability than only yttria-doped zirconia coatings at given dopant concentrations. In fact, within the optimum low-conductivity region of 6-13 mol% dopant concentration, significant coating life improvements (in some cases, coating lives comparable to those of zirconia-4.55 mol%yttria) have been observed for the initially processed (no processing optimization) oxide cluster coatings as compared to the yttria-stabilized zirconia coatings. Moderate coating life increases were also observed by coating composition, microstructure and bond coat modifications[23]. Further life improvements will be expected by utilizing advanced coatings architecture design, dopant type and composition optimization, and improved processing techniques.

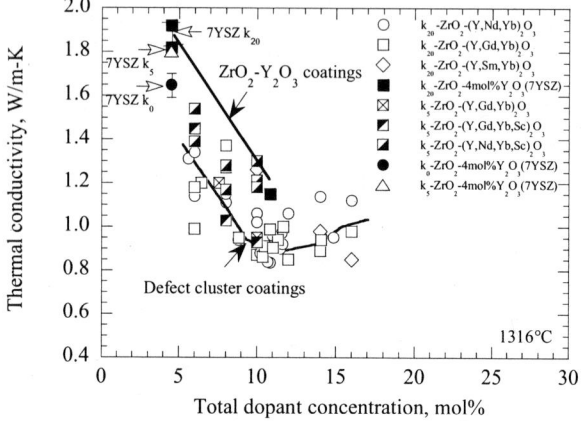

(a)

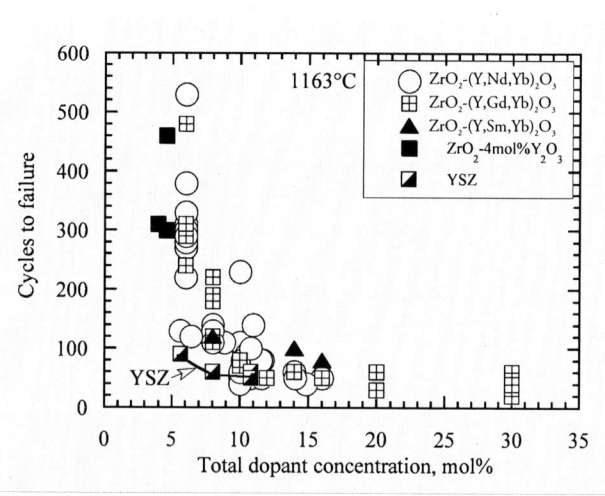

(b)

Fig. 7. *Furnace cyclic test results for (a) plasma-sprayed and (b) EB-PVD multi-component oxide defect-cluster coatings. The coating specimens were cyclic tested in a tubular or a box furnace at 1163°C with 45-min hot time cycles[23]. The coating cyclic life generally decreased with increasing total dopant concentration. The oxide defect-cluster coatings showed promise to achieve significantly better cyclic durability than only yttria-doped zirconia coatings at given dopant concentrations.*

Discussion

The intrinsic thermal conductivity of a ceramic coating is closely related to its lattice structure and lattice defects. The interactions between lattice phonon waves, and scattering of the lattice phonon and radiative photon waves by various-length scale defects will greatly affect the thermal conductivity behavior[24]. As mentioned earlier, the multi-dopant oxides were incorporated into the ZrO_2-Y_2O_3 system by considering their interatomic and chemical potentials, lattice elastic strain energy (ionic size effect), polarization and electro-neutrality within the oxides[15]. The defect-cluster design approach, using the high-stability, paired dopant oxides of distinctively different ionic sizes, will effectively produce lattice distortion in the oxide solid solutions and facilitate local ionic segregation, and thus defect clustering. Oxide defect clusters with appropriate sizes can effectively attenuate and scatter lattice phonon waves as well as radiative photon waves at a wide spectrum of frequencies. Therefore, by promoting the creation of thermodynamically stable, highly defective lattice structures with controlled defect-cluster sizes, one can expect a reduced oxide intrinsic lattice and radiation thermal conductivity for theses coatings.

The measured thermal conductivity for plasma-sprayed and EB-PVD TBCs include both the contributions from the intrinsic coating conductivity and from the micro-structural (such as coating porosity) effect. The coating thermal conductivity can be greatly reduced by the presence of micro-cracks and micro-porosity within the ceramic coatings. However, the conductivity reduction achieved by micro-porosity may not persist at high temperatures. The laser thermal conductivity test data for both the plasma-sprayed and EB-PVD TBCs showed a significant coating conductivity increase with time. The increase in measured coating thermal conductivity has been attributed to ceramic sintering and densification[20-21]. The advanced cluster oxide TBCs showed reduced conductivity increase rates and thus improved sintering resistance due to the addition of the dopant oxides.

The added cluster dopant oxides can facilitate the formation of defective oxide lattice structures with essentially immobile defect clusters and/or nanoscale-ordered phases, which improves the coating sintering resistance. The defect-clustering phenomena have been observed in the multi-component rare-earth oxide doped TBCs using high-resolution transmission electron microscopy[17]. As exemplified for the ZrO_2- or HfO_2-based oxide systems, the following reactions, describing the defect clustering and dissociation, can be written using Kröger-Vink notation[25]:

$$\left[\left(Mf_M^{'}(Mf^{3+})\right)V_O^{\cdot\cdot}\right]^{\cdot} \Leftrightarrow Mf_M^{'}(Mf^{3+}) + V_O^{\cdot\cdot}$$
(for a two-member defect complex case)

$$\left[\left(Mf_M^{'}(Mf^{3+})\right)V_O^{\cdot\cdot}\left(Mf_M^{'}(Mf^{3+})\right)\right]^{X} \Leftrightarrow 2Mf_M^{'}(Mf^{3+}) + V_O^{\cdot\cdot}$$
(for a three-member defect complex case)

$$\left[\text{defect clusters, sub-microdomains}\right]^{n} \Leftrightarrow \text{single point defects}$$
(for a general defect cluster case)

where $\left(Mf_M^{'}(Mf^{3+})\right)$ is a valence defect for a dopant cation (valence +3) at the Zr (valence +4) site, and $V_O^{\cdot\cdot}$ is the anion oxygen vacancy. Because the defect clusters are in dynamic equilibrium with the single point defects at high temperatures, the sophisticated oxide cluster design may suppress the cluster dissociation reactions at extremely high temperatures. By reducing the mobile defect concentrations through the defect clustering, the atomic (both cationic and anionic) mobility and mass transport within the oxides can be greatly reduced. This can explain why the oxide cluster TBCs exhibited a lower conductivity rate of increase, and thus sintering resistance, than the baseline yttria partially stabilized zirconia coatings.

The thermal conductivity and the conductivity rate increase showed a clear minimum for the oxide cluster TBCs at about 10 mol% total dopant concentration. This composition approximately corresponds to the phase boundary between the tetragonal phase zirconia (for the partially stabilized zirconia, t′ phase) and cubic phase zirconia at the testing temperatures. It is possible that the oxide defect clustering occurs most extensively near this phase boundary, thus showing the maximum conductivity reductions and the minimum conductivity rates of increase. The sophisticated compositional design approach will expect to significantly reduce thermal conductivity, improve sintering-creep resistance and other mechanical properties (such as fracture toughness) at high temperatures, for both the ZrO_2-based cubic and tetragonal phase coating systems.

Conclusions

Advanced multi-component, low-conductivity oxide TBCs have been developed based on an oxide defect-clustering design approach and using a laser high-heat-flux thermal conductivity technique. The laser test approach emphasizes real-time monitoring of the coating conductivity at high temperatures in order to assess the overall coating thermal conductivity performance under engine-like heat-flux and thermal gradient conditions.

The durability of the advanced low-conductivity coatings was evaluated using cyclic furnace tests. Although the advanced oxide cluster coatings followed a similar trend as the pseudo-binary ZrO_2-Y_2O_3 coatings in the furnace cyclic behavior where the coating cyclic life generally decreases with increasing the total dopant concentration, the oxide cluster coatings showed promise to have significantly better cyclic durability (comparable to that of zirconia-4.55mol%yttria) than the binary ZrO_2-Y_2O_3 coatings with equivalent dopant concentrations. Further life improvements will be expected by utilizing advanced coating architecture design, dopant type and composition optimization, and improved processing techniques. The advanced low-conductivity TBCs will be developed with long-term cyclic durability and stability at very high temperatures which are far beyond the current ZrO_2-Y_2O_3 coating capabilities.

Acknowledgments

This work was supported by NASA Ultra-Efficient Engine Technology (UEET) Program. The authors gratefully acknowledge the help of their NASA colleagues, Narottam Bansal, for hot-processed specimen processing, Jeffrey I. Eldridge, for emittance measurements, James A. Nesbitt and Charles A. Barrett, for cyclic furnace testing of the coating materials. The authors are also grateful to George W. Leissler at QSS group, Inc. at the NASA Glenn Research Center, for his assistance in the preparation of plasma-sprayed thermal barrier coatings, and to Robert W. Bruce at General Electric Aircraft Engines and Kenneth S. Murphy at Howmet Research Corporation for EB-PVD coating processing.

References

1. Dongming Zhu and Robert A. Miller, "Defect Cluster Design Considerations in Advanced Thermal Barrier Coatings," *NASA UEET Presentation*, NASA Glenn Research Center, Cleveland, Ohio, April 1999.
2. D. D. Hass, A. J. Slifka, and H. N. G. Wadley, "Low Thermal Conductivity Vapor Deposited Zirconia Microstructures," *Acta Materialia*, 49 973-983 (2001).
3. S. Gu, T. J. Lu, D. D. Haas, and H. N. G. Wadley, "Thermal Conductivity of Zirconia Coatings with Zig-Zag Pore Microstructures," *Acta Materialia*, 49 2539-2547 (2001).
4. Tian J. Lu, Carlos G. Levi, Haydn N. G. Wadley, Anthony G. Evans, "Distributed Porosity as a Control Parameter for Oxide Thermal Barriers by Physical Vapor Deposition," *Journal of the American Ceramic Society*, 84 2937-2946 (2001).
5. J. R. Nicholls, K. J. Lawson, A. Johnstone, D. S. Rickerby, "Method to reduce the thermal conductivity of EB-PVD TBCs," *Surface and Coatings Technology*, 383-391 151-152 (2002).
6. A. Kulkarni, Z. Wang, T. Nakamura, S. Sampath, A. Goland, H. Herman, J. Allen, J. Ilavsky, G. Long, J. Frahm, R. W. Steinbrech, "Comprehensive Microstructural Characterization and Predictive Property Modeling of Plasma-Sprayed Zirconia Coatings," *Acta Materialia*, 51 2457-2475 (2003).
7. M. J. Malony, "Thermal Barrier Coating Systems and Materials," *US Pat. No. 6,284,323*, September 2001.
8. R. Subramanian, "Thermal Barrier Coating Having High Phase Stability," *US Pat. No. 6,387,539*, May 2002.
9. Jie Wu, Xuezheng Wei, Nitin P. Padture, Paul G. Klemens, Maurice Gell, Eugenio García, Pilar Miranzo, and Maria I. Osendi, "Low-Thermal-Conductivity Rare-Earth Zirconates for Potential Thermal-Barrier-Coating Applications," *Journal of the American Ceramic Society*, 85 3031-3035 (2002).
10. Henry Lehmann, Dieter Pitzer, Gerhard Pracht, Robert Vassen, and Detlef Stöver, "Thermal Conductivity and Thermal Expansion Coefficients of the Lanthanum Rare-Earth-Element Zirconate System," *Journal of the American Ceramic Society*, 86 1338-1344 (2003).
11. Ulrich Bast and Eckart Schumann, "Development of Novel Oxide Materials for TBCs," *Ceramic Eng. Sci. Proc.*, 23 525-532 (2002).
12. R. Gadow, and M. Lischka, "Lanthanum Hexaaluminate-Novel Thermal Barrier Coatings for Gas Turbine Applications—Materials and Process Development," *Surface and Coatings Technology*, 151-152 392-399 (2002).
13. Dongming Zhu, Narottam P. Bansal, and Robert A. Miller, "Thermal Conductivity and Stability of Hafnia- and Zirconate-Based Materials for 1650°C Thermal/Environmental Barrier Coating Applications," *Ceramic Transactions*, 153 331-343 (2003). Also NASA TM-212544, 2003.
14. Dongming Zhu and Robert A. Miller, "Sintering and Creep Behavior of Plasma-Sprayed Zirconia and HfO2-Based Thermal Barrier Coatings," *Surface and Coatings Technology*, 108-109 114-120 (1998).
15. Dongming Zhu and Robert A. Miller, "Low Conductivity and Sintering Resistant Thermal Barrier Coatings," US Provisional Patent Application Serial No. 60/263,257, USA; US Patent Application Serial No. 09/904,084, USA, January 2001.
16. Dongming Zhu and Robert A. Miller, "Thermal Conductivity and Sintering Behavior of Advanced Thermal Barrier Coatings," *Ceramic Eng. Sci. Proc.*, 23, 457-468 (2002).
17. Dongming Zhu, Yuan L. Chen and Robert A. Miller, "Defect Clustering and Nano-phase Structure Characterization of Multicomponent Rare-Earth Oxide Doped Zirconia-Yttria Thermal Barrier Coatings," *Ceramic Eng. Sci. Proc.*, 24 525-534 (2003). Also NASA TM-212480.
18. Dongming Zhu and Robert A. Miller, "Hafnia-Based Materials Development for Advanced Thermal/Environmental Barrier Coating Applications," in *Research and Technology 2003*, NASA Glenn Research Center, NASA TM, in press.
19. Dongming Zhu and Robert A. Miller, "Thermal Conductivity Change Kinetics of Ceramic Thermal Barrier Coatings Determined by the Steady-State Laser Heat Flux Technique," in Research and Technology 1999, NASA Glenn Research Center, NASA TM-209639, 29-31, March 2000.
20. Dongming Zhu and Robert A. Miller, "Thermal Conductivity and Elastic Modulus Evolution of Thermal Barrier Coatings Under High Heat Flux Conditions," *Journal of Thermal Spray Technology*, 9 175-180 (2000).
21. Dongming Zhu, Robert A. Miller, Ben A. Nagaraj, and Robert W. Bruce, "Thermal Conductivity of EB-PVD Thermal Barrier Coatings Evaluated by a Steady-State Laser Heat Flux Technique," *Surface and Coatings Technology*, 138 1-8 (2001).
22. Dongming Zhu, Narottam P. Bansal, Kang N. Lee and Robert A. Miller, "Thermal Conductivity of Ceramic Coating Materials Determined by a Laser Heat Flux Technique," *High Temperature*

Ceramic Matrix Composites IV, Proc. the 4th High Temperature Ceramic Matrix Composites Conference (HT-CMC 4), Munich, Germany, Oct. 1-3 (2001), edited by W. Krenkel, R. Naslain, and H. Schneider, Wiley-VCH, Verlag GmBH, Germany, 262-267 (2001). Also NASA TM-211122, NASA Glenn Research Center, Cleveland, 2001.

23.   Dongming Zhu, James A. Nesbitt, Terry R. McCue, Charles A. Barrett, and Robert A. Miller, "Furnace Cyclic Behavior of Plasma-Sprayed Zirconia-Yttria and Multi-Component Rare Earth Doped Thermal Barrier Coatings," *Ceram. Eng. Sci. Proc.*, 23 533-546 (2002).

24.   P. G. Klemens and M. Gell, "Thermal Conductivity of Thermal Barrier Coatings," *Materials Science and Engineering*, A245 143-149 (1998).

25.   F. A. Kröger, *The Chemistry of Imperfect Crystals*, Amsterdam: North-Holland, 1964.

Developments in Processing

PROCESS AND EQUIPMENT FOR ADVANCED THERMAL BARRIER COATINGS

Albert Feuerstein, Neil Hitchman, Thomas A. Taylor, Don Lemen
Praxair Surface Technologies, Inc.
Indianapolis, Indiana, USA

ABSTRACT

State of the art advanced thermal barrier coating (TBC) systems for aircraft engine and power generation hot section components consist of EB-PVD applied yttria stabilized zirconia and platinum modified diffusion aluminide bond coating. Thermally-sprayed ceramic coatings are still extensively used for combustors and power generation blades and vanes. This paper highlights the key features of plasma spray and EB-PVD coating processes for TBC. The process and coating characteristics of APS low density and dense vertically cracked (DVC) Zircoat™ TBC as well as EB-PVD coatings are described. The most important bondcoat processes are touched. The major coating cost elements such as material, equipment and processing are explained for the different technologies. New trends in TBC development such as ultra pure Zirconia for improved sintering resistance and low conductivity compositions are addressed.

INTRODUCTION

Thermal barrier coating systems (TBC's) are widely used in modern gas turbine engines to lower the metal surface temperature in combustor and turbine section hardware and to meet increasing demands for greater fuel efficiency, lower NOx emissions, and higher power and thrust. The engine components exposed to the most extreme temperatures are the combustor and the initial rotor blades and nozzle guide vanes of the high pressure turbine. Metal temperature reductions of up to 165°C are possible when TBC's are used in conjunction with external film cooling and internal component air cooling[1]. A diagram of the relative temperature reduction achieved using both TBC and cooling air technologies on hot section hardware is shown in Figure 1.

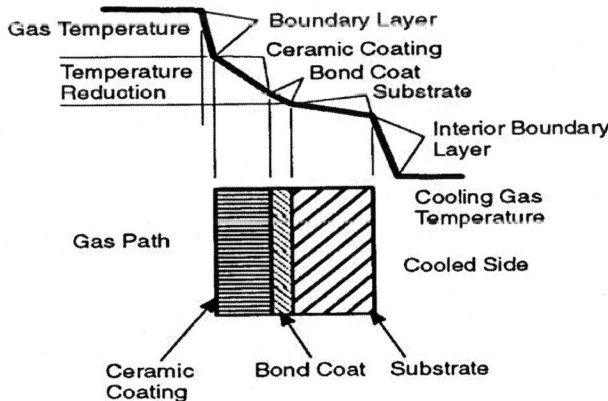

Figure 1. Schematic of a TBC on an air-cooled gas turbine engine component [2]. The thermal barrier acts as a heat flow resistor and is only efficient in air cooled engine components.

A typical thermal barrier coating system consists of two key layers: an oxidation resistant bondcoat such as diffusion aluminide or overlay MCrAlY bond coating, and a plasma sprayed or electron beam evaporated [3] ceramic top layer, typically 7-8 wt.% Y_2O_3-stablilized ZrO_2 (7YSZ), to reduce the heat flux into the component.

Significant effort is going into the development of new ceramic compositions with lower thermal conductivity. Concepts used are advanced multicomponent zirconia (ZrO_2)-based TBC's using an oxide defect clustering design with and materials with a pyrochlore structure [4,5]. A promising innovation consists of doping partially-stabilized YSZ with paired-cluster rare-earth oxides. [6,7,8,9]

An aluminum-enriched bond coat composition is used to provide a slow growing, adherent aluminum oxide film, otherwise know as a thermally grown oxide, or TGO [10]. This alumina scale is an ideal oxygen diffusion barrier, since it has one of the lowest oxygen diffusion rates of all protective oxide films [11]. Bondcoat oxidation is the primary cause of TBC failure [12]. Thermal cycling testing has shown the modes of failure to include the growth of a delamination crack in the zirconia layer just above the TGO and bondcoat, cracking within the TGO, and at the TGO / bondcoat interface [13,14,15].

This paper concentrates on thermal barrier coating systems for selected high temperature applications in a modern aircraft and industrial gas turbine engines. Specifically, the key features of each layer which affect performance, common deposition technologies, and the economic factors are addressed. More recent development at Praxair Surface Technologies, Inc., (PST), towards more sintering resistant TBC are highlighted. The feasibility of low conductivity ceramic compositions as a substitute for 7YSZ as low density plasma TBC and DVC are investigated.

7YSZ CERAMIC THERMAL BARRIER COATINGS

State of the art 7YSZ ceramic top layers are applied either by plasma spray or by electron beam physical vapor deposition (EB-PVD) [2,3]. Alternate thermal spray technologies using liquid precursors show promising results and potential but need still optimization work [16].

Plasma Sprayed Ceramic Thermal Barrier Coatings

A ceramic top layer deposited by air plasma spray (APS) consists of either of two morphologies. A "low density" coating exhibits an even spacing of pores and voids ranging from 20 µm to nano-sized, with sub-critical horizontal micro-cracking between individual splat layers. The coating density usually ranges between 80 and 86 percent of the theoretical value. Figure 2 shows optical and SEM images of a typical low density coating in cross-section.

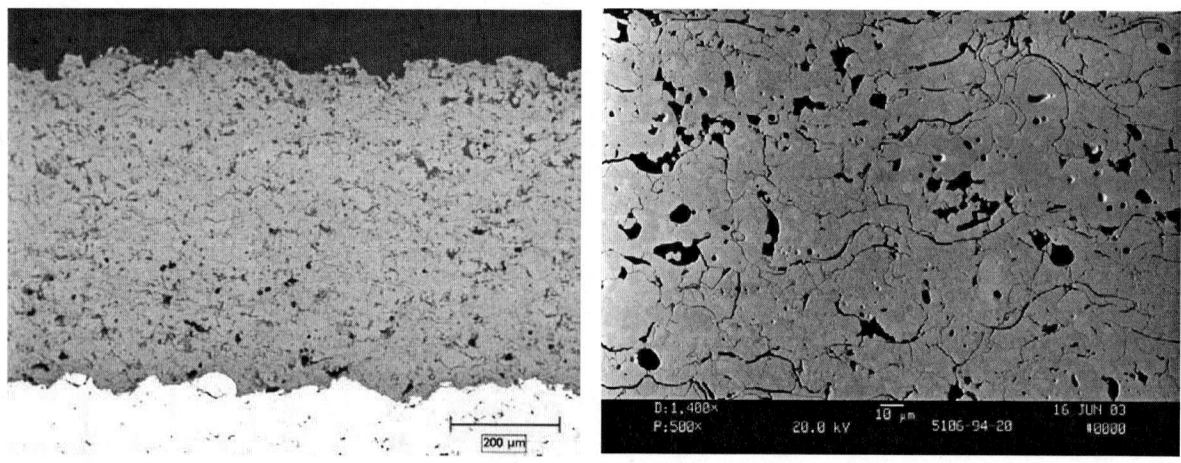

Figure 2: Cross-section of APS low density TBC, containing approximately 15% porosity. Two different magnifications are shown to reveal the "macro" and "micro" structure of the coating.

Alternatively, dense, vertically segmented coatings with improved tolerance of the ceramic layer to the strain caused by the CTE mismatch of ceramic and bondcoat., such as Zircoat™ [17] shown in Figure 3, are successfully used in both aircraft and land-based gas turbine engines [18,19].

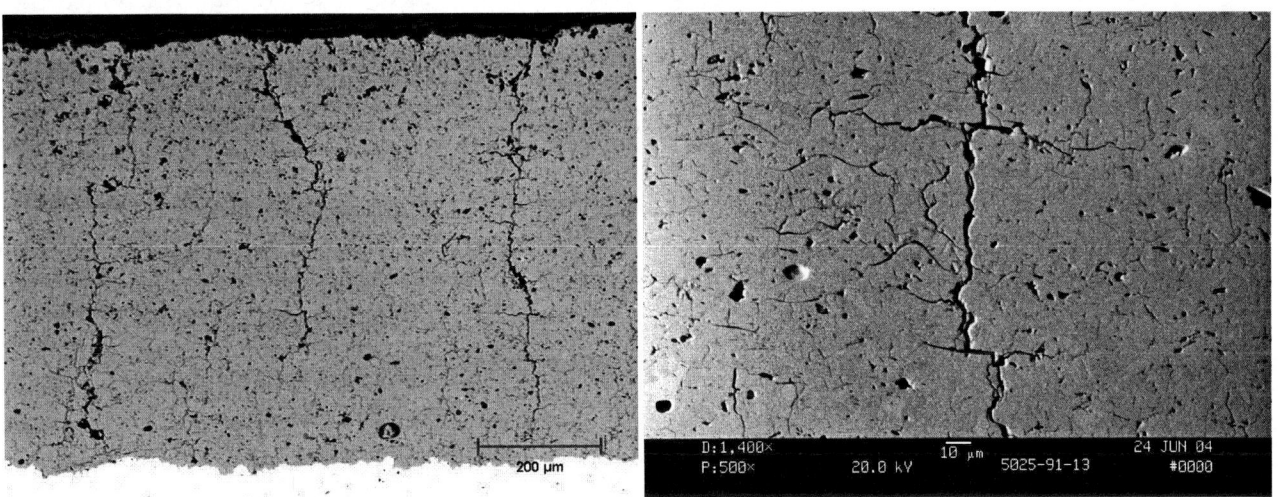

Figure 3: Cross-section of APS Zircoat™ TBC, containing approximately 16-24 cracks per linear centimeter. Two different magnifications are shown to reveal the "macro" and "micro" structure of the coating.

EB-PVD Ceramic Thermal Barrier Coatings

The 7YSZ EB-PVD TBC polished cross-section in Figure 4 shows a plurality of fine columnar grains nucleating on top of an aluminide bond coat. These subsequently increase in size during the vapor deposition process due to competitive growth. The loosely bonded columnar grains provide a high degree of mechanical compliance. However, the lack of large splat boundaries and other features normal to the heat flow direction ensures that 7YSZ EB-PVD TBC's will have relatively higher thermal conductivity values versus their plasma-sprayed counterparts of the same composition [13].

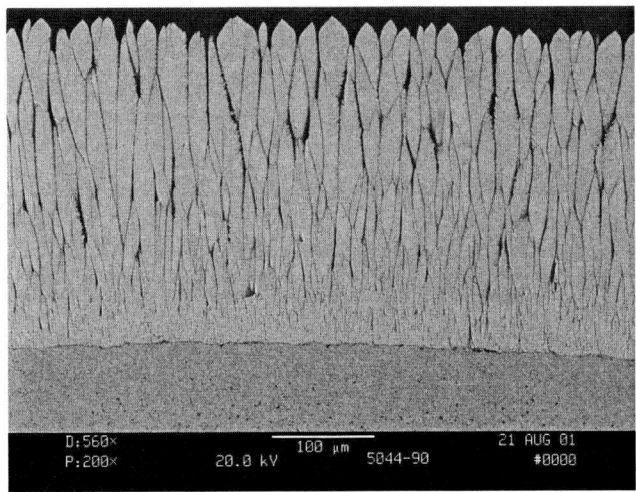

Figure 4: EB-PVD TBC, featuring a plurality of vertical, loosely-bonded columnar grains.

THERMAL CONDUCTIVITY OF APS AND EB-PVD TBC

In order to calculate thermal conductivity, one must know the specific heat of that particular ceramic composition, the density, and the thermal diffusivity as a function of temperature (measured using the laser flash technique [20]). More specific details on these techniques are published elsewhere [21,22]. TBC coatings deposited by APS offer a thermal conductivity significantly lower (0.8-0.9W/mK), than that of fully dense 7YSZ (> 2W/mK). The thermal conductivity of dense vertically cracked TBC and EB-PVD TBC is substantially higher than that of APS coatings, see Figure 5. The nominal densities of plasma sprayed TBC's range from 92% (Zircoat™) to about 85% ("Low Density"). EB-PVD coatings are close to 100% dense in direction of the column growth, however have significant inter-columnar porosity.

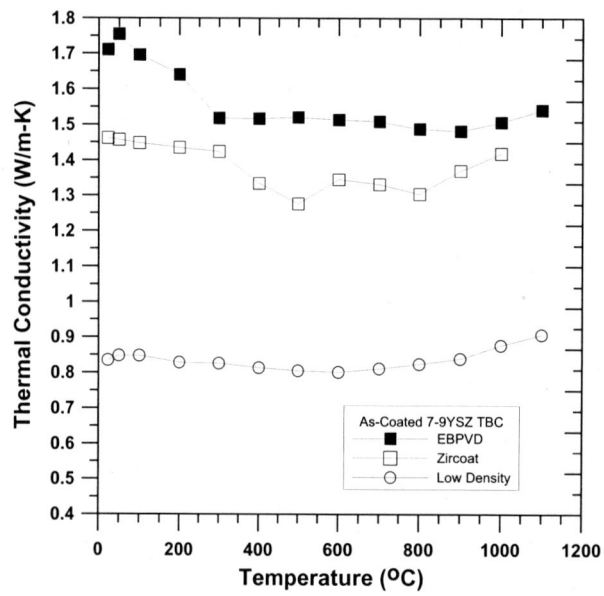

Figure 5: Temperature dependence of thermal conductivity for 7YSZ TBC's.

The thermal conductivity of APS coatings typically increases with increasing pressure, as it is the case in the engine, whereas EB-PVD thermal conductivity is almost pressure independent. The thermal conductivity typically also increases with long time thermal ageing of the coatings (APS as well as EB-PVD coatings) due to sintering effects and phase transformations in 7YSZ. Actually this is a main issue which justifies the today search for new ceramic coatings with low conductivity AND increased stability [23, 24].

ADVANCED THERMAL BARRIER COATINGS AT PST

In the next generation of engines, increased demand for greater efficiency, higher horsepower, and lower NOx emissions will require higher combustion temperatures, lighter weight materials, and less fan air diversion for cooling [4]. At Praxair we are investigating two routes for ceramic coatings with higher temperature capability. We are looking into improvements / enhancements of the current 7YSZ coatings with regard to sintering resistance and also investigate new low conductivity ceramic compositions with rare-earth oxides doped partially stabilized YSZ [6,7,8].

High Purity Sinter Resistant 7YSZ

The improvement comes from using special, highly pure YSZ powders, and spraying with either the standard PST plasma torch or a newly developed plasma torch capable of long standoff coatings of the high density, vertically crack segmented structure. The composition of the new powder compared to the standard present purity material is at least 100 time more pure in the critical alumina and silica analyses. The coatings are called Zircoat-HP™ to make this distinction [25]. Zircoat-HP™ is more thermal shock resistant than even the first version of Zircoat™. It has the same low thermal conductivity, better finishability for smoothness, and has remarkable resistance to density change upon high temperature exposure. The density of standard Zircoat™ after exposures up to 100 hours and up to 1400°C can decrease by nearly 20% due to fine porosity agglomeration. Also, the segmentation cracks begin to close at 1400°C. In contrast, Zircoat-HP™ density does not change (Figure 6) and the segmentation remains after the same exposures [26].

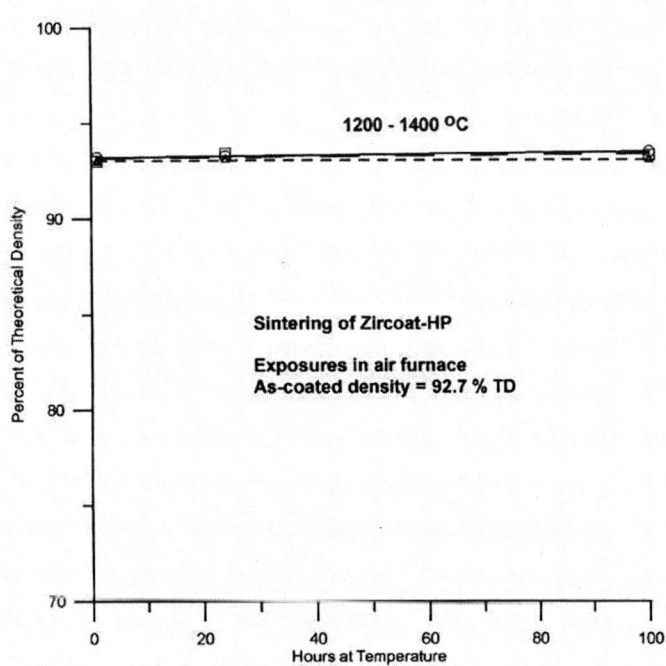

Figure 6: Density of Zircoat-HP™ vesus long term exposure at 1200 and 1400 C is demonstrates the stability of the coating against sintering.

The new long standoff plasma torch developed by PST [27] allows either Zircoat™ or Zircoat-HP™ to be coated on large IGT blades and vanes.

Low Conductivity TBC

In a baseline study we have investigated the feasibility of utilizing lower thermal conductivity coatings of a rare-earth oxide doped YSZ composition in two APS coating morphologies – as low density coating and as dense, vertically macro-cracked (DVM) coating. The APS coatings were generated from spray-dried and sintered powder of a rare-earth oxide doped TBC consisting of Bal. ZrO_2 – 9.23 wt.% Y_2O_3 – 5.12 wt.% GdO_2 – 5.56 wt.% Yb_2O_3 [9]. The micrographs of polished TBC cross-sections shown for the rare-earth oxide doped YSZ samples in Figure 7, exhibit the typical features that are found in APS 7YSZ coatings. The dense vertically macrocracked coatings contain a controlled population of vertical segments between 16 to 24 cracks per linear centimeter.

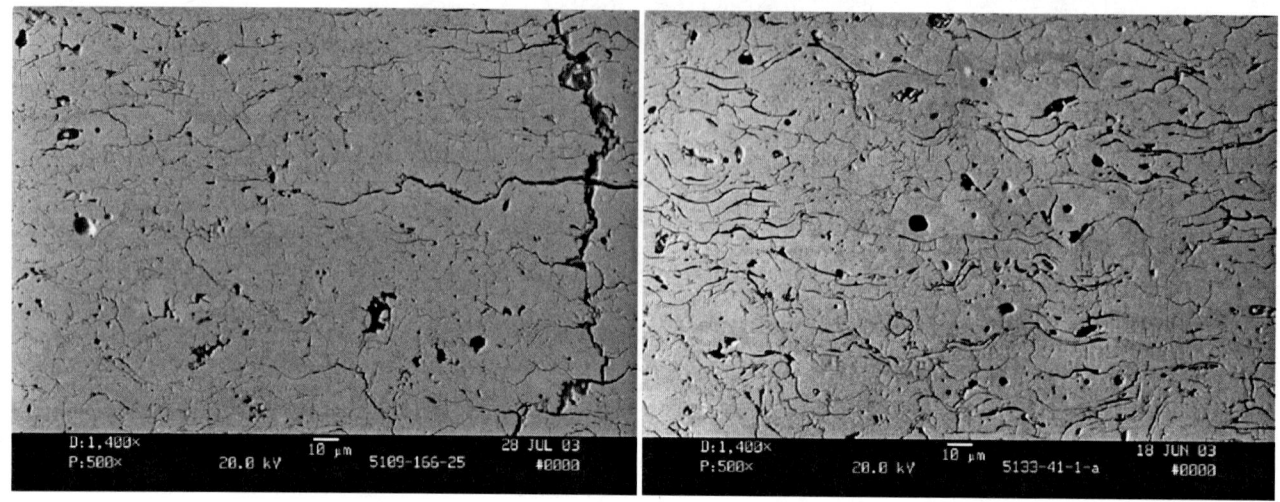

Figure 7: SEM images of rare-earth oxide doped YSZ TBC metallographic cross-sections - APS vertically macrocracked (left), nominally 92% dense. APS low density (right), nominally 85% dense.

The thermal conductivity of the REO-doped YSZ coatings measured on heating is shown in Figure 8.

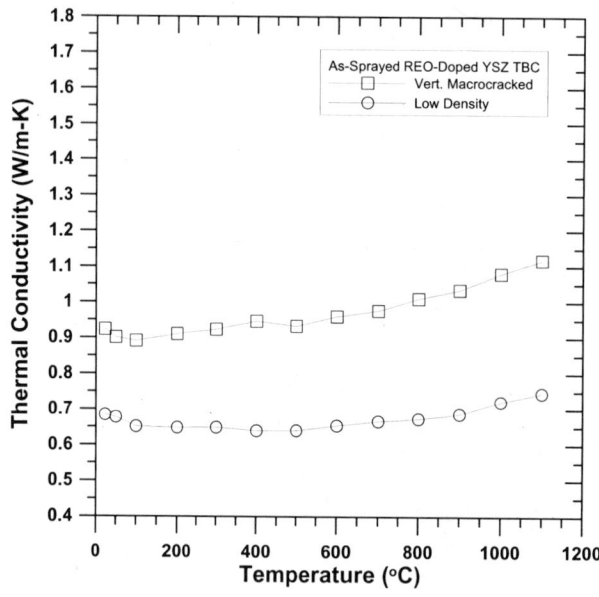

Figure 8: Temperature dependence of thermal conductivity for REO-doped YSZ TBC's

Figure 9 depicts the differences in "net benefit," or percentage reduction in thermal conductivity for each APS coating morphology, resulting from the substitution of rare-earth oxide doped YSZ for 7-8 wt.% YSZ. At room temperature, the dense, vertically segmented TBC's realize the greater "net benefit" from utilizing REO-doped YSZ, with a 38% reduction in thermal conductivity. At 1000°C, this benefit falls to a 24% reduction.

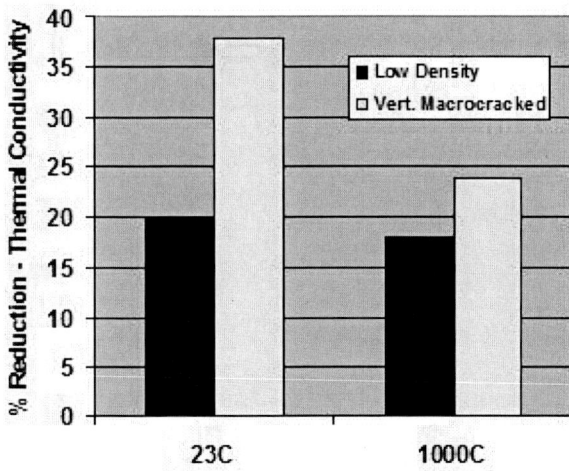

Figure 9: Plot showing differences in percentage reduction in thermal conductivity at 23°C and 1000°C in APS low density and dense vertically macrocracked TBC's when rare-earth doped YSZ is substituted for 7-8 wt. % YSZ.

There is still an issue with the thermal shock resistance of these advanced coating compositions. We observed premature spallation of the coating at the interface to the MCrAlY bondcoat. A solution to this problem is the deposition of a two layer ceramic coating, a first thin 7YSZ adherence layer followed by a second low conductivity thermal barrier layer. This approach combines the excellent thermal stability and shock resistance of the 7YSZ with the better thermal barrier properties of the low conductivity composition. To minimize the stress at the ceramic / ceramic interface of this dual ceramic layer, one must carefully adjust the density of both ceramic layers. Furnace cycle testing showed substantial improvement of the thermal shock resistance of the ceramic / ceramic interface when the density was matched (Figure 10).

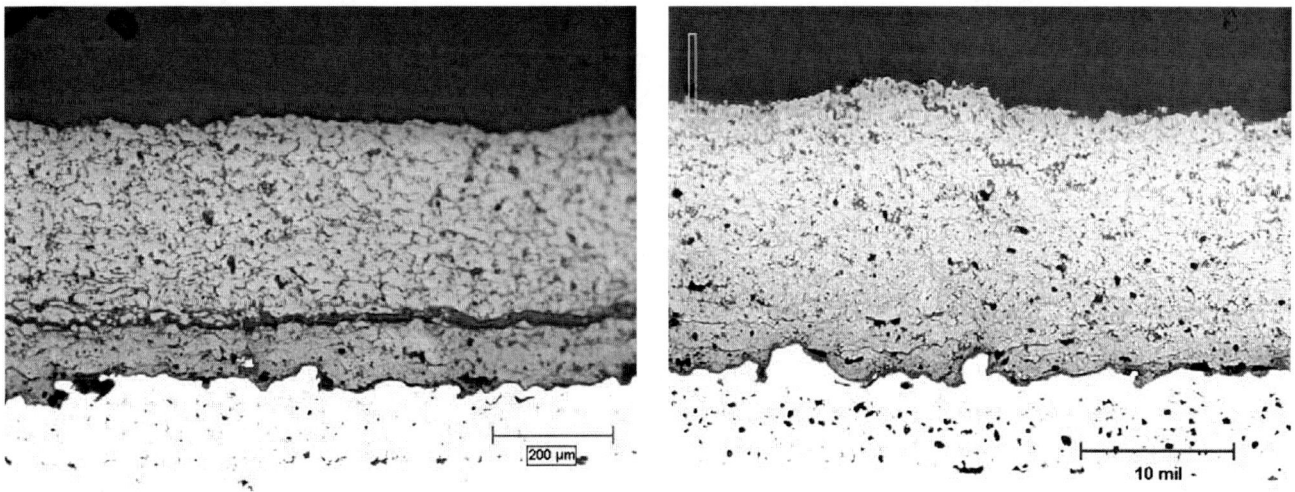

Fig 10: Thermal shock testing of two layer ceramic systems. Bond layer is 7YSZ, top layer is low conductivity TBC.
Left - two layer TBC system – dissimilar density – after 4 FCT cycles.
Right - Improved two layer TBC system –matching density – after 10 FCT cycles.

BONDCOAT

State of the art bond coatings are essentially MCrAlY composition type overlay coatings or diffusion aluminide type coatings. Initially, MCrAlY coatings were applied by electron beam physical vapor deposition. Due to cost considerations, coatings manufacturers soon developed techniques such as air, vacuum, and low-pressure plasma spraying (APS, VPS, and LPPS) and high velocity oxy-fuel (HVOF) deposition. Union Carbide (now PST) developed an inert gas shrouded plasma spray process for MCrAlY and other oxygen –reactive coatings [28] which maintained oxygen pickup to below 0.1 wt. percent. More recently, MCrAlY coatings have also been produced by Tribomet™ [29], a Ni or Co electroplating process with CrAlY powder entrapment and subsequent diffusion heat treatment.

Diffusion aluminide coatings are based on the intermetallic compound β-NiAl. Pack cementation is a commonly used process, because it is relatively inexpensive and capable of coating many small parts in one batch. Coating takes place at temperatures between 800-1000°C. Aluminum halides react on the surface of the part and deposit aluminum. More advanced processes consist of "over the pack" vapor phase aluminizing (VPA) or chemical vapor deposition (CVD). Depending on the activity of the aluminum and the coating temperature one can achieve two coating microstructures [30]. The low activity - high temperature process (1050-1100°C), forms NiAl by outward diffusion of nickel. In the high activity - low temperature process (700-950°C), Ni_2Al_3 and possibly β-NiAl forms by inward diffusion of aluminum. Typically a diffusion heat treatment is applied to form a fully homogeneous β-NiAl layer. The addition of platinum to the diffusion aluminide coating system enhances the diffusion of aluminum [31] into the substrate alloy during the diffusion aluminizing process and drastically improves the oxidation properties of the aluminide coating [32]. Typically 5-10 μm Pt is deposited by electroplating, followed by a diffusion aluminide process.

CURRENT TBC COATING PROCESSES

In the following, we concentrate on three important TBC systems as applied at Praxair Surface Technologies for aero and industrial power gas turbines, their related deposition processes and the key elements determining the coating cost (Table 1).

Table 1: Selected examples of TBC systems for various gas turbine applications.

Example	Bondcoat	Topcoat
Aerospace Combustor	NiCoCrAlY via Shrouded Plasma Spray	Zircoat™ via Air via Plasma Spray
IGT HPT Blade	NiCoCrAlY via HVOF	Low density YSZ via Air Plasma Spray
Aerospace HPT Blade	Platinum-Aluminide via Diffusion	Columnar YSZ via EB-PVD

PLASMA SPRAY COATING OF COMBUSTORS, BLADES, AND VANES

A plasma-spray booth for applying TBCs onto aerospace and IGT combustors consists not only of the physical enclosure and the equipment contained therein, such as torches, gas and power service lines, tooling, multi-axis robots for gun and/or part manipulation, and ventilation systems, but also external process control components such as control panels for power, gas flow regulation, and powder feeder settings. Other necessary capital equipment includes grit blasting and cleaning/inspection facilities for component pre-coating preparation, and may also incorporate heat treatment furnaces and stripping tanks. Both MCrAlY and YSZ ceramic layers are typically applied in the same booth.

For the coating of most annular combustors, a simple two axis manipulator is sufficient to move the plasma torch and rotate the component. Combustors with more complex geometry or with features far off-angle to the torch centerline require the use of robots and turntables or other manipulation equipment with multiple axes of movement.

BOND COATING DEPOSITION PROCESS

Two bond coat deposition techniques are reviewed here, the Shrouded Plasma system and HVOF. Both systems have similarities in their requirements to provide an acceptable coating. In the Shrouded Plasma torch, the effluent is surrounded by an argon shroud to minimize the pickup of oxygen during flight to the component. With this proprietary design, oxygen content in argon-shrouded plasma spray deposited MCrAlY has been shown to contain an order of magnitude lower oxygen content in comparison to conventional air plasma sprayed or HVOF-deposited MCrAlY coatings.

In order to provide optimum adherence for a TBC topcoat, a surface roughness of the MCrAlY bondcoat at least approximately 10 µm Ra is desirable so as to mechanically anchor the layers together. Accordingly, PST has historically applied a two-layer bondcoat system [33] onto combustors and other aerospace and IGT components via gas-shrouded plasma spray [28], in conjunction with vacuum heat treatment. The inner layer is sprayed from a finer cut of powder, and sinters to at least 95 percent theoretical density during heat treatment. This layer has an average roughness of 5 µm. The outer layer of bondcoat is consists of the same composition but of coarser powder, so as to produce a final roughness of at least 10 µm on the surface.

The High Velocity Oxy-Fuel (HVOF) deposition system comprises an oxygen and fuel mixture consisting of either kerosene, propylene, propane, natural gas or hydrogen. The mixture of oxygen and fuel is injected into the combustion chamber and is ignited. The powder is injected internally into the upper stream of the combustion flame, either axially or radially. The ignited gases form a circular flame, which surrounds the required coating powder as it flows through the nozzle. The combustion temperatures can exceed 2800°C depending on gun operating parameters and the fuel type. The flame configuration shapes the powder stream to provide uniform heating and acceleration of the powder particles. Similar to the plasma spray process the selection of gun parameters is based on providing the optimum heating and acceleration of the powder particles by the flame. Typical gas velocities are 1000 to 1200 m/s and can exceed 1500 m/s, depending on which hardware and spray parameters are utilized. To achieve the required bond coat density and surface roughness, MCrAlY applied with this technology can be either a single or dual layer. The schematic in Figure 11 demonstrates the working principle of the Praxair TAFA JP5000 HVOF gun.

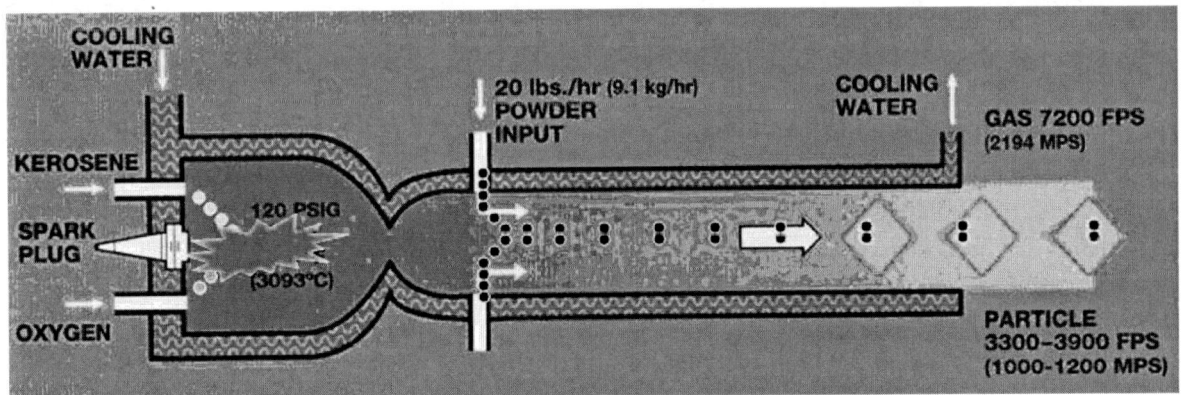

Figure 11: Working principle of the Praxair Tafa JP5000 HVOF gun.

THERMAL BARRIER COATING DEPOSITION PROCESS

The TBC is applied using plasma spray technology, with the resulting coating and its properties significantly influenced by the torch settings. YSZ ceramic powder is injected into the plasma effluent, heated, and accelerated toward the combustor surface. Where this injection takes place, in conjunction with other processing factors, dictates the TBC microstructure. Low density TBC's are best obtained from plasma torches with powder injection ports externally positioned to the exit nozzle. Higher density, vertically segmented TBCs require a more complete treatment of the injected powder. This may be done through external injection with increased torch power, or through internal injection within the torch body.

TBC Systems for Combustors

Figure 12 shows an example of a TBC coating system for a combustor, consisting of a NiCrAlY bond coating and Zircoat ceramic top layer. Generally, the thickness of MCrAlY bond coatings ranges from 125 to 250 μm. The thickness and type of thermal barrier layer is often dependent on the desired morphology, as well as the expected conditions in service. Low density YSZ coatings typically range from 250 to 500 μm, although thicknesses up to 1000 μm are not uncommon in IGT combustors. Dense vertically segmented coatings are generally thicker, due to their higher thermal conductivity. Thicknesses above 500 μm are often recommended for good thermal insulation.

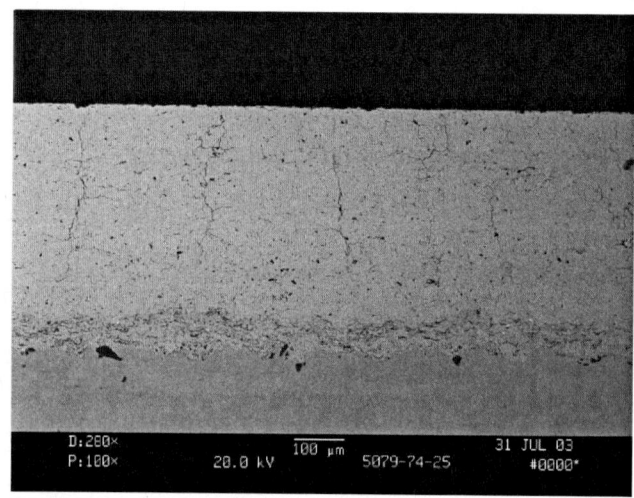

Figure 12: Combustor TBC system consisting of Zircoat™ and NiCrAlY bondcoat. Total TBC thickness appr. 500 μm.

TBC Systems for IGT Blades and Vanes.

The total coating thickness for IGT components ranges between 400 to 500 μm. Typically, HVOF is used at PST to apply the bond coating due to the large part size and long stand-off distances required during coating. A low density TBC topcoat provides maximum thermal protection and good thermal shock resistance over extended time at high operation temperatures. Plasma-sprayed ceramic thermal barriers are almost always employed. More recently, the Tribomet® process has been applied to complex 3D airfoil geometries which utilize the non line of sight features of the powder entrapment plating process. Figure 13 shows a TBC system for a typical IGT component with a low density thermal barrier coating on a HVOF bondcoat and the same system on a Tribomet® bondcoat. Figure 14 shows a large industrial vane receiving a HVOF bond coating.

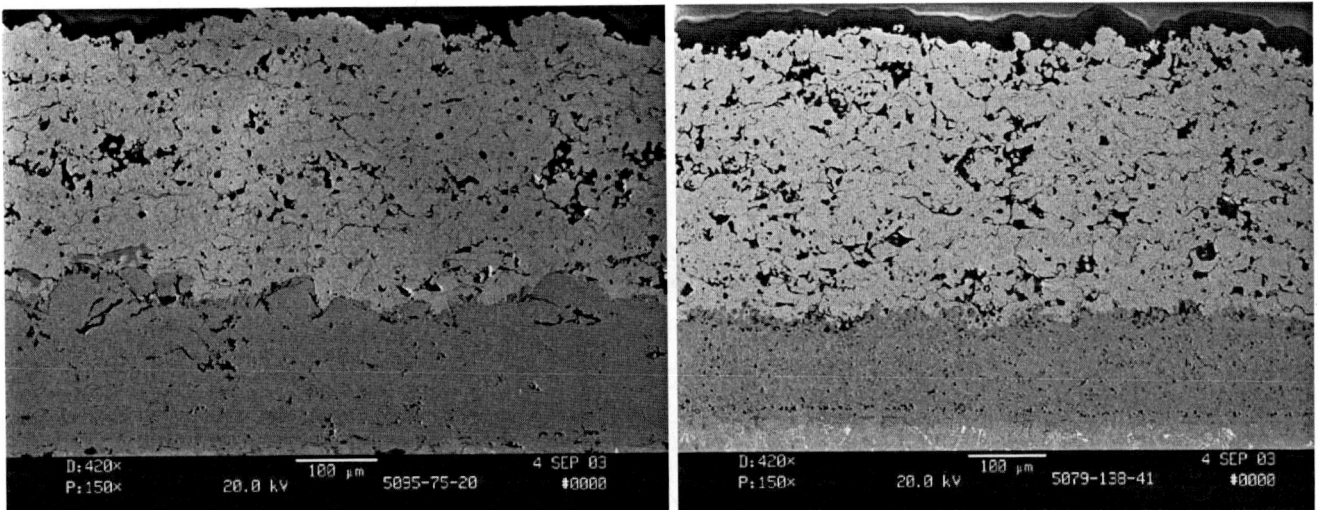

Figure 13: Left - IGT TBC coating system with two-layer HVOF bondcoat and low density TBC. Right - Low density TBC on Tribomet® NiCoCrAlY

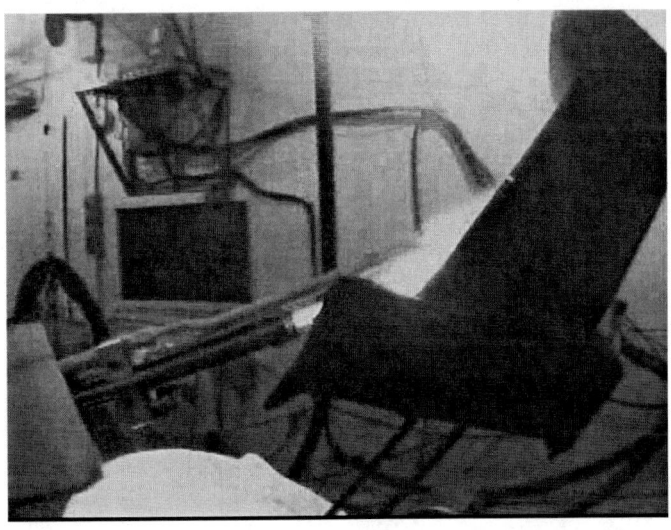

Figure 14: HVOF torch coating the T/E of a large IGT vane

PLATINUM ALUMINIDE / EB-PVD TBC SYSTEM

The platinum aluminide / EB-PVD TBC system can be applied to a variety of modern aircraft and IGT parts. The only limitation is whether the engine hardware can fit into and be easily handled within the various processing chambers. This TBC system requires three separate coating processes: electroplating, aluminizing, and EB-PVD.

Platinum Aluminide Deposition.

Platinum plating and diffusion aluminide are basically batch processes. Depending on the part size, ten to one hundred parts are processed simultaneously. Parts are cleaned in a vapor degreaser; grit blasted, and weighed prior to platinum electroplating (Figure 15a). The electroplated platinum is then diffused in a vacuum furnace prior to the aluminum diffusion. Platinum plated parts are then loaded in a retort containing the aluminum donor and activator for vapor phase aluminizing (Figure 15b). Areas of parts not receiving coating, such as the roots of turbine blades, are masked by nickel paste or tape. Retorts are then placed in the hot zone of a furnace.

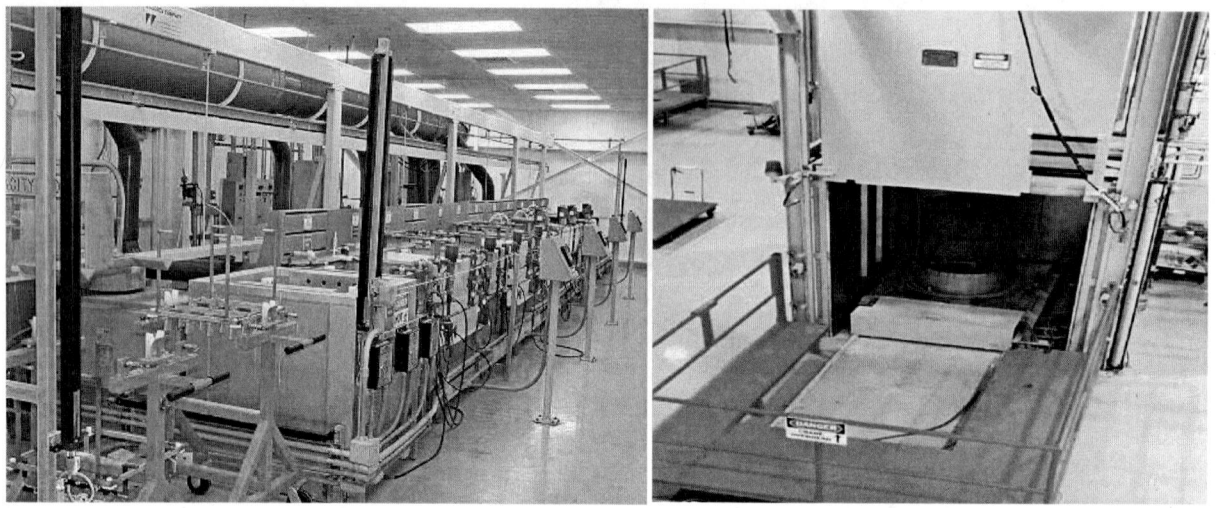

Figure 15: Pt Electroplating line (a),and VPA Aluminizing Furnace (b)

EB-PVD TBC

EB-PVD TBC coatings are produced by vacuum deposition of YSZ in a reactive atmosphere at elevated temperatures (app. 1000°C). 7YSZ ingot is evaporated by electron beams in vacuum and deposited onto the preheated parts. By a combination of rotation and tilting, a uniform coating over the airfoil surface is accomplished. To compensate for some oxygen loss during coating, a minor amount of oxygen is added to the process. Typical deposition speeds are 2 - 6 μm per minute. A highly efficient EB-PVD production coater is shown in Figure 16.

Figure 16: EB-PVD TBC 4-chamber production coater.

Progress in Thermal Barrier Coatings

COST STRUCTURE FOR SELECTED EXAMPLES

The cost factors can be grouped into three cost categories:

A- **One time cost factors** such as application development, tooling development and design, process qualification and approval, production process documentation.

B- **Direct coating related cost factors** such as direct materials (powder, ingot, plating salt, aluminizing donor alloy), auxiliary materials (grit, cleaning agent, gloves); labor for incoming inspection, cleaning and surface preparation, labor for fixturing and masking, post coating processing equipment amortization, energy, process gases.

C- **Indirect cost factors** (materials and services) such as material preparation and recycling, preventive maintenance and spare parts, strip and rework in case of nonconformance, tooling cleaning and rework, tooling replacement, packing and shipping.

Thus, the cost to coat an actual part is very much dependant on the part volume and the production life cycle of a part. With the variability in the cost elements, it can be misleading to state actual coating costs with precision. For each of the coating example given above, it is more useful to state a typical range of costs related to various processing factors. Figure 17 exhibits the main cost groups B and C, as previously mentioned. For this cost breakdown, the following assumptions hold:

- Material costs between $30 and $70 per kg for ceramic powder and ingot, and $50-100 per kg for MCrAlY powder. The price for 1 oz Pt is above $1000.

- Energy costs are essentially site related, and are based on US averages.

- Gas prices of approximately $0.1-0.2 per kg for CO_2, $0.5 per m3 for argon and $2 per liter for kerosene.

- For the investment, an APS plasma cell for approximately $0.4-0.8 million was considered. Adding an automated parts handling system can add another $0.5 to 0.8 million. A PtAl diffusion coating facility consisting of a platinum plating line, vacuum heat treat furnaces, and VPA diffusion furnaces costs between $3 million and $6 million. Equipment cost is highest for EB-PVD. Depending on the machine capacity, an EB-PVD TBC facility costs between $15-30 million. An equipment amortization scheme of 10 years depreciation with a 10% annual interest rate is assumed.

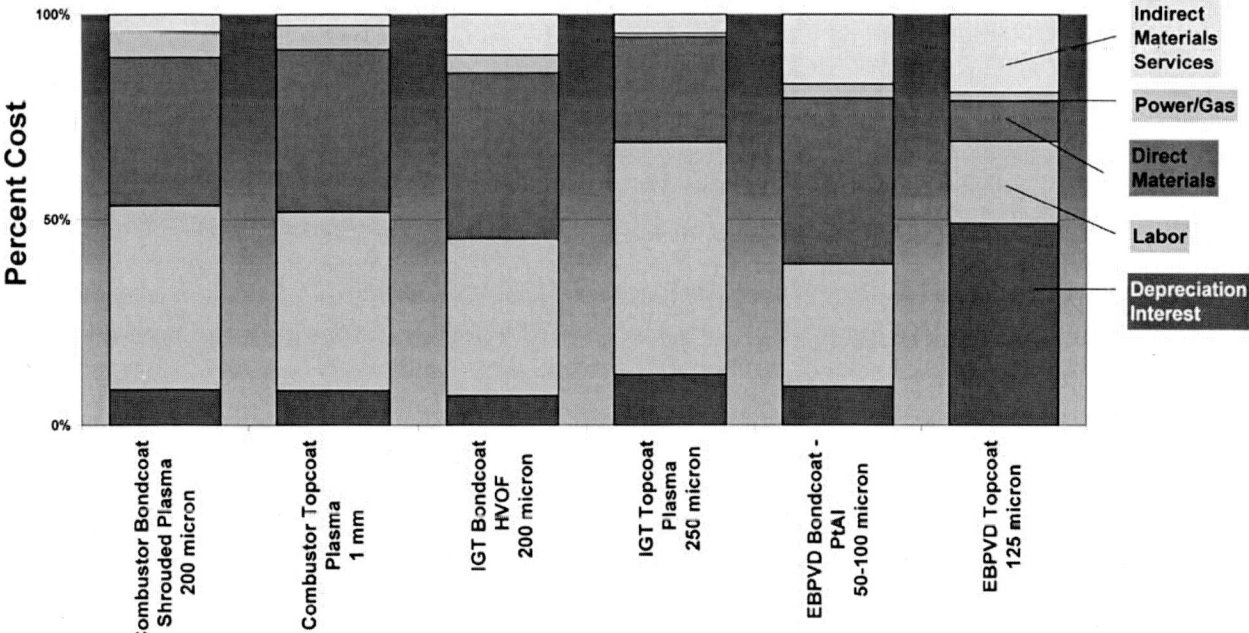

Figure 17: Comparative cost structure for thermal barrier topcoat of selected coating applications.

Discussion of Cost Elements

The cost of the thermal spray process, is essentially driven by the material (powder) cost and labor cost. With APS TBC application, the relatively low deposition rates and comparatively low material expense, labor becomes the most significant portion of the cost. For IGT parts, dependent on the blade or vane size, costs can range between several hundred and several thousand dollar per part. The material cost aspect is even more extreme in the case of platinum aluminide, where platinum can count for several 10$ per part. In the case of EB-PVD with the substantial equipment cost, the equipment depreciation calls for close to 50% of the overall production cost. For EB-PVD TBC only a guideline for the coating cost is for $120-200 per part. For larger aero parts with thicker coating requirements such as nozzle guide vanes, where only 4-6 parts can be coated in one batch, this cost easily triples.

The relative contribution of these cost factors also provides guidance for application development for process and product optimization. Processes with a high percentage of material cost demand improvement of the material utilization. In the case of thermal spray processes, optimized torch parameters can increase the material deposition efficiency substantially and need to be evaluated for each new application. In the case of platinum electroplating, the application development tends towards improving the uniformity and reproducibility of the platinum distribution. This allows adjusting the overall thickness in such a way that the specified thickness requirement is met even at critical locations without having to deposit an excessive thickness on uncritical locations.

Equipment amortization cost dictates the utilization of the facility. In the case of thermal spray processes, where the labor and material cost are the major cost factors, the equipment amortization is moderate; the equipment can be utilized in a one shift or two shift operations. In the case of expensive capital equipment such as EB-PVD and platinum aluminide, 3 shift operation is mandatory. In the case of EB-PVD, development tends toward effective utilization of the expensive capital equipment.

SUMMARY AND CONCLUSION

The key features of state of the art processes for the deposition of TBC systems were outlined and compared. An outlook into PST's new TBC developments such as sintering resistant high purity 7YSZ coatings and low conductivity ceramic compositions was given. The main processes, the equipment and the coating cost elements have been investigated and compared on an index basis for selected examples. The rough order of magnitude cost figures given for selected examples serve as a guideline; each application needs its own detailed evaluation considering all the elements as mentioned before. In general, all presented technologies are considered to be mature. Still, substantial efforts are underway to improve quality and reduce cost by Lean Manufacturing and 6 Sigma programs.

ACKNOWLEDGEMENTS

The authors would like to thank Ann Bolcavage, Adil Ashary, James Knapp, Dan Fillenwarth, Dan Helm and Daming Wang for their input and valuable discussions.

REFERENCES

[1] S.M Meier, D.K. Gupta, and K.D. Sheffler, Ceramic Thermal Barrier Coatings for Commercial Gas Turbine Engines, J. of Metals, 43 (3), (1991), pp. 50-53.

[2] A. Maricocchi, A. Barz, and D. Wortman, PVD TBC Experience on GE Aircraft Engines, Thermal Barrier Coating Workshop, NASA Lewis Research Center, Cleveland, OH, March 27-29, NASA Conference Publication 3312, (1995), pp. 79-90

[3] D.V. Rigney, R. Viguie, D.J. Wortman, D.W. Skelly, PVD Thermal Barrier Coating Applications and Process Development for Aircraft Engines, Journal of Thermal Spray Technology Vol. 6 (2) June 1997 p. 167

[4] D. Zhu, J.A Nesbitt, C.A. Barrett, T.R. McCue, R.A.Miller, Furnace cyclic oxidation behavior of multicomponent low conductivity thermal barrier coatings, Journal of Thermal Spray Technology, Volume 13, Number 1, March 2004, pp. 84-92

[5] D. Stöver; G. Pracht, H. Lehmann, M. Dietrich, J-E. Döring, R.Vaßen , New material concepts for the next generation of plasma-sprayed thermal barrier coatings, Journal of Thermal Spray Technology, Volume 13, Number 1, March 2004, pp. 76-83

[6] J.R. Nicholls, K.J. Lawson, A. Johnstone, and D.S. Rickerby, "Methods to Reduce the Thermal Conductivity of EB-PVD TBCs," Surf. Coat. Technol., 151/152, (2002), pp. 383-391.

[7] D. Zhu and R.A. Miller, "Thermal Conductivity and Sintering Behavior of Advanced Thermal Barrier Coatings," NASA/TM 2002-211481, (2002).

[8] D. Zhu and R.A. Miller, "Low Conductivity and Sintering-Resistant Thermal Barrier Coatings," US Patent 6,812,176 B1, issued Nov. 2, 2004.

[9] D. Zhu and R.A. Miller, "Low Conductivity and Sintering-Resistant Thermal Barrier Coatings," US Patent 6,812,176 B1, issued Nov. 2, 2004.

[10] F.H. Stott, Elevated Temperature Coatings, Science and Technology 11,The Minerals, Metals and Materials Society 1996, pp. 151-161

[11] The Oxide Handbook, Editor G.V Samsonov, IFI / Plenum, 1982

[12] P.K. Wright, A.G. Evans, Mechanisms governing the performance of thermal barrier coatings , Current Opinions in Solid State and Materials Science 4, (1999), pp. 255-265

[13] S. Alperine, M. Derrien, Y. Jaslier, R. Mevrel , Thermal Barrier Coatings – The Thermal Conductivity Challenge, AGARD Report 823 "Thermal Barrier Coatings", 15-16 October 1997, pp. 1.1 -1.10

[14] V. Teixeira, M. Andritschky, H. Gruhn, W. Maliener, H.P. Buchkremer, D., Stoever, Failure of Physically Vapor Deposition / Plasma-Sprayed Thermal Barrier Coatings During ThermalCycling, Journal of Thermal Spray Technology, Vol. 9(2) June 2000 – pp. 191-197

[15] J.A. Haynes, M.K. Ferber, W.D. Porter, Thermal cycling behavior of plasma-sprayed thermal barrier coatings with various MCrAlX bond coats, Journal of Thermal Spray Technology, Volume 9, Number 1, March 2000, pp. 38-48(11)

[16] E.H. Jordan, L. Xie, M. Gell, N.P. Padture, B. Cetegen, A. Ozturk, J. Roth, T.D. Xiao, P.E.C. Bryant, Superior thermal barrier coatings using solution precursor plasma spray, Journal of Thermal Spray Technology, Volume 13, Number 1, March 2004, pp. 57-65(9)

[17] Thomas A. Taylor, US Patent 5,073,433, Dec 17, 1991

[18] T.A. Taylor, D.L. Appleby, A.E. Weatherill, and J. Griffiths, Surf. Coat. Technol., Plasma Sprayed Yttria-Stabilized Zirconia Coatings: Structure-Property Relationships, Surf. Coat. Technol., 43/44, (1990), pp. 470-480.

[19] T. A. Taylor, Dense Vertically Segmented Thermally Sprayed YSZ for TBC and other High Temperature Applications, Proceedings 2nd International Surface Engineering Congress, ASM International, Sept. 2003, Indianapolis.

[20] A. Feuerstein, A. Bolcavage, Thermal Conductivity of Plasma and EB-PVD Thermal Barrier Coatings, Proceedings ASM International Surface Engineering Congress, 2004, Orlando, Florida

[21] R.E. Taylor, Thermal Transport Property and and Contact Conductance Measurements of Coatings and Thin Films, Inter. Journal of Thermophysics, 19 (3), (1998), pp. 931-940.

[22] W. Chi, S Sampath, H.Wang, Ambient and High-Temperature Thermal Conductivity of Thermal Sprayed Coatings, Journal of Thermal Spray Technology, Volume 15, Number 4, December 2006, pp. 773-778

[23] Cernuschi et al., Studies of the sintering kinetics of thick TBC's by thermal diffusivity measurements, J. Europ. Ceram. Soc., 25 (2005), pp. 393-400

[24] Flores Renteria et al., Effect of morphology on thermal conductivity of EB-PVD PYSZ TBC's, Surf. Coat. Technol., 201 (2006), pp. 2611-20

[25] Thomas Taylor, US Patent 5,073,433 and others pending.

[26] Thomas Taylor, Neil Hitchman and Albert Feuerstein, Zircoat-HPTM, A New High Purity Segmented YSZ Coating, ICMCTF Session A2, San Diego, May 2007.

[27] Thomas Taylor and John Jackson, US Patent 7,045,172.

[28] Union Carbide Patent US 3,470,347, Sep 30, 1969

[29] Foster et al, US patent Nos. 5,558,758, Sep 24, 1996 and 5,824,205, Oct 20, 1998

[30] F.S Petit, G.W. Goward, Oxidation – Corrosion-Erosion Mechanisms of Environmental degradation of High Temperature Materials, in Coatings for High Temperature Processes, Editor E. Lang, Applied Science Publishers, 1985

[31] R. Bouchet, R. Mevrel, Influence of platinum and palladium on diffusion in beta-NiAl phase, Defect Diffusion Forum, 237-240 (2005), pp. 238-245

[32] D.K Das, Vakil Singh, S.V. Joshi, The yclic Oxidation Performance of Aluminide and Pt-Aluminide Coatings on Cast i-Based Superalloy CM-247, JOM-e, 52 (1), 2000

[33] Weatherly and Tucker, Union Carbide, U.S. Patent 4,095,003, Jun 13, 1978

INFLUENCE OF POROSITY ON THERMAL CONDUCTIVITY AND SINTERING IN SUSPENSION PLASMA SPRAYED THERMAL BARRIER COATINGS

H. Kaßner, A. Stuke, M. Rödig, R. Vaßen, D. Stöver,
Forschungszentrum Jülich, Institut für Energieforschung 1, IEF-1
Jülich, NRW, Germany

ABSTRACT

Suspension plasma spraying (SPS) was investigated as a potential manufacturing route for thermal barrier coatings. In this process powders with a particle size typically between a few up to several hundred nanometres are dispersed and stabilized in a suspension and then injected into the plasma. So, the SPS process makes it possible to directly feed nano scaled particles into the plasma plume, in contrast to the standard APS process, in which powders with a particle size above 10 μm have to be used. The direct processing of nano particles by the SPS process leads to new microstructures and properties. The size range of the porosity is shifted to lower values and also the porosity levels can be increased easily by the use of the SPS technology.

Free-standing coatings made of yttria partially stabilized zirconia (YSZ) with different porosity levels were produced and the pore size distribution was measured by mercury porosimetry. The thermal conductivity was measured for different porosity levels. Additionally, the effect of these different porosity levels on the sintering was investigated by dilatometric measurements during annealing at high temperatures. Furthermore the thermal expansion coefficient was determined from expansion during heating. For comparison, the results were compared to those of standard APS coatings.

Keywords: suspension plasma spraying, thermal barrier coating, yttria stabilized zirconia, thermal conductivity, thermal expansion coefficient, sintering

INTRODUCTION

The increase of the efficiency and performance of gas turbines is a great challenge in ecological as well as in economical perceptions. Both are related to the inlet temperature and the heat loss of the gas turbine.[1-3] For increasing typically thermal barrier coatings (TBC's) generated by atmospheric plasma spraying (APS) comprising a 0.25 to 0.5 mm thick yttria stabilized zirconia (YSZ) combined with an air cooling are widely used. The heat flux and hence the heat loss is mainly controlled by the thermal conductivity and the temperature gradient in the ceramic layer. The thermal conductivity mainly depends on the grain size, the phase composition, morphology and porosity.[4-8] Here the suspension plasma spraying (SPS) process offers new possibilities to adjust new microstructures with a low conductivity. By establishing the nanotechnology with its improved physical, mechanical and thermal properties also new ways for processing these small particles in the APS process were focused.[9-17] One possibility is their agglomeration to a particle size which enables a sufficient flowability. Nevertheless the semi-molten or molten droplets are still in the upper μm scale. Another possibility is suspension plasma spraying (SPS). Therefore a heterogeneous mixture of nano-particles together with a fluid is injected into the plasma plume. This enables the directly process of nano-particles. Thus molten droplets with a diameter of a few hundred nanometres to a few micrometers can be generated. Thereby also the generated splats are much smaller compared to conventional APS splats. Also the grain size of the ceramic is much lower. Hence new and often improved coating structures combined with a wide band of porosity levels could be generated. So the SPS process seems to be a promising technique for increasing the performance of conventional YSZ APS coatings.

In this work the influence of different porosity levels on the thermal conductivity, the thermal expansion coefficient and the sintering behaviour of different SPS coatings is investigated.

EXPERIMENTAL

For the experiments four different SPS coatings with different porosity levels and a thickness of about 300µm were produced. For comparison also an APS coating with a thickness of 300µm was sprayed. The coatings were produced with a Triplex II APS plasma gun supplied by Sulzer Metco AG, Wohlen, Switzerland. Square steel disks with a side length of 50 mm and a thickness of 2mm served as substrate. All substrates were sand blasted before coating. For the suspension preparation yttria stabilized zirconia (5YSZ) from Tosoh Corporation, Tokyo, Japan with a particle diameter of $d_{50}=300$ nm was used. The powder was dispersed in an ethanol based suspension. The mass content varied between 10 to 30 wt%. To achieve the required particle size all powders were dispersed in ethanol and ball-milled for at least 24h. As grinding stock zirconia or alumina balls with a diameter from 2 to 5 mm were used. The suspensions have been stabilized by the addition of 1.5 wt.% of a dispersant. For the APS standard coatings a 5YSZ from Sulzer Metco AG, Wohlen, Switzerland with a diameter of $d_{50}=50$µm was used. Particle size and distribution was measured with an Acoustic Spectrometer DT-1200 (Dispersion Technology Inc., Bedford Hills, USA) or an Analysette 22 (Fritsch GmbH, Idar Oberstein, Germany) using Fraunhofer diffraction. The microstructure was inspected in cross-sections or free standing coatings with a scanning electron microscopy (SEM), Ultra 55 (Carl Zeiss NTS AG, Germany). Dilatometric measurements were performed in a high-temperature dilatometer, Setsys 16/18, Setaram, Kep Technologies, France. The samples were cut from the freestanding layers. The sample length was 25 mm. A heating rate of 3 K/min up to 1200°C and dwell times at this temperature of 10 h was used. Pore size distributions in the layers were determined by Pascal 140 and 440 mercury porosimeters made by CE-Instruments, Milan, Italy, operating in a pressure range between 0.008 and 400 MPa, corresponding to pore diameters between 3.6 nm and 90 µm. Thermal diffusivity experiments were conducted using a laser flash device (Model: THETA, Netzsch, Germany) on disk-shaped specimens 12 mm in diameter and 1.5 mm in height. Measurements were performed at six different temperatures in the range 20–1200 °C, and repeated fives times at each temperature for statistical purposes.

RESLUTS AND DISCUSSION

Porosity

In sum four SPS and one APS coating with different porosity levels were generated. The pore size distribution of the SPS as well as the APS coating is typical bimodal. All SPS coatings have a higher number of pores all over the pore size distribution compared to the APS coating. Thereby the much higher porosity level of the SPS coatings are mainly induced due to the increased fine pore volume lower than 1 µm. Meanwhile fine pores radii with a size smaller than 1 µm are attributed to micro cracks and micro pores, the pore radii larger 1 µm are globular pores. The pore size distribution of all coatings can be seen in figure 1.

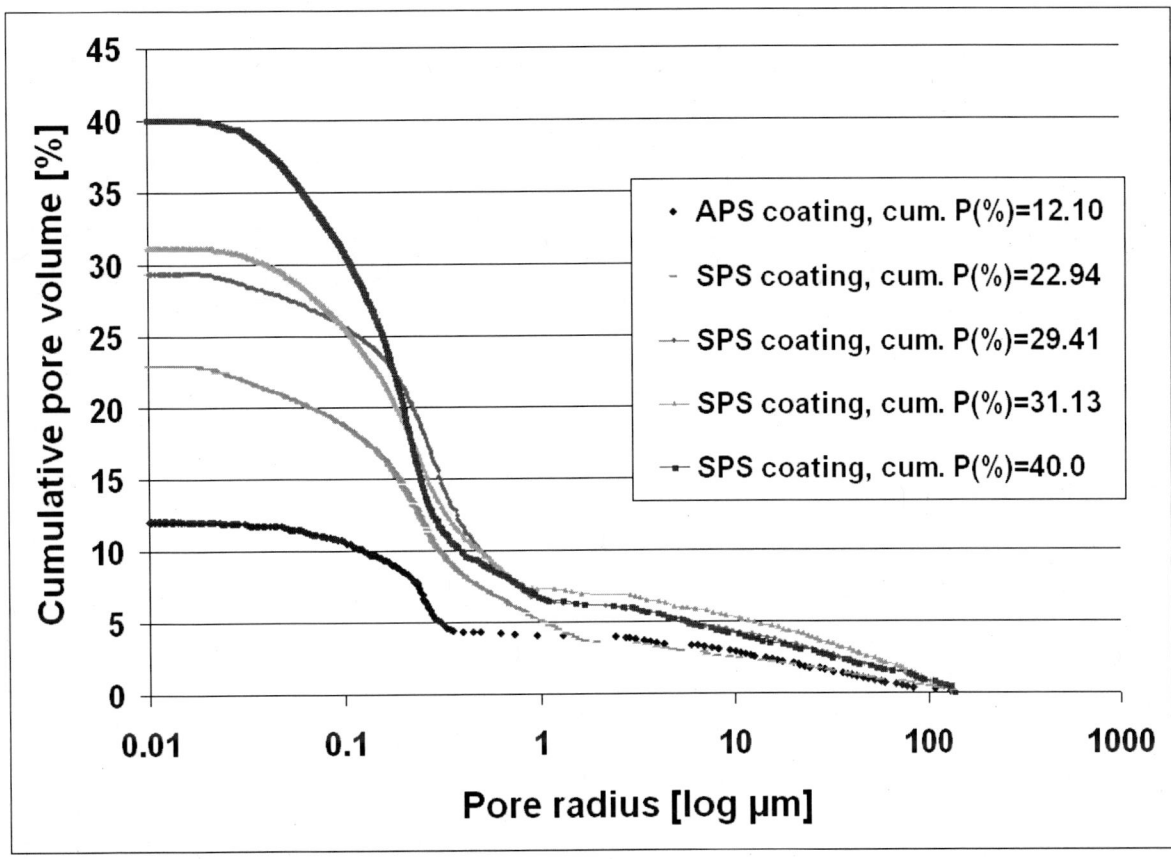

Figure 1. Porosity distribution of the SPS and APS coating as a result of Hg-porosimetry measurements.

The porosity level of the APS coating is at about 12.10%, the SPS coatings have porosity levels of up to 40.0%. The exact porosity levels of the different SPS coatings are 22.94%, 29.41%, 31.13% and 40.0%. The difference in the microstructure and the porosity levels of the SPS coatings can be attributed to different processing conditions like mass content of the suspension, plasma power and spraying. It has to draw out that there are different effects on the porosity levels for SPS and APS coatings. The pores lower 1 μm of the APS coating belong to intralamella and intersplat cracks. The pores of the SPS coatings are split into two areas. The pore radii between 0.2 μm to 1 μm are caused by segmentation cracks. Pores lower 0.2 μm mainly belong to micropores and embedded clusters of unmolten primary particles. Figure 2 points out the different curve progression of an APS and SPS coating.

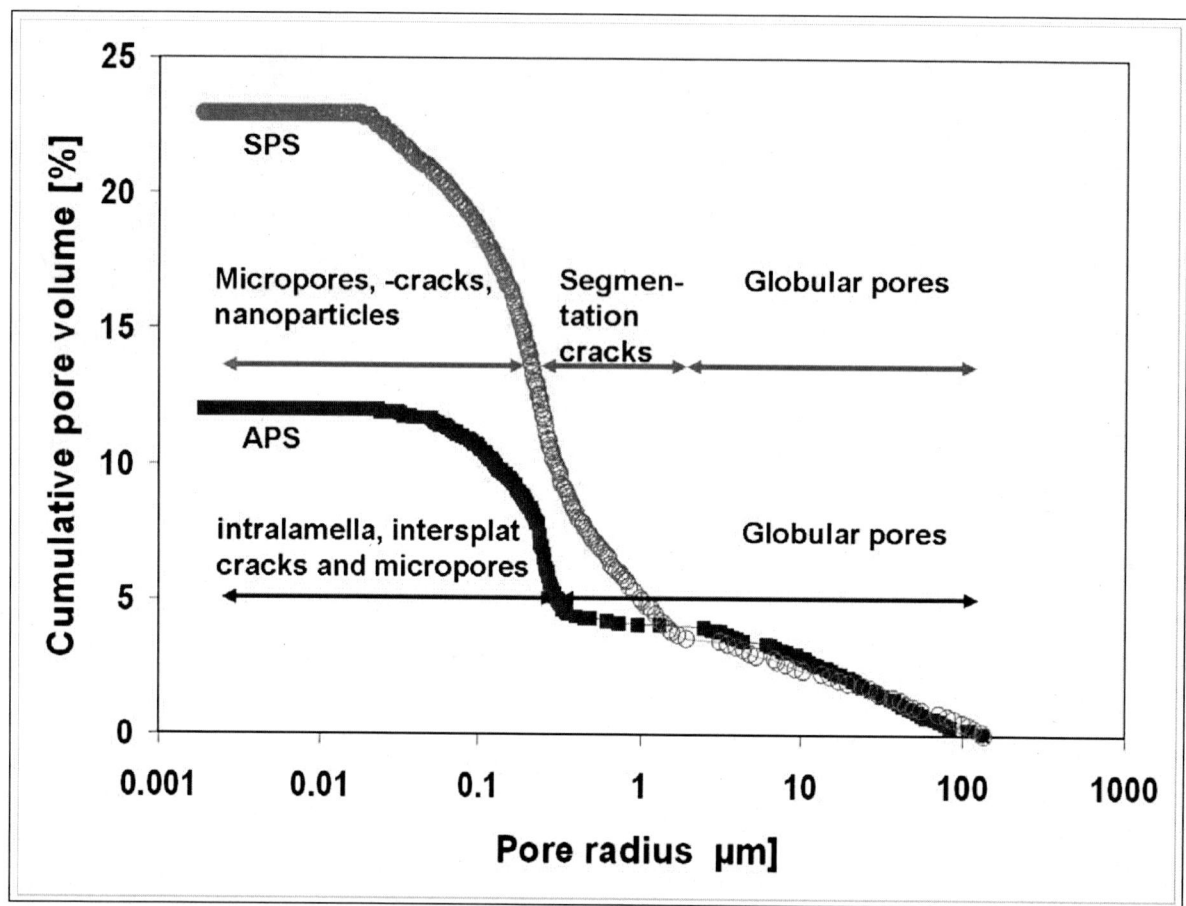

Figure 2: Curve progression of an APS and SPS coating.

Also the ratio between pores bigger than 1 μm to pores lower than 1 μm is different. The ratio of the APS coating is at about 0.5. For SPS coatings it is lower than 0.3. This means that SPS coatings contain a much higher amount of micro pores and micro cracks. Figure 3 depicts a cross section of the SPS coating with an open porosity of 22.94%. The visible small pores are isolated and homogenously distributed. Also clusters of unmolten nanoparticles can be seen. The big vertical crack belongs to a segmentation crack. Compared to the APS coating all SPS coatings show also a number of segmentation cracks.

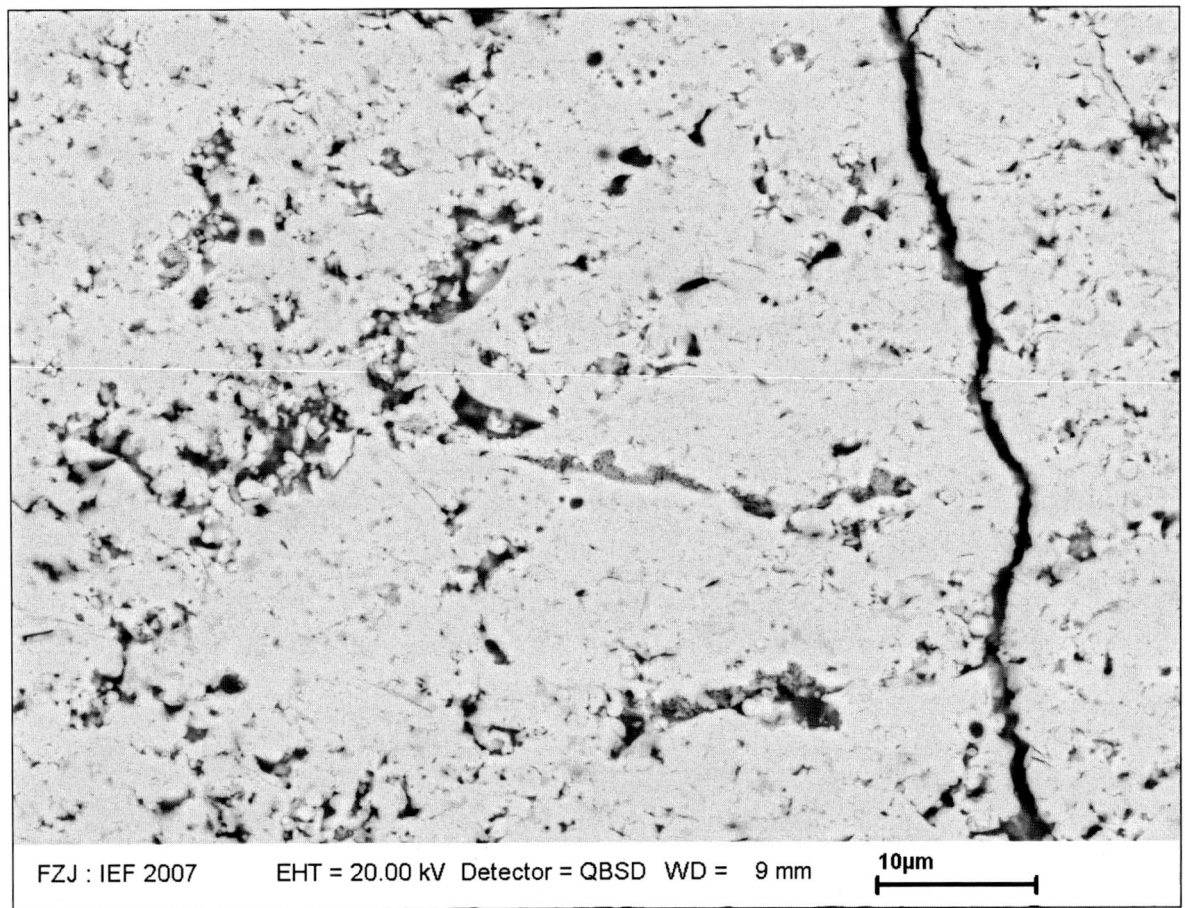

FZJ : IEF 2007   EHT = 20.00 kV Detector = QBSD WD =   9 mm   10µm

Figure 3: Cross section of the SPS coating with a porosity of 22.94%.

Sintering

Figure 4 illustrates the results of dilatometric measurements of the SPS sample with a porosity of 22.94%. After reaching the dewll temperature of 1200°C the sintering rate is rather high but decreases fast with time. The period at the end of the heating phase and at the beginning of the dwell time is shown in figure 5. The deviation from a straight line at the end of the heating at about 1100°C is a result of the onset of sintering. This effect is observed for all SPS and APS coatings.

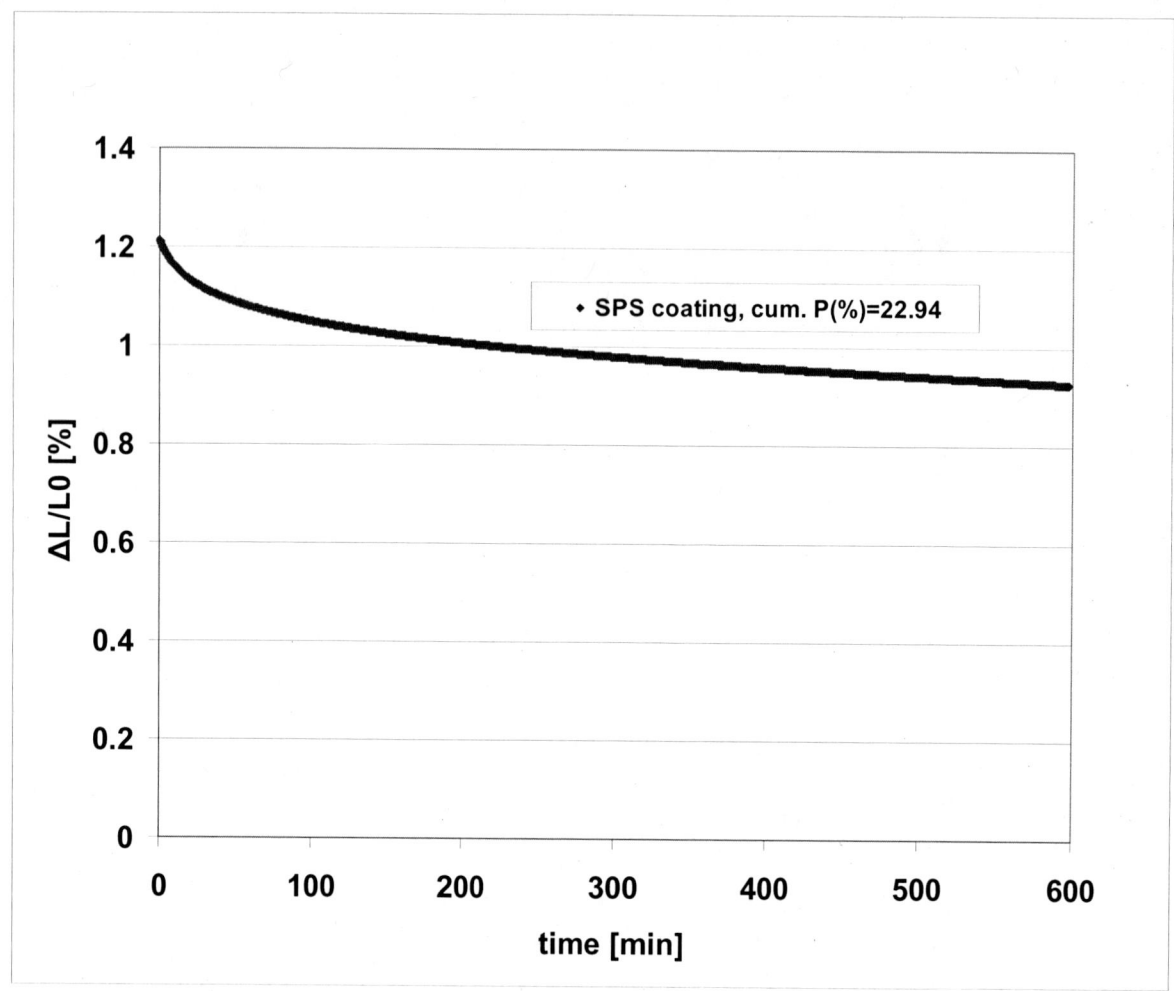

Figure 4: Dilatometric curve for the freestanding SPS coating with a porosity of 22.94%

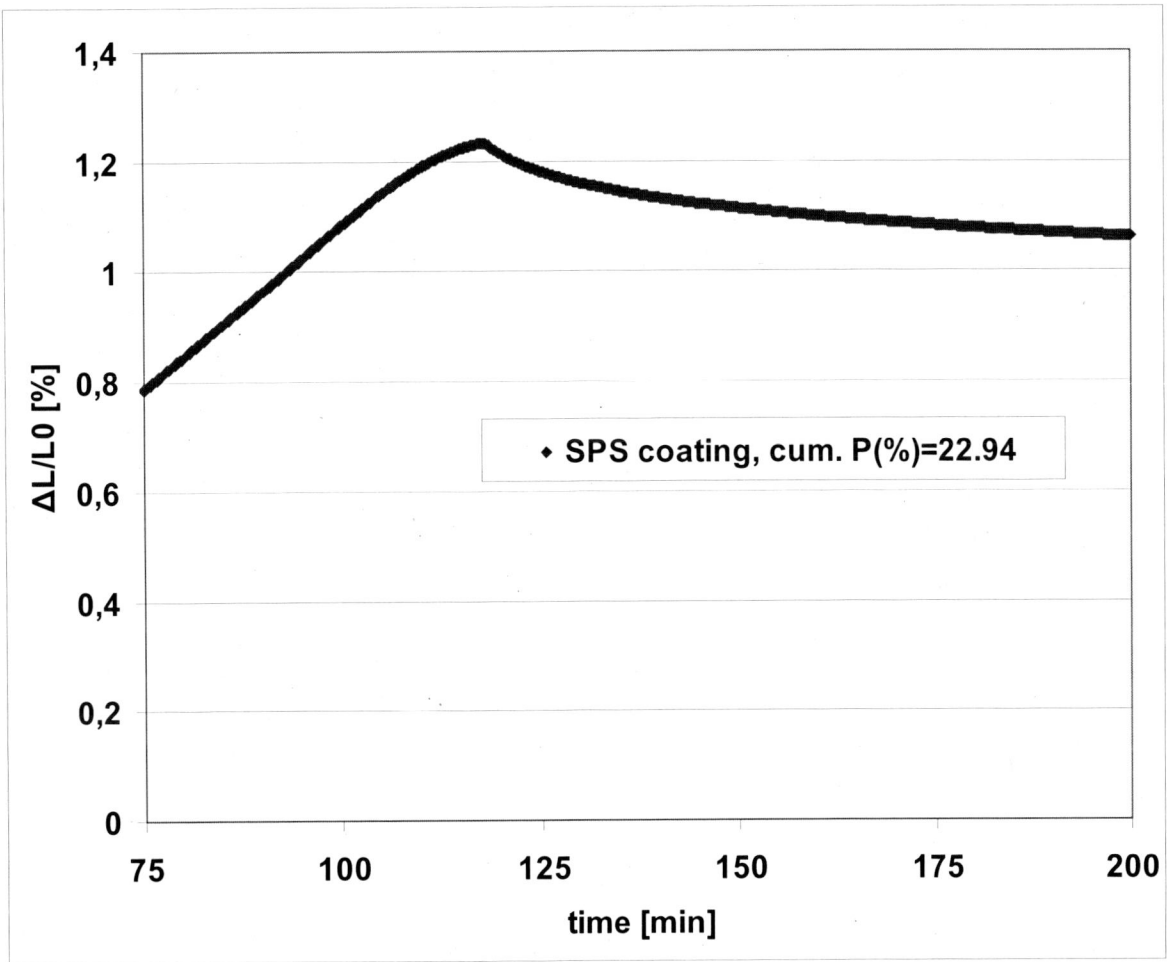

Figure 5: Period of the last minutes of heating phase and at the beginning of the dwell time

The effect of the porosity on the sintering of the SPS coatings after 10h is shown in figure 6. The sintering of the SPS coatings depends directly on the porosity level. Meanwhile the coating with a the lowest porosity of 22.94% features a sintering rate of 0.27% the coating containing a porosity of 40.0% has a sintering of 0.42% compared to the initial length. The APS coating shows a rather low sintering of 0.02%. Reasons for the higher sintering of the SPS coatings could be seen in the high amount of un-molten nanoscaled primary particles and the big number of micro- and nanopores in the SPS coatings. Both factors lead to an increased sintering especially at the beginning of the annealing. Another influencing parameter could be the higher silicon content in the nanopowder that serves as sintering aid.

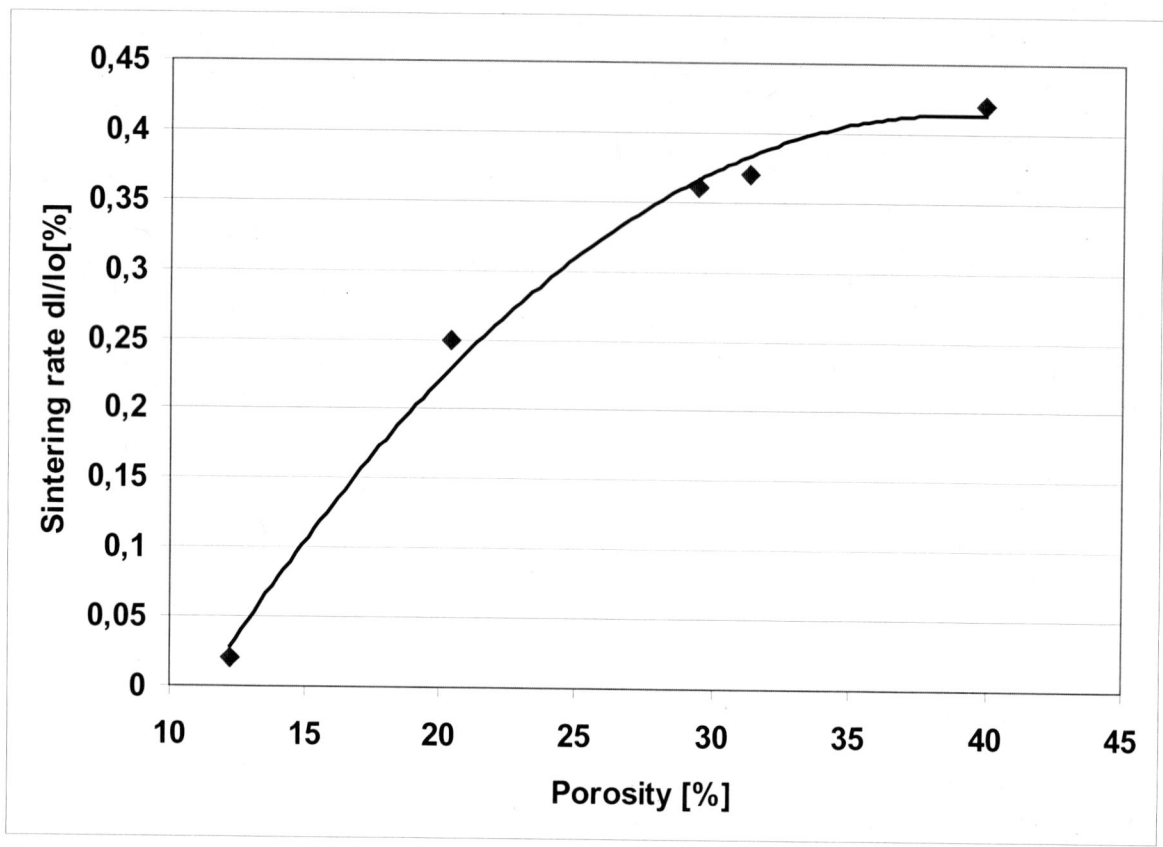

Figure 6: Shrinkage within 10h annealing at 1200°C.

Thermal expansion coefficient

As remarked the beginning of the sintering leads to a deviation of the straight line of the thermal expansion above 1100°C. The thermal expansion between room temperature and 1200 °C is almost linear for all coatings. The average value of thermal expansion coefficient (TEC) for the coating with 29.41% is determined to be 10.5×10^{-6} K^{-1}. Also the other porosity levels show a TEC between 10.5 to 10.9×10^{-6} K^{-1} independent from the porosity levels. The SPS coating with a density of 20.4% shows a thermal expansion coefficient of 10.6×10^{-6} K^{-1}. The TEC of the SPS coating with a porosity level of 31.3% is determined to be 10.9×10^{-6} K^{-1}. In contrast the APS coating reaches 11.2×10^{-6} K^{-1}. The thermal expansion data of the SPS coating with a porosity of 29.41% is shown in figure 7.

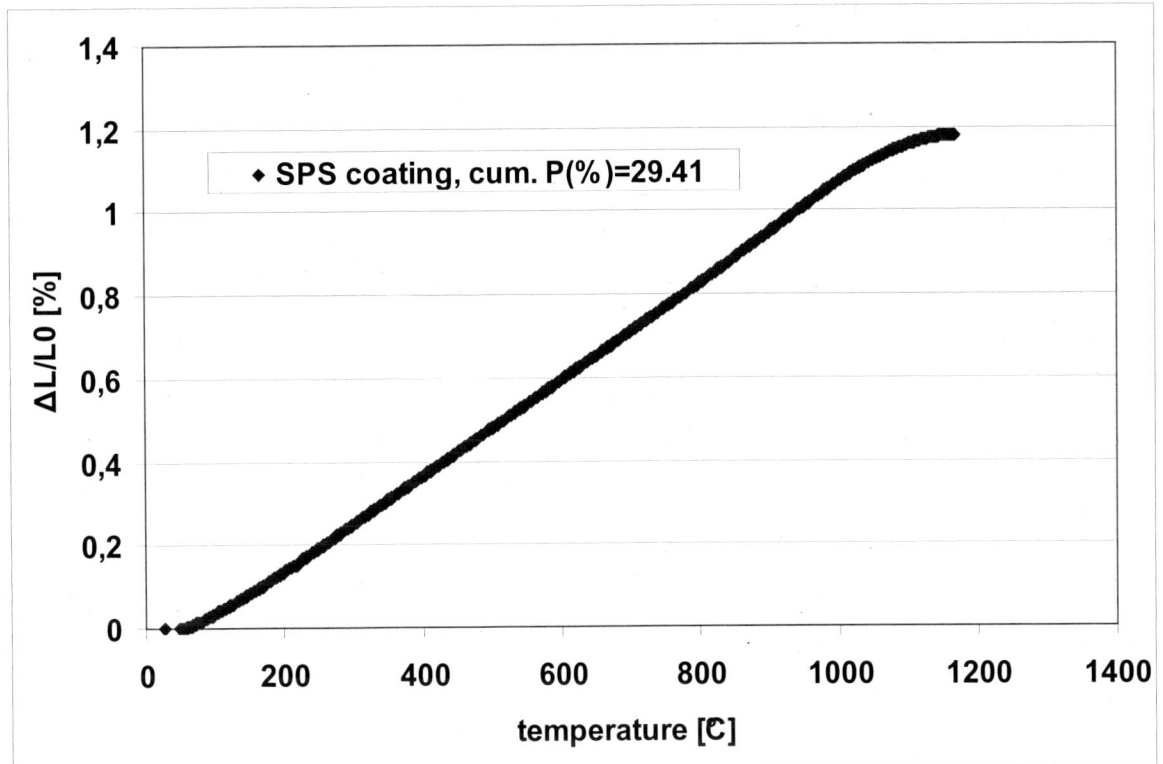

Figure 7: Thermal expansion of the SPS coating as a function of temperature

Thermal conductivity

The thermal conductivity was calculated with the equation $\lambda(T)=\alpha(T)\cdot c_p(T)\cdot \rho(T)$. Thereby the density $\rho(T)$ is the density of the porous material. The thermal diffusivity measurements of the SPS coatings reveal rather low values compared to the APS coating. The SPS coating with a porosity of 20.41% ranges from 0.0027 to 0.003 cm²/s (from 25 to 1200°C) the APS coating has 0.0038 to 0.0035 cm²/s (from 25 to 1200°C). Hence the calculated thermal conductivity of the SPS coatings is much lower as the standard APS coating. Figure 8 points out the influence of the temperature on the thermal conductivity for these two coatings as a function of the temperature.

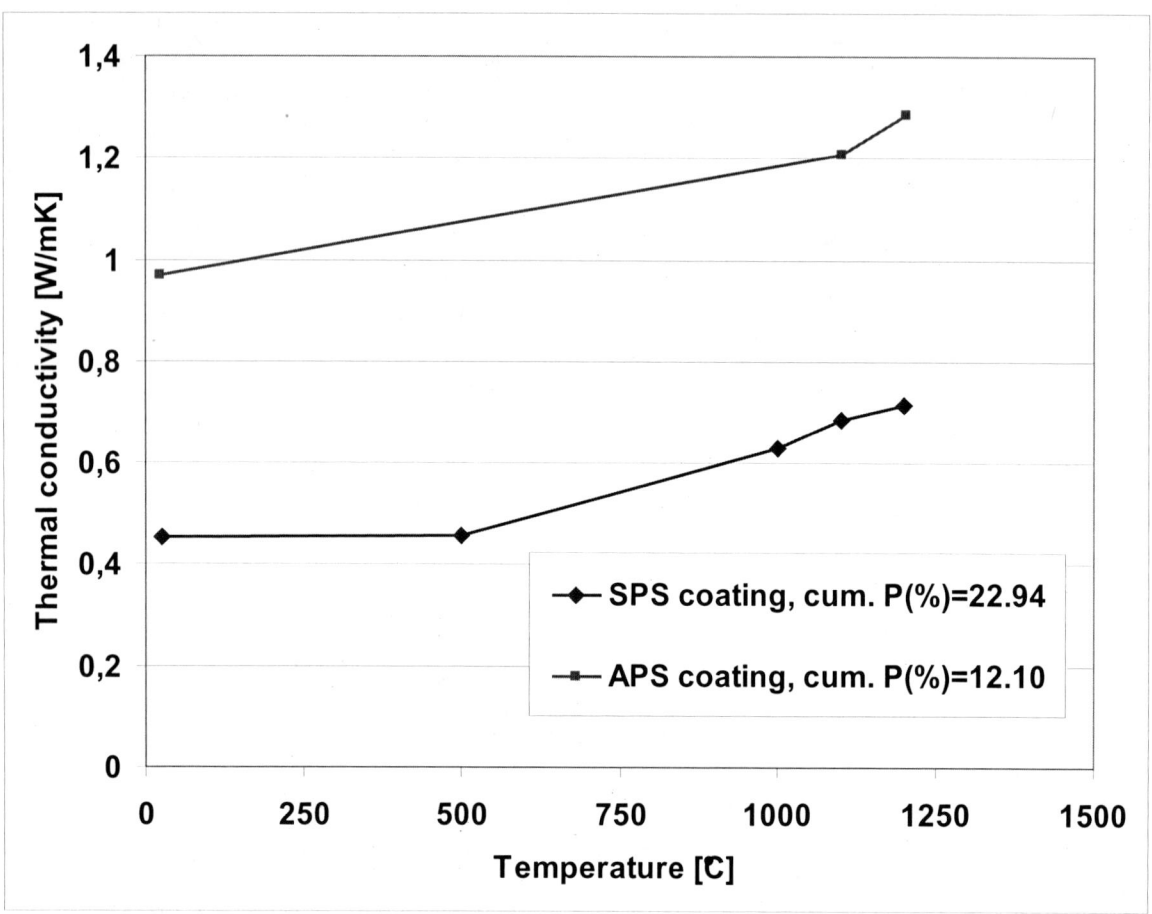

Figure 8: Thermal conductivity of an APS and SPS coating as a function of the temperature

The shape of the curves is for both coatings similar. Beginning from a low level at room temperature the thermal conductivity rises continuously with increasing the temperature. Nevertheless the values of the SPS coating are up to 50% less than that ones of the APS coating. The thermal conductivity at the technical relevant temperature of 1200°C for all coatings is given in table I.

Table I: Thermal conductivity at 1200°C

	porostiy level [%]	thermal conductivity [W/mK]
SPS	22.94	0.92
SPS	29.41	0.58
SPS	31.31	0.50
SPS	39.90	0.42
APS	12.12	1.28

There is a correlation between the porosity level and the thermal conductivity (figure 9). An increased porosity level leads to a decreased thermal conductivity. The thermal conductivity is 0.92 W/mK for the coating with a porosity of 22.94%. The coating with 39.9% only has a thermal conductivity of 0.42 W/mK. So a doubling of the porosity leads to a reduction of the thermal conductivity of 50%. The 1.3 W/mK at 1200°C of the APS coating is nearly 3 times higher.

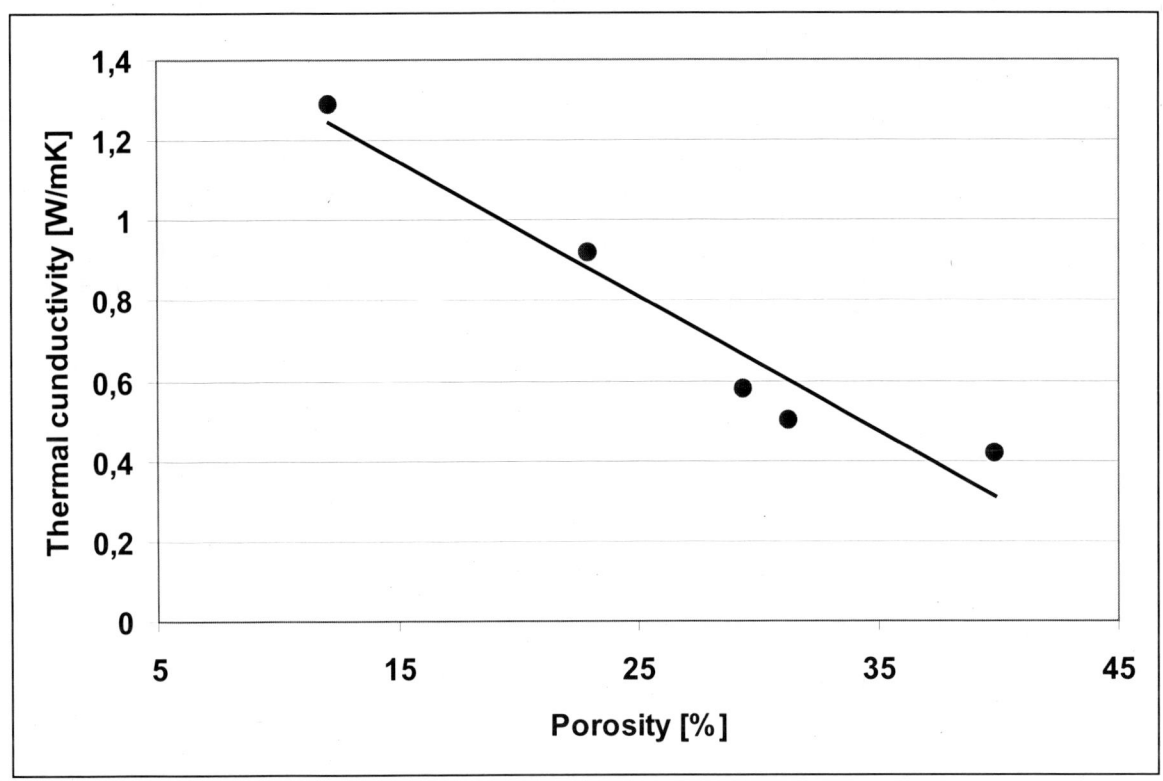

Figure 9: Influence of the porosity on the thermal conductivity.

CONCLUSIONS

As the results show, the SPS process allows improving the microstructure and the thermo physical properties of conventional coatings significantly.

Compared to the conventional APS process the SPS process enables a much wider band of porosity levels in thermal barrier coatings. Meanwhile conventional APS coatings normally have porosity levels of less than 15% the defined changing of the process parameters allows the SPS process to generate coatings with a porosity of up to 40%. Especially the amount of porosity with a pore radii lower than 1 μm induced by micro cracks, micro pores and some embedded un-molten nano particles is much higher compared to APS coatings.

The sintering curve characteristics are similar to conventional ones. The results also show that the sintering rate correlates directly with the porosity of the SPS coatings. So the coating with the lowest porosity level of 22.94% features a sintering rate of 0.27%, the coating a porosity level of 40.0% has a sintering rate of 0.42%. Reasons for the higher sintering of the SPS coatings could be seen in the high micro porosity and embedded unmolten primary particles in the coating. Also the higher silicon content in the nanopowder might enhance the sintering.

The thermal expansion and the thermal expansion coefficient are not influenced by the porosity levels. The thermal expansion coefficient varies between 10.6 to 10.9 $\times 10^{-6}$ K^{-1}. This is a little bit lower than that one of the APS coating with 11.2$\times 10^{-6}$ K^{-1}.

As expected the high porosity levels in the SPS coatings lead to a decrease of the thermal conductivity by a factor of 3. Values of 0.42 W/mK can be obtained at the technical relevant temperature of 1200°C.

REFERENCES

[1] R.A. Miller., Current status of thermal barrier coatings - an overview, *Surf. Coat. Technol.,* **30,** 1-11, (1987).

[2] S.M. Meier, D.K. Gupta and K.D. Sheffler, Ceramic thermal barrier coatings for commercial gas turbine engines, *J. Me.t,* **43,** 50-53, (1991).

[3] A.G. Evans, D.R. Mumm, J.W. Hutchinson, G.H. Meier and F.S. Pettit, Mechanisms controlling the durability of thermal barrier coatings, *Prog. Mater. Sci.,* **46,** 505-553, (2001).

[4] F. Cernuschi, P. Bianchi, M. Leoni and P. Scardi, Thermal diffusivity/microstructure relationship in Y-PSZ thermal barrier coatings, *Journal of Thermal Spray Technology,* **8,** 102-109, (1999).

[5] H. B. Guo, R. Vaßen and D. Stöver, Thermophysical properties and thermal cycling behavior of plasma sprayed thick thermal barrier coatings, *Surface and Coatings Technology,* **192,** 48-56, (2005).

[6] V.P. Swaminathan and N.S. Cheruvu, Gas Turbine Hot-Section Materials and Coatings in Electric Utility Applications, *Advanced Materials and Coatings for Combustion Turbines,* ASM International, (1994).

[7] M.G. Hocking, V. Vasantaree, and P.S. Sidky, *Metallic & Ceramic Coatings: Production, High Temperature Properties & Applications,* Longman Scientific & Technical, Harlow, UK, (1989)

[8] P. Scardi, Matteo Leoni, Microstructure and Heat Transfer Phenomena in Ceramic Thermal Barrier Coatings, J. *Am. Ceram. Soc.,* **84** [4], 827–35, (2001).

[9] J. Oberste Berghaus, S. Bouaricha, J.-G. Legoux, C. Moreau, Injection conditions and in-flight states in suspension plasma spraying of alumina and zirconia nano-ceramics, *Thermal Spray 2005: Explore its surface potential!,* ASM International, 512 – 518, (2005)

[10] C. Delbos, J. Fazilleau, V. Rat, J.F. Coudert, P. Fauchais, B. Pateyron, Phenomena Involved in Suspension Plasma Spraying, Part 1: Suspension Injection and Behaviour, *Plasma Chem. And Plasma Processing,* **26**(4), 371-391, (2006)

[11] C. Delbos, J. Fazilleau, V. Rat, J.F. Coudert, P. Fauchais, B. Pateyron, Phenomena Involved in Suspension Plasma Spraying, Part 2: Zirconia particle treatment and coating formation, *Plasma Chem. And Plasma Processing,* **26**(4), 393-414, (2006)

[12] H. Gleiter, Nanostructured Materials: Basic concepts and microstructure, *Acta Materialia,* **48,** 1-29, (2000).

[13] D. Vollath, D.V. Szabó, Nanocoated Particles: A Special Type of Ceramic Powders, *Nanostructured Materials,* 4(8), 927-938, (1994).

[14] R.S. Lima, A. Kucuk and C.C. Berndt, Evaluation of microhardness and elastic modulus of thermally sprayed nanostructured zirconia coatings. *Surf. Coat. Technol,.* **135,** 166–172 (2001).

[15] Y. Zeng, S.W. Lee, L. Gao and C.X. Ding, Atmospheric plasma sprayed coatings of nanostructured zirconia. *J. Eur. Ceram. Soc.,* **22,** 347–351, (2002).

[16] M. Gell, Application opportunities for nanostructured materials and coatings. *Mater. Sci. Eng.,* **A204,** 246–251, (1995).

[17] B.H. Kear and G. Skaudan, Thermal spray processing of nanoscale materials. *Nanostruct. Mater.,* **8,** 765–769(1997).

THERMAL AND MECHANICAL PROPERTIES OF ZIRCONIA/MONAZITE-TYPE LaPO$_4$ NANOCOMPOSITES FABRICATED BY PECS

Seung-Ho Kim, Tohru Sekino, and Takafumi Kusunose
The Institute of Scientific and Industrial Research, Osaka University
8-1 Mihogaoka, Ibaraki, Osaka 567-0047, Japan

Ari T. Hirvonen
Materials Science and Engineering, Helsinki University of Technology
P.O. Box 6200, FI-02015 TKK, Finland

ABSTRACT

Thermal barrier coatings (TBC's) perform the important function of insulating components such as gas turbine parts that operate at elevated temperatures. The most commonly applied TBC material is yttria stabilized zirconia (3YSZ), because it has a coefficient of thermal expansion similar to that of substrate metals. In this study, 3YSZ/monazite-type LaPO$_4$ nanocomposites were prepared by the pulse electric current sintering (PECS) method. The amount of LaPO$_4$ added to 3YSZ was varied from 0 to 40 vol.% and thermal and mechanical properties of these nanocomposites were investigated. The XRD results of the 3YSZ/LaPO$_4$ nanocomposites demonstrated differences in the crystalline phase of as-sintered and annealed zirconia. The phase transformation of as-sintered specimens was related to the amount of monazite-type LaPO$_4$, but this phenomenon was not observed in annealed specimens. The density of 3YSZ/LaPO$_4$ nanocomposites after annealing was decreased and the porosity was increased. Also, mechanical properties of 3YSZ/LaPO$_4$ nanocomposites decreased with increasing dispersion of monazite-type LaPO$_4$ particles. It was caused by low mechanical properties of LaPO$_4$ and weak bonding between 3YSZ. Thermal conductivity of 3YSZ/LaPO$_4$ nanocomposites was lower than 3YSZ. The difference of thermal conductivity between 3YSZ and 3YSZ/LaPO$_4$ nanocomposites at high temperatures was higher than that at low temperatures.

INTRODUCTION

The continual development of high-tech industries, such as aerospace engineering and power plants etc., has required the development of new materials to replace traditional turbine materials, which have reached the limits of their temperature capabilities[1,2]. Thermal barrier coating (TBC) was proposed to improve the thermal efficiency of gas turbines due to the protective effect of the metal substrate. The protection materials of hot-path parts of gas turbine engines or jet engines were mainly developed using ceramics which have excellent thermal properties, chemical stability and heat insulation properties, etc[1,3,4]. The efficiency of gas turbines using ceramic TBC can be increased by 5 - 8 %. A ceramic coating of 300 μm thickness yields a temperature difference of up to 200 °C between the top coat and the metallic substrate.

The selection of materials has been very important for the development of TBC. Some basic requirements of TBC materials is as follows: high melting point, no phase transformation for a variation of temperatures, low thermal conductivity, chemical stability, low mismatch of thermal expansion with the metallic substrate, good adherence to the metallic substrate and low sintering rate of the porous microstructure[1,5,6]. Specially, TBC materials must have low thermal conductivity and a high thermal expansion coefficient. TBC layers in the gas turbine should be considered for thermal stress, thermodynamic affinity between the bond- and top-coat, and protection materials at high temperature, etc. The materials that can be applied to TBC have yttria stabilized zirconia (YSZ), a mixture of rare earth oxide, $La_2Zr_2O_7$, and zirconium phosphate, etc. In particular, the most commonly applied TBC material is YSZ because these materials have a coefficient of thermal expansion similar to that of the metallic substrate. The operation temperature of YSZ was limited below 1200 °C, because it undergoes a volume change as a result of a phase transformation from a monoclinic to tetragonal form at this temperature. To overcome this problem of YSZ, many researchers have studied how to improve its thermal properties such as high temperature stability, low thermal conductivity and high thermal expansion coefficient.

In this study, to improve the high temperature stability and low thermal conductivity of YSZ, monazite-type $LaPO_4$ was composed with YSZ to add properties such as weak bonding between oxides. 3YSZ/monazite-type $LaPO_4$ nanocomposites were fabricated by a pulse electric current sintering (PECS) method using composite powders, which consisted of chemically precipitated $LaPO_4$ on zirconia powder surfaces. The amount of monazite-type $LaPO_4$ added to 3YSZ varied from 0 to 40 vol.% and thermal and mechanical properties of 3YSZ/monazite-type $LaPO_4$ nanocomposites were investigated.

EXPERIMENTAL PRECEDURES

In this study, the composite powders were prepared by chemical precipitation[7,8] of monazite-type $LaPO_4$ on zirconia powder surfaces. The chemical precipitation of monazite-type $LaPO_4$ on zirconia powders was performed as follows. The raw material to make the Monazite-type $LaPO_4$ powder was La_2O_3 powder (99.9%, Shin-Etsu Chem. Co. Ltd., Japan). La_2O_3 powder was completely dissolved in HCl (6mol/l, Wako Pure Chem. Ind. Ltd., Japan). 3YSZ (TZ-3YE, Tosoh Corp., Japan) was ball-milled in a $LaCl_2$ solution for 12 h. The aqueous solution of H_3PO_4 (Wako Pure Chem. Ind. Ltd., Japan) was added on the slurry when the molar ratio of La to P was 1:1, and then the slurry was further mixed for 6 h. To precipitate monazite, the slurry was added to ammonia water (28%, Ishizu Seiyaku Ltd., Japan). This slurry was ball-milled for 6 h in order to homogenously distribute precipitation materials. After ball milling, the slurry was washed several times with de-ionized water and acetone. The washed powder was dried for 24 h in a dry oven at 55 °C. Dry powders were calcined at 700 °C for 2 h in atmospheric conditions. Calcined powders were ball-milled in a pot for 24 h. The specimens were fabricated by PECSed methods at 1300 °C, 30 MPa for 5 min in an Ar gas flow environment. The elevated temperature was 100 °C/min. The sintered body of composites was analyzed for thermal conductivity and its microstructure observed.

Density of the PECSed samples was measured by the Archimedes immersion method in toluene. Hardness and fracture toughness of sintered samples was measured by an indentation method. The indentation test of specimens was carried out by using a Vickers hardness tester (AVK-C2, Akashi Co. Ltd., Japan). Test conditions were made at the applied load of 98 N and the duration time of 15 seconds. The value of fracture toughness was calculated from the Niihara equation[9]. Flexural strength specimens of 3 x 4 x 36 mm in dimension were cut and ground from the sintered body, and the three point bending test was performed with conditions for crosshead speed of 0.5 mm/min with spans of 30 mm.

To confirm the microstructure of specimens, scanning electron microscopy (SEM, model S-5000, Hitachi Co. Ltd., Japan) was used. Specimens were fixed on the holder for SEM and coated with a thin evaporated gold film to avoid charging under the electron beam. The thermal conductivity (κ) was calculated from the equation,

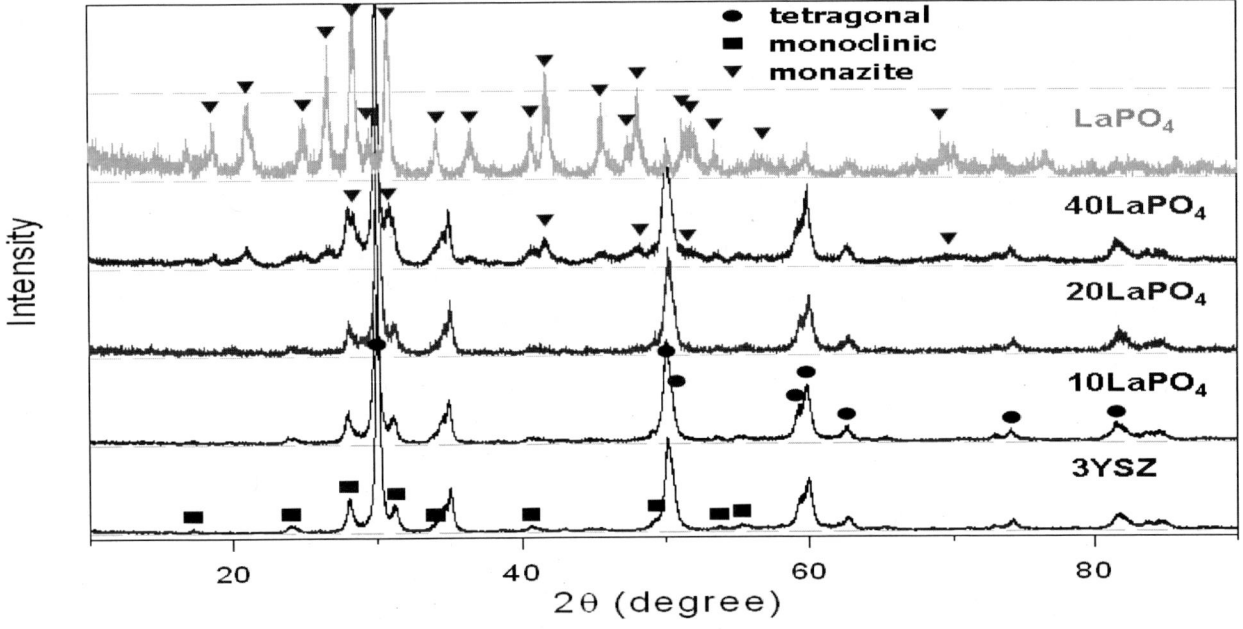

Figure 1. XRD patterns of 3YSZ/LaPO$_4$ compound powders.

$$\kappa = C_P \lambda \rho \qquad (1)$$

where C_P is the specific heat, λ is the thermal diffusivity and ρ is density. Bulk density of the sintering specimens was measured by the Archimedes immersion method using toluene solvent at room temperature, and the relative density was calculated with the theoretical density of each powder by the mixing rule. The thermal diffusivity was measured using the laser flash thermal constant analyzer (TC-7000, Ulvac-Riko, Japan). The specific heat was measured using the differential scanning calorimeter (DSC404C, NETZSCH, Germany).

RESULTS AND DISCUSSION

The XRD results of 3YSZ/LaPO$_4$ compound powder are shown in figure 1. 3YSZ had a mixture of the peak of tetragonal- and monoclinic-phases. Chemically precipitated LaPO$_4$ powder had a peak similar to that of the monazite-type. Figure 2 shows the XRD results of as-sintered and annealed 3YSZ/LaPO$_4$ nanocomposites. The XRD results of 3YSZ/LaPO$_4$ nanocomposites were different before (figure 2(a)) and after annealing (figure 2(b)) at 1500 $^{\circ}$C. When LaPO$_4$ was added up to 30 vol.%, the XRD results of as-sintered nanocomposites showed an increase in the monoclinic phase, but that of nanocomposites with 40 vol.% LaPO$_4$ only showed a tetragonal-phase. The zirconia phase was affected by the amount of LaPO$_4$. When sintered nanocomposites were annealed at 1500 $^{\circ}$C for 1 h, the XRD results of the annealed nanocomposites were different as compared to that of sintered nanocomposites. The XRD results of annealed nanocomposites were similar to all compositions. The XRD results of annealed nanocomposites showed a peak of a mixture of tetragonal- and monoclinic-phases, as shown in figure 2(b). The peaks after annealing were sharper than as-sintered specimens. The phase transformation of zirconia to the amount of LaPO$_4$ was not observed in the annealed specimen.

Figure 3 shows the density and porosity of 3YSZ/LaPO$_4$ nanocomposites before and after annealing. Density and porosity was measured by the Archimedes immersion method using toluene solvent at room temperature. As the 3YSZ/LaPO$_4$ nanocomposites annealed, the

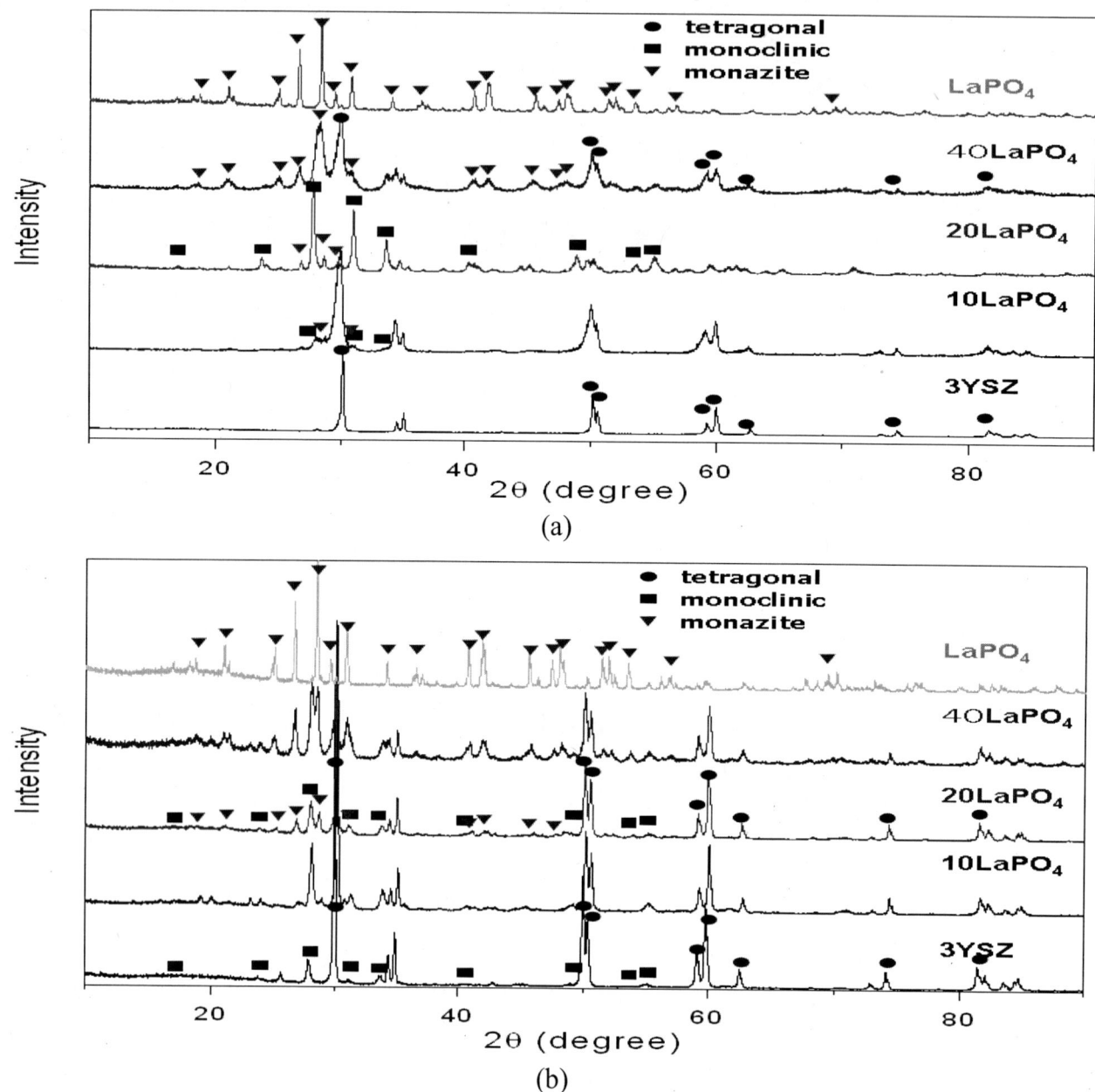

Figure 2. XRD patterns of 3YSZ/LaPO$_4$ nanocomposites (a) before- (as-sintered) and (b) after annealing at 1500 $^\circ$C for 1 h in atmosphere.

experimental density decreased, but porosity increased. However, the density and porosity of monolithic LaPO$_4$ was increased. The decreasing density of 3YSZ/LaPO$_4$ nanocomposites after annealing was related to the phase transformation shown in figure 2. It suggested that the decreasing density of 3YSZ/LaPO$_4$ nanocomposites was caused by an increasing porosity as a result of a volume change resulting from a phase transformation from a monoclinic- to tetragonal-phase. In annealed 3YSZ/LaPO$_4$ nanocomposites, the relative density decreased from 97.9 – 99.3 % of theoretical density to 95.3 – 96.6 %. On the other hand, the density of monolithic LaPO$_4$ increased from 97.0 to 99.6 %. Figure 4 shows the mechanical properties of 3YSZ/LaPO$_4$ nanocomposites as a function of the amounts of monazite-type LaPO$_4$ particles. As the amount of monazite-type LaPO$_4$ particles increased, mechanical properties of 3YSZ/LaPO$_4$

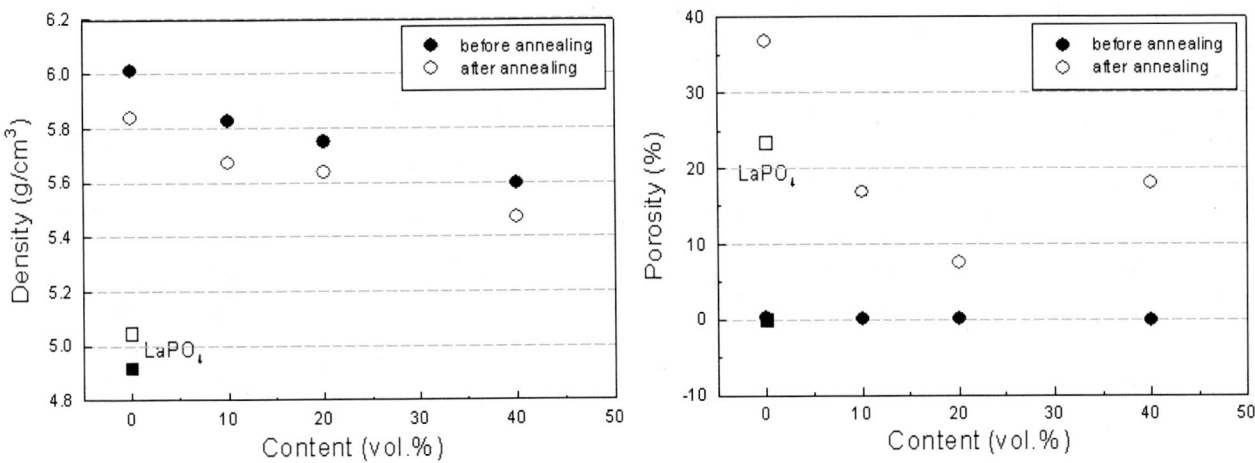

Figure 3. Density and porosity of 3YSZ/LaPO$_4$ nanocomposites with/without annealing at 1500 ^{o}C for 1 h in atmosphere.

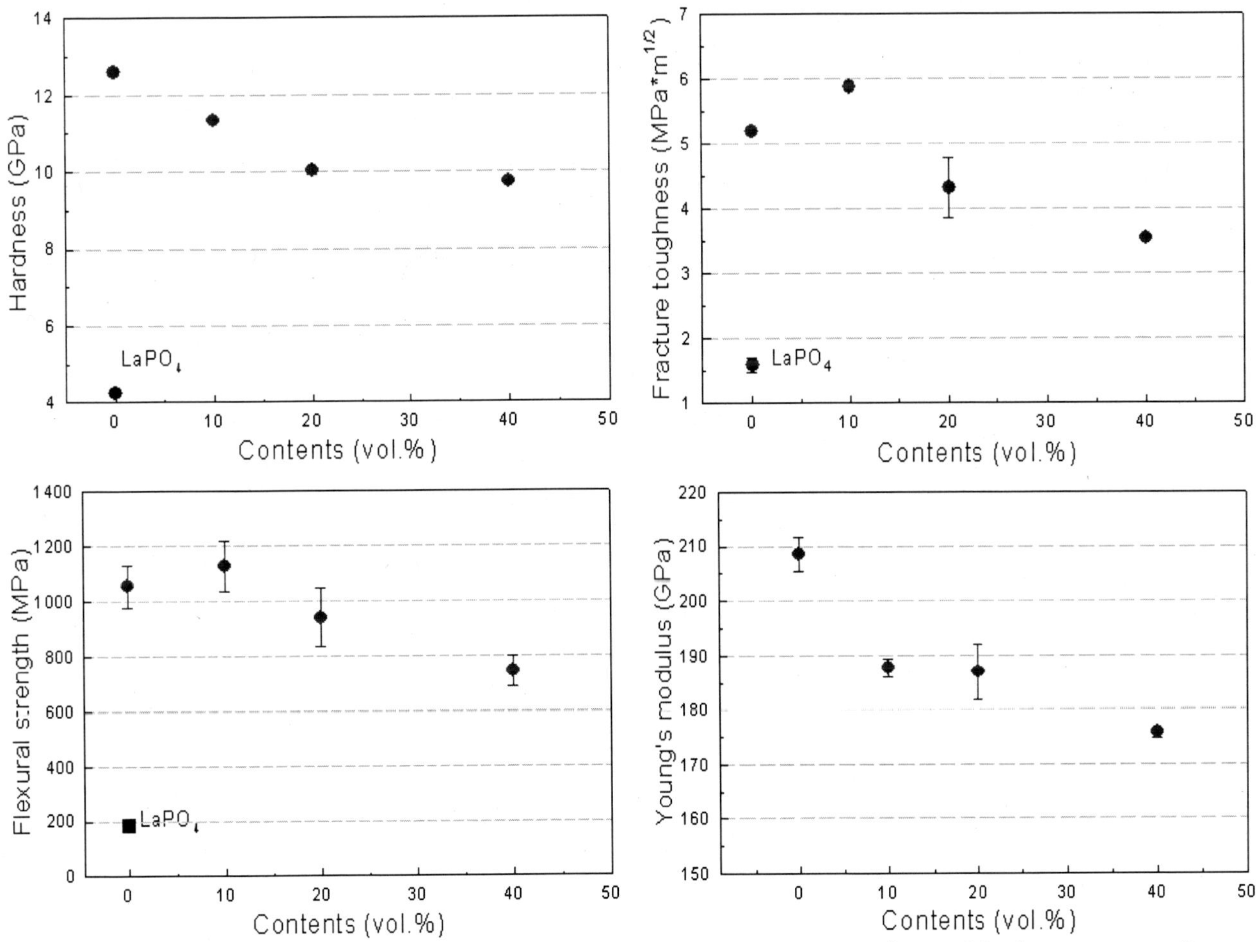

Figure 4. Mechanical properties of 3YSZ/LaPO$_4$ nanocomposites with the amount of monazite-type LaPO$_4$.

nanocomposites decreased. The degradation of mechanical properties was caused by dispersed monazite-type LaPO$_4$ particles. Monazite-type LaPO$_4$ materials have poor mechanical properties and weak bonding between oxides[10]. Actually, mechanical properties of PECSed monolithic

Figure 5. SEM micrographs of the fracture surfaces of 3YSZ/LaPO₄ nanocomposites with the varying amounts of monazite-type LaPO₄; (a) 3YSZ, (b) 10 vol.% LaPO₄, (c) 20 vol.% LaPO₄ and (d) 40 vol.% LaPO₄.

LaPO$_4$ have a hardness of 4.26 GPa, 1.60 MPa· m$^{1/2}$ of fracture toughness and 174.9 MPa of flexural strength. Young's modulus can not be measured by range error. Mechanical properties of 3YSZ/LaPO$_4$ nanocomposites were decreased due to dispersed monazite-type LaPO$_4$ particles. Mechanical properties of 3YSZ/LaPO$_4$ nanocomposites were lower than 3YSZ. In 3YSZ/10 vol.% LaPO$_4$ nanocomposites, flexural strength was higher (~1127 MPa) than other nanocomposites. When 40 vol.% LaPO$_4$ particles was added, flexural strength of 3YSZ/LaPO$_4$ nanocomposites increased to more than 700 MPa.

The fracture behaviors of 3YSZ and 3YSZ/LaPO$_4$ nanocomposites were compared as shown in figure 5. As the amount of monazite-type LaPO$_4$ increased, the grain size of 3YSZ/LaPO$_4$ nanocomposites gradually decreased. The grain size of 3YSZ and 3YSZ/40 vol.% LaPO$_4$ nanocomposites were about 200-300 nm and about 100-150 nm, respectively. This effect was likely due to the addition of dispersed monazite-type LaPO$_4$ particles, which likely suppressed grain growth in 3YSZ. In general, decreasing grain size increases fracture strength due to a Hall-Petch relationship. However, in these systems, the flexural strength of 3YSZ/LaPO$_4$ nanocomposites by dispersed monazite-type LaPO$_4$ decreased in spite of

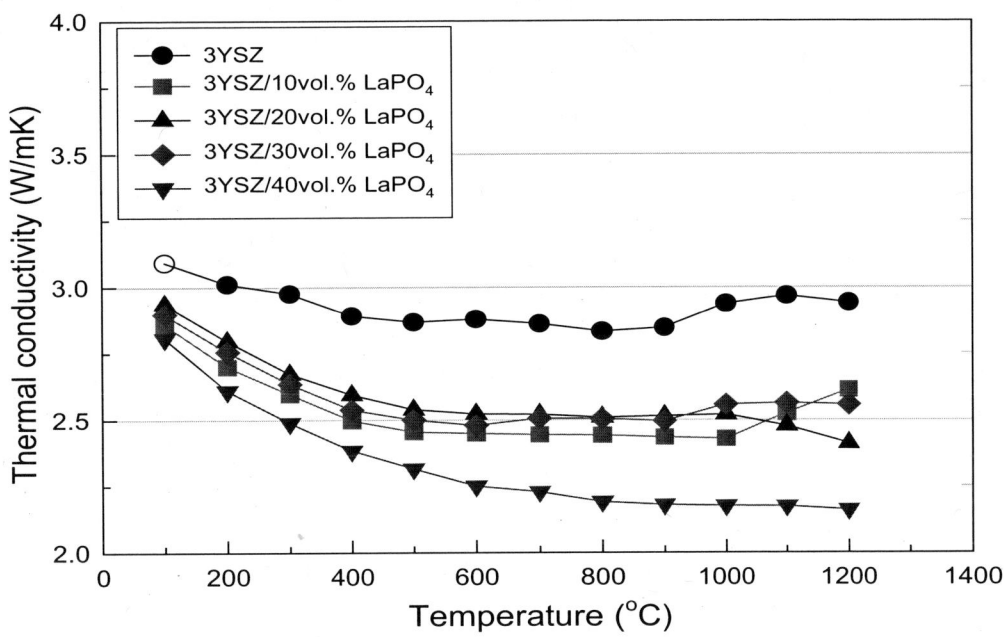

Figure 6. Thermal conductivity of 3YSZ/ LaPO$_4$ nanocomposites with the amount of monazite-type LaPO$_4$ and temperature.

decreasing grain size. The degradation of flexural strength was due to weak bonding between 3YSZ and LaPO$_4$. For this reason, fracture behaviors were dominated by intergranular-type fracture along the grain boundaries, as shown in figure 5.

Figure 6 shows the thermal conductivity of PECSed 3YSZ/monazite-type LaPO$_4$ nanocomposites with monazite-type LaPO$_4$ contents. Thermal conductivity of 3YSZ/monazite-type LaPO$_4$ nanocomposites was lower than the 3YSZ monolith. As the amount of monazite-type LaPO$_4$ increased, thermal conductivity of 3YSZ/LaPO$_4$ nanocomposites decreased. Also, the stabilization range of thermal conductivity of specimens was wider than monolithic 3YSZ. Thermal conductivity of monolithic 3YSZ slightly decreased or was constant up to 900 $^{\circ}$C, and increased above 900 $^{\circ}$C. In the case of nanocomposites, these phenomena were observed at higher temperatures than the monolith. The temperature of nanocomposites observed in these phenomena was above 1000 $^{\circ}$C or more. These phenomena were related to grain size and the amount of LaPO$_4$. As the amount of LaPO$_4$ increased, weak boundaries between 3YSZ and LaPO$_4$ increased. Thermal conductivity was related to phonon diffraction generated by grain boundaries such as 3YSZ/3YSZ and 3YSZ/LaPO$_4$. The effect of phonon diffraction in weak boundaries was higher than that of strong interfaces. Therefore, thermal conductivity was likely caused by increasing phonon diffraction due to increments in areas of grain boundaries. The stable range of nanocomposites was longer than the monolith, as shown in figure 6. The difference of thermal conductivity between 3YSZ and 3YSZ/LaPO$_4$ nanocomposites increased with elevating temperature and the amount of LaPO$_4$. In 3YSZ/40 vol.% LaPO$_4$ nanocomposites, thermal conductivity was reduced to 9 % at 100 $^{\circ}$C and 27 % at 1200 $^{\circ}$C , respectively, when compared with 3YSZ. This indicated that the high temperature stability of monazite-type LaPO$_4$ added to 3YSZ improved as compared to the monolith.

CONCLUSIONS

In order for compounds to be considered as TBC materials, they must have high temperature stability and low thermal conductivity, etc., In order to improve high temperature stability and low thermal conductivity of YSZ, monazite-type $LaPO_4$ was composed with 3YSZ to add properties such as weak bonding between zirconia. Monazite-type $LaPO_4$ was prepared by chemical precipitation methods. The XRD results of $3YSZ/LaPO_4$ nanocomposites demonstrated differences in the crystalline phases between as-sintered and annealed zirconia. The phase transformation of as-sintered specimens was related to the amount of monazite-type $LaPO_4$, but annealed specimens were not observed to be affected by the amount of monazite-type $LaPO_4$ on phase transformation. The density of $3YSZ/LaPO_4$ nanocomposites after annealing was decreased and porosity was increased. In addition, mechanical properties of $3YSZ/LaPO_4$ nanocomposites decreased with increasing dispersion of monazite-type $LaPO_4$ particles. This was likely caused by low mechanical properties of $LaPO_4$ and weak bonding between 3YSZ. Thermal conductivity of $3YSZ/LaPO_4$ nanocomposites was lower than 3YSZ. The difference in thermal conductivity between 3YSZ and $3YSZ/LaPO_4$ nanocomposites at high temperatures was higher than that at low temperatures.

ACKNOWLEDGMENTS

This work was carried out as a part of the "Nanostructure Coating Project (Nano-Coating Functions, Structural Design and Control Technique)" under the Nanotechnology Materials Program supported by The New Energy and Industrial Technology Development Organization (NEDO), Japan.

REFERENCES

[1]X.Q. Cao, R. Vassen, and D. Stoever, "Ceramic Materials for Thermal Barrier Coatings," *J. Euro. Ceram. Soc.*, **24**, 1-10 (2004).

[2]D.R. Clarke, and S.R. Phillpot, "Thermal Barrier Coating Materials," *Materialstoday*, 22-29 (2005).

[3]U. Schulz, B. Saruhan, K. Fritscher, and C. Leyens, "Review on Advanced EB-PVD Ceramic Topcoats for TBC Applications," *Int. J. Appl. Ceram. Technol.*, **1**, 302-15 (2004).

[4]D. Zhu, and R.A. Miller, "Thermal Barrier Coatings for Advanced Gas Turbine and Diesel Engines," *NASA Glenn Research Center, Cleveland, Ohio, NASA/TM-209453*, 1-12 (1999).

[5]X. Cao, R.Vassen, W. Fischer, F. Tietz, W. Jungen, and Detlev Stöver, "Lanthanum-Cerium Oxide as a Thermal Barrier-Coating Material for High-Temperature Applications," *Adv. Mater.*, **15**, 1438-42 (2003).

[6]F. Cernuschi, P. Bianchi, M. Leoni, and P. Scardi, "Thermal Diffusivity/Microstructure Relationship in Y-PSZ Thermal Barrier Coatings," *J. Therm. Spray Technol.*, **8**, 102-109 (1999).

[7]W. Min, K. Daimon, T. Matsubara, and Y. Hikichi, "Thermal and Mechanical Properties of Sintered Machinable $LaPO_4$-ZrO_2 Composites," *Mater. Res. Bull.*, **37**, 1107-15 (2002)

[8]W. Min, D. Miyahara, K. Yokoi, T. Yamaguchi, K. Daimon, Y. Hikichi, T. Matsubara, and T. Ota, "Thermal and Mechanical Properties of Sintered $LaPO_4$-Al_2O_3 Composites," *Mater. Res. Bull.*, **36**, 936-45 (2001)

[9]K. Niihara, "Indentation Microfracture of Ceramics - Its Application and Problems," *Ceramics Japan*, **20**, 12-18 (1985).

[10]J.B. Davis, D.B. Marshall, R.M. Housley, and P.E.D. Morgan, "Machinable Ceramics Containing Rare-Earth Phosphates," *J. Am. Ceram. Soc.*, **81**, 2169-75 (1998).

Dense Alumina–Zirconia Coatings Using the Solution Precursor Plasma Spray Process

Dianying Chen,[‡] Eric H. Jordan,[†,§] Maurice Gell,[‡] and Xinqing Ma[¶]

[‡]Materials Science and Engineering Program, Institute of Materials Science, University of Connecticut, Storrs, Connecticut 06269

[§]Department of Mechanical Engineering, University of Connecticut, Storrs, Connecticut 06269

[¶]Inframat Corporation, Farmington, Connecticut 06032

For the first time, dense coatings have been made by the solution precursor plasma spray (SPPS) process. The conditions are described for the deposition of dense Al_2O_3–40 wt% 7YSZ (yttria-stabilized zirconia) coatings; the coatings are characterized and their thermal stability is evaluated. X-ray diffraction analysis shows that the as-sprayed coating is composed of α-Al_2O_3 and tetragonal ZrO_2 phases with grain sizes of 72 and 56 nm, respectively. The as-sprayed coating has a 95.6% density and consists of ultrafine splats (1–5 μm) and unmelted spherical particles (<0.5 μm). The lamellar structure, typical of conventional plasma-sprayed coatings, is absent at the same scale in the SPPS coating. The formation of a dense Al_2O_3–40 wt% 7YSZ coating is favored by the lower melting point of the eutectic composition, and resultant superheating of the molten particles. Phase and microstructural thermal stabilities were investigated by heat treatment of the as-sprayed coating at temperatures of 1000°–1500°C. No phase transformation occurs, and the grain size is still in the nanometer range after the 1500°C exposure for 2 h. The coating hardness increases from 11.8 GPa in the as-coated condition to 15.8 GPa following 1500°C exposure due to a decrease in coating porosity.

I. Introduction

BULK nanostructured materials have exhibited excellent properties such as hardness, strength, and wear resistance over conventional, micrograined counterparts.[1] Recently, the technology for deposition of nanostructured ceramics coatings using the plasma spraying process has been developed.[2–5] It is reported that nanostructured coatings, with a duplex microstructure, have improved mechanical properties compared with that observed in commercial coatings. Duplex nanostructured Al_2O_3–13 wt% TiO_2 coatings have superior mechanical properties, such as a two times increase in adhesion strength, a three times increase in abrasive wear resistance, and a much better spallation resistance in bend and cup tests, in comparison with conventional coatings.[2–4] Plasma-sprayed nanostructured zirconia coatings also possess better wear resistance than traditional coarser-grained coatings.[6]

In air plasma spray (APS), individual nanoparticles cannot be thermally sprayed because of their low mass and the resultant inability to be carried in a moving gas stream and deposited on a substrate.[3] To overcome this, reconstitution of individual nanoparticles into spherical micrometer-sized granules is necessary.

Recently, a solution precursor plasma spray (SPPS) process was developed for the deposition of durable, low-thermal-conductivity 7YSZ (yttria-stabilized zirconia) thermal barrier coatings.[7,8] In the SPPS process, liquid-precursor solutions are injected directly into the plasma jet. The atomized droplets undergo a series of physical and chemical reactions before deposition on the substrate as a coating. The SPPS process for the deposition of ceramic coatings offers several advantages over the conventional plasma spray method, such as circumvention of the powder-feedstock preparation step, better control over the chemistry of the deposit, the ability to deposit compositionally graded coatings with ease, the ability to deposit coatings that are inherently nanostructured (nanometer-scale grain sizes), and processing versatility. These advantages and the potential to deposit a wide range of ceramics (oxides and nonoxides) make the SPPS method attractive.

Both alumina and zirconia have gained wide applications as structural ceramics or protective coatings due to their excellent mechanical and thermal properties. In the Al_2O_3–ZrO_2 binary system, there is a eutectic with the composition of Al_2O_3–40 wt% ZrO_2 at 1880°C. Because the eutectic temperature is lower than the melting points of pure zirconia (~2700°C) and pure alumina (~2100°C), the droplets with this eutectic composition will be easily melted compared with the pure zirconia and alumina. Because a dense coating using the SPPS process is dependent on generating exclusively ultra-fine splats,[9] it was hypothesized that this would be a favorable composition for demonstrating the ability of the SPPS process to make dense coatings.

II. Experimental Procedure

(1) Precursor Preparation

The precursor is an aqueous solution containing aluminum, yttrium, and zirconium salts that are mixed based on molar volumes to produce a ceramic composition of Al_2O_3–40 wt% 7YSZ (7 wt% Y_2O_3).

(2) Precursor Characterization

The precursor solution was dried on a hot plate at ~100°C. To study the phase evolution of the solution precursor, the dried precursor powders were then heated to various temperatures (800°–1200°C) at a heating rate of 10°C/min, and then held for 1 h.

(3) Plasma Spray Deposition

Figure 1 shows the SPPS process, where an atomizing nozzle, attached to the plasma torch, injects solution precursor mist into the plasma jet. The direct current (DC) plasma torch used here is the Metco 9MB (Sulzer—Metco, Westbury, NY), which is attached to a six-axis robotic arm. Argon and hydrogen are used

W. Mullins—contributing editor

Manuscript No. 22628. Received December 26, 2006; approved July 20, 2007.
This work was supported by the U.S. Office of Naval Research under Grant No. N00014-02-1-0171 managed by Dr. Lawrence Kabacoff.
[†]Author to whom correspondence should be addressed. e-mail: jordan@engr.uconn.edu

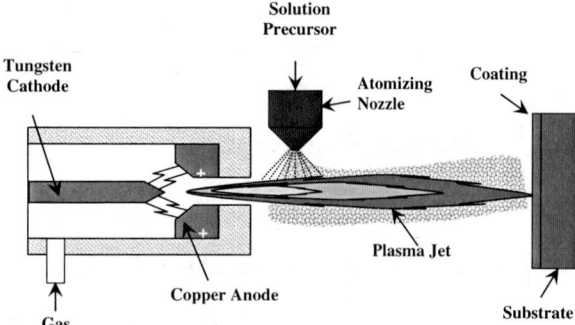

Fig. 1. Schematic illustration of the solution precursor plasma spray process.

as the primary and the secondary plasma gases, respectively. The coating was deposited on Type 304 stainless steel substrates (disks 25 mm diameter, 3 mm thickness) with an APS NiAl bond coat (Metco 450NS, Ni5Al, ∼100 μm thickness).

To investigate phase and microstructural stability, the as-sprayed coatings were detached by immersion of the specimen in a bath of hydrochloric acid. The acid attacked the substrate/top coat interface and the coating become detached after 1 week. The detached specimens were then heated in a furnace at temperatures of 1100°, 1200°, 1300°, 1400°, and 1500°C for 2 h.

(4) Characterization

The crystalline phase composition of all samples was determined using X-ray diffraction, (XRD, CuKα radiation; D5005, Bruker AXS, Karlsruhe, Germany). The XRD patterns were collected in a 2θ range from 20° to 80° at a scanning rate of 2°/min. The average crystallite size was estimated based on XRD peak broadening using the Scherrer formula[10]

$$D_{hkl} = \frac{0.9\lambda}{\beta_{hkl}\cos\theta}$$

where D_{hkl} is the average dimension of crystallites, λ is the wavelength of the X-ray radiation ($\lambda = 0.15405$ nm), and θ is the Bragg angle of reflection of a specific crystalline plane, and β_{hkl} is the full-width of half-maximum (FWHM) of the peak intensity after correction for the instrumental line broadening. For tetragonal ZrO_2, a (111) reflection was used and for α-Al_2O_3, a (012) reflection was used. There are several ways to take the instrumental line broadening into account. One suggested by Taylor[11] and used in this study is

$$\beta_{hkl} = \sqrt{B^2 - b^2}$$

where B is the FWHM of the peak and b is the instrumental line broadening. The instrumental line broadening was obtained from FWHM of peaks in the same angular region for polycrystalline α-Al_2O_3 and ZrO_2 samples with a crystal size > 1 μm.

The chemical composition and the binding state of elements in the deposited coating surface were characterized by X-ray photoelectron spectroscopy (XPS, VG Scientific ESCALAB Mark II) using a MgKα X-ray source and a pass energy of 60 eV. All the spectra were calibrated at the Carbon 1s binding energy (284.8 eV). Survey scans were performed up to 1100 eV at 1 eV/s. The data analysis was performed with CasaXPS software. The relative sensitivity factors were selected from the Scofield element library in CasaXPS.

An environmental scanning electron microscope (ESEM 2020, Philips Electron Optics, Eindhoven, the Netherlands) and a JEOL JSM-6335F field emission scanning electron microscope (FESEM) were used to characterize the coating microstructure. Coating porosity was measured on the polished cross section (×2500 magnification) by image analysis. The Vickers hardness of the as-sprayed coatings was measured on

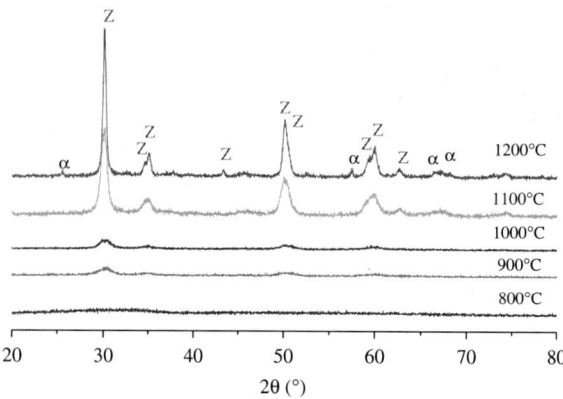

Fig. 2. X-ray diffraction patterns of as-calcined powders: (Z: Zirconia, α: α-Al_2O_3).

the polished cross section with a 0.98 N normal load and a dwell time of 15 s. The hardness value for each sample is the average of 10 measurements.

III. Experimental Results

(1) Solution Precursor Phase Evolution

The XRD patterns of the Al_2O_3–40 wt% 7YSZ composite powders heated in the lab furnace at various temperatures are displayed in Fig. 2. It can be seen that at 800°C, the composite powders are still amorphous. The zirconia crystalline peak in the composite powders begins to appear at 900°C, which is much higher than the crystallization temperature (∼400°C) of zirconia from the pure zirconium salt precursor.[12] The increased crystallization temperature of zirconia may result because Al_2O_3 suppresses zirconia crystallization.[13] α-Al_2O_3 was formed at ∼1200°C, without experiencing any other intermediate phases, such as, γ, δ, θ, in the calcined samples.

(2) Deposition of Dense SPPS Al_2O_3–7YSZ Coatings

Figure 3 shows the representative surface morphology of the as-sprayed Al_2O_3–40 wt% 7YSZ coating. The coating is composed of ultrafine splats (1–5 μm) and dense fine spheres. No gel-like unpyrolyzed precursor appears on the coating surface. The ultrafine splats are quite similar to what is observed in APS coatings; however, the diameters of the splats in the SPPS coating (1–5 μm) are much smaller than that in a typical APS coating (100–150 μm).[5,14] These splats and spherical particles indicate that melting and solidification take place during SPPS coating formation.

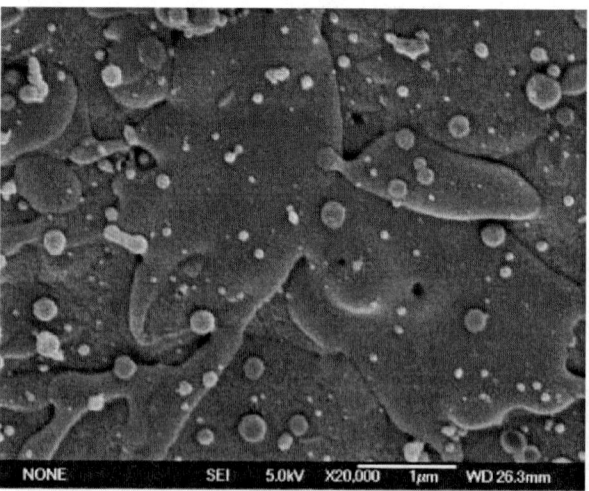

Fig. 3. Typical coating surface morphology.

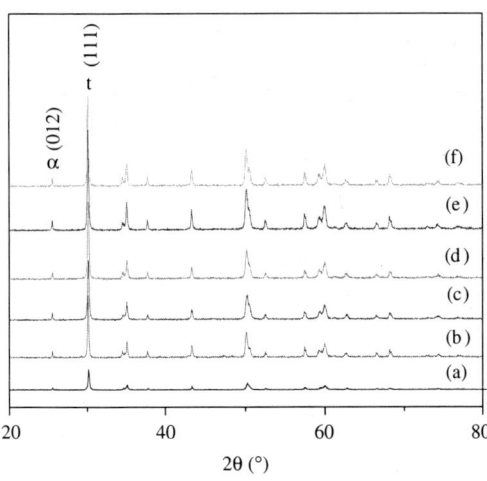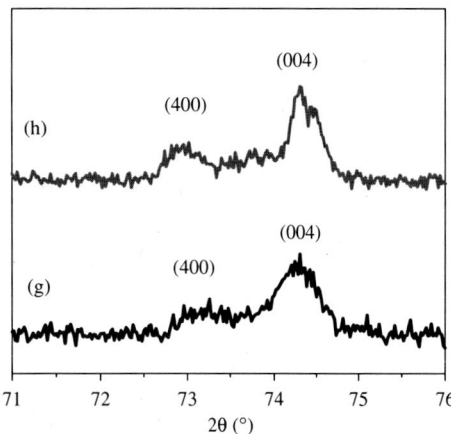

Fig. 4. X-ray diffraction of the as-sprayed (a, g) and heat-treated coatings at temperatures of (b) 1100°C, (c) 1200°C, (d) 1300°C, (e) 1400°C, and (f) and (h) 1500°C. (t: tetragonal zirconia; α: α-alumina).

The XRD pattern (Fig. 4(a)) shows that the coating is composed of two distinct phases: α-Al_2O_3 and ZrO_2. The split of (400) and (004) peaks shown in Fig. 4(g) indicates that the zirconia phase is metastable tetragonal (t'). The grain sizes of α-Al_2O_3 and metastable tetragonal ZrO_2 phases determined by the Scherrer equation are 72 and 56 nm, respectively.

XPS investigation of the as-sprayed coating surface has been conducted (Fig. 5). The atomic ratio of Al and Zr elements was calculated by considering the sensitivity factor using CasaXPS software from the integrated area under the assigned element peak in the XPS spectra. It shows the molar ratio of Al over Zr is 3.65, which is very close to the theoretical molar ratio of 3.61. The results indicate that the coating composition can be controlled from the precursor chemistry.

Polished cross sections of the as-sprayed coating are presented in Figs. 6(a) and (b). The thickness of the coating is approximately 50 μm. A lamellar structure, which is a common characteristic in conventional plasma-sprayed coatings, cannot be seen at this magnification in the present SPPS coating. The coating is quite dense, with a measured porosity of only 4.4%, based on image analysis. The average hardness of the as-sprayed coating is 11.8 GPa. The typical Vickers indentation shape is shown in Fig. 6(b). Figure 6(c) shows a SEM micrograph of the fracture surface of an as-deposited coating. Spherical particles are observed in the fracture surface (as indicated by the arrow), which are trapped and cemented into the coating by subsequent deposits.

(3) Heat Treatment of the As-Sprayed Coatings

(A) Phase Stability and Grain Growth Activation Energy: The coating phase and microstructure stability are investigated by heat treatment for 2 h at temperatures between 1100° and 1500°C. Figure 4 shows the typical XRD patterns of the heat-treated coatings. It can be seen that both the Al_2O_3 and ZrO_2 peak intensities after heat treatment are stronger than in the as-received coating. Again, the split of (400) and (004) peaks (Fig. 4(h)) of the 7YSZ phase indicates that the coating still retains the metastable tetragonal zirconia phase after 1500°C heat treatment; no phase transformation occurred during heat treatment. The average grain size of Al_2O_3 and ZrO_2 determined by the Scherrer equation is plotted in Fig. 7. Up to 1400°C, both the Al_2O_3 and ZrO_2 grain sizes increase very slowly. At 1500°C, the Al_2O_3 grain size coarsens more quickly than 7YSZ grains, but is still within the nanometer range. Owing to the rapid Al_2O_3 grains' growth at 1500°C, in the following part, the calculation of the activation energy for Al_2O_3 grain growth is based on the grain size data at heat treatments below 1500°C.

The activation energies of grain growth for Al_2O_3 and ZrO_2 are calculated by the following equation[15]

$$d^2 - d_0^2 = kt$$

where d is the grain size, d_0 is the initial grain size, and t is the time. The term k is sensitive to temperature and is usually written as

$$k = k_0 \exp\left(\frac{-Q}{RT}\right)$$

The grain size as a function of temperature can also be expressed as

$$\ln(d^2 - d_0^2) = \frac{-Q}{RT} + k'_0$$

where k_0 and k'_0 are constant. By plotting $\ln(d^2 - d_0^2)$ as a function of the inverse temperature, the activation energy can be calculated from the slope. Figure 8 shows the plot of $\ln(d^2 - d_0^2)$ as a function of the inverse temperature for Al_2O_3 and ZrO_2, respectively. From the slope of Fig. 8, the activation energies for Al_2O_3 and ZrO_2 grain growth were determined to be Q (Al_2O_3) = 150.5 ± 18.5 kJ/mol and Q (ZrO_2) = 154.6 ± 36.5 kJ/mol, respectively.

(B) Microstructural Evolution: Figure 9 illustrates the SEM secondary electron images of coatings at various heat-

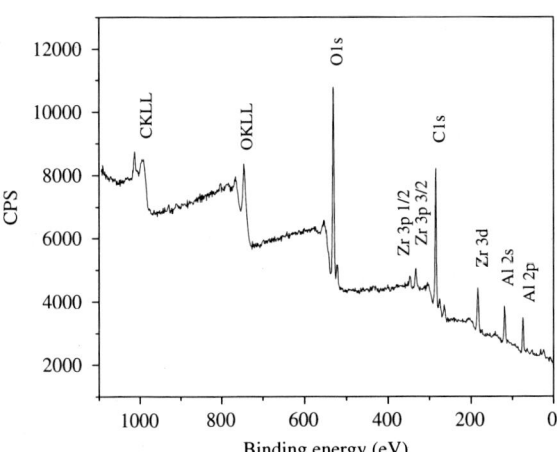

Fig. 5. X-ray photoelectron spectroscopy of the as-sprayed coating.

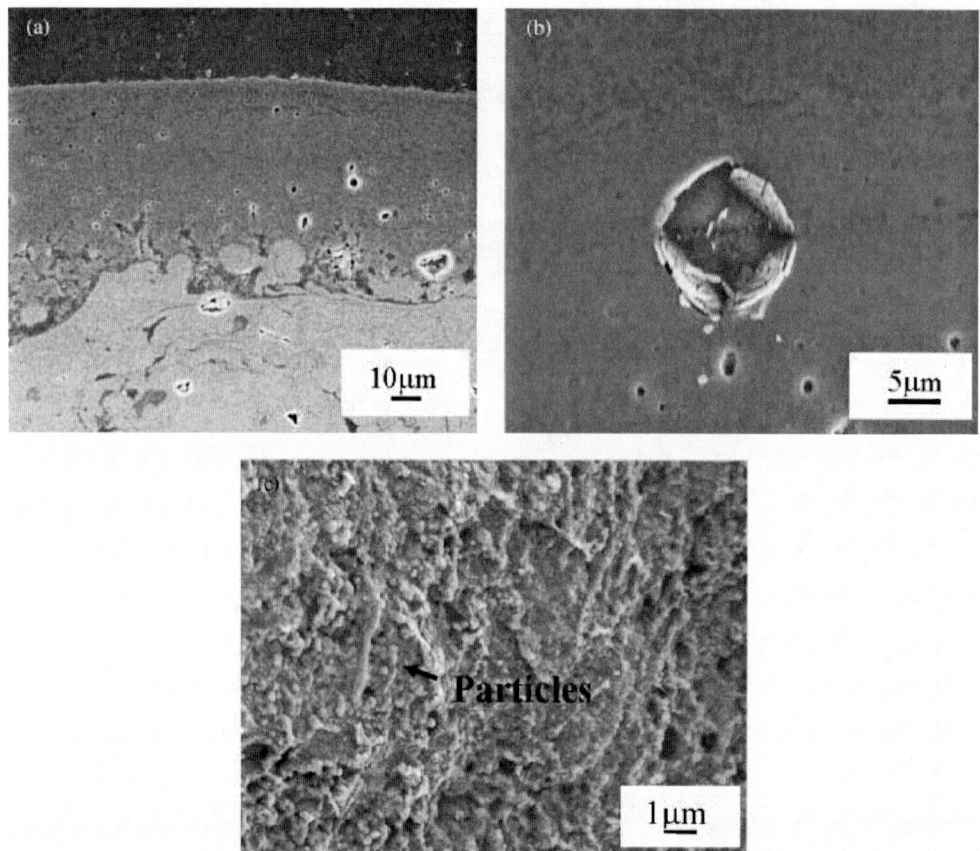

Fig. 6. Microstructure of the as-sprayed coating: (a) polished cross section, (b) Vickers hardness indent shape, and (c) fractured surface.

treatment temperatures. It can be seen that the coating surface morphology heat treated at 1100°C (Fig. 9(b)) is similar to that of the as-received sample (Fig. 9(a)). Ultrafine spheres and splats could still be observed. The splats after heat treatment at 1200°C are almost invisible. At 1300°C, the splats disappear and gradually evolve to grains. The grains become more delineated when the heat-treatment temperature increases to 1400° and 1500°C. The back-scattered electron image of coatings heat treated at 1500°C for 2 h (Fig. 10) shows the Al_2O_3 (black phase) and ZrO_2 (bright phase) with grain sizes around 350 and 170 nm, respectively, which are slightly higher than the grain size values determined by the Scherrer equation.

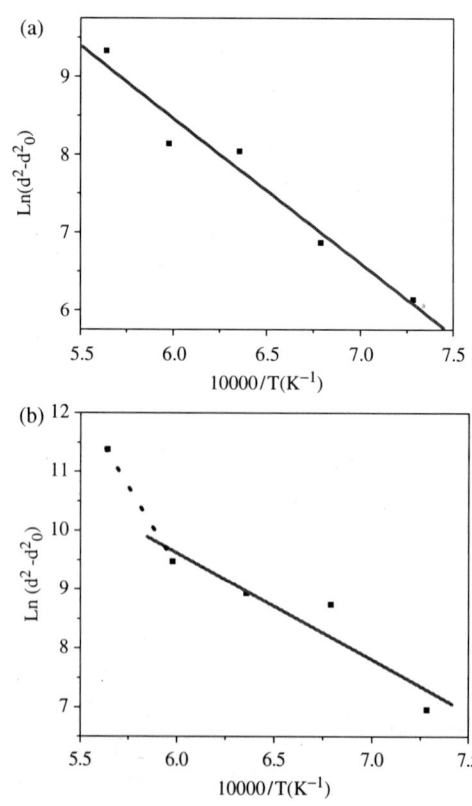

Fig. 8. Plot of $\ln(d^2 - d_0^2)$ as a function of reverse temperature: (a) ZrO_2, (b) Al_2O_3.

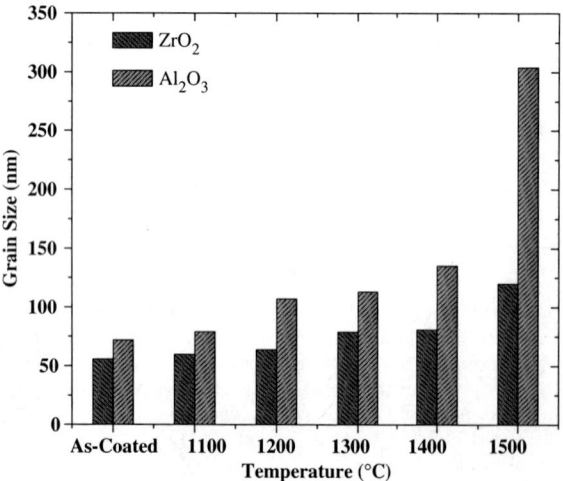

Fig. 7. Plot of grain size as a function of heat treatment temperature.

Progress in Thermal Barrier Coatings

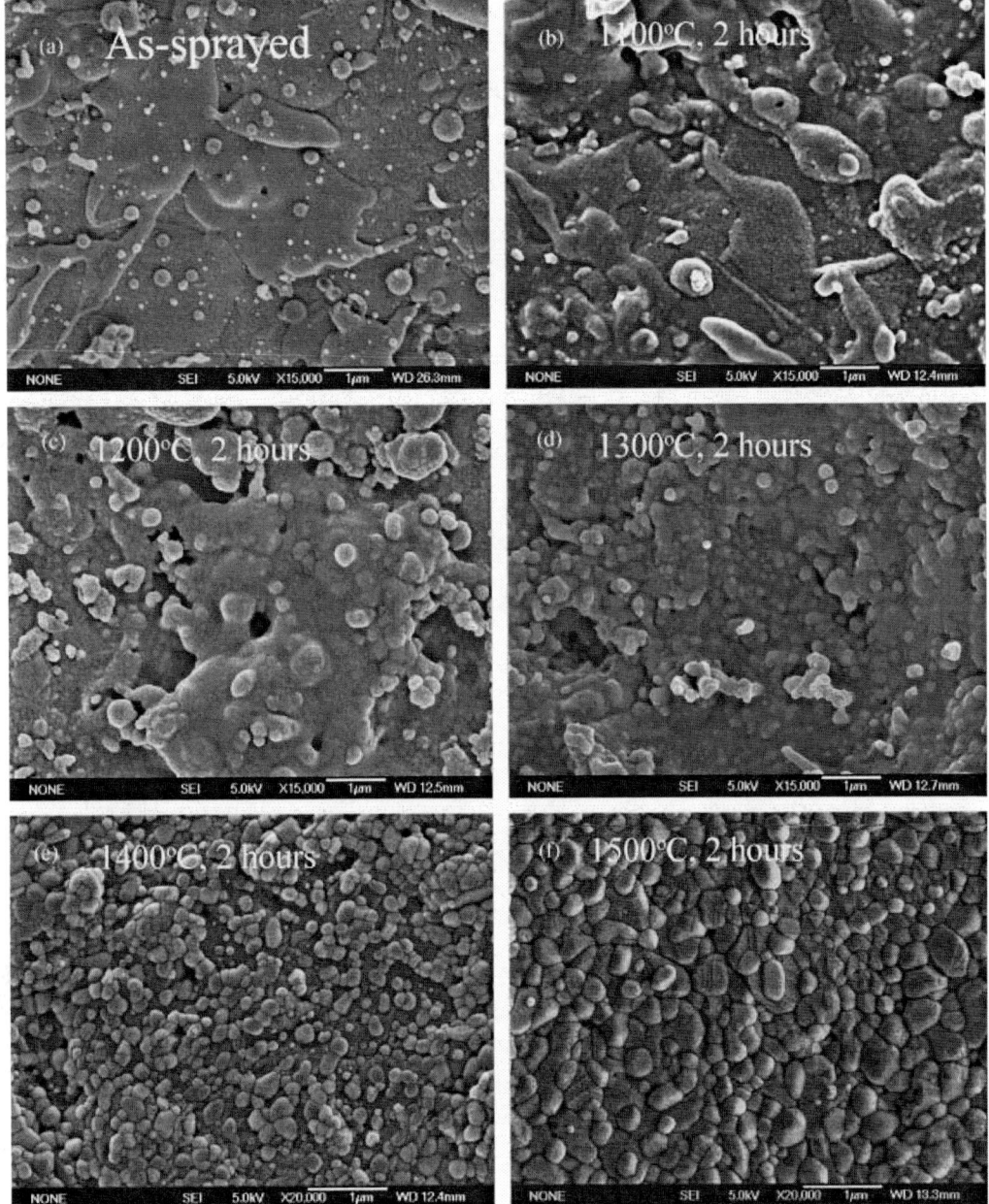

Fig. 9. Secondary electron images of coatings heat treated at various temperatures for 2 h showing the microstructure evolution: (a) as-received, (b) 1100°C, (c) 1200°C, (d) 1300°C, (e) 1400°C, and (f) 1500°C.

Figure 11 shows the SEM micrograph of a coating fracture surface following heat treatment at 1500°C. Comparing Figs. 11 and 6(c), it can be seen that the spherical particles have disappeared and evolved to grains.

(C) Hardness and Porosity: Figure 12 shows the coating hardness as a function of heat-treatment temperature. Each data point is the average of 10 readings, and the error bars are the standard deviation of the readings made on each sample. Coating hardness increases from 11.8 GPa before heat treatment to around 13.0 GPa at 1300°C, and then increases to 15.8 GPa at 1500°C. In addition, the coating porosity decreases from 4.4% in an as-sprayed condition to 1% after 1500°C exposure due to sintering. Kim and Khalil[16] studied the relationship of sintering density and the mechanical behavior of Al_2O_3–27.8 wt% 3YSZ (3mol% YSZ) ceramics and reported that the ceramics have a hardness of ~ 12.0 GPa with 96% density and ~ 17.0 GPa in fully densified ceramics. The decrease of the porosity accompa-

nied by hardness increase is very similar to the behavior observed in the present experiment.

IV. Discussion

(1) Phase Formation

When the precursor powders are heat treated, crystallizations for zirconia and alumina are observed to form at temperatures of 900° and 1200°C, respectively (Fig. 2). During coating deposition, the substrate temperature (~ 500°C) is lower than the crystallization temperature of alumina and zirconia; however, both crystalline alumina and zirconia phases are present in the as-sprayed coatings, which is due to the high plasma temperature (over 10 000 K), where the atomized droplets will undergo rapid solvent evaporation, solute precipitation, decomposition, melting, solidification, and crystallization. In addition to the predominant crystalline alumina and zirconia phases in the

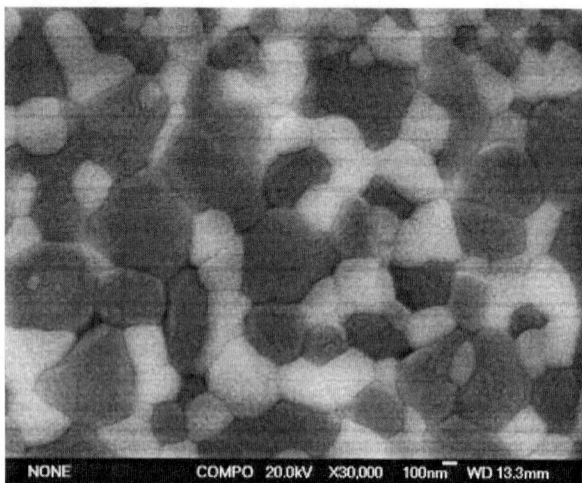

Fig. 10. Back-scattered electron image of coating heat treated at 1500°C for 2 h showing the Al_2O_3 (black) and 7 wt% yttria stabilized zirconia (white) grains.

coating, some amorphous phase is also present. When the as-sprayed coatings are heat treated, the amorphous phases crystallize, as can be seen by the increased XRD peak intensity of heat-treated samples in comparison with the as-sprayed coating (Fig. 4).

(2) Splats and Dense Coating Formation

In conventional plasma spraying, powder particles injected into a plasma jet are melted and propelled onto a substrate, where they spread upon impact and rapidly solidify. The flattened and solidified particles, called splats, represent the building blocks of the coating, along with voids and unmelted particles. The dynamics of the splat formation involves flattening (spreading) of a molten droplet driven by the droplet kinetic energy, surface tension, heat transfer, and rapid solidification processes.[17] The diameter of splats in conventional powder spray is generally in the range of 100–150 μm.

In the SPPS process, the coating is built up in a similar way as that of the APS process. However, the splats' diameter ranges from 1 to 5 μm, which is about 50 times smaller than the splats in APS coatings. The fine splats are formed because the droplets undergo a break-up in the plasma.[18] Because there are no obvious splat boundaries observed on the fracture surface of Fig. 6, it can be concluded that the contact between splats is improved, likely resulting from the ultrafine splat diameter and

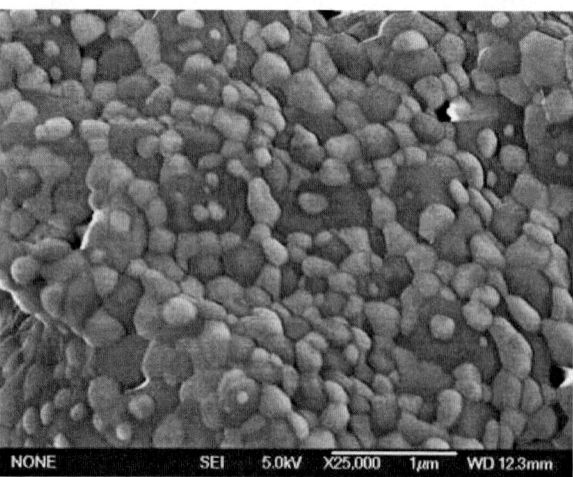

Fig. 11. Fractured surface morphologies of coating heat treated at 1500°C for 2h.

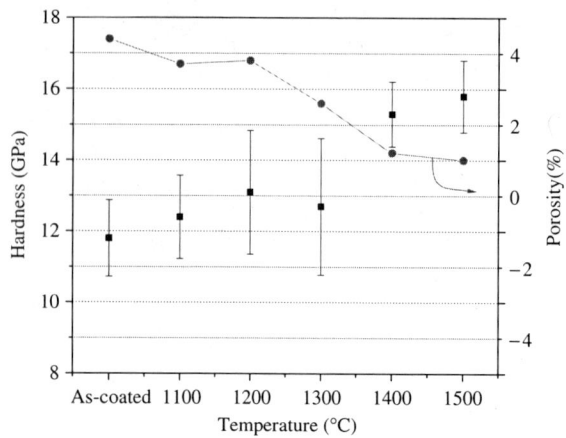

Fig. 12. Coating hardness as a function of heat treatment temperature.

the resultant, reduced out-of-plane distortion upon solidification and cooling. Normal splats, produced in the APS process, might be viewed as taking on the shape of a potato chip, because of varied temperature gradients and cooling rates across the splat area. With the much smaller splat area (1/2500) of the SPPS splats compared with the APS splats, the temperature gradients across the splat area are reduced and the splat solidifies and cools to a thin disk shape.

In previous research on SPPS thermal barrier coatings,[8,9] the as-deposited coating was relatively porous: about 80% theoretical density. In the present research, dense Al_2O_3–40 wt% 7YSZ coatings with a high hardness were deposited by the SPPS process, and the as-sprayed coating is mainly composed of fine splats and a small amount of spherical particles.

The deposition of a dense coating for Al_2O_3–40 wt% 7YSZ and the deposition of a porous coating for 7YSZ are due to the difference in the melting temperatures of the solid particles in a plasma jet. In Al_2O_3–7YSZ composites, a eutectic composition of the coating, which has a much lower liquid-formation temperature (1880°C) compared with the pure ZrO_2 (2700°C), was chosen. The relatively low liquid-formation temperature ensures melting of most of the solid particles in the plasma. These liquid spheres form ultrafine splats when impacting the substrate and produce a dense coating with high hardness. It can be predicted that a material with a similar low melting point, such as TiO_2 (1840°C), can also be SPPS deposited with a high density. These experiments have been successfully carried out and will be the subject of another publication.

Solid spherical particles formed in the SPPS process are considered to arise from the two sources: (1) melted particles resolidify; and (2) periphery of the plasma jet,[9] where the plasma temperature is relatively low; the pyrolyzed precursor may crystallize or even sinter in this low plasma temperature region, but the thermal energy is insufficient for melting the solid particles. These fine solid particles will be trapped and cemented into the coating by subsequent deposits.

(3) Phase and Microstructural Stability

α-Al_2O_3 is the thermodynamically stable phase of crystalline aluminum oxide at standard temperature and pressure. A phase transformation will not occur for the stable α-Al_2O_3 after a high-temperature heat treatment. Zirconia has three polymorphs that are stable at different temperatures. Pure ZrO_2 has a monoclinic crystal structure at room temperature and transitions to tetragonal and cubic at increasing temperatures. The volume expansion caused by the cubic to tetragonal to monoclinic transformation induces very large stresses, and will cause pure ZrO_2 to crack upon cooling from high temperatures. Therefore, a dopant, such as Y_2O_3, is added to zirconia to stabilize the tetragonal and/or cubic phases. The 7 wt% Y_2O_3 content

Fig. 13. 7 wt% yttria stabilized zirconia coating microstructure heat treated at 1500°C for 2 h.

relative to the ZrO_2 was chosen for the current composition to stabilize the zirconia. Thus, the as-sprayed and heat-treated coatings have the desired metastable tetragonal ZrO_2 phase.

Up to 1400°C, both the Al_2O_3 and ZrO_2 grain sizes increase slowly and the two phases' grain sizes are very close (Fig. 7), which are consistent with their near-identical activation energy values. At 1500°C, the Al_2O_3 grain size coarsens more quickly than 7YSZ grains (Fig. 7); this suggests that the mechanism for grain coarsening for the Al_2O_3 above 1400°C may be different from that at lower temperatures. It is worth comparing the grain growth behavior in single- and dual-phase materials. The average grain size for the single- phase 7YSZ coating heat treated at 1500°C for 2 h is $\sim$ 500 nm (Fig. 13), which is about three times larger than that in the Al_2O_3–40 wt% 7YSZ dual-phase coating. The limited grain growth of zirconia in the Al_2O_3–40 wt% ZrO_2 nanocomposite can be attributed to the limited solid solubility of the two phases and the increased diffusion path relative to the single-phase oxides. According to the phase diagram,[19] the mutual solubility of Al_2O_3 and ZrO_2 is very limited. Grain growth will, therefore, be controlled by grain boundary and interphase-boundary diffusion. Because there is 40 wt% 7YSZ in the composite coating, the 7YSZ and Al_2O_3 grains are interpenetrated (Fig. 10). It is expected that the grain growth occurs mainly by diffusion of species along interphase boundaries, rather than grain boundaries. Therefore, the diffusion path will be increased. As a result, grain growth behavior is inhibited. The results presented here are consistent with the previous work on bulk ceramic sintering of Al_2O_3–58.7 wt% ZrO_2,[20] where coarsening is severely retarded by a dual-phase interpenetrating microstructure. The growth rates in the duplex composition Al_2O_3–58.7 wt% ZrO_2 are 160 and 3500 times lower that that for single-phase Al_2O_3 and ZrO_2, respectively.[20]

V. Conclusions

Dense Al_2O_3–7YSZ nanocomposite coatings with high hardness have been deposited using the SPPS process. The coating is composed of nanograined α-alumina and tetragonal zirconia phases. The deposition of a dense coating is attributed to the formation of ultra-fine splats, which are generated from the low liquid formation temperature of the eutectic composition. XPS quantitative analyses indicate that the coating composition can be accurately controlled by the precursor chemistry composition. High-temperature heat treatments of the coating show that both the phase and nanograin structure are very stable. Grain growth is suppressed in the dual-phase Al_2O_3–7YSZ compared with the single-phase 7YSZ coating.

References

[1] J. R. Weertman, D. Farkas, K. Hemker, H. Kung, M. Mayo, R. Mitra, and H. Van Swygenhoven, "Structure and Mechanical Behavior of Bulk Nanocrystalline Materials," *MRS Bull.*, **24** [2] 44–50 (1999).

[2] E. H. Jordan, M. Gell, Y. H. Sohn, D. Goberman, L. Shaw, S. Jiang, M. Wang, T. D. Xiao, Y. Wang, and P. Strutt, "Fabrication and Evaluation of Plasma Sprayed Nanostructured Alumina–Titania Coatings with Superior Properties," *Mater. Sci. Engg. A—Struct. Mater. Properties Microstruct. Process.*, **301** [1] 80–9 (2001).

[3] L. L. Shaw, D. Goberman, R. M. Ren, M. Gell, S. Jiang, Y. Wang, T. D. Xiao, and P. R. Strutt, "The Dependency of Microstructure and Properties of Nanostructured Coatings on Plasma Spray Conditions," *Surface Coatings Technol.*, **130** [1] 1–8 (2000).

[4] Y. Wang, S. Jiang, M. D. Wang, S. H. Wang, T. D. Xiao, and P. R. Strutt, "Abrasive Wear Characteristics of Plasma Sprayed Nanostructured Alumina/Titania Coatings," *Wear*, **237** [2] 176–85 (2000).

[5] P. Fauchais, V. Rat, U. Delbos, J. F. Coudert, T. Chartier, and L. Bianchi, "Understanding of Suspension DC Plasma Spraying of Finely Structured Coatings for SOFC," *IEEE Trans. Plasma Sci.*, **33** [2] 920–30 (2005).

[6] H. Chen, C. X. Ding, P. Y. Zhang, P. Q. La, and S. W. Lee, "Wear of Plasma-Sprayed Nanostructured Zirconia Coatings Against Stainless Steel Under Distilled-Water Conditions," *Surf. Coatings Technol.*, **173** [2–3] 144–9 (2003).

[7] M. Gell, L. D. Xie, X. Q. Ma, E. H. Jordan, and N. P. Padture, "Highly Durable Thermal Barrier Coatings Made by the Solution Precursor Plasma Spray Process," *Surf. Coatings Technol.*, **177**, 97–102 (2004).

[8] N. P. Padture, K. W. Schlichting, T. Bhatia, A. Ozturk, B. Cetegen, E. H. Jordan, M. Gell, S. Jiang, T. D. Xiao, P. R. Strutt, E. Garcia, P. Miranzo, and M. I. Osendi, "Towards Durable Thermal Barrier Coatings with Novel Microstructures Deposited by Solution-Precursor Plasma Spray," *Acta Mater.*, **49** [12] 2251–7 (2001).

[9] L. D. Xie, X. Q. Ma, E. H. Jordan, N. P. Padture, D. T. Xiao, and M. Gell, "Deposition of Thermal Barrier Coatings Using the Solution Precursor Plasma Spray Process," *J. Mater. Sci.*, **39** [5] 1639–46 (2004).

[10] H. P. Klug and L. E. Alexander, *X-Ray Diffraction Procedures for Polycrystalline and Amorphous Materials*, pp. 491–4. John Wiley & Sons Inc., London, 1954.

[11] A. Taylor, *An Introduction to X-ray Metallography*. Chapman & Hall Ltd., London, 1952.

[12] D. Chen, E. H. Jordan, and M. Gell, "Thermal and Crystallization Behavior of Zirconia Precursor Used in the Solution Precursor Plasma Spray Process," *J. Mater. Sci.*, **42** [14] 5576–80 (2007).

[13] S. Bhattacharyya, S. K. Pratihar, R. K. Sinha, R. C. Behera, and R. I. Ganguly, "Preparation of Alumin-Zirconia Microcomposite by Combined Gel Precipitation," *Mater. Lett.*, **53**, 425–31 (2002).

[14] S. Sampath, X. Y. Jiang, J. Matejicek, A. C. Leger, and A. Vardelle, "Substrate Temperature Effects on Splat Formation, Microstructure Development and Properties of Plasma Sprayed Coatings Part I: Case Study for Partially Stabilized Zirconia," *Mater. Sci. Engg. A—Struct. Mater. Properties Microstruct. Process.*, **272** [1] 181–8 (1999).

[15] H. V. Atkinson, "Theories of Normal Grain Growth in Pure Single Phase Systems," *Acta Metallurg.*, **36** [3] 469–91 (1988).

[16] S. W. Kim and K. A. R. Khalil, "High-Frequency Induction Heat Sintering of Mechanically Alloyed Alumina–Yttria-Stabilized Zirconia Nano-Bioceramics," *J. Am. Ceram. Soc.*, **89** [4] 1280–5 (2006).

[17] A. Vardelle, C. Moreau, and P. Fauchais, "The Dynamics of Deposit Formation in Thermal-Spray Processes," *MRS Bull.*, **25** [7] 32–7 (2000).

[18] L. D. Xie, X. Q. Ma, E. H. Jordan, N. P. Padture, D. T. Xiao, and M. Gell, "Identification of Coating Deposition Mechanisms in the Solution-Precursor Plasma-Spray Process Using Model Spray Experiments," *Mater. Sci. Engg. A—Struct. Mater. Properties Microstruct. Process.*, **362** [1–2] 204–12 (2003).

[19] "Phase Diagrams for Ceramics"; National Institute of Standards and Technology, The American Ceramic Society Inc., Al_2O_3–ZrO_2 Fig. 4377, 4378, 6452.

[20] D. Joanthan, M. P. H. French, M. C. Helen, and G. A. Miller, "Coarsening-Resistant Dual-Phase Interpenetrating Microstructures," *J. Am. Ceram. Soc.*, **73** [8] 2508–10 (1990). □

Thermal Stability of Air Plasma Spray and Solution Precursor Plasma Spray Thermal Barrier Coatings

Dianying Chen and Maurice Gell[†]

Materials Science and Engineering Program, Institute of Materials Science, University of Connecticut, Storrs, Connecticut 06269, USA

Eric H. Jordan

Department of Mechanical Engineering, University of Connecticut, Storrs, Connecticut 06269, USA

Eric Cao

Gillette Corporation, Waterbury, Connecticut, USA

Xinqing Ma

Inframat Corporation, Farmington, Connecticut 06032, USA

Yttria-stabilized zirconia (7YSZ) thermal barrier coatings (TBCs) were produced by conventional air plasma spray (APS) and solution precursor plasma spray (SPPS) processes. Both TBCs were isothermally heat treated from 1200° to 1500°C for 100 h. Changes in the phase content, microstructure, and hardness were investigated. The nontransformable tetragonal (t') phase is the predominant phase in both the as-sprayed APS and SPPS TBCs. APS and SPPS coatings exhibit similar thermal stability behavior such as densification rate, hardness increase, and grain coarsening rate. Both the as-received and heat-treated APS and SPPS TBCs show a bimodal pore size distribution with nano- and micro-size pores. After 1400°C/100 h heat treatment, equiaxed grains replace the columnar structure in APS TBCs and the splat structure disappears. Vertical cracks remain after the 1500°C/100 h exposure in SPPS TBCs. The monoclinic phase appears in APS TBCs after a 1400°C/100 h exposure and in SPPS coatings after a 1500°C/100 h exposure.

I. Introduction

THERMAL barrier coatings (TBCs) are widely used in aircraft engines, marine propulsion, and industrial gas turbines.[1–4] A TBC system usually consists of four layers: (1) a metal substrate providing structural strength; (2) a bond coat providing oxidation resistance; (3) a ceramic top coat providing insulation; and (4) a thermally grown oxide (TGO) formed between the ceramic top coat and the bond coat due to high-temperature oxidation of the bond coat. ZrO_2–7 wt% Y_2O_3 (7YSZ) is the choice for ceramic top coat, primarily because it has a thermal expansion coefficient ($\sim 10^{-5}$ °C)$^{-1}$ closer to that of the metallic substrate, and it has a low high-temperature thermal conductivity (2.5 W·(m·K)$^{-1}$ at 1000°C for a dense polycrystalline ceramic).[5]

The air plasma spray (APS) process is widely used for the deposition of TBCs. In this process, ceramic powder ZrO_2–7wt% Y_2O_3 (7YSZ), is injected into the high temperature, high-velocity plasma jet. The powder is melted and propelled toward the substrate. Upon impact, the molten particles solidify and form "splats." The accumulation of splats results in the buildup of the ceramic coating. The deposited coating is highly defective, containing porosity, and microcracks, which contribute to the low-thermal conductivity of APS TBCs. The splat interface contains considerable porosity, has low toughness, and is the site for crack initiation.

Recently, a solution precursor plasma spray (SPPS) process has been developed to deposit various ceramic coatings.[3,5–8] It has been demonstrated that the SPPS process can produce highly durable, low-thermal conductivity 7YSZ TBCs.[3,5] In the SPPS process, an aqueous chemical precursor feedstock is injected into the plasma jet. The droplets undergo a series of physical and chemical reactions before deposition on the substrate as a 7YSZ coating. The SPPS TBC has a unique microstructure with vertical cracks in a porous matrix and the absence of coarse splats. The matrix of the coating consists of ultra-fine splats and limited amount of unmelted particles that are formed in the plasma jet from the solution precursor. The unmelted particles, the porosity, and the through-thickness cracks all impart strain tolerance to the TBC, while the porosity helps reduce the thermal conductivity.

Zirconia exists in three crystallographic phases: the low-temperature monoclinic phase; the intermediate temperature tetragonal (t) phase; and the high-temperature cubic (c) phase. The phase transformation of tetragonal to monoclinic (m) phase is accompanied by significant volume expansion (approximately 3–5 vol%).[9] During service, TBCs are subject to high-temperature cyclic exposure that can result in phase transformation and sintering of the ceramic topcoat. Sintering is of practical importance in several ways, first the shrinkage and loss of strain tolerance is a critical element in coating failure in the presence of thermal gradients and second sintering can lead dramatic loss of insulating abilities with thermal conductivity increasing by over a factor of 2.[10] We note that the SPPS TBC has initially favorable thermal insulating properties.[11] Finally, in addition to sintering, the phase change from the tetragonal to monoclinic and its associated volume change is generally catastrophic mechanically for TBCs. These can lead to an increase in the driving force for spallation failure of TBCs. In this study, the phase and

J. Smialek—contributing editor

Manuscript No. 22990. Received March 27, 2007; approved May 8, 2007.
This work is supported by U.S. Office of Naval Research under Grant No. N00014-02-1-0171 managed by Dr. Lawrence Kabacoff.
[†]Author to whom correspondence should be addressed. e-mail: jordan@engr.uconn.edu

microstructural stability of APS and SPPS 7YSZ TBCs are evaluated.

II. Experimental Procedures

(1) APS Coating Preparation

Commercial, reconstituted ZrO_2–7.0 wt% Y_2O_3 (7YSZ) powder with average grain size of ~200 nm (Metco 204NS, Sulzer Metco, Westbury, NY) is used as a feedstock to deposit YSZ TBCs with a Metco 9MB plasma torch (Sulzer Metco, Westbury, NY), and a six-axis robotic arm. Argon and hydrogen are used as the primary and the secondary plasma gases, respectively. The coatings were deposited on type 304 stainless steel substrates (disks 25 mm diameter, 3 mm thickness).

(2) SPPS Coating Preparation

The deposition of 7YSZ coatings by the SPPS process is similar to APS process, except that the powder feedstock is replaced by droplets of an aqueous solution containing zirconium and yttrium salts to produce a solid solution of 93 wt% ZrO_2 and 7 wt% Y_2O_3 (7YSZ). The detailed processing conditions can be found in previous publications.[3,5,7,12–18]

(3) Heat Treatment of Stand-Alone Ceramics

All the as-sprayed APS and SPPS coatings were detached by immersion of the specimen in a bath of hydrochloric acid. The acid attacked the substrate/top coat interface and the coating become detached after 1 week. The detached samples were soaked in water for 24 h and rinsed a second time to remove acid residue before heat treatment. The specimens were then heat treated in air at temperatures in the range 1200°–1500°C for 100 h with heating rate of 15°C/min.

(4) Coating Characterization

The crystalline phase composition of as-sprayed and heat-treated samples was determined using X-ray diffraction (XRD, Cu$K\alpha$ radiation; D5005, Bruker AXS, Karlsruhe, Germany). The XRD patterns were collected in a 2θ range from 20° to 80° with a scanning rate of 2°/min and a slow scan rate of 0.1°/min in the 2θ range from 72° to 76°. The average crystallite size was estimated based on XRD peak broadening using the Scherrer formula. The grain size of heat-treated APS and SPPS TBCs was determined from SEM images using a rectangular intercept procedure.[19] The average grain size, D, is then given by

$$D = \sqrt{\frac{4A}{\pi(n_i + n_0/2)}}$$

where A is the area of rectangular, n_i and n_0 are the grain numbers in the rectangular and on the rectangular boundary, respectively.

All as-sprayed and heat-treated APS and SPPS TBCs were cut to prepare the cross sections. These cross sections were then polished to a 1 μm finish using routine metallographic methods. The polished cross sections of the APS and SPPS TBCs were characterized using a scanning electron microscope (SEM) (ESEM 2020, Philips Electron Optics, Eindhoven, the Netherlands). A JEOL (Japan) JSM-6335F field emission scanning electron microscope (FESEM) was used to characterize the coating surface and fracture surface microstructure. Hardness measurements (Vickers indenter, Leco Corporation, St. Joseph, MI, 100 g load) were performed at random locations on the polished cross section of the TBCs. The reported hardness value for each specimen is an average of 10 measurements. Coating porosity was measured on the polished cross section (×1000 magnification) by image analysis. Coating pore size distribution was measured by a mercury intrusion porosimetry method. The pore volume distribution was obtained from the derivative curve of the cumulative intruded pore volume as a function of pore

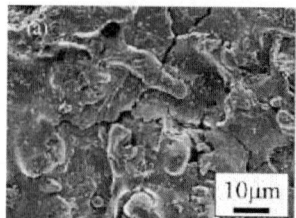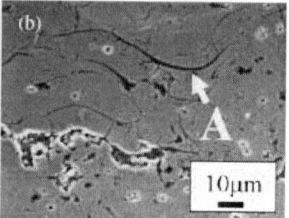

Fig. 1. Microstructure of as-sprayed air plasma spray coating: (a) surface morphology; (b) polished cross section.

diameter. This latter parameter is related to the measured pressure according to the Washburn equation.[20]

III. Experimental Results and Discussion

(1) Microstructure and Phase Composition of As-Sprayed Coatings

Figures 1(a) and (b) show the typical surface morphology and polished cross section of the as-sprayed APS coating. "Splat" boundaries (e.g., position A in Fig. 1) in the cross section can be clearly seen. The SPPS coating shows ultra-fine splats (1–5 μm) and a unique microstructure with evenly spaced through-thickness vertical cracks and lack of horizontal large-scale "splat" boundaries (Figs. 2(a) and (b)). The inset in Fig. 2(b) presents the uniformly distributed porosity. Vertical cracks formation in SPPS TBCs is predominately caused by shrinkage strains associated with the pyrolysis of unpyrolyzed material in the deposited coating.[8] The unmelted particles, the porosity, and the through-thickness cracks all impart strain tolerance to SPPS TBCs, while the porosity reduces thermal conductivity. The lack of large-scale "splat" boundaries toughens the TBC.[3]

XRD spectra of the as-sprayed APS and SPPS coatings are shown in Fig. 3. Both coatings are composed of tetragonal and/or cubic zirconia phases because of the similarities in tetragonal and cubic zirconia XRD patterns. However, the peak splitting (inset in Fig. 3) in the range of 72°–76° indicates that both APS and SPPS as-sprayed coatings are composed of nontransformable tetragonal phase (henceforth called t' to distinguish it from the transformable tetragonal t phase). The nontransformable tetragonal phase (t') in the as-sprayed coatings contains the same yttria concentration as the starting powders or solution precursor and its formation is the result of rapid cooling of molten particles upon impact on the substrate.[21] The high quenching rate during solidification in the plasma spray process causes the diffusionless transformation from the high-temperature cubic phase to the nontransformable tetragonal phase (t') without a composition change.[21] Because this nontransformable tetragonal phase (t') contains a much higher nonequilibrium amount of yttria (7.0 wt%) than that of the equilibrium tetragonal YSZ (~4.0 wt%), it is unstable with respect to heat treatment.

(2) Thermal Stability of APS and SPPS Coatings

(A) Surface Morphologies: Figures 4 and 5 illustrate the SEM secondary electron images of APS and SPPS coatings at

Fig. 2. Microstructure of as-sprayed solution precursor plasma spray coating: (a) surface morphology; (b) polished cross section.

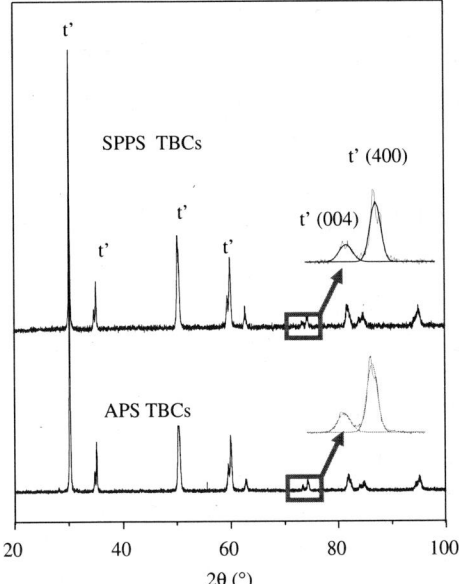

Fig. 3. XRD of as-sprayed solution precursor plasma spray (SPPS) and air plasma spray (APS) thermal barrier coatings (TBCs).

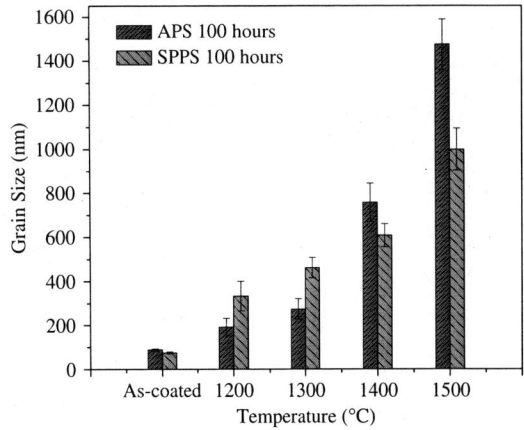

Fig. 6. Grain size of air plasma spray (APS) and solution precursor plasma spray (SPPS) coatings as a function of temperature.

heat treatment temperatures of 1200°–1500°C for 100 h. For APS coatings, with increasing temperature, the splats disappear and grains grow gradually. Grain growth in SPPS coatings is similar. The splats and spherical particles gradually evolve to grains. The calculated grain size based on the image analysis is plotted in Fig. 6. The grain size in the as-sprayed APS and SPPS coating is 89 and 75 nm, respectively. APS and SPPS TBCs have similar coarsening rates up to 1300°C. Above 1300°C, APS grains coarsen at a greater rate. After 1500°C/100 h exposure, the APS and SPPS grain size are 1474 and 998 nm, respectively, which are ∼16 and ∼13 times larger than that in the as-sprayed coatings. The higher grain-coarsening rate of APS TBCs at temperature above 1300°C compared with SPPS TBCs may be due to the higher impurity content of 204NS 7YSZ powders. The reduction of impurity content will significantly improve sintering resistance of 7YSZ TBCs.

(B) Fracture Surface Morphologies: The fracture surfaces of APS TBCs following various thermal exposures are illustrated in Fig. 7. The splats with a columnar structure still exist

up to 1300°C/100 h exposure. After that, the columnar grains disappear and are replaced by the equiaxed grains. Grain growth across splat boundaries is clearly visible in the fracture surface of the 1300°C/100 h heat-treated sample (circle indicated in Fig. 7(b)). Bonding and coherence between splats boundary increases with increasing temperature. Because of the improved bond strength, a transgranular fracture mode is observed in the APS coatings after the 1500°C/100 h exposure.

The SPPS TBCs show a totally different fracture mode, as is illustrated in Fig. 8. Splats and a columnar grain structure are not observed on the fracture surface. This is likely because the coating is fractured along the already-existing vertical cracks or the weak interface between splats. It is noted that intergranular fracture, rather than transgranular fracture, is observed in the SPPS coating after a 1500°C/100 h exposure.

(C) Polished Cross Section: Figure 9 shows a polished cross section of a freestanding SPPS sample heat treated at 1500°C/100 h. The vertical cracks are well-retained, or reformed on cooling, after the 1500°C exposure. Moreover, the through-thickness crack spacing is very uniform following heat treatment, with the vertical crack spacing varying between 100 and 160 μm (Fig. 10).

Figures 11 and 12 show the high-magnification microstructure of polished cross sections of APS and SPPS TBCs after 100

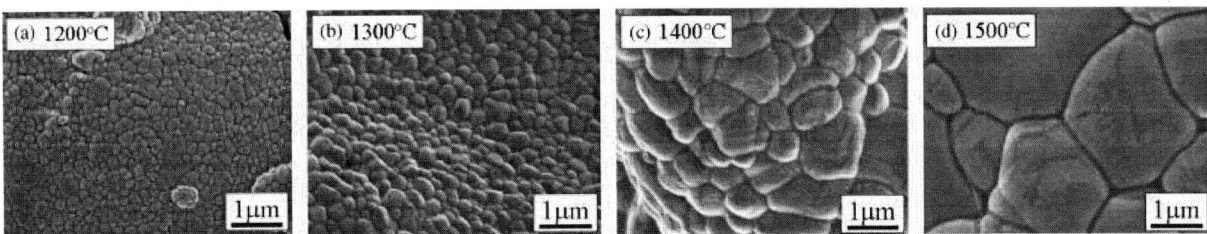

Fig. 4. Surface morphologies evolution of air plasma spray coatings as a function of temperature.

Fig. 5. Surface morphologies evolution of solution precursor plasma spray coatings as a function of temperature.

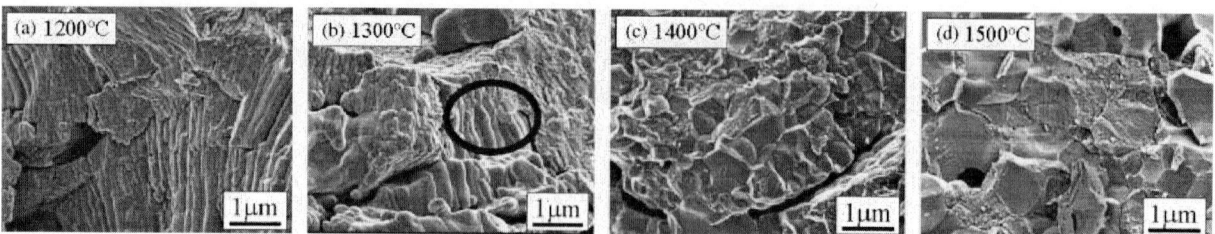

Fig. 7.   Fracture surface of air plasma spray coatings as a function of temperature.

Fig. 8.   Fracture surface of SPPS coatings as a function of temperature.

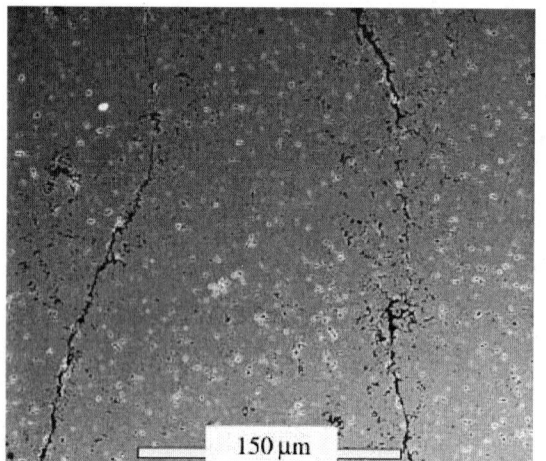

Fig. 9.   Microstructure of 1500°C/100 h heat-treated solution precursor plasma spray coating showing vertical cracks remaining.

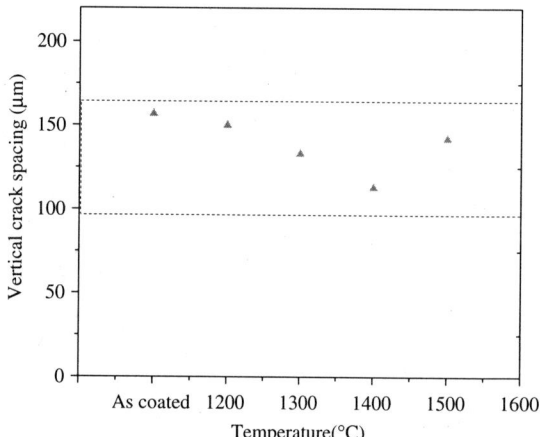

Fig. 10.   Vertical crack spacing of solution precursor plasma spray coatings as a function of temperature.

h heat treatments. In both APS and SPPS coatings, total porosity decreases gradually with increasing temperature. After 1500°C/100 h exposure, the APS coating total porosity has decreased from 21.1% to 9.3%. SPPS coatings have the same tendency as the APS coating, the total porosity decrease from 25% in as-sprayed state to 8.9% after 1500°C/100 h exposure. The coating total porosity and Vickers hardness relationship for APS and SPPS coatings is shown in Fig. 13. As expected, the Vickers hardness increases gradually with the increase of temperature due to the decreased porosity.

(D)   Pore Size Distribution: The coating pore size distributions before and after a 1400°C/100 h heat treatment was measured by mercury intrusion and is shown in Fig. 14. Both APS and SPPS coatings before and after 1400°C/100 h heat treatment show a bi-modal pore size distribution consisting of nano- and micron-sized pores. The average nano- and micron-pore size are 0.45 and 42.6 μm in APS TBCs and 0.38 and 34.6 μm in SPPS TBCs, respectively. Comparing the nano-sized pore distribution before and after heat treatment, small pores, with diameters below 0.1 m, have sintered and disappeared after heat treatment in both APS and SPPS coatings. These results are in agreement with the observations, described above, on polished cross sections. Large pores (∼100 μm) are revealed by mercury intrusion after high-temperature and longtime sintering in both APS and SPPS coatings.

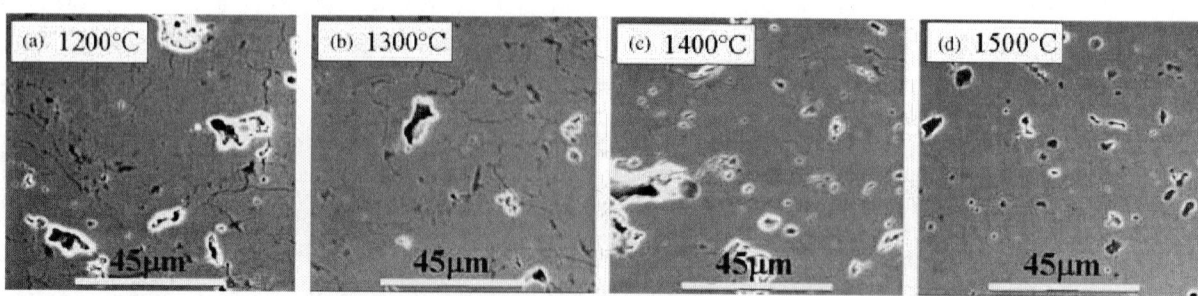

Fig. 11.   Polished cross section of air plasma spray coatings as a function of temperature.

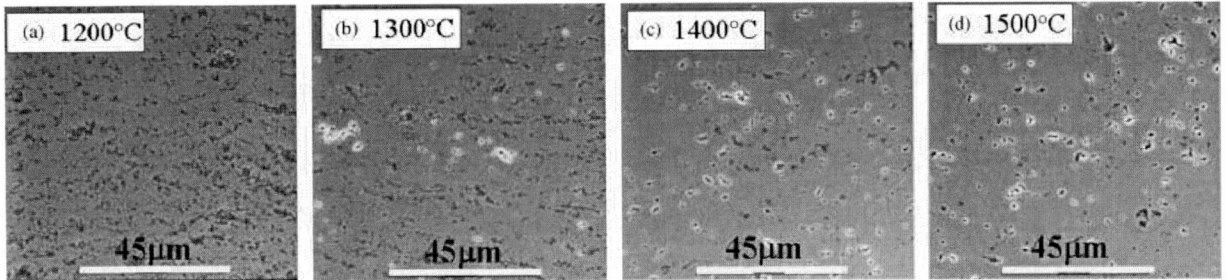

Fig. 12. Polished cross section of solution precursor plasma spray coatings as a function of temperature.

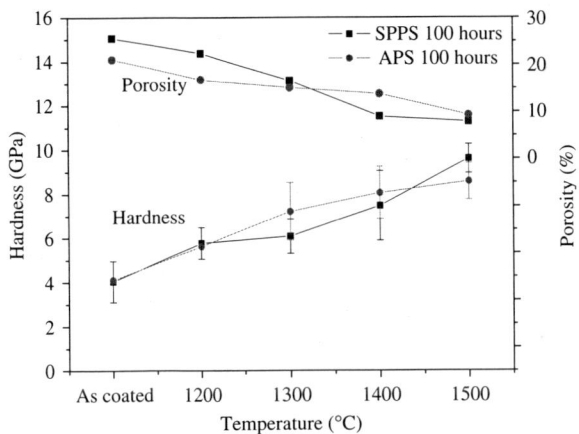

Fig. 13. Coating porosity and hardness as a function of temperature. APS, air plasma spray; SPPS, solution precursor plasma spray.

The relative volume fractions of the nano- and micron-sized pores are shown in Fig. 14. Nano-sized pores in as-sprayed APS TBCs occupy ~41.6%, whereas it is ~84.9% in as-sprayed SPPS TBCs. After heat treatment, the relative amount of nano-sized pores decreases to 24.9% in APS TBCs and 70.2% in SPPS TBCs due to the pore sintering. The relative micron porosity level increases with the high-temperature heat treatment in both APS and SPPS TBCs even in the face of a reduction in overall porosity (Fig. 13). The average micron-sized large pore size in SPPS coating is smaller than that of APS coatings (Fig. 15) before and after heat treatment.

(3) Phase Stability of APS and SPPS TBCs

XRD patterns of all heat-treated APS and SPPS coatings were collected in the ranges of 20–70° and 72–76°. The XRD patterns of heat-treated samples at various temperatures for 100 h are shown in Fig. 16.

It is noted that the diffraction peak for the metastable tetragonal (t') (400) plane has shifted to lower angle compared with that of t (400) plane after heat treatment at 1300°C and above in

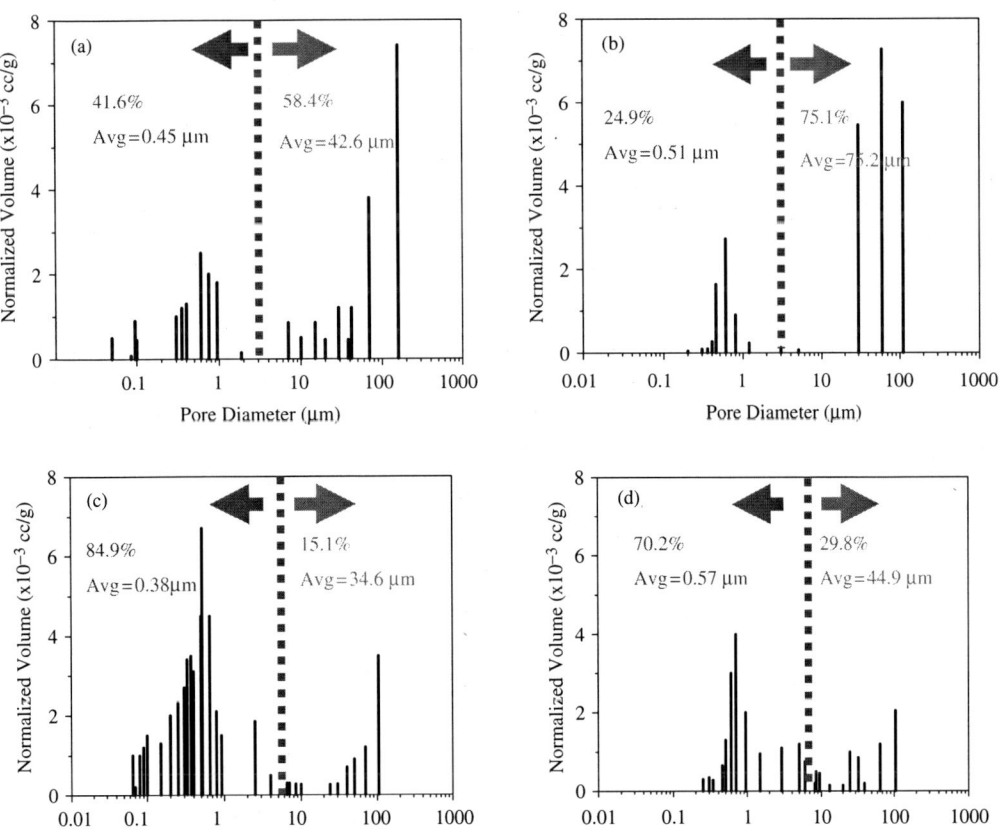

Fig. 14. Bimodal pore size distribution of air plasma spray (APS) and solution precursor plasma spray coating: (a) as-sprayed APS thermal barrier coatings (TBCs); (b) APS TBCs at 1400°C/100 h; (c) as-sprayed SPPS TBCs; (d) SPPS TBCs at 1400°C/100 h.

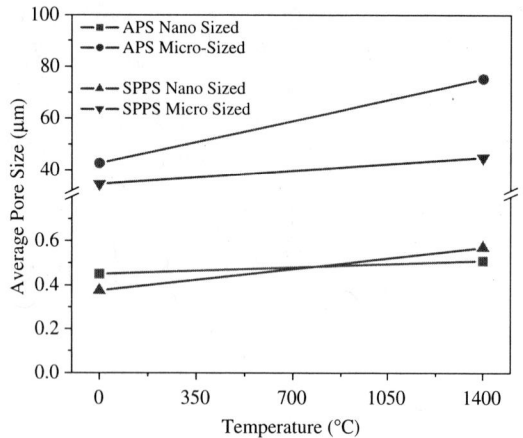

Fig. 15. Average pore size of air plasma spray (APS) and solution precursor plasma spray (SPPS) coating as a function of temperature for 100 h exposure.

both APS and SPPS coatings (Figs. 16(a) and (b)), which is caused by the excess Y^{3+} ions in the t' phase.[10] Upon heat treatment, yttria diffuses out of the metastable zirconia phase until it reaches its equilibrium concentration (~4.0 wt%), which results in the diffraction peak shifting to higher 2θ angles.

Up to 1300°C, both APS and SPPS TBCs exhibit similar phase changes. The t' phase decomposes to two high-temperature equilibrium tetragonal (t) and cubic (c) phases. When the temperature is increased to 1400°C, the tetragonal t (400) peak in APS coating becomes very weak and the cubic peak becomes very strong in the range of 72°–76° (Fig. 16(a)). At the same

time, the monoclinic peaks become apparent at 1400°C (Fig. 16(c)), which indicates the APS coating has partially transformed to m-ZrO_2 and c-ZrO_2. In contrast, the t (400) peak in SPPS coating (Fig. 16(b)) is still very strong and there is no monoclinic phase identified at 1400°C (Fig. 16(d)). After 1500°C/100 h exposure, the tetragonal peaks t (400) in both APS and SPPS coatings disappear and only the cubic peaks exist (Figs. 16(a) and (b)). Both APS and SPPS coatings have completely transformed to monoclinic and cubic zirconia at 1500°C.

The above XRD analysis indicates that the monoclinic phase appears at a higher temperature (1500°C/100 h) in SPPS coatings than that in APS coatings (1400°C/100 h). The tetragonal to monoclinic ($t \rightarrow m$) transformation is a martensitic transformation. The transformation strongly depends on grain size, i.e., transformation does not take place in ceramics smaller than a minimum grain size. Moon et al.[22] studied the effects of heat treatment on the phase transformation behavior of plasma-sprayed zirconia coatings and indicated that the transformation of tetragonal to monoclinic phase is greater for larger grains, but the critical grain size value was not reported. For the APS and SPPS coatings heat treated at 1400°C/100 h, the grains size is 757 and 608 nm (Fig. 6), respectively. So the effect of grains size is a possible explanation for this difference in temperature for the phase change.

Despite APS and SPPS TBCs having very different starting microstructures, many thermal stability characteristics of the two TBCs are similar. These characteristics include densification and hardness increase and, to a lesser extent, grain growth and phase stability. The likely explanation is that these thermal behaviors like grain growth and densification are controlled by diffusion processes at the atomic scale, which is similar to that in zirconia ceramic sintering process.[23] Despite the difference in as-coated microstructures, the two TBCs have the same composition and compositional homogeneity, and, therefore, respond similarly to thermal exposure.

IV. Conclusions

Effects of heat treatment on the APS and SPPS TBCs have been investigated. Both APS and SPPS coatings exhibit similar thermal behaviors such as densification rate, increase in hardness and grain coarsening rate. Both the as-received and heat-treated APS and SPPS TBCs show a bimodal pore size distribution with nano- and micrometer-size pores. After 1400°C/100 h exposure, equiaxed grains replaced the long columnar structure in APS TBCs and at the same time the splat structure disappears. The vertical cracks remain after 1500°C/100 h exposure in SPPS TBCs, which may be one of the reasons why these SPPS coatings have demonstrated superior thermal cycling resistance levels when compared with those of APS coatings. The APS TBCs exhibit the monoclinic phase after 1400°C/100 h exposure; in contrast, SPPS TBCs exhibit the monoclinic phase at 1500°C/100 h.

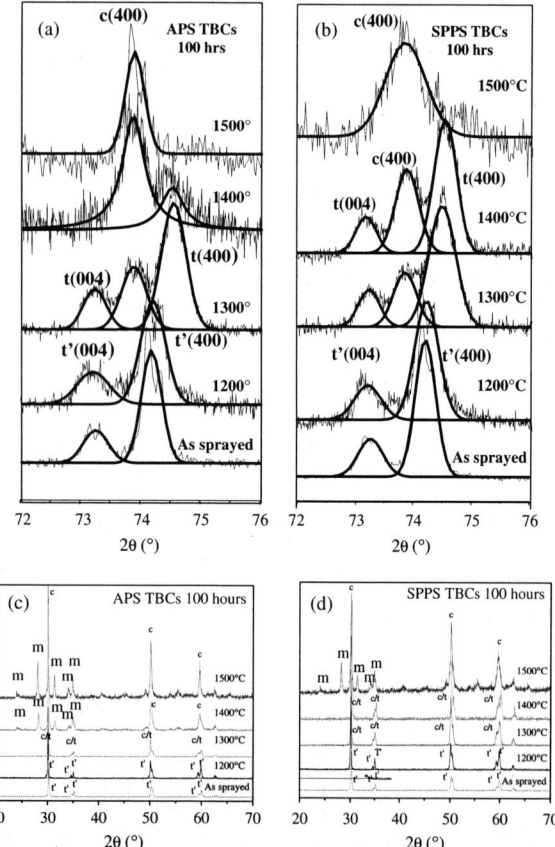

Fig. 16. XRD of air plasma spray (APS) and solution precursor plasma spray (SPPS) coatings after 100 h heat treatment: (a, c) APS thermal barrier coatings (TBCs); (b, d) SPPS TBCs.

References

[1]A. M. Robert, "Current Status of Thermal Barrier Coatings—An Overview," *Surf. Coat. Technol.*, **30** [1] 1–11 (1987).

[2]N. P. Padture, M. Gell, and E. H. Jordan, "Thermal Barrier Coatings for Gas–Turbine Engine Applications," *Science (Washington, DC, United States)*, **296** [5566] 280–4 (2002).

[3]M Gell, L. D. Xie, X. Q. Ma, E. H. Jordan, and N. P. Padture, "Highly Durable Thermal Barrier Coatings Made by the Solution Precursor Plasma Spray Process," *Surf. Coat. Technol.*, **177**, 97–102 (2004).

[4]A. G. Evans, D. R. Mumm, J. W. Hutchinson, G. H. Meier, and F. S. Pettit, "Mechanisms Controlling the Durability of Thermal Barrier Coatings," *Prog. Mater. Sci.*, **46** [5] 505–53 (2001).

[5]N. P. Padture, K. W. Schlichting, T. Bhatia, A. Ozturk, B. Cetegen, E. H. Jordan, M. Gell, S. Jiang, T. D. Xiao, P. R. Strutt, E. Garcia, P. Miranzo, and M. I. Osendi, "Towards Durable Thermal Barrier Coatings With Novel Microstructures Deposited by Solution-Precursor Plasma Spray," *Acta Mater.*, **49** [12] 2251–7 (2001).

[6]A. Jadhav, N. P. Padture, F. Wu, E. H. Jordan, and M. Gell, "Thick Ceramic Thermal Barrier Coatings With High Durability Deposited Using Solution-Pre-

cursor Plasma Spray," *Mater. Sci. Eng. A—Struct. Mater. Properties Microstruct. Process.*, **405** [1–2] 313–20 (2005).

[7]E. H. Jordan, L. Xie, X. Ma, M. Gell, N. P. Padture, B. Cetegen, A. Ozturk, J. Roth, T. D. Xiao, and P. E. C. Bryant, "Superior Thermal Barrier Coatings Using Solution Precursor Plasma Spray," *J. Thermal Spray Technol.*, **13** [1] 57–65 (2004).

[8]L. Xie, D. Chen, E. H. Jordan, A. Ozturk, F. Wu, X. Ma, B. M. Cetegen, and M. Gell, "Formation of Vertical Cracks in Solution-Precursor Plasma-Sprayed Thermal Barrier Coatings," *Surf. Coat. Technol.*, **201** [3-4] 1058–64 (2006).

[9]A. H. Heuer, "Transformation Toughening in ZrO_2-Containing Ceramics," *J. Am. Ceram. Soc.*, **70** [10] 689–98 (1987).

[10]R. W. Trice, Y. Jennifer Su, J. R. Mawdsley, and K. T. Faber, "Effect of Heat Treatment on Phase Stability, Microstructure, and Thermal Conductivity of Plasma-Sprayed YSZ," *J. Mater. Sci.*, **37**, 2359–65 (2002).

[11]A. D. Jadhav, N. P. Padture, E. H. Jordan, M. Gell, P. Miranzo, and E. R. Fuller, "Low-Thermal-Conductivity Plasma-Sprayed Thermal Barrier Coatings With Engineered Microstructures," *Acta Mater.*, **54** [12] 3343–9 (2006).

[12]L. D. Xie, X. Q. Ma, A. Ozturk, E. H. Jordan, N. P. Padture, B. M. Cetegen, D. T. Xiao, and M. Gell, "Processing Parameter Effects on Solution Precursor Plasma Spray Process Spray Patterns," *Surf. Coat. Technol.*, **183** [1] 51–61 (2004).

[13]L. D. Xie, X. Q. Ma, E. H. Jordan, N. P. Padture, D. T. Xiao, and M. Gell, "Deposition Mechanisms of Thermal Barrier Coatings in the Solution Precursor Plasma Spray Process," *Surf. Coat. Technol.*, **177**, 103–7 (2004).

[14]L. D. Xie, X. Q. Ma, E. H. Jordan, N. P. Padture, D. T. Xiao, and M. Gell, "Deposition of Thermal Barrier Coatings Using the Solution Precursor Plasma Spray Process," *J. Mater. Sci.*, **39** [5] 1639–46 (2004).

[15]L. D. Xie, X. Q. Ma, E. H. Jordan, N. P. Padture, D. T. Xiao, and M. Gell, "Identification of Coating Deposition Mechanisms in the Solution-Precursor Plasma-Spray Process Using Model Spray Experiments," *Mater. Sci. Eng. A—Struct. Mater. Properties Microstruct. Process.*, **362** [1–2] 204–12 (2003).

[16]L. D. Xie, E. H. Jordan, N. P. Padture, and M. Gell, "Phase and Microstructural Stability of Solution Precursor Plasma Sprayed Thermal Barrier Coatings," *Mater. Sci. Eng. A—Struct. Mater. Properties Microstruct. Process.*, **381** [1–2] 189–95 (2004).

[17]M. Gell, L. D. Xie, E. H. Jordan, and N. P. Padture, "Mechanisms of Spallation of Solution Precursor Plasma Spray Thermal Barrier Coatings," *Surf. Coat.Technol.*, **188–89**, 101–6 (2004).

[18]T. Bhatia, A. Ozturk, L. D. Xie, E. H. Jordan, B. M. Cetegen, M. Gell, X. Q. Ma, and N. P. Padture, "Mechanisms of Ceramic Coating Deposition in Solution-Precursor Plasma Spray," *J. Mater. Res.*, **17** [9] 2363–72 (2002).

[19]J. Luo, S. Adak, and R. Stevens, "Microstructure Evolution and Grain Growth in the Sintering of 3Y-TZP Ceramics," *J. Mater. Sci.*, **33** [22] 5301–9 (1998).

[20]E. W. Washburn, "A Method of Determining the Distribution of Pore Sizes in a Porous Material," *Proc. Natl Acad. Sci.*, **7**, 115 (1921).

[21]L. Lelait, S. Alperine, C. Diot, and M. Mevert, "Thermal Barrier Coatings: Microstructural Investigation after Annealing," *Mater. Sci. Eng. A*, **A121**, 475–82 (1989).

[22]J. Moon, H. Choi, H. Kim, and C. Lee, "The Effects of Heat Treatment on the Phase Transformation Behavior of Plasma-Sprayed Stabilized ZrO_2 Coatings," *Surf. Coat. Technol.*, **155**, 1–10 (2002).

[23]K. R. Venkatachari, D. Huang, S. P. Ostrander, W. A. Schulze, and G. C. Stangle, "Preparation of Nanocrystalline Yttria-Stabilized Zirconia," *J. Mater. Res.*, **10** [3] 756–61 (1995). ☐

Mechanical Design for Accommodating Thermal Expansion Mismatch in Multilayer Coatings for Environmental Protection at Ultrahigh Temperatures

Jie Bai, Kurt Maute, Sandeep R. Shah, and Rishi Raj[†]

Ultrahigh Temperature Materials Laboratory, Department of Mechanical Engineering, University of Colorado, Boulder, Colorado 80302

The design of coatings is like designing a system. Every coating has one or more specific functions that determine the choice of materials, and its architecture. In the case of environmental barrier coatings the topcoat must be chemically inert to the atmosphere. In high-temperature applications the stresses arising from thermal expansion mismatch between the topcoat and the substrate must be ameliorated. In this article we consider the design of an intermediate layer of a multilayer coating system with the explicit objective of managing thermal expansion difference between the topcoat and the substrate. The design is based upon a columnar architecture where the columns serve as flexible beams to accommodate relative displacement without fracture. The value of the maximum stresses in the beam and in the topcoat are calculated and used to develop a map with fail and safe regimes. The safe region is defined by the prevention of fracture in the beams, since their fracture would precipitate delamination of the topcoat. As a rule of thumb the topcoat thickness should be less than the width of the columns for safe operation (this condition changes somewhat with the aspect ratio of the columns). A larger aspect ratio of the columns also promotes safe design. We further consider how the tractions induced by the thermal stresses on the surface of the substrate may influence the intrinsic fracture strength of the substrate. The stresses in the coating are predicted to have an insignificant effect on the intrinsic fracture strength of the substrate.

I. Introduction

THE gas turbine epitomizes the significance and the need for high temperature coatings for structural applications. The highest temperatures and the most severe corrosive environments in the gas turbine are experienced by nozzles, linings, and most of all, by the rotating turbine blade. The push for higher combustion temperatures is creating a need for multifunctional coatings that can withstand thermal shock, adhere well to the substrate, provide thermal insulation, and protect from environmental corrosion. The state-of-the-art blade materials are metallic superalloys. Zirconia-based thermal barrier coatings (TBCs) for superalloys have been in use for over a decade. The TBCs were developed by intuition and experience,[1] yet they have laid the foundation for the conceptual design of high-temperature coatings. As described in Strangman and Schienle[2] they were based upon a reactive metallic bond coat for adherence between zirconia and the superalloy, a columnar, strain tolerant zirconia layer, and a dense zirconia topcoat.

The next generation gas turbines are slated to contain ceramic components made from silicon nitrite (Si_3N_4). However, silicon nitride suffers from erosion in the streaming humid environment of the gas turbine. The weight loss can be severe and is caused by the volatilization of the passivating silica scale.[3–5] Example of such a result from our laboratory given in Fig. 1, which shows increasing weight loss with higher streaming velocity. These data reach up to a relative velocity of 35 cm/s[6]; the velocities encountered in the gas turbine are an order of magnitude higher which would be clearly intolerable for Si_3N_4. The objective of environmental barrier coatings (EBCs) is to protect the load bearing silicon-nitride structure from corrosive weight loss in the high velocity, high temperature, and humid environment of the gas turbine.

The design of EBCs is constrained by at least three criteria: (a) the topcoat of the EBC must be able to survive in the gas-turbine environment, (b) the topcoat must be securely bonded to the Si_3N_4 substrate, and (c) the coating architecture must have good thermal shock resistance.

The choice of the optimum material for the topcoat is often juxtaposed against the issue of thermal shock, as matching thermal expansion and, at the same time providing chemical durability, can pose a challenge to the materials engineer. However, this approach has been successfully used to develop coatings for siliconcarbide-based ceramic structures.[7] In the present work we consider another approach, one where the topcoat is chosen entirely for its corrosion resistance, and then the thermal shock is managed by adding a compliant intermediate layer which accommodates the difference in the coefficients of thermal expansion of the topcoat and the substrate. For example, zirconia has a proven record of durability as thermal barrier coatings, making it a good choice as the material for the topcoat. However, its thermal expansion is much larger than that of Si_3N_4 (10 ppm vs about 3.5 ppm/K) which would cause it to spall. A coating design which can ameliorate the thermal stresses is illustrated in Fig. 2. It consists of a topcoat, a compliant intermediate coat that accommodates thermal strains, and a bond coat that secures the upper layers to the substrate. The mechanical design of the compliant intermediate coat is the main subject of this paper.

The principal purpose of the above coating architecture is to prevent the *high velocity* humid environment from impinging directly on to the surface of Si_3N_4. The thermal expansion of the topcoat can be expected to produce "periodically spaced" cracks,[8] which will allow the humid environment to seep into the coating. Therefore the function of this EBC is merely to subdue the velocity of the environment. The question then arises whether oxidation of Si_3N_4 under *static* humid environments can be acceptable.[9] Work to be published in a companion paper by Shah and Raj[10] shows that a special bond coat that is made from polymer derived siliconcarbonitride is effective against oxidation at high temperatures in static humidity (but not under high-velocity conditions). Thus a combination of the polymer-derived coating[10] and the thermal stress management approach described in this paper can be used to

C.-H. Hsueh—contributing editor

Manuscript No. 21601. Received March 17, 2006; approved August 24, 2006.
This research was supported by the MEANS program at the Air Force Office of Scientific Research under the direction of Dr. Joan Fuller. The Grant number is F49620-01-1-052. The experimental part of this research was supported under the Power and Energy CTA Program at Honeywell Inc., Phoenix, AZ, under the direction of Laura Lindberg.
[†]Author to whom correspondence should be addressed. e-mail: rishi.raj@colorado.edu

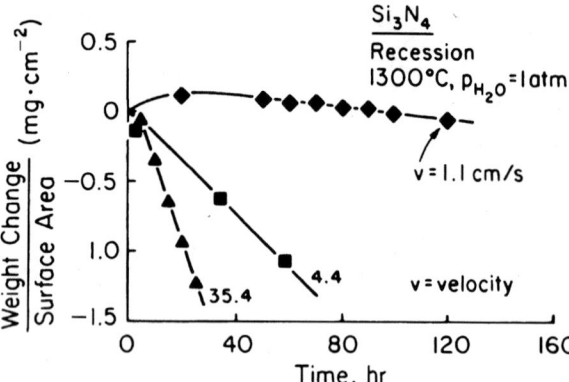

Fig. 1. Influence of streaming water-vapor velocity on weight loss in coated silicon nitride at 1300°C.

produce effective EBCs for high-temperature, silicon-based structural ceramics.

The first objective of this paper is to develop design guidelines for the geometrical structure of the columnar intermediate layer. The performance goal is to *prevent* the delamination of the top-coat—this condition is achieved if the maximum stress in the topcoat is greater than in the columnar structure, that is, if the columns resist fracture better than the topcoat. The second problem analyzed in this article pertains to the degradation in the fracture strength of the substrate as a result of the coating. We consider increased loading on a surface flaw in the substrate exerted by the forces of the thermal strains; a formal derivation shows this effect to be relatively inconsequential.

II. Mechanical Design of the Columnar Interlayer and the Topcoat

(1) Analysis

The problem is analyzed in two dimensions, that is, the schematic in Fig. 2 represents the cross-section of the EBC, which does not change in the normal direction. For volumetric quantities, e.g. the strain energy per column, we assume a unit depth normal to the paper. The geometrical parameters of the multilayer coating are:

(i)   the thickness, or the height, of the topcoat and of the columnar interlayer, given by h_{TC} and h_{BC}, respectively,

(ii)   the aspect ratio of the columns is $A_r = h_{BC}/W$, where W is the width of the columns,

(iii)   the spacing between the columns, called L, and

(iv)   the relative density of the columnar interlayer, ρ, which is simply $\rho = W/L$.

The elastic moduli of the two materials, the topcoat, and the material used to construct the columns are written as E_{TC} and E_{BC}. As we shall see later on it is likely to be a good practice to use the same material for both, in which case the ratio of the two moduli becomes equal to unity. The difference between the coefficient of thermal expansion between the topcoat and the substrate is written as $\Delta\alpha$.

The difference in the thermal expansion between the topcoat and the substrate will lead to periodic cracks in the topcoat (assuming that the in-plane stress given by $\Delta\alpha\Delta TE_{TC}$ is greater than the fracture stress of the topcoat). The spacing of these periodic cracks is likely to be similar to the description of the interfacial strength of thin films under uniform loading by shear-lag models.[8] In the present instance the interfacial stresses are accommodated by the flexure of the columns. The mechanics of deformation may therefore be illustrated as in Fig. 3. The periodic crack spacing in the topcoat, λ, is equivalent to n columns, or to a length of nL as L is the column spacing. If the topcoat is sufficiently thin then the in-plane stress in the film may be assumed to be uniform in the z, or the out of plane direction. The stress in the topcoat will then be symmetrically distributed about the center-line of the spacing between adjacent cracks in the topcoat, while the bending displacements of the columns, parallel to the interface are antisymmetric. These bending displacements in the columns are called u_j, where $j = 0, n$. The topcoat is also described by discrete elements, each of length L, such that they are synchronous with the columns. The stress in these elements are written as σ_{TCj}, where $j = 1, n$. The strains in these elements are described by the difference between the displacements of the two columns on either edge of the element (after compensating for the thermal expansion strain). Thus the first element is stretched by $(u_1 - u_0)$, and the jth element by $(u_j - u_{j-1})$ and so on. In this way we have n elements in the topcoat, and $(n+1)$ columns which flex to accommodate the strain in the topcoat. The objective of the analysis is to solve for the shear displacements in the columns, i.e. u_j, $j = 0, n$.

The analysis is based on the principle of minimum potential energy. The potential energy is the sum of the elastic strains in the columns and in the elements of the topcoat. (The substrate being much thicker than the topcoat and the columnar layer, may be safely assumed to be a rigid body; as such the strain

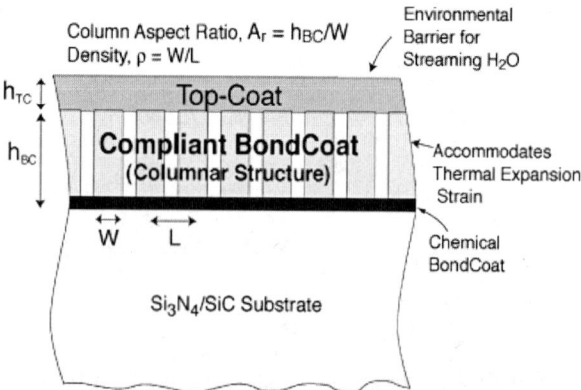

Fig. 2. The three elements of the architecture of an environmental barrier coatings. The topcoat provides environmental protection, the compliant columnar interlayer accommodates the thermal strains, and the chemical bond coat helps adherence of the upper layers to the substrate.

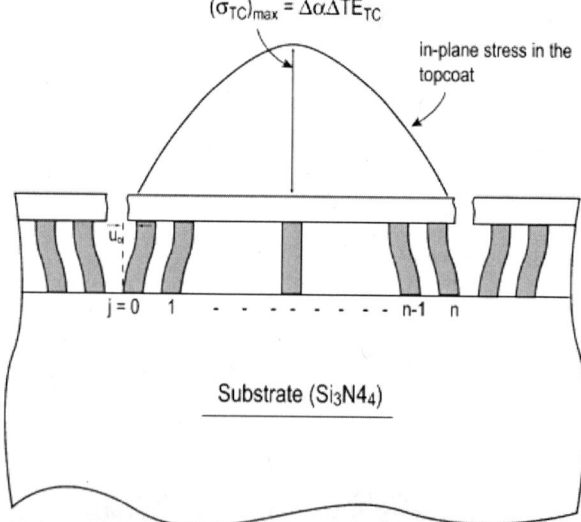

Fig. 3. The symmetrical in-plane stress in the topcoat, and the antisymmetric flexure displacements in the columns between two adjacent fractures in the topcoat. The model is conceptually equivalent to the shear-lag models of interfacial tractions between thin films and rigid substrates.

energy in the substrate can be ignored. The chemical bond coat, on the other hand, is extremely thin, relative to the topcoat and the columnar layer; therefore its volume is negligible in comparison and the strain energy in it as well can be ignored.) First, we consider the latter. The strain energy in the jth element of the topcoat is given by:

$$\prod_{j}^{TC} = \frac{1}{2}E_{TC}\left(\frac{u_j - u_{j-1}}{L} - \Delta\alpha\Delta T\right)^2 h_{TC}L \qquad (1)$$

where $j = 1, n$. The terms within the brackets represent the net strain in the jth element of the topcoat. The elastic strain in this element is equal to the difference between displacement in the element, divided by the length of the element L, and the thermal expansion strain. The strain energy being equal to the square of the net displacement is insensitive to the sign of this difference. The quadratic term in Eq. (1) when expanded gives the following equation for the strain energy in the jth element of the topcoat:

$$\prod_{j}^{TC} = \frac{h_{TC}E_{TC}}{2L}(u_j - u_{j-1})^2 - E_{TC}h_{TC}\Delta\alpha\Delta T(u_j - u_{j-1})$$
$$+ \frac{1}{2}(\Delta\alpha\Delta T)^2 E_{TC}h_{TC}L \qquad (2)$$

The strain energy in the columns is simply related to the bending displacement, u_j, to h_{BC} and to the moment of inertia of the columns, I, by beam theory which gives[11]:

$$\prod_{j}^{beam} = \frac{6E_{BC}Iu_j^2}{h_{BC}^3} \qquad (3)$$

where $j = 0, n$. The total potential energy is equal to the sum of Eqs. (2) and (3) which leads to the following result after proper summation:

$$\prod = \sum_{j=0}^{n}\frac{6E_{BC}Iu_j^2}{h_{BC}^3} + \sum_{j=1}^{n}\frac{h_{TC}E_{TC}}{2L}(u_j - u_{j-1})^2$$
$$- \sum_{j=1}^{n}E_{TC}h_{TC}\Delta\alpha\Delta T(u_j - u_{j-1}) \qquad (4)$$
$$+ \frac{n}{2}(\Delta\alpha\Delta T)^2 E_{TC}h_{TC}L$$

Using principle of minimum potential energy, we obtain the governing set of equations for u_j:

$$\frac{\partial\prod}{\partial u_0} = \frac{12E_{BC}Iu_0}{h_{BC}^3} + \frac{E_{TC}h_{TC}}{L}(u_0 - u_1) + E_{TC}h_{TC}\Delta\alpha\Delta T \qquad (5)$$
$$= 0$$

$$\frac{\partial\prod}{\partial u_k} = \frac{12E_{BC}Iu_k}{h_{BC}^3} + \frac{E_{TC}h_{TC}}{L}(u_k - u_{k+1})$$
$$+ \frac{E_{TC}h_{TC}}{L} \times (u_k - u_{k-1}) = 0, \ k \neq 0, n \qquad (6)$$

$$\frac{\partial\prod}{\partial u_n} = \frac{12E_{BC}Iu_n}{h_{BC}^3} + \frac{E_{TC}h_{TC}}{L}(u_n - u_{n-1})$$
$$- E_{TC}h_{TC}\Delta\alpha\Delta T = 0 \qquad (7)$$

After some arrangement, the above equations can be written in matrix form and in terms of one normalized parameter c,

$$\begin{bmatrix} 1+c & -1 & & & & \\ -1 & 2+c & -1 & & & \\ & & \cdot & & & \\ & & & \cdot & & \\ & & -1 & 2+c & -1 & \\ & & & -1 & 1+c \end{bmatrix}\begin{bmatrix} \bar{u}_0 \\ \cdot \\ \cdot \\ \cdot \\ \cdot \\ \bar{u}_n \end{bmatrix} = \begin{bmatrix} -1 \\ 0 \\ 0 \\ \cdot \\ \cdot \\ 0 \\ 1 \end{bmatrix} \qquad (8)$$

where

$$\bar{u}_j = \frac{u_j}{L\Delta\alpha\Delta T} \quad j = 0, n \qquad (9)$$

Here $\bar{u}_j$ is the normalized value of the displacement, as a fraction of the displacement to be expected from "free" thermal expansion. The results of the analysis can now be described in terms of the non-dimensional parameter, c, which embodies within it the elastic properties of the topcoat and the compliant interlayer, and is given by

$$c = 12\frac{E_{BC}}{E_{TC}}\frac{IL}{h_{TC}h_{BC}^3} \qquad (10)$$

Equations in the matrix in (8) now are solved numerically for $\bar{u}_j$ in terms of c and n. Note that n is a measure of the spacing between the cracks in the topcoat as $\lambda = nL$.

It remains to write down the expressions for the stresses in the topcoat and in the columns, which are explicitly related to $\bar{u}_j$. The stress in the jth element of the topcoat, σ_{TCj}, is normalized with respect to the thermal expansion stress, that is the unrelaxed thermal stress in the topcoat, so that

$$\bar{\sigma}_{TCj} = \frac{\sigma_{TCj}}{E_{TC}\Delta\alpha\Delta T} \qquad (11)$$

Combining Eq. (11) with Eq. (9) and recognizing that:

$$\sigma_{TCj} = E_{TC}\left(\frac{u_j - u_{j-1}}{L} - \Delta\alpha\Delta T\right)$$

we obtain the following relationship between the stress in the topcoat elements and the displacements:

$$\bar{\sigma}_{TCj} = \bar{u}_{j+1} - \bar{u}_j - 1 \quad \text{for } j = 1, n \qquad (12)$$

The stresses in the columns of the interlayer arise from bending. We are interested in the maximum value of the stress arising from the bending; we call this stress σ_{BCj} and normalize it in the same way as the stress in the topcoat, as given by Eq. (11), and denote it with a bar. Using the equation for maximum stress from beam theory, the following result is obtained[11]:

$$\bar{\sigma}_{BCj} = \frac{\sigma_{BCj}}{E_{TC}\Delta\alpha\Delta T} = \frac{3E_{BC}Wu_j}{h_{BC}^2\Delta\alpha\Delta TE_{TC}} \quad \text{where } j = 0, n \qquad (13)$$

Note that Eqs. (11)–(13) describe the in plane stress in the topcoat, parallel to the interface, and the maximum stress in the beam produced by flexure, which is normal to the interface. This description deviates from the description of stresses in continuous films on substrates that are dealt by shear lag models; in these models the shear stresses in the interface are equilibrated against the in-plane stress in the film.[8]

The calculation of the displacements from Eq. (8) and substituting them into Eqs. (12) and (13) for obtaining the stress in the topcoat and in the columns completes the solution to the problem. However, Eq. (8) must be solved numerically. The nature of the results depends on the non-dimensional parameter, c, given by Eq. (10). Substituting for I in Eq. (10), and recognizing that the aspect ratio of the columns, $A_r = h_{BC}/W$, gives:

$$c = \frac{E_{BC}}{E_{TC}} \frac{(L/h_{TC})}{A_r^3} \qquad (14)$$

Note that c pulls together the principal materials parameters, the elastic moduli of the top coat and the columnar layer, as well as the geometrical parameters of the columnar structure, and the thickness of the topcoat. The stresses and displacements in the EBC can be described in terms of this non-dimensional parameter, adding to the generality of the results. Later we shall find that $A_r \approx 3$ lies in the transition region for safe design of the EBC. Assuming the elastic moduli of the topcoat and the columns to be nearly equal, and the spacing of the columns, L, to be about the same as the thickness of the topcoat, h_{TC}, we note that this condition corresponds to $c \approx 0.05$. This overview gives the range of the values for c that should be explored in the results from the analysis.

III. Results

(1) Distribution of Stresses and Displacements

The physically interesting result that can be obtained from the analysis is how the stresses in the topcoat and the bending displacements in the columns vary from the free edge of a crack in the topcoat, that is, how these quantities vary for $j = 0, 1, 2, 3.$ ... and so on. The stresses are symmetric and the displacements antisymmetric between two neighboring cracks in the topcoat. Both quantities of course will depend on c.

The stresses and displacements are plotted in Fig. 4 for two values of $c = 0.04$ and 0.1. The spacing between the adjacent cracks in the topcoat is held constant (at $n = 60$) to show how the decay of these quantities depends on c. A smaller value of c leads to a more gradual decay, that is, the stresses are spread over a larger distance. This result can be qualitatively understood from Eq. (14), as a smaller value of c is obtained for a larger aspect ratio of the columns. A larger aspect ratio means that the columns are more compliant and therefore the relaxation of the stress in the topcoat is spread over a larger distance.

The variation of the decay length of the stress from a free edge in the topcoat, with respect to c, is plotted in Fig. 5. The decay length, expressed in terms of the number of units, n, of column spacing varies from about 50 at $c = 0.01$ to approximately 20 at $c = 0.1$ and $n = 10$ at $c = 0.3$.

(2) Map for Safe Design

The objective of the mechanical design is that the topcoat should not delaminate under the influence of thermal strains. The cri-

terion for safe design, therefore, is that the fracture in the topcoat should accommodate the thermal strains, in preference to a fracture in the columnar beams. This "safe" criterion is expressed by the following equation:

$$\frac{\hat{\sigma}_{BC}}{\hat{\sigma}_{TC}} < 1 \qquad (15)$$

where the hat signifies the maximum value of the stresses in the topcoat, $\hat{\sigma}_{TC}$, and in the columnar beams, $\hat{\sigma}_{BC}$. The maximum stress in the topcoat is simply given by $E_{TC}\Delta\alpha\Delta T$, while $\hat{\sigma}_{BC}$ is given by $[\sigma_{BCj}]_{j=0}$. From Eq. (13), since, as seen in Fig. 3 the first beam at the edge of the crack in the topcoat suffers the greatest elastic strain. Substituting these expressions into Eq. (15) gives the following result for "safe" design:

$$3\frac{E_{BC}}{E_{TC}} \frac{\bar{u}_0}{A_r^2 \rho} < 1 \qquad (16)$$

where $\bar{u}_0$ is the bending displacement in the first column. The value for $\bar{u}_0$ was computed various values of c, which varies with h_{BC}/W. For simplicity, it was assumed that $E_{BC}/E_{TC} = 1$. With this assumption the map depends on three geometrical parameters: the thickness of the topcoat relative to the width of the columns, h_{TC}/W, the aspect ratio of the columns, A_r, and the relative density of the columnar structure which is given by $\rho = W/L$.

The map showing the fail (where the topcoat is likely to delaminate), and the safe region (where the topcoat will develop periodic cracks but will remain attached to the substrate via the columnar structure) is given in Fig. 6. The design space is charted in a field described by h_{TC}/W and A_r. The "fail" and "safe" regimes for two values of $\rho = 0.25$ and 0.5 are shown. A higher density of the columnar interlayer enlarges the safe region, and it becomes, therefore, more forgiving. This observation can be physically explained by the greater load bearing capacity of the columns without increasing the maximum stress experienced by the bending.

The map shows two regimes for safe design, one to the left of the minimum in the boundary separating the two regions, and the other to the right of the minimum. The left hand regime bears theoretical uncertainty as the present analysis is based upon Bernoulli beam analysis[11] where the edge effects of the beam are neglected. For small aspect ratio this assumption would be inaccurate. In any event the left hand region is likely to be narrow and therefore may not offer the same degree of latitude in design and manufacturing as the safe regime on the right-hand side. On this side we find that a larger aspect ratio of the columns is safer, as is a thin topcoat relative to the width of

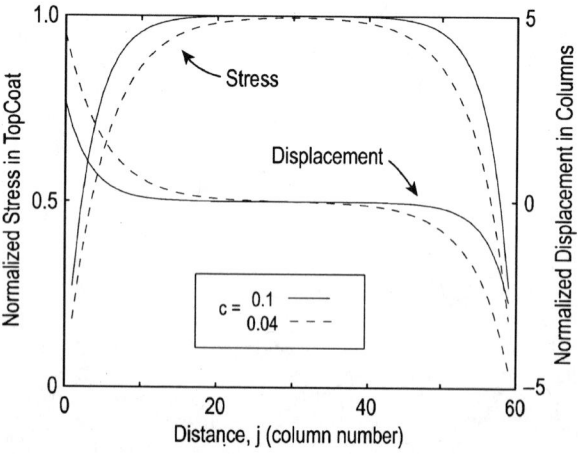

Fig. 4. Results for the in-plane stress in the topcoat and the flexure displacements in the columns for two values of the non-dimensional parameter, c.

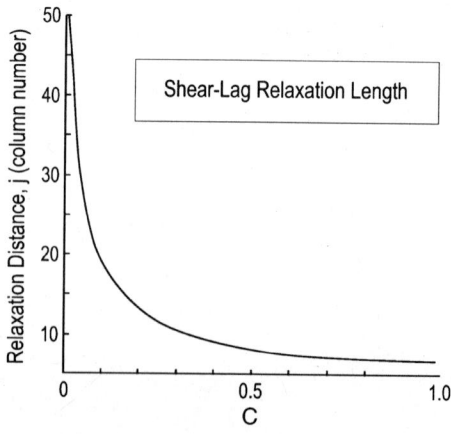

Fig. 5. The decay distance of the in-plane stress next to a free edge of the topcoat as a function of c.

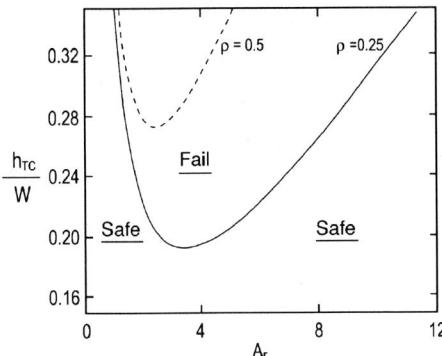

Fig. 6. The design-map for choosing the aspect ratio of the columns and the thickness of the topcoat in order to preclude delamination of the topcoat.

the columns (which should typically be less than about 1/3). In summary, the design map suggests that a thinner topcoat, a larger aspect ratio of the columns and a higher packing density in the columnar interlayer, favor the probability of avoiding delamination of the topcoat due to thermal strains.

IV. Influence of the EBC on the Fracture Strength of a Brittle Substrate

A thermally strained coating exerts surface tractions on the substrate. These tractions can increase the loading on the flaws near the surface of the substrate material thereby having a negative impact on its fracture strength. Such effect has been reported for EBCs made from BAS for silicon carbide ceramics.[12] The question arises to what extent the compliant interlayer architecture of the present EBC can influence the fracture behavior of the substrate.

The problem is approached analytically by assuming the *worst-case* scenario where the location of the flaw in the substrate coincides with the position where the highest shear traction is exerted on the substrate by the bending of the beams. These tractions are highest near the free edge of a crack in the topcoat, that is next to $j = 0, 1, 2, 3 \ldots$ and so on. The analytical procedure then, as illustrated in Fig. 7, is to calculate the additional loading on a flaw of size a placed near $j = 0$, by adding up the incremental stress-intensity produced at the crack tip by the periodically occurring shear tractions exerted by the bending of the columns at $j = 0, 1, 2 \ldots$ on one side and $j = n, (n-1), (n-2), \ldots$ on the other side. Being antisymmetric both sets of tractions increase the loading on the flaw.

The intrinsic flaw size in the substrate is estimated from the fracture toughness, K_{IC} and the fracture strength of the material, using the equation:

$$K_{IC} = \sigma_f \sqrt{\pi a} \tag{17}$$

where σ_f is the fracture strength and a is the flaw size. Typical numbers for the fracture toughness and the fracture strength of Si_3N_4 are 5 MPa·m$^{1/2}$ and 1 GPa, respectively, which yields a flaw size of approximately 10 μm.

The incremental stress intensity, ΔK, exerted by the coating is calculated by considering a point force acting from each of the columns on the crack and summing up the effect from all columns. Forces from the $(n+1)$ columns, from each side of the crack, will exert a pull force on the crack of size a. The stress intensity exerted by one of these symmetrically placed pair of forces, each of strength P_j at a distance y_j from the crack, as shown schematically in Fig. 7, is given by[13]

$$\Delta K_j = \frac{2P_j}{\sqrt{\pi a}} \Omega_j \tag{18}$$

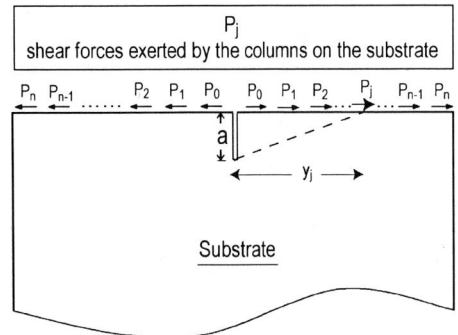

Fig. 7. The tractions exerted by the environmental barrier coatings on a flaw in the substrate.

where

$$\Omega_j = \frac{1 + 2Y_j^2}{(1 + Y_j^2)^{1.5}} \left[1.3 - 0.3 \left(\frac{Y_j}{\sqrt{1 + Y_j^2}} \right)^{5/4} \left(0.665 - 0.267 \left(\frac{Y_j}{\sqrt{1 + Y_j^2}} \right)^{5/4} \left(\frac{Y_j}{\sqrt{1 + Y_j^2}} - 0.73 \right) \right) \right]$$

Here

$$Y_j = \frac{y_j}{a} \tag{19}$$

The total increase in the loading of the crack due to shear tractions induced by the coating is given by summing over all ΔK_j:

$$\Delta K = \sum_{j=0}^{n} \Delta K_j = \frac{2}{\sqrt{\pi a}} \sum_{j=0}^{n} P_j \Omega_j \tag{20}$$

The fracture strength decreases because the critical stress intensity factor given in Eq. (17) is effectively reduced by ΔK given above in Eq. (20). The ratio of the fracture strength with the coating, σ_f', with respect to the intrinsic fracture strength of the substrate, σ_f is then given by

$$\frac{\sigma_f'}{\sigma_f} = 1 - \frac{\Delta K}{K_{IC}} \tag{21}$$

We normalize the intrinsic fracture toughness, K_{IC}, in the following way:

$$\bar{K}_{IC} = \frac{K_{IC}}{E_{TC} \Delta \alpha \Delta T \sqrt{\pi a}} \tag{22}$$

To get an order of magnitude for $\bar{K}_{IC}$ following values for the terms in the denominator: $E_{TC} = 300$ GPa, $\Delta \alpha = 7 \times 10^{-6}$, and $a = 10$ μm. Substituting these values the denominator becomes equal to 3.7 MPa·m$^{1/2}$. Assuming K_{IC} for silicon-nitride of 5.1 MPa·m$^{1/2}$ we note that $\bar{K}_{IC}$: 2.

The expression for P_j in Eq. (18) is related to the displacements in the columns parallel to the interface, u_j, by the well-known force–displacement equation for the constrained bending of a beam with rectangular cross-section, which translates into the following relationship with the present nomenclature[11]:

$$P_j = u_j \frac{E_{BC}}{2A_r^3} \tag{23}$$

Introducing the normalization for the displacements u_j in the above equation according to Eq. (9), and substituting this equation for P_j into Eq. (20) for ΔK and again inserting this expression into (21) gives the following final result for the degradation in the fracture strength due to thermal stresses induced by the coating:

$$\frac{\sigma'_f}{\sigma_f} = 1 - \frac{L}{a}\frac{E_{BC}}{E_{TC}}\frac{1}{\pi A_r^3 \bar{K}_{1C}}\sum_{j=0}^{n}\bar{u}_j\Omega_j \tag{24}$$

A special case of Eq. (24) can be reduced to a physically important result. For this purpose assume that

$$\frac{E_{BC}}{E_{TC}} \approx 1, \text{ and, } \frac{L}{a} \approx 1 \tag{25}$$

which states that the elastic modulus for the topcoat and the column materials is the same and that the periodic spacing of the columns is about equal to the flaw size (which is approximately equal to 10 μm). In this case Eq. (24) reduces to the following expression:

$$\frac{\sigma'_f}{\sigma_f} = 1 - \frac{1}{\pi A_r^3 \bar{K}_{1C}}\sum_{j=0}^{n}\bar{u}_j\Omega(n_j) \tag{26}$$

Note that $\Omega_j = \Omega(n_j)$ as it follows from Eq. (19) that $Y_j = n_j$ where $n_j = 1,2, \ldots, n+1$ with $j = 0, 1,\ldots, n$ when $L \approx a$. Equation (26) can be expressed in another form by substituting for A_r^3 from Eq. (14). Assuming again that $E_{BC} \approx E_{TC}$ and that $(L/h_{TC}) \approx 1$, we obtain

$$\frac{\Delta\sigma_f}{\sigma_f} = \frac{\sigma_f - \sigma'_f}{\sigma_f} = \frac{f(c)}{12\pi\bar{K}_{1C}} \tag{27}$$

where

$$f(c) = c\sum_{j=0}^{n}\bar{u}_j\Omega(n_j) \tag{28}$$

Note that since $f(c)$, in Eq. (28), depends only on c, the degradation in strength, which is given by Eq. (27), also depends primarily on c, except for the additional effect of $\bar{K}_{1C}$. Therefore, the degradation in strength can be estimated by plotting $f(c)$, as a function of c. This plot is given in Fig. 8. The result is that $f(c)$ is of the order of unity, varying from approximately 0.3 to 1.1, over a wide range of values for c. Taking the near-highest value of this range, that is $c = 1$, and assuming that $\bar{K}_{1C} \approx 2$ as discussed earlier (for silicon nitride), we have that the degradation in strength is approximately equal to:

$$\frac{\Delta\sigma_f}{\sigma_f} : \frac{1}{24\pi} \tag{29}$$

that is, less than 2%. The conclusion to be drawn from the above analysis is that the columnar structure of the bond coat will reduce the influence of the topcoat on the fracture strength of the substrate to just a few percent.

V. Experiments

Detailed experiments for the survivability EBCs prepared according to the guidelines developed in this paper are being reported separately,[10] where the processing of the multilayer coatings on silicon-nitride is also described. Here we report on the concept for creating the columnar structure, and study whether or not the design-map in Fig. 6 gives a credible prediction. Note that the safe design space depends on the thickness of the topcoat relative to the width of the columns, and upon the aspect ratio of the columns.

The method for controlling the above design parameters is illustrated in Fig. 9. The effective width of the columns and the aspect ratio is controlled by introducing latex spheres into a

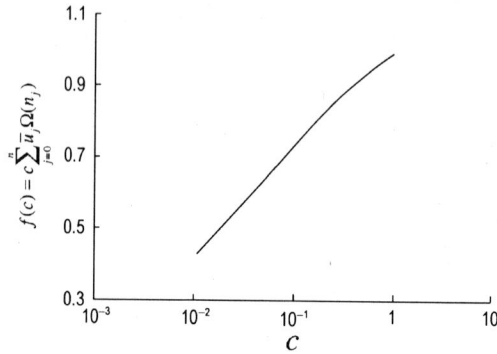

Fig. 8. A plot of $f(c)$, as given by Eq. (28) as a function of c, for estimating the degradation in the fracture strength.

slurry of zirconia which is spun coated on to the substrate. This procedure results in a type of structure shown schematically in Fig. 9. The height of the columns is equal to the diameter of the latex spheres, while width of the columns is varied by changing the volume fraction of the latex spheres in the slurry. Simplified geometrical analysis gives the following relationship:

$$A_r = \frac{h_{BC}}{W} = \sqrt{\frac{6v_f}{\pi}} \tag{30}$$

where v_f is the volume fraction of the latex spheres. The result obtained with latex spheres having a diameter of 25 μm and volume fraction $v_f = 0.5$ is shown in the micrograph in Fig. 9. As expected from Eq. (30) this constitution gives an aspect ratio $A_r \approx 1$ and $W \approx 25$ μm. In this example the columnar structure is made from particles of zirconia while the continuous topcoat is made from electron-beam physical-vapor deposited HfO_2.

According to the design map in Fig. 4, the aspect ratio of one falls to the left hand side of the minimum. Since in the present case $\rho \approx 0.5$, we note that for "safe" operation of the coating, it is necessary that $(h_{TC}/W) \leq 0.4$. Since $W \approx 25$ μm, this condition means that h_{TC} should be less than about 10 μm.

The results from two EBCs, one with $h_{TC} \approx 40$ μm and the other with $h_{TC} \approx 10$ μm after exposing to streaming humid environment at 1250°C for 30 h are shown in Fig. 10. While the thick topcoat delaminates, the thin topcoat, which meets the safe

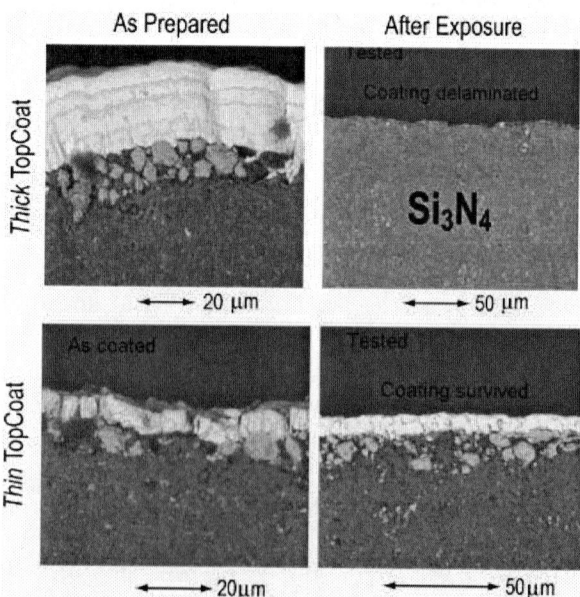

Fig. 9. An experimental strategy for creating a columnar structure for accommodating the thermal strain in the topcoat.

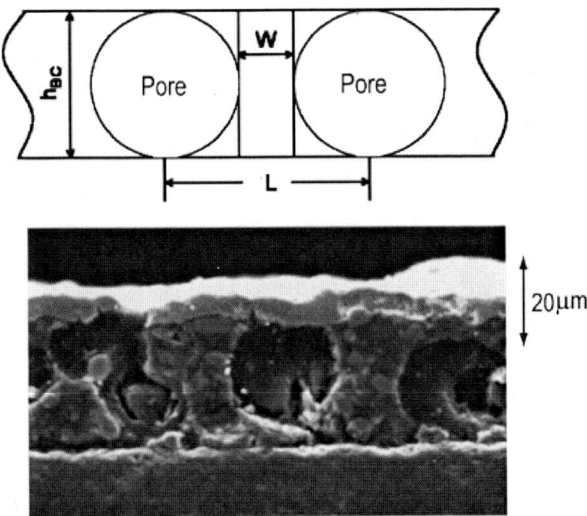

Fig. 10. A thick topcoat delaminates while a thin topcoat does not, in agreement with the design-map in Fig. 6.

design criterion, does indeed survive the exposure. These observations are in approximate agreement with the design guidelines developed in this article.

VI. Summary

High-temperature EBCs are multifunctional; they not only protect the substrate from environmental attack but also must be designed to withstand thermal shock. In general, silicon containing ceramics are unstable in streaming water-vapor environment at high temperatures due to volatilization of the silica scale which otherwise is protective in static oxidation conditions. It is therefore likely that oxides, especially those that do not convert to hydroxides in humid conditions, such as zirconia or hafnia, are a natural choice for the topcoat in EBCs. However, these oxides also have a much larger coefficient of thermal expansion relative to silicon-based ceramics. The accommodation of this thermal expansion strains by employing a compliant interlayer is the main topic of this article.

The compliant interlayer is assumed to be constructed from a columnar structure of "beams" which can flex to accommodate the thermal strain, without fracture. Indeed the safe design of the columnar structure and the dense topcoat deposited on to it, lies in the condition that the maximum value of the stress in the columns, due to flexure, must be less than the stress in the topcoat. In this case, if both the columns and the topcoat are made from the same material, then the columns would *not* fracture, thereby precluding delamination of the topcoat.

The analysis in the article focuses on the analysis of the stresses in the topcoat and the flexure displacements in the columns. The approach is conceptually similar to the shear lag models for the interplay between the in-plane stress in thin films and the shear tractions at the interface when the film is mechanically loaded by thermal strain. The difference in the present problem is that the shear tractions in the interfacial layer are borne by the flexure of an array of discreetly distributed columns.

The analysis is formulated in terms of a non-dimensional parameter, c, given by Eq. (14). The displacements in the columns are given by the set of difference equations in Eqs. (8) and (9). These equations lead to the full solution to the problem. Figure 4 gives the stress distribution in the topcoat, and the displacements in the columns, as a function of the distance from a free edge of a crack in the topcoat. The effective decay distance of the displacements is plotted as a function of c in Fig. 5. The criterion for safe design leads to the map in Fig. 6. The map shows two "safe" regimes, one at low aspect ratio of the columns and the other for the high aspect ratio of the columns. Experi-

ments that are apparently in agreement with the design map in Fig. 6 are described; these results are presented in Figs. 9 and 10.

Finally, the concern that stresses in the EBC can degrade the intrinsic fracture strength of the substrate is addressed quantitatively. The general result is given by Eqs. (27) and (28), while the approximate result, which should suffice for most applications, is given by Eq. (29). The most notable feature of Eq. (29) is that the columnar structure of the bond coat reduces the fracture-stress penalty to less than 5%.

VII. Conclusions

A general architecture for an environmental barrier coating is presented. It consists of a topcoat material which resists corrosion, a columnar bond coat that is strain tolerant, and the chemical bond coat that helps the adhesion of the columnar structure and the topcoat to the substrate. The function of the compliant bond coat is to accommodate the mismatch between the thermal expansion coefficients of the topcoat and the substrate.

The stresses and displacements in the topcoat and columnar structure are analyzed. The fracture criterion in the columns is related to the maximum bending stress produced in them. The "safe" design of the coating is based upon the maximum stress in the columns being less than the maximum stress in the topcoat, as this would prevent delamination of the coating. The analysis then leads to the design map shown in Fig. 6. Experiments reported in Figs. 9 and 10 are agreement with the prediction from this map. It should be noted that the topcoat cannot be expected to provide hermetic isolation between the environment and the substrate. The topcoat will necessarily develop periodic cracks if the thermal stress in it is greater than its fracture strength; however, the spacing between such cracks in the top coat can be controlled by the design of the columnar structure (the half spacing between the cracks in the topcoat will be equal to the decay distance shown along the y-axis in Fig. 5). The main purpose of the topcoat is to subdue the velocity of the environment at the substrate interface.

Finally, formal analysis shows that the tractions exerted by the above architecture of the coating on the substrate surface will have an insignificant influence on the intrinsic fracture strength of the substrate.

References

[1]T. E. Strangman, "Columnar Grain Ceramic Thermal Barrier Coatings"; U.S. Patent No. 4,321,331, 1982.
[2]T. E. Strangman and J. L. Schienle, "Tailoring Zirconia Coatings for Performance in Marine Gas Turbine Environment," *J. Engin. Gas Turbines Power*, **112**, 531–5 (1990).
[3]J. Opila, "Variation of the Oxidation Rate of Silicon Nitride with Vapor Pressure," *J. Am. Ceram. Soc.*, **82** [3] 625–36 (1999).
[4]N. S. Jacobson, "Corrosion of Silicon-Based Ceramics in Combustion Environment," *J. Am. Ceram. Soc.*, **76** [1] 3–28 (1993).
[5]K. N. Lee, "Current status of EBCs for Si Based Ceramics," *Surf. Coat. Tech.*, **133–134**, 1–7 (2000).
[6]B. Sudhir and R. Raj, "Effect of Steam Velocity on the Hydrothermal Oxidation/Volatilization of Silicon Nitride," *J. Am. Ceram. Soc.*, **89** [9] 1380–7 (2006).
[7]T. Bhatia, H. Eaton, J. Holowczak, E. Sun, and V. Vedula, "Development and Evaluation of Environmental Barrier Coatings for Silicon Nitride"; United Technologies Research Center, East Hartford, CT, DOE-EBC Workshop, Nashville, TN, November 18, 2003.
[8]D. C. Agrawal and R. Raj, "Measurement of the Ultimate Shear Strength of a Metal Ceramic Interface," *Acta Met. Mater.*, **37** [4] 1265–70 (1989).
[9]K. More, unpublished work, Oak Ridge National Laboratory, Knoxville, TN.
[10]S. R. Shah and R. Raj, "Multilayer Design and Evaluation of a High Temperature Environmental Coating for Si-Based Ceramics," *J. Am. Ceram. Soc.*, (2006), in press.
[11]F. P. Beer, E. R. , Jr. Johnston, and J. T. DeWolf (eds), *Mechanics of Materials*, 4th edition, McGraw Hill, New York 2006.
[12]K. Sharma, P. S. Shankar, and J. P. Singh, "Mechanical Behavior of Si_3N_4 Substrates with Environmental Barrier Coatings"; *Proceedings of the Symposium on Innovative Processing and Synthesis of Ceramics, Glasses, and Composites at the 105th American Ceramic Society Annual Meeting and Exposition*, Nashville, TN, April 27–30, 2003.
[13]H. Tada, P. C. Paris, and G. R. Irwin (eds), *The Stress Analysis of Cracks Handbook*, 3rd edition, ASME Press, New York, 2000. ☐

Grain-Boundary Grooving of Plasma-Sprayed Yttria-Stabilized Zirconia Thermal Barrier Coatings

Kendra A. Erk,* Christophe Deschaseaux, and Rodney W. Trice*,†

Purdue University, School of Materials Engineering, West Lafayette, Indiana 47907

The focus of this study was to determine the mechanisms responsible for the microstructural changes of plasma-sprayed 7 wt% Y_2O_3–ZrO_2 thermal barrier coatings with annealing from 800° to 1400°C. Mullins's thermal grooving theories have been applied to plasma-sprayed TBCs to determine the dominant mass transport mechanism at various temperatures. Grain-boundary groove widths were measured as a function of annealing time and temperature using atomic force microscopy (AFM). The same collection of grains was analyzed after progressive heat treatments. Surface diffusion was found to be the dominant diffusion mechanism at 1000°C, corresponding to the disappearance of intralamellar cracks at that temperature. At 1100°C, both surface and volume diffusion were active. Volume diffusion, found to be the dominant diffusion mechanism at 1200°C and above, was responsible for the sintering of interlamellar pores observed from AFM analysis of a single, progressively heat-treated interlamellar boundary. Surface roughening was observed to coarsen with increased annealing time and disappear with increased annealing temperature.

I. Introduction

PLASMA-SPRAYED ceramic thermal barrier coatings (TBCs) are widely used to protect the metallic components of gas turbine engines and improve operating efficiencies by permitting higher operating temperatures and reduced cooling requirements. TBC systems are comprised of a yttria-stabilized zirconia (YSZ) top coat with a composition of 6–8 wt% Y_2O_3–ZrO_2 plasma sprayed over a metallic bond coat. A MCrAlY (M being either Ni, Co, or Fe) bond coat is often used to facilitate strong adhesion between the YSZ and metallic component (such as a turbine blade).[1] The temperature experienced at the surface of the structure has been shown to decrease by 189°C for a YSZ coating 127 μm thick.[2]

Plasma spraying of YSZ powders results in a complex microstructure.[1,3,4] As droplets of molten YSZ strike the substrate at high velocity, they spread to form flattened discs. Coatings are made of multiple stacked discs, termed lamellae. The large thermal gradient between the relatively cool substrate and the newly sprayed lamellae causes directional solidification at a cooling rate in the range of 10^4–10^6 K/s,[3,4] resulting in a columnar grain structure within each lamella. The columnar grains, oriented perpendicular to the substrate, are clearly resolved on transmission electron micrographs.[5,6] Intralamellar cracking occurs perpendicular to the substrate upon cooling to relieve thermal stresses.[7] Total porosity of plasma-sprayed ceramic coatings is typically between 3% and 20%; high porosity is advantageous as it acts to reduce thermal conductivity of the coating.[1] Total

porosity is comprised of two main types: porosity between lamellae (i.e., interlamellar pores) and porosity within lamellae (i.e., intralamellar pores).[3] The greatest amount of porosity is attributed to interlamellar pores, which are oriented perpendicular to the spray direction.[4]

The rapid cooling rates achieved during plasma spraying prevent the diffusion of yttrium cations (Y^{3+}) during solidification, resulting in the formation of a non-transformable yttrium-rich tetragonal phase designated here as t'-ZrO_2.[8,9] Non-transformable zirconia is a non-equilibrium phase as the concentration of yttrium cations in the zirconia lattice is greater than the concentration predicted by the ZrO_2–Y_2O_3 phase diagram.[10] Thus, t'-ZrO_2 is unstable at high temperatures.[9]

As plasma-sprayed YSZ coatings are exposed to temperatures simulating the service conditions in an engine (800°–1300°C), numerous microstructural changes occur that greatly affect the material properties of the coatings.[11] Morphological changes lead to increases in thermal conductivity. Ilavsky et al.[12] used small angle neutron scattering (SANS) to analyze the shape of microstructural defects present in YSZ coatings as a function of temperature. Healing of intralamellar microcracks was found to begin around 800°C. Microcracks continued to close with increasing temperature and were fully closed at 1000°C. Additionally, at temperatures above 1000°C, surface area attributed to interlamellar porosity began to decrease. The different sintering regimes were believed to be related to the size and shape of the voids.[13] Trice et al.[6] presented visual confirmation of intralamellar microcrack and interlamellar pore disappearance (after 1000°C/50 h and 1200°/50 h heat treatments, respectively) which was linked to the observed increase in thermal conductivity of the coatings.

Although the microstructural evolution adversely affecting the properties of thermal barrier coatings during service at high temperatures is well documented, the mechanisms responsible for these microstructural changes remain unclear. The purpose of the present study is to investigate the mechanisms responsible for the microstructural changes of plasma-sprayed YSZ coatings during heat treatment. Atomic force microscopy (AFM) is used to observe grain-boundary grooving, observe interlamellar porosity evolution, and determine the dominant diffusion mechanism at certain temperatures by Mullins's thermal grooving theories.[13,14] The shrinkage behavior of cylindrical stand-alone coatings, indicative of coating densification, is investigated using dilatometry to confirm the AFM dominant diffusion mechanism results.

II. Experimental Procedure

(1) Sample Preparation

The YSZ powder used in the processing of all plasma-sprayed coatings in this study had a nominal composition of 7 wt% Y_2O_3 and 93 wt% ZrO_2, with an average particle size of 22 μm (Amperit 825.0, H.C. Starck, Newton, MA). It was a fused and crushed powder. The coatings were plasma sprayed at two different locations, though similar spraying parameters were used.[15]

G. Rohrer—contributing editor

Manuscript No. 20837. Received August 3, 2005; approved December 12, 2005.
This work was supported by the National Science Foundation under contract no. 0134286-DMR.
*Member, American Ceramic Society.
†Author to whom correspondence should be addressed. e-mail: rtrice@purdue.edu

The YSZ specimen used for AFM analysis was plasma sprayed at Ames National Laboratory with a SG-100 plasma spray gun (Praxair, Danbury, CT) mounted to a two-way stage. The YSZ was sprayed directly onto a flat copper substrate to a thickness of approximately 2 mm. The coating surface was ground and sectioned into samples roughly 10 mm long and 5 mm wide using a diamond-coated blade. The copper substrate was dissolved by immersion in a strong nitric acid solution. The sample surface to be analyzed was coarse ground on a 14 μm SiC polishing wheel and progressed through finer polishing media until a final polishing suspension of 0.05 μm chrome oxide particles was used. The coating specimen used for the AFM studies had a bulk density of 5.5 g/cm^3 with average total porosity of 9.4%.[15,16]

Specimens used for dilatometry were sprayed at Northwestern University in Evanston, IL, using a small particle plasma-spray injector.[17] A 100–200 μm thick aluminum layer was sprayed uniformly onto a rotating alumina tube (300 mm long and 13 mm in diameter), on top of which YSZ was sprayed to a thickness of approximately 200–300 μm. Each sprayed rod was cut into approximately 15 mm long samples. The edges of the sample were machined on a lathe using a diamond tool, so that the faces of each section were parallel. The intermediate aluminum layer was dissolved in a weak solution of hydrochloric acid, releasing the YSZ coating from the substrate. The final dimensions of each sample were nominally 13 mm tall with a wall thickness of 250 μm. The bulk density of the tubes was 5.0 g/cm^3 with an average total porosity of approximately 17.1%, calculated assuming the theoretical density of YSZ to be 6.08 g/cm^3.[15,16]

(2) AFM Analysis and Theory

The topography and microstructural evolution of YSZ coatings was investigated as a function of heat-treatment temperature and time using a MultiMode AFM (Digital Instruments, Santa Barbara, CA) in tapping mode. PointProbe® Plus silicon AFM probes (Nanosensors™, Neuchatel, Switzerland) having a probe tip radius <10 nm were used. Polished coating samples were scanned to quantify initial surface roughness before heating. Coatings were isothermally etched in air for 1–100 h at temperatures ranging from 800° to 1400°C in a programmable Lindberg box furnace. Heating and cooling rates were 600° and 300°C/h, respectively. To provide a barrier to furnace contaminants, the coatings were contained in a covered alumina crucible. After each heat treatment, AFM area scans 4, 9, or 25 μm^2 in size were taken, corresponding to lateral resolutions of ±4, ±6, or ±10 nm, respectively. Two separate studies were performed: grain-boundary widths were measured from plane-view coating scans, and interlamellar porosity and grain structure evolution were studied from cross-sectional coating scans.

In the first AFM study, grain-boundary groove profiles were analyzed from plane-view-oriented YSZ coatings progressively heat treated at 1000°, 1100°, and 1200°C. The same collection of grains was analyzed after every heat treatment for each of the three coatings. Peak-to-peak groove width measurements were obtained from cross-sectional topography profiles oriented perpendicular to the grain boundaries. Severely asymmetric grooves, polishing scratches, newly nucleated grains, and grains bordering pores or cracks were avoided in the measurement. Expected error for the groove-width measurements is dependent solely on image pixel resolution, as there is no error associated with tip geometry when measuring groove peak positions.[18,19]

Mullins's theories of grain-boundary grooving were used to determine the dominant diffusion mechanism at 1000°, 1100°, and 1200°C.[13,14] For volume, surface, and evaporation/condensation diffusion, the theories prove the resulting groove profiles have groove widths w proportional to annealing time $t^{1/3}$, $t^{1/4}$, and $t^{1/2}$, respectively. Klinger[20] showed the changes in groove width with annealing time in a two-component system still follow the $t^{1/4}$ law for surface diffusion, as Mullins's original theories were for single component systems. Robertson[21] and Zhang and Schneibel[22] have further extended the theories to include finite surface slopes beyond Mullins's assumption that the slope of the surface was small compared with unity.[13,14]

For the second AFM study, a cross-sectional-oriented YSZ coating was progressively heat treated from 1000° to 1400°C for varying lengths of time to observe the evolution of interlamellar porosity and grain structure. The same lamella boundary was analyzed after each heat treatment. The AFM fast scan direction was perpendicular to the boundary to preserve height data after zero-order plane fitting was performed to remove the z-offset associated with AFM scans.

(3) Dilatometric Study

The shrinkage behavior in the axial direction of the YSZ coating tubes during 10 h isothermal holds from 800° to 1400°C was investigated using a horizontal Orton® 1600D dilatometer (Westerville, OH). A heating rate of 10°C/min was used for all tests. Length variations of the coatings were recorded using a linear variable displacement transducer (LVDT) capable of a resolution of ±1 μm; however, the noise in the measurements because of temperature fluctuation was estimated to be ±2 μm. The percent linear change recorded by the dilatometer reflects the thermal expansion of the apparatus, the thermal expansion of the coating, and the shrinkage of the coating. After isolating the data specific to the YSZ coating, the thermal expansion of the coating being tested was calculated as a function of temperature and subtracted from the raw data in order to obtain the data relative to the shrinkage of the YSZ. This step adds additional uncertainty to the final shrinkage measurements. Overall, the multiple sources of uncertainty limit the linear shrinkage resolution to approximately ±0.1% for a 15 mm long sample.

III. Results

(1) AFM Analysis of Plane-View-Oriented Samples

The YSZ coating heat treated at 800°C for 10 h did not display any observable faceting or grain-boundary grooving. Average surface roughness for the measured images, calculated as a ratio of surface area to projected area, did not increase with heating. Notable surface features were limited to surface pores, surface cracks, and residual polishing scratches (not shown).

The 900°C/10 h treated coating exhibited signs of early grain grooving. Grooving was non-uniform across the surface. Regions existed where grain boundary definition was insufficient to clearly outline entire grains. Some of the defined grain surfaces were more extensively roughened than others. Surface cracks and pores were observed.

Grain-boundary profiles of the same 9 μm^2 area of the plane-view-oriented coating were analyzed after progressive heat treatments of 1, 5, 10, 50, and 100 h at 1000°C. Grain-boundary grooves were sufficiently defined after the 1 h heat treatment to allow for accurate width measurements. As the 1 and 5 h images varied merely in contrast, only the area scans taken after the 10, 50, and 100 h treatments are displayed in Fig. 1(a). The previously polished surfaces of the grains became rough after annealing (see grain A, outlined in the first image of Fig. 1(a)). Surface roughening resulted in the formation of hills and valleys on the surfaces of the grains. Sometimes the roughened surfaces displayed clear striations (see grain B of Fig. 1(a)). The majority of grains exhibited various extents of surface roughening which coarsened with further annealing. Most grain-boundary groove profiles were also asymmetric.

Grains that displayed extensive surface roughening tended to groove at varying rates. Variable groove rates resulted in a range of groove widths observed within single grain boundaries and from one measured boundary to another; this will be subsequently referred to as non-uniform grooving. After the 1000°C/50 h heat treatment, grains less than 60 nm in size, such as those identified by the box in the middle image of Fig. 1(a), seemed to disappear. A wide intralamellar microcrack spanning the 9 μm^2 analyzed area was exposed after the 100 h treatment. A larger

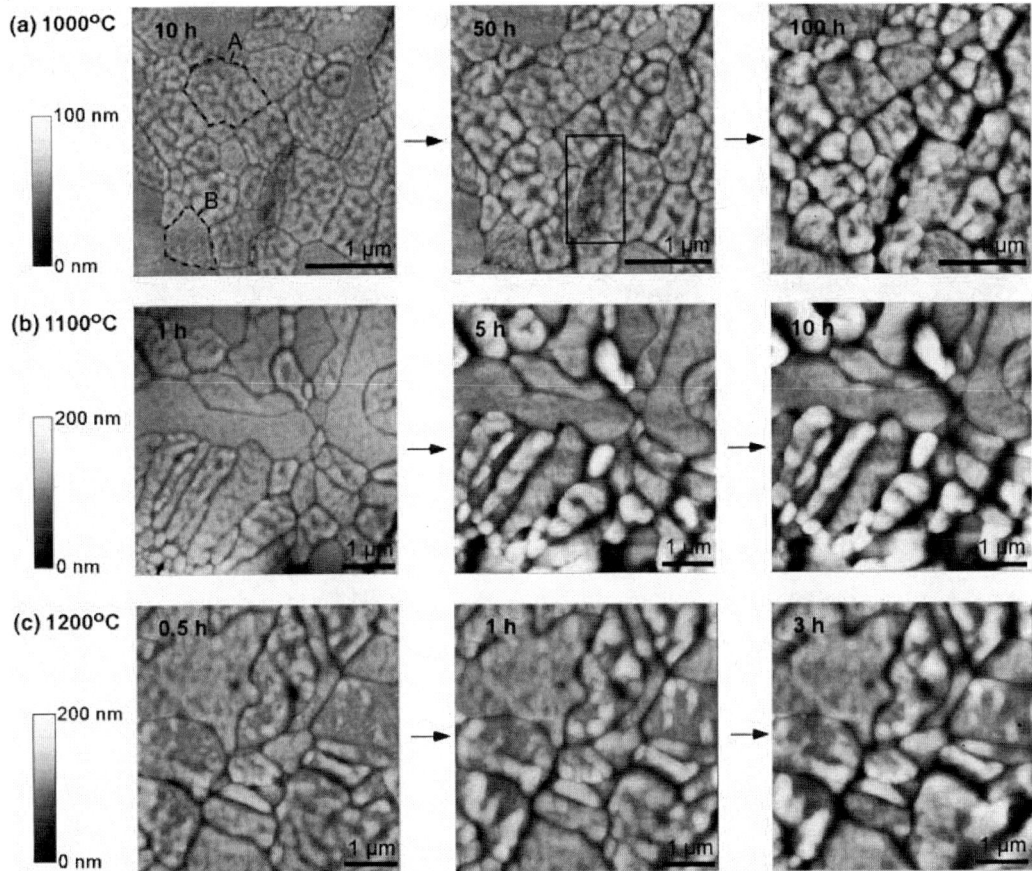

(a) 1000°C 10 h A B 50 h 100 h 1 µm

(b) 1100°C 1 h 5 h 10 h 1 µm

(c) 1200°C 0.5 h 1 h 3 h 1 µm

Fig. 1. Atomic force microscope area scans of plane-view oriented yttria-stabilized zirconia coatings annealed for different lengths of time at (a) 1000°C, (b) 1100°C, and (c) 1200°C illustrating the evolution of grain grooving, surface roughening, and grain growth/shrinkage. Contrast changes between the images at each temperature indicate changes in height. An example of the small grain disappearance discussed in the text is shown in (a): many grains present in the box of the 1000°C/50 h image seem to be absent from the 100 h image.

area scan of the coating revealed the crack extended beyond the 9 µm² area and was clearly defined outside the analyzed area after the initial 1000°C/1 h heat treatment.

Figure 1(b) displays the plane-view 25 µm² area scans of a coating progressively heat treated at 1100°C for 1, 5, and 10 h. As shown in the lower left-hand corner of the images, after 1 h, grains less than 200 nm in diameter began to decrease in size and coalesce with the surrounding grains. Surface roughening coarsened with annealing time. Non-uniform grain-boundary grooving and groove asymmetry were more prevalent at annealing temperatures of 1100°C than at 1000°C as the images in Fig. 1 demonstrate. Some grain boundaries grooved extensively between the 1 and 5 h treatments at 1100°C, while other boundaries did not appear to change and therefore, were not measured. As non-uniform grooving amplified with time, analysis of the 50 and 100 h heat treatments was disregarded because of increasingly sporadic grooving.

Finally, grain grooving was studied from isothermal holds at 1200°C for 0.5, 1, and 3 h (analyzed scans shown in Fig. 1(c)). Decreased treatment times were used as the heightened temperature caused more extensive grooving in smaller periods of time than compared with the grooving observed at 1000° and 1100°C. Surface roughening coarsened with annealing time. Non-uniform grooving again led to a range of width measurements.

With annealing at all three temperatures, the average width of the grain-boundary grooves measured between each heat treatment increased with annealing time as shown by the log w–log t plot in Fig. 2. Error bars indicated the range of observed groove widths. The 1000°C data in the plot displayed a linear curve fit with a slope similar to 1/4, illustrating that surface diffusion was the dominant mechanism. The log w–log t data from the 1100°C heat treatments had a curve-fit slope of 0.28. When both surface

and volume diffusion are taking place simultaneously, multiple works have concluded the slope corresponding to the log w–log t plot will be between 1/4 (dominant surface diffusion) and 1/3 (dominant volume diffusion).[23,24] Thus, both surface and volume diffusion were active at 1100°C. The linear curve fit of the 1200°C log w–log t data exhibited a slope equal to 1/3, indicating

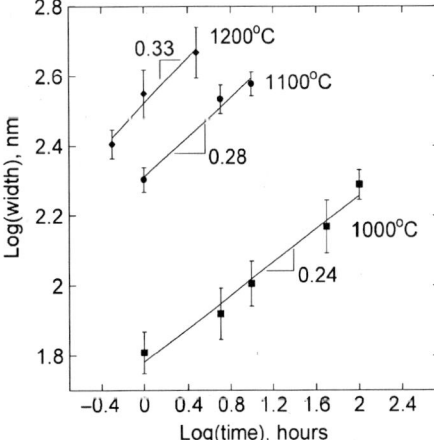

Fig. 2. Plot expressing the relationship between the grain-boundary groove widths as a function of annealing time at 1000°, 1100°, and 1200°C. The 0.24 slope of the linear curve fit for the 1000°C data indicates the dominant mass transport mechanism is surface diffusion. The 0.28 curve-fit slope of the 1100°C data indicates both volume and surface diffusion are active. The 0.33 curve-fit slope of the 1200°C data indicates volume diffusion is dominant.

volume diffusion was the dominant transport mechanism at 1200°C.

(2) AFM Analysis of Cross-Sectional-Oriented Samples

Figures 3(a)–(f) show the evolution of a single lamellar boundary during a progressive heat treatment at temperatures from

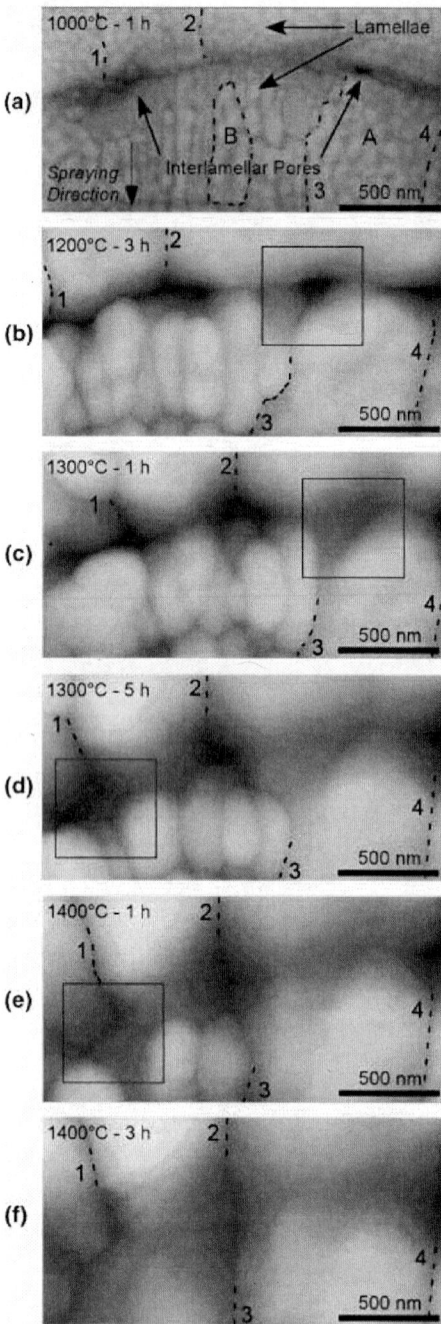

Fig. 3. Atomic force microscope area scans of one lamella boundary showing the evolution of the interlamellar porosity and grain structure of the boundary as the cross-sectional yttria-stabilized zirconia coating was progressively heat treated from 1200° to 1400°C. The same area was scanned each time. The same four grain boundaries are indicated by dotted lines and numbered 1–4 in each image to aid the reader in following the microstructural evolution. With increased annealing at higher temperatures, a number of changes occurred: the columnar grain structure grooved and coarsened; porosity became more spheroidal and segregated; and grains below the surface bridged the pore and ultimately sintered (as shown in the boxed areas). Contrast changes from black to white indicate a positive height change of 150 nm in image (a) and 400 nm in all other images.

1000° to 1400°C. The same four-grain boundaries are indicated with dotted lines and numbered in all the images to aid the reader in observing the evolution of the region. Heat treating commenced at 1000°C for 1 h to allow for uniform grain definition across the coating.

Figure 3(a) shows the single lamella boundary targeted throughout the analysis after the 1000°C/1 h heat treatment. The area between grain boundary #3 and #4 comprises one columnar grain approximately 600 nm in width and labeled grain A. The surface of grain A displayed extensive surface roughening. A number of smaller columnar grains (roughly 100–200 nm in width) existed to the left of grain A (see grain B in Fig. 3(a)). These grains also displayed surface roughening, though due to the decreased size of the grains, the roughening was more difficult to resolve than the roughening present on grain A. The structure of the columnar grains coarsened[25] with further annealing at temperatures of 1200°C and above as shown in Figs. 3(b–f): existing grains enlarged at the expense of smaller grains during each heat treatment. The surface roughening observed after the 1000°C/1 h treatment in Fig. 3(a) was also observed to decrease at temperatures of 1200°C and above.

After 1 h at 1300°C (Fig. 3(c)), sintering across the interlamellar pores was observed which caused the pores to become more spheroidal and segregated. This densification resulted from grain grooving directed across the pore at depths below the exposed surface grains. The bridging grain (outlined by the boxes in Figs. 3(b) and (c)) existed at a depth below the surface grains. Bridging from below was further verified from the 1400°C/1 h scan: as indicated by the boxes in Figs. 3(d) and (e), a surface grain merged with the grain beneath, which was responsible for closing the pore. The interlamellar pores further healed after the 1400°C/3 h treatment (Fig. 3(f)).

(3) Dilatometer Studies

Figure 4 is a plot of the shrinkage behavior of YSZ coatings as a function of time at different test temperatures. No shrinkage of the YSZ was detected after 10 h at 800° or 900°C. Approximately 0.08% linear shrinkage was measured after 1000°C/10 h trial, which was believed to be bordering on the resolution limits of the apparatus. Greater linear shrinkages were observed for similar isothermal trials at 1100°, 1200°, 1300°, and 1400°C. The amount of shrinkage after the 1200°C/10 h trial doubled compared with the 1100°C/10 h trial. Similarly, after 10 h at 1300°C, the amount of measured shrinkage was twice that of samples heat treated for 10 h at 1200°C. Some shrinkage occurs during heat-up of the coating at testing temperatures greater than 1100°C; thus, some linear shrinkage of the coating has already occurred when the isothermal hold begins.

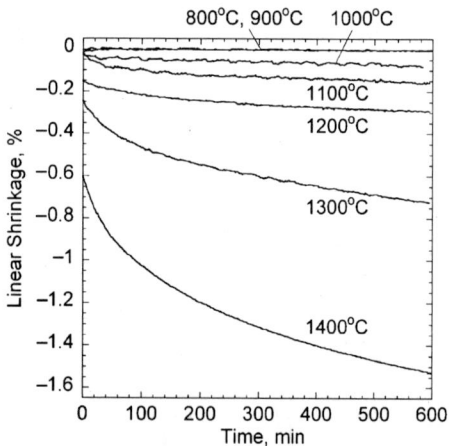

Fig. 4. A plot of the linear dilatometry response of yttria-stabilized zirconia coatings tested at different temperatures showing the shrinkage behavior of the coatings as a function of time.

IV. Discussion

(1) Microstructural Changes from 800° to 1000°C

As discussed in the introduction, Trice et al.[6] and Ilavsky et al.[12] observed the disappearance of intralamellar cracks in fused and crushed YSZ coatings annealed at 1000°C. Here, surface diffusion was found to be the dominant diffusion mechanism at 1000°C, indicating the crack disappearance reported in the literature was the result of a non-densifying mass transport mechanism.[26] Dilatometry results further confirmed this conclusion as no appreciable shrinkage was detected until the 10 h isothermal hold at 1100°C.

There is evidence in the literature supporting surface diffusion as the dominant mass transport mechanism in zirconia at relatively low temperatures. In a study using pore smoothing observations and particle neck growth measurements by Akash and Mayo,[27] surface diffusion was shown to be the dominant mass transport mechanism in tetragonal 3 mol% Y_2O_3–ZrO_2. Evidence of Zr^{4+} ion transport to the neck between spherical powders was observed at temperatures from 870° to 1050°C. In situ transmission electron microscope experiments at 890°C also noted surface diffusion was the likely transport mechanism for pore smoothing between nanocrystalline grains of zirconia.[28]

Since surface diffusion is a non-densifying transport mechanism, this mechanism can only change the shape of an isolated intralamellar crack. However, narrow intralamellar cracks connected to interlamellar pores would completely heal; atoms present on the surface of the pore would diffuse in the direction of decreased chemical potential[25] and fill the crack. As densification is not occurring, the decrease in volume of the crack would be balanced by an increase in volume of the pore. If the intralamellar crack is not coupled with an interlamellar pore, the crack could not heal in the traditional sense but become spheroidal. Allen et al.[4] suggested intralamellar cracks in annealed coatings tended to form globular or irregular-shaped pores during a study of plasma-sprayed YSZ coatings made with fused and crushed 7–8 wt% YSZ powder. Evidence of spheroidal intralamellar cracks was not observed on AFM images, but the small volume of such defects in the process of healing might make observation difficult.

Another noteworthy microstructural change was the development of surface roughening on the exposed grains. AFM analysis of the plane-view coatings showed different extents of surface roughening existed on the surfaces of the grains after annealing. Roughening coarsened with increased annealing time during the 1000°C progressive heat treatment. Surface roughening was also observed on the cross-sectional coating after the 1000°C/1 h heat treatment.

Non-uniform grooving and asymmetric grain-boundary grooves were frequently observed from the AFM analysis of the annealed coatings. Robertson and Chang[29] concluded variations of the surface diffusion coefficient with orientation will lead to non-uniform grooving and the formation of asymmetric grooves. Multiple studies support this conclusion.[30–32] Thus, the surface roughening observed during the AFM analysis of the annealed YSZ coatings was most likely due to surface diffusivity anisotropy. The diffusion variations may be induced by crystallographic faceting from surface energy anisotropy[31,33] of the plasma-sprayed coating or a heterogeneous composition across the surface of the coating.[34] The distribution of yttrium atoms within the lamellar microstructure depends on the interface morphology adopted during solidification.[35] By modeling the heat transfer and solidification of plasma-sprayed YSZ splats, Wang et al.[35] showed that poor heat transfer during solidification results in low solidification interface velocity which can lead to microsegregation of yttria at grain or dendrite boundaries. Regarding the validity of Mullins's surface diffusion theories in the presence of surface roughening and faceting, Sachenko et al.[31] showed unfaceted grains and faceted grains that coarsened with further annealing remained in qualitative agreement with Mullins's theories. The validity of the theories was also unaffected by the presence of asymmetric grain-boundary grooves.

The driving force behind the exposure of the wide microcrack after the plane-view 1000°C/100 h heat treatment (shown in Fig. 1(a)) remains to be understood. No evidence of coating densification was observed from the AFM plane-view images presented in Fig. 1. The dilatometer was also unable to detect significant shrinkage with annealing at 1000°C. As the plane-view region analyzed was small relative to the bulk volume of the annealed coating, it was believed the morphological evolution at the surface was effectively constrained by the interior grains. Sudre and Lange[36] studied the breakup of grain bridges within polycrystalline 8 mol% Y_2O_3–ZrO_2 compacts resulting in the formation of "large cracklike voids." When the distance between particle centers remains constant in a three-grain bridge (similar to the small grains present between the larger grains in the box of Fig. 1(a)), it is possible for the necking region to become unstable, causing the larger grains to desinter from the smaller, center grains in order to reduce the free energy of the system. Rankin and Boatner[37] also observed the desintering of single-crystal MgO particles when surface diffusion was the likely mass transport mechanism in the system. As the AFM images alone do not yield enough information about the factors which may have caused desintering by destabilizing the coating (such as the existence of a porosity gradient within the coating or the coarsening of smaller grains within the neck[37]), a more in-depth analysis is needed.

(2) Microstructural Changes from 1100° to 1400°C

As shown in the AFM images of Fig. 3, thermal grooving resulted in columnar grains assuming a rounded appearance at the pore interface, able to bridge interlamellar pores. Grain bridging was favorable at temperatures of 1200°C and above. The images also revealed the bridging grains would ultimately come into contact and sinter during heat treatments of 1300° and 1400°C, healing the interlamellar pores.

As observed in the cross-sectional AFM study, the interlamellar pores appeared to enlarge between the 1000°C/1 h and 1200°C/3 h heat treatments (Figs. 3(a) and (b), respectively). This pore widening was due to the grooving of surrounding grains. Consistent with the plane-view AFM study and thermal grooving theories, with increased annealing the columnar grains exposed at the polished surface grooved rapidly and increased in height above the original surface height of the coating. Thus, mass was being removed from the upper walls of the pores and redistributed, creating the illusion of pore expansion at the surface. Below the surface, thermal grooving experienced by grains would only have taken place along the pore interface as the surrounding grains would have constrained growth in all other directions. This idea of constrained, directed grooving explains why the bridging and healing of the pores initiated below the outer surface grains as observed in the cross-sectional study.

Densification of the coating was driven by volume diffusion, shown to be the competing and dominant transport mechanism in the coatings at 1100° and 1200°C, respectively. At temperatures above 1200°C, volume diffusion remained the dominant transport mechanism as the reduction in porosity and resulting shrinkages detected by the dilatometer at 1300° and 1400°C could not result from either surface or evaporation/condensation diffusion mechanisms. Diffusion along the grain boundaries (limited by the slowest diffusing species: Zr^{+4} ions[27]) was most likely occurring as opposed to lattice diffusion due to the large grain-boundary area relative to the bulk volume of the coatings.[26] The rate of shrinkage (represented by the first derivative of the curves in Fig. 4) decreased with time as the interlamellar pores became spheroidal and/or healed, reducing the total interfacial energy of the system and thus, reducing the driving force for sintering.[38] Shrinkage rates remained non-zero for the duration of the test for coatings treated at or above 1100°C, indicating densification would continue at these temperatures.

Surface roughening coarsened with increased annealing time (as seen in Figs. 1(a)–(c) from the AFM plane-view analysis) and disappeared with increased annealing temperature (as seen in

Figs. 3(a)–(f) from the cross-sectional analysis). Surface roughening was previously described as being a collection of hills and valleys. Similar to the driving force behind grain grooving, the hill-and-valley morphology created chemical potential gradients across the surface of the grain.[25] Atomic diffusion became more active with heightened annealing temperatures, allowing for the redistribution of atoms from the hills (high chemical potential) to the valleys (low chemical potential), eliminating the surface roughening and lowering the related interfacial energy.[38]

The results of this present work enable the TBC scientific community to take a closer look at the phenomena behind coating densification, leading to decreased durability and performance of TBCs. As mentioned earlier,[12] previous findings link the low- and high-temperature sintering regimes to the size and shape differences of the voids. Though the atomic migration causing sintering does depend on size and shape differences (leading to chemical potential gradients, for example), this study finds that the dominance of two distinctly different diffusion mechanisms is responsible for the void reductions at the various temperatures. By directly identifying the dominant mechanisms, future research can investigate ways of impeding or controlling the mechanisms leading to coating densification. For example, the temperature at which volume diffusion and densification becomes significant may be increased by the addition of impurities or altered plasma-spraying parameters and feedstock (leading to different microstructural characteristics).

V. Summary

Plasma-sprayed TBCs have low thermal conductivity, owed in part to the highly porous microstructure. It has been recognized previously that the densification of the coatings with prolonged exposure to high temperatures will lead to increases in thermal conductivity. This research is aimed at improving coating durability and performance by a thorough investigation of the mechanisms responsible for the microstructural changes causing the conductivity increase.

With increased annealing temperatures and times, the microstructure of the plasma-sprayed YSZ coatings was observed to coarsen and sinter. The surface roughening, non-uniform grain grooving, and the formation of asymmetric grooves observed at temperatures of 900°C and above was most likely because of surface diffusivity anisotropy. Surface roughening coarsened with increased annealing time and disappeared with increased annealing temperatures. Changes in porosity began with the disappearance of narrow intralamellar microcracks at 1000°C. Application of Mullins's thermal grooving theories and AFM measurements of the same collection of grains after successive heat treatments revealed surface diffusion was the dominant diffusion mechanism at 1000°C. Thus, intralamellar cracks were eliminated via a non-densifying mass transport mechanism, either by atoms of a connected pore filling the crack or by the shape of the crack becoming spheroidal. Overall, the total volume attributed to voids in the coatings would not change.

Interlamellar pores closed at annealing temperatures of 1200°C and above. Surrounding columnar grains were able to bridge the widths of the voids and come into contact, sintering with other bridging grains. Volume diffusion was found to be the dominant mass transport mechanism at 1200°C. This result was confirmed by the appreciable shrinkages observed by the dilatometer, indicating volume diffusion dominated above 1200°C as well.

References

[1]R. B. Heimann, *Plasma-Spray Coating: Principles and Applications*, pp. 164–165, 209–23. VCH, Weinheim, Germany, 1996.

[2]R. A. Miller, "Current Status of Thermal Barrier Coatings — An Overview," *Surf. Coat. Technol.*, 30, 1–11 (1987).

[3]R. McPherson, "A Review of Microstructure and Properties of Plasma-Sprayed Ceramic Coatings," *Surf. Coat. Tech.*, 39/40, 173–81 (1989).

[4]A. J. Allen, J. Ilavsky, G. G. Long, J. S. Wallace, C. C. Berndt, and H. Herman, "Microstructure Characterization of Yttria-Stabilized Zirconia Plasma-Sprayed Deposits Using Multiple Small-Angle Neutron Scattering," *Acta Mater.*, 49, 1661–75 (2001).

[5]T. Chraska and A. H. King, "Growth of Columnar Grains During Zirconia-Yttria Splat Solidification," *J. Mater. Sci. Lett.*, 18, 1517–9 (1999).

[6]R. W. Trice, Y. J. Su, J. R. Mawdsley, K. T. Faber, A. R. De Arellano-López, H. Wang, and W. D. Porter, "Effect of Heat Treatment on Phase Stability, Microstructure, and Thermal Conductivity of Plasma-Sprayed YSZ," *J. Mater. Sci.*, 37, 2359–65 (2002).

[7]R. McPherson, "Relationship Between the Mechanism of Formation, Microstructure and Properties of Plasma-Sprayed Coatings," *Thin Solid Films*, 83 [3] 297–310 (1981).

[8]L. Lelait, S. Alperine, C. Diot, and M. Mevrel, "Thermal Barrier Coatings: Microstructural Investigation After Annealing," *Mater. Sci. Eng. A*, A121, 475–82 (1989).

[9]K. Muraleedharan, J. Subrahmanyam, and S. B. Bhaduri, "Identification of t' Phase in ZrO_2–7.5 wt% Y_2O_3 Thermal-Barrier Coatings," *J. Am. Ceram. Soc.*, 71 [5] C226–7 (1988).

[10]H. G. Scott, "Phase Relationships in the Zirconia–Yttria Systems," *J. Mater. Sci.*, 10 [9] 1527–35 (1975).

[11]S. R. Choi, D. Zhu, and R. A. Miller, *Effects of Sintering on Mechanicals and Physical Properties of Plasma-Sprayed Thermal Barrier Coatings*. John H. Glenn Research Center, Cleveland, OH, 2004.

[12]J. Ilavski, G. G. Long, A. J. Allen, and C. C. Berndt, "Evolution of the Void Structure in Plasma-Sprayed YSZ Deposits During Heating," *Mater. Sci. Eng.*, A272, 215–21 (1999).

[13]W. W. Mullins, "Theory of Thermal Grooving," *J. Appl. Phys.*, 28 [3] 333–9 (1957).

[14]W. W. Mullins, "Grain Boundary Grooving by Volume Diffusion," *AIME Trans.*, 218, 354–61 (1960).

[15]G. R. Dickinson, C. Petorak, K. Bowman, and R. W. Trice, "Stress-Relaxation of Compression Loaded Plasma-Sprayed 7 wt% Y_2O_3–ZrO_2 Stand-Alone Coatings," *J. Am. Ceram. Soc.*, 88 [8] 2202–8 (2005).

[16]C. Deschaseaux, "A Sintering Study of Plasma-Sprayed Yttria Stabilized Zirconia Thermal Barrier Coatings Using Stand-Alone Coating Tests." Master's Thesis, Purdue University, West Lafayette, IN, 2002.

[17]T. F. Bernecki and D. R. Marron, "Small Particle Plasma Spray Apparatus, Method and Coated Article." U.S. Patent No. 5,744,777, April 28, 1998.

[18]D. M. Saylor and G. S. Rohrer, "Measuring the Influence of Grain-Boundary Misorientation on Thermal Groove Geometry in Ceramic Polycrystals," *J. Am. Ceram. Soc.*, 82 [6] 1529–36 (1999).

[19]E. Saiz, R. M. Cannon, and A. P. Tomsia, "Energetics and Atomic Transport at Liquid Metal/Al_2O_3 Interfaces," *Acta Mater.*, 47 [15] 4209–20 (1999).

[20]L. Klinger, "Surface Evolution in Two-Component System," *Acta Mater.*, 50, 3385–95 (2002).

[21]W. M. Robertson, "Grain-Boundary Grooving by Surface Diffusion for Finite Surface Slopes," *J. Appl. Phys.*, 42 [1] 463–7 (1971).

[22]W. Zhang and J. H. Schneibel, "Numerical Simulation of Grain-Boundary Grooving by Surface Diffusion," *Comp. Mater. Sci.*, 3, 347–58 (1995).

[23]W. M. Robertson and S. R. Srinivasan, "Interpretation of Grain Boundary Grooving Data for Combined Surface and Volume Diffusion," *Met. Trans. A.*, 6A, 1653–4 (1975).

[24]W. M. Robertson, "Thermal Etching and Grain-Boundary Grooving of Silicon Ceramics," *J. Am. Ceram. Soc.*, 64 [1] 9–13 (1981).

[25]Y.-M. Chiang, D. P. Birnie III, and W. D. Kingery, *Physical Ceramics: Principles for Ceramic Science and Engineering*, pp. 357–88. John Wiley & Sons Inc., New York, 1997.

[26]S-J. L. Kang, *Sintering: Densification, Grain Growth & Microstructure*, pp. 37–87. Elsevier Butterworth-Heinemann, Burlington, MA, 2005.

[27]Akash and M. J. Mayo, "Zr Surface Diffusion in Tetragonal Yttria Stabilized Zirconia," *J. Mater. Sci.*, 35 [2] 437–42 (2000).

[28]J. Rankin and B. W. Sheldon, "In situ TEM Sintering of Nano-Sized ZrO_2 Particles," *Mater. Sci. Eng.*, A204, 48–53 (1995).

[29]W. M. Robertson and R. Chang, "Chapter 4: The Kinetics of Grain-Boundary Groove Growth on Alumina Surfaces"; pp. 49–60 in *Materials Science Research, Vol. 3: The Role of Grain Boundaries and Surfaces in Ceramics*, Edited by W. W. Kriegel and H. Palmour III. Plenum Press, New York, 1966.

[30]E. Rabkin and L. Klinger, "The Fascination of Grain Boundary Grooves," *Adv. Eng. Mater.*, 3 [5] 277–82 (2001).

[31]P. Sachenko, J. H. Schneibel, and W. Zhang, "Effect of Faceting on the Thermal Grain-Boundary Grooving of Tungsten," *Philos. Mag. A.*, 82 [4] 815–29 (2002).

[32]E. Rabkin, Y. Amouyal, and L. Klinger, "Scanning Probe Microscopy Study of Grain Boundary Migration in NiAl," *Acta Mater.*, 52, 4953–9 (2004).

[33]C. A. Handwerker, J. M. Dynys, R. M. Cannon, and R. L. Coble, "Dihedral Angles in Magnesia and Alumina: Distributions from Surface Thermal Grooves," *J. Am. Ceram. Soc.*, 73 [5] 1371–7 (1990).

[34]Y.-M. Chiang, D. P. III Birnie, and W. D. Kingery, *Physical Ceramics: Principles for Ceramic Science and Engineering*, p. 186. John Wiley & Sons, Inc., New York, 1997.

[35]G.-X. Wang, R. Goswami, S. Sampath, and V. Prasad, "Understanding the Heat Transfer and Solidification of Plasma-Sprayed Yttria-Partially Stabilized Zirconia Coatings," *Mater. Manuf. Processes*, 19 [2] 259–72 (2004).

[36]O. Sudre and F. F. Lange, "The Effect of Inclusions on Densification: III, The Desintering Phenomenon," *J. Am. Ceram. Soc.*, 75 [12] 3241–51 (1992).

[37]J. Rankin and L. A. Boatner, "Unstable Neck Formation During Initial-Stage Sintering," *J. Am. Ceram. Soc.*, 77 [8] 1987–90 (1994).

[38]S-J. L. Kang, *Sintering: Densification, Grain Growth & Microstructure*, pp. 1–36. Elsevier Butterworth-Heinemann, Burlington, MA, 2005. □

Novel Deposition of Columnar $Y_3Al_5O_{12}$ Coatings by Electrostatic Spray-Assisted Vapor Deposition

Yiquan Wu

Department of Materials, Imperial College London, London SW7 2AZ, U.K.

Jing Du and Kwang-Leong Choy[†]

School of Mechanical, Materials Manufacturing Engineering and Management, University of Nottingham, Nottingham NG7 2RD, U.K.

A novel and cost-effective electrostatic spray-assisted vapor deposition (ESAVD) was used to deposit $Y_3Al_5O_{12}$ (YAG) coatings. Polycrystalline single-phase $Y_3Al_5O_{12}$ coatings were synthesized using the ESAVD method in an open atmosphere at 650°C, and then annealed at 700°–900°C for 1 h. The ESAVD process involves the decomposition and chemical reactions of charged aerosol in vapor phase. The low-temperature coating deposition characteristics of the ESAVD process using a suitable sol precursor decreases the reaction and crystallization temperatures for forming $Y_3Al_5O_{12}$ coatings. The microstructure of the $Y_3Al_5O_{12}$ coating prepared using the ESAVD method is columnar and such strain-resistance microstructure could be useful for thermal barrier coating applications.

I. Introduction

FROM the Y_2O_3–Al_2O_3 phase diagram, besides the alumina and yttria, there are three crystallographically defined compounds known to exist: $Y_3Al_5O_{12}$ (YAG) (m.p. = 1942°C), $Y_4Al_2O_9$ (YAM) (m.p. = 1977°C), and $YAlO_3$ (YAP) (m.p. = 1917°C). $Y_3Al_5O_{12}$ is the most important material of these compositions and has a rather complex cubic structure with 160 atoms per unit cell ($Z = 8$). $Y_3Al_5O_{12}$ has the chemical formula of $A_3B_2(CO_4)_3$, where the A cation (Y^{3+}) is in a dodecahedral coordination site, the B cation (Al^{3+}) is in an octahedral site, and the C cation (Al^{3+}) is in a tetrahedral site.[1] $Y_3Al_5O_{12}$ possesses low creep rate, high oxidation resistance, low electrical conductivity, no birefringence effects, and isotropic thermal expansion. Therefore, it has wide engineering applications in electronic, optical, and magneto-optic devices. For example, $Y_3Al_5O_{12}$ in the form of coating or film can be used as an optical scintillator, and it is also a promising candidate as thermal barrier coatings, or as oxidation and erosion-resistant materials, which can be used in gas turbine engines, solid-state lasers and various display applications.[2–4]

Usually, lengthy heat treatment at high temperatures is required to obtain pure $Y_3Al_5O_{12}$ phase through traditional solid-state oxide reaction. Thus, the microstructure, grain size, and grain size distribution cannot be readily controlled in the solid state processing conditions. Liquid-phase and soft-chemical routes have also been developed to achieve low temperature processes.[5–8] However, most of these methods produce only powder, which requires subsequent materials processing such as thermal spraying to produce coatings. Therefore, there is limited reported work in the preparation of $Y_3Al_5O_{12}$ coating. The conventional physical vapor deposition method would have a problem controlling the coating stoichiometry of multicomponent oxides. Similarly, conventional chemical vapor deposition (CVD) would require high deposition temperature besides the expensive plasma-assisted CVD equipment. Moreover, the vaporization of chemical precursor at different rate in the conventional CVD process would also lead to the difficulty in controlling the stoichiometry of the multicomponent oxides.

Recently, Choy[9,10] and Bai[9] have developed a novel and cost-effective deposition technique called electrostatic spray-assisted vapor deposition (ESAVD) to synthesize various coatings and films in the form of single layer, multilayer, and compositionally gradient layer for a variety of applications, which include catalytic coatings, solid oxide fuel cell components, ceramic membranes for selective gas separation, thermal barrier coatings, bioactive coatings, optical films, and ferroelectric films for sensors and memory devices. ESAVD involves atomizing and spraying precursor droplets under an electric field, where the charged droplets undergo decomposition and/or chemical reactions in the vapor phase in the vicinity of the heated substrate. This produces stable, solid coatings and films with excellent adhesion to the substrate with high deposition efficiency (>90%), in an open atmosphere, without involving the use of any sophisticated reactor or vacuum systems, making ESAVD a simple and cost-effective deposition technique.

In this paper, we report the investigation of $Y_3Al_5O_{12}$ coatings prepared using ESAVD for the first time. ESAVD offers a viable route for coating on metal or ceramic substrates at relatively low temperature in order to produce high-temperature thermal insulation and oxidation resistance coatings with a controllable microstructure.

II. Experimental Procedure

The precursor solution was prepared from a stoichiometric mixture of acetylacetonates of yttrium and aluminum. In the ESAVD process, a high voltage in the range of 5–25 kV was applied to the nozzle and the electric field-induced charges into the precursor at the tip of the nozzle. Under the electrostatic field, a droplet at the tip of the nozzle is supposed to jet into a conical shape. Once the electric field strength exceeded the critical value of electrostatic repulsion needed to overcome the surface tension of the precursor, fine liquid jets were formed from the tip of the nozzle. The electrostatically charged jets were attracted towards the substrate by a Coulombic force under a gradient temperature field. A stable electrospraying for preparing the coatings can be formed above a minimum electric field strength. As a comparison, spray pyrolysis deposition without

L. Klein—contributing editor

Manuscript No. 10663. Received November 6, 2003; approved June 3, 2004.
Supported by the Overseas Research Scholarship from the U.K. government to Y. Wu and J. Du.
[†]Author to whom correspondence should be addressed. e-mail: kwang-leong.choy@nottingham.ac.uk

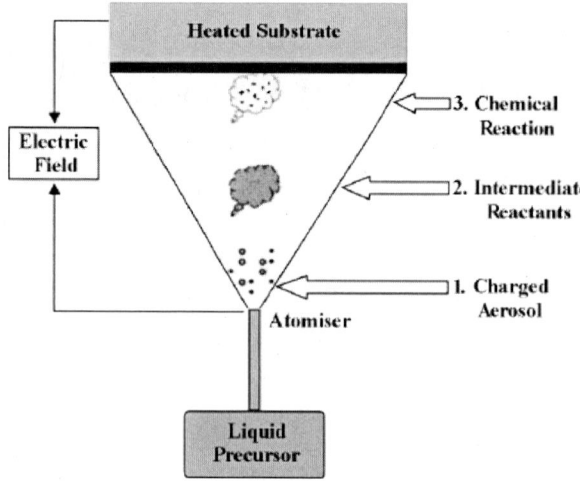

Fig. 1. Schematic diagram of the electrostatic spray-assisted vapor deposition process.

an electric field was also used to prepare the coatings. The same precursor was vaporized with an ultrasonic nebulizer and delivered onto the heating substrate by argon carrier gas through a pipe. Both of the depositions were performed at 650°C on silicon substrates. The coated samples were annealed in the temperature range of 700°–900°C for 1 h. The schematic diagram of the ESAVD process is shown in Fig. 1. The detailed information about the ESAVD method is described by Choy.[11]

A Netzsch 404 (Selb, Germany) differential scanning calorimetry (DSC) was used to characterize the thermal decomposition of precursors. The DSC thermal analysis was performed at a heating rate of 10°C/min from 20° to 1200°C in air on powder materials and α-Al_2O_3 was used as a reference. A Philips PW 1710 (Almelo, the Netherlands) X-ray diffraction (XRD) spectrometer with Cu$K\alpha$ radiation was employed to study the crystallinity of the coatings. JEOL 220T (Hiba, Japan) and Philips XL 30 scanning electron microscope (SEM) were used to analyze the microstructure of the coatings.

III. Results and Discussion

Figure 2 shows the XRD patterns of the as-deposited coatings prepared by the ESAVD and annealed coatings at different temperatures. The peaks in the as-deposited coatings at 650°C were identified as $Y_3Al_5O_{12}$, $Y_4Al_2O_9$, $YAlO_3$, and Y_2O_3. In addition to the $Y_3Al_5O_{12}$ phase, some intermediate phases also appeared in the as-deposited coatings. This might be due to the low deposition temperature and short electrospraying deposition time. The coatings annealed at 850°C showed less $YAlO_3$ phase and there was an increase in the $Y_4Al_2O_9$ content. Polycrystalline $Y_3Al_5O_{12}$ phase was completely formed without any other observed intermediate phases after heating at 900°C for 1 h. However, pure phase $Y_3Al_5O_{12}$ cannot be obtained by annealing the coatings at 900°C for 1 h, in which the coatings were prepared by the spray pyrolysis deposition without an electric field using the same precursor. The XRD patterns of the annealed coatings prepared by the spray pyrolysis deposition are shown in Fig. 3. At 850°C, more $Y_4Al_2O_9$ phase was identified whereas less $Y_3Al_5O_{12}$ phase as compared with 900°C.

The DSC curves of the thermal decomposition behaviour of the Y–Al–O gel powder are shown in Fig. 4. Four main peaks were observed at 180°–190°C, 380°–410°C, 425°–430°C, and approximately 920°C. The first two peaks were due to the evaporation of organic solvents and the combustion of metal–organic matters. The third sharp endothermic peak was due to the formation of intermediate inorganic oxide phases and the last exothermic peak was attributed to the crystallization of $Y_3Al_5O_{12}$. The DSC results showed that crystallized $Y_3Al_5O_{12}$

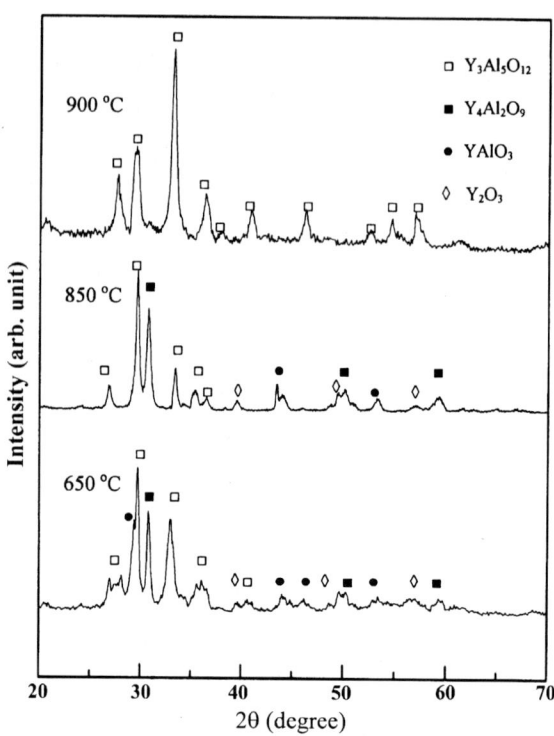

Fig. 2. X-ray diffraction patterns of as-deposited coatings prepared by the electrostatic spray-assisted vapor deposition at 650°C, and coatings annealed at 850° and 900°C.

phase powder could be formed at approximately 920°C, which was proved by a separated XRD analysis. These results demonstrated that pure-phase $Y_3Al_5O_{12}$ coating can be produced by the ESAVD method at the temperature lower than that needed to form the $Y_3Al_5O_{12}$ powder above 920°C using the same Y–Al–O precursor. It has been noticed that ESAVD can produce nanodroplets and better chemical mixture during the deposition, which could result in decreasing the temperature of crystallization of $Y_3Al_5O_{12}$ coatings.[10] Importantly, the main advantage of ESAVD process is that $Y_3Al_5O_{12}$ coating, not the $Y_3Al_5O_{12}$ powder, can be deposited at relatively low temperature using a sol precursor. Compared with some other processes for preparing coatings, the ESAVD process atomized the precursor solution containing Y and Al ions mixed at molecular scale under a high electric field and such mixing has provided a better reactivity of the mixture and allowed a relative shorter reaction time and a lower reaction temperature during the coating deposition

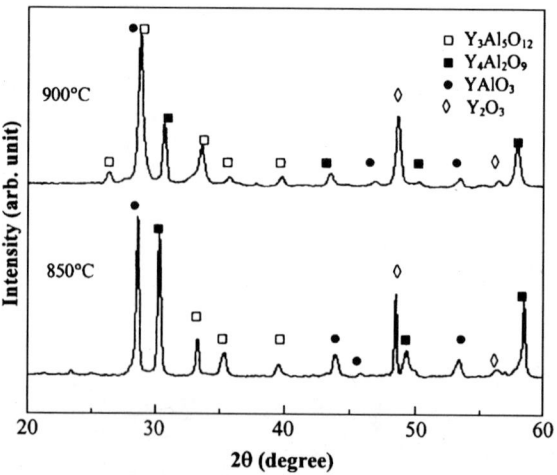

Fig. 3. X-ray diffraction patterns of the annealed coatings at 850° and 900°C with spray pyrolysis deposition.

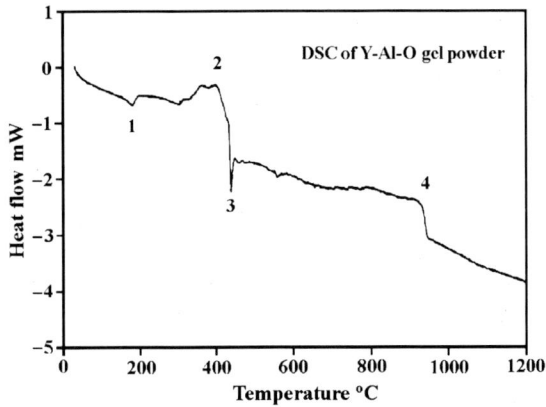

Fig. 4. Differential scanning calorimetry curve of the Y–Al–O gel powder.

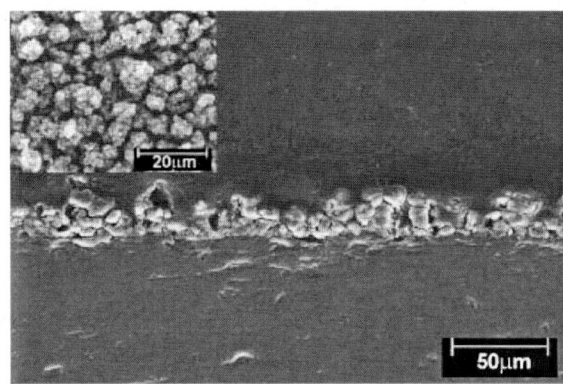

Fig. 6. Cross-section and surface scanning electron microscope micrographs of the coating annealed at 900°C with the spray pyrolysis deposition.

process. For example, the solid-state route would normally require a relatively high processing temperature and the plasma spraying also involves a high deposition temperature for preparing coatings.[12,13] Although the sol–gel method can produce $Y_3Al_5O_{12}$ at fairly low temperature, it is limited to thin film deposition and the process has to be repeated to achieve the desired thickness, and the use of binder also tends to cause crack during annealing.

The SEM micrographs of the cross-section and the surface morphology (inset in Fig. 5) of the annealed $Y_3Al_5O_{12}$ prepared by the ESAVD are shown in Fig. 5. A dense and uniform $Y_3Al_5O_{12}$ coating with columnar microstructure has been deposited using the ESAVD process. The uniform $Y_3Al_5O_{12}$ coating with a thickness of about 30 μm consisted of the grain clusters with an average size of 5–10 μm and the average crystal size estimated from the Scherrer equation is 100 nm. The microstructure of $Y_3Al_5O_{12}$ coatings exhibited the desirable columnar growth features with micro-porosity. The spacing between the columns in the $Y_3Al_5O_{12}$ coatings was uniformly distributed and the average spacing between the columns is about 2 μm. Such distributed spacing in the columnar microstructure of thermal barrier coatings is a key factor in determining the coating compliance, and consequently its resistance to spallation, as well as its thermal conductivity.[14] The cross-section and surface SEM micrograph of the coatings prepared by the spray pyrolysis deposition without an electric field in this work revealed a porous and non-columnar microstructure with particles of approximately 1 μm in diameter being agglomerated to form cluster of approximately 10 μm in diameter, as shown in Fig. 6. The columnar $Y_3Al_5O_{12}$ coatings prepared by the ESAVD could have potential applications as strain-resistance coating materials and prospective phosphor host materials for the thermal barrier coating when doped with rare-earth elements because of the en-

durance of thermal stress through the unique columnar microstructure, which is a unique desirable property for thermal barrier coating.[15]

IV. Conclusions

Columnar $Y_3Al_5O_{12}$ coatings could be successfully prepared by spraying stoichiometric yttrium and aluminium acetylacetonate precursor at relative low temperature in an open atmosphere using a rapid, simple and cost-effective ESAVD method. The deposition of columnar $Y_3Al_5O_{12}$ coatings by the ESAVD method was attributed to the atomized charged aerosol droplets, which underwent a sequence of atomization, evaporation, reactions, nucleation, and growth under a high electric field on the substrate. The $Y_3Al_5O_{12}$ coatings with a columnar microstructure have potential application as high-temperature engineering materials for the strain-resistance and insulating smart coatings having temperature-sensitive luminescent properties.

References

[1]P. Florian, M. Gervais, A. Douy, D. Massiot, and J. P. Coutures, "A Multi-Nuclear Multiple-Field Nuclear Magnetic Resonance Study of the Y_2O_3–Al_2O_3 Phase Diagram," *J. Phys. Chem. B*, **105**, 379–91 (2001).

[2]J. L. Kennedy and N. Djei, "Operation of Yb: YAG Fiber-Optic Temperature Sensor up to 1600°C," *Sens. Act. A*, **100** [9] 187–91 (2002).

[3]D. Ravichandran, R. Roy, A. G. Chakhovskoi, C. E. Hunt, W. B. White, and S. Erdei, "Fabricataion of $Y_3Al_5O_{12}$: Eu Thin Films and Powders for Field Emission Display Application," *J. Lumin.*, **71** [4] 291–7 (1997).

[4]T. A. Parthasarathy, T. Mah, and L. E. Matson, "Deformation Behavior of an Al_2O_3–$Y_3Al_5O_{12}$ Eutectic Composite in Comparison with Sapphire and YAG," *J. Am. Ceram. Soc.*, **76** [1] 29–32 (1993).

[5]Q. M. Lu, W. S. Dong, H. Wang, and X. K. Wang, "A Novel Way to Synthesize Yttrium Aluminum Garnet from Metal–Inorganic Precursors," *J. Am. Ceram. Soc.*, **85** [2] 490–2 (2002).

[6]K. R. Han, H. J. Koo, and C. S. Lim, "A Simple Way to Synthesize Yttrium Aluminium Garnet by Dissolving Yttria Powder in Alumina Sol," *J. Am. Ceram. Soc.*, **82** [6] 1598–600 (1999).

[7]M. Inoue, H. Otsu, H. Kominami, and T. Inui, "Synthesis of Yttrium Aluminum Garnet by the Glycothermal Method," *J. Am. Ceram. Soc.*, **74** [6] 1452–4 (1991).

[8]Y. Liu, Z. F. Zhang, B. King, J. Halloran, and R. M. Laine, "Synthesis of Yttrium Aluminum Garnet from Yttrium and Aluminum Isobutyrate Precursors," *J. Am. Ceram. Soc.*, **79** [2] 385–94 (1996).

[9]K. L. Choy and W. Bai. British Patent, 95 2550551995.

[10]K. L. Choy, *Innovation Processing of Films and Nanocrystalline Powder*. Imperial College Press, UK, 2002.

[11]K. L. Choy, "Chemical Vapour Deposition of Coatings," *Prog. Mater. Sci.*, **48** [2] 57–170 (2003).

[12]M. Mymam, J. Caruso, M. J. H. Smith, and T. T. Kodas, "Comparison of Solid-State and Spray-Pyrolysis Synthesis of Yttrium Aluminate Powders," *J. Am. Ceram. Soc.*, **80** [5] 1231–8 (1997).

[13]S. D. Parukuttyamma, J. Morgolis, H. Liu, C. P. Grey, S. Sampath, H. Herman, and J. B. Parise, "Yttrium Aluminum Garnet (YAG) Films Through a Precursor Plasma Spraying Technique," *J. Am. Ceram. Soc.*, **84** [8] 1906–8 (2001).

[14]T. Nakamura, G. Qian, and C. C. Berndt, "Effects of Pores on Mechanical Properties of Plasma-Sprayed Ceramic Coatings," *J. Am. Ceram. Soc.*, **83** [3] 578–84 (2000).

[15]I. Matsubara, M. Paranthaman, S. W. Allison, M. R. Cates, D. L. Beshears, and D. E. Holcomb., "Preparation of Cr-Doped $Y_3Al_5O_{12}$ Phosphors by Heterogeneous Precipitation Methods and their Luminescent Properties," *Mater. Res. Bull.*, **35** [2] 217–24 (2000). □

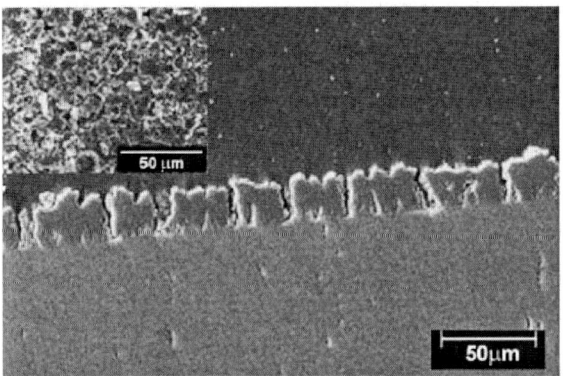

Fig. 5. Cross-section and surface SEM micrographs of the $Y_3Al_5O_{12}$ coating annealed at 900°C with the electrostatic spray-assisted vapor deposition.

Testing and Characterization

MONITORING THE PHASE EVOLUTION OF YTTRIA STABILIZED ZIRCONIA IN THERMAL BARRIER COATINGS USING THE RIETVELD METHOD

G. Witz, V. Shklover, W. Steurer
Laboratory of Crystallography, ETH Zürich,
Wolfgang-Pauli-Str. 10
Zürich, 8093 Zürich, Switzerland

S. Bachegowda, H.-P. Bossmann
Alstom (Schweiz) AG
Brown Bovery Str. 7
Baden, 5401 Baden, Switzerland

ABSTRACT

Plasma sprayed thermal barrier coating composed of tetragonal Yttria Stabilized Zirconia (YSZ) have a limited lifetime and after a certain operating time they fail, usually by spallation. One of the proposed failure mechanisms is the transformation of metastable tetragonal YSZ phase into its monoclinic polymorph, which can lead to destabilization of coating structure and increase of thermal conductivity. Study of the monoclinic content in YSZ is usually performed by X-Ray diffraction using evaluation of the intensities of a few diffraction peaks for each YSZ phase. But this method is missing some important information that can be further extracted from the X-Ray diffraction pattern using Rietveld method. Using the Rietveld method, one can estimate phase content more accurately, take into account some phases present at level even below 1%, gain information on the grain size and strains. By applying Rietveld method, we can observe that during the first stage of the coating ageing process (a) small grains of cubic YSZ are crystallizing and (b) yttria content within the tetragonal phase is lowering. The tetragonal phase can be described as a mixture of two tetragonal phases; one (t') with typical for YSZ c/a ratio and unchanged yttria content and second one (t) with an increased c/a ratio, corresponding to a lower yttria content. These observations are used for modeling the decomposition of the tetragonal YSZ, considering the segregation of yttria at YSZ phase boundaries, leading to the formation of YSZ domains with the cubic structure.

INTRODUCTION

Thermal barrier coatings (TBC) are used in gas turbines to reduce the temperature of blades and hot engine components, they allow operating them at higher temperature, increase the components lifetime and reduce the cooling needs. A TBC system typically consists of a MCrAlY bond coat and an yttria stabilized zirconia (YSZ) topcoat. The topcoat is in general composed of metastable tetragonal YSZ containing 6-8 wt% of yttria. As can be seen in Figure 1, by varying the yttria content, other polymorphs of YSZ can be obtained.[1] Upon annealing at temperature above 1000 °C, the metastable tetragonal YSZ phase is expected to decompose into a mixture of cubic YSZ having a high yttria content and tetragonal YSZ having a low yttria content. The low yttria content tetragonal phase is stable only at high temperature and it undergoes a martensitic transformation to a monoclinic phase at a temperature between 600 °C and 1000 °C depending on the yttria content accompanied by a change of the unite cell volume by about 4%, which can result in cracking and coating failure.

Further improvement of engine efficiency and reduction of operating costs are expected by increasing the operating temperature and maintenance intervals.[2] Both of these goals require a better understanding of the mechanisms leading to YSZ coating failure. Decomposition of the metastable tetragonal polymorph of YSZ into cubic and monoclinic polymorphs has been considered as a potential mechanism for coating failure.[3] A spinodal decomposition model proposed by Katamura et al. dicates that yttrium diffuses to the domain boundaries, which have a cubic structure.[4] This leads to the formation of yttria depleted domains., which at some point of the decomposition process are sufficiently depleted in yttria to undergo the tetragonal-monoclinic transition upon cooling leading to volume change and potentially to coating failure.

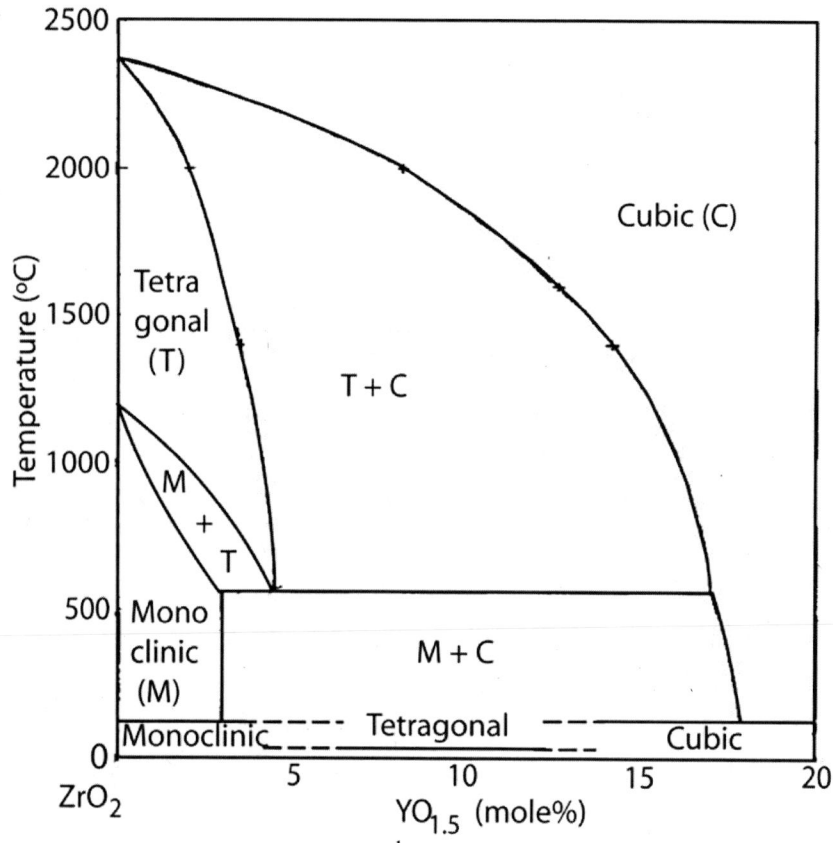

Fig. 1. Phase diagram of YSZ[1]

X-ray diffraction (XRD) combined with Rietveld refinement is an efficient technique to determine microstructural parameters and phase content of multiphase compounds. The Rietveld method uses a least square approach to fit a measured XRD pattern with a theoretical one. Microstructural and instrumental parameters, like cell dimensions, atomic positions and occupancies, phase composition, peak shapes including microstrains and grain size contributions, texture, polarizations effects, 2θ offset, sample displacement, absorption factors, atomic displacement factors and background contribution can be refined. Nevertheless, the Rietveld refinement of multiphase systems can be difficult if there are overlapping peaks like in YSZ polymorphs, in spite of using a monochromatic radiation that generally increases the quality of the collected spectra and reduces the peak overlaps. The X-rays scattering factors of Zr and Y are too close to allow refining directly their relative content in the unit cell. But since the lattice parameters of the tetragonal and cubic phases depend on the yttria content,[1,5] it is possible to determine the yttria content in both phases and study its evolution.

A previous neutron scattering study by Ilavsky et al. of the phase evolution in YSZ coatings during annealing shows that metastable tetragonal YSZ decomposes into monoclinic and cubic phases while the yttria content of the tetragonal phase is reduced.[5] XRD studies on plasma sprayed coatings by Brandon and Taylor and on EB-PVD coatings by Schulz indicate that at 1300 °C and 1400 °C the metastable tetragonal phase decomposes into a new low-yttria content tetragonal phase and a high-yttria content cubic or tetragonal phase.[6,7] Computer simulation, combined with TEM studies has shown that the domain boundaries have a cubic-like structure and that yttrium ions tend to concentrate in these boundaries.[4]

In the present work, the microstructure evolution in plasma sprayed YSZ-TBC during annealing at temperature ranging from 1100 °C to 1400 °C is studied. Lattice parameters, phases content, strains and crystallite size of as sprayed and annealed coatings are determined using Rietveld refinement of XRD patterns.

EXPERIMETAL PROCEDURE

YSZ coatings were produced by an atmospheric plasma spraying system (APS). The feedstock material is a zirconia with 7.8 wt% yttria powder (204NS, Sulzer-Metco, Westbury, NY). Samples were mechanically removed from the substrate before annealing at temperatures ranging from 1100 °C to 1400 °C. After annealing samples were furnace cooled down to room temperature such that the transformation from tetragonal to monoclinic YSZ can occur during cooling. XRD patterns were collected using an X-ray powder diffractometer (X'Pert Pro, PANalytical, Almelo, The Netherlands) with monochromatic CuKα_1 radiation, within 2θ=20 to 100° range in 0.01° steps using standard θ–2θ Bragg–Brentano geometry. Rietveld refinement was performed using the General Structure Analysis System (GSAS) package and the EXPGUI interface.[8,9] The Rietveld refinement was processed following the Rietveld refinement guidelines formulated by the International Union of Crystallography Commission on Powder Diffraction.[10] The background was fitted using Chebyschev polynomials and the peak profiles were fitted using a convolution of a pseudo-Voigt and asymmetry function together with the microstrain broadening description of P. Stephens.[11-13] The pseudo-Voigt is defined as a linear combination of a Lorentzian and a Gaussian. The Gaussian variance of the peak, σ^2, varies with 2θ as:

$$\sigma^2 = U \tan^2 \theta + V \tan \theta + W + \frac{P}{\cos^2 \theta} \tag{1}$$

where U, V and W are the coefficients described by Cagliotti et al. and P is the Scherrer coefficient for Gaussian broadening.[14] The Lorentzian coefficient, γ, varies as:

$$\gamma = \frac{X}{\cos \theta} + Y \tan \theta \tag{2}$$

The P and X coefficients are related to the Scherrer broadening and gives information about crystallite dimensions. The U and Y coefficients are related to the strain broadening and allow estimating micro-strains within crystallites:

$$\varepsilon_G(\%) = \frac{\pi}{180} \sqrt{8 \ln 2 (U - U_i)} \tag{3}$$

$$\varepsilon_L(\%) = \frac{\pi}{180}(Y - Y_i) \tag{4}$$

The surface roughness and porosity of the coating leads to absorption of the X-Ray beam and reduction of the diffracted intensities at low angles. If this effect is not taken into account during Rietveld refinement, the atomic displacement parameters are reduced and can even become negative having then no more physical meaning. To take into account the absorption effect, we used the function described by Pitschke et al.[15] Since atomic displacement parameters and absorption factors modify the profile intensities with a similar 2θ dependency and refining all of them together usually leads to unrealistic results, we fixed the atomic displacement parameters to standard values and refined only the absorption factors. The number of refined parameters during the last step of the refinement process and R_p factors indicating the quality of the fit compared to the experimental XRD pattern are listed in Table I, the R_p values are calculated after removal of the background contribution.

Table I. Data on annealing conditions and Rietveld refinement

Temperature °C	Annealing time h	Number of refined parameters	R_p (without background)
-	-	27	7.8%
1400	1	33	6.9%
1400	10	33	6.5%
1300	1	27	7.4%
1300	10	33	6.4%
1300	100	33	5.7%
1300	1000	32	7.3%
1200	1	27	7.6%
1200	10	27	8.3%
1200	1000	33	7.6%
1100	1	27	6.7%
1100	24	27	6.8%
1100	100	27	6.0%
1100	250	33	6.3%
1100	450	33	6.1%
1100	650	33	6.1%
1100	850	33	6.6%
1100	1400	33	7.7%

Four YSZ phases were used for interpretation of the XRD patterns: two tetragonal phases with different unit cell dimensions corresponding to different yttria content of 7-8 wt% (t′-YSZ) and 4-5 wt% (t-YSZ), one monoclinic (m-YSZ) and one cubic phase having 13-15 wt% yttria (c-YSZ). For the coatings annealed at 1400 °C, the pattern of the phase with high yttria content (11-13 wt% yttria) was fitted using a tetragonal phase with a low c/a ratio, we will refer to this phase as c-YSZ to be consistent with the phase definitions used for lower annealing temperature. In some patterns, the intensities of peaks of some phases are too low to allow a structural refinement. In such cases, the unit cell dimensions are fixed by setting them to a defined yttria content. The peak shapes parameters are then either fixed or the number of refined parameters is reduced to 1 or 2. The yttria content of the cubic phase is calculated using the formula:

$$YO_{1.5}(mol\%) = (a - 5.1159)/0.001547 \qquad (5)$$

derived by Ilavsky[5] from the data of Scott[1], where a is the unit cell dimension in angstroms. The yttria content of the tetragonal phase is calculated using the formula:

$$YO_{1.5}(mol\%) = \frac{1.0225 - \dfrac{c}{a\sqrt{2}}}{0.0016} \qquad (6)$$

where a and c are the unit cell dimensions expressed in angstroms. This formula was derived from data of Scott[1] and empirically corrected by Ilavsky[16] to obtain a better fit with data coming from various samples.

RESULTS AND DISCUSSION

XRD patterns of the YSZ coating annealed at 1300 °C for various times are displayed in Figure 2. The pattern of the as-deposited coating contains only peaks belonging to t′-YSZ. After 10 hours of annealing, shoulders appear in Figure 2b at lower angles for the (004) peak at 2θ=73.2° and to higher angles for the (220) peak at 2θ=74.2°, they belong to the t-YSZ which has a higher c/a ratio. A new peak also appears between the (004) and (220) peaks of t′-YSZ indicating the presence of c-YSZ. After 100 hours of annealing, peaks of t-YSZ and c-YSZ grow when those of t′-YSZ are getting smaller, and after 1000 hours of annealing the peaks of t′-YSZ completely disappeared. In Figure 2a, peaks attributed to m-YSZ phase also appear after 1000 hours of annealing. The peak present at 2θ=30.2° is a combination of the (101) peaks of both tetragonal YSZ phases and the (111) peak of c-YSZ. This peak appears at the lowest angle for c-YSZ and at the highest angle for t′-YSZ, explaining its asymmetric shape.

These observations show that fitting XRD patterns using a combination of two tetragonal YSZ phases having different yttria content and unit cell dimensions is a good way to obtain a realistic description of the YSZ sample composition. Using only one unique tetragonal YSZ phase as was done by Ilavsky et al.[5] is not a good solution, since as can be observed in Figure 2b, peaks of the tetragonal phase are not shifted during annealing due to a steady loss of yttria, but a new set of peaks appears while the peaks of the as-deposited YSZ disappear. Using more than two tetragonal YSZ phases did not improve the quality of the fit, the phase contents obtained for phases having intermediate yttria contents were always close to 0 wt%. This indicates that we do not observe YSZ phases having intermediate yttria content, but only the as-deposited phase or the decomposition products, as was already observed by Brandon and Taylor in plasma sprayed coatings annealed at 1300 °C and higher temperatures.[6]

Evolution of the phase composition of the coatings during annealing at different temperatures is displayed in Figure 3 and the phase composition, cell parameters and yttria content of each phase are listed in Table II and III. We always observe decomposition of t′-YSZ into c-YSZ and t-YSZ with a decomposition rate increasing by around one order of magnitude when the temperature is increased by 130 °C. After 1000 hours of annealing at 1300 °C the m-YSZ appears; at the same time the peaks of t′-YSZ are no more visible and the amount of the t-YSZ starts to decrease.

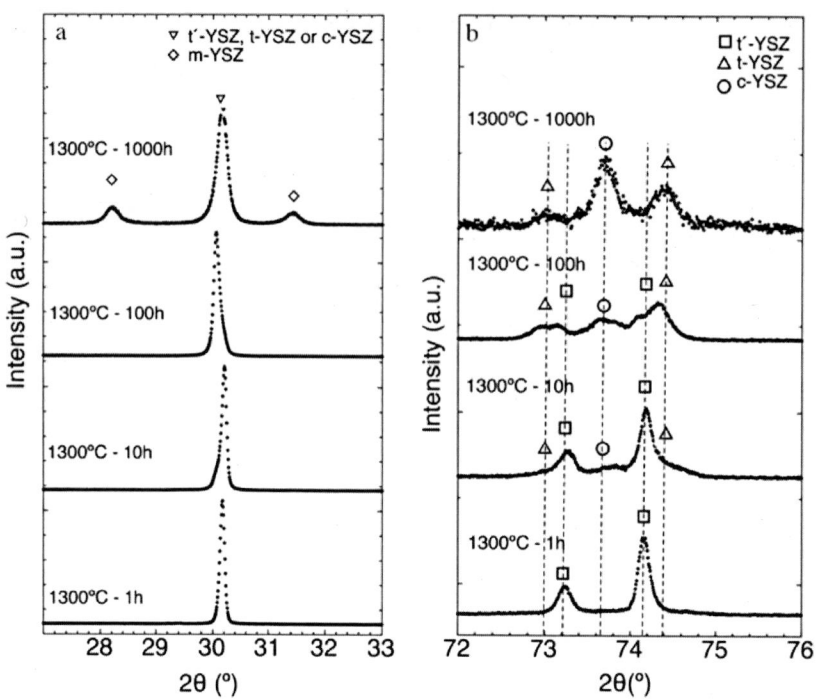

Fig. 2. a) XRD patterns of YSZ coatings, annealed for various times at 1300 °C, for a 2θ range between 27° and 33°. b) XRD patterns of YSZ coatings, annealed for various times at 1300 °C, for a 2θ range between 72° and 76°.

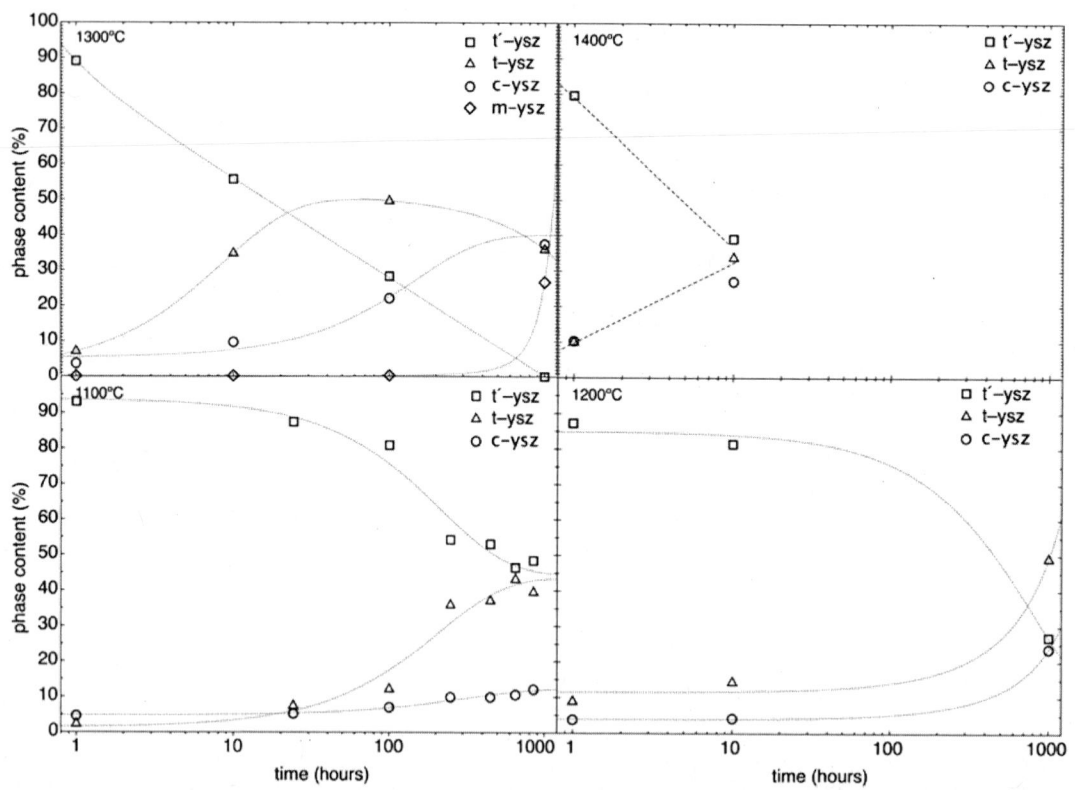

Fig. 3. Evolution of phase composition of YSZ coatings during annealing at different temperatures

Table II. Data on YSZ phases obtained from Rietveld refinement

Tempe rature (°C}	Annea ling time (h)	t'-YSZ wt %	t-YSZ wt %	c-YSZ wt %	m-YSZ wt %	a t'-YSZ (Å)	c t'-YSZ (Å)	a t-YSZ (Å)	c t-YSZ (Å)	c c-YSZ (Å)
-	-	88.8	7.5	3.7	-	3.6150	5.1617	3.6044	5.1709	5.1325
1400	1	79.6	10.0	10.4	-	3.6132	5.1654	3.6006	5.1780	5.1467/ 3.6246
1400	10	39.3	33.5	27.2	-	3.6109	5.1634	3.6005	5.1783	5.1456/ 3.6248
1300	1	89.2	7.1	3.7	-	3.6144	5.1664	3.6045	5.1709	5.1353
1300	10	52.3	38.4	9.3	-	3.6134	5.1647	3.6057	5.1687	5.1353
1300	100	28.3	49.6	22.1	-	3.6138	5.1665	3.6029	5.1741	5.1361
1300	1000	-	35.9	37.4	26.7	-	-	3.6066	5.1774	5.1395
1200	1	87.2	8.9	3.8	-	3.6142	5.1669	3.6045	5.1709	5.1353
1200	10	81.5	14.3	4.2	-	3.6142	5.1671	3.6045	5.1709	5.1353
1200	1000	27.1	49.2	23.7	-	3.6143	5.1685	3.6031	5.1823	5.1403
1100	1	93.1	2.3	4.6	-	3.6121	5.1638	3.6041	5.1689	5.1373
1100	24	87.4	7.4	5.2	-	3.6125	5.1643	3.6041	5.1689	5.1373
1100	100	80.9	12.1	7.0	-	3.6118	5.1647	3.6041	5.1689	5.1373
1100	250	54.2	35.9	9.9	-	3.6116	5.1645	3.6041	5.1689	5.1373
1100	450	53.0	37.1	9.9	-	3.6117	5.1650	3.6041	5.1689	5.1373
1100	650	46.4	43.1	10.6	-	3.6108	5.1632	3.6041	5.1689	5.1373
1100	850	48.4	39.5	12.1	-	3.6117	5.1657	3.6041	5.1689	5.1373
1100	1400	42.3	45.0	12.7	-	3.6102	5.1638	3.6032	5.1672	5.1373

Curves describing the time evolution of t'-YSZ decomposition level can be scaled such to fall on a master curve. To do this, time scale was modified with an equation of the form:

$$t^* = At \cdot e^{-\frac{B}{T}} \tag{7}$$

where t is the annealing time and T is the temperature in Kelvin. The master curve is displayed in Figure 4. The time scaling used allow that all data fall well on the same curve and it allows to estimate how fast t'-YSZ will decompose at a given temperature.

Information about micro-strains and grain size was extracted from the function describing the Gaussian part of the diffraction peak shape. It indicates that peak broadening is mainly governed by micro-strains within the crystallites. The micro-strain level in t'-YSZ is around 0.2-0.4% for short annealing times and increases up to 1% upon further annealing as can be seen in Figure 5. After 1 hour of annealing the micro-strains in the coatings annealed at 1300 °C and 1400 °C are lower than in the coating annealed at lower temperatures, this can be explained by the high level of micro-strains of 0.48% in the as-deposited coating that are released during the first hours of annealing at high temperature. The general subsequent increase in micro-strains can be related to the reduction in t'-YSZ phase content leading to an increase in straining due to the other phases present in the coating. In the c-YSZ and t-YSZ phases, micro-strains vary between 0.4% and 0.7% and decrease upon annealing. The micro-strain level that is reached in the t'-YSZ phase upon annealing at high temperature is high enough to allow cracks nucleation and their subsequent propagation. But it is probably overestimated because the micro-strain broadening is described by a distribution of cell dimensions; in the case of YSZ, this distribution can also come from a non-homogeneous yttria distribution within the grains or between the grains. Therefore, the increase of the observed micro-strains could also be interpreted as an effect of yttria diffusion, which leads to a broader distribution of the yttria content within t′-YSZ crystallites.

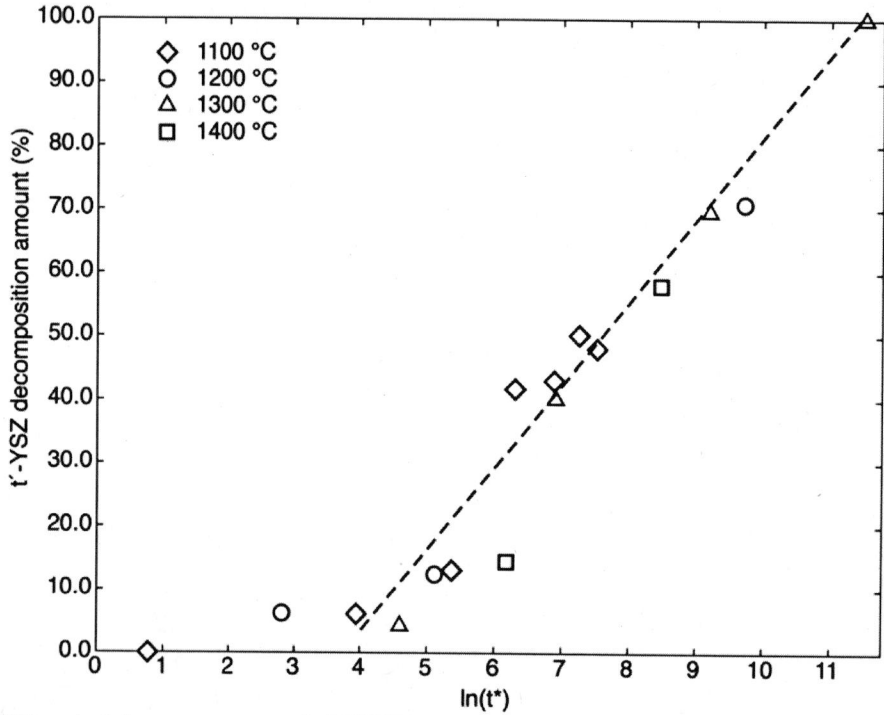

Fig. 4. Master curve of t′-YSZ content as a a function of equivalent annealing time at 1100 °C.

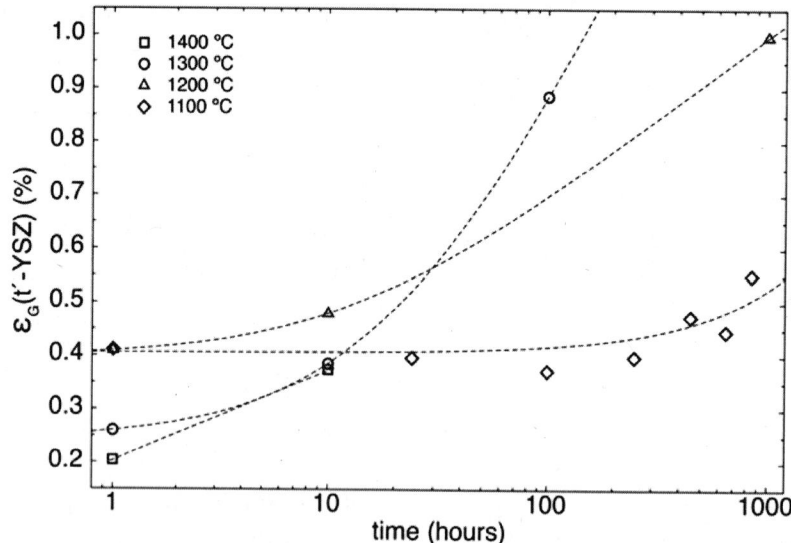

Fig. 5. Evolution of micro-strains in the t′-YSZ phase as a function of annealing time.

The absence of t′-YSZ undergoing the decomposition process indicates that the diffusion of yttrium should happen on a short time scale. This observation together with the diffraction peak profiles showing that the broadening comes mainly from a high level of strains or a non-homogeneous yttria distribution in the particles, are in agreement with previous TEM studies by Shibata et al.[17] and the spinodal decomposition model proposed by Katamura et al.[4] According to this model, yttrium segregates at domain boundaries producing grains composed of domains having the cubic structure with high yttria content and other domains having the tetragonal structure and a reduced yttria content. Only when the size of yttria-depleted domains reaches some critical size, they undergo the martensitic transformation to the monoclinic YSZ structure.[18] One can write the decomposition process as:

$$t′\text{-YSZ} \to t\text{-YSZ} + c\text{-YSZ} \to m\text{-YSZ} + c\text{-YSZ} \tag{8}$$

It is still not clear if the remaining traces of t′-YSZ come from grains having a slower decomposition rate or if they are equally distributed in grains undergoing the decomposition process. It is also not yet clear if the m-YSZ is forming only after the t′-YSZ is fully decomposed. The answer to this question can be important for the consideration of the stabilization mechanism; it can gives information if t-YSZ domains can be stabilized by the presence of adjacent t′-YSZ domains or if the transformation from tetragonal to monoclinic YSZ is only governed by the size of the t-YSZ domains.

Table III. Data on YSZ phases obtained from Rietveld refinement

Temperature (°C)	Annealing time (h)	Yttria content t′-YSZ	Yttria content t-YSZ	Yttria content c-YSZ	U t′-YSZ	ε t′-YSZ
-	-	8.0%	5.0%	10.7%	135	0.48%
1400	1	7.3%	3.4%	13.2%	24.6	0.20%
1400	10	7.1%	3.4%	12.9%	82.8	0.38%
1300	1	7.4%	5.1%	12.5%	40.1	0.26%
1300	10	7.4%	5.0%	12.5%	87.9	0.39%
1300	100	7.2%	4.2%	13.1%	465	0.89%
1300	1000	-	4.6%	15.3%	-	-
1200	1	7.3%	5.0%	12.5%	99.2	0.41%
1200	10	7.2%	5.0%	12.5%	136	0.48%
1200	1000	7.1%	3.4%	15.8%	590	1.00%
1100	1	7.3%	5.3%	13.8%	101	0.42%
1100	24	7.3%	5.3%	13.8%	92.8	0.41%
1100	100	7.1%	5.3%	13.8%	82.1	0.40%
1100	250	7.1%	5.3%	13.8%	94.3	0.37%
1100	450	7.1%	5.3%	13.8%	133	0.40%
1100	650	7.1%	5.3%	13.8%	118	0.48%
1100	850	7.0%	5.3%	13.8%	180	0.55%
1100	1400	6.9%	5.3%	13.8%	191	0.57%

CONCLUSIONS

X-Ray diffraction combined with Rietveld refinement of full patterns allowed monitoring of evolution of phase composition in YSZ plasma sprayed coatings. When the coating is annealed at temperature ranging between 1100 °C and 1400 °C, the as-deposited tetragonal YSZ phase containing 7-8 wt% of yttria decomposes into a low yttria content tetragonal YSZ phase with 4-5 wt% of yttria and a high yttria content YSZ phase. This high yttria content YSZ phase is either cubic and contains 13-15 wt% of yttria when the annealing is performed at 1300 °C and lower temperatures or is tetragonal and contains 11-13 wt% of yttria when the annealing is performed at 1400 °C. When the domins of the low yttria content tetragonal YSZ phase have grown enough in size they transform into monoclinic YSZ upon colling. The kinetic of the phase transformation depends on the annealing temperature and is increased by around one order of magnitude when the temperature is increased by 130 °C. The decomposition of the as-deposited metastable YSZ phase comes together with an increase of its micro-strains having values high enough to initiate cracks that after propagation would lead to the coating failure.

REFERENCES

[1] H. G. Scott, "Phase relationships in the zirconia-yttria system", *J. Mater. Sci.*, **10**, 1527-35 (1975)

[2] D. R. Clarke and C. G. Levi, "Material Design for the Next Generation Thermal Barrier Coatings", *Annu. Rev. Mater. Res.*, **33**, 383-417 (2003)

[3] R. A. Miller, J. L. Smialek, and R. G. Garlick, "Phase Stability in Plasma-Sprayed, Partially Stabilized Zirconia–Yttria"; pp. 241–53 in Advances in Ceramics, Vol. 3, *Science and Technology of Zirconia I*. Edited by A. H. Heuer and L. W. Hobbs. American Ceramic Society, Columbus, OH, 1981

[4] J. Katamura and T. Sakuma, "Computer Simulation of the Microstructural Evolution during the Diffusionless Cubic-to-Tetragonal Transition in the System $ZrO_2-Y_2O_3$", *Acta mater.*, **46** [5], 1569-75 (1998)

[5] J. Ilavsky, J. K. Stalick, and J. Wallace, "Thermal Spray Yttria-stabilized Zirconia Phase Changes during Annealing", *J. Therm. Spray Technol.*, **10** [3], 497-501 (2001)

[6] J. R. Brandon and R. Taylor, "Phase stability of zirconia-based thermal barrier coatings Part I, Zirconia-yttria alloys", *Surf. Coat. Technol.*, **46**, 75-90 (1991)

[7] U. Schulz, "Phase Transformation in EB-PVD Yttria Partially Stabilized Zirconia Thermal Barrier Coatings during Annealing", *J. Am. Ceram. Soc.*, **83** [4], 904 –10 (2000)

[8] A. C. Larson and R. B. Von Dreele, "General Structure Analysis System (GSAS)", *Los Alamos National Laboratory Report* LAUR 86-748 (2000)

[9] B. H. Toby, "EXPGUI, a graphical user interface for GSAS", *J. Appl. Cryst.*, **34**, 210-13 (2001)

[10] L. B. McCusker, R. B. Von Dreele, D. E. Cox, D. Louër, and P. Scardi, "Rietveld refinement guidelines", *J. Appl. Cryst.* **32**, 36-50 (1999)

[11] P. Thompson, D. E. Cox and J. B. Hastings, J. Appl. Cryst., "Rietveld refinement of Debye-Scherrer synchrotron X-ray data from Al_2O_3", *J. Appl. Cryst.*, **20**, 79-83 (1987)

[12] L. W. Finger, D. E. Cox and A. P. Jephcoat, "A correction for powder diffraction peak asymmetry due to axial divergence", *J. Appl. Cryst.*, **27**, 892-900 (1994)

[13] P. Stephens, "Phenomenological model of anisotropic peak broadening in powder diffraction", *J. Appl. Cryst.*, **32**, 281-89 (1999).

[14] G. Caglioti, A. Paoletti and F. P. Ricci, "Choice of collimators for a crystal spectrometer for neutron diffraction", *Nucl. Instrum.*, **3**, 223-28 (1958)

[15]W. Pitschke, H. Hermann, and N. Mattern, "The influence of surface roughness on diffracted X-ray intensities in Bragg-Brentano geometry and its effect on the structure determination by means of Rietveld analysis", *Powder Diffr.*, **8**, 74-83 (1993)

[16]J. Ilavsky and J. K. Stalick, "Phase composition and its changes during annealing of plasma-sprayed YSZ", *Surf. Coat. Technol.*, **127** [2-3], 120-29 (2000)

[17]N. Shibata, J. Katamura, A. Kuwabara, Y. Ikuhara, T. Sakuma, "The instability and resulting phase transition of cubic zirconia", *Mater. Sci. Eng.*, **A312**, 90–8 (2001)

[18]T. K. Gupta, F. F. Lange, J. H. Bechtold, "Effect of stress-induced phase transformation on the properties of polycrystalline zirconia containing metastable tetragonal phase", *J. Mat. Sci.* **13**, 1464-70 (1978)

THERMAL IMAGING CHARACTERIZATION OF THERMAL BARRIER COATINGS

J. G. Sun
Argonne National Laboratory
Argonne, IL 60439

In a three-layer thermal barrier coating (TBC) system consisting of a ceramic TBC topcoat, a bond coat, and a metallic substrate, a large disparity in thermal conductivity exists between the TBC and the substrate and, when TBC is debonded and air fills the gap, between the TBC and the air. For TBC system characterization, flash thermal imaging is effective because it involves nondestructive measurement of thermal properties. This paper describes a new thermal-imaging method for multilayer TBC characterization and imaging to simultaneously determine the TBC thickness, conductivity, and optical absorptance. This method directly accounts for the TBC translucency that has been a major issue in thermal imaging application for TBCs. Results from theoretical analyses and experimental measurements are presented and discussed.

INTRODUCTION

Thermal barrier coatings (TBCs) are being extensively used for improving the performance and extending the life of combustor and gas turbine components. In this application, a thermally insulating ceramic topcoat (the TBC) is bonded to a thin oxidation-resistant metal coating (the bond coat) on a metal substrate. Because TBCs play critical role in protecting the substrate components, their failure (spallation) may lead to unplanned outage or safety threatening conditions. Therefore, it is important to inspect and monitor the TBC condition to assure its quality and reliability.

Most TBCs consist of yttria-stabilized zirconia (YSZ). They are usually applied by electron-beam physical vapor deposition (EB-PVD) or air plasma spay (APS) to thicknesses ranging from 0.1 to >2 mm. TBC failure during high-temperature operations typically starts from initiation of small cracks at the TBC/bond coat interface. These cracks then grow and link together to form delaminations under the TBC which will eventually cause the TBC spallation. To monitor this TBC failure process and detect TBC delamination, several nondestructive evaluation (NDE) methods have been developed, most are based on optical principles because TBCs are typically either semi-transparent (most EB-PVD TBCs) or translucent (APS TBCs). These optical NDE methods include mid-infrared reflectance [Eldridge et al., 2006], luminescence spectroscopy [Tolpygo et al., 2004], and elastic optical scattering [Ellingson et al., 2006]. These methods have been demonstrated to be capable of detecting TBC degradation and pre-spall conditions. However, they can only be used for semi-transparent EB-PVD TBCs or thin APS TBCs (<0.4 mm) because of the limited optical penetration depth of the lights used in these methods. In addition, these methods are qualitative and may become useless once a TBC is coated or infiltrated by "dirty" contaminants.

In the three-layer TBC system consisting of a TBC topcoat, a bond coat, and a metal substrate, a large disparity in thermal conductivity exists between the TBC and the substrate and, when TBC is delaminated and air fills the gap, between the TBC and the air. Therefore, pulsed (or flash) thermal imaging is effective for TBC system characterization because it involves nondestructive measurement of thermal properties. Thermal imaging has been widely used to detect TBC delamination [e.g., Chen et al., 2001]. Recently, it has also been extended for estimation of TBC thickness and thermal conductivity [Shepard et al., 2005; Ringermacher, 2004]. However, because TBC system is multilayer and its top layer (TBC) is translucent, conventional methods for pulsed thermal imaging cannot be used directly to analyze the TBC system. This paper describes a new thermal-imaging technology for multilayer TBC characterization and imaging. Results from experimental and theoretical analyses are presented and discussed.

PULSED THERMAL IMAGING METHODS FOR SINGLE- AND MULTI-LAYER MATERIALS

Pulsed thermal imaging is based on monitoring the temperature decay on a specimen surface after it is applied with a pulsed thermal energy that is gradually transferred inside the specimen. The premise is that the heat transfer from the surface (or surface temperature/time response) is affected by internal material structures and properties and the presence of flaws such as cracks [Sun, 2006a]. A schematic one-sided pulsed-thermal-imaging setup for testing a 3-layer material system is illustrated in Fig. 1. Theoretical development for analyzing material properties from thermal imaging data is described below.

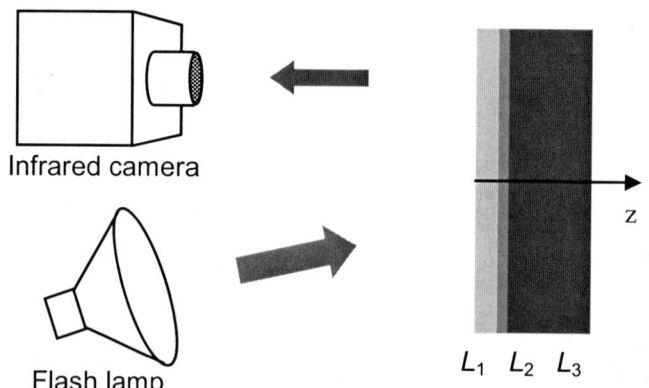

Fig. 1. Schematics of pulsed thermal imaging of a 3-layer material system.

Because thermal imaging is inherently a 2D method (it images the temperature of the 2D x-y specimen surface), theoretical development is usually carried out in 1D (the depth or z direction) models. The temperature/time response at a surface position (a pixel in a 2D image) is related to the depth variation of material properties under that surface position. By analyzing the surface temperature/time response, the material property and depth of various subsurface layers under that pixel can be determined. The final thermal imaging results for all pixels are usually presented in image forms; the value at each pixel represents a particular physical parameter such as thermal conductivity or depth that was determined from the analysis. To understand the thermal responses due to material thermal and optical properties and depth, heat conduction theory is examined first.

The 1D governing equation for heat conduction in a solid material is:

$$\rho c \frac{\partial T}{\partial t} = \frac{\partial}{\partial z}\left(k \frac{\partial T}{\partial z} \right),$$

(1)

where $T(z, t)$ is temperature, ρ is density, c is specific heat, k is thermal conductivity, t is time, z is coordinate in the depth direction, and $z = 0$ is the surface that receives pulsed heating. It is noted that Eq. (1) contains two independent thermal parameters, the heat capacity ρc and the thermal conductivity k, both are normally assumed constant in each material layer.

During flash thermal imaging, an impulse energy is applied on surface $z = 0$ at $t = 0$. Under ideal thermal imaging conditions which assumed (1) flash is instantaneous or flash duration is zero and (2) flash heat is absorbed on surface or heat-absorption depth is zero (for opaque materials), analytical solution of Eq. (1) for single-layer materials has been obtained by Parker et al. [1961]. Analytical solutions of Eq. (1) for single-layer materials under finite flash duration and finite heat-absorption depth (for translucent materials) were also obtained [Sun & Benz, 2004; Sun, 2006b]. These theories have been directly used for thermal imaging analysis of single-layer materials [Sun, 2006a, 2007].

Thermal imaging analysis for multilayer materials is more complex. For multilayer materials, parameters in each layer include: conductivity k, heat capacity ρc, layer thickness L, and, for translucent materials, the absorption coefficient a. In comparison, only one parameter α/L^2 ($\alpha = k/\rho c$) controls the entire heat transfer process in single-layer materials. Despite of the complexity, surface temperature decay for multilayer materials under pulsed thermography conditions has been well understood. For a 2-layer opaque material, depending on the ratio of heat conductivities between the first and second layers, k_1/k_2, the expected surface temperature decay is illustrated in Fig. 2 (in log-log scale). In the early time period, flash heat absorbed on the surface propagates within the first layer, and the surface temperature decay follows the -0.5 slope in log-log scale. When heat approaches the interface between the first and the second layer, the temperature decay rate deviates from the -0.5 slope if $k_1/k_2 \neq 1$. The temperature decay within this intermediate time period is therefore indicative of the interface condition. In later times, heat propagation proceeds in the second layer so the temperature decay rate is determined by the conductivity ratio k_1/k_2: when $k_1/k_2 < 1$ the (absolute) magnitude of the slope is >0.5, and when $k_1/k_2 > 1$ the (absolute) slope amplitude is <0.5. The surface temperature decay rate will eventually approaches to zero when the temperature of the entire specimen becomes equalized.

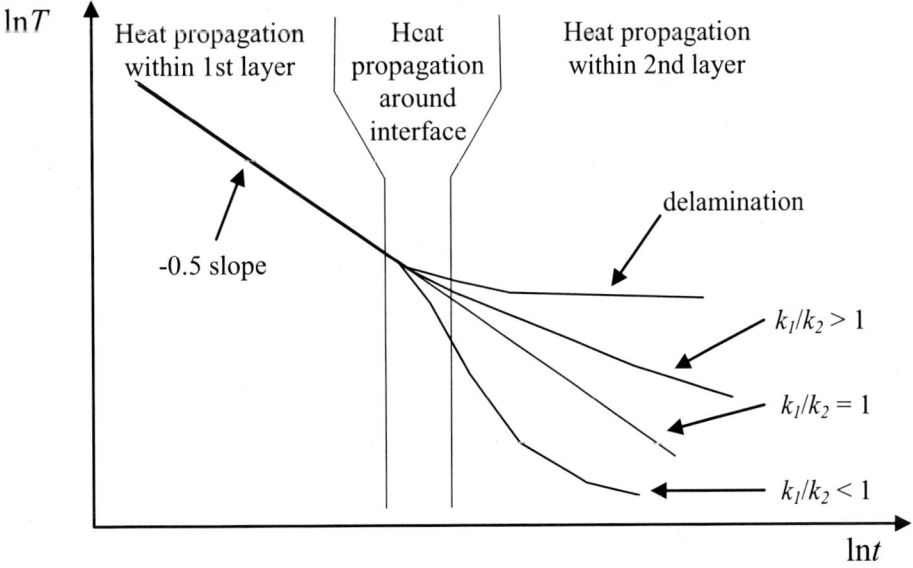

Fig. 2. Illustration of surface temperature decay from pulsed thermal imaging for a 2-layer material system.

For multilayer materials, direct solution of Eq. (1) is possible. Balageas et al. [1986] derived analytical solutions for two- and three-layer materials. However, these solutions are difficult to be used for general applications because a large number of eigenfunctions are involved. New robust and efficient methods are needed for thermal imaging analysis of multilayer material systems.

A general method and numerical algorithm has been developed for automated analysis of thermal imaging data for multilayer materials [Sun, 2006b]. It is based on a theoretical model of the material system which is solved numerically. The numerical formulation also incorporates finite flash duration and finite heat absorption depth effects. The numerical solutions are fitted with the experimental data by least-square minimization to determine unknown parameters in the multilayer material system. Multiple parameters in one or several layers can be determined simultaneously. For a TBC system, the most important parameters are the thickness, thermal conductivity, and absorption coefficient of the TBC in the first layer.

The least-square fitting is carried out for each pixel within the 2D thermal imaging data, and each fitted parameter is expressed in a 2D image. This data analysis process has been fully automated to simultaneously determine the distributions (images) of TBC thickness, conductivity, and absorption coefficient [Sun, 2006b]. Typical results for TBC analysis are presented below.

THERMAL IMAGING ANALYSIS OF TBC MATERIALS

For a TBC system, because the bond coat is typically thin and has thermal properties comparable to those in the substrate, thermal imaging analysis can be carried out for a two-layer material consisting of a TBC and a substrate. In this system, the conductivity of the TBC is typically much lower than that of the substrate, i.e., $k_1/k_2 < 1$. Therefore, the surface temperature decay should follow the curve for $k_1/k_2 < 1$ in Fig. 2 for flash thermal imaging of opaque TBCs. However, TBCs are typically translucent at levels determined by the amount of contamination. Because TBC translucency affects the heat absorption and infrared detection during a thermal imaging test, it must be determined explicitly in order for accurate prediction of other TBC parameters.

The numerical thermal imaging method described above is used to analyze a 2-layer TBC system to demonstrate its sensitivity and accuracy for determining TBC thickness, conductivity, and optical absorption coefficient. The results will be compared with thermal imaging data for a TBC specimen presented in the next section. In this 2-layer TBC system, the substrate is assumed to have constant thermal properties: $k_2 = 8$ W/m-K, $\rho c = 4$ J/cm^3-K, and thickness $L_2 = 2.5$ mm. The TBC has generic properties: $k_1 = 1.3$ W/m-K, $\rho c = 3$ J/cm^3-K, $L_1 = 0.62$ mm, and optical absorption coefficient $a = 4$ mm^{-1}.

Figure 3 shows the calculated results for TBCs with different optical absorption coefficient a. The TBC translucency (i.e., a finite a) can significantly reduce the temperature decay rate $d(\ln T)/d(\ln t)$ in the early times. When the absorption coefficient approaches infinity (for opaque TBCs), the initial surface-temperature slope becomes -0.5. Because TBC absorption depends on TBC material composition and structure as well as TBC surface conditions (such as contamination), this thermal imaging method can be used to analyze as-processed TBCs that have uniform optical property as well as used TBCs that may have various levels of surface contamination.

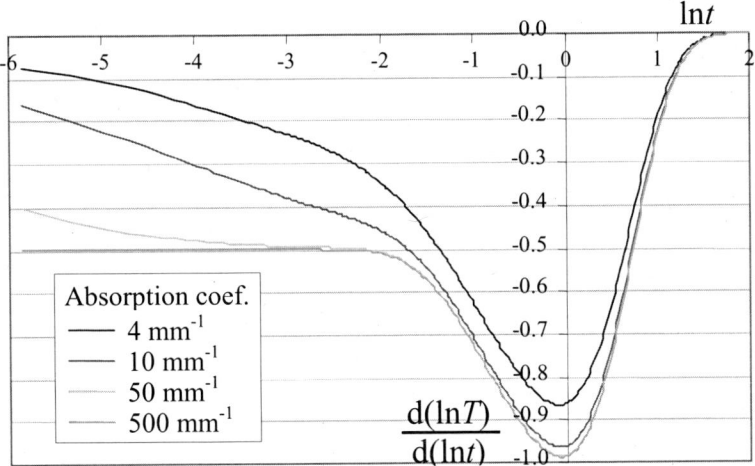

Fig. 3. Calculated surface-temperature slope as function of time (in log-log scale) for TBCs with various optical absorption coefficients.

Figure 4 shows calculated results for TBCs of different thicknesses. The surface-temperature slope $d(\ln T)/d(\ln t)$ initially follows approximately straight lines with (absolute) magnitudes below 0.5; the slope magnitude becomes larger than 0.5 when heat transfer reaches the substrate. The time when the slope change occurs is related to the TBC thickness.

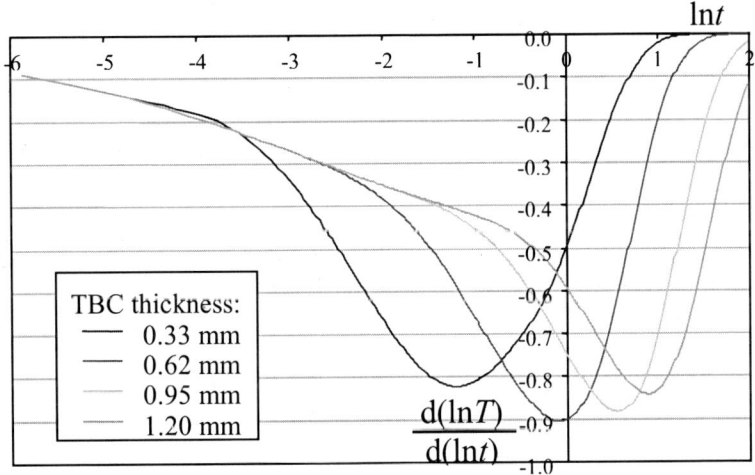

Fig. 4. Calculated surface-temperature slope as function of time (in log-log scale) for TBCs of different thicknesses.

Figure 5 shows calculated results for TBCs with different thermal conductivities. It is seen that the magnitude of the surface-temperature slope is very sensitive to the change of thermal conductivity of the TBC layer; a lower TBC conductivity will result in a higher peak magnitude of the slope.

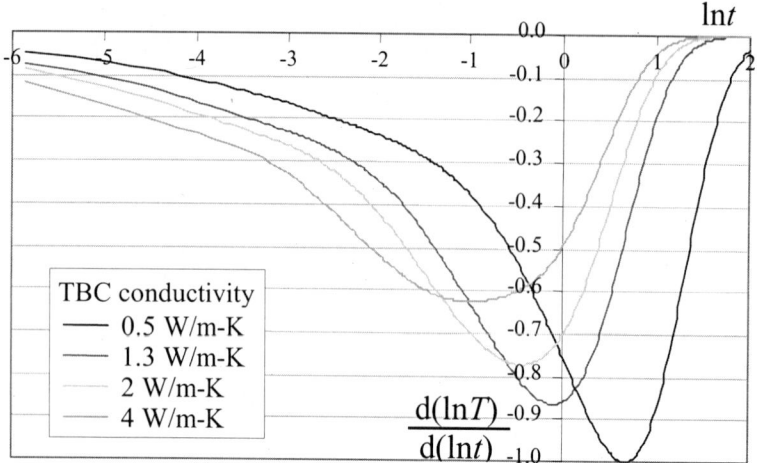

Fig. 5.   Calculated surface-temperature slope as function of time (in log-log scale) for TBCs with different conductivities.

THERMAL IMAGING RESULTS FOR A TBC SPECIMEN

Pulsed thermal imaging test was conducted for an as-processed TBC specimen shown in Fig. 6a. It consists of a nickel-based substrate of 2.5 mm thick and a TBC layer with its surface being divided into 4 sections having nominal thicknesses 0.33, 0.62, 0.95, and 1.2 mm. Because this TBC specimen is as-processed, its thermal conductivity and optical absorption coefficient are expected to be uniform. Pulsed thermal imaging data was obtained for a total duration of 13 seconds at an imaging speed of 145 Hz. A typical thermal image is shown in Fig. 6b. Figure 7 shows measured surface-temperature slopes from the 4 thickness sections of this TBC specimen. Compared with the theoretical results in Fig. 4, the experimental data in Fig. 7 clearly indicate the difference of TBC thickness in these 4 sections.

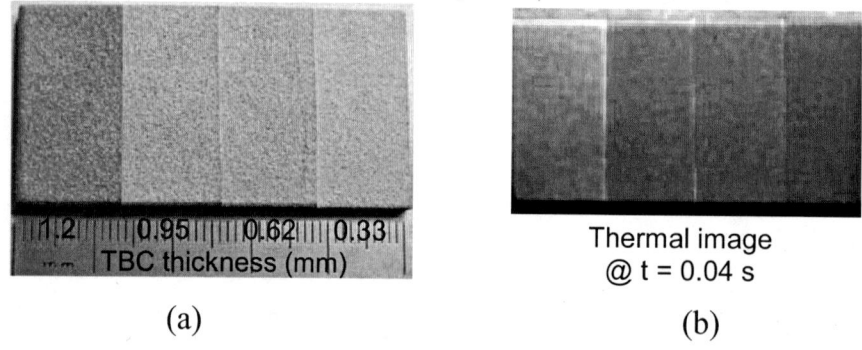

(a)   (b)

Fig. 6.   (a) Photograph and (b) thermal image of a TBC specimen with 4 sections of thicknesses.

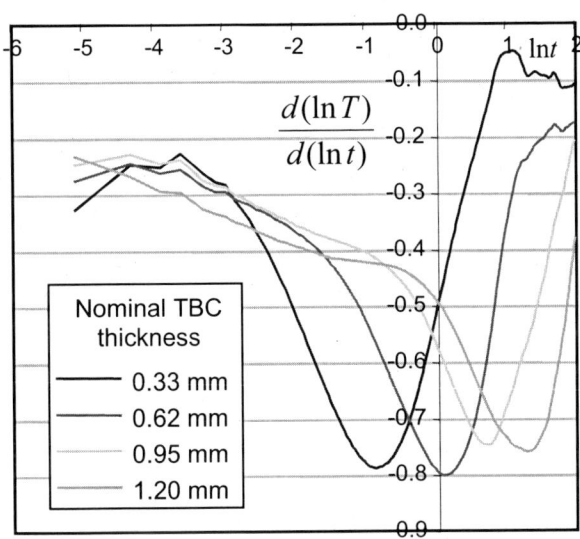

Fig. 7.   Measured surface-temperature-slope data for TBCs of different thicknesses.

CONCLUSION

A new thermal-imaging method for analyzing TBCs was developed. It is based on a theoretical model of the multilayer material system which is solved numerically. The numerical solution is fitted with experimental data by least-square minimization to determine unknown parameters. The method was used to analyze a 2-layer TBC system to demonstrate its sensitivity and accuracy for determining TBC thickness, conductivity, and optical absorption coefficient. The theoretical results agree well with thermal imaging data for a TBC specimen. This method has been fully automated to image TBC thickness, conductivity, and absorption coefficient distributions [Sun, 2006b]. Because it can accurately determine TBC thickness and conductivity variation, this method may be applied for health monitoring of TBC materials.

ACKNOWLEDGMENT

Work sponsored by the U.S. Department of Energy, Energy Efficiency and Renewable Energy, Office of Industrial Technologies, Office of Power Technologies, under Contract W-31-109-ENG-38.

REFERENCES

D. L. Balageas, J. C. Krapez, and P. Cielo, 1986, "Pulsed Photothermal Modeling of Layered Metarials," J. Appl. Phys., Vol. 59, pp. 348-357.

X. Chen, G. Newaz, and X. Han, 2001, "Damage Assessment in Thermal Barrier Coatings Using Thermal Wave Image Technique", Proc. 2001 ASME Int. Mech. Eng. Congress Expo., Nov. 11-16, 2001, New York, NY, paper no. IMECE2001/AD-25323.

J. I. Eldridge, C. M. Spuckler, and R. E. Martin, 2006, "Monitoring Delamination Progression in Thermal Barrier Coatings by Mid-Infrared Reflectance Imaging," Int. J. Appl. Ceram. Technol., Vol. 3, pp. 94-104.

W. A. Ellingson, R. J. Visher, R. S. Lipanovich, and C. M. Deemer, 2006, "Optical NDE Methods for Ceramic Thermal Barrier Coatings," Materials Evaluation, Vol. 64, pp. 45-51.

W. J. Parker, R. J. Jenkins, C. P. Butler, and G. L. Abbott, 1961, "Flash Method of Determining Thermal Diffusivity, Heat Capacity, and Thermal Conductivity," J. Appl. Phys., Vol. 32, pp. 1679-1684.

H. I. Ringermacher, 2004, "Coating Thickness and Thermal Conductivity Evaluation Using Flash IR Imaging," presented in Review of Progress in Quantitative NDE, Golden, CO, July 25-30, 2004.

S. M. Shepard, Y. L. Hou, J. R. Lhota, D. Wang, and T. Ahmed, 2005, "Thermographic Measurement of Thermal Barrier Coating Thickness," in Proc. SPIE, Vol. 5782, Thermosense XXVII, 2005, pp. 407-410.

J. G. Sun, 2006a, "Analysis of Pulsed Thermography Methods for Defect Depth Prediction," J. Heat Transfer, Vol. 128, pp. 329-338.

J. G. Sun, 2006b, "Method for Analyzing Multi-Layer Materials from One-Sided Pulsed Thermal Imaging," Argonne National Laboratory Invention ANL-IN-05-121, US patent pending.

J. G. Sun, 2007, "Evaluation of Ceramic Matrix Composites by Thermal Diffusivity Imaging," Int. J. Appl. Ceram. Technol., in press.

J. G. Sun and J. Benz, 2004, "Flash Duration Effect in One-Sided Thermal Imaging," in Review of Progress in Quantitative Nondestructive Evaluation, eds. D.O. Thompson and D.E. Chimenti, Vol. 24, pp. 650-654.

V. K. Tolpygo, D. R. Clarke, and K. S. Murphy, 2004, "Evaluation of Interface Degradation during Cyclic Oxidation of EB-PVD Thermal Barrier Coatings and Correlation with TGO Luminescence," Surf. Coat. Technol., Vol. 188-189, pp. 62-70.

EXAMINATION ON MICROSTRUCTURAL CHANGE OF A BOND COAT IN A THERMAL BARRIER COATING FOR TEMPERATURE ESTIMATION AND ALUMINUM-CONTENT PREDICTION

Mitstutoshi Okada and Tohru Hisamatsu
Central Research Institute of Electric Power Industry
2-6-1 Nagasaka
Yokosuka, 240-0196, Japan

Takayuki Kitamura
Kyoto University
Yoshida-honmachi
Kyoto, 606-8501, Japan

ABSTRACT

Specimens of superalloy with thermal barrier coating (TBC) are exposed to high-temperature atmosphere in order to develop a prediction method for local temperature and Al-content at bond coat (BC). The Al-content measured by means of an electron probe microanalyzer decreases as the test time passes. It is due to the Al transport induced by the oxidation of BC and the interdiffusion between BC and substrate. The Al-decreased layer (ADL) is formed at the boundary between BC and top coat since Al diffuses to the BC surface for the oxidation. Its thickness increases in proportion to the square root of test time, and the growth rate follows the Arrhenius relationship. Based on this relation, the local temperature of an in-service blade can be estimated by the ADL thickness if the operation period is known. The decrease of Al-content is also in proportion to the square root of test time, and Arrhenius relationship is established for the decrease rate. The prediction method of the Al-content is presented.

INTRODUCTION

In order to increase thermal efficiency of a gas turbine for electric generation, its turbine inlet temperature (TIT) has reached 1500°C at present [1]. Particularly, hot-gas-path parts such as combustors, vanes and blades are exposed to combustion gas flow, which is critical environment. For their reliability and reduction of maintenance cost, development of life evaluation method is inevitable.

Thermal barrier coating (TBC) as well as the internal cooling plays an important role as the gas temperature increases. Since it is difficult to measure the surface temperature of the hot-gas-path parts, the accurate estimation of the temperature distribution is important for life evaluation. Several estimation methods based on the microstructural change of substrates or coatings have been proposed. The temperature estimation method based on the diameter of γ' precipitate was presented for Ni-base superalloys, which are widely used for turbine blades [2-4]. However its growth rate changes due to the coalescence with neighboring γ', the influence on the temperature estimation has not been examined. On the other hand, an estimation method based on the microstructural change of coating (corrosion-resistant coating, MCrAlY) was proposed [2, 5]. Although the material for BC is often almost same as that for the corrosion-resistant coating, the estimation method by means of its microstructural change has not been developed. For the TBC, the methods focused on the oxide of BC (TGO) [6, 7] and top coat porosity [8] have been

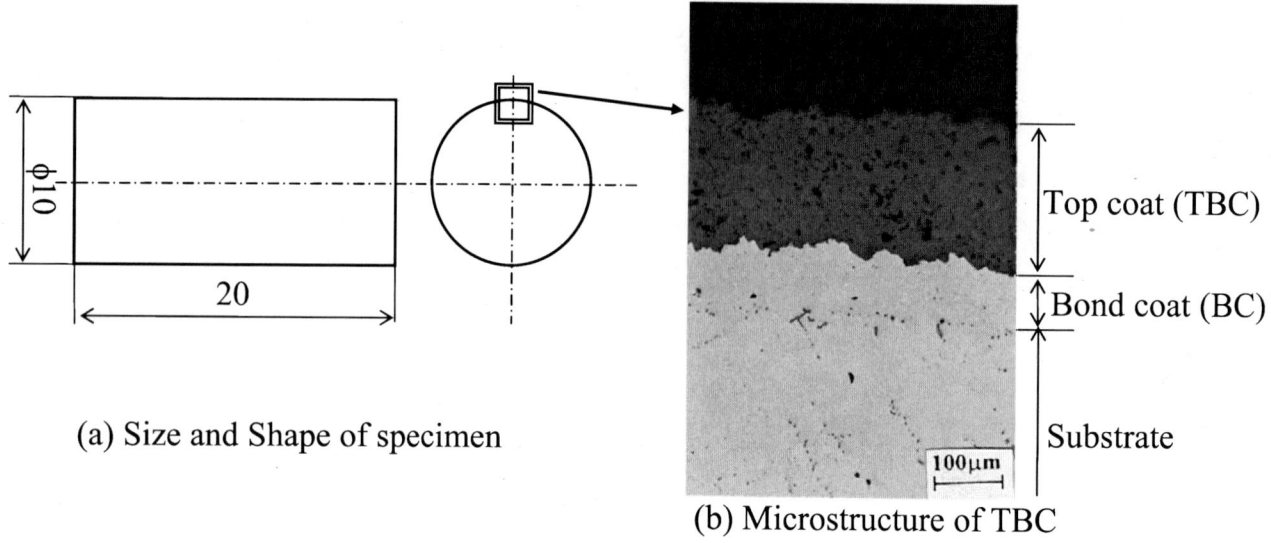

(a) Size and Shape of specimen

(b) Microstructure of TBC

Figure 1. Schematic representation of specimen.

Table I. Chemical composition of Inconel738LC

wt%

C	Si	Mn	P	S	Ni	Cr	Mo	Co
0.09	0.02	0.01	<0.005	0.001	Bal.	16.00	1.70	8.48

W	Al	Ti	Fe	Ta	Cu	Ag	Bi
2.54	3.52	3.45	0.06	1.74	<0.01	<0.5 ppm	<0.1 ppm

reported. Both of them, however, have not been examined in terms of applicable limits and accuracy.

The BC oxidation causes TBC delamination under high-temperature environment [9-12]. Particularly, the decrease of Al-content in BC accelerates the delamination since it promotes the oxidation. There are few researches on the microstructural change due to Al diffusion caused by the oxidation [13-15].

This paper clarifies the microstructural change of BC, and proposes an estimation method of local temperature. The decrease of Al-content in BC, moreover, is examined.

EXPERIMENTAL PROCEDURE
Specimen

Figure 1(a) shows the size and shape of specimen. The cylindrical substrate, 10mm in diameter and 20mm in length, is made of an Inconel738LC, which is a typical material for gas turbine blade. Table I indicates its chemical composition. Figure 1(b) represents the microstructure of TBC. BC of CoNiCrAlY (Co-32Ni-21Cr- 8Al-0.5Y (wt %)) with the thickness of 100μm is formed on the substrate by the low pressurized plasma spraying (LPPS). The heat treatment is carried out after the spraying at 1393K × 2h and 1118K × 24h in a vacuum. Then, top coat (TBC) of yttria partially-stabilized zirconia (YSZ, 8wt%Y$_2$O$_3$-ZrO$_2$) with the thickness of 200μm is deposited by the air plasma spraying (APS).

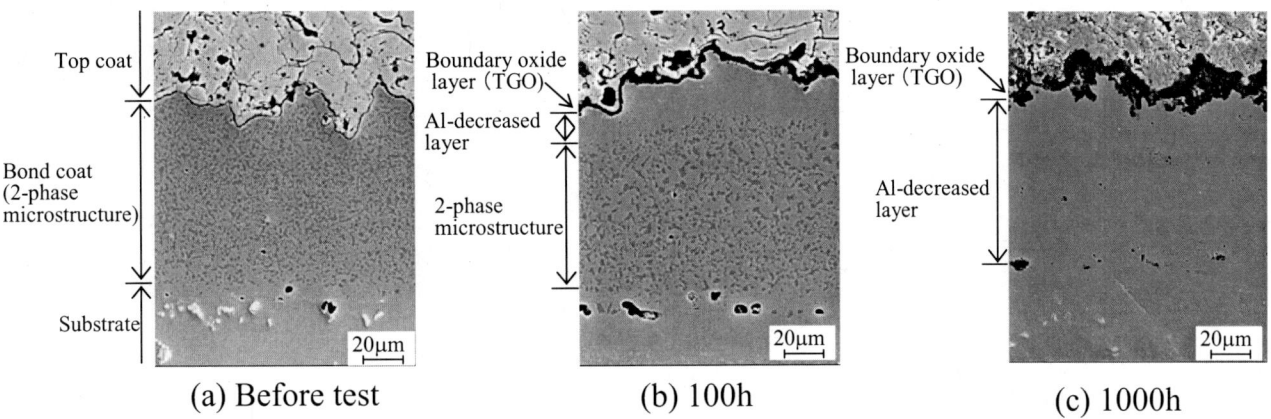

(a) Before test	(b) 100h	(c) 1000h

Figure 2. Microstructural change of bond coat in TBC specimens at 1273K in air.

Experimental procedure

The test is carried out at constant temperatures of 1173K, 1223K, 1273K and 1323K in an air by means of an electric tube furnace with the internal diameter of 70mm. The specimen is heated at 200K/h before the test, and the temperature fluctuation in the test section is kept in about 2~3K during the test. After the test, the specimen is cooled to about 773K at the rate of 200K/h and to the room temperature inside the furnace without the temperature control. Then, the microstructure of specimen is examined by means of an optical microscope, a scanning electron microscope (SEM) and an electron probe microanalyzer (EPMA).

TEMPERATURE ESTIMATION METHOD BY MICROSTURUCTURAL CHANGE OF BOND COAT

Figure 2 shows the microstructural change of BC at 1273K. The boundary between TBC and BC has the asperity of about 10μm, and the oxide (boundary oxide layer) grows along it. The oxide thickness increases as the test time passes.

The BC originally consists of 2-phases; dark dots in bright mother-phase as shown in Figure 2(a). The microstructure disappears from the BC surface near the TBC/BC boundary as shown in Figures 2(b) and (c). On the other hand, no significant microstructural change is observed at BC/substrate boundary. Figure 3 indicates the distribution of elements around the TBC/BC boundary before the test and after 1223Kx500h observed by an EPMA. It clearly points out that the 2 phases are β-(Ni, Al) and γ-(Co, Cr), and aluminum oxide is formed at the boundary. The lower Al-content region "Al-decreased layer (ADL)", which is caused by the diffusion, is observed on the boundary between the oxide layer and 2-phase microstructure in the BC. The Al-content in ADL is about 4wt%.

Figure 4 shows the relationship between the test time and the ADL thickness, which is the average of 12 data points in each specimen. The average squared root of the unbiased variance of the thickness is smaller than 10μm in all the test conditions. The ADL thickness increases monotonously according to the test time and temperature. The ADL grows to the whole bond coat after 500h at 1273K and 200h at 1323K. The relationship between the ADL thickness, l (μm) and the test time t (h) is in the form;

$$l = kt^{\frac{1}{2}}$$

(1)

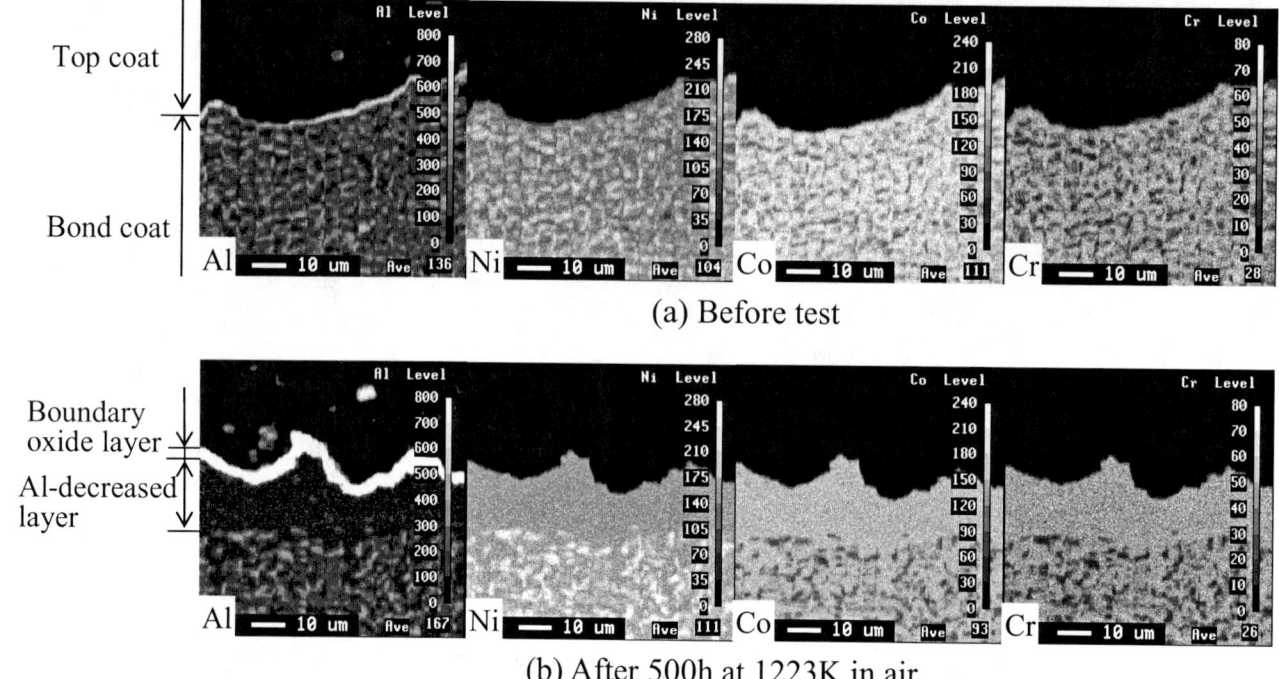

Top coat

Bond coat

(a) Before test

Boundary oxide layer

Al-decreased layer

(b) After 500h at 1223K in air

Figure 3. Distribution of elements around the boundary between top coat and bond coat by means of EPMA.

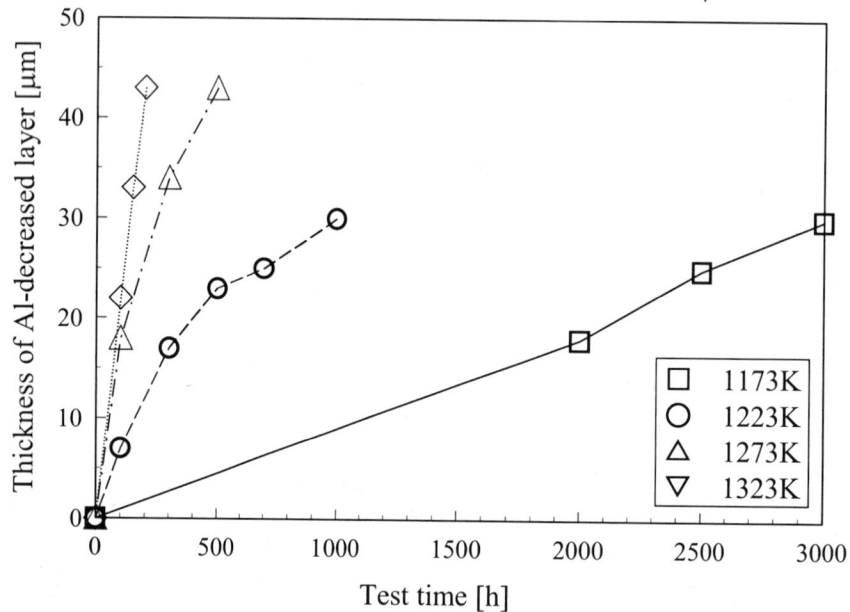

Figure 4. Relationship between thickness of Al-decreased layer and test time.

where k is the constant representing the growth rate. The Arrhenius plot shown in Figure 5 indicates

$$k = 2.96 \times 10^6 \exp\left(-\frac{152 \times 10^3}{RT}\right)$$
(2)

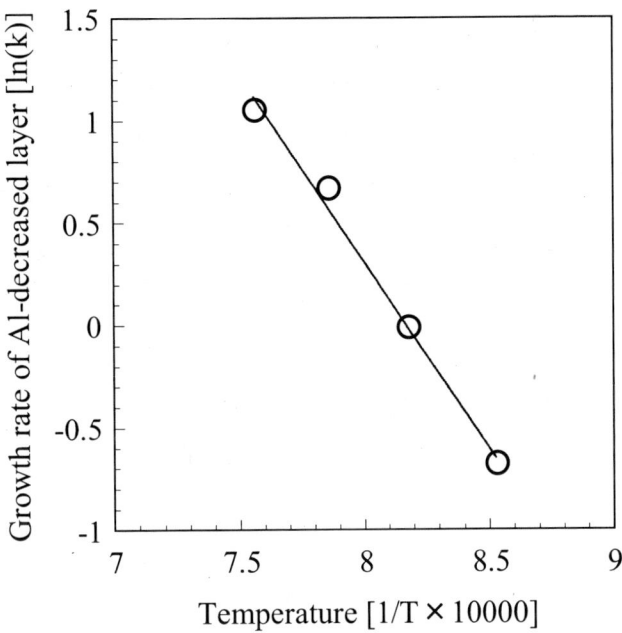

Figure 5.   Arrhenius plot growth rate of Al-decreased layer.

where T and R are temperature (K) and gas constant (8.31J/(mol·K)), respectively. The activation energy for the layer growth is given by 152kJ/mol. Then, T is described as follows from the equations (1) and (2).

$$T = \varphi(l,t) = -\frac{152 \times 10^3}{R} \cdot \frac{1}{\ln \dfrac{l}{2.96 \times 10^6 t^{\frac{1}{2}}}} \qquad (3)$$

Equation 3 enables us to estimate the temperature at the vicinity of the BC surface by the measurement of ADL thickness at the hot section of in-service component if the operation time is known.

The detection limit of ADL by an optical microscope is about 10μm. It is the minimum thickness to be recognized as layer since the thickness has scatter. When the thickness is larger than about 50μm, the thickness cannot be distinguished. The ADL larger than 50μm coalescences with another ADL growing at vicinity of the boundary between the BC and the substrate due to the interdiffusion. Thus, the ADL thicknesses from 10μm to 50μm are the applicable condition of the method. As the applicable limits depend on the temperature, the applicable operation time varies as shown in Figure 6. The scatter of data caused by the asperity at the BC surface as shown in Figure 2 affects the accuracy as well.

Assuming that the error of operation time is neglected, the unbiased variance of temperature u_T^2 is expressed as follows [16].

$$u_T^2 = \left(\frac{\partial \varphi}{\partial l}\right)_0^2 \frac{u_l^2}{n} \qquad (4)$$

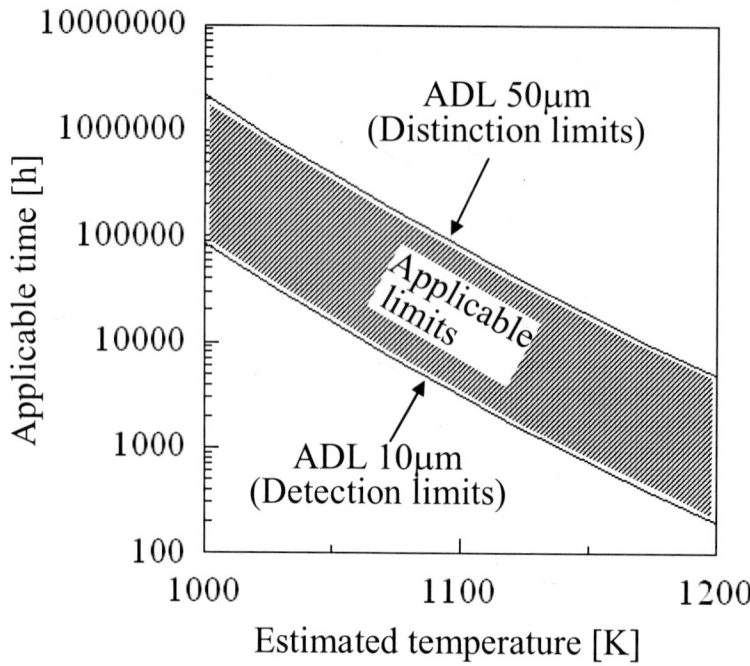

Figure 6.   Applicable limits of the temperature estimation method.

Here, $\left(\frac{\partial \varphi}{\partial l}\right)_0$ is the differential of equation (3) by the ADL thickness l. u_l and n are the the square root of unbiased variance of ADL thickness and the number of the measurement, respectively. Let $u_l = 10$, $n=10$, u_T is calculated. u_T varies with temperature and time since the ADL thickness is their function. Figure 7 shows the relationship between the estimated temperature and its error $\lambda \frac{u_T}{\sqrt{n}}$ (confidence coefficient 99%). The estimation error is smaller than 20K in almost all applicable limits.

Al-CONTENT PREDICTION METHOD

Figure 8 shows the relationship between the decrease of Al-content in BC, Δc, and square root of oxidation time. The Al-content is average of the whole bond coat. When it reaches about 4wt%, the diffusion saturates. Thus, these data are removed from Figure 8. The relationship is in the form;

$$\Delta c = c_0 - c = k'\sqrt{t} \tag{5}$$

The Arrhenius plot shown in Figure 9 indicates

$$k' = 8.93 \times 10^4 \exp\left(-\frac{142 \times 10^3}{RT}\right) \tag{6}$$

Thus, the time t is formulated as,

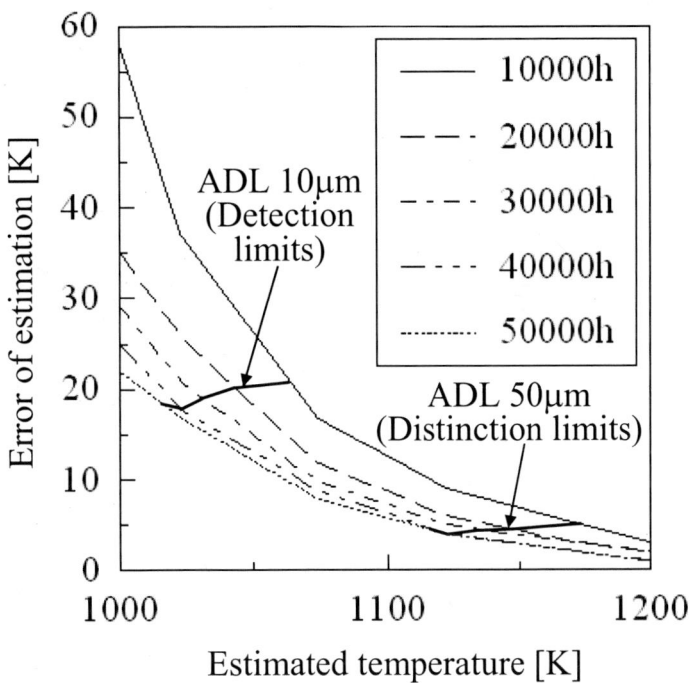

Figure 7. Error of estimated temperature by means of Al-decreased layer (Confidence coordinate 99%).

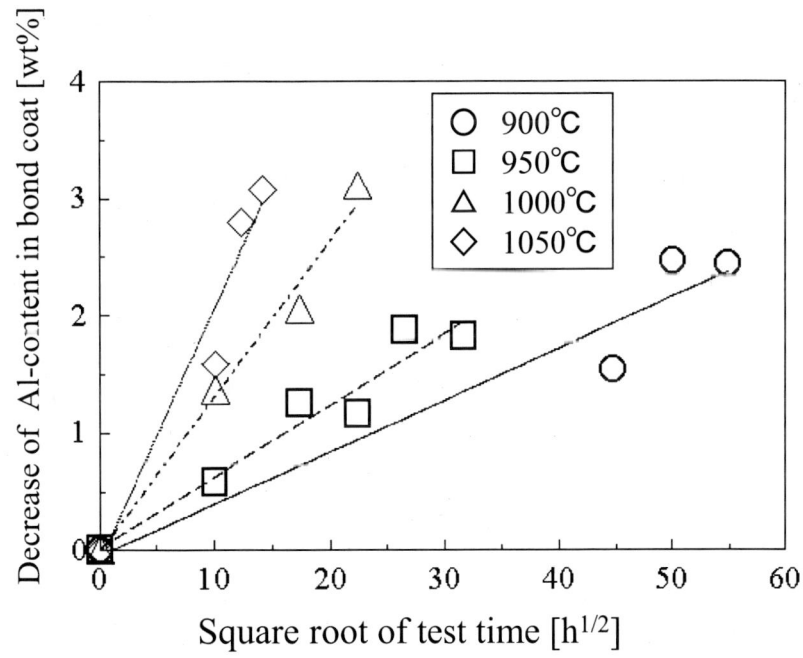

Figure 8. Relationship between the decrease of Al-content in bond coat and square root of test time

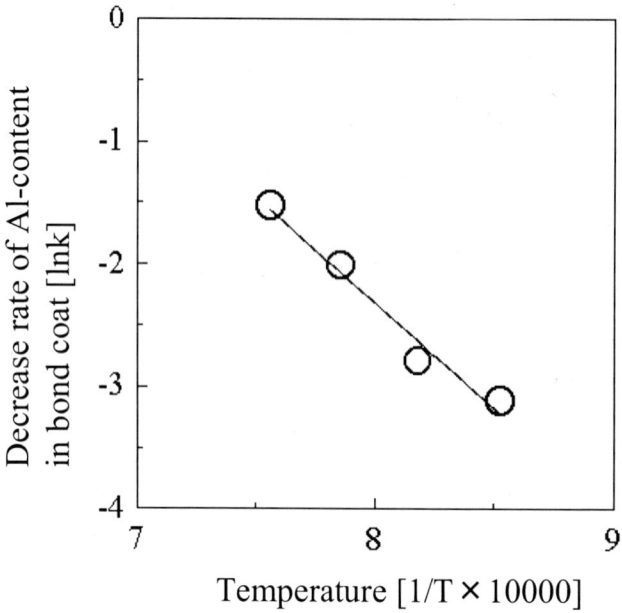

Figure 9. Arrhenius plot of Al-content decrease rate.

$$t = \left(\frac{c_0 - c}{8.93 \times 10^4 \exp\left(-\dfrac{142 \times 10^3}{RT}\right)} \right)^2 \qquad (7)$$

Substituting the estimated temperature T evaluated by the ADL thickness into the equation (7), the time when the Al-content reaches an arbitrary value c is predicted. This method, however, can be used for the Al-content from about 8wt% to 4wt %.

CONCLUSION

The Al-decreased layer is formed at the vicinity of the bond coat surface in the TBC. Its thickness increases in proportion to square root of the test time. The operation temperature can be estimated by measuring the thickness of the Al-decreased layer in a blade of in-service gas turbine on the basis of the relation among the thickness, time and temperature. The applicable condition due to the thickness of BC and detection limit of ADL is given in Figure 6.

The Al-content of the bond coat is measured by means of an electron probe microanalyzer. The decrease of the Al-content is also in proportion to square root of the time. Using the relationship, the time to reach arbitrary Al-content can be predicted.

REFERENCES

[1]T. Okubo, "1500°C Class Steam Cooled Gas Turbine Combined Cycle Technology", Journal of the Gas Turbine Society of Japan", **31**, 161-166 (2003)

[2]V. Srinivasan, N. S. Cheruvu, T. J. Carr and C. M. O'Brien, "Degradation of MCrAlY Coating and Substrate Superalloy During Long Term Thermal Exposure", Materials and Manufacturing Process, **10**, 955-969 (1995)

[3]Y. Yomei, N. Okabe, D. Saito, K. Fujiyama and T. Okamura, "Service Temperature Estimation of Gas Turbine Buckets Based on Microstructural Change", Journal of the Society of Materials Science , Japan, **45**, 699-704 (1996)

[4]A. Nomoto, M. Yaguchi and T. Ogata, "Evaluation of Creep Properties of Directionally Solidified Nickel Base Superalloy for Gas Turbine Blades Based on Microstructures", Central Research Institute of Electric Power Industry Report, T99094 (2000)

[5]M. Okada, Y. Etori, T. Hisamatsu and T. Takahashi, "Temperature estimation and prediction of Aluminum –content by means of microstructural change in gas turbine coatings", Journal of the Society of Materials Science , Japan, **54**, 257-264 (2005)

[6]M. Arai and U. Iwata, "Temperature estimation of gas turbine combustor based on thermally grown oxidation measurement in thermal barrier coating", The Thermal and Nuclear Power, **54**, 1064-1069 (2003)

[7]T. Torigoe, S. Aoki, I. Okada and H. Koguma, "Metal temperature estimation of high temperature components", JP.2003-4548 (2003).

[8]T. Fujii and T. Takahashi, "Development of Operating Temperature Prediction Method Using Thermophysical Properties Change of Thermal Barrier Coatings", Journal of Engineering for Gas Turbine and Power, **126**, 102-106 (2004)

[9]R. A. Miller, "Oxidation-Based Model for Thermal Barrier Coating Life", Journal of the American Ceramic Society, **67**, 517-521 (1984)

[10]S. Bose and J. DeMasi-Marcin, "Thermal Barrier Coating Experience in Gas Turbine Engines at Pratt & Whitney", Journal of Thermal Spray Technology, **6**, 99-104 (1997)

[11]A. Rabiei and G. Evans, Failure Mechanism Associated with the Thermally Grown Oxide in Plasuma-sprayed Thermal Barrier Coatings, Acta materialia., **48**, 3963-3976 (2000)

[12]S. Takahashi, M. Yoshiba and Y. Harada, "Nano-Characterization of Ceramic Top-coat/ Metallic Bond-Coat Interface for Thermal Barrier Coating Systems by Plasma Spraying", Materials Transactions, **44**, 1181-1189 (2003)

[13]E.Berghof-Hasselächer, H.Echsler, P.Gawenda, M. Schorr and M. Schütze, Time and Temperature Dependent Development of Physical Defects in Thermal Barrier Coating Systems, Prackt Metallogr, **40**, 219-231 (2003)

[14]H. Echsler, D. Renusch and M. Schütze, Bond coat oxidation and its significance for life expectancy of thermal barrier coating systems, Materials Science and Technology, **20**, 307-318 (2004)

[15]M. Hasegawa and Y. Kagawa, "Microstructural and Mechanical Properties Changes of a NiCoCrAlY Bond Coat with Heat Exposure Time in Air Plasma-Sprayed Y2O3-ZrO2 TBC systems", International Journal of Applied Ceramic Technology, **3**, 293-301 (2006)

[16]Y. Yoshizawa, "New theory of error", 157-161 (1989) Kyoritsu-shuppan.

QUANTATIVE MICROSTRUCTURAL ANALYSIS OF THERMAL BARRIER COATINGS PRODUCED BY ELECTRON BEAM PHYSICAL VAPOR DEPOSITION

Matthew Kelly[1], Jogender Singh[1], Judith Todd[2], Steven Copley[1], Douglas Wolfe[1]
[1]The Applied Research Laboratory
[2]Engineering Sciences and Mechanics Department
Pennsylvania State University
University Park, Pa, 16802

ABSTRACT

Thermal Barrier Coatings (TBC) produced by Electron Beam Physical Vapor Deposition (EB-PVD) are used primarily for system critical components of power turbines. The performance of coatings is highly dependant on micro and nano structural features. This paper proposes and demonstrates quantitative microstructural analysis of TBC produced by the EB-PVD. Metallographic techniques were applied to surfaces parallel and perpendicular to the columnar growth direction. Multiple levels perpendicular to the columnar growth direction were prepared to identify descriptive statistical values for coating microstructure throughout coating thickness. Metallographic surfaces were imaged with Scanning Electron Microscopy (SEM) and evaluated with commercial image analysis software. Samples produced with different vapor incidence angles were evaluated by the proposed method and show distinct quantitative differences of column grain size, inter-columnar porosity, and levels of re-nucleation. Microstructural results will be presented as a function of coating thickness and vapor incidence angle.

INTRODUCTION

Thermal Barrier Coatings (TBC) produced by Electron Beam Physical Vapor Deposition (EB-PVD) are used primarily for system critical components of power turbines due to the increased coating life compared to thermal sprayed coatings and ability to coat components with active air cooling without sealing external cooling passages [1]. Under current and expected future operating temperatures up to 1400C , failure of the coating will result in rapid metal structural component degradation, turbine performance decreases, and eventual system failure [2]. PVD coatings exhibit what has been described as columnar morphology, high aspect ratio grains oriented normal to the substrate separated by voids, which offer mechanical compliance parallel to the component surface [1, 3-5]. This structural compliance aids in reducing stress in the ceramic coating during thermal cycling and thickening of the Thermally Grown Oxide (TGO) due to high temperature oxidation when compared to fully dense structures [6].

Accurate, cost-effective characterization of coating structure is an essential tool for coating development and quality control. There are two main areas of material structure that should be evaluated: physical structure and atomic structure. The evaluation of atomic structure, which includes determination of crystallographic information and chemical composition, is handled well by bulk techniques such as X-Ray Diffraction (XRD) and Spectroscopy techniques as discussed elsewhere and is routinely reported [7, 8]. Evaluating and quantifying physical structure of Physical Vapor Deposited (PVD) coatings is often ignored in nearly all publications related to TBC due to the lack of reliable cost-effective techniques. The fact that physical structure is not commonly evaluated for PVD coatings is not surprising, considering the broad range of possible parameters needed to describe the complex range of geometric morphologies shown in classic Structural Zone Models (SMZ) [4]. However, this lack of information has created a void in relating structure to both process parameters and coating performance, causing many studies to relate machine specific parameters to testing based performance. While development based on the relation between machine specific process parameters and material performance is very beneficial to industrial scale development, it prohibits implementation of outside developed knowledge. Moreover, the lack of coating evaluation prevents accurate quality control of simple structural parameters such as grain size or porosity. Without quality control and standardization, coating performance can be unreliable and not ideal for the environmental use.

The physical structure of TBC varies throughout the thickness and is determined by the process parameters of the equipment used to produce it. Qualitatively it can be observed that columns are often thinner at the "root" or area closest to the substrate. Ballistic deposition models have explained that columns nucleate from initial particles, grow in size rapidly, compete during growth, merge, reach a maximum size for the given surface mobility, and eventually nucleation begins again on the deposited column surface [9, 10]. The rate at which changes take place and geometry of structure during the growth is process dependant. Additionally, three types of porosity can also be observed that coatings produced on substrates rotating in the vapor cloud. Most easily observed is the void space between columns (Type I inter-columnar porosity), which has been described to be ribbon-like and believed to be directly related to coating compliance [11, 12]. The second type of inter-columnar porosity has been described as the open "feathery" surface of columns (Type II intercolumnar porosity) and can be observed under high magnification of coatings produced under high rotation rates fractured through thickness [13]. Type II intercolumnar porosity is believed to affect the low temperature phonon thermal transmission [14]. Finally the third type of porosity is trapped voids on the sub-micron scale (intra-columnar) porosity, which reduces high temperature electro-magnetic thermal transfer [15-17].

These previously described structural changes with respect to thickness imply that neither a single technique nor position evaluation will be adequate in complete structural characterization. Useful measurements must include sets of measured parameters with respect to position during growth or detailed knowledge on growth dynamics of the material system. The following research summarizes and demonstrates metallographic techniques discussed previously on a variety of TBC produced by EB-PVD [18]. The described technique readily provides micron scale information on Type I inter-columnar porosity and columnar grain size with respect to coating thickness and is only a step in the direction of characterization of EB-PVD TBC physical structure.

EXPERIMENTAL
Sample Production

Samples were generated to evaluate the processing parameter effect of Vapor Incidence Angle (VIA) for an industrial scale prototype deposition system to simulate the range of structure seen on a complex geometry component similar to a turbine blade. Substrate material was nickel alloy 625 cut into 1.9 cm diameter buttons 0.476 cm thick. Buttons were polished smooth using successive steps of wet grinding with silicon carbide paper to an 800 grit finish. Substrates were "heat tinted" in air at 700C for 15 minutes to form an oxide layer. The surface and the side of each test specimen were grit blasted in a Unihone brand grit blaster using high purity 400 micron-size aluminum oxide particles. The distance from the edge of the nozzle to the surface of the samples was approximately 38 cm, with a pressure of 200 kPa. The angle of the nozzle with respect to the sample surface was 45° to minimize the amount of embedded Al_2O_3 particles incorporated into the substrate surface. The grit blast time on each sample varied between 10–15 seconds and was performed until a uniform matte finish was obtained. Grit blasted substrates were ultrasonically cleaned in acetone for 20 minutes, rinsed with methanol, ultrasonically clean again in methanol for 20 minutes, and dried with nitrogen gas.

Prepared substrates were then tack welded to strips of stainless steel foil that were tack welded to an 8.64 cm diameter mandrel having wedges milled at angles 0°, 15°, 30°, 45°, 60°, 75°, and 90° with respect to the primary vapor flux direction. The mandrel was loaded into an industrial prototype Sciaky Inc. EB-PVD unit consisting of six EB-guns and a three-ingot continuous feeding system described in previous papers [19]. The deposition chamber and gas feed lines were then evacuated to a pressure of 10^{-3} Pa. Two electron beams were then used on the graphite heater assembly to bring the heating surfaces to ~1200° for 20 minutes before samples were positioned 28 cm above the center of a 4.93 cm diameter 7%wt Yttria Stabilized Zirconia (7YSZ) ingot lot number 1297726, provided by Trans Tech Inc. of Adamstown, MD. Samples were then set to rotate at 12 RPM and allowed to soak at an average temperature of 1020°C for 20 minutes. During the soak period oxygen was introduced near the samples at 100 sccm to grow a thin uniform oxide layer (TGO). Following the TGO period, samples were exposed to an established 7YSZ vapor at an average chamber pressure of 1.7×10^{-3} torr. Evaporation rate was established at 5.7 grams per minute, resulting in a deposition rate of 2 microns per minute. Deposition was continued for a period of 90 minutes, producing a 177± 4 micron-thick 7YSZ coating for the sample with normal incidence angle (0° VIA). Samples were left to cool under vacuum and 200 sccm oxygen flow for 10 minutes before venting to atmospheric conditions. Coatings exhibited what would be considered typical range of morphologies for TBC deposited at VIA in the 0° to 90° range and are shown in Figure 1.

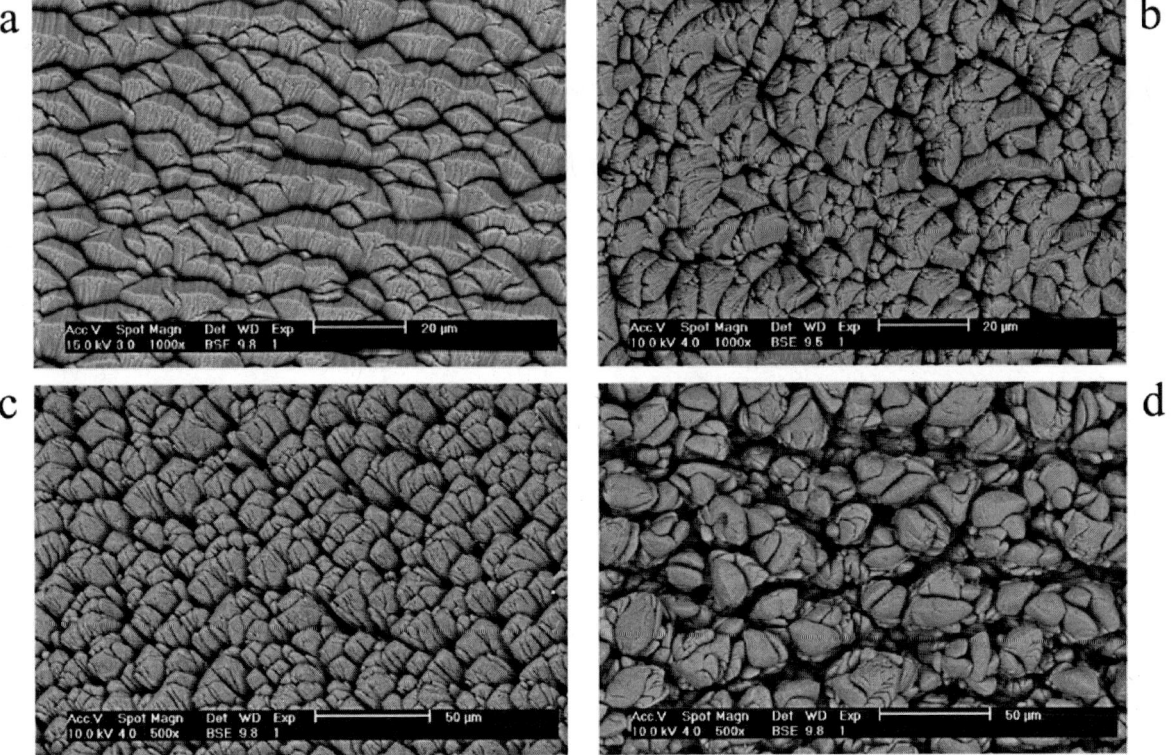

Figure 1. Typical range of morphologies shown a, b, c, and d for TBC deposited at Vapor Incidence Angles (VIA) of 0°, 30°, 60°, 90° respectively.

Sample Preparation

Samples were sectioned with a diamond waffering saw through the center of each disk along the line normal to the thickness variation, producing indistinguishable halves. One half was vacuum mounted in metallographic epoxy on a 50 micron sacrificial shim at taper inclination angles greater than 1°, yet less than 3° to minimize the magnification correction factor. Mounted samples were polished and imaged using Back Scattered Scanning Electron Microscopy (BSE) [18]. Image position with respect to Thermally Grown Oxide (TGO) was recorded as described by a previous article [18].

Image Analysis

Image analysis was conducted using commercial software produced by Clemex Technologies Inc. of Longueuil, Canada. Microstructural phase selection was determined using grey scale threshold limits. Voids were determined to have grey scale levels of the metallographic epoxy and lower. Porosity was measured as an area fraction based on the ratio of selected pixels to total image pixels. Grains were determined as areas with greater grey scale levels than the lightest fully developed grain boundaries. Fully developed boundaries are those that completely bisect column regions. Grains were separated using software built in functions that perform low level binary operations and verified by an operator for accuracy. The area of each separated grain was measured and reported as a diameter of a circle representing equivalent area.

RESULTS

Surfaces exposed using the described Transcolumnar metallographic method provided a surface that can be used to detail the growth of the coating through the entire thickness. Features such as intercolumnar porosity and columnar grain size, that are difficult to measure using the more common metallographic cross, section become very apparent with the Transcolumnar surfaces. Metallographic surfaces perpendicular to the growth direction eliminate the depth of view difficulties observed with conventional imaging techniques applied to cross sectional surfaces. These "Transcolumnar" surfaces provide additional contrast due to increased visible depth of pores and provide additional surface area available for evaluation at different positions with respect to distance from TGO. An example image of a thick coating displaying features of notable interest is shown in Figure 2. Figure 2a displays the most common measurable features visible with this evaluation technique; Type I inter-columnar porosity and columnar grains with defined boundaries. Highlighted in Figure 2b is Type II "feather like" inter-columnar porosity, developed re-nucleation, and co-competitive grain boundaries. As columnar re-nucleation becomes well defined by void like boundaries, it is classified as an independent grain. A select group of images with corresponding relative positions for Vapor Incidence Angles (VIA) 0°, 30°, 60°, and 90° are shown in Figure 3.

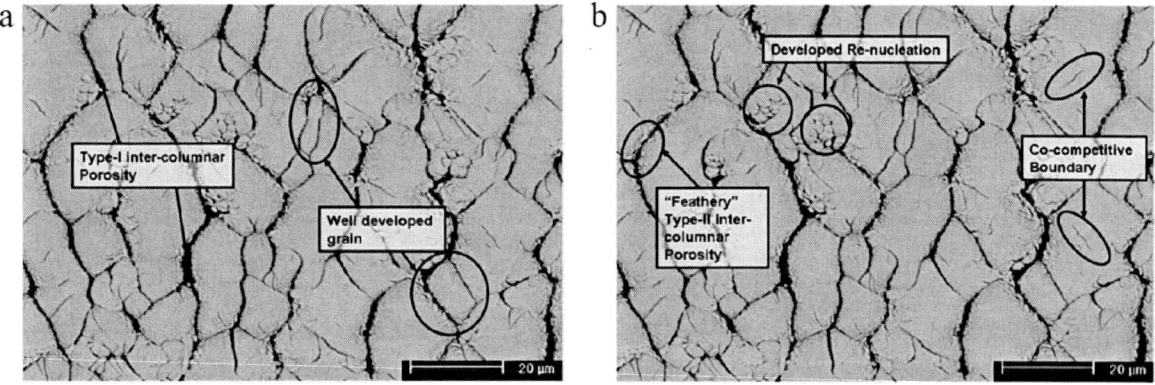

Figure 2. Shown in 2a are measured microstructural features of Type I inter-columnar porosity and developed columnar grains. Shown in 2b are notable features of observation; Region of Type II "feathery" inter-columnar porosity, region of developed re-nucleation, and a co-competitive grain boundary.

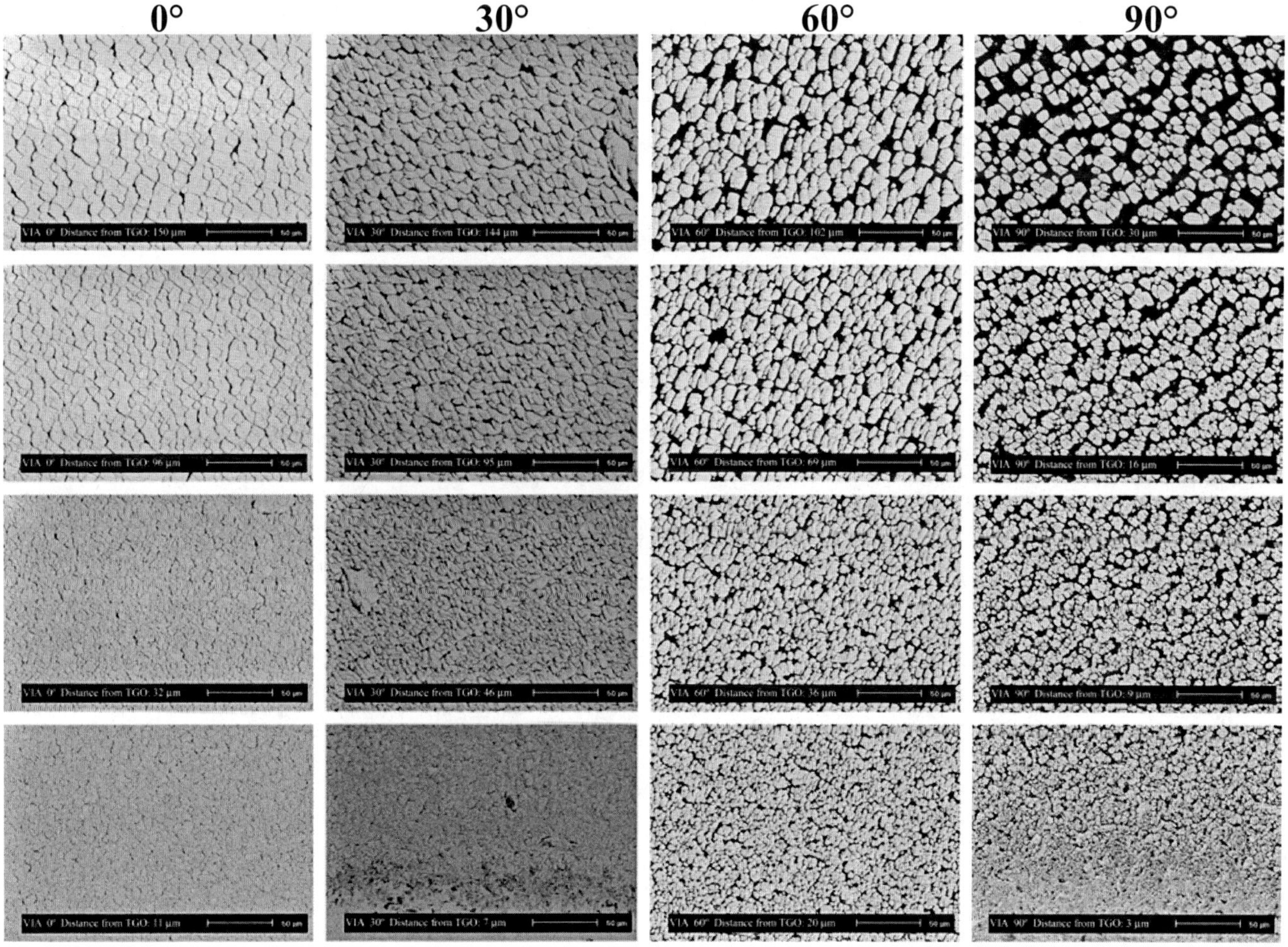

Figure 3. Sample images at consistent magnification for coatings produced at VIA of 0°, 30°, 60°, 90° shown in columns from left to right respectively. Position with respect to distance from TGO layer in microns beginning with 0° VIA top image; 150, 96, 32, 11 micron. Position with respect to distance from TGO layer in microns for 30° VIA top image; 144, 95, 46, 7 micron. Position with respect to distance from TGO layer in microns for 60° VIA top image; 102, 69, 36, 20 micron. Position with respect to distance from TGO layer in microns for 90° VIA top image; 30, 16, 9, 3 micron.

Samples produced with different VIA were evaluated with the Transcolumnar metallographic method. It is clear that microstructural differences exist from visual inspection of SEM images shown in Figure 3. The most apparent visual difference is the changes of the void space between columns with both thickness and VIA. When quantitative image analysis was applied to the images for Type-I inter-columnar porosity and grain size the differences can be compared mathematically and graphically. Values of Type-I inter-columnar porosity are shown with respect to distance through thickness from the TGO layer graphically in Figure 4 for the different VIA. For the VIA 30° the porosity remains below 6% and somewhat constant with undulations. As VIA is increased above 30°, for this deposition process, porosity increases greatly. The amount of porosity becomes more dependent on thickness. Columnar grain size distributions were measured and the median grain size at each imaged position through the thickness is presented for the different VIA in Figure 5. Analysis of the mean grain size suggests that the minimum median grain size for all coating thickness under these deposition parameters should be found between 15° and 30° VIA. Additionally, the dependence of grain size on thickness suggests columns are growing divergently, or columnar grains are coalescing. Grain size distribution variance was plotted in Figure 6 to verify that re-nucleation and column divergence was the dominate trend. Parent-child image analysis could also be conducted with columns-columnar grains to make the same conclusion. One interesting observation when analyzing the grain size variance is that 15° VIA samples appear to have the tightest grain size distribution. Competitive growth may begin to dominate above 140 microns in thickness for these deposition parameters.

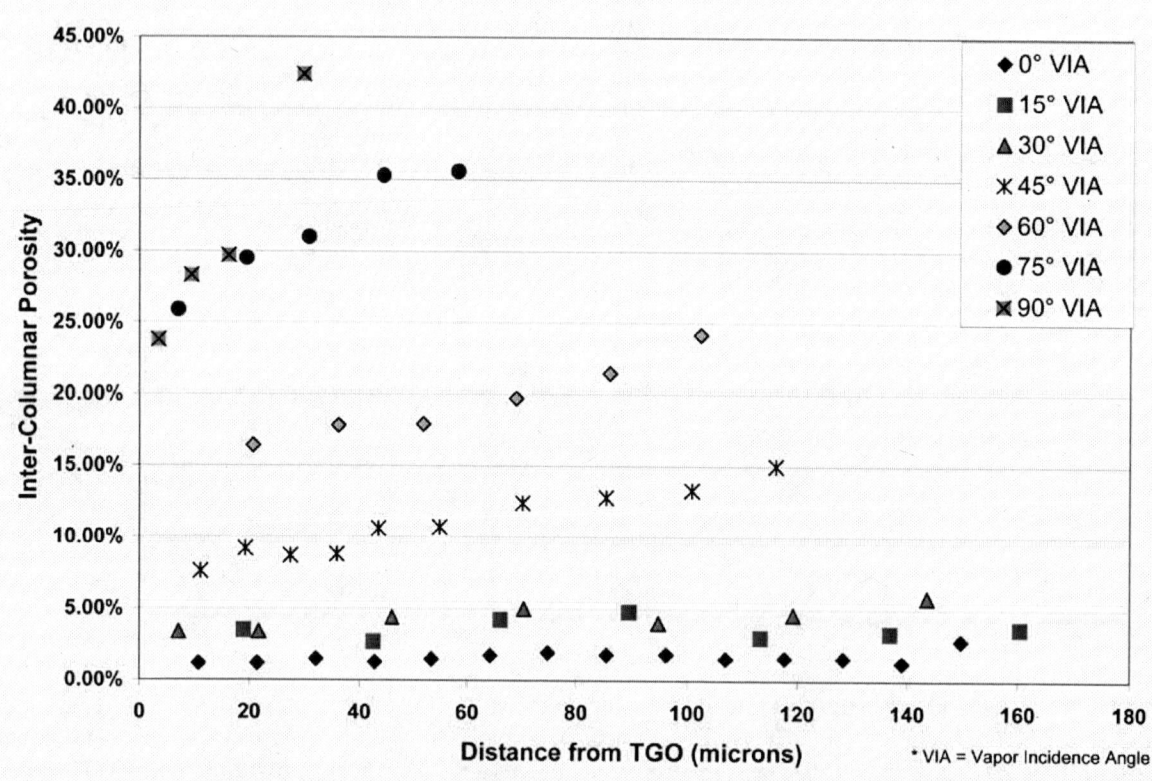

Figure 4. Measured values of Type-I inter-columnar porosity are shown with respect to distance through thickness from the TGO layer.

Progress in Thermal Barrier Coatings

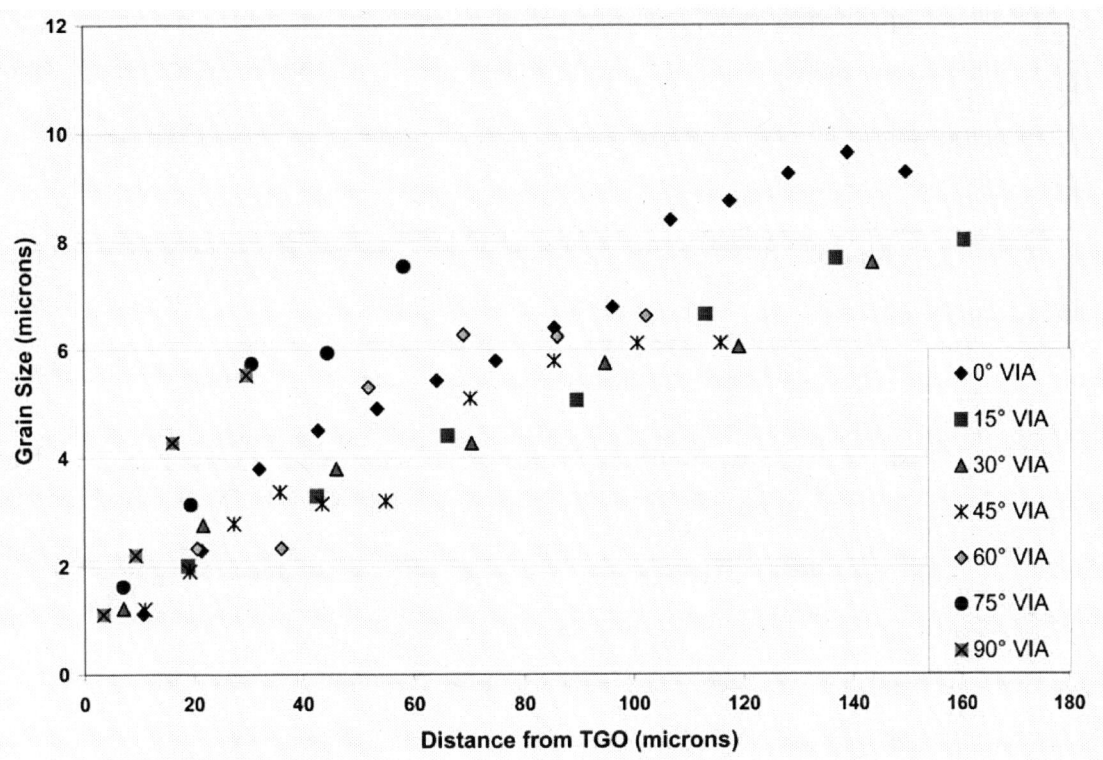

Figure 5. Measured values of median grain size are shown with respect to distance through thickness from the TGO layer.

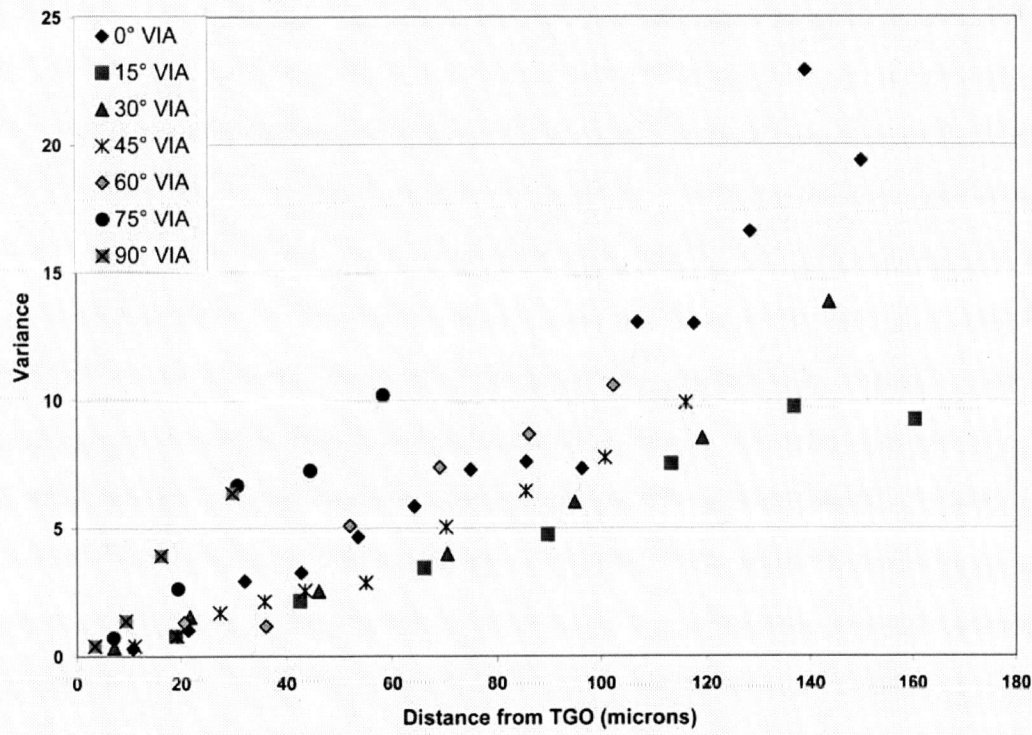

Figure 6. Calculated variance of grain size distribution is shown with respect to distance through thickness from the TGO layer.

Additional to the clear microstructural differences measured, trends on the growth mechanics can be observed. The most notable trend is the relation between grain size variance and porosity. Re-nucleation creates many small grains that grow rapidly, creating self-shadowing of nearby areas and increased porosity. This growth phenomenon can be observed by sharp changes in the slope of the graphs shown in Figures 4 and 6, and visually shown within in the images shown in Figure 3. This result agrees with accepted ballistic models for PVD and may offer insight into many of the unknown distribution functions needed for more complex growth modeling.

CONCLUSION

Thermal Barrier Coatings were applied to test coupons at discrete Vapor Incidence Angles on a rotating mandrill to simulate microstructures observed on complex components such as turbine blades. Samples were evaluated using a developed quantitative metallographic technique to evaluate microstructure at multiple levels through the thickness for Type-I inter-columnar porosity and columnar grain size. A ten fold increase of Type-I inter-columnar porosity was observed between normal incidence and samples parallel to vapor flux. Relative porosity and columnar grain size dependence was measured with respect to thickness. For the fixed process parameters, both porosity and columnar grain size were found to be dependant with coating thickness. Microstructural information corresponding to position with respect to thickness for a specified set of deposition parameters provides new knowledge into coating growth dynamics not obtainable through evaluation of historically evaluated surfaces. While research for this study was limited to the evaluation of Type-I inter-columnar porosity and columnar grain size distribution analysis, additional descriptive parameters can be measured from the same prepared samples.

Knowledge of microstructural information with respect to coating thickness for fixed deposition parameters is directly applicable for quality control. Exposed surfaces of deposited coatings can be imaged and analyized rapidly for columnar grain size and exposed porosity. Using predetermined charts as shown in figures 4-6 and measured thickness of the coating being evaluated for quality control purposes, an expected microstructure can be compared to the evaluated sample. Rapid verification of microstructure in this manner would provide an added level of assurance that coated components should perform as the tested samples during coating development.

Combinations of additional techniques and further development of metallographic microstructural evaluation techniques are required to achieve the level of resolution and microstructural information needed for true materials engineering of TBC coatings produced by EB-PVD. It is know that different microstructures of materials will perform differently under the same environmental usage. It was shown in this research that microstructure varies significantly with respect to geometry. Using the information retained from this study, it should be possible to design deposition parameters to increase microstructural uniformity on complex geometry substrates thereby increasing uniformity of coating performance. Incorporation of more advanced emerging techniques that accurately characterize the submicron levels of structure could provide the ability to design idealized material structures to maximize desired performance for individual applications.

ACKNOWLEDGEMENT

Distribution Statement A: Approved for public release. Distribution unlimited.
This research was sponsored by the United States Navy Manufacturing Technology (ManTech) Program, Office of Naval Research, under Navy Contract N00024-02-D-6604. Any opinions, findings, conclusions, or recommendations expressed in this material are those of the authors and do not necessarily reflect the views of the U.S. Navy.

REFERENCES

1. Peters, M., et al., *Design and Properties of Thermal Barrier Coatings for Advanced Turbine Engines*. Materialwissenschaft und Werkstofftechnik, 1997. **28**: p. 357-362.

2. Layne, A.W., *Advanced Turbine Systems*, N.E.T. Laboratory, Editor. 2000, U.S. Department of Energy.

3. Strangman, T.E., *Thermal barrier coatings for turbine airfoils*. Thin Solid Films, 1985. **127**(1-2): p. 93.

4. Thornton, J.A., *High Rate Thick Film Growth*. Annual Review Materials Science, 1977. **7**: p. 239-260.

5. Nicholls, J.R., M.J. Deakin, and D.S. Rickerby, *A comparison between the erosion behaviour of thermal spray and electron beam physical vapour deposition thermal barrier coatings*. Wear, 1999. **233-235**: p. 352-361.

6. Evans, A.G., et al., *Mechanisms controlling the durability of thermal barrier coatings*. Progress in Materials Science, 2001. **46**(5): p. 505.

7. Almeida, D.S., et al., *EB-PVD TBCs of zirconia co-doped with yttria and niobia, a microstructural investigation*. Surface and Coatings Technology, 2006. **200**(8): p. 2827.

8. Bernier, J.S., et al., *Crystallographic texture of EB-PVD TBCs deposited on stationary flat surfaces in a multiple ingot coating chamber as a function of chamber position*. Surface and Coatings Technology, 2003. **163-164**: p. 95.

9. Smith, D.L., *Thin-Film Deposition; Principles and Practice*. 1995: Mc Graw Hill. 616.

10. Hill, R.J., *Physical Vapor Deposition*. 1986: Temescal.

11. Clarke, D.R. and S.R. Phillpot, *Thermal barrier coating materials*. Materials Today, 2005. **8**(6): p. 22-29.

12. Terry, S.G., J.R. Litty, and C.G. Levi, *Evolution of Porosity and Texture in Thermal Barrier Coatings Grown by EB-PVD*. Elevated Temperature Coatings: Science and Technology III, 1999: p. 13-26.

13. Cho, J., et al., *A kinetic Monte Carlo simulation of film growth by physical vapor deposition on rotating substrates*. Materials Science and Engineering, 2005. **391**(1-2): p. 390-401.

14. Zhao, X., X. Wang, and P. Xiao, *Sintering and failure behaviour of EB-PVD thermal barrier coating after isothermal treatment*. Surface and Coatings Technology, 2005(Issues 20-21): p. 5946-5955.

15. Zhu, D., et al., *Thermal conductivity of EB-PVD thermal barrier coatings evaluated by a steady-state laser heat flux technique*. Surface and Coatings Technology, 2001. **138**(1): p. 1.

16. Saruhan, B., et al., *Liquid-phase-infiltration of EB-PVD-TBCs with ageing inhibitor.* Journal of the European Ceramic Society, 2006. **26**(1-2): p. 49.

17. Zhu, D. and R.A. Miller, *Thermophysical and Thermomechanical Properties of Thermal Barrier Coating Systems*, G.R.C. NASA, Editor. 2000, NASA Center for Aerospace Information. p. 22.

18. Kelly, M.J., et al., *Metallographic techniques for evaluation of thermal barrier coatings produced by EB-PVD.* Materials Characterization, In Press.

19. Wolfe, D.E., et al., *Tailored microstructure of EB-PVD 8YSZ thermal barrier coatings with low thermal conductivity and high thermal reflectivity for turbine applications.* Surface & Coatings Technology, 2005(190): p. 132-149.

INVESTIGATION OF DAMAGE PREDICTION OF THERMAL BARRIER COATING

Y. Ohtake
Ishikawajima-Harima Heavy Industries Co., Ltd.
1, Shin-Nakahara-Cho, Isogo-ku,
Yokohama-shi, Kanagawa 235-8501, Japan

ABSTRACT

Thermal barrier coating (top coating) for protecting turbine blades in airplane engines causes delaminalion by cyclic thermal loading. The delamination depends on the growth of thermal growth oxidation (TGO) layer at the interface when the coating system consists of top coating over environment barrier coating (bond coating). The growth behavior of TGO layer had examined by testing at constant temperature in furnace, but it didn't clear by testing of cyclic thermal loading. This paper investigates to examine the growth behavior of TGO layer for cyclic thermal loading. The burner rig testing is conducted to the behavior of TGO layer in cyclic thermal loading. Testing equipment is also developed to test four circular plate specimens at same time by burner rig. The thickness of TGO layer was measured from the observation at the interface in the specimen after the testing. The thickness increased as the number of cyclic thermal loading. The growth of TGO layer also denoted same tendency of the results of heating testing in furnace. It was found that the growth of TGO layer could predict by an equation. The equation was proposed in the relationships between the thickness of TGO layer and heating time when heating time was adopted for total holding times at maximum temperature in cyclic thermal loading.

INTRODUCTION

A typical coating system consists of thermal barrier coating (top coating) over environment barrier coating (bond coating). The coating system is applied for protecting engine parts, for example turbine blades, in airplane engines. The coating system fractures in top coating when the part is given cyclic thermal loading. The fracture mechanism is classified into two types. One is vertical crack in the normal direction of thickness of top coating. The cause depends on thermal expansion difference between top coating and base metal (or bond coating). Another is delamination of top coating at the interface over bond coating. The vertical crack can easily detect by regular inspection to appear on the surface of the engine parts, but the delamination can not detect in the first stage of the fracture because the crack propagates in top coating near the interface. Ohtake et al. had examined the damage of rectangular plate specimen with top coating by burner rig testing [1]-[4]. The delamination occurred in top coating from the observation of the specimen after the testing. The fracture mechanism was examined in both finite element analysis and burner rig testing. It was found that the delamination was caused by the growth of thermal growth oxidation (TGO) layer, thermal stress, the shape of interface at bond coating, the pore in top coating etc.

A simple life prediction model of top coating had been proposed from those results in previous paper [4]. The model is composed of three damage parameters. The growth of TGO layer is one of parameters in the model. The growth law of TGO layer had been examined in the results of the testing at constant temperature in furnace, but it didn't clear for cyclic thermal loading. This paper investigates to examine the growth behavior of TGO layer for cyclic thermal loading. The grown behavior of TGO layer was examined in burner rig testing. The testing takes in a lot of times even if it gives cyclic thermal loading to one specimen. The designer may also changes the material of base metal or the compositions (thickness, material and manufacturing process) of top coating. Thus, testing equipment was developed to test four circular plate specimens at same time by burner rig.

EXPERIMENTAL PROCEDURE

The specimen is composed of a typical coating system and base metal. The coating system consists of top coating over bond coating. Base metal is single crystal CMSX-2 substrate of nickel base superalloy. Bond coating applies CoNiCrAlY that is manufactured by low pressure plasma spray (LPPS). Top coating applies 8 wt. percent yttria stabilized zirconia (YSZ) that is manufactured by air plasma spray (APS).

Figure 1 shows the appearance in burner rig testing. The testing equipment was developed can test many specimens in short time. The equipment can test four specimens at same time by burner. Those specimens are circular plate and the size of the plate is diameter 20mm and thickness 3mm. Thickness of top coat of the specimen is 0.5m and bond coat 0.125mm. The surface of the specimen is heated by high temperature gas and the back surface is cooled by air. The thermal history of one cycle is total time 3 min, 20s heating time, 60s holding time and 100s cooling time. The maximum temperatures on surface of top coating are about 1473K, and then the temperature of the cooling surface are 1173K. The thermal loading is repeated until 1000, 2000 and 3000 cycles by burner rig. Those specimens are cut by diamond saw after the testing. TGO layer at interface in all specimens is observed by scanning electron microscope (SEM).

Holder   Burner

Four test specimens

The thickness of TGO layer is measured at ten points for one specimen and the average value is adopted for thickness of TGO layer of the specimen. Heating time is defined as the sum of holding time at maximum temperature in burner rig testing. The growth behaviors of TGO layer is examined from the relationship between thickness of TGO layer and heating time after cyclic thermal loading by burner rig testing.

EXPERIMENTAL RESULTS

Figure 2 shows the relationship between heating time and thickness of TGO layer, where heating time is total time at maximum temperature and the thickness of TGO layer is measured in the specimen at heating time in burner rig testing. TGO layer grows as the increase of heating time as shown in Fig.2 and the thickness increases as the number of cyclic thermal loading. The grown behaviors don't change for the testing at constant temperature in furnace. Thus, it was found that the grown of TGO layer didn't depend on testing method and heating time was important factor to predict the grown of TGO layer. Equation (1) was also proposed for the testing in furnace in previous paper [1]-[4]. The equation is expressed in terms of thickness of TGO layer w, heating time t and two constants k and n.

$$w = kt^n \tag{1}$$

The line in Fig.2 is k=0.03 and n=0.3 in Eq.(1). The data of burner rig testing could predict by using the line of Eq.(1) in Fig.2. It was found that the growth of TGO layer in cyclic thermal loading could predict by Eq.(1).

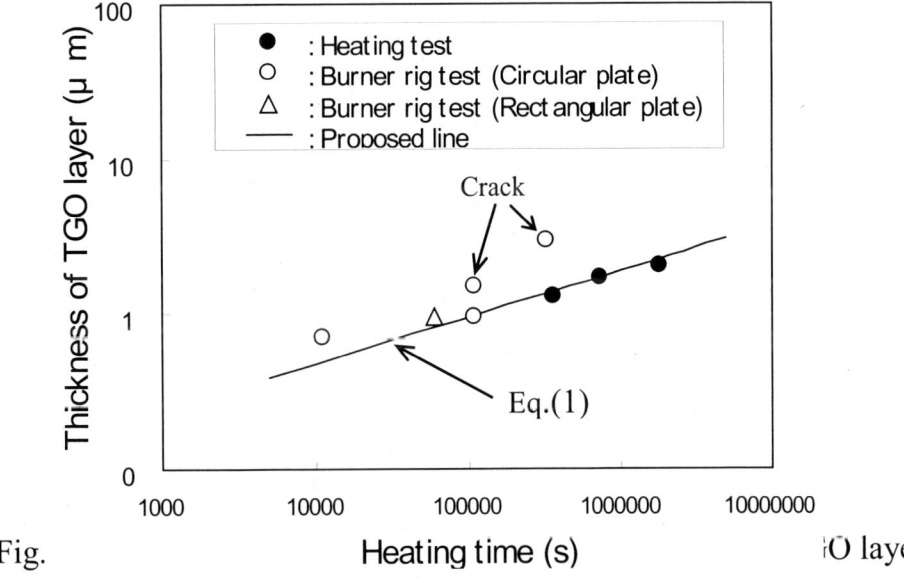

Fig.   Heating time (s)   O layer

Crack was detected in top coat of the specimen after burner rig testing as shown in Fig.2 when the number of cycle reached more over 1000 cycles. The cracks existed in parallel direction of top coating surface from pore near the interface. It was considered that the delamination of top coating was occurred in the progress and the combination with a lot of cracks in top coating near the interface when TGO layer was increased with heating time.

CONCLUSIONS

This paper investigated growth behavior of TGO layer for cyclic thermal loading. The burner rig testing was used to examine the behavior of TGO layer in cyclic thermal loading. Testing equipment was also developed to test four circular plate specimens at same time by burner rig. The thickness of TGO layer after the testing increased as the number of cyclic thermal loading. The growth of TGO layer also denoted same tendency for the results of heating testing in furnace. It was found that the growth of TGO layer could predict by an equation. The equation was proposed in the relationships between the thickness of TGO layer and heating time when heating time was adopted for total holding times at maximum temperature in cyclic thermal loading.

REFERENCES

[1]Y. Ohtake, N. Nakamura, N. Suzumura and T. Natsumura, "Evaluation for Thermal Cycle Damage of Thermal Barrier Coating," *Ceramic Engineering and Science Proceedings*, 24(3) 561-566 (2003).

[2]Y. Ohtake, T. Natsumura, "Investigation of Thermal Fatigue Life of Thermal Barrier Coating," *Ceramic Engineering and Science Proceedings*, 25(4) 357-362 (2004).

[3]Y. Ohtake, T. Natsumura, K.Miyazawa, "Investigation of Thermal Fatigue Life Prediction of Thermal Barrier Coating," *Ceramic Engineering and Science Proceedings*, 26(3) 89-93 (2005).

[4]Y. Ohtake, "Damage Prediction of Thermal Barrier Coating," *Ceramic Engineering and Science Proceedings*, 27(3) (2006).

[5]R. A. Miller, "Oxidation-Based Model for Thermal Barrier Coating Life," Journal of the American Ceramic Society, 67 [8] 517-21 (1984).

[6]R. A. Miller, "Thermal Barrier Coatings for Aircraft Engines History and Directions," *Journal of Thermal Spray Technology*, 6 [1] 35-42 (1997).

[7]R. A. Miller, "Life Modeling of Thermal Barrier Coatings for Aircraft Gas Turbine Engines," *Journal of Engineering for Gas Turbines and Power*, 111 301-05 (1989).

[8]A. G. Evans, M. Y. He and J. W. Hutchinson, "Mechanics-based scaling laws for the durability of thermal barrier coatings," *Progress in Materials Science*, 46 249-271 (2001).

[9]A. G. Evans, D. R. Mumm, J. W. Hutchinson, G. H. Meier and F.S. Pettit, "Mechanisms controlling the durability of thermal barrier coating," *Progress in Materials Science*, 46 505-553 (2001).

[10]T. A. Cruse, S. E. Stewart and M. Ortiz, "Thermal Barrier Coating Life Prediction Model Development," Journal of Engineering for Gas Turbines and Power, 110 610-616 (1988).

[11]S. M. Meier, D. M. Nissley, K. D. Sheffler and T. A. Cruse, "Thermal Barrier Coating Life Prediction Model Development," *Journal of Engineering for Gas Turbines and Power*, 114 258-263 (1992).

CORROSION RIG TESTING OF THERMAL BARRIER COATING SYSTEMS

Robert Vaßen, Doris Sebold, Gerhard Pracht, Detlev Stöver
Institut für Werkstoffe und Verfahren der Energietechnik
Forschungszentrum Jülich GmbH
Jülich, 52425, Germany

ABSTRACT

A burner rig test facility was built up which allows the testing of protective coatings as thermal and environmental barrier coatings under harsh environmental conditions. In the experimental set-up liquids as salt solutions or kerosene are injected into a gas burner flame, accelerated and heated or burned in the combustion gases and finally deposited on the sample surface. As the samples are cooled by compressed air a thermal gradient can be established across the coatings allowing typically a surface temperature of 1150-1250 °C and a bond coat temperature of about 950 -1000°C.

YSZ based thermal barrier coating systems have been cycled in the new rig using different corrosive liquids as pure water, NaCl and Na_2SO_4 solutions. Depending on the concentration of the corrosive media the lifetime of the coatings were reduced up to a factor of more than 100. Microstructural evaluation as well as X-ray diffraction analysis will be presented and the reasons for the early failure of the coatings discussed.

INTRODUCTION

Thermal barrier coating systems (TBCs) are widely used in land-based or aero gas turbines to improve their performance. The isolative layer can provide a significant reduction of the temperature of the metallic substrate which results in an improved component durability. Alternatively, an increase of efficiency can be achieved by allowing an increase of the turbine inlet temperatures [1].

TBC systems consist typically of two layers, a so-called bond coat layer, and an isolative, ceramic topcoat. The bond coat is in most cases an alloy and has two major functions. It improves the bonding between the substrate and the topcoat and it protects the substrate from corrosion and oxidation. Two types of bond coats are frequently used, a (platinum-) aluminide based one and a so-called MCrAlY with M being Ni and/or Co. The choice of the adequate bond coat depends on the used deposition technique for the topcoat. Electron beam physical vapour deposition (EB PVD) and atmospheric plasma spraying (APS) are the most frequently used techniques. The meanwhile as standard material established oxide is the 6-8 wt.-% Y_2O_3 stabilised zirconia (YSZ, [2]) which is frequently used in aero and stationary gas turbines from the beginning of the eighties [3, 4, 5, 6]. The stabilizing agent is essential for the performance of the TBCs to avoid the phase transition from tetragonal to monoclinic which is observed in pure zirconia.

All of the different components, substrate, bond coat, and topcoat, interact with each other or the environment to a more or less extent and/or they undergo detrimental changes due to thermo-mechanical treatments during operation.

This study is focused on the influence of corrosive media on the performance of TBCs. Corrosive media can be introduced from the environment e.g. in aviation engines from sand particles during landing and take-off. In addition, also the fuel contains corrosive constituents, the most harmful are vanadium, sulphur and sodium. The concentration can vary in wide ranges. While for aviation fuel the sulphur content is limited to 0.05 wt.%, it may reach 4 wt.% in heavy oil for stationary gas turbines [7].

A large number of investigations have been performed in the past to study the influence of the corrosive media in laboratory tests. An older review of the results has been made by Bürgel and Kvernes [8] and a more recent one by Jones [9], showing the most important reactions of the corrosive media with the TBCs. Under severe corrosive environments it is found that the stabilizing agent reacts with the corrosive media leading to a destabilizing of the coating. Two examples are:

$$ZrO_2(Y_2O_3) + V_2O_5 \longrightarrow ZrO_2 \text{ (monoclinic)} + 2YVO_4 \qquad (1)$$

$$ZrO_2(Y_2O_3) + 3SO_3(+Na_2SO_4) \rightarrow ZrO_2 \text{ (monoclinic)} + Y_2(SO_4)_3 \qquad (2)$$
$$\text{(in Na}_2\text{SO}_4 \text{ solution)}$$

Also new stabilizing agents as Sc_2O_3 or new ceramics have been studied [10, 11]. A general improvement of the corrosion behaviour of these materials compared to YSZ was not found. To the authors knowledge no material was identified up to now which shows largely improved corrosion resistance compared to YSZ for most of the relevant testing conditions and corrosive media.

The distinct influence of the testing conditions also raises the question under which conditions corrosion tests should be performed. Typically, the testing is performed under isothermal conditions in a furnace with testing temperatures up to about 1100°C, typically between 800 and 1000°C. In earlier investigations sometimes also burner rigs are used for corrosion testing [12, 13]. The surface temperature in these studies was also well below 1100°C at about 982°C, the substrate was at about 842°C. In this study a major influence of the bond coat composition on the performance of the TBC systems was found. Typically higher Cr and Al contents improved the lifetime of the coatings. These results show that the corrosion of the bond coat and the attack of the formed thermally grown oxide play a major role. Typically in the temperature range above 800°C type I hot corrosion is expected. Reactions of the formed Al_2O_3 scale with SO_3 containing atmospheres can be as follows [14], with (3) being dominant at high temperatures (hot corrosion type I):

$$Al_2O_3 + O^{2-} \longrightarrow 2\,AlO_2^- \quad \text{for low SO}_3 \text{ pressures} \qquad (3)$$

$$Al_2O_3 + 3SO_3 \rightarrow Al_2(SO_4)_3 \quad \text{for high SO}_3 \text{ pressures} \qquad (4)$$

In addition, under type I conditions, the sulphide formation of metals (Cr, Ni) at the slag/metal interface play an important role:

$$M + SO_2 \longrightarrow \text{M-oxide} + \text{M-sulphide} \qquad (5)$$

In the present investigation corrosive media are injected into the flame of a burner rig. With the used setup temperature profiles can be adjusted which are considered as relevant for modern gas turbines (i.e. bond coat temperatures in the range of 900 to 1000°C and surface temperature of about 1200°). Hence a rather realistic testing of TBC systems under corrosive conditions should be possible.

EXPERIMENTAL

The investigated thermal barrier coating systems have been produced by plasma spraying with two Sulzer Metco plasma-spray units. Vacuum plasma spraying with a F4 gun was used to deposit a 150 μm NiCo21Cr17Al13Y0.6 bond coat (Ni 192-8 powder by Praxair Surface Technologies Inc., Indianapolis, IN) on disk shaped IN738 superalloy substrates. The diameters of the substrates used for thermal cycling tests were 30 mm, the thickness 3 mm. At the outer edge a radius of curvature of 1.5 mm was machined to reduce the stress level.

The ceramic top coats with a thickness of about 350 μm were produced by atmospheric plasma spraying (APS) using a Triplex I gun. During the manufacture of the thermal cycling specimens also steel substrates were coated. These coatings were used to characterize the as-sprayed condition. The porosity level was measured using free - standing coatings with mercury porosimetry given values of about 12 vol.%. The Argon and Helium plasma gas flow rates were 20 and 13 standard liter per minute (slpm), the plasma current was 300 A at a power of 20 kW.

Corrosion tests were performed in a gas burner rig setup, in which the disk shaped specimens were periodically heated up to the desired surface temperature of about 1200 °C in approximately 1 min by a natural gas/ oxygen burner. After 5 min heating the specimens were cooled for 2 min from both sides of the specimens by compressed air. The surface temperature was measured with an infrared pyrometer operating at a wavelength of 9.6 - 11.5 μm and a spot size of 5 mm, while the substrate temperature was measured using a NiCr/ Ni thermocouple inside a hole drilled to the middle of the substrate. More details on the standard thermal cycling rigs can be found in [15]. In this investigation the central gas nozzle was used to introduce atomized liquids containing the corrosive media into the flame. In this paper only the results on water based salt solutions will be presented. Other media as kerosene are also possible.

In Figure 1 a photo of the rig in operation is shown. The injection of water droplets into the flame led to a reduce stability of the flame and therefore larger variations in the temperature profile. In order to evaluate this effect also a pure water injection (without additional corrosive media) was examined.

Fig. 1 Flame of the corrosion rig during injection of Na_2SO_4 solution.

The test conditions used with respect to the corrosive media are given in Table 1. The concentrations of the corrosive media have been calculated from the injected quantities of water-based solutions and the mean methane and oxygen gas flows during thermal cycling. It turned out that the gas flows of the burner gases had to be varied in rather wide ranges to adjust similar temperature profiles. This fact led to the rather large differences in the corrosive media concentration in Table 1 column 4. As the injection nozzle is located in the center of the flame the profile of the corrosive media concentration has a maximum in the centre. This concentration profile has not been considered for the calculation of the concentration in Table 1.

The test was stopped when obvious degradation of the coatings (large delamination) occurred during the thermal cycling. This definition of failure results in an uncertainty in the lifetime data especially if partial delamination of the coatings occurs. However, as seen below the resulting error does not effect the interpretation of the results.

Sprayed specimens were vacuum impregnated with epoxy, and then sectioned, ground and polished. As for the preparation water was used as a media, water-soluble corrosion products might be removed by the preparation process. Cross-sections of the coatings were examined by optical microscopy and scanning electron microscopy (Ultra 55, Zeiss). The surface of the tested samples were analysed by X-ray diffraction using a Siemens D5000 facility at a wavelength of 1.5406 Å.

Sample #	Corrosive media	Mass flow of corrosive solution [g/min]	Appr. concentration of corrosive media (Cl/S) in relation to CH_4 [%]
A	Non	-	-
B	Water	1.4 - 2.8	-
C	1% NaCl in water	1.4	0.433
D1	0.1 % Na_2SO_4 in water	1.4	0.010
D2a	1.0 % Na_2SO_4 in water	1.4	0.108
D2b	1.0 % Na_2SO_4 in water	1.4	0.122
D2c	1.0 % Na_2SO_4 in water	1.4	0.039
D2d	1.0 % Na_2SO_4 in water	1.4	0.098
D3	10 % Na_2SO_4 in water	1.4	0.588

Table 1 Corrosive testing conditions.

RESULTS AND DISCUSSION

In Table 2 and Figure 1 the temperatures and the cycles to failure of the thermal cycling tests are given. The listed temperatures are mean values calculated from the temperature recordings during the heating phase. The standard deviation of the bond coat temperature is about 50 K, for the surface temperature it is about 75 K. For the used test conditions the bond coat temperature is about 40 K higher than the substrate temperature. The results refer to single samples, D2a to D2d show results of samples cycled under similar conditions.

Sample #	Substrate temperature [°C]	Surface temperature [°C]	Cycles to failure	Time at temperature [h]
A	about 950	about 1200	about 5000	333
B	965	1253	4402	293
C	944	1225	936	62
D1	960	1242	1243	83
D2a	938	1241	547	36
D2b	923	1162	614	41
D2c	981	1195	150	10
D2d	948	1235	843	56
D3	942	1156	21	1.4

Table 2 Results of thermal cycling with corrosive media (conditions of samples A to D are explained in Table 1).

Progress in Thermal Barrier Coatings

The lifetime of the sample cycled with pure water injection (sample B) is similar to the results of coatings cycled without additional injection (sample A). The stated last value is a mean value taking into account several experiments in the indicated temperature regime. More details are given in [16]. Obviously the use of water does not significantly influence the lifetime of the coatings. In contrast, the cycling with corrosive media led to a significant reduction of lifetime for all investigated conditions.

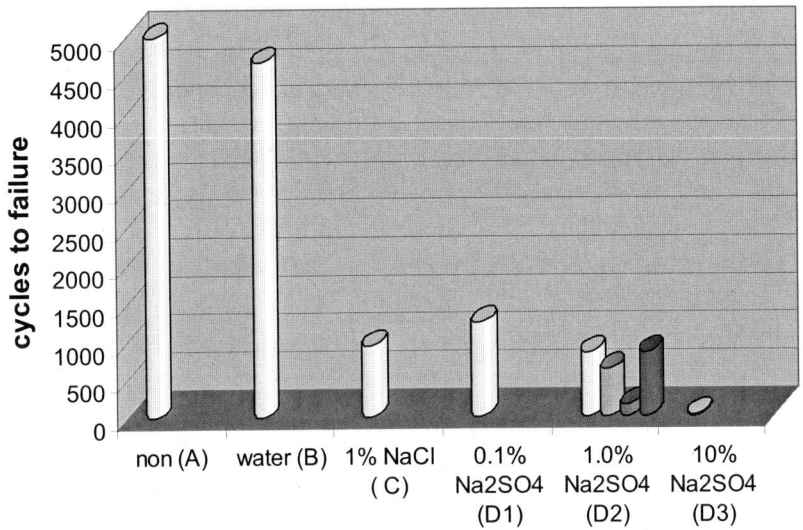

Figure 1 Cycles to failure for the used corrosion conditions (Table 2).

Cycling with NaCl media (sample C)
Sample C cycled with 1% NaCl injection showed an early failure with a lifetime reduction of about a factor of 5 compared to thermal cycling without corrosive media. The sample shows a massive spallation of the coating and also a yellowish and greenish appearance (Figure 2). In contrast, sample B cycled with water appears still white.
The XRD pattern (Figure 3) shows peaks corresponding to following three major phases, the tetragonal YSZ phase, the fcc phase of the Ni-base superalloy and α− alumina phase. There is no indication of the formation of significant amounts of monoclinic phase. These results correspond to the results of Mitamura et al. who did not find a reaction between 3YSZ and NaCl + KCl in the temperature range between 850 and 1050°C [17].

Figure 2 Photos of samples after thermal cycling, left with water injection (sample B in Table 1+2), right with 1% NaCl injection (sample C).

SEM-images of the metallographic cross-section of the sample are shown in Figure 4 a,b). In the left image a long crack in the TBC is found which runs close to the bond coat / TBC interface. In the right micrograph the interface is shown with a higher magnification. At three locations EDX analysis has been performed (Fig. 4, c-e). At location 1 mainly the elements Al, Y, Cr, Co, Ni and Cr were found. At location 2 mainly Al, Cr and less Y showed up. At the third location peaks from Al, Ni, Co, Zr, relatively much Cr and clear indications of Na and Cl were visible. Although at other locations clear evidence of Na and Cl were found, it is assumed that the alumina scale is attacked by the liquid NaCl (melting point 800°C) which can easily penetrate through the interconnected cracks to the thermally grown oxide. Although it is known that Al_2O_3 scales are relatively resistant to Cl-containing gases [18] the given conditions obviously significantly degraded the alumina scale. Once the alumina scale is removed elements as Cr, Ni and Co can easily react with the Cl-species.

In addition to the changes of the TGO also at some locations an infiltration of the micro cracks by Cr-species has been detected. An explanation of this finding can be the formation of liquid or even volatile $CrCl_2$ species which are transported by capillary forces or via the gas phase through the crack network.

A more detailed analysis and discussion of the reaction will be performed in the future.

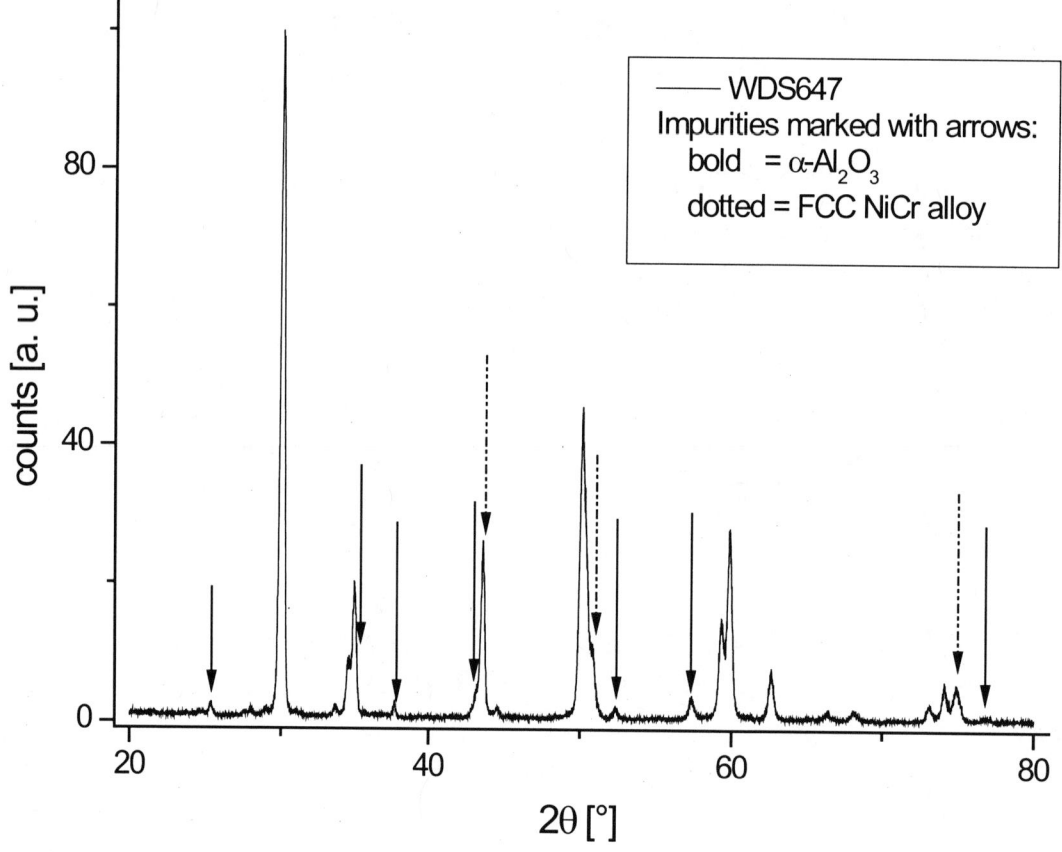

Figure 3 X-ray diffraction patterns of samples cycled with NaCl injection (sample C, Table 1). Peaks not indexed can be ascribed to YSZ (tetragonal phase).

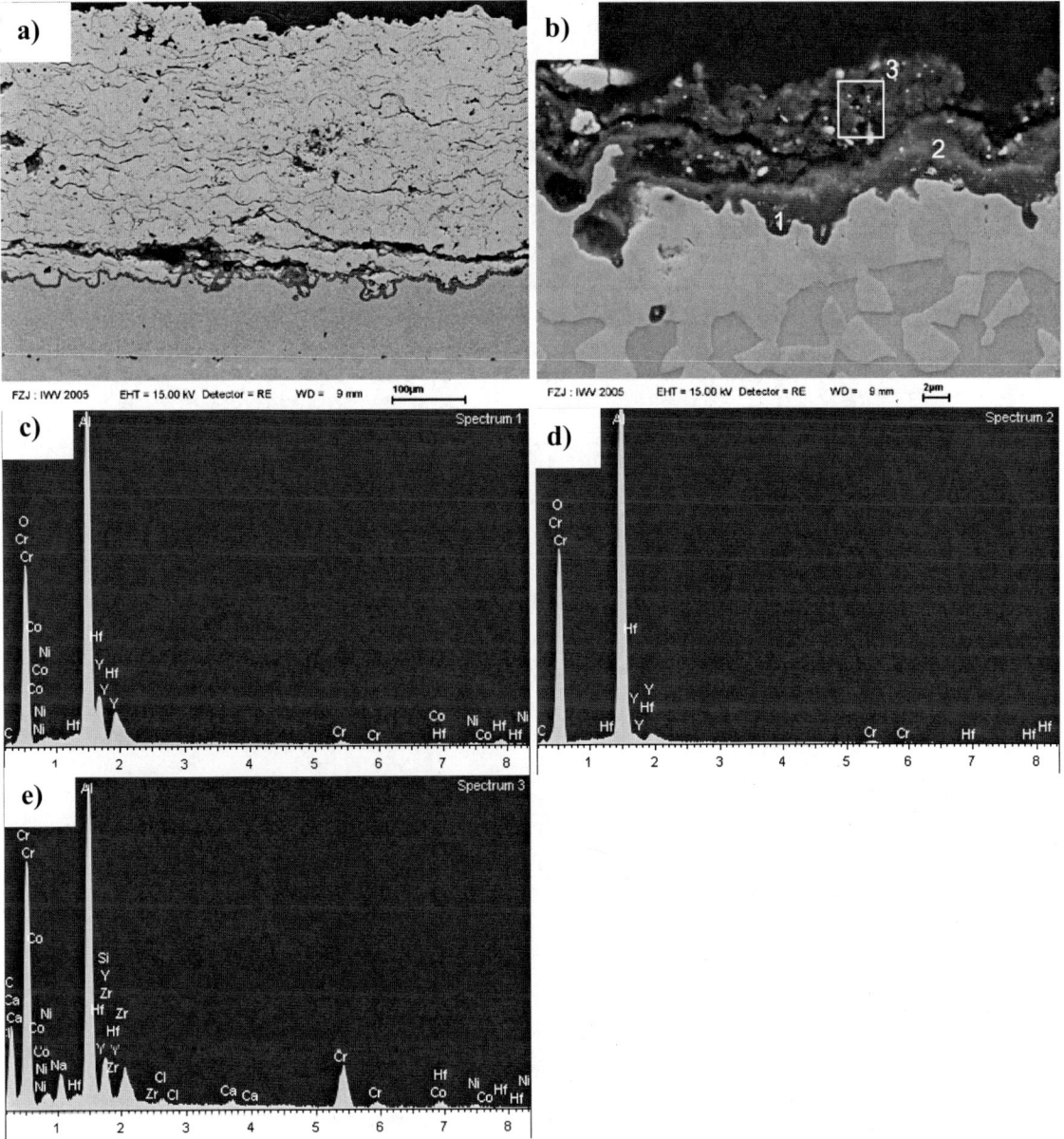

Figure 4 SEM images (a, b) of metallographic cross-sections as well as EDX results (c, d, e) at the different locations given in b) of the sample cycled with NaCl addition.

Cycling with Na_2SO_4 media (samples D)

Similar to the results of the thermal cycling experiment with NaCl addition also the Na_2SO_4 injection led to an early failure of the samples for all investigated concentrations (see Table 2). Photos of the samples after cycling are shown in Figure 4. Similar to the NaCl injection also here different colourations of the surface of the samples are visible. The circular spall pattern of sample D1 can be attributed to the rather circular concentration profile of the corrosive media due to the central injection.

For the Na_2SO_4 three different concentrations have been used. Clearly, an increased concentration of corrosive media led to a reduction of lifetime. Especially the use of solutions with the highest concentration (10%) resulted in a very early failure of the coatings after about 21 cycles (D3).

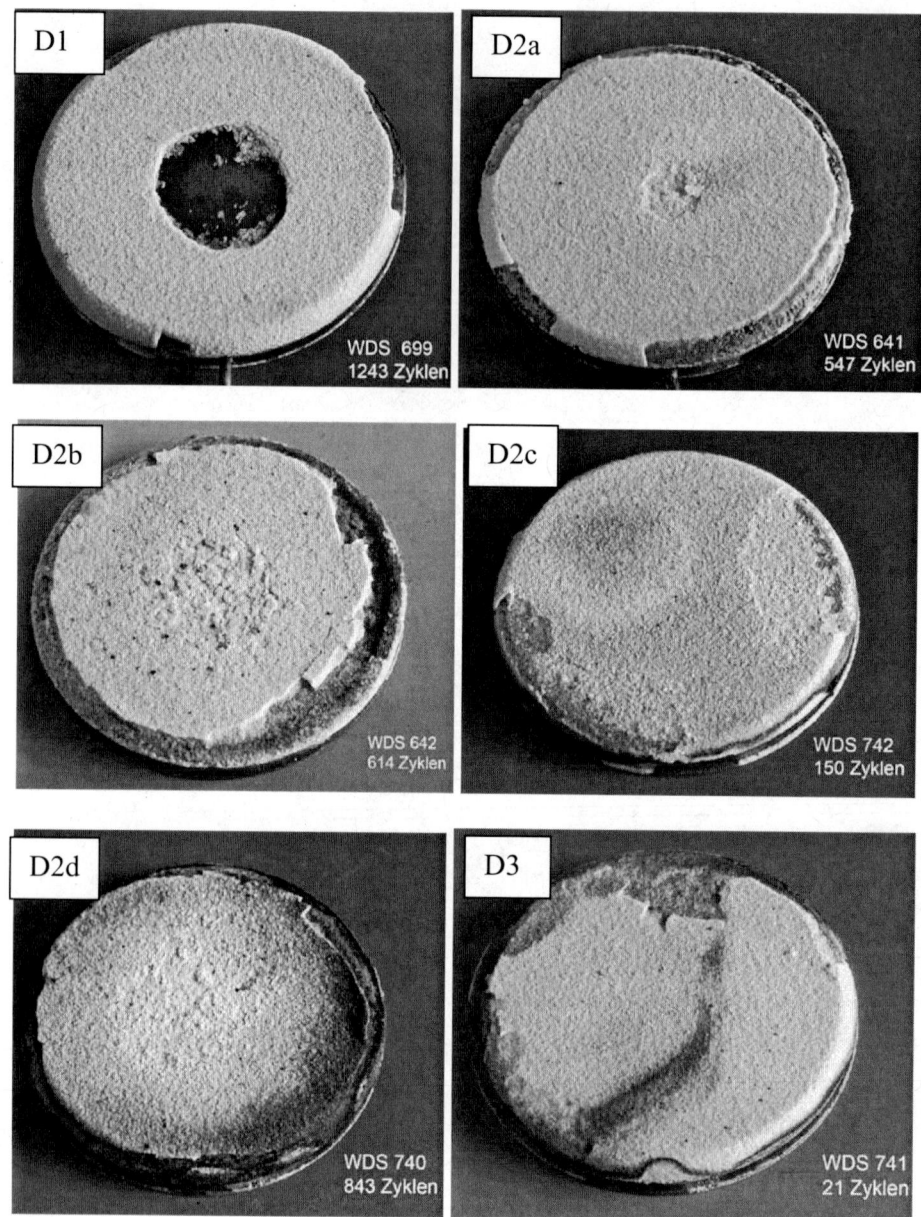

Figure 5 Photos of the samples after thermal cycling with Na_2SO_4 injection corresponding to the samples D1, D2a – d, and D3 from top left to bottom right.

In Fig. 6 the XRD patterns for samples cycled with the 0.1 and 1% Na_2SO_4 injection are plotted. All patterns reveal that the tetragonal YSZ phase is the dominant phase. There are only minor peaks at about $2\theta = 28°$ which can be attributed to the existence of monoclinic ZrO_2. The corresponding amount of monoclinic phase is similar to the one in the as-sprayed condition [19], i.e. the cycling under the investigated conditions did not significantly destabilize the YSZ as might be expected according to reactions as given in (2). Also in [11] YSZ underwent little change in a sulphur containing environment (SO_2 + Na_2SO_4 + $MgSO_4$) even after treatment for 360 h at 900°C. Obviously, the higher testing temperature (>1150°C for the TBC surface) used in the present investigation did not significantly accelerate the reaction of sulphur containing species and YSZ.

Besides the YSZ phase also additional phases can be identified by the XRD patterns. For the samples D2a and D2b Na_2SO_4 is clearly visible, which is expected due to its injection in the gas flame. For D1 further phases are found. Looking at Figure 5 it is visible that a part of the coating spalled off completely. Hence, the layers formed on the bond coat are contributing to the XRD patterns. Consequently, the thermally grown oxide consisting of spinel and alumina layers are found. In addition, also the nickel substrate is visible.

In addition to the XRD analysis metallographic sections have been prepared and analysed by optical and in the case of sample D2a by electron microscopy. Two examples of optical micrographs are shown in Figure 7.

In both micrographs cracks are found within the TBC close to the bond coat. While in D2a the crack is still rather short, a long crack is observed in D3. In addition, also large dark areas are found in the TBC, which will be discussed below.

A better insight in the development of the TBC system during thermal cycling in corrosive media is possible with a SEM analysis of the coatings. Figure 8 shows SEM images of sample D2a (1% Na_2SO_4). In the two top images it is obvious that the dark "TGO" in the central part is much thicker than at the outer rim of the sample. This can be correlated to the central injection of the corrosive media. In fact, the "TGO" is not simply an alumina scale but much more complex as seen in Fig 8 c.

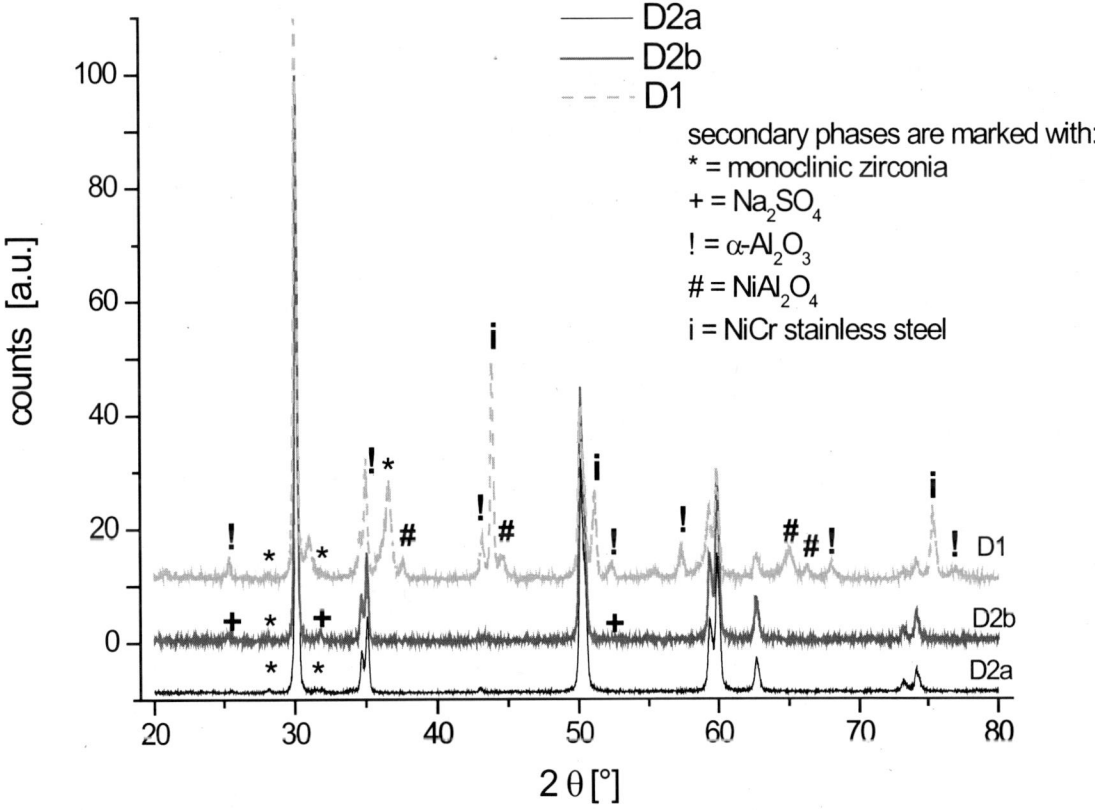

Figure 6 X-ray diffraction patterns of samples cycled with Na_2SO_4 injection (samples D1, D2a, D2b, Table 1). Peaks not indexed can be ascribed to YSZ (tetragonal phase).

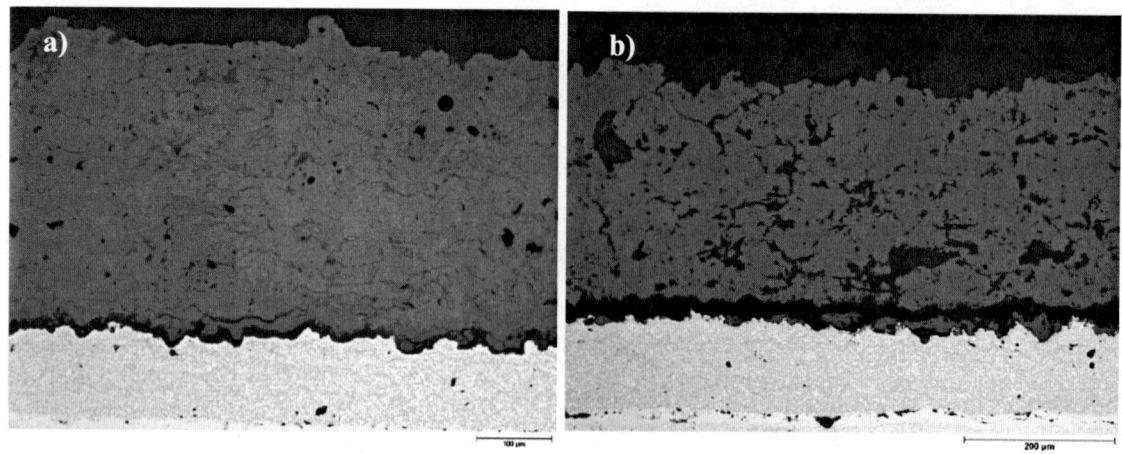

Figure 7 Optical micrographs of sample cycled with the injection of 1% (D2a, a) and 10 % Na$_2$SO$_4$ (D3, b).

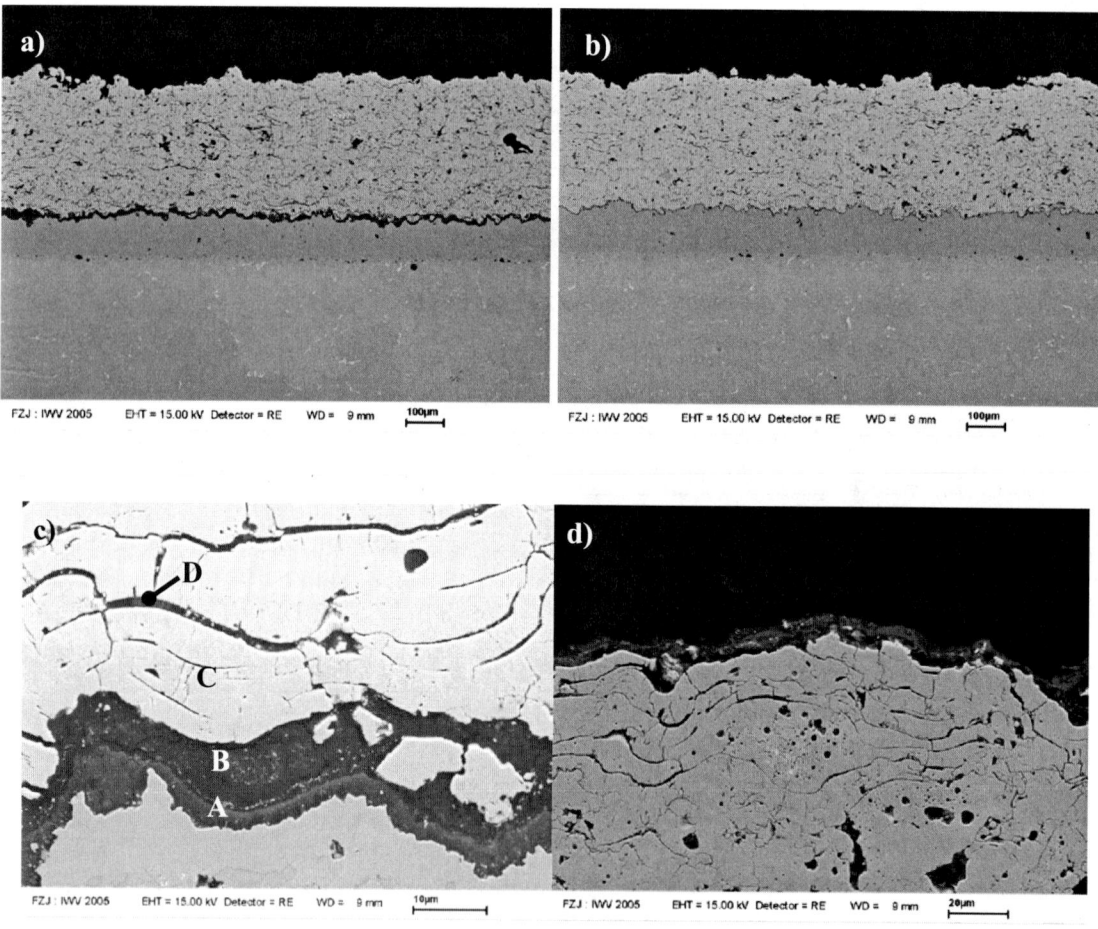

Figure 8 SEM images of sample cycled with the injection of 1% (D2a), central part (a), area close to the outer rim (b), central part at bond coat TBC interface (c), surface region of the TBC (d).

Location	Al	Cr	Ni, Co	Na, S	Zr, Y
A	++				
B	++	++	+	++	
C					++
D	++	++		++	++

Table 3 Qualitative results of an EDX analysis of the micrograph in Fig. 8c at the given locations A-D. Only major elements are shown. Oxygen is present at all locations.

A qualitative EDX analysis of the regions in Figure 8c is shown in Table 3. Close to the bond coat still a part of the alumina scale is found. On top of this scale, complex corrosive products have been formed (region B in Fig. 8c) which consist of Al/Cr and Na containing compounds with some additions of Ni and Co. According to equation (3) the formation of $NaAlO_2$ can lead to the basic fluxing of the alumina scale. After removal of the protective scale at different locations the reaction of Na_2SO_4 and oxygen with the Ni base bond coat will start leading to the observed corrosion products. Both oxygen and sulfur is observed in the EDX analysis. So, it can not be determined by the present investigation whether mixed oxide, sulfates or sulfides of the various elements as Al, Cr, Ni are present. In [20] complex oxides have been found together with some remaining alumina scale after hot corrosion tests with Na_2SO_4 at 950°C. However, in this investigation only low levels of sulfur were found in the oxide areas. This might be related to the use of pure oxygen in this investigation.
An indication of internal sulfide formation of the bond coat, often found in type I hot corrosion, according to reaction (5) was not found in the samples.
At location C only YSZ could be detected, no indication of TBC decomposition was found. Within the micro cracks (location D) often corrosion products have been observed. Probably capillary forces led to the filling of the micro cracks at high temperatures by the liquid corrosion products.
Sample D2c showed a significantly lower lifetime than the other samples cycled with the same Na_2SO_4 concentration. This might be related to the relatively high substrate temperature which can lead to an increased melt formation and penetration of corrosive products into the TBC. A detailed analysis of the influence of the temperature profile will be made in the future.
Finally, the sample tested with the highest amount of Na_2SO_4 (D3c) will be analyzed by SEM. In Figure 7 b already large dark areas have been detected in this sample. In the SEM image this areas can be identified as pores (Figure 9a). It is possible that these pores have been filled with water soluble corrosion products which have been removed during metallographic sample preparation. Clearly the size and the amount of the pores is much larger than in the as-sprayed condition. In addition, the appearance of the pore surface is unusual for plasma-sprayed coatings (Figure 9 b). The surface is highly fragmented indicating severe attack of the former rather straight surfaces. The described findings suggest a considerable attack of the YSZ by Na_2SO_4 at the given high concentration. Under this conditions failure took place within the ceramic and not by fluxing of the oxide layer on the bond coat. The oxide layer seems to be intact and rather thin (Figure 9a) corresponding to the short time at high temperature.

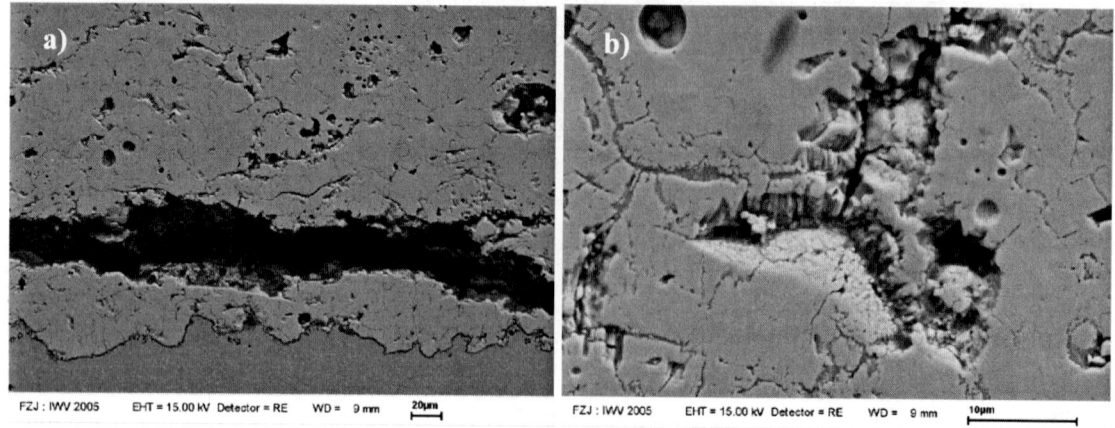

FZJ : IWV 2005   EHT = 15.00 kV Detector = RE   WD = 9 mm   20µm

FZJ : IWV 2005   EHT = 15.00 kV Detector = RE   WD = 9 mm   10µm

Figure 9 SEM-images of a metallographic section of a sample cycled with 10% Na_2SO_4 (D3c).

Further work will be performed in the near future to achieve a better insight in the corrosion mechanisms. In addition, additional corrosive media as kerosene and V containing compounds will be tested.

CONCLUSIONS

The results of a new burner rig allowing corrosion testing of TBC systems under rather realistic conditions have been presented. The injection of NaCl and Na_2SO_4 led to a reduction of lifetime by more than a factor of 100 for the highest concentration (0.6 % S). Failure of the coatings was in most cases related to fluxing of the TGO oxide by the corrosive species. Only for the highest S-concentration failure of the ceramic topcoat has been observed.

ACKNOWLEDGEMENT

The authors would like to thank Mr. K.H. Rauwald and Mr. R. Laufs (both IWV1, FZ Jülich) for the manufacture of the plasma-sprayed coatings and Mrs. A. Lemmens for the thermal cycling of the specimens. The authors also gratefully acknowledge the work of Mrs. H. Moitroux (IWV1), Mr. P. Lersch (IWV2), Mrs. S. Schwartz-Lückge (IWV1) and Mr. M. Kappertz (IWV1) who supported the characterization of the samples by photography, XRD, optical microscopy, and sample preparation.

REFERENCES

1 P. Hancock, and M. Malik, Materials for Advanced Power Engineering Part 1, D. Coutsouradis et al. (eds.), Kluwer Academic Publishers, Dordrecht, 1994, 658-704.
2 S. Stecura, Advanced Ceramic Materials, 1 [1] (1986) 68-76.
3 S. Bose, J. DeMasi-Marcin, J. of Thermal Spray Technology 6 [1] (1997) 99-104.
4 W.A. Nelson, R.M. Orenstein, J. of Thermal Spray Technology, 176 (1997) 176-80.
5 J. Wigren, L. Pejryd, Proc. of the 15th Int. Thermal Spray Conf., ASM International, Ohio, USA, 1998, 1531-1541.
6 D.R. Clarke and C.G. Levi, Annu. Rev. Mater. Res. 33 (2003) 383-417.
7 B.R. Marple, J. Voyer, C. Moreau, D.R. Navy, "Corrosion of Thermal Barrier Coatings by Vanadium and Sulfur Components," Materials at High Temperatures, 17 (3) (2000) 397-412.

8 R. Bürgel, I. Kvernes, "Thermal Barrier Coatings," High Temperature Alloys for GasTurbines and Other Applications 1986, W. Betz et al. , ed. D. Reidel Publiching, 1986, p.327-356.

9 R.L. Jones, "Some Aspects of the Hot Corrosion of Thermal Barrier Coatings," J. of Thermal Spray Technology, 6 (1) (1997) 77-84.

10 M. Yoshiba, K. Abe, T. Arami, Y. Harada, "High-Temperature Oxidation and Hot Corrosion Behavior of Two Kinds of Thermal Barrier Coating Systems for Advanced Gas Turbines," J. of Thermal Spray Technology, 5 (3) (1996) 259-68.

11 B.R. Marple, J. Voyer, M. Thibodeau, D.R. Nagy, R. Vassen, "Hot Corrosion of Lanthanum Zirconate and Partially Stabilized Zirconia Thermal Barrier Coatings," J. of Engineering for Gas Turbines and Power, July 2005 Vol 127, p1-9.

12 P.E. Hodge, R.A. Miller, M.A. Gedwill, "Evaluation of the Hot Corrosion Behaviour of Thermal Barrier Coatings," Thin Solid Films, 73 (1980) 447-453.

13 I. Zaplatynsky, "Performance of Laser-Glazed Zirconia Thermal Barrier Coatings in Clyclic Oxidation and Corrosion Burner Rig Tests," Thin Solid Films, 95 (1982) 275-284.

14 P. Kofstad, High Temperature Corrosion, Elsevier Appl. Sci., London, 1988.

15 F. Traeger, R. Vaßen, K.-H. Rauwald, D. Stöver, "A Thermal Cycling Setup for Thermal Barrier Coatings," Adv. Eng. Mats., 5, 6 (2003) 429-32.

16 F. Traeger, M. Ahrens, R. Vaßen, D. Stöver, "A Life Time Model for Ceramic Thermal Barrier Coatings," Materials Science and Engineering, A358 (2003) 255-65.

17 T. Mitamura, E. Kogure, F. Noguchi, T. Iida, T. Mori, Y. Matsumoto, "Stability of Tetragonal Zirconia in Molten Fluoride Salts," Advancces in Ceramics, Vol 24°, Science and Technolgy of Zirconia III, S. Somiya, N. Yamamoto, H. Yanagida, eds., Amercan Ceramic Society, 1988, p. 109-118.

18 R. Bürgel, Handbuch Hochtemperatur-Werkstofftechnik, Friederich Vieweg-Verlagsgesellschaft, Braunschweig, 1998.

19 J.-E. Döring, R. Vaßen, D. Stöver, R.G. Castro, „Particle Properties Tailor Coating Microstructure, Porosity and Phase Composition," Proceedings of the 2003 International Thermal Spray Conference, Orlando, FL, USA, 5-8 May 2003, edited by B.R. Marple, C. Moreau, ASM International, Materials Park, OH, 2003, pp. 1197-1204.

20 C. Leyens, I.G. Wright, B.A. Pint, „Hot Corrosion of an EB-PVD Thermal Barrier Coating System at 950°C," Oxidation of Metals, 54, 5/6 (2000) 401-424.

OXIDATION BEHAVIOR AND MAIN CAUSES FOR ACCELERATED OXIDATION IN PLASMA SPRAYED THERMAL BARRIER COATINGS

Hideyuki Arikawa, Yoshitaka Kojima
Hitachi Research Laboratory, Hitachi, Ltd
Hitachi, 317-8511, Japan

Mitsutoshi Okada, Takayuki Yoshioka and Tohru Hisamatsu
Materials Science Research Laboratory, Central Research Institute of Electric Power Industry
Yokosuka, 240-0196, Japan

ABSTRACT

The growth of thermally grown oxide (TGO) can influence the durability of thermal barrier coatings (TBCs). In order to clarify its main causes and to find a method to restrain it, we examined an influence of surface treatment on a bond coat and its yttrium (Y) content on TGO growth. Shot peening was applied on the bond coat surface of TBC specimens, and high temperature oxidation behavior in air was compared with that of the TBC specimens without shot peening. As a result, in the specimen without shot peening, wart-like oxide was observed on a continuous oxide layer, which was formed between the bond coat and top coat. The wart-like oxide was caused due to unmelted particles of the bond coat. And in the continuous oxide layer, Y-rich granular microstructure was found. On the other hand, by means of shot peening on the bond coat surface, unmelted particles were removed, and the formation of the wart-like oxide was restrained. The growth of the continuous oxide layer, moreover, was also decreased because the Y-rich microstructure in the oxide was reduced. And the TBC specimen with the bond coat eliminating Y-content (Y-free bond coat) was prepared, and its oxidation behavior was examined by the high-temperature oxidation test in air. In the TBC specimen with Y-free bond coat, whether or not shot peening was applied, the growth of the continuous oxide was restrained since Y did not exist in the oxide. The possibility that Y-free bond coat could restrain TGO growth was found.

INTRODUCTION

Turbine inlet temperatures (TIT) of gas turbines have been increased in order to improve thermal efficiency of combined cycles using them, and the temperatures have reached around 1500°C recently. In order to withstand such critical thermal conditions, Co-base or Ni-base superalloys with excellent high-temperature strength are employed in hot-gas-path parts, and, moreover, thermal barrier coatings (TBCs) are applied for better heat resistance.

In general, TBC consists of top coat (TC) and bond coat (BC). Yttria-partially stabilized zirconia (YSZ) is used for TC, and MCrAlY metals (M is cobalt and/or nickel), which are also used for corrosion-resistant coatings, are employed for BC[1].

TBC is exposed in high-temperature oxidizing environment during the operation of gas turbines, and the oxide layer at the boundary between TC and BC, which is called thermally grown oxide, TGO, grows due to the oxidation of BC. It is considered that the growth of the oxide layer may accelerate the delamination of TBC[2-6]. Thus, it is one of important issues to restrain the growth of the oxide layer in order to improve the reliability of the hot-gas-path parts and to increase their lives.

It is reported that the oxide layer formed in TBC grows faster than that on MCrAlY coating[2]. It is considered that the oxide growth is influenced by the formation of wart-like oxide consisting of BC compositions such as Ni and Co[2] and/or the segregation of yttrium in the oxide layer[3, 4]. As for the formation of the wart-like oxide, it is reported that they are caused by unmelted particles formed in spraying process, and that surface polish and chromate processing can restrain its growth. On the other hand, as for the segregation of yttrium, its influence on the accelerated oxidation has not been clarified yet. It can be concluded that the cause of the faster growth of the oxide layer in TBC has not been fully elucidated yet.

In this paper, the authors examined the main causes for the growth of the oxide layer at the boundary between TC and BC in TBC in order to find a method to restrain its growth. The oxidation behavior was observed when the unmelted particles were removed from the BC surface and when yttrium was eliminated from the chemical composition of BC. And, in each case, the observation was also performed when top coat was not applied. And then, we examined the effects to remove the unmelted particles and to eliminate yttrium on the restraint of the oxide growth at the boundary.

EXPERIMENTAL PROCEDURE

Figure 1 illustrates schematic representation of specimen shape and coatings. Substrate material was Inconel738LC, and its size and shape was 20mm×70mm×3mm. 100μm thick of CoNiCrAlY bond coat (Co-32.0Ni-21.2Cr-8.0Al-0.5Y (wt %)) was applied on the substrate plates by means of low pressurized plasma spraying, and after that, diffusion heat treatment was carried out. In order to examine the influence of surface treatment on BC, any treatment was not performed on one specimen, and shot peening was applied on the other. And 200μm thick of YSZ top coat (ZrO_2-8wt%Y_2O_3) was applied on one side of each specimen, and top coat was not

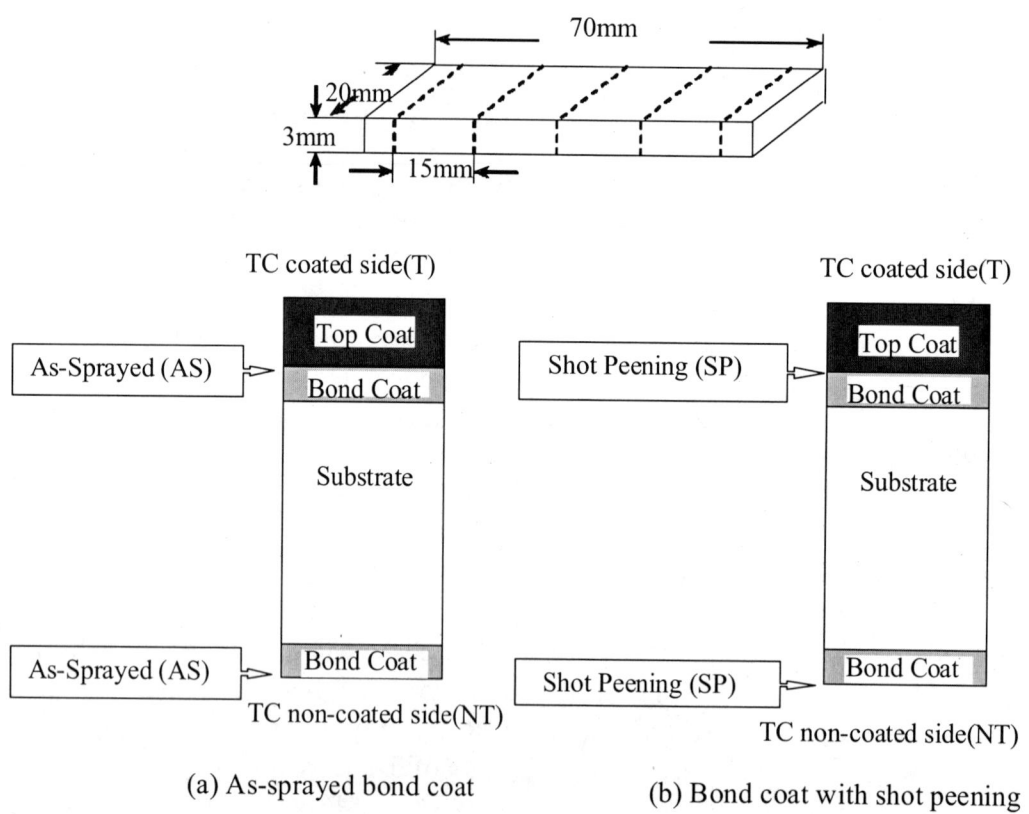

(a) As-sprayed bond coat (b) Bond coat with shot peening

Fig. 1 Schematic representation of specimen shape and coatings

overlaid on the other. The influence of TC, therefore, was examined by comparing the sides with TC and without TC. In this paper, "AS" and "SP" represent as-sprayed BC and BC with shot peening, respectively. And "T" and "NT" represent the side with TC and the one without TC, respectively. For example, the TC coated side of the specimen with as-sprayed BC is described as "AS-T", and the TC non-coated side of the specimen with shot-peened BC was described as "SP-NT".

The CoNiCrAl metal powder was prepared in order to examine the influence of yttrium in BC on the oxidation behavior. The powder consists of the chemical composition where yttrium was eliminated from CoNiCrAlY metal, and the analysis of the powder showed its chemical composition of Co-31.7Ni-20.7Cr-7.8Al (wt%). The coated specimens were produced in the same condition as above. "YF" indicates that the BC without yttrium (Y-free BC) was applied. According to the analysis by an electron probe micro-analyzer (EPMA), the bond coat did not contain yttrium.

One plate was cut into four specimens after coatings were applied, and then they were used for oxidation tests.

The oxidation tests were carried out in air at 950°C. A test time was defined as the one held at the test temperature. The specimens were taken out from a furnace after certain oxidation time. And they were cut and polished, and then its microstructure in the cross section was analyzed. For the microstructural analysis, scanning electronic microscopy (SEM) and electron probe micro-analyzer (EPMA) were used.

The heat cycle test was conducted in air by means of an electric furnace. One heat cycle consists of a rapid heat-up from room temperature to 1100°C for 15minutes, a 10hours-hold at 1100°C, and a 15 minutes cooling to room temperature by a fan. TBC failure was defined as the cycle when 20% of whole area of top coat was delaminated, and then, the thermal cycling was stopped.

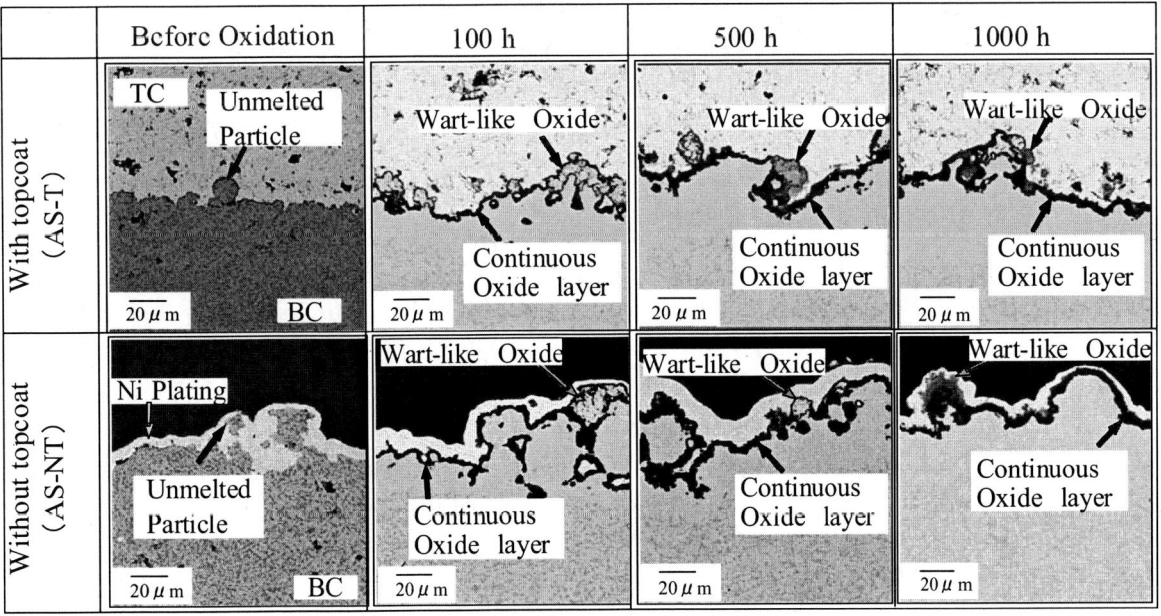

Fig. 2 Cross-sectional morphologies of the specimens with as-sprayed bondcoat after oxidation test at 950°C in air

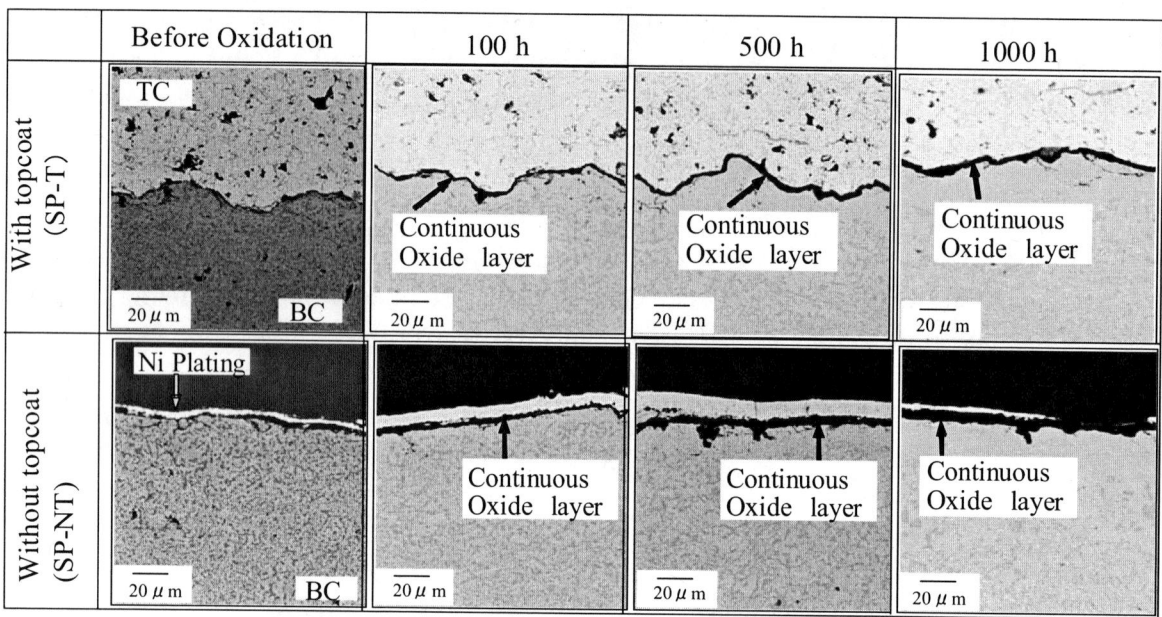

Fig. 3 Cross-sectional morphologies of the specimens with shot peening treatment on bondcoat surface after oxidation test at 950°C in air

RESULTS AND DISCUSSION

Influence of surface treatment for bond coat on oxide morphology

Figure 2 indicates cross-sectional morphologies of the specimens with as-sprayed bond coat. The upper side and the bottom side in the figure show the specimen with TC (AS-T) and the one without TC (AS-NT), respectively. In the specimen without TC, Ni-plating was performed so as not to damage the oxide layer in the process of cutting and grinding. In the specimen with TC, an oxide layer was observed along the BC/TC boundary, and it is defined as continuous oxide layer. The continuous oxide layer consisted mainly of alumina according to EPMA analysis. On the continuous oxide layer, moreover, wart-like oxides were formed locally. It has been reported that the wart-like oxide was found particularly in TBC. It has been also pointed out that the oxide results from unmelted particles left on the BC/TC boundary in spraying process. The unmelted particles were found also in our specimen with as-sprayed BC before the test. According to EPMA analysis, the wart-like oxide consisted of the oxides of aluminum and chromium in the external scale of the particle and of the metals of cobalt and nickel in its center after 100hours of the oxidation test. After 500hours, the metals in the center almost disappeared and the whole particle changed into the mixed oxide of aluminum, chromium, cobalt and nickel. The above analysis ensures that the wart-like oxide results from the unmelted particle of BC formed in coating spraying. Its formation process can be described as follows. At first, the unmelted particle, which was not contacted with BC, was oxidized, and mainly alumina was formed. And then, since aluminum in the particle was consumed for oxidation, the mixed oxide of cobalt and nickel was formed. In the specimen without TC, the continuous oxide layer and wart-like oxide were also observed on as-sprayed BC surface.

Figure 3 indicates cross-sectional morphologies of the specimens with shot peening on BC surface. The upper side and the bottom side in the figure show the specimen with TC (SP-T) and the one without TC (SP-NT), respectively. In the specimen without TC, Ni-plating was performed as well as in Figure 2. The BC surface treatment restrained the formation of the wart-like oxide with TC as well as without TC, and only the continuous oxide layer was formed. Even after 1000hours of the oxidation test, the wart-like oxide was not observed.

320

Whether or not TC was applied, the wart-like oxide was formed in the specimen with as-sprayed BC, and it was restrained in the specimen with BC surface treatment. That is because the BC surface treatment removes the unmelted particle that is the cause of the wart-like oxide.

Influence of top coat on the thickness of continuous oxide layer

In order to examine the influence of TC on the oxidation behavior, the continuous oxide layer and the wart-like oxide were distinguished, and only the thickness of the former was considered. That is because the formation mechanism of the wart-like oxide is different from that of the continuous oxide layer.

The thickness of the continuous oxide layer is defined as follows. The area of the continuous oxide layer was measured from the cross-sectional micrograph. And, dividing the area by the length of the BC/TC boundary, the thickness was obtained. The average of 10 fields of view with a magnitude of 500times was used as a thickness value of certain test condition. The average square root of the unbiased variance of the thickness was about 0.4μm in all the specimens. The oxide was observed in the specimen even before the oxidation test. It was formed during the thermal treatment after spraying process. Its thickness, however, was too small to be measured by the above method.

Figure 4 indicates the relationship between the thickness of continuous oxide layer and time. Close and open symbols represent the specimens with TC and those without TC, respectively. As shown in the figure, when the surface condition was same, significant difference in the thickness of the continuous layer was not found between the specimens with TC and without TC. This result, therefore, reveals that the chemical reaction of BC with TC and the change of the partial pressure of oxygen due to overlaid TC do not have an influence on the thickness growth of the continuous oxide layer to the extent of these oxidation tests.

Influence of surface treatment for bond coat on the growth of continuous oxide layer

The influence of the surface treatment for BC on the thickness of the continuous oxide

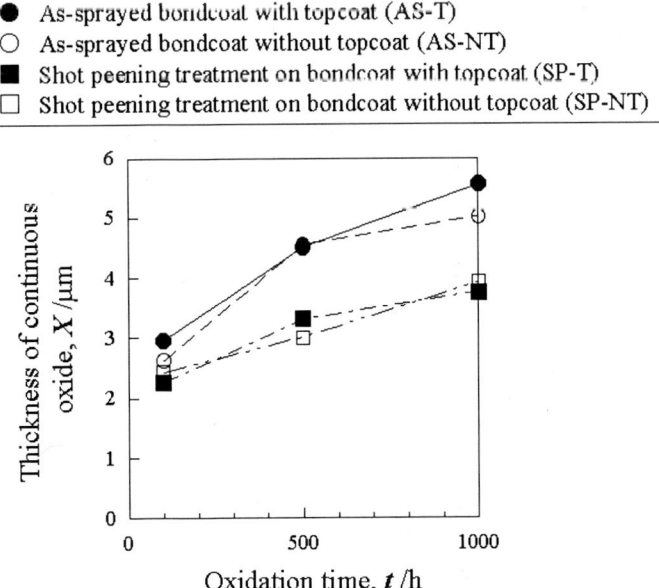

Fig. 4   Relationship between thickness of continuous oxide layer and time in oxidation tests at 950°C in air

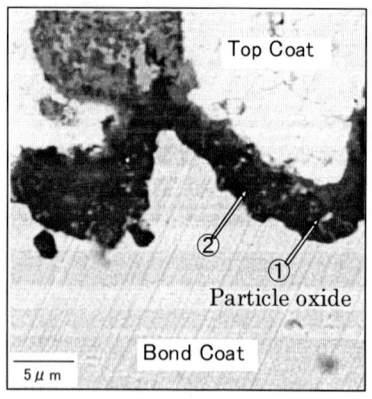

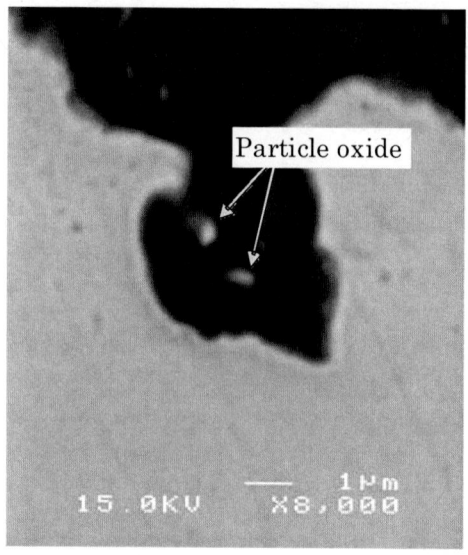

	Concentrations (mass%)						
	Al	O	Y	Co	Ni	Cr	Zr
①	43.8	45.5	9.1	0.6	0.6	0.4	0.01
②	47.5	46.0	1.0	2.0	2.1	1.3	0.04

(a) Cross-sectional micrograph and results   (b) Morphology of the boundary between
    of quantitative analysis by EPMA   continuous oxide layer and bond coat

Fig. 5   Analytical results of the continuous oxide layer in TBC specimen with as-sprayed bond coat
       (AS-T) after oxidation test at 950°C for 1000h in air

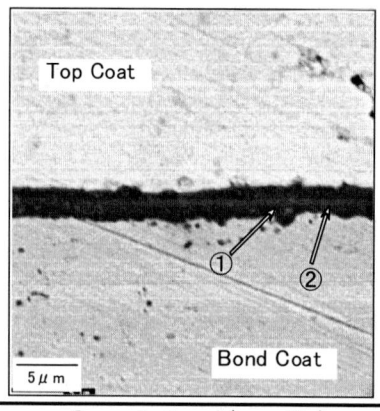

	Concentrations (mass%)						
	Al	O	Y	Co	Ni	Cr	Zr
①	47.5	40.9	3.0	1.7	1.7	2.0	3.1
②	43.1	36.1	2.4	5.6	5.6	3.5	3.7

Fig. 6   Analytical results of the continuous oxide layer in TBC specimen with shot peening treatment
       on bond coat surface (SP-T) after oxidation test at 950°C for 1000h in air

layer was also shown in Figure 4. Whether or not TC is applied, the thickness of the continuous oxide layer in the specimen with the BC surface treatment was thinner than that in the specimen with as-sprayed BC. This result clarifies that the BC surface treatment not only restrains the formation of the wart-like oxide as described above, but also reduces the thickness of the continuous oxide layer.

      The continuous oxide layer was observed in more detail and analyzed by EPMA in order to examine the cause of the difference in the thickness of the continuous oxide layer due to the BC surface treatment. Figures 5 and 6 show analytical results of the continuous oxide layer in the specimen with as-sprayed BC and of that with the BC surface treatment, respectively. Both specimens were oxidized at 950°C for 1000h in air.

White granular microstructure was scattered inside the continuous oxide layer in the specimen with as-sprayed BC (AS-T) as shown in Figure 5. The chemical compositions were analyzed by EPMA as for the granular microstructure (① in Figure 5 (a)) and the other part in the continuous oxide layer (② in Figure 5(a)). And the microstructure of ① was identified as yttrium-rich oxide. As shown in Figure 5(b), in some local parts, the continuous oxide layer surrounded the granular microstructure and the layer grew inside BC.

On the other hand, the white granular microstructure was hardly observed in the specimen with the BC surface treatment (SP-T) as shown in Figure 6. EPMA analysis could not find the Y-rich microstructure indicated in Figure 5.

Zirconium was detected in the EPMA analysis of figures 5 and 6. The authors estimate that this results from the chemical composition of TC.

The BC surface treatment causes the difference in the microstructure of the continuous oxide layer, and this leads to the difference in its thickness. It is suggested that yttrium-rich granular oxide is the origin of the oxidation and that it results in the growth of the boundary oxide layer since there is a possibility that the diffusion rate is comparatively high in the yttrium-rich oxide or in the boundary between the Y-rich oxide and alumina[5,7]. Our research also ensures that a number of Y-rich granular oxides were included in the continuous oxide layer of the specimen with as-sprayed BC and that this granular oxide was the origin of oxidation. The acceleration mechanism of oxidation due to yttrium such as how yttrium is distributed in bond coat and how it forms granular oxide, however, has not been clarified yet, and it is necessary to be investigated more in detail. On the other hand, in the specimen with the BC surface treatment, the formation of alumina with smaller amount of impurities such as Y-rich oxide restrains the growth of the continuous oxide layer. As for its cause, it is pointed out that the surface treatment produces lattice defects at the vicinity of surface and that it accelerates the oxidation of aluminum and the growth of protective alumina[6], but it is necessary to be examined further.

Influence of Yttrium in bond coat on the growth of continuous oxide layer

Figure 7 indicates cross-sectional morphologies of TBC specimens with Y-free BC after

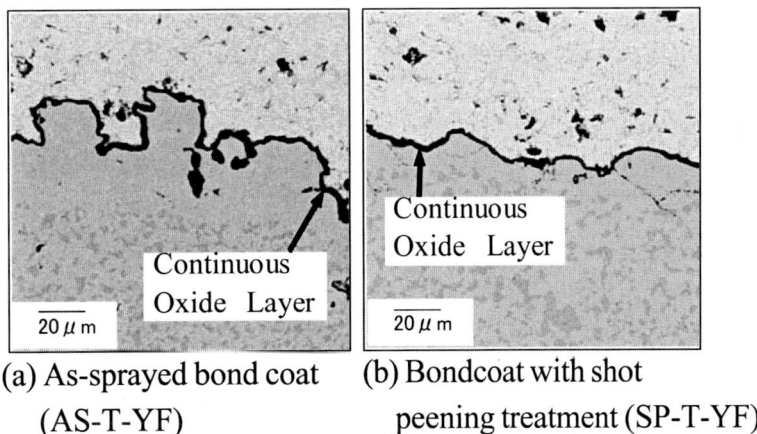

(a) As-sprayed bond coat   (b) Bondcoat with shot
    (AS-T-YF)   peening treatment (SP-T-YF)

Fig. 7   Cross-sectional morphologies of TBC specimens with Y-free bondcoat after oxidation test
    at 950°C for 1000h in air

oxidation test at 950°C for 1000h in air. Figures 7(a) and (b) show the TBC specimen with as-sprayed BC (AS-T-YF) and the one with BC surface treatment (SP-T-YF), respectively. In the Y-free BC specimen with the surface treatment (SP-T-YF), only the continuous oxide layer of alumina was formed along the boundary of BC/TC as well as in the Y-included BC (CoNiCrAlY coated) specimen with the surface treatment (SP-T) shown in Figure 3. In the specimen with as-sprayed BC (AS-T-YF), even though some parts were estimated to be unmelted particles, the continuous oxide layer was formed along the rough boundary, and the wart-like oxide including cobalt or nickel was not observed.

The relationship between the thickness of continuous oxide layer and time in TBC specimens with Y-free BC was indicated in Figure 8. In the specimens with Y-free BC (AS-T-YF

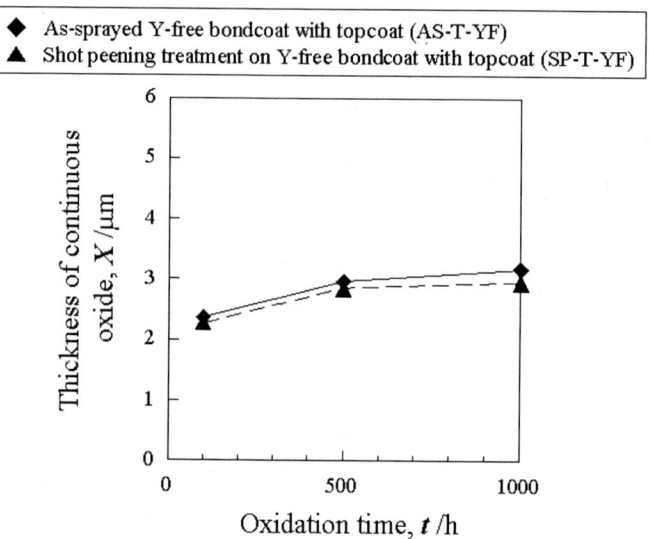

Fig. 8 Relationship between thickness of continuous oxide layer and time in TBC specimens with Y-free bondcoat in oxidation test at 950°C in air

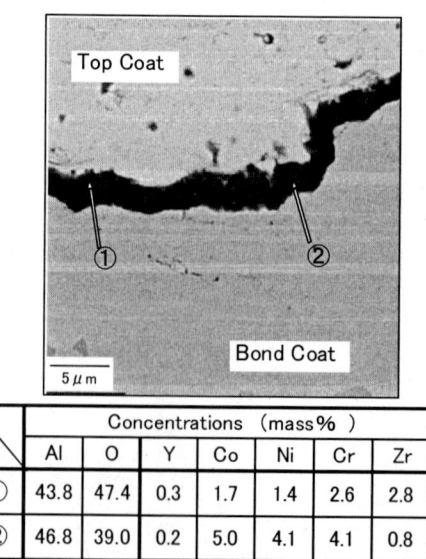

	Concentrations (mass%)						
	Al	O	Y	Co	Ni	Cr	Zr
①	43.8	47.4	0.3	1.7	1.4	2.6	2.8
②	46.8	39.0	0.2	5.0	4.1	4.1	0.8

Fig. 9 Analytical results of the continuous oxide layer in TBC specimen with as-sprayed Y-free bond coat (AS-T-YF) after oxidation test at 950°C in air

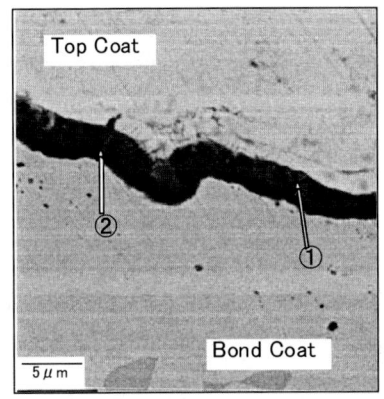

	Concentrations (mass%)						
	Al	O	Y	Co	Ni	Cr	Zr
①	47.4	44.6	0.3	1.7	1.3	1.8	3.0
②	47.8	46.0	0.02	1.7	1.4	2.4	0.6

Fig. 10 Analytical results of the continuous oxide layer in TBC specimen with Y-free bond coat and shot peening treatment (SP-T-YF) after oxidation test at 950°C in air

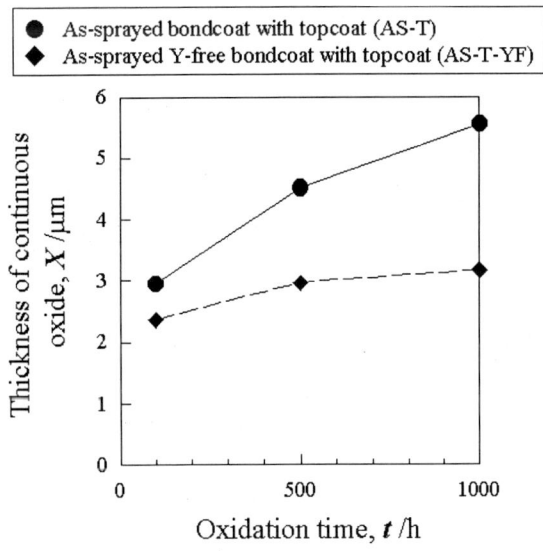

Fig.11   Relationship between thickness of continuous oxide layer and time in TBC specimens with CoNiCrAlY bondcoat and Y-free CoNiCrAl bondcoat in oxidation test at 950°C in air. Bond coat is as-sprayed.

and SP-T-YF), whether or not the BC surface treatment was carried out, the thicknesses of the continuous oxide layer were almost same. Figures 9 and 10 show analytical results of the continuous oxide layer in the specimen with as-sprayed Y-free BC and in the one with the surface treatment on Y-free BC, respectively. Whether or not the BC surface treatment was applied, granular microstructure was not included in the continuous oxide layer as observed in Figure 5(b). In both specimens, the shape of the boundary between the continuous oxide layer and BC was relatively flat, and the local growth of the oxide into BC shown in Figure 5(b) was not found. According to EPMA analysis on the specimens, moreover, the continuous oxide layer was alumina, and Y-rich microstructure was not found. Subtle content of yttrium was detected in the continuous oxide layer in Figures 9 and 10. That is because the thickness of the oxide layer was so small that the electronic beam of EPMA detected the yttrium of top coat or because yttrium might diffuse from TC. The EPMA analysis also comfirmed that Y-free BC did not contain yttrium even after the oxidation test.

Figure 11 shows the relationship between the thicknesses of the continuous oxide layer and time in the specimens with as-sprayed Y-included BC (AS-T) and with as-sprayed Y-free BC (AS-T-YF). In the specimen with Y-included BC, since the continuous oxide layer including yttrium was formed as described in the previous paragraph, thicker oxide layer grew. On the other hand, in the specimen with since Y-free BC, alumina with relatively smaller amount of impurities was formed, and its growth rate was smaller.

Figure 12 indicates the relationship between the thicknesses of continuous oxide layer and time in the specimens with Y-included bond coat and (SP-T) and with Y-free one (SP-T-YF) where the BC surface treatment was performed. Since the continuous oxide layer with smaller content of yttrium was formed in the specimens with the BC surface treatment, its thickness was almost same in the specimen with Y-included BC as in the specimen with Y-free BC. In the long-term oxidation test, the thickness was thinner in the specimen with Y-free BC than in the one with Y-included BC.

In the specimen with Y-free BC, whether or not the BC surface treatment was applied, yttrium was not included in the continuous oxide layer and its thickness was thinner. This clarifies that yttrium accelerates the growth of the continuous oxide layer. Since yttrium was not included by eliminating it from the chemical composition of BC, it was supplied from BC and the yttrium in TC does not have an influence on the oxidation.

Figure 13 summarizes the factors of the accelerated oxide growth in TBC. The oxidation at the boundary of TBC is accelerated by following factors; ① the formation of the wart-like oxide, and ② yttrium in bond coat. The formation of the wart-like oxide results from unmelted particles on BC surface, and the surface treatment can remove them. But, since the rough boundary is indispensable to adhere BC and TC, it is necessary to develop other surface treatment that not only

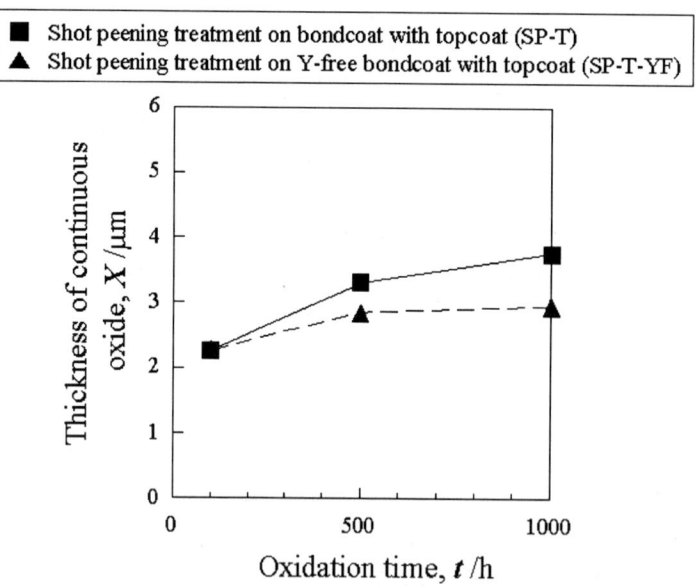

Fig. 12   Relationship between thickness of continuous oxide layer and time in TBC specimens with CoNiCrAlY bond coat and Y-free CoNiCrAl bond coat in oxidation test at 950°C in air. Shot peening was carried out on bond coat surface.

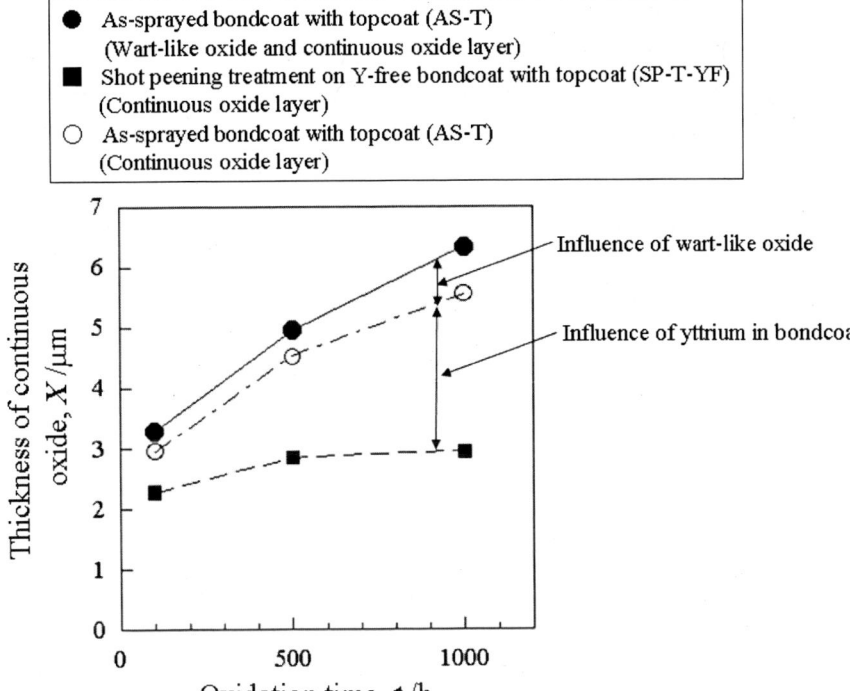

● As-sprayed bondcoat with topcoat (AS-T)
  (Wart-like oxide and continuous oxide layer)
■ Shot peening treatment on Y-free bondcoat with topcoat (SP-T-YF)
  (Continuous oxide layer)
○ As-sprayed bondcoat with topcoat (AS-T)
  (Continuous oxide layer)

Figure 13 Factors of oxide growth acceleration in TBC

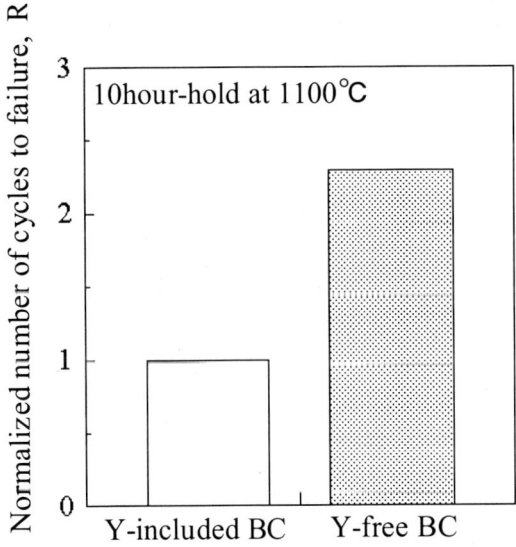

Figure 14 Normalized number of cycles to failure in thermal cycling test

can remove the unmelted particles and but can maintain the rough boundary.

On the other hand, yttrium in bond coat is included in the boundary oxide, and it accelerates its growth. The surface treatment and/or BC without yttrium can restrain the growth of the continuous oxide layer. Yttrium in CoNiCrAlY coating is added in order to improve the adhesion between the protective oxide layer of alumina and the coating[8]. It is, therefore, necessary to examine the influence of yttrium on the resistance to delamination.

Resistance to thermal cycling of TBC with Y-free bond coat

The thermal cycling test was performed by means of the specimens with Y-included BC and Y-free BC. Figure 14 shows the normalized number of cycles to failure of each specimen. The normalized number of heat cycles to failure of the specimen with Y-free BC was about twice larger. These results reveal that the CoNiCrAl (Y-free) bond coat is effective in order to improve the durability of TBC. One of the reasons is that it restrains the growth of continuous oxide layer, but the further analysis is necessary.

SUMMARY

The TBC specimen with shot-peening treatment on bond coat surface was prepared in order to clarify the influence of the surface treatment on the growth of the oxide layer at the boundary between bond coat (BC) and top coat (TC) in thermal barrier coating (TBC). Its high-temperature oxidation in air was compared with that of the specimen with as-sprayed BC. The high-temperature oxidation test, moreover, was carried out by means of the specimen with Y-free BC, and the influence of yttrium on the oxide layer growth was also examined. The results are as follows.

(1) The continuous oxide layer was formed at the BC/TC boundary of the specimen with as-sprayed BC, and the wart-like oxide, which resulted from unmelted particle of BC, was observed. In the continuous oxide layer, Y-rich granular microstructure was found.

(2) The shot-peening treatment on BC surface removed the unmelted particles, and the formation of the wart-like oxide was restrained. The content of yttrium, moreover, was decreased in the continuous oxide layer, and its growth was also restrained.

(3) In the specimen with Y-free BC, yttrium did not exist in the continuous oxide layer. And, whether or not the surface treatment was applied, its growth was restrained. There is a possibility that Y-free BC can restrain the growth of the continuous oxide layer.

REFERENCES

[1]H. Arikawa and Y. Kojima, "Heat Resistant Coatings for Gas Turbine Materials (in Japanese)," J. the Surface Finishing Soc. Japan, 52, 11-15 (2001).

[2]M.Narumi, Z.Yu, K.Taumi and T.Narita, "Oxidation Behavior of Plasma Sprayed CoNiCrAlY/YSZ Film at 1173K and 1273K in Air (in Japanese)," Zairyo-to-Kankyo, 50, 466-471 (2001).

[3]A.Rabiei and A.G.Evans, "Failure Mechanism Associated with the Thermally Grown Oxide in Plasma-sprayed Thermal Barrier Coatings," Acta mater., 48, 3963-3976 (2000).

[4]S.Takahashi, M.Yoshiba and Y.Harada, "Nano-Characterization of Ceramic Top-Coat/ Metallic Bond-Coat Interface for Thermal Barrier Coating Systems by Plasma Spraying," Materials Transactions, 44, 1181-1189 (2003).

[5]L.Lelait, S.Alpérine and R.Mévre, "Alumina scale growth at zirconia-MCrAlY interface: a microstructural study," Journal of Materials Science, 27, 5-12 (1992).

[6]T.Teratani, T.Suidzu, K. Tani and Y.Harada, "Formation of Alumina Protective Layer on MCrAlY Atmospheric Plasma Sprayed Coating by Chromate Processing (in Japanese)," J. High Temperature Soc. 29, 247-252 (2003).

[7]J. Klöwer, "Factors affecting the oxidation behaviour of thin Fe-Cr-Al foils Part II The effect of alloying elements: Overdoping," Materials and Corrosion. 51, 373-385 (2000).

[8]K. Shimotori and T. Aisaka, "The Trend of MCrAlX Alloys for High-temperature-Protective Coatings –On the Effects of Alloy Compositions- (in Japanese)," Tetstu-to-Hagane, 69, 1229-1241 (1983).

CRACK GROWTH AND DELAMINATION OF AIR PLASMA-SPRAYED Y$_2$O$_3$-ZrO$_2$ TBC AFTER FORMATION OF TGO LAYER

Makoto Hasegawa[1], Yu-Fu Liu[1] and Yutaka Kagawa[1,2]

[1]Research Center for Advanced Science and Technology, The University of Tokyo,
4-6-1 Komaba, Meguro-ku,
Tokyo, 153-8904, Japan

[2]Center for Collaborative Research, The University of Tokyo,
4-6-1 Komaba, Meguro-ku,
Tokyo, 153-8904, Japan

ABSTRACT

Failure behavior in an air plasma-sprayed Y$_2$O$_3$-ZrO$_2$ thermal barrier coating systems (TBCs) under tensile loading has been examined. The coating system is heat exposed in ambient air at 1423 K for 10, 50 and 100 h. Tensile testing of the system is conducted to observe cracking behavior in the TBC layer. Periodic cracking in the TBC layer, which is located perpendicular to loading direction, is observed after tensile loading. Density of the cracks increases with the increase in heat exposure time.

INTRODUCTION

Thermal barrier coating systems (TBCs) are widely used to protect the hot section components in gas turbine blades and vanes from high temperature [1, 2]. They usually consist of outer ceramic TBC layer and inner intermetallic bond coat (BC) layer for protecting the nickel-base superalloy substrate from the high temperature and from the oxidation. Failure of TBCs certainly shortens the service life of components and hence understanding of the failure behavior of the TBCs is important for improving the performance of components.

Failure of TBCs under unexpected over-load under service leads damage in the TBC systems. Therefore, it is important to understand damage of TBC systems under tensile force. Studies on the damage under a tensile load have been reported [3, 4]. However, the reports do not reveal the effect of heat exposure on the failure behavior. The properties of constitutes and entire system of a TBC system strongly depends on a history of heat exposure, as reported elsewhere [5-8]. However, the cracking behavior of the TBC layer under in-plane tensile force is not well known. The present study has been focused on the experimental observation of the failure behavior in TBC layer under tensile loading.

EXPERIMENTAL PROCEDURE

Thermal barrier coating layer (TBC) of an Y_2O_3 (7~8 mass%)-partially-stabilized-ZrO_2 was deposited on bond coat using an atmospheric plasma spray process. Bond coat (BC) material was a NiCoCrAlY alloy (Co22 Cr17 Al12.5 Y0.6 and balance Ni in mass%) that was coated on a nickel-base superalloy substrate (Inconel 738, Co8.5 Cr16 Al3.5 Ti3.5 W2.6 Mo1.8 Nb0.9 and balance Ni in mass%) by a low-pressure plasma spraying process. The thickness of TBC and BC layers, respectively, was ~200 and ~100 μm. The coated materials were heat exposed in ambient air at 1423 K for 10, 50 and 100 h. Microstructural characterizations on as-sprayed and heat-exposed TBC systems were performed on polished transverse plane of the TBC system using optical microscope (OM) and scanning electron microscope (SEM).

Specimens for tensile test were prepared from as-sprayed and heat-exposed TBC systems by diamond-saw cutting and polishing processes. **Figure 1** shows testing configurations of the specimen. Specimen size was 40 mm in length, ℓ, 2 mm in width, w, and 3.4 mm in thickness, t, which included TBC and BC layers. Side surfaces of the specimen were polished up to 0.5 μm diamond paste finish to allow direct observation of the failure behavior.

Tensile tests were performed in ambient air atmosphere at 295 K by using a screw-driven type testing machine. Tests were done with a crosshead speed of 0.1 mm/min. Tensile strain was measured with a strain gage (effective gage area: 2 × 0.84 mm). The strain gage was attached to the center of the substrates bottom. After tests, fracture behavior of the TBC systems was observed by OM and SEM.

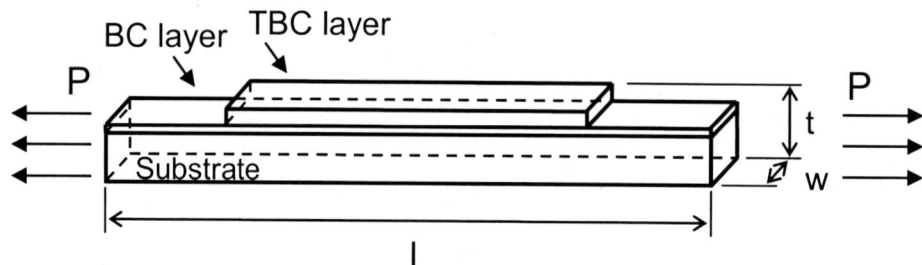

Fig. 1 Shape of tensile test specimen (P: applied load).

RESULTS AND DISCUSSION

Figure 2 shows entire view of the polished sections of the thermal barrier coating systems before and after heat exposure. Hereafter, respective layers are designated as thermal barrier coating (TBC) layer, bond coat (BC) layer and thermally grown oxide layer (TGO). Defects such as pores and inter-splat boundaries are observed in the polished cross section of the TBC layer as black spots and black lines, respectively. Voids and inter-splat boundaries are observed in as-sprayed state (Fig. 2(a)) and the numbers of them seem to decrease with the increase in heat exposure time (Fig. 2 (b) ~ (d)). Such the change in microstructure is caused by

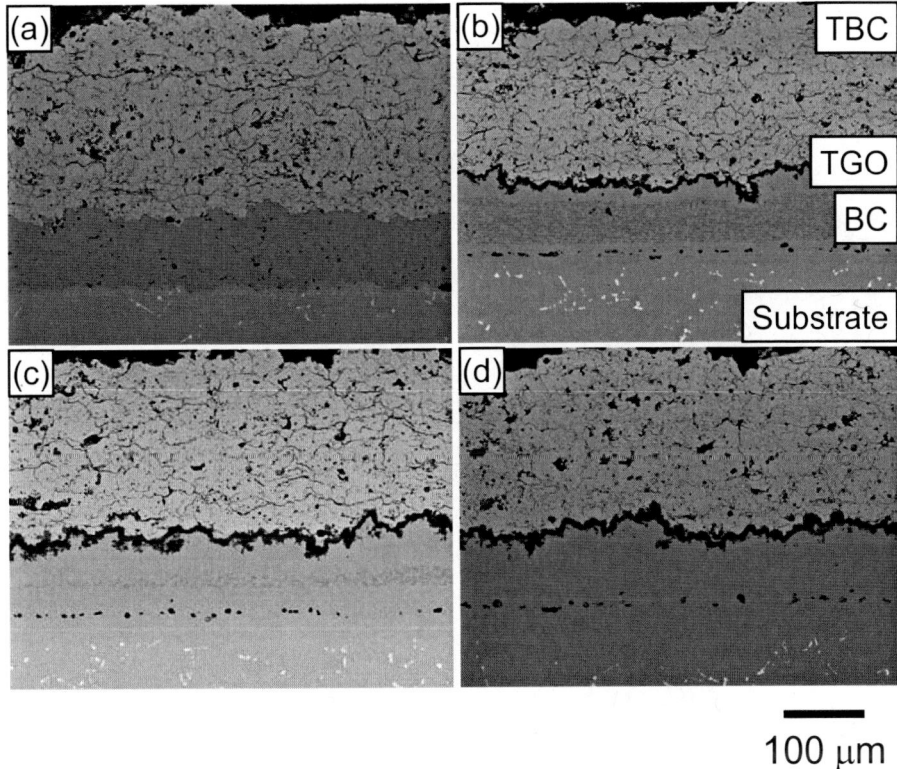

Fig. 2 SEM micrographs of polished transverse section of TBC system: (a) as-sprayed, (b), (c) and (d) after heat exposure for 10, 50 and 100 h, respectively.

shrinkage of TBC layer by progress of sintering. More detail sintering behaviors of the same material are reported elsewhere [5, 6].

The stress-strain curves obtained by tensile test of as-sprayed and heat-exposed TBC specimens are shown in **Fig. 3**. The specimens were loaded at a strain $\varepsilon = 0.01$. For all curves derived from TBC system specimens, starts from a linear regime and then shows nonlinear behavior. This behavior is observed independent of heat exposure time. In the as-sprayed specimen, a maximum tensile stress was ~900 MPa and the maximum stress tends to decrease with the increase in the heat exposure time. The range of the maximum stress in the heat-exposed specimen was 650~620 MPa. Transverse cracking of TBC is observed when the stress reaches ~ 600 MPa and the cracks are formed with and equal spacing. **Figure 4** shows typical optical micrographs of the

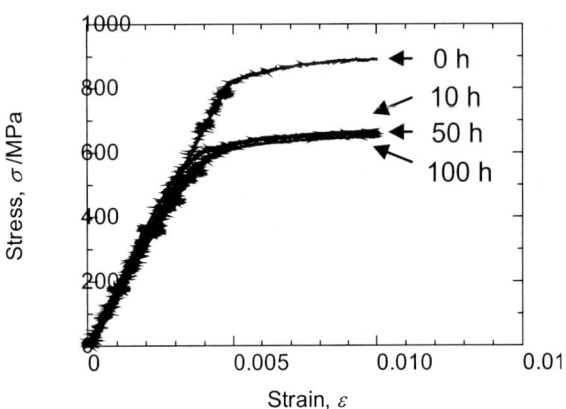

Fig. 3 Tensile stress-strain curves of as-sprayed specimen and heat-exposed specimens with heat exposure time.

transverse cracks in the as-sprayed and heat-exposed (50 h) specimens. The transverse cracks are formed perpendicular to loading direction with an almost equal spacing. However, the crack spacing depends on heat exposure time. The average spacing of transverse cracks in as-sprayed state is the largest, and the spacing decreases with the increase in heat exposure time. This means that the density of transverse cracks under tensile loading in the present coating system tends to increase after heat exposure.

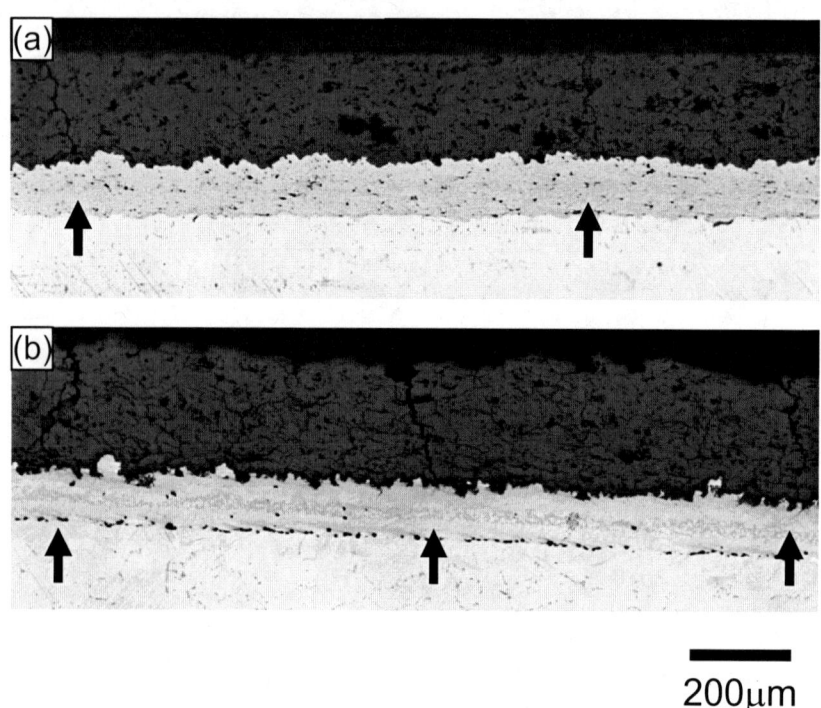

200μm

Fig. 4 Optical micrographs showing transverse cracks in (a) as-sprayed specimen and in the specimens subjected to heat exposure for (b) 50 h.

CONCLUSIONS

Change in transverse cracking behavior in air plasma-sprayed TBC layer under tensile load has been observed. The transverse crack density is highly dependent on heat exposure time of TBC systems. The density increases with the increase in heat exposure time. Full understanding of this behavior needs more detail understanding of the change in constitute properties, however.

Acknowledgement

A part of this study was performed by Nanostructure Coating Project carried out by New Energy and Industrial Technology Development Organization (NEDO).

REFERENCES

[1] S. M. Meier and D. K. Gupta, " The Evolution of Thermal Barrier Coatings in Gas Turbine Engine Applications," *Trans. ASME J. Eng. Gas Turbines Power* **116**, 250-257 (1994).

[2] A. G. Evans, D. R. Mumm, J. W. Hutchinson, G. H. Meier and F. S. Pettit, " Mechanisms Controlling the Durability of Thermal Barrier Coatings," *Prog. Mater. Sci.* **46**, 505-553 (2001).

[3] Y. C. Zhou, T. Tonomori, A. Yoshida, L. Liu, G. Bignall and T. Hashida, " Fracture Characteristics of Thermal Barrier Coatings after Tensile and Bending Test," *Surf. Coat. Technol.* **157**, 118-127 (2002).

[4] L. Qian, S. Zhu, Y. Kagawa and T. Kubo, " Tensile Damage Evolution Behavior in Plasma-Sprayed Thermal Barrier Coating System," *Surf. Coat. Technol.* **173**, 178-184 (2003).

[5] S. Guo and Y. Kagawa, " Young's Moduli of Zirconia Top-Coat and Thermally Grown Oxide in a Plasma-Sprayed Thermal Barrier Coating System," *Script. Mater.* **50**, 1401-1406 (2004).

[6] A. Shinmi, M. Hasegawa, Y. Kagawa, M. Kawamura and T. Suemitsu, " Change in Microstructure of Plasma Sprayed Thermal Barrier Coating by High Temperature Isothermal Heat Exposure," *J. Jap. Inst. Metals* **69**, 67-72 (2005).

[7] M. Tanaka, M. Hasegawa , A.F. Dericioglu and Y. Kagawa, " Measurement of Residual Stress in Air Plasma-Sprayed Y_2O_3-ZrO_2 Thermal Barrier Coating System using Micro-Raman Spectroscopy," Accepted in *Mater. Sci. Eng. A*

[8] M. Hasegawa and Y. Kagawa, "Microstructural and Mechanical Properties Changes of A NiCoCrAlY Bond Coat with Heat Ecposure Time in Air Plasma-Sprayed Y_2O_3-ZrO_2 TBC Systems," Submitted in *Inter. J. Appl. Ceram. Technol.*

CHARACTERISATION OF CRACKS IN THERMAL BARRIER COATINGS USING IMPEDANCE SPECTROSCOPY

Lifen Deng, Xiaofeng Zhao, Ping Xiao[*]
Materials Science Centre, School of Materials, Grosvenor Street, University of Manchester, Manchester, M1 7HS, UK

ABSTRACT

Finite element method has been developed to calculate impedance spectra of a model thermal barrier coating (TBC) with presence of cracks. Impedance spectra have been calculated to examine the effect of the crack size, total length of cracks, crack thickness, and location of the cracks on impedance spectra of TBCs. The calculated results indicate that both crack size and total length of cracks have significant effect on the impedance spectra, whereas the crack thickness and crack location have little influence on the impedance spectra. The initial impedance measurement results have confirmed some of the calculated results.

INTRODUCTION

Crack formation and propagation in thermal barrier coatings (TBCs) due to thermal treatment leads to spallation of TBCs[1, 2]. However, failure mechanisms of TBCs are not clear yet although there have been extensive studies on this important topic[3-6]. Therefore, monitoring crack propagation in TBCs is important in predicting lifetime of TBCs in service and will become critical in further development of TBCs for wide applications. Experimental studies of TBC degradation suggested that cracks often occur at either TGO/BC[7] or YSZ/TGO[6] interfaces, and the accumulation of cracked regions led to a final spallation of the TBCs. Several techniques such as ultrasonic imaging coupled with acoustic emission[8, 9] and thermal wave[10] have been used to directly monitor the crack propagation in TBCs. Fluorescence spectroscopy has been used to examine the stress distribution in TGO, and used the non-uniformity of stresses in cracked region to detect crack formation[11, 12]. However, other factors than cracks could affect the stress distribution, so that uncertainty in measurements would be introduced. Previous works tried to link impedance measurements of TBCs to degradation and crack propagation of TBCs[13, 14]. However, it is difficult to quantify the crack size and location based on measured impedance spectra. We have done extensive studies on impedance spectra of TBCs[14, 15], and studied the TGO growth[16], TBC sintering and phase transformation[17, 18] using impedance spectroscopy.

The present study is to simulate impedance spectra of TBCs based on the finite element method and to examine the effect of the cracks on the impedance spectra. A two dimensional model of TBC structure has been established to calculate impedance spectra of the TBCs. The

---

[*] To Whom correspondence should be addressed (ping.xiao@manchester.ac.uk)

calculated Bode plots indicate that the presence of cracks contributes to the increase and widening of the peak at frequency of ~10^2Hz, which is also attributed to the presence of the TGO layer. Impedance measurements of TBCs with and without cracks further confirm the overlapping effect of the cracks and TGO. Further modelling results suggest that impedance spectra of TBCs change significantly with an increase of total crack length, ie. the sum of crack sizes of all cracks, while the impedance spectra varies moderately with the variation of crack size when the total crack length is constant. However, little change appears in impedance spectra with the change of crack thickness and location of cracks. This study provides a further understanding of impedance spectra of TBCs and indicates that cracks in TBCs can be characterised using impedance measurements.

IMPEDANCE MODELLING

Figure 1(a) shows that the EB-PVD TBC system consists of three layers, YSZ top coat with a columnar microstructure, a thermally grown oxide (TGO) layer and a metallic bond coat (BC) layer. Fig. 1(b) shows a two dimensional schematic structure of the TBC prepared by electron beam physical vapor deposition (EB-PVD). The gb_1 represents vertical gap between columns, and the gb_2 represents grain boundaries inside the columns. The sample size is 5 mm width while the YSZ column width was set as 250 μm to save computer memory and computing time. Although the approximation of the column size would lead to some error in calculated impedance spectra at a frequency higher than 10^3 Hz, the effect of cracks on impedance spectra at a frequency lower than 10^3 Hz would be very similar. The YSZ layer thickness is 150 μm and the TGO layer thickness is 5 μm. The width of grain boundaries is set as 2 μm. The electrode size is 1mm. Fig. 2 shows the presence of cracks at either the YSZ/TGO or TGO/BC interface. The parameters of crack numbers, size and thickness are given in Table 1. Seven series of parameters have been used for calculation. At first, a total length of cracks is constant as 4000μm, and individual crack size decreased and the crack number increased in order to investigate the effect of crack size distribution on impedance spectra (No.1-3). Secondly, the total crack length was reduced by decreasing the number of uniform cracks with a fixed crack size of 200 μm to study the influence of crack total length on impedance spectra (No. 4-6). With the same crack size and number of cracks, crack thickness decreased from 5μm to 2μm (No.3 and No.7 series of parameters) or change crack location from TGO/BC to YSZ/TGO interface, to identify the effect of crack thickness and location on impedance spectra.

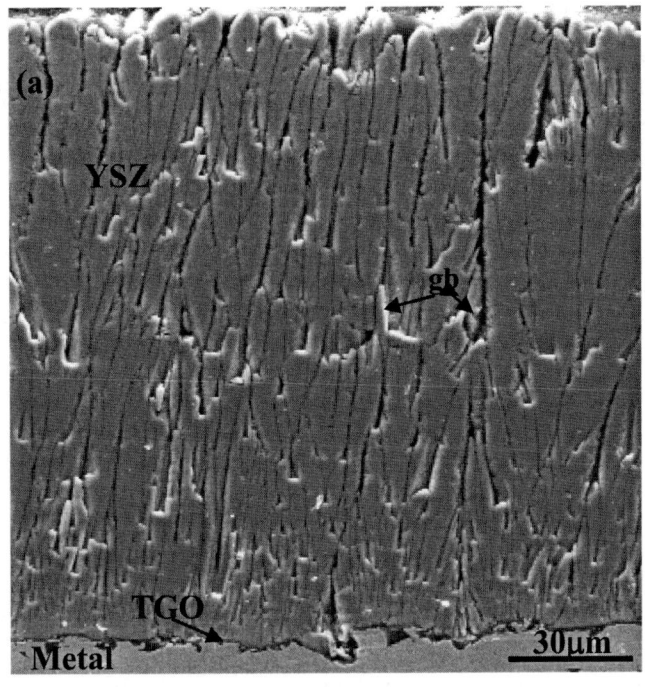

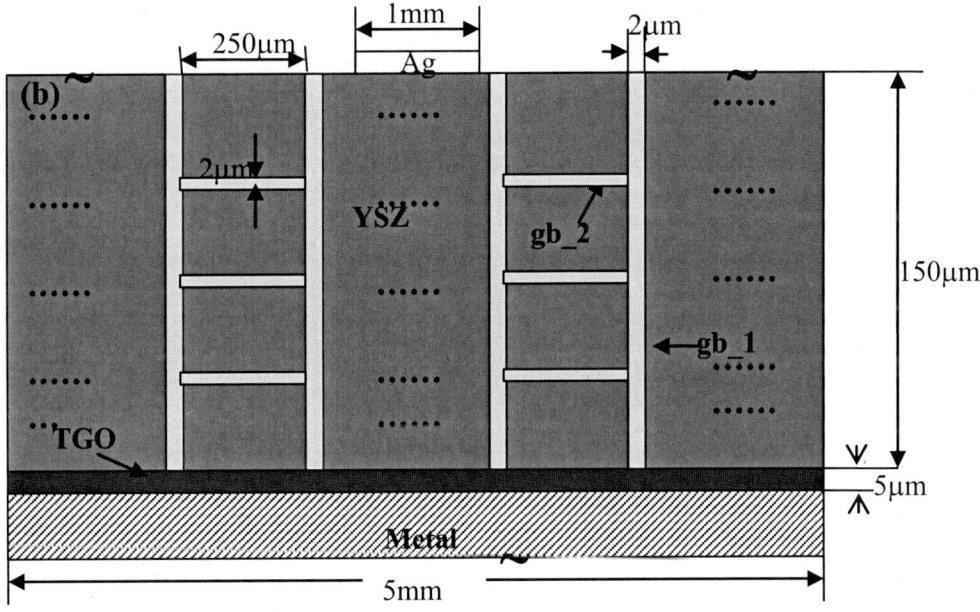

Figure 1. (a) SEM image of a EB-PVD TBC cross section; (b) A schematic figure of cross section of EB-PVD TBCs.

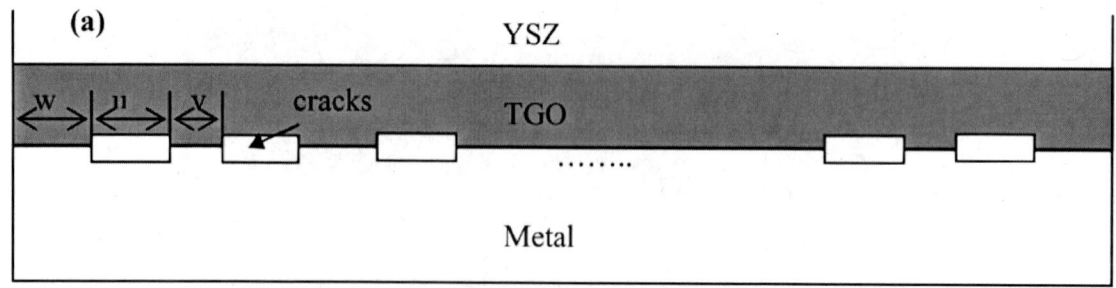

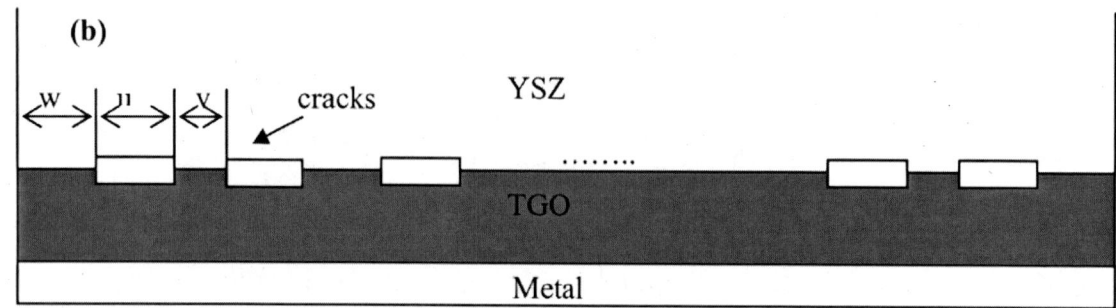

Figure 2. Schematic figures of TBC showing cracks at (a) YSZ/TGO interface and (b) TGO/BC interface

Table 1. The geometric parameters of cracks at TGO/BC interface in models.

No.	Thickness /μm	Size u/μm	Number of cracks	Separation distance μm	Total length of cracks μm
1.	5	500	8	200	8X500=4000
2.	5	400	10	100	10X400=4000
3.	5	200	20	40	20X200=4000
4.	5	200	15	100	15X200=3000
5.	5	200	10	300	10X200=2000
6.	5	200	5	800	5X200=1000
7.	2	200	20	40	20X200=4000

The electric potential in TBC satisfied the Poisson's equation, as stated by Fleig [19],

$$\text{div} \cdot \text{grad}\, \widehat{\Phi}(r,t) = -\frac{1}{\varepsilon}\rho(r,t) \quad (1)$$

where $\widehat{\Phi}$ is AC electric potential, ρ is charge density, ε is the absolute dielectric constant and ρ is the charge density, r and t are space and time respectively. There is no applied magnetic field during impedance measurements and the diffuse charge on the surface and interface can be ignored, therefore, the charge density in a uniform domain of TBC samples is equal to zero,

$\rho(r,t)=0$, thus Eq(1) can be reduced to Laplace's equation

$$\frac{\partial^2 \hat{\Phi}}{\partial x^2} + \frac{\partial^2 \hat{\Phi}}{\partial y^2} = 0 \tag{2}$$

Three boundary conditions were used to solve Laplace's equation (2): i) There is no potential drop occurred at the electrode/electrolyte interface; ii) No current goes out of the sample, therefore, the normal current density at free edges of the model is equal to zero: grad $\hat{\varphi} \bullet n = 0$, where n represents a normal vector of relevant side; iii) At interfaces such as the Ag/YSZ interface, YSZ/gb1 interface, YSZ/gb2 interface, YSZ/TGO interface, TGO/gb1 interface and TGO/BC interface, the normal component of complex current density is continuous. For example, at the interface between the 1st sub-domain and the 2nd sub-domain: $\hat{\kappa}_1 grad \hat{\varphi} \bullet n = \hat{\kappa}_2 grad \hat{\varphi} \bullet n$.

Here, complex conductivity $\hat{\kappa} = \sigma + i\omega\varepsilon$, σ is the electric conductivity, ε is the absolute dielectric constant. With the three boundary conditions, the equation (2) has been solved, and then the complex current density was calculated according to

$$\hat{j} = -\hat{\kappa} grad \hat{\varphi} \tag{3}$$

Further, the current at the electrode/electrolyte interface was obtained by integrating the current density along the electrode boundary:

$$I = \int_s j \bullet ds = \int_s \hat{\kappa} grad \hat{\varphi} \bullet ds \tag{4}$$

With applied voltage U_0 and the gained current I, impedance can be gained by:

$$\hat{Z} = \frac{U_0}{I} = \frac{U_0}{\int_s \hat{\kappa} grad \hat{\varphi} \bullet dA} . \tag{5}$$

Table 2 shows material properties used for calculations. The electrical properties of YSZ were calculated from the experimental data by measuring YSZ bulk samples with standard shape and size at 400 °C. The conductivity of grain boundary is much less than that of bulk and assumed as $\sigma_{gb} \approx 10^{-3}\sigma_{YSZ}$[20], while the permittivity of YSZ grain boundary can be regarded as the same with that of YSZ bulk, $\varepsilon_{gb} = \varepsilon_{bulk}$[21]. TGO layer is mainly composed of Al_2O_3, so that set its electric conductivity $\sigma_{TGO} = 1.0 \times 10^{-8}$ /Ω and permittivity $\varepsilon_{,TGO} = 10$ [22]. Cracks are filled with air, electric conductivity of air $\sigma_{air}=0$, and relative permittivity of air $\varepsilon_{r\,air}=1$. The electric properties of Ag and substrate metal electrodes are set as $\sigma=1.0\times10^7$ /Ω, $\varepsilon_r=1$. For the two dimensional model, the thickness of the 3rd dimension is assumed as infinite. The unit of electric conductivity is Ω^{-1}. The calculation was carried out using FEMLAB with MATLAB software.

Table 2. The electric properties of materials used in models.

	σ /Ω^{-1}	ε_r
YSZ grain	1e-3	28
YSZ Grain boundary	1e-6	28
TGO	1e-8	10
air	0	1
Ag / Metal	1e+7	1

EXPERIMENTAL

EB-PVD TBCs, provided by Rolls-Royces, Derby, UK, consisted of Ni-superalloy, Pt diffusion bond coat and 8 wt.% YSZ topcoats. 5×5 mm^2 plate samples were cut from engine components and isothermally heated at 1200°C for 60 hours and 90 hours.

Impedance measurements were carried out using a Solartron SI 1255 HF frequency response analyser (FRA) coupled with a computer controlled Solartron 1296 dielectric interface, with a horizontal tube furnace at 400 °C. During measurements, an a.c. amplitude of 100 mV was used, over a frequency range of 10^6–10^{-2} Hz. 1 mm^2 silver (Ag) was painted on the YSZ surface and then fired at 690 °C for 30 min. The fired Ag acted as an electrode. For TBCs served at temperature higher than 1000°C, microstructure change of TBCs caused by impedance spectroscopy measurement 400 °C and 690 °C for 30 min Ag annealing can be ignored. The polished base of the superalloy substrate acted as the other electrode. Samples were placed in a jig between Al$_2$O$_3$ plates; Pt wires was attached the electrodes for impedance measurements.

SEM/EDX (JEOL 6300 and Philips 525M) was used to examine the cross-section of the TBC samples. The samples were prepared using a precision diamond wheel saw (Struers Accutom-50) at low speed to minimise mechanical damage to the coating. The samples were placed in epoxy resin, before grinding and polishing using successively finer SiC paper. Polished samples were carbon coated (Edwards E306A coater) for SEM examination.

RESULTS

Modelling results

Figure 3 shows a Bode plot (minus phase angle (θ) vs. frequency (Log (f)) and a Nyquist plot (imaginary impedance (Zi) vs. real impedance (Zr)) of TBCs with cracks (using No. 2 parameters

Progress in Thermal Barrier Coatings

in table 1) and without crack. The Bode plot shows three peaks, corresponding to three relaxation processes in TBCs, YSZ grain (g, approx. $10^{6\sim7}$Hz), YSZ grain boundary (gb, approx. 10^4Hz), and TGO (approx. $10^{2\sim3}$Hz) which were confirmed experimentally in our previous study without consideration of cracks[17]. The Nyquist plots show large semicircles of the TGO layer, whereas the semicircles for g and gb at high frequencies are too small to be seen in figure 3b. In comparison with the TBC without cracks, the cracks in the TBC induced a significant change in impedance spectra, i.e. the presence of cracks at the TGO/BC interfaces led to higher and broader "TGO" peak in the Bode plot. The semicircle corresponding to the "TGO" in the Nyquist plot became much larger with the presence of cracks.

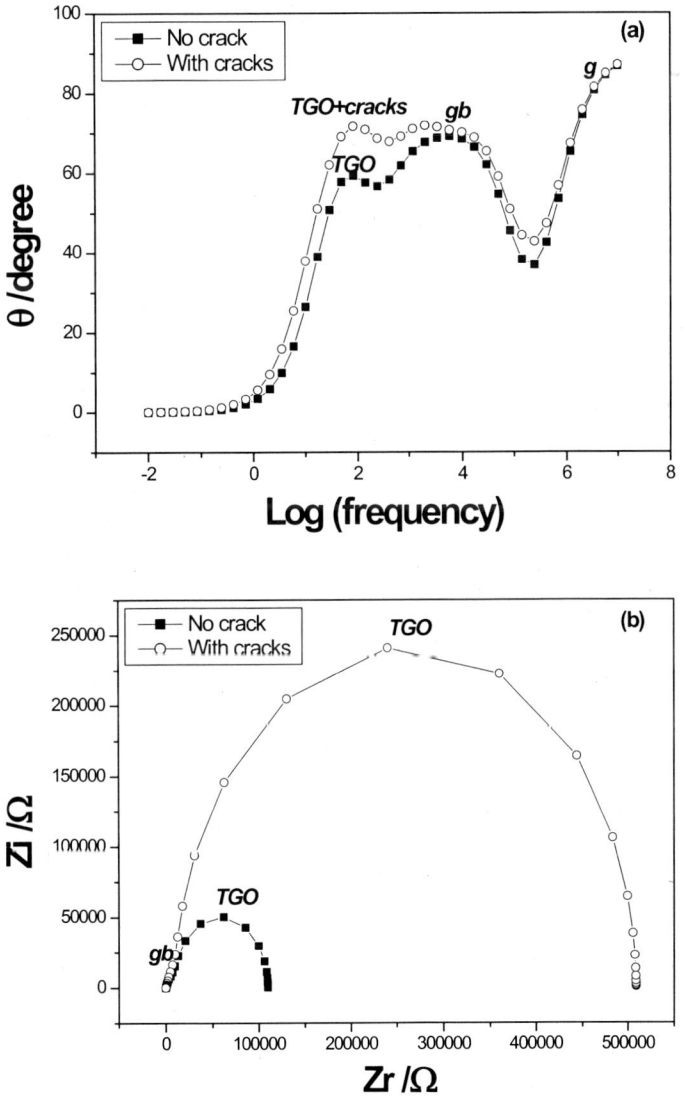

Figure 3. (a) Bode plots (minus θ vs. Log(f)) and (b) Nyquist plots (Zi vs. Zr) of models with cracks (No 2 in table 1) and without crack.

Experimental Results

Fig. 4 shows the cross section of EB-PVD TBC samples after isothermal treatments at 1200 °C for 60 hours and 90 hours. With the treatment time increasing from 60 hr to 90 hr, there is little change in the TGO thickness, but a large crack appeared at the TGO/BC interface in the 90 hr treated sample whereas no apparent crack appears in the 60 hr treated sample. Although the TGO is separated from BC metal in Fig. 4b, the TGO layer is still attached to the BC in other region. The large crack in Fig. 4b was partially induced by polishing during sample preparation for SEM, but difference in thermal treatment certainly lead to difference in crack size in the two samples.

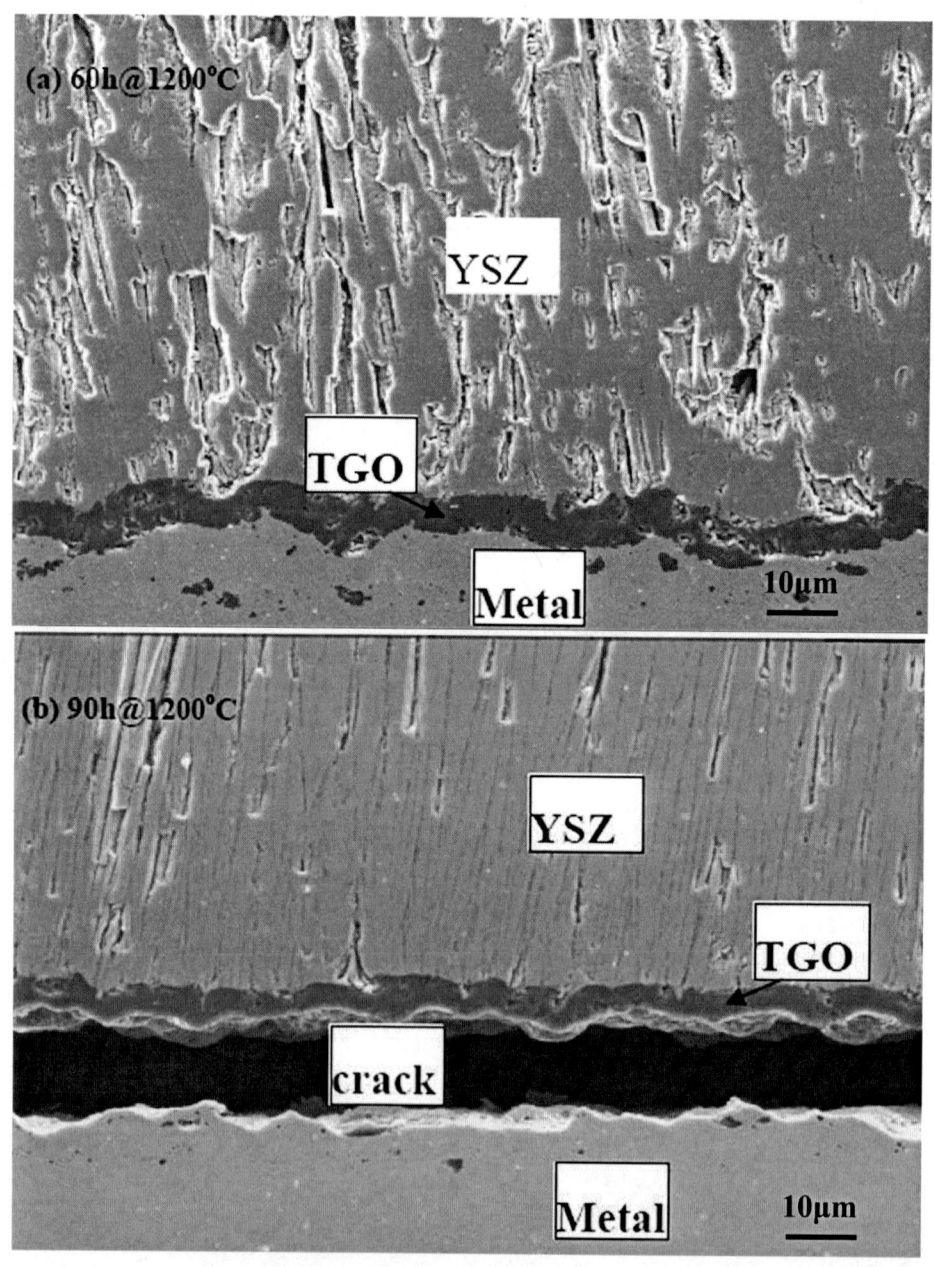

Figure 4. SEM images of cross section of EB-PVD TBC samples isothermally heated at 1200 °C for (a) 60 hours and (b) 90 hours.

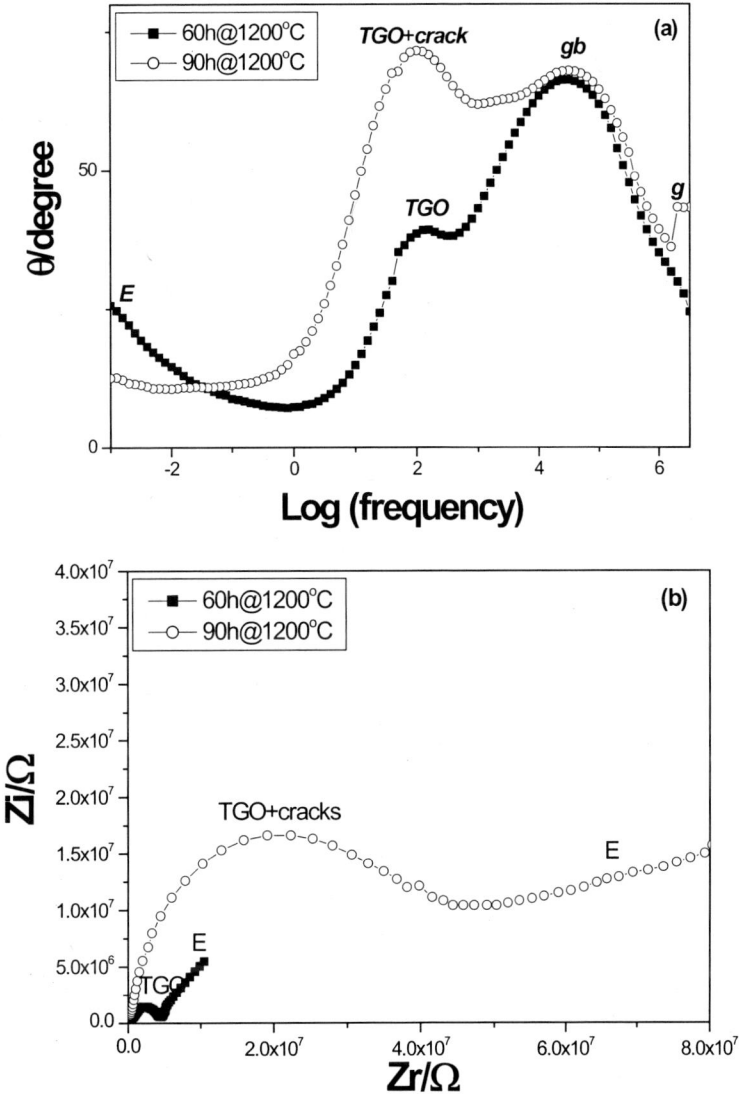

Figure 5. The Bode (a) and Nyquist (b) plots measured from EB-PVD TBCs after isothermal treatments at 1200 °C for 60 hours and 90 hours.

Fig. 5 shows impedance spectra of EB-PVD TBC samples with microstructure shown in figure 4. The Bode plots indicate grain effect (approx. $10^{6\sim7}$ Hz), grain boundary effect (approx. 10^4 Hz), TGO effect (approx. $10^{2\sim3}$ Hz) and electrode effect (E, approx. $10^{-2\sim3}$ Hz). Because the spectrum measured at frequencies higher than 10^6 Hz is not reliable[23], the grain effect, ie. g peak in the Bode plot is not clear in figure 5. The Nyquist plots show a semicircle contributed by TGO+cracks, and a tail corresponding to electrode effect (E). However, the significant increase in the "TGO" effect in both the Bode plot and the Nyquist plot confirmed the effect of the cracks. In addition, the crack effect on the Bode plot and the Nyquist plot is the same as in the calculated spectra (Fig. 3). It should be noted that there is no consideration of electrode effect in calculation of the impedance spectra.

Further Modelling

Experimental measurements (Fig. 5) are in good agreement with modelling results (Fig. 3) about the effect of cracks at the TGO/BC interfaces on the impedance spectra of TBCs. Then, the effects of crack size, the total length of cracks, crack location and crack thickness have been examined and discussed as follows:

Total Crack Length Effect

The size of each crack was fixed as 200μm, the number of cracks changed from 20 to 15, 10, and 5 (No.4-6 in table 1). Fig. 6 shows calculated impedance spectra where the Bode plots show that the TGO+cracks peak near 10^2Hz increased with increasing total crack length. In addition, semicircles in Nyquist plots become larger with an increase in the total crack length. When the total crack length equal to or less than 5×200 μm=1000 μm (No.6 parameters in table 1), it is difficult to differentiate impedance spectra with cracks from those without crack. This suggests that for the 1000μm Ag electrode, the minimum total length of cracks which can be monitored is about 1000μm. Measurements with smaller Ag electrode size might increase sensitivity of impedance spectra to the total crack length. Further investigation need to be carried out to examine this issue.

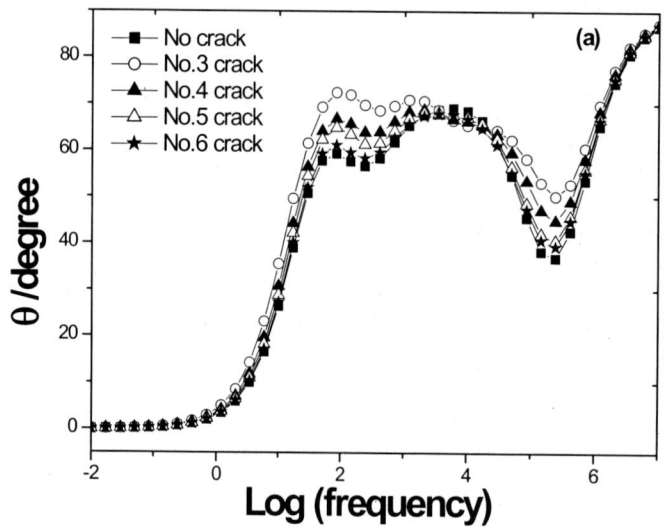

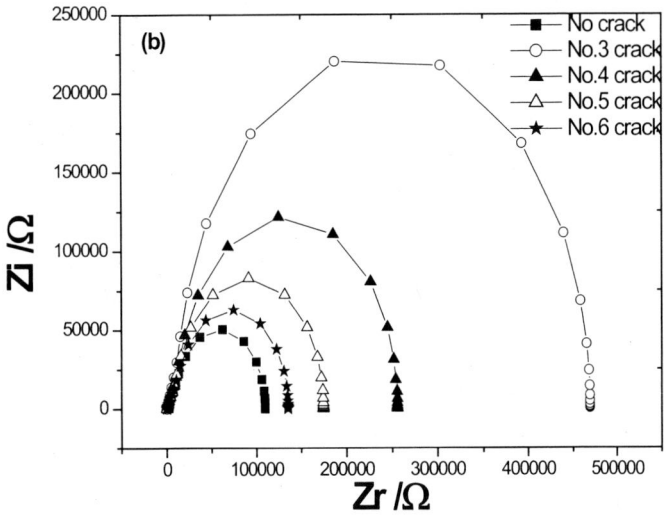

Figure 6. Calculated Bode and Nyquist plots of model with No.3, No.4, No.5 No.6 cracks in table 1 (the crack size is constant while the number of cracks changes) and without crack .

Crack Size

With the total length of cracks fixed, e.g. 4000 μm, the size of individual crack changing from 1000 μm, to 500 μm, to 200 μm, (No.1 to No.3 parameters in table 1), the calculated impedance spectra are shown in Fig. 7. The Bode plots indicate that the "TGO+cracks" peak near 10^2 Hz shows little change with the change of crack size. However, the semicircles on Nyquist plots increased slightly with the increase of crack size.

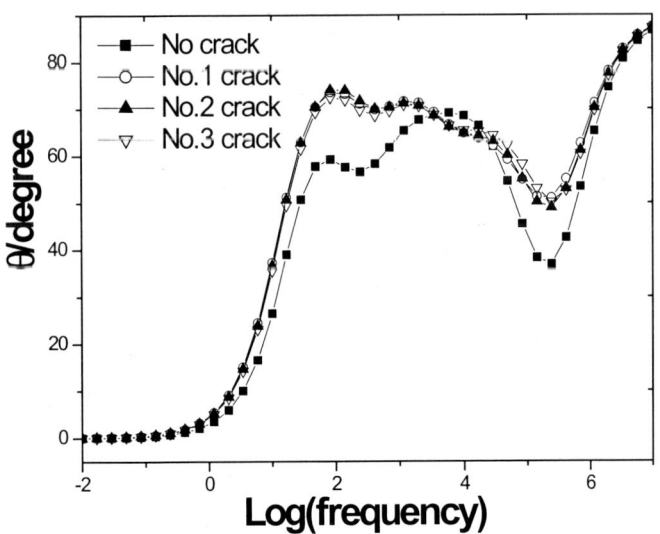

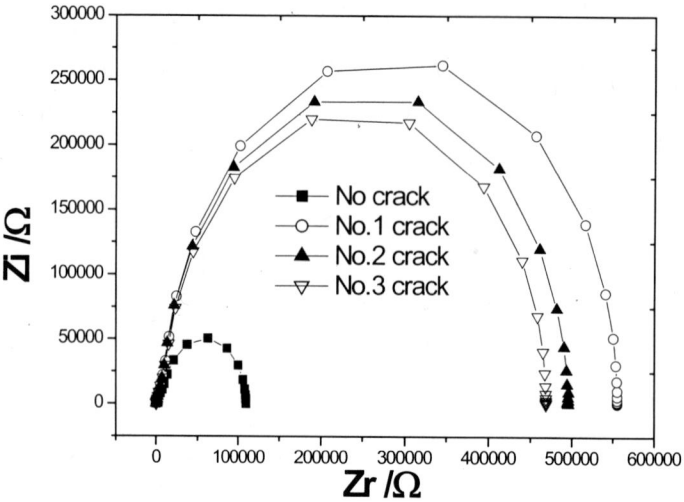

Figure 7. Calculated Bode and Nyquist plots of model with No.1, No.2 and No.3 cracks in table 1 (the total length of cracks is same as 4000um, the crack size is different) and without crack.

Crack thickness and location

For TBC degradation, cracks appeared sometimes at the TGO/BC interfaces[7] and sometimes occurred at the YSZ/TGO interfaces[24]. Fig. 2 shows the cracks at different locations in TBCs. With No. 2 cracks in Table 1, Fig. 8 shows impedance spectra of TBCs with different crack locations. No visible difference appears on impedance spectra due to difference in the crack location.

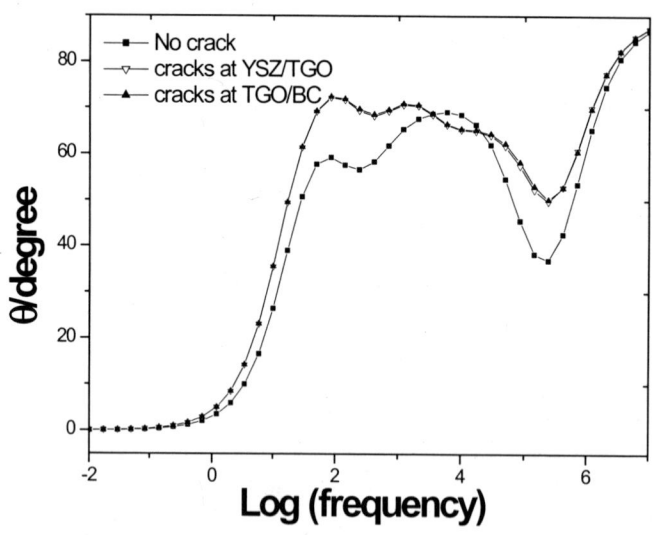

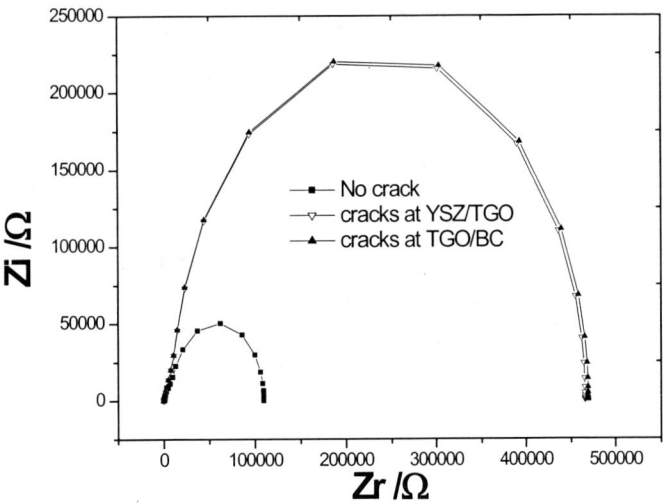

Figure 8. Impedance spectra calculated from models with cracks (No.2 crack parameter in the table 1) at the YSZ/TGO or TGO/BC interface and without cracks.

With the thickness of cracks changing from 5 μm to 2 μm, calculated impedance spectra in Fig. 9 show no apparent difference. It should be noted that the calculation was made with arrangement of 1mm Ag electrode at surface of TBC and the metallic substrate as the other electrode. The crack thickness effect and crack location effect on impedance spectra may be different with different arrangements of the electrodes. Further research is required to study the effects of crack thickness and location on impedance spectra of TBCs.

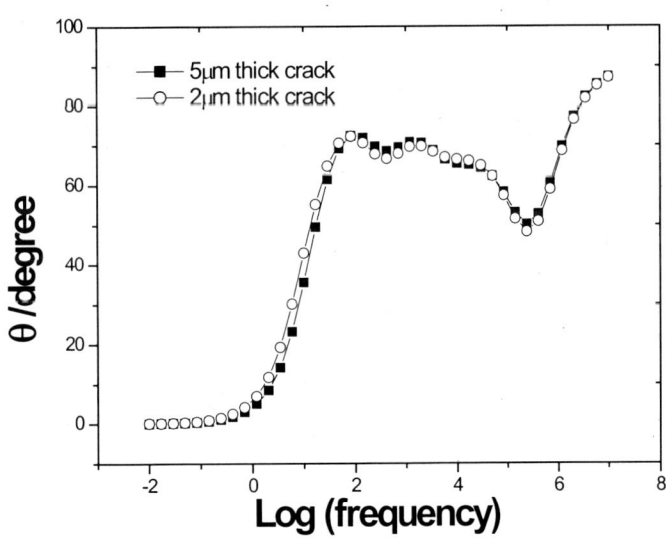

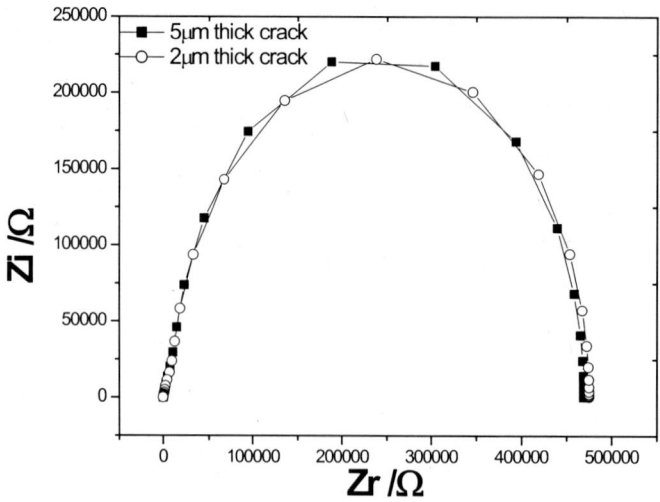

Figure 9 Bode and Nyquist plots calculated from models when the thickness of cracks was changed from 5μm to 2μm (No.3 and No.7 cracks in table 1).

CONCLUSION

A 2-dimensional model of TBC has been built to calculate impedance spectra of the TBCs using the finite element method. The effect of cracks in TBCs has been investigated by modelling and experimental results. Both results show that the higher TGO peak near 10^2 Hz in the Bode plot and larger semicircles on Nyquist plots were induced with crack formation. An increase in the total length of cracks led to significant change in both the Nyquit plot and the Bode plot. With the total length of cracks being constant, an increase in crack size induced change in the Nyquist plot but little change in the Bode plot. The crack thickness change and change in crack location has negligible effect on impedance spectra with arrangement of 1mm Ag electrode at surface of TBC and the metallic substrate as the other electrode.

REFERENCES

[1.] S. Nusier,G. Newaz, "*Analysis of interfacial cracks in a TBC/superalloy system under thermal loading,*" Eng. Fract. Mech. **60**, 577-81 (1998).

[2.] D. R. Clarke, R. J. Christensen,V. Tolpygo, "*The evolution of oxidation stresses in zirconia thermal barrier coated superalloy leading to spalling failure,*" Surf. Coat Tech. **94-95**, 89-93 (1997).

[3.] A. G. Evans, D. R. Mumm, J. W. Hutchinson, G. H. Meier,F. S. Pettit, "*Mechanisms controlling the durability of thermal barrier coatings,*" Prog. Mater. Sci. **46**, 505-53 (2001).

[4.] D. R. Mumm,A. G. Evans, "*On the role of imperfections in the failure of a thermal barrier coating made by electron beam deposition,*" Acta Mater. **48**, 1815-27 (2000).

[5.] G. Qian, T. Nakamura,C. C. Berndt, "*Effects of thermal gradient and residual stresses on thermal barrier coating fracture,*" Mech. of Mater. **27**, 91-110 (1998).

[6.] A. Rabiei,A. G. Evans, "*Failure mechanisms associated with the thermally grown oxide in plasma-sprayed thermal barrier coatings,*" Acta Mater. **48**, 3963-76 (2000).

[7.] D. R. Clarke,S. R. Phillpot, "*Thermal barrier coating materials,*" Mater. Today. **8**, 22-29 (2005).

[8.] Y. C. Zhou,T. Hashida, "*Thermal fatigue failure induced by delamination in thermal barrier coating,*" Int. J. Fatigue. **24**, 407-17 (2002).

[9.] N. Mesrati, Q. Saif, D. Treheux, A. Moughil, G. Fantozzi,A. Vincent, "*Characterization of thermal fatigue damage of thermal barrier produced by atmospheric plasma spraying,*" Surf.& Coat. Tech. **187**, 185-93 (2004).

[10.] G. Newaz,X. Chen, "*Progressive damage assessment in thermal barrier coatings using thermal wave imaging technique,*" Surf Coat Tech. **190**, 7-14 (2005).

[11.] M. M. Gentleman,D. R. Clarke, "*Concepts for luminescence sensing of thermal barrier coatings,*" Surf Coat Tech. **188-189**, 93-100 (2004).

[12.] V. K. Tolpygo, D. R. Clarke,K. S. Murphy, "*Evaluation of interface degradation during cyclic oxidation of EB-PVD thermal barrier coatings and correlation with TGO luminescence,*" Surf. Coat. Tech. **188-189**, 62-70 (2004).

[13.] B. Jayaraj, S. Vishweswaraiah, V. H. Desai,Y. H. Sohn, "*Electrochemical impedance spectroscopy of thermal barrier coatings as a function of isothermal and cyclic thermal exposure,*" Surf Coat Tech. **177-178**, 140-51 (2004).

[14.] M. S. Ali, *A novel technique for evaluating the degradation of engine components non-destructivtely*, in *Department of Mechanical Engineering*. 2002, Brunel University: London.

[15.] M. S. Ali, S. Song, P. Xiao, "*Evaluation of degradation of thermal barrier coatings using impedance spectroscopy,*" J. Eur. Ceram. Soc. **22**, 101-07 (2002).

[16.] X. Wang, J. Mei, P. Xiao, "*Determining oxide growth in thermal barrier coatings (TBCs) non-destructively using impedance spectroscopy,*" J. Mater. Sci. Lett. **20**, 47-49 (2001).

[17.] P. S. Anderson, X. Wang,P. Xiao, "*Impedance spectroscopy study of plasma sprayed and EB-PVD thermal barrier coatings,*" Surf. Coat. Tech. **185**, 106-19 (2004).

[18.] P. S. Anderson, X. Wang, P. Xiao, "*Effect of isothermal heat treatment on plasma-sprayed yttria-stabilized Zirconia studied by impedance spectroscopy,*" J. Am.Ceram.Soc. **88**, 324-30 (2005).

[19.] J. Fleig,J. Maier, " *Finite element calculations of impedance effects at point contacts,*" Electrochemica Acta. **41**, 1003-09 (1995).

[20] J. Fleig, J. Maier, "*The impedance of ceramics with highly resistive grain boundaries: validity and limits of the brick layer model,*" J. Eur. Ceram. Soc. **19**, 693-96 (1999).

[21] J.-S. Lee, J. R. Fleig, J. Maier, "*Conventional and microcontact impedance studies of Mn–Zn ferrite ceramics,*" J. Mater. Res. **19**, 864-71 (2004).

[22] D. R. Lide, *CRC Handbook of Chemistry and Physics.* 85th ed. 2004: CRC press LLC. 12-52.

[23] X. Wang, *Non-Destructive Characterisation of Structural Ceramics Using Impedance Spectroscopy*, in *Department of Mechanical Engineering.* 2001, Brunel University: London. p. 8.

[24] M. Okazaki, H. Yamano, "*Mechanisms and mechanics of early growth of debonding crack in an APSed Ni-base superalloy TBCs under cyclic load,*" Int. J. Fatigue. **27**, 1613-22 (2005).

NONDESTRUCTIVE EVALUATION METHODS FOR HIGH TEMPERATURE CERAMIC COATINGS

William A. Ellingson, Rachel Lipanovich, Stacie Hopson, Robert Visher
Argonne National Laboratory
9700 S. Cass Avenue
Argonne, IL 60439

ABSTRACT

Various high temperature ceramic coatings are under development for components for the hot gas path of gas turbine engines. These coatings are being developed for both metal and ceramic substrates. Regardless of the coating system, environmental barrier coatings for ceramics or thermal barrier coatings for metals, there is a need to determine the condition of the coating as well as determine physical parameters such as thickness. Two laser-based, non-contact, nondestructive methods are being developed that provide information regarding the condition or "health" of the coating, including thickness variations. The elastic optical scatter (EOS) method has been demonstrated to correlate to spall conditions for thermal barrier coatings however there has been a question about sensitivity to surface features. The laser-based optical coherence tomography (OCT) method has been demonstrated to provide thickness measurements for thermal and environmental barrier coatings. Several well-controlled sample sets have been examined with these two NDE methods and extensive data have been acquired. The test methods, test samples, and results obtained to date showing correlations with destructive measurements are presented.

INTRODUCTION

Two very different high temperature coating systems are being addressed by NDE technology: 1)- Environmental barrier coatings mainly for ceramic composites being used in high temperature oxygen-rich environments and 2)-thermal barrier coatings mainly for metals used in the hot gas flow path of turbine engines. For gas-turbine combustor liners made of ceramic matrix composites-or any high temperature application where oxygen is present, there is a need to reduce oxidation-induced recession in silicon-based ceramic composites [1] and a need to reduce the operating temperature to reduce creep in oxide-based composites [2]. These two negative effects on composite performance can be mitigated by application of environmental barrier coatings (EBC) [2]. For SiC/SiC materials, present EBCs, typically 100-200 micron thick, are composed of some form of barium-strotium-alumino-silicate (BSAS). For oxide-based composites, a special oxide-based functionally graded insulator (FGI) has recently been developed and patented by Siemens-Westinghouse [1]. For either of these EBCs, what is desired to be measured (determined) by NDE technology are: uniformity of thickness, adherence quality and detection of any flaws in the CMC under the EBC. Several processing steps are necessary to fully fabricate a CMC liner with an EBC. It is possible that initial flaws within the CMC material could affect the EBC adhesion or provide an increased potential to spall. Thermal barrier coatings (TBCs), on the other hand, are necessary to reduce the operating temperature of metal components in the hot gas path of gas turbines. Components such as the first stage blades and vanes down stream of the combustor as well as the combustor itself are examples of hot gas path components. Failure of a TBC could lead to a costly unplanned outage, and could lead to catastrophic events. It is therefore necessary to monitor the condition of the TBC so as to avoid such failures - and, if possible, provide pre-cursor information that would suggest a spallation is immanent. Significant work has been done by many investigators [3-6] and this work has shown that failure of TBCs depend upon the coating type EB-PVD or APS. Use of NDE technology for detection of regions where spallation might occur in the future therefore depends upon the coating type because the failure mechanism is different. There is also a need to develop NDE technology that can determine changes in the thermal conductivity of these TBC materials. Changes in thermal conductivity can arise from a diffusion of combustion constituents into the coating, especially for turbines burning "dirty" fuels; the conductivity can also change from microstructural changes in the TBC such as in-situ sintering. From the NDE point of view, the microstructural differences between an EB-PVD and an APS TBC are significant. An EB-PVD TBC has a better defined columnar microstructure [7] and is less optically scattering than the "splat" microstructure of an APS. NDE technologies that can accommodate these microstructural differences and measure the desired TBC parameters are important for long-term reliable operation of advanced, high efficiency and low-emission gas turbines.

TEST METHODS

Two optical NDE test methods have been under development for application to these coating systems. One is optical coherence tomography (OCT) [8] and the other is elastic optical back scatter (EOS)

(⁹⁻¹⁰). OCT is a fairly new NDE technique that can be used to take cross-sectional images of various materials providing that the material is optically translucent. Our work on OCT has been to investigate this as a tool for evaluating the thickness of TBCs and EBCs as well as perhaps detecting any disbond. The OCT technique is described more fully in reference (⁸) and a block diagram of the Argonne OCT system is shown in Figure 1.   In an OCT set up, light from an optical source is split into two paths, a sample path and a reference path. Light in the reference path is reflected from a fixed-plane mirror whereas light in the sample path is reflected from surface and subsurface features of the ceramic sample. The reflected light from the sample path will only be detected if it travels a distance that closely matches the distance traveled by the light in the reference path; this constraint incorporates depth resolution into the technique. Thus, data can be obtained from a cross-sectional plane perpendicular or parallel to the surface of the sample.

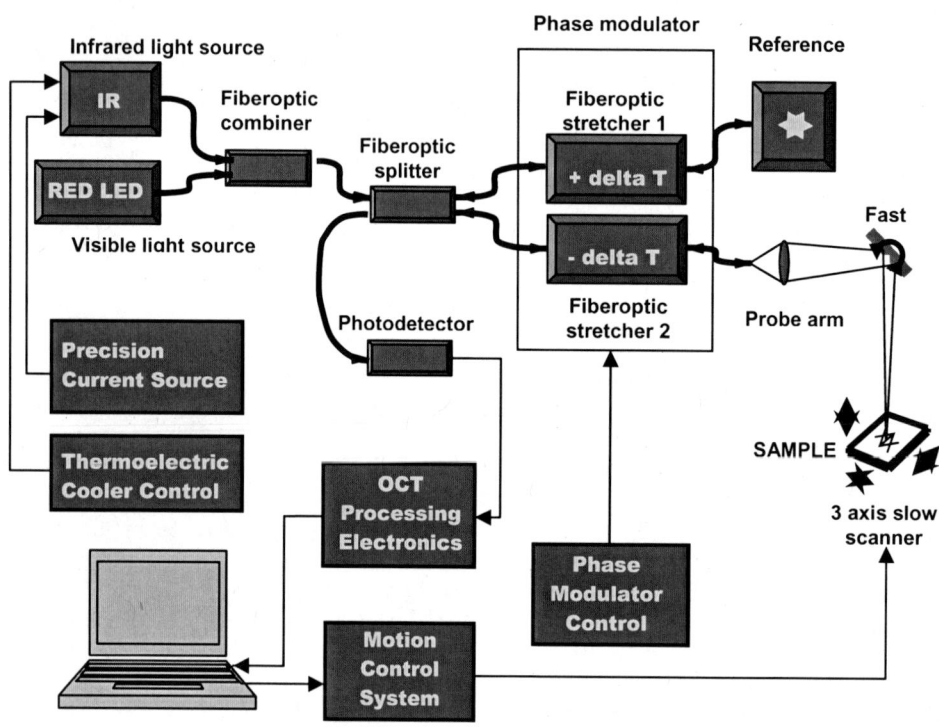

Figure 1: Block diagram of ANL OCT system.

The EOS or laser backscatter, technique (⁹⁻¹⁰), has been extensively discussed previously. In this method, polarized laser light is used to investigate the surface and subsurface characteristics ceramic coatings such as TBCs and EBCs.   The underlying physical principle behind laser backscatter is that light incident on the surface of the coating will be partially reflected and partially transmitted. The polarization state of any surface- reflected light will not change, whereas the polarization state of a portion of the transmitted light that is subsequently reflected will change. The characteristics of the reflected light can be used to distinguish between light reflected from the surface and light reflected from a sub-surface feature. A two-detector EOS system is shown in Figure 2. In this system, the sample under investigation is mounted on the translational motion stages. The laser beam is incident on the sample at a given coordinate and a measurement is taken. The sample is moved so that the laser beam is now incident on a new location, adjacent to the previous location, and the measurement is repeated. This process continues until the area of the sample being studied has been covered. Typically, the locations are separated by a distance on the order of 5-10 microns. The two-dimensional array of collected measurements is then normalized with respect to the minimum and maximum measured values to create a gray-scale image.

Progress in Thermal Barrier Coatings

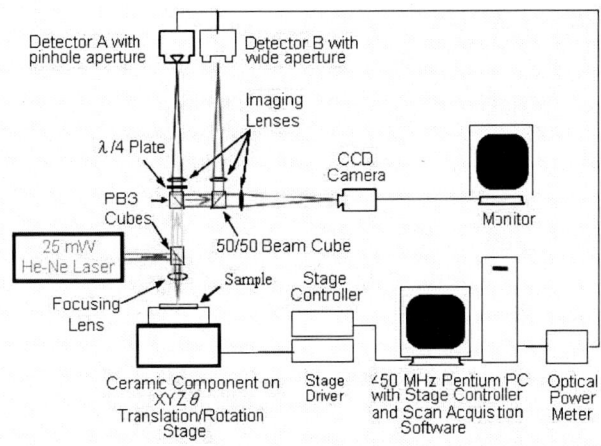

Figure 2: Schematic representation of ANL-developed EOS system.

RESULTS

Thermal Barrier Coatings

Previous data has resulted in empirically correlated results between the laser back scatter and spallation of both EB-PVD and APS TBC sample sets [9]. An example of what the EOS data look like for an EB-PVD coating and one example of the correlation with spallation is given in Figure 3.

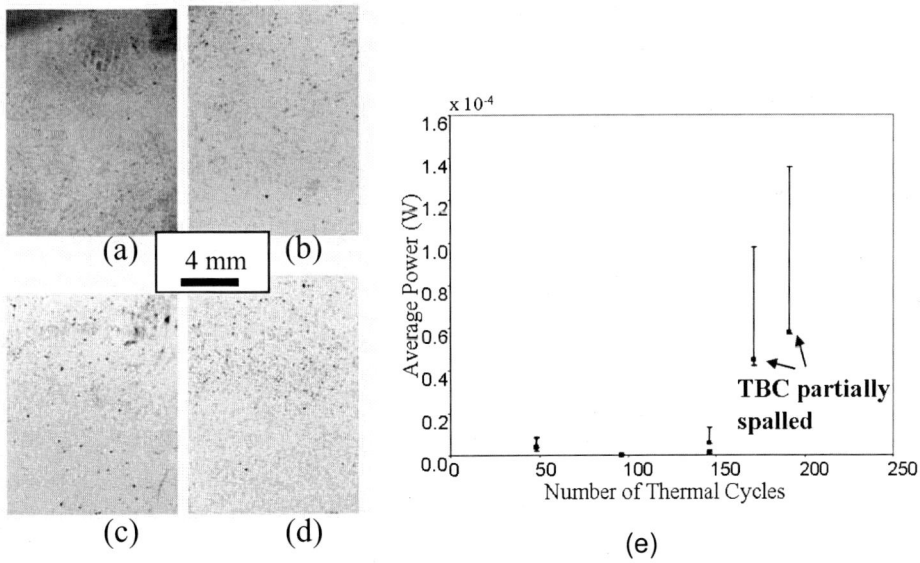

Figure 3: Laser backscatter sum images for: EB-PVD TBC sample after exposure to (a) 48, (b) 146, (c) 171, and (d) 191 thermal cycles and (e) average detected power of back scattered light as a function of the number of thermal cycles

However, there has been some concern over the sensitivity of this method to small features on the surface of the samples. To address this issue a special TBC sample was prepared and then polished. An

EB-PVD TBC sample was prepared at the German Aerospace Institute in Koln, Germany, DLR. The test sample, 10 mm by 28 mm, consisted of an EB-PVD TBC on a metal substrate. For reference purposes, three indents were placed on the sample. First, the sample was examined for surface features using a microscope and recording the images. Then data were acquired with the laser backscatter system. In these laser scatter tests the sample was raster scanned in 10 µm steps at a velocity of 20 mm/s. Using the two detectors in the set up, the first detector detected the light directly reflected by the sample, while the second detected the scattered light. One half of the sample was then polished to reduce surface effects and was scanned using the same data acquisition parameters. The total reflected light is the sum of detector 1 and detector 2. Figure 4 shows the resulting sum images obtained before and after polishing.

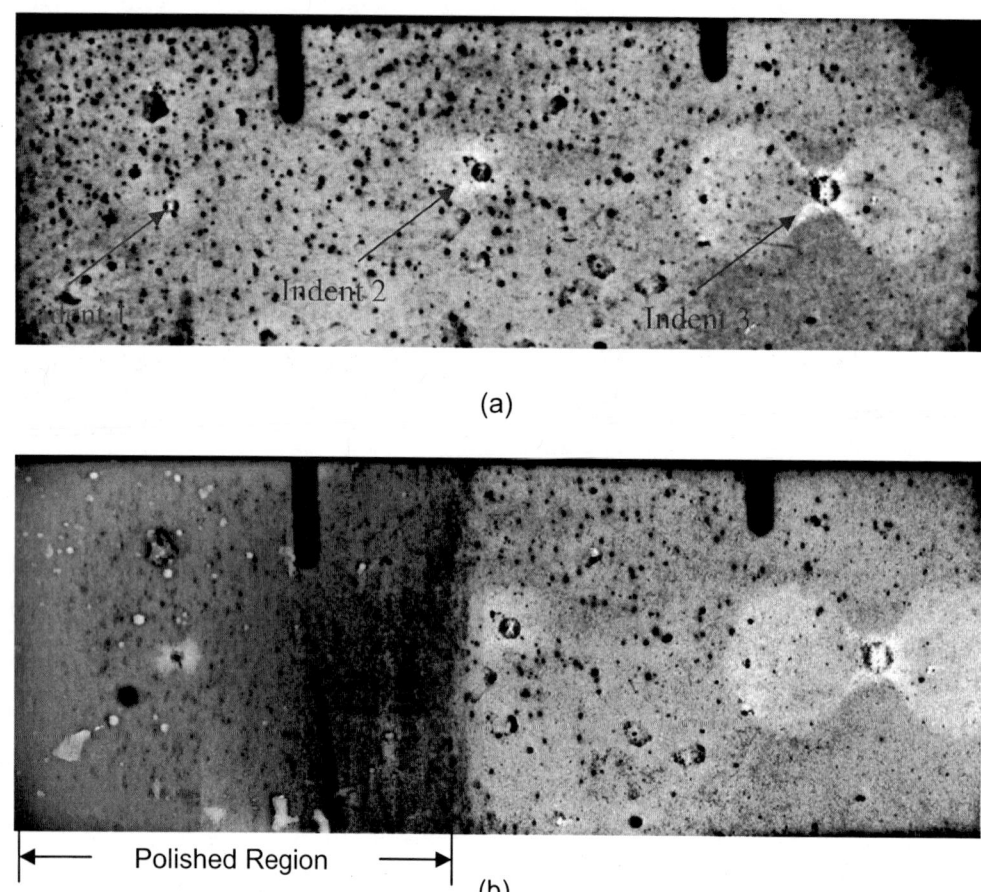

(a)

(b)

Figure 4: EOS sum image data

(a) before polishing and (b) after polishing.

Examination of Figure 4b reveals that there are fewer "dark spots" in the EOS data in the polished area after polishing than before polishing. This suggests that some "dark spots" that were present prior to polishing in fact are a result of surface features. The correlation between "dark spots" before and after polishing was confirmed by comparing microscope images and scan images. The digital microscope images were overlaid on the digital EOS images. The opacity of the microscope image was digitally adjusted from zero to 100 percent, where zero shows only the EOS data scan and 100 shows only the microscope image. Results of these overlays with opacities of 0%, 50%, and 100% are shown in Figure 5, with corresponding "dark spots" labeled A through G. The "'butterfly" region to the right side is a point where an indent was placed.

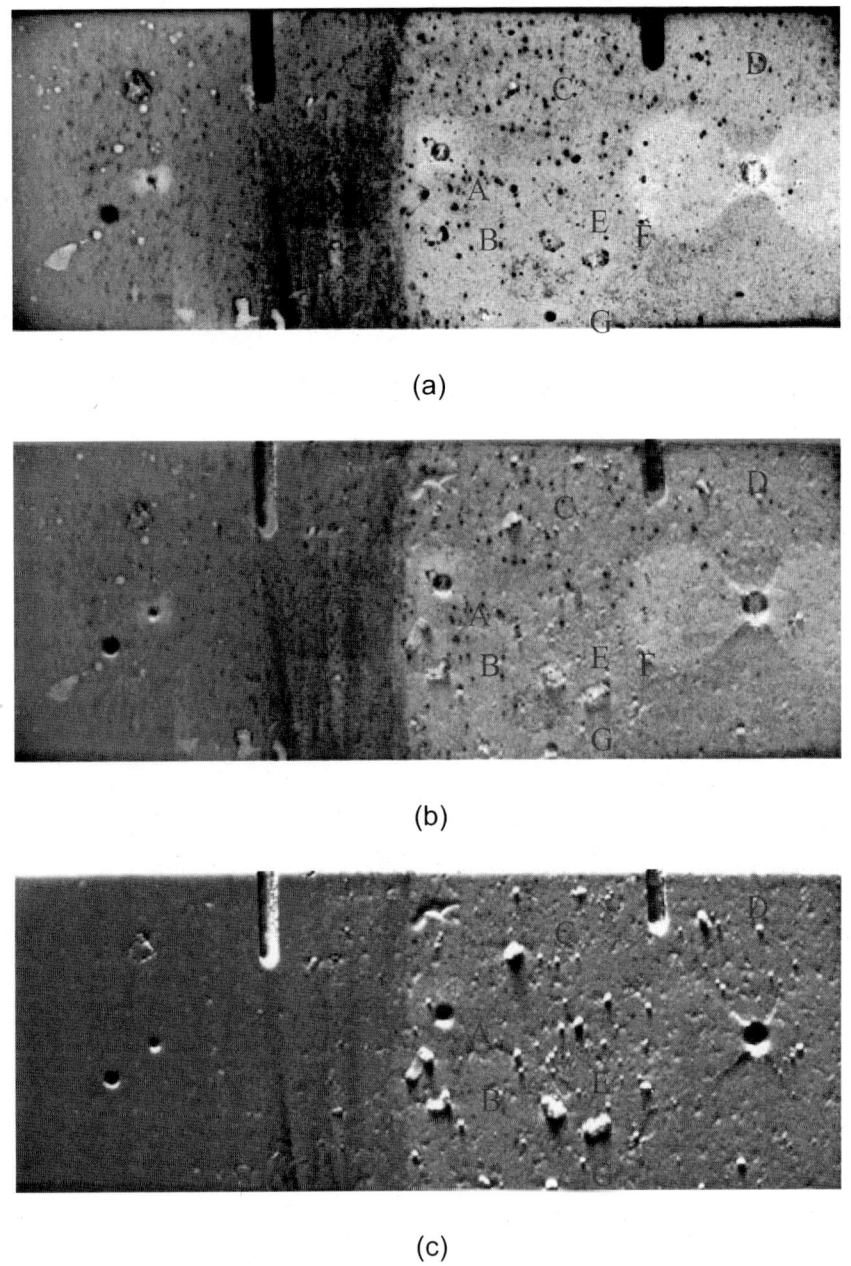

(a)

(b)

(c)

Figure 5: Results of overlaying digital microscope images over EOS images with opacities of (a) 0%, (b) 50%, and (c) 100%.

Figure 5 however shows that the lighter areas around the three indents are not surface features. Instead, these sections are areas of probable spallation as a result of the indentations.

Further analysis of TBCs has been done using OCT to obtain direct thickness measurements on an actual airfoil section. An example is shown in Figure 6.

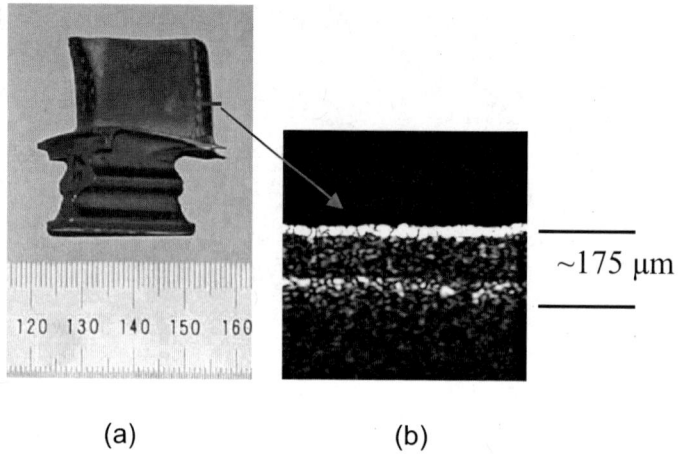

(a)   (b)

Figure 6: (a) Photograph of turbine blade with TBC and (b) OCT cross sectional image showing TBC thickness of ~ 175 um

Environmental Barrier Coatings

Use of OCT to measure thickness of EBCs has recently been demonstrated on both plasma sprayed BSAS EBCs and slurry applied EBCs made of a proprietary oxide material. The first example to be shown is of a BSAS EBC applied on a melt infiltrated (MI) SiC/SiC. Figures 7(a) shows a diagram where several OCT cross sections were acquired from the sample shown in Figure 7(b). Figure 7(c) show resulting cross sectional images that correspond to the various locations noted in 7a. Clearly the OCT data provide reasonable thickness data within a few 10s of microns.

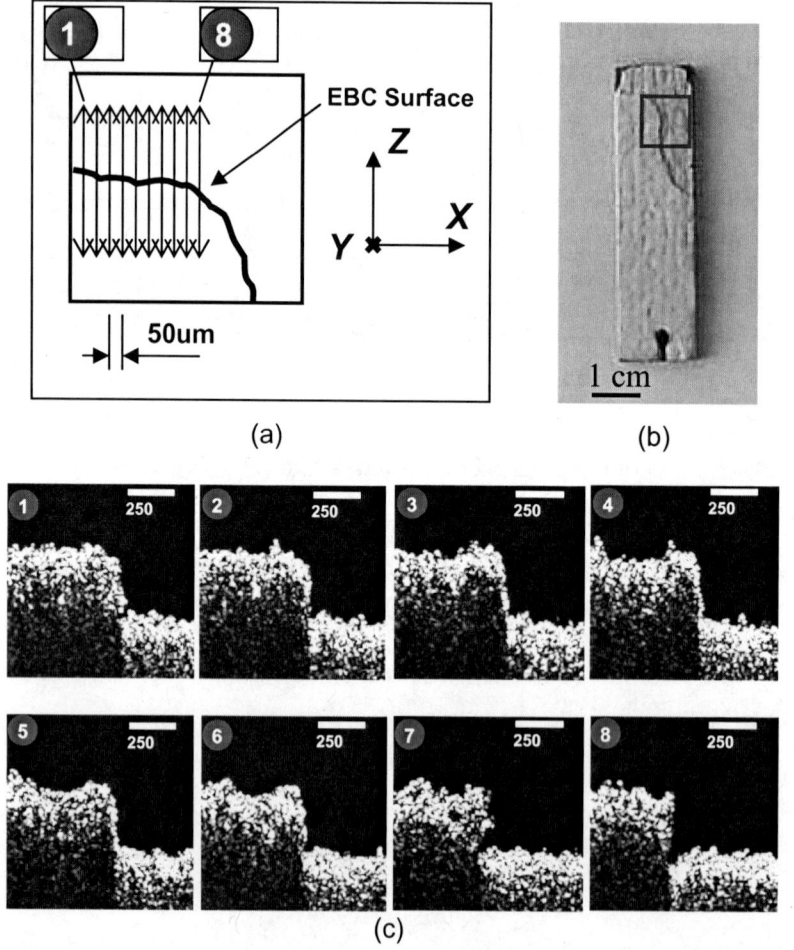

The second example of the use of OCT to measure EBC thickness of a new slurry dipped coating. Optical transmission characteristics for the slurry coating in the "green" state suggested that there were poor optical properties. However, at 1038 um of the OCT there was small transmission and therefore attempts were made to use the OCT. A special sample was prepared by using a standard quartz microscope slide so that the slurry coating could be readily observed. Figure 8(a) shows a photograph of the slide with the slurry coating applied only on one end. Figure 8(b) shows a resulting OCT cross sectional image that clearly shows the thickness of the slurry coated EBC on the top as well as the bottom. By focusing in on the top section, see Figure 8(c), one can estimate the thickness of the coating very well. Data from the OCT using a calibrated setting, suggests that the coating is 42um um thick. By placing the sample under the microscope the actual thickness was estimated to be 40 um . Thus, the comparison is very good.

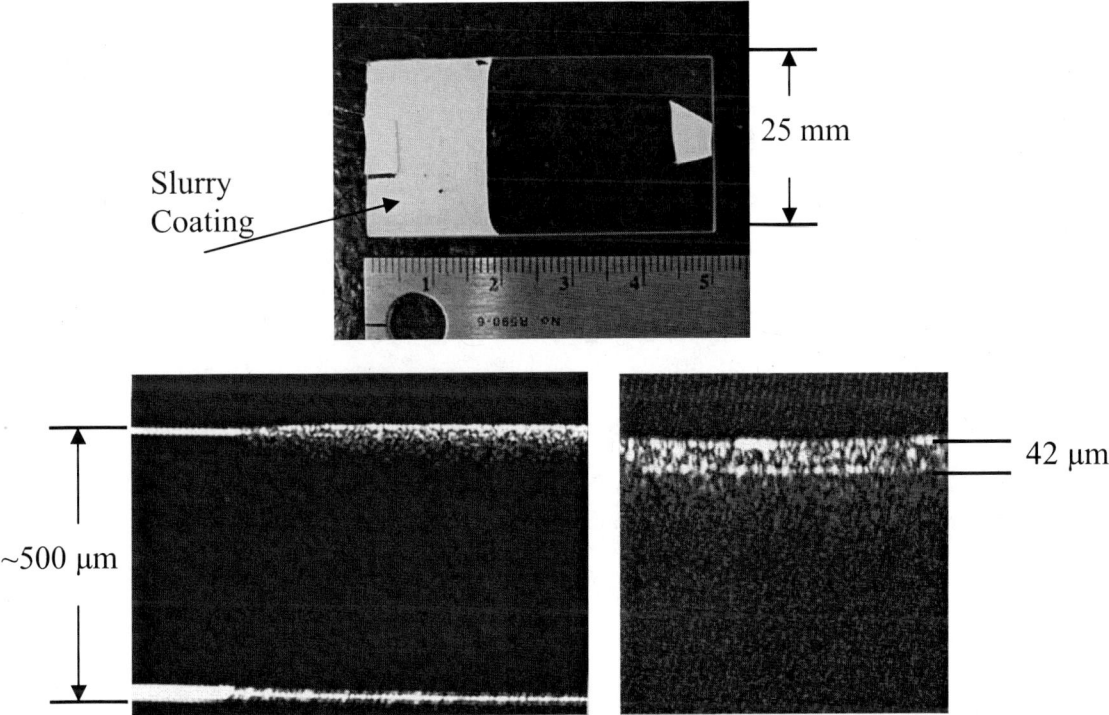

Figure 8: (a) Photograph of a slurry coating on a quartz slide and (b, c) the resulting cross-sectional images

CONCLUSIONS

Two optical NDE methods have been studied for characterizing high temperature ceramic coatings: environmental barrier coatings and thermal barrier coatings. Optical coherence tomography (OCT) has been shown to be able to determine thickness of both TBCs and EBCs including green state slurry coated EBCs. Elastic Optical Scatter (EOS) has been shown to provide one possible way to estimate spallation of TBCs and that, indeed, while there is more sensitivity to surface features than initially considered, there is a dominant effect of other factors that out weigh the effects of surface features.

REFERENCES

1. A. Szweda, T. E. Easler, R. A. Jurf and S. C. Butner, "Ceramic Matrix Composites for Gas Turbine Engines", in Ceramic Gas Turbine Component Development and Characterization, eds., M. van Roode, M. K. Ferber and D. W. Richerson, ASME press, 2003, pgs 277-289.

2. K. N. Lee, H. Fritze and Y. Ogura, "Coatings for Engineering Ceramics", in Ceramic Gas Turbine Component Development and Characterization, eds., M. van Roode, M. K. Ferber and D. W. Richerson, ASME press, 2003, pgs 641-664.

3. Karlsson. A. M., Hutchinson, J. W., and Evans, A. G. "The Displacement of the Thermally Grown Oxide in Thermal Barrier Coating Systems upon Temperature Cycling", in Mat. Sci. and Eng. A, 351 1-2, pgs 244-257, 2003

4. Karlsson. A. M., Hutchinson, J. W., and Evans, A. G. "A Fundamental Model of Cyclic Instabilities in Thermal Barrier Systems", Journal of the Mechanics and Physics of Solids, 50, 1565-89.

5. Mumm, D. R., and Evans, A. G., 2000, "On the Roles of Imperfections in the Failure of a Thermal Barrier Coating Made by Electron-Beam Physical Vapor Deposition," Acta Materiala, 48, 1815-27.

6. D. R. Clarke, J. R. Christensen, V. Tolpygo, "The Evolution of Oxidation Stresses in Zirconia Thermal Barrier Coated Superalloy Leading to Spalling Failure," *Surf. Coat. TechnoL*, **94/95**, 89-93 (1997).

7. Advisory Group for Aerospace Research and Development (AGARD)/NATO, "Thermal Barrier Coatings", 1997, AGARD report 823, 1997, North Atlantic Treaty Organization

8. B. E. Bouma and G. J. Tearney, Handbook of Optical Coherence Tomography, Marcel Dekker, New York (2002).

9. W. A. Ellingson, R. J. Visher, R. S. Lipanovich and C. M. Deemer, "Optical NDT Techniques for Ceramic Thermal Barrier Coatings", in Mat. Eval. Vol. 64, No. 1, pgs 45-51, 2006

10. Visher, R. J., Gast, L., Ellingson, W. A. and Feuerstein, A. "Health Monitoring and Life Prediction of Thermal Barrier Coatings using a Laser-Based Method", ASME paper GT2005-68252, 2005

Phase Evolution in Yttria-Stabilized Zirconia Thermal Barrier Coatings Studied by Rietveld Refinement of X-Ray Powder Diffraction Patterns

Grégoire Witz,[†] Valery Shklover, and Walter Steurer

Laboratory of Crystallography, Department of Materials, ETH Zurich, 8093 Zurich, Switzerland

Sharath Bachegowda and Hans-Peter Bossmann

Alstom (Schweiz) AG, 5401 Baden, Switzerland

One failure mechanism of thermal barrier coatings composed of yttria-stabilized zirconia (YSZ) has been proposed to be caused, in part, by the transformation of the tetragonal phase of YSZ into its monoclinic phase. Normally, studies of phase evolution are performed by X-ray diffraction (XRD) and by evaluating the intensities of a few diffraction peaks for each phase. However, this method misses some important information that can be obtained with the Rietveld method. Using Rietveld's refinement of XRD patterns, we observed, upon annealing of YSZ coatings, an increase of cubic phase content, a reduction in as-deposited tetragonal phase content, and the appearance of a new tetragonal phase having a lower yttria content that coexists with the as-deposited tetragonal phase of YSZ.

I. Introduction

THERMAL barrier coatings are used in cooled parts of gas turbines to reduce the temperature of blades and hot engine components, allowing higher operating temperatures and yielding increased efficiency. The most commonly used material for plasma-sprayed TBCs is yttria-stabilized zirconia (YSZ), which is produced as a tetragonal metastable polymorph containing around 7%–8% weight of yttria. As shown in Fig. 1, other polymorphs of YSZ can be obtained by varying the yttria content.[1] YSZ is stabilized into its cubic form by a high yttria content; monoclinic YSZ, on the other hand, forms with low yttria contents. The monoclinic polymorph is stable only at low temperatures and undergoes a martensitic transformation to a tetragonal phase around 1000°C. The monoclinic-to-tetragonal transformation causes a volume change in the unit cell by about 4%, which can result in cracking and coating failure. For this reason, it is commonly believed that the monoclinic phase in coatings should be avoided.

It is expected that engine efficiency will improve, and operating costs will decrease with increased operating temperatures and less frequent maintenance.[2] Both of these aims require a better understanding of the failure mechanisms of YSZ coatings. The decomposition of the metastable tetragonal polymorph of YSZ into cubic and monoclinic polymorphs has been discussed as a mechanism contributing to coating failure.[3] For example, the spinodal decomposition model proposed by Katamura et al.[4] suggests diffusion of yttrium- and zirconium-forming domains with a cubic structure as well as yttria-depleted tetragonal domains. At some point in the decomposition process, part

of the coating can become sufficiently depleted in yttrium to allow the martensitic tetragonal–monoclinic transition upon cooling, leading to volume change of the coating and failure.

Rietveld refinement and X-ray diffraction (XRD) together are efficient techniques to determine the microstructural parameters and phase content of a multiphase compound. However, Rietveld's refinement of multiphase systems can be difficult if there are overlapping peaks, as is the case for YSZ polymorphs, in spite of using a monochromatic radiation that generally decreases the noise of the collected spectra and reduces the peak overlaps. Rietveld refinement of peak shape parameters can give information about microstructural parameters like crystallite size and strains. As zirconium and yttrium are separated in the periodic table by only one atomic number, their X-rays scattering factors are close, and it is not possible to determine their relative content in the unit cell. As the lattice parameters of the tetragonal and cubic phases depend on the yttria content, it is therefore possible to determine the yttria concentration of both phases and study its evolution.

A previous XRD study by Miller et al.[3] and another neutron scattering study by Ilavsky et al.[5] of the phase evolution in plasma-sprayed YSZ coatings during annealing have shown that metastable tetragonal YSZ decomposes into monoclinic and cubic phases as the yttria content of the tetragonal phase is reduced. XRD studies on plasma-sprayed coatings by Brandon and Taylor and on EB-PVD coatings by Schulz indicate that between 1300° and 1400°C, the metastable tetragonal phase decomposes into a low-yttria tetragonal phase and a high-yttria cubic or tetragonal phase.[6,7] Another XRD, combined with a transmission electron microscopy (TEM) study by Azzopardi et al.[8] on EB-PVD coatings heat treated at temperatures ranging from 1100° to 1500°C shows that the metastable tetragonal YSZ phase decomposes in a low-yttria tetragonal phase, a high-yttria tetragonal phase, and a high-yttria cubic phase; their TEM images show alternating layers having high-yttria and low-yttria contents. Computer simulations, combined with TEM studies, have shown that the domain boundaries have a cubic-like structure and a relatively high concentration of yttrium ions.[4]

In the present work, we investigate the microstructure evolution in plasma-sprayed YSZ–TBC during annealing at 1100°–1400°C. The lattice parameters, phase content, strains, and crystallite size of as-sprayed and annealed coatings have been determined using Rietveld's refinement of XRD patterns.

II. Experimental Procedure

YSZ coatings were produced with an atmospheric plasma spraying system (APS) using parameters typical for preparing TBCs. The feedstock material was zirconia with 7.8 wt% yttria powder (204NS, Sulzer-Metco, Westbury, NY). Samples were mechanically removed from the substrate before annealing at temperatures ranging from 1100° to 1400°C. After annealing, samples were furnace cooled down to room temperature such that the

J. Smialek—contributing editor

Manuscript No. 22603. Received December 20, 2006; approved April 4, 2007.
Supported by the Innovation Promotion Agency, Federal Office for Professional Education and Technology, Switzerland under Grant No. 7820.3 EPRP-IW.
[†]Author to whom correspondence should be addressed. e-mail: gregoire.witz@mat.ethz.ch

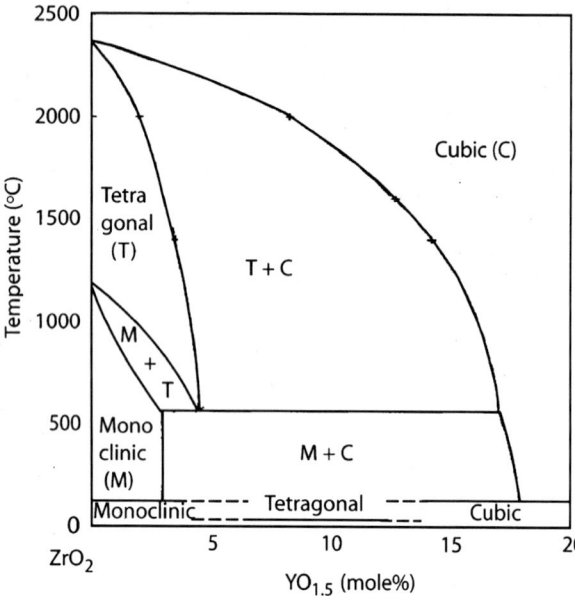

Fig. 1. Phase diagram of yttria-stabilized zirconia.

transformation from tetragonal to monoclinic YSZ can occur during cooling. The porosity content of the coatings was from 17% to 21% after plasma spraying, and declined to 12%–18% upon annealing.

XRD patterns were obtained using an X-ray powder diffractometer (X'Pert Pro, PANalytical, Almelo, the Netherlands) with monochromatic CuK$^{\alpha 1}$ radiation. Diffraction patterns were obtained within the $2\theta = 20$–$100°$ range in $0.01°$ steps using standard θ–2θ Bragg—Brentano geometry. Rietveld's refinement was performed using the general structure analysis system (GSAS) package and the EXPGUI interface.[9,10] The Rietveld refinement was processed following the Rietveld refinement guidelines formulated by the International Union of Crystallography Commission on powder diffraction.[11] The background was fitted using Chebyschev polynomials and the peak profiles were fitted using a convolution of a pseudo-Voigt and asymmetry function, together with the microstrain broadening description of Thompson and colleagues.[12–14] The pseudo-Voigt is defined as a linear combination of a Lorentzian and a Gaussian. The Gaussian variance of the peak, σ^2, varies with 2θ as

$$\sigma^2 = U \tan^2 \theta + V \tan \theta + W + P/\cos^2 \theta \qquad (1)$$

where U, V, and W are the coefficients described by Cagliotti et al.[15] and P is the Scherrer coefficient for Gaussian broadening. The Lorentzian coefficient$^{©}$ varies as:

$$\gamma = \frac{X}{\cos \theta} + Y \tan \theta \qquad (2)$$

The P and X coefficients are related to the Scherrer broadening and gives information about crystallite dimensions. The U and Y coefficients are related to the strain broadening and allow estimating microstrain within crystallites

$$\varepsilon_G (\%) = \frac{\pi}{180} \sqrt{8 \ln 2 (U - U_i)} \qquad (3)$$

$$\varepsilon_L (\%) = \frac{\pi}{180} (Y - Y_i) \qquad (4)$$

The surface roughness of plasma-sprayed coatings leads to absorption and reduction of the diffracted intensities at low

angles. If this effect is not taken into account during Rietveld's refinement, the atomic displacement parameters are reduced and can even become negative, having no physical meaning. To take into account the absorption effect, we used a normalized form of the function described by Pitschke et al.[16] Atomic displacement parameters and absorption factors modify the profile intensities with a similar 2θ dependency, and refining all of them together usually leads to unrealistic results. We have therefore fixed the atomic displacement parameters to standard values for each atom and have refined only the absorption factors. As all YSZ phases have a very close microstructure, atomic displacement parameters are expected to be very close for all phases and fixing them should not influence the quality of the fit. The number of refined parameters during the last step of the refinement process and R_p factors indicating the quality of the fit compared with the experimental XRD pattern are listed in Table I, and the R_p values were calculated after removing the background contribution.

The yttria content of the cubic phase can be calculated using the formula

$$YO_{1.5}(mol\%) = (a - 5.1159)/0.001547 \qquad (5)$$

derived by Ilavsky[5] from Scott's data on phase analysis of YZ powders with different yttria contents,[1] where a is the unit cell dimension in angstroms. The yttria content of the tetragonal phase can be calculated using the formula

$$YO_{1.5}(mol\%) = \frac{1.0225 - \frac{c}{a\sqrt{2}}}{0.0016} \qquad (6)$$

where a and c are the unit cell dimensions expressed in angstroms. This formula was derived from Scott's data[1] and empirically corrected by Ilavsky and Stalick[17] to obtain a better fit with data for a wide range of samples. Four YSZ phases were used for interpretation of the XRD patterns: two tetragonal phases with different unit cell dimensions corresponding to different yttria contents of 7–8 wt% (t'-YSZ) and 4–5 wt% (t-YSZ), one monoclinic (m-YSZ) and one cubic phase having cell dimensions corresponding to a 13–15 wt% yttria content (c-YSZ). For the coatings annealed at 1400°C, the pattern of the phase with a high yttria content (11–13 wt% yttria) was fitted using a tetragonal phase with a low c/a ratio. In this paper, we will refer to this phase as c-YSZ to be consistent with the phase definitions used for lower annealing temperatures. For some of the collected patterns, the intensities of peaks of some phases are too low to allow a structural refinement. In such cases, the

Table I. Annealing Conditions and Rietveld's Refinement

Temperature (°C)	Annealing time (h)	No. refined parameters	R_p (%)
—	As deposited	27	7.8
1100	1	27	6.7
1100	24	27	6.8
1100	100	27	6.0
1100	250	33	6.3
1100	450	33	6.1
1100	650	33	6.1
1100	850	33	6.6
1100	1400	33	7.7
1200	1	27	7.6
1200	10	27	8.3
1200	1000	33	7.6
1300	1	27	7.4
1300	10	33	6.4
1300	100	33	5.7
1300	1000	32	7.3
1400	1	33	6.9
1400	10	33	6.5

unit-cell dimensions are fixed by setting them to correspond to a chosen yttria content. Either the peak shape parameters were then fixed, or the number of refined parameters is reduced to 1 or 2. When peak shape parameters were fixed, they were set to the values obtained for the closest annealing time at the same temperature where peak shape parameters could be refined. As we observed that peaks are broader for reduced phase content, fixing peak shape could lead to a slight underestimation of the corresponding phase content.

III. Results and Discussion

The XRD patterns of the YSZ coating that had been annealed at 1300°C for various times are given in Fig. 2. The pattern of the as-deposited coating contains only the peaks of the tetragonal t'-YSZ phase. After 10 h of annealing, shoulders appear at lower angles for the (004) peak and at higher angles for the (220) peak. These shoulders correspond to the presence of low peaks belonging to a tetragonal t-YSZ phase having a higher c/a ratio. A new peak also appears between the (004) and (220) peaks of the tetragonal phases, indicating the presence of the cubic c-YSZ phase. After 100 h of heat treatment, the peaks of the t-YSZ and c-YSZ continue to grow, while those of the t'-YSZ became smaller. After 1000 h of annealing, the t'-YSZ peaks completely disappeared and were replaced by those of the c-YSZ and t-YSZ.

After 1000 h of annealing, peaks attributed to the monoclinic m-YSZ phase appeared, as shown in Fig. 3. The peak present at $2\theta = 30.2°$ may have arisen from the (101) peaks of both tetragonal YSZ phases and the (111) peak of the c-YSZ. For t'-YSZ, this peak appears at a higher angle than that of t-YSZ and for c-YSZ it appears at the lowest angle. Depending on the annealing time, this peak has different shapes due to the presence of different proportions of these three phases.

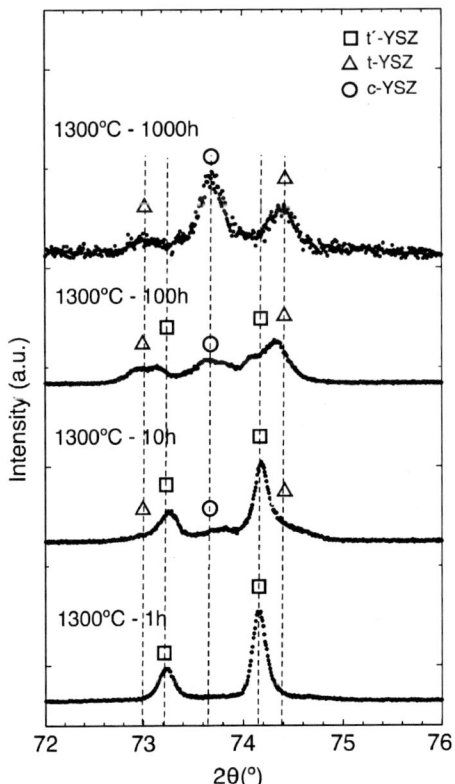

Fig. 2. X-ray diffraction patterns of yttria-stabilized zirconia (YSZ) coatings, annealed for various times at 1300°C, with 2θ between 72° and 76°. The intensities of the peaks of the t-YSZ decrease with annealing time, while new peaks of the cubic c-YSZ phase and tetragonal t-YSZ phase appear, having a low yttria content.

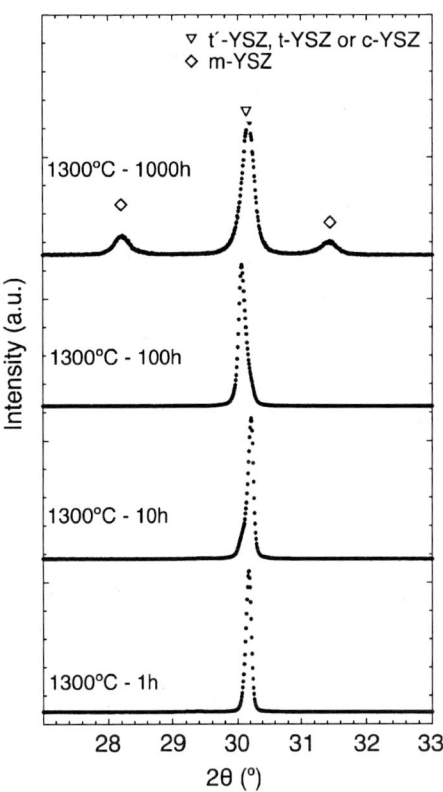

Fig. 3. X-ray diffraction patterns of yttria-stabilized zirconia (YSZ) coatings, annealed for various times at 1300°C, with 2θ range 27° and 33°. Peaks of monoclinic m-YSZ appear after 1000 h of annealing.

These observations show that fitting XRD patterns using a combination of two tetragonal YSZ phases having different yttria contents and unit cell dimensions is a good way to obtain a realistic description of the YSZ sample composition. Using only one unique tetragonal YSZ phase is not a good solution because, as shown in Fig. 2, peaks of the tetragonal phase are no longer symmetric and do not shift due to a steady loss of yttria during annealing; instead, a new set of peaks appears and the peaks of the as-deposited YSZ disappear. From this, it appears that no phases with intermediate yttria content can be seen. Instead, only the as-deposited phase or the decomposition products can be observed, as reported by Brandon and Taylor[6] in coatings annealed at 1300°C and higher.

Evolution of the phase composition of the coatings during annealing at different temperatures is displayed in Fig. 4, and the phase composition, cell parameters, and yttria content of each phase are listed in Table II. At all temperatures, there is a decrease in the t'-YSZ content and an increase in the amounts of c-YSZ and t-YSZ. This means that the main difference between the results of the heat-treatment temperature is the decomposition rate of the t'-YSZ. The yttria content of c-YSZ also increases above 15 wt% after 1000 h of heat treatment at 1200° and 1300°C.

The decomposition rate of the t'-YSZ increases by about one order of magnitude as the temperature is increased by 100°C. For instance, the phase composition after a 1000-h anneal at 1200°C is very close to the composition after an anneal of 100 h at 1300°C as well as 10 h at 1400°C. After 1000 h of annealing at 1300°, the m-YSZ appears. At the same time, the t'-YSZ peaks are no longer visible and the amount of t-YSZ begins to decrease.

Information about microstrain and grain size was obtained from the function describing the Gaussian part of the diffraction peak shape. This function shows that the peak broadening is mainly governed by microstrain within the crystallites. The microstrain level in the as-deposited t'-YSZ is around 0.2%–0.4%

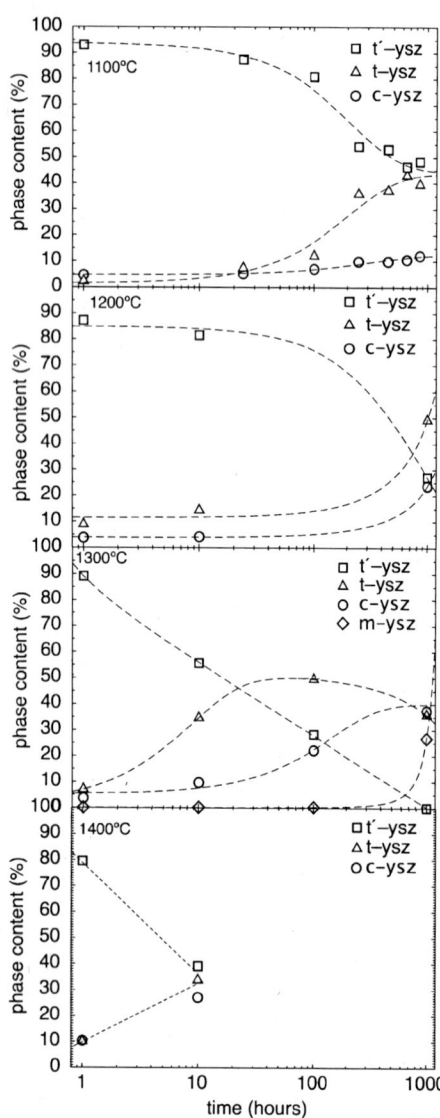

Fig. 4. Evolution of the phase composition of yttria-stabilized zirconia coatings during annealing at different temperatures.

and can increase up to 1% upon annealing, as can be seen in Fig. 5. After 1 h of annealing at 1300°C or 1400°C, the microstrain in the coatings is lower than in the coating annealed at lower temperatures. This can be explained by the high level (0.48%) of microstrain in the as-deposited coating that is released during annealing at a high temperature. The general subsequent increase in microstrain can be related to the reduction in the t'-YSZ phase content, leading to an increase in straining due to the other phases present in the coating. In the c-YSZ and t-YSZ phases, microstrain varies between 0.4% and 0.7% and decreases upon annealing. The microstrain level that is reached in the t'-YSZ phase upon annealing at a high temperature is high enough to allow cracks to nucleate and propagate. The microstrain levels can be overestimated because the microstrain broadening is described by a distribution of cell dimensions; in the case of YSZ, this distribution can also arise from a non-homogeneous yttria distribution within the grains or between the grains. Therefore, the increase of the observed microstrain can also be interpreted as an effect of yttria diffusion, which leads to a broader distribution of the yttria content within t'-YSZ crystallites.

In every XRD patterns, the absence of YSZ phases having an intermediate yttria content (in the range 5–7 wt% or 8–13 wt%) indicates that the diffusion of yttrium is relatively fast. This

Table II. YSZ Phases Content and Characteristics Obtained from Rietveld Refinement

Temperature (°C)	Annealing time (h)	t'-YSZ (wt%)	t-YSZ (wt%)	c-YSZ (wt%)	m-YSZ (wt%)	$a(t'\text{-YSZ})$ (Å)	$c(t'\text{-YSZ})$ (Å)	Yttria content (t'-YSZ) (%)	$a(t\text{-YSZ})$ (Å)	$c(t\text{-YSZ})$ (Å)	Yttria content (t-YSZ) (%)	$c(c\text{-YSZ})$ (Å)	Yttria content c-YSZ (%)	ε (t'-YSZ) (%)
	As-deposited	88.8	7.5	3.7	—	3.6150	5.1617	8.0	3.6044	5.1709	5.0	5.1325	10.7	0.48
1100	1	93.1	2.3	4.6	—	3.6121	5.1638	7.3	3.6041	5.1689	5.3	5.1373	13.8	0.42
1100	24	87.4	7.4	5.2	—	3.6125	5.1643	7.3	3.6041	5.1689	5.3	5.1373	13.8	0.41
1100	100	80.9	12.1	7.0	—	3.6118	5.1647	7.1	3.6041	5.1689	5.3	5.1373	13.8	0.40
1100	250	54.2	35.9	9.9	—	3.6116	5.1645	7.1	3.6041	5.1689	5.3	5.1373	13.8	0.37
1100	450	53.0	37.1	9.9	—	3.6117	5.1650	7.1	3.6041	5.1689	5.3	5.1373	13.8	0.40
1100	650	46.4	43.1	10.6	—	3.6108	5.1632	7.1	3.6041	5.1689	5.3	5.1373	13.8	0.48
1100	850	48.4	39.5	12.1	—	3.6117	5.1657	7.0	3.6041	5.1689	5.3	5.1373	13.8	0.55
1100	1400	42.3	45.0	12.7	—	3.6102	5.1638	6.9	3.6032	5.1672	5.3	5.1373	13.8	0.57
1200	1	87.2	8.9	3.8	—	3.6142	5.1669	7.3	3.6045	5.1709	5.0	5.1353	12.5	0.41
1200	10	81.5	14.3	4.2	—	3.6142	5.1671	7.2	3.6045	5.1709	5.0	5.1353	12.5	0.48
1200	1000	27.1	49.2	23.7	—	3.6143	5.1685	7.1	3.6031	5.1823	3.4	5.1403	15.8	1.00
1300	1	89.2	7.1	3.7	—	3.6144	5.1664	7.4	3.6045	5.1709	5.1	5.1353	12.5	0.26
1300	10	52.3	38.4	9.3	—	3.6134	5.1647	7.4	3.6057	5.1687	5.0	5.1353	12.5	0.39
1300	100	28.3	49.6	22.1	—	3.6138	5.1665	7.2	3.6029	5.1741	4.2	5.1361	13.1	0.89
1300	1000	—	35.9	37.4	26.7	—	—	—	3.6066	5.1774	4.6	5.1395	15.3	—
1400	1	79.6	10.0	10.4	—	3.6132	5.1654	7.3	3.6006	5.1780	3.4	5.1467/3.6246	13.2	0.20
1400	10	39.3	33.5	27.2	—	3.6109	5.1634	7.1	3.6005	5.1783	3.4	5.1456/3.6248	12.9	0.38

YSZ, yttria-stabilized zirconia.

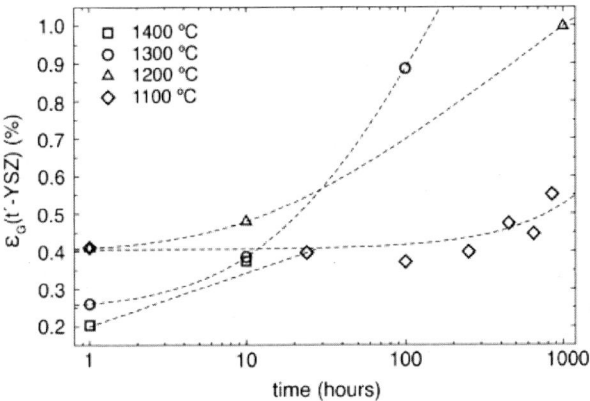

Fig. 5. Evolution of microstrain in the t'-yttria-stabilized zirconia phase as a function of annealing time.

observation, together with the diffraction peak profiles showing that the broadening arises mainly from a high level of strains or a nonhomogeneous yttria distribution in the particles, is in agreement with previous TEM studies by Lanteri et al.,[18] Shibata et al.,[19] and by Azzopardi et al.[8] and the spinodal decomposition model proposed by Katamura et al.[4] This allows the formulation of the following model for decomposition of the metastable t'-YSZ, which is schematically presented in Fig. 6. According to this model, yttrium segregates at domain boundaries, yielding grains having a modulated structure composed of cubic, high-yttria domains, and tetragonal, low-yttria domains.

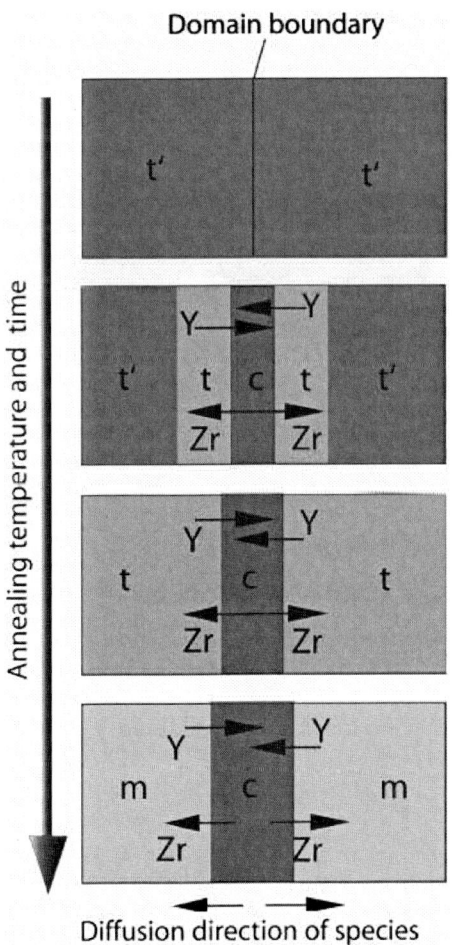

Domain boundary

Annealing temperature and time

Diffusion direction of species

Fig. 6. Schematic diagram of the yttria-stabilized zirconia phase distribution and its evolution upon annealing (see comments in text).

Only when the size of yttria-depleted domains reaches a critical size do they undergo martensitic transformation to a monoclinic YSZ structure.[20] The decomposition process can thus be written as

$$t'\text{-YSZ} \rightarrow t\text{-YSZ} + c\text{-YSZ} \rightarrow m\text{-YSZ} + c\text{-YSZ}$$

In the latter stage of the decomposition process, not only is the content of cubic phase increasing but also the yttria content of the cubic phase.

This model for the decomposition of t'-YSZ can also be applied to results obtained in previous XRD or neutron studies. In the studies by Miller et al.[3] and by Suhr et al.,[22] the absence of t-YSZ can be explained by the broad peaks of XRD patterns and the difficulty in resolving the overlapping peaks' pattern in the 72°–76° 2θ range using the relative intensity ratio (RIR) method; the c-YSZ content increases faster upon annealing than in the present study, probably because of the lower t'-YSZ content after plasma spraying and the absence of t-YSZ in the analysis of XRD patterns.[21] The tetragonal peak at 74.5° of the coating annealed for 1 h at 1400°C, in the study by Ilavsky and Stalick[17], has a shoulder indicating a possible presence of t-YSZ. The study by Brandon and Taylor[6] agrees with the decomposition model of t'-YSZ when considering that the phases T' and T'_2 are in fact describing the same phase (t'-YSZ); the phase compositions after 10 h at 1400°C and 100 h at 1300°C obtained in this study and in the present study are similar. The study by Schulz on EB-PVD coatings can also be interpreted using the decomposition model of t'-YSZ if the t2 phase, when having a high yttria content, is identified with t'-YSZ and the t1 phase and the t2 phase, when having a low yttria content, are identified with t-YSZ; the decomposition of t'-YSZ is faster in this study than in the present one, which could have been due to the difficulty to fit accurately the XRD pattern in the 72°–76° 2θ range with a combination of 10 peaks describing the CuKα1 and CuKα2 contributions of two tetragonal and one cubic YSZ phase.[7] The description of t'-YSZ decomposition by Azzopardi et al.[8] uses a model similar to the one we proposed, the only noticeable difference being the two coexisting high yttria content phases (one tetragonal and one cubic); the t'-YSZ content after 100 h at 1100°C or 100 h at 1300°C is similar to the content observed in our coatings.

It is still not clear whether, after 1000 h at 1300°C, the remaining traces of t-YSZ arise from grains having a slower decomposition rate or whether they are equally distributed in the grains undergoing the decomposition process. It is also not apparent whether the m-YSZ only forms after the t-YSZ is fully decomposed. The answers to these questions will suggest whether t'-YSZ is stabilized by the presence of the t-YSZ, or whether the transformation from tetragonal into monoclinic YSZ is governed by another mechanism.

IV. Conclusions

The XRD study, combined with Rietveld refinement of full patterns, allows monitoring of the evolution of phase composition in YSZ plasma-sprayed coatings; four phases were unambiguously distinguished and their respective proportions were refined even when they were as low as 3 wt%. This level of accuracy cannot be obtained using the RIR method, because of the difficulty in resolving the overlapping peaks of the different YSZ phases. First, at annealing temperatures between 1100° and 1300°C, the as-deposited tetragonal metastable YSZ phase decomposes into two phases: a high yttria content cubic YSZ phase and a new low yttria content tetragonal YSZ phase. The low yttria content tetragonal YSZ transforms into a monoclinic polymorph upon further annealing. In the coating annealed at 1400°C, the as-deposited tetragonal metastable YSZ phase decomposes into a high yttria content tetragonal YSZ phase and another low yttria content tetragonal YSZ phase. The kinetics of the phase transformation depend on the annealing temperature and proceeds around one order of magnitude faster when the temperature is increased by 100°C. The decomposition of the as-deposited metastable YSZ phase comes together with an increase of its microstrain having values high enough to initiate cracks that, after propagation, lead to the coating failure.

References

[1]H. G. Scott, "Phase Relationships in the Zirconia–Yttria System," *J. Mater. Sci.*, **10**, 1527–35 (1975).

[2]D. R. Clarke and C. G. Levi, "Material Design for the Next Generation Thermal Barrier Coatings," *Annu. Rev. Mater. Res.*, **33**, 383–417 (2003).

[3]R. A. Miller, J. L. Smialek, and R. G. Garlick, "Phase Stability in Plasma-Sprayed, Partially Stabilized Zirconia–Yttria"; pp. 241–53 in *Advances in Ceramics*, **Vol. 3**, *Science and Technology of Zirconia I*. Edited by A. H. Heuer, and L. W. Hobbs. American Ceramic Society, Columbus, 1981.

[4]J. Katamura and T. Sakuma, "Computer Simulation of the Microstructural Evolution during the Diffusionless Cubic-to-Tetragonal Transition in the System ZrO_2–Y_2O_3," *Acta Mater.*, **46** [5] 1569–75 (1998).

[5]J. Ilavsky, J. K. Stalick, and J. Wallace, "Thermal Spray Yttria-stabilized Zirconia Phase Changes during Annealing," *J. Therm. Spray Technol.*, **10** [3] 497–501 (2001).

[6]J. R. Brandon and R. Taylor, "Phase Stability of Zirconia-Based Thermal Barrier Coatings Part I, Zirconia–Yttria Alloys," *Surf. Coat. Technol.*, **46**, 75–90 (1991).

[7]U. Schulz, "Phase Transformation in EB-PVD Yttria Partially Stabilized Zirconia Thermal Barrier Coatings during Annealing," *J. Am. Ceram. Soc.*, **83** [4] 904–10 (2000).

[8]A. Azzopardi, R. Mévrel, B. Saint-Ramond, E. Olson, and K. Stiller, "Influence of Aging on Structure and Thermal Conductivity of Y-PSZ and Y–FSZ EB-PVD Coatings," *Surf. Coat. Technol.*, **177–178**, 131–9 (2004).

[9]A. C. Larson and R. B. Von Dreele, "*General Structure Analysis System (GSAS)*"; Los Alamos National Laboratory Report LAUR 86–748 (2000).

[10]B. H. Toby, "EXPGUI, a Graphical User Interface for GSAS," *J. Appl. Cryst.*, **34**, 210–13 (2001).

[11]L. B. McCusker, R. B. Von Dreele, D. E. Cox, D. Louër, and P. Scardi, "Rietveld Refinement Guidelines," *J. Appl. Cryst.*, **32**, 36–50 (1999).

[12]P. Thompson, D. E. Cox, and J. B. Hastings, "Rietveld Refinement of Debye–Scherrer Synchrotron X-Ray Data from Al_2O_3," *J. Appl. Cryst.*, **20**, 79–83 (1987).

[13]L. W. Finger, D. E. Cox, and A. P. Jephcoat, "A Correction for Powder Diffraction Peak Asymmetry Due to Axial Divergence," *J. Appl. Cryst.*, **27**, 892–900 (1994).

[14]P. Stephens, "Phenomenological Model of Anisotropic Peak Broadening in Powder Diffraction," *J. Appl. Cryst.*, **32**, 281–9 (1999).

[15]G. Caglioti, A. Paoletti, and F. P. Ricci, "Choice of Collimators for a Crystal Spectrometer for Neutron Diffraction," *Nucl. Instrum.*, **3**, 223–8 (1958).

[16]W. Pitschke, H. Hermann, and N. Mattern, "The Influence of Surface Roughness on Diffracted X-Ray Intensities in Bragg–Brentano Geometry and its Effect on the Structure Determination by Means of Rietveld Analysis," *Powder Diffr.*, **8**, 74–83 (1993).

[17]J. Ilavsky and J. K. Stalick, "Phase Composition and its Changes during Annealing of Plasma-Sprayed YSZ," *Surf. Coat. Technol.*, **127** [2–3] 120–9 (2000).

[18]V. Lanteri, A. H. Heuer, and T. Mitchell, "Tetragonal Phase in the System ZrO_2–Y_2O_3"; pp. 118–30 in *Advances in Ceramics*, **Vol. 12**, *Science and Technology of Zirconia II*. Edited by N. Claussen, M. Ruhle, and A. Heuer. American Ceramic Society, Columbus, 1984.

[19]N. Shibata, J. Katamura, A. Kuwabara, Y. Ikuhara, and T. Sakuma, "The Instability and Resulting Phase Transition of Cubic Zirconia," *Mater. Sci. Eng.*, **A312**, 90–8 (2001).

[20]T. K. Gupta, F. F. Lange, and J. H. Bechtold, "Effect of Stress-Induced Phase Transformation on the Properties of Polycrystalline Zirconia Containing Metastable Tetragonal Phase," *J. Mat. Sci.*, **13**, 1464–70 (1978).

[21]R. A. Miller, R. G. Garlick, and J. L. Smialek, "Phase Distributions in Plasma-Sprayed Zirconia–Yttria," *Am. Ceram. Soc. Bull.*, **62** [12] 1355–8 (1983).

[22]D. S. Suhr, T. E. Mitchel, and R. J. Keller, "Microstructure and Durability of Zirconia Thermal Barrier Coatings"; pp. 503–17 in *Advances in Ceramics*, **Vol. 12**, *Science and Technology of Zirconia II*. Edited by N. Claussen, M. Ruhle, and A. Heuer. American Ceramic Society, Columbus, OH, 1984. □

Characterization of Chemical Vapor-Deposited (CVD) Mullite+CVD Alumina+Plasma-Sprayed Tantalum Oxide Coatings on Silicon Nitride Vanes After an Industrial Gas Turbine Engine Field Test

J. A. Haynes,[†] S. M. Zemskova, H. T. Lin, and M. K. Ferber

Oak Ridge National Laboratory, Oak Ridge, Tennesse 37831-6063

W. Westphal

Rolls-Royce Allison, Indianapolis, Indiana 46206-0420

Silicon nitride ceramic vanes coated with chemical vapor-deposited (CVD) mullite, CVD alumina, and plasma-sprayed tantalum oxide were exposed to field tests in an industrial gas turbine engine. Results varied due to expected non-uniformities in the CVD coating microstructures, but dense CVD mullite/alumina showed excellent stability and protective capacity after 1148 h of engine testing. Surfaces without CVD coatings experienced massive intragranular subsurface oxidation and/or rapid recession of the ceramic substrate due to volatilization of silica species formed by oxidation. These results suggest that thin (<5 μm), dense, high-purity CVD mullite and CVD alumina are viable components for an environmental barrier coating system to protect structural ceramics in combustion environments.

I. Introduction

SILICON-based ceramics, such as Si_3N_4 and SiC, are being considered for use as materials for high-performance gas turbine engines due to their high strength and creep resistance. However, the presence of H_2O in the combustion environment significantly degrades the environmental response of silica-forming ceramics by accelerating the rate of oxidation and by volatilizing the silica through formation of gaseous silicon hydroxides.[1–5] Recession of silica is further accelerated by the high-velocity gas stream of a turbine engine, likely due to the increased flux of volatile silicon species as the thickness of the gas boundary layer is reduced.[2]

Effective options for protecting Si-based ceramics are currently limited. Thick (∼300 μm) plasma-sprayed (PS) coatings based on the barium–strontium–alumino-silicate (BSAS) system have shown promise on SiC/SiC when tested in simulated and actual combustion environments.[6] However, BSAS coatings are likely to show less success on Si_3N_4 due to the greater thermal expansion difference between BSAS and Si_3N_4 (as compared with SiC) and the significant thickness of PS coatings. Tantalum oxide (Ta_2O_5) coatings have been considered as potential protective coatings for Si-based ceramics, as Ta_2O_5 is stable up to 1360°C and its coefficient of thermal expansion (Table I) is compatible with that of Si_3N_4.[7,8] However, it is not clear whether Ta_2O_5 is a permeability barrier to oxygen and

H_2O, and hot corrosion of Ta_2O_5 may be a problem in some combustion environments.[7]

Thermally grown Al_2O_3 has been used for decades as the primary oxidation barrier for high-temperature metallic turbine components.[9] However, the relatively high thermal expansion of alumina, as compared with Si_3N_4 (Table I), makes it non-ideal as a single-layer coating to protect Si_3N_4. It is possible that a thin layer of alumina could remain thermo-mechanically stable as a protective coating on Si_3N_4 if it was layered with other lower expansion coating materials.

Mullite ($3Al_2O_3 \cdot 2SiO_2$) has also been considered as a protective coating for Si_3N_4 due to its low thermal expansion. Most prior research has applied mullite via PS,[6,10,11] resulting in thick, porous coatings with marginal protective capacity and limited mechanical stability. The stability of PS mullite in combustion environments is questionable. During engine testing, PS mullite was partially volatilized,[6,10] possibly due to decomposition to Al_2O_3 and volatile SiO_2 in the presence of high-velocity H_2O.

Dense mullite can be deposited via chemical vapor deposition (CVD).[12] Thin layers (3 μm) of CVD mullite on SiC were extremely protective in high-pressure (10 atm), static water vapor at 1200°C,[13] indicating that mullite has excellent potential as a high-temperature permeability barrier to oxygen and H_2O. The practical advantages of CVD mullite over PS mullite include higher purity, greater crystallinity, higher density, excellent adherence, non-line-of-sight deposition, and a much thinner coating (1–10 μm). However, it has not yet been demonstrated whether CVD mullite will be resistant to volatilization in a turbine engine.

At present, there is no ideal single material to serve as an environmental barrier for Si_3N_4; thus, the present study tested a multi-layer coating concept. Gas turbine engine field tests were used to test our hypothesis that thin layers of dense CVD mullite would be protective in an actual combustion environment, if it were combined with other stable ceramic overlayers (alumina and Ta_2O_5 in this study) that functioned as a volatility barrier.

II. Experimental Procedure

Ceramic vanes were fabricated from AS800 silicon nitride (Honeywell Ceramic Components, Torrance, CA), densified by gas pressure sintering. The crystalline phases in the as-received material were β-Si_3N_4 and a modified H phase ($Sr_2La_4Y_4(SiO_4)_6O_2$). The surfaces of these vanes were oxidized along the grain boundaries to a depth of 4–6 μm.

The vanes were coated with a three-layer coating system consisting of: (1) CVD mullite (intended as a permeability barrier to oxygen and H_2O vapor), (2) CVD alumina (intended as a permeability barrier and a stable overlayer to protect mullite from volatility), and (3) PS Ta_2O_5 (intended as a stable overlayer to

E. Opila—contributing editor

Manuscript No. 20882. Received August 15, 2005; approved June 5, 2006.
This work was funded by the U.S. Department of Energy, Office of Distributed Energy, Microturbine Materials Program, under Contract DE-AC05-00OR22725 with UT-Battelle, LLC.
[†]Author to whom correspondence should be addressed. e-mail: haynesa@ornl.gov

Table I. **Thermal Expansion Comparisons of Coating Materials for Si-Based Ceramics**

Material	SiC[18]	Si_3N_4[18]	SiO_2[19]	Mullite[18]	Al_2O_3[18]	Ta_2O_5[18]
CTE ($\times 10^{-6}$/°C)	5.5	3.0	6.0†	5.7	8.0	3.6

†Cristobalite.

protect mullite from volatility if the alumina cracked due to thermal expansion mismatch). These three materials were selected due to their combination of stability, protective capacity, and thermal expansion compatibility (Table I).

Mullite coatings (0.5–5 µm thick with an Al:Si ratio of 2.8–3.0) were deposited directly onto the as-received Si_3N_4 vanes at 1050°C and 10.6 KPa (80 Torr) for 120 min, using a laboratory-scale, horizontal, hot-wall CVD reactor flowing $SiCl_4$, $AlCl_3$, H_2, and CO_2.[13–15] X-ray diffraction of control coupons indicated crystalline mullite coatings with a 001 texture. A thin layer of CVD alumina (0.2–1.5 µm) was deposited over the CVD mullite by ceasing $SiCl_4$ flow and increasing the reactor temperature to 1100°C for 60 min. The CVD coatings were then overlaid with PS Ta_2O_5 (150–200 µm) by a commercial coating source (Honeywell Ceramic Components).

It must be emphasized that uniform, high-quality CVD coatings were not expected on most regions of the vane surfaces. The small, laboratory-scale CVD reactor was designed to deposit thin, dense mullite coatings on small ceramic coupons (1 cm × 1 cm × 0.3 cm). The ceramic vanes (approximately 4.0 cm × 4.0 cm × 2.0 cm) barely fit within the CVD reactor. The CVD coatings were not expected to be uniform due to severe gradients in gas flow and temperature within the reactor hot zone. However, it was anticipated that some regions of CVD alumina/mullite on the vane surfaces would be dense. The protective behavior of these dense areas could then be compared after engine testing with the areas with no alumina/mullite coating or with inferior microstructures.

These first-stage ceramic vanes were designed for retrofit into a Model 501-K turbine (Rolls-Royce, Indianapolis, IN). The vane assembly, which included numerous Si_3N_4 vanes with various coating types and configurations, was mounted in a 501-K turbine at a commercial site in Mobile, AL.

During the engine tests, the average temperature and pressure of gas impinging on the vanes were approximately 1066°C (1950°F) and 8.9 atm (128 psia), respectively. However, due to the combustor temperature pattern, it was estimated that the mid-span gas temperature could have been as high as 1260°C (2300°F) at the "hot spot." The inlet gas velocity at vane mid-span was approximately 162 m/s (530 ft/s), with the gas accelerating to approximately 573 m/s (1880 ft/s) at the vane exit. The engine was exposed to the extremely humid conditions of the Gulf of Mexico region, with a mole fraction of water vapor for the gas entering the vanes calculated at 0.10.[16] The engine sustained full power and was subjected to the normal rigors of commercial operation including emergency shutdown and full-power water wash. The engine was periodically shut down (at 200, 624, 815, and 1148 h) for pre-planned inspections and removal/replacement of selected vanes for subsequent component evaluations. The vanes described in this study were removed for metallographic analysis after 624 and 1148 h of field testing.

Tested vanes were encapsulated in epoxy, sectioned along the mid-span (highest temperature region), and metallographically polished. Cross sections were examined via field-emission gun scanning electron microscopy (FEG-SEM: S4700, Hitachi Ltd., Tokyo, Japan) and energy-dispersive spectroscopy (Hitachi Ltd.).

III. Results

Characterization of non-coated Si_3N_4 vanes removed from the engine test at various time intervals confirmed that rapid oxidation, volatilization, and recession of the Si_3N_4 occurred, as

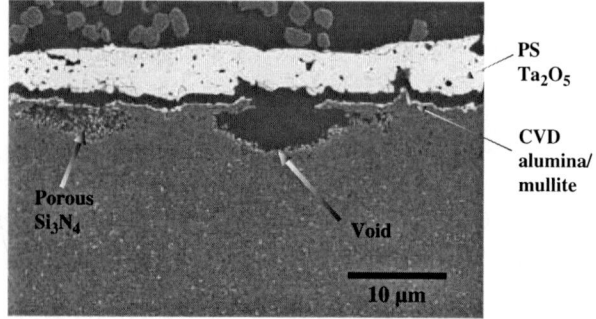

Fig. 1. Scanning electron microscopy of a cross section from the surface of a coated silicon nitride vane after 624 h of engine exposure. Damage to the alumina/mullite resulted in a severe localized substrate attack.

expected.[17] The extent of AS800 recession on trailing edge surfaces of non-coated vanes was measured at 400 µm after 624 h, as determined by a coordinate measuring machine.[16]

Two vanes with CVD mullite+CVD alumina+PS Ta_2O_5 were engine tested for 624 and 1148 h, respectively. Characterization of the mid-span of both vanes showed that although the coatings completely spalled in some areas, there were numerous areas where all three coating layers (mullite/alumina/Ta_2O_5) survived. Most of these intact areas had mullite coatings with low levels of porosity, as determined by analysis of metallographic cross sections. On surfaces where no coating remained, there was severe recession of Si_3N_4. Further, there were no SiO_2 corrosion products remaining on the bare Si_3N_4 suction surfaces (convex side of the vane), and only localized pockets of porous silica remaining on the bare pressure surfaces (concave side of the vane), suggesting high rates of silica evaporation.

There were also regions of intact Ta_2O_5 coating with localized subsurface Si_3N_4 oxidation and volatilization that left either pockets of very porous Si_3N_4 material or large voids in the Si_3N_4 beneath the surface of the coatings (see Fig. 1). These pockets of subsurface damage were associated with high levels of porosity or through-cracks in the alumina/mullite layer. The localized damage in the alumina/mullite layers may be the result of flaws in the CVD coatings or cracking in the alumina, due to its higher modulus and coefficient of thermal expansion as compared with mullite and Si_3N_4. The significant sub-surface attack under the intact Ta_2O_5 in Fig. 1 demonstrates that although thick PS Ta_2O_5 coatings can survive in a combustion environment, PS Ta_2O_5 did not serve as an effective barrier to oxygen or H_2O permeation. The large void in Fig. 1 suggests that the volatility of silica is not prevented by the presence of PS Ta_2O_5, although volatility may be reduced (as other non-coated surfaces showed up to 400 µm of recession). The intact, overhanging segments of alumina/mullite coating around the void in Fig. 1 confirm that CVD alumina/mullite is much less volatile than SiO_2 (as expected).

Figure 2 shows a cross section from a region of porous mullite coating after 1148 h. There was no discrete alumina layer visible in this region, most likely due to lack of coating formation during the brief period of CVD alumina deposition. Although the porous mullite coating apparently offered some protection, sub-surface oxidation of the Si_3N_4 is visible beneath the coating. Beneath the intact mullite, there were typically two layers: (1) a layer of silica (0.2–8 µm thick, depending on mullite porosity) formed by oxidation of the Si_3N_4, and (2) a subsurface damage zone (SDZ) where silicon nitride grains were partially oxidized at the grain boundaries to an additional depth of 4–15 µm. The thickness of the SDZ varied, based on the microstructure and thickness of the mullite and alumina coatings. The depth of the SDZ on the pressure side of the vane was typically greater than that on the suction side.

Figure 3 shows a region where the dense alumina/mullite coating was intact and there was little sign of oxidation damage

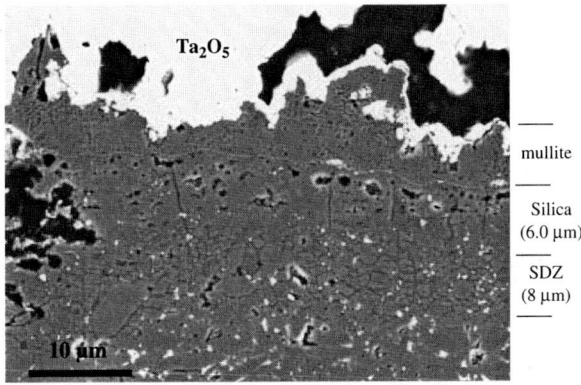

Fig. 2. Scanning electron microscopy of cross sections from the surface of a coated silicon nitride vane after 1148 h of engine exposure. The 6.0 μm thick silica layer, which is likely cristobalite, is vertically cracked.

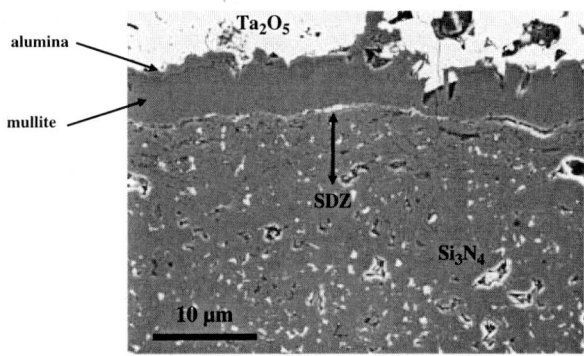

Fig. 4. Scanning electron microscopy of cross sections from the surface of a coated silicon nitride vane after 1148 h of engine exposure.

after 1148 h. Figure 4 shows a region with a very thin layer of alumina (<0.5 μm) and dense, crack-free mullite (indicating good thermo-mechanical stability on Si_3N_4). In this case, the underlying silica layer is <0.2 μm thick (as measured by FEG-SEM) and the total SDZ is 6 μm thick or less, which is very similar to the SDZ thickness on the as-received Si_3N_4 vanes. There was very little porosity in the mullite or silica in Fig. 4, suggesting little oxidation or volatilization after 1148 h. The surface condition of the vane cross section in this region looked very similar to the as-received vanes before engine testing.

IV. Discussion

These results clearly demonstrate the potential for mullite-based coating systems, by demonstrating that it is possible for thin films of CVD mullite to not only survive multiple duty cycles in a gas turbine engine, but in combination with alumina and PS Ta_2O_5 overlayers, to also confer effective protection to Si_3N_4 components. Comparison of the 6 μm SDZ after 1148 h (Fig. 4) with the uncoated specimens that experienced over 400 μm of Si_3N_4 recession after 624 h emphasizes the potential of these coatings. Within the boundaries of this study, there is no way to quantify the relative effects of the mullite and alumina to suppress oxidation of the Si_3N_4. It could be argued that the CVD alumina was the layer that prevented oxidation of the Si_3N_4, as alumina is the most well-known protective oxide layer for metallic materials in gas turbine engines.[9] However, previous laboratory furnace studies confirmed that CVD mullite alone drastically reduces oxidation rates of silica-forming ceramics in

nearly stagnant, high-pressure water vapor at 1200°C.[13] Further, alumina would not be an effective protective barrier by itself, due to its poor thermal expansion match with Si_3N_4 (Table I). Mullite has a much lower modulus of elasticity than alumina and a much closer thermal expansion match with Si_3N_4 (Table I) and thus has a much better chance of long-term survival. Mullite may also serve as an intermediate thermal expansion layer that allows a thin alumina film to remain thermo-mechanically stable during multiple duty cycles. The protective capacity of this coating system was attributed to three factors: (1) the good thermo-mechanical stability of mullite on Si_3N_4; (2) the excellent permeation barrier characteristics of dense, high-purity CVD mullite and CVD alumina; (3) and the reduction in gas velocity provided by the CVD alumina (and possibly the PS Ta_2O_5) overlayers to further reduce volatility effects on the CVD mullite.

V. Conclusions

(1) Thin (<5.0 μm) coatings of dense CVD mullite overlaid with CVD Al_2O_3 (<1.0 μm) and PS Ta_2O_5 (150–200 μm) provided excellent localized protection to Si_3N_4 vanes after 1148 h in a gas turbine engine.

(2) Stable ceramic top coatings of CVD Al_2O_3 appear to reduce the volatility of silica and mullite, possibly by reducing the influence of gas velocity.

(3) Thin layers (~1 μm) of CVD Al_2O_3 showed good thermo-mechanical stability on Si_3N_4 when deposited over CVD mullite.

(4) The performance of thin coatings of CVD mullite/alumina demonstrates the potential of high-purity, dense mullite to remain stable in a turbine engine environment and to serve as a viable permeation barrier for protecting Si_3N_4 in combustion environments.

(5) PS Ta_2O_5 was permeable to oxygen and water vapor.

Acknowledgments

The authors would like to thank Peter Tortorelli, Michael Lance, and Dave Stinton of ORNL for reviewing this paper.

References

[1]E. J. Opila and R. E. Hann, "Paralinear Oxidation of CVD SiC in Water Vapor," *J. Am. Ceram. Soc.*, **80** [1] 197–205 (1997).
[2]E. J. Opila, J. L. Smialek, R. C. Robinson, D. S. Fox, and N. S. Jacobson, "SiC Recession Caused by SiO₂ Scale Volatility Under Combustion Conditions: II, Thermodynamics and Gaseous Diffusion Model," *J. Am. Ceram. Soc.*, **82** [7] 1826–2834 (1999).
[3]K. L. More, P. F. Tortorelli, M. K. Ferber, L. Walker, J. R. Keiser, N. Miriyala, W. D. Brentnall, and J. R. Price, "Exposure of Ceramics and Ceramic Matrix Composites in a Simulated Combustion Environment"; ASME Paper 99-GT-292, 1999.
[4]J. Kimmel, J. Price, K. L. More, P. F. Tortorelli, E. Sun, and G. Linsey, "The Evaluation of CFCC Liners After Field Testing in a Gas Turbine"; GT-2003-38920, *Proceedings of ASME Turbo Expo*, 2003. June 15–19, 2003, Atlanta, GA.

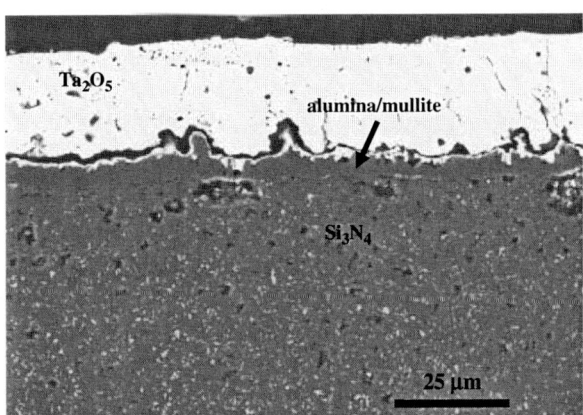

Fig. 3. Scanning electron microscopy of a cross section from the surface of a coated silicon nitride vane after 1148 h of engine exposure. The dense mullite/alumina coating with a plasma-sprayed tantalum oxide top coat provided excellent protection.

[5]R. Krishnamurthy, B. W. Sheldon, and J. A. Haynes, "Stability of Mullite Protective Coatings for Silicon-Based Ceramics," *J. Am. Ceram. Soc.*, **88** [5] 1099–107 (2005).

[6]K. L. More, P. F. Tortorelli, L. R. Walker, J. B. Kimmel, N. Miriyala, J. R. Price, H. E. Eaton, E. Y. Sun, and G. D. Linsey, "Evaluating Environmental Barrier Coatings on Ceramic Matrix Composites after Engine and Laboratory Exposures"; GT-2002-30630, *Proceedings of ASME Turbo Expo*, 2002. June 2–6, 2002, Amsterdam.

[7]W. Y. Lee, Y. W. Bae, and D. P. Stinton, "Na_2SO_4-Induced Corrosion of Si_3N_4 Coated with Chemically Vapor Deposited Ta_2O_5," *J. Am. Ceram. Soc.*, **78** [7] 1927–30 (1995).

[8]M. Moldovan, C. M. Weyant, D. L. Johnson, and K. T. Faber, "Tantalum Oxide Coatings as Candidate Environmental Barriers," *J. Thermal Spray Technol.*, **13** [1] 51–6 (2004).

[9]J. Stringer and I. G. Wright, "Current Limitations of High-Temperature Alloys in Practical Applications," *Oxid. Met.*, **44** [1/2] 265–308 (1995).

[10]K. N. Lee and R. A. Miller, "Development and Environmental Durability of Mullite and Mullite/YSZ Dual Layer Coatings for SiC and Si_3N_4 Ceramics," *Surf. Coat. Technol.*, **86–87**, 142–8 (1996).

[11]K. N. Lee, "Contamination Effects on Interfacial Porosity During Cyclic Oxidation of Mullite-Coated Silicon Carbide," *J. Am. Ceram. Soc.*, **81** [12] 3329–32 (1998).

[12]R. P. Mulpuri and V. K. Sarin, "Synthesis of Mullite Coatings by Chemical Vapor Deposition," *J. Mater. Res.*, **11** [6] 1315–24 (1996).

[13]J. A. Haynes, M. J. Lance, M. K. Ferber, K. M. Cooley, R. A. Lowden, and D. P. Stinton, "CVD Mullite Coatings in High Temperature, High Pressure Water Vapor," *J. Am. Ceram. Soc.*, **83** [3] 657–9 (2000).

[14]J. A. Haynes, K. M. Cooley, D. P. Stinton, R. A. Lowden, and W. Y. Lee, "Corrosion Resistant CVD Mullite Coatings for Si_3N_4," *Ceram. Eng. Sci. Proc.*, **20** [4] 355–62 (1999).

[15]S. M. Zemskova, C. Y. Jones, K. M. Cooley, and J. A. Haynes, "Optimization of CVD Parameters for Fabrication of Oxidation-Resistant Mullite Coatings on Silicon Nitride," *J. Am. Ceram. Soc.*, **87** [12] 2201–7 (2004).

[16]H. T. Lin, M. K. Ferber, W. Westphal, and F. Macri, "Evaluation of Mechanical Reliability of Silicon Nitride Vanes After Field Tests in an Industrial Gas Turbine"; ASME GT2002-30629, presented and published in *Proceedings of at TURBO EXPO Land, Sea, & Air 2002*, June 3–6, Amsterdam, the Netherlands.

[17]H. T. Lin and M. K. Ferber, "Mechanical Reliability Evaluation of Silicon Nitride Ceramic Components After Exposure in Industrial Gas Turbines," *J. Eur. Ceram. Soc.*, **22**, 2789–97 (2002).

[18]D. P. Stinton, J. C. McLaughlin, and L. Riester, "Fabrication and Testing of Corrosion Resistant Coatings"; pp. 1146–53 in *4th International Symposium on Ceramic Materials and Components for Engines*, Edited by R. Carlsson, T. Johansson, and L. Kahlman, Swedish Ceramic Society, Goteborg, Sweden, Elsevier Applied Science, London and New York, 1992.

[19]Y. Fei, "Thermal Expansion"; pp. 29–44 in *Mineral Physics and Crystallography, A Handbook of Physical Constants*, Edited by T.J. Ahrens, Washington, DC: American Geophysical Union 1995. □

International Journal of
Applied Ceramic Technology
Ceramic Product Development and Commercialization

Monitoring Delamination Progression in Thermal Barrier Coatings by Mid-Infrared Reflectance Imaging

Jeffrey I. Eldridge* and Charles M. Spuckler

NASA Glenn Research Center, Cleveland, Ohio 44135

Richard E. Martin

Cleveland State University, Department of Civil and Environmental Engineering, Cleveland, Ohio 44115

Mid-infrared (MIR) reflectance imaging is shown to be a reliable diagnostic tool for monitoring delamination progression in thermal barrier coatings (TBCs). MIR reflectance imaging utilizes the maximum transparency of TBCs in the 3–6 μm wavelength region to probe below-surface delamination crack propagation that is typically hidden from visible wavelength inspection. The image contrast that identifies delamination progression arises from the increased reflectance produced by a large component of total internal reflection at the TBC/buried-crack interface. Imaging was performed at a wavelength of 4 μm to take advantage of the relatively high transmittance of plasma-sprayed 8 wt% yttria-stabilized zirconia (8YSZ) TBCs along with a desirable relative insensitivity to potentially interfering absorptions by atmospheric constituents at that wavelength. A key advantage of MIR reflectance imaging over competing techniques is that it is sensitive to delamination progression even at very early stages before delamination cracks start linking together; therefore, TBC health assessment can be achieved throughout the life of the TBC well before TBC failure is imminent. Examples are presented to demonstrate monitoring delamination progression by MIR reflectance imaging in 8YSZ TBC-coated specimens subjected to furnace cycling to 1163°C. The experimental results were in good agreement with reflectance values predicted by a four-flux Kulbelka–Munk approximation applied to the extreme cases of a completely adherent and a completely detached TBC. Practical considerations, including potential interfering effects from surface contamination, sintering, and erosion are discussed.

Introduction

While thermal barrier coatings (TBCs) provide beneficial thermal protection to turbine engine components, TBCs are susceptible to spallation, which can re-

sult in performance- or safety-threatening conditions. Therefore, reliable diagnostic tools that are practical to implement are needed to identify the location and severity of degradation in TBCs to protect against premature TBC failure. TBC replacement can then be based on an informed assessment of TBC damage rather than by a fixed timetable. TBC failure typically progresses by buried cracks propagating both within the TBC near the interface with the thermally grown oxide (TGO) as well as within the TGO.[1,2] These cracks grow

This work was financially supported by the NASA Independent Research & Development Program.

*jeffrey.i.eldridge@nasa.gov

and link together to produce TBC delamination and eventual TBC spallation. Because these cracks are not visible at the TBC surface, any successful TBC damage-inspection tool must probe through the thickness of the TBC to detect this buried damage. TBC translucency allows the application of optical probes to evaluate subsurface TBC damage. For example, piezospectroscopy (Cr^{3+} luminescence) has been utilized to monitor the stress state of the TGO that forms beneath the TBC, and impending TBC failure is signified by TGO stress relaxation.[3-5] Unfortunately, this stress relaxation often occurs immediately preceding failure, and therefore, does not provide sufficient warning to take preventative action. Infrared (IR) thermography[6-9] relies on the discontinuity in heat flow presented by coating delamination to produce "hot spots" on the coating surface that can be thermally imaged; however, as with piezospectroscopy, there may not be adequate warning of impending TBC failure since the discontinuity in heat flow needed to produce image contrast often does not occur until the later stages of TBC damage progression when there is significant crack linkage. Elastic optical (laser) scattering[10] and the technique of optical coherence tomography[10] both rely on increased elastic scattering of visible to near-infrared wavelength light by subsurface defects to monitor TBC failure progression, and have been demonstrated for low-scattering electron-beam physical-vapor-deposited (EB-PVD) TBCs. However, these two methods showed much more limited depth penetration into the more highly scattering plasma-sprayed TBCs. This paper presents a new approach using mid-infrared (MIR) reflectance imaging as a diagnostic tool that overcomes many of the obstacles encountered by these other optical techniques for evaluating remaining TBC life. This approach offers the advantage of working at wavelengths at which the TBC has much greater transmittance than for visible light,[11,12] and therefore can be applied to highly scattering (compared with EB-PVD) plasma-sprayed TBCs that have proven difficult to probe by either piezospectroscopy or elastic (visible) light scattering. A hemispherical transmittance measurement (Fig. 1) for a freestanding plasma-sprayed 8 wt% yttria-stabilized zirconia (8YSZ) TBC clearly shows that the maximum TBC transparency is in the MIR wavelength range, specifically in the 3–6 μm wavelength range. MIR reflectance imaging takes advantage of this optical "window" to achieve much greater probe depth than visible wavelengths and can therefore better detect the buried crack

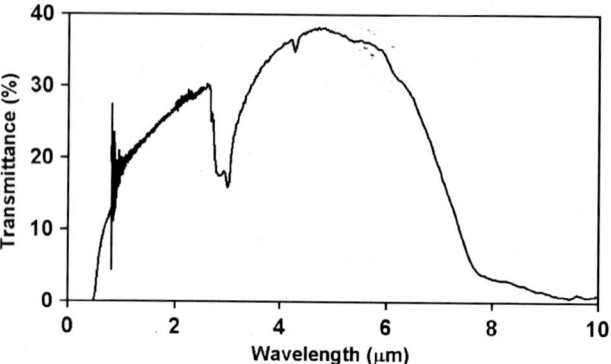

Fig. 1. *Hemispherical transmittance spectrum of 170-μm-thick freestanding plasma-sprayed 8YSZ coating.*

progression that ultimately produces TBC failure. In addition, in contrast to piezospectroscopy and IR thermography, MIR reflectance imaging will be shown to be sensitive to the unlinked cracks that are present in the early stages of TBC damage progression so that TBC health assessment can be achieved throughout the life of the TBC well before TBC failure is imminent.

While damage sensitivity and spatial resolution are essential attributes for a TBC damage inspection tool, additional constraints must be met to achieve practical implementation. In particular, spot analysis as typically performed in the laboratory will be ineffective for monitoring damage over significant areas; therefore, an imaging-based method is preferred over either spot- or raster-based methods to achieve greater inspection coverage at shorter acquisition times. In addition, the measurements should be simple to interpret and feasible to implement in an engine environment. In this regard, the need for a flash heat source and the analysis of thermal transients makes IR thermography difficult to apply in an engine environment. Similarly, electrical measurements such as eddy current[13] and impedance[14] are not easily implemented in difficult-to-access engine environments. As presented in this paper, many of these limitations have been overcome by applying MIR reflectance as an imaging technique that functions as a practical, easily interpreted TBC health-monitoring tool.

Spectral Basis for MIR Reflectance Imaging of TBC Damage Progression

The selection of a wavelength range for performing MIR reflectance imaging was based on spectral analysis of the IR reflectance from TBC-coated specimens furnace-cycled to various stages of TBC life as well as on transmittance of freestanding TBCs. Because TBCs are scattering materials,[11] hemispherical detection was necessary to fully capture the reflected (backscattered) and transmitted (forward scattered) radiation. Room-temperature spectral hemispherical reflectance spectra were obtained for the TBC specimens using a Nicolet 760 Fourier transform infrared (FTIR) spectrometer (Thermo Nicolet, Madison, WI) with a Labsphere integrating sphere accessory. The TBC specimens were prepared by atmospheric plasma-spraying a top coat of 8YSZ onto 25.4 mm diameter Rene N5 superalloy disks that were precoated with a ~120-μm-thick NiCrAlY bond coat by low-pressure plasma-spraying. The freestanding TBCs used for transmittance and reflectance measurements were produced by spraying onto sacrificial carbon substrates that were later heat treated in air at 800°C to burn off the carbon. Furnace cycling tests were performed in a tube furnace in air, with each cycle consisting of a 45-min hold at 1163°C followed by a 15 min cooling period to ~120°C.

While high transmittance (Fig. 1) is a key criterion favoring the selection of MIR wavelengths to probe buried damage in TBCs by reflectance, it is also essential that the incoming radiation interact with buried cracks to produce a change in reflectance that can be monitored. Figure 2 displays the increase in MIR hemispherical reflectance in the high-transmittance 3–5 μm wavelength range that accompanies furnace cycling for a 165-μm-thick TBC-coated specimen. This reflectance increase associated with furnace cycling has been observed consistently, with reflectance values among specimens with the same TBC thickness tracking closely

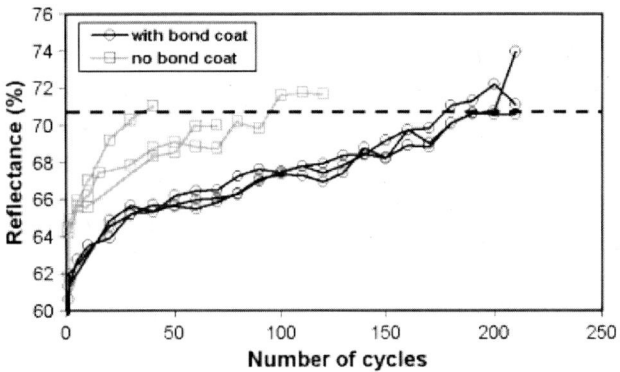

Fig. 3. *Effect of furnace cycling on hemispherical reflectance at λ = 4.0 μm for ~ 200-μm-thick TBC-coated specimens with and without bond coats.*

together as a function of number of furnace cycles until reaching a reflectance value that is associated with TBC failure for that specific TBC thickness.[12] Reflectance increases due to furnace cycling among different TBC-coated specimens can be more easily compared by plotting reflectance at a single wavelength in the wavelength range displayed in Fig. 2. For this purpose, a wavelength of 4.0 μm was selected for comparing hemispherical reflectance values. Figure 3 plots the relationship between hemispherical reflectance measured at a wavelength of 4.0 μm and the number of furnace cycles for one set of 200-μm-thick TBC-coated specimens with a bond coat beneath the TBC and another set with no bond coat. The specimens were furnace cycled until TBC failure occurred. The three specimens with bond coats showed failure between 200 and 210 cycles and reflectance values tracked very closely together with number of furnace cycles. In contrast, the three TBC-coated specimens without bond coats failed at much fewer numbers of cycles and with much higher scatter among cycles-to-failure (failure observed at 40, 70, and 120 cycles). Note that the shorter-lived specimens (without bond coats) showed a more rapid increase in hemispherical reflectance than the longer-lived specimens (with bond coats), and that there is a reflectance associated with TBC failure of about 71% for all specimens. In all cases, the reflectance increases smoothly from the beginning of TBC life and appears to be sensitive to even the very early stages of TBC failure progression. The association between reflectance increases and TBC failure progression was supported by observing furnace-cycling-induced damage propagation by SEM examination of cross-sections of TBCs that had been cycled to different stages of cyclic

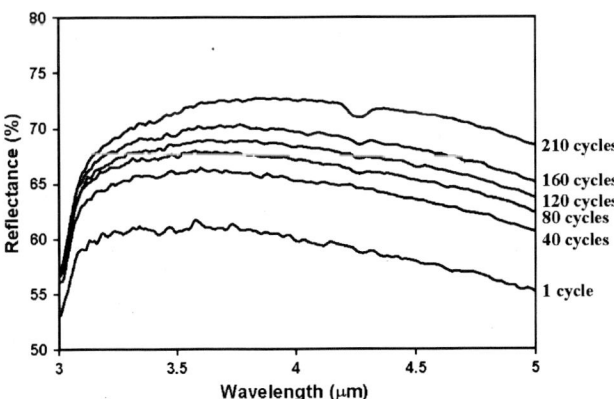

Fig. 2. *Effect of furnace cycling of TBC-coated specimen (165-μm-thick 8YSZ) on hemispherical reflectance spectra in mid-infrared (MIR) wavelength range.*

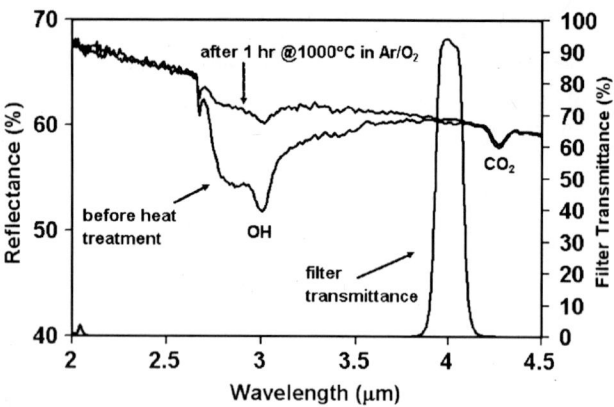

Fig. 4. *Hemispherical reflectance spectra of 82-μm-thick freestanding plasma-sprayed 8YSZ coating before and after 1 h heat treatment at 1000°C in Ar/O₂ mixture showing reduction in OH absorption after heat treatment. Transmittance spectrum for bandpass filter centered at 4.0 μm wavelength is displayed to illustrate desirable relative insensitivity to OH content and to atmospheric CO₂ absorption at this wavelength.*

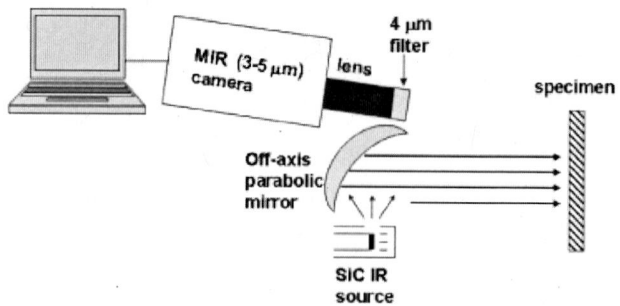

Fig. 5. *Mid-infrared (MIR) reflectance imaging system.*

life (see Fig. 8 for examples), suggesting that hemispherical reflectance at a wavelength of 4 μm can be used as an indication of accumulated TBC damage that culminates in a reflectance value associated with TBC failure.

Selection of a wavelength range for MIR reflectance imaging must also take into account potential interferences by absorptions from atmospheric constituents as well as absorptions by the TBC itself. In particular, absorption from atmospheric CO_2 at 4.25 μm and a broad absorption at 3 μm by OH incorporated in the TBC, both unrelated to the TBC damage state, can affect reflectance measurements in the vicinity of those wavelengths (Fig. 4). Imaging at a wavelength of 4.0 μm is then advantageous, not only for the high transmittance and sensitivity to damage at this wavelength, but also for being sufficiently distant from (and therefore insensitive to) these absorptions as well as from TBC intrinsic absorption that occurs at wavelengths greater than 5 μm. Figure 4 displays a transmittance curve for the bandpass filter, centered at 4.0 μm, used for MIR reflectance imaging, showing the relative insensitivity within this narrow wavelength range to the potentially interfering absorptions.

MIR Reflectance Imaging Experimental Procedures

Figure 5 illustrates the configuration of the MIR normal directional reflectance imaging system. All im-

ages were obtained with a Phoenix (Indigo Systems, Santa Barbara, CA) MIR camera with a LN_2 cooled 320×256 indium antimonide (InSb) focal plane array and a 25 mm f/2.3 lens. Imaging was restricted to the desired wavelength range (Fig. 4) by mounting the narrow 4.0 μm wavelength bandpass filter (full-width at half-maximum = 0.15 μm) onto the lens assembly. The camera was positioned 120 mm from the specimen. MIR illumination was provided by a SiC IR emitter operated at 50 W. The illumination from the source was collimated with a 63.5 mm diameter off-axis parabolic mirror that illuminated the specimens without shadowing by the source or loss of intensity with distance. It is important to note that while the off-axis parabolic mirror provides collimated, reproducible illumination, the illumination intensity is non-uniform; the intensity of the beam is higher where it reflects from areas of the mirror closer to the source. A flat detector response over the entire illumination phase of data collection has shown that the IR illumination source does not significantly heat the sample. Each image represents an average of 800 frames captured at 30 Hz rate. Additionally, a detector integration time of 4.5 ms was used for each frame to maximize signal level. Custom software was developed to perform the necessary image acquisition, processing, and analysis routines.

A sequence of image processing steps was performed using the customized data acquisition software to produce a true MIR reflectance image. These steps are necessary because the untreated image, $I_{\text{raw image}}$, includes thermal radiation emitted by the specimen and reflected background radiation as well as a dependence on illumination and pixel-to-pixel sensitivity variations:

$$
\begin{aligned}
I_{\text{raw image}}(x, y) = S(x, y)[I_0(x, y)R_{\text{specimen}}(x, y) \\
+ I_{\text{thermal}}(x, y)]
\end{aligned}
\tag{1}
$$

where x and y are the image coordinates, $S(x,y)$ is the individual pixel sensitivity, $I_0(x,y)$ is the position-dependent field-of-view illumination intensity, $R_{\text{specimen}}(x,y)$ is the specimen reflectance, and $I_{\text{thermal}}(x,y)$ is the thermal radiation emanating from the specimen (sum of thermal emission by specimen and reflected background radiation). The thermal radiation term in Eq. (1) is eliminated by subtracting the image obtained with the illumination off, $I_{\text{image}}^{\text{no illumination}}(x,y) = S(x,y)I_{\text{thermal}}(x,y)$, from the image obtained with the illumination on:

$$
\begin{aligned}
I_{\text{reflectance image}}(x,y) &= I_{\text{raw image}}(x,y) \\
&\quad - I_{\text{image}}^{\text{no illumination}}(x,y) \\
&= S(x,y)I_0(x,y)R_{\text{specimen}}(x,y)
\end{aligned}
\tag{2}
$$

A flatfield correction is then made to remove the dependence on illumination and pixel-to-pixel variations. This is performed by normalizing to a reflectance image, $I_{\text{standard}}(x,y)$, collected from a 250-μm-thick TBC standard with measured hemispherical reflectance, $R_{\text{standard}}(x,y) = R_{\text{standard}} = 0.703$, that fills the field of view. Here, the hemispherical reflectance (measured by FTIR spectrometer with integrating sphere) of the TBC standard is assumed to be equal to the normal directional reflectance (measured by reflectance imaging), due to the diffuse nature of reflectance from plasma-sprayed TBCs. This assumption was validated by the excellent agreement over a wide range of reflectance values for reflectance measurements obtained by both methods, where all imaging reflectance values were obtained using the single calibration point at $R_{\text{standard}} = 0.703$. The normalized reflectance image is then scaled to the reflectance of the standard to produce an image where the pixel intensity is equal to the specimen reflectance:

$$
\begin{aligned}
I_{\text{final}}(x,y) &= I_{\text{reflectance image}}^{\text{normalized}}(x,y)R_{\text{standard}} \\
&= \frac{I_{\text{reflectance image}}(x,y)}{I_{\text{standard}}(x,y)}R_{\text{standard}} \\
&= \frac{S(x,y)I_0(x,y)R_{\text{specimen}}(x,y)}{S(x,y)I_0(x,y)R_{\text{standard}}}R_{\text{standard}} \\
&= R_{\text{specimen}}(x,y)
\end{aligned}
\tag{3}
$$

Results

The primary objective for developing MIR reflectance imaging was to monitor the progression of TBC delamination. To demonstrate the spatial resolution of the imaging approach and to determine the range of thicknesses over which the progress of TBC delamination could be distinguished, MIR reflectance images were acquired from freestanding TBCs that had been backside-coated over half their back surfaces (half-backed) with 2-μm-thick NiAl by sputter deposition to mimic an adherent substrate (the unbacked half mimics a delaminated TBC). Because no heat treatment was applied to produce "delamination," any differences in MIR reflectance could be attributed to the presence or absence of the adherent backside NiAl coating. Figure 6 displays a sequence of 73-, 125-, and 295-μm-thick half-backed TBCs that show a significantly lower reflectance over the well-delineated half-backed section for the two thinner TBCs, with the reflectance difference decreasing with increased TBC thickness until the contrast virtually disappears at 295 μm thickness, at which point the reflectance contrast can only be discerned when viewed using an expanded contrast scale. Reflectance values for a line scan across the specimens are also shown beneath the images (Fig. 6). These results indicate MIR reflectance should be capable of discerning TBC delamination with good spatial resolution, but with decreasing contrast as the TBC thickness increases.

While the half-backed specimens essentially simulated a side-by-side completely adherent versus completely detached condition, the next step was to

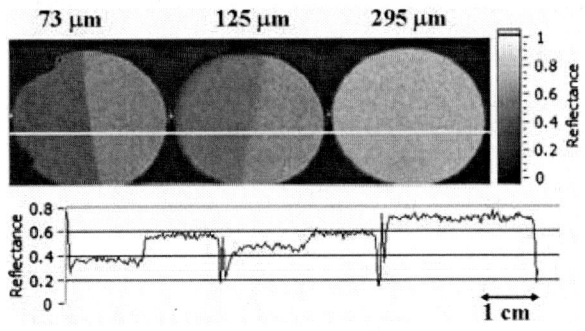

Fig. 6. *Mid-infrared (MIR) reflectance images of freestanding 73, 125, and 295 μm thick 8YSZ coatings that were backside-coated over half their back surfaces with NiAl to mimic adherent substrate. As viewed, NiAl backside coating is below left half of each TBC. Reflectance values from line scan (white solid line) across images are plotted below images.*

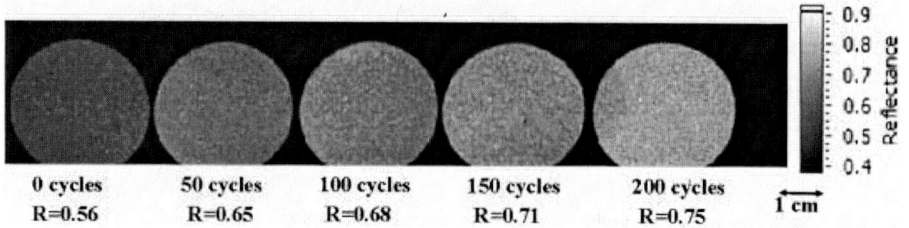

	0 cycles	50 cycles	100 cycles	150 cycles	200 cycles	
	R=0.56	R=0.65	R=0.68	R=0.71	R=0.75	1 cm

Fig. 7. Mid-infrared (MIR) reflectance images of 165-μm-thick 8YSZ TBC-coated specimens after 0, 50, 100, 150, and 200 furnace cycles. Average reflectance for each specimen is also indicated. Visible TBC buckling was observed at 200 cycles.

demonstrate that MIR reflectance imaging can also monitor the gradual and more complex delamination progression associated with furnace cycling of TBC-coated substrates. Figure 7 compares MIR reflectance images of 165-μm-thick TBC-coated specimens after 0, 50, 100, 150, and 200 furnace cycles (visible buckling was observed at 200 cycles). While there was no noticeable contrast between visible light images of these specimens, there was a continuous reflectance intensity increase in MIR reflectance images (4-μm wavelength) from 0.56 to 0.75 that is associated with delamination progression. SEM inspection (Fig. 8) of the cross-sections of the same furnace-cycled specimens confirmed a TBC failure progression similar to that observed in pre-

vious studies:[1,2] the formation and growth of a delamination crack network along the bottom of the TBC, mostly above the thermally grown oxide (TGO) that grows on top of the bond coat. Backscatter electron detection provided superior contrast for cracks compared with secondary electron detection. Figure 8 shows that the crack network grows by lengthening and widening of individual cracks as well as increasing crack interconnectedness.

Because delamination progression proceeded uniformly over the surfaces of the furnace-cycled specimens imaged in Fig. 7, another set of specimens was prepared for which the bond coat was applied to only half of the substrate top surface before the 8YSZ top coat deposition. Because TBCs without bond coats fail much earlier and with a higher degree of scatter than TBCs with bond coats (Fig. 3), these test specimens were imaged to demonstrate spatially resolved monitoring of localized delamination. Figure 9 shows MIR reflectance images of one 175-μm-thick 8YSZ-coated specimen after 0, 10, 20, and 30 furnace cycles. The left half of the specimen has a bond coat while the right half does not. While the bond coat half shows a uniform increase in reflectance (consistent with Fig. 7), the no-bond-coat half shows both a greater increase in overall reflectance along with localized bright regions that grow from the edge, eventually evolving into visible edge-initiated delamination.

Discussion

The two observable phenomena that could be identified as mechanisms for the observed increase in MIR reflectance with furnace cycling are the propagation of delamination cracks and TGO growth (Figs. 8 and 10a). However, measurements of TGO thickness (determined by SEM inspection of prepared cross-sections) versus number of cycles (Fig. 10b) show very little in-

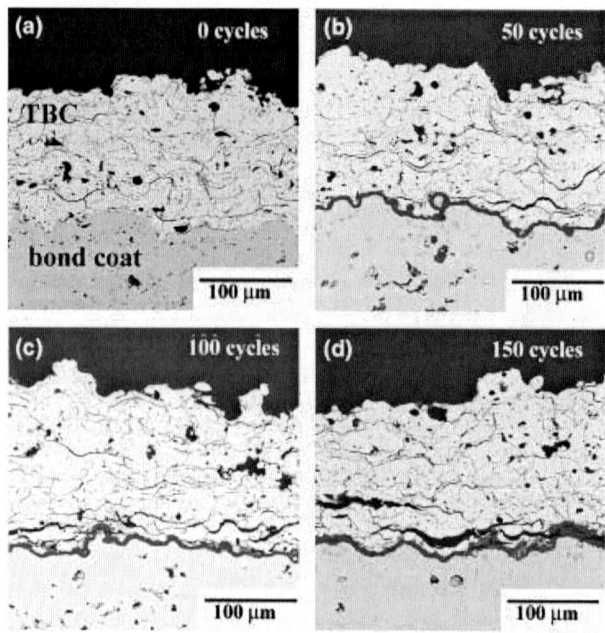

Fig. 8. Backscatter electron images of cross-sections of 165-μm-thick 8YSZ TBC-coated specimens after (a) 0, (b) 50, (c) 100, and (d) 150 furnace cycles.

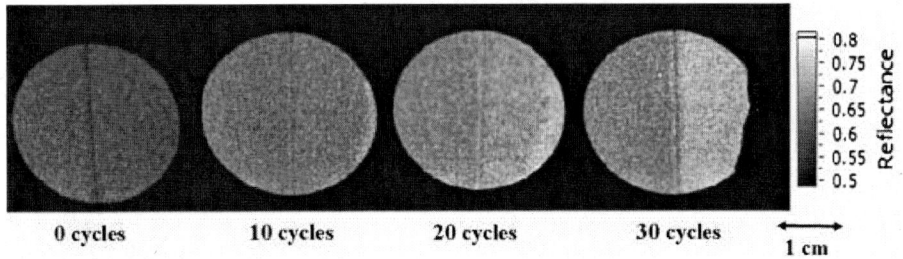

Fig. 9. Mid-infrared (MIR) reflectance images of a 175-μm-thick 8YSZ TBC-coated specimen after 0, 10, 20, and 30 furnace cycles. Bond coat was applied only to left half of specimen before 8YSZ top coat deposition.

crease in TGO thickness beyond 50 cycles, consistent with the expected parabolic growth of the TGO. Because the reflectance continues to increase at a nearly linear rate beyond 50 cycles, it is proposed that the reflectance increases beyond 50 cycles are not due to TGO growth but are primarily associated with the observed progression of buried delamination cracks. In order to further isolate the effects of TGO thickness on reflectance by avoiding the effects of TBC crack propagation generated by furnace cycling, a set of hemispherical re-

flectance measurements was obtained from TBCs where the bond coat was pre-oxidized in air at 1163°C to produce a range of TGO thicknesses prior to TBC deposition (Fig. 10c). Pre-oxidation times of 0.075, 0.75, 7.5, 37.5, and 75 h were selected to match time-at-temperature for 0.1, 1, 10, 50, and 100 cycles, respectively, and no furnace cycling was performed after TBC deposition. The plot of measured TGO thickness and hemispherical reflectance (after TBC deposition) versus pre-oxidation time revealed virtually no change in reflect-

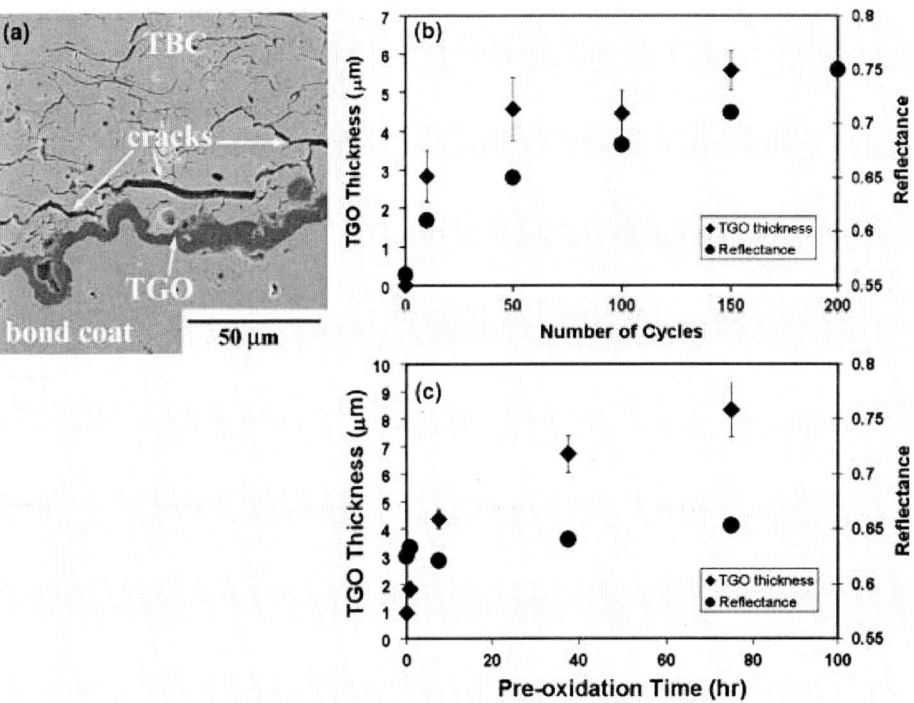

Fig. 10. (a) SEM micrograph of TBC/bond coat interfacial region after 100 furnace cycles showing early-stage delamination cracks and thermally grown oxide (TGO) formation. (b) TGO thickness and hemispherical reflectance at 4 μm wavelength versus number of furnace cycles. (c) TGO thickness (produced by pre-oxidation exposure) and hemispherical reflectance at 4 μm wavelength versus pre-oxidation exposure time (no furnace cycling).

ance despite large increases in TGO thickness. Even the small increase in reflectance for the pre-oxidation times of 37.5 and 75 h could be due to the cracks that are produced within the TGO for these longer pre-oxidation times. These results further indicate that TGO growth does not significantly affect hemispherical reflectance of TBC-coated specimens and that the reflectance increases observed during furnace cycling (Figs. 2, 3, 7, and 9) are associated with delamination crack progression.

The change in reflectance with the introduction of cracks into the TBC was modeled to check whether it was reasonable to attribute the observed reflectance increases to crack propagation. The driving force behind the increase in reflectance is the introduction of coating/air-gap interfaces where cracks are introduced into the TBC. Richmond[15] showed that at an internal interface, such as the TBC/crack interface, the total internal reflection of the fraction of incident radiation with an angle of incidence above the critical angle has a large effect and results in a diffuse internal reflectance, ρ_i, that can be determined from the relation

$$\rho_i(n) = 1 - \frac{1}{n^2}[1 - \rho_o(n)] \qquad (4)$$

where ρ_i is the diffuse internal reflectivity at the TBC/crack interface, ρ_o is the external reflectivity at the air/TBC interface that can be determined from the Fresnel equation,[16] and n is the index of refraction of the TBC ($n = 1.0$ in air gap). Using the value of $n = 2.1$ at the 4.0 μm wavelength for YSZ,[17] Eq. (4) predicts a high diffuse internal reflectivity of 0.81 at the TBC/crack interface. In terms of optical interaction, the contrast-producing mechanism responsible for the sensitivity of MIR reflectance to the presence of buried cracks is that crack generation produces internal interfaces with high diffuse internal reflectivity that will increase the overall reflectance of the TBC-coated specimen. In order to predict the range of TBC thicknesses over which delamination progression could be monitored by MIR reflectance at 4 μm wavelength, a modified four-flux Kulbelka–Munk approximation[18] was applied to the extreme cases of a completely adherent and a completely detached TBC. In the four-flux Kulbelka–Munk approximation, the total light flux is represented by the sum of forwards and backwards traveling collimated and diffuse light fluxes, and each of these four light fluxes is scattered and absorbed within the TBC and partially reflected at each interface encountered.[19] Boundary con-

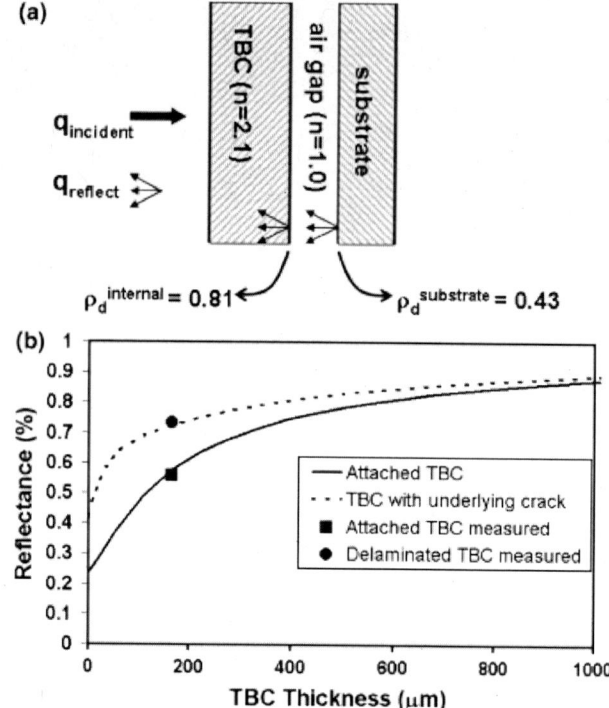

Fig. 11. (a) Model for predicting effect of underlying crack (air gap) on hemispherical reflectance of TBC-coated specimen. (b) Predicted effect of introduction of crack (air gap) at plasma-sprayed 8YSZ TBC/substrate interface on hemispherical reflectance at 4 μm wavelength as a function of TBC thickness. Measured reflectance values are also included for a 165-μm-thick TBC-coated specimen after 0 furnace cycles (attached) and 200 furnace cycles (delaminated).

ditions are set up for both an attached and detached TBC and the equations are solved for the total reflected light flux, $q_{reflect}$. The detached TBC was modeled simply by a uniform air gap between the TBC and the substrate (Fig. 11a). In this model, the substrate (bond coat) was assigned a diffuse reflectivity of 0.43 (obtained from reflectance measurements of bare bond coat), and the TBC/air-gap interface was assigned the diffuse reflectivity of 0.81 as described above. The total reflectance, R, was calculated from the ratio of the reflected radiative flux to the incident radiative flux, $R = q_{reflect}/q_{incident}$, as a function of TBC optical thickness. TBC optical thickness was then converted to actual thickness by plugging in the scattering and absorption coefficients at 4.0 μm wavelength for plasma-sprayed 8YSZ. The scattering and absorption coefficients, 359 and 0.035 cm^{-1}, respectively, were determined from a set of reflectance/transmittance versus thickness measurements from freestanding 8YSZ TBCs. The predicted effect of introducing the air gap (crack) on the reflectance at 4 μm as a function of TBC thickness is plotted in Fig. 11b. Two experimentally measured points for a 165-μm-thick TBC-coated specimen after 0 and 200 cycles, representing an attached and detached TBC, respectively, have been added to Fig. 11b, illustrating good agreement with the prediction. This plot has important implications for the TBC thickness range where MIR reflectance imaging will be viable for commercial application. The plot clearly shows that the difference in reflectance between an adherent and completely detached TBC (which produces the delamination-sensitive contrast) decreases with increasing TBC thickness until the reflectance difference is experimentally indistinguishable, at which point MIR reflectance will no longer be useful for monitoring delamination progression. For example, at a TBC thickness of 1000 μm, the reflectance is predicted to barely increase from 0.88 to 0.89 when the TBC becomes detached, a change that cannot be discerned within experimental scatter. It should be noted that these results are specific for plasma-sprayed TBCs and that MIR reflectance will probe much deeper into electron-beam physical vapor deposited (EB-PVD) coatings, due to their much lower scattering coefficients, especially at shorter wavelengths; indeed, TBC delamination progression in EB-PVD coatings is often evident at visible wavelengths.[20]

The prospect of using MIR reflectance imaging as a useful health-monitoring tool depends on the gradual nature of the TBC failure process. If TBC failure proceeded by sudden catastrophic propagation of singular flaws, then crack detection would provide insufficient advance warning. However, previous work[1,2] has shown that plasma-sprayed TBCs degrade by the gradual accumulation and linking of numerous cracks produced along the bottom of the TBC and that this process initiates very early in TBC life. Therefore, the extent of crack network progression, if it can be monitored, provides a good indication of the fraction of TBC life remaining. SEM inspection (Fig. 8) and the agreement between reflectance predictions and measurements (Fig. 11) indicate that the MIR reflectance increase that correlates with continued furnace cycling (Figs. 2, 3, and 7) is associated with the gradual progression of the delamination cracks that propagate along the bottom of the TBC, and that the reflectance increases until it reaches a value associated with TBC failure. Because the reflectance increases smoothly (associated with the gradual

progression of the buried crack network), one can confidently select a threshold reflectance that is safely below the reflectance associated with failure, at which the TBC should be replaced. The distinguishing feature of MIR reflectance imaging over other methods is that it is sensitive to delamination progression, even in its earlier stages, whereas other health monitoring approaches are primarily sensitive only to later stages. For example, the presence of a buried crack network as exhibited in Figs. 8 and 10a strongly affects MIR reflectance, but would not disrupt heat flow sufficiently to be observed by IR thermography, and the early stages of crack development will not result in significant relaxation of TGO stresses and therefore would not be effectively monitored by piezospectroscopy.

For the successful real-world application of MIR reflectance imaging, influences on MIR reflectance other than delamination progression must be considered to avoid misinterpretation of MIR reflectance changes. Most notably, MIR reflectance is sensitive to TBC thickness, as predicted in Fig. 11b and measured over the 3–5 μm wavelength range in Fig. 12a for freestanding 8YSZ TBCs. Consequently, MIR reflectance imaging exhibits TBC thickness sensitivity as displayed in

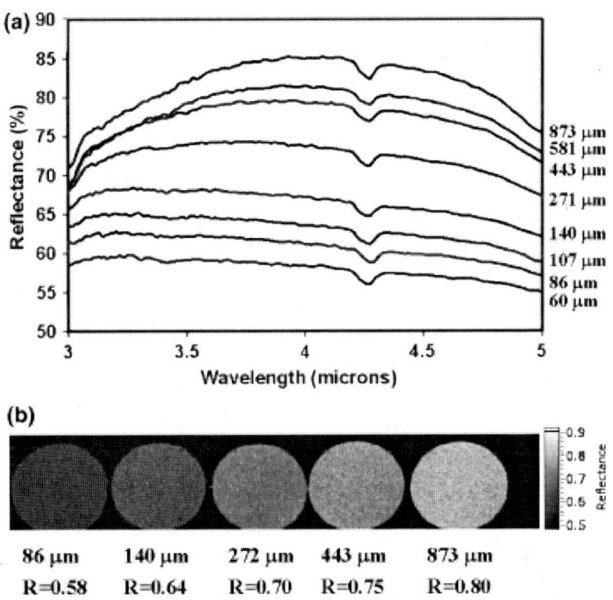

Fig. 12. (a) Hemispherical reflectance spectra of freestanding plasma-sprayed 8YSZ TBCs showing reflectance increase with increased TBC thickness. (b) MIR reflectance images (at 4 μm wavelength) of same set of freestanding plasma-sprayed 8YSZ TBCs showing increased reflectance intensity with increased TBC thickness.

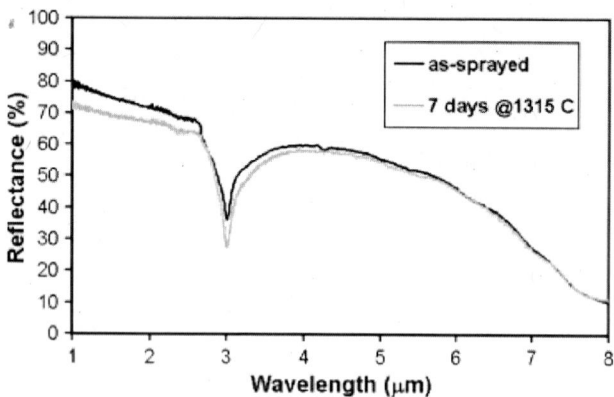

Fig. 13. *Hemispherical reflectance spectra of 160-μm-thick 8YSZ TBC before and after 7-day exposure at 1315°C.*

MIR reflectance images of the same set of freestanding TBCs (Fig. 12b). While this sensitivity to thickness can be utilized for monitoring TBC erosion,[21] thickness dependence must be accounted for when monitoring delamination progression. Therefore, reflectance values associated with TBC failure and threshold reflectance values selected to prompt TBC replacement are specific to the TBC thickness. It should be noted that TBC erosion will always decrease reflectance and therefore will not produce false positive signs of delamination (which increases reflectance) but can increase the risk of false-negative indication by offsetting reflectance increases associated with TBC delamination.

TBC sintering can also result in reflectance changes. Figure 13 shows the change in hemispherical reflectance caused by subjecting a 160-μm-thick freestanding 8YSZ coating to 7 days at 1315°C. While there is a noticeable decrease in reflectance due to sintering, the reflectance decrease occurs preferentially at shorter wavelengths, with only a minimal decrease in reflectance at the 4 μm wavelength used for imaging. This decrease in reflectance due to sintering is too small to significantly offset the reflectance increases associated with delamination progression. In addition, unlike the reflectance changes observed in Fig. 13, lower temperature heat treatments up to 7 days at 1163°C produced no change in specimen reflectance over the same wavelength range.

The most difficult influence on reflectance to address is the effect of surface contamination, which will generally result in decreased reflectance. While not a problem in clean-burning environments, surface contamination can be severe; for example, at temperatures above 1240°C, calcium-magnesium-alumino-silicate (CMAS) can deposit onto and infiltrate into TBCs.[22] Severe contamination, such as coverage by opaque soot,[23] will completely prevent MIR reflectance from probing delamination progression. As with TBC erosion and sintering, contamination will decrease reflectance and therefore cannot produce false-positive indications of delamination but can produce a false-negative indication if the contamination-induced reflectance decrease compensates for the delamination-induced reflectance increase.

Future Work

Multi-wavelength reflectance imaging will be developed to address component service environments where there are influences on reflectance simultaneously competing with delamination progression, such as erosion, sintering, or contamination. Multi-wavelength reflectance imaging may be capable of distinguishing effects of different types of TBC degradation by their differing spectral dependencies. For example, the sensitivity of reflectance to TBC delamination decreases more rapidly than the sensitivity to TBC thickness as the wavelength decreases so that adding near-infrared to MIR reflectance imaging may aid the separation of thickness and delamination effects. The problematic influence of modest levels of translucent contamination may be addressed by multi-wavelength imaging. Multi-wavelength imaging can help distinguish the effects of contamination from delamination because, unlike delamination, contamination will also decrease reflectance at visible wavelengths and may exhibit specific absorption peaks associated with the contaminating species. Future effort will also be directed towards establishing a more quantitative relationship between delamination crack network growth and MIR reflectance, using image analysis of crack networking observed in SEM micrographs of TBCs at various stages of damage.

Conclusions

Spatially resolved MIR reflectance imaging has been developed to monitor delamination progression in TBC-coated materials. Imaging at a wavelength of 4 μm takes advantage of the relatively high transmittance of plasma-sprayed 8YSZ at this wavelength along with a desirable relative insensitivity to potentially interfering absorptions by atmospheric constituents or to

OH incorporated within the TBC. The image contrast that indicates delamination progression arises from the increased reflectance due to a significant component of total internal reflection at the TBC/buried-crack interface. Predictions of the overall reflectance increase based on the high-diffuse reflectivity at the TBC/crack interface agreed well with reflectance measurements. The contrast between attached and delaminated TBC regions was both predicted and demonstrated to diminish with increasing TBC thickness to the point where attached and delaminated regions are experimentally indistinguishable, and identification of delaminated regions was shown to already be difficult at a TBC thickness of $300\,\mu m$. Therefore, industrial applications of MIR reflectance imaging for monitoring delamination progression of plasma-sprayed TBCs will be restricted to thinner TBCs. While TGO growth may be the driving force behind TBC failure progression, TGO thickness increases were shown to have minimal effects on reflectance at $4\,\mu m$ wavelength.

MIR reflectance imaging shows great potential as a diagnostic tool for investigating TBC delamination progression in the laboratory where it will aid the development of longer-lived TBCs. In addition, the technique is well suited to between-flight engine inspection for early-stage delamination in non-surface-fouling engine environments.

Acknowledgments

The authors thank George Leissler, Sandy Leissler, and Gary Kostyak for coating deposition, Chuck Barrett for furnace cycling, and Luke Hertert for assistance in developing the image acquisition software.

References

1. J. T. De Masi-Marcin, K. D. Sheffler, and S. Bose, "Mechanisms of Degradation and Failure in a Plasma-Deposited Thermal Barrier Coating," *ASME J. Eng. Gas Turbines Power*, 112 521–526 (1990).
2. I. T. Spitsberg, D. R. Mumm, and A. G. Evans, "On the Failure Mechanisms of Thermal Barrier Coatings with Diffusion Aluminide Bond Coats," *Mater. Sci. Eng. A*, 394 176–191 (2005).
3. X. Peng and D. R. Clarke, "Piezospectroscopic Analysis of Interface Debonding in Thermal Barrier Coatings," *J. Am. Ceram. Soc.*, 83 [5] 1165–1170 (2000).
4. V. K. Tolpygo, D. R. Clarke, and K. S. Murphy, "Evaluation of Interface Degradation During Cyclic Oxidation of EB-PVD Thermal Barrier Coatings and Correlation with TGO Luminescence," *Surf. Coat. Technol.*, 188–189 62–70 (2004).
5. M. Gell, S. Sridharan, M. Wen, and E. H. Jordan, "Photoluminescence Piezospectroscopy: A Multi-Purpose Quality Control and NDI Technique for Thermal Barrier Coatings," *Int. J. Appl. Ceram. Technol.*, 1 [4] 316–329 (2004).
6. M. K. Ferber, A. A. Wereszczak, M. Lance, J. A. Hanes, and M. A. Antelo, "Application of Infrared Imaging to the Study of Controlled Failure of Thermal Barrier Coatings," *J. Mater. Sci.*, 35 2643–2651 (2000).
7. S. Marinetti, V. Vavilov, P. G. Bison, E. Grinzato, and F. Cernuschi, "Quantitative Infrared Thermographic Nondestructive Testing of Thermal Barrier Coatings," *Mater. Eval.*, 61 [6] 773–780 (2003).
8. G. Newaz and X. Chen, "Progressive Damage Assessment in Thermal Barrier Coatings Using Thermal Wave Imaging Technique," *Surf. Coat. Technol.*, 190 7–14 (2005).
9. B. Franke, Y. H. Sohn, X. Chen, J. R. Price, and Z. Mutasim, "Thermal Wave Imaging Application in Thermal Barrier Coatings," *Ceram. Eng. Sci. Proc.*, 26 [3] 113–119 (2005).
10. R. J. Visher, W. A. Ellingson, M. D. Shields, and A. Feuerstein, "Laser-Based Inspection of Thermal Barrier Coatings," Surface Engineering—Proceedings of 3rd International Surface Engineering Congress, 299–306, 2004.
11. J. I. Eldridge, C. M. Spuckler, K. W. Street, and J. R. Markham, "Infrared Radiative Properties of Yttria-Stabilized Zirconia Thermal Barrier Coatings," *Ceram. Eng. Sci. Proc.*, 23 [4] 417–430 (2002).
12. J. I. Eldridge, C. M. Spuckler, J. A. Nesbitt, and K. W. Street, "Health Monitoring of Thermal Barrier Coatings by Mid-Infrared Reflectance," *Ceram. Eng. Sci. Proc.*, 24 [3] 511–516 (2003).
13. P. Crowther, "Nondestructive Evaluation of Coatings for Land Based Gas Turbines Using a Multi-Frequency Eddy Current Technique," *Insight*, 46 [9] 547–549 (2004).
14. N. W. Wu, K. Ogawa, M. Chyu, and S. X. Mao, "Failure Detection of Thermal Barrier Coatings Using Impedance Spectroscopy," *Thin Solid Films*, 457 301–306 (2004).
15. J. C. Richmond, "Relation of Emittance to Other Optical Properties," *J. Res. Nat. Bureau Stand.*, 67C [3] 217–226 (1963).
16. R. Siegel and J. R. Howell, *Thermal Radiation Heat Transfer*, 4th ed., Taylor & Francis, New York, 2002, p. 87.
17. D. L. Wood, K. Nassau, and T. Y. Kometani, "Refractive Index of Y_2O_3 Stabilized Zirconia: Variation with Composition and Wavelength," *Appl. Optics*, 29 [16] 2485–2488 (1990).
18. T. Makino, T. Kunitomo, I. Sakai, and H. Kinoshita, "Thermal Radiation Properties of Ceramic Materials," *Heat Transfer—Jpn. Res.*, 13 [4] 33–50 (1984).
19. A. Ishimaru, *Wave Propagation and Scattering in Random Media*, Vol. 1. Academic Press, New York, 1978, p. 198.
20. J. I. Eldridge, T. J. Bencic, C. M. Spuckler, J. Singh, and D. E. Wolfe, "Delamination-Indicating Thermal Barrier Coatings Using YSZ:Eu Sublayers," *J. Am. Ceram. Soc.*, submitted.
21. J. I. Eldridge, C. M. Spuckler, J. A. Nesbitt, and R. E. Martin, "Nondestructive Evaluation of Thermal Barrier Coatings by Mid-Infrared Reflectance Imaging," *Ceram. Eng. Sci. Proc.*, 26 [3] 121–128 (2005).
22. C. Mercer, S. Faulhaber, A. G. Evans, and R. Darolia, "A Delamination Mechanism for Thermal Barrier Coatings Subject to Calcium-Magnesium-Alumino-Silicate (CMAS) Infiltration," *Acta Mater.*, 53 1029–1039 (2005).
23. R. Siegel and C. M. Spuckler, "Analysis of Thermal Radiation Effects on Temperatures in Turbine Engine Thermal Barrier Coatings," *Mater. Sci. Eng. A*, 245 [2] 150–159 (1998).

International Journal of
Applied Ceramic Technology
Ceramic Product Development and Commercialization

Noncontact Methods for Measuring Thermal Barrier Coating Temperatures

Molly M. Gentleman, Vanni Lughi, John A. Nychka,* and David R. Clarke[†]

Materials Department, College of Engineering, University of California, Santa Barbara, California 93106-5050

Three noncontact, optical methods for measuring temperature are reviewed with an emphasis on their application to the measurement of temperatures of thermal barrier coatings (TBCs). The methods are: infrared pyrometry, Raman spectroscopy, and photo-stimulated luminescence from lanthanide-doped coatings. Although each has the capability of measuring temperatures pertinent to monitoring TBCs, the finite thickness of typical coatings together with the optical properties of zirconia place severe restrictions on the depth from which the temperature sensing can be obtained. Some of these limitations can be circumvented using photo-stimulated luminescence with coatings containing dopants at specific locations. To illustrate this, it is demonstrated that by depositing coatings with a lanthanide dopant, such as Eu^{3+}, at specific locations, for instance in contact with the metallic alloy, temperature sensing can be performed with much higher spatial resolution.

Introduction

In situ, noncontact monitoring of hot-section turbine components has long been sought because the life of these components often limits the time between maintenance intervals and, in some extreme cases, the onset of failure. Of all the parameters that influence the life and reliability of a component, however, the principal one is temperature. This is largely because the degradation mechanisms

that determine life, such as creep and oxidation, are thermally activated processes and hence are notionally exponentially dependent on temperature. Temperature measurement in an engine poses many practical challenges, not least of which is access to the pertinent components, whether they are vanes, blades, transition parts, or combustor liners. For these reasons, our focus is on optical methods as they are not only intrinsically noncontact but are also compatible with the limited access available for monitoring the hot-section of turbines. In this contribution, we discuss some of the advantages and disadvantages of three noncontact methods being developed for measuring the temperature of stationary and rotating gas turbine components. The three methods are optical pyrometry in the infrared, Raman spectroscopy, and luminescence spectroscopy.

In addition to the requirement that any temperature measurement method be capable of providing accurate

The work of Molly Gentleman has been supported by DOE-NETL and the NSF through its TTTP program. The work of Professor John Nychka was supported by the Office of Naval Research while he was a graduate student at UCSB. The work of Vanni Lughi has also been supported by the Office of Naval Research but through the MURI program based at UCSB.

*Now at Department of Chemical and Materials Engineering, University of Kentucky, Lexington, Kentucky 40506.

[†]clarke@engineering.ucsb.edu

values of temperature, there are other requirements. For instance, measuring the temperature distributions across a component requires high lateral spatial resolution. For rotating components, there is the additional requirement that the measurement be sufficiently fast that individual blades can be monitored in real time. In addition, for coated components, it is also desirable that the temperature at the bottom of coating, in contact with the alloy, as well as at its outer surface be measurable.

Infrared Pyrometry

Optical pyrometry is the established noncontact method of measuring temperature of materials and structures at high-temperatures.[1] The method relies on knowledge of the emissivity as a function of temperature at the wavelengths to which the pyrometer is sensitive. By the use of the technique of "two-color" pyrometry, in which the pyrometry signal at two different wavelengths is compared, the effect of uncertainty in the emissivity can, in principle, be removed. With the advent of high-speed infrared detectors, the method has recently been extended to "multi-spectral" pyrometry.[2,3] Nevertheless, there remain uncertainties as to the actual surface temperature, especially in practical situations where, for instance, absorption in the hot-gas occurs and contamination layers form on the surface and alter the emissivity.

Measuring the temperature of thermal barrier coatings (TBCs) poses particular complications not usually considered important in most high temperature measurement situations. Probably the most important is that thermal gradients exist through the thickness of the coating and so surface temperature measurements require that the selected pyrometry signal originates from the surface rather than be distributed through the coating thickness. Taking as an example estimates of the current performance of TBCs,[4] they operate under a temperature gradient of ~0.5°C/μm (13°C/mil) implying that the pyrometry penetration depth must be less than ~4 μm for measurements to be accurate to within ±2°C. Also important is that the reflectivity should be as low as possible to avoid significant reflection of thermal radiation from other radiating parts of the engine off the TBC surface. Lastly, the coatings are relatively thin, in some cases less than or at least commensurate with the optical absorption length at many wavelengths and so finite thickness values for the emissivity must be used rather

than those for bulk materials. In these cases, the optical properties of the underlying thermally grown oxide and the alloy must also be taken into account.

These requirements and restrictions place constraints on the optimum wavelengths for accurate pyrometry. The majority of pyrometer measurements are carried out at or near the peak in the black-body radiation curve as the radiation signal is generally highest at these wavelengths. However, in this range zirconia has low emissivity and is essentially transparent with only scattering from porosity limiting the depth from which the signal is obtained.[5–8] An alternative approach to selecting the optimum wavelengths for pyrometry is to consider the optical properties of the TBC materials themselves. The optical properties of cubic zirconia are well established[6] but current TBC coatings have a different composition, typically 8YSZ, as well as a different crystal structure, namely tetragonal-prime zirconia.[9] Fortunately, detailed measurements of the transmission and reflectivity of plasma-sprayed 8YSZ coatings have been reported by NASA Glenn.[7] In addition, more recently, measurements of the reflectivity as a function of temperature up to 1400°C have been reported on thick slabs of fully dense, tetragonal-prime 8YSZ material.[8] These latter complement the NASA results as optical scattering from porosity and thin sample effects are circumvented.

The measurements indicate that there are no abrupt changes in reflectance with temperature up to 1400°C but rather a smooth decrease in reflectance at all wavelengths other than the wavelength of the minimum reflectance, at 12 μm (Fig. 1). This wavelength, sometimes erroneously referred to as the Christiansen wavelength, is the only wavelength at which the emissivity is unity at all temperatures. It corresponds to the wavelength at which the combination of the optical constants, n and k, in the equation for reflectivity, R,[10]

$$R = \frac{(n-1)^2 + k^2}{(n+1)^2 + k^2}$$

is such that R equals 0. At other wavelengths, this singular condition is not satisfied and the reflectance, and hence emissivity, of bulk materials changes with temperature. Analysis of the reflectance data indicates that the optical constants, n and k, themselves are relatively insensitive to temperature for wavelengths shorter than 12 μm. Using the optical constant data in reference 8, the optical penetration depth as a function of wavelength is shown in Fig. 2, where the emitted intensity as

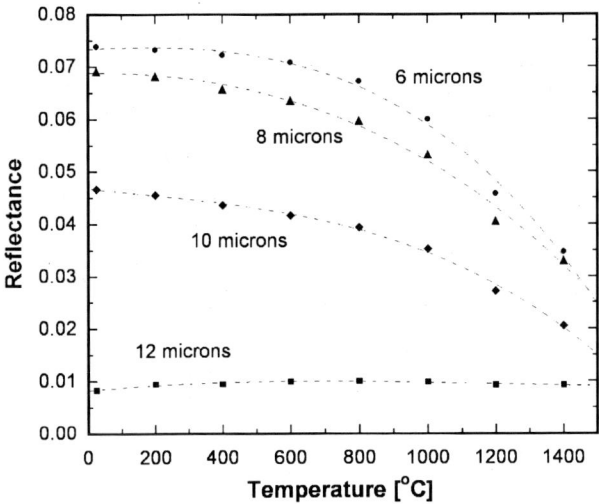

Fig. 1. *The reflectance as a function of temperature at the indicated wavelengths for a thick slab of dense 8YSZ material having the tetragonal-prime structure.*

a function of depth is given by

$$I = I_O e^{-\alpha x} = I_O e^{-x/L_P}$$

The penetration depth, L_P, is defined here as the reciprocal of the absorption coefficient, α.

In closing this section, it is worth emphasizing that the relationship often used in the literature to convert measured values of reflectance, $R(\lambda)$, to emissivity, $\varepsilon(\lambda)$, known as Kirchhoff's law,

$$\varepsilon(\lambda) = 1 - R(\lambda)$$

is not always held unless proper account is taken of scattering, especially in porous or particulate materials.[11,12] This requires care in deriving emissity values for coatings from dense, bulk and single crystal data.

Raman Spectroscopy

Zirconia exhibits one of the strongest Raman signals of all materials reported in the literature[13] and consequently it is tempting to consider using the characteristic features of Raman spectra as a noncontact method of temperature measurement. It is known that temperature affects Raman spectra in three distinct ways, each of which provides the basis for measuring temperature. One is the relative strength of the so-called Stokes and anti-Stokes lines.[14] A second is the variation in the broadening of the Raman lines with temperature.[15] The third is the variation in the position of the Raman lines with temperature. In our work we have recently calibrated the shift in the principal Raman lines with temperatures up to 1200°C for a zirconia TBC having a composition of 8YSZ.[16] Typical Raman spectra, recorded at different temperatures, from a zirconia coating is illustrated in Fig. 3. Analysis of the spectra indicates that the peak positions vary linearly with temperature indicating the potential for a straightforward measurement of temperature.

However, in common with other measurement methods, including pyrometry, the biggest difficulty is in obtaining absolute values of the measurement varia-

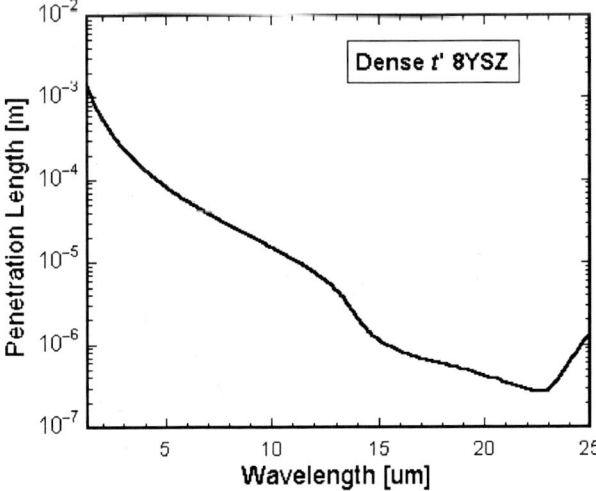

Fig. 2. *The optical penetration length as a function of wavelength for 8YSZ computed from the optical constants, n and k.*

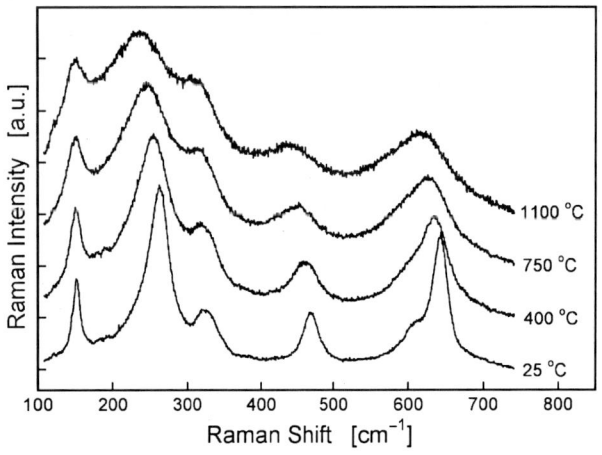

Fig. 3. *Examples of the Raman spectra of an electron-beam evaporated 8YSZ thermal barrier coating recorded at the indicated temperatures.*

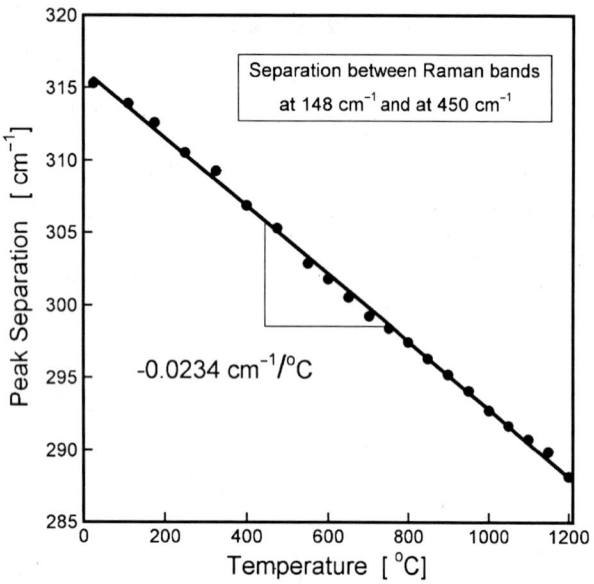

Fig. 4. *The separation of the 148/cm¹ and 450/cm¹ Raman peaks as a function of temperature for a 8YSZ thermal barrier coating.*

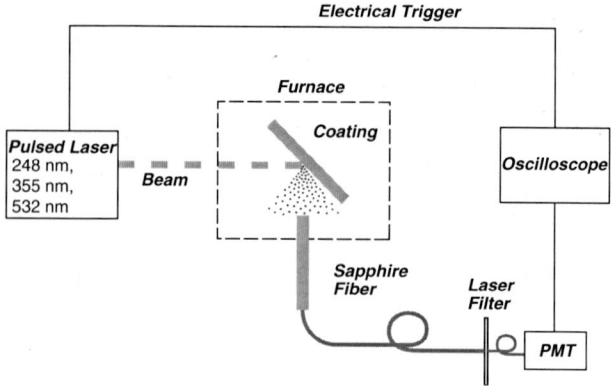

Fig. 5. *The experimental arrangement used to measure the luminescence lifetime of a coating at temperature.*

ble. For this reason, the ratio of the intensities of the Raman lines in the Stokes and anti-Stokes regions of the spectra are often used in measuring temperature, as only a measurement of a ratio is required. This becomes increasingly difficult as the temperature increases because of the low intensity of the anti-Stokes Raman signal. Alternatively, it is possible to use a frequency calibration, for instance by recording simultaneously the emission spectra from a gas discharge lamp, to determine the absolute value of the wavelength of the Raman lines. To circumvent the difficulties associated with this type of calibration, we propose using the *separation* of two of the Raman lines.[17] As indicated by the results in Fig. 4, the separation of the lines at 148 and 450/cm⁻¹ decrease linearly with increasing temperature.

Luminescence Spectroscopy

The third method we review relies on the characteristic luminescence from ions incorporated within the crystal structure of the TBC itself. As with the Raman spectroscopy reviewed above, the luminescence is excited by a strong source of light, such as a laser beam. The technique itself has a long history, dating back to the early 1950s, when luminescent phosphors were painted onto the wings of a wind-tunnel model.[18] The distinction between this, more traditional method and what we

describe here is that the luminescent ions are incorporated directly into the crystal structure of the coating, rather than as a separately applied "paint" on top of the coating. This provides two advantages: (i) the coating material itself can be layered with different luminescent ions and information at different depths probed and (ii) the measurement is not dependent on the adherence of a paint film that is unlikely to withstand prolonged operation in a turbine environment.

Traditionally, three properties of the luminescence spectrum have been used to measure temperature: the relative intensities of the peaks; the peak position and the time decay.[19,20] The first two are used for measurement at relatively low temperatures but are ill-suited for the highest temperatures because of extensive peak broadening and decreased intensity. In our work, we have utilized the luminescence lifetime decay method, in large part because it requires simpler electronics and is less susceptible to uncertainties introduced by both background variations and thermally induced line broadening. In the luminescence lifetime method, the coating is illuminated with a very short pulse of light and the intensity of the luminescence recorded as a function of time after the end of the pulse (Fig. 5). In simplest cases, the logarithm of the intensity decreases linearly with time and a decay time can be identified as the slope of the decay curve (Fig. 6). For some coating compositions, the intensity decreases with a more complex function, characterized by a sum of exponential decays. In addition, the presence of gradients in the concentration of luminescence ions or in the temperature, can cause nonexponential decays, often referred to as "stretched exponentials," to arise.

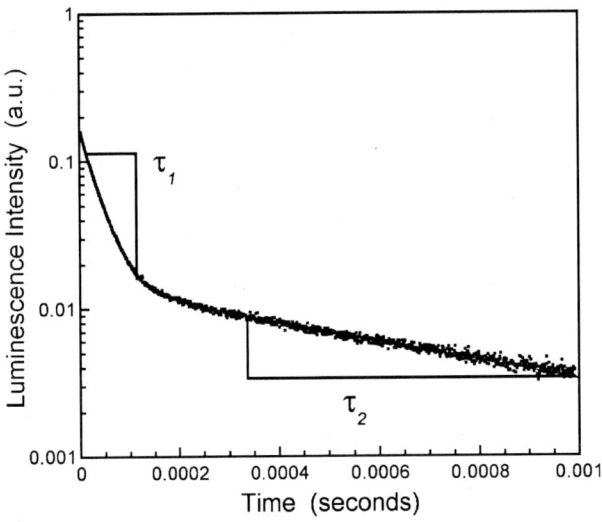

Fig. 6. *Typical luminescence decay from a Eu-doped YSZ coating measured at room temperature illustrating a bi-exponential decay with a fast time constants, τ_1, and a slower decay, τ_2.*

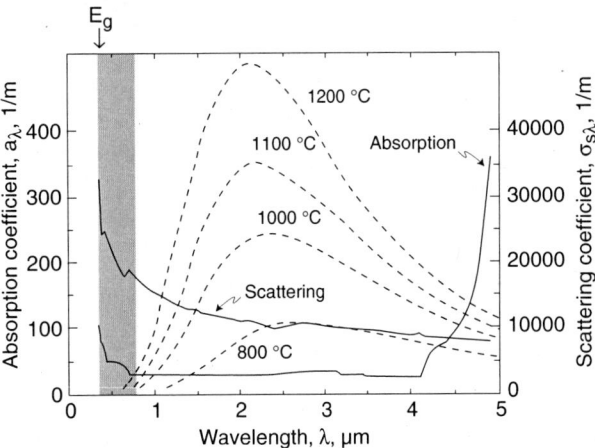

Fig. 7. *Wavelength "window", shown as a gray band, for temperature sensing of zirconia lying between its absorption edge at $\sim 250\,nm$ and the onset of the black body thermal radiation, estimated to be $\sim 650\,nm$, together with curves for the optical absorption and scattering of a thermal barrier coating. Superimposed are the calculated radiation curves (dashed) for the emission from a black body at the temperatures indicated.*

For reasons described in detail in reference,[21] there is a narrow range of wavelengths over which temperature sensing can be performed at high temperatures while avoiding the infrared emission radiation peak characteristic of hot bodies (Fig. 7). At the short wavelength boundary, the range is limited by the ultraviolet (UV) absorption edge of zirconia itself, which occurs at $\sim 250\,nm$.[22] At this wavelength, the optical penetration depth is of the order of a few unit cells. The longer wavelength boundary is less well defined but in practice is dependent on the ratio of the luminescence signal to the infrared radiation at the same wavelength. This is proving to be $\sim 650\,nm$. Thus, to measure temperatures, both the excitation and emission wavelengths must be between these two wavelengths (The same considerations apply to the Raman method outlined above). There are two practical considerations that limit the maximum temperature that can be sensed. One is the decreasing luminescence signal relative to the increasing background with increasing temperature. This depends on both the concentration of luminescence ions and the magnitude of the thermal radiation. The other is the bandwidth of the luminescence optical detector which determines the minimum lifetime detectable with the particular electronics used to collect the signal. With our current system, we can measure lifetimes as short as ~ 20 ns. For the Eu-doped YSZ

coatings this corresponds to a maximum temperature of $\sim 1150°C$.[23]

The preferred luminescence ions that can be incorporated into yttria-stabilized zirconia are the trivalent lanthanide ions. They have the same notional valence as the Y^{3+} stabilizer and similar size so they can be substituted, in part, for these stabilizer ions without destabilizing the tetragonal-prime phase crucial to the long-term cyclic life of TBCs.[9] Furthermore, their optical transitions are dependent on shielded $4f$ electrons that yield intense and sharp luminescence emission lines, many of which can be excited as well as emitted in the visible portion of the spectrum. Of the rare-earth ions, we have focused principally on Eu^{3+} as the luminescent ions for temperature measurement, although we have also studied other ions, such as Sm^{3+} and Er^{3+}.[21] The luminescence spectrum of Eu^{3+} ions in YSZ is shown in Fig. 8. In contrast to Eu^{3+} ions in Y_2O_3, where the luminescence consists of a single, sharp line at 606 nm, the luminescence consists of several lines. This is a consequence of the different site symmetries of the Eu^{3+} in the two materials. These multiple peaks are also seen in $Eu_2Zr_2O_7$ where the D_{3d} symmetry of the Eu^{3+} site in the pyrochlore structure splits the $^7F_1$ level into a singlet and a doublet as

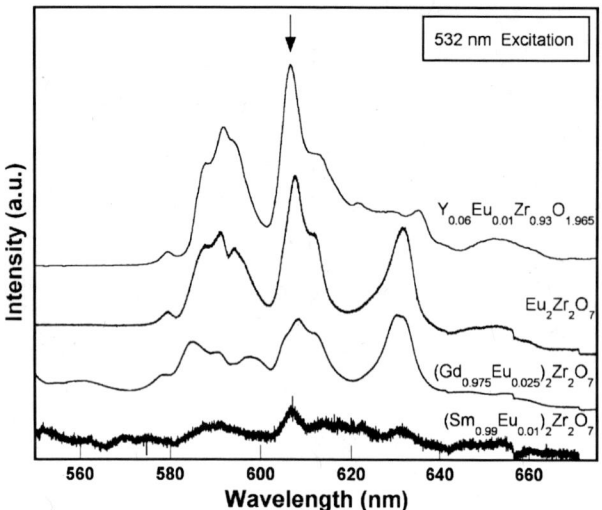

Fig. 8. *The luminescence associated with Eu^{3+} ions incorporated in several materials, including 8YSZ and $Eu_2Zr_2O_7$, all excited with a laser at 532 nm.*

indicated by the presence of the three peaks at approximately 592 nm.

The temperature dependence of the luminescence lifetime is shown in Fig. 9 for Eu^{3+}-doped YSZ (YSZ:Eu) and $Eu_2Zr_2O_7$.[23] In the case of YSZ:Eu the luminescence intensity as a function of time can be characterized by two distinct exponential decays (Fig. 5). The faster of the two decays is likely because of energy migration to defects such as Y^{3+} ions and Eu^{3+} ions in sites adjacent to grain boundaries or other defects. This fast decay is not useful in temperature measurements because it drops below the lifetime detection

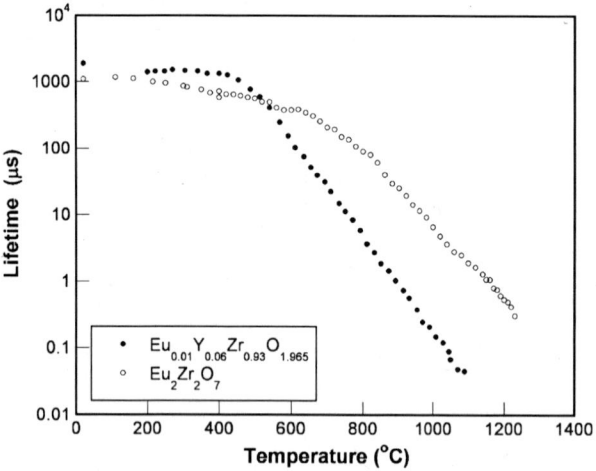

Fig. 9. *Comparison of the temperature dependence of the luminescence lifetime for Eu-doped YSZ and $Eu_2Zr_2O_7$.*

limit at relatively low temperatures ($\sim 900^\circ$C). The longer decay, τ_2, in YSZ:Eu^{3+}, on the other hand, allows temperature sensitivity and measurability up to $\sim 1100^\circ$C. The reason that the $Eu_2Zr_2O_7$ displays a significantly longer lifetime at high temperatures than the Eu^{3+}-doped YSZ is not completely understood at present but is very reproducible from sample to sample and in bulk as well as in coatings. The longer decays also make it feasible to make temperature measurements up to higher temperatures. Currently, we have been able to measure up to $\sim 1280^\circ$C and are limited by signal intensity, not lifetime measurement limitations as is the case for the Eu^{3+}-doped YSZ.

Discussion

All three of the optically based methods reviewed have been demonstrated to be capable of noncontact measurement in air of temperatures typical of those to which TBCs are exposed in an engine. In contrast to other methods of measuring temperature, all three rely on intrinsic properties of the coatings materials themselves and so can be expected to give reliable measurements of the actual temperatures as there are no attachment and contact problems such as those associated with thermometers. Apart from very real practical issues, such as signal to noise ratios, signal intensity and cost considerations, there are important distinctions to be made concerning the attainable spatial resolution and speed of each method.

We distinguish between two spatial resolutions: the lateral resolution, namely distinguishing the spatial distribution of temperature across a component, and the depth resolution, namely the depth into the coating from which the measurement signal comes. This latter is especially important for measuring TBCs as their function is to provide a thermal gradient between the outside and the inside of the coating and hence depth discrimination is essential. In all three methods, the optical penetration depth, and its wavelength dependence, plays a key role in determining spatial resolution. For instance, in optical pyrometry the penetration length determines the depth into the coating from which the emissivity signal originates. As remarked earlier, in many respects the ideal wavelength corresponds to the Christiansen wavelength, since at this wavelength the emissivity is unity at all temperatures. However, even at this wavelength, which is at 12 μm for zirconia,[7,8] the penetration distance is finite and has a value of ~6 μm (Fig. 2). At shorter wavelengths, the penetration depth is greater and the emissivity is also temperature dependent. For instance, at 8 μm, a commonly used pyrometry wavelength for high-temperature materials, the penetration depth is 300 μm, greater than the thickness of many aerospace coatings (In practice, multiple scattering because of the presence of porosity limits the penetration length and so these numbers represent an upper limit on the penetration depths). The lateral spatial resolution is more difficult to define as it also depends on the optics used to collect the signal but is unlikely to be less than the penetration length unless small numerical aperture optics are used. Indeed, as in other optical systems a compromise will probably have to be made between maximizing signal and lateral resolution in determining temperature distributions.

In Raman spectroscopy and photo-luminescence, the optical penetration dictates both the depth into the coating that can be probed as well as the distance from which the signal can be collected. Thus, in the visible, a laser can probe through the whole coating and consequently the Raman signal also originates from throughout the coating thickness. Being a photo-stimulated signal, the lateral spatial resolution depends primarily on the diameter of the laser used to excite the Raman signal. As the wavelength of the Raman signal is simply shifted from that of the laser by a small amount (the Raman shift), the only means of restricting the depth is to use a shorter wavelength laser, such as an excimer laser in the UV, close to the absorption edge of zirconia at ~250 nm. Under such conditions, the Raman signal would be localized to a few nanometers of the surface and a true surface temperature measurement could be realized. The disadvantage, however, is that as the Raman signal is proportional to the volume of material probed, the signal would necessarily be weak.

Consideration of the optics associated with the photo-luminescence method of measuring temperature would suggest that its spatial resolutions would be the same as for Raman spectroscopy, with the lateral resolution determined by the laser diameter and the depth resolution determined by the optical penetration distance. However, the fact that the luminescence originates from only the doped regions of the coating provides an opportunity to localize the temperature sensing to where the dopant has been placed during the deposition of the coating. For instance, by localizing the doped sensor layer to the vicinity of the bond-coat during deposition and using a laser

wavelength with a long penetration length, the temperature at the bond-coat can be determined. Similarly, if the doped layer is applied to the outer layer of the coating, the signal comes from that layer irrespective of the absorption length of the laser in the zirconia. An example of the former is shown in Fig. 9 where a 10 μm thick Eu^{3+} doped layer was deposited directly on a bond-coat and then covered with 125 μm of undoped 8YSZ.[24] The coating was then probed at temperature using a frequency-doubled Nd:YAG laser at 532 nm, which can penetrate through the entire coating. Despite this the luminescence originates from the 10 μm region at the TBC/alloy interface. This is an illustration of the versatility provided by combining the optical properties of the coating with the ability to place the dopant in the coating where it is required to provide the temperature measurement.

Of all the challenges in temperature measurement in turbines, the most difficult is measuring the temperature of rotating components, especially as a function of position on a blade surface. The measurement has to be made before the blade rotates out of the field of view of the measurement system and so the speed of data acquisition and the signal response time are critical and both need to be much faster than the time in which the blade is in view. This calls for synchronous excitation and detection of the luminescent, or Raman signal, coincident with the same position on the blade. An estimate of the time window can be made as follows. Assuming that the measurement system is fixed to the turbine casing, the measurement on a single blade has to be made before the next blade comes into view. This time, t_{max}, in seconds, is given by the relationship

$$t_{max} = \frac{60}{rm}$$

where r is the rotational speed of the turbine in revolutions per minute and m is the number of blades per row. Thus, for an industrial, utility power generating turbine that operates at 3600 rpm (for 60 Hz, and correspondingly 3000 rpm for 50 Hz electric power) and has 72 blades per row, the maximum time for measurement is 230 μs. In comparison, for an aeroturbine that rotates at 16,000 rpm and has 36 blades, the time window is only 104 μs. In practice, the measurement has to be made even faster because the field of view changes very rapidly with the angle of illumination and collection away from normal incidence. Nevertheless, these estimates provide guidance as to the times involved. Indeed, researchers at Siemens have recently demonstrated that *in situ* pyrometry of individual blades, using a very fast infrared CCD detector array, is feasible.[25] In addition to using a fast detector, the measurement was facilitated by the fact that the intervening hot gases had a low emissivity and at that the wavelengths chosen the signal probably originated from a substantial thickness of the coating. Comparison of these estimated times with the lifetimes shown in Fig. 9 also indicate that the luminescence lifetime method has the speed necessary for high temperature measurements of rotating blades.

Finally, during operation the surfaces of blades and vanes can become discolored, presumably by deposition of, and reaction with, various impurities. In such cases, even though the contamination is only superficial it may nevertheless adversely affect optical sensing. Fortunately, the discoloration generally appears most pronounced on the cooler portions of the blades. Despite this, further studies are needed to quantify the effect in detail as it is likely that each method described here will be affected differently depending on the emissivity and optical absorption of the contamination.

Acknowledgements

The research summarized in this contribution has been drawn from the PhD theses of David Clarke's co-authors. The authors are especially grateful to Dr. K. Murphy of Howmet Corporation for providing the electron-beam deposited TBC samples we have studied, including the Eu^{3+} doped coatings.

References

1. D. R. DeWitt and G. D. Nutter, *Theory and Practice of Radiation Thermometry*. Wiley, New York, 1988.

2. D. Ng and G. Fralick, "Use of Multiwavelength Pyrometry in Several Elevated Temperature Aerospace Applications," *Rev. Scientific Instrum.*, 72 [2] 1522 (2001).

3. M. E. Thomas and M. J. Linevsky, "Multispectral Pyrometry for Insulating Material Emissometry," *Optical Diagnostic Methods for Inorganic Transmissive Materials*. Eds., R. V. Datla and L. M. Hanssen, Proc. SPIE, 3425 126–133 (1998).

4. R. Darolia, private communication.

5. D. Yajima, A. Ohnishi, and Y. Nagasaka, "Simultaneous Measurement Method of Normal Spectral Emissivity and Optical Constants at High Temperature," *Proceedings 15th Symposium on Thermophysical Properties*, Boulder, CO, 2003.

6. D. L. Wood and K. Nassau, "Refractive Index of Cubic Zirconia Stabilized with Yttria," *Appl. Optics*, 21 [16] 2978 (1982).

7. J. I. Eldridge, C. M. Spuckler, K. W. Street, and J. R. Markham, "Infrared Radiative Properties of Yttria-Stabilized Zirconia Thermal Barrier Coatings," *Proc. Ceramic Sci. Eng.*, 23 [4] 417–430 (2002).

8. J. A. Nychka, T. Naganuma, M. R. Winter, Y. Kagawa, and D. R. Clarke, "Temperature Dependent Optical Reflectivity of Tetragonal-Prime Yttria-stabilized Zirconia," *J. Am. Ceram. Soc.*, 89 908–913 (2006).

9. D. R. Clarke and C. G. Levi, "Materials Design for the Next Generation Thermal Barrier Coatings," *Ann. Rev. Mater. Res.*, 33 383 (2003).

10. M. Born and E. Wolf, *Principles of Optics*. Pergamon Press, Oxford, 1970.

11. B. Hapke, *Theory of Reflectance and Emittance Spectroscopy*. Cambridge University Press, Cambridge, 1993.

12. B. G. Henderson, P. G. Lucey, and B. M. Jakosky, "New Laboratory Measurements of Mid-IR Emission Spectra of Simulated Planetary Surfaces," *Journal of Geophysical Research*, 101 [E6] 14969 (1996).

13. D. R. Clarke and F. Adar, "Measurement of the Crystallographically Transformed Zone Produced by Fracture in Tetragonal Zirconia Containing Ceramics," *Journal of the American Ceramic Society*, 65 284 (1982).

14. W. Hayes and R. Loudon, *Scattering of Lights by Crystals*. Wiley, New York, 1978.

15. D. A. Long, *Raman Spectroscopy*. McGraw-Hill, New York, 1977.

16. V. Lughi and D. R. Clarke, to be published.

17. Suggested by V. K. Tolpygo, UCSB, private communication.

18. L. C. Bradley, "A Temperature Sensitive Phosphor Used to Measure Surface Temperatures in Aerodynamics," *Review of Scientific Instruments*, 24 [3] 219 (1953).

19. S. A. Allison and G. T. Gillies, "Remote Thermometry with Thermographic Phosphors: Instrumentation and Applications," *Review of Scientific Instruments*, 7 2615 (1997).

20. K. T. V. Grattan and Z. Y. Zhang, *Fiber Optic Fluorescence Thermometry*. Chapman and Hall, 1995.

21. M. M. Gentleman and D. R. Clarke, *Surface and Coatings Technology*, 188–189 93–100 (2004).

22. D. C. Harris, *Materials for Infrared Windows and Domes*. SPIE Optical Engineering Press, 1999.

23. M. M. Gentleman and D. R. Clarke, "Luminescence Sensing of Temperature in Pyrochlore Zirconate Materials for Thermal Barrier Coatings," *Surface and Coating Technology*, 200 1264–1269 (2005).

24. M. M. Gentleman, J. L. Eldridge, D. Zhu, and D. R. Clarke, in preparation.

25. Siemens Corporation Magazine for Research and Innovation. (2005) (www.photonics.com/spectra/applications/)

Modeling the Influence of Reactive Elements on the Work of Adhesion between Oxides and Metal Alloys

I. J. Bennett

Netherlands Institute of Metals Research, Delft, The Netherlands

J. M. Kranenburg and W. G. Sloof[†]

Department of Materials Science and Technology, Delft University of Technology, Delft, The Netherlands

A method is presented that allows determination of the work of adhesion between oxides and metals using a macroscopic atom model. The method allows complex interfaces to be modeled easily and the influence of additives and impurities to be assessed. The model is used to study the work of adhesion between α-Al₂O₃ and β-NiAl. This interface is of importance for the performance of thermal barrier coatings as applied to jet turbines. The model shows that the work of adhesion is not significantly altered by so-called reactive element additions. Reactive elements are known to improve the durability at the alloy/oxide interface and include elements such as Zr, Y, and Hf added in concentrations of less than 1 at.%. A significant weakening of the interface is predicted when impurities such as sulfur and carbon are present. The model also predicts a large interaction enthalpy between the reactive elements and impurities. It is proposed that the primary effect of reactive element additions is impurity scavenging. The impurities are fixed in the bulk of the alloy by the reactive elements and cannot diffuse to the oxide/metal interface to weaken it.

I. Introduction

THE work of adhesion between ceramics and metals is of importance in a number of advanced material applications.[1] These include metal–matrix composites, coatings, microelectronics, and photovoltaic devices, sensors, and magnetic nanoparticles. In electronic components, good metal–ceramic joints are essential for thermal management and electrical isolation.[2] Good adhesion is crucial for metal brazes used for joining two ceramic components, for example, silver-based brazes used for joining sapphire to alumina.[3] The physical, chemical, mechanical, and optical properties of sapphire are of interest to the electronics industry, and the joining of sapphire to other ceramics is the key to its practical application. Further metal/ceramic interfaces that have been investigated include copper/alumina[2] and niobium/alumina.[4]

The growth of oxide layers on high-temperature-resistant alloys is another field where the work of adhesion between an oxide and a metal alloy is important.[5–8] The alloys are usually Fe or Ni based with Cr and/or Al as the oxide former. The oxide acts as a diffusion barrier, limiting the diffusion of oxygen through the oxide to the metal/oxide interface. For temperatures up to 900°C, Cr₂O₃ is the preferred oxide. Above 900°C,

α-Al₂O₃ is desired. Other alumina phases that are not as protective may also be formed and to limit this, the oxidation temperature should ideally be above 1100°C.

Modeling of the work of adhesion between oxides and metals has been performed primarily by *ab initio* calculations, often using the Hartree–Fock method or density functional theory (DFT).[1] Examples of these calculations include adhesion energies between various metals and oxides.[9–11] Calculations have also been made for simple nickel/alumina interfaces.[12–14] Although DFT calculations are in principle exact, they do require the use of powerful computers and long calculation times.

A combination of first-principle calculations and thermodynamic data has been used to obtain surface and interfacial energies as a function of oxygen partial pressure and temperature for Nb/Al₂O₃ interfaces.[15] The work of separation is calculated using first-principle calculations. The results of these calculations are then combined with thermodynamic data to obtain surface energies and interfacial energies. This method avoids the difficulties associated with calculating the absolute interface energy for nonstoichiometric interfaces.

An alternative route is offered by the macroscopic atom model, originally developed to predict whether an alloy of two metals was thermodynamically possible.[16] To avoid problems of defining what type of bonding (ionic, metallic, or covalent) was dominant in a particular alloy system, a model based on empirical quantities was constructed. The model has been shown to be valid for metals and inter-metallics, but the exclusion of a specific type of bonding should make the model suitable for other systems. The model also allows the calculation of interaction and surface energies, which in turn can be used to calculate the work of adhesion between two metallic elements. Examples of the development and application of the macroscopic atom model can be found in Jeurgens *et al.*[17] and Benedictus *et al.*[18] In this work, the macroscopic atom model has been extended to estimate the work of adhesion between two compounds, with both metallic and nonmetallic components (Table I). Specifically, it has been used to estimate the work of adhesion between α-Al₂O₃ and β-NiAl and to investigate the influence of reactive elements and impurities on this work of adhesion. The reactive elements considered are yttrium, zirconium, and hafnium. Impurities include sulfur and carbon, which may be introduced during materials processing. (See Tolpygo and Grabke,[19] Smialek,[20] and Smialek and Pint[21] for examples of impurity content in nickel aluminum alloys.) Segregation of carbon at an Al₂O₃/FeCrAl interface has been reported,[22] although a reduction in carbon surface concentration measured with X-ray photoemission spectroscopy (XPS) was seen on heating of alloys to temperatures above 650°C.[23]

The interface between α-Al₂O₃ and β-NiAl is the critical interface in modern high-temperature coating systems as applied, for example, in turbine jet engines. Gas temperatures in these engines can reach in excess of 1600°C, with surface temperatures of the turbines blades reaching 1200°C. To protect the underly-

---

J. Smialek—contributing editor

Manuscript No. 20055. Received September 20, 2004; approved March 7, 2005.
Supported by the Strategic Research Programme of the Netherlands Institute for Metals Research in the Netherlands (www.nimr.nl) under Project No. MC7.00080 and also by Innovation-driven Research Programme (IOP) on Surface Technology in the Netherlands (www.iop.nl) under Project No. IO100004A.
†Author to whom correspondence should be addressed. e-mail: w.g.sloof@tnw.tudelft.nl

Table I. Work of Adhesion Calculated using the Macroscopic Atom Model

Interface	Work of adhesion (J/m²)
Al_2O_3–Hf	6.02
Al_2O_3–Zr	5.86
Al_2O_3–NiAl	4.84
Al_2O_3–Y	4.75
Al_2O_3–S	2.50
Al_2O_3–C	2.21
Al_2O_3–HfO_2	5.81
Al_2O_3–ZrO_2	5.40
Al_2O_3–Y_2O_3	4.67
Al_2O_3–Cr_3C	3.88
Al_2O_3–Y_2S_3	3.72
NiAl–HfO_2	5.70
NiAl–ZrO_2	5.61
NiAl–Y_2O_3	4.56
NiAl–Cr_3C	3.85
NiAl–Y_2S_3	4.08
NiAl–Hf	4.88
NiAl–Zr	4.73
NiAl–Y	3.56
NiAl–C	3.30
NiAl–S	3.07

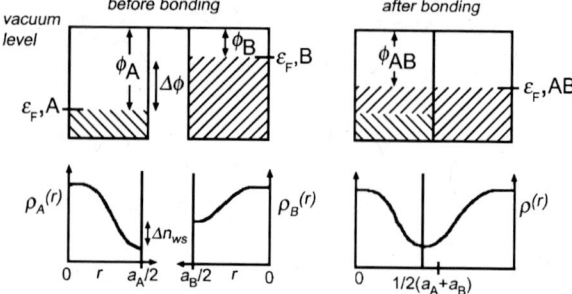

Fig. 1. Schematic plots showing how the electronegativity and electron density change at an interface between A and B with respect to the values for the components. The left two plots show the difference in electronegativity, $\Delta\phi$, and electron density, Δn_{WS}, between the two elements. On formation of an alloy, these differences can no longer exist. The energy required to compensate for changes in electronegativity and electron density on alloy formation determine the enthalpy of formation of the alloy (from Rohrer[65]).

ing substrate material, a thermal barrier system is used.[24,25] This consists of a thermal barrier coating fixed to the substrate with a bond coat. The thermal barrier coating is usually yttria-stabilized zirconia with a thickness of about 150 μm. This, in combination with internal cooling of the turbine blades, provides a drop of 150°C as compared with the gas temperature. In a modern high-temperature coating system, the bond coat is made of a β-NiAl-type alloy with additions of Cr (or Pt) and a reactive element. In service, the bond coat forms a thermally grown oxide (TGO) at the interface between itself and the thermal barrier coating. This oxide layer acts as a diffusion barrier, preventing oxygen from reaching and oxidizing the substrate. The oxide layer also enhances the adhesion of the thermal barrier coating to the bond coat. Ideally, the TGO consists of α-Al_2O_3, which, at these operating temperatures, is the most protective oxide.

One of the two most probable modes of failure will be delamination of the TGO from the bond coat, which will result in loss of the thermal barrier coating. Delamination is enhanced by the presence of impurities such as sulfur in the coating alloy and also by the environment in which the turbine is operated. The addition of reactive elements to the bond coat has been found to dramatically improve the adhesion of the TGO to the bond coat.[14,26–35] This reactive element effect is attributed to a number of mechanisms, including the direct promotion of chemical bonding between the oxide and the bond coat alloy. Other mechanisms include impurity scavenging, in particular, scavenging of sulfur,[36] reduction of cation diffusion from the bond coat to the oxide, and enhanced nucleation of α-Al_2O_3. Reactive elements readily form oxides with a high melting point and include yttrium, zirconium, and hafnium.

II. Macroscopic Atom Model as Applied to Metal/Oxide Interfaces

To model the influence of impurities and reactive elements on the work of adhesion between an oxide and a metal alloy, a macroscopic atom model is used. The term macroscopic atom implies that there is no difference in the interfacial energy between two blocks of metal and the enthalpy of formation of an inter-metallics alloy.[16] This suggests a relationship between the surface energy of a solid or liquid metal and its enthalpy of vaporization, which is also found in experimental data. The model was originally designed to predict the formation of an alloy between two metallic elements. Atoms of each of the components

of the alloy are treated as units of metal atoms with the same properties as the pure metals. The pure metals are described in terms of the electron density at the boundary of their Wigner–Seitz cells and their chemical potential for electronic charge. This is shown schematically in Fig. 1, where r is the distance from the center of the Wigner–Seitz cell, ρ is the electron density of the cells A and B at r, a is the diameter of the respective Wigner–Seitz cells, and ε_F is the Fermi level. The difference in electron density at the boundary of the Wigner–Seitz cell is given as Δn_{WS} and the difference in chemical potential as $\Delta\phi$. The chemical potential or electronegativity used in the model is called the Miedema electronegativity and takes the symbol ϕ^*.[16] It is derived from work functions of the pure metals and adjusted using available experimental data for the enthalpies of formation. Using these values, the enthalpy of formation of an alloy between A and B, ΔH^{for}, can be calculated using

$$\Delta H^{for} \propto \left[-P(\Delta\phi^*)^2 + Q(\Delta n_{WS}^{1/3})^2 \right] \qquad (1)$$

where P and Q are experimentally determined constants (where $Q/P = 9.4$ V/(density unit)²; see De Boer et al.[16]). The sign of this enthalpy predicts whether an alloy is thermodynamically possible or not. A negative enthalpy of formation indicates that an alloy may be formed.

The macroscopic atom model has been extended to other applications including the calculation of surface and interfacial energies.[16] In this work, the model is adopted to calculate the work of adhesion between two compounds or between an element and a compound. This is done by summation of the interactions between the components of the compounds on each side of the interface (or between components of the compounds on one side and an element on the other side of the interface), taking into account the extra atomic volume, and so surface area, occupied by the other components (see Fig. 2). Because of the empirical nature of the model, it should be applicable to systems other than the transition metals for which it was designed. In this work, it has been used to describe the work of adhesion between a metallic alloy and a TGO. Most parameters required for the model are readily available (ϕ^*, n_{WS}, and molar volumes for transition metals from De Boer et al.[16] and for group 16 atoms from Neuhausen and Eichler[37]) and others, such as the surface energy of compounds (see Eq. (11)), can be estimated.

To calculate the work of adhesion between two dissimilar compounds (e.g. between compounds A_kX_l and B_mY_n), the interaction energy between the components of each compound needs to be calculated. For the interaction between components A and B, this is done using the following equation:

$$\gamma_{A-B}^{T=0,\,interaction} = \frac{\Delta \overline{H}_{A\,in\,B}^{\circ interface}}{c_0 V_A^{2/3}} \qquad (2)$$

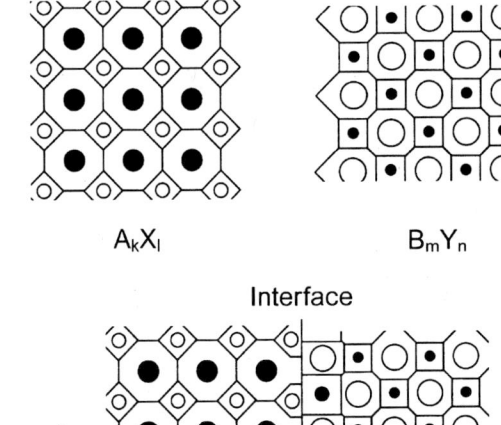

A_kX_l B_mY_n

Interface

Fig. 2. Schematic diagram showing the formation of an interface between compounds A_kX_l and B_mY_n and the deformation of the unit cells of the surface atoms of each compound required to achieve this interface. This deformation is associated with a change in the electronegativity and density of electrons at the cell boundary for the components of each compound.

where $\Delta \overline{H}_{A \, in \, B}^{\circ interface}$ is the interface enthalpy for an atomic cell of A fully surrounded by atomic cells of B, V_A is the molar volume of A as it is in pure A, and c_0 is a proportionality constant linking the atomic volume to the atomic surface area for an atomic cell. A similar equation can be written for $\gamma_{B-A}^{T=0, \, interaction}$, using $\Delta \overline{H}_{B \, in \, A}^{\circ interface}$ and $c_0 V_B^{2/3}$. This is later described in this section as looking up at the oxide (or impurity or reactive element) surface and looking down at the metal surface. Equations can also be written for the other possible interactions across the interface, i.e. the interactions between A–X, B–Y, and X–Y. As the interface enthalpy for an atomic cell surrounded by atomic cells of the same element is zero, the interactions between similar components across the interface are also zero. For the interface between A_kX_l and B_mX_n only the interactions A–B, A–X, and B–X need to be considered.

The surface area fractions of the components i on one side of the interface, C_i^S, are calculated from the ratio of their concentrations and corrected molar volumes. The use of component concentrations to calculate surface area fractions implies that crystal orientations on either side of the interface are not considered in this work. Crystal orientations can be included by determining the component concentrations of a particular crystal plane and termination type, and substituting these values for C_i^S. The corrected molar volumes are the molar volumes for the compound components after charge transfer associated with the formation of the compound. This is an iterative calculation, and the surface area fractions are initially calculated using the uncorrected molar volumes. For A_kX_l, the surface concentration of component $i = X$ is given by

$$C_X^S = \frac{C_X V_X^{2/3}}{C_X V_X^{2/3} + C_A V_A^{2/3}} \qquad (3)$$

where

$$C_A^S = 1 - C_X^S \qquad (4)$$

where C_X and C_A are the respective atomic concentrations of components X and A in A_kX_l and V_X is the molar volume of components X. The degree to which the X-component atoms are

surrounded by A-component atoms is denoted as f_A^X and is calculated for A_kX_l using the following empirical expression applicable to compounds:

$$f_A^X = C_A^S \left[1 + 8(C_X^S C_A^S)^2 \right] \qquad (5)$$

The corrected molar volume for component X is given by

$$\left(V_X^{2/3} \right)_{alloy} = \left(V_X^{2/3} \right)_{pure \, X} \left[1 + a f_A^X (\phi_X^* - \phi_A^*) \right] \qquad (6)$$

where ϕ^* is the electronegativity of the respective components. The constant a has a value dependent on the nature of the component for which the molar volume is being corrected.[16] For noble and trivalent metals, including aluminum, $a = 0.07$; for divalent metals, including nickel, $a = 0.10$; and for other metals $a = 0.04$. For oxygen and sulfur $a = 0.04$ is also applied.[37]

The corrected molar volumes are then used to recalculate the surface area fractions of the compound components using Eq. (3).

The interaction energy between components A and B in A_kX_l and B_mY_n, including surface area fractions of the components, is given by

$$\gamma_{A-B}^{T=0, \, interaction} = (C_A^S C_B^S) \frac{\Delta \overline{H}_{A \, in \, B}^{\circ interface}}{c_0 V_A^{2/3}} \qquad (7)$$

As previously mentioned, the calculations can be carried out looking both up at the oxide (or impurity or reactive element) surface and down on the metal surface, so that the interaction energy is dependent on the molar volume of both components of the interface couple. The interaction between B and A, the inverse of the orientation for Eq. (7), is given by

$$\gamma_{B-A}^{T=0, \, interaction} = (C_B^S C_A^S) \frac{\Delta \overline{H}_{B \, in \, A}^{\circ interface}}{c_0 V_B^{2/3}} \qquad (8)$$

The interaction energies for all component combinations are added for a specific interface orientation to give the total interaction energy for that interface and that orientation. For A_kX_l and B_mY_n, looking from A_kX_l to B_mY_n, this gives

$$\begin{aligned} \gamma_{A_MX_N-B_RY_S}^{interaction, \, T=0} = & \, \gamma_{A-B}^{interaction, \, T=0} + \gamma_{A-Y}^{interaction, \, T=0} \\ & + \gamma_{X-Y}^{interaction, \, T=0} + \gamma_{X-B}^{interaction, \, T=0} \end{aligned} \qquad (9)$$

Where the interaction enthalpy is not tabulated, it can be calculated using

$$\Delta \overline{H}_{A \, in \, B}^{\circ interface} = \frac{V_A^{2/3}}{\left(n_{WS}^{-1/3} \right)_{av}} \left[-P(\Delta \phi^*)^2 + Q(\Delta n_{WS}^{1/3})^2 \right] \qquad (10)$$

where $(n_{WS}^{-1/3})_{av}$ is the inverse of the average $n_{WS}^{-1/3}$ values for atoms A and atoms B. The above equation can be used to calculate the interaction energy for an atom of A surrounded by B atoms or an atom B surrounded by A atoms. The accuracy with which the equation can predict the interaction enthalpy is highest for two transition metal atoms. For nontransition metal atoms, a structure-dependent term has to be added. This term has also been applied in this work for interactions between metals and oxygen and sulfur. It has also been noted that the quadratic terms $\Delta \phi^*$ and Δn_{WS} represent the first terms in a series expansion. For alloys and compounds with a significant charge transfer and a significant change in atomic electron configuration, the large $\Delta \phi^*$ and Δn_{WS} values may not fit this approximation. In this work, for interfaces between oxygen and a reactive element, $\Delta \phi^*$ approaches 4 V, whereas for interactions between inter-metallics, $\Delta\phi^*$ is always less than 2 V. Similar variations are seen for Δn_{WS}.

As a result of the above, a difference in the work of adhesion is found when calculated using $\Delta\overline{H}_{A\,in\,B}^{\circ interface}$ and $\Delta\overline{H}_{B\,in\,A}^{\circ interface}$. The problem is most significant for the interactions between the reactive elements and oxygen. These combinations have the largest values of $\Delta\phi^*$ and Δn_{WS} and a large difference in atomic volume. The anomaly in the work of adhesion is reduced by increasing the term describing the degree to which an atom is surrounded by others, f_A^X, in Eq. (5). This equation is derived from experimental results for the formation of alloys between two transition metals. As stated above, the values of $\Delta\phi^*$ and Δn_{WS} are generally smaller for the transition metals, as are the differences in atomic volume. By increasing f_A^X, the volume corrections for the interacting atoms are enhanced. For atoms where an increase in volume is predicted on interaction, the increase is greater. Where a reduction in volume is predicted, the reduction is also greater. This results in values for the work of adhesion at the interface with a smaller anomaly. One solution would be to derive an alternative equation for the calculation of f_A^X suitable for atoms other than the transition metals. This would involve looking at the packing of the large atoms relative to the small atoms and vice versa. It may be necessary to derive separate equations for the large and small atoms. This is beyond the scope of this article and, as an increase in f_A^X causes the value for the work of adhesion calculated looking up at the oxide (or impurity or reactive element) surface and looking down at the metal surface to approach each other, it was considered sufficient to use an average work of adhesion.

The above parameters are also needed for determination of the surface energy of the substrate and the oxide. For A_kX_l, the surface energy is given by

$$\gamma_{A_mX_n}^0 = \frac{C_A^S \Delta H_A^{surf}}{f_{vacuum}^A c_0 V_A^{2/3}} + \frac{C_X^S \Delta H_X^{surf}}{f_{vacuum}^X c_0 V_X^{2/3}} \quad (11)$$

where ΔH^{surf} is the surface enthalpy of the respective atoms and f_{vacuum} is the degree to which the respective surface atoms are surrounded by vacuum (an average value of 0.31 is used[16]). The surface energy for each of the components is summed in the ratio of the surface area fraction for that component. The surface enthalpy for oxygen, ΔH_O^{surf}, is not tabulated and is calculated using a known value of the surface energy for α-Al$_2$O$_3$[37] in combination with ΔH_{Al}^{surf}. A literature value is taken for the surface energy of sulfur,[38] carbon,[39] and other elements.[16]

The surface energies are used in combination with the interaction energy to obtain the work of adhesion. The work of adhesion between A_kX_l and B_mY_n is given by

$$W_{A_kX_l-B_mY_n}^{ad,\,T=0} = -(\gamma_{A_kX_l}^{T=0} + \gamma_{B_mY_n}^{T=0}) + \gamma_{A_kX_l-B_mY_n}^{interaction,\,T=0} \quad (12)$$

Calculations were performed for combinations of β-NiAl and α-Al$_2$O$_3$, α-Al$_2$O$_3$ and the reactive element metals and oxides, and β-NiAl and the reactive element metals and oxides. In addition to this, the work of adhesion for the above metals, alloys, and oxides is calculated when forming an interface with sulfur and carbon.

All the calculations in this work are performed at zero Kelvin. Investigation of the change in enthalpy with temperature shows that at 298 K, interaction energies of 1% or 2% lower can be expected. This is balanced by the change in surface energy with temperature. Thus, it is assumed that the influence of temperature on the work of adhesion can be neglected (see also Zhang et al.[12]).

III. Calculated Values for Work of Adhesion for Combinations of Oxides and Metal Alloys

The average work of adhesion between undoped β-NiAl and α-Al$_2$O$_3$ is estimated as 4.8 J/m^2 (see Figs. 3 and 4), which corresponds well with literature values. The calculations show that the reactive elements Zr and Hf (except for Zr in combination with NiAl, which has a value of 4.7 J/m^2) have a positive effect of the work of adhesion. For all possible interface configurations with Hf or its oxide, an increase in the work of adhesion is predicted. The highest work of adhesion is found for the combination of Hf and Al$_2$O$_3$ with a value of 6.0 J/m^2. Interfaces with Y and its oxide all result in a reduction in the work of adhesion. The largest reduction is seen for Y in combination with β-NiAl with a value of 3.6 J/m^2. This is in conflict with reported improvement in oxide adhesion due to yttrium additions.[14,40] Differences may be because of the method used to define the structure of interfaces. The calculations presented for the macroscopic atom model are based on an average element, oxide, and substrate compositions with no account taken of the type of termination or lattice plane. For all reactive elements, the largest work of adhesion is calculated for the interface between the reactive element and α-Al$_2$O$_3$, with the weakest interface between the reactive element and β-NiAl.

Figures 3 and 4 also show that the work of adhesion between β-NiAl and α-Al$_2$O$_3$ is weaker than between α-Al$_2$O$_3$ and ZrO$_2$. This is consistent with the observation that, within a thermal coating system where the thermal barrier is usually ZrO$_2$, the β-NiAl/α-Al$_2$O$_3$ interface is the critical interface.[41–43] The system is more likely to fail by delamination of the α-Al$_2$O$_3$ layer from the alloy, than by delamination of the thermal barrier from the α-Al$_2$O$_3$ layer.

With sulfur as one of the components of the interface and either β-NiAl or α-Al$_2$O$_3$ as the other, the calculated work of adhesion shows a large drop relative to the reference interface (see Figs. 3 and 4). The work of adhesion drops from 4.8 J/m^2 for the interface β-NiAl/α-Al$_2$O$_3$, to 2.3 J/m^2 for the interface between α-Al$_2$O$_3$ and S. The largest contributing factor to this drop is the low surface energy of sulfur. The work of adhesion between β-NiAl and S is calculated to be 3 J/m^2. Again, this is because of the low surface energy of the sulfur as the interaction energy between sulfur and β-NiAl and α-Al$_2$O$_3$ is similar to that of the reference interface. A similar effect is found with carbon as one component of an interface. For the interface between α-Al$_2$O$_3$ and C, the work of adhesion is approximately 2.2 J/m^2. For this interface, the interaction energy is higher than for the reference interface, but the low surface energy of carbon results in a work of adhesion closer to that for sulfur.

A combination of Y and S as Y$_2$S$_3$ does increase the work of adhesion with both β-NiAl and α-Al$_2$O$_3$ relative to sulfur alone.

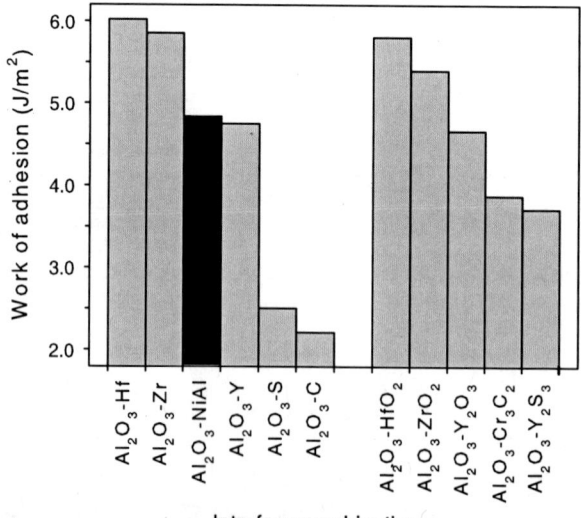

Fig. 3. Work of adhesion calculated using the macroscopic atom model for combinations of α-Al$_2$O$_3$ with reactive elements, impurities, and compounds with the work of adhesion for an ideal α-Al$_2$O$_3$/β–NiAl interface as reference (in black).

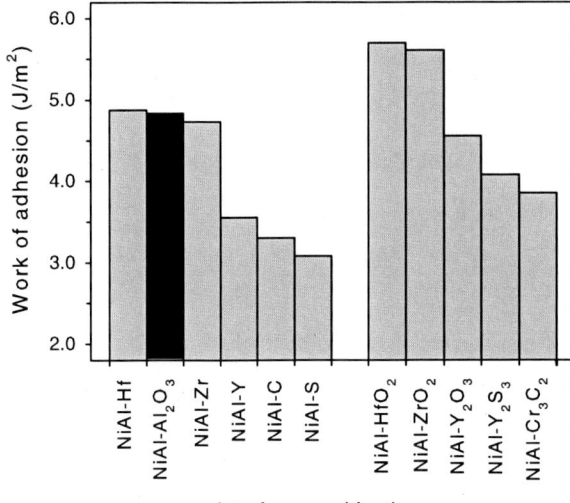

Fig. 4. Work of adhesion calculated using the macroscopic atom model for combinations of β-NiAl with reactive elements, impurities, and compounds with the work of adhesion for an ideal β-NiAl/α–Al₂O₃ interface as reference (in black).

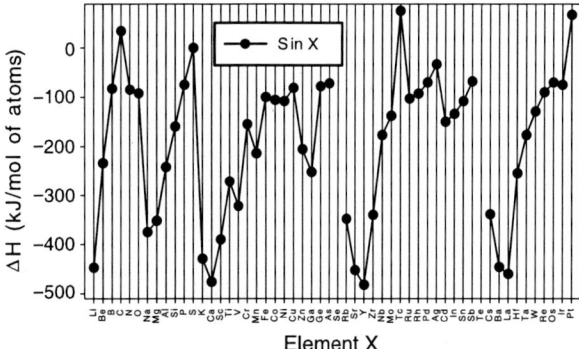

Fig. 5. Interaction enthalpy for metallic elements and sulfur where the sulfur is completely surrounded by metal atoms (S in X).

For the α-Al₂O₃–Y₂S₃ interface, the work of adhesion is 3.6 J/m². The surface energy of Y₂S₃ is higher than just sulfur because of the Y atoms, but the work of adhesion is still lower than for the reference interface between β-NiAl and α-Al₂O₃.

Similarly, the work of adhesion between Cr₃C₂ and α-Al₂O₃ and β-NiAl is higher than for just C. For both combinations, the average work of adhesion is approximately 3.8 J/m². Again, it is the increase in surface energy of the impurity in going from C to Cr₃C₂ that results in the higher work of adhesion.

In summary, the calculated values show that the work of adhesion is greatly influenced by the surface energy of one or both of the components of an interface. A lower interaction energy will also result in a lower work of adhesion, but for the interfaces calculated in this study, the surface energy has a greater effect. The calculation of the work of adhesion for the β-NiAl and α-Al₂O₃ interface assumes a clean interface with no contamination. Calculations for combinations of either of these two forming an interface with an impurity show a significant reduction in the work of adhesion. For sulfur, the work of adhesion is reduced from 4.8 to 2.3 J/m²; for carbon, it is reduced to 2.2 J/m². Although some gains are calculated for work of adhesion when a reactive element is present at the interface, yttrium, which is known to improve the performance of the TGO, shows a small drop. For both hafnium and zirconium, a rise in the work of adhesion is calculated for most interfaces, but the rise is not as significant as the drop seen with impurities present.

IV. Choice of Reactive Element

Although a rise in the work of adhesion is seen for some reactive elements and some interface configurations, the largest gain would appear to be obtained by preventing the diffusion of impurities to that interface. This can be achieved either by the use of pure starting materials and the formation of the interface under controlled conditions or by the use of reactive elements as scavengers for impurities, fixing them in the bulk of the alloy.

A plot of the interaction enthalpy for a range of metals with sulfur and carbon shows that the reactive elements considered in this work are among those metals with the most negative interaction enthalpies (see, for S, Fig. 5 and, for C, Fig. 6). Interaction enthalpies can be calculated for the metal surrounded by the impurity (X in S and X in C where X is the metal) or for the impurity surrounded by the metal (S in X and C in X). As the concentration of reactive elements is many times higher than

the concentration of impurities, the S in X and C in X values are considered in this work. Highly negative interaction enthalpies indicate that the reactive elements will readily form an interface with sulfur and carbon. It is proposed that the value of interaction enthalpy is a measure of the scavenging efficiency of a metal. The more negative the value, the more efficient the metal will be at scavenging that particular impurity. In this work, calculations are restricted to interactions between metallic elements and S or C. Other interactions, such as S–O, are not included. Data in Neuhausen and Eichler[37] indicate that enthalpy of formation of metallic sulfides is larger than for possible S–O compounds. The plots also show that scavenging may further be improved by using more than one reactive element. Certain reactive elements are better at scavenging for sulfur whereas others are preferred for scavenging carbon impurities. Considering the C in X plot for carbon, Zr and Hf have the most negative value. For S in X for sulfur, Y would appear to be the best candidate, whereas Zr and Hf, although still strongly negative, have values less negative than, for example, Ba and Cs. A combination of Y and Zr or Hf would cover both S and C impurities. Alternatives to the three reactive elements considered in this work can also be derived from the interaction enthalpy plots, but their influence on other parameters, such as the work of adhesion, need to be assessed. For example, Ca may appear to be a candidate for S scavenging, but it had been shown to have a significant negative effect on the work of adhesion when present at the interface between the TGO and the alloy.

Previous work[44] has also shown a difference in the scavenging efficiency of group IIIB and group IVB elements, with the former having a preference for S and the latter for C. The scavenging efficiency was determined by comparison of the standard free energies of formation of carbides and sulfides. Y is found to react more strongly than Hf or Zr with S, forming YS, whereas Hf forms the most stable carbide, with no reaction expected between Y and C. Another article[45] suggests co-doping of FeCrAl

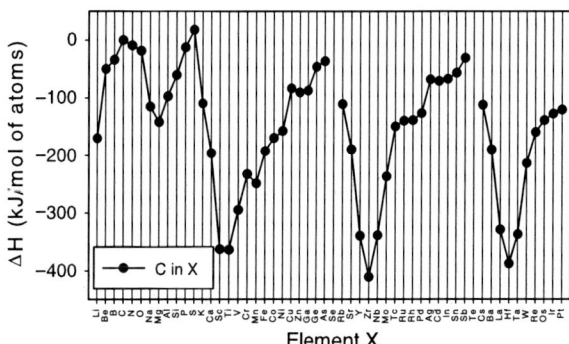

Fig. 6. Interaction enthalpy for metallic elements and carbon where the carbon is completely surrounded by metal atoms (C in X).

alloys with Hf and Y. The Y:S and Hf:C ratios are found to be a determining factor in the performance of these alloys during thermal cycling.

V. Comparison with Other Models

First-principle calculations using DFT have been performed for systems similar to that studied in this work.[13] For the interface between Ni and Al_2O_3, an increase in the work of adhesion was calculated when the interface was doped with a reactive element, including Y and Zr. The work of adhesion increased from 1.88 J/m^2 with a half monolayer of Ni at the interface to 3.24 and 3.21 J/m^2 with Y and Zr, respectively. The increase in the work of adhesion is attributed to modifications in the electronic structure of the interface. This is similar in magnitude to the increase seen when comparing the β-NiAl/α-Al_2O_3 interface with the Zr/α-Al_2O_3 and Hf/α-Al_2O_3 interfaces calculated using the macroscopic atom model. The differences between the models with Y at the interface and the negative influence of Al on the work of adhesion using DFT may be the result of the choices made for atomic volumes and bond lengths. For the macroscopic atom model, the large atomic volume for Y results in a lowering of the work of adhesion. In the DFT model, the bond length appears to correlate with the work of adhesion. For both models, small adjustments to these values can have a large effect on the work of adhesion and consequently on the conclusions that can be drawn from their values. The works of adhesion for the DFT model are generally lower than for the macroscopic atom model. Again, this is possibly because of the choice of atomic volume and bond length. The DFT model is also used to model a specific Ni and Al_2O_3 orientation, whereas the macroscopic atom model calculations ignore specific crystal orientations.

Calculations for the work of adhesion between Ni and the oxides Al_2O_3 and ZrO_2 using the same DFT method are also lower than calculated with the macroscopic atom model.[46,47] The interface Ni/Al_2O_3 has a maximum calculated work of adhesion of 0.94 J/m^2. The Ni/ZrO_2 and ZrO_2/Al_2O_3 interfaces have maximum works of adhesion of 2.01 and 1.26 J/m^2, respectively. A lower work of adhesion with increasing thickness of the oxide beyond two atomic layers is also predicted. Calculations are limited to the Ni(111), Al_2O_3(0001), and ZrO_2(111) planes.

Another study, also using a DFT approach, looked at the connection between the calculations and interface adhesion measurements on metals/oxide systems including Ni/Al_2O_3.[12] They used the calculations to explain the differences seen in experimentally determined works of adhesion between pure Ni and Ni alloys containing small amounts of Al, predicting that the Ni alloys were Al terminated. Works of separation are found in the range of 1–7 J/m^2 depending on the type of interface. These values are of the same order as those found with the macroscopic atom model for the work of adhesion. The influence of reactive elements and impurities was not discussed.

Similar calculations have been used to assess the possibility of using SiO_2 instead of Al_2O_3.[48] The work of adhesion at the Ni/SiO_2 interface is predicted to be three times greater than for the Ni/Al_2O_3 interface. This is attributed to the more covalent nature of SiO_2. Using the macroscopic atom model, a work of adhesion for the interface β-NiAl/SiO_2 of 5.9 J/m^2 was calculated, which is only slightly higher than the 4.8 J/m^2 calculated for β-NiAl/α-Al_2O_3. The differences in values obtained between the two models do not imply that one model is more accurate than the other, but that the sensitivity of the models to variations in the nature of atoms included at an interface is dissimilar. The advantage of the macroscopic atom model is that the calculations are fast, which allows easier analysis of the sensitivity of the result to variations in the input parameters and for many interfaces to be studied.

A model for the wetting and adhesion behavior of ceramics with liquid metals using a combination of the Gibbs free energy change per unit area of interface formed and bulk contributions to the work of adhesion has been developed.[49] This model shows some similarities to the macroscopic atom model using an enthalpy of mixing term when calculating the Gibbs free energy change for the reaction between the metal and the ceramic. The work of adhesion values are generally lower than those found in this work, but of a similar magnitude. The work of adhesion is calculated at elevated temperatures ($<900°C$) in all cases.

VI. Comparison with Experimental Results

Experimental values for the work of adhesion for the NiAl–Al_2O_3 interface are not easy to find and their worth in the evaluation of theoretical work of adhesion is questionable. Determination of the work of adhesion of a brittle ceramic on a metallic substrate is far from simple, with difficulties including the determination of when delamination occurs, where delamination occurs, and the amount of energy directly involved in delamination as opposed to deformation of the substrate. Evidence for the detrimental effect of a high sulfur content in the alloy on oxide adherence is readily available.

The work of adhesion for the interfaces Al/Al_2O_3 and Ni/Al_2O_3 has been calculated from experimentally determined surface energy values extrapolated to room temperature.[50] For the interface Al/Al_2O_3, the work of adhesion is given as 1.2 J/m^2. For Ni/Al_2O_3, a value of 1.13 J/m^2 is given. Earlier work using a modified sessile drop method resulted in a work of adhesion of 0.65 J/m^2 for the Ni/Al_2O_3 interface.[51] A higher value of 3 J/m^2 was also found.[52] Variations in the work of adhesion may be the result of differences in materials and measurement techniques. All the results are for an interface as it would be found during experimentation, so the presence of impurities at the interface, even at a very low concentration, cannot be excluded. This would give a measured value for the work of adhesion lower than theoretically possible. Values calculated for the uncontaminated β-NiAl/α-Al_2O_3 interface using the macroscopic atom model can be seen as the upper limit of the work of adhesion. It can be expected that all experimental values will be lower. Works of adhesion measured on other ceramic/metal interfaces show values similar to those cited above.[53]

A number of publications show observations related to the presence of sulfur at the metal/oxide interface. Cyclic oxidation results of an as-received alloy with a sulfur content of 20 ppm show extensive oxide spallation and a drop in mass after approximately 300 cycles, whereas a similar alloy annealed in hydrogen with an estimated sulfur content of less than 0.1 ppm only starts to lose mass after 1000 cycles.[54] Similar results were found with Zr-doped alloys, and a relationship between the sulfur content and the amount of reactive element needed to improve adhesion was found.

A number of articles describe experimental evidence for the segregation of sulfur to an oxide/alloy interface resulting in weak bonding and a nonadherent scale.[22,55–58] Evidence is also presented for an additional effect above that of sulfur scavenging for Zr-doped Fe_3Al alloys. The oxide on the Zr-doped alloys performed better in a scratch test than a similar H_2-annealed alloy. This could correspond to the slight improvement in the work of adhesion seen in the calculation using the macroscopic atom model with Zr present.

Cyclic oxidation of a number of alumina-forming alloys showed that scale adhesion was drastically improved by the reduction of the sulfur content to below 1 ppmw.[59] An improvement in scale adhesion was also seen on addition of yttrium to a single crystal of an Ni-based alloy. Combined H_2 annealing and doping with yttrium saw a further improvement in scale adhesion, but without a reduction in sulfur content. It was proposed that this improvement may be because of decarburization.

Other studies of the influence of sulfur on oxide/alloy adhesion also conclude that sulfur weakens the interface, either by the formation of cavities[60,61] or by interface embrittlement.[62]

Direct observation of sulfur at yttrium-rich particles in the bulk of a Y-doped NiAl alloy after oxidation at 1200°C provides evidence of impurity scavenging.[63] Analysis was performed using scanning transmission electron microscopy/energy-dispersive spectroscopy. Yttrium-rich oxide particles were found underneath the oxide scale, no voids were seen at the oxide/metal interface, and sulfur was detected in some of the yttrium-rich particles. These observations led to the conclusion that yttrium changes the mechanism of scale growth, thus eliminating interfacial voids, and traps sulfur as predicted using the macroscopic atom model.

Our own experimental observations show that cycles of annealing and sputtering of the surface of a β-NiAl sample reduces the amount of sulfur segregated in the subsequent annealing step.[64] Oxidation of a sample subjected to a number of annealing and cycling steps resulted in an adherent oxide with delamination levels similar to untreated alloys containing reactive elements. Also, the degree of sulfur segregation measured with XPS after annealing was found to be lower for alloys containing reactive elements than for an undoped alloy. Of the reactive elements tested, Y was the most efficient at reducing sulfur segregation, followed by Zr and Hf.

VII. Conclusions

A method is presented to determine the work of adhesion between oxides and metal alloys using a macroscopic atom approach. The advantage of this method when compared with first-principle calculations using DFT is that complex interfaces can readily be studied. The effects of reactive elements and impurities can be easily calculated.

The model presented has been used to determine the work of adhesion at the interface between Al_2O_3 and NiAl. It has been found that the work of adhesion is not directly affected by reactive element additions to the alloy. The work of adhesion is not significantly increased with reactive elements present and for some situations, the work of adhesion may even be reduced relative to an undoped interface.

A significant weakening of the Al_2O_3/NiAl interface is predicted when the impurities such as sulfur and carbon are present. A large interaction enthalpy is calculated between the reactive elements and the impurities. These results support the hypothesis that the primary action of the reactive elements is to scavenge the impurities, preventing them from diffusing to the oxide/metal interface.

Using the macroscopic atom model, alternative reactive elements for scavenging of sulfur, carbon, and other possible impurities and the effect of these reactive elements on the work of adhesion are assessed. A reactive element must form a stable compound with the impurities and may itself have no negative effect on the work of adhesion between the oxide and the alloy.

References

[1]G. Pacchioni, "Metal/Oxide Adhesion Energies from First-Principles," *Surf. Sci.*, **520**, 3–5 (2002).

[2]X.-G. Wang, J. R. Smith, and M. Scheffler, "Adhesion of Copper and Alumina from First Principles," *J. Am. Ceram. Soc.*, **86** [4] 696–700 (2003).

[3]H. Ning, Z. Geng, J. Ma, F. Hung, Z. Qian, and Z. Han, "Joining of Sapphire and Hot Pressed Al_2O_3 using $Ag_{70.5}Cu_{27.5}Ti_2$ Brazing Filler Metal," *Ceram. Int.*, **29**, 689–94 (2003).

[4]J. D. Sugar, J. T. McKeown, R. A. Marks, and A. M. Glaeser, "Liquid-Film-Assisted Formation of Alumina/Niobium Interfaces," *J. Am. Ceram. Soc.*, **85** [10] 2823–30 (2002).

[5]G. D. Oxx Jr., "Which Coating at High Temperature?," *Prod. Eng.*, **29**, 61–64 (1958).

[6]H. Hindam and D. P. Whittle, "Microstructure, Adhesion and Growth Kinetics of Protective Scales on Metals and Alloys," *Oxid. Met.*, **18** [5–6] 245–84 (1982).

[7]M. Schütze and D. R. Holmes, *Protective Oxide Scales and the Breakdown*. Wiley, New York, 1997.

[8]A. S. Khanna, *Introduction to High Temperature Oxidation and Corrosion*. ASM International, Materials Park, OH, 2002.

[9]A. E. Mattsson and D. R. Jennison, "Computing Accurate Surface Energies and the Importance of Self-Energy in Metal/Metal–Oxide Adhesion," *Surf. Sci.*, **520**, L611–8 (2002).

[10]E. A. Kotomin, J. Maier, Y. F. Zhukovskii, D. Fuks, and S. Dorfman, "Ab Initio Modelling of Silver Adhesion on the Corundum (0001) Surface," *Mater. Sci. Eng. C*, **23**, 247–52 (2003).

[11]Y. F. Zhukovskii, E. A. Kotomin, B. Herschend, K. Hermansson, and P. W. M. Jacobs, "The Adhesion Properties of the $Ag/\alpha–Al_2O_3(0001)$ Interface: An Ab Initio Study," *Surf. Sci.*, **513**, 434–58 (2002).

[12]W. Zhang, J. R. Smith, and A. G. Evans, "The Connection Between Ab Initio Calculations and Interface Adhesion Measurements on Metal/Oxide Systems: Ni/Al_2O_3 and Cu/Al_2O_3," *Acta Mater.*, **50**, 3803–16 (2002).

[13]E. A. Jarvis and E. A. Carter, "An Atomic Perspective of a Doped Metal–Oxide Interface," *J. Phys. Chem. B*, **106**, 7995–8004 (2002).

[14]A. B. Anderson, S. P. Mehandru, and J. L. Smialek, "Dopant Effect of Yttrium and the Growth and Adherence of Alumina on Nickel–Aluminium Alloys," *J. Electrochem. Soc.*, **132** [7] 1695–701 (1985).

[15]I. G. Batyrev, A. Alavi, and M. W. Finnis, "Equilibrium and Adhesion of Nb/Sapphire: The Effect of Oxygen Partial Pressure," *Phy. Rev. B*, **62** [7] 4698–706 (2000).

[16]F. R. De Boer, R. Boom, W. C. M. Mattens, A. R. Miedema, and A. K. Niessen, *Cohesion in Metals, Transition Metal Alloys*. North-Holland, Amsterdam, 1988.

[17]L. P. H. Jeurgens, W. G. Sloof, F. D. Tichelaar, and E. J. Mittemeijer, "Thermodynamic Stability of Amorphous Oxide Films on Metals: Application to Aluminium Oxides Films on Alumium Substrates," *Phys. Rev. B*, **62** [7] 4707–19 (2000).

[18]R. Benedictus, A. Böttger, and E. J. Mittemeijer, "Thermodynamic Model for Solid-State Amorphization in Binary Systems at Interfaces and Grain Boundaries," *Phys. Rev. B*, **54** [13] 9109–25 (1996).

[19]V. K. Tolpygo and H. J. Grabke, "The Effect of Impurities of Scale Growth: An Alternative View," *Scr. Mater.*, **38** [1] 123–9 (1998).

[20]J. L. Smialek, "The Effect of Hydrogen Annealing on the Impurity Content of Alumina-Forming Alloys," *Oxid. Met.*, **55** [1/2] 75–86 (2001).

[21]J. L. Smialek and B. A. Pint, "Optimizing Scale Adhesion for Single Crystal Superalloys," *Mat. Sci. Forum*, **369–372**, 459–66 (2001).

[22]P. Y. Hou, "Impurity Effects on Alumina Scale Growth," *J. Am. Ceram. Soc.*, **86** [4] 660–8 (2003).

[23]D. T. Jayne and J. L. Smialek, "A Sulphur Segregation Study of PWA 1480, NiCrAl, and NiAl Using X-ray Photoelectron Spectroscopy With *In Situ* Sample Heating"; pp. 183–96 in *Microscopy of Oxidation 2*, Edited by S. B. Newcomb and M. J. Bennett. Institute of Materials, London, 1993.

[24]M. Peters, C. Leyens, U. Schultz, and W. A. Kaysser, "EB-PVD Thermal Barrier Coatings for Aeroengines and Gas Turbines," *Adv. Eng. Mater.*, **3** [4] 193–204 (2001).

[25]P. Kofstad, *High Temperature Corrosion*. Elsevier Applied Science, Amsterdam, 1988.

[26]B. A. Pint, J. A. Haynes, K. L. More, I. G. Wright, and C. Leyens, "Compositional Effects on Aluminide Oxidation Performance: Objectives for Improved Bond Coats"; pp. 629–38 in *Superalloys 2000*, Edited by T. M. Pollock, R. D. Kissinger, R. R. Bowman, K. A. Green, M. McLean, S. Olsen, and J. J. Schirra. TMS, Warrendale, PA, 2000.

[27]J. Nowok, "Formation Mechanisms of Keying or Pegging Yttrium Oxide and Increased Plasticity of Alumina Scale on FeCrAlY," *Oxid. Met.*, **18** [1–2] 1–17 (1982).

[28]E. Roszczynialska, J. G. A. Jedlinski, and M. Danielewski, "The Influence of Yttrium on the Oxidation Behaviour of Ni–23Co–19Cr–12Al Alloy at High Temperatures," *Werkst. Korros.*, **43**, 124–30 (1992).

[29]M. K. Loudjani, A. M. Huntz, and R. Cortès, "Influence of Yttrium on Microstructure and Point Defects in Alpha-Alumina in Relation to Oxidation," *J. Mater. Sci.*, **28**, 6466–73 (1993).

[30]M. K. Loudjani and C. Haut, "Influence of the Oxygen Pressure on the Chemical State of Yttrium in Polycrystalline Alpha-Alumina," *J. Eur. Ceram. Soc.*, **16**, 1099–106 (1996).

[31]N. Czech, F. Schmitz, and W. Stamm, "Improvement of MCrAlY Coatings by Addition of Rhenium," *Surf. Coat. Technol.*, **68–69**, 17–21 (1994).

[32]N. Czech, V. Kolarik, W. J. Quadakkers, and W. Stamm, "Oxide Layer Phase Structure of MCrAlY Coatings," *Surf. Eng.*, **13** [5] 384–8 (1997).

[33]E. Schumann, J. C. Yang, M. J. Graham, and M. Rühle, "The Effect of Y and Zr on the Oxidation of NiAl," *Mat. Corr.*, **47**, 631–2 (1996).

[34]I. M. Allam, D. P. Whittle, and J. Stringer, "Improvements in Oxidation Resistance by Dispersed Oxide Addition: Al_2O_3-Forming Alloys," *Oxid. Met.*, **13** [4] 381–401 (1979).

[35]J. L. Cocking, J. A. Sprague, and J. R. Reed, "Oxidation Behaviour of Ion-Implanted NiCrAl," *Surf. Coat. Tech.*, **36**, 133–42 (1988).

[36]A. W. Funkenbusch, J. G. Smeggil, and N. S. Bornstein, "Reactive Element–Sulfur Interaction and Oxide Scale Adherence," *Metall. Trans. A*, **16A** [6] 1164–6 (1985).

[37]J. Neuhausen and B. Eichler, "Extension of Miedema's Macroscopic Atom Model to the Elements of Group 16 (O, S, Se, Te, Po)," PSI Report No. 03-13, 2003.

[38]R. F. Cook, "Crack Propagation Thresholds: A Measure of Surface Energy," *J. Mater. Res.*, **1** [6] 852–60 (1986).

[39]W. N. Reynolds, *Physical Properties of Graphite*. Elsevier, Amsterdam, 1968.

[40]J. L. Smialek and R. Browning, "Current Viewpoints on Oxide Adherence Mechanisms"; in *Proceedings of the Symposium in High Temperature Materials Chemistry III*, Edited by Z. A. Munir and D. Cubicciotti. Electrochemical Society, Pennington, NJ, 1986.

[41]A. G. Evans, D. R. Mumm, J. W. Hutchinson, G. H. Meier, and F. S. Pettit, "Mechanisms Controlling the Durability of Thermal Barrier Coatings," *Mater. Sci.*, **46**, 505–53 (2001).

[42]N. M. Yanar, G. Kim, S. Hamano, F. S. Pettit, and G. H. Meier, "Microstructural Characterization of the Failures of Thermal Barrier Coatings on Ni-base Superalloys," *Mater. High Temp.*, **20** [4] 495–506 (2003).

[43]D. R. Mumm and A. G. Evans, "On the Role of Imperfections in the Failure of a Thermal Barrier Coating Made by Electron Beam Deposition," *Acta Mater.*, **48**, 1815–27 (2000).

[44]D. T. Sigler, "Aluminium Oxide Adherence on Fe–Cr–Al Alloys Modified with Groups IIIB, IVB, VB and VIB Elements," *Oxide. Met.*, **32** [5/6] 337–54 (1989).

[45]B. A. Pint, "Optimization of Reactive-Element Additions to Improve Oxidation Performance of Alumina-Forming Alloys," *J. Am. Ceram. Soc.*, **86** [4] 686–95 (2003).

[46]E. A. A. Jarvis, A. Christensen, and E. A. Carter, "Weak Bonding of Alumina Coatings on Ni(111)," *Surf. Sci.*, **487**, 55–76 (2001).

[47]E. A. Jarvis and E. A. Carter, "The Role of Reactive Elements in Thermal Barrier Coatings," *Comput. Sci. Eng.*, **4** [2] 33–41 (2002).

[48]E. A. A. Jarvis and E. A. Carter, "Exploiting Covalency to Enhance Metal–Oxide and Oxide–Oxide Adhesion at Heterogeneous Interfaces," *J. Am. Ceram. Soc.*, **86** [3] 373–86 (2003).

[49]J. Chen, M. Gu, and F. Pan, "A Model of Work of Adhesion for Reactive Metal/Ceramic Systems," *Metal. Mater. Trans. A*, **32A**, 2033–8 (2001).

[50]D. M. Lipkin, J. N. Israelachvili, and D. R. Clarke, "Estimating the Metal–Ceramic van der Waals Adhesion Energy," *Philos. Mag. A*, **76** [4] 715–28 (1997).

[51]R. M. Pilliar and J. Nutting, "Solid–Solid Interfacial Energy Determinations in Metal–Ceramic Systems," *Philos. Mag.*, **16** [139] 181–8 (1967).

[52]M. Ohring, *Materials Science of Thin Films*. Academic Press, New York, 2002.

[53]M. W. Finnis, "The Theory of Metal–Ceramic Interfaces," *J. Phys. Condens. Matter*, **8** [32] 2811–36 (1996).

[54]C. Sarioglu, C. Stinner, J. R. Blanchere, N. Birks, F. S. Pettit, G. H. Meier, and J. L. Smialek, "The Control of Sulfur Content in Nickel-Base, Single Crystal Superalloys and its Effect in Cyclic Oxidation Resistance"; pp. 71–80 in *Superalloys 1996*, Edited by R. D. Kissinger, D. J. Deye, D. L. Anton, A. D. Cetel, M. V. Nathal, T. M. Pollock, and D. A. Woodford. TMS, Warrendale, PA, 1996.

[55]P. Y. Hou, "Beyond the Sulfur Effect," *Oxid. Met.*, **52** [3/4] 337–51 (1999).

[56]P. Y. Hou, K. Prüßner, D. H. Fairbrother, J. G. Roberts, and K. B. Alexander, "Sulfur Segregation to Deposited Al_2O_3 Film/Alloy Interface at 1000°C," *Scr. Mater.*, **40** [2] 241–7 (1999).

[57]P. Y. Hou, "Sulfur Segregation to Growing Al_2O_3/Alloy Interfaces," *J. Mater. Sci. Lett.*, **19**, 577–8 (2000).

[58]P. Y. Hou and J. Moskito, "Sulfur Segregation to Al_2O_3–FeAl Interfaces Studied by Field Emission-Auger Electron Spectroscopy," *Oxid. Met.*, **59** [5/6] 559–74 (2003).

[59]J. L. Smialek, "Scale Adhesion, Sulfur Content, and TBC Failure on Single Crystal Superalloys"; pp. 485–95 in *26th Annual Conference on Composites, Advanced Ceramics, Materials and Structures: B*, Edited by H. -T. Lin, and M. Singh. American Ceramics Society, Westerville, OH, 2002.

[60]L. Rivoland, V. Maurice, P. Josso, M.-P. Bacos, and P. Marcus, "The Effect of Sulfur Segregation on the Adherence of the Thermally-Grown Oxide on NiAl-I: Sulfur Segregation on the Metallic Surface of NiAl(001) Single-Crystals and at NiAl(001)/Al_2O_3 Interfaces," *Oxid. Met.*, **60** [1/2] 137–57 (2003).

[61]L. Rivoland, V. Maurice, P. Josso, M.-P. Bacos, and P. Marcus, "The Effect of Sulfur Segregation on the Adherence of the Thermally-Grown Oxide on NiAl-II: The Oxidation Behavior at 900°C of Standard, Desulfurised or Sulfur-Doped NiAl(001) Single-Crystals," *Oxid. Met.*, **60** [1/2] 159–78 (2003).

[62]J. D. Kiely, T. Yeh, and D. A. Bonnell, "Evidence for the Segregation of Sulfur to Ni–Alumina Interfaces," *Surf. Sci.*, **393**, L126–30 (1997).

[63]E. Schumann, J. C. Yang, and M. J. Graham, "Direct Observation of the Interaction of Yttrium and Sulfur in Oxidized NiAl," *Scr. Mater.*, **34** [9] 1365–70 (1996).

[64]Own work, to be published.

[65]G. S. Rohrer, *Structure and Bonding in Crystalline Materials*. Cambridge University Press , Cambridge, 2001. □

Hot Corrosion Mechanism of Composite Alumina/Yttria-Stabilized Zirconia Coating in Molten Sulfate–Vanadate Salt

Nianqiang Wu,[1,2] Zheng Chen,[1,3] and Scott X. Mao[†,1]

[1]Department of Mechanical Engineering, University of Pittsburgh, Pittsburgh, Pennsylvania 15261

[2]Keck Interdisciplinary Surface Science Center, NU*ANCE*, Northwestern University, Evanston, Illinois 60208-3108

[3]Department of Materials Science and Engineering, East China Shipbuilding Institute, Zhenjiang 212003, China

In order to improve the hot corrosion resistance of yttria-stabilized zirconia (YSZ), an Al_2O_3 overlay has been deposited on the surface of YSZ by electron-beam physical vapor deposition. Hot corrosion tests have been performed on the YSZ coatings with and without an Al_2O_3 overlay in the molten salt mixture (Na_2SO_4+0–15 wt% V_2O_5) at 950°C. The presence of V_2O_5 in the molten salt exacerbates degradation of both the monolithic YSZ coating and the composite YSZ/Al_2O_3 system. The formation of a low-melting Na_2O–V_2O_5–Al_2O_3 liquid phase is responsible for degradation of the Al_2O_3 overlay. The Al_2O_3 overlay acts as a barrier against the infiltration of the molten salt into the YSZ coating during exposure to the molten salt mixture with <5 wt% vanadate.

I. Introduction

THERMAL-BARRIER COATINGS (TBCs), which consist of an yttria-stabilized zirconia (YSZ) top coating and an intermediate MCrAlY (M = Ni, Co, Fe) bond coating, are extensively used in gas turbines.[1–5] The application of TBCs can improve the durability of components and enhance the engine efficiency by increasing the turbine inlet combustion temperature. The common failure mode of TBCs used in aviation gas turbines is that a thermally growth oxide (TGO) forms and continuously grows between the top coating and the bond coat. Because of the thermal expansion mismatch between the TGO and the bond coat, thermal cycling results in cracking, and even spalling of TBCs.

TBCs are also finding increasing application in land-based industrial engines and sea engines that are usually operated with low-quality fuels containing sulfur and vanadium.[5] In this case, another failure mode, hot corrosion, becomes predominant and crucial to the lifetime of TBCs. During service, molten sulfate and vanadate salts condense on the TBCs at the temperature of 600°–1000°C.[6,7] Although zirconia itself shows good resistance to the attack of the molten sulfate or vanadate compounds arising from fuel impurities, yttria is leached out of zirconia by the reaction with V_2O_5 or $NaVO_3$ to form YVO_4, causing the structural destabilization of ZrO_2 (i.e., transformation of zirconia from the tetragonal and/or cubic to the monoclinic phase). The structural destabilization of ZrO_2 is accompanied by a large destructive volume change, leading to large stresses within YSZ, which eventually results in delamination and spalling of coatings.[8–13]

N. S. Jacobson—contributing editor

Manuscript No. 10540. Received September 18, 2003; approved July 27, 2004.
[†]Author to whom correspondence should be addressed. e-mail: smao@engr.pitt.edu

Many methods have been developed to improve the hot corrosion resistance of TBCs to such harsh environments containing sulfate–vanadate deposits. For instance, based on the Lewis acid–base concept, zirconia stabilized with oxides such as india (In_2O_3),[14,15] scandia (Sc_2O_3),[16] ceria (CeO_2),[11,17] tantalic oxide (Ta_2O_5),[9,18] and $YTaO_4$[18] have been evaluated in terms of hot corrosion resistance. On the other hand, an attempt has been made to seal the surface of TBCs by laser glazing and arc lamp[19–21] or by various "seal coats"[21–25] to prevent the penetration of molten deposits into the porous YSZ coating.

In the present work, a high-purity Al_2O_3 overlay is deposited on the surface of YSZ coating by means of an electron-beam physical vapor deposition (EB-PVD) technique in order to improve the hot corrosion resistance of TBCs. Alumina has a high melting point and it is stable without showing a phase transition at high-temperature-like ZrO_2. Al_2O_3 has a small solubility particularly in molten salts and is expected to show excellent corrosion resistance.[26] Hot corrosion testing of TiAl with an Al_2O_3 coating in the sulfate melt at 900°C has shown that the Al_2O_3 coating itself is very stable in the sulfate melt and effectively prevents intermetallic TiAl from hot corrosion.[27] Chen et al.'s experiment[28] demonstrated that the Al_2O_3 coating could resist hot corrosion attack of molten Na_2SO_4 salt for a longer time than the YSZ coating. In addition, Al_2O_3–ZrO_2 composite coatings have been explored as thermal barrier applications, showing better resistance in NaCl melt than YSZ.[29] This suggests the potential application of Al_2O_3 in gas turbines.

In our study, both the monolithic YSZ coating and the composite YSZ/Al_2O_3 system will be exposed to the molten salt (Na_2SO_4+0–15 wt% V_2O_5) at 950°C. The hot corrosion mechanism of the composite YSZ/Al_2O_3 system will be studied. The effect of the V_2O_5 content on the hot corrosion behavior of coatings will be investigated. The role of Al_2O_3 in the hot corrosion environment will be explored.

II. Experimental Procedure

The TBC system used in the present work consisted of an IN-CONEL 601 nickel-based superalloy substrate, a CoNiCrAlY alloy bond coat as well as a zirconia–8 wt% yttira (YSZ) ceramic top coating. The 100 μm thick bond coat and the 200 μm thick YSZ coating were produced by low-pressure plasma spray and air plasma spray, respectively.

The Al_2O_3 overlay was deposited by an EB-PVD unit. Prior to deposition, the 1.5″ × 1.5″ coupons were ultrasonically cleaned and dried. The vacuum unit was pumped down to a base pressure of 7.5×10^{-6} Torr with the oxygen gas lines being evacuated. The samples were preheated to ~1000°C and allowed to soak at 1000°C for 20 min with the electron beam. During evaporation of aluminum oxide, ~150 sccm of oxygen was flowed into the chamber to maintain the oxygen stoichiometry of the condensing coating (chamber pressure $\sim 1 \times 10^{-3}$ Torr). The average condensation rate was 0.88 μm/min. At the end of the desired deposition time, the samples were retracted into the load lock chamber and allowed to cool for 10 min with ~200 sccm of oxygen flow before venting to atmosphere. The thickness of the Al_2O_3 coating was approximately 25 μm.

Hot corrosion tests were carried out on the TBCs with and without an Al_2O_3 coating. The TBC plates were coated with a 150 mg/cm² salt mixture (Na_2SO_4+0-15 wt% V_2O_5) by dipping into an aqueous slurry of salt (1000 g/L), then placed carefully into a static air furnace, and isothermally held at 950°C for 10 h. Upon heating, the salt mixture was melted, leading to the formation of a thin liquid film on the surface of specimens. After exposure, the specimens were cooled down to room temperature in the furnace. The exposed specimens were cleaned in de-ionized water, rinsed in isoprypol alcohol, and then dried. The Philips PW1700 X-ray diffractometer (Philips, Eindhoven, Netherlands) was then used to analyze the corrosion products and the structure of ZrO_2 in the exposed samples.

After the exposed samples were cleaned and then dried, X-ray photoelectron spectroscopy (XPS) analyses were performed on the top surface of coatings using an Omicron ESCA Probe (Omicron, Nanotechnology, Taunusstein, Germany), which was equipped with an EA125 Energy Analyzer (Omicron, Nanotechnology). Photoemission was stimulated by monochromated $AlK\alpha$ radiation (1486.6 eV) with an operating power of 300 W. A low-energy electron flood gun was used for charge neutralization. Binding energies of XPS spectra were referenced to the C 1s binding energy set at 284.8 eV.

After the exposed samples were cleaned in de-ionized water and then dried, the TBC samples were embedded into the epoxy resin, and then cut by a diamond saw to obtain the cross-section specimens. The cross-section specimens were polished by a diamond paste. Time-of-flight secondary ion mass spectrometry (ToF-SIMS) analyses were performed on the cross-section specimens with a Physical Electronics PHI TRIFT III (Physical Electronics, Inc., Eden Prairie, MN), which is equipped with a pulse Ga^+ liquid ion gun operated at 15 kV. The ion source was operated with a current of 600 pA. The charging effect was neutralized by a pulsed low-energy electron flood gun. The secondary ions were accelerated to ±3 kV by applying a bias on the sample. Spectra were acquired for positive secondary ions over a mass range of $m/z = 0$–1850. Positive spectra were calibrated using CH_3^+, $C_2H_5^+$, and $C_3H_7^+$ peaks.

The microstructure, the composition of the coating surface, and the cross-section were also examined using a PHILIPS XL30 scanning electron microscope (FE-SEM) equipped with an energy-dispersive spectrometer (EDS).

III. Results

(1) X-Ray Diffraction (XRD) Measurement

XRD analysis was performed on the coatings before and after hot corrosion testing. Pattern A in Fig. 1 demonstrates that the T-phase of ZrO_2 was predominant in the monolithic YSZ coating before corrosion testing. After the monolithic YSZ coating was exposed to pure Na_2SO_4 melt at 950°C for 10 h, only a small amount of M-phase ZrO_2 was detected (Pattern B in Fig. 1), and no chemical reaction was found. However, after exposure to the mixed molten salt of Na_2SO_4+5 wt% V_2O_5, YVO_4 was formed (Pattern C in Fig. 1), implying the leaching of Y_2O_3 from YSZ by the reaction of Y_2O_3 with V_2O_5. As a result, the intensity of T-phase decreased remarkably, and a considerable amount of M-phase was formed because of the leaching of Y_2O_3 from YSZ. As the V_2O_5 content in the molten salt mixture increased to 15 wt%, much more M-phase ZrO_2 occurred in the coating after hot corrosion testing (Pattern D in Fig. 1).

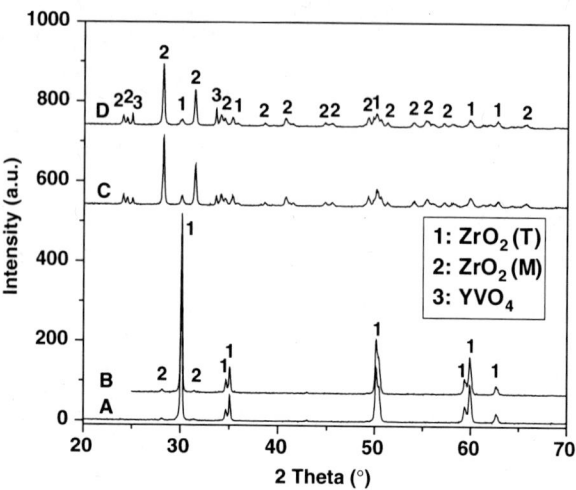

Fig. 1. X-ray diffraction patterns taken from the surface of the monolithic yttria-stabilized zirconia coating before and after 10 h of hot corrosion testing at 950°C (Pattern A, before corrosion testing; Patterns B–D, after corrosion testing in Na_2SO_4, Na_2SO_4+5 wt% V_2O_5, and Na_2SO_4+15 wt% V_2O_5, respectively).

Figure 2 shows the XRD patterns of the composite YSZ/Al_2O_3 system before and after hot corrosion testing. The as-deposited Al_2O_3 overlay showed the γ-phase structure (Pattern A in Fig. 2). As shown in Pattern B in Fig. 2, there were no changes in the structure of both the Al_2O_3 overlay and the YSZ layer after exposure to pure Na_2SO_4 melt. In contrast, part of γ-Al_2O_3 phase was transformed to α-Al_2O_3 phase after exposure to the mixed molten salt of Na_2SO_4+5 wt% V_2O_5 (Pattern C in Fig. 2). Moreover, the entire γ-Al_2O_3 phase was transformed to α-Al_2O_3 phase after exposure to the mixed molten salt with 15 wt% V_2O_5 (Pattern D in Fig. 2). It is well known that α-Al_2O_3 can only be produced by heating pure Al_2O_3 in the air above 1000°C.[30] This indicated that V_2O_5 played an important role in the phase transformation process from γ-Al_2O_3 to α-Al_2O_3 at 950°C. However, no evidence from the XRD patterns was found that the chemical reaction between the Al_2O_3 overlay and the molten salt had taken place. The results also showed that no YVO_4 peaks were present because of its low content, which was below the detection limit of the XRD. The T-phase of ZrO_2 in the YSZ coating was still predominant and only a small amount

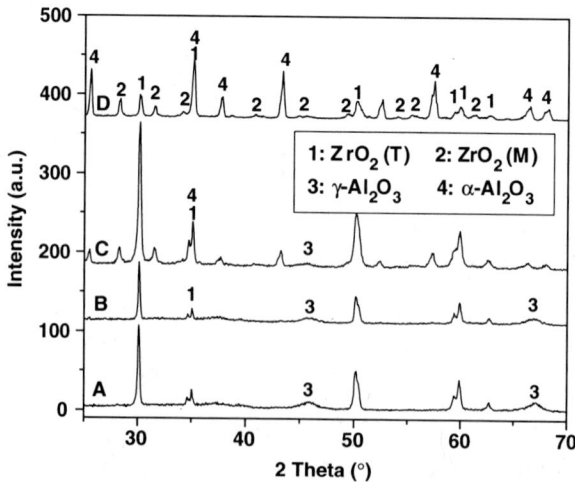

Fig. 2. X-ray diffraction patterns taken from the surface of composite yttria-stabilized zirconia/Al_2O_3 coating before and after 10 h of hot corrosion testing at 950°C (Pattern A, before corrosion testing; Patterns B–D, after corrosion testing in Na_2SO_4, Na_2SO_4+5 wt% V_2O_5, and Na_2SO_4+15 wt% V_2O_5, respectively).

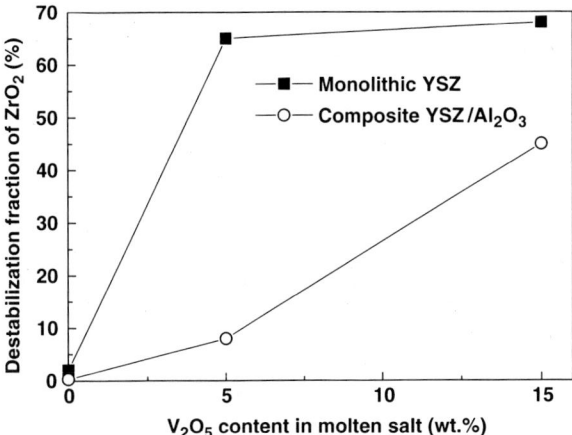

Fig. 3. Destabilization fraction of zirconia in the yttria-stabilized zirconia (YSZ) layer as a function of V_2O_5 content in the molten salt after hot corrosion testing at 950°C for 10 h.

of the M-phase of ZrO_2 existed in the YSZ coating after exposure to the molten salt of Na_2SO_4+5 wt% V_2O_5. As demonstrated in Pattern D in Fig. 2, however, the amount of M-phase of ZrO_2 in the YSZ remarkably increased when the V_2O_5 content in the salt melt reached 15 wt%.

In order to evaluate the hot corrosion resistance of the TBCs with and without an Al_2O_3 overlay, the extent of destabilization (D) of zirconia was estimated by[8]

$$D(\%) = \frac{M}{T+M} \times 100 \qquad (1)$$

where T is the height of the ZrO_2 tetragonal (111) peak, and M is the height of the ZrO_2 monoclinic ($11\bar{1}$) peak in XRD tests. Figure 3 shows the destabilization fraction of ZrO_2 ($D\%$) after exposure to the molten salts at 950°C for 10 h. It can be seen from Fig. 3 that both the monolithic YSZ and the composite YSZ/Al_2O_3 system exhibited excellent resistance to hot corrosion in pure Na_2SO_4 melt, whereas the T-phase ZrO_2 in both the monolithic YSZ and the composite YSZ/Al_2O_3 system became destabilized during exposure to the mixed sulfate–vanadate salt. The destabilization fraction of ZrO_2 ($D\%$) increased with an increase in the V_2O_5 content in the molten salt mixture. The Al_2O_3 overlay had a significant effect on the destabilization fraction of ZrO_2. The destabilization fraction of ZrO_2 in the composite YSZ/Al_2O_3 system was about 8% after exposure to the mixed sulfate–vanadate salt with 5 wt% V_2O_5. In contrast, the destabilization fraction of ZrO_2 in the monolithic YSZ coating reached up to 65%. As the V_2O_5 content in the salt increased to 15 wt%, the destabilization fraction of ZrO_2 in the composite YSZ/Al_2O_3 coating sharply reached up to 45%. This indicated that the Al_2O_3 overlay was no longer a protective coating when exposed to the molten salt mixture containing a high V_2O_5 content.

(2) XPS Analysis

In order to identify the corrosion product in the composite YSZ/Al_2O_3 coating, XPS analysis was performed on the surface of coatings before and after corrosion testing in the melt of Na_2SO_4+5 wt% V_2O_5. XPS analysis showed that the surface of composite YSZ/Al_2O_3 coating was composed of Al, V, Na, and O after corrosion testing, but no sulfur was detected. Figure 4 shows the XPS spectra of Al 2p core level. The Al 2p peak with a binding energy of 74.1 eV was attributed to γ-Al_2O_3 phase.[31] The peak with a binding energy of 73.7 eV was corresponding to α-Al_2O_3 phase.[31] Figure 5 reveals the XPS spectra of V $2p_{3/2}$ core level. The V $2p_{3/2}$ spectrum, which was obtained from the surface of composite YSZ/Al_2O_3 coating after corrosion testing,

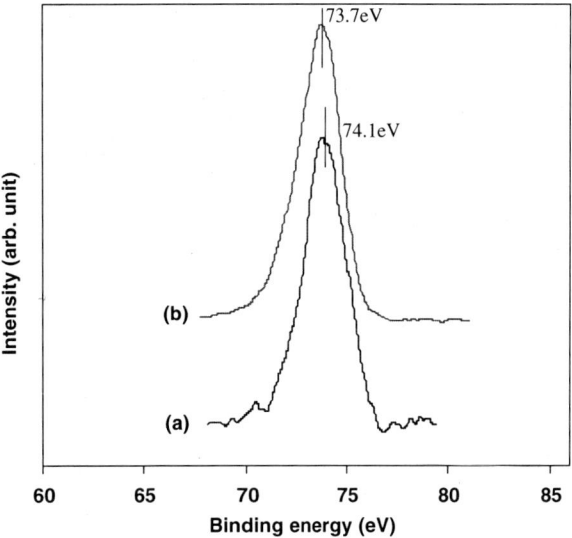

Fig. 4. X-ray photoelectron spectroscopy spectra of Al 2p core level: (a) obtained from the composite yttria-stabilized zirconia (YSZ)/Al_2O_3 coating before corrosion testing, (b) obtained from the composite YSZ/Al_2O_3 coating after corrosion testing.

exhibited a peak at 517.6 eV. This value was consistent with that obtained from the reactant of V_2O_5 before corrosion testing. This indicated that the corrosion product on the surface of composite coating was α-Al_2O_3 and V_2O_5, and no other chemical substance containing V or Al was detected after corrosion exposure. In contrast, the V $2p_{3/2}$ spectrum, which was obtained from the surface of monolithic YSZ coating after corrosion testing, illustrates that the V $2p_{3/2}$ peak chemically shifted to 516.5 eV from 517.6 eV (as shown in Fig. 5). This was attributed to the transition to YVO_4 from V_2O_5 induced by corrosion exposure.

(3) SEM Observation

SEM examination was carried out on the surface of monolithic YSZ coating after 10 h of exposure to the melt of Na_2SO_4+5 wt% V_2O_5, and revealed many pyramid-like particles (Fig. 6(a)). EDS analysis confirmed that the particles were rich in yttrium (40.53 at.%) and vanadium (36.31 at.%) and contained no zirconium. Keeping the XRD pattern in mind, these particles were identified to be YVO_4. The cross-section of the YSZ coating was

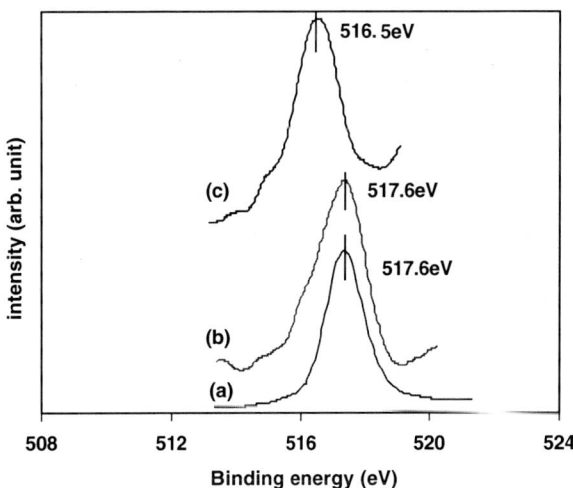

Fig. 5. X-ray photoelectron spectroscopy spectra of V $2p_{3/2}$ core level: (a) obtained from the reactant of V_2O_5 before corrosion testing, (b) obtained from the surface of composite yttria-stabilized zirconia (YSZ)/Al_2O_3 coating after corrosion testing, (c) obtained from the surface of monolithic YSZ coating after corrosion testing.

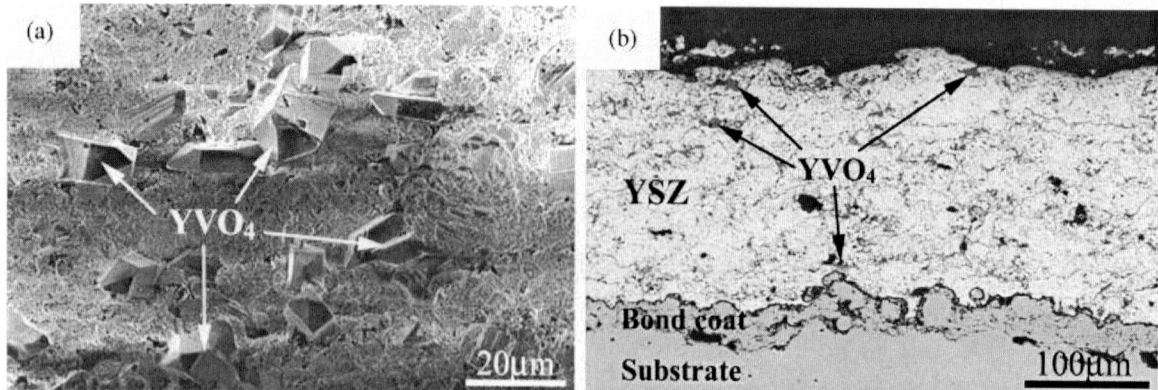

Fig. 6. Scanning electron microscope images taken from the corroded thermal-barrier coating without Al_2O_3 overlay ((a) surface image, (b) cross-section image), showing the formation of YVO_4 after 10 h of corrosion testing at 950°C in salt melt of $Na_2SO_4 + 5$ wt% V_2O_5.

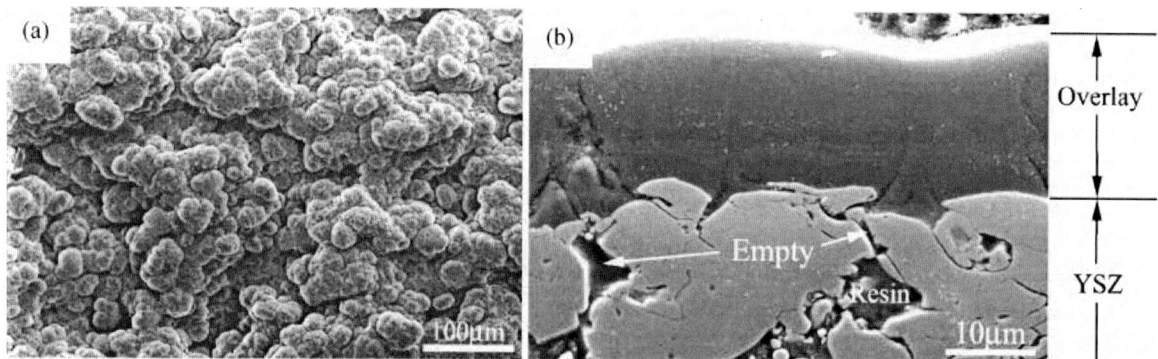

Fig. 7. Scanning electron microscope photographs showing the surface of the Al_2O_3 overlay (a), and the cross-section of the yttria-stabilized zirconia (YSZ)/Al_2O_3 coating (b).

examined as shown in Fig. 6(b). YVO_4 was detected not only near the surface of YSZ but also in the area adjacent to the bond coat, implying penetration of the molten salt into the inner YSZ coating.

The surface of as-deposited Al_2O_3 overlay revealed a "cauliflower" type of morphology (Fig. 7(a)). It can be clearly seen from the SEM image of the cross-section that the Al_2O_3 overlay was dense and adherent to the porous YSZ coating (Fig. 7(b)). The thickness of the Al_2O_3 coating was measured to be about 25 μm. After exposure to pure Na_2SO_4 melt, the surface of the composite YSZ/Al_2O_3 system became dark green. But the morphologies of both the top surface and the cross-section showed little variation after exposure (Fig. 8). The Al_2O_3 overlay was still dense and continuous.

After exposure to the molten salt mixture containing V_2O_5, visual examination showed that the surface became brown from white in color. SEM examination results are shown in Fig. 9. Comparing Fig. 9 with Figs. 7 and 8, it can be found that a significant change in the surface morphology of the Al_2O_3 overlay took place after 10 h of exposure to the molten salt containing V_2O_5. Coarse acicular-shaped α-Al_2O_3 crystals and faceted crystals were present, and the orientation of the crystals varied from region to region. Also, the preferential dissolution of grain boundaries occurred. Observation under a quantitative microscope showed that about 97% of the surface was still covered by the Al_2O_3 overlay after exposure to molten $Na_2SO_4 + 5$ wt% V_2O_5 salt (Fig. 9(a)). However, after exposure to the molten $Na_2SO_4 + 15$ wt% V_2O_5 salt, the YSZ was clearly visible on the surface, where Al_2O_3 was removed during corrosion exposure (Fig. 9(b)).

Figure 10(a) shows the cross-section of a composite YSZ/Al_2O_3 system after 10 h of exposure to the molten salt of

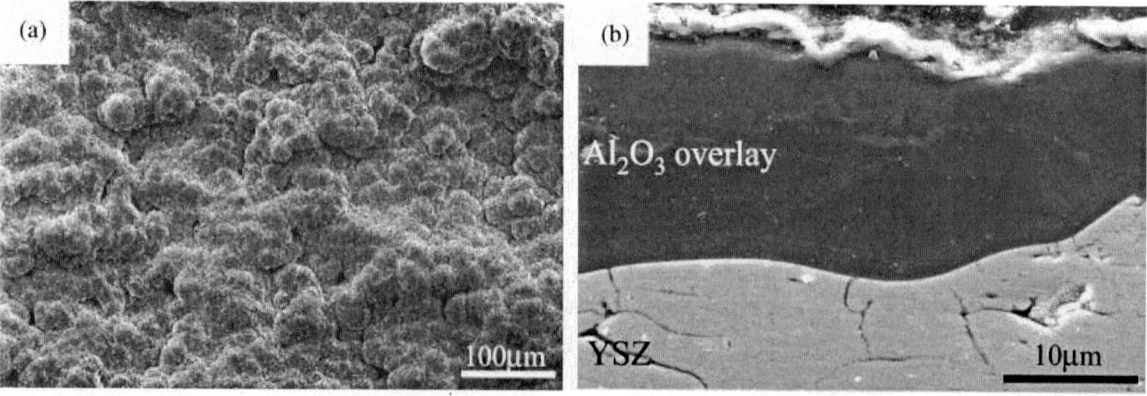

Fig. 8. Scanning electron microscope images of yttria-stabilized zirconia (YSZ) with Al_2O_3 overlay after hot corrosion in molten Na_2SO_4 ((a) surface image, (b) cross-section image).

Progress in Thermal Barrier Coatings

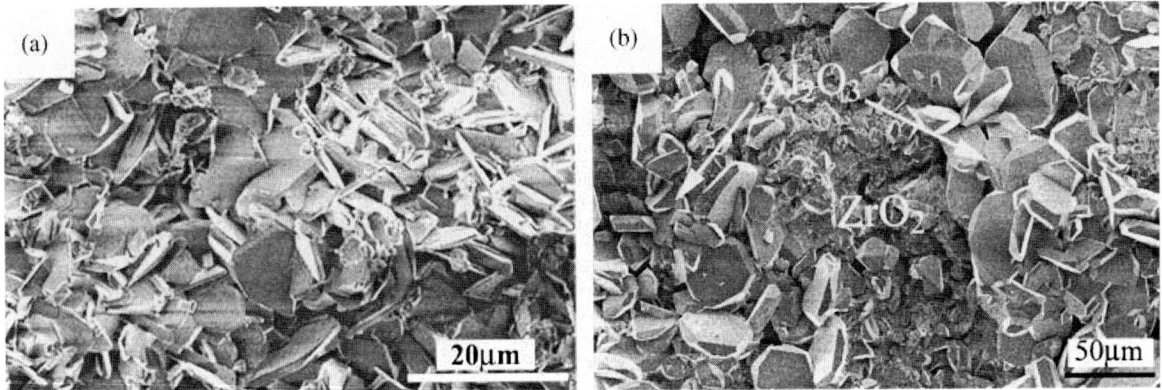

Fig. 9. Scanning electron microscope photographs taken from the surface of composite yttria-stabilized zirconia/Al_2O_3 coating after 10 h exposure to the molten salts of (a) Na_2SO_4+5 wt% V_2O_5 and (b) Na_2SO_4+15 wt% V_2O_5.

Na_2SO_4+5 wt% V_2O_5. The Al_2O_3 overlay can be divided into two different regions. (i) The region close to the outer surface marked by "A" was bright and loose, which was corresponding to the acicular-shaped α-Al_2O_3 crystals and faceted crystals (Fig. 9(a)). (ii) The region adjacent to YSZ marked by "B" was gray and relatively dense as compared with region "A". Because of the presence of dense layer "B", the attack of YSZ by the molten salt was arrested considerably (Fig. 3). But some grooves existed within the region "B". It is worth noting that the pores and cracks within the YSZ coating were filled with Al_2O_3 (indicated by arrows), which was confirmed by the EDS analysis as shown in Fig. 10(b). And YVO_4 was also found in some pores and cracks within YSZ near the surface (Figs. 10(c) and (d)). In contrast, the pores and cracks within the YSZ coating before hot corrosion testing were empty (Fig. 7(b)). Therefore, it is inferred that Al_2O_3 was absent in the YSZ layer before hot corrosion testing, and the presence of Al_2O_3 in the defects was caused by hot corrosion rather than by the EB-PVD process.

Figure 11 illustrates the cross-section of a composite YSZ/Al_2O_3 system after exposure to the molten salt of Na_2SO_4+15 wt% V_2O_5. The integrated and dense Al_2O_3 overlay was no longer present, leaving separate α-Al_2O_3 fragments on the surface of YSZ, which was consistent with the surface micrograph (Fig. 9(b)). Large Al_2O_3 pieces were "embedded" into the YSZ layer, leading to the interlacing of the Al_2O_3 layer with the YSZ layer at the interface. Owing to the absence of a compact Al_2O_3 overlay on the surface of the YSZ, the YSZ layer was directly exposed to the molten salt. As a result, serious destabilization of ZrO_2 occurred during hot corrosion exposure (Fig. 3).

(4) SIMS Characterization

After hot corrosion testing, SIMS analysis was performed on the cross-section of TBCs as shown in Fig. 12. Sodium and vanadium were infiltrated into the YSZ and existed over the YSZ layer. For the composite YSZ/Al_2O_3 coating, sodium was

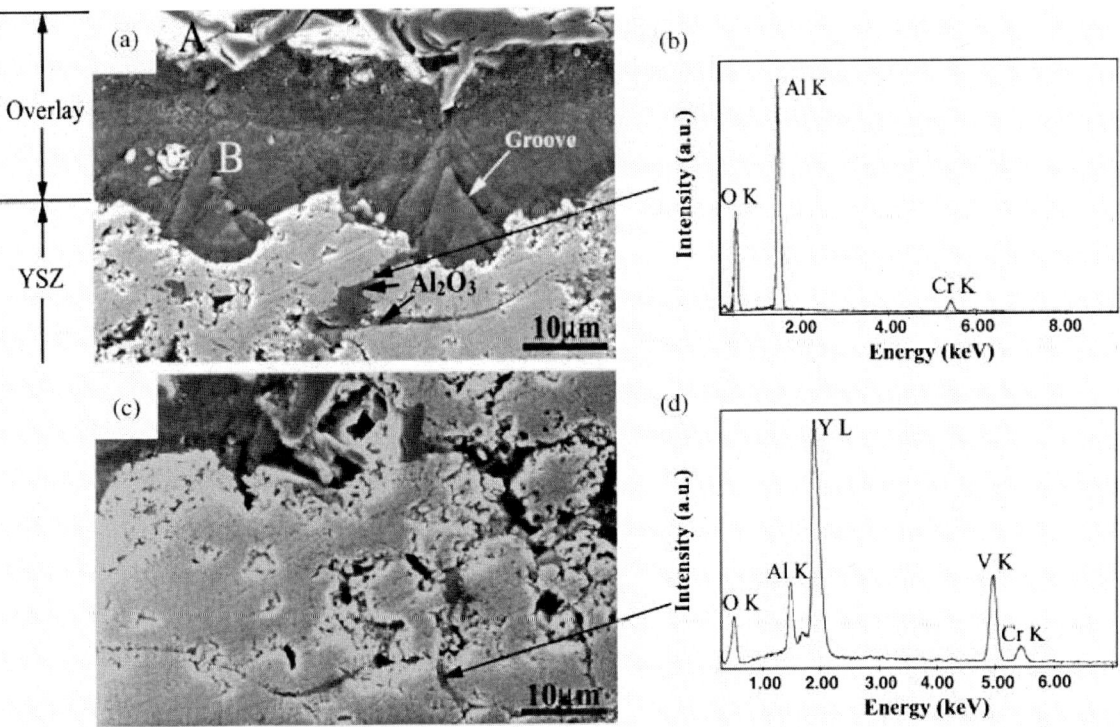

Fig. 10. Scanning electron microscope (SEM) photographs taken from the cross-section of yttria-stabilized zirconia (YSZ)/Al_2O_3 overlay system. (a) The microstructure of composite YSZ/Al_2O_3 coating after 10 h exposure to the molten salt of Na_2SO_4+5 wt% V_2O_5; (b) the energy dispersive spectrometer (EDS) spectrum taken from the pores that are indicated in (a); (c) SEM photograph showing the filled cracks and pores; (d) the EDS spectrum taken from pores in (c).

Testing and Characterization

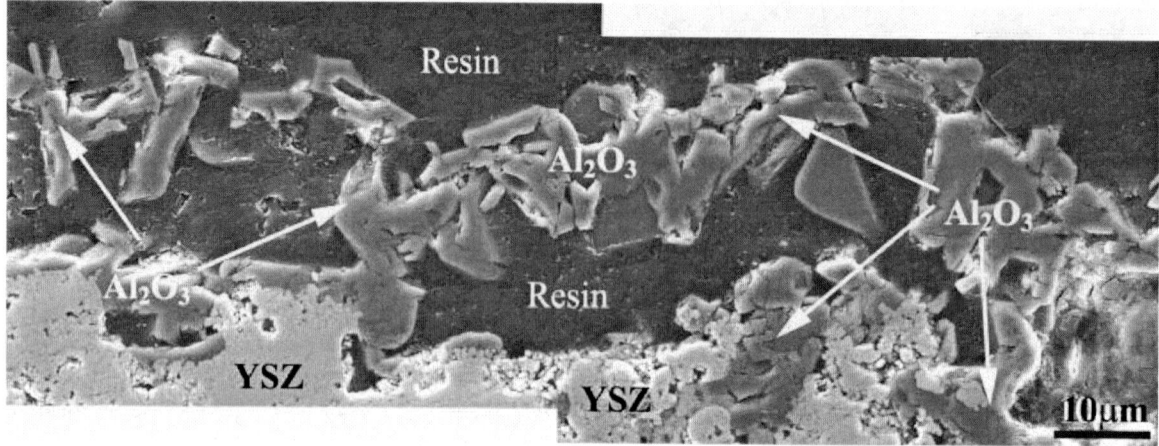

Fig. 11. Cross-section scanning electron microscope photographs of yttria-stabilized zirconia (YSZ)/Al$_2$O$_3$ coating after 10 h of exposure to the molten salt of Na$_2$SO$_4$+15 wt% V$_2$O$_5$.

limited within the YSZ layer. In contrast, sodium was penetrated into the bond coat of the monolithic YSZ coating. In order to evaluate the content of infiltrated sodium and vanadium in the YSZ layer, SIMS analysis was focused on the YSZ layer. The SIMS spectrum was obtained from the spot with a raster size of 100 μm × 100 μm. The intensity ratio of the impurity peak to the ^{90}Zr peak, I_M/I_{Zr} (M = ^{51}V, ^{23}Na, ^{89}Y, ^{56}Fe, ^{52}Cr), was used to evaluate the extent of infiltration of sodium and vanadium into the YSZ layer. Totally 15 random spots were analyzed and the average value is listed in Table I. It is evident that the contents of sodium and vanadium in the composite YSZ/Al$_2$O$_3$ coating were much lower than those in the monolithic YSZ coating without the Al$_2$O$_3$ overlay. This indicated that the Al$_2$O$_3$ overlay acted as a barrier against the infiltration of sodium and vanadium into the YSZ layer. During corrosion exposure, chromium and iron diffused into the YSZ layer from the bond coat. The Cr and Fe contents in the YSZ layer with the Al$_2$O$_3$ overlay were also lower than those in the YSZ layer without the Al$_2$O$_3$ overlay. However, the Y content in the YSZ layer exhibited no difference between the composite YSZ/Al$_2$O$_3$ coating and monolithic YSZ coating. This indicated that yttria still remained in the YSZ layer despite the transition of Y$_2$O$_3$ to YVO$_4$.

IV. Discussion

In agreement with previous studies,[3–11] the present work showed that YSZ was susceptible to the attack of sulfate–vanadate salt. Failure of the YSZ coating was ascribed to the infiltration of molten salt into the YSZ coating along pores and cracks in the YSZ and the subsequent reaction of molten salt with Y$_2$O$_3$, leading to the destabilization of YSZ coating (Figs. 3 and 6). The degradation mechanism of YSZ was addressed in the previous investigations.[3–11] Therefore, the present study is focused on the hot corrosion mechanism of the composite YSZ/Al$_2$O$_3$ system.

(1) Hot Corrosion in Pure Sulfate Melt

Lawson *et al.*[26] have reported that intergranular corrosion can occur if an alumina-silicate phase existed along the grain boundaries of Al$_2$O$_3$ grains. The preferential dissolution of grain boundary phase can be excluded in the present study because of the absence of the impurities along the grain boundaries. Chen *et al.*[28] have studied the hot corrosion behavior of the plasma-sprayed Al$_2$O$_3$ coating when exposed in the molten Na$_2$SO$_4$ at 850°C. They have found that Al$_2$O$_3$ reacted with molten Na$_2$SO$_4$ to form NaAlO$_2$. In the present investigation, however, no new corrosion products were detected by XRD and XPS after hot corrosion testing of the Al$_2$O$_3$ overlay in the molten salt of pure Na$_2$SO$_4$. In addition, the Al$_2$O$_3$ overlay did not

show phase transformation and dissolution after exposure (Figs. 1 and 8).

At 950°C, there exists equilibrium for Na$_2$SO$_4$ melt as expressed by[32]

$$Na_2SO_4 \rightarrow Na_2O + SO_3 \qquad (2)$$

$$\log a_{Na_2O} + \log P_{SO_3} = \frac{\Delta G°(l)}{2.303RT} = -15.0 \quad (\text{at } 950°C)$$

This melt exhibits acid/base chemistry, with the basic component Na$_2$O and the acidic component SO$_3$. As shown by Eq. (1), as the activity of Na$_2$O, a_{Na_2O}, increases, the pressure of SO$_3$, P_{SO_3}, decreases and vice versa. During hot corrosion, the activity of Na$_2$O or the pressure of SO$_3$ in the molten salt determines the type and extent of reaction.[33–35] Hot corrosion may involve fluxing of oxides as either basic or acidic solutes in the molten salt. In case of the high Na$_2$O activity, alumina can react with Na$_2$O and dissolve in the molten sulfate by basic fluxing,[33] which can be given by

$$Na_2SO_4(l) + Al_2O_3(s) \rightarrow 2NaAlO_2(l) + SO_3(g) \qquad (3)$$

In case of the low Na$_2$O activity and the correspondingly high P_{SO_3}, alumina can dissolve by acidic fluxing,[33] which is shown by

$$Na_2SO_4(l) + Al_2O_3(s) \rightarrow Al_2(SO_4)_3(l) + SO_3(g) \qquad (4)$$

If the activity of Na$_2$O and the pressure of SO$_3$ are within the intermediate range, alumina is stable and negligible solubility in the molten sulfate occurs. Keeping this in mind, it is not surprising that in our experiment no evident corrosion took place in the Al$_2$O$_3$ overlay.

(2) Hot Corrosion in Sulfate–Vanadate Melt

For the mixed sulfate–vanadate salt, V$_2$O$_5$ would be first melted upon heating because of its lower melting point (690°C). NaVO$_3$ could be formed by the reaction of molten Na$_2$SO$_4$ with V$_2$O$_5$ at a testing temperature of 950°C

$$Na_2SO_4(l) + V_2O_5(l) \rightarrow 2NaVO_3(l) + SO_3(g)$$
$$(\Delta G° = -11.9 \text{ kJ/mol at } 950°C) \qquad (5)$$

As a result, when the YSZ/Al$_2$O$_3$ system was exposed to the Na$_2$SO$_4$+V$_2$O$_5$ salt melt, the NaVO$_3$–Na$_2$SO$_4$ melt covered the surface of the Al$_2$O$_3$ overlay. As reported in the previous literature,[34,36] the presence of NaVO$_3$ increases the acidic solubility. Consequently, it is possible that metal oxides, such as Al$_2$O$_3$ and

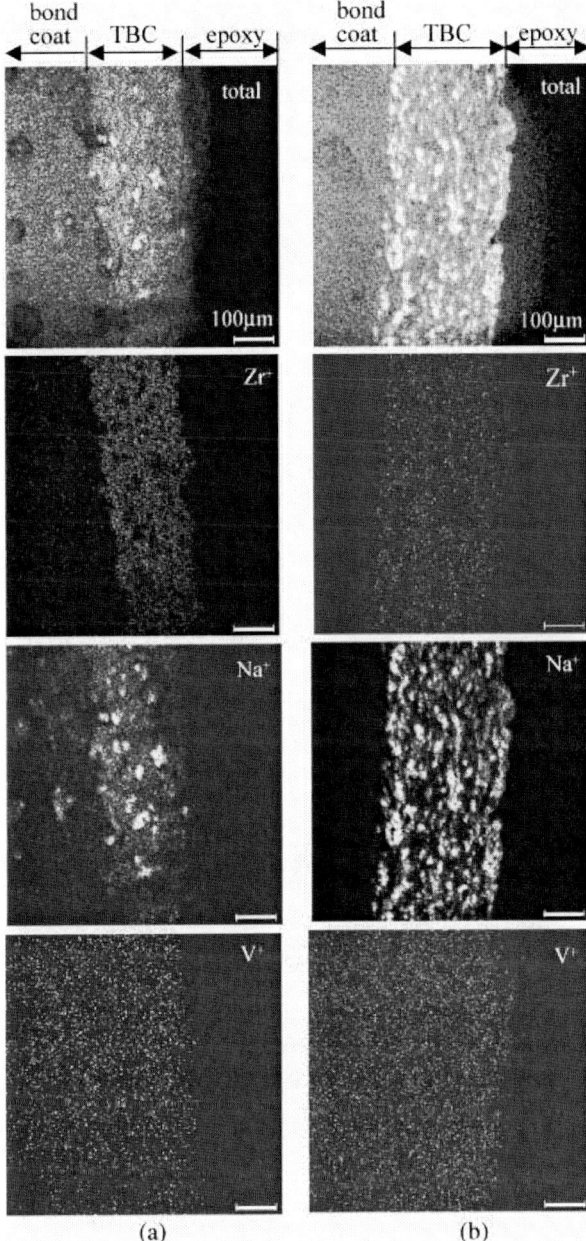

Fig. 12. Secondary ion mass spectrometry images of the cross-section of thermal-barrier coating coatings after exposure to mixed salt melt of $Na_2SO_4 + 5\ wt\%V_2O_5$: (a) the monolithic yttria-stabilized zirconia (YSZ) coating, (b) the composite YSZ/Al_2O_3 coating.

Y_2O_3, react with $NaVO_3$ to be dissolved by acidic flux, which is expressed by

$$Al_2O_3(s) + 2NaVO_3(l) \rightarrow 2AlVO_4(l) + Na_2O(l) \qquad (6)$$

$$Y_2O_3(s) + 2NaVO_3(l) \rightarrow 2YVO_4(s) + Na_2O(l) \qquad (7)$$

Table I.   Quantitative Comparison of the Content of Elements in the YSZ Layer

	I_V/I_{Zr}	I_{Na}/I_{Zr}	I_Y/I_{Zr}	I_{Cr}/I_{Zr}	I_{Fe}/I_{Zr}
YSZ without overlay	0.10	3.54	0.21	2.72	0.25
Composite YSZ/Al_2O_3	0.016	0.53	0.21	0.02	0.017

YSZ, yttria-stabilized zirconia.

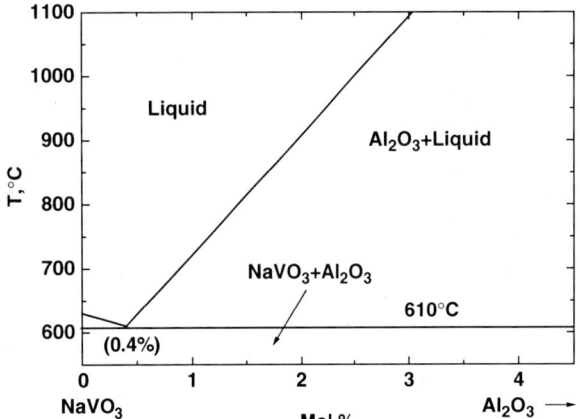

Fig. 13. $NaVO_3$–Al_2O_3 Phase diagram showing the formation of liquid phase at a temperature above 610°C (after Klinkova and Ukshe[37]).

However, $AlVO_4$ was not found after the composite Al_2O_3/YSZ coating had been exposed to the molten sulfate–vanadate salt for 10 h. This suggests that the degradation mechanism of the composite Al_2O_3/YSZ coating cannot be interpreted by Eqs. (6) and (7).

If we take a look at the phase diagram of $NaVO_3$–Al_2O_3 as shown in Fig. 13,[37] we may find the reason for degradation of Al_2O_3 during exposure to the molten sulfate–vanadate salt. The $NaVO_3$–Al_2O_3 phase diagram demonstrates that a liquid phase containing Al, Na, V, and O will be formed when the temperature exceeds 610°C. Therefore, it is deduced that the formation of low-melting liquid phase was responsible for the failure of the Al_2O_3 overlay during hot corrosion testing. This notion can be confirmed by the further evidence in the present study. For instance, the surface of Al_2O_3 overlay exhibited a "cauliflower" type of morphology before hot corrosion testing (Fig. 7(a)), whereas the surface after hot corrosion testing was characteristic of acicular-shaped crystals and faceted crystals (Fig. 9). In addition, a gap between the crystals was observed (Fig. 9). In particular, the pores and cracks within the YSZ coating were filled with Al_2O_3 (Fig. 10). This indicates that a liquid melt flowed along the pores and cracks.

When pure Al_2O_3 co-existed with $NaVO_3$ at 950°C, a low-melting liquid phase containing Al, Na, V, and O was formed. The continuous formation of the liquid phase can result in loss of integrity of the surface. Upon the cooling process, the liquid phase was then decomposed and Al_2O_3 crystallized to form acicular-shaped crystals on the surface of the Al_2O_3 overlay (Fig. 9). With this in mind, it was not surprising that the surface morphology of the overlay (region "A" in Fig. 10(a)) was different from the subsurface region (region "B" in Fig. 10(a)). Further attack by molten salt etched the subsurface region and caused the grooves. Then the liquid phase was infiltrated along the grooves and penetrated into pores and cracks within the YSZ coating. During the cooling process, Al_2O_3 was crystallized from liquid phase and trapped in the pores and cracks within the YSZ coating. This was confirmed by SEM observation and EDS analysis as shown in Fig. 10.

The solubility of Al_2O_3 in solvent $NaVO_3$ was estimated to be about 2.2 mol% according to the phase diagrams.[37] When the molten salt contained only 5 wt% V_2O_5, the amount of alumina that could be dissolved by molten salt was limited because of limited V_2O_5 content. When the V_2O_5 content in the molten salt increased to 15 wt%, however, more Al_2O_3 could be dissolved. Consequently, continuous formation of the liquid phase eventually resulted in loss of integrity of the surface and removal of dense alumina overlay. The outermost surface of YSZ could be fully covered by the liquid phase at 950°C. When this occurred, the Al_2O_3 overlay was no longer able to protect the inner YSZ layer. The liquid phase containing molten salts penetrated freely into the YSZ layer. As a result, a significant M-phase of ZrO_2

was formed in this case. After cooling, Al_2O_3 was crystallized to form large α-Al_2O_3 acicular-shaped crystals (Fig. 9(b)).

In short, before corrosion exposure alumina exhibited the γ-phase structure, which was thermodynamically metastable. During corrosion exposure, γ-Al_2O_3 reacted with vanadate to form a low-melting liquid eutectic phase. After corrosion exposure, α-Al_2O_3, which was thermodynamically stable, was crystallized from the liquid phase upon cooling (Fig. 2).

Dissolution of Al_2O_3 in the molten salt depended on the V_2O_5 content in the mixture. In the current hot corrosion testing, the V_2O_5 content is much higher than that in real fuel (usually, the V_2O_5 content in fuels is in the <100 ppm range.[38] After combustion, the V_2O_5 content in the ash deposit is expected to be much higher than 100 ppm, which is more relevant to corrosion). Although part of the Al_2O_3 overlay was degraded by the formation of low melting-point liquid phase, the Al_2O_3 overlay in the composite Al_2O_3/YSZ system acted as a barrier layer against the infiltration of the molten salt into the YSZ coating (as shown in Table I). Destabilization of ZrO_2 was considerably restrained when the amount of V_2O_5 in the molten salt was less than 5 wt% (Fig. 3).

V. Conclusions

(1) Both the monolithic YSZ coating and the composite YSZ/Al_2O_3 system exhibited good hot corrosion resistance in the molten salt of pure Na_2SO_4.

(2) During exposure to the molten sulfate–vanadate salt mixture, the monolithic YSZ coating reacted with V_2O_5 to form YVO_4, leading to a significant structural destabilization of ZrO_2.

(3) During exposure to the molten sulfate–vanadate salt mixture, γ-Al_2O_3 reacted with vanadate to form a low-melting liquid eutectic phase, leading to dissolution of Al_2O_3 in molten $NaVO_3$. Degradation of the Al_2O_3 overlay deteriorated with an increase in the V_2O_5 content in the molten salt mixture.

(4) The Al_2O_3 overlay acted as a barrier against the infiltration of the molten salt into the YSZ coating when the V_2O_5 content in salt was lower than about 5 wt%. This considerably restrained the destabilization of ZrO_2.

Acknowledgment

We are grateful to Keck-II, NU*ANCE* at Northwestern University for XPS and SIMS measurements.

References

[1]M. J. Stiger, N. M. Yanar, M. G. Topping, F. S. Pettit, and G. H. Meier, "Thermal Barrier Coatings for the 21st Century," *Z. Metallkd.*, **90** [12] 1069–78 (1999).

[2]L. Singheiser, R. Steinbrech, W. J. Quadakkers, and R. Herzog, "Failure Aspects of Thermal Barrier Coatings," *Mat. High Temp.*, **18** [4] 249–59 (2001).

[3]I. Gurrappa, "Hot Corrosion of Protective Coatings," *Mat. Manuf. Process*, **15** [5] 761–73 (2000).

[4]I. Gurrappa, "Thermal Barrier Coating for Hot Corrosion Resistance of CM 247 LC Superalloy," *J. Mater. Sci. Lett.*, **17**, 1267–9 (1998).

[5]R. L. Jones, "Thermogravimetric Study of the 800 Degree Reaction of Zirconia Stabilizing Oxides with SO_3–$NaVO_3$," *J. Electrochem. Soc.*, **139**, 2794–9 (1992).

[6]K. L. Luthra and H. S. Spacil, "Impurity Deposits in Gas-Turbines from Fuels Containing Sodium and Vanadium," *J. Electrochem. Soc.*, **129** [3] 649–56 (1982).

[7]N. S. Bornstein and W. P. Allen, "The Chemistry of Sulfidation Corrosion—Revisited," *Mater. Sci. Forum*, **127**, 251–4 (1997).

[8]A. S. Nagelberg, "Destabilization of Yttria-Stabilized Zirconia Induced by Molten Sodium Vanadate–Sodium Sulfate Melts," *J. Electrochem. Soc.*, **132** [10] 2502–7 (1985).

[9]R. L. Jones, C. E. Williams, and S. R. Jones, "Reaction of Vanadium Compounds with Ceramic Oxides," *J. Electrochem. Soc.*, **133** [1] 227–30 (1986).

[10]R. L. Jones, "High Temperature Vanadate Corrosion of Yttria-Stabilized Zirconia Coatings on Mild Steel," *Surf. Coat. Technol.*, **37**, 271–84 (1989).

[11]R. L. Jones and C. E. Williams, "Hot Corrosion Studies of Zirconia Ceramics," *Surf. Coat. Technol.*, **32**, 349–58 (1987).

[12]D. W. Susnitzky, W. Hertl, and C. B. Carter, "Destabilization of Zirconia Thermal Barriers in the Presence of V_2O_5," *J. Am. Ceram. Soc.*, **71** [11] 992–1004 (1988).

[13]R. A. Miller and C. E. Lowell, "Failure Mechanism of Thermal Barrier Coatings Exposed to Elevated Temperature," *Thin Solid Films*, **95**, 265–73 (1982).

[14]R. L. Jones, "India as a Hot Corrosion-Resistant Stabilizer for Zirconia," *J. Am. Ceram. Soc.*, **75**, 1818–21 (1992).

[15]R. L. Jones and R. F. Reidy, "Vanadate Hot Corrosion Behavior of India, Yttria-Stabilized Zirconia," *J. Am. Ceram. Soc.*, **76** [10] 2660–2 (1993).

[16]R. L. Jones, "Scandia-Stabilized Zirconia for Resistance to Molten Vanadate–Sulfate Corrosion," *Surf. Coat. Technol.*, **39/40**, 89–96 (1989).

[17]S. A. Muqtader, R. K. Sidhu, E. Nagabhushan, K. Muzaffaruddin, and S. G. Samdani, "Destabilization Behavior of Ceria-Stabilized Tetragonal Zirconia Polycrystals by Sodium Sulphate and Vanadium Oxide Melts," *J. Mater. Sci. Lett.*, **12**, 831–3 (1993).

[18]S. Raghavan and M. J. Mayo, "The Hot Corrosion Resistance of 20 mol% $YTaO_4$ Stabilized Tetragonal Zirconia and 14 mol% Ta_2O_5 Stabilized Orthorhombic Zirconia for Thermal Barrier Coating Applications," *Surf. Coat. Technol.*, **160**, 187–96 (2002).

[19]A. Petitbon, L. Boquet, and D. Delsart, "Laser Surface Sealing and Strengthening of Zirconia Coatings," *Surf. Coat. Technol.*, **49**, 57–61 (1991).

[20]Z. Liu, "Crack-Free Surface Sealing of Plasma Sprayed Ceramic Coating Using an Excimer Laser," *Appl. Surf. Sci.*, **186**, 135–9 (2002).

[21]S. Ahmaniemi, P. Vuoristo, and T. Mantyla, "Improved Sealing Treatment for Thick Thermal Barrier Coatings," *Surf. Coat. Technol.*, **151–152**, 412–7 (2002).

[22]T. Mantyla, P. Vuoristo, and P. Kettunen, "Chemical Vapor Deposition Densification of Plasma-Sprayed Oxide Coatings," *Thin Solid Films*, **118**, 437–44 (1984).

[23]I. Berezin and T. Troczynski, "Surface Modification of Zirconia Thermal Barrier Coatings," *J. Mater. Sci. Lett.*, **15**, 214–8 (1996).

[24]T. Troczynski, Q. Yang, and G. John, "Post-Deposition Treatment of Zirconia Thermal Barrier Coatings Using Sol–Gel Alumina," *J. Therm. Spray Technol.*, **8** [2] 229–34 (1999).

[25]M. Vippola, P. Vuorinen, T. Vuoristo, J. Lepisto, and T. Mantyla, "Thermal Analysis of Plasma Sprayed Oxide Coatings Sealed with Aluminum Phosphate," *J. Euro. Ceram. Soc.*, **22**, 1937–46 (2002).

[26]M. G. Lawson, F. S. Pettit, and J. R. Blachere, "Hot Corrosion of Al_2O_3," *J. Mater. Res.*, **8**, 1964–71 (1993).

[27]Z. Tang, F. Wang, and W. Wu, "Effect of Al_2O_3 and Enamel Coatings on 900°C Oxidation and Hot Corrosion Behaviors of Gamma-TiAl," *Mater. Sci. Eng. A*, **276**, 70–5 (2000).

[28]H. C. Chen, Z. Y. Liu, and Y. C. Chuang, "Degradation of Plasma-Sprayed Alumina and Zirconia Coatings on Stainless Steel During Thermal Cycling and Hot Corrosion," *Thin Solid Films*, **223**, 56–64 (1992).

[29]P. Ramaswamy, S. Seetharamu, K. B. R. Varma, and K. J. Rao, "Al_2O_3–ZrO_2 Composite Coatings for Thermal Barrier Applications," *Comp. Sci. Technol.*, **57**, 81–9 (1997).

[30]I. Levin and D. D. Brandon, "Metastable Alumina Polymorphs: Crystal Structures and Transition Sequences," *J. Am. Ceram. Soc.*, **81** [8] 1995–2012 (1998).

[31]J. Moulder, W. Stickle, P. Sobel, and E. Bomben, *Handbook of X-Ray Photoelectron Spectroscopy*. Physical Electronics, Eden Prairie, MN, 1995.

[32] "JANAF Thermochemical Tables. 3rd edition," *J. Phys. Chem. Ref. Data*, [Suppl. 1] 14 (1985).

[33]P. D. Jose, D. K. Gupta, and R. Rapp, "Solubility of α-Al_2O_3 in Fused Na_2SO_4 at 1200 K," *J. Electrochem. Soc.*, **132** [3] 735–7 (1985).

[34]R. A. Rapp and Y. S. Zhang, "Hot Corrosion of Materials: Fundamental Studies," *Journal of Metals*, **46** [12] 47–55 (1994).

[35]M. G. Lawson, H. R. Kim, F. S. Pettit, and J. R. Blachere, "Hot Corrosion of Silica," *J. Am. Ceram. Soc.*, **73** [4] 989–95 (1990).

[36]Y. S. Hwang and R. R. Rapp, "Thermochemistry and Solubilities of Oxides in Sodium Sulfate–Vanadate Solutions," *Corrosion*, **45** [1] 933–7 (1989).

[37]L. A. Klinkova and E. A. Ukshe, "Solution of Corundum in Fused Vanadates," *Russ. J. Inorg. Chem.*, **20** [2] 799–803 (1975).

[38]N. S. Jacobson, "Corrosion of Silicon-Based Ceramics in Combustion Environments," *J. Am. Ceram. Soc.*, **76** [1] 3–28 (1993). □

Microstructure–Property Correlations in Industrial Thermal Barrier Coatings

Anand A. Kulkarni,* Allen Goland, and Herbert Herman

Department of Materials Science and Engineering, State University of New York, Stony Brook, New York   11794

Andrew J. Allen,* Jan Ilavsky,* and Gabrielle G. Long*

National Institute of Standards and Technology, Gaithersburg, Maryland   20899

Curtis A. Johnson* and Jim A. Ruud*

General Electric Corporate Research Division, Schenectady, New York   12309

This paper describes the results from multidisciplinary characterization/scattering techniques used for the quantitative characterization of industrial thermal barrier coating (TBC) systems used in advanced gas turbines. While past requirements for TBCs primarily addressed the function of insulation/life extension of the metallic components, new demands necessitate a requirement for spallation resistance/strain tolerance, i.e., prime reliance, on the part of the TBC. In an extensive effort to incorporate these TBCs, a design-of-experiment approach was undertaken to develop tailored coating properties by processing under varied conditions. Efforts focusing on achieving durable/high-performance coatings led to dense vertically cracked (DVC) TBCs, exhibiting quasi-columnar microstructures approximating electron-beam physical-vapor-deposited (EB-PVD) coatings. Quantitative representation of the microstructural features in these vastly different coatings is obtained, in terms of porosity, opening dimensions, orientation, morphologies, and pore size distribution, by means of small-angle neutron scattering (SANS) and ultra-small-angle X-ray scattering (USAXS) studies. Such comprehensive characterization, coupled with elastic modulus and thermal conductivity measurements of the coatings, help establish relationships between microstructure and properties in a systematic manner.

I. Introduction

COMPREHENSIVE efforts have been under way to incorporate prime-reliant and energy-efficient ceramic thermal barrier coatings (TBCs) into advanced gas turbine and diesel engine components. TBCs provide insulation to metallic structures in the hot section of land-based/aero-turbine engines and offer three important benefits: (1) increased operating temperature of the engine and therefore enhanced efficiency; (2) enhanced durability and extended life of metallic components subjected to high temperatures and high stresses; and (3) reduced cooling requirements to metallic components.[1–3] Present-day TBCs are comprised of a two-layer coating system on a superalloy turbine blade substrate. The materials of interest for such systems are MCrAlY (where M is Ni, Co, etc.) alloys or Pt–Al-based oxidation-resistant bond-coats followed by an yttria-stabilized zirconia (YSZ) TBC. The bond-coat is typically deposited using atmospheric or low-pressure plasma spray while the topcoat is deposited using either electron-beam physical vapor deposition (EB-PVD) or plasma spray deposition.[4] Each technique has merits for TBC applications depending on size scale, performance requirements and cost. The EB-PVD process produces a unique columnar microstructure with wide intercolumnar spacing, thus providing superior strain tolerance and thermal shock resistance; hence it is used on rotating airfoils for significant lifetime enhancements. Because plasma-sprayed TBCs produce splat-based layered structures and offer advantages in terms of thermal insulation and economics, they are widely used in combustion chambers. However, such coatings are not considered durable on rotating airfoils due to lack of in-plane compliance that leads to premature delamination failure during thermomechanical cycling under oxidative conditions.[5] This is mainly due to a myriad array of process-related defects in the form of interlamellar pores, cracks, and gas porosity. These imperfections can, to a certain extent, offer beneficial attributes, such as compliance to the coating, enabling high-temperature thermal cycling and reduction of thermal conduction due to phonon scattering, etc.[6] Understanding the characteristics of these defects and their control is critical for the enhancement of the system's performance and reliability.

In plasma spraying, feedstock material is melted and accelerated to high velocities. The resultant melt impinges on the substrate and rapidly solidifies to form a "splat" (a flattened particle). The deposit develops by successive impingement and interbonding among the splats. The microstructure consists of the splats, separated by interlamellar pores resulting from rapid solidification of the lamellae, very fine voids formed by incomplete intersplat contact or around unmelted particles, and cracks due to thermal stresses and tensile quenching stress relaxation.[7,8] Taken together, these imperfections introduce a measurable porosity in the coatings. The cracks increase the compliance of the coating and hence enhance the thermal shock resistance. This microcrack-related feature of YSZ might be exploited to advantage through generation of varying stress states during deposition. Such a microstructure with controlled micro/macrocracks can yield a compliant TBC coating, which is considered to be beneficial relative to strain tolerance, spallation resistance, and component life during service.[9] Recent efforts on processing/deposition methodologies have

R. Hannink—contributing editor

Manuscript No. 10162. Received April 29, 3004; approved January 24, 2004.
This research was supported by the National Science Foundation (NSF) MRSEC program at the SUNY Stony Brook under Grant No. DMR-0080021. This work used facilities supported in part by the National Science Foundation under Agreement No. DMR-9986442. The UNICAT facility at the Advanced Photon Source (APS) is supported by the University of Illinois at Urbana-Champaign, Materials Research Laboratory (U.S. Department of Energy (DOE), the State of Illinois-IBHE-HECA, and the NSF), the Oak Ridge National Laboratory (U.S. DOE under contract with UT-Battelle LLC), the National Institute of Standards and Technology (U.S. Department of Commerce), and UOP LLC. The APS is supported by the U.S. DOE, Basic Energy Sciences, Office of Science under contract No. W-31-109-ENG-38.
*Member, American Ceramic Society.

allowed the development of plasma spray processes that produce dense vertically cracked (DVC) TBC microstructures. While the details of the DVC microstructures remain proprietary,[10] studies have clarified that these unique microstructures can be achieved under conditions involving plasma spraying at high power, high powder flow rates, carefully controlled spray distances, and relatively high substrate temperatures (e.g., >400°C).[11–13]

In this paper, advanced scattering techniques to characterize and to compare microstructures of conventional plasma-sprayed coatings with DVCs are explored. Small-angle neutron scattering (SANS) methods have been previously used to characterize and quantify the anisotropic nature of the thermal sprayed ceramic coatings.[14,15] Ultra-small-angle X-ray scattering (USAXS) studies have been conducted in concert to provide quantitative microstructure maps as a function of size of scatterers. This information obtained on porosity, pore size distributions, and pore orientations provides an enhanced understanding that can lead to improvements in coating behavior during service.

II. Experimental Procedure

(1) Deposit Properties

Four coatings were studied, each deposited under incrementally different spray conditions onto a NiCrAlY bond-coated superalloy (IN 718) (General Electric, Schenectady, NY).[†] The coatings[1] are labeled GE-1 to GE-4, going from conventional layered structure to DVC, respectively. The coating thickness was ~450 μm in each case. Free-standing coatings were used for porosity determinations, thermal conductivity measurements, and SANS studies. Microstructural evaluation using optical and scanning electron microscopy (SEM) were also conducted. Surface-connected porosity was measured by mercury intrusion porosimetry (MIP) using an Autoscan 33 porosimeter (Quantachrome Corp., Bayton Beach, FL). The total porosity content was determined using the precision density (PD) method, where mass-over-volume ratios were obtained for a cut rectilinear specimen. The technique gives a fractional density (or porosity) uncertainty of standard deviation, $\pm 1\%$, based on the average of 10 measured identical specimens and an assumed theoretical density of 6 g/cm^3. Thermal conductivity measurements were conducted on a 12.5 mm (0.5 in.) diameter disk, coated with carbon, using a laser flash thermal diffusivity instrument (Netzsch Corp., Boston, MA). Elastic modulus measurements and USAXS studies were conducted on the coatings bonded to the substrate. Elastic modulus measurements using depth-sensitive indentation studies were conducted with a Nanotest 600 instrument (Micromaterials, Inc., Cambridge, U.K.) with a 1/16 in. WC–Co spherical indenter with a maximum load of 10 N. The instrument enables a basic load/displacement curve to be obtained, or multiple partial load/unload cycles to be performed. This allows hardness and elastic modulus values to be measured as a function of the load/contact stress. The indentation procedure used usually consists of 10–15 loading/unloading cycles. Also, modulus measurements were conducted on polished top-surface (out-of-plane) and cross sections (in-plane) to examine the anisotropy in the coatings.

(2) Small-Angle Neutron Scattering

SANS studies were conducted on the NIST NSF 30 m NG3 SANS instrument at the Cold Neutron Research Facility at the National Institute of Standards and Technology, Gaithersburg, MD. The use of large sample-to-detector distances on this instrument provides a powerful opportunity to measure the microstructure of thin coatings, which are more representative of industrial applications. This will be the case, provided enough multiple scattering exists for the multiple small-angle neutron scattering (MSANS) formalism to apply. A monochromatic beam of cold long-wavelength neutrons passes through the specimen in transmission geometry and the scattered neutrons are recorded on a two-dimensional detector. The details of the experiment are described elsewhere.[14,15] The scattering occurs at the void–grain interface due to differences in scattering-length density between the material and the pores. The experiment involved two types of measurements, the first being anisotropic Porod scattering. It is advantageous to measure the Porod scattering since it amplifies the microstructural anisotropies. However, on orientational averaging of the Porod scattering from the sample, one can obtain the total void surface area per unit sample volume, independent of the precise void morphology. The fine features in the microstructure are major contributors to this deduced surface area. The second type of measurement is the anisotropic MSANS, which involves a measurement of the beam-broadening due to anisotropic multiple scattering at long neutron wavelengths (1–1.8 nm). The multiple scattering usually arises from the coarse features in the microstructure. The MSANS beam broadening versus wavelength for two sample orientations, with the incident beam out-of-plane (in the spray direction) and in-plane (i.e., in the substrate plane), yields information on microstructural anisotropy. The sector-averaged anisotropic MSANS data also provide microstructural orientation information, as discussed in detail elsewhere.[15,16]

The goal is to obtain quantitative information on each population of the porosity. This can be obtained by combining MSANS measurements for different sample orientations, anisotropic Porod surface area distributions, and the total porosity found from precision density measurements. To acquire a quantitative delineation of the three void components (interlamellar pores, intrasplat cracks, and globular pores) in terms of their porosity contributions, dimensionality, and orientation distribution, the following four constraints are imposed in the MSANS analysis:

(1) The component porosities are consistent with the total porosity obtained using precision density measurements.

(2) The component surface areas are consistent with the total surface area obtained from anisotropic Porod scattering experiments.

(3) The circularly averaged MSANS beam broadening versus wavelength model predictions are consistent with the experimental data for both orientations: out-of-plane (spray direction) and in-plane (orthogonal direction).

(4) The predicted MSANS anisotropy (perpendicular to the substrate) is consistent with that observed experimentally.

With these constraints, it is possible to determine the volume-weighted mean-opening dimensions of the intrasplat cracks and interlamellar pores, their orientation distributions with respect to the spray direction, and the diameters of the globular pores. Porosity and surface area contributions may also be distinguished.

(3) Ultra-Small-Angle X-ray Scattering

Ultra-small-angle X-ray scattering (USAXS) studies were conducted on the UNICAT beam line 33-ID at the Advanced Photon Source, Argonne National Laboratory, Argonne, IL. This instrument uses Bonse–Hart double-crystal optics[17] to extend the range of SAXS to low-scattering vectors, Q, where $Q = (4\pi/\lambda)(\sin\theta)$ and 2θ is the scattering angle. In small-angle scattering (SANS and USAXS) studies, the anisotropic structural information is measured along the direction of Q. In a modified form of the standard USAXS experiment, a finely collimated and highly monochromatic X-ray beam (prepared using horizontally and vertically diffracting crystals) is incident on the specimen in transmission geometry and scattered intensity is measured. Use of orthogonal diffracting crystals removes the intrinsic slit-smeared geometry of the standard USAXS experiment. The details of the experiment are described elsewhere.[18] The X-ray energy was 17 keV to penetrate through highly absorbing YSZ coatings. Two methods are usually combined. In the first, the scattered intensity is measured as a function of Q for each orientation of the azimuthal angle, α. Alternatively, the scattered intensity at a particular Q is measured as a function of α by rotating the sample in the beam. The different anisotropies in the scattering at different Q are related to the

---

[†]Information on commercial products is given for completeness and does not constitute or imply their endorsement by the National Institute of Standards and Technology.

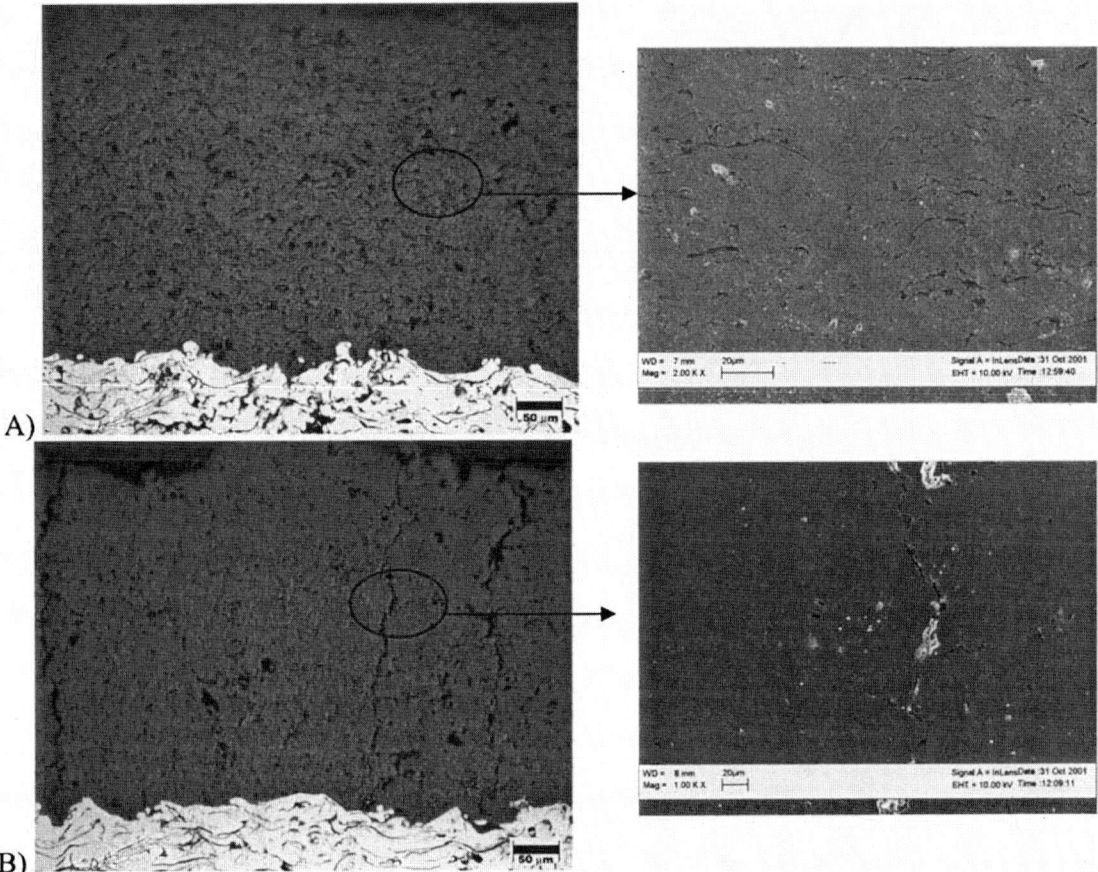

Fig. 1. Optical microscopy images of coatings with high-magnification SEM micrographs (on the right) showing detailed microstructure: (A) GE-1 and (B) GE-4.

different anisotropies of the microstructure at different length scales, thus giving a quantitative map as a function of the sizes of the scattering populations. However, the **Q** resolution of the instrument limits detection of sizes to those below about 1.5 μm in diameter.

III. Results and Discussion

The cross-sectional micrographs of the coatings sprayed at the endpoints of incrementally varied conditions (GE-1 and GE-4) are shown in Fig. 1. Figure 1(A) shows a typical plasma-sprayed coating with a layered structure. The interlamellar porosity, which results from poor adhesion between splats, is evident in the inset. Figure 1(B) shows a significantly different structure with vertical macrocracks. These cracks may be beneficial from the point of view of strain tolerance and component life during service. Also, a dense coating structure is observed in the magnified image on the right.

The measured coating properties for the four samples are presented in Table I: density from the precision density method;

porosity from precision density; porosity from mercury intrusion porosimetry (MIP); thermal diffusivity; and conductivity values. The density increases from GE-1 to GE-4 and the MIP surface-connected porosity decreases from GE-1 to GE-3 and then increases for the DVC microstructures (GE-4). This is due to macrocracks being accounted as the surface-connected porosity. The thermal diffusivity and conductivity values measured using the laser flash technique show an inverse relationship with porosity. The elastic modulus measured using depth-sensitive indentation is presented in Fig. 2. In-plane and out-of-plane measurements show anisotropy in these coatings. It is observed that the out-of-plane (top surface) modulus increases consistently similar to the trend of thermal conductivity of the coatings. The in-plane (cross section) modulus, which is sensitive to the crack networks, increases except for the DVC case.

(1) SANS Results

SANS results along with MSANS model fits are presented in this section. The anisotropic surface area derived from the Porod scattering regime (Fig. 3) shows the orientation dependence of the

Table I. Coating Property Measurements

Coating	Density (g/cm³)	PD porosity (%)	MIP porosity (%)	Thermal conductivity (W/(m·K))	Thermal diffusivity (cm²/s)
GE-1	5.25 ± 0.11	13.3 ± 0.7	14 ± 0.3	1 ± 0.1	0.005 ± 0.0006
GE-2	5.38 ± 0.05	11 ± 0.7	10 ± 0.4	1.2 ± 0.04	0.009 ± 0.0004
GE-3	5.42 ± 0.18	10.3 ± 0.6	7 ± 0.6	1.7 ± 0.03	0.012 ± 0.0004
GE-4	5.48 ± 0.28	9 ± 0.5	9 ± 0.5	1.9 ± 0.06	0.015 ± 0.0007

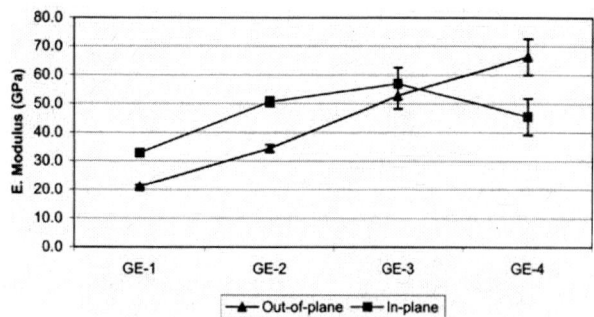

Fig. 2. Comparison of in-plane and out-of-plane elastic properties showing anisotropy in the coatings. Error bars indicate standard deviations for each measurement.

scatterers. Since the scattering data are represented in Fourier space, the contribution from horizontal components is observed vertically and vice versa. It is seen in Fig. 3(A) that the scattering is dominated by horizontal elements (interlamellar pores) in the traditionally plasma-sprayed GE-1 coating. The scattering appears to be crack-dominated in Fig. 3(B) for the GE-4 (DVC) coating.

Since the coatings are deposited by plasma spraying, the microstructure develops by splat–splat layering, suggesting the presence of interlamellar pores in the coating, even though they are not well resolved in the SEM micrograph of Fig. 1(B). The total surface area from the Porod scattering, combined with density/porosity measurements from precision density measurements and MSANS model parameters, together quantify the many details of the microstructure.

Using the constraints in the MSANS model, the results for porosity contributions and mean opening dimensions obtained, are summarized in Table II. Estimated uncertainties are given in parentheses. The incremental changes in the anisotropic orientation distribution of the intrasplat cracks and the interlamellar pores (which are significantly different in the two extreme cases of GE-1 and GE-4) are also shown below in Table II. The orientation distributions for the interlamellar-pore 1/5 aspect ratio spheroidal elements and the intrasplat-crack 1/10 aspect ratio elements are separately parameterized in terms of the relative probabilities of finding the normal to these elements within the range 0–30° from the spray direction, 30–60° from the spray direction, and 60–90° from the spray direction. To obtain good MSANS model fits that satisfy all the constraints, intrasplat cracks are found to be predominantly perpendicular to the substrate (spheroidal-elements normals 60–90° from the spray direction), and the interlamellar pores are found to be predominantly parallel to the substrate

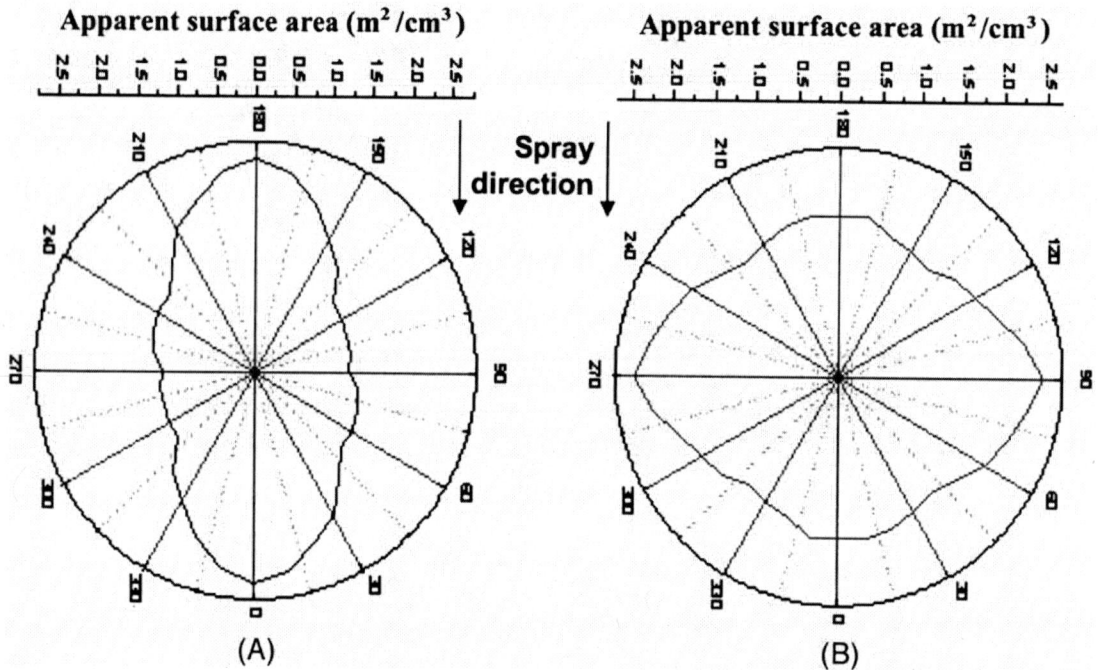

Fig. 3. Porod surface area anisotropy plots for (A) GE-1 and (B) GE-4 coatings.

Table II. Quantitative MSANS Model Results along with Orientation Information

Material	Porosity (%)	Component porosities (%)			Mean opening dimensions (μm)	Globular pore diameter (μm)
		Interlamellar pores	Intrasplat cracks	Globular pores		
GE-1	13.3 ± 0.3	9.9(8)	1.4(3)	1.9(9)	0.067	0.43(8)
GE-2	11 ± 0.2	7.4(7)	1.8(7)	1.8(6)	0.06	0.39(2)
GE-3	10.3 ± 0.2	5.1(5)	2.1(6)	3.1(2)	0.071	0.46(2)
GE-4	9 ± 0.2	3.6(1)	3.1(5)	2.2(5)	0.066	0.42(9)

	GE-1		GE-4	
	Cracks (%)	Pores (%)	Cracks (%)	Pores (%)
(0°–30°)	4.5	95.2	9.3	87.6
(30°–60°)	12.2	3.5	12.7	11.5
(60°–90°)	83.3	1.3	78	0.9

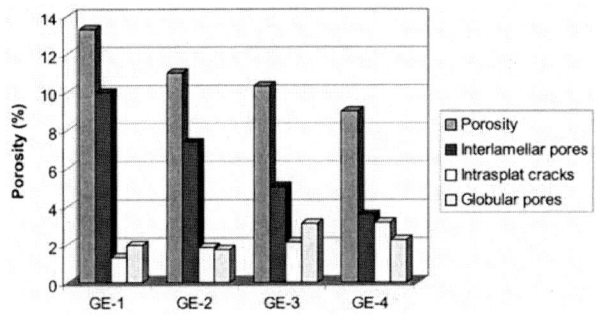

Fig. 4. Quantitative separation of the total porosity, obtained from MSANS analysis, for each of the four samples.

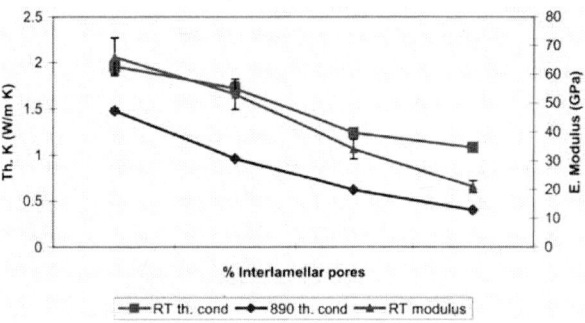

Fig. 5. Thermal conductivity (left ordinate) and out-of-plane elastic modulus (right ordinate) correlated with interlamellar porosity derived from MSANS. The vertical bars are the standard deviations on the mean of 10 measurements of each sample.

(spheroidal-elements normals 0–30° from the spray direction). The quantitative separation of the coating microstructure into its components, obtained from the MSANS model, is shown in Fig. 4. It is seen that the component porosities of interlamellar pores decrease, from GE-1 to GE-4, thus enabling a correlation with the thermal conductivity and elastic modulus of the coatings. The component porosities of intrasplat cracks increase monotonically from GE-1 to GE-4, thereby explaining the decrease in cross-sectional modulus for the DVC coating. The porosity–thermal conductivity and porosity–elastic modulus correlations can be better understood in terms of the percentage of interlamellar pores in the coatings, as shown in Fig. 5. The behavior is very similar to the thermal conductivity values measured both at room temperature and at 890°C, where the lower value at 890°C occurs due to greater phonon scattering occurring at high temperature than at room temperature.

The USAXS measurements giving the pore size distributions for the two extreme cases (GE-1 and GE-4) are presented in Fig. 6. The scattering intensity as a function of azimuthal angle for the wave vector $\mathbf{Q} = 0.00038$ A^{-1} is presented for both cases and compared with the observed microstructural features. The plot for sample GE-1 shows dominant scattering from interlamellar pores in the coating. The intensity for GE-4 shows contributions from both vertical and horizontal components of the porosity in the coating. The combination of scattering data collected as a function of $\mathbf{Q}$ for each orientation and as a function of orientation at a particular $\mathbf{Q}$ provides quantitative maps of the anisotropy of the microstructure as a function of sizes within the scattering population for the two coatings, as shown in Fig. 7. For each sample, the maximum entropy size distribution (MAXENT) routine gives size

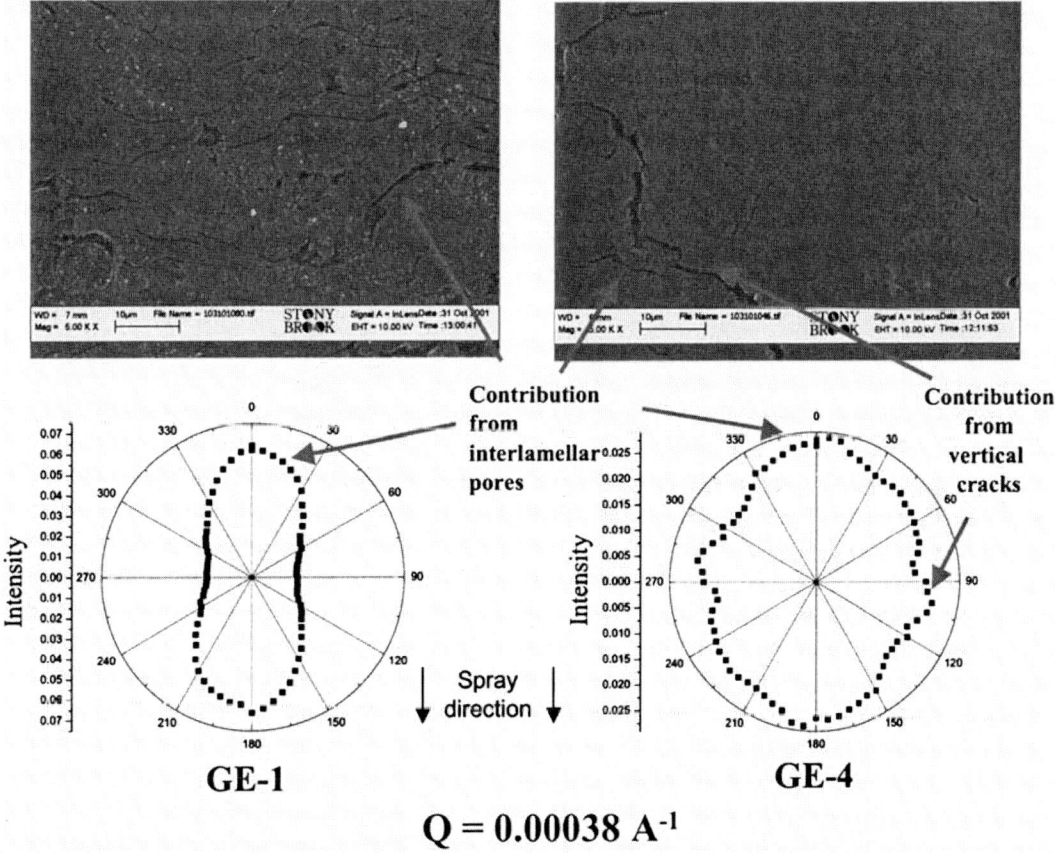

Fig. 6. Scattering intensity as a function of azimuthal angle α for the GE-1 and GE-4 coating at $\mathbf{Q} = 0.00038$ Å^{-1}. The errors are within the sizes of the data points.

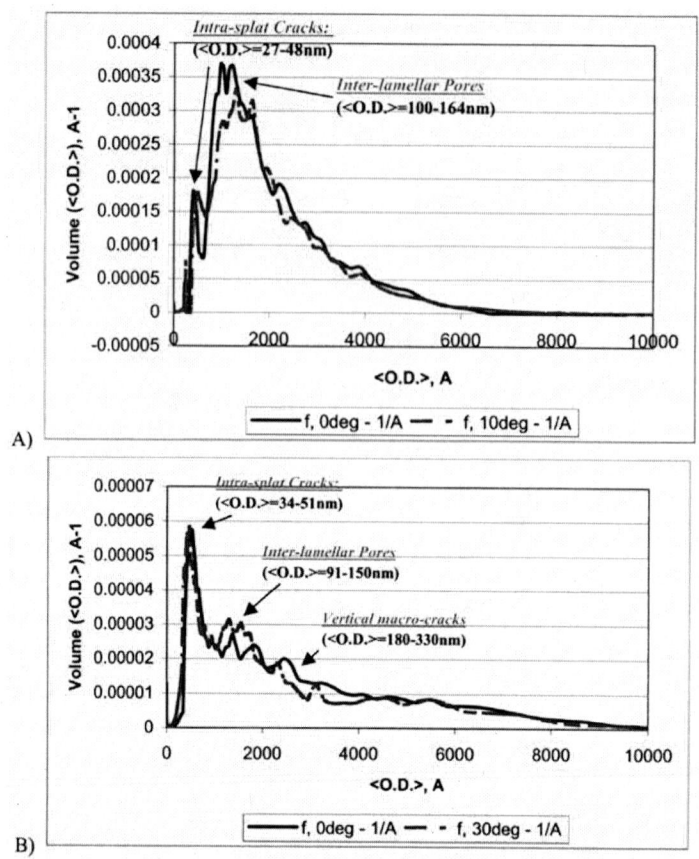

A)

B)

Fig. 7. MAXENT volume fraction size distributions of the coatings, for two orientations (α) with respect to the spray direction: (A) GE-1 coating showing interlamellar pores as the dominant system and (B) GE-4 coating showing a crack-dominant system.

distributions (no volume fraction contributions) of the void components that vary for different orientations of **Q** with respect to the spray direction. These differences arise from the anisotropy in the scattering and result from an assumption of random orientation distributions in the MAXENT routine.[19,20] It is envisaged that future work will extend the MAXENT routine to deal with nonrandom orientation distributions. However, in the present study, these discrepancies with orientation are not sufficient to detract from a comparison of the size distributions from different samples. It is seen that different features dominate the anisotropy at different magnitudes of **Q**. The results show an interlamellar pore dominant system for the GE-1 case in Fig. 7(A) as opposed to a crack dominant system for the GE-4 case in Fig. 7(B). The size ranges of the intrasplat cracks and interlamellar pores were similar in both systems. The analysis shows the opening dimension for interlamellar pores to be between 90 and 150 nm and that for the intrasplat cracks to be between 27 and 50 nm. Also observed are vertical macrocracks with opening dimensions between 180 and 330 nm in the GE-4 (DVC) case. The aspect ratio was assumed as 1/10 for both the interlamellar pores and intrasplat cracks, similar to the SANS model assumption. While this is not the true aspect ratio of macrocracks seen in the DVC micrographs, this spheroidal shape assumption gives us a way to deduce the volume fractions, opening dimensions, and orientation distributions. The assumption works because the scattering in this range is sensitive to the short area dimensions, and not particularly sensitive to the large area dimensions.

IV. Conclusions

The drive to develop these prime-reliant TBCs has fostered growing interest in comprehensive materials characterization for establishing processing–structure–property relationships. It has been successfully demonstrated that SANS, in combination with USAXS and microscopy, has provided a quantitative representation of the different void components within plasma-sprayed deposits. The paper presents valid results of scientific characterization techniques to thin coatings of engineering importance. Complete microstructural information of constituent porosities, opening dimensions, along with orientation information of void morphologies sought using SANS, complemented by void size distribution obtained using USAXS, have led to rational porosity–property correlations. The studies show DVCs with the dominant vertical crack network system, depicting low in-plane modulus, to fall between traditional plasma sprayed (layered structures) and EB-PVD (columnar structures), thus offering optimization of processing economics, appropriateness, and performance.

Acknowledgments

The authors wish to thank Dr. Sanjay Sampath of SUNY Stony Brook for valuable discussions, Dr. Boualem Hammouda of the NIST Center for Neutron Research, and Dr. Pete Jemian of the Advanced Photon Source for scientific and technical support. We acknowledge the support of the National Institute of Standards and Technology, U.S. Department of Commerce, in providing the neutron research facilities used in this work.

References

[1]R. A. Miller, "Current Status of Thermal Barrier Coatings—An Overview," *Surf. Coat. Technol.*, **30**, 1–11 (1987).
[2]W. J. Brindley and R. A. Miller, "TBCs for Better Engine Efficiency," *Adv. Mater. Proc.*, **8**, 29–33 (1989).
[3]S. M. Meier and D. K. Gupta, "The Evolution of Thermal Barrier Coatings in Gas Turbine Applications," *J. Eng. Gas Turbines Power*, **116**, 250–57 (1994).

[4]R. L. Jones, "Thermal Barrier Coatings"; p. 194 in *Metallurgical and Protective Coatings*. Edited by K. H. Stern. Chapman and Hall, London, U.K., 1996.

[5]K. D. Sheffler and D. K. Gupta, "Current Status and Future Trends in Turbine Application of Thermal Barrier Coatings," *J. Eng. Gas Turbines Power*, **110**, 605–609 (1988).

[6]W. Mannsmann and H. W. Grunling, "Plasma Sprayed TBC for Industrial Gas Turbines: Morphology, Processing, and Properties," *J. Phys. IV*, **3**, 903–12 (1993).

[7]H. Herman, "Plasma Sprayed Coatings," *Sci. Am.*, **259** [3] 112–15 (1988).

[8]R. McPherson, "The Relationship Between the Mechanisms of Formation, Microstructure, and Properties of Plasma Sprayed Coatings," *Thin Solid Films*, **83**, 297–303 (1981).

[9]P. Bengtsson, T. Ericsson, and J. Wigren, "Thermal Shock Testing of Burner Cans Coated with a Thick Thermal Barrier Coating," *J. Therm. Spray Technol.*, **7** [3] 340–48 (1998).

[10]Gray, D. M.; Lau, Y. C., Johnson, C. A., Borom, M. P., Nelson, W. A (General Electric Co.). Thermal Barrier Coatings Having an Improved Columnar Microstructure. U.S. Patent 6,180,184, October 24, 1997.

[11]A. Kulkarni, A. Vaidya, A. Goland, S. Sampath, and H. Herman, "Processing Effects on Porosity–Thermal Conductivity Correlations in Plasma Sprayed Yttria-Stabilized Zirconia Coatings," *Mater. Sci. Eng., A*, **359**, 100–111 (2003).

[12]S. Sampath and X. Jiang, "Splat Formation and Microstructure Development During Plasma Spraying: Deposition Temperature Effects," *Mater. Sci. Eng. A*, **304–306**, 144–50 (2001).

[13]S. Sampath, X. Y. Jiang, J. Matejicek, A. C. Leger, and A. Vardelle, "Substrate Temperature Effects on Splat Formation, Microstructure Development and Properties of Plasma Sprayed Coatings: Part I, Case Study for Partially Stabilized Zirconia," *Mater. Sci. Eng. A*, **272** [1] 181–88 (1999).

[14]J. Ilavsky, A. J. Allen, G. G. Long, S. Krueger, C. C. Berndt, and H. Herman, "Influence of Spray Angle on the Pore and Crack Microstructure of Plasma Sprayed Deposits," *J. Am. Ceram. Soc.*, **80**, 733–42 (1997).

[15]A. J. Allen, J. Ilavsky, G. G. Long, J. S. Wallace, C. C. Berndt, and H. Herman, "Microstructural Characterization of Yttria-stabilized Zirconia Plasma-Sprayed Deposits Using Multiple Small-Angle Neutron Scattering," *Acta Mater.*, **49**, 1661–75 (2001).

[16]A. J. Allen and N. F. Berk, "Analysis of Small-Angle Scattering Data Dominated by Multiple Scattering for Systems Containing Eccentrically Shaped Particles or Pores," *J. Appl. Crystallogr.*, **27**, 878–91 (1994).

[17]U. Bonse and M. Hart, "A New Tool for Small-Angle X-ray Scattering and X-ray Spectroscopy: The Multiple Reflection Diffractometer," *Appl. Phys. Lett.*, **7**, 238–40 (1965).

[18]J. Ilavsky, A. J. Allen, G. G. Long, and P. R. Jemian, "Effective Pinhole-Collimated Ultrasmall-Angle X-ray Scattering Instrument for Measuring Anisotropic Microstructures," *Rev. Sci. Instrum.*, **73** [3] 1660–62 (2002).

[19]J. A. Potton, G. J. Daniell, and B. D. Rainford, "Particle Size Distribution from SANS Data Using the Maximun Entropy Method," *J. Appl. Crystallogr.*, **21**, 663–68 (1988).

[20]G. G. Long, S. Krueger, P. R. Jemian, D. R. Black, H. E. Burdette, J. P. Cline, and R. A. Gerhardt, "Small-Angle Scattering Determination of the Microstructure of Porous Silica Precursor Bodies," *J. Appl. Crystallogr.*, **23**, 535–44 (1990). □

J.I. Eldridge, C.M. Spuckler, J.A. Nesbitt and
K.W. Street
NASA Glenn Research Center, Cleveland, Ohio

TBC
INTEGRITY

Although thermal barrier coatings (TBCs) provide thermal protection for turbine engine components, the risk of TBC spallation severely restricts their use by either forcing extreme safety margins to guide TBC replacement or limiting TBC application to engine temperatures at which an unprotected component can survive.

Because TBC failure results from crack/flaw propagation near the TBC/bond coat interface, a health-monitoring tool must monitor damage evolution beneath the overlying TBC. One approach is to take advantage of TBC translucency, as demonstrated by the successful application of piezospectroscopy (Cr^{3+} luminescence) to monitor the stress state of the thermally grown oxide (TGO) that forms beneath the TBC.[1-3] Unfortunately, the TGO stress state does not provide a good indication of remaining TBC life, because the indication of impending TBC failure by TGO stress relaxation tends to occur immediately preceding failure.

We present a new approach that uses mid-infrared (MIR) reflectance as a diagnostic tool for evaluating the fraction of TBC lifetime remaining. The approach correlates MIR reflectance with the progress of the buried TBC delamination crack network that ultimately produces TBC failure. This approach offers the advantage of working at wavelengths where the TBC has much greater transmittance than it has for visible light.[4] Therefore, this approach can be applied to highly attenuating plasma-sprayed TBCs that are difficult to probe using piezospectroscopy.[2]

A hemispherical transmittance measurement (Fig. 1) for a freestanding plasma-sprayed 8-wt%-yttria-stabilized zirconia (8YSZ) TBC shows that the maximum

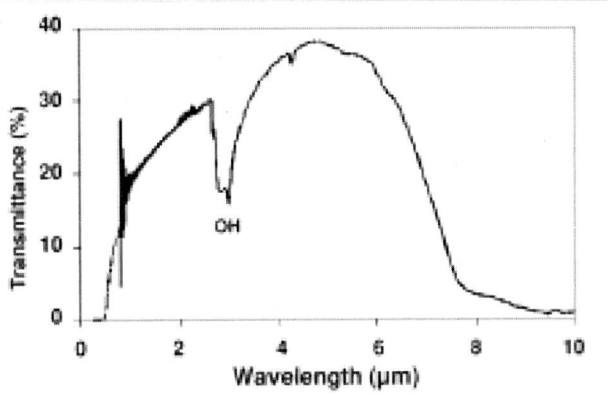

Fig. I. Hemispherical transmittance of 172 mm thick freestanding plasma-sprayed 8YSZ.

TBC transparency is in the MIR wavelength region (peak transmittance at 5 μ m), which is substantially greater than at visible wavelengths (<0.8 μ m).

TBC Preparation and Testing

Two types of TBC specimens were tested. The first type consisted of a ~200 μ m thick top coat of 8YSZ deposited by atmospheric plasma-spraying on top of a ~120 μ m thick NiCrAlY bond coat that had been deposited by low-pressure plasma-spraying onto a 25.4 mm diameter nickel-based superalloy René N5 substrate. The second type consisted of identical TBC specimens, except there was no NiCrAlY bond coat between the TBC and the substrate.

Furnace cycling tests were performed in a tube furnace; each cycle consisted of a 45 min interval at 1163°C followed by a 15 min cooling period to ~120°C. Furnace cycling was interrupted after the first cycle and after every tenth cycle to obtain hemispherical reflectance measurements and to inspect the specimen for TBC failure. TBC failure was judged to occur when buckled or spalled regions exceeded 20% of the total coating area. All specimens were eventually cycled to failure, except for one specimen for which cycling was terminated after 150 cycles. This specimen was examined using cross-sectional SEM (along with an uncycled control specimen) for evidence of buried crack propagation that occurred before external signs of TBC

> *A four-flux pure-scattering model is used to show that MIR reflectance reproducibly tracks the progression of YSZ TBC delamination produced by repeated thermal cycling of plasma-sprayed TBCs coated on superalloy substrates.*

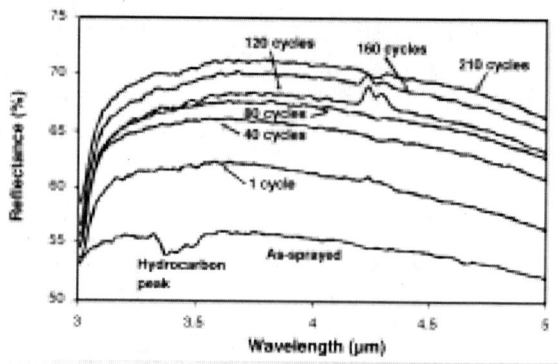

Fig. 2. *Effect of furnace cycling of TBC-coated specimen (200 μ m thick plasma-sprayed 8YSZ on René N5 with bond coat) on MIR spectral hemispherical reflectance.*

damage.

Because TBCs are scattering materials, hemispherical detection was necessary to fully capture the reflected radiation. Room-temperature spectral directional-hemispherical reflectance spectra were obtained for the TBC specimens using a FTIR spectrometer (Model 760, Nicolet Instrument Corp., Madison, Wis.) equipped with a integrating sphere accessory (Model RSA-NI-550-ID, Labsphere, North Sutton, N.H.).

Effect of Furnace Cycling

MIR hemispherical reflectance in the 3–5 μ m wavelength range of a TBC-coated specimen with a bond coat increased with furnace cycling (Fig. 2). At higher and lower wavelengths outside the 3–5 μ m wavelength range, the differences in reflectance decreased until they were no longer distinguishable. Much of the increase in reflectance observed after one cycle was due to the removal of oxygen deficiency present in the as-sprayed TBC during heat treatment that was accompanied by a visible "whitening" of the TBC. All other specimens showed similar reflectance increases, and the reflectance increases were larger for the early stages of furnace cycling.

To facilitate comparison of the effect of furnace cycling on MIR reflectance of all specimens, a single wavelength at 3.8 μ m was selected, because it offered good discrimination and was sufficiently distant from potential interference by OH and CO_2 absorptions at 3 and 4.25 μ m, respectively (as well as from hydrocarbon adsorption at 3.4 μ m that disappears upon heat treatment).

Hemispherical reflectance at 3.8 μ m and the number of furnace cycles for TBC specimens with and without a bond coat was determined (Fig. 3). Three TBC-coated specimens with bond coats were tested until TBC failure; all showed failure between 200 and 210 cycles. A fourth specimen with bond coat was terminated after 150 furnace cycles, before TBC failure was evident. The increased hemispherical reflectance with number of furnace cycles tracked closely for all four specimens. In contrast, the three TBC-coated specimens without bond coats failed much earlier and with a much higher degree of scatter between specimens (failure was observed after 40, 70, and 120 cycles).

The hemispherical reflectance for the TBC-coated specimens without bond coats increased more rapidly than for the specimens with bond coats. Among the specimens without bond coats, the shorter-lived specimens showed a more rapid increase in hemispherical reflectance with furnace cycling. There also was reflectance

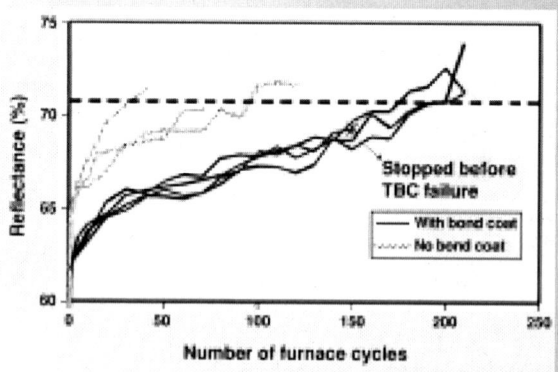

Fig. 3. *Effect of furnace cycling on hemispherical reflectance at λ = 3.8 μ m for ~200 μ m thick TBC-coated specimens with and without bond coats. Dashed line shows reflectance for all specimens that seems to be associated with final TBC failure.*

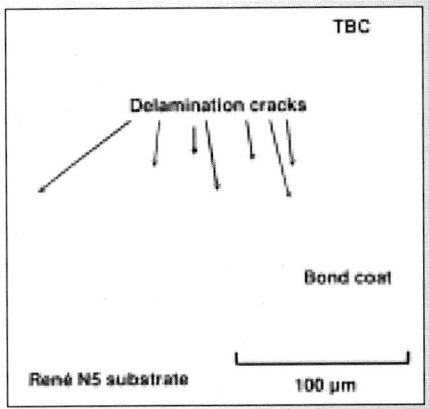

Fig. 4. SEM micrograph of TBC/bond coat interfacial region for TBC-coated specimen after 150
furnace cycles showing formation of delamination crack network.

for all specimens (with and without bond coat) of ~71%
that was associated with final TBC failure.

SEM inspection of the cross-section of the TBC-coated
specimen with bond coat that had undergone 150 furnace cycles revealed the formation of a network of
delamination cracks along the bottom of the TBC, mostly
above the bond coat and TGO (Fig. 4). Many of these
cracks occurred along splat boundaries. The degree of
crack interconnectedness after 150 cycles was insufficient to produce large-scale separation between the TBC
and substrate. These delamination cracks were absent in
a control specimen that had not been furnace cycled.

Crack/Reflectance Model

Realative transparency is not sufficient to determine
delamination without an effect on reflectance produced
by the introduction of cracks. The effect on reflectance
caused by introducing a crack at the bottom of the TBC has been modeled (Fig. 5).

In the MIR wavelength range <4 μ m, except for the OH absorption, the
TBC hemispherical reflectance can be effectively predicted by a zero-
absorption (pure scattering) four-flux model.[5] When this model is applied,
the substrate is assigned a diffuse reflectance of 0.43 (the measured bond
coat hemispherical reflectance at 3.8 μ m).
The underlying crack is modeled by an air
gap between the TBC and substrate. It is
assumed that the air gap is sufficiently
wide to prevent significant radiative tun-
neling across the gap. The introduction of
this air gap increases reflectance because
of the large index of refraction change
across the TBC/air gap interface (from n =
2.1 to n = 1.0).

This change produces a high diffuse
reflectance at the TBC/air gap interface of
81% because of total internal reflection.
The overall effect of introducing this high
internal reflectance interface at the bottom of the TBC has been illustrated by
plotting predicted hemispherical
reflectance versus optical thickness for
TBC-coated specimens (Fig. 5). Significantly higher reflectance is predicted
for a TBC with an underlying crack, although the effect decreases and
eventually disappears at high optical thicknesses.

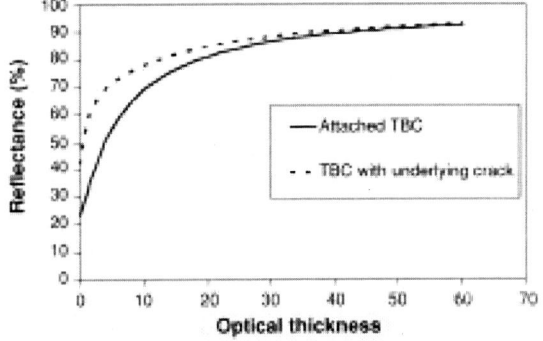

Fig. 5. Predicted effect of introduction of TBC/bond coat interface on hemispherical reflectance (based on zero-absorption
four-flux model).

The eventual loss of discrimination for the presence of an underlying
crack is expected, because the fraction of incoming radiation that reaches
the "buried" crack and then is reflected through the TBC decreases with
increasing optical thickness. This explains why visual inspection, which uses
visible wavelength reflectance (with higher associated optical thickness),
reveals no evidence of the development of a buried crack network.
Fortunately, at MIR wavelengths, even the highly scattering plasma-sprayed
TBCs tested have optical thicknesses sufficiently low to easily observe the
increase in reflectance produced by the introduction of underlying cracks.

MIR reflectance as a useful health-monitoring tool depends on the gradual nature of the TBC failure process. If TBC failure proceeds by sudden catastrophic propagation of singular flaws, then crack detection provides insufficient advance warning. Plasma-sprayed TBCs degrade by the gradual accumulation and linking of numerous cracks produced along the bottom of the TBC and this process initiates early in TBC life.[6]

Therefore, the extent of crack network progression, if it can be monitored, provides a good indication of the fraction of TBC life that remains. We propose that the MIR reflectance increase that correlates with continued furnace cycling (Figs. 2 and 3) is associated with the gradual progression of the delamination crack network buried near the bottom of the TBC and that the reflectance increases until it reaches a value associated with TBC failure. Reflectance increases smoothly (associated with the gradual progression of the buried crack network). Therefore, a threshold reflectance can be selected that is safely below the reflectance associated with failure, at which the TBC should be replaced.

MIR reflectance is sensitive to the progressive development of this crack network, even in its earlier stages, whereas other health-monitoring approaches are sensitive only to later stages. For example, the presence of a buried crack network (Fig. 4) strongly affects MIR reflectance, but does not significantly affect TGO stresses and, therefore, is not effectively monitored using piezospectroscopy.

MIR reflectance is affected by factors other than buried crack growth. Thicker TBCs exhibit higher reflectance;[4] therefore, threshold and failure levels for reflectance need to be effectively normalized to TBC thickness. TBC sintering appears to decrease reflectance. Although TGO growth on a bare substrate decreases reflectance, the effect of buried TGO growth is unclear. TBC erosion and sintering have the opposite effect of crack network growth (decreases instead of increases reflectance) and do not produce false positive signs of impending TBC failure, but must be considered for potential masking effects. ∎

Acknowledgments

The authors wish to thank C.A. Barrett for performing the furnace cycling tests and G.W. Leissler and S.L. Leissler for producing the TBC specimens.

Editor's Note:

This article is adapted from a paper presented at the 27th International Cocoa Beach Conference on Advanced Ceramics and Composites and published in *Ceramic Engineering & Science Proceedings*, Vol. 24, No. 3, 2003, "Health Monitoring of Thermal Barrier Coatings by Mid-Infrared Reflectance," pp 511–16. Single copies of the full text paper are available for a fee. To purchase the book, click here, or contact Customer Service at 614-794-5892.

References

[1]X. Peng and D.R. Clarke, "Piezospectroscopic Analysis of Interface Debonding in Thermal Barrier Coatings," *J. Am. Ceram. Soc.*, **83** [5] 1165-70 (2000).

[2]K.W. Schlichting, K. Vaidyanathan, Y.H. Sohn, E.H. Jordan, M. Gell and N.P. Padture, "Application of Cr^{3+} Photoluminescence Piezo-Spectroscopy to Plasma-Sprayed Thermal Barrier Coatings for Residual Stress Measurement," *Mater. Sci. Eng. A*, **291**, 68-77 (2000).

[3]A. Selcuk and A. Atkinson, "Analysis of the Cr^{3+} Luminescence Spectra from Thermally Grown Oxide in Thermal Barrier Coatings," *Mater. Sci. Eng. A*, **335**, 147-56 (2002).

[4]J.I. Eldridge, C.M. Spuckler, K.W. Street and J.R. Markham, "Infrared Radiative Properties of Yttria-Stabilized Zirconia Thermal Barrier Coatings," *Ceram. Eng. Sci. Proc.*, **23** [4] 417-30 (2002).

[5]T. Makino, K. Kunitomo, I. Sakai and H. Kinoshita, "Thermal Radiation Properties of Ceramic Materials," *Heat Transfer-Jpn. Res.*, **13** [4] 33-50 (1984).

[6]J.T. DeMasi-Marcin, K.D. Sheffler and S. Bose, "Mechanisms of Degradation and Failure in a Plasma-Deposited Thermal Barrier Coating," *ASME J. Eng. Gas Turbines Power*, **112**, 521-26 (1990).

International Journal of
Applied Ceramic Technology
Ceramic Product Development and Commercialization

Photoluminescence Piezospectroscopy: A Multi-Purpose Quality Control and NDI Technique for Thermal Barrier Coatings

Maurice Gell*, Swetha Sridharan, Mei Wen

Department of Metallurgy and Materials Engineering, University of Connecticut, Storrs, CT 06269

Eric H. Jordan

Department of Mechanical Engineering, University of Connecticut, Storrs, CT 06269

Many thermal barrier coating (TBC) applications have been hampered by the lack of accurate quality control, non-destructive inspection, remaining life assessment, and lifetime prediction tools for TBCs. Photoluminescence piezospectroscopy (PLPS) has the potential to provide these capabilities. The PLPS technique provides for the measurement of stress in the thermally grown oxide (TGO) and for the determination of TGO type (α, θ, γ alumina) and distribution. The capability of PLPS for these various applications will be described based on specimen and engine component measurements.

Introduction

Premature spallation of thermal barrier coatings (TBCs) during engine service continues to impact component durability and limits the more aggressive use of prime reliable TBCs for performance. Non-destructive techniques that can assess coating quality and determine TBC damage prior to spallation would contribute to the more effective use of TBCs. Three NDI techniques are presently under development: Mid-Infrared Reflectance (MIR),[1] Thermal Wave Imaging (TWI),[2] and Photoluminescence Piezospectroscopy (PLPS).[3-11]

The MIR and TWI techniques depend on the partial debonding of the ceramic from the bond-coated substrate. Debonds larger than 1 mm can be detected. PLPS can detect debonds as small as ~1-2 μm and can potentially assess coating quality and bond coat and TGO damage prior to debonding.

The PLPS technique, invented by Paton, Murphy, and Clarke for TBCs (US Patent #6, 072, 568), and further developed by Clarke, et al.,[12-16] and Gell, Jordan, et al.,[7-11] has been successful in non-destructively measuring the stress in the TGO under the ceramic. The technique is based on the frequency shift in the Cr^{3+} (present in the TGO) photo stimulated R-luminescence (Fig. 1). This technique has been applied to both electron beam physical vapor-deposited (EB-PVD) and air plasma-sprayed (APS) coatings.[7-11] While the columnar microstructure of the EB-PVD coatings acts as optical wave guides for the lasers and ensures that the optical signals are not attenuated, the presence of microstructural defects, such as porosity and splat boundaries in APS coat-

*Corresponding author: Maurice Gell
Tel: +1-860-486-3514
Fax: +1-860-486-4745
Email: mgell@mail.ims.uconn.edu)

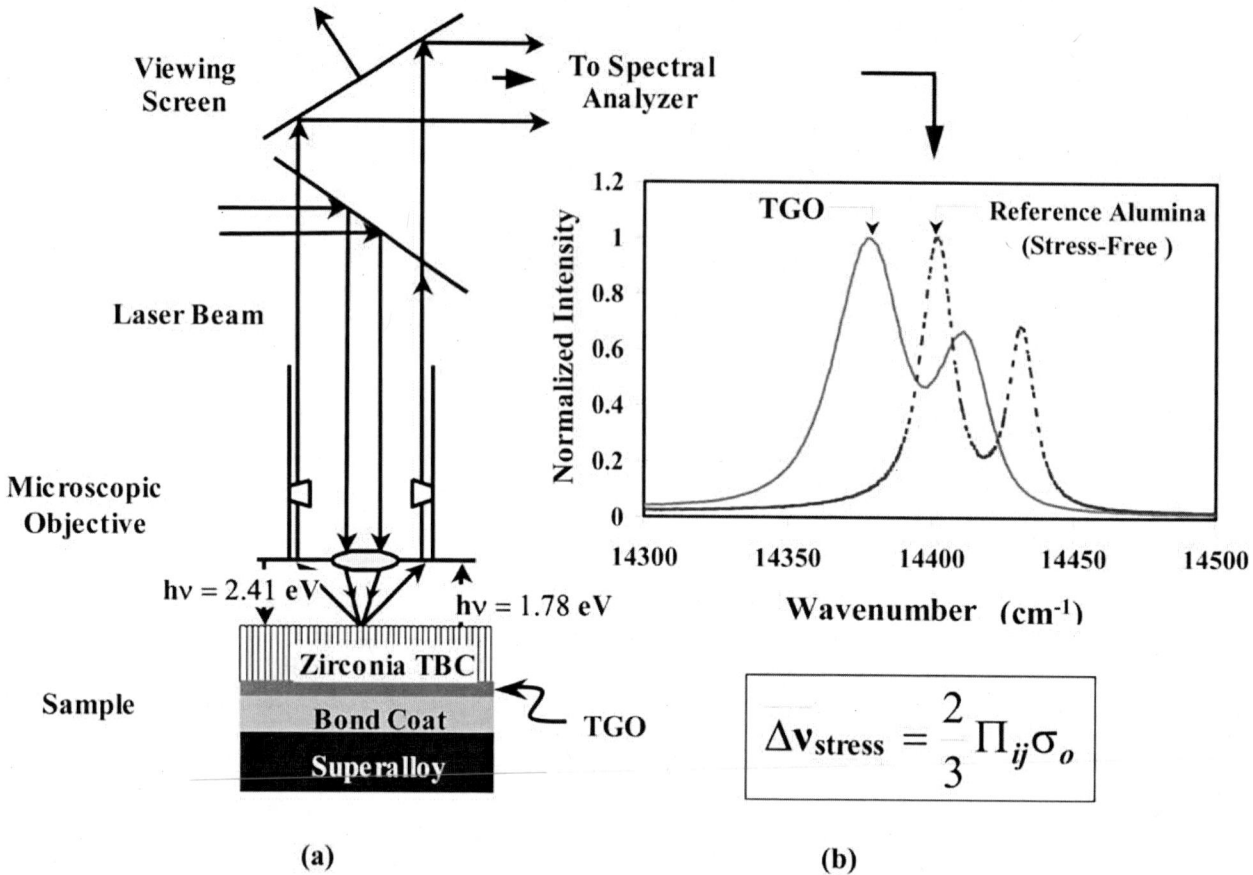

Fig. 1. (a) Schematic illustration of the photo-stimulated luminescence piezospectroscopy technique; (b) Typical R_1/R_2 fluorescence spectra for Cr-containing stress-free and stressed α-Al_2O_3.

$$\Delta v_{stress} = \frac{2}{3} \Pi_{ij} \sigma_o$$

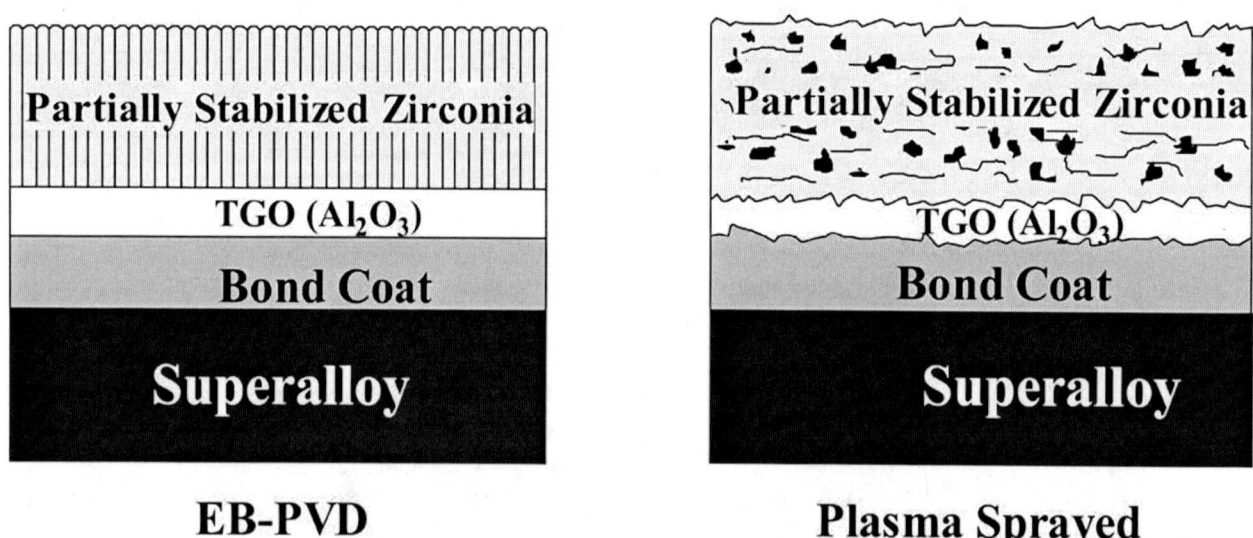

Fig. 2. Schematic of EB-PVD and APS TBCs.

ings, results in attenuation of the optical signal (Fig. 2). PLPS can be used for full-thickness EB-PVD coatings (≥300 μm),[8-11] but, at present, can be used for APS coatings of only ≤75 μm.[7,17] Thicker APS TBCs can be examined if the pores and splat boundaries are filled with a liquid whose refractive index more closely matches that of the YSZ.[7,17]

Experimental Details

Disk-shaped coupons coated with five production TBCs have been evaluated in this work. The composition and thickness of the layers of the TBC systems are given in Table I. Types I, II, and III TBCs have platinum aluminide bond coats, with Types II and III being grit blasted prior to deposition of the ceramic layer, while Types IV and V TBCs have MCrAlY bond coats. Type II and Type III TBCs are nominally the same, but produced by two different coating suppliers.

Stress measurements were made using a Renishaw™ Ramascope™ 2000 (Renishaw, Gloucestershire, UK) system in conjunction with a DM/LM light microscope (Leica™). The schematic of the setup is shown in Fig. 1. The setup utilizes an Ar-ion 514-nm-wavelength laser beam operating at 35 mW for exciting the chromium impurity (Cr^{3+}) in the TGO.[7,8] Two distinct fluorescence transitions corresponding to the R_1 and R_2 fluorescence doublets occur at frequencies of 14402 cm^{-1} and 14432 cm^{-1}, respectively, for the stress-free sapphire disk (Fig. 1). A systematic shift in the position of these peaks, when the crystal is strained, is related to the bi-axial compressive stress in the TGO.[4-6] At a magnification of 20X, the spot size for these stress measurements is typically 5-7 μm at the TGO, in the absence of the YSZ. Due to scattering within the YSZ, the effective spot size for these samples is ~20 μm. Detailed information on the experi-

mental setup and the procedure used to make the TGO stress measurements can be found elsewhere,[7] as well as experimental details about thermal cycling testing.[8-11]

Experimental Results and Discussion

PLPS as a Quality Control Tool

The ability of PLPS to non-destructively determine the oxide type and its uniformity makes it a very valuable quality control tool. Fig. 3 shows the effects of bond coat surface finish and heat treatment on TBC performance. TBC durability increases as a function of pre-oxidation treatment. The surface finishes and heat treatments that result in the formation of α-alumina have been shown to yield TBCs with longer failure lives than those that form non-α-alumina (γ,θ) or mixtures of α-, γ-, and θ-alumina. Thus, the presence of a thin layer of continuous TGO comprising α-alumina was found to be favorable for producing durable TBCs.

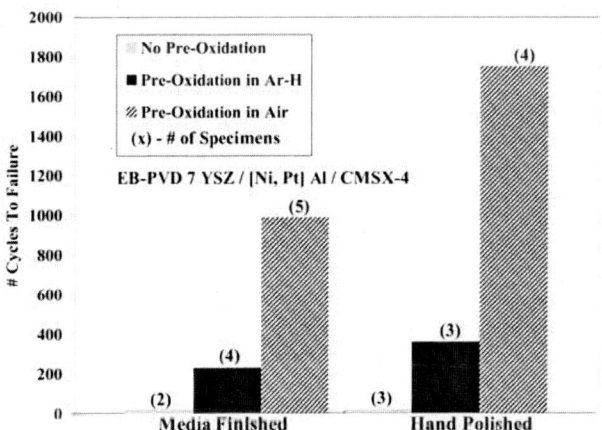

Fig. 3. Effect of surface finish and heat treatment atmosphere on TBC performance for platinum aluminide bond-coated samples.

Table I. Composition and Thickness of the Five TBCs

Type	Superalloy Substrate	Bond Coat		Ceramic (7 YSZ)	
		Composition (weight percent)	Thickness (μm)	Type	Thickness (μm)
I	Single Crystal Rene N5	[(Ni, Pt) Al] - Ni-21 Al-20 Pt	65	EB-PVD	125
II	Single Crystal CMSX-4	Grit Blasted - [(Ni, Pt) Al] - Ni-21Al-20 Pt	75	EB-PVD	150
III	Single Crystal CMSX-4	Grit Blasted - [(Ni, Pt) Al] - Ni-21 Al-20 Pt	50	EB-PVD	140
IV	Single Crystal CMSX-4	Ni-20Co-18 Cr-12.5 Al-0.6 Y-0.4 Si-0.25 Hf	100	EB-PVD	145
V	Polycrystalline IN-738	Ni-20 Co-20 Cr-8 Al-0.5 Y	150	EB-PVD	300

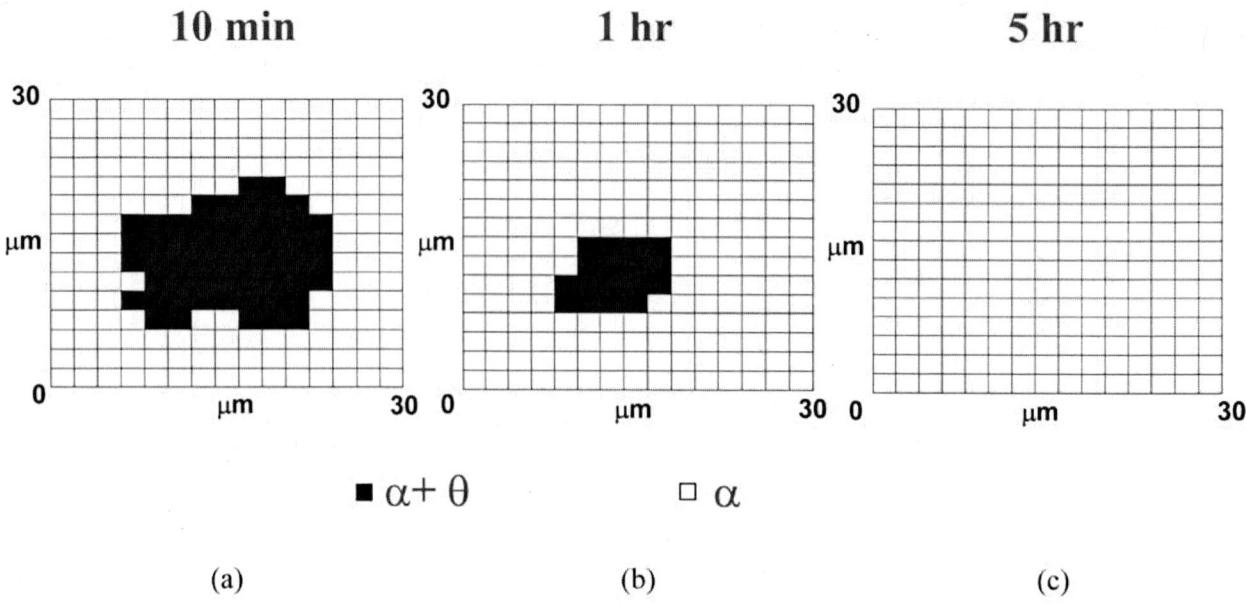

Fig. 4. Oxide-type distribution maps showing the distribution of α- and θ-alumina in Type III bare bond-coated TBC sample after (a) 5 min, (b) 1 h, and (c) 5 h at 1151°C.

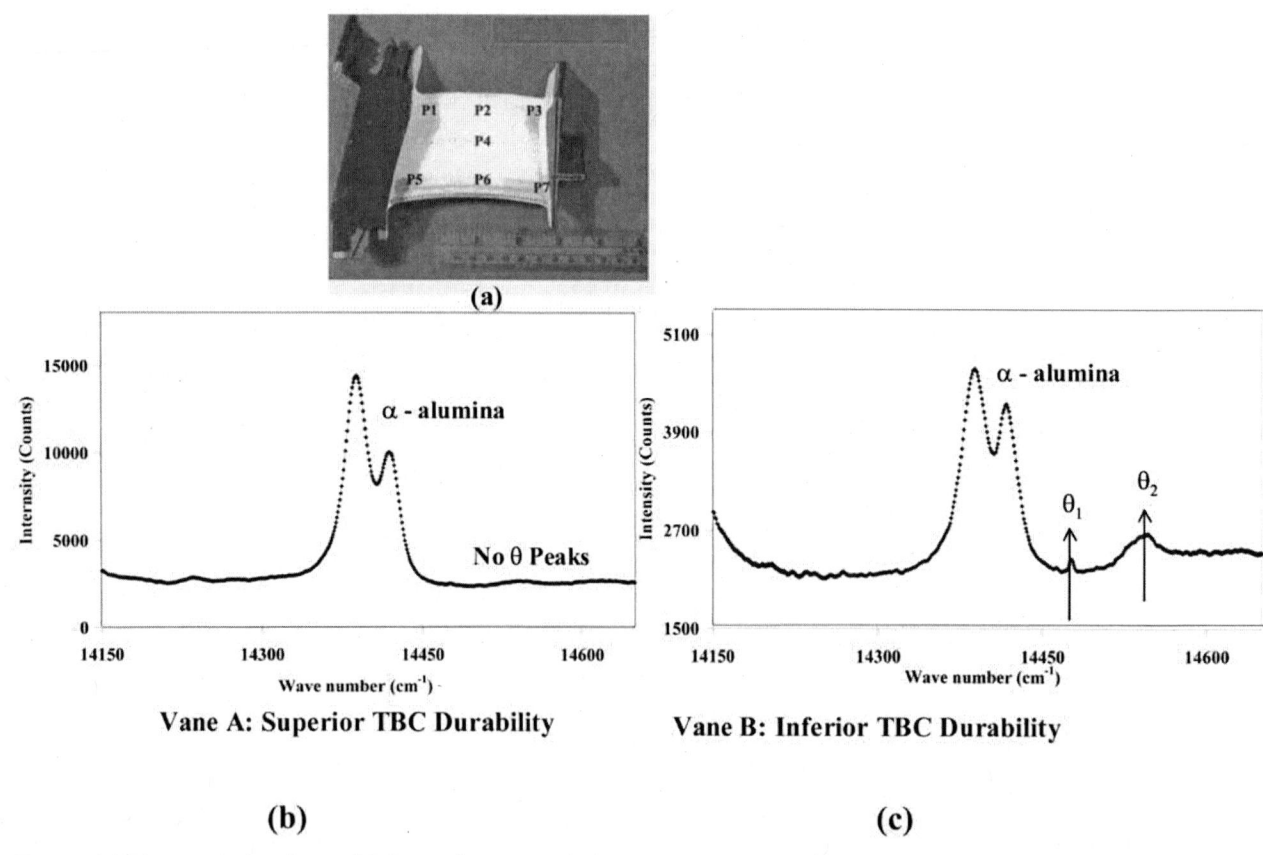

Fig. 5. (a) Photograph describing a TBC coated vane on which measurements were made. (b) and (c) Photoluminescence spectra obtained from TBC coated vanes.

Fig. 4 is an oxide-type distribution area map indicating the distribution of alumina (α and non-α), obtained using PLPS from a bare bond-coated Type III TBC specimen after short times at 1151°C in air. The figure shows the initial presence of an island of θ-alumina (R_2 = 14613 cm^{-1}) surrounded by α-alumina (R_2 = 14420 cm^{-1}). The island is reduced in size by the transformation of θ- to α-alumina, after 1-h exposure, and disappears after 5 h. Based on the result of Fig. 3, PLPS can be used to verify a continuous α-alumina layer prior to deposition of the ceramic topcoat.

PLPS measurements were made on as-coated vanes received from an engine-manufacturer (Fig. 5a). The vanes came from two distinct populations: one showing long life and the other showing short life. PLPS measurements showed that the long-life vanes had a TGO consisting only of α-alumina, while the short life vanes had a TGO consisting of both α- and θ-alumina (Figs. 5b and c). The additional stress and new zirconia to TGO interfaces created when θ transforms into α could account for the reduced life.

PLPS For NDI Applications

For PLPS spectral characteristics to be useful in detecting early damage, they should exhibit a monotonic change with thermal cycling. It has been found that the particular spectral characteristic that shows a monotonic change may vary from TBC to TBC, consistent with the damage mechanism. The operative TBC damage mechanism and associated spectral changes with thermal cycling should be determined for each TBC of interest, under thermal cycling conditions that match or approximate those of engine service.

Most EB-PVD TBCs on platinum-aluminide bond coats showed a consistent and monotonic decline in TGO stress and a systematic increase in its standard deviation with thermal cycling, as shown in Figs. 6 and 7.[8,10,11] Figs. 6a and b show the effects of peak temperature and cycle time on the evolution of TGO stress and its standard deviation for Type II TBCs. The TGO stress evolution, relative to the TBC life fraction (on a relative basis, normalized to TBC lives), is seen to be insensitive to cycling temperature but dependent on hold time (Fig. 6a). No bimodal stress states were observed at any point in the TBC life for the Type II TBCs. The observed stress change for this system is possibly due to TGO rumpling and localized separations at the TBC/TGO interface. Type III TBCs also showed a similar monotonic decrease in stress

vs. life fraction, which was found to be independent of temperature and a monotonically increasing standard deviation. This temperature independence of TGO stress vs. life fraction data, if repeatable, will greatly simplify the use of PLPS as an NDI tool for turbine coatings. Fig. 7 shows a plot of TGO stress vs. cycles for Type I TBCs. As seen in Fig. 7, there is no significant stress decrease after 600 cycles. Thus, the stress-drop with cycling is more evident in Type II TBC (Fig. 6a) than in Type I TBC (Fig. 7). This difference is presumably due to the greater extent of rumpling in Type II TBCs,[18] which have a rougher initial bond coat surface than Type I TBCs.[11]

PLPS measurements on EB-PVD TBCs with MCrAlY bond coats (Type IV and Type V) show a very different behavior. In Type V TBCs, there is a monotonic increase in TGO stress with cycling followed by a sharp decline, as shown in Fig. 8a. The observed stress change for this system is due to an increase in the bimodal stress states, associated with intact TBC and locally damaged regions. The standard deviation of stress increases with cycling (Fig. 8b). All spectra collected from Type IV and Type V TBCs were bimodal. The bimodal fraction (stress-free component) for Type V TBCs, calculated based on the intensity of R_2 photoluminescence, is presented in Fig. 8c.[9] This bimodal fraction is seen to initially increase sharply, then gradually decrease and increase again close to TBC failure.[9] Fig. 9 shows the stress evolution vs. cycles for Type IV TBCs. As seen, only two of the six TBC specimens tested showed a slight decline in the TGO stress before failure, while the stresses for the remaining four specimens either remained constant after an initial increase or changed randomly. The stress evolution vs. cycles for Type IV TBCs did not provide any useful means of detecting early TBC damage and its impending failure. Thus, while there is detectable stress-drop in Type V TBCs (Fig. 8a), it is almost absent in Type IV TBCs (Fig. 9). The standard deviation of measured TGO stress increases, in Figs. 6b and 8b, due to an increase in localized areas of damage and an increase in the number of low-stress regions and possibly bimodal peaks. The standard deviation can be reduced by measuring the stress over a larger area of the sample or, more specifically, by making more stress measurements over a given sample area.

The TGO stress values at failure in the failed Type II[10] TBC specimens were found to be quite constant (~1.2 GPa), irrespective of the cycling temperatures (1100°C, 1121°C, and 1151°C) and cycle times (1 h and 24 h). Type III TBC specimens also showed a constant stress at failure (0.9 GPa) after 1-h cycling at 1121°C and 1151°C.

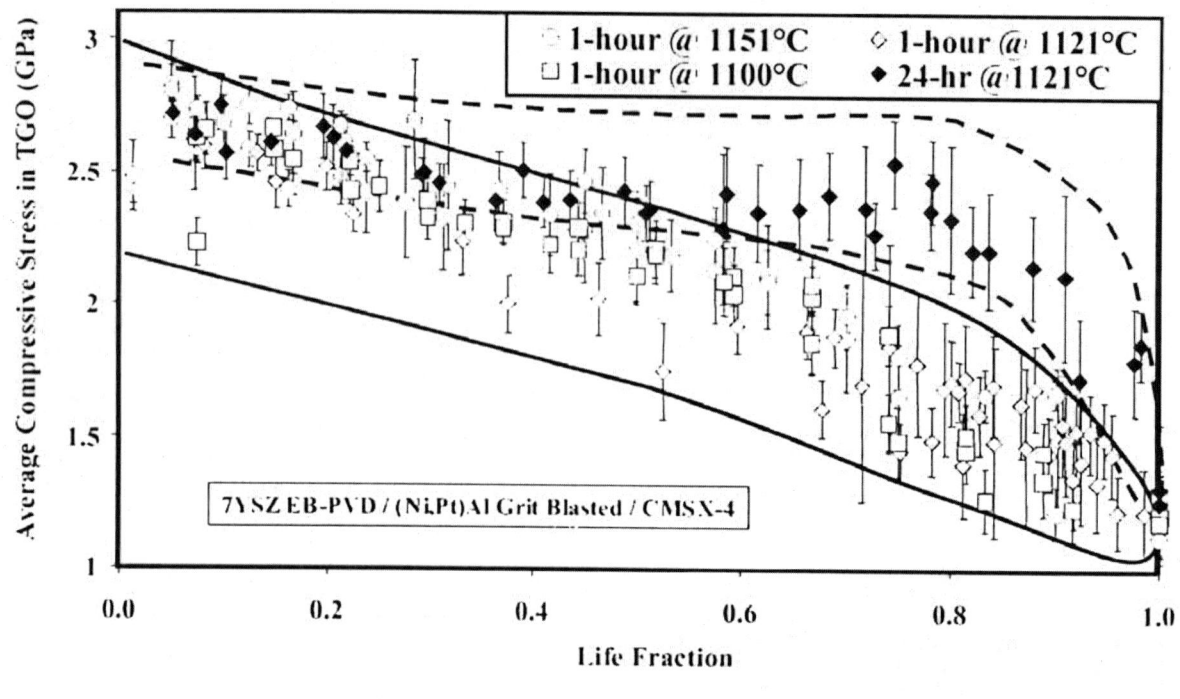

(a)

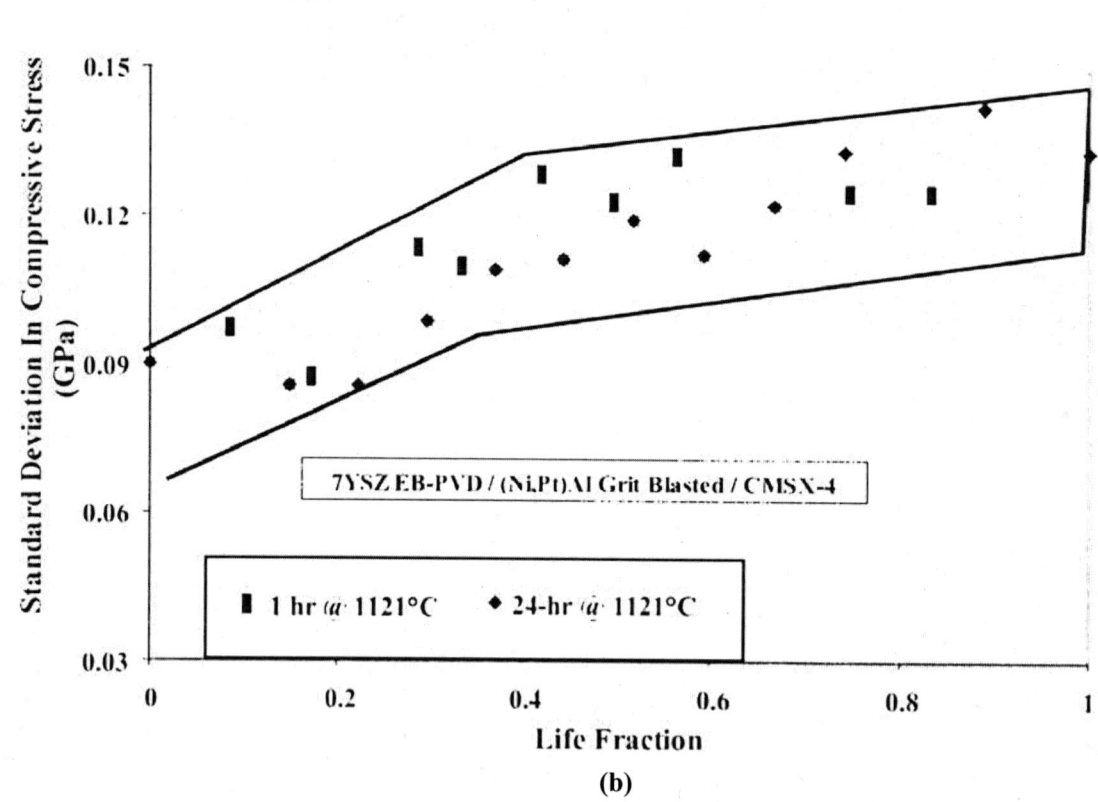

(b)

Fig. 6. Effect of peak temperature and hold time on (a) TGO stress and (b) standard deviation vs. life fraction for Type II TBC specimens.

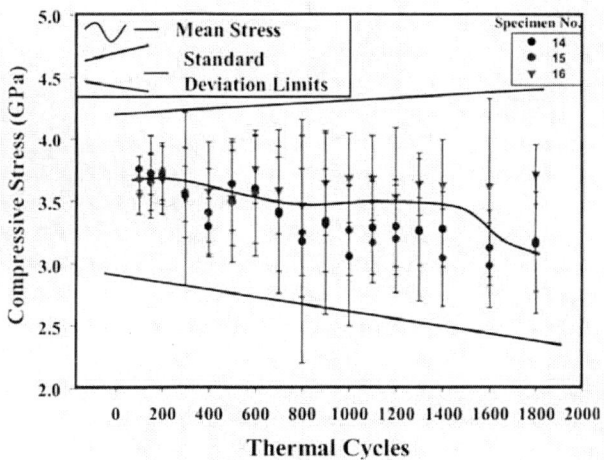

Fig. 7. Evolution of TGO compressive residual stress as a function of thermal cycling for Type I TBC specimens.

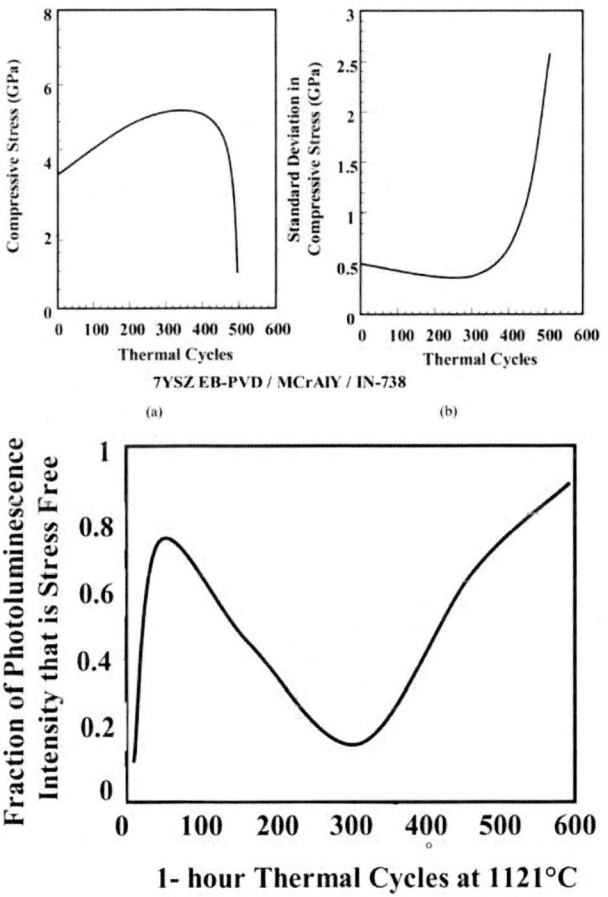

Fig. 8. Evolution of (a) TGO compressive residual stress, (b) its standard deviation, and (c) bimodal fraction as a function of thermal cycling for Type V TBC specimens. (Derived from Ref. [9].)

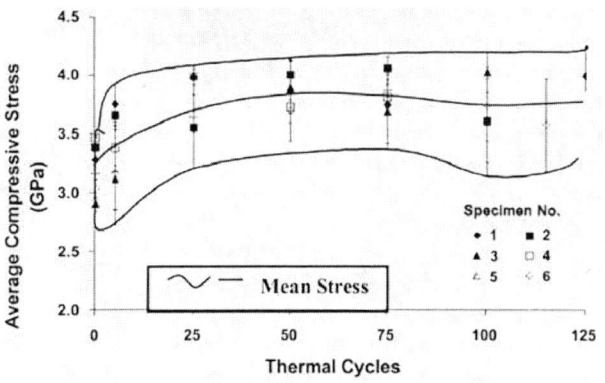

Fig. 9. Evolution of TGO compressive residual stress as a function of thermal cycling for Type IV TBC specimens.

This critical TGO stress value has been found useful in developing reliable life prediction models.[19] This TGO stress associated with failure offers great potential to detect impending failure in Type II and Type III TBCs, since the consistent values of the critical stress at failure provide sufficient warning of imminent TBC failure well in advance.

Early detection of TBC damage and progression can be obtained using filled area maps for bimodal spectra, composed of two R_1/R_2 doublets, corresponding to two different residual stress levels in the analyzed volume.[20-22] A typical example of a bimodal spectrum illustrating the stressed (shifted) and stress-free (not-shifted) α-Al_2O_3 is shown in Figs. 10a and b. The high stress peaks (5 GPa) are from the α-alumina attached to the bond coat and the low stress peaks (0 GPa) are from the flakes of α-alumina lying on the spallation surface. These bimodal spectra have been deconvoluted to give the two contributing spectra, using the recently developed physically based curve-fitting procedures developed by Nychka, et al.,[23] and Atkinson, et al.[20] For all those TBC systems (Type I, Type IV, and Type V), for which bimodal stress states were observed, the TGO stress plotted corresponds to the stressed component of the bimodal spectra.

Figs. 11a and b show the stress maps obtained for Type III TBCs after 27 and 470 1-h cycles at 1121°C, respectively. The low stress components, attributed to localized TBC damage, obtained from the bimodal spectra progressively increase with thermal cycling. A visual record of the number and size of the damage sites and their progression with cycling can be obtained with this technique.

Other spectral characteristics evaluated for these TBCs, such as peak width, R_1/R_2 area ratio, and Lorentzian to Gaussian ratio, did not show systematic

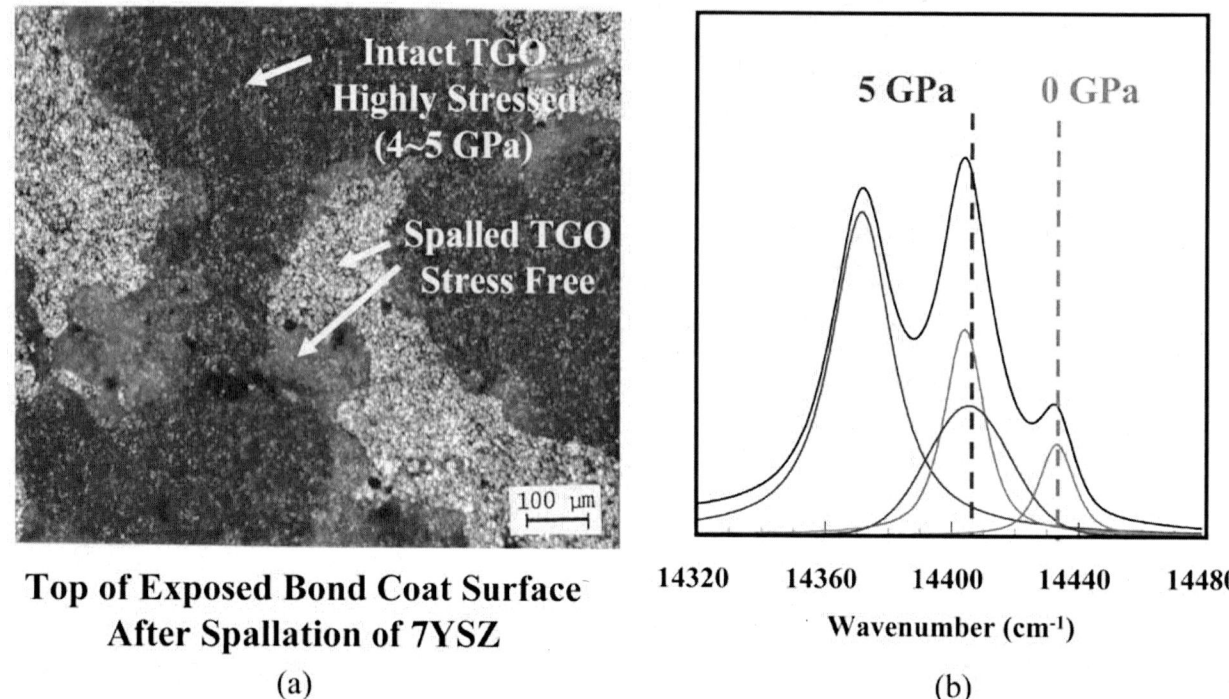

Top of Exposed Bond Coat Surface
After Spallation of 7YSZ

(a)

5 GPa   0 GPa

14320   14360   14400   14440   14480

Wavenumber (cm⁻¹)

(b)

Fig. 10. Bimodal stress state associated with spallation of TBC. (a) Micrograph showing the top of the bond coat surface after TBC spallation. (b) Typical bimodal luminescence spectra showing two R_1–R_2 doublets from intact and locally damaged regions of the TBC.

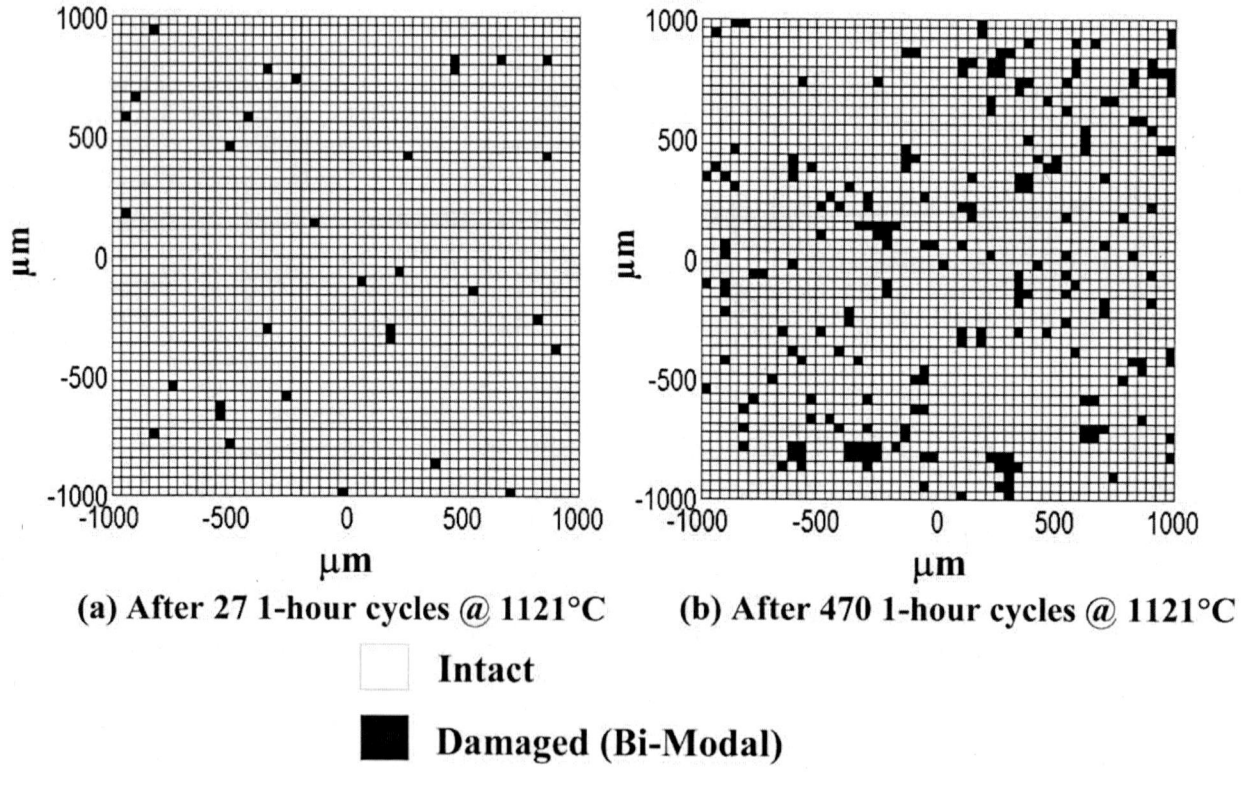

(a) After 27 1-hour cycles @ 1121°C

(b) After 470 1-hour cycles @ 1121°C

☐ Intact

■ Damaged (Bi-Modal)

Fig. 11. TGO Mapping Distribution on Type III TBC after (a) 27 cycles and (b) 470 cycles at 1121°C.

trends with cycling in this work. However, Atkinson, *et al.*,[20] and Nychka, *et al.*,[23] have demonstrated systematic changes in these spectral features with thermal cycling. Thus, all these features need to be studied on a system-by-system basis for possible use for NDI, since these results are highly system specific.

PLPS has also been demonstrated on actual turbine parts,[17] where the change in stress levels with cycles was found to follow trends similar to those seen in laboratory samples. TGO stress measurements were made on both EB-PVD and plasma-sprayed turbine blades using a fiberoptic probe that could be connected to the laboratory Raman instrument.[17] This setup is shown in Figs. 12a and b.

PLPS measurements were made on paired vanes (see Fig. 13a) on the pressure and suction sides and along the leading edge and the trailing edge. These vanes were made available in the as-coated condition and after 2000 cycles of developmental engine testing. Fig. 13b shows that the TGO stress values were found to decrease by about a factor of two on engine testing. Also, the stress profiles obtained across the vanes from the leading edge to the trailing edge, on both the pressure and suction sides, showed systematic changes as a function of cycling and location on the vane (Fig. 13b) that can be associated with the thermal history of the particular position on the vane.

PLPS for Assessment of Life Remaining

When TBC coated parts are examined following field service, it would be desirable to evaluate the condition of the coating to determine whether it can be reliably used

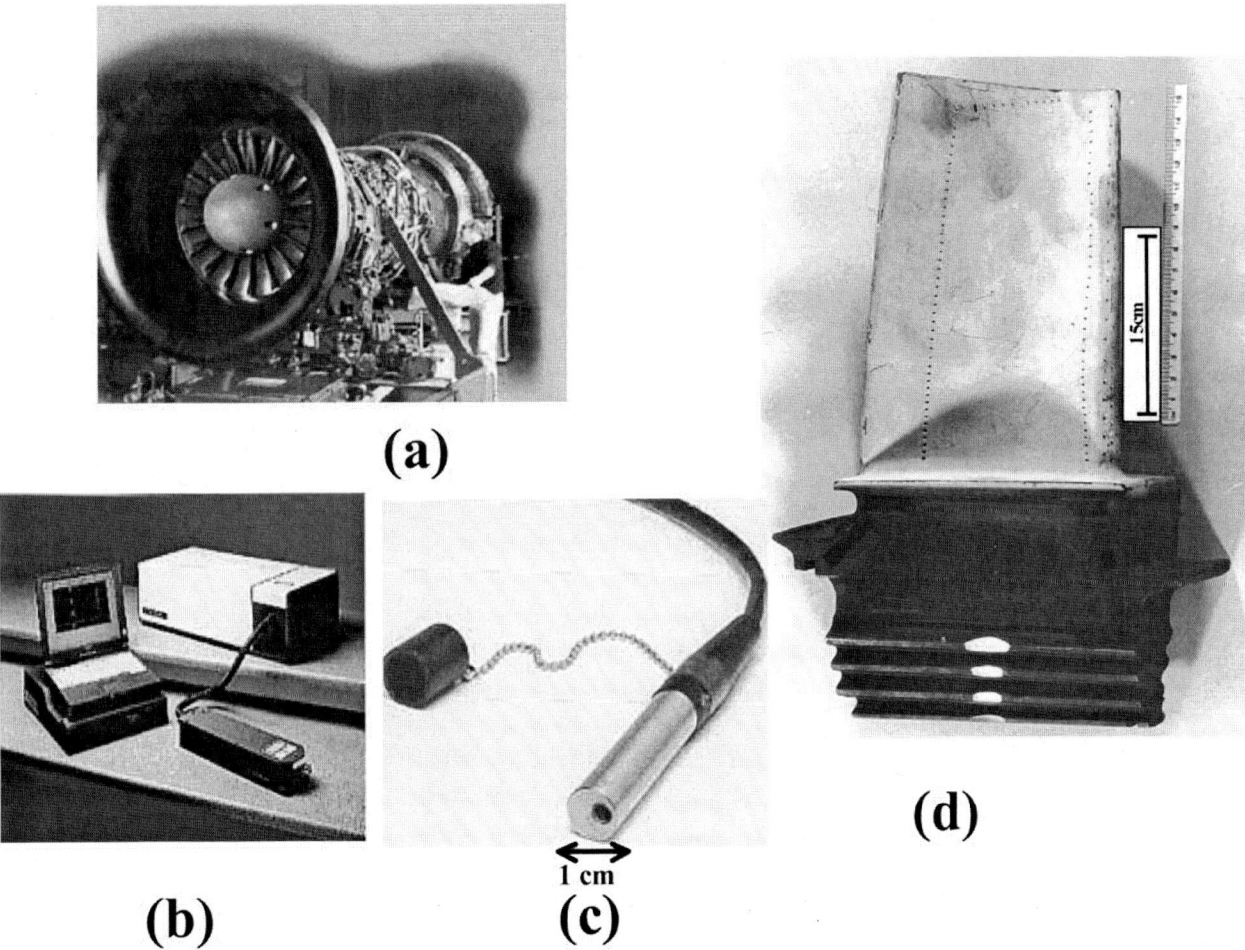

(a)

(b)

(c)

1 cm

(d)

Fig. 12. Photographs showing (a) a gas turbine engine, (b) portable PLPS instrument, (c) a fiberoptic probe, and (d) an EB-PVD TBC coated turbine blade.

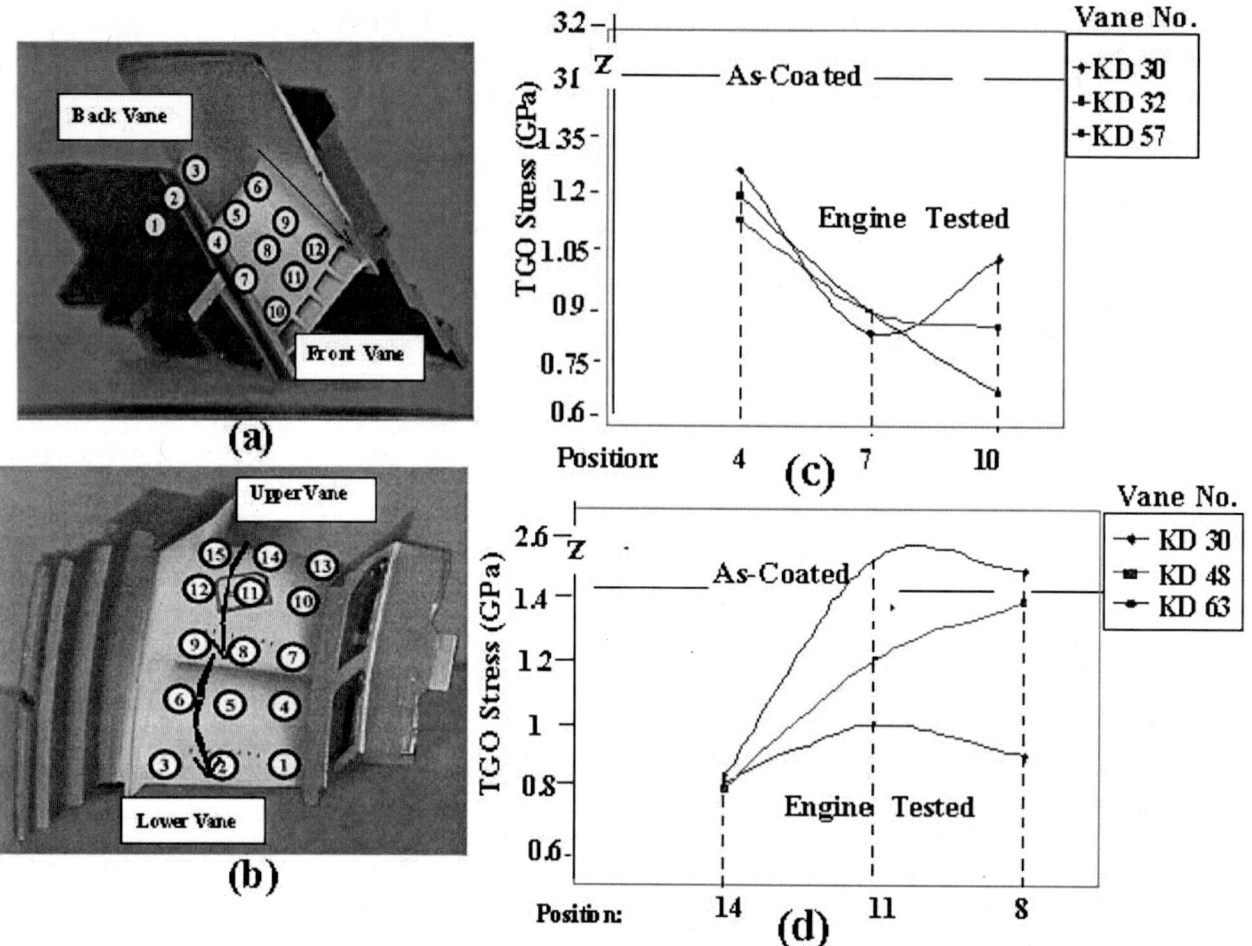

Fig. 13. (a) Pressure side and (b) suction side of a paired vane used for making PLPS measurements; stress profiles across the engine-tested vanes from the leading edge to trailing edge on the (c) pressure side and (d) suction side.

until the next planned inspection. This section describes preliminary studies for making reliable remaining life assessment using PLPS spectral characteristics that show a monotonic change with cycling, including TGO stress, standard deviation, and bimodal fractions. Before continuing to describe the methods used for making these remaining life assessments, the terms Predicted Life and Expected Life should be defined. Predicted Life is based on measured TGO stress at failure and the TGO thickness values; Expected Life is the experimentally observed average TBC life with the standard deviation.

A straightforward regression approach, based on the measured TGO stress, for Type II TBCs, and the rate of change of stress for Type V TBCs, and a method based on training neural networks have been examined. Two dif-

ferent neural network models have been used for prediction purposes.[19] For each measurement point, the stress, number of cycles to failure, first-order derivative of stress, standard deviation of measurements, and second-order derivative of stress were available from PLPS measurements.

For Type II TBCs, the measured average stress and standard deviation vs. cycles were curve-fit using a quadratic polynomial function, and regression calculations were completed. The error in predicted life was found to be less than 7% of the expected life (Fig. 14). For Type V TBCs, the slope of the stress vs. cycles curve was used for regression analysis. The error in predicted life was found to be about 8% of the expected life.

The standard deviation of the predicted TBC life for Type II TBCs, calculated using the generalized regression

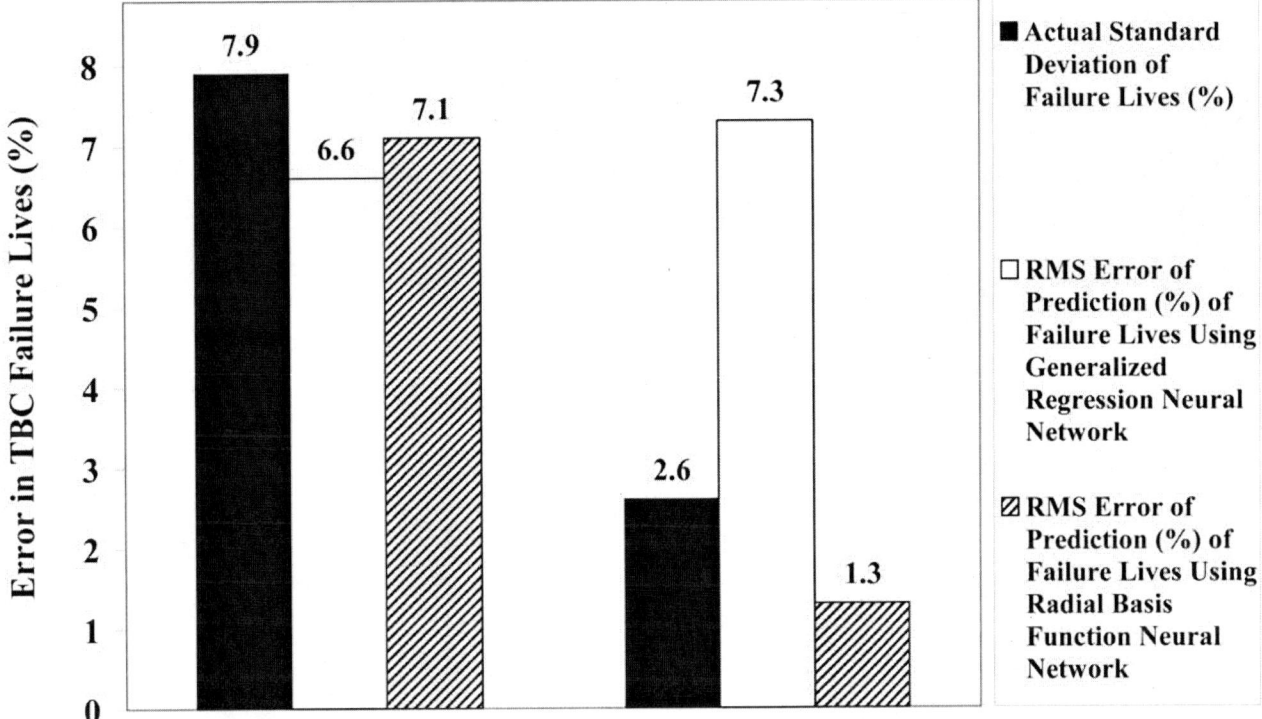

neural network (GRNN) method, was found to be 6.6% (Fig. 14). The standard deviation of the predicted TBC life for Type V TBCs, determined using radial basis function network (RBF), was found to be 1.3%.

Thus, satisfactory remaining life assessments are possible for these TBCs. Neural network-based predictions were found to be substantially better than those based on simple regression for Type V TBCs where multiple useful data features were available. However, for Type II TBCs, the simple regression-based results were found to be better than those obtained using neural network methods.

PLPS for Life Prediction

Since TBCs and internal cooling allow gas turbine engines to operate at gas temperatures well above the melting temperature of metallic components, coating spallation can lead to premature component and system failure. Current TBC life prediction methods, based on the constitutive properties of the different TBC layers and fracture mechanics approaches, can predict life only to a factor of about two times. It is thought that a simple, more accurate life prediction methodology can be em-

ployed with the non-destructive measurement of three critical TBC properties: initial bond coat surface geometry, TGO stress, and TGO thickness.

Thermal cyclic tests were conducted on a Type IV TBC system at 1121°C. Spallation of the coating occurred predominantly at the interface of TGO layer and bond coat. The proposed failure mechanism is that damage initiates at localized debonding region at the TGO/bond coat interface due to increasing out-of-plane tensile stress and the final spallation of coating is driven by the strain energy stored in the TGO layer.[24,25] The out-of-plane tensile stress across the TGO/bond coat interface may be estimated from the TGO stress, the TGO thickness, and the radius of curvature at asperities.[24,26] Strain energy is also related to the TGO stress and the TGO thickness.[27] Based on this failure mechanism, a life prediction method has been developed and experimentally validated by measurement of these three TBC parameters.

Surface geometry has been quantified using interferometric surface profilometry.[25] The average roughness of the bond coat surface (Fig. 15) before barrel finishing is higher than that after finishing by a factor of three, but the curvature (Fig. 16) before finishing is smaller than

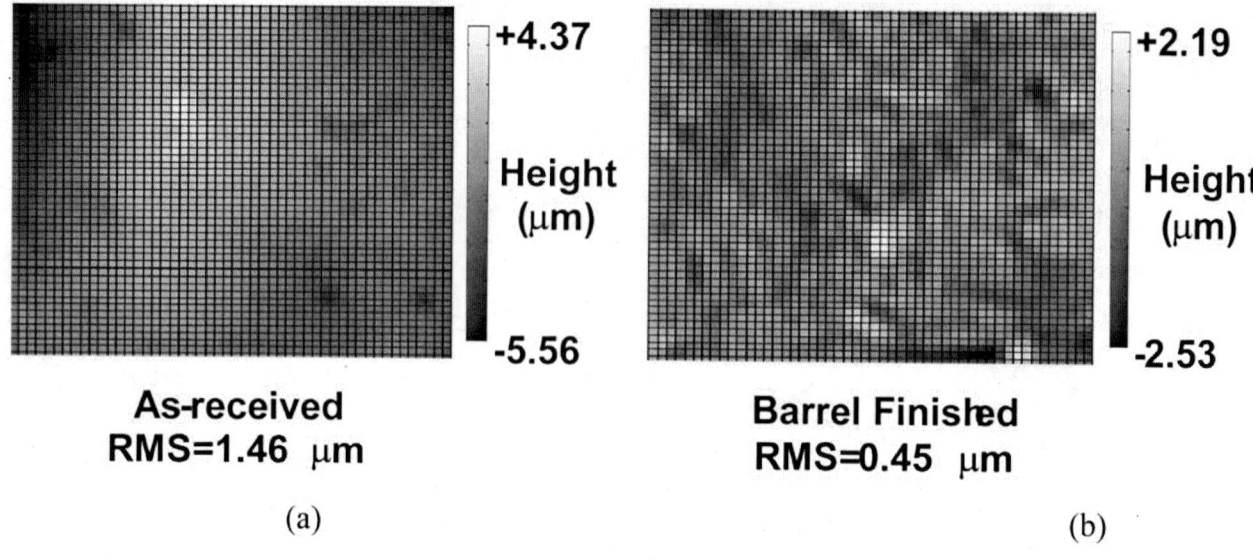

Fig. 15. *Bond coat surface geometry of Type IV TBCs (a) before and (b) after barrel finishing.*

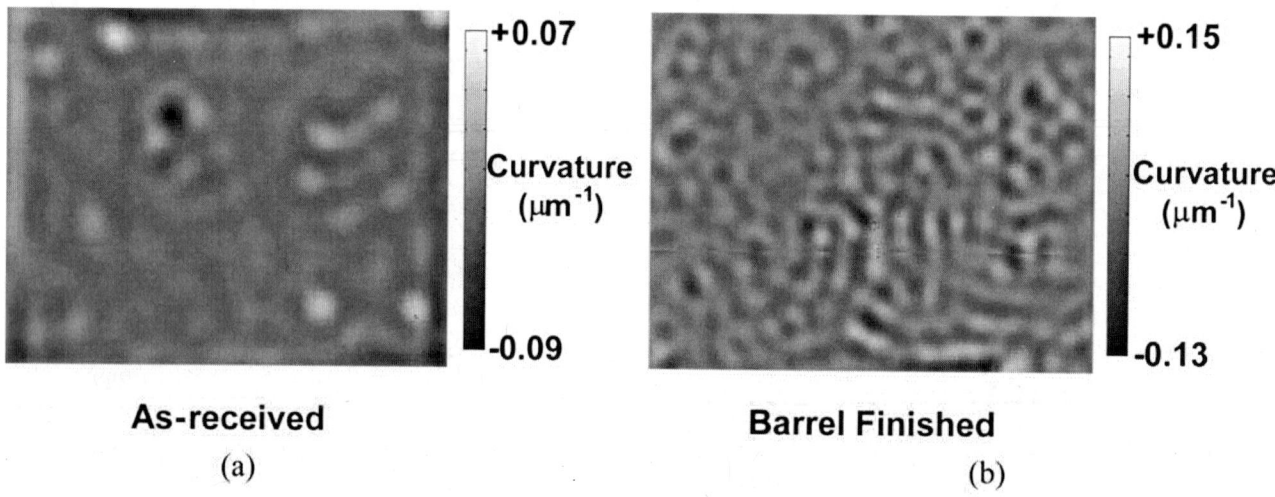

Fig. 16. *Bond coat curvature maps of Type IV TBCs (a) before and (b) after barrel finishing.*

that after finishing, which indicates that the asperity of the barrel-finished surface is more severe.

Fig. 8 shows the TGO stress vs. cycling with the stress increasing from 2.8 to about 4.0 GPa initially and then remaining constant. TGO thickness as a function of cycles was measured using metallography, but could be measured non-destructively using AC impedance.[28] The oxide growth rate was obtained by fitting the thickness-time data according to parabolic law. Thus, the TGO strain energy can be calculated accurately.

Under the mechanism of damage initiation, the debond region size at the TGO/bond coat interface can be obtained from the curvature map, assuming the asperity debonding occurs at a fixed value of out-of-plane normal stress.[24] The critical debond region size, and correspondingly the minimum TGO thickness that leads to the final spallation, can be determined by fracture mechanics.[24,25] Finally, spallation life is predicted based on the minimum TGO thickness criterion using the parabolic TGO growth rate obtained from the TGO growth rate equation. Fig. 17 shows that the predicted and actual lives agree within 15%.

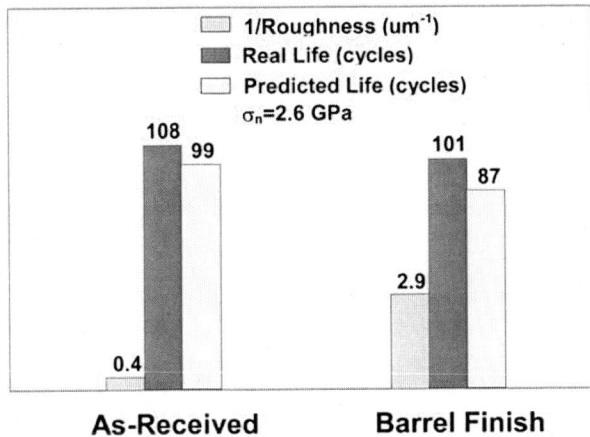

Fig. 17. Comparison of actual life and predicted life for Type IV TBCs after different surface finishes.

Concluding Remarks

The potential for PLPS to support a number of EB-PVD TBC applications has been demonstrated, including (a) quality control, (b) NDI, (c) assessment of life-remaining, and (d) input to simplified, accurate lifetime prediction methodologies.

The basis for using PLPS as a quality control instrument is to ensure that the bond coat has a uniform layer of α-alumina, and no θ- or γ-alumina present, prior to the deposition of the ceramic layer. Used in this way, PLPS should provide TBCs that have both longer and more consistent lives.

The key to using PLPS for NDI is that one or more of the spectral characteristics exhibits a monotonic change with thermal cycling. In these studies, the TGO stress and the standard deviation of the TGO stress showed systematic changes with thermal cycling for the EB-PVD TBCs on Pt-Al bond coats. In addition, the bimodal fraction was applicable for EB-PVD TBCs on MCrAlY bond coats.

Bimodal spectra, with one R_1/R_2 doublet indicating zero stress, are excellent indications of early damage. The number and size of the damage sites can be systematically examined using filled area maps.

Life-remaining assessment of service components and specimens can be assessed using the monotonic trends of the PLPS spectra as a function of thermal cycling. Improved predictability is obtained using neural network analyses where two or more spectral characteristics showing monotonic changes are included. Additional accuracy

in life assessments is associated with those TBCs that show spallation at a constant value of the TGO stress.

In all the TBCs studied, spallation failure initiates at bond coat to TGO interface at peak asperities. Assuming the spallation occurs at a constant value of the localized out-of-plane normal stress, then simplified, accurate life predictions can be made on the basis of three critical TGO characteristics that can be measured non-destructively: initial bond coat surface curvature, TGO stress, and TGO thickness.

There are a number of failure mechanisms that TBCs can exhibit depending on bond coat composition and processing, and service history variables that include time, temperature, and thermal transient effects. It is important in using PLPS as NDI, remaining life assessment, and life prediction tools, that laboratory testing duplicates service spallation modes and that knowledge of the specific spallation mechanism(s) be used in interpreting the PLPS spectra.

Acknowledgments

Supported by the U.S. Department of Energy/Advanced Gas Turbine System Research Program, under Contract #s 95-01-SR030, 99-01-SR073, 00-01-SR081, 00-01-SR091, 02-01-SR097.

References

1. J.I. Eldridge, C.M. Spuckler, J.A. Nesbitt, K.W. Street, "Health Monitoring of Thermal Barrier Coatings by Mid-Infrared Reflectance," *Ceramic Eng. Sc. Proc.*, 24 511-516 (2003).
2. X. Chen, G. Newaz, X. Han, "Damage Assessment in Thermal Barrier Coatings using Thermal Wave Image Technique," *Proceedings of 2001 ASME Int. Mech. Eng. Congress and Exposition*, Nov. 11-16 2001, New York, NY, Paper # IMECE2001/AD-25323
3. D.M. Lipkin, D.R. Clarke, "Measurement of the Stress in Oxide Scales Formed by Oxidation of Alumina-Forming Alloys," *Oxid. Met.*, 45 [3-4] 267-280 (1996).
4. Q. Ma, D.R. Clarke, "Piezo-Spectroscopic Determination of Residual Stresses in Polycrystalline Alumina," *J. Am. Ceram. Soc.*, 77 298-302 (1994).
5. J. He, D.R. Clarke, "Determination of the Piezospectroscopic Coefficients for Chromium-Doped Sapphire," *J. Am. Ceram. Soc.*, 78 [5] 1347-1353 (1995).
6. Q. Ma, D.R. Clarke, "Stress Measurement in Single Crystal and Polycrystalline Ceramics Using Their Optical Fluorescence," *J. Am. Ceram. Soc.*, 76 1433-1440 (1993).
7. K.W. Schlichting, K. Vaidyanathan, Y.H. Sohn, E.H. Jordan, M. Gell, N. P. Padture, "Application of Cr³⁺ Photoluminescence Piezo-Spectroscopy to Plasma-Sprayed Thermal Barrier Coatings for Residual Stress Measurement," *Materials Science and Engineering*, A 291 [1-2] 68-77 (2000).
8. M. Gell, E.H. Jordan, K. Vaidyanathan, K. McCarron, B. Barber, Y.H. Sohn, V.K. Tolpygo, "Bond Strength, Bond Stress and Spallation Mechanisms of Thermal Barrier Coatings," *Surface and Coatings Technology*, 120-121 53-60 (1999).
9. Y.H. Sohn, K. Vaidyanathan, M. Ronski, E.H. Jordan, M. Gell, "Thermal Cycling of EB-PVD/MCrAlY Thermal Barrier Coatings: II. Evolution of Photo-Stimulated Luminescence," *Surface and Coatings Technology*, 146-147 102-109 (2001).

10. S. Sridharan, L. Xie, E. H. Jordan, M. Gell, "Stress Variation with Thermal Cycling in the Thermally Grown Oxide of an EB-PVD Thermal Barrier Coating," *Surface and Coatings Technology*, 179 [2-3] 286-296 (2004).

11. L.D. Xie, Y.H. Sohn, E.H. Jordan, M. Gell, "The Effect of Bond Coat Grit Blasting on the Durability and Thermally Grown Oxide Stress in an Electron Beam Physical Vapor Deposited Thermal Barrier Coating," *Surface and Coatings Technology*, 176 [1] 57-66 (2003).

12. D.M. Lipkin, D.R. Clarke, "Sample-Probe Interactions in Spectroscopy: Sampling Microscopic Property Gradients," *J. Appl. Phys.*, 77 [5] 1855-1863 (1995).

13. R.J. Christensen, D.M. Lipkin, D.R. Clarke, K.S. Murphy, "Nondestructive Evaluation of the Oxidation Stresses Through Thermal Barrier Coatings using Cr^{3+} Piezospectroscopy," *Applied Phys. Lett.*, 69 [24] 3754-3756 (1996).

14. D.M. Lipkin, D.R. Clarke, M. Hollatz, M. Bobeth, W. Pompe, "Stress Development in Alumina Scales Formed Upon Oxidation of (111) NiAl Single Crystals," *Corrosion Science*, 39 [2] 231-242 (1997).

15. D.R. Clarke, R.J. Christensen, V.K. Tolpygo, "The Evolution of Oxidation Stresses in Zirconia Thermal Barrier Coated Superalloy Leading to Spalling Failure," *Surf. Coat. Technology*, 94-95 89-93 (1997).

16. X. Peng, D.R. Clarke, "Piezospectroscopy Analysis of Interface Debonding in Thermal Barrier Coatings," *J. Am. Ceramic Soc.*, 83 [5] 1165-1170 (2000).

17. Y.H. Sohn, K. Schlichting, K. Vaidyanathan, E. Jordan, M. Gell, "Communication: Non-Destructive Evaluation of Residual Stress in Thermal Barrier Coated Turbine Blades by Cr^{3+} Photoluminescence Piezo-Spectroscopy," *Metallurgical Mater. Trans.*, 31A 2388-2391 (2000).

18. E.H. Jordan, Swetha Sridharan, L.D. Xie, M. Gell, "Damage Evolution in an Electron Beam Physical Vapor Deposited Thermal Barrier Coating as a Function of Cycle Temperature and Time" Submitted to *Materials Science & Eng. A.*

19. E.H. Jordan, Y.H. Sohn, W. Xie, M. Gell, L.D. Xie, F. Tu, K.R. Pattipati, P. Willet, "Residual Stress Measurement of Thermal Barrier Coatings using Laser Fluorescence Technique and Their Life Prediction," in *Autotestcon*, Valley Forge, PA, 2001.

20. A. Selcuk, A. Atkinson, "Analysis of the Cr^{3+} Luminescence Spectra from Thermally Grown Oxide in Thermal Barrier Coatings," *Materials Science & Eng.*, A 335 [1-2] 147-156 (2002).

21. A. Selcuk, A. Atkinson, "The Evolution of Residual Stress in the Thermally Grown Oxide on Pt Diffusion Bond Coats in TBCs," *Acta Materialia*, 51 [2] 535-549 (2003).

22. A. Atkinson, A. Selcuk, S.J. Webb, "Variability of Stress in Alumina Corrosion Layers Formed in Thermal-Barrier Coatings," *Oxidation of Metals*, 54 [5-6] 371-384 (2000).

23. J.A. Nychka, D.R. Clarke, "Damage Quantification in TBCs by Photo-Stimulated Luminescence Spectroscopy," *Surface and Coatings Technology*, 146-147 110-116 (2001).

24. K. K. Vaidyanathan, E. H. Jordan, M. Gell, "Surface Geometry and Strain Energy Effects in the Failure of a (Ni, Pt) Al/EB-PVD Thermal Barrier Coating," *Acta Materialia*, 52 1107-1115 (2004).

25. M. Wen, E.H. Jordan, M. Gell, University of Connecticut, unpublished results.

26. X.Y. Gong, D.R. Clarke, "On the Measurement of Strain in Coatings Formed on a Wrinkled Elastic Substrate," *Oxid. Met.*, 50 355-376 (1998).

27. J.W. Hutchinson, Z. Suo, "Mixed Mode Cracking in Layer Materials," in *Adv. Appl. Mech.*, eds. J.W. Hutchinson, YY Wu, 29 63-191 (1992).

28. N. Goldfine, D. Schlicker, Y. Sheiretov, "Conformable Eddy-Current Sensors and Arrays for Fleet Wide Gas Turbine Component Quality Assessment," *Monitoring and Diagnostics Applications*, ASME Turbo Expo Land, Sea & Air, 1-9, June 2001.

Mechanical Properties

ELASTIC AND INELASTIC DEFORMATION PROPERTIES OF FREE STANDING CERAMIC EB-PVD COATINGS

Marion Bartsch and Uwe Fuchs
Institute of Materials Research, German Aerospace Center (DLR)
Linder Hoehe
D-51103 Cologne, Germany

Jianmin Xu
Rolls-Royce Deutschland Ltd & Co KG
D-15827 Blankenfelde-Mahlow, Germany

ABSTRACT

Thermal barrier coatings (TBC) processed by electron beam physical vapor deposition (EB-PVD) show complex deformation behavior due to their columnar microstructure and multi-scale porosity. Limited information is available regarding the inelastic deformation behavior. Furthermore, the existing data for the elastic properties of EB-PVD coatings varies in a wide range, depending on the measurement method. In most cases, the fragile EB-PVD coatings are tested on the substrate, onto which they were deposited. Since the ceramic coatings are deposited onto metallic substrates, which have about 1000°C, compressive residual stresses develop in the coating after cooling down. Since the deformation behavior of EB-PVD coatings is stress dependent, the residual stresses cannot be calculated without knowledge of the stress-strain response, and without knowledge of the residual stress the elastic properties cannot be measured correctly. To overcome this dilemma, tubular free standing EB-PVD coating samples with a thickness of 250μm have been processed and tested in compression at room temperature. For introducing the mechanical load safely into the gauge length a soft clamping has been developed, and the strain was measured using a contact-free laser extensometer. Several subsequent loading and unloading cycles were performed on specimens with different thermal pre heat treatments. Non-linear stress-strain behavior was observed, showing increasing stiffness with increasing compressive stress. Cycle by cycle the stiffness increased, indicating inelastic deformation, which has also been observed in creep experiments in the same test configuration. Long term thermally aged specimens showed higher stiffness than as processed specimens.

INTRODUCTION

The knowledge of elastic and inelastic deformation properties is important to understand and model the damage and failure behavior of EB-PVD thermal barrier coatings. The existing data on EB-PVD TBCs are limited and varies in a wide range, depending on the testing method. Several authors observed evidence for stress dependence of the elastic properties of EB-PVD coatings[1,2,3,4]. This is important, since EB-PVD coatings develop high compressive residual stresses after deposition on metallic substrates after cooling down due to the thermal mismatch between substrate and coating. Thus, it is necessary to know the residual stress state when determining the elastic properties of the EB-PVD coating. Since a free standing EB-PVD TBC is nearly stress free, it would be ideal to perform mechanical tests on it, provided that it would not be so extremely fragile. The deposition of significantly thicker TBCs does not help solving the problem since the mechanical properties of EB-PVD coatings depend on the thickness due to a porosity gradient over the thickness, which corresponds to the difference between near-to-interface and distant-to-interface columnar microstructures. For example, finite element calculations, considering the measured inter-columnar porosity over the thickness of an EB-PVD coating, predict a decrease in elastic modulus of about 30% from the near-to-interface to a 60 μm distant-to-interface microstructure [5]. The porosity gradient has also to be considered when testing flat free standing specimens in bending. The challenge is to achieve free standing coatings like for industrial applications with a thickness of about 200 to 300 μm with sufficient stability and to test them in-plane. Some advantages concerning stability provide tubular specimens, which have been successfully processed in the Institute of Materials Research of the German Aerospace Center in Cologne. A technique has been developed by the authors to test these tubular specimens in compression, using soft fixtures to clamp the specimens and a laser extensometer for contact-free strain measurement. Tests have been performed on specimens, which have been mechanically stabilized by thermal pre-treatment and on specimens, which have been additionally aged at high temperature, in order to study the effect on changes of the mechanical properties due to the thermal ageing. The data obtained on the free standing tubular specimens are compared to data reported in the literature on the same coating system but using different testing methods.

EXPERIMENTAL
Specimens

EB-PVD coatings were deposited onto metallic cylinder-shaped substrates using standard coating conditions with rotating the cylinders around their length axis. The coating was subsequently removed from the substrate so that as free standing tubes with a thickness of 250μm were produced, which is typical in standard coatings for industrial application. Tubes with 2 different diameters - 12, and 19 mm - were processed. Specimens of about 20mm length were cut from the tubes by laser cutting. The cut edges were plan-parallel but small parts of the coatings always broke from the edges, see fig. 1.

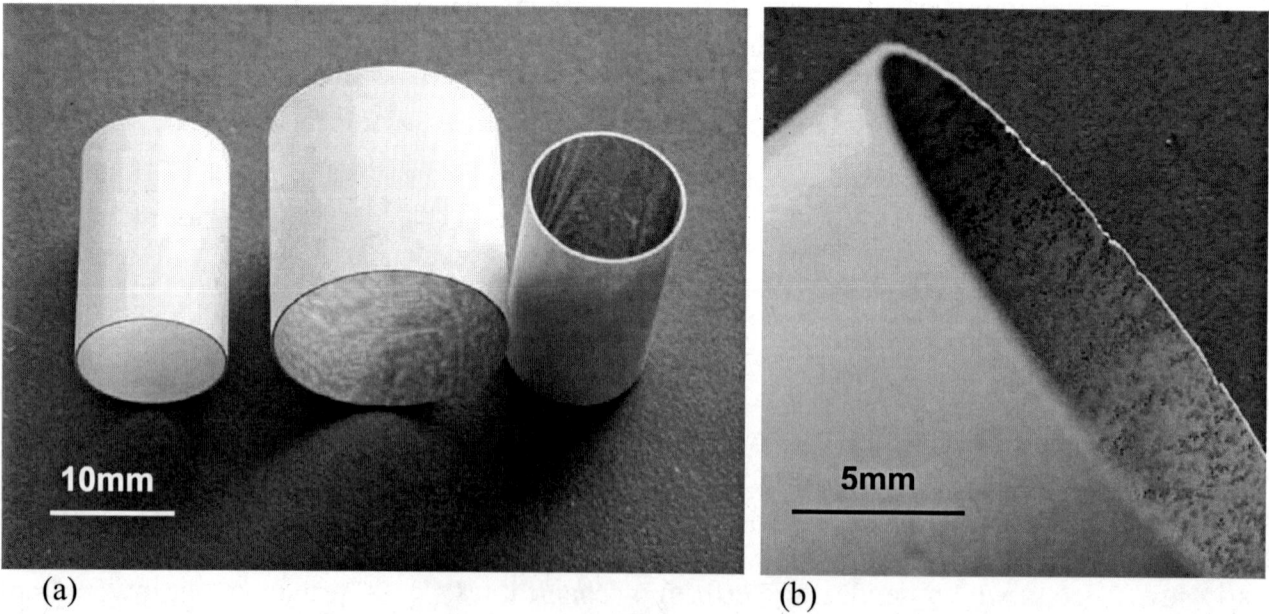

(a)   (b)

Fig.1. Free standing tubular specimens after laser cutting (a) and detail of a laser cut edge (b).

The ceramic coatings were made from zirconia, partially stabilized with 7-8 wt % yttria. The microstructure of the coatings was identical to standard coatings as used for gas turbine blades and showed the typical columnar structure with higher density at the inner surface toward the (later removed) substrate than at the outer surface (fig. 2).

(a)   (b)

Fig. 2. Free standing EB-PVD coating, overview (a) and enlarged detail (b)

Due to their columnar microstructure, the specimens were extremely fragile in the as coated condition. In order to strengthen the connections between the columns and thus to avoid fracture by handling, all specimens were thermally pre treated according to industrial standard. Some specimens have been additionally thermally aged for long term in air in order to investigate the effect of microstructural changes, such as sintering, on the mechanical properties. The long term thermal ageing resulted in a slight reduction of the diameter (12.65 mm to 12.5 mm) and often in damage of the specimens by the formation of cracks at the edges, so that only one intact specimen was available for further testing. An overview of the tested specimens is given in Table I.

Table I: Overview over tested specimens

Specimen Nr.	outer diameter [mm]	thermal ageing in air	comment	performed tests
1	19,5	-		compr. load cycles, creep
2	19,5	-		compr. load cycles, creep
3	19.5	-		1 compressive load cycle
4	12.5	1130°C/ 201h		compr. load cycles
5	12.5	1130°C/ 238h	partially cracked at edge	compressive load cycles

Testing procedure

Before testing in compression, soft fixtures were attached to both ends of the specimens in order to even out slivered parts of the edges, avoid stress concentrations, and stabilize the free standing tubes. The fixtures consist of polycarbonate caps, which were laid out with a layer of flexible glue. The specimens were aligned, the ends fixed to the caps, and glued into the cap. Finally, the specimens were furnished with parallel stripes, which give a good contrast for contact–free strain measurements by means of a laser extensometer in further compression testing. Fig. 3 shows a schematic of the fixture and a photograph of a specimen in testing configuration.

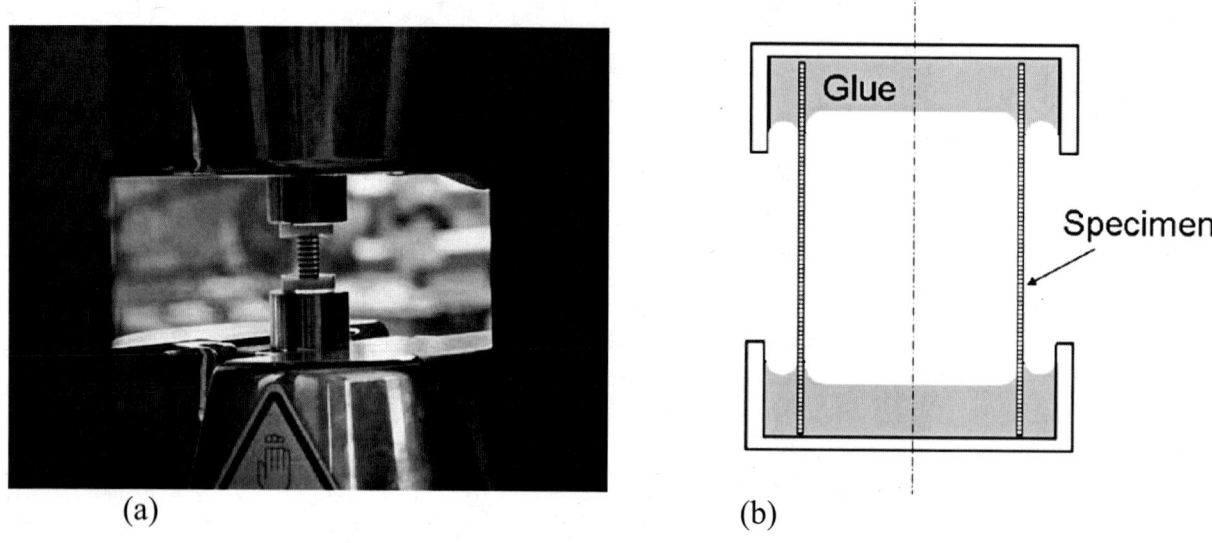

(a)   (b)

Fig.3. Photograph of a specimen in testing configuration (a) and schematic of the specimen fixture (a)

Compressive loading and unloading tests and some creep tests have been performed in the same test configuration with an electro-mechanic spindle testing machine. The specimens were loaded in several subsequent test cycles until a defined load and unloaded until a load of about 10N. The load was measured by means of a load cell and controlled manually. The displacement rate of the cross head was in the first experiments selected to 0.01mm/min and changed to 0.1mm/min in order to save testing time. The strain was measured by means of a laser extensometer (Fiedler Measurement Systems), which emits a laser beam scanning the specimen. The laser signals, which were reflected from the contrast-rich stripes on the specimens, are utilized to measure the strain; changes between the distances of the stripes due to deformation of the specimen result in changes between the time distances of the maxima of the reflected laser signal. The precision of the measurement depends on the accuracy of the stripe pattern on the specimen. On the first specimens, the stripe pattern was painted manually using a microscope, resulting in stripes with some arbitrary irregularities. Later, it was recognized that the fixture stabilizes the specimens sufficiently to press on the stripe pattern provided by the manufacturer of the laser extensometer. Using the press-on stripes instead of painting the stripes resulted in significantly less scatter of the monitored strain data.

RESULTS

Several test sets with subsequent compressive loading and unloading cycles have been performed on the specimens, see Table I. Specimens without ageing showed non-linear deformation behavior with an increasing 'elastic modulus' with increasing mechanical load. Between loading and unloading a hysteresis occurred. After the first loading/unloading cycle the slope of the load-strain curve became steeper and the hysteresis loop became narrower. After about 5 cycles a steady state was reached. However, when the test set was repeated after some minutes of total unloading, the specimens recovered and showed almost the same load-strain behavior like in the preceding test set. Fig. 4a displays the load-strain data of an exemplary test set and Fig. 4b gives an overview of measured values for the elastic modulus as a function of the applied compressive stress.

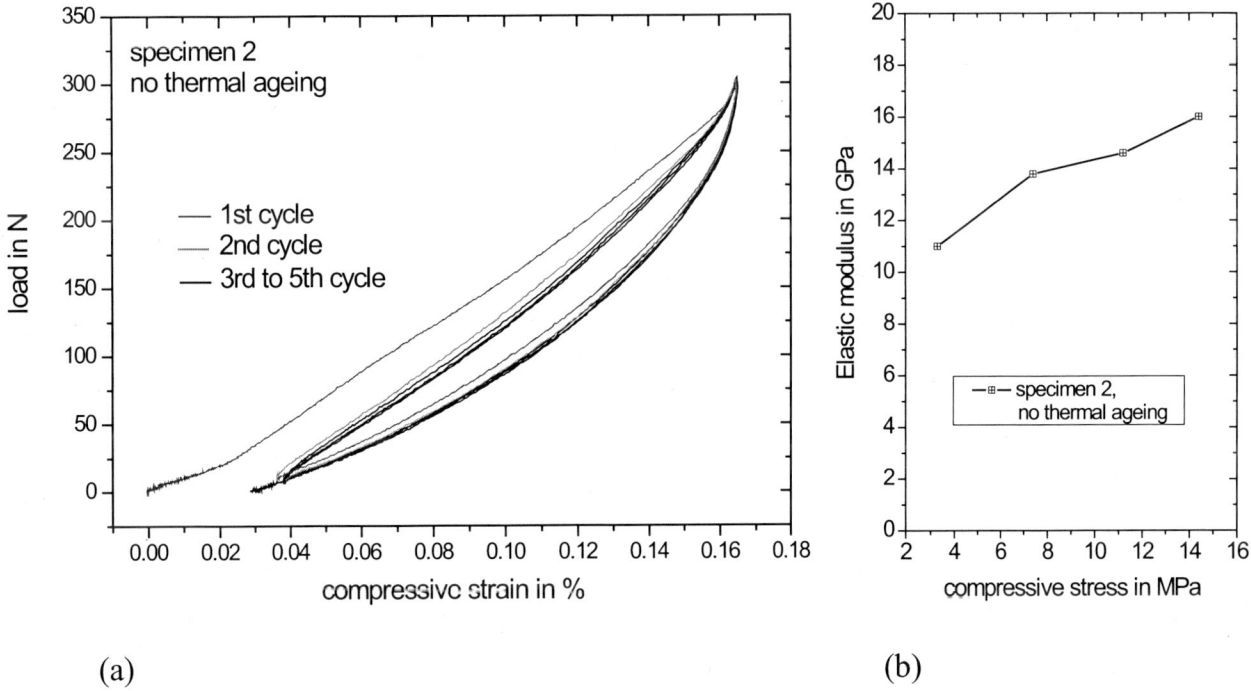

(a)   (b)

Fig. 4. Load – strain behavior of a specimen (a) and overview of measured elastic-moduli as function of compressive stress (b)

Specimens, which have been thermally aged, did not show an expressed non-linear behavior and no hysteresis, but the elastic modulus of the specimens increased in subsequent loading-unloading cycles. Exemplary loading-unloading cycles of specimen 4 and an overview of the measured elastic moduli are displayed in Fig. 5. The specimen 5 developed cracks at the edges during the long term thermal treatment and showed at higher loads untypical non-linear behavior. However, at loads between 50 and 100N the load-strain behavior was almost linear and a value of 66GPa has been measured, which agrees well with the data on the intact thermally aged specimen.

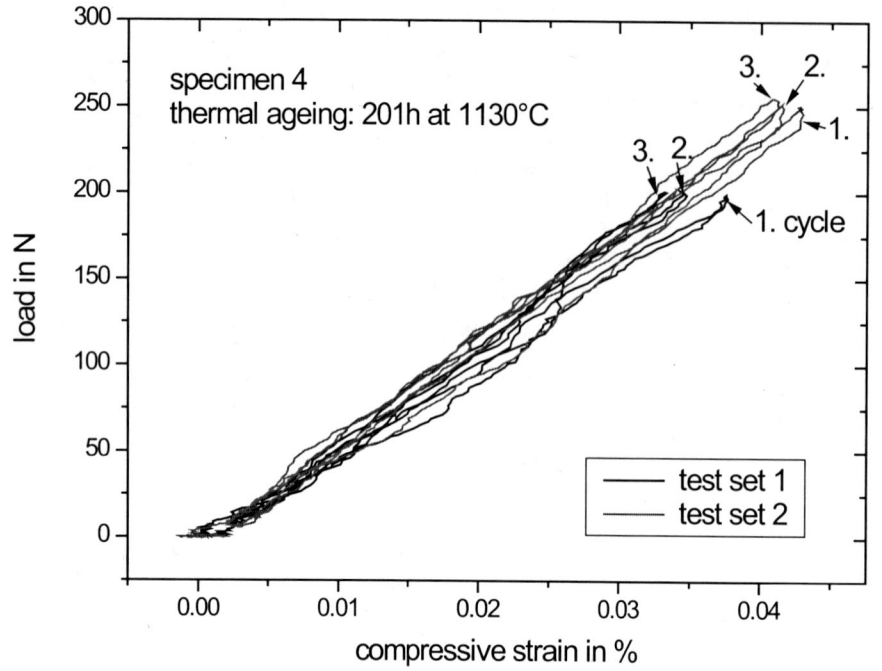

Fig. 5. Load-strain behavior of a specimen after thermal ageing

cycle	F_{max} [N]	E [GPa]
1st		55
2nd	200	59
3rd		61
1st		58
2nd	250	60
3rd		62

Some creep tests have been performed on non aged specimens in the same test configuration like the loading-unloading cycles. The specimens displayed time dependent deformation behavior, specimen Nr. 2 fractured during the creep test after 4900 seconds. The test data are displayed in Fig. 6.

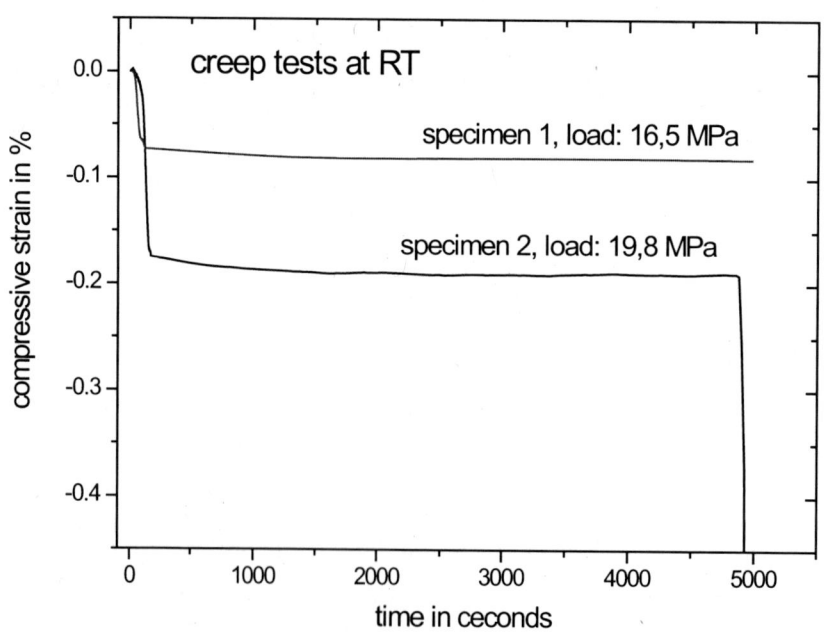

Fig. 6. Compressive strain as function of time under constant load for two specimens without ageing

DISCUSSION

Non linear stress-strain behavior was observed in compressive tests on free standing EB-PVD zirconia coatings. The non linear behavior depends on the stress-level, the loading history, and on the thermal history of the specimens. The stress dependence of EB-PVD coatings is thought to depend on micro-contacts between the columns of the coating, following the micro-contact model discussed by Szücs[2]. In stress free condition micro-contacts between columns ensure the integrity of the coating. With increasing compressive stress, existing micro-contacts deform and subsequently more micro-contacts can form and bridge the gaps between single columns, so that each further deformation increment requires a higher compressive stress increment. This model can be used to explain the non linear load-strain behavior. Assuming that many contacts are not rigid, dissipation by local friction can occur and led to the observed hysteresis between loading and unloading. Thermally aged specimens did not show expressed non-linearity or hystereses, which may be attributed to the formation of many rigid micro-contacts due to sintering. The formation of sinter-contacts between the columns of EB-PVD coatings have been observed by several authors[6,7] and can explain also the higher values obtained for the elastic modulus of thermally aged specimens compared to non aged ones. The micro-contact model may also be used to explain irreversible strain fractions, as indicated by the increase of elastic moduli in subsequent cycles, especially after the first cycle by assuming micro-fracture at the inter-columnar contacts. However, it does not provide a sound explication for the observed recovery effects. Also not clear is the mechanism behind the creep behavior at room temperature. Further experiments and microstructural investigations are planned in order to elucidate the creep mechanisms.

The data for elastic moduli obtained on the free standing tubes vary in the case of the non aged specimens between 10 and 18 GPa in the first load cycle at low stresses of about 4 MPa and between 14 and 26 GPa at about 12 MPa in repeated cycling. The differences between the individual specimens may be attributed to differences in the coating conditions, because the specimens were cut from tubes of about 80 mm length, which are subjected to different coating conditions depending on the location in the vapor cloud during deposition. The obtained values are comparable to the value of 16 GPa, reported by Szücs[2] on 250 µm thick flat free standing specimens in dynamic mechanical bending tests. Data reported on EB-PVD coatings on metallic substrates vary between 15 and 116 GPa, depending on the residual stress in the coating system and the testing method.[1,2,4,8] The thermal ageing led to a significant increase of the elastic modulus of the free standing tubes to values between 55 and 66 GPa. Dynamic mechanical bending tests on 250 µm thick flat free standing specimens gave an increase of the elastic modulus from 16 GPa before to 30.5 GPa after 4h ageing at 1000°C in air[2], and own measurements on 220µm thick EB-PVD coatings on IN 625 substrates resulted in an increase of the elastic modulus from about 35 GPa before to 72 GPa after 1000h ageing at 1000°C in air.[9] In the variety of data the obtained values on the free standing TBC tubes are reasonable.

CONCLUSIONS

Compressive testing of free standing tubular EB-PVD coatings gave reasonable results for the elastic and inelastic deformation behavior. The observed non-linear behavior and loading-unloading hystereses observed on the specimens can be explained by a micro-contact model, assuming the formation of increased inter-columnar contacts during compressive loading.

ACKNOWLEDGEMENTS

This investigation was funded by the German Bundesministerium für Wirtschaft und Technologie on the research program MARCKO with FKZ 03266888C. The authors thank Klaus Kröder and Jörg Brien for manufacturing the free standing tubular coatings, Dan Renusch (DECHEMA e.V) for performing the laser cutting of the specimens, and Liudmila Chernova (German Aerospace Center, DLR) for taking the micrographs.

REFERENCES

[1]C.A, Johnson, J.A. Ruud, R. Bruce, D.Wortmann, "Relationship between residual stress, microstructure and mechanical properties of electron beam – physical vapor deposition thermal barrier coatings", Surf. & Coat. Technology **108-109**, 80-85 (1998).

[2]F. Szücs, "Thermomechanische Analyse und Modellierung plasmagespritzter und EB-PVD aufgedampfter Wärmedämmschicht –Systeme für Gasturbinen", PhD-Thesis Technical University Berlin, in German (1997).

[3]U. Schulz, K. Fritscher, C. Leyens, M. Peters, "Influence of processing on microstructure and performance of electron beam physical vapor deposition (EB-PVD) thermal barrier coatings", Jour. Engineering for Gas Turbines and Power, **124**, 229-234 (2002).

[4]X. Zhao, P. Xiao, "Residual stresses in thermal barrier coatings measured by photoluminescence piezospectroscopy and indentation technique", Surf. & Coat. Technology **201**, 1124-1131 (2006).

[5]M. R. Locatelli, E. R. Fuller, Jr., "Using OOF to model mechanical behaviour of thermal barrier coatings", presentation given at the int. conf. on adv. ceramics and composites, Cocoa Beach, FL (2002). http://www.ctcms.nist.gov/~fuller/PRESENTATIONS/index.html

[6]X. Zhao, X. Wang, P. Xiao, "Sintering and failure behaviour of EB-PVD thermal barrier coating after isothermal treatment", Surf. & Coat. Technology **200**, 5946-5955 (2005).

[7]C. Leyens, U. Schulz, B.A. Pint, I.G. Wright, "Influence of electron beam physical vapor deposited thermal barrier coating microstructure on thermal barrier coating system performance under cyclic oxidation conditions", Surf. & Coat. Techn. **120-121**, 68-76 (1999).

[8]T. Lauwagie, K. Lambrinou, I. Mircea, M. Bartsch, W. Heylen, O. Van der Biest, "Determining the elastic moduli of the individual component layers of cylindrical thermal barrier coatings by means of a mixed numerical-experimental technique", Materials Science Forum **492-493**, 653-658 (2005).

[9]M. Bartsch, U.Schulz, B. Saruhan, "EB-PVD thermal barrier coatings for gas turbines, part I: processing", Proceedings of the summer school of the European Research Training Network SICMAC – HRPN-CT-2002-00203, 11th-16th June 2006, Maó, Spain, 191-200.

CREEP BEHAVIOUR OF PLASMA SPRAYED THERMAL BARRIER COATINGS

Reza Soltani, Thomas W. Coyle, Javad Mostaghimi
Centre for Advanced Coating Technologies, University of Toronto
40 St George Street
Toronto, Ontario, Canada

ABSTRACT

Engineers have always been concerned about the creep/sintering properties of thermal barrier materials, especially thermally sprayed zirconia coatings, which in the past few years have received exceptional attention. Under service conditions, the surface of these coatings is under compressive stress as a result of the high surface temperatures and steep thermal gradient. Stress relaxation occurs through mechanisms such as creep and sintering. Upon cooling, a new stress distribution develops, which may introduce cracks and reduce service life. In this work coatings from two different zirconia feedstocks (nanostructured and hollow sphere particles, HOSP^{TM}) with 6-8Wt% Y_2O_3, as stabilizer, were prepared by air plasma spraying. By altering the process parameters of spraying, different amounts of porosity and non-melted particles were incorporated into the deposits. The creep strain and creep rate of these coatings were measured in four point flexure under a range of load and temperature levels. Results show that in spite of having almost the same rate of creep in the secondary stage (equal stress exponents, n), the total creep strain of coatings produced from nanostructured feedstock is lower than the coatings produced by the hollow sphere feedstock.

1. INTRODUCTION

Plasma sprayed thermal barrier coatings for diesel engines and turbine blades are being developed, however the scenario for diesel engines is different than that for turbines. As operating temperatures approach 1000°C the presence of relatively high compressive stresses at the coating surface lead to stress relaxation by creep and/or sintering. Shrinkage at high temperatures followed by crack initiation and propagation during cooling will cause spallation of the coating, consequently shortening the lifetime of the deposit. Creep/sintering behaviour of zirconia and its plasma sprayed form have been investigated by several researchers.[1-10]

At service temperatures sintering may heal microcracks and eliminate some porosity, creating a denser structure which could be an advantage in reducing oxygen transport to the bond coat. Unfortunately upon cooling micro cracks reappear. Eventually the coating becomes sufficiently damaged that spalling occurs. The thickness of the coating is another important factor; nowadays depositions of more than 1mm in thickness are required in some applications to increase protection and reduce heat transfer. The thicker the coating, the greater the thermal gradients present during service, which results in higher stresses within the cross section of the material.[3]

Therefore the development of a new top coat structure which shows more resistance to sintering is a topic of active research. Bimodal structured thermal barrier coatings can provide the improved performance of an alternative material without the necessity of redesigning the entire component. Researchers have shown that bimodal nanostructured coatings can exhibit better properties than conventional coatings of the same composition.[11-13]

In the current work, partially stabilized zirconia (PSZ) coatings were deposited by air plasma spraying. Two different types of powder feedstock were employed, a powder consisting of hollow spherical particles and a nanostructured particle powder. The influence of the powder type on the thermo-mechanical behaviour of the coatings was investigated.

2. EXPERIMENTAL PROCEDURE

Two commercially available 6-8 wt% Y_2O_3-PSZ powders were used to deposit coatings by air plasma spraying. A nano-structured powder (Nanox S4007, Inframat Corp., CT, USA) was deposited using the SG-100 (Praxair, Concord, NH, USA) plasma torch. This powder consists of porous agglomerated particles 15-150 μm in diameter made up of crystallites on the order of 200 nm. A powder consisting of hollow spherical particles (204B-NS, Sulzer-Metco, NY, USA) was deposited with the same gun. The particle size of this HOSP[TM] powder (hollow oven spherical particles)was 45-75 μm.

A range of deposition parameters was used to produce coatings to get structures with different levels of porosity and non-melted particles. These parameters as well as mean velocity and temperature of in flight particles have been given elsewhere.[19] Substrates were carbon steel plates, with dimensions of 50x50x4 mm (WxLxH). The thickness of the coating was more than 3 mm. Samples then were cut to 3x4x40 mm (HxWxL) and then substrates were detached from the coatings. The tensile and compressive surfaces of the specimens, where the upper and lower rollers of flexure fixture contact, then were ground to assure the surfaces were flat, smooth, and parallel.

The powder and coating microstructures were examined by scanning electron microscopy (SEM Hitachi S-4500, Japan) using low voltage. The chemical composition of the powders was obtained from flame photometry (FP) and inductively coupled plasma (ICP). Image analysis (Clemex Vision Professional, Clemex, QC, Canada) was used to measure the average porosity of each sample; details were published elsewhere.[19] Creep tests were conducted employing a SiC four-point bend test fixture (with 20 and 40 mm inner and outer spans, respectively) inside a box furnace at elevated temperatures and under a range of applied loads.

3. RESULTS

Microstructure: prior to conducting creep tests, SEM investigations of feedstock particles and coatings deposited using 204B-NS and Nanox powders was conducted, Figs.1 and 2. Thin shell particles increase the probability of having fully melted particles resulting in very thin final splats in coating.

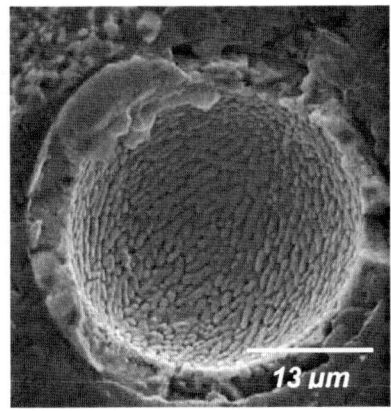 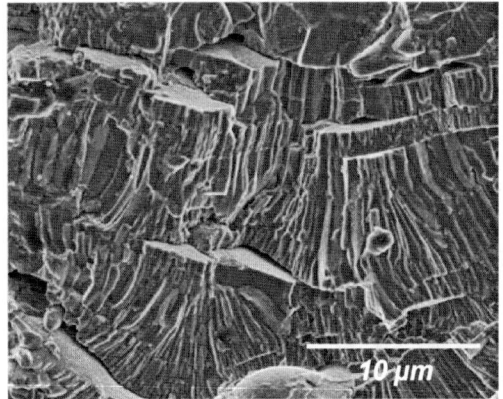

Fig. 1. Typical SEM image of 204B-NS; left: feedstock particle and right: polished and etched cross section of 204B-NS coating.

Feedstock of Nanox particles is an agglomeration of nano particles in spherical shapes; some of these nano species will be retained in coating as is clear in Figure 2.

Porosity and non-molten nano particle percentages in deposits were investigated by image analysis, Table I. The most significant process parameters were argon and hydrogen flow rate, current and type of powder feeder (internal/external).

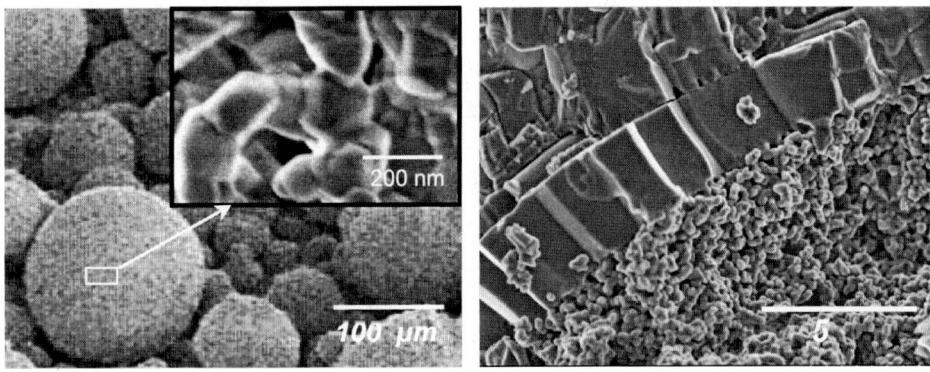

Fig.2. SEM image of Nanox S4007; left: feedstock particles and right: fractured cross section showing nano particles next to big columnar grains of an adjacent splat.

In this work the Nanox sample which showed the lowest thermal diffusivity is compared with the 204B-NS deposit.

Table I: Porosity and non molten percentages of Nanox and 204B-NS coatings.

Powder	Plasma power (KW)	Feeder	Velocity m/sec	Temp. °C	Thermal diffusivity m^2/Sec	Image Analysis %	
						Porosity	Non molten
Nanox	17	Ext.	130	2533	5.09E-7	17 ±1	4.5 ±1
204B-NS	25	Int.	144	2809	6.12E-7	11 ±1	---

Int.: Internal, Ext.:External

Creep tests were performed at a range of temperatures (800,1000 and 1200 ºC) and stresses for 2 days. Creep strain and strain rate of samples were calculated and plotted versus time, details can be found elsewhere[14]. At least two specimens of each sample were tested and results showed a very good repeatability.

Fracture images of 204B-NS samples, under 30 N applied force at 1000 ºC, illustrate phenomena such as crack healing and inter/intra splat grain growth due to material transport by diffusion, Fig.3. Under these circumstances, SEM investigations show that inter-splat grain growth is more significant than intra-splat joining of grains. It may be because of a preferential growth direction of the columnar grains created during solidification and the stable geometrical shape of grains in a splat.

Figure 4 shows SEM images of some nano particles after two and five days under 30 N bending load at 1000ºC. Little sintering is apparent after two days but after five days extensive sintering could be observed in many areas.

Figure 5 shows the results of creep tests conducted under a range of stresses for Nanox and 204B-NS coatings at 1000 ºC. With increasing applied stress, creep strain rises but there is not any remarkable difference in strain rates after the first day of testing, showing little dependency of strain rate on stress in the quasi-stationary stage.

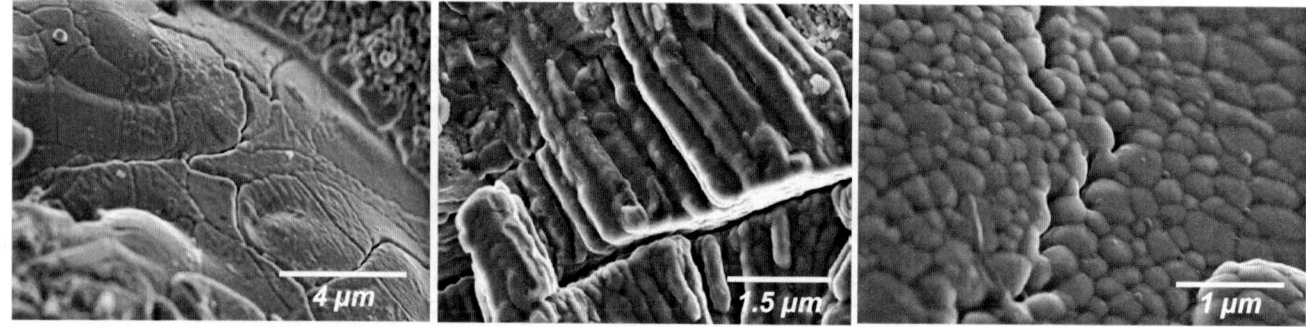

Fig. 3. SEM image of 204B-NS splats after two days at 1000 ºC under 30 N flexure load; left: surface of splat shows intra splat grain growth, middle: inter splats grain growth, right: micro-crack healing.

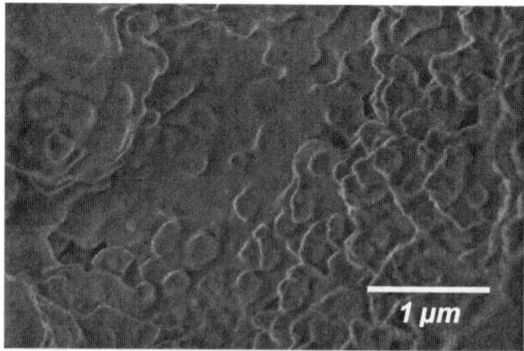

Fig. 4. SEM image of non molten nano particles after 5 days at 1000 ºC under 30 N flexure

Figure 5 shows the results of creep tests conducted under a range of stresses for Nanox and 204B-NS coatings at 1000 ºC. With increasing applied stress, creep strain rises but there is not any remarkable difference in strain rates after the first day of testing, showing little dependency of strain rate on stress in the quasi-stationary stage.

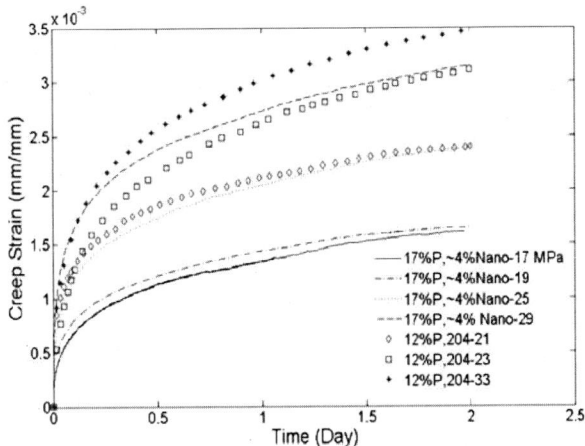

Fig. 5. Creep strain of 204B-NS and Nanox under different stresses at 1000 °C

Plotting strain rate against stress in a log-log scale will provide the stress exponent in a power law equation, Fig. 6. Creep exponents of Nanox and 204B-NS samples are 1.1 and 1.3, respectively, very close to 1.

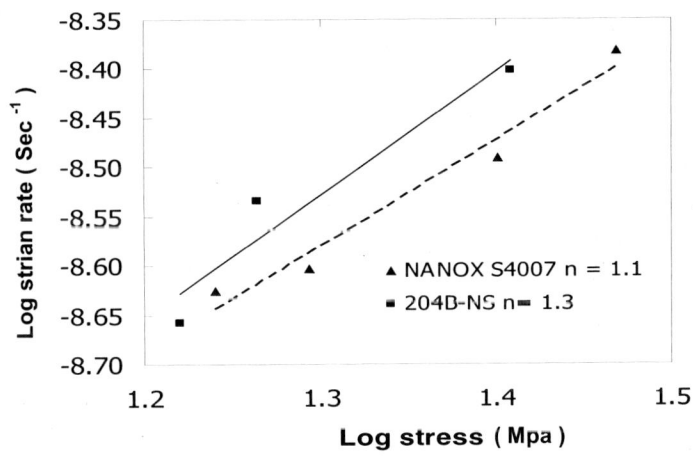

Fig.6. Stress versus strain rate of 204B-NS and Nanox samples at 1000 °C .

Activation energy is another concern in creep properties; lower activation energy generally means a lower resistance to creep.

Thermal barrier coatings experience compression stresses at high temperatures. If relaxation mechanisms such as creep and/or sintering eliminate this compressive stress, upon cooling to room temperature tensile stresses will develop in the coating which could in turn lead to cracking of the coating; consequent delamination reduces the service life of the deposit. Therefore a high activation energy in thermal barrier deposits is advantageous. In order to determine the apparent activation energy, creep tests were conducted for both materials under the same stress and at temperatures 800, 1000 and 1100°C, Fig.7.

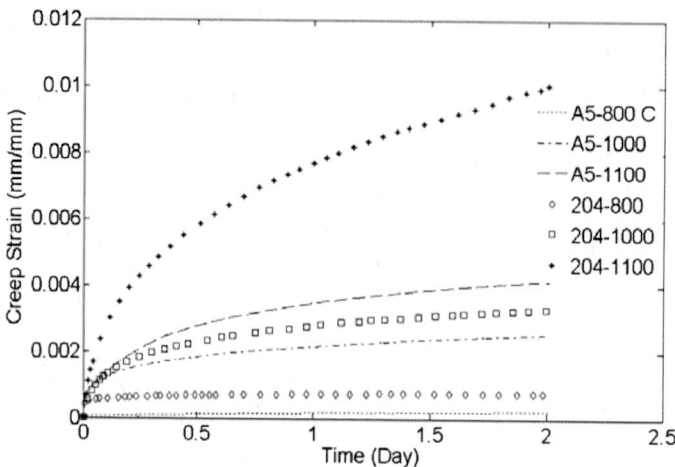

Fig. 7. Creep strain of 204B-NS samples under 25 MPa flexure stress at different temperatures.

Plasma sprayed thermal barrier coatings posses a long primary creep stage which is not only a function of temperature and stress but also is time dependent. Generally the strain rate of these samples can be expressed by following equation :

$$\dot{\varepsilon} = A \left\{ \frac{\sigma^n}{D^p} \right\} \exp \left\{ \frac{-Q}{RT} \right\} t^{-s} \qquad (1)$$

Equation (1) was used to plot natural log of strain rate versus 1/T, where $\dot{\varepsilon}$ denotes creep rate (Sec[-1]), A, a material constant (Sec[-1]), σ the normalized applied stress, n the stress exponent, D grain size, p grain size exponent, Q the apparent activation energy of creep (J/mol), R the gas constant (J/mol K) and T the absolute temperature (K). Slope of the fitted line will give Q. Figure 8 shows results for two samples of Nanox and 204B-NS, having apparent activation energies of 154 and 190 KJ/mol, respectively.

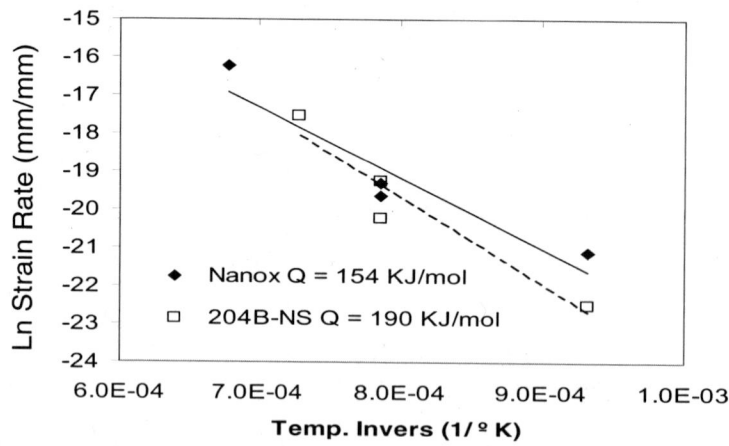

Fig. 8. Activation energy of 204B-NS and Nanox under 30 MPa flexure stress.

4. DISCUSSION

The amount of deformation in a creep test,^{TM}, is a function of stress σ, time t , temperature T and structure S[15]:

$$^{TM} = f(\sigma, t, T, S) \tag{2}$$

Since deformation is strongly dependent on the constitution of the material, a structure term is required which involves both macro and micro structural features .[26] In this work the effect of some of these structural features on creep behaviour of zirconia coatings has been examined.

Compositional analysis was performed to investigate the levels of several common impurities in the two feedstocks. No major differences between the two materials were found (Table II). Particular components such as silica and alumina, which may create a glassy phase at grain boundaries, were present in almost the same amounts in both powders. Therefore compositional differences do not account for the observed differences in creep behavior.

The area percentage of non-melted nano-particles does not exceed 5% in any sample, excluding porosity from the calculated area, Table I. Therefore the discrete areas of nano particles could not have a significant effect on the creep/sintering behaviour.

Considering the lamella structure of thermally sprayed coatings, similar composition and a negligible content of non melted nano particles, the thickness of splats and the location and shape of porosity seem to be the main features influencing creep/sintering processes.

Table II: Chemical composition of the two feedstock powders; nanostructured and 204B-NS.

	Nanox(%)	204B-NS(%)	Technique
Al_2O_3	0.66	0.66	ICP
CuO	0.007	0.008	ICP
CaO	0.025	0.026	ICP
Fe_2O_3	0.006	0.009	ICP
HfO_2	1.74	1.52	ICP
P_2O_5	0.04	0.04	ICP
SiO_2	0.12	0.11	ICP
TiO_2	0.06	0.14	ICP
Y_2O_3	6.43	7.30	ICP
$Na2O$	0.03	0.02	FP

There are several parameters which affect the final thickness of splats. The velocity and temperature of the in-flight particles are particularly important. Table I shows velocity and temperature of in-flight particles for each coating. The following equations show correlations between the flattening ratio and these parameters. Temperature appears through the kinematic viscosity, which is exponentially dependant on temperatue[16]:

$$R_f = \alpha \, (\, V_p \cdot D_p / \, \nu \,)\beta$$

(3)

$$\nu = A \exp (B / KT)$$

(4)

Where R_f is flattening ratio, V_p Velocity, D_p particle diameter, ν kinematic viscosity, K Boltzmann's constant, T absolute temperature, α, β, A and B are all constants. Based on equations 3 and 4, by increasing the velocity, temperature, and mean diameter of in-flight particles, the flattening ratio increases and therefore the thickness of splats decreases. The effect of particle size is the least significant among these. Thinner splats result in a higher density of splat interfaces in a given thickness of coating.

Table I shows that the 204B-NS sample exhibited a higher velocity and temperature of in-flight particles than the Nanox coating. Empirical constants for equation 3 obtained by Yoshida[17], Madejski[18] and Solonenko[19] indicate that the average thickness of 204B-NS splats would be more than 20% less than in the Nanox deposit. Extensive SEM examination of splat thickness in chemically etched samples (H_3PO_3 at 250°C) , see Fig. 9, showed good agreement with these predictions.

The lower thickness of splats in 204B–NS coatings results in an increase of about 25% in the linear density of splat interfaces in a given thickness. The inter-splat boundaries and their associated laminar pores are well recognized as a mechanical weak link in thermal spray coatings. Sliding along splat boundaries has been suggested to play a significant role in the inelastic deformation of thermal spray coatings.[1] According to Fig. 5, 204B-NS coatings show a creep strain which is more than 50% higher than the Nanox coatings, supporting the premise that the inter-splat boundaries play a controlling role in the creep behavior.

A large primary creep strain and a low creep activation energy for plasma sprayed zirconia coatings were observed (Figs.5 and 7). The creep rate continuously decreases during the test though no steady-state regime was observed. The large decrease of creep rate with time can be explained by microstructural changes in the coating during the creep test. At the beginning of the creep test the total contact area between splats is very small.[20] At the elevated temperatures experienced during the creep test, the flat surfaces of the columnar grains at the splat surface gradually develop spherical caps as grain boundary grooves form by surface diffusion. Wherever columnar grains in adjacent splats are separated by a small distance, the spherical caps can touch and start necking, as shown schematically in Figure 10. These inter-splat bridges pin the splats together, increasing the contact area between splats.

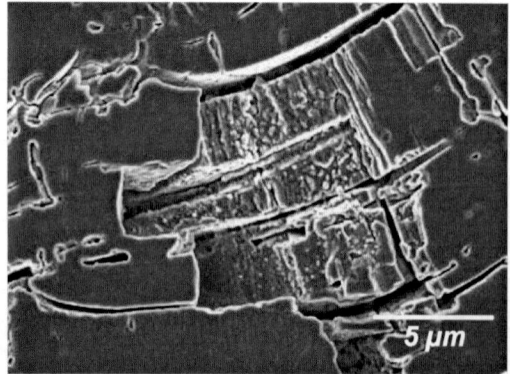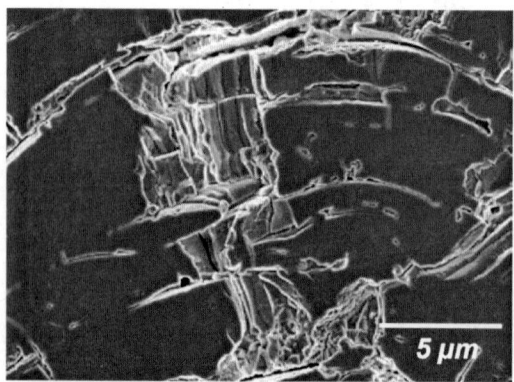

Fig. 9. Typical SEM images of Nanox (left) and 204B-NS (right) showing thickness of splats.

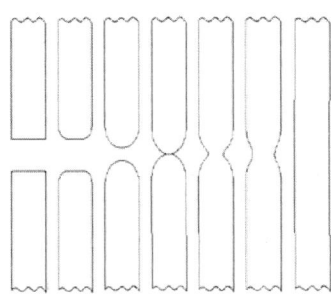 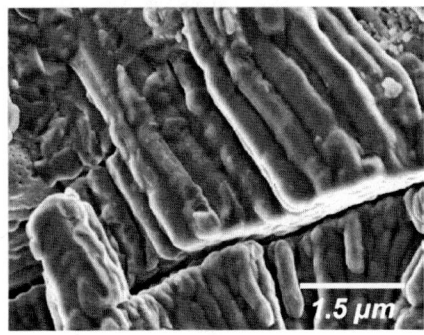

Fig. 10. Schematic and SEM images of inter-splat grain bridging.

5. CONCLUSION

At all temperatures and stress levels the bimodal coating showed lower creep strain and strain rate than 204B-NS coatings. Stress exponents were close to one, showing very little dependency of creep rate on applied stress. The apparent activation energy was 154 KJ/mol for Nanox samples and 190 KJ/mol for 204B-NS. The splat interfaces were found to be the key factor affecting the creep behavior of these coatings. The formation of bridges across splat interfaces was observed during the creep tests, and is believed to account for the steady decrease in creep rate with time.

6. REFERENCES

[1] D.Zhu and Miller R. A. , "Determination of creep behaviour of thermal barrier coatings under laser imposed temperature and stress gradients"; NASA Technical Memorandum 113169, ARL-TR-1556(1997).

[2] D.Zhu and Miller R.A., "Sintering and creep behaviour of plasma-sprayed zirconia –and hafnia-based thermal barrier coatings"; Surface and Coatings Technology, 108-109,114-120 (1998).

[3] Ed. F. Rejda, D.F. Socie and T. Itoh, "Deformation behavior of plasma-sprayed thick thermal barrier coatings"; Surface and Coatings Technology, 113,218-226 (1999).

[4] A. H. Chokshi , "Diffusion creep in oxide ceramics", Journal of the European Ceramic Society 22, 2469-2478 (2002).

[5] A. H. Chokshi, "Diffusion, diffusion creep and grain growth characteristics of nanocrystalline and fine-grained monoclinic, tetragonal and cubic zriconia"; Scripta Materialia, 48,791-796, (2003).

[6] R.Schaller, M. Daraktciev, "Mechanical spectroscopy of creep appearance in fine-grained yttria-stabilized zirconia"; Journal of the European Ceramic Society, 22, 2461-2467,(2002).

[7] M.J. Adnrews, M. K. Ferber and E. Lara-Curzio, "Mechanical properties of zirconia –based ceramics as functions of temperature" ; Journal of the European Ceramic Society, 22, 2633-2639,(2002).

[8] K.Kokini, Y. R. Takeuchi and B.D. Choules, "Surface thermal cracking of thermal barrier coatings owing to stress relaxation: zirconia vs. Mullite"; Surface and Coating Technology, 82, 77-82 (1996).

[9]L. Vasylkiv, Y. Skka, and V. V .Skorokhod, "Low-temperature processing and mechanical properties of zirconia and zirconia-alumina nanoceramics"; J. Am. Ceram. Soc., 86[2],299-304 (2003).

[10]M.Daraktchiev and R. Schaller, "High-temperature mechanical loss behaviour of 3 mol% yttria-stabilized tetragona zirconia polycrystals (3Y-TZP)"; Journal of Phys. Stat. Sol. 2, 293-304 (2003).

[11]Shaw,L., Goerman, D., Ren,R. and Gell, M., "The dependency of microstructure and properties of nanostructured coatings on plasma spray conditions"; Surface and Coating Technology, 130, 1-8, (2000).

[12]Gell, M., Jordan, E.H., Sohn, Y.H. and Gberman, D., "Development and implementation of plasma sprayed nanostructured ceramic coatings"; Surface and Coating Technology, 146-147,48-54, (2001).

[13]Chen,H. and Ding , C.X., "Nanostructured zirconia coating prepared by atmospheric plasma spraying", Surface and Coatings Technology, 150, 31-36, (2002).
sprayed nanostructured zirconia coatings", Surface and coating technology, 135, 166-172, (2001).

[14]R.Soltani, T.W.Coyle, J.Mostaghimi, "Thermo-physical property of bimodal structured plasma sprayed TBCs", To be published.

[15] W.D.Kingery, H.K.Bowen and D.R.Uhlmann, "Introduction to Ceramics",2nd Ed., John Wiley &Sons, New York,1975.

[16] P.Fauchais, M.Fukumoto, A.Vardelle, M.Vardelle, "Knowledge concerning splat formation: An invited review", Journal of Thermal Spray Technology, 13 (3), 2004, p. 337-360.

[17] J. Madejski, "Solidification of droplets on a cold surface", Int. J.Heat Mass Transfer, 19, 1976, p. 1009-1013.

[18] T. Yoshida, T. Okada, H. Hamatani, and H. Kumaoka, "Integrated fabrication process for solid oxide fuel cells using novel plasma sprayingv" , Plasma Sources Sci. Technol., 1(3), 1992, p. 195-201.

[19] O.P. Solonenko, A.V. Smirnov, V.A. Klimenov, Y.G. Butov, and Y.F.Ivanov, "Role of interfaces in splat and coatings structure formation", Physical Mesomechanics, 2(1-2), 1999, p. 113-129.

[20] R. McPherson, "A Review of microstructure and properties of plasma sprayed ceramic coatings", Journal of Surface and Coatings Technology, 39/40 ,1989, p. 173- 181

SIMULATION OF STRESS DEVELOPMENT AND CRACK FORMATION IN APS-TBCS FOR CYCLIC OXIDATION LOADING AND COMPARISON WITH EXPERIMENTAL OBSERVATIONS

R. Herzog, P. Bednarz, E. Trunova, V. Shemet, R.W. Steinbrech, F. Schubert, and L. Singheiser

Institute of Materials and Processes in Energy Systems 2,
Research Centre Juelich, 52425 Juelich, Germany

ABSTRACT

Oxidation induced spallation of plasma-sprayed thermal barrier coatings (APS-TBCs) is regarded as one major failure mode of ceramic coated gas turbine components. A failure crack path, which is located partly in the thermally grown oxide (TGO) and partly in the TBC is typical for this kind of failure (grey failure). Recent investigations have shown that the related damage evolution starts within the first 10% of life by the formation of micro cracks at the TGO and by opening of pre-existing micro cracks in the TBC. Crack growth and linking of these cracks along the interface lead to final spallation. However, parameters, which govern the kinetics and thus the life-time are not sufficiently known. Finite element simulations of the stress response near the TGO at micrometer scale were conducted corresponding to cyclic furnace tests with identical loading. The load cycle consisted of thermal cycling between 20°C and 1050°C and a dwell-time of 2 h at 1050°C. Continued TGO growth was considered (thickness increase and lateral growth). To include realistic material data, the deformation properties of both the actual NiCoCrAlY bond coat and the plasma-sprayed TBC (ZrO_2 with 7-8 wt. % Y_2O_3) as well as the oxidation kinetics have been experimentally determined and implemented in the FE code. The stress calculations showed two distinct features: (i) a fast development of high tensile stresses in the bond coat with a maximum value directly at the interface bond coat / TGO below a roughness peak, which occurred during the cooling stage and which were maximum at the lowest cycle temperature, and (ii) a development of a lateral region of larger tensile stresses alongside the roughness peak over roughness valleys. The simulation of crack formation at the interface bond coat / TGO using cohesive elements resulted in an early formation of a micro crack at the roughness peak .

INTRODUCTION

Thermal barrier coatings (TBCs) offer a considerable potential for a further increase of the turbine inlet temperature (TIT) and thus for a further increase of the efficiency of gas turbines and combined cycle power plants. Even so TBCs are still applied predominantly as a wrapping against unexpected overheating and for an increase of component life without increasing the TIT, because basic aspects of damage and spalling of thermal barrier coatings have not been understood as yet, and the time or number-of-cycles to failure cannot be reliably predicted.

Finite element analysis (FEA) can contribute to understand the degradation and damage processes by providing a tool for analysing the stress response in thermal barrier systems and for simulating the evolution of stresses and the formation and growth of cracks during cyclic and high temperature loading. By doing so, finite element analysis principally allows to separate the parameters, which affect the stress response and to assess their respective impact.

A large number of research groups have reported results from numerical simulations of the stress response at the metal/ceramic interface of TBCs ([1,2,3,4,5,6,7,8,9,10,11,12]). Different approaches have been chosen in the past years. Some workings comprise parametrical studies with respect to the thickness of the thermally grown oxide (TGO) by implementing various but constant thickness values. In these cases TGO growth stresses were neglected, because bond coat oxidation was not considered and implemented as a continuous process at high temperature. Some include oxidation as a continuous process, but did not consider creep and high temperature stress relaxation. Others again comprise plastic and partly creep and stress relaxation properties of the coatings, but generally those properties were taken from third publications and it was not possible to conclude whether the material data were representative for a real TBC composite or not. Thus, the demand for more realistic FE simulations for thermal barrier coatings based on more realistic material properties for the coatings increased in the last years. The present work aims at a further step towards FE simulations for APS TBCs with improved significance. The simulations were conducted with material data, which have been determined predominantly on actual coatings used in corresponding life-time experiments. The data comprise plastic deformation and creep/stress relaxation at high temperature. The numerical calculations include furthermore the simulation of crack formation at the bond coat / TBC interface. Both, simulations of the stress response and crack formation are compared with experimental observations.

EXPERIMENTAL OBSERVATIONS

The thermal barrier composite, which was experimentally investigated, consisted of the single-crystalline Ni-alloy CMSX-4, a NiCoCrAlY bond coat made by vacuum plasma-spraying and an APS TBC of zirconia doped with about 7-8 wt% yttria. Micrographs of the materials are displayed in Fig. 1.

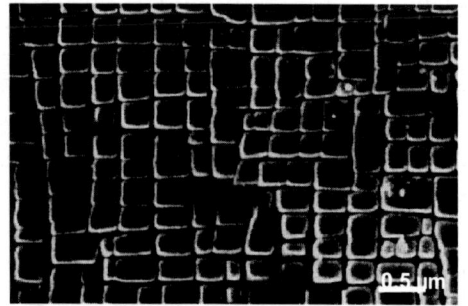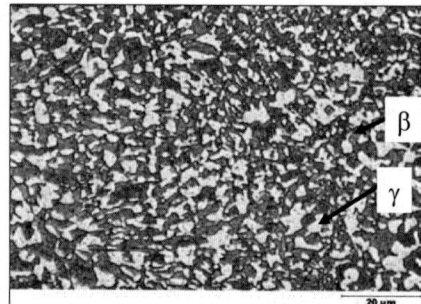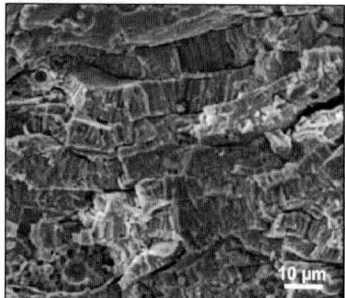

Fig. 1: Left: base material CMSX-4, center: NiCoCrAlY bond coat; right: APS TBC of zirconia doped with about 7-8 wt% yttria.

The thermal barrier composite was cyclically exposed between 60°C (approx.) and 1050°C with a dwell-time of 2 hours at 1050°C. The material was exposed until macroscopic spallation of the TBC was observed. The resulting failure mode is represented in Fig. 2. The failure crack path was located partly in the TBC and partly in the TGO. The mean TGO thickness was >10μm after failure.

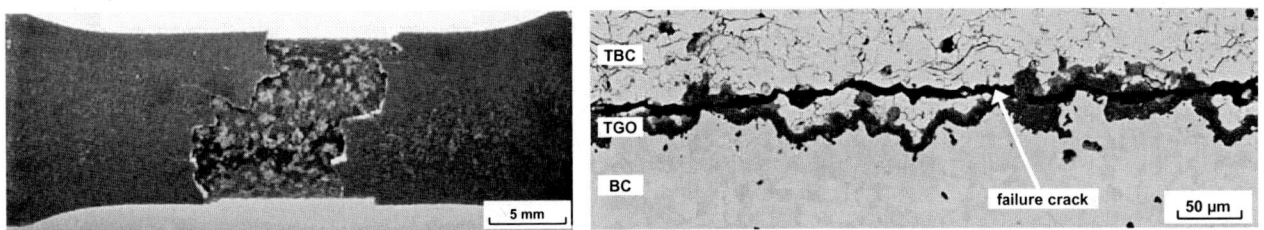

Fig. 2: left: observed failure of the plasma-sprayed TBC after cyclic thermal tests; right: metallographically prepared cross section directly next to the spalled area revealing a failure crack path which is located partly in the TBC and partly in the TGO.

Additional experiments were conducted up to selected fractions of life. The specimens were metallographically prepared to document the damage state. Fig. 3 shows examples of crack pattern observed at early stages of exposure (up to about 30% of time-to-failure). Frequently, micro cracks were observed at or near the interface bond coat / TGO at roughness peaks. Some were directly located at and along the interface downwards both sides of a roughness peak. They partly crossed the TGO towards the TBC and the crack tips were sometimes located within the TBC. Other micro cracks showed a similar shape, but were located partly in the TGO with some distance to the bond coat / TGO interface. Those types followed also the shape of the roughness peak, crossed the TGO with their tips partly penetrating the TBC above roughness valleys.

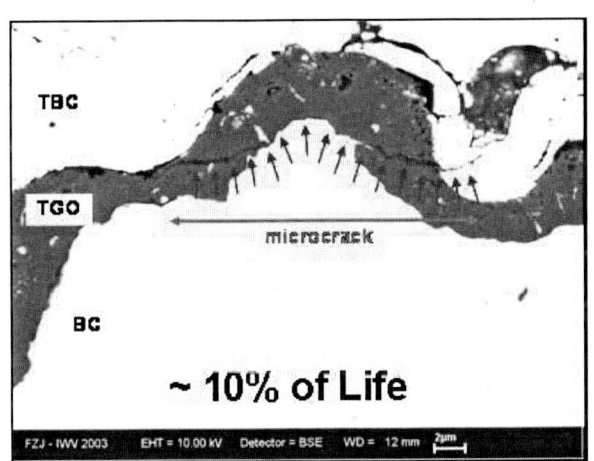
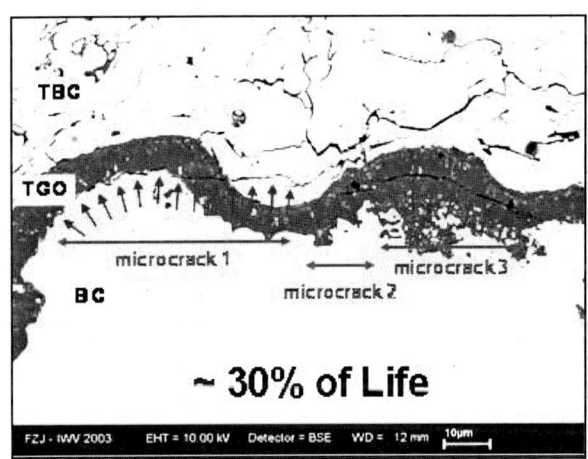

Fig. 3: Frequently observed crack pattern at early stages of exposure indicating weak points for crack formation and initial crack growth.

It has to be assumed that crack pattern which were frequently observed at early stages of exposure indicate weak points for crack formation and initial crack growth. The present results suggest that for the material investigated and the type of loading the interface bond coat / TGO at roughness peaks or the area directly above the bond coat / TGO interface within the TGO have to be considered as weak points for crack formation. Further weak points, which are indicated by the observed crack pattern, are the TGO at roughness flanks, the interface TGO / TBC at roughness valleys and the area above roughness valleys within the TBC. It has to be further taken into account that single micrographs taken from selected fractions of life are snapshots of a continuous damage evolution. Thus, cracks which on one micrograph are located in the TGO above the bond coat / TGO interface at roughness peaks might have been formed directly at the interface bond coat / TGO at an earlier stage of exposure. Prolonged bond coat oxidation and growth of the TGO would let them appear as if being formed and located within the TGO. See also [7] for a similar and more detailed discussion.

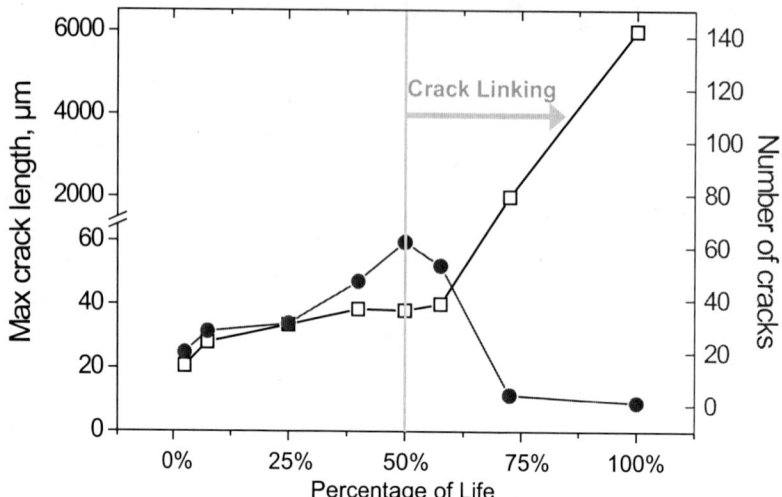

Fig. 4: Maximum crack length and number (density) of micro cracks, which have been observed in or near the TGO, for selected exposure times, here plotted against the fraction of life. The data covers the range from early stages of exposure to macroscopic spalling. At about 50% of life, micro cracks started to form links resulting in accelerated crack growth.

Similar micrographs were taken from each sample, which had been exposed to a certain fraction of life. To analyse and quantify the damage state, an area of the cross section, constant for all samples, was defined. All micro cracks from this area associated with the TGO were counted and characterized with respect to their length and location. From these crack data, the maximum crack length and the number (density) of micro cracks were determined for each exposure time and plotted against the fraction of life. The results are displayed in Fig. 4. Both curves reflect the evolution of damage at the interface of the investigated thermal barrier composite with respect to the applied load type.

The number density of micro cracks increased within the first 50% of life indicating that more and more micro cracks were formed along the interface. The maximum crack length increased also within the first 50% of life, but less steeply. It is worth mentioning that the shape of the crack growth curve indicates decelerated crack growth, that is the crack increment per time or number of cycles was decreasing within the first 50% of life. The maximum crack length was restricted to values below 50µm. Above 50% of life both curves reflect a different behaviour. The number density of cracks was significantly reduced indicating that individual micro cracks started to form links. At the same time the evolution of the maximum crack length was significantly accelerated until final macroscopic failure occurred.

The crack growth curve as well as the number density of micro cracks vs. exposure time in Fig. 4 highlight spalling failure of TBCs as a consequence of a continued evolution and accumulation of damage. A damage process, which starts below 10% of life with the formation of micro cracks along the bond coat / TBC interface and ends with macroscopic spalling of the TBC. Thus, failure and life time of this type of plasma-sprayed thermal barrier composite under cyclic furnace testing appears to be determined by the kinetics of the whole and quite slow damage process. More detailed results about the evolution of damage in APS TBCs were presented for instance by Echsler and Trunova [13,14]. The Finite Element Analysis (FEA), which is described in the next section, aims at simulating the local stress response, the evolution of local stresses and the crack formation near the TGO during the initial stage of exposure to obtain more information about the kinetics of the observed damage evolution.

FINITE ELEMENT SIMULATION

Mesh and boundary conditions

A cylindrical geometry had been chosen for the FE model, which corresponds to the specimen geometry used in the experiment. It consists of four layers or material constituents: base material, bond coat, thermally grown oxide and TBC. The base material had an outer diameter of 10 mm. The thickness of the bond coat was 150 µm and of the TBC 300 µm. The initial thickness of the TGO was 0.5 µm. The undulations (or roughness profile) at the bond coat / TBC interface have been approximated as sinusoidal. The sinus function has been parameterized by an amplitude of 15 µm and a wavelength of 60 µm. The mesh consists of 4 node Generalized Plane Strain elements with reduced Gauss integration (CPEG4R, HKS/ABAQUS1 FEA). Geometry and mesh are shown in Fig. 5. Regarding a cylindrical co-ordinate system, the nodes which are lying on the edges of the segment (Fig. 5: right) has been constrained normal to the edges.

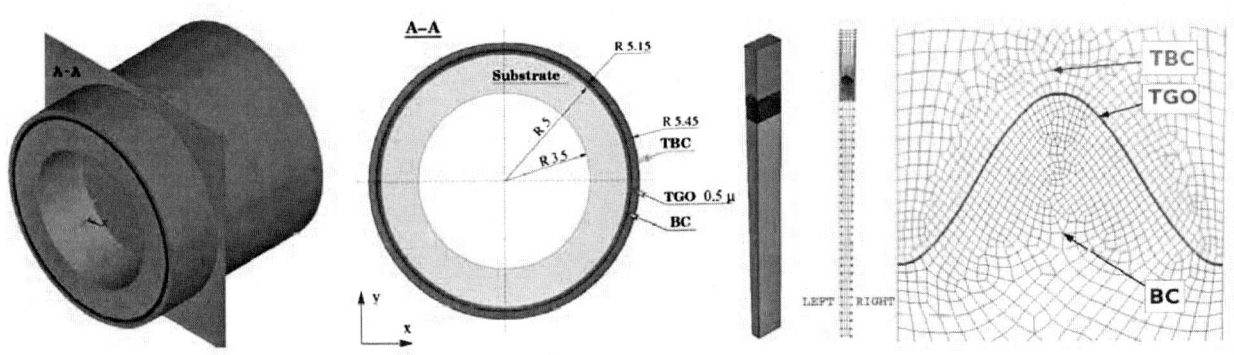

Fig. 5: Geometry and mesh of the FE-model

Material data and bond coat oxidation

The base material (CMSX-4) was treated as an entirely elastic material, because of the low stresses which occur during pure thermal cycling without additional creep or fatigue loading. The bond coat was considered as elastic-visco-plastic and the TGO as elasto-plastic. Elastic and creep data has been considered for TBC deformation. All material properties were temperature dependent. Fig. 6 shows the data for the coefficient of thermal expansion (CTE) and the Young's modulus. Fig. 7 displays creep data for the TBC determined by compression creep tests with stand-alone coatings [15]. Creep properties of the bond coat were determined by compression creep tests with stand-alone coatings and shear deformation experiments on TBC composites

---

[1] distributed by Habbit, Karlsson & Sorensen

[16]. Creep of bond coat and TBC was generally considered for T ≥ 750°C, whereas bond coat creep was substantial at 750°C, but significant primary creep rates of the TBC occur only above 950°C. Primary and secondary creep stages were taken into account. The data were implemented using Eq. (1):

$$\dot{\varepsilon} = A' \cdot \sigma^{n'} \cdot e^{-\frac{\varepsilon}{\varepsilon'}} + A'' \cdot \sigma^{n''} \cdot e^{-\frac{\varepsilon}{\varepsilon''}} + A \cdot \sigma^{n} \qquad (1)$$

where $\dot{\varepsilon}$ is the deformation rate, A, A', A'' are pre-factors, n, n', n'' are stress exponents, ε is the strain, ε' and ε'' are model parameters for the primary creep stage and σ is the stress. The right hand term covers steady state creep, the first two cover primary creep. The model parameters are temperature dependent.

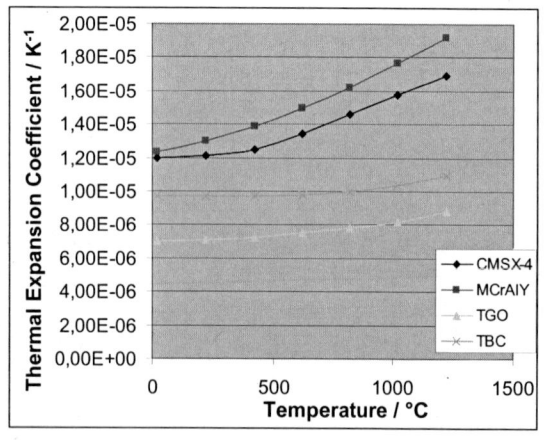 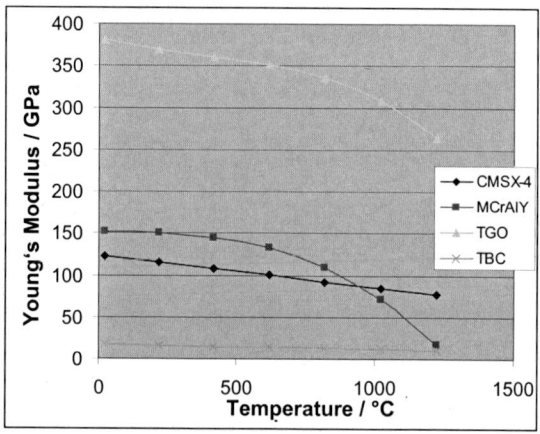

Fig. 6: Data for CTE and Young's modulus, which have been used for the simulations.

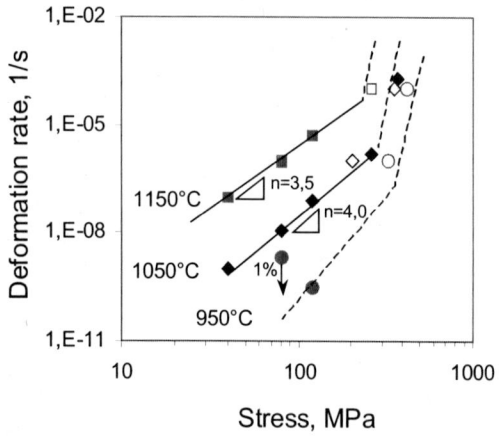 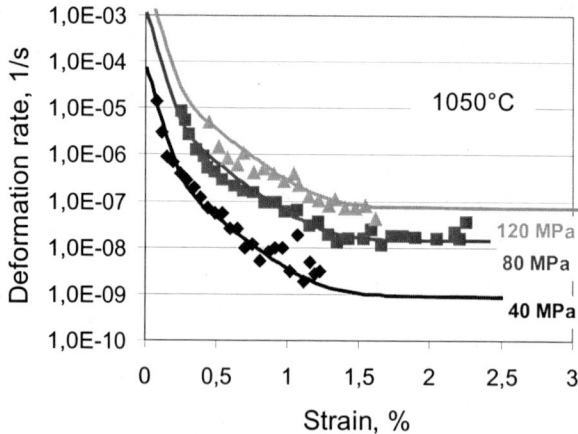

Fig. 7: Creep data for the APS TBC; left: Norton plot (minimum deformation rate vs. stress); right: deformation rate vs. strain and model curves with primary and secondary creep stages as an example for 1050°C at 3 different stress levels.

Growing of the alumina scale at high temperature due to bond coat oxidation was simulated as a continued process using the swelling option in ABAQUS. It was modeled as an orthotropic swelling strain of the TGO, whereby lateral TGO growth (length increase) was considered as a constant amount of thickness growth (generally 5%). The initial thickness of TGO was defined as 0.5 µm. The oxidation kinetics of this thermal barrier composite has been experimentally determined for 3 different temperatures (950°C, 1000°C, 1050°C) [13] and were implemented for this temperature range using a parabolic time law.

Load parameters

The simulated load cycle consists of thermal cycling and high temperature exposure corresponding to the experiments. It consists of four steps (Fig. 3): (i) heating from 20°C to 1050°C in 103 s (10°C/s), (ii) dwell-time at 1050°C for 2 hours, (iii) cooling from 1050°C to 20°C in 103 s (10°C/s), (iv) dwell-time at low temperature (20°C) for 15 min. A temperature of 200°C was selected at which the TBC composite was initially stress free. It matches approximately with the material temperature during the air-plasma-spraying process. The simulations comprised generally 160 load cycles, what amounted to about 30% of life. The TGO thickness was about 5.7 µm after the last cycle.

Results

All presented results from the numerical calculations comprise the stress response near the TGO at room temperature after the 160th load cycle. The displayed stresses are radial stresses.

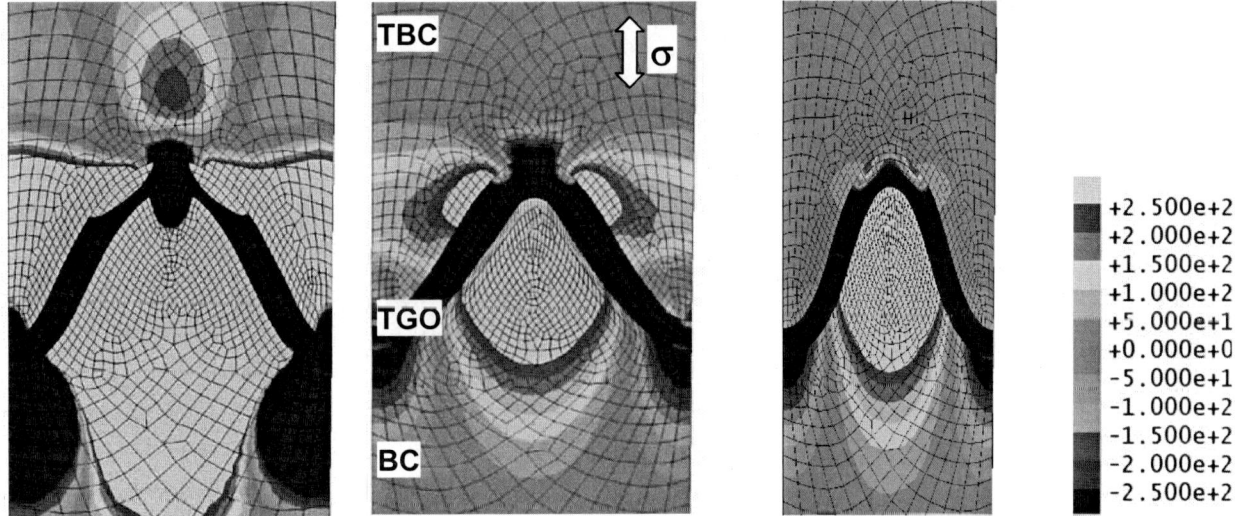

Fig. 8: Simulated stress response near the TGO after 160 load cycles at 20°C with continued TGO growth (here: TGO thickness = 5.7 µm); 1. case (left): all materials elastic; 2. case (center): like 1. case, but BC and TGO additionally with plastic properties; 3. case (right): like 2. case, but BC and TBC additionally with creep properties; light grey regions = tensile > 250 MPa, dark grey regions = compression < 250 MPa

Fig. 8 represents three different stress distributions, whereby the applied material properties were different for each case. The stress response at the left hand side (1. case) is a result from an entirely elastic calculation but with continued TGO growth (thickness increase and 5% lateral increase) to get information about the undisturbed effect of growth stresses in combination with thermal cycling. Noticeable are rather high and localized compressive and tensile stresses. High tensile stresses (light grey regions) were present in the TBC and in the bond coat as well as in some smaller regions in the TGO (off-peak). Compressive stresses were developed primarily in the TGO and in the BC below roughness valleys, and also within smaller regions in the BC directly below the peak and in the TBC directly above the valley. The absolute stress values were quite high (> 10 GPa and <-10 GPa) and probably not realistic.

The second stress distribution (Fig. 8, center) results from a calculation for which the plastic deformation properties of bond coat and TGO have been additionally taken into account in contrast to the first case. The main effects were an overall stress decrease and some redistributions of local stresses. The largest tensile stress was 1040 MPa and was located directly at the interface bond coat / TGO at the roughness peak. At this position, the stresses were changed from compression to tension compared to the first case. The maximum compressive stress occurred directly above the largest tensile stress in the TGO with appr. 2400 MPa. The third case (Fig. 8, right) comprised additionally creep in bond coat and TBC and thus the possibility of stress relaxation. By comparing Fig. 8 (center) and Fig. 8 (right), at first, a decrease of the high tensile stresses alongside the roughness peak in the TBC becomes apparent, and secondly a shape change of the interface. The curvature at the peak became less sharp. This effect was directly due to stress relaxation, which relaxed the entire structure. In contrast, the tensile stresses at the bond coat / TGO interface were decreased only slightly by less than 10%. However, this region showed the largest tensile stresses.

The material properties used in the last case were taken as a reference parameter set. It includes plastic and creep properties corresponding to the experimentally investigated MCrAlY bond coat, plastic properties of the TGO and creep properties of the experimentally investigated plasma-sprayed TBC. One of the first questions was how the stress response is developing with increasing number of load cycles. In particular the stress distribution at 20°C was of interest, because the largest stresses appeared at the lowest cycle temperature. Fig. 9 displays the stress response after selected load cycles at 20°C. Two remarkable features characterize the simulated stress response. At first, high tensile stresses occurred at the bond coat / TGO interface even after the first load cycle. Thus, early crack formation at this site appears quite likely depending of course on actual interface shape, material properties (deformation properties as well as resistance against crack formation at the interface) and load parameters. The corresponding cyclic furnace tests revealed crack formation at the bond coat / TGO interface within the first 10% of life (about 50 cycles). For comparison see also Fig. 3.

Secondly, a coherent lateral region of tensile stresses was developing in the TBC at both sides of the roughness peak indicating higher loaded regions. According to the cyclic tests, the regions in between roughness peaks and over valleys showed cracking. However, crack formation and propagation in the TBC would be critically affected by pre-existing splat boundaries, micro cracks and pores. Thus, the numerical simulations provide here only a tool for merely rough estimations of the load situation directly in the TBC.

An additional result is indicated by Fig. 9. Up to the 10th cycle the tensile stresses alongside the roughness peak in the TBC increased due to the initially fast oxidation and thus large growth stresses. Afterwards the tensile stresses decreased in this region, because the oxidation rate decreased (parabolic behavior) and the stress relaxation got relatively more influence on the stress response. Then again, a narrow zone of large tensile stresses was emerging directly from the TGO and was growing in lateral direction. This particular result indicates the fairly complex interaction of growth stresses, thermal stresses and stress redistributions due to plastic deformation and even more due to stress relaxation.

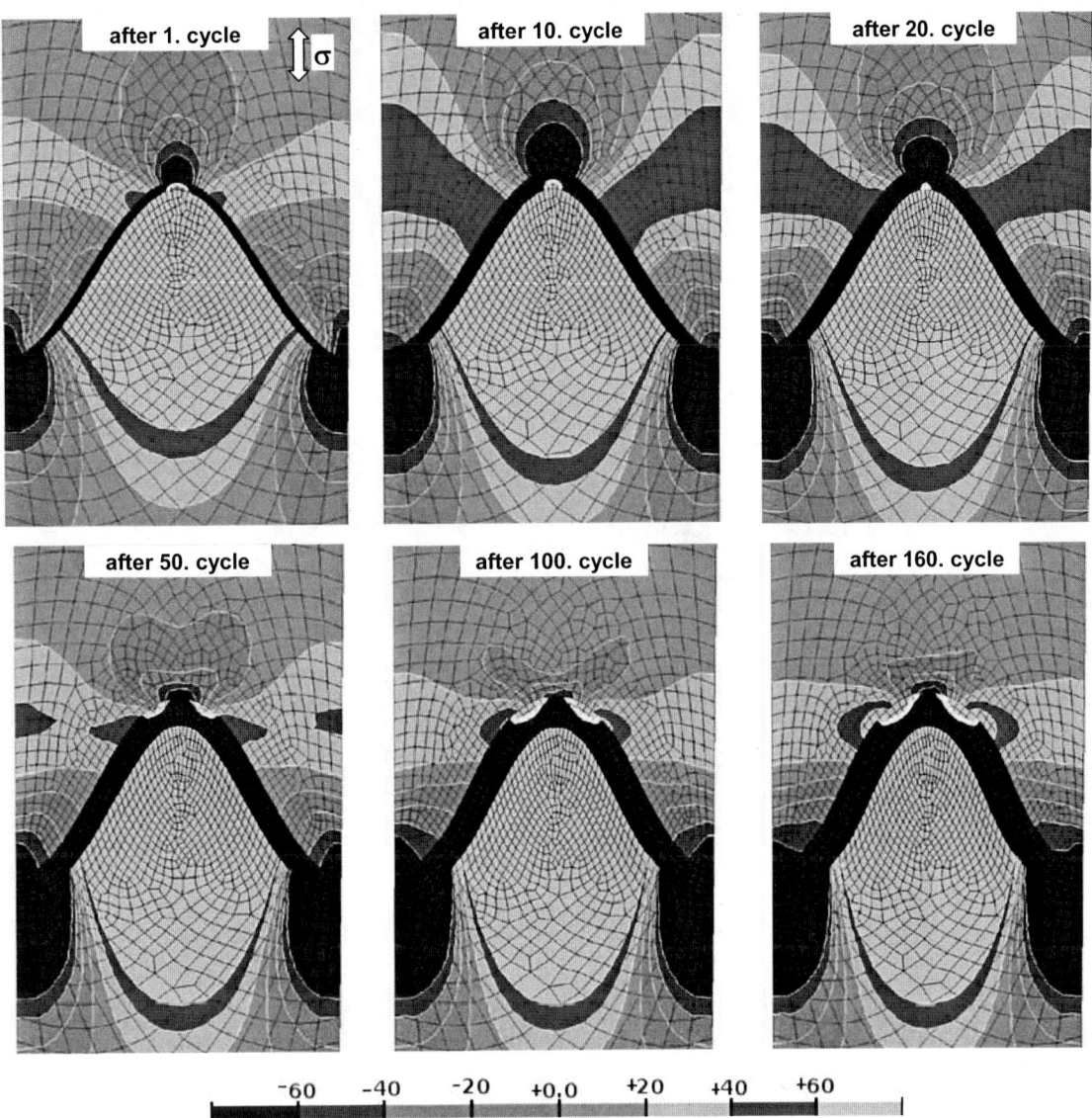

Fig. 9: Simulated stress response near the TGO after selected load cycles at 20°C with continued TGO growth using the reference parameter set, which is described in the text.

Starting from the reference simulation a couple of systematic variations of certain material parameters, such as CTE, Young's modulus and creep rate of the materials were conducted to analyze their influence. Here, only the influence of the stiffness variation of the thermal barrier coating should be described exemplarily. Fig. 10 displays the stress response associated with the reference parameter set (Fig. 10, center) as well as simulation results with a 50% higher (Fig. 10, left) and lower (Fig. 10, right) Young's modulus of the TBC (reference value: 17.5 GPa). The images show the stress distribution at 20°C after the 160th cycle.

The stress plots indicate that an increase of the stiffness in the TBC increases predominantly the tensile stresses in the TBC except the small regions of compressive stresses directly above the roughness peak and directly above the roughness valley. Particularly the lateral region of tensile stresses alongside the roughness peak in the TBC increased by the stiffness increase of the TBC. The stresses in the lateral tensile zone directly at the boundary of the unit cell are increased approximately linearly, the stresses directly at the interface TGO / TBC in the TBC are affected more than linearly. Beyond that, the tensile stresses at the bond coat / TGO interface were increased. On the other hand, the tensile stresses in the TBC were generally decreased by decreasing the stiffness and above the lateral region of tensile stresses they were even changed into small compressive stresses.

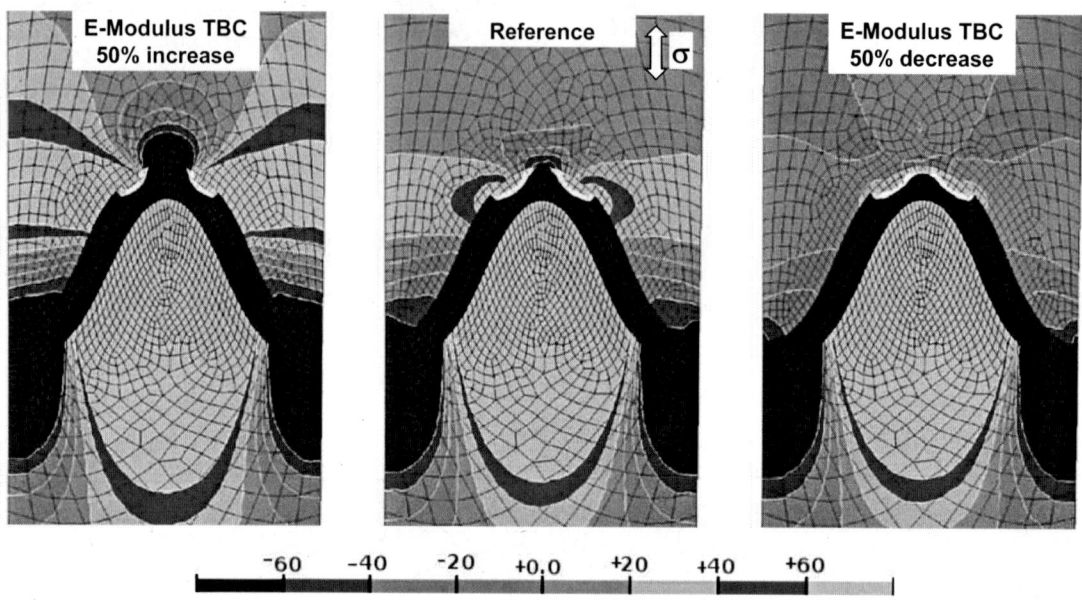

Fig. 10: Simulated stress response near the TGO after 160 load cycles at 20°C with continued TGO growth; the Young's modulus of the TBC was modified by ± 50% with respect to the reference parameter set.

One could use the calculation with the higher stiffness value also as a rough impression of how the stress response would change for the case high temperature exposure causes a time dependent stiffness increase of the TBC.

In addition to the simulation of the stress response, simulations of crack formation and initial crack growth have been conducted using cohesive elements [17]. Due to the fact that the highest local tensile stresses were developing at the interface bond coat / TGO, cohesive elements were implemented directly at the interface. Critical stress values of 600 MPa (normal) and 1200 MPa (shear) have been applied for crack formation. For crack opening a critical strain energy release rate of 20 N/m has been taken into account for normal and shear loading. Fig. 11 represents the stress state after two selected cycles taking into account the reference parameter set. After the 19th load cycle the maximum tensile stress exceeded the critical stress value of 600 MPa at the roughness peak and a crack was formed during cooling (Fig. 11, left).

The formation of the micro crack at the interface affected the stress field substantially. The tensile stress region directly below the crack was relaxed. In contrast, small regions with high tensile stresses occurred in the TGO close to the crack tips at both sides. After 160 cycles the crack was elongated downwards the roughness profile along the interface at both sides of the peak. However, between the 19th and the 160th cycle the crack was not growing steadily. As a result of the interaction between thermal stresses, growth stresses and stress relaxation the crack was after certain cycles partly closed and opened again. No tendency was found for the crack to propagate further downwards the roughness profile up to the 160th cycle. In contrast, the crack tip saw high tensile stresses in the adjacent TGO indicating a potentially bending and a penetration of the crack into the TGO. This behavior would correspond to the frequently observed crack pattern, which are exemplarily shown in Fig. 3.

Further development of the simulations are planned including the prediction of crack growth direction as well as dynamical re-meshing.

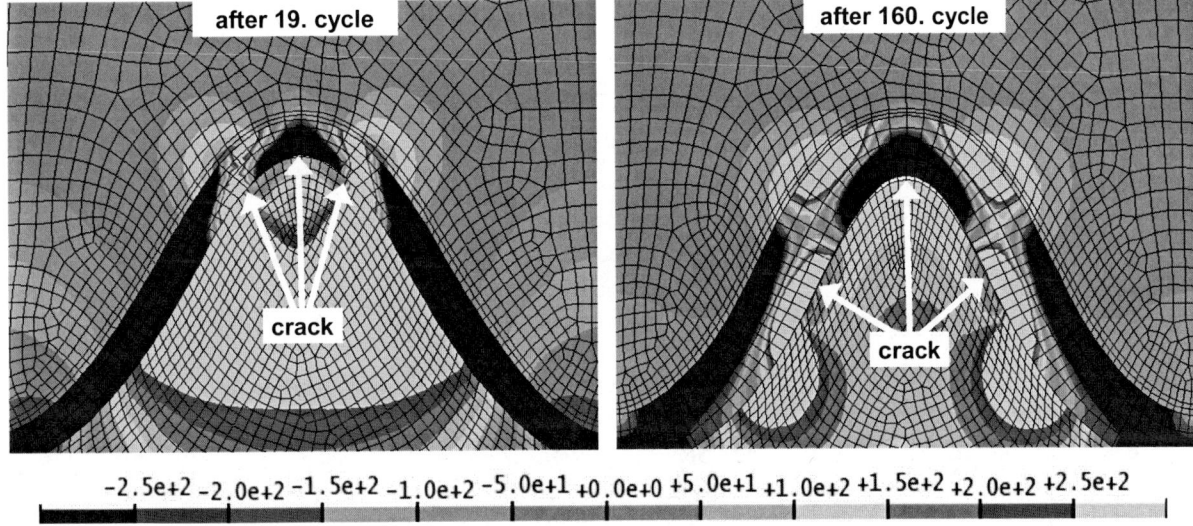

Fig. 11: Simulated stress response near the TGO and formation of a micro crack at the bond coat / TGO interface after 19 cycles and initial growth after 160 cycles at 20°C. It shall be noticed, that the scale of the stress distribution is different with respect to the other stress plots.

CONCLUSIONS

Simulations of the response in a plasma-sprayed thermal barrier system were presented and compared with corresponding cyclic thermal tests including a dwell time of 2 hours at high temperature (cyclic furnace tests). The material properties of the MCrAlY bond coat and the ceramic thermal barrier coating were determined in order to obtain more realistic deformation properties, particular creep and thus stress relaxation properties as input data for the FE simulations. The simulation results showed that considering a continued oxidation process by simulating a continued TGO growth at high temperatures was required to cover the influence of growth stresses on the stress response. In general, the stress response in and near the TGO at the rough bond coat / TBC interface and its evolution during prolonged cyclic loading, was the consequence of the complex interaction of thermally induced stresses, oxidation induced growth stresses and redistribution of stresses due to plastic deformation and even more due to stress relaxation as a result of local creep processes. Pure stress calculations showed two distinct features: (i) the fast development of high tensile stresses in the bond coat with a maximum value directly at the interface bond coat / TGO below a roughness peak, which occurred during the cooling stage and which were maximum at the lowest cycle temperature, and (ii) the development of a lateral region of larger tensile stresses alongside the roughness peak over roughness valleys. The simulation of crack formation at the interface bond coat / TGO using cohesive elements resulted in an early formation of a micro crack at the roughness peak (after the 19th cycle). Up to 160 cycles the crack was elongated downwards the roughness profile, but was also partly closed and opened again and no tendency was observed for the crack to propagate further downwards. Instead, high tensile stresses in the adjacent TGO would suggest a penetration of the crack into the TGO in the upper half of the sinusoidal interface, but this process was not covered by the simulations, at this time. The results of the numerical simulations were in agreement with the experimental observations of crack pattern obtained within the first 30% of life from the corresponding cyclic furnace tests.

ACKNOWLEDGMENTS

The German Science Foundation (DFG) is greatly acknowledged for the financial support of the work within the Sonderforschungsbereich 370. Furthermore, many thanks are due to Priv.-Doz. Dr. R. Vaßen for manufacturing the coatings at the Institute of Materials and Processes in Energy Systems (IWV 1) of Research Centre Juelich.

REFERENCES

1 G.C. Chang, W. Phucharoen, R.A. Miller: *Surf. Coat. Technol.*, 1987, *30*, 13.

2 R. Vaßen, G. Kerkhoff, D. Stöver: *Mater. Sci. Eng. A*, 2001, *303*, 100.

3 G. Kerkhoff, R. Vaßen, C. Funke, D. Stöver: *Proceedings of the 6th Liege Conference on Materials for Advanced Power Enineering 1998*, 1998, 1669.

4 M. Ahrens, R. Vaßen, D. Stöver: *Surf. Coat. Technol.*, 2002, *161*, 26.

5 A. M. Freborg, B.L. Ferguson, W.J. Brindley, G.J. Petrus: *Mater. Sci. Eng. A*, 1998, *245*, 182.

6 A.G. Evans, D.R. Mumm, J.W. Hutchinson, G.H. Meier, F.S. Pettit, in *Prog. Mater. Sci.*, 2001, *46*, 505.

7 J. Cheng, E.H. Jordan, B. Barber, M. Gell: *Acta Mater.*, 1998, *46*, 5839.

8 J. Rösler, M. Bäker, M. Volgmann: *Acta Mater.*, 2001, *49*, 3659.

9 J. Rösler, M. Bäker, K. Aufzug: *Acta Mater.*, 2004, *52*, 4809.

10 E.P. Busso, J. Lin and S. Sakurai: *Acta Mater.*, 2001, *49*, 1529.

11 K. Sfar, J. Aktaa, D. Munz: *Mater. Sci. Eng. A*, 2002, *333*, 351.

12 P. Bednarz, R. Herzog, E. Trunova, R.W. Steinbrech, L. Singheiser: *Proceedings of the 29th International Conference on Advanced Ceramics and Composites 2005*, Ceramic Engineering and Science Proceedings, 2005, *26, 3*, 55.

13 H. Echsler: *PhD thesis, RWTH Aachen University*, 2003, ISBN 3-8322-1895-5.

14 E. Trunova: *PhD thesis, submitted to RWTH Aachen University*, Berichte des Forschungs-zentrums Jülich, Reihe Energietechnik/Energy Technology, to be published in 2006.

15 R. Herzog, E. Trunova, R.W. Steinbrech, E. Wessel, R. Vaßen, F. Schubert, L. Singheiser: Proceed. of the International Conference "Creep and Fracture in High Temperature Components – Design & Life Assessment Issues", 12-14 September 2005, Institution of Mechanical Engineers, Central London, UK

16 P. Majerus, R.W. Steinbrech, R. Herzog, F. Schubert: Proceedings of the 7th Liege Conference 2002, 30. September - 2. October 2002, Liege, Belgium, Materials for Advanced Power Engineering 2002, ISBN 3-89336-213-2

17 M. Cliez, J.-L. Chaboche, F. Feyel, S. Kruch: *Acta Mater.*, 2003, 52, 1133-1141

NUMERICAL SIMULATION OF CRACK GROWTH MECHANISMS OCCURRING NEAR THE BONDCOAT SURFACE IN AIR PLASMA SPRAYED THERMAL BARRIER COATINGS

A. Casu, J.-L. Marqués, R. Vaßen, D. Stöver
Institut für Werkstoffe und Verfahren der Energietechnik (IWV1), Forschungszentrum Jülich GmbH, Jülich/Germany

ABSTRACT

Under thermal cycling, the failure of an air plasma sprayed thermal barrier coating (TBC) on a metallic bondcoat (BC) usually occurs near the interface between both coatings. The local curvature of such an interface is responsible for the stress components which lead to the growth of micro-cracks already produced during the plasma spraying. The growth of oxide scales (TGO) between BC and TBC at high temperatures determines the stress level near the TGO-TBC interface during cooling, where the main stress source is the mismatch in thermal expansion.

A failure mechanism based on finite-element calculations of thermal stress within the TBC is presented, which models the TGO-TBC interface as a sinusoidal profile. Assuming the coating system has completely relaxed its stresses during the thermal cycling hot phase, sub-critical crack growth after cooling to room temperature is calculated for horizontal cracks starting at every hill of the curved TGO-TBC interface profile. Failure is assumed when the growing cracks cover one whole profile wavelength. In a second step, an extension of the presented model is discussed where crack growth follows the path where the energy release rate becomes maximum. Finally, the crack path is implemented directly in the finite-element mesh. The conclusions drawn from the numerical calculations are compared to crack configurations near the TGO-TBC interface, taken from micrographs of thermally cycled samples.

INTRODUCTION

The increasing temperatures inside a gas turbine require the metallic components to be protected against the high thermal load. This is carried out by depositing a thermally insulating ceramic coating, the thermal barrier coating (TBC), on top of the metallic component. Due to the different thermal expansion coefficient of both materials, high stress levels develop in the ceramic coating during temperature changes, which lead to crack growth of previously existing micro-cracks and, eventually, to the failure of the ceramic coating. A reliable prediction method for the lifetime of the metallic-ceramic system is, nevertheless, still to be achieved.

The most used material as TBC is yttria partially stabilized zirconia (YSZ), and one of the most extended techniques for depositing it on a metallic component is using atmospheric plasma spraying (APS). In this technique the ceramic TBC material, injected into the plasma jet as particles of size 10-100μm, is molten and the fluid particles are flattened on the metallic component to build up the thermal barrier coating. In order to improve the adhesion between the metallic substrate and the ceramic layer, a metallic bondcoat (BC) is sprayed first on the metallic component before depositing of the TBC. The roughness of the resulting BC, sprayed under very low gas pressure to avoid its oxidation, ensures the interlocking of the BC with the ceramic TBC deposited on it. Additionally, the aluminium content within the BC material, usually MCrAlY (M=Ni, Co), eventually forms an alumina scale, impeding the oxidation of the metallic component below the BC. This oxide scale produced during the exposure of the system to high temperatures is called the thermally grown oxide (TGO). The whole system consisting of a metallic component, the metallic bondcoat and the ceramic thermal barrier coating is denoted TBC system. The main aim of the present work is to develop a better understanding of the crack growth mechanism for such a system under thermal cycling, in order to improve the lifetime estimation of the TBC system.

The paper is organized as follows: the next section presents a new technique to calculate the most relevant length scales for the BC-TBC interface profile. These length scales determine the out-of-plane components of the thermal stress, responsible for the crack growth within the TBC near the BC-TBC interface. In the third section a simplified lifetime model will be discussed, which estimates the horizontal growth of TBC cracks based on the thermal stress distribution calculated by means of the Finite Element method (FEM), but without including explicitly the crack into the FEM mesh. This approach is improved in a second step by determining at every growth step the crack path for the maximum opening stress. The results collected from these two steps are used in the final section to develop a full FEM lifetime model by tracking explicitly within the FEM mesh the crack propagation through the TBC for the maximum energy release rate. The paper is concluded with the discussion of the results obtained.

CHARACTERIZATION OF THE BC-TBC INTERFACE PROFILE

The failure of the TBC system under thermal cycling, with a high temperature above 1000 °C and below 1300°C, usually takes place near the interface between the bondcoat and the thermal barrier coating. This is due to the growth of delamination cracks which run approximately parallel to the BC-TBC profile.[1] For a perfectly flat BC, the compression/tension stress developed within the TBC during the thermal cycling has only in-plane components, parallel to the BC-TBC interface, which therefore cannot lead to the just mentioned delamination crack growth. In order to produce stress components which would eventually open those cracks, a local curvature in the BC-TBC profile is required. Such condition fully applies to plasma sprayed BCs (and TBCs), where their inherent disordered character and fragmentation of the sprayed molten particles lead to an irregular, and thus curved, BC surface.

In a first approximation for the description of TBC failure, the BC-TBC interface will be considered periodic, with a certain characteristic wavelength λ_0 still to be determined. The idea behind this assumption is to consider that the failure crack growth is a local process, not influenced by other cracks outside the length scale λ_0, and taking place more or less simultaneously at similar places along the whole BC-TBC interface. The segment of the BC-TBC profile where this representative single crack growth occurs will be taken as a cosine function

$$y = A_0 \cos\left(\frac{2\pi}{\lambda_0}x\right) \tag{1}$$

The amplitude A_0 is determined through the measured average roughness of the BC profile, either the mean roughness R_a or the root mean square roughness R_q

$$R_a = \frac{1}{L}\int_0^L |h(x)|dx \quad \text{or} \quad R_q = \sqrt{\frac{1}{L}\int_0^L (h(x))^2 dx} \tag{2}$$

where L is the sampling length along the BC profile and $h(x)$ denotes the profile height (or deviation) with respect to the line of averaged height. Assuming that the periodic function in equation (1), with $L=\lambda_0$, describes effectively the BC surface profile whose experimental roughness is given by R_a or R_q, the amplitude A_0 becomes

$$A_0 = \frac{\pi}{2} R_a \quad \text{or} \quad A_0 = \sqrt{2} R_q \qquad (3)$$

For a typical 150μm thick MCrAlY bondcoat sprayed under low pressure conditions and injecting particles with diameter about 40μm, the resulting roughness is $R_a \approx 6$μm and $R_q \approx 7$μm, which in both cases leads to an amplitude $A_0 \approx 10$μm.

The other characteristic length scale necessary to describe locally the BC-TBC profile is λ_0. The plasma sprayed BC (as well as the TBC) is formed through the flattening of impinging molten particles, called splats, and the subsequent disordered stacking of these splats into lamellae. The surface profile resulting from such process is of course non-periodic and thus a simple Fourier analysis of the BC profile cannot lead to a reliable estimation of λ_0 as the effective longitudinal scale of the profile. On the contrary, one could consider the sprayed profile to be completely random. In such case the BC surface would not contain any kind of preferred length scale and would result to be self-similar. A double logarithmic representation of the Fourier components as a function of the corresponding wavelengths would yield one straight line, with the slope related to the fractal dimension of the BC profile.

The surface of plasma sprayed coatings contains many irregularities with size scales ranging from 1μm to 1mm and is therefore rough. Nevertheless, the Fourier components for the profile of such coatings, when represented double logarithmically, are not arranged along one single straight line as it would correspond to a pure random profile. For plasma sprayed coatings, an inflection point characteristically separates two branches of different slope, one of higher inclination for small length scales (small wavelength λ or equivalently large wave vector $k=2\pi/\lambda$) and second one of lower inclination for the large length scales corresponding to near macroscopic undulations. The length scale for this inflection denotes the smallest size at which the Fourier components deviate from a pure random signal. Hence the BC profile, although apparently rough and random, does contain a characteristic longitudinal scale. Further, the length value of this inflection point is well correlated to the size of the particle injected into the plasma jet to produce the coating.[2]

The existence of a length scale in the BC profile, manifested as the just mentioned inflection, should actually be expected from the very nature of the deposition technique: it is the footprint of the size of the flattened particles (splats) building the plasma sprayed coating. This interpretation is further supported from the possibility to derive the existence of the inflection for the Fourier spectrum and the observed values for the inclination of both branches from a simple stochastic model. Such model describes the splat flattening as a diffusion process under the additional influence of a spatially correlated noise corresponding effectively to processes at very small length scales (splat fragmentation, local flattening hindrance).[2]

Figure 1 shows a typical example for the Fourier transformation of a MCrAlY bondcoat surface profile, where the diameter of the sprayed particles ranges from 25μm to 55μm. A perthometer Mahr M2 rasters the height profile of the BC surface along a sampling length of $L=4.1$mm with a diamond head of radius 2μm, and this profile is Fourier transformed. The double logarithmic representation of the squared Fourier transformation components $\tilde{h}(k)$, $\tilde{D}(k) = 2\pi \left| \tilde{h}(k) \right|^2$, averaged over five different measured profile lines of the same sample, is represented as a function of the length scale λ described by the wave vector $k=2\pi/\lambda$; the wavelength is measured in micrometer. The Fourier components are grouped into two different branches: the red line corresponding to the linear fitting of the range of large length scales (small wave vector k) and the green line to those of very small length. The inflection point in the Fourier spectrum for this bondcoat sample corresponds to a length scale of $70\pm10\mu m$, which is equated to the wavelength λ_0 for the effective periodic function (1) describing locally the BC-TBC interface. This is the smallest length scale incorporated in the BC profile not being pure random noise and thus characteristic for the BC surface. It is also the length yielding the strongest curvature (the curvature of a cosine is a decreasing function of the wavelength) and therefore the highest stress level leading to the delamination of the ceramic TBC deposited on the BC. Hence, such λ_0 is the length scale determining the shortest lifetime of the TBC system.

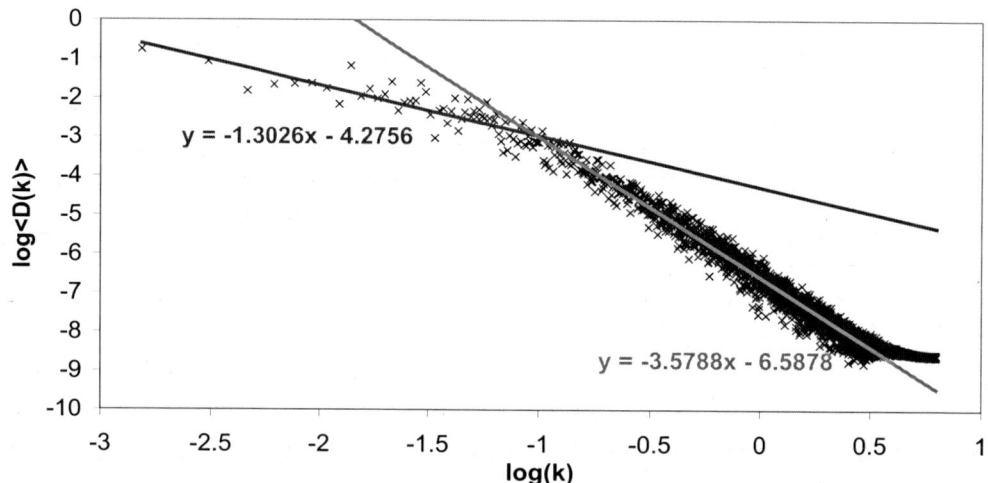

Figure 1. Double logarithmic representation of the averaged squared Fourier components for a plasma sprayed bondcoat, displaying the characteristic inflection point.

SIMPLIFIED LIFETIME MODEL FOR HORIZONTAL AND INCLINED CRACKS

For the following discussion, the TBC system to be investigated consists of a flat metallic substrate 3mm thick made of IN738, a 150μm thick vacuum plasma sprayed (VPS) MCrAlY bondcoat with a resulting surface roughness of $R_a \approx 6\mu m$, and a 300μm thick atmospheric plasma sprayed (APS) thermal barrier coating made of YSZ.[3] The thermal cycling consists, firstly, of heating up the TBC surface for 5min by means of a gas burner and cooling simultaneously the back side of the metallic substrate, in order to achieve a controlled BC-TBC interface temperature about 1050°C; and secondly, removing the burner and, by means of compressed air, cooling rapidly the TBC surface down to a temperature slightly above 30°C. The sample is maintained at the latter temperature for 2min, before repeating the whole process.[4] During the thermal cycling, the TBC surface temperature is scanned with a pyrometer, and the temperature in the metallic substrate with a thermocouple. Under thermal cycling, the main stress sources within the TBC system are the thermal stress at heating up or cooling down, as well as the stress created by the growth of the TGO scale at the BC-TBC interface. Nevertheless, since both the thermal stress during the rapid heating and the growth stress occur at a high temperature, the TBC system, particularly the sprayed materials at the highest temperatures, is able to relax these stresses in a short time, without initializing crack growth. Only during the cooling down the temperature near to the BC-TBC interface goes down very rapidly and no relaxation process can be efficiently activated to reduce the resulting thermal stress. Hence crack growth, resulting in the TBC system failure, takes place only during the cooling down. This is confirmed by the measurement of acoustic emission (AE), which is directly correlated to crack growth, during the thermal cycling. As shown in Fig. 2, AE signals arriving at sensors located at three different positions on the back side of the TBC system mainly occur during the whole cooling down phase, not during the heating up or the hot phase.

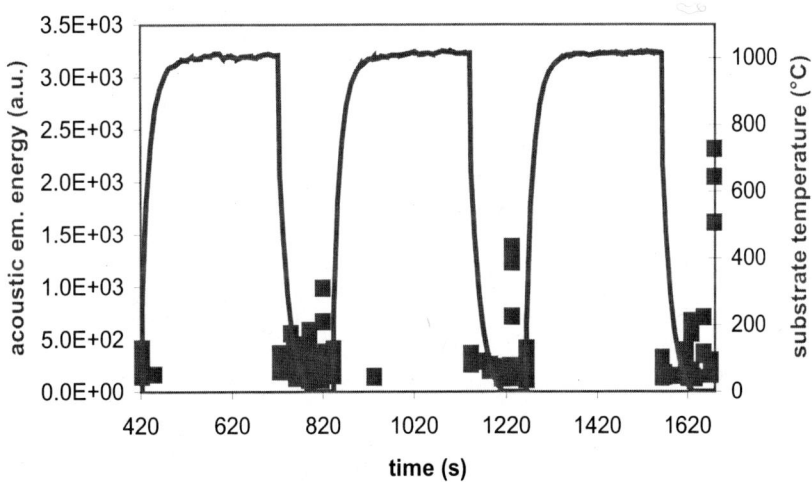

Figure 2. Acoustic emission energy during 3 cycles, with the corresponding temperature in the metallic substrate, after 1173 cycles.

Table I. Thermo-mechanical parameters for the TBC system materials at room temperature.[7]

	metallic substrate, IN378	metallic BC, MCrAlY	oxide TGO, Al$_2$O$_3$	ceramic TBC, YSZ
thermal expansion coeff. α	$15.8 \times 10^{-6} K^{-1}$	$17.5 \times 10^{-6} K^{-1}$	$8.0 \times 10^{-6} K^{-1}$	$10.7 \times 10^{-6} K^{-1}$
elastic modulus E	191GPa	140GPa	360GPa	25GPa
Poisson number v	0.3	0.3	0.22	0.22

Let us discuss qualitatively the evolution of the thermal stress, firstly for the initial TBC system state, without an oxide scale separating the TBC from the BC. The thermal expansion coefficient of the ceramic TBC is lower than that of the underlying metallic layer (see Table I). A local convex curvature (or "hill") along the BC-TBC interface can be approximated as a small BC lump surrounded by a quite extended TBC. A simple way to estimate the thermal stress in the neighborhood of the BC hill is to consider the surrounding TBC as an inert background and to subtract its thermal expansion coefficient from that of the materials locally in contact with the TBC. To the BC hill, thus, a positive effective thermal expansion coefficient $\alpha_{BC}-\alpha_{TBC}>0$ is assigned. During the cooling down, for a negative temperature change about $\Delta T=-1000K$ near the BC-TBC interface, the top of the BC hill tries to contract but it is held by the inert TBC: the TBC above the hill wants to be pulled apart in radial direction. This TBC part is put under radial tension and horizontal TBC cracks situated near such a hill can grow due to the tensile stress. On the other hand, if the oxide scale separating the BC from the TBC has grown to a thickness high enough to screen the BC from the TBC, the previous reasoning is reversed. Now the small hill is made of a material with a lower α than the surrounding extended TBC and can be taken as behaving with a negative thermal expansion coefficient $\alpha_{TGO}-\alpha_{TBC}<0$ against an inert TBC. During the cooling down, the TGO hill tries now to expand but it is hindered by the TBC: the TBC region above the hill is now put under radial compression and cracks located there will become closed. The previous discussion applies analogously for a concave curvature (a "valley") in the BC-TBC interface, by interchanging the role of tensile and compressive stresses.

Summarizing, at the start of thermal cycling, those cracks running parallel to the BC-TBC interface which are placed near to convex curvatures (hill) are able to grow in the cooling phase. But only until they penetrate in a TBC region of concave curvature (valley) where they become temporarily halted. With the increasing oxidation of the BC during thermal cycling, the radial compressive stress around the BC-TBC valley gets progressively converted into tensile: the previously stopped delamination can continue growing. Assuming that such a process occurs simultaneously at every hill-valley, the failure of the TBC system takes place when the single crack vertex reaches a location above the valley middle point.[3]

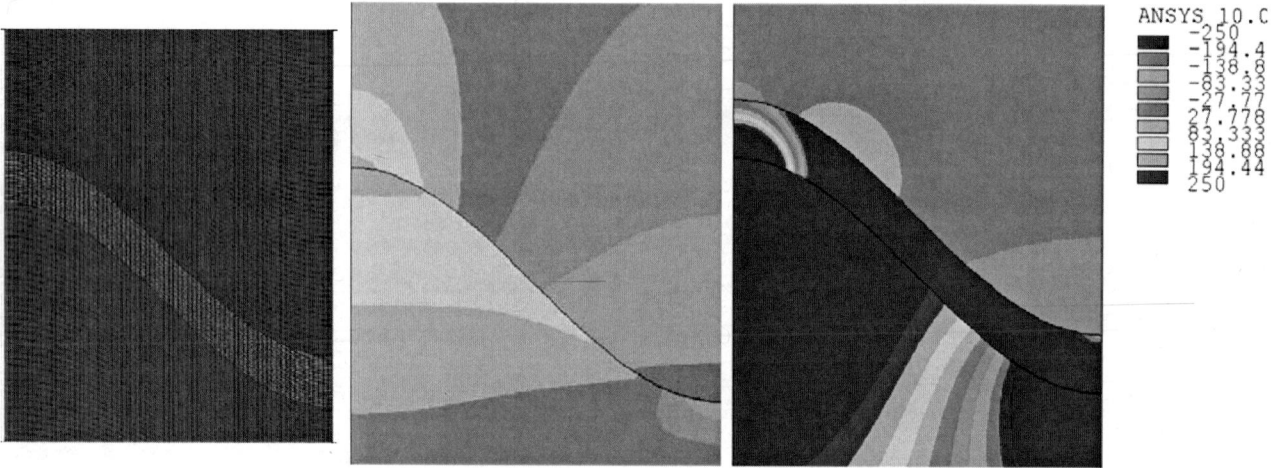

Figure 3. Left, FEM mesh of the upper BC part, a 5μm TGO and the lower TBC part, for a cosine profile with $A_0=10$μm and $\lambda_0=70$μm. Right, distribution of the thermal stress vertical component (range –250MPa to +250MPa) at the BC-TBC interface without oxide scale (middle) and at the TGO-TBC interface for a 5μm TGO (right).

As the next step, this qualitative failure mechanism is numerically implemented by means of the Finite Element method (FEM) using the software ANSYS (Ansys Inc., Canonburgh, Pittsburgh, PA, USA). With the already mentioned dimensions for the TBC system, and assuming the hill-valley structure of the BC-TBC profile to be the equation (1) with $A_0=10$μm and $\lambda_0=70$μm, half wavelength of such profile is generated with a fine meshing of approximately square cells of 0.5μm side for the region neighboring the BC-TBC interface. A plane strain state is considered. At one vertical flank a periodic boundary condition is fixed and at the opposite side a free displacement in horizontal direction for the whole flank is chosen. As shown in Figure 3, within the TBC and when no TGO is present, the vertical stress component σ_{yy} is positive (tensile) above the hill, able to open horizontal cracks, but negative (compressive) above the valley. For an intermediate state in the TGO growth, the tensile region above the hill has disappeared, the compressive region above the valley has been reduced and the positive vertical stress has moved to the lateral flank of TGO-TBC profile. As already discussed, only the thermal stress during the cooling down will be considered here. Actually, a free standing plasma sprayed TBC is also able to display creep even at room temperature,[5] although such effect, regarding the crack growth at low temperatures in the TBC, will be not considered. The reason is that TBC creep is mainly a collective effect of all the sprayed lamellae when they slightly slide over each other. For a thermal cycling high temperature phase below 1300°C, the crack growth leading to the TBC delamination occurs only within the first two deposited lamellae, which are quite good adhered to each other and thus partially impeded to participate in the collective creep relaxation. Furthermore, the brittle TBC coating is assumed to behave elastically. The effect of plastic stress relaxation very near to the crack vertex is included effectively in the TBC fracture toughness value, being the calculation pure elastic for thermal stress at distances above 0.5μm from vertex.

Crack growth inside the TBC is sub-critical and proceeds slowly. Let a denote the current crack length: the sub-critical crack growth under mode I (crack opening mode) is described by the Paris-Erdogan law as function of the stress intensity factor $K_I(\sigma,a)$ for mode I

$$\frac{da}{dt} = v_0 \left(\frac{K_I(\sigma,a)}{K_{I,crit}} \right)^m \tag{4}$$

with $K_{I,crit}$ the fracture toughness for mode I, approximately equal to 1MPa m$^{1/2}$ for plasma sprayed YSZ;[6] v_0 is the crack velocity at critical conditions, equal to 7.6×10^{-5}m/s for YSZ, and $m=18$.[7] For a simplified formulation of the lifetime model, the thermal stress is obtained from the numerical FEM calculations (as in Fig. 3), without including explicitly the crack in the FEM mesh. And the stress intensity factor is approximated by that of a crack in an infinite plane under uniform stress, $K_I(\sigma,a) \approx \sigma\sqrt{\pi a/2}$, being now σ the thermal stress component perpendicular to the flank of a fictive crack (which grows without modifying the stress distribution) evaluated at the vertex location of such crack. It is clear that this can only lead to an estimation of the lifetime prediction, since the crack does locally modify the stress distribution. However it will be used as a first approximation to discuss later the most rapid form of crack growth.

The TGO growth determines the sign of the thermal stress and thus whether a crack is able to increase its length or not. Additionally, another time dependent effect strongly contributes to accelerate the crack growth: the sintering of the plasma sprayed TBC due to its relative high porosity (about 10%). The crack growth evolution is therefore controlled by the current TGO thickness and sintering of the TBC elastic modulus, both of them described by a diffusion law $\sim t^{1/2}$ with a temperature dependent activation factor

$$d_{TGO} = A_{TGO}e^{-E_{TGO}/k_B T}\sqrt{t} \quad \text{and} \quad E_{TBC}(t) = E_{TBC}(t=0) + A_{sint}e^{-E_{sint}/k_B T}\sqrt{t} \tag{5}$$

with $A_{TGO}=3.715 \times 10^{-3}m/s^{1/2}$ and $A_{sint}=2.49 \times 10^{19}Pa/s^{1/2}$, and activation energies $E_{TGO}=1.435$eV and $E_{sint}=3.10$eV;[7] T is to the temperature at the BC-TBC interface during the hot temperature phase. The elastic modulus for YSZ in as-sprayed state is taken as $E_{TBC}(t=0)=25 \times 10^9$Pa.

Now the lifetime of the TBC system is estimated as follows: the stress distribution is numerically calculated for a homogeneous temperature change of $\Delta T=-1000$K and for different

TGO thickness between 0μm and 30μm. An initial horizontal crack of length 20μm, assumed to be already produced during the solidification of the sprayed molten lamellae, is situated (fictively, without being included in the FEM mesh) and centered over the BC-TBC hill. At every thermal cycle, during the 5min of the hot phase the two processes described by equation (5) take place. They determine the thermal stress which arises during the subsequent cooling down phase of 2min duration at which the crack grows according to equation (4). The thermal stress distribution is obtained by interpolation between the two corresponding FEM simulations nearest to the current TGO thickness. The process is repeated until the crack has covered a complete wavelength of the BC-TBC profile. For a crack running horizontally, opened only by a positive vertical stress component σ_{yy}, the evolution of crack length as a function of the increasing TGO is represented in Fig. 4 (black line, left diagram).

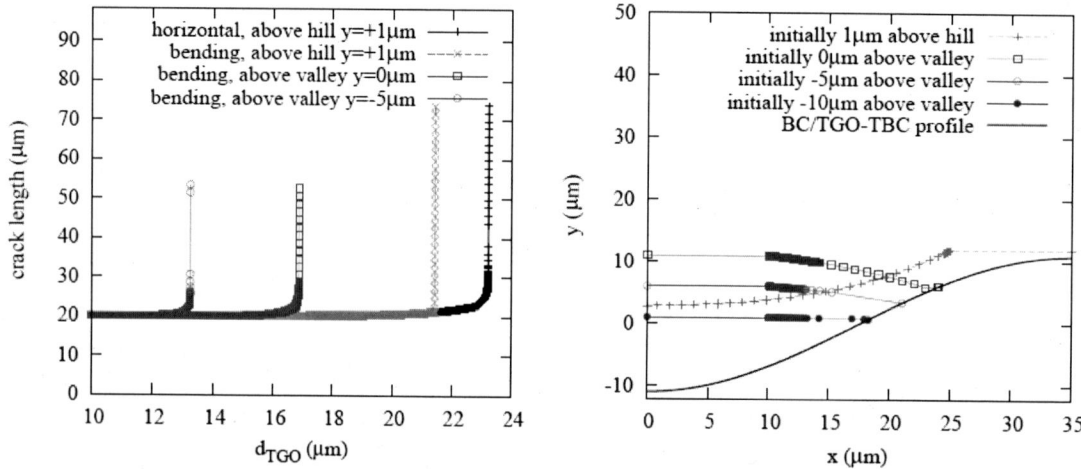

Figure 4. Crack length as function of TGO thickness and crack path, for a horizontal crack (black) and bending cracks, starting centered from the BC-TBC hill (green) or from the valley (red). The initial crack length is 20μm. Growth for valley cracks stopped when reaching TGO.

An improvement to the previous lifetime model is achieved if the crack is now allowed to bend and choose at each growth step the direction at which the crack vertex is "seeing" the maximum opening stress. Hence a faster crack growth results. Let σ_{xx}, σ_{zy} and σ_{yy} be the local stress distribution at the point where the fictive crack vertex is placed, then the azimuthal or hoop tension and the shear component along an inclination θ with the horizontal axis are

$$\sigma_{\theta\theta} = \sigma_{xx} \sin^2 \theta - \sigma_{xy} \sin(2\theta) + \sigma_{yy} \cos^2 \theta, \quad \sigma_{r\theta} = -0.5\left(\sigma_{xx} - \sigma_{yy}\right)\sin(2\theta) + \sigma_{xy} \cos(2\theta) \quad (6)$$

The inclination $\theta=\beta$ at which the hoop stress is maximized is given by

$$\tan(2\beta) = 2\sigma_{xy} / \left(\sigma_{xx} - \sigma_{yy}\right) \quad (7)$$

together with the additional condition $\sigma_{xx} < \sigma_{yy}$ for ensuring $\left.\dfrac{d^2\sigma_{\theta\theta}}{d\theta^2}\right|_{\theta=\beta} < 0$ at $\beta \leq 45°$. Such

inclination also corresponds to one local principal stress direction due to relation $\dfrac{d\sigma_{\theta\theta}}{d\theta} = -2\sigma_{r\theta}$.

From the simulated local distribution of thermal stress for the current location of the (fictive) crack tip, the bending angle β is determined from equation (7), and the corresponding maximum stress component which makes the crack grow under mode I from

$$\sigma_{open} = \sigma_{\theta\theta}(\theta = \beta) = \frac{1}{2}\left[\left(\sigma_{xx} + \sigma_{yy}\right) + \sqrt{\left(\sigma_{xx} - \sigma_{yy}\right)^2 + 4\sigma_{xy}^2}\right] \qquad (8)$$

The stress intensity factor for the crack growth equation (4) is taken, in analogy to circularly curved cracks in an infinite plane under homogeneous stress[8], as $K_I \approx \sigma\sqrt{\pi a'/2}$, being a' the crack length projected onto the horizontal direction.

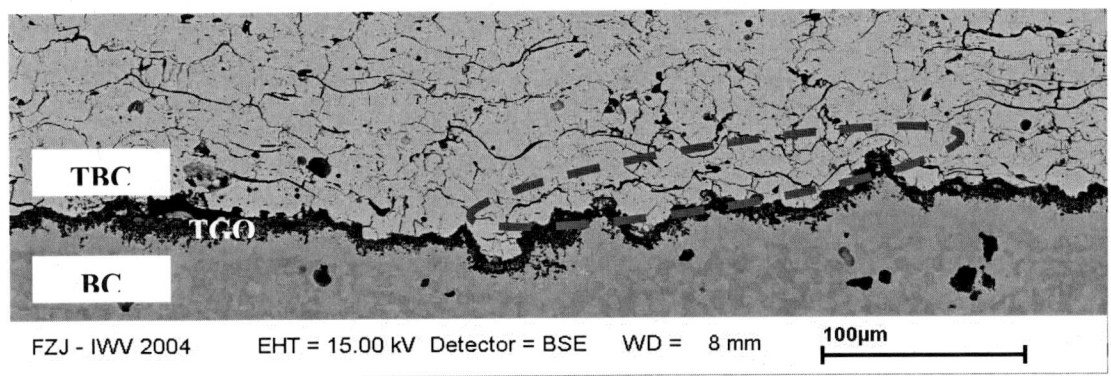

FZJ - IWV 2004   EHT = 15.00 kV Detector = BSE   WD =   8 mm   100μm

Figure 5. Micrograph of a TBC system after 1500 cycles with a temperature at the TBC surface of 1250°C and of 1020°C in the metallic substrate. The intermediate growth state of a delamination crack parallel to the TGO-TBC interface, and covering several hill-valley groups, is visible. The TGO is about 6μm thick..

The crack growth evolution thus achieved is shown in Fig. 4. Cracks starting from the BC-TBC hill, when the TGO reaches a thickness high enough, bend downwards toward the valley, where for the growing TGO the opening stress becomes increasingly larger. Such a crack, however, does not penetrate deep into the valley since a very large TGO thickness would be required to yield a positive opening stress. The crack growth now proceeds more rapidly than when the growth direction was fixed to be horizontal. The downward trajectory is also observed experimentally (see Fig. 5). Further, and according to the simplified lifetime model, cracks starting centered over the BC-TBC valley grow even faster than when starting over the hill. Here, the crack is initially not able to grow since it is located in the valley region of vertical compressive stress; only after a thick enough TGO has pulled down the compression region, the crack starts to grow rapidly, even for a TGO thickness lower than for hill cracks. The growth direction is slightly downwards, pointing to the interior of the TGO flank. In the latter case, failure is assumed when the crack reaches the TGO. For an inclined crack there exists in the TGO flank a stress level high enough to drive crack growth. This growth is additionally enhanced through the TGO elastic modulus, higher by one order of magnitude than that of the plasma sprayed YSZ, for quite similar fracture toughness values in both layers. Hence crack growth in the TGO can be assumed in a first approximation to be near critical and thus very fast.

However, the result on the starting point for fastest crack growth can only be provisional as discussed above. For the crack bending and growth only the stress distribution at the current crack tip location is used for deriving the stress intensity factor, without considering whether the remaining crack part is still open or closed. Hence, for a crack centered over the valley, the vertex can soon "feel" the concentration of opening stress being developed on the TGO-TBC flank (see Fig. 3, right), different from the crack starting over the hill, where its vertex is more distant from such a flank. The latter, nevertheless, extends for most of its length within a non-compression region, whereas a large part of the valley crack is under compression.

EXTENDED LIFETIME MODEL: CRACK PATH IMPLEMENTED IN THE FEM MESH

In the previous simplified model, although the crack growth direction has been correctly reproduced, the necessary TGO thickness at failure is unrealistic high (above 13μm), when compared to actual failed TBC systems. Further, neglecting the extension and presence of the crack itself in the FEM calculation leads to a conclusion about which initial crack location is more relevant for the lifetime which can be wrong. Hence the model will be extended by explicitly tracking the crack growth in the TBC.

The accurate calculation of the stress intensity factor in equation (4) has to be carried out through a path integration around the crack tip, which requires locally a very fine mesh and, moreover, to re-adapt the mesh every time the crack grows. This path integral can be avoided by considering the change in elastic potential energy of the TBC system when the crack grows. Herein, the crack will be modeled by giving very soft properties (elastic modulus of 100Pa, negligible compared to that of the other materials, vanishing thermal expansion coefficient) to the mesh cells occupied by the crack. This can be easily implemented in the FEM simulation. At every crack growth step, the elastic energy U is calculated for the current crack state and 9 further extensions: case 1, crack extended one cell ahead; case 2, one cell up; case 3, one cell down; cases 4/5, one cell up/down, one cell ahead; cases 6/7, one cell up/down, two cells ahead; cases 8/9, one cell up/down, three cells ahead (see Fig. 6, left). Then the energy release rate for each of the 9 cases is calculated, defined as the relative energy change between the current crack state of length a and each crack extension, $G(\sigma,a) = \dfrac{U(\sigma,a) - U(\sigma, a+\Delta a)}{\Delta a}$, and the maximum value is selected. Since G is equal, up to a material dependent parameter, to the squared stress intensity factor, the sub-critical growth equation (4) can be re-formulated as

$$G_{I(,crit)}(\sigma,a) = \frac{1-v_{TBC}^2}{E_{TBC}}\left(K_{I(,crit)}(\sigma,a)\right)^2 \quad \Rightarrow \quad \frac{da}{dt} = v_0\left(\frac{G_I(\sigma,a)}{G_{I,crit}}\right)^{m/2} \tag{9}$$

Nevertheless, one point should be considered. The energy release rate is always a positive scalar and thus the softening of a cell ahead of the crack vertex always reduces the energy, even if that cell was under compression and thus the crack would not be able to grow there. Therefore, the maximum energy release rate for crack growth has to be found out but only among the cases where the opening stress on the crack tip is positive. Hence at every growth step and for the current crack state, the stress distribution averaged over the cell ahead of the crack vertex is read (cell average to avoid the stress singularity just at the crack tip). As next step, angle β is determined according to equation (7) and then the maximum energy release rate for mode I, G_I, is selected for only those two extension cases corresponding to the two directions nearest to the just calculated β. Such G_I and crack extension direction are used to calculate the current growth velocity (eq. (9)). Since angle β is referred to the horizontal, whereas the 9 crack extensions are defined on the local mesh (Fig. 6, left), angle φ has to be used to convert β into the corresponding crack extension case.

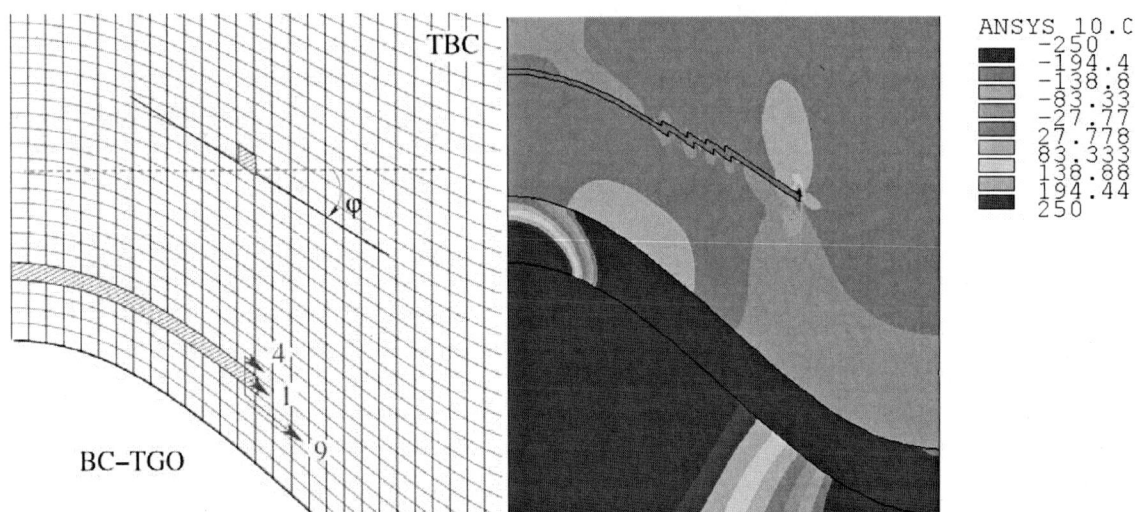

Figure 6. Left, TBC crack extension cases 1-4-9 and angle φ between horizontal and crack tip cell. Right, vertical stress σ_{yy} (range –250MPa to +250MPa) at an intermediate growth step for a TGO thickness (middle layer) of 5.2µm. Initial 20µm crack centered 10µm above hill.

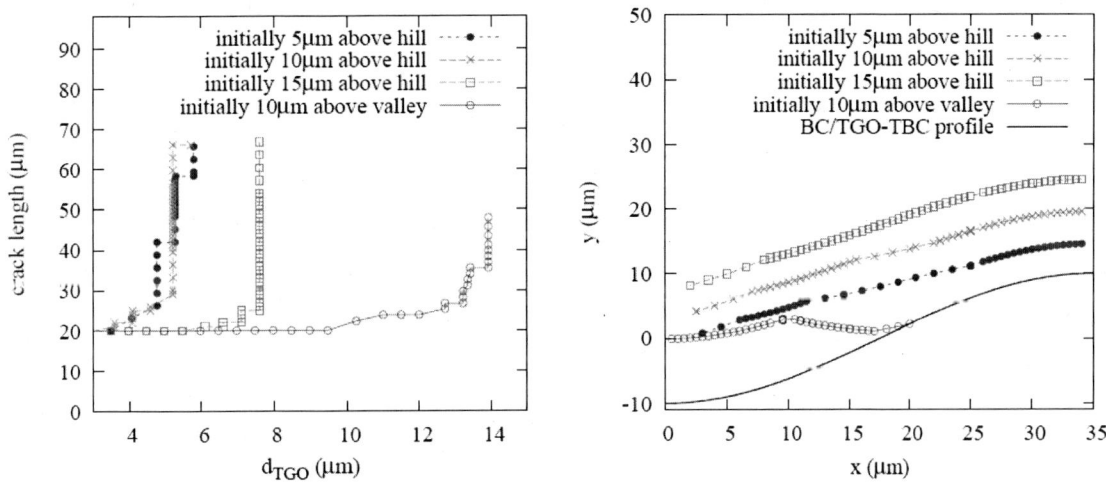

Figure 7. Extended lifetime model: crack length as function of TGO thickness and path for a crack starting centered from the BC-TBC hill (green and blue) or from the valley (red). The initial crack length is 20µm. Growth for valley crack stopped when reaching TGO.

Now the growth for initial 20µm cracks is simulated, both starting from the BC-TBC hill and valley. Since the typical thickness of the first sprayed TBC lamella is about 10µm, this is the reference height above the BC-TBC interface for the initial crack. In Figure 6 (right) the vertical stress distribution at an intermediate growth state for a hill crack is shown when the TGO has reached a thickness of 5.2µm. The crack path is similar to that in the simplified model of the previous section (Fig. 4, right) but now a quite lower TGO is required for the crack to cover a whole wavelength of BC-TBC hill-valley. Figure 7 displays the crack length and growth path for different initial crack locations. The shortest growth time corresponds to cracks starting from the hill (not from the valley as in the simplified model), particularly that initially placed 10µm above, which requires a 5.7μm TGO at failure (actually below 5.3μm for most of the crack growth). This value lies well within the values measured experimentally for delamination cracks (see Fig. 5). Also the crack path, following approximately parallel the hill-valley structure, is the correct one. A crack closer to the BC-TBC interface (Fig. 7, blue line) starts earlier but then grows more slowly across the valley.

It should be noted that cracks initially centered over the valley need a much thicker TGO to be able to start growing. The path followed then bends towards, and eventually penetrates into, the TGO. It might be that if the cooling down phase would start after a very long hot phase, such that the TGO has already grown above, let's say, 12μm, then valley cracks, being already in a positive tension region closer to the TGO-TBC interface and "feeling" thus a higher stress, would grow faster than hill cracks. This could yield an explanation why the failure for long isothermal oxidation occurs rather inside the TGO, in contrast to the thermal cycling where failure characteristically takes place within the first sprayed TBC lamellae.

CONCLUSIONS

A lifetime model has been developed, which tracks directly in the FEM mesh the sub-critical growth of delamination cracks near the curved BC-TBC interface for an increasing TGO growth during thermal cycling. The force driving the crack growth is the thermal stress during the cooling down phase. The crack growth velocity is determined through a combination of trajectory for maximal opening stress and maximum energy release rate, and the resulting TGO thickness at failure well corresponds to experimental values.

ACKNOWLEDGMENTS

The authors thank Mr K.-H. Rauwald and Mr R. Laufs for the sprayed samples, Mr M. Kappertz for the cross-section preparation and Dr D. Sebold for the careful micrographs.

REFERENCES

[1] R. Vaβen, F. Traeger, D. Stöver, "Correlation Between Spraying Conditions and Microcrack Density and Their Influence on Thermal Cycling Life of Thermal Barrier Coatings," *J. Thermal Spray Technol.* **13**, 396-404 (2003).

[2] S. Giesen, "Characterization of plasma sprayed coatings by means of Fourier analysis and stochastic equations," diploma thesis, Fachhochschule Aachen/Jülich (2005), in German.

[3] R. Vaβen, G. Kerhoff, D. Stöver, "Development of a micromechanical life prediction model for plasma sprayed thermal barrier coatings," *Mater. Sci. Eng. A* **303**, 100-109 (2001).

[4] F. Traeger, R. Vaβen, K.-H. Rauwald, D. Stöver, "Thermal Cycling Setup for Testing Thermal Barrier Coatings," *Adv. Eng. Mater.* **5**, 429-432 (2003).

[5] M. Ahrens, S. Lampenshcerf, R. Vaβen, D. Stöver, "Sintering and Creep Processes in Plasma-Sprayed Thermal Barrier Coatings," *J. Thermal Spray Technol.* **13**, 432-442 (2003).

[6] S.R. Choi, D. Zhu, R. Miller, "Mechanical Properties/Database of Plasma-Sprayed ZrO_2-8wt% Y_2O_3 Thermal Barrier Coatings," *Int. J. Appl. Ceram. Tehcnol.* **1**, 330-342 (2004).

[7] F. Traeger, M. Ahrens, R. Vaβen, D. Stöver, "A life time model for ceramic thermal barrier coatings," *Mater. Sci. Eng. A* **358**, 255-265 (2003).

[8] H. Tada, P.C. Paris, G.R. Irwin, "The Stress Analysis of Cracks Handbook," Del Research Co., Pennsylvania, section 21 (1973).

DAMAGE PREDICTION OF THERMAL BARRIER COATING

Y. Ohtake
Ishikawajima-Harima Heavy Industries Co., Ltd.
1, Shin-Nakahara-Cho, Isogo-ku,
Yokohama-shi, Kanagawa 235-8501, Japan

ABSTRACT

Thermal barrier coatings are applied to many high temperature airplane engine and gas turbine hot section parts. The durability of the thermal barrier coating determines the life of the parts. Thus, the development of a thermal fatigue life prediction method of thermal barrier coating is paramount to the design of the parts. The fracture mechanism of thermal barrier coating by thermal cycle test and furnace heating test are examined here. All specimens fractured in delamination of the thermal barrier coating. The delamination driving forces were calculated using the finite element method. The stress in thermal barrier coating was calculated using two models; a full scale 3D model and a subscale 2D model to capture the waviness of the BC/TC interface. It was found that the delamination was due to the growth of thermal growth oxidation layer, thermal stress, the shape of the part, the configuration of the interface and the mismatch strain at the interface. Thus, this paper proposed a simplified method to predict the thermal fatigue life of thermal barrier coating based on the present results. A damage factor was proposed to evaluate the damage of the thermal barrier coating. The factor was constructed using three parameters: a) thickness of thermal growth oxidation layer, b) stress, and c) mismatch strain at the interface. These parameters are determined from a combination of the experiment and analytical results. The damage prediction method is very simple to apply and hence it is effective in the design of the high temperature coated parts.

INTRODUCTION

Thermal barrier coating (top coat ; TC) is applied to the parts of the airplane engine and gas turbine. Top coat is used for a coating system together environment barrier coating (bond coat ; BC). Fracture of the top coat occurs by either delamination or the crack in the normal direction to the thickness of the top coat when the coating system is heated. The delamination of top coat occurs near the BC/TC interface by the growth of thermal growth oxidation (TGO) layer at the interface and other causes [1]-[3]. The vertical crack in top coat is due to the difference of thermal expansion of the materials. We had examined the damage of a plate specimen with a typical coating system by burner rig test [1]. The damage was the delamination of top coat at the interface [1]. The delamination of thermal barrier coating had been investigated by a lot of researchers until now [4]-[10]. This paper proposed a simplified method of thermal fatigue life to predict the delamination of top coat. The method should be used to predict thermal fatigue life of thermal barrier coating in the design and it may be used for the decision of the shape of application part and the compositions (thickness, material, layer number) of top coat in the design for the application of coating system to the part.

Fig.1   Appearance of thermal cycle test

EXPERIMENTAL PROCEDURE

The specimen was made of a coating system that was applied to single crystal CMSX-2 substrate of nickel base superalloys. The coating system consisted of bond coat and top coat. The bond coat was CoNiCrAlY that was deposited by low pressure plasma spray (LPPS). The top coat was 8 wt. percent yttria stabilized zirconia (YSZ) that was deposited by air plasma spray (APS). Thermal fatigue damage of top coat was examined in thermal cycle test by burner rig [1]. A plate specimen with corner edges was used for the test. The dimension of substrate of the specimen is length 500mm, width 500mm and thickness 3mm. Thickness of top coat and bond coat of the specimen are 0.5mm and 0.125mm, respectively. The total cycle time is 5 min with 3 min of holding time at maximum temperature for a total of 1000 cycles. Figure 1 shows a photograph of the burner rig test. The burner rig test heats on the surface of top coat of specimen by heating gas and cools on the back surface by air, and then the specimen has a temperature distribution in the direction of the thickness in heating. The distributions produce stress in the specimens because the thermal expansions of top and bond coats are different from one of substrate. The specimen was removed to examine the damage of top coat after the thermal cycling test. A second test was performed to examine the damage of top coat and the growth of TGO layer [2][3]. The test specimen consisted of small circular plate used in the furnace test. The dimension of substrate of the specimen is diameter 20mm and thickness 3mm. Two coating thickness were used 0.3mm and 0.5mm, with the same bond coat thickness of 0.125mm. The specimens were tested at 800℃, 900℃ and 1200℃ [2][3]. The heating intervals were 100h, 200h and 500h. The specimen was examined for the damage of top coat and to measure the thickness of the TGO layer. An experimental relation was deduced based on the heating time and the thickness of TGO layer to determine the TGO growth rates at temperatures.

ANALYTICAL PROCEDURE

The stresses in the thermal fatigue specimens were calculated using the finite element method [1]. Figure 2 shows the mesh of the analytical model (Specimen model). The model consists of 8-nodes three-dimensional solid elements. The effect of coating waviness was examined using a two-dimensional model to examine the stress in top coat near the interface between bond coat and top coat after the growth of TGO layer. Figurer 3 shows the mesh of the analytical model. The model consists of 4-nodes generalized plane strain elements. A TGO layer is inserted in the model and it grows with heating time. The BC/TC interface shape is modeled as cosine wave with three different amplitudes 5 μ m, 10 μ m and 20 μ m. The furnace temperature is set at 900℃ uniformly. The TGO growth rate is 1.5 μ m/s and the total heating time is 6000 second. The results of burner rig test [1] were used to determine the relationship between the temperature and the growth rate of TGO layer. Table 1 denotes material properties of the coatings that were used to make the FEM analysis. Those properties of substrate are taken from [2] [11]. The material property of TGO layer was adopted from [12]. The bond coat was supposed for elastic-ideal plastic material [12] and is added creep property in combination with the Norton law [13]. The ABAQUS program [14] was used for the finite element calculations.

Table.1   Material property

	Temperature (K)	Bond coat CoNiCrAlY	Top coat APS YSZ	Top coat EB-PVD YSZ
Thermal conductivity (W/mK)	373	10.10	0.95	1.80
	773	19.30	0.96	1.80
	1273	25.32	0.97	1.84
	1473	29.87	0.98	1.84
Specific heat (J/kg·K)	373	429	608	498
	773	430	610	505
	1273	410	600	500
	1473	415	605	510
Density (kg/m³)	298	7078	5723	3780
Young's modulus (GPa)	298	177.5	38.0	44.0
	873	151.3	-	-
Poission's ratio (-)	298	0.3	0.3	0.3
Thermal expansion coefficient (×10⁻⁵ /K)	1273	15.70	11.10	8.12

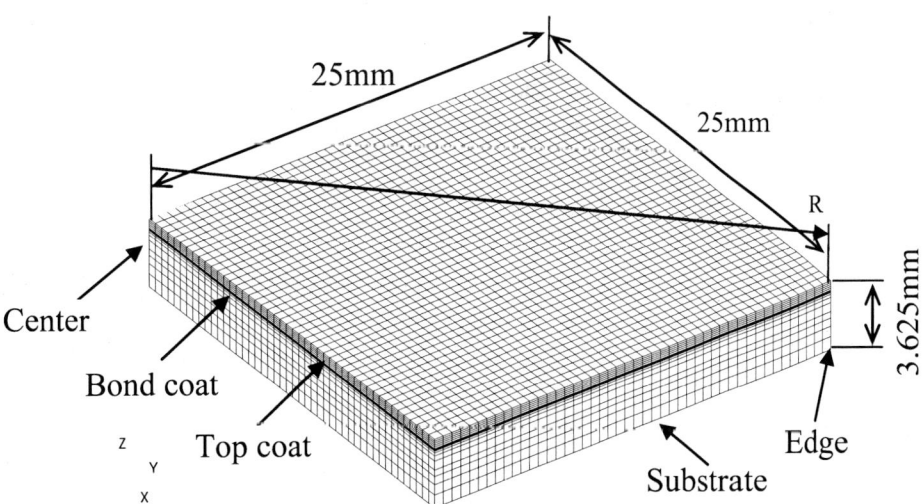

Fig.2   Modeling and division of mesh element for FEM analysis

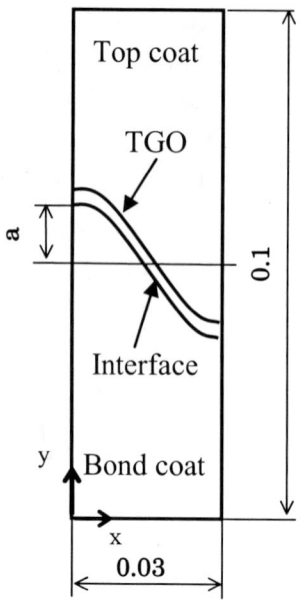

Fig.3 Modeling for FEM analysis

EXPERIMENTAL RESULTS

Figure 4 shows two photographs of the microstructure of the plate specimen after thermal cycle test after 1000 cycles. The observed damage as seen in Fig.4 was delamination of the top coat near the BC/TC interface. The delamination was detected at two locations at the center and the edges of the specimen. Many small cracks were observed in top coat near the interface. The cracks initiated from the pores in top coat. Thus, it is found that the delamination of top coating occurred by linking of the micro-cracks in the top coat near the interface. Moreover, a TGO layer was observed at BC/TC interface. The delamination of top coat near the interface may be affected by the growth of the TGO layer. Figure 5 shows the relationship between heating time and the thickness of TGO layer at the interface.

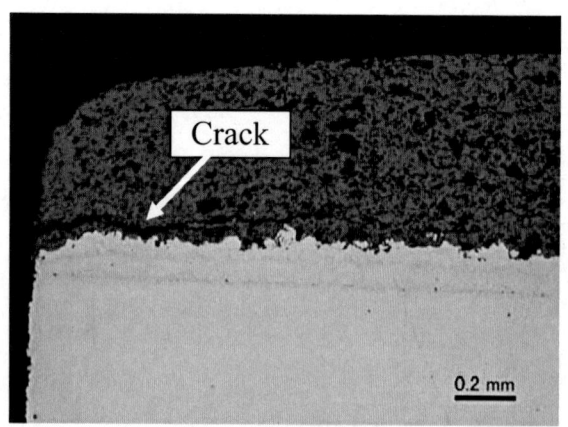

(i) At edge area   (ii) At center area

Fig. 4 Microstructure of specimen at 1000 cycles

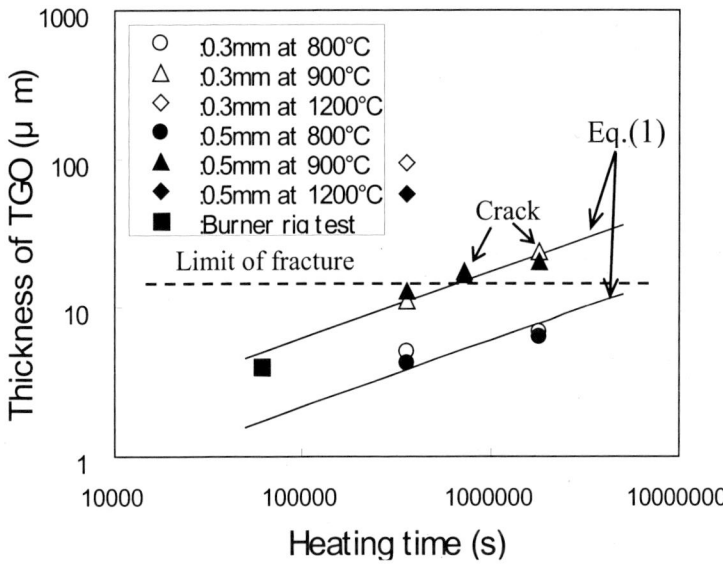

Fig. 5   Relationship between heating time and thickness of TGO layer

The results of heating test and thermal cycle test were denoted in Fig.5. It was found that the thickness of TGO layer increased with test temperature and heating time. The growth of TGO layer was almost independent of the initial thickness of the top coat. The growth of TGO layer can be easily modeled as a relationship in terms of thickness of TGO layer w, heating time t and two constants k and n.

$$w = kt^n \tag{1}$$

The line at 900℃ in Fig.5 was k=0.035 and n=0.45 and the line at 800 was k=0.013 and n=0.45. The results of burner rig test could be predicted by using the 900℃ curve of Fig. 5. Small cracks are detected in top coat of those specimens after 200h and 500h at 900℃. The cracks propagated parallel to the interface direction of top coating surface from pore near the interface. The delamination of top coat of the specimen occurred after 100h at 1200℃. Thus, it was found that TGO layer was grown as increasing test temperature and heating time, so that the failure occurred by delamination of top coat near the BC/TC interface.

ANALYTICAL RESULTS

Figure 6 shows the calculation results of specimen at 20s after heating start in burner rig tests. Figure 6(i) shows the relationships between distance R and the stresses in full scale model in Fig.2. Those stresses are normal stress σ_z and shear stress τ_{zx} in mesh of top coat at BC/TC interface. The maximum values of those stresses were generated at the corner edge of the model by the effects of both the shape of the specimen and thermal stress. It resulted the delamination of top coat at the edge area of the specimen in Fig.4(i). Figure 6(ii) shows the relationship between distance R in Fig.2 and mismatch strain. The mismatch strain had been proposed by Miller [4]-[6]. The maximum value of the mismatch strain occurred at the center of the full scale model. It was found that the mismatch strain caused the delamination of top coat at the center of the specimen in Fig.4(ii).

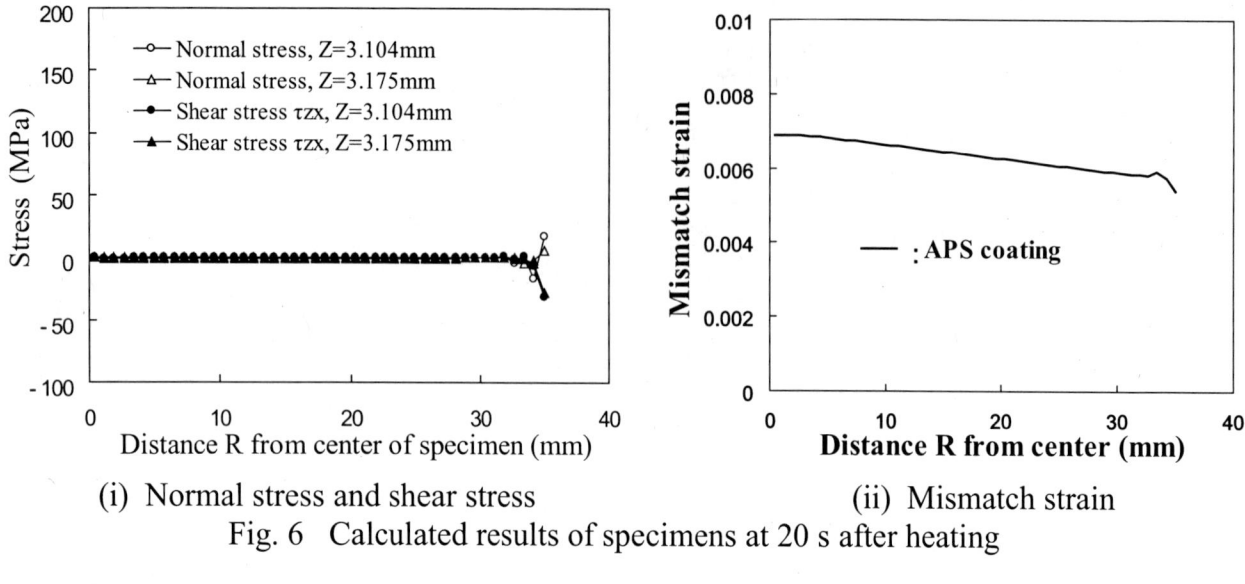

(i) Normal stress and shear stress   (ii) Mismatch strain

Fig. 6   Calculated results of specimens at 20 s after heating

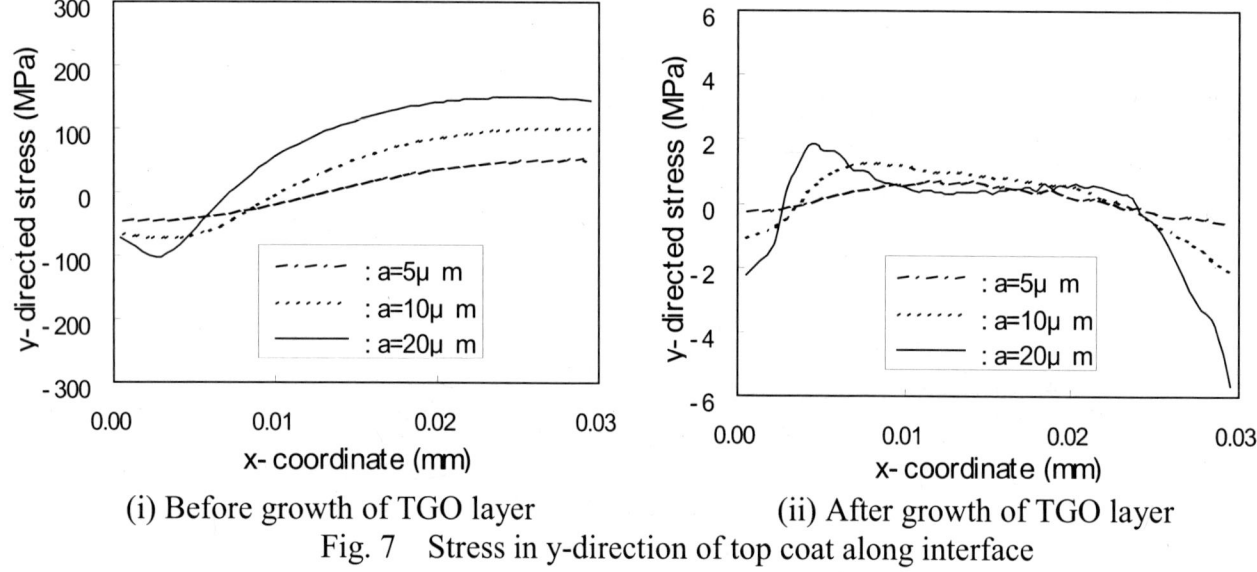

(i) Before growth of TGO layer   (ii) After growth of TGO layer

Fig. 7   Stress in y-direction of top coat along interface

Figure 7 shows normal stress σ_y in the y-direction in the mesh of top coat along BC/TC interface for x-coordinate of a subscale 2D model in Fig.3. The maximum values of the normal stress σ_y in all models were increased as TGO layer glows with heating. The location of the maximum value moved to the convexity (x=0mm) from the concavity (x=0.3mm) of the BC/TC interface of the model in Fig.3. The simulation results coincided with the locations of the occurrence of the delamination near the convexity of the interface.

THERMAL FATIGUE LIFE PROCEDURE

Thermal cycle test was performed by using the plate specimen with a coating system. The damage was the delamination of top coat near BC/TC interface at those locations of the center and the edge of the plate specimen after 1000 cycles. It was found that the delamination of top coat was due to the growth of TGO layer, thermal stress, the shape of the parts that applied top coat, the configuration of the interface and mismatch strain at the interface. Those fracture mechanisms were cleared from thermal cycle test, heating test and thermal stress analysis. Thus, this paper was investigated a simplified method to predict thermal fatigue life of top coat from those examinations. Equation (2) was proposed to predict the thermal fatigue life of top coat. The delamination of top coat was caused when the damage factor D in Eq.(2) was reached to one. The factor D was composed of three kinds of parameters ε_o, ε_m ε_c in Eq.(2).

$$D = \frac{\varepsilon_o}{\varepsilon_{oc}} + \frac{\varepsilon_m}{\varepsilon_{mc}} + \frac{\varepsilon_c}{\varepsilon_{cc}} \tag{2}$$

The strain ε_o is determined from the calculation of finite element analysis of the growth of TGO layer. The mismatch strain ε_m and the strain ε_c are decided by the calculation of the analysis model of the part, where ε_c implies the strain that is caused by the effect of the shape of part. Three kinds of parameters ε_{oc}, ε_{mc} and ε_{cc} in Eq.(2) denotes limit values of those strains when the delamination of top coat happens at BC/TC interface. Two parameters ε_{oc}, ε_{mc} are determined by the heating test of small specimen. The test is very simple in comparison with thermal cycle test. it was found that parameter ε_{oc} of first term in Eq.(2) was main parameter for the thermal fatigue damage of top coat from the results until now [1]-[3] and it may be set the value of a dotted line in Fig.5. On the other hand, parameter ε_{cc} is determined from several mechanical tests of the specimen. However, it can solve the effect of the parameter ε_{cc} in the design change of the shape of the parts.

CONCLUSIONS

This paper examined the fracture mechanism of top coat in the experiments both thermal cycle test and heating test. Those specimens fracture in delamination near the BC/TC interface. The stress was calculated in both micro-model at the interface and specimen model in finite element analysis. As a results, it was found that the delamination is due to the growth of TGO layer, thermal stress, the shape of engine parts, the configuration of the interface and mismatch strain at the interface from the examinations of those fracture mechanisms. Thus, this paper investigated a simplified method to predict thermal fatigue life of top coat from those examinations. The method proposes an accumulated damage parameter D to predict the occurrence of delamination of top coat. The damage parameter D is composed three terms of the thickness of TGO layer, mismatch strain at BC/TC interface and the effect of the shape of part. Those parameters are determined both the calculations of finite element analysis and some specimens. A dotted line in Fig.5 shows a main parameter for the thickness of TGO layer. The proposed method in this paper is very simple, and also it is an effective method in the design of thermal fatigue life of top coat on engine part.

REFERENCES
[1]Y. Ohtake, N. Nakamura, N. Suzumura and T. Natsumura, "Evaluation for Thermal Cycle Damage of Thermal Barrier Coating," *Ceramic Engineering and Science Proceedings*, 24(3) 561-566 (2003).

[2]Y. Ohtake, T. Natsumura, "Investigation of Thermal Fatigue Life of Thermal Barrier Coating," *Ceramic Engineering and Science Proceedings*, 25(4) 357-362 (2004).

[3]Y. Ohtake, T. Natsumura, K.Miyazawa, "Investigation of Thermal Fatigue Life of Thermal Barrier Coating," *Ceramic Engineering and Science Proceedings*, 26(3) 89-93 (2005).

[4]R. A. Miller, "Oxidation-Based Model for Thermal Barrier Coating Life," Journal of the American Ceramic Society, 67 [8] 517-21 (1984).

[5]R. A. Miller, "Thermal Barrier Coatings for Aircraft Engines History and Directions," *Journal of Thermal Spray Technology*, 6 [1] 35-42 (1997).

[6]R. A. Miller, "Life Modeling of Thermal Barrier Coatings for Aircraft Gas Turbine Engines," *Journal of Engineering for Gas Turbines and Power*, 111 301-05 (1989).

[7]A. G. Evans, M. Y. He and J. W. Hutchinson, "Mechanics-based scaling laws for the durability of thermal barrier coatings," *Progress in Materials Science*, 46 249-271 (2001).

[8]A. G. Evans, D. R. Mumm, J. W. Hutchinson, G. H. Meier and F.S. Pettit, "Mechanisms controlling the durability of thermal barrier coating," *Progress in Materials Science*, 46 505-553 (2001).

[9]T. A. Cruse, S. E. Stewart and M. Ortiz, "Thermal Barrier Coating Life Prediction Model Development," Journal of Engineering for Gas Turbines and Power, 110 610-616 (1988).

[10]S. M. Meier, D. M. Nissley, K. D. Sheffler and T. A. Cruse, "Thermal Barrier Coating Life Prediction Model Development," *Journal of Engineering for Gas Turbines and Power*, 114 258-263 (1992).

[11]K. Nishimoto, K. Saida, D. Kim, Y. Nakao, "Transient Liquid Phase Bonding of Ni-base Singe Crystal Superalloy, CMSX-2," *The Iron and Steel Insitute of Japan*, **35[10]** 1298-1306 (1995).

[12]K. Sfar, J. Aktaa and M. F. Kanninen, "Analysing the Failure Behaviour of Thermal Barrier Coating Using the Finite Element Method," *Ceramic Engineering and Science Procedings*, 21(3) 203-211 (2000).

[13]R. G. Kerkhoff, R. Vaβen and D. Stover, "Numerlcally calculated thermal stresses in thermal barrier coatings on cylindrical substrates," *Proc. United Thermal Spray Conference*, 787-792 (1999).

[14]ABAQUS, "ABAQUS/Standard, Version5.8, " Hibbitt, Karlsson & Sorensen, Inc. (1999).

Creep Behavior of Plasma-Sprayed Zirconia Thermal Barrier Coatings

Reza Soltani,[*,†] Thomas W. Coyle,[*,‡] and Javad Mostaghimi[§]

Centre for Advanced Coating Technologies, University of Toronto, Toronto, Ontario, Canada

Thermally sprayed ceramic coatings deposited from nano-structured feedstock powder have often demonstrated improved properties relative to coatings produced from conventional powders. This type of coating has been reported to exhibit better wear resistance and higher adhesion strength compared with conventional deposits. Powder consisting of hollow spherical particles has been reported to produce coating with lower unmelted particles and lower thermal conductivity. In this study, the thermo-mechanical properties of plasma-sprayed yttria-stabilized zirconia coatings deposited using each of these types of powder were investigated. Creep strain and creep rate were measured using free-standing thick coatings loaded in a four-point bend configuration at temperatures ranging from 800° to 1200°C in air under a range of loads. The creep exponent and activation energy were determined.

I. Introduction

PLASMA-sprayed thermal barrier coatings have been used or considered for protection of a variety of components in aircraft and power generation turbines and in diesel engines. As operating temperatures approach 1000°C, the presence of relatively high compressive stresses at the coating surface lead to stress relaxation by creep and/or sintering. Shrinkage at high temperatures followed by crack initiation and propagation during cooling may cause spallation of the coating, consequently shortening the lifetime of the deposit. The thickness of the coating is an important factor; depositions of more than 1 mm in thickness may be required to achieve the desired level of protection and reduction in heat loss.

The thicker the coating, the greater the thermal gradients present during service, which results in larger stresses within the cross section of the material. The high-temperature behavior of monolithic zirconia and its plasma-sprayed form have been investigated by several researchers.[1–10]

In the current work, yttria partially stabilized zirconia (PSZ) coatings were deposited by air plasma spraying using two different types of powder feed stock. The influence of the powder type on the microstructure and thermo-mechanical behavior of the coatings was investigated.

II. Experimental Procedure

(1) Coating Deposition and Characterization

Two commercially available 6–8 wt% Y_2O_3–PSZ powders were used to deposit coatings by air plasma spraying. A nano-structured powder (Nanox S4007; Inframat Corp., Willington, CT) was deposited using the F4-MB (Sulzer-Metco, Westbury, NY) torch. This powder consists of porous agglomerates 15–150 μm in diameter made up of particles on the order of 200 nm in diameter. A powder consisting of hollow spherical particles (204B-NS, HOSP; Sulzer-Metco) was deposited with an SG-100 (Praxair, Concord, NH) plasma torch. The particle size of this HOSP powder was 45–75 μm.

Diffraction patterns of Nanox and HOSP powders using an X-ray diffractometer (D5000; Siemens, Madison, WI) indicated a tetragonal structure for both feedstocks. The chemical composition of the starting powders was analyzed by inductively coupled plasma (ICP) atomic emission spectroscopy and flame photometry (FP); the results did not show any major differences in feedstock powder compositions.

A range of deposition parameters was used to produce coatings, resulting in different levels of porosity and nonmelted particles, as described elsewhere.[11] An image analyzer (Clemex Vision Professional, Clemex, Canada) was used to determine the average porosity of each sample. The deposition conditions and image analyses results for the coatings selected for this study are given in Table I. The velocity, temperature, and average particle size of in-flight particles were obtained by using a Tecnar DPV 2000 in-flight particle diagnostics instrument (Tecnar Automation, St. Hubert, Canada) at the point of impacting the substrate. By air cooling the coating and the substrate, the temperature was maintained at about 140°C during spraying. After deposition the coating was detached from the substrate by dissolving in nitric acid.

Nanox samples A3 and A4 have the same porosity and almost the same nonmolten nano particle percent. Nanox sample A5 has almost the same nonmolten nano particle content as A3 but a higher porosity level. Finally, HOSP sample A8 has a total porosity close to A3 and A5. Selection was in such a way that the effect of porosity and nonmolten particles on creep/sintering properties of the Nanox and HOSP coatings could be investigated under comparable conditions.

The thickness of the coatings was more than 3 mm. Samples were cut into dimensions of 3 mm × 4 mm × 40 mm ($H \times W \times L$), and then the carbon steel substrates were detached from the coatings. The surfaces of the specimens, where the upper and lower rollers of the flexure fixture made contact, were then ground with diamond disks to assure that the surfaces were flat, smooth, and parallel.

Phase analysis of powders and coatings was carried out by X-ray diffraction (Philips PW2273, Eindhoven, the Netherlands) and the results showed that all coatings contained only tetragonal phase ZrO_2. The powder and coating microstructures were examined by scanning electron microscopy (SEM; Hitachi S-4500, Tokyo, Japan).

(2) Flexure Creep Testing

Potential problems associated with gripping of tensile samples at elevated temperatures and difficulties in measuring displacement at these temperatures have motivated researchers to utilize the four-point bend test method to measure the creep strain and creep rate of ceramics. The simple shape of the sample and its small size minimizes the expenses of specimen preparation. Both tensile and compressive stresses are introduced into the same sample simultaneously. It is known that ceramic creep rates under these two different stress states may vary significantly if any damage occurs during the test. Although this complicates the analysis somewhat,

T. Troczynski—contributing editor

Manuscript No. 22670. Received January 16, 2007; approved April 18, 2007.
Based in part on the thesis submitted by R. Soltani for the Ph.D. degree in Materials Science and Engineering, University of Toronto, Toronto, ON, Canada, 2006.
*Member of American Ceramic Society.
†Author to whom correspondence should be addressed. e-mail: rsoltani@mie.utoronto.ca
‡Department of Materials Science and Engineering.
§Department of Mechanical and Industrial Engineering.

Table I.   In-Flight Properties of Particles, Porosity, and Nonmolten Percent of Nanox and HOSP Coatings

Sample	Powder	Plasma power (KW)	Feeder type	Velocity m/s	Particle diameter (im)	Temperature (°C)	Image analysis % Porosity	Image analysis % Nonmolten
A3	Nanox	27	Int.	179	56.51	3021	12±1	4±2
A4	Nanox	25	Int.	148	52.82	2958	12±1	1.5±1
A5	Nanox	17	Ext.	130	48.01	2533	17±1	4.5±1
A8	HOSP	25	Int.	144	56.06	2809	11±1	—

Int., internal; and Ext., external.

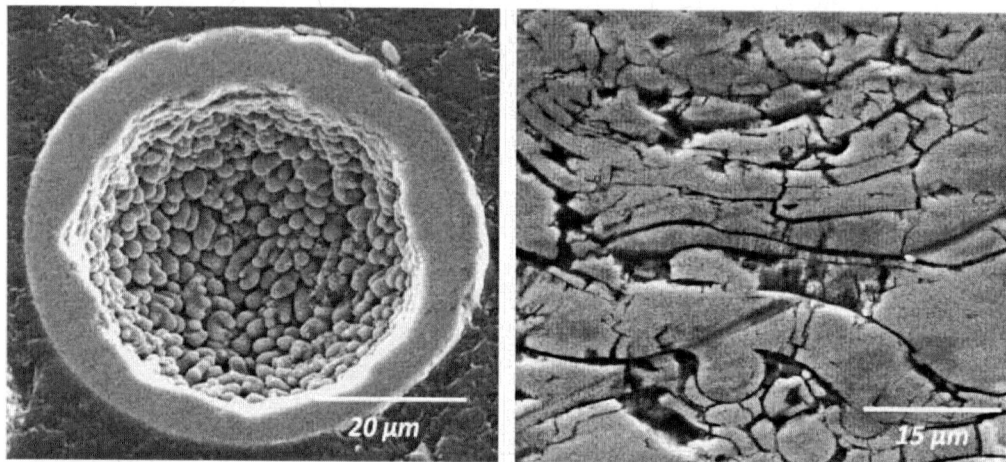

Fig. 1.   Typical scanning electron microscopy image of HOSP; left: feedstock particle and right: polished and etched cross section of coating A8.

it is possible to determine both tensile and compressive creep rates simultaneously in one sample with this approach.[12]

To perform the creep test, a SiC four-point bend test fixture, with inner and outer spans of 20 and 40 mm, respectively, was used.[13] Creep tests were performed at three different temperatures (800°, 1000°, and 1200°C) under a range of stresses for 2 days. The *in situ* dynamic beam profiling method[14] was adopted to measure the creep properties of samples.

III.   Results

(1)   Microstructure of Feedstocks and Coatings

SEM examinations of feedstock powder particles and of the coatings deposited using HOSP and Nanox powders were conducted. Figure 1 shows a hollow spherical HOSP feedstock particle next to a cross section of the corresponding coating. The thin shell of the particles increases the probability of having fully

melted particles and results in a high flattening ratio and very thin final splats in the coating. The morphologies of Nanox particles and the microstructure of the corresponding deposit are illustrated in Fig. 2. The feedstock is an agglomeration of nano particles in spherical shapes; some of these nano particles are retained in the coating as is shown in Fig. 2. The maximum amount of nonmolten nano particles in deposited coatings varied from 1.5% (for A4 sample) to 5% (for A5).

SEM investigations after the creep test were performed on specimens. SEM images from fractured surfaces of samples illustrated different sintering phenomena such as crack healing and intrasplat grain growth and intersplat joining of columnar grains in both types of coatings (Fig. 3).

(2)   Creep Results

Creep tests were performed at a range of temperatures (800°–1200°C) under various stresses for 2 days to make sure

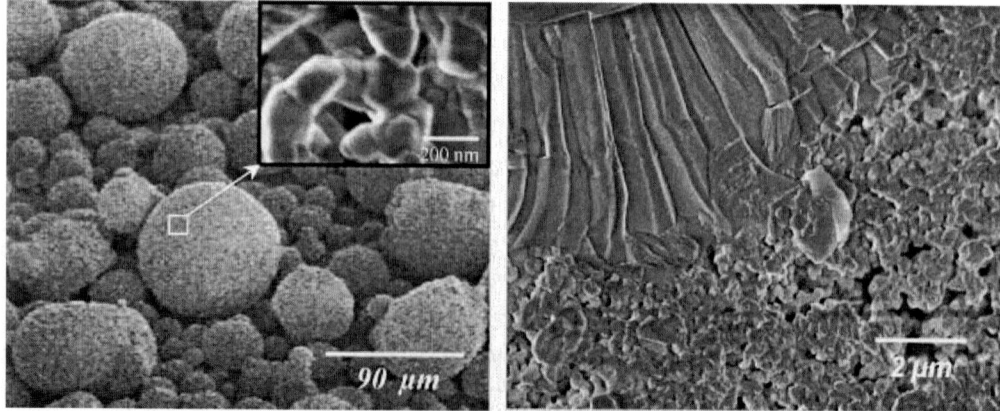

Fig. 2.   Scanning electron microscopy image of Nanox S4007; left: feedstock particles and right: fractured cross section of coating A5 showing nano particles next to large columnar grains of an adjacent splat.

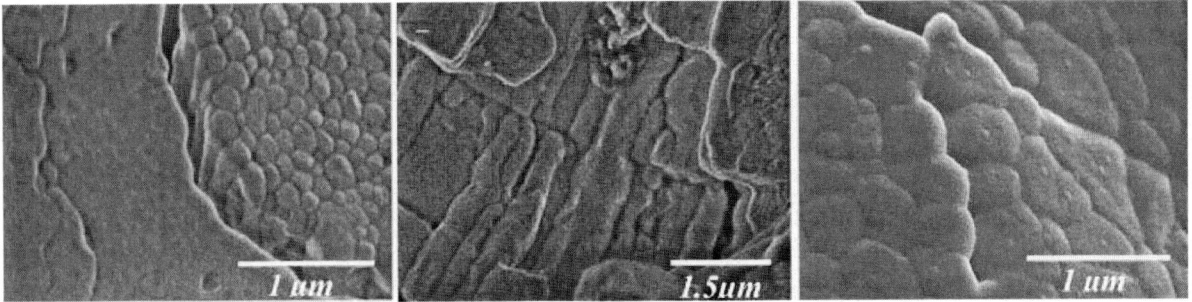

Fig. 3. Scanning electron microscopy image of HOSP splats after 2 days at 1000°C under a 30 N flexure load; left, the surface of the splat shows intrasplat grain growth; middle, intersplat grain growth; right, micro crack healing.

that the samples entered the steady-state creep region. Using the beam profiling method, the strain-time graphs were calculated from the displacement-time data. A Matlab program (The MathWorks Inc., Natick, MA) was prepared to perform all the calculations to obtain creep strain for any given time based on the relative displacement of the center point to the loading points of the beam. The program was modified to calculate the strain rate and displacement of the neutral axis as well.

The creep strains of samples A3, A4, A5, and A8 are shown in Fig. 4. The preliminary results illustrate that samples show the primary and steady-state stages. The creep strains are larger for higher stresses, but there is no significant difference in the stain rates after the first day of testing, showing little dependency of strain rate on applied stress. In these graphs, the elastic portion of the tests was removed and therefore they just represent the creep information.

Assuming that the creep graphs show a steady-state behavior, the stress exponent, n, and preexponential constant, A, were obtained from an equation[15] of the following form:

$$\dot{\varepsilon} = A\sigma^n \tag{1}$$

The creep test was performed on a range of stresses for both Nanox and HOSP samples at 1000°C (Fig. 5) and then the strain rate at the steady-state part of the graphs was calculated. Plotting strain rate versus applied stress, in a logarithmic scale, yielded stress exponents of 1.1 and 1.3, for the Nanox and HOSP samples, respectively (Fig. 6).

$$\log \dot{\varepsilon} = \log A + n \log \sigma \tag{2}$$

Creep activation energies of Nanox and HOSP deposits were calculated by using Arrhenius type of Eq. (1).[15] Taking the nat-

ural logarithm of the equation yielded the strain rate as a function of the inverse temperature

$$\mathrm{Ln}\,\dot{\varepsilon} = \mathrm{Ln}\,A + \left(\frac{-Q}{R}\right)\left(\frac{1}{T}\right) \tag{3}$$

Therefore, creep of the Nanox and HOSP coatings in a range of temperatures was performed. Figure 7 shows the creep strain graphs for the free-standing A5 and A8 coatings in the temperature range of 800°–1100°C.

Based on these graphs, the apparent activation energies of coatings obtained from Eq. (3) were 154 and 190 kJ/mol for Nanox and HOSP deposits, respectively. Considering the large standard deviation (about ±50 kJ/mol), this difference is not significant. When performing a creep test in the four-point bend configuration, the creep exponent for the tension side may be different from that for the compression side due to cavitation damage under tensile loading. This results in the neutral axis being displaced from its original position, causing stress redistribution and consequently different creep rates in tension and in compression. The dynamic beam profile method determines the cross-section rotational center displacement in order to calculate the strain in the compressive and tensile outer fibers of the beam. At relatively low stresses and temperatures of 800°–1100°C, no significant displacement of the neutral axis occurred.

To examine whether samples enter the steady-state stage (having a constant creep rate) after 2 days or not, the creep rates of coatings were plotted versus time in Fig. 9. No constant creep rate was observed, although the creep rate showed a quasi power law behavior. A couple of samples from both materials

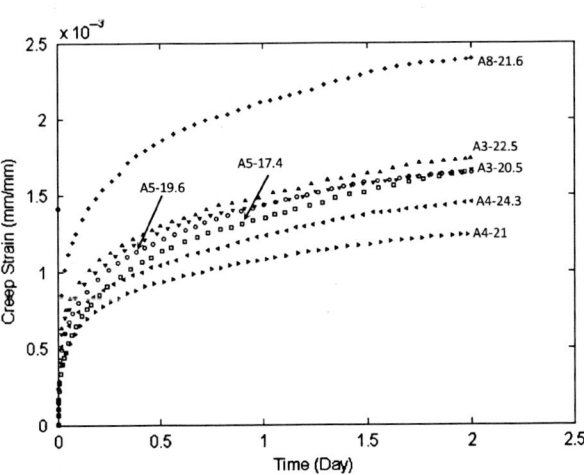

Fig. 4. Creep strain of Nanox and HOSP samples under a range of stresses at 1000°C for 2 days.

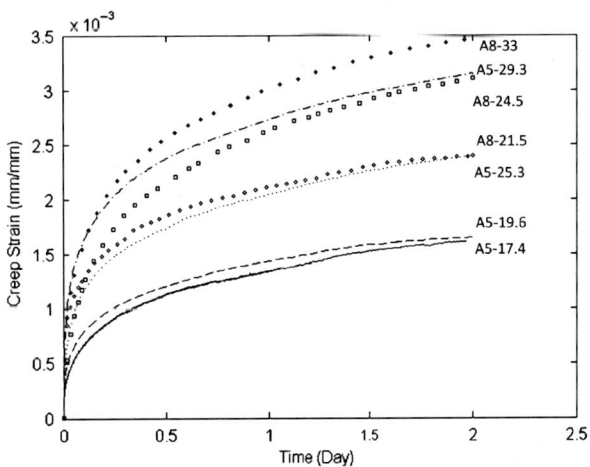

Fig. 5. Comparison of the creep strain of Nanox and HOSP specimens under a range of stresses at 1000°C.

Mechanical Properties

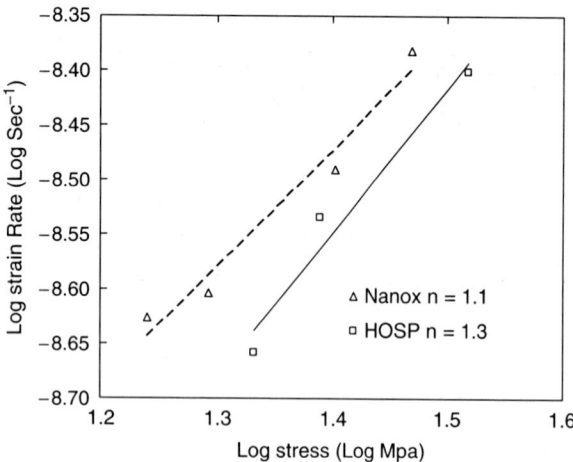

Fig. 6. Plot of strain rate versus applied stress for Nanox and HOSP samples.

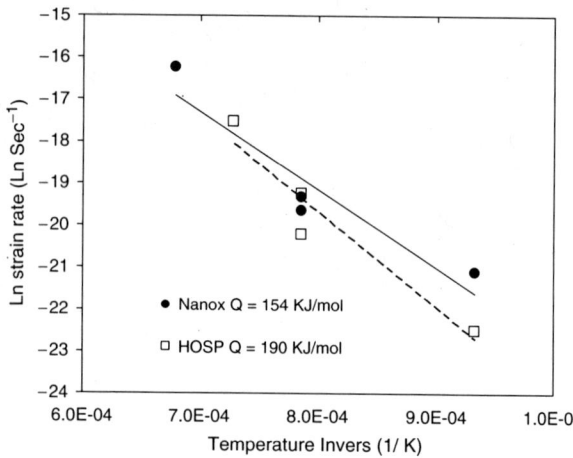

Fig. 8. Natural logarithm of the strain rate against the inverse temperature.

were kept under the same conditions for more than 4 days but still no stationary creep was observed. Increasing the temperature to 1200°C for the Nanox coating, A5, resulted in sample failure after 2 days without entering the steady-state stage.

Stress relaxation and redistribution are the phenomena that are generally observed in creep experiments. The change of stress during the creep test may affect the results obtained from a constant stress assumption. The analytical method using applied moments,[16–18] instead of stresses, to measure the creep constants was used here to calculate these constants and compare them with the dynamic beam profiling results. One of these results for the Nanox sample A5 is shown in Fig. 10.

As is clear, the stress exponent of Nanox A5 sample is in good agreement with that obtained from the steady-state creep rate equation (very close to one); however, there was an insignificant decrease in its value as the test time increased, showing lower dependency of the creep rate on the applied stress.

IV. Discussion

A large primary creep strain and a low creep activation energy were observed for both types of coatings. The creep rate continuously decreased during the test, although no steady-state regime was observed. The large primary creep phase and low apparent creep activation energy for the coatings have been attributed to stress-induced mechanical sliding, and diffusion through the splat boundaries.[1] The large decrease in creep rate

with time can explained by sintering and microstructural changes. A two-stage stiffening mechanism has been proposed for sintering[19]: the first stage involves the healing of interlamellar porosity, where the separation between splats is very small, and the second stage involves is microcrack healing at higher temperatures and longer times.

At the beginning of a creep test, the total contact area among splats is very small.[20] The small contact areas result in high local stresses and short paths for diffusion, resulting in a high initial creep rate. At the temperature of the creep tests, microstructural changes occur due to the applied stress and local curvature differences.

The flat surfaces of splats gradually change as grooves form along the boundaries of the columnar grains, producing spherical caps. When the surfaces of adjacent splats are sufficiently close, the caps may touch and form a neck, bridging the gap between splats, as shown in Fig. 11. Microcracks within splats can heal through a similar process when asperities are brought into contact as a result of elastic or creep strains (Fig. 12). The greater contact area leads to lower stress concentrations and longer diffusion lengths, both of which may reduce the creep strain rate dramatically.

In all creep tests, under similar conditions, the Nanox samples showed a higher creep resistance than the HOSP coatings. To investigate the differences between creep strains of the coatings, the compositions and microstructural features of both coatings were studied.

(1) Glassy Phases

The presence of glassy phases at elevated temperatures has a major influence on increasing deformation of ceramics in case

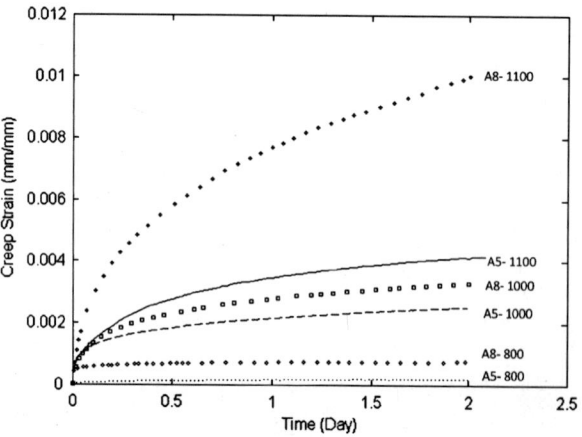

Fig. 7. Creep strain of Nanox and HOSP coatings in a range of temperatures under a 21 MPa applied stress.

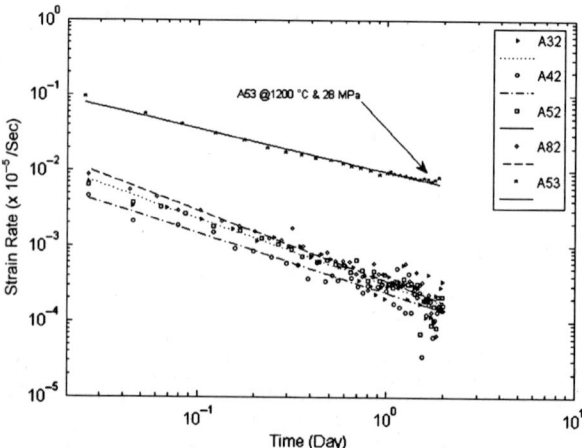

Fig. 9. Logarithmic scale of creep rate versus time for Nanox deposits.

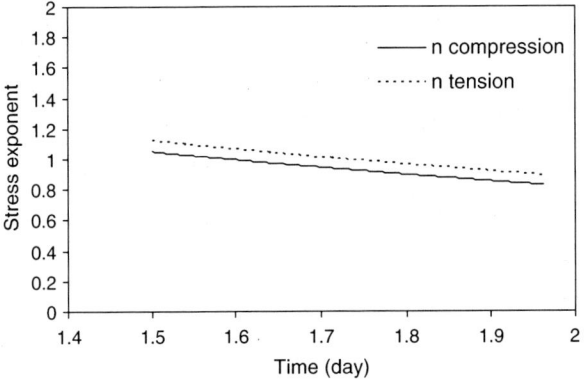

Fig. 10. Stress exponent variation versus time for Nanox coatings obtained from analytical models.

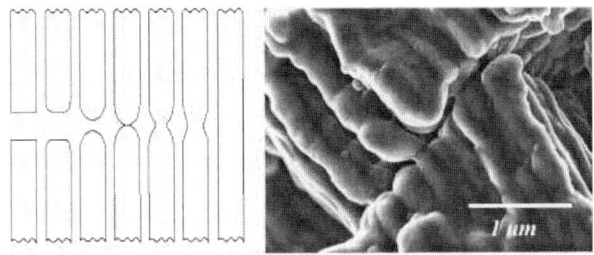

Fig. 11. Schematic and scanning electron microscopy images of intersplat grain growth.

they could wet the material surfaces.[15] Compositional analysis was performed to investigate the levels of several common impurities in the two feedstocks. No major differences between the two materials were found. Particular components such as silica, alumina, and titania, which may create a glassy phase at grain boundaries, were present in almost the same amounts in both powders. Therefore, if these constituents have any effect on the creep/sintering behavior of the samples, it is expected that it would be similar for both, and a comparison based on compositional considerations should be valid.

(2) Porosity

A proposed relationship between creep rate and porosity is based on the assumption that the porosity, P, reduces the cross-sectional area available to resist creep[15]

$$\dot{\varepsilon} \propto (1 - P^{2/3})^{-1} \tag{4}$$

Therefore, a higher porosity will result in a higher creep strain. In this study, Nanox samples showed the same trend.

Nanox A5 with 17% porosity showed a higher creep strain compared with the creep strain of Nanox A3 and A4 specimens, but HOSP sample A8, with a lower porosity level, showed a higher creep strain than the Nanox A5 coating. Therefore, only the total amount of porosity alone cannot be used to explain the creep behavior. The shape of the pores will affect the creep resistance of coatings in the primary stage as well; however, this effect is not as large as other parameters in stationary creep,[21] consistent with this study.

(3) Splat Boundaries

Intersplat boundaries can act as a perfect surface passage for material transfer leading to splat sliding. HOSP, A8, shows a creep strain that is more than 50% higher than the Nanox coating, A5. This sample has a good fraction of hollow sphere particles (Fig. 2), which means that a smaller amount of material has to be spread compared to a solid core particle and therefore results in thinner splats than Nanox deposits.

This reduction in thickness causes a consequent increase in the linear density of splats interfaces in a given thickness. These boundaries play an important role in controlling the creep/sintering behavior of thermally sprayed coatings, especially in the primary creep stage. Because splats can slide under stress at high temperatures, the more splat interfaces, the easier the sliding, leading to a higher creep strain. This behavior was observed in HOSP samples compared with Nanox coatings.

Empirical equations,[22–25] accommodating the properties of in-flight particles, show that by increasing the velocity and temperature of particles, the flattening factor increases, resulting in thinner splats.

$$R_f = \alpha(V_p.D_p/\nu)^\beta \tag{5}$$

$$\nu = A\exp(B/KT) \tag{6}$$

where R_f is the flattening factor; V_p and D_p are the velocity and diameter of in-flight particles, respectively; ν is the kinematic viscosity, α, β, A, and B are empirical constants; K is Boltzmann's constant; and T is the absolute temperature.

In Nanox samples, on increasing the velocity and temperature of the particles, the creep strain increases too; however, the A5 coating, due to 5% higher porosity, shows a creep strain similar to the A3 specimen. HOSP sample A8 with a relatively high in-flight temperature for particles exhibits a higher creep strain in spite of having a lower velocity. It seems that the presence of hollow sphere particles helps to produce thinner splats in this coating.

V. Summary

Different process parameters were used in this study to produce coatings. Samples with a thickness of 3 mm were prepared for creep tests. Creep experiments on partially stabilized zirconia (ZrO_2–7%Y_2O_3) with two different structures were conducted. At a range of temperatures and stresses, creep strain and creep rate graphs were plotted for both structures. Under similar

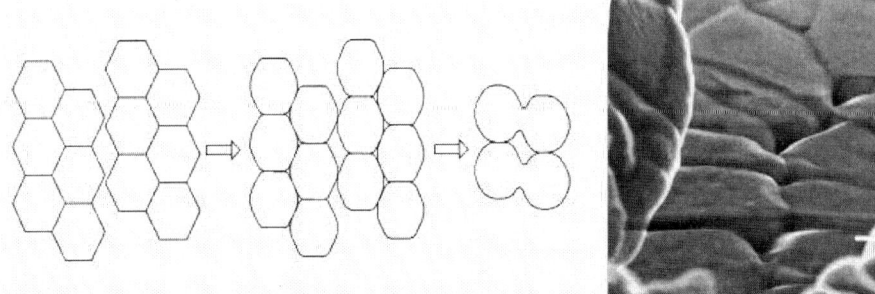

Fig. 12. Schematic and scanning electron microscopy images of microcrack healing.

Mechanical Properties

conditions, all the Nanox coatings showed a lower creep strain than the HOSP coatings. Lower creep strain may result in lower stress relaxation caused by creep mechanisms, which may lead to less microcracking in thermal contraction and increased lifetime of the coating. Stress exponents were calculated and for both samples were close to one, showing very little dependency of the stationary creep rate on the applied stresses. Activation energy was measured and both coatings showed similar values considering the standard deviation. The authors recommend that the thickness of splats in plasma-sprayed coatings be considered as one of the important parameters to control the creep strain while all other factors are being kept the same. Increasing the splat thickness may lead to a reduction of creep strain, especially at the primary stage but it will increase the thermal conductivity of the deposit as well. It was shown that the splat interface is an important parameter controlling the thermal diffusivity of coatings.[11] Therefore, there should be an optimized splat thickness to satisfy creep and thermal diffusivity expectations of plasma-sprayed thermal barrier coatings.

References

[1]D. Zhu and R. A. Miller, "Determination of Creep Behaviour of Thermal Barrier Coatings Under Laser Imposed Temperature and Stress Gradients," NASA Technical Memorandum 113169, ARL-TR-1556 (1997).

[2]D. Zhu and R. A. Miller, "Sintering and Creep Behaviour of Plasma-Sprayed Zirconia-and Hafnia-Based Thermal Barrier Coatings," *J. Surf. Coat. Tech.*, **108–109**, 114–20 (1998).

[3]E. F. R. Fejda, D. F. Socie, and T. Itoh, "Deformation Behaviour of Plasma-Sprayed Thick Thermal Barrier Coatings," *J. Surf. Coat. Tech.*, **113**, 218–26 (1999).

[4]A. H. Chokshi, "Diffusion Creep in Oxide Ceramics," *J. Eur. Ceram. Soc.*, **22**, 2469–78 (2002).

[5]A. H. Chokshi, "Diffusion, Diffusion Creep and Grain Growth Characteristics of Nanocrystalline and Fine-Grained Monoclinic, Tetragonal and Cubic Zirconia," *Scripta Mater.*, **48**, 791–6 (2003).

[6]R. Schaller and M. Daraktciev, "Mechanical Spectroscopy of Creep Appearance in Fine-Grained Yttria-Stabilized Zirconia," *J. Eur. Ceram. Soc.*, **22**, 2461–7 (2002).

[7]M. J. Adnrews, M. K. Ferber, and E. Lara-Curzio, "Mechanical Properties of Zirconia-Based Ceramics as Functions of Temperature," *J. Eur. Ceram. Soc.*, **22**, 2433–9 (2002).

[8]K. Kokini, Y. R. Takeuchi, and B. D. Choules, "Surface Thermal Cracking of Thermal Barrier Coatings Owing to Stress Relaxation: Zirconia vs. Mullite," *J. Surf. Coat. Tech.*, **82**, 77–82 (1996).

[9]L. Vasylkiv, Y. Skka, and V. V. Skorokhod, "Low-Temperature Processing and Mechanical Properties of Zirconia and Zirconia–Alumina Nanoceramics," *J. Am. Ceram. Soc.*, **86** [2] 299–304 (2003).

[10]M. Daraktchiev and R. Schaller, "High-Temperature Mechanical Loss Behaviour of 3 mol% Yttria-Stabilized Tetragona Zirconia Polycrystals (3Y-TZP)," *J. Phys. Stat. Sol.*, **2**, 293–304 (2003).

[11]R. Soltani, T. W. Coyle, and J. Mostaghimi, "Thermo-Physical Property of Bimodal Structured Plasma Sprayed TBCs," *J. Surf. Coat. Tech.* (to be published).

[12]D. C. Cranmer and D. W. Richerson, *Mechanical Testing, Methodology for Ceramic Design and Reliability*. Marcel Dekker Inc, NY, 1998.

[13]R. Soltani, T. W. Coyle, and J. Mostaghimi, *Thermo-Mechanical Behavior of Bimodal Structured Thermal Barrier Coatings, Thermal Spray 2004: Advances in Technology and Application, ASM International, May 10–12, 2004*. ASM International, Osaka, Japan, 1129pp.

[14]C. A. Costa and J. A. Todd, "Creep Behavior of Si$_3$N$_4$- Whisker-Reinforced Si$_3$N$_4$-Matrix Composites Using an In Situ Dynamic Beam Profiling Method," *J. Am. Ceram. Soc.*, **82** [3] 607–15 (1999).

[15]W. D. Kingery, H. K. Bowen, and D. R. Uhlmann, *Introduction to Ceramics*, 2nd ed, John Wiley & Sons, New York, 1975.

[16]C. F. Chen and T. J. Chuang, "Improved Analysis for Flexural Creep with Application to Sialon Ceramics," *J. Am. Ceram. Soc.*, **73** [8] 2366–73 (1990).

[17]T. J. Chuang, "Estimation of Power-Law Creep Parameters from Bend Test Data," *J. Mater. Sci.*, **21**, 165–75 (1986).

[18]C. F. Chen and T. J. Chuang, "High Temperature Mechanical Properties of SiAlON Ceramic: Creep Characterization," *Ceram. Eng. Sci. Proc.*, **8**, 7–8 (1987).

[19]T. W. Clyne, J. T. Klocker, and J. A. Thompson, "Sintering the Top Coat in Thermal Spray TBC Systems Under Service Conditions"; pp. 685–9 in *Super Alloys 2000*, Edited by K. Green, R. Kissinger, and T. Pollock. TMS, Seven Springs, PA, 2000.

[20]R. McPherson, "A Review of Microstructure and Properties of Plasma Sprayed Ceramic Coatings," *J. Surf. Coat. Tech.*, **39–40**, 173–81 (1989).

[21]R. F. Firestone, W. R. Logan, J. W. Adams, and R. C. Jr. Bull, "Creep of Plasma-Sprayed-ZrO$_2$ Thermal Barrier Coatings," *Ceram. Eng. Sci. Proc.*, **13**, 758–71 (1982).

[22]J. Madejski, "Solidification of Droplets on a Cold Surface," *Int. J. Heat Mass Transfer*, **19**, 1009–13 (1976).

[23]T. Yoshida, T. Okada, H. Hamatani, and H. Kumaoka, "Integrated Fabrication Process for Solid Oxide Fuel Cells Using Novel Plasma Spraying," *Plasma Sources Sci. Technol.*, **1** [3] 195–201 (1992).

[24]O. P. Solonenko, A. V. Smirnov, V. A. Klimenov, Y. G. Butov, and Y. F. Ivanov, "Role of Interfaces in Splat and Coatings Structure Formation," *Physical Mesomechanics*, **2** [1–2] 113–29 (1999).

[25]P. Fauchais, M. Fukumoto, A. Vardelle, and M. Vardelle, "Knowledge Concerning Splat Formation: An Invited Review," *J. Therm. Spray Tech.*, **13** [3] 337–60 (2004). □

Application of Hertzian Tests to Measure Stress–Strain Characteristics of Ceramics at Elevated Temperatures

Estíbaliz Sánchez-González, Juan J. Meléndez-Martínez, and Antonia Pajares[†]

Departamento de Física, Facultad de Ciencias, Universidad de Extremadura, Badajoz 06071, Spain

Pedro Miranda and Fernando Guiberteau

Departamento de Electrónica e Ingeniería Electromecánica, Escuela de Ingenierías Industriales, Universidad de Extremadura, Badajoz 06071, Spain

Brian R. Lawn

Materials Science and Engineering Laboratory, National Institute of Standards and Technology, Gaithersburg, Maryland 20899-8500

A new method for evaluating the elastic–plastic properties of ceramics from room temperature up to the onset of creep based on Hertzian indentation testing is proposed. Indentation stress–strain curves are compiled for representative alumina and zirconia ceramics at prescribed temperatures. Deconvolution of the indentation stress–strain curves for each material provides a measure of Young's modulus, yield stress, and work-hardening coefficient as a function of temperature, enabling construction of true stress–strain curves. The stress–strain curves flatten out with increasing temperature, in accordance with an expected increased plastic response at elevated temperatures.

I. Introduction

DESIRABLE MECHANICAL properties of advanced ceramics include high modulus and hardness, and high wear and chemical resistance. Such properties enable ceramics to supplant metals in applications such as bearings, cutting tools, seal valves, and heat exchangers. Any increase of temperature above ambient can significantly affect performance, particularly under contact conditions where local stress levels are uncommonly high. Accordingly, there is a need for a fundamental understanding of the contact properties of ceramics above room temperature, particularly elastic–plastic responses. Indentation studies with spheres are ideally suited to meet this need, because of their experimental simplicity, their amenability to analysis, and (especially) their unique capacity to determine the full elastic–plastic response without premature fracture.[1] Such Hertzian studies have been conducted at room temperature on a wide range of ceramics including alumina, zirconia, silicon nitride, silicon carbide, and dental materials,[2–7] as well as on thermal barrier coatings and other layer systems.[8,9] One of the advantages of Hertzian tests is that it covers a range of stresses (from 1 to tens of GPa) and testing size scale (from tens of micrometer to millimeter) not accessible by the more conventional Vickers indentation (stresses above 10 GPa and a scale of micrometer to tens of micrometer) or uniaxial tests (stresses from 1 MPa to 1 GPa and a scale above 1 mm). However, an extension of this testing methodology to high temperatures has not been carried out. Hertzian tests have been conducted above room temperature for the determination of the temperature dependence of toughness in silicate materials.[10] Routine hot hardness Vickers indentation measurements have also been performed on a wide variety of ceramics.[11–13] Extensive uniaxial testing and impression creep tests[14–16] have been conducted on the creep properties of ceramics at very high temperatures. None of these other studies provides a full description of the temperature dependence of the elastic–plastic stress–strain response in the pre-creep region.

The current paper seeks to redress this deficiency. Our aim here is to expand the existing Hertzian testing methodology to deconvolute true stress–strain responses of ceramics from indentation data as a function of temperature. A furnace enables *in situ* indentation testing above room temperature. Commercial polycrystalline alumina and zirconia are used as test materials, with sphere indenters made from the same materials. The deconvolution is carried out using basic elastic–plastic relations in conjunction with finite-element modeling (FEM). Our current focus will be on the methodology, with a more detailed description of material properties deferred to later reports.

II. Experimental Procedure

(1) Materials

Alumina and zirconia were chosen as test materials because their properties have been comprehensively studied in the ceramics literature. The alumina was a commercial polycrystalline material with a grain size of about 6 μm and a porosity of 3% (Goodfellow, Cambridge, U.K.). The zirconia was a commercial Y–TZP containing 5 mol% Y_2O_3 (Imetra, Elmsford, NY). Both materials were supplied as spheres of 3 and 9 mm radius. The spheres were cut in half and used as indenters. Plate specimens for testing were cut from the center regions of the larger spheres to 8 mm thickness and were polished to a 1 μm finish. This procedure ensured that specimen and indenter were always similar materials.

(2) Indentation Tests

Hertzian contact tests were performed using a universal testing machine (Model AG-IS 100 kN, Shimadzu, Kyoto, Japan). A vertical split furnace was incorporated into the testing machine as shown in Fig. 1. The furnace consisted of a cylindrical chamber with a frontal aperture to facilitate specimen access. Upper and lower alumina push rods (40 mm diameter and 350 mm length) were used to support the specimen and deliver the load,

N. Padture—contributing editor

Manuscript No. 21959. Received June 29, 2006; approved August 17, 2006.
This work was funded by the Ministerio de Ciencia y Tecnología (Government of Spain) and the Fondo Social Europeo through grant MAT2003-05584.
[†]Author to whom correspondence should be addressed. e-mail: apajares@unex.es

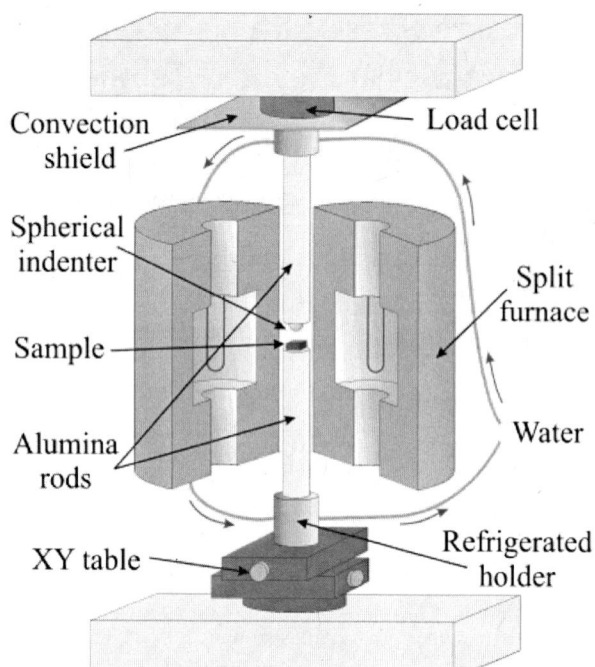

Fig. 1. Schematic of the experimental setup used for the Hertzian tests at elevated temperatures.

with the indenter and specimen at the center of the furnace chamber. The half-sphere indenter and the specimen were, respectively, bonded to the upper and lower push rods using alumina paste (Ceramabond 569, Aremco Products Inc, NY). External push rod holders were cooled with circulating water to protect the load cell. Before bonding the specimen to the rod, a metal film of 50 nm thickness was sputter coated onto the top surface (Polaron SC7640, Quorum Technologies Ltd., New Haven, U.K.), with different metals at different temperatures to provide an optimum imprint of the ensuing indentations—gold at room temperature, rhodium-palladium up to 600°C, and platinum at higher temperatures. The bottom push rod holder was placed on an X–Y table to allow at least 15 tests to be conducted on any specimen at any selected temperature. Indentations were made at a constant crosshead displacement rate of 0.05 mm/min.

Indentation sequences were made at peak loads up to 5000 N in air at temperatures in the range 25°–1200°C for alumina and 25°–1000°C for zirconia. At higher temperatures, creep was observed—data in this region were discarded. The specimens were heated at a rate of 6°C/min, held for 1 h at peak temperature before indentation, and then allowed to cool over several hours by switching off the furnace.

After cooling, contact radius a at each peak load P and indenter radius r were measured by optical microscopy at each indentation site from the contact imprint. Plots of indentation stress ($p_0 = P/\pi a^2$) versus indentation strain (a/r) for each temperature were thereby obtained for each material.

(3) Analysis

Young's modulus E for each prescribed temperature was determined from the linear region of the indentation stress–strain curve using the Hertzian relation for elastic contacts with similar indenters

$$p_0 = [2E/3\pi(1-v^2)]a/r \quad (p_0 < 1.1Y) \tag{1}$$

where v is Poisson's ratio (generically taken as 0.22 for our materials) and Y is the yield stress.[1,17,18]

FEM was used to determine yield stress and work-hardening coefficients from the indentation stress–strain curves using AB-AQUS/Standard software (Hibbitt, Karlsson & Sorensen Inc, Pawtucket, RI).[19] The algorithm models a half-sphere indenter of 3 mm radius in axisymmetric contact with a flat specimen, incrementally loaded to prescribed peak loads, with 1 μm minimum dimension square elements in the near-contact region. Deformation in both the indenter and the specimen is assumed to occur in accordance with a Lüdwig constitutive stress–strain relation[20]

$$\sigma = E\varepsilon \quad (\sigma < Y) \tag{2a}$$

$$\sigma = Y(E/Y)^n\varepsilon^n \quad (\sigma > Y) \tag{2b}$$

with n being a dimensionless strain-hardening coefficient of value between 0 (fully plastic) and 1 (fully elastic). This model provides a more realistic strain hardening behavior than other simple bilinear models used in previous work,[19,21–23] without increasing the number of adjustable parameters. For each material at any given temperature, given E from Eq. (1), Y and n are iteratively adjusted to fit the indentation stress–strain data using the algorithm.

III. Results

Figure 2 shows micrographs of the indentation-induced surface damage for a 1500 N load at room temperature and 1000°C for alumina (Fig. 2(a)) and zirconia (Fig. 2(b)). In both materials, the residual impression is markedly larger at the higher temperature, indicating greater deformation. Some ring and radial cracks are observed, most apparent in the zirconia at the higher temperature, but these are considered subsidiary to the greater deformation under the present loading conditions. Some plastic deformation was also observed in the indenter after the tests, especially at the higher temperatures.

Indentation stress–strain curves at temperatures up to 1200°C for alumina and 1000°C for zirconia are shown in Fig. 3. Each point in these curves represents a single indentation performed at a prescribed peak load and temperature. The solid curves through the experimental data are FEM best fits, with appropriately adjusted parameters E, Y, and n for both the test material and the indenter (the latter to allow for observed yield in the indenter). The curves move substantially downward with increasing temperature, as expected. In principle, asymptotic plateaus of these curves at high strain would yield Meyer's hardness of the material at any given temperature. However, in the

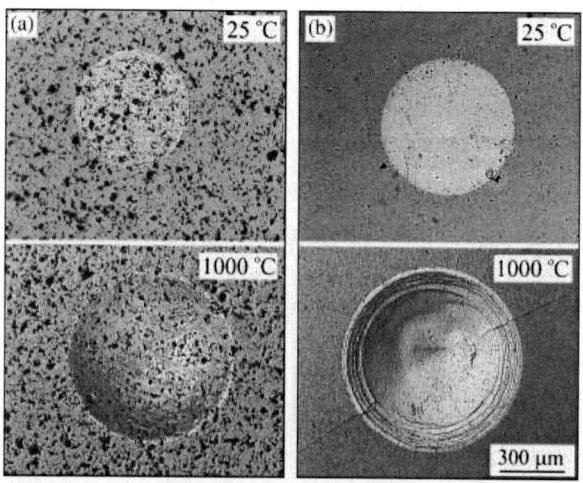

Fig. 2. Optical micrographs showing surface Hertzian damage in (a) alumina and (b) zirconia generated with 3 mm radius indenters of similar materials for 1500 N peak load at room temperature and 1000°C.

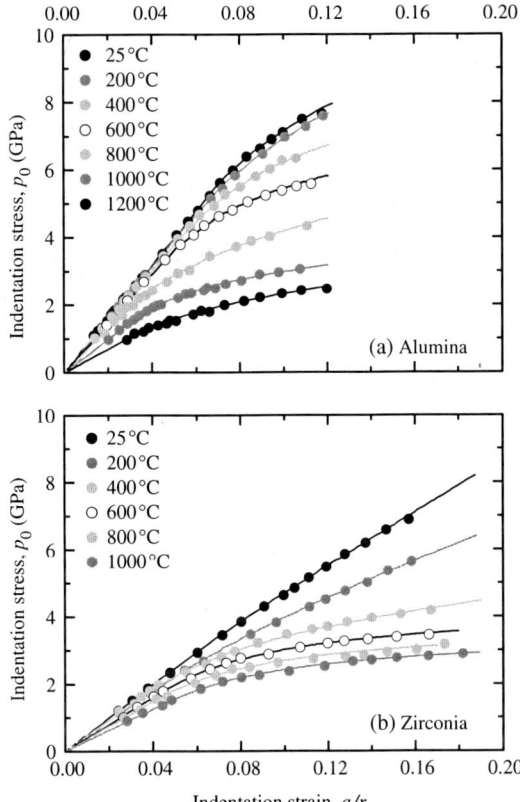

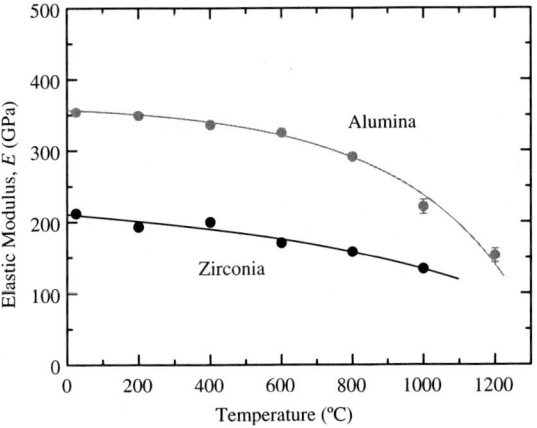

Fig. 3. Indentation stress–strain data for (a) alumina and (b) zirconia at designated temperatures. Hertzian tests were performed with indenters of similar materials of 3 and 9 mm radius (not distinguished here). The solid curves indicate finite-element modeling best fits.

present experiments, we could not reach these plateaus without breaking either the sample or the indenter.

Young's moduli E obtained from the slopes of the linear region of these curves using Eq. (1) are plotted versus temperature for alumina and zirconia in Fig. 4. Error bars are calculated from the uncertainties in the estimation of the slopes from linear regressions and in most cases are smaller than the size of the symbols, except at higher temperatures, where the number of data points within the elastic region is limited and impression visibility is less clear (Fig. 2). The solid lines through the data points indicate empirical fits. The reduction in E over the tem-

Fig. 4. Young's modulus E versus temperature for alumina and zirconia. Data points evaluated from slope of the initial linear region of stress–strain curves of Fig. 3 in conjunction with Eq. (1). The solid lines through data indicate empirical fits.

perature range is apparent for both materials, but is especially so for alumina above about 600°C.

Corresponding yield stresses Y and strain-hardening coefficients n obtained from the FEM analyses are plotted as a function of temperature for alumina in Fig. 5(a) and for zirconia in Fig. 5(b). Error bars are estimated uncertainties in the trial FEM calculations, and the solid lines through the data indicate empirical fits. Whereas Y for zirconia diminished relatively slowly and continuously over the temperature range, that for alumina showed an abrupt drop at around 600°C. Closer inspection of the indentation sites revealed some enhanced damage, possibly from enhanced grain boundary degradation in this particular alumina, but this aspect was not investigated in depth. The trends in n show even greater disparities in the two materials, remaining consistently low in the alumina but falling rapidly from near unity in the zirconia. These trends in Y and n, taken together, indicate significant differences in the elastic–plastic responses in the two materials.

IV. Discussion

In this work, we propose a simple Hertzian test methodology to determine the elastic–plastic behavior of ceramics at temperatures up to the onset of creep. Young's modulus E and yield parameters Y and n are deconvoluted from indentation stress–strain curves using Eqs. (1) and (2), in conjunction with FEM. This information can be used to reconstruct the true stress–strain curve for any given ceramic. Accordingly, in Fig. 6, we plot stress σ versus strain ε for our alumina and zirconia by inserting E, Y, and n from Figs. 4 and 5 into Eq. (2). The entire deformation evolution in each material at each temperature, from initial elastic to fully plastic, is now apparent.

These results provide some insights into the deformation processes in the two test ceramics. For the alumina, the curves in the plastic region are relatively flat, consistent with a small, rela-

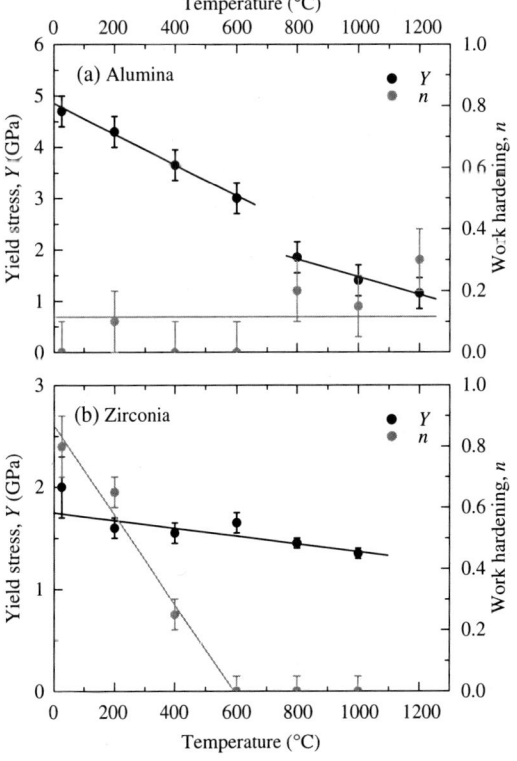

Fig. 5. Yield stress Y and strain-hardening coefficient n versus temperature for (a) alumina and (b) zirconia. Points evaluated by iterative finite element modeling analysis in conjunction with Eq. (2) from stress–strain data of Fig. 3.

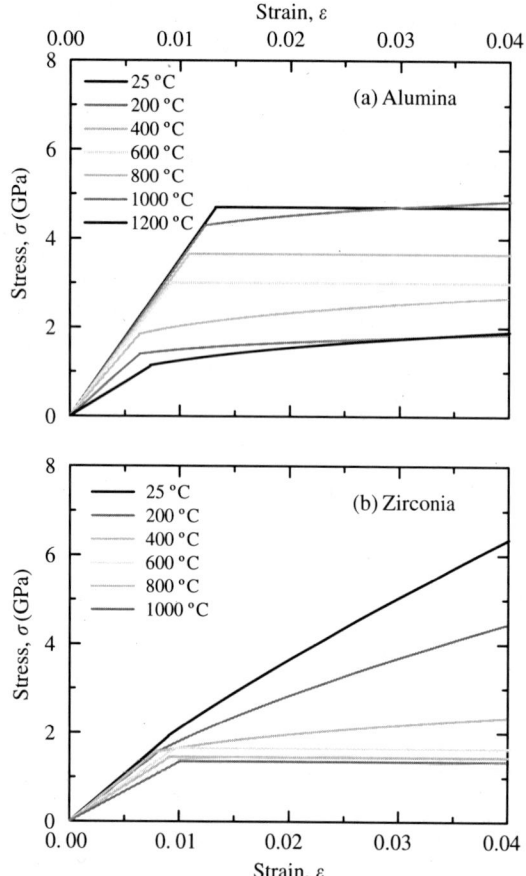

Fig. 6. True stress–strain curves for (a) alumina and (b) zirconia, obtained from data in Figs. 4 and 5 in conjunction with Eq. (2).

tively invariant work-hardening coefficient n in Fig. 5(a). In this case, increasing plasticity is associated with a fast degradation of yield stress. For the zirconia, the curves in the plastic region are much steeper at lower temperatures, leveling out at higher temperatures, reflecting a strongly diminishing n in Fig. 5(b). In this case, the yield stress is more slowly varying. These results suggest basic differences in the underlying plasticity processes in the two materials.

These results raise many interesting issues concerning materials science aspects. However, as already mentioned, the aim of the present paper is to present a methodology to measure mechanical properties at temperatures below the onset of creep. A consideration of underlying material features will be given elsewhere.

Apart from enabling determination of the entire stress–strain curve, the Hertzian methodology is relatively straightforward and economical. Using microscopy to determine contact radii directly in the construction of indentation stress–strain curves (Fig. 3) eliminates errors from thermal or electronic drift that may occur with more instrumented procedures. Measurement errors from thermal contraction of higher temperature indents are small, amounting to less than 1% for materials with expansion coefficients $\approx 10 \times 10^{-6}$ K^{-1} and for temperature ranges of $\approx 1000^{\circ}$C. It may be argued that the method is cumbersome because it is necessary to perform many individual contacts to produce each full indentation stress–strain curve. However, this can be used to advantage, as it allows one to study the full evolution of contact damage with increasing contact pressure at any given temperature.

Other methods for measuring elastic and plastic properties in ceramics below the creep temperature have been described in the literature. Dynamic testing techniques (e.g., natural resonance

frequency, sound velocity, etc.) are simple, but measure only elastic properties.[24,25] Uniaxial compression tests are also simple in principle, but are not feasible in most ceramics at lower temperatures because of premature fracture (unless a lateral confinement is imposed).[26] Also, uniaxial strain measurements are not easy to make in the elastic region. High-temperature hardness testing machines using Vickers or other sharp indenters are commercially available, enabling routine measurements of hardness H over the same temperature range covered here.[11–13] However, measurements provide no information on the contribution of elastic modulus E and work-hardening coefficient n to the elastic–plastic response. Instrumented indentation techniques with blunt indenters offer the prospect of determining true stress–strain curves from load–displacement data using a transformation algorithm to compute contact radii,[27] but require precision high-temperature displacement measurements, which are subject to thermal and electronic drift.

Apart from its clear application to ceramic materials in general, the indentation methodology technique described here may be extended to more complex high-temperature material systems such as thermal barrier coatings. In such layer systems, it is possible, using FEM, to predict composite responses from stress–strain functions of the individual component materials.[28]

References

[1]B. R. Lawn, "Indentation of Ceramics with Spheres: A Century After Hertz," *J. Am. Ceram. Soc.*, **81** [8] 1977–94 (1998).

[2]A. C. FischerCripps and B. R. Lawn, "Stress Analysis of Contact Deformation in Quasi-Plastic Ceramics," *J. Am. Ceram. Soc.*, **79** [10] 2609–18 (1996).

[3]F. Guiberteau, N. P. Padture, H. Cai, and B. R. Lawn, "Indentation Fatigue— A Simple Cyclic Hertzian Test for Measuring Damage Accumulation in Polycrystalline Ceramics," *Philos. Mag. A—Phys. Condens. Matter Struct. Defect Mech. Prop.*, **68** [5] 1003–16 (1993).

[4]A. Pajares, F. Guiberteau, and B. R. Lawn, "Hertzian Contact Damage in Magnesia-Partially-Stabilized Zirconia," *J. Am. Ceram. Soc.*, **78** [4] 1083–6 (1995).

[5]S. K. Lee, S. Wuttiphan, and B. R. Lawn, "Role of Microstructure in Hertzian Contact Damage in Silicon Nitride. 1. Mechanical Characterization," *J. Am. Ceram. Soc.*, **80** [9] 2367–81 (1997).

[6]B. R. Lawn, N. P. Padture, H. D. Cai, and F. Guiberteau, "Making Ceramics Ductile," *Science*, **263** [5150] 1114–6 (1994).

[7]I. M. Peterson, A. Pajares, B. R. Lawn, V. P. Thompson, and E. D. Rekow, "Mechanical Characterization of Dental Ceramics by Hertzian Contacts," *J. Dent. Res.*, **77** [4] 589–602 (1998).

[8]A. Pajares, L. H. Wei, B. R. Lawn, N. P. Padture, and C. C. Berndt, "Mechanical Characterization of Plasma Sprayed Ceramic Coatings on Metal Substrates by Contact Testing," *Mater. Sci. Eng. A—Struct. Mater. Prop. Microstruct. Process.*, **208** [2] 158–65 (1996).

[9]H. Y. Liu, B. R. Lawn, and S. M. Hsu, "Hertzian Contact Response of Tailored Silicon Nitride Multilayers," *J. Am. Ceram. Soc.*, **79** [4] 1009–14 (1996).

[10]M. V. Swain, J. S. Williams, B. R. Lawn, and J. J. H. Beek, "Comparative Study of Fracture of Various Silica Modifications Using Hertzian Test," *J. Mater. Sci.*, **8** [8] 1153–64 (1973).

[11]C. P. Alpert, H. M. Chan, S. J. Bennison, and B. R. Lawn, "Temperature-Dependence of Hardness of Alumina-Based Ceramics," *J. Am. Ceram. Soc.*, **71** [8] C371–3 (1988).

[12]V. Tikare and A. H. Heuer, "Temperature-Dependent Indentation Behavior of Transformation-Toughened Zirconia-Based Ceramics," *J. Am. Ceram. Soc.*, **74** [3] 593–7 (1991).

[13]V. V. Milman, S. I. Chugunova, I. V. Goncharova, T. Chudoba, W. Lojkowski, and W. Gooch, "Temperature Dependence of Hardness in Silicon-Carbide Ceramics with Different Porosity," *Int. J. Refract. Met. Hard Mat.*, **17** [5] 361–8 (1999).

[14]S. N. G. Chu and J. C. M. Li, "Impression Creep–New Creep Test," *J. Mater. Sci.*, **12** [11] 2200–8 (1977).

[15]C. H. Hsueh, P. Miranda, and P. F. Becher, "An Improved Correlation Between Impression and Uniaxial Creep," *J. Appl. Phys.*, **99** [11] (2006).

[16]J. C. M. Li, "Impression Creep and Other Localized Tests," *Mater. Sci. Eng. A—Struct. Mater. Prop. Microstruct. Process.*, 113513, **322** [1–2] 23–42 (2002).

[17]M. V. Swain and B. R. Lawn, "A Study of Dislocation Arrays at Spherical Indentations in Life as a Function of Indentation Stress and Strain," *Physica Status Solidi*, **35** [2] 909–23 (1969).

[18]K. L. Johnson, *Contact Mechanics*. Cambridge University Press, Cambridge, U.K., 1985.

[19]P. Miranda, A. Pajares, F. Guiberteau, F. L. Cumbrera, and B. R. Lawn, "Contact Fracture of Brittle Bilayer Coatings on Soft Substrates," *J. Mater. Res.*, **16** [1] 115–26 (2001).

[20]P. Ludwig, *Element der Technologischen Mechanik*, pp. 32–44. Springer, Berlin, 1909.

[21]H. Zhao, P. Miranda, B. R. Lawn, and X. Z. Hu, "Cracking in Ceramic/Metal/Polymer Trilayer Systems," *J. Mater. Res.*, **17** [5] 1102–11 (2002).

[22]P. Miranda, A. Pajares, F. Guiberteau, Y. Deng, and B. R. Lawn, "Designing Damage-Resistant Brittle-Coating Structures: I. Bilayers," *Acta Mater.*, **51** [14] 4347–56 (2003).

[23]P. Miranda, A. Pajares, F. Guiberteau, Y. Deng, H. Zhao, and B. R. Lawn, "Designing Damage-Resistant Brittle-Coating Structures: II. Trilayers," *Acta Mater.*, **51** [14] 4357–65 (2003).

[24]J. B. Wachtman and D. G. Lam, "Young Modulus of Various Refractory Materials as a Function of Temperature," *J. Am. Ceram. Soc.*, **42** [5] 254–60 (1959).

[25]M. Fukuhara and I. Yamauchi, "Temperature-Dependence of the Elastic-Moduli, Dilational and Shear Internal Frictions and Acoustic-Wave Velocity for Alumina, (Y)TZP and Beta'-Sialon Ceramics," *J. Mater. Sci.*, **28** [17] 4681–8 (1993).

[26]W. Chen and G. Ravichandran, "An Experimental Technique for Imposing Dynamic Multiaxial-Compression with Mechanical Confinement," *Exp. Mech.*, **36** [2] 155–8 (1996).

[27]J. S. Field and M. V. Swain, "A Simple Predictive Model for Spherical Indentation," *J. Mater. Res.*, **8** [2] 297–306 (1993).

[28]S. Wuttiphan, A. Pajares, B. R. Lawn, and C. C. Berndt, "Effect of Substrate and Bond Coat on Contact Damage in Zirconia-Based Plasma-Sprayed Coatings," *Thin Solid Films*, **293** [1–2] 251–60 (1997).   □

Effect of Sintering on Mechanical Properties of Plasma-Sprayed Zirconia-Based Thermal Barrier Coatings

Sung R. Choi,[†] Dongming Zhu, and Robert A. Miller

National Aeronautics and Space Administration, Glenn Research Center, Cleveland, Ohio 44135

The effect of sintering on the mechanical properties of free-standing, plasma-sprayed ZrO_2–8 wt% Y_2O_3 thick thermal barrier coatings (TBCs) was determined after annealing at 1316°C in air. Mechanical properties of the TBCs, including flexure strength, modes I and II fracture toughnesses, constitutive relations, elastic modulus, and microhardness, were determined at ambient temperature as a function of annealing time ranging from 0 to 500 h. In addition, some physical properties such as density and phase stability were also determined. Mechanical and physical properties increased significantly in 5–100 h and then reached a plateau above 100 h. An exception to this was the monoclinic phase that increased monotonically without forming a plateau. Annealing resulted in healing of microcracks and pores, and in grain growth, accompanying densification of the TBC's body because of the sintering effect. However, an inevitable adverse effect also occurred such that the desired lower thermal conductivity and good strain-tolerant capability, which makes the TBCs unique in thermal barrier applications, were degraded upon annealing. A phenomenological model was proposed to assess and quantify all the property variables in response to annealing in a normalized scheme.

I. Introduction

THERMAL barrier coatings (TBCs) have attracted ever-increasing attention for advanced aero- and land-based engine applications because of their ability to provide thermal insulation to engine components.[1–3] The merits of using ceramic TBCs are well recognized and include a potential increase in engine operating temperature with reduced cooling requirements, resulting in significant improvements in thermal efficiency, performance, and reliability. Plasma-sprayed zirconia-based ceramics are one of the most important coating materials in light of their low thermal conductivity, relatively high strain tolerance, and unique microstructure as a result of the plasma-spraying process. However, the durability of TBCs under severe thermal and mechanical loading conditions encountered in heat engines remains one of the major problems. As a result, the development of TBCs requires a better understanding of both mechanical and thermal behavior of the coating materials to ensure life and reliability of the related components.

Because of their unique application process, plasma-sprayed TBCs exhibit a complicated but unique microstructure with numerous macro- and microcracks, porosity, and lamellar-like splat morphology. As a result, plasma-sprayed TBCs, when heated to elevated temperatures, have exhibited a sintering effect through crack healing, inter- and/or transsplat bonding, and grain growth with their degree of sintering depending on heating

temperature and time.[4–9] The sintering effect has been well known to change the mechanical and physical properties of TBCs through the processes mentioned above. It has been shown that elastic modulus,[4–9] strength,[4,5] and work of fracture[5] of as-sprayed TBCs all increased upon annealing at elevated temperatures. Thermal conductivity has been shown to increase as well.[4,6] Sintering could result in an increase in mechanical properties but at the same time deteriorate thermal conductivity and strain-tolerant capability, thereby degrading the important feature of the coatings as a thermal barrier.

Therefore, it is important to quantify the effect of sintering on the mechanical and physical properties of TBCs to ensure reliability and life of the components in service conditions. Various efforts have been concentrated primarily to determine one or two mechanical properties (such as elastic modulus and/or strength), although a recent study by Thurn et al.[5] expanded more to include elastic modulus, strength, and work of fracture. No systematic sintering study, aimed at a particular service temperature as a function of time up to 500 h from the same batch of TBCs, has been carried out to include a wide variety of mechanical and physical properties such as strength, fracture toughness, constitutive relations, elastic modulus, microhardness, thermal conductivity, and phase stability, etc. This work, as a consequence, was focused on determining as completely and systematically as possible the effect of sintering on the mechanical properties (and some physical properties as well) of free-standing, plasma-sprayed TBCs that were from the same batch of the material. Test coupons of as-sprayed TBCs, with their specimen configurations depending on the test matrix, were first annealed in air at a prospective aeroengine service temperature of 1316°C as a function of time ranging from 0 to 500 h, and their respective properties were then determined at ambient temperature. The properties thus determined included flexure strength, modes I and II fracture toughnesses, deformation, elastic modulus, microhardness, density, and phase stability. A generalized, phenomenological model is proposed and discussed to better quantify mechanical and physical responses of TBCs to sintering. The detailed descriptions regarding this sintering-associated work can be found from a previous report.[10]

II. Experimental Procedures

(1) Material

ZrO_2–8 wt% Y_2O_3 powder with an average particle size of 60 μm, fabricated by sintering and crushing, was first plasma sprayed on a rectangular graphite substrate measuring $150 \times 100 \times 6.5$ mm to a thickness of about 6 mm, using a Sulzer Metco ATC-1 plasma-coating system (Sulzer Metco Inc., Westbury, NY) with an industrial robot. The plasma-spray conditions can be found elsewhere.[11] A free-standing, plasma-sprayed ceramic billet was then obtained by slowly burning out the graphite substrate at 680°C in air for 24 h. The billets were machined into the configurations of test specimens with a final finish using a #500 diamond grinding wheel according to the test matrix described below. Two test-specimen configurations were machined from the billets oriented as shown in Fig. 1. The two configurations made were rectangular cross-section bars used in

K. T. Faber—contributing editor

Manuscript No. 20190. Received October 21, 2004; approved April 6, 2005.
Supported by the Ultra-Efficient Engine Technology (UEET) Program, NASA Glenn Research Center, Cleveland, Ohio.
[†]Author to whom correspondence should be addressed. e-mail: sung.r.choi@grc.nasa.gov

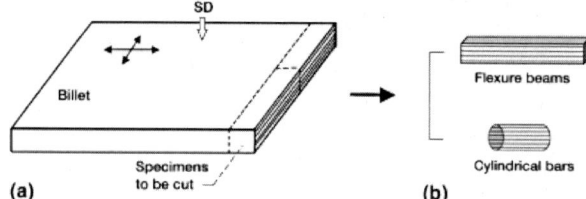

Fig. 1. Thermal barrier coating billet and test specimen configurations. (a) Billet showing plasma-spray direction (SD) with specimens to be cut, (b) test specimen configurations of flexure and cylindrical bars machined from billet. Layers are indicated.

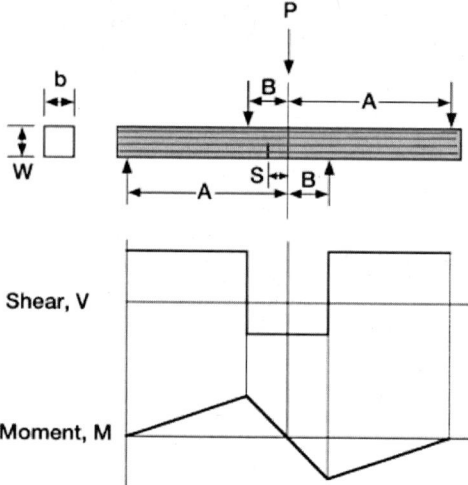

Fig. 2. Schematics of asymmetric four-point flexure specimen geometry with accompanying shear force and bending moment diagrams used in mode II fracture toughness testing for plasma-sprayed ZrO_2–8 wt% Y_2O_3 thermal barrier coatings.

flexure tests and cylindrical bars used in compression tests. It has been found that the coatings primarily consist of the tetragonal t' phase[12] and have a porosity of about 10%. Despite their prevailing lamellar-like (layered), microcracked and porous, and platelet structure, the coatings exhibited a somewhat insignificant effect of material direction on mechanical properties, termed directionality, when measured via fracture toughness, microhardness, and elastic modulus.[10,13] The TBCs used in this work can be categorized in size as thick TBCs (TTBCs), which impart a significant simplicity and convenience in conducting many of the mechanical testings, compared with thin coatings, which would provide a great challenge in both testing techniques and interpretation of the results. However, this does not necessarily guarantee the applicability of TTBCs' properties to thin TBC systems in which interface, processing, and size effects would all be involved in controlling the overall mechanical and physical properties.

(2) Annealing and Mechanical Testing

(A) Annealing: Annealing was conducted at 1316°C in air using either rectangular flexure bars or cylindrical bars depending on the type of testing. The annealing temperature chosen in this work was one of the target service temperatures from the ongoing advanced aeroengine programs at the NASA Glenn. Test specimens were heated at a heating rate of about 20°C/min, held for a specified annealing time, and then furnace cooled. After annealing, each of the mechanical or physical properties was determined as a function of annealing time in accordance with the procedures described below. Typically, a total of five different annealing times of 0, 5, 20, 100, and 500 h were used for a given property evaluation.

(B) Flexure Strength: Flexure test specimens measuring 4, 3, and 25 mm, respectively, in width, depth, and length were annealed at 1316°C in air for 0, 5, 20, 100, and 500 h. A total of five to ten test specimens were annealed for each duration. The corresponding strength of annealed test specimens was determined at ambient temperature in four-point flexure with 10 mm inner and 20 mm outer spans using an actuator speed of 0.0083 mm/s in an electromechanical test frame (Model 8562, Instron, Canton, MA). One of two 4 mm wide faces of each test specimen that was perpendicular to the plasma-spray direction was subjected to tension in flexure loading. In other words, the prospective fracture plane was normal with respect to the layered structure (or lamellae). The coatings have exhibited negligible slow crack growth at either ambient or elevated (800°C) temperature with slow-crack-growth parameter n greater than 100;[14] hence, the test rate used here, 0.0083 mm/s, was considered sufficiently fast to minimize slow crack growth effects during strength testing. Flexure testing and specimen preparation, in general, were in accordance with ASTM C 1611.[15]

(C) Modes I and II Fracture Toughnesses: Flexure test specimens with measurements of 3 and 4 mm in width and depth and a measurement of either 25 or 50 mm in length were used for determination of modes I and II fracture toughnesses after annealing at 1316°C in air for 5, 10, 20, and 500 h. For each mode of fracture toughness, a total of four to five test coupons

were tested at ambient temperature after annealing as described below.

Sharp V notches were introduced in test specimens using the single-edge-V-notched-beam (SEVNB) method.[16] This method utilizes a razor blade with diamond paste to introduce a sharp root radius by tapering a saw cut. A starter straight-through notch about 1.2 mm deep and 0.026 mm wide was made on the 3 mm wide face of test coupons. The final notch depth and root radius were finished with a steel razor blade with diamond paste to be about 2.0 mm[‡] and 20–50 μm, respectively. Details of this procedure can be found elsewhere.[14,19] The coating material had less sharpness in its root radius compared with typical dense ceramics because of its porous and microcracked nature. However, it has been observed that the sharpness ranging from 20 to 50 μm is sufficient to give a consistent and accurate value of fracture toughness of the coating material.[14,19] Other methods used to generate sharp cracks to estimate fracture toughness, such as the single-edge-precracked-beam (SEPB) and the surface crack in flexure (SCF) techniques by indentation (ASTM C 1421[20]), were not feasible for the coating material, since the indentation response was very poor because of the material's significant porosity, microcracks, and tendency to densify upon localized indentation.[19] Note that the through-the-thickness sharp notches thus prepared were aligned perpendicular with respect to the layered structure.

In mode I fracture toughness testing, each test specimen with a sharp V notch located at its center was loaded in symmetric four-point flexure with 10 mm inner and 20 mm outer spans for the 25 –mm long test specimens or with 20 mm inner and 40 mm outer spans for the 50 mm long test specimens. In mode II testing, the sharp V-notched test specimens were loaded in asymmetric four-point flexure as shown in Fig. 2. The precrack was centered with respect to the loading point ($s = 0$) so that the precrack was subjected to pure mode II loading. Of course, the ratio of mode I to mode II loading can be varied by varying the distance of the precrack from the center plane, s, as previously done for mixed-mode fracture behavior of TBCs at both 25° and 1316°C.[19] The values of $A = 10$ mm and $B = 5$ mm were used in mode II fracture toughness testing. Both modes I and II fracture toughness tests were conducted at ambient temperature in air using the same actuator speed (0.0083 mm/s) and test frame that

---

[‡]It has been reported that plasma-sprayed TBCs exhibit rising R-curve behavior in fracture resistance with short cracks <1 mm.[5,9,17] The crack size of 2 mm used in this work may represent a long crack that would give rise to a steady-state, asymptotic value of fracture toughness or may represent a short crack since an actual crack size with a small crack opening displacement that is responsible for toughening would be small around <1 mm. The difference in fracture toughness between SEPB and SEVNB methods has been observed in advanced dense ceramics with a rising R-curve.[18]

were used in flexure strength testing. After testing, the crack size of each specimen tested was determined optically from its fracture surface based on the three-point measurements in accordance with the test standard ASTM C 1421.[20] Fracture toughnesses K_{Ic} and K_{IIc} were calculated using the following typical relations:

$$K_I = \sigma(\pi a)^{1/2} F_I\left(\frac{a}{W}\right) \qquad (1)$$

$$K_{II} = \tau(\pi a)^{1/2} F_{II}\left(\frac{a}{W}\right) \qquad (2)$$

where σ is the applied (remote) normal stress, τ is the applied shear stress, a is the crack size, and W is the specimen depth (4 mm). F_I and F_{II} are crack geometry factors in modes I and II, respectively. The shear stress in pure mode II is given from the elementary beam theory by

$$\tau = \frac{A-B}{A+B}\frac{P}{bW} \qquad (3)$$

where b is the specimen width (3 mm), and P is the applied force. Several different expressions of F_{II} were suggested by Suresh et al.,[21] Wang et al.,[22] and He and Hutchinson[23] for the case of a through-the-thickness crack, and their respective results of F_{II} do not yield any significant difference, particularly when $a/W = 0.35$–0.50. The solution by He and Hutchinson[23] which provides a convenient polynomial expression, is used here as follows:

$$
\begin{aligned}
F_{II}\left(\frac{a}{W}\right) = {} & \frac{a/W}{\sqrt{\pi(1-a/W)}} \\
& \times \left[7.264 - 9.37\left(\frac{a}{W}\right) + 2.74\left(\frac{a}{W}\right)^2 \right. \\
& \left. + 1.87\left(\frac{a}{W}\right)^3 - 1.04\left(\frac{a}{W}\right)^4\right] \\
& \text{for } 0 \le \frac{a}{W} \le 1
\end{aligned} \qquad (4)
$$

The solution for F_I is taken from Murakami,[24] which is almost identical to Srawly and Gross'[25] solution for $a/W \le 0.7$:

$$
\begin{aligned}
F_I\left(\frac{a}{W}\right) = {} & 1.122 - 1.121\left(\frac{a}{W}\right) + 3.740\left(\frac{a}{W}\right)^2 \\
& + 3.873\left(\frac{a}{W}\right)^3 - 19.05\left(\frac{a}{W}\right)^4 + 22.55\left(\frac{a}{W}\right)^5
\end{aligned} \qquad (5)
$$

(D) Deformation and Elastic Modulus: Cylindrical rod test specimens (see Fig. 1) were used to determine the constitutive relations and elastic modulus in compression as a function of annealing time. The nominal dimensions of test specimens were 5 and 10 mm, respectively, in diameter and length. The plasma-spraying direction was normal to the specimen's longitudinal axis. Five different annealing times at 1316°C, $t = 0$, 5, 20, 100, and 500 h, were used. Two specimens were annealed at each specified time. After annealing, each test specimen was strain gauged and subjected to a monotonic loading and unloading cycle in compression to its longitudinal direction. A maximum compression force of 900 N (46 MPa) was applied using a specially designed test fixture. Resultant stress–strain curves were obtained, and the elastic modulus was determined from their slopes. The test frame was the same as that used in flexure strength and fracture toughness testing.

The impulse excitation of the vibration method was also used to determine the elastic modulus of annealed flexure test specimens measuring $4 \times 3 \times 50$ mm in width, depth, and length, respectively, in accordance with ASTM C 1259.[26] Although this method is primarily for dense, homogeneous materials, this technique was used to find its applicability for porous, less

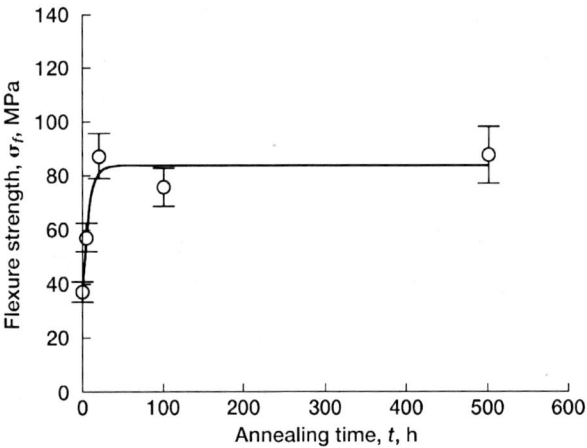

Fig. 3. Flexure strength as a function of annealing time for plasma-sprayed ZrO_2–8 wt% Y_2O_3 thermal barrier coatings annealed at 1316°C in air. Error bars indicate ± 1.0 standard deviation.

dense TBCs that typically exhibit nonlinear elastic behavior, and to compare with the strain gauge method. Two flexure test specimens were continuously used throughout testing such that each individual elastic modulus was determined at each of the five annealing times, starting from $t = 0$–500 h. Impulse excitation was applied to the 4 mm wide face of each flexure beam specimen, where the direction of vibration of a beam was parallel to the plasma-spray direction (or normal to the layered structure or lamellae). At each step of elastic modulus measurements, the density was subsequently determined at the corresponding annealing time using the mass–volume method. The impulse excitation technique has been applied to determine the elastic modulus of thin TBCs on metal substrates.[27]

(E) Microhardness, Phase Stability, and Microstructure: Vickers microhardness was determined at ambient temperature with the remaining flexure specimen halves from mode I fracture toughness testing, chosen at different annealing times of 0, 5, 100, and 500 h. The flexure halves, measuring approximately $3 \times 4 \times 12$ mm, were mounted, and the 4 mm wide face of each piece was polished. Two indentation loads P_i of 9.8 and 19.6 N were applied using a type 3212 Zwick indenter (Zwick Roell AG, Ulm, Germany) on the polished surface that was parallel to the plasma-spray direction. In general, there were five indents made at each load for a given specimen annealed for a given time. The test procedure for hardness measurements was followed in accordance with ASTM C 1327.[28]

Fracture surfaces and polished surface morphologies of test specimens were characterized as a function of annealing time using limited optical microscopy and scanning electron microscopy (SEM).

The amount of monoclinic phase was also determined as a function of annealing time by X-ray diffractometry[29] using the remaining specimens broken in the mode I fracture toughness testing.

III. Results and Discussion

(1) Flexure Strength

The results of flexure strength as a function of annealing time are presented in Fig. 3. The increase in flexure strength upon annealing was significant, occurring in a short period of time of $t = 20$ h. Flexure strength reached a plateau at $t \ge 20$ h with an approximated value of $\sigma_f = 85$ MPa, a 130% increase with respect to the as-sprayed ($t = 0$) strength of $\sigma_f = 37 \pm 4$ MPa.§

§Average strength (σ_f) and Weibull modulus (m) determined previously with earlier vintages of coating material were $\sigma_f = 33 \pm 7$ MPa and $m = 6$ using a total of 30 flexure test specimens (3 mm $\times$ 4 mm $\times$ 25 or 50 mm) in four-point flexure at ambient temperature.[13] This shows that the flexure strength of TBCs is shown to be almost unchanged regardless of vintage over a period of time, indicative of consistency in the plasma-spraying process and raw coating materials. The same was observed to be true for fracture toughness.

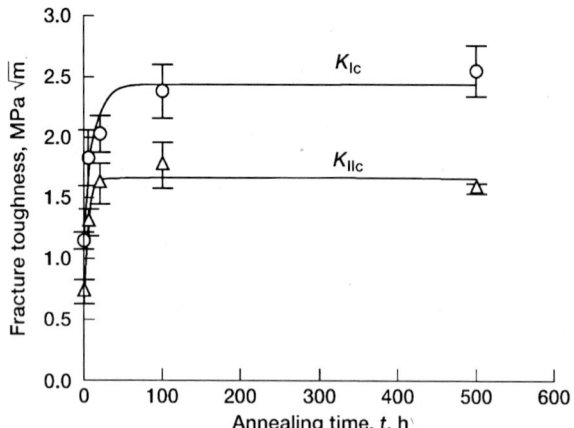

Fig. 4. Modes I and II fracture toughnesses (K_{Ic} and K_{IIc}) as a function of annealing time for plasma-sprayed ZrO_2–8 wt% Y_2O_3 thermal barrier coatings annealed at 1316°C in air. Error bars indicate ±1.0 standard deviation.

Despite the limited number of test specimens used in this work, the results showed a clear trend in strength upon annealing with a steady-state condition occurring within 100 h. However, the flexure strength (85 MPa) of TBCs annealed up to 500 h was still much lower than that (270 MPa) of hot-pressed 10 mol% yttria-stabilized zirconia (YSZ) at ambient temperature.[30] This indicates that the basic open structure inherent in TBCs remained mostly unchanged even after annealing for 500 h. A similar order of strength increase by 170%–180%, as compared with the as-sprayed counterpart, was also observed for ZrO_2–7% Y_2O_3 TBCs by Eaton and Novak[4] after annealing at 1290°C for 24 h and by Thurn et al.[5] at 1300°C for 10 h. Their data were only for 10 or 24 h, so that a direct comparison in strength trend in terms of annealing time cannot be made between this work and those previous studies. The type of strength-controlling flaws, fracture origins, and their natures were barely able to be determined from fracture surfaces because of problems associated with the open, porous, microcracked, and splat structure of TBCs in either the as-sprayed or annealed condition. Because of this, fractography in plasma-sprayed TBC material—unlike in dense ceramics—is an enormous challenge.[14] The microstructural changes of TBCs upon annealing will be described later in Section III(7).

(2) Modes I and II Fracture Toughnesses

Figure 4 shows the results of modes I and II fracture toughness testing for TBCs annealed at 1316°C in air. Previously determined modes I and II fracture toughnesses of as-sprayed TBCs, $K_{Ic} = 1.15 \pm 0.07$ MPa·m$^{1/2}$ and $K_{IIc} = 0.73 \pm 0.10$ MPa·m$^{1/2}$,[19] were also included. Mode I fracture toughness, upon annealing, increased significantly up to $t = 100$ h and then reached a plateau thereafter with $K_{Ic} = 2.6 \pm 0.2$ MPa·m$^{1/2}$ at $t = 500$ h, resulting in a 120% increase, as compared with the as-sprayed K_{Ic}. It is noted that unlike strength, the fracture toughness of TBCs at the plateau region was greater than that ($K_{Ic} = 1.7$ MPa·m$^{1/2}$) of hot-pressed 10 mol% YSZ.[30] Similar to the case in mode I, fracture toughness in mode II increased rapidly within 20 h, and then reached a plateau at $t \geq 100$ h, with an approximate 140% increase with respect to the as-sprayed value at $t = 0$ h. This increase in K_{IIc} compares well with that ($\approx 120\%$) in K_{Ic} as well as that ($\approx 130\%$) in strength. Thurn et al.[5] also observed a 130 % increase in fracture resistance (K_{Ir}) of a ZrO_2–7% Y_2O_3 TBCs system after heat treating at 1300°C for 10 h.

The ratio K_{IIc}/K_{Ic} for a given annealing time did not vary significantly, ranging from 0.63 to 0.80, with a resulting average of $K_{IIc}/K_{Ic} = 0.70 \pm 0.08$ for $t = 0$–500 h. This contrasts with some other observations that K_{IIc} ($= 3$–5 MPa·m$^{1/2}$) of as-sprayed TBCs was greater than K_{Ic} ($= 1$ MPa·m$^{1/2}$), because of the strong influence of crack face friction.[17] For some dense

ceramics, K_{IIc} was also reported to be greater than K_{Ic}, again attributed to the frictional interaction between the two crack planes.[31] However, the previous studies on advanced ceramics including silicon nitrides, alumina, and zirconia,[32,33] using naturally sharp (SEPB) precracks, showed a different result in which K_{IIc} was almost identical to K_{Ic}, indicative of insignificant frictional effects on K_{IIc}, regardless of grain structure/size. The coating material in this work yielded lower K_{IIc} than K_{Ic} values in both as-sprayed and annealed conditions, implying that the effect of crack face friction would be negligible.

The critical crack size at failure was estimated as a function of annealing time using the already determined values of respective flexure strength and fracture toughness, based on an assumption of half-penny-shaped crack configuration. The stress intensity factor related to such a crack configuration in an infinite body¶ is typically described by

$$K_I = Y\sigma_f\sqrt{\pi c_f} \tag{6}$$

where Y is the crack geometry factor ($Y = 2/\pi$) and c_f is the critical crack size. The critical crack sizes‖ based on Eq. (6) were estimated to be 760, 800, 420, 780, and 660 μm, respectively, for $t = 0$, 5, 20, 100, and 500 h. Except for the value at $t = 20$ h, the estimated crack sizes were not significantly different. Use of more test specimens in strength testing, of course, can yield more statistically reliable crack-size data. The current crack-size data, nonetheless, provide the insight that strength-controlling cracks or flaws in as-sprayed TBCs would remain reasonably unchanged in size, regardless of the sintering duration. This, from a fracture mechanics point of view, leads to the conclusion that fracture toughness is the only variable to affect the magnitude of strength of either the as-sprayed or annealed body of the current TBC system. This conclusion might be contrary to the notion that sintering would result in a decrease in crack and flaw size via healing or grain growth, thus leading to an increase in strength. Although micro- or macrocracks and pores might be healed via the annealing–sintering process, which gives rise to an increase in fracture toughness, major strength-controlling cracks or flaws having significant crack (or flaw) openings would be little affected in their overall physical sizes upon annealing. This suggests a necessity of simultaneous evaluation of both strength and fracture toughness of TBCs as sprayed or annealed; otherwise, a misleading interpretation could be attained regarding crack or flaw size if only one property, say strength, is used. This is particularly important in TBCs for which fractography, used to determine crack and flaw sizes as well as their configurations, is extremely difficult.

(3) Deformation and Elastic Modulus

The stress–strain curves of TBCs in compression determined for different annealing times by strain gauges are shown in Fig. 5. The as-sprayed cylindrical rod test specimen exhibited unique nonlinearity in both loading and unloading curves with significant hysteresis, primarily because of internal friction from the loosely connected open structure.[7,14,35] This type of nonlinear behavior has also been observed in sandstone and mortar, which exhibited an open microstructure similar to as-sprayed TBCs.[36] Despite their nonlinearity, the coatings have shown almost

---

¶The stress intensity factor (SIF) for a half-penny-shaped crack in a finite plate under flexure is expressed as[34]

$$K_I = H\sigma\sqrt{\pi\frac{c}{Q}}F\left(\frac{c}{W},\frac{c}{b},\phi\right)$$

where H, Q, and F are crack-geometry and specimen-dimension-dependent functions, σ is the remote outer-fiber flexure stress, c is the crack size, and ϕ is an angle of a particular point in crack front measured from specimen surface. The maximum SIF occurs at $\phi = 0$; i.e., at the specimen surface. It can also be found that the maximum SIF varies by only a few percent (about 6%) even though c/W varies between 0.2 and 0.5. Therefore, a closed form, infinite body solution without considering the complexity associated with crack and specimen size corrections in functions H, Q, and F can be used, which leads to Eq. (6).

‖The exact configuration of critical cracks whether they were either in single cracks or in multiple cracks connected to each other was not certain, because of difficulty in fractography. Hence, the term "critical crack" can include a meaning of "apparent" or "resultant" critical cracks in size.

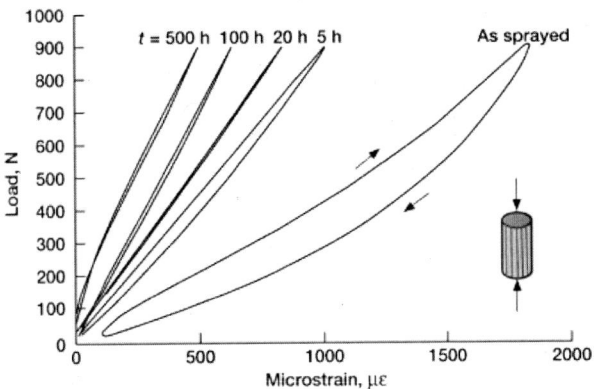

Fig. 5. Stress versus strain responses for different annealing times t of plasma-sprayed ZrO_2–8 wt% Y_2O_3 thermal barrier coatings annealed at 1316°C in air, determined in compression with cylindrical rod specimens by strain gauging.

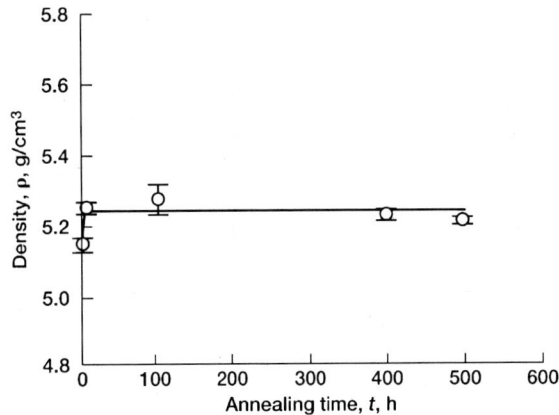

Fig. 7. Density as a function of annealing time for plasma-sprayed ZrO_2–8 wt% Y_2O_3 thermal barrier coatings annealed at 1316°C in air. Error bars indicate ± 1.0 standard deviation.

completely elastic behavior with little plastic deformation. This nonlinear elastic behavior was found to be independent of the loading rate, number of cycles, level of applied force, and type of loading configurations (flexure, tension, or compression).[14] As annealing time increased, the nonlinearity and hysteresis decreased, at the same time losing strain-tolerant capability. At $t \geq 20$ h, the annealed test specimens exhibited reasonably well-developed linear constitutive relations with little hysteresis.

Elastic modulus, E, determined from the slope of each stress–strain curve in Fig. 5, was plotted as a function of annealing time in Fig. 6. It should be noted that the elastic modulus of each test specimen as sprayed or annealed at $t = 5$ h was determined, for comparison, from the slope of a straight line drawn from zero to peak stress, although a single value of elastic modulus can never be defined particularly in as-sprayed TBCs.[7,13] Elastic modulus, as seen in the figure, increased quickly from $E = 25$ at $t = 0$–55 GPa within a short period of time $t = 20$ h and reached a plateau at $t \geq 100$ h with $E = 75$ GPa. Hence, a significant increase in elastic modulus by about 200% with respect to the as-sprayed value resulted from annealing at $t \geq 100$ h, much greater than the 120%–130% increase in strength and fracture toughness. The elastic modulus in the plateau, however, was considerably lower than that ($E = 219$ GPa) of dense 10 mol% YSZ,[30] indicating that the annealed TBC body was still of an open, loose structure, compared with the dense ceramic, even after a long sintering of $t = 500$ h. The elastic modulus has been observed to

increase from the as-sprayed counterpart by 200%–500% for plasma-sprayed yttria–zirconia coatings[4,5,8,37] and by about 200% for plasma-sprayed alumina coatings,[7] upon annealing at 1300°C in air. However, in general, a large scatter in the elastic-modulus data is typified, depending on material, type of measurements, and even on investigators. Poisson's ratio was also observed to change from an ill-defined to a well-defined status with a value of 0.2 when as-sprayed TBCs were annealed at 1316°C for 500 h.[13]

Figure 6 also includes the elastic modulus determined with flexure bars by the impulse excitation of the vibration method.[26] Fundamental resonant frequency was invariably defined either in as-sprayed or in annealed test specimens. The general trend of elastic modulus with respect to annealing time was almost identical to that determined by strain gauges. However, the magnitude of elastic modulus for a given annealing time was consistently greater by 64%–68% in the excitation method than in the strain-gauge method, as also observed previously by the sound velocity measurements.[38] This contrasts with general observations that most dense, homogeneous materials exhibit negligible or almost no difference in elastic modulus between the two methods. The reason for the elastic-modulus difference in TBCs between the two methods is not clear yet. Difference in the mode and/or magnitude of deformation—relatively large strain in uniaxial compression (in the strain-gauge method) versus very small strain in uniaxial vibratory flexure (in the impulse excitation of the vibration method)—incorporated with microscopic inhomogeneity of the open, loose splat structure of TBCs might be a plausible reason.

(4) Density

Figure 7 shows the change in density as a function of annealing time. The overall trend in density was similar to that in strength, fracture toughness, and elastic modulus, except for the degree in increase and the time to reach a plateau. The density increased from the as-sprayed value of 5.15 ± 0.02 g/cm^3 by 2.0%, 2.5%, 1.6%, and 1.3%, respectively, at $t = 5$, 100, 400, and 500 h. This increase in density was observed as a result of dimensional change (contraction) of flexure test specimens upon annealing. The plateau in density was formed at an earlier time of $t = 5$ h, compared with that formed at $t = 20$ or 100 h in strength, fracture toughness, or elastic modulus. A small increase (2%) in density because of annealing–sintering resulted in a significant change in mechanical properties such as strength, fracture toughness, and elastic modulus. It might be argued that annealing for a long time (e.g., 500 h) at 1316°C in air might have changed the weight of test specimens, adding an inherent error in the density measurement; however, the weight change was found to be negligible, with only a 0.03% decrease from the as-sprayed counterpart.

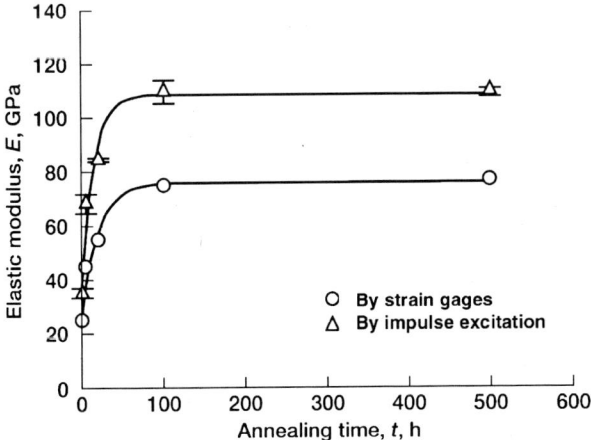

Fig. 6. Elastic modulus as a function of annealing time for plasma-sprayed ZrO_2–8 wt% Y_2O_3 thermal barrier coatings annealed at 1316°C in air. Elastic modulus determined by both strain gauging in compression and impulse excitation of vibration in flexure is included. Error bars indicate ± 1.0 standard deviation.

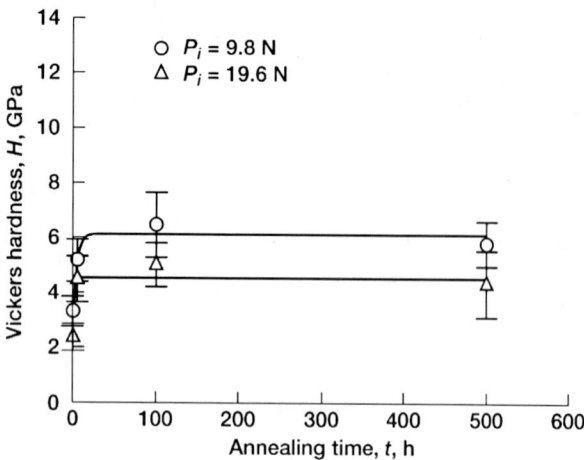

Fig. 8. Vickers hardness as a function of annealing time for plasma-sprayed ZrO_2–8 wt% Y_2O_3 thermal barrier coatings annealed at 1316°C in air, determined at two different indentation loads P_i of 9.8 and 19.6 N. Error bars indicate ±1.0 standard deviation.

(5) Hardness

Because of the presence of microcracks and porosity, Vickers microhardness impressions in TBCs were not as easily and clearly defined as those in dense ceramics. Surface cracks and pores obscured the impression measurements, yielding increased uncertainty. Although hardness impressions seemed a little better defined in their configuration for annealed TBCs, the overall uncertainty in impression measurements was about the same as that for the as-sprayed specimens. The results of Vickers hardness measurements as a function of annealing time, determined at indentation loads P_i of 9.8 and 19.6 N, are depicted in Fig. 8. As in some of the mechanical properties and density, hardness increased quickly at $t = 5$ h, seemingly reaching a plateau at $t \leq 100$ h. The hardness values at 9.8 N were about 25% greater than those at 19.6 N, attributed to the localized densification (plastic deformation) that occurs upon indentation: less localized densification would occur at a lower P_i and vice versa.

Unlike dense ceramics, TBCs exhibited significant scatter in microhardness; the coefficient of variation ranged from 12% to 20%, independent of annealing time or indentation load. This consistent but significant scatter was primarily because of the loosely connected open structure of TBCs, making the material microscopically inhomogeneous in response to microindentation. In addition, pore or flaws in some cases were comparable

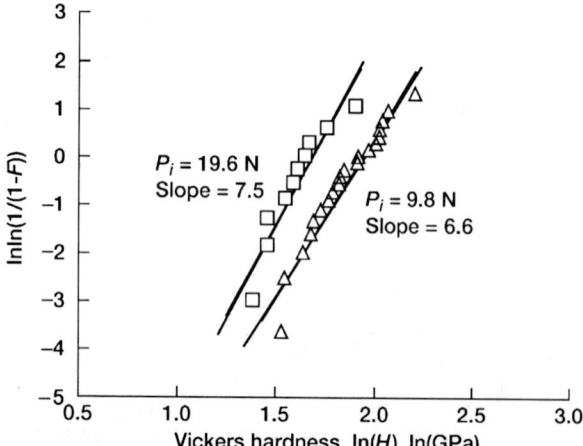

Fig. 9. Typical Weibull microhardness distributions of plasma-sprayed ZrO_2–8 wt% Y_2O_3 thermal barrier coatings annealed at 1316°C in air, determined by Vickers indenter at indentation loads P_i of 9.8 and 19.6 N.

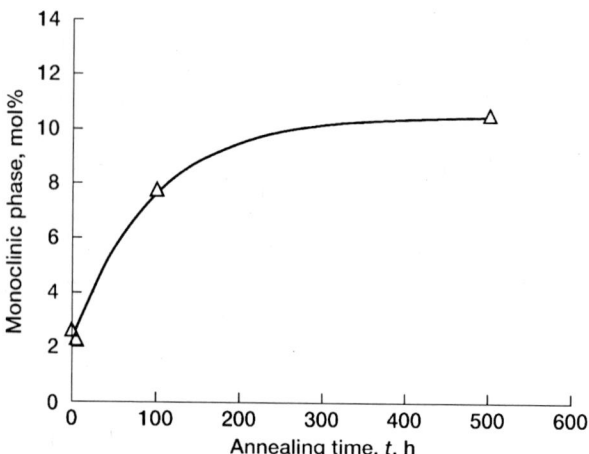

Fig. 10. Percent of monoclinic phase as a function of annealing time of plasma-sprayed ZrO_2–8 wt% Y_2O_3 thermal barrier coatings annealed at 1316°C in air.

with the impression sizes of 50–100 μm, which made the impression size measurements obscure, thereby increasing uncertainty in hardness measurements. For dense 10 mol% YSZ, microhardness is around $H = 14$ GPa with insignificant scatter, having a coefficient of variation of 4%.[30] Typical Weibull hardness distributions of the 100 h annealed TBCs with both 9.8 and 19.6 N are presented in Fig. 9. A relatively low Weibull modulus (slope) of around 7–8 was observed at both loads and was thus characterized as a unique response of plasma-sprayed TBCs to elastic–plastic microindentation deformation. For dense ceramics such as 10 mol% YSZ, the Weibull modulus in microhardness distribution was estimated to be 30.[30]

(6) Phase Stability and Thermal Conductivity

(A) Phase Stability: The plasma-sprayed, partially stabilized ZrO_2–8 wt% Y_2O_3 coatings can degrade during high-temperature exposure because of the nontransformable tetragonal t' phase separation into a low-yttria, transformable tetragonal and a high-yttria cubic phases. The low-yttria tetragonal phase can transform to the lower temperature monoclinic phase upon cooling or thermal cycling. Figure 10 shows the amount of the monoclinic phase as a function of annealing time at 1316°C. It can be seen that the amount of monoclinic phase showed a considerable increase with increasing time under the furnace heating and cooling condition. The monoclinic phase amount remained at 2–3 mol% in the first 5 h, but increased rapidly to 7–8 mol% after 100 h. The monoclinic phase amount seemed to saturate and reach a final value around 10–11 mol% after 500 h. The slow furnace cooling rate in the present study seemed to produce more monoclinic phase for the plasma-sprayed coatings than some of the more rapid cooling cases examined previously.[39] Although a general increasing trend is observed for the amount of monoclinic phase formation with annealing, the monoclinic phase increase showed a slower response with respect to the annealing time as compared with the mechanical properties, density, and thermal conductivity. This is because the monoclinic phase formation is mainly associated with the volume diffusion-related partitioning of some of the metastable tetragonal t' phase and the subsequent transformation into the monoclinic phase.

(B) Thermal Conductivity: Although sintering, from a mechanical-property viewpoint, gives favorable results such as increased strength and fracture toughness, it also gives an adverse effect stemming from a change in microstructure from open and loosely connected to more tightly closed, resulting in a more densified and stiffer network of the TBC body. This microstructural change was observed to have a considerable

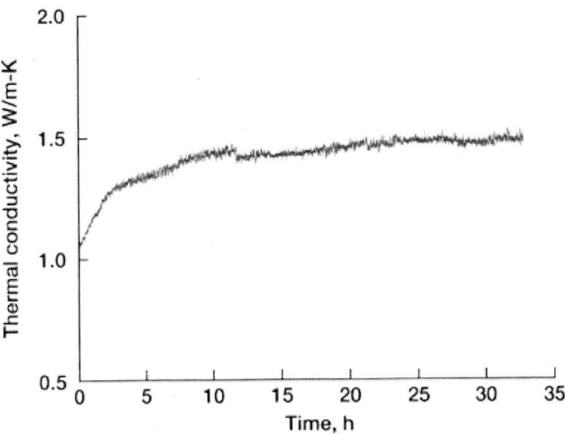

Fig. 11. Thermal conductivity as a function of time for plasma-sprayed ZrO_2–8 wt% Y_2O_3 thermal barrier coatings annealed at a surface temperature of 1320°C in air under high-heat-flux conditions by laser.[6]

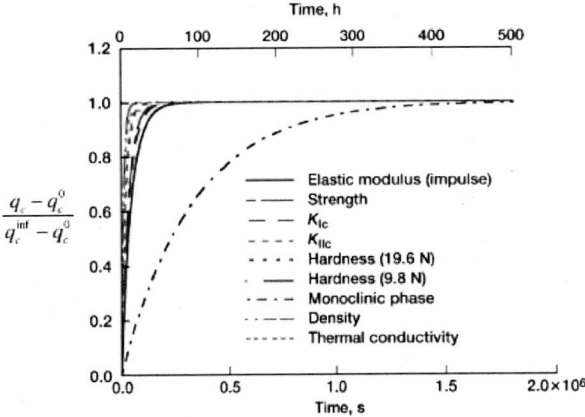

Fig. 13. Results of application of model (Eq. (7)) to various property values determined for plasma-sprayed ZrO_2–8 wt% Y_2O_3 thermal barrier coatings annealed at 1316°C in air.

influence on the inherently unique function of plasma-sprayed TBCs by increasing thermal conductivity[4,6,40] and reducing strain tolerance. An example of a change in thermal conductivity as a function of time[6] is shown in Fig. 11, in which testing was conducted in air under high heat flux conditions by laser while maintaining a surface temperature of 1320°C. A similarity in behavior between mechanical properties and thermal conductivity is manifest from the figure such that a significant thermal conductivity increase occurred at $t \leq 5$–10 h, followed by a plateau thereafter. Resistance to sintering, while maintaining a low thermal conductivity, particularly at higher temperatures, is of prime importance in the performance and life of a TBC system. Various efforts toward development of more sinter-resistant and higher-temperature TBCs are under way.

(7) Microstructures

SEM of polished surfaces of TBCs did provide some information regarding microcrack and/or pore healing as a result of sintering.[10] Somewhat large pores and macrocracks were observed from surfaces of either as-sprayed or annealed specimens. The 100 and 500 h annealed specimens showed an overall decrease in pore number and size, and microcrack density. The surfaces of the annealed specimens were also much smoother than the as-sprayed counterparts. A change in color from dull yellow to a somewhat shiny white-yellow was also noted for the specimens annealed particularly at longer times.

Figure 12 shows typical fracture surfaces with respect to annealing time for specimens tested in mode I fracture toughness testing. For specimens annealed at $t = 20$ and 100 h, both

initiation and development of grain growth because of the sintering effect were evident, with grain growth being more dominant for the specimen annealed for a longer time of $t = 100$ h. Note that a crack particularly located at the splat boundaries still exhibited an appreciable crack opening displacement of approximately ≤ 0.5 μm even after annealing for 20 h (Fig. 12(b)). The pores, gaps, and/or microcracks were significantly diminished for the specimen annealed at $t = 100$ h (see Fig. 12(c)) by the increase in contact areas between the splats and between splat particles, attributed to bonding via grain growth, as also noted previously.[4,5,8,9]

(8) Modeling: Normalization of Property Quantities

A simplified model to quantify the time-dependent coating property behavior at 1316°C is proposed here by normalizing the property-value increases at a given time and then fitting them into an exponential growth formulation as follows:

$$\frac{q_c - q_c^0}{q_c^{\text{inf}} - q_c^0} = C \exp\left[(1 - \exp\left(-\frac{t}{\gamma}\right)\right] \tag{7}$$

where q_c is the value of any property at a given annealing time t, q_c^0 and q_c^{inf} are the corresponding values at the initial time and at an infinite long time, respectively, γ is the 'relaxation' time, and C is a constant related to temperature. It should be mentioned that in all cases $C = 1$ because q_c^{inf} is taken as the final value of the property at the only test temperature of 1316°C (instead of using the material intrinsic value). This model has been used to

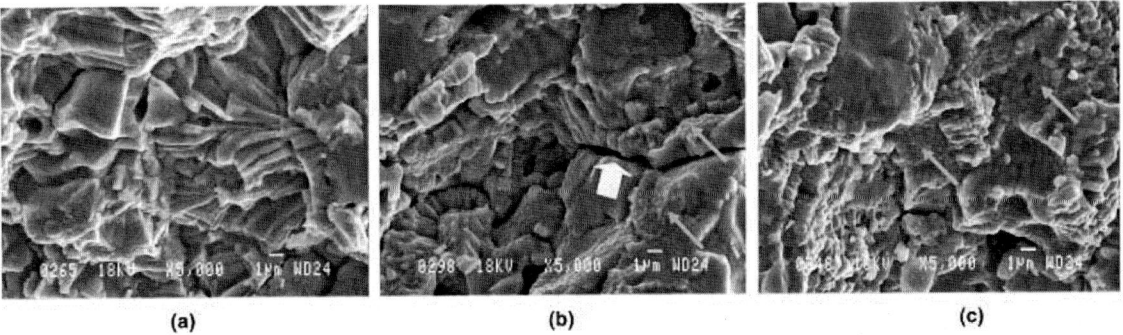

Fig. 12. Fracture surfaces of mode I fracture toughness specimens of plasma-sprayed ZrO_2–8 wt% Y_2O_3 thermal barrier coatings annealed at 1316°C in air at different times, t: (a) $t = 0$ h (as sprayed); (b) $t = 20$ h; and (c) $t = 100$ h. Arrows in (b) and (c) indicate regions of grain growth. A crack with a significant crack opening displacement is indicated with an open arrow in (b).

Mechanical Properties

describe the elastic modulus evolution of TBCs under a high heat flux condition by a laser.[6]

The results based on the model using constants found from the data collected are shown in Fig. 13. The figure shows that most of the properties, except for the monoclinic phase content, reached their final values within a relatively short time and are characterized by a relatively short relaxation time. The relatively short relaxation time for these properties suggests that they may be strongly related to the coating microcracks and splat boundaries, which tend to be sintered away quickly because of the fast surface and boundary diffusion.[41] The t' phase partitioning (later resulting in the monoclinic phase formation) would be related to the volume diffusion of cations and anions. Therefore, the phase transformation does have a much longer relaxation time than other properties that would be more sensitive to the surface-diffusion-related microcrack sintering.

Although the model was not based on a rigorous physical and theoretical foundation, it nevertheless allows one to quantify and assess the phenomenological effect of sintering on various property quantities in a convenient and expedient manner.

IV. Summary

The mechanical and physical properties of free-standing plasma-sprayed ZrO_2–8 wt% Y_2O_3 TBCs annealed at 1316°C in air were determined at ambient temperature as a function of annealing time ranging from 0 to 500 h. Flexure strength, modes I and II fracture toughnesses, elastic modulus, density, and microhardness all increased significantly in 5–100 h and then reached a plateau above 100 h; the percentage of monoclinic phase increased monotonically without forming a plateau. Annealing, which provided a sintering effect, resulted in microcracks and micropores healing, grain growth, and densification so that a loosely connected, open structure was changed into a more closely connected, densified structure. However, an inevitable adverse effect also occurred such that the relatively low thermal conductivity and good strain tolerance—which make the TBCs unique in thermal barrier applications—were degraded upon annealing because of the formation of a densified network of the TBC body. The proposed model quantified all related property variables as a function of annealing time in a simple normalized fashion.

Acknowledgments

The authors are grateful to R. Pawlik for experimental work and to G. Leissler for processing TBCs.

References

[1]R. A. Miller, "Current Status of Thermal Barrier Coatings—An Overview," *Surf. Coat. Technol.*, **30**, 1–11 (1987).

[2]R. A. Miller, "Thermal Barrier Coatings for Aircraft Engines—History and Direction;" pp. 17–34 in *NASA CP-3312*, Edited by W. J. Brindley. National Aeronautics and Space Administration, Glenn Research Center, Cleveland, OH, 1995.

[3]T. M. Yonushonis, "Thick Thermal Barrier Coatings for Diesel Components;" NASA CR-187111, National Aeronautics and Space Administration, Glenn Research Center, Cleveland, OH, 1991.

[4]H. E. Eaton and R. C. Novak, "Sintering Studies of Plasma-Sprayed Zirconia," *Surf. Coat. Technol.*, **32**, 227–36 (1987).

[5]G. Thurn, G. A. Schneider, H. A. Bahr, and F. Aldinger, "Toughness Anisotropy and Behavior of Plasma Sprayed ZrO_2 Thermal Barrier Coatings," *Surf. Coat. Technol.*, **123**, 147–58 (2000).

[6]D. Zhu and R. A. Miller, "Thermal Conductivity and Elastic Modulus Evolution of Thermal Barrier Coatings Under High Heat Flux Conditions," *J. Thermal Spray Technol.*, **9** [2] 175–80 (2000).

[7]R. W. Trice, D. W. Prine, and K. T. Faber, "Deformation Mechanisms in Compressed-Loaded, Stand-Alone Plasma-Sprayed Alumina Coatings," *J. Am. Ceram. Soc.*, **83** [12] 3057–64 (2000).

[8]J. A. Thompson and T. W. Clyne, "The Effect of Heat Treatment on the Stiffness of Zirconia Top Coats in Plasma-Sprayed TBCs," *Acta Mater.*, **49**, 1565–75 (2001).

[9]R. W. Steinbrech, "Thermomechanical Behavior of Plasma-Sprayed Thermal Barrier Coatings," *Ceram. Eng. Sci. Proc.*, **23** [4] 397–408 (2002).

[10]S. R. Choi, D. Zhu, and R. A. Miller, "Effect of Sintering on Mechanical and Physical Properties of Plasma-Sprayed Thermal Barrier Coatings;" NASA/TM-2004-212625, National Aeronautics and Space Administration, Glenn Research Center, Cleveland, OH, 2004.

[11]D. Zhu and R. A. Miller, "Influence of High Cycle Thermal Loads on Thermal Fatigue Behavior of Thick Thermal Barrier Coatings;" NASA/TP 3676, National Aeronautics and Space Administration, Glenn Research Center, Cleveland, OH, 1997.

[12]D. Zhu and R. A. Miller, "Sintering and Creep Behavior of Plasma-Sprayed Zirconia- and Hafnia-Based Thermal Barrier Coatings," *Surf. Coat. Technol.*, **108–109**, 114–20 (1998).

[13]S. R. Choi, D. Zhu, and R. A. Miller, "Mechanical Properties/Database of Plasma-Sprayed ZrO_2–8wt% Y_2O_3 Thermal Barrier Coatings," *Int. J. Appl. Ceram. Technol.*, **1** [4] 330–42 (2004).

[14]S. R. Choi, D. Zhu, and R. A. Miller, "(a) High-Temperature Slow Crack Growth, Fracture Toughness and Room-Temperature Deformation Behavior of Plasma-Sprayed ZrO_2–8 wt% Y_2O_3," *Ceram. Eng. Sci. Proc.*, **19** [4] 293–301 (1998); (b) "Deformation and Strength Behavior of Plasma-Sprayed ZrO_2-8 wt% Y_2O_3 Thermal Barrier Coatings in Biaxial Flexure and Trans-Thickness Tension," *Ceram. Eng. Sci. Proc.*, **21** [4] 653–61 (2000); (c) "Deformation and Tensile Cycle Fatigue of Plasma-Sprayed ZrO_2-8 wt% Y_2O_3 Thermal Barrier Coatings," *Ceram. Eng. Sci. Proc.*, **22** [4] 427–34 (2001).

[15]ASTM C 1161 "Test Method for Flexural Strength of Advanced Ceramics at Ambient Temperature;" in *Annual Book of ASTM Standards*, Vol. 15.01. American Society for Testing and Materials, West Conshohocken, PA, 2003.

[16]J. Kübler, "(a) Fracture Toughness of Ceramics Using the SEVNB Method: Preliminary Results," *Ceram. Eng. Sci. Proc.*, **18** [4] 155–62 (1997) (b) "Fracture Toughness of Ceramics Using the SEVNB Method; Round Robin," VAMAS Report No. 37, EMPA, Swiss Federal Laboratories for Materials Testing & Research, Dübendorf, Switzerland, 1999.

[17]A. G. Evans, D. R. Mumm, J. W. Hutchinson, G. H. Meier, and F. S. Pettit, "Mechanisms Controlling the Durability of Thermal Barrier Coatings," *Prog. Mater. Sci.*, **46**, 505–53 (2001).

[18]S. R. Choi and J. P. Gyekenyesi, "Fracture Toughness in Advanced Monolithic Ceramics–SEPB vs. SEVNB Methods;" Proceedings of the 11th International Conference on Fracture (ICF11), Turin, Italy, March 20–25, 2005.

[19]S. R. Choi, D. Zhu, and R. A. Miller, "Fracture Behavior under Mixed-Mode Loading of Ceramic Plasma-Sprayed Thermal Barrier Coatings at Ambient and Elevated Temperatures," *Eng. Fract. Mech.*, in press.

[20]ASTM C 1421 "Test Method for Determination of Fracture Toughness of Advanced Ceramics at Ambient Temperature;" in *Annual Book of ASTM Standards*, Vol. 15.01. American Society for Testing and Materials, West Conshohocken, PA, 2002.

[21]S. Suresh, C. F. Shih, A. Morrone, and N. P. O'Dowd, "Mixed-Mode Fracture Toughness of Ceramic Materials," *J. Am. Ceram. Soc.*, **73** [5] 1257–67 (1990).

[22]K. J. Wang, H. C. Lin, and K. Hua, "Calculation of Stress Intensity Factors for Combined Mode Bend Specimens;" pp. 123–33 in *Advances in Research on the Strength and Fracture of Materials*, Vol. 4, Edited by M. D. R. Taplin. ICF4, Waterloo, Canada, 1977.

[23]M. Y. He and J. W. Hutchinson, "Asymmetric Four-Point Crack Specimen," *J. Appl. Mech.*, **67**, 207–9 (2000).

[24]Y. Murakami (ed.), *Stress Intensity Factors Handbook*, Vol. 1, p. 16. Pergamon Press, New York, 1987.

[25]J. E. Srawley and B. Gross, "Side-Cracked Plates Subjected to Combined Direct and Bending Forces;" pp. 559–79 in *Cracks and Fracture, ASTM STP 601*. American Society for Testing and Materials, Philadelphia, 1976.

[26]ASTM C 1259 "Test Method for Dynamic Young's Modulus, Shear Modulus, and Poisson's Ratio for Advanced Ceramics by Impulse Excitation of Vibration;" in *Annual Book of ASTM Standards*, Vol. 15.01. American Society for Testing and Materials, West Conshohocken, PA, 2003.

[27]C. A. Johnson, J. A. Rudd, A. C. Kaya, and H. G. deLorenzi, "A Method for Measuring Non-Linear Elastic Properties of Thermal Barrier Coatings;" pp. 415–20 Proceedings of the 8th National Thermal Spray Conference, September 11–15, Houston TX, 1995.

[28]ASTM C 1327 "Test Method for Vickers Indentation Hardness of Advanced Ceramics;" in *Annual Book of ASTM Standards*, Vol. 15.01. American Society for Testing and Materials, West Conshohocken, PA, 2003.

[29]R. A. Miller, R. G. Garlick, and J. L. Smialek, "Phase Distributions in Plasma-Sprayed Zirconia–Yttria," *Am. Ceram. Soc. Bull.*, **62** [12] 1355–8 (1983).

[30]S. R. Choi and N. P. Bansal, "Mechanical Behavior of Zirconia/Alumina Composites," *Ceram. Int.*, **31**, 39–46 (2005).

[31]D. K. Shetty, A. R. Rosenfield, and W. H. Duckworth, "Mixed-Mode Fracture of Ceramics in Diametral Compression," *J. Am. Ceram. Soc.*, **69** [6] 437–43 (1986).

[32]V. Tikare and S. R. Choi, "Combined Mode I and Mode II Fracture of Monolithic Ceramics," *J. Am. Ceram. Soc.*, **76** [9] 2265–72 (1993).

[33]V. Tikare and S. R. Choi, "Combined Mode I–Mode II Fracture of 12-Mol-%-Ceria-Doped Tetragonal Zirconia Polycrystalline Ceramic," *J. Am. Ceram. Soc.*, **80** [6] 1624–6 (1997).

[34]I. S. Raju and J. C. Newman, "Stress-Intensity Factors for a Wide Range of Semi-Elliptical Surface Cracks in Finite–Thickness Plates," *Eng. Fract. Mech.*, **11**, 817–29 (1979).

[35]K. F. Wesling, D. F. Socie, and B. Beardsley, "Fatigue of Thick Thermal Barrier Coatings," *J. Am. Ceram. Soc.*, **77** [7] 1863–8 (1994).

[36]S. R. Choi, NASA Glenn Research Center, Cleveland, OH, 2004, unpublished work.

[37]A. J. Allen, G. G. Long, Y. Wallace, Y. Llavsky, C. C. Berndt, and H. Hermann, "Microstructural Changes in YSZ Deposits During Annealing;" pp. 228–33 in *Proceedings of the Unified Thermal Spray Conference*, Edited by E. Lugscheider, and P. A. Kammer. DVS, Germany, 1999.

[38]J. I. Eldridge, G. N. Morscher, and S. R. Choi, "Quasistatic vs. Dynamic Modulus Measurements of Plasma-Sprayed Thermal Barrier Coatings," *Ceram. Eng. Sci. Proc.*, **23** [4] 371–8 (2002).

[39]J. Moon, H. Choi, H. Kim and C. Lee, "The Effect of Heat Treatment on the Phase Transformation Behavior of Plasma-Sprayed Stabilized ZrO$_2$ Coatings," *Surface and Coatings Technology*, **155**, 1–10 (2002).

[40]D. Zhu and R. A. Miller, "Thermal Barrier Coatings for Advanced Gas-Turbine Engines," *MRS Bull.*, **25** [7] 43–7 (2000).

[41]D. Zhu and R. A. Miller, "Determination of Creep Behavior of Thermal Barrier Coatings Under Laser Imposed High Thermal and Stress Gradient Conditions," *J. Mater. Res.*, **14** [1] 146–61 (1999). ☐

The Measurement of Residual Strains Within Thermal Barrier Coatings Using High-Energy X-Ray Diffraction

J. Thornton[†] and S. Slater

Defence Science and Technology Organisation, Port Melbourne, Vic. 3207, Australia

J. Almer

Advanced Photon Source, Argonne National Laboratory, Argonne, Illinois

The residual strains through the entire thickness of the zirconia layer of pristine and heat-treated thermal barrier coatings (TBCs) were mapped to help elucidate the failure mechanisms of TBCs. The strains were measured using 80.72 keV synchrotron radiation and a transmission geometry. The heat-treated TBC showed that a compressive strain formed in the zirconia layer of the TBC on cooling but this strain was diluted and reversed by the oxidation-driven expansion of the underlying metals. It also showed large (0.0024) out-of-plane tensile strains in the zirconia layer just above its interface with a thick underlying oxide layer.

I. Introduction

THERMAL barrier coatings (TBCs) typically consist of a 0.3 mm layer of partially stabilized zirconia over a 0.2 mm layer of NiCoCrAlY applied to a metal substrate. The NiCoCrAlY layer is commonly known as the bond coat. TBCs are applied to components that are exposed to hot combustion gases within gas turbine and piston engines—combustion liners for example. These coatings thermally insulate the components and therefore reduce the components' temperatures. TBCs are only effective when there is a temperature gradient through the skin of the coated component, which is the case for the air-cooled components in the hot ends of gas turbine engines. The temperature reduction claimed for TBCs varies. For example, one source estimates reductions from 110° to 170°C[1] and another measured values of 55°–80°C.[2] Nevertheless, the reduction in temperature improves the durability of the components. Alternatively, the application of TBCs can allow an increase in the temperature of the combustion gases, which then enables the engine to produce greater specific thrust.

While the use of TBCs is now widespread, the lifetime of TBCs cannot be predicted reliably.[2,3] After a long exposure to the high temperatures and oxidizing environments in heat engines, TBCs can fail (substantial portions of the zirconia layer delaminate). Thus, engines are not designed to run hotter by the full-temperature reduction enabled by TBCs because the loss of these coatings could result in rapid degradation of the component.[2,3] The ability to reliably predict the lifetimes of TBCs will therefore enable a greater utilization of the temperature reduction provided by these coatings. This ability will require a full understanding of the failure mechanisms of TBCs.

There are two main methods for depositing TBCs: plasma spray (PS)[4] and electron beam physical vapor deposition (EB-PVD).[5] In the PS technique, molten droplets of the coating material impact on the target and flatten to form what are commonly known as "splats." The outlines of the splat edges can be discerned in the micrographs shown in Section III(2) (e.g. Fig. 8). EB-PVD creates a different microstructure of vertical columns and wedges.[5] NiCoCrAlY bond coats are used with EB-PVD TBCs but platinum aluminide bond coats are increasingly popular. Early thermal cycling tests established that PS TBCs fail not on heating but on cooling, and that oxidation of the bond coat influences failure.[6] Mechanical tests showed that placing the coating in compression caused failure, while equivalent stresses in tension did not.[7] The loci of failures were in the ceramic layer just above the bond coat layer. It was proposed that at high temperatures, plastic deformation gradually relieved strains, but on cooling, the ceramic was placed in compression because of the thermal expansion mismatch with the bond coat[7] a cooling stress. The degree of compression is dependent on the rate of cooling and the ability of the bond coat to plastically deform quickly. Rapid cooling occurs in the combustion and turbine section of gas turbine engines on shutdown. It was also proposed that bond coat oxidation reduced the ability of the bond coat to plastically deform on cooling and thereby its ability to reduce the compression of the ceramic layer.[7] The compressive stresses were therefore expected to grow with bond coat oxidation until large enough to cause the ceramic to delaminate. An additional proposal was that bond coat oxidation promoted cracking at the bond coat to zirconia interface, and that failure was the result of slow crack growth and micro-crack linkage.[4] A compressive cooling stress in the ceramic layer[7] could still drive the failure but in the interfacial-cracking proposal,[4] delamination would occur at lower stresses because of the weakening of the interface.

More recent work has focused on EB-PVD TBCs,[8] for which the locus of failure is usually at the interface between the bond coat oxide and the bond coat.[1] The present consensus[8] is that the compressive stresses (3–5 GPa) generated within the bond coat oxide upon cooling drive the failure. The stresses that arise during the growth of the oxide layer are smaller (< 1 GPa).[8] The topology of the interface between the bond coat and the zirconia is an important factor in TBC failure. For both EB-PVD and PS TBCs, it has been proposed that rough interfaces between the bond coat and the zirconia[4,8,9] result in localized tensile strains that cause cracking and an overall weakening of the adhesion of the zirconia layer. The strain distributions expected around bond coat asperities have been calculated for real bond coat profiles.[10] Changes in the topography of the bond coat surface during oxidation have also been linked to failure of EB-PVD TBCs.[8,11]

S. J. Glass—contributing editor

Manuscript No. 11096. Received July 9, 2004; approved December 31, 2004.
This work was supported by the Australian Synchrotron Research Program, which is funded by the Commonwealth of Australia under the Major National Research Facilities Program. Use of the Advanced Photon Source was supported by the U.S. Department of Energy, Basic Energy Sciences, Office of Energy Research, under Contract No. W-31-109-Eng-38.
Originally presented at the 2004 Annual Meeting in the symposium in honor of Ed Fuller.
[†]Author to whom correspondence should be addressed. e-mail: john.thornton@dsto.defence.gov.au

The importance of residual strains in the failure of TBCs led to X-ray measurements of the strain in the zirconia layer.[12] However, these measurements were limited to the top 50 μm of the zirconia layer, and the samples were not subjected to the high temperatures that lead to TBC failure. Subsequent neutron measurements on a thick (1 mm) zirconia layer on a copper substrate did provide depth profiles of strain through the zirconia layer,[13] but were also limited because the sample had not experienced simulated engine conditions. Measurements on heat-treated and as-sprayed samples, using higher energy X-rays to penetrate deeper into the zirconia, showed large strains in the heat-treated samples.[14]

An aim of this present work was to measure the depth distribution of strain through the zirconia layers in heat-treated and as-sprayed samples, where the zirconia layer thickness was typical of gas turbine engine standards (0.3 mm). Determining the depth distribution of strain through the zirconia layer in heat-treated samples is important, as different depth distributions of strain may produce the same curvature of a simple coated strip, but may have significantly different propensities to cause delamination. Furthermore, given that the locus of failure in PS TBCs is in the zirconia layer just above the bond coat layer,[1,6] it is important to measure the strains of the zirconia at the interface between the bond coat and the zirconia. The strains in the bond coat oxide layer have been extensively studied using spectroscopy.[15] The second aim of this present work was to compare the depth profile of the strain in regions where different amounts of bond coat oxidation were occurring, and thus determine the effects of changing topography on the strain distribution in the zirconia layer. The strains in an underlying oxide layer were also measured in this work.

II. Experimental Procedure

(1) Experimental Outline

The work required a method to map strain over at least two dimensions (depth and a transverse direction) over the thickness of a TBC (0.5 mm). To meet this requirement, TBCs were cross-sectioned and high-energy (80.72 keV) collimated X-rays were used to produce transmission diffraction patterns from rod-shaped volumes through the cross-section (Fig. 1). Complete rings of the diffraction patterns were recorded on an area detector. The transmission technique has been applied to layered[16] and bulk[17] materials and thin (1–10 μm) coatings.[18] A practical concern for thick coatings, such as TBCs, was that sectioning might introduce damage or strain relaxation to coatings, and that this damage would extend through a significant portion of the illuminated volume. The range of the damage or relaxation was anticipated to be of the order of the thickness of the coating.

The alternative techniques of neutron diffraction and high-energy X-ray diffraction with reflection geometry were considered unsuitable for TBCs or other thick coatings. Neutron diffraction can be used to depth profile strain[13] but the spatial resolution (0.3 mm) is insufficient when plausible accumulation times are used. The use of high-energy X-rays in reflection was considered unsuitable because the diffraction of high-energy X-rays is biased toward smaller angles, and this reduces the actual depth sampled.[19] Intense beams and fine exit beam collimation of high-energy X-rays[20] can enable depth profiling but require more beam time than the transmission method used.

(2) Sample Preparation

Five centimeter diameter discs were cut from a 2 mm thick Hastelloy-X sheet (Haynes International, Kokomo, IN) using a laser process. Commercially available equipment (Metco 9 MB gun, Sulzer-Metco Holding AG, Winterthur, Switzerland) was used to apply the plasma-sprayed coatings. The deposition parameters are shown in Table I. A 0.2 mm layer of bond coat was applied to the discs. With two of the discs, an open mask with two parallel wires was used during the deposition of 0.2 mm of bond coat. The mask was then removed and a layer of 0.3 mm of

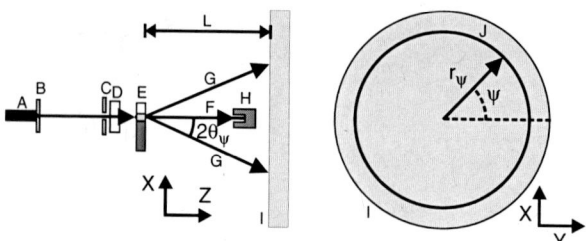

Fig. 1. The left-hand side shows a schematic of the experimental apparatus (side view): A, monochromatic X-ray beam; B, primary slits; C, guard slits; D, ion chamber to measure incident beam flux I_0; E, sample; F, transmitted beam; G, diffracted beam; H, beam stop with photodiode to measure transmitted flux I_F; and I, the area detector (MAR345). The three shades on E represent the nickel alloy substrate (dark gray), the NiCoCrAlY alloy TBC layer (light gray), and the zirconia TBC layer (white). The X and Z arrows define the co-ordinate axis; the Y-axis is into the page. $2\theta_\psi$ is twice the Bragg angle for the diffracted beam for azimuthal angle ψ (rotation about z-axis) and L is the distance between the sample and the area detector. The right-hand side shows a schematic of I, the area detector, and J a diffraction ring viewed from along the beam. The radius, r_ψ, of the diffraction ring at a particular azimuthal angle, ψ, is shown.

zirconia was applied. The TBCs produced therefore contained two linear gaps in the bond coat, which exposed the Hastelloy-X to oxidation during heat treatment.

One disc was heat-treated and the second was left in the as-sprayed condition. The heat treatment consisted of two stages:

1. 20 h in air at 1150°C followed by slow cooling (hours) in the furnace, and

2. 1 h in air at 1000°C followed by a rapid quenching in water.

The first stage was to age the sample by producing bond coat oxidation as well as enabling some sintering of the zirconia layer. The second stage was to produce large cooling strains.

The preparation of the samples for the strain mapping is shown in Fig. 2. Rectangular segments (2 cm × 1.2 cm) were cut from each disc. The segments were prepared as cross-sectional SEM specimens: mounted in cylindrical pots with cold-setting epoxy with the linear gaps vertical. The top surfaces of the cylinders produced were cut back by 2 mm and polished. SEM images of the exposed surface of the TBCs were taken. The cylinders were then cut 2.5 mm back from this exposed surface to produce the 2.5 mm thick sections (the samples) for the strain mapping.

The main purpose of the epoxy was to prevent spallation of the zirconia layer during sectioning. In order to ascertain whether the samples sustained any cutting damage, one section was further sectioned perpendicular to the first cut in order to inspect in cross-section. The commercially available epoxy (Struers SpeciFix-20, Struers Inc, Westlake, OH) is formulated to introduce no strain to the sample.

(3) Strain Mapping by High-Energy X-Ray Diffraction

A schematic diagram of the transmission method for the strain mapping is shown in Fig. 1, together with the co-ordinate

Table I. Thermal Barrier Coating (TBC) Deposition Parameters

Layer	Powder	V (V)	I (A)	f (mm)
Zirconia (7 wt% yttria stabilizer)	Metco 204NS (Sulzer-Metco Holding AG, Winterthur, Switzerland)	70	600	100
Bond coat (NiCrCoAlY)	Amdry 365-II (Sulzer-Metco)	70	500	100

V, potential difference; I, current; f, spray distance.

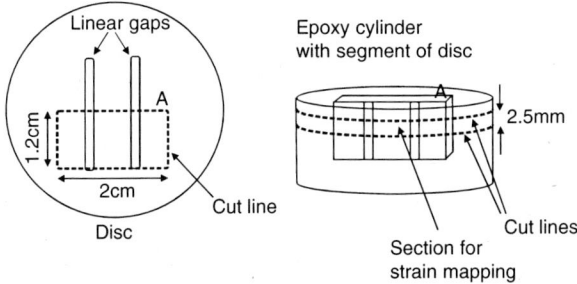

Fig. 2. A schematic diagram showing the preparation of the sections for strain mapping. The left-hand side shows the segment cut from each disc and the right-hand side shows the mounting of the segment in a cylinder of epoxy and the sectioning of that cylinder.

system. The diffraction rings produced were distorted from circles by the strains in the illuminated volume; for example, compression in the plane of the zirconia layer (y direction) would cause a widening of the rings in the y direction.

Prior to the mounting of a sample, the X-ray beam was aligned through the center of rotation (COR) of a goniometer using a steel pin, and the COR position was recorded on a television screen with two orthogonal CCD cameras. Samples were then mounted on the center of the goniometer with reference to the camera positions. The goniometer was equipped with three orthogonal translations and rotations, all remotely controlled.

To avoid beam degradation of spatial resolution, the faces of the sample had to be aligned with the co-ordinate system. The degradation caused by the top surface of the TBC not being parallel to the beam along the z direction was the most important as this affected the depth resolution strongly, and the most precise spatial measurements were required in depth. Any errors in sample parallelism with respect to the beam along the z-axis (α) would cause a degradation in depth resolution given by

$$\Delta x = \alpha \tau \tag{1}$$

where α is in radians and τ is the thickness of the sample in the z direction.

To minimize α, the top surface of the TBC was placed in the beam and the transmitted intensity was monitored while the sample was rotated about the y-axis. The estimated error in this alignment was 0.1°; this error degraded the depth resolution by 4 μm. This error is acceptable as the smallest step size used in the x direction was 10 μm.

The other sample surfaces were made parallel to the y and x directions optically with reference to a CCD camera, with an estimated error of 1°. Thus, the spatial resolution in the y direction was degraded by 40 μm; however, this is small in comparison with the dimension of the edge of the gap in the bond coat, 0.1 mm, and the distances between the depth scans >0.5 mm.

A monochromatic 80.72 keV X-ray beam was created using an undulator and double-Laue optics,[21] providing a flux of approximately 10^{12} ph/s/mm². The primary slits defining the transverse beam size were set to 25 [10] μm in x and 50 [100] μm in y for medium [high] depth-resolved measurements, respectively.

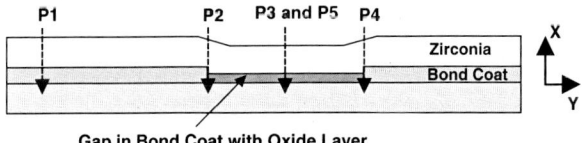

P1   P2   P3 and P5   P4

Gap in Bond Coat with Oxide Layer

Fig. 3. A schematic diagram showing the location of the depth profiles on a cross-section of a heat-treated sample with a gap in the bond coat. Depth profile P1 is 1.5 mm from P2 at the edge of the gap. The labels for the depth profiles of the as-sprayed sample followed the same pattern.

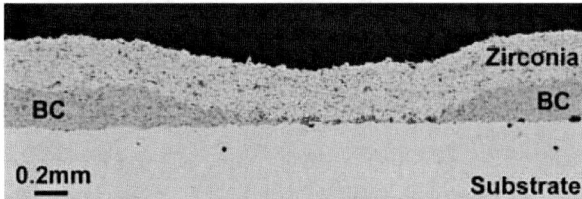

Fig. 4. A cross-section of an as-sprayed TBC showing the gap in the bond coat end-on. BC, bond coat. The magnification is lower than that for Fig. 5 in order to include the gap's shallow edge on the left-hand side.

Guard slits were placed near the sample to minimize the scattered radiation reaching the sample. A tungsten beam stop, containing a Si photodiode, was placed between the sample and detector, allowing for continuous monitoring of transmitted flux. The transmission of the high-energy X-rays through the 2.5 mm thickness of the TBC zirconia was approximately 17%.

The illuminated volume of the specimen was rod like in shape, with the cross-section of the X-ray beam and with a length equal to the sample thickness (25 μm × 50 μm × 2500 μm). The diffraction pattern produced from the illuminated volume was detected using an on-line image plate detector (MAR345, Marresearch GmBH, Norderstredt, Germany) at 974 mm from the sample. The detector was a circle of 345 mm diameter and an area of 0.093 m², with a spatial resolution of 100 μm in both directions. The detector to sample distance was chosen to balance the need for large distances (to reduce systematic strain errors because of sample displacement and to increase strain resolution) and the need to include a range of diffraction rings (to show any variation in strain sensitivity with diffraction plane).

The layout and labelling of the strain mapping locations are depicted in Fig. 3. Four depth profiles (P1, P2, P3, and P4) were produced by stepping the sample in the x direction, and taking a diffraction pattern at each (50 μm) step. The layout of these depth profiles was similar for both the as-sprayed and heat-treated sections. To obtain a higher depth resolution, a depth profile with smaller steps (10 μm) was also performed in the center of the gap in the heat-treated section (P5), using the high depth resolution slit settings.

Diffraction patterns were collected from diffraction standard cerium dioxide (held in a 100 μm diameter capillary) between the strain mapping runs. The capillary was located at the same position along the beam as the sample (determined optically with an accuracy of 100 μm). These diffraction patterns enabled the calibration of the beam center, sample–detector distance, and detector tilt with respect to the beam.

III. Results

(1) Secondary Electron Microscopy

All the images shown were taken with a backscattered detector and 20 keV electrons, unless otherwise specified in the figure caption. The backscattered electron images of the polished surfaces of the as-sprayed and heat-treated sections are shown in

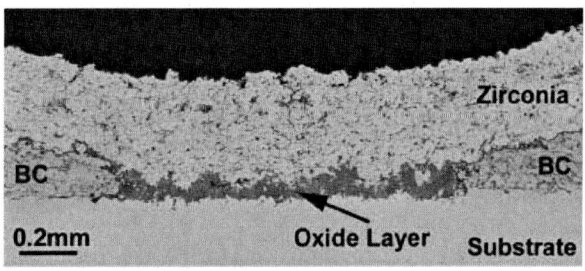

Fig. 5. A cross-section of a heat-treated TBC showing the gap in the bond coat end-on. BC, bond coat.

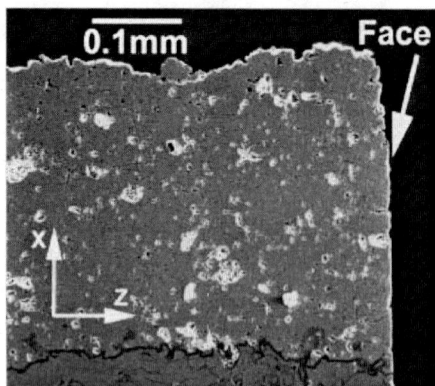

Fig. 6. A cross-section of a section for the X-ray strain measurements. Any cracking because of cutting is limited to 10 μm of the face. The image was taken in an SEM with 25 keV secondary electrons to accentuate cracking.

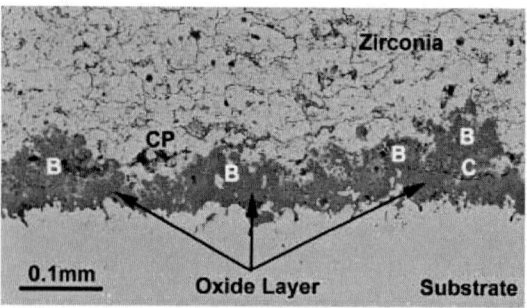

Fig. 7. A cross-section of the heat-treated sample viewed at the center of the gap in the bond coat (P5 in Fig. 3). Some cracking can be identified in the oxide layer (C). There is also a flaw that extends from the oxide layer and into the zirconia (CP); whether this is a crack from which more material has been pulled during polishing or purely "pull-out" during polishing is not clear. The label B indicates where the oxide has locally grown into the zirconia to form "balloons" of oxide.

Figs. 4 and 5, with the x direction vertical and the y direction horizontal as in Fig. 3. In this view, the gaps in the bond coat run perpendicular to the page in the z direction. Similarly, this is also the direction in which the X-ray beam enters the specimens.

Figure 6 shows the cross-section from a different sample. This view provides a perspective in which cutting damage could be seen; however, no significant damage was observed. It appears that the epoxy protected the coating from sectioning-induced spallation. Relaxation because of cutting damage is limited in the z direction to 10 μm from the surface. It is expected that polishing damage is also limited to a shallow region and does not affect the bulk of the diffracting material.

X-ray spectroscopy (EDS) showed that oxidation of the bond coat produced alumina and mixed (Ni, Cr, Co, and Al) oxides, while the bulk of the oxide formed on the Hastelloy-X substrate was chromium oxide. The backscattered technique shows these oxides as dark regions because of their low electron density, with alumina being the darkest. The as-sprayed TBC (Fig. 4) shows little oxidation of the bond coat or in the gap. The heat-treated TBCs (Fig. 5) exhibited some oxidation on the bond coat surface, and some ribbons of oxidation within its bulk; however, a much larger thickness of oxide had formed within the gap. Some cracking was also seen within this large body of oxide. These cracks ran parallel to the substrate surface, and rose at the edge with the bond coat. An example of this cracking can be seen in Fig. 7(C). Whether this cracking was formed during heat treatment or as a result of polishing is unknown.

Figures 7, 8, and 9 show the cross-section of the heat-treated sample viewed at the positions corresponding to strain depth profiles P5 (and part of P3), P2, and P1, respectively (Fig. 3). The two cross-sections of the X-ray beam (medium and high depth resolution) are shown superimposed on the micrograph of Fig. 8 for comparison with the typical microstructural features.

(2) Analysis of Diffraction Data

Diffraction lines at greater Bragg angles are generally more sensitive to strain. To measure strain in the zirconia layer, the diffraction line corresponding to the (213) atomic planes of tetragonal zirconia (JCPDS[‡] 42-1164) was used, because it was the strongest and most isolated line available at large Bragg angles. The actual d-spacings were slightly different from the standard pattern because the zirconia in the coating is doped with yttrium. To measure the strain in the layer of buried oxide, the diffraction line corresponding to the (104) atomic planes of chromium oxide (JCPDS 38-1479[‡]) was used. The large grain size in the substrate and bond coat produced spotty diffraction rings unsuitable for strain measurement. A substantial increase in the X-ray beam cross-section, and hence degradation of the

spatial resolution, was required to produce continuous rings from the substrate and bond coat.

Strain is the change in length divided by the strain-free length. Thus, the relationship between strain, ε_ψ, at a particular azimuthal angle, ψ (Fig. 1), and the change in d-spacings is

$$\varepsilon_\psi = (d_\psi - d_0)/d_0 \qquad (2)$$

where for strain measurements, d_0 is the d-spacing of the selected lattice planes when the material is free of strain, and d_ψ is the d-spacing of the selected lattice planes, in the strained material, at the azimuthal angle ψ (Fig. 1).

The program FIT2d[22] was used to integrate the ring-like diffraction patterns (right-hand side of Fig. 1), transforming them to linear plots of azimuthal angle, ψ, versus radius, r_ψ.[17] The experimental geometry (Fig. 1) combined with the Bragg equation enabled d_ψ to be determined from the radius versus azimuthal data:

$$d_\psi = \lambda/\{2\sin[0.5\arctan(r_\psi/L)]\} \qquad (3)$$

where λ is the X-ray wavelength, r_ψ is the radius of the diffraction line centroid at a given azimuthal angle, and L is the detector to sample distance.

The centroids of the diffraction "lines" in the linear plots were then found by a least-squares routine for each azimuthal position.[18] In common with many of the strong lines from the

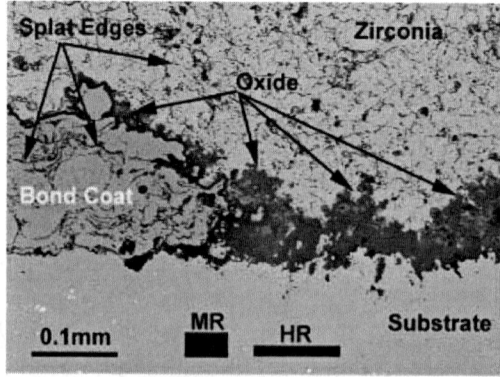

Fig. 8. A cross-section of the heat-treated sample viewed at the edge of the gap in the bond coat (P2 in Fig. 3). The thicker oxide with the medium shade of gray is predominantly chromia and the thinner black line that defines the surface of the bond coat is predominantly alumina. The block denoted MR indicates the size of the X-ray beam cross-section for medium depth resolution, and similarly the block denoted HR indicates the size of the X-ray beam cross-section used for high depth resolution.

[‡]Joint Committee on Powder Diffraction Standards (JCPDS), Swarthmore, PA (now International Centre for Diffraction Data (ICDD), Newtown Square, PA).

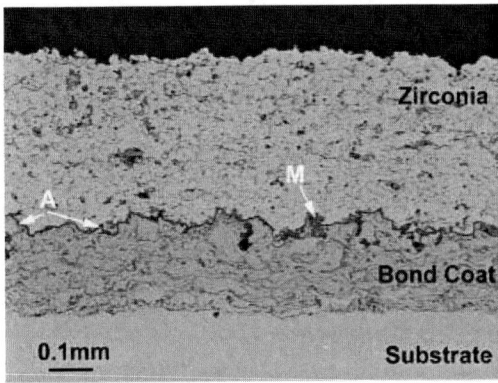

Fig. 9. A cross-section of the heat-treated sample viewed 1.5 mm away from the edge of the gap in the bond coat (P1 in Fig. 3). A thin (3 μm) band of alumina (A) is seen covering the top surface of the bond coat; it appears black in this image. A few patches of mixed oxide (M) are also observed, appearing as dark gray. The metallic components of the mixed oxide are Cr, Ni, Co, or Al.

tetragonal zirconia, the (213) line was part of a doublet. The least-squares routine used fitted the two peaks in each doublet, providing the centroids for each. The largest d-spacing (0.1178 nm) of the doublet was used. Similarly, Eq. (3) was used to determine d_0 from the centroid of the radius for strain-free material as described below. The values of d_0 and d_ψ were then used in Eq. (2) to find the strain, ε_ψ, as a function of ψ.

Figure 10 relates to the heat-treated sample and shows d_ψ as a function of azimuthal angle at two depths in depth profile P3 (Fig. 3). The data are grouped 1° bins and were smoothed using a 10-point moving average. The d_ψ values of the 0.35 mm data were offset by 0.0002 nm for clarity. The magnitude of the strain at the substrate to zirconia interface (0.35 mm) is much larger than the strain at the surface of the zirconia (0 mm).

Equation (2) is for the strain at a particular azimuthal angle, ε_ψ. Provided the strain distribution was two dimensional, then ε_ψ could be resolved into the strains in both the x and y directions (ε_x and ε_y)—as well as an additional shear component (ε_{xy}) if the principal stresses were not parallel to the x and y directions. This resolution was accomplished by fitting the ε_ψ versus ψ data to the function

$$\varepsilon_\psi = \varepsilon_x \sin^2 \psi + \varepsilon_y \cos^2 \psi + \varepsilon_{xy} \sin 2\psi \quad (4)$$

This relationship was derived from the more general three-dimensional case.[23] The depth profiles of ε_x, ε_y, and ε_{xy} within the zirconia layer are shown in Figs. 11 and 12. Figure 13 shows ε_x

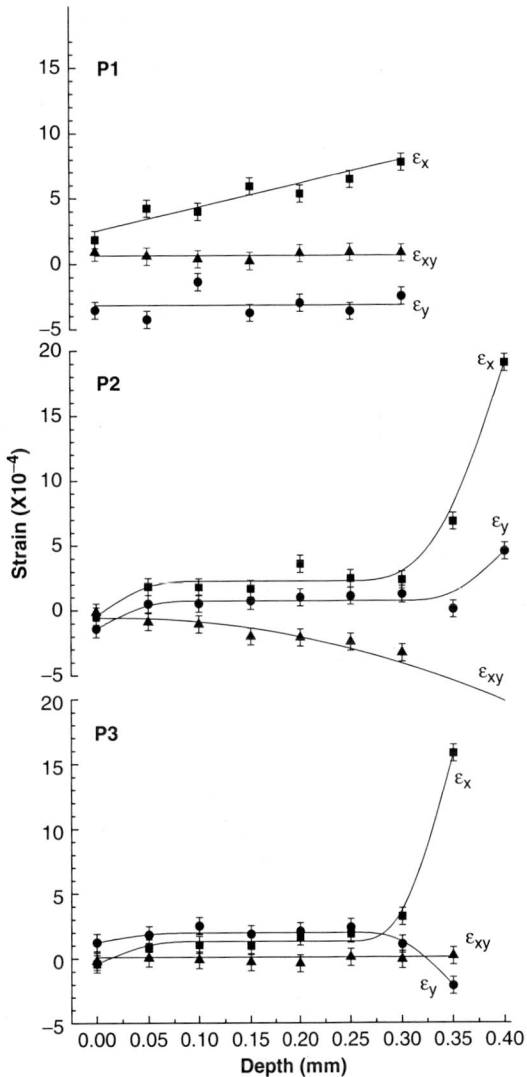

Fig. 11. Depth profile of strain in the zirconia for the heat-treated sample at P1, P2, and P3 in Fig. 3.

and ε_y for the zirconia and the chromia layers at the center of the bond coat gap in the heat-treated sample.

As the normals to the diffracting planes are tilted away from the x–y plane by the Bragg angle (θ), then the direction of the

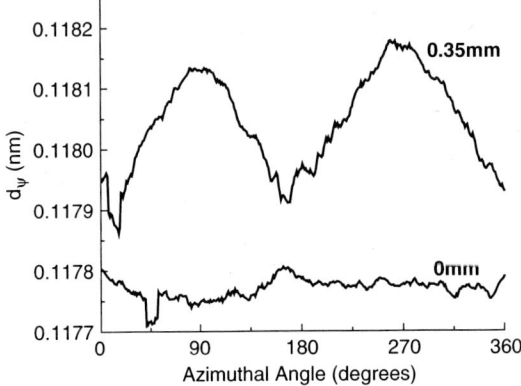

Fig. 10. d-spacing, d_ψ, as a function of azimuthal angle, ψ, at the top surface of the zirconia (0 mm) and at the substrate to zirconia interface (0.35 mm). The measurements were made on the heat-treated sample at the center of the gap in the bond coat (P3 in Fig. 3).

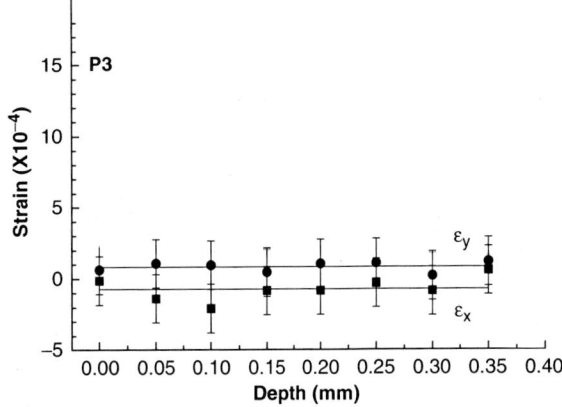

Fig. 12. Depth profile of strain in the zirconia for the as-sprayed sample at P3 in Fig. 3.

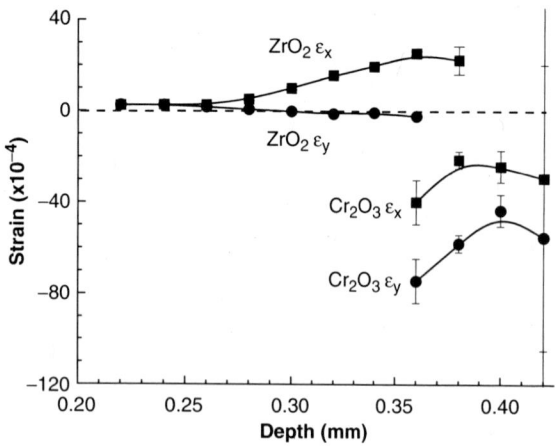

Fig. 13. A depth profile of strain measured with high depth resolution at the interface between the zirconia and the substrate oxide taken from the heat-treated sample at P5 in Fig. 3. All the shear values for the zirconia were close to zero.

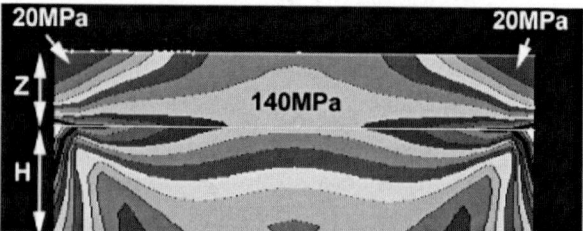

Fig. 14. The 2-D von Mises stress distribution from a finite element calculation of the cooling to 0° from 1000°C of a 2 mm wide section of 0.3 mm thick plasma-sprayed zirconia (Z) on a 2.2 mm thick Hastelloy-X substrate (H).

measured strains are not actually in the x–y plane either but at an angle θ to it. However, as the Bragg angles used here are small (3.75°), the strains are a good approximation to those in the x–y plane. The error produced by this approximation can be estimated from the general version of Eq. (4).[23] Errors could affect the $\sin^2 \psi$ and the $\sin 2\psi$ terms, but only if the shear strains ε_{zx} and ε_{zy} were comparable with ε_x and ε_{xy}; however, provided the sample alignment placed the beam perpendicular to the face of the cross-section, then ε_{zx} and ε_{zy} were small. The largest error was in the $\sin^2 \psi$ term that modifies the derived ε_x by $0.13\varepsilon_{zx}$ for the Bragg angles used. Therefore, the measured strains were treated as if they were in the x–y plane.

Two methods to determine the reference radius corresponding to a strain-free state were attempted. Neither proved entirely satisfactory. In the first method, diffraction patterns were collected from freestanding pieces of the zirconia chipped from the off-cuts of the discs. The d-spacings varied from the different pieces varied significantly enough to bias the smaller strains observed. The variation was probably because of slight spatial variations in the concentration of yttria in the zirconia. To reduce this variation, strain-free d-spacings closer to the measurement locations were needed. The second method, described below, partially achieved this aim and was therefore used in this work.

At free surfaces, there is no normal stress, therefore ε_x should be just the sum of the Poisson contraction (or expansion) due to ε_y and ε_z. Also, away from the gap in the bond coat, the sample appears symmetric in y and z and hence $\varepsilon_y \sim \varepsilon_z$. Combining these two statements algebraically produces

$$\varepsilon_x = -2\upsilon/(1-\upsilon)\varepsilon_y \sim -0.857\,\varepsilon_y \qquad (5)$$

where a Poisson's ratio, υ, of 0.3 has been used. Thus, for small surface strains, the d-spacing for zero strain lay approximately half-way (1.857 $\sim$ 2) between the d-spacings in the x and y directions. Thus, the strain-free d-spacing, d_0, was approximately equal to the mean of d over the whole 360° of ψ.

For the heat-treated samples, d_0 was obtained from the diffraction pattern at the surface position of P1 (Fig. 3). To produce d_0 for the as-sprayed samples, the d-spacing for each x and y position was averaged over ψ and then these were also averaged with those from the surface regions excluded. The surface values were removed from the mean because both ε_y and ε_x rose by 3.5×10^{-4} at P1 and 1.45×10^{-4} at P4. These two rises were likely to be artifacts caused by either part of the illuminated volume extending above the zirconia surface, or by local surface contamination. Partial illumination of the surface can produce false shifts in the diffraction rings, and local contamination can produce a local change in d_0.

The statistical scatter in d_ψ versus ψ data only produced a small error in ε_x, ε_y, and ε_{xy} (mean absolute deviation/$\sqrt{\text{number}}$ of data points $= 1.5 \times 10^{-5}$). However, the error bars shown in Figs. 11 and 12 are the standard deviation in d_0 calculated from all the surface positions on a sample divided by the mean d_0 (7×10^{-5} heat-treated, 1.7×10^{-4} as sprayed). Shifts of this degree were likely to be because of variation in d_0, especially if ε_x and ε_y shifted in the same sense. The error bars are similar to the 1σ error in strain obtained with conventional diffraction (10^{-4}).[24] For some data points in the interface region (Fig. 13), the statistical scatter became the dominant error because only small amounts of the diffracting material were illuminated.

Although cutting of the specimens was shown (see Fig. 6) to cause no identifiable changes in the z direction, with the exception of some shallow cracking at the cut surfaces, the cutting will allow the top edge to relax some of the stress it contains. A finite element simulation of a 2 mm wide section on cooling from a relaxed state at 1000°C was performed using the material properties from Koolloos[9] and CINTAS[25]; the von Mises stresses generated are shown in Fig. 14. The substrate material replaced the bond coat in this calculation because the material properties of the substrate were similar but known more accurately. Young's modulus used for the zirconia layer was 27.6 GPa[9]; thus, the 140 MPa shown corresponds to a strain of 5×10^{-3}. Without pressure from surrounding material, the zirconia in the top edges was not compressed as much as the zirconia next to the substrate or in the center. However, the average cooling stress was still predicted to be significant throughout the bulk of the zirconia layer, and substantial (>100 MPa) for more than 50% of the surface. Therefore, while the use of a limited length of the sample in the z direction is likely to modify the strain distribution from that of the full disc, it is not expected to completely remove the strain. A gradual tapering of strain toward the surface (as found for ε_x in Fig. 11, P1) was the expected consequence of the limited length in the z direction. It cannot account for the absence of cooling strains throughout the bulk of the zirconia layer (as found in Fig. 11, P3). It was therefore assumed that the greater 2.5 mm thickness of the samples was large enough to preserve the majority of the strains in the bulk of the zirconia layer.

The oxide in the center of the gap on the substrate surface in the heat-treated specimens was identified as hexagonal chromium oxide (Cr_2O_3 JCPDF file number 38-1479). A strain-free oxide standard was obtained by collecting oxide spalled from pieces of heated (1150°C 20 h) substrate material. However, the laboratory-based diffractometer system available was not sufficiently accurate to obtain d_0, as the errors in the derived lattice parameters were comparable with the strain amplitudes observed in the coating. The lattice parameters at the lower end of this error range agreed with the lattice parameters in the JCPDF file. Thus, the JCPDF values were used to calculate the d-spacings for the strain-free standard used here.

(3) Depth Profiles of Strain

The layout and labelling of the depth profiles are depicted in Fig. 3. Figures 11–13 show the corresponding depth profiles for

Table II. Strain Amplitude $\Delta\varepsilon$ in the Topcoat at a Depth of 0.05 mm from the Surface	
Sample and position	$\Delta\varepsilon = \varepsilon_x - \varepsilon_y \ (\times 10^{-5})$
G position P3	9 ± 1
G position P2	16 ± 1
G position P4	10 ± 1
G position P1 (over the bond coat)	66 ± 1
W, over the bond coat	62 ± 1

Table III. Thermal Expansion Coefficients (TEC) of the Materials Within Thermal Barrier Coatings		
Material	TEC $(\times 10^{-6}/°C)$	References
PS zirconia	9 (25°C)	Hillery et al.[26]
	9 (1073°C)	
NiCoCrAlY bond coat	11 (25°C)	Hillery et al.[26]
(low pressure	18 (1000°C)	
plasma sprayed)	14.5 (average)	
Substrate material	14.0 (25°C)	25
(Hastelloy-X)	16.6 (1000°C)	
Chromia (Cr_2O_3 hexagonal)	7.5	Lackey et al.[27]
Alumina (Al_2O_3 hexagonal)	7.2–8.6	Lackey et al.[27]

strain in the zirconia layer, with zero depth corresponding to the top of the zirconia layer in Fig. 3. It is important to note that because of the rod-like shape of the illuminated volume, the strain values are averaged in the z direction over the 2.5 mm thickness of the sections. Also, the depth of the zirconia varied slightly from P1 to P4, and thus the depth profiles in Fig. 11 end at slightly different depths.

The depth profiles of the heat-treated samples can be divided into three zones: the surface (0–0.05 mm); the bulk (0.05–0.2 mm or 0.25 mm); and the interface region (0.2 or 0.25–0.35 or 0.4 mm).

From a comparison of the strain depth profiles of the heat-treated sample (Fig. 11), it can be seen that the strain in the bulk of the zirconia layer was larger when the zirconia was over the bond coat (P1). To confirm this trend, the strain data from a similar sample (W) but without the gap were examined. It had been coated at the same time as the gap sample (G) and given the same heat treatment. A comparison of the strains is shown in Table II. To remove the confusion of the slight d_0 variation from sample to sample and from location to location, the strain amplitudes ($\varepsilon_x - \varepsilon_y$) at a depth of 0.05 mm are listed. Thus, the error because of the scatter in d_0 does not apply here and the appropriate error in the table values is the statistical error when fitting to the centroid data (mean scatter/$\sqrt{360}$). A depth of 0.05 mm was chosen for the comparison, because many of the depth plots showed that the strain in the bulk was approximately constant from 0.05 to 0.2 mm but the strain at the surface was sometimes different. The values of strain amplitude for sample W confirmed that the strain in the bulk of the zirconia layer for heat-treated samples was greater over the bond coat than over the substrate material alone.

IV. Discussion

The discussion of the strain data is divided into four cases, and is then followed by a general discussion.

(1) As Sprayed (Fig. 12)

It was expected that the strain in the as-sprayed samples would be low and dominated by a biaxial in-plane tensile strain generated during deposition. This strain was expected to originate when the splats solidify and attach to the substrate, and then subsequently cool. This cooling was expected to generate tensile strains within the splats as they tried to contract with respect to the underlying material. Furthermore, the direction of curvature of long thin strips after coating suggested an in-plane tensile strain in the zirconia layer. The strain depth profiles were compatible with these expectations. The center of the bond coat gap (P3) is a typical profile and is shown in Fig. 12; here, the average strains are $\varepsilon_y = 8 \pm 3 \times 10^{-5}$ (a tensile in-plane strain) and $\varepsilon_x = -7 \pm 8 \times 10^{-5}$ (Poisson's contraction given the biaxial tensile strains). It should be noted that the symmetry of these results (ε_y tension and ε_x compression) is highly dependent on the choice of d_0. Also, the strains are only slightly different from each other and zero.

(2) Heat-Treated Over Bond Coat (P1 in Fig. 11)

It was expected that during heating, the metallic bond coat and substrate would plastically deform to enable the relaxation of

any residual stresses between it and the zirconia layer. Quenching was then expected to generate compressive strains within the zirconia layer as the substrate and bond coat both underwent a greater contraction—a cooling strain. The cooling rate during the water quench was expected to be too rapid to allow full relaxation of the compressive strains by plastic deformation. The thermal expansion coefficients of the materials in a TBC, including the chromia and alumina in the oxidation layers, are given in Table III.

With no plastic deformation on cooling through a temperature drop of ΔT, the upper limit on the cooling strain in the zirconia was expected to be

$$\varepsilon_y = (\alpha_{zirconia} - \alpha_{bond\ coat\ average})\Delta T = -5.5 \times 10^{-3} \quad (6)$$

Plot P1 in Fig. 11 shows that the in-plane strain ($\varepsilon_y = -3.1 \pm 0.7 \times 10^{-4}$) of the zirconia over the bond coat and after heat treatment was compressive, in agreement with Eq. (5); however, this strain is an order of magnitude lower than the upper limit. With temperature, the bulk of such a strain can be relaxed in 25 s with some bond coat formulations.[28] However, 90% of the cooling occurred in less than 10 s (boiling of the quench water stops in less than 5 s), and so a large fraction of the -5.5×10^{-3} cooling strain was expected to remain. A more plausible explanation is that bond coat oxidation may have caused the lateral expansion of the bond coat and the overlying zirconia during heating. Thus, the zirconia may have had an in-plane tensile strain before cooling.

Cermet layers or porous bond coats show this lateral expansion effect dramatically by bending thin substrates during heating and causing vertical cracking in the zirconia layer.[29,30] The curvature of the bending indicated strains of the order of 10^{-2}. For the samples in the present study, the amount of bond coat oxidation was small in comparison; however, the tensile strains required to counteract the cooling strain ($\sim 10^{-3}$) were also much smaller and hence oxidation-induced tension may still have reduced the cooling-induced compression.

(3) Heat-Treated Over Gap in Bond Coat (P3 in Figs. 11 and 13)

The strain in the zirconia layer reached high levels ($\varepsilon_x = 0.0024$, Fig. 13); however, this highly strained zirconia was located close to the interface amid substantial (50 μm thick) and uneven oxidation (Fig. 7). Away from the interface in the bulk of the zirconia layer, the strain was actually less than over the bond coat. Using the lines drawn through the data points in Fig. 11, at a depth of 0.1 mm, $\varepsilon_x - \varepsilon_y$ was -7.1×10^{-5} over the gap (P3) and 7.4×10^{-4} over the bond coat (P1). Also, over the gap, ε_y was more tensile than ε_x. It is possible and likely that the appropriate d_0 for this region (P3, 0.05–0.25 mm deep) would place the lines for ε_y and ε_x almost symmetrically about, and almost on, the zero strain line. It is therefore likely that the bulk of the zirconia layer above the gap was strain free.

The differences in the strains observed in the bulk of the zirconia layer above the gap (P3 in Fig. 11) and above the bond coat (P1 in Fig. 11) may also be explained by the lateral expansion because of oxidation. Over the gap, there is a much greater amount of oxidation (Fig. 7 compared with Fig. 9) and hence there will be a greater amount of lateral expansion. Therefore, the contraction of the substrate upon cooling will not be sufficient to produce an in-plane compression in the bulk of the zirconia layer at P3.

The strains at the interface between the zirconia and the oxide layer are shown in Fig. 13 with high depth resolution. From about 0.28 mm below the surface of the zirconia, the vertical strain, ε_x, rose, becoming a large tensile strain of 0.0023 at a depth of 0.38 mm. Here, the presence of both zirconia and chromium oxide diffraction rings indicated that the zirconia interpenetrated the layer of uneven substrate chromium oxide, as shown with microscopy in Fig. 7.

The oxide (labelled Cr_2O_3 in Fig. 13) appears to be in hydrostatic compression as both ε_x and ε_y are negative. This is plausible, as the oxide will occupy a larger volume than the metal it consumed and the interface region is confined both laterally by the bond coat, and vertically by the bridge of zirconia. If the large growth of oxide were not confined to a gap, then it would still be confined laterally by its substrate; however, it would be generally free to expand upward as then the surrounding zirconia would also be raised by the growth of a similar amount of oxide. In this situation, the zero strain line would be expected to cut between the ε_x and ε_y lines. An error in d_0 may have also made both ε_x and ε_y negative.

A schematic diagram illustrating how the uneven growth of an oxide layer will place the interpenetrating zirconia into tension vertically (ε_x) is shown in Fig. 15. As oxidation occurs at a greater rate in some areas, balloons of oxide (B) push upward into the zirconia, compressing (C) the zirconia directly above them and stretching (T) the zirconia protrusions beside them. This stretching will continue above the height of the oxide balloons as the compressed zirconia will also stretch the neighboring zirconia. In this way, columns of zirconia about twice the height of the dips in the oxide will be placed in tension. From Figs. 7 and 8, these columns were 0.09 mm high in the sample studied. This is compatible with the depth over which ε_x was strongly tensile (0.1 mm).

This explanation is appropriate for the zirconia between the oxide balloons; however, the length of the diffracting volume in the z direction was 2.5 mm and would therefore have averaged over the compressed and tensile regions when the beam was

above the peaks of the oxide balloons (0.35 mm deep). The average strain should therefore have fallen to zero above 0.35 mm. A plausible extension to the mechanism is that the center of the gap was itself a region with a lower oxidation rate compared with the two adjacent sides. Thus, as the sides rose, pushed by the faster growing oxide, the center was placed in tension. A higher oxidation rate at the sides may have been caused by the extra surface area introduced by the faint spray of bond coat that penetrated a small way behind the masking, or by the extra surface area introduced by the edge of the bond coat.

It is difficult to attribute the cracking shown on a cross-section to intrinsic failure because it may have resulted from polishing damage. Thus, features like CP or C in Fig. 7 cannot be relied upon. However, it is clear that the large tensile out-of-plane strains generated at the bottom of the zirconia (ε_x in P3 of Fig. 11 and in Fig. 13) would have weakened the bonding of the zirconia layer to the substrate. The strain measurements in this paper therefore confirm many of the proposed failure mechanisms for TBCs based on localized or uneven swelling of the material underlying the zirconia.[8,11]

(4) The Edges of the Gap (P2 in Fig. 11)

At the edges of the gaps, the strains in the bulk of the zirconia layers were low for both the as-sprayed and heat-treated samples. However, the heat-treated edges did show large strains below 0.25 mm.

The similar amount of oxidation shown in Figs. 7 and 8 leads to the expectation that for the heat-treated sample, the edges would behave similar to the center of the gap. That is, the compressive in-plane cooling strains in the bulk of the zirconia were counteracted by the expansion of the oxide, and the uneven oxidation producd large strains in the interface region. The strain plot of P2 in Fig. 11 shows these basic features.

The other major feature of the heat-treated edge plots is the rise in magnitude of the shear component of the strains, ε_{xy}. This is a consequence of the geometry. At the edges of the gap, the mean surface was not parallel to the y direction, or the x direction but lay in between. The in-plane compression of the oxide layer and the push of the oxide balloons were also not along either axis. The principal strain directions were thus not close to x or y and consequently, there were large shear components. A rotation of the co-ordinate system could be made to eliminate them. As the edge at P4 faced in the opposite y direction, its ε_{xy} rose with increasing depth. The shear components for the as-sprayed case were close to zero for the whole range of depths. The shear strain is usually a fraction of ε_x and ε_y. As ε_x and ε_y were low, the shear components were effectively zero.

(5) General Discussion

In general, localized swelling, as shown schematically in Fig. 15, could be caused by a local high rate of oxidation. The oxidation could be of the bond coat surface, the internal porosity of the bond coat, or, in the case of a gap in the bond coat, of the substrate. Substrate materials in gas turbine engines generally have poorer oxidation resistance than a bond coat.

The comparison of cases IV 3 and IV 2 indicates that an in-plane compressive strain was generated in the zirconia layer on rapid cooling, but this strain was reduced and eventually replaced by the in-plane expansion, driven by oxidation of the underlying metal. An implication is that a greater fraction of the full compressive cooling strain (5×10^{-3}) could be produced when less oxidation of the bond coat occurs—provided sufficient time is spent at high temperatures for any deposition strains to be annealed out. Less oxidation will occur for shorter heat treatments and when bond coats with greater oxidation resistance are used. Thus, thermal cycling may initially generate large compressive strains but these will reduce as the bond coat oxidizes and laterally expands.

Delamination of the zirconia layer may have two mechanisms:

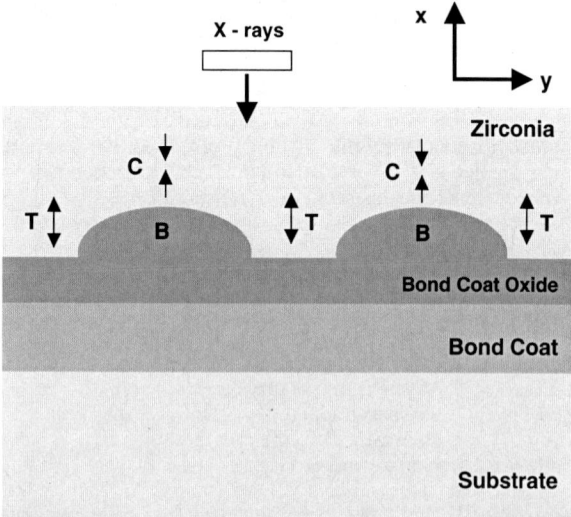

Fig. 15. A schematic diagram illustrating how balloons of oxidation (B) can cause the deepest parts of the zirconia layer (lowest on page) to be placed in tension (T) in the x direction. In the layer just above the balloons, the zirconia is alternately in tension, between the balloons, and compression (C), directly above the balloons. The X-ray beam is also shown in cross-section scanning along the x-axis.

1. Delamination driven by the compressive strains in the zirconia layer occurring after a low number of thermal cycles.

2. Delamination caused by the weakening of the zirconia to bond coat interface by uneven oxidation of the bond coat, and driven by the compressive strains in the oxide layer formed after a high number of thermal cycles or a long time at high temperature.

The presence of both low and high cycle mechanism for delamination would help to explain the observed large spread in TBC lifetimes.[3,31] Although the interface weakening mechanism may not directly drive delamination, the zirconia would delaminate more readily when subject to accidental loads.

The sensitivity of the strain measurement could not be fully utilized because of the variation in strain-free d-spacing (d_0) across the sample. An improved methodology might consist of locating the regions scanned to within 0.1 mm after the strain scanning; then cutting small (0.1 mm $\times$ 0.1 mm $\times$ 0.1 mm) volumes from these small regions; and then obtaining d_0 from these free standing, and therefore strain free, volumes with the same experimental setup.

V. Conclusions

The use of high-energy X-rays in transmission has enabled the mapping of residual strains within the zirconia layer of TBCs in two orthogonal directions. A depth resolution of 20 μm has been achieved. The technique required the sample to be cross-sectioned into thick (>2 mm) slices and is therefore destructive. Furthermore, the use of epoxy resin to support the coating during cross-sectioning was necessary to prevent spallation of the coating. However, the technique still provides strain and texture information over a depth range that is normally inaccessible. The depth range covers many other thick coatings and is suitable for the study of crack initiation and propagation—especially through residual strain fields. The sensitivity of the technique may improve if small (0.1 mm) pieces of material were cut from the regions of interest, post-measurement, for use as the strain-free references.

In general, the results showed that oxidation changed the magnitude and distribution of the strains. In particular, the results showed:

1. a small in-plane tensile strain in the zirconia layer of the as-sprayed TBC;

2. a larger in-plane compressive cooling strain in the zirconia layer of the heated and quenched TBC;

3. the dilution and reversal of this compressive strain by lateral expansion driven by oxidation of the metals underlying the zirconia; and

4. the generation of out-of-plane tensile strains in the zirconia, within and just above the rough interface between the zirconia and the metal oxide.

It was proposed that the out-of-plane tensile strains were the effect of uneven oxidation, and a mechanism was described to show how this could occur. The observations of these out-of-plane tensile strains at and above the interface supports many of the proposed mechanisms for TBC failure based on localized swelling of the material underlying the zirconia.

Acknowledgments

The authors would like to thank D. Weaven, R. Hussein, and E. Pescott for sample preparation, and Dr. D. Cookson for discussions about strain measurement.

References

[1]S. M. Meier and D. K. Gupta, "The Evolution of Thermal Barrier Coatings in Gas Turbine Engine Applications," *Trans. ASME*, **116**, 250–7 (1994).

[2]A. Maricocchi, A. Bartz, and D. Wortman, "PVD TBC Experience on GE Aircraft Engines," *J. Thermal Spray Technol.*, **6**, 193–8 (1997).

[3]R. V. Hillery, *Coatings for High-Temperature Structural Materials*. Committee on Coatings for High-Temperature Structural Materials, National Research Council, National Academy Press, Washington, DC, USA, 1996.

[4]R. A. Miller, "Current Status of Thermal Barrier Coatings—An Overview," *Surf. Coat. Technol.*, **30**, 1–11 (1987).

[5]D. V. Rigney, R. Viguie, D. J. Wortman, and D. W. Skelly, "PVD Thermal Barrier Coating Applications and Process Development for Aircraft Engines," *J. Thermal Spray Technol.*, **6**, 167–75 (1997).

[6]R. A. Miller and C. E. Lowell, "NASA Tech"; Memo, 82905, 1982.

[7]A. Bennett, "Properties of Thermal Barrier Coatings," *Mater. Sci. Technol.*, **2**, 257–61 (1986).

[8]P. K. Wright and A. G. Evans, "Mechanisms Governing the Performance of Thermal Barrier Coatings," *Curr. Opin. Solid State Mater. Sci.*, **4**, 255–65 (1999).

[9]M. F. J. Koolloos, "Behaviour of Low Porosity Microcracked Thermal Barrier Coatings Under Thermal Loading;" Doctoral Thesis, Technische Universiteit, Eindhoven, 2001.

[10]C.-H. Hsueh and E. R. Fuller Jr., "Residual Stresses in Thermal Barrier Coatings: Effects of Interface Asperity Curvature/Height and Oxide Thickness," *Mater. Sci. Eng.*, **A283**, 46–55 (2000).

[11]V. K. Tolpygo and D. R. Clarke, "Damage Induced by Thermal Cycling of Thermal Barrier Coatings"; pp. 93–108 in *Proceedings of Elevated Temperature Coatings IV, Louisiana, USA*, Edited by N. B. Dahotre, J. M. Hampikian, and J. E. Morral. The Mineral, Metals and Materials Society, Warrendale, PA, USA, 2001.

[12]P. Scardi, M. Leoni, L. Bertini, and L. Bertamini, "Residual Stress in Partially-Stabilised-Zirconia TBCs: Experimental Measurement and Modelling," *Surf. Coat. Technol.*, **94/95**, 82–8 (1997).

[13]P. Scardi, M. Leoni, L. Bertini, L. Bertamini, and F. Cernuschi, "Strain Gradients in Plasma-Sprayed Zirconia Thermal Barrier Coatings," *Surf. Coat. Technol.*, **108–109**, 93–8 (1998).

[14]J. Thornton, D. Cookson, and E. Pescott, "The Measurement of Strains Within the Bulk of Aged and As-Sprayed Thermal Barrier Coatings Using Synchrotron Radiation," *Surf. Coat. Technol.*, **120–121**, 96–102 (1999).

[15]Q. Ma and D. R. Clarke, "Stress Measurement in Single-Crystal and Polycrystalline Ceramics Using Their Optical Fluorescence," *J. Am. Ceram. Soc.*, **76**, 1433 (1993).

[16]U. Lienert, C. Schulze, V. Honkima, Th. Tschentscher, S. Garbe, O. Hignette, A. Horsewell, M. Lingham, H. F. Poulsen, N. B. Thomsen, and E. Ziegler, "Focusing Optics for High-Energy X-Ray Diffraction," *J. Synchrotron Rad.*, **5**, 226–31 (1998).

[17]A. Wanner and D. C. Dunand, "Synchrotron X-Ray Study of Bulk Lattice Strains in Externally Loaded Cu-Mo Composites," *Met. Mater. Trans.*, **31A**, 2949–62 (2000).

[18]J. Almer, U. Lienert, R. L. Peng, C. Schlauer, and M. Oden, "Strain and Texture Analysis of Coatings Using High-Energy X-rays," *J. Appl. Phys.*, **94**, 697–702 (2003).

[19]P. J. Withers and P. J. Webster, "Neutron and Synchrotron X-ray Strain Scanning," *Strain*, **37**, 19–33 (2001).

[20]J. Thornton, D. Cookson, and S. Slater, "A Method for Measuring the Depth Profile of Strain Through a Coating of Intermediate (0.05 mm to 0.2 mm) Thickness"; Advanced Photon Source Activity Report (2000), Argonne National Laboratory, Argonne, IL; ANL-01/03, December 2001.

[21]S. D. Shastri, K. Fezzaa, A. Mashayekhi, W.-K. Lee, P. B. Fernandez, and P. L. Lee, "Cryogenically Cooled Bent Double-Laue Monochromator for High-Energy Undulator X-rays," *J. Synchrotron Radiat.*, **9**, 317–22 (2002).

[22]A. P. Hammersley, *European Synchrotron Research Facility (ESRF), Internal Report. ESRF98HA01T, FIT2D V9.129 Reference Manual V3.1*. ESRF, Grenoble, France, 1998.

[23]I. C. Noyan and J. B. Cohen, *Residual Stress*. Springer-Verlag Inc., New York, 1987.

[24]B. D. Cullity, *Elements of X-Ray Diffraction*, 2nd edition, p. 458 Addison-Wesley, Reading, MA, 1978.

[25]*Aerospace Structural Metals Handbook*, 1992 edition, CINDAS/Purdue University, West Lafayette, IN, USA, 1992.

[26]R. V. Hillery, B. H. Pilsner, R. L. McKnight, T. S. Cook, and M. S. Hartle, "Thermal Barrier Coating Life Prediction Model Development"; NASA-CR-180807, 1988.

[27]W. J. Lackey, D. P. Stinton, G. A. Cerny, A. C. Schaffhauser, and L. L. Fehrenbacher, "Ceramic Coatings for Advanced Heat Engines—A Review and Projection," *Adv. Ceram. Mat.*, **2**, 24–30 (1987).

[28]W. J. Brindley, "Properties of Plasma Sprayed Bond Coats"; pp. 189–202 in *Thermal Barrier Coating Workshop, Cleveland, OH, USA, 1995*, **Vol. 3312**. NASA-AQ, 1995.

[29]J. Thornton and S. Slater, "The Evolution of Thermal Barrier Coating Design"; in *Proceedings of ISABE 2003, Cleveland, OH, USA, Paper Number ISABE-2003-1138*. AIAA, Reston, VA, USA.

[30]J. Thornton, N. Ryan, and G. Stocks, "The Production of Stresses in Thermal Barrier Coating Systems by High Temperature Oxidation"; pp. 633–6 in *Proceedings of the National Thermal Spray Conference: Thermal Spray Industrial Applications, MA, USA, 1994*, Edited by C. C. Berndt, and S. Sampath. ASM International, Materials Park, OH, USA.

[31]Y. Jaslier and S. Alperine, "EB-PVD TBCs: A Comparative Evaluation of Competing Deposition Technologies"; pp. 8-1–10 in *Proceedings of AGARD SMP Meeting on "Thermal Barrier Coatings," Aalborg, Denmark, 1997, AGARD Report 823*. NATO, 1998. □

Stress Relaxation of Compression Loaded Plasma-Sprayed 7 Wt% Y₂O₃–ZrO₂ Stand-Alone Coatings

Graeme R. Dickinson, Chris Petorak, Keith Bowman, and Rodney W. Trice[†]

School of Materials Engineering, Purdue University, West Lafayette, Indiana 47907

Stand-alone plasma-sprayed tubes of 7 wt% Y_2O_3–ZrO_2 made from the same starting powder but at two different sites were subject to stress-relaxation testing in axial compression at temperatures of 25°, 1000°, 1050°, 1100°, and 1200°C and at an initial stress of 10–80 MPa. A time-dependent stress response was observed for both coatings at all temperatures. For example, a 20 MPa stress applied at 1050°C relaxed to ~3 MPa in 180 min. When the same initial stress was applied at 1200°C, the coating fully relaxed in 32 min. For all experimental conditions evaluated, an initial fast stress-relaxation regime was observed (< 10 min), followed by a slower second stress-relaxation regime at later times (> 10 min). Coatings with higher as-sprayed densities exhibited a lengthened fast relaxation regime as compared with less dense coatings. A Maxwell model was modified in order to provide an accurate fit to the experimental stress-relaxation curves. From scanning electron microscopy experiments and mechanical data, the mechanism for stress relaxation from 25°C through 1200°C, particularly during fast relaxation, was proposed to be the formation of cracks parallel with respect to the applied load. In addition to this mechanism, stress relaxation that occurred in specimens tested at 1000°C through 1200°C was proposed to be due to partial or complete closure of cracks oriented perpendicular to the applied stress.

I. Introduction

PLASMA-SPRAYED thermal barrier coatings (TBCs) are used as protective coatings in high-temperature environments such as gas-turbine engines.[1-4] TBCs typically consist of a metallic substrate coated with an ~100 μm MCrAlY (where M stands for Ni, Co, or Fe) or PtAl bondcoat, followed by an ~200 μm topcoat with a composition of 7 wt% Y_2O_3–93 wt% ZrO_2 (YSZ). Plasma-sprayed coatings of YSZ have a layered lamellar microstructure that contains a significant volume fraction of cracks and pores, differing in both orientation and morphology, and porosity as high as 20%.[5] Lenticular-shaped interlamellar pores are oriented approximately parallel to the substrate and form due to air entrapment during rapid solidification of the molten particles.[5,6] Intralamellar cracks are oriented approximately perpendicular to the substrate and form due to thermally generated tensile stresses that occur upon cooling.[5,6] A columnar grain structure also forms within each lamella upon cooling due to the temperature gradient that develops between the cooler substrate and solidifying molten droplet.[5]

Significant morphological changes such as intralamellar crack and pore closure and loss of the columnar grain structure can occur in plasma-sprayed YSZ as a result of exposure to high temperatures. Ilavsky et al.[7] have related the shapes of defects in plasma-sprayed YSZ to their respective behavior at elevated temperatures. In their study, in situ small-angle neutron scattering (SANS) was used to monitor the decrease in the void-specific surface area of both interlamellar pores and intralamellar cracks during heating to 1400°C. It was shown that a decrease in the specific surface area of intralamellar cracks was visible after short times at 800°C, below the normal operating temperature of a gas turbine engine. These cracks were observed to be closed completely at 1000°C, while the interlamellar pores began to shrink at temperatures above 1000°C.

Thermomechanical stresses also develop within these coatings during service. Due to the low thermal conductivity of the YSZ as compared with the metallic structure, surface temperatures can increase rapidly during engine start-up, resulting in a large thermal gradient in the TBC. This gradient results in thermal expansion differences through the thickness of the TBC, and due to the cooler and stiffer metallic substrate ($E \sim 200$ GPa for Ni-based superalloys), can result in an effective compressive stress in the YSZ.[8] Kokini et al.[9] have found that the compressive stress in a 1 mm thick YSZ on a steel substrate under high heat flux laser heating reached 112 MPa after only 1.1 s. At ~800°C, complex time-dependent deformation consisting of sintering and compaction causes densification of the coating; it has been proposed that the time-dependent deformation observed in these coatings is a result of stress-induced mechanical sliding, as well as a thermal- and stress-activated diffusion process.[10] Sintering of the coating is enhanced by the thermally generated compressive stress state; this stress will decrease or "relax" with coating densification. For example, the maximum compressive stress of 112 MPa previously cited[9] decreased under thermal load to a value of ~30 MPa via stress-relaxation mechanisms.

Upon cooling, the TBC system will contract. While the substrate will return to its original dimensions, the densification of the topcoat during service will cause residual tensile stress to develop in the coating, as high as 124 MPa, as reported by Kokini et al.[9] The porous ceramic microstructure will not sustain this tensile stress, resulting in crack initiation and propagation through the thickness of the YSZ[9] proportional to the amount of coating relaxation. Thus, the goal of the current work is to investigate the stress-relaxation behavior of YSZ coatings by simulating the stress state developed during engine start-up. In the present work, this is simulated using stress-relaxation testing at elevated temperatures on stand-alone tubes of YSZ loaded in compression. Despite the fact that stress relaxation has similarities to the loading conditions experienced in service, it is still important to realize that it is a simplification of the stress state actually experienced by the coating. A uniaxial compressive stress-relaxation test will produce a uniaxial compressive stress, while a coating in service will experience biaxial compressive stresses. Regardless, the stress-relaxation behavior of a coating in service has a direct impact on crack growth, coating life, and failure of that coating. The amount that the compressive stresses decrease over time due to coating relaxation will determine the amount of tensile stress a coating will experience upon cooling.[11,12] This stress-relaxation behavior will therefore dictate the

G. Scherer—contributing editor

Manuscript No. 20179. Received July 29, 2004; approved March 4, 2005.
This work was supported by the National Science Foundation through DMR-0134286 and HRD-0120794.
Based in part on the thesis submitted by G. Dickinson for the M.S. Degree in Materials Engineering, Purdue University, West Lafayette, Indiana, 2004.
The plasma arc spraying work at Site 2 was performed by the Materials Preparation Center at the Ames Laboratory, which is supported by U.S. Department of Energy, Office of Science through Iowa State University under Contract No. W-7405-ENG-82.
[†]Author to whom correspondence should be addressed. e-mail: rtrice@purdue.edu

severity and density of cracking due to tensile stresses upon cooling.

II. Experimental Procedure

(1) Sample Fabrication and Physical Testing

Cylindrical stand-alone coatings were prepared using a yttria-stabilized zirconia powder (H.C. Starck, Amperit 825.0 powder, Newton, MA) with a composition of 7 wt.% Y_2O_3–93 wt.% ZrO_2 (YSZ), consisting of a particle size of ~ 22.5 μm. Coatings were produced at two different sites; the parameters used for each are listed in Table I. For Site 1 (designated as S1), stand-alone coatings were fabricated by first spraying aluminum powder (Sulzer Metco, 54NS-1, Winterthur, Switzerland) onto a 300 mm long alumina rod. Each alumina rod was 13 mm in diameter, and was rotated at a rate of 200 rpm during plasma-spray processing, with air cooling through the center of the tube to minimize its heating. The plasma-spray gun rastered along the length of the rod, depositing a thin ~ 200 μm layer of aluminum, which was subsequently lightly sanded and cleaned in acetone. Next, an ~ 300 μm layer of YSZ was sprayed onto the aluminum-coated alumina substrate. For Site 2 (designated as S2) coatings, the YSZ was sprayed directly on 13 mm diameter copper rods.

Rods sprayed at either site were next sectioned into multiple ~ 15–20 mm long cylinders using a diamond-coated blade. The two faces of each cylinder were machined parallel using a diamond tool on a lathe. The outer diameter of each YSZ tube was measured using digital calipers with a resolution of ± 10 μm. This measurement was performed before removal from the alumina substrate due to handling issues inherent to the fragile stand-alone coatings. After machining, the samples were immersed in a weak HCl solution to dissolve either the intermediate aluminum layer (S1 coatings) or the copper substrate (S2 coatings) and released the YSZ layer. Approximately 10–12 samples were produced from each plasma-sprayed rod. The height and thickness of the coatings were measured after removal from the substrate, from which the inner diameter and cross-sectional area of each sample were calculated.

Bulk density, open, closed, and total porosity were measured from coatings made at both sites using the Archimedes Method, using 6.08 g/cm^3 as the theoretical density.[13] YSZ tubes were soaked in water at least 12 h prior to measuring suspended and saturated masses.

(2) Stress-Relaxation Testing

High-temperature compression testing was performed using a servo-hydraulic load frame (MTS 810 load frame) equipped with hydraulic collet grips, an alignment fixture (MTS 609 alignment fixture), a 100 kN force transducer, SiC pushrods, and a high-temperature furnace (Applied Test Systems, Inc., Butler, PA). Strain was measured with a high-temperature extensometer (MTS 632.70H-01) with a resolution of ± 1 μm. The design of

the extensometer was such that an alumina pushrod extended vertically through the center of the lower SiC pushrod, through a small hole in the lower SiC compression platen, through the hollow stand-alone YSZ sample, and to the surface of the upper SiC pushrod. Strain was measured as the extensometer recorded any displacement between the stationary upper compression platen and the cantilever supporting the pushrod. The distance between the upper platen and the cantilever under an applied load was governed by the material response of the stand-alone coating between the platens.

Alignment of the load frame was adjusted prior to testing the coatings to ensure that load was distributed equally around the circumference of the sample. Alignment was achieved by elastically loading a cylindrical aluminum specimen that had four strain gauges (Vishay Measurements Group, Raleigh, NC) positioned 90° from each other on the outside surface. The alignment fixture was then adjusted until identical strain readings were obtained from the strain gauges adhered to the aluminum sample.

Stress-relaxation tests were performed on the stand-alone YSZ samples at temperatures ranging from 1000° to 1200°C, and compressive stresses ranging from 10 to 80 MPa. Each sample was heated to the desired temperature at a rate of 10°C/min under no applied load. After the target temperature was reached and stabilized for 15 min, the test was initiated by monotonically increasing the stress applied to the YSZ tube at a rate of 20 N/s to a pre-determined initial stress. Once the initial stress was reached, the control system was rapidly changed from force to strain feedback, and the strain was held constant for 3 h. The material response was measured by observing the stress reduction with time. The noise in the load signal represents ± 1 MPa uncertainty. This uncertainty value, in part, was determined by hanging weights from the load cell; the uncertainty in the load cell measurement was within ± 1.5 N for loads from 0 to 90 N. Replicate testing of coatings made at Site 2 were used to establish variability in sample behavior as a function of test temperature and stress. Thermal expansion of the YSZ tubes and the load frame did not influence the material response, as no load was applied during heating or during the 15-min stabilization period. Inspection of samples after testing to stresses as high as 80 MPa shows no evidence of barreling.

The modulus values of YSZ S1 coatings tested at initial stress levels of 10 and 20 MPa at 1000°C were measured during stress-relaxation tests to evaluate the change in the elastic properties as a function of time relaxed. Modulus measurements were made by unloading YSZ tubes at specific time intervals (2, 5, 10, 60, and 180 min) during a stress-relaxation test, followed immediately by reloading to the strain level held constant throughout the test. The initial linear region of the stress-strain response was used to calculate modulus, and the compliance of the load frame was taken into account.

(3) Stress-Relaxation Modeling

A Maxwell element, consisting of an elastic spring and viscous dashpot in series, is often used to model the stress-relaxation behavior of polymeric materials.[14–17] For the case where there are multiple relaxation phenomena, individual Maxwell elements are arranged in parallel to better fit the experimental data. In general, the experimental data can be approximated by a series of Maxwell elements arranged in parallel. This is given by

$$\sigma(t) = \sigma_0 \sum_{k=1}^{n} w_k \exp\left(-\frac{t}{\tau_k}\right) \tag{1}$$

$$\sum_{k=1}^{n} w_k = 1 \tag{2}$$

where σ_0 is the initial stress, n is the number of Maxwell elements in parallel, w_k is a weighting factor that is summed to unity, and t is the time. The time constant for each Maxwell

Table I. Comparison of Spray Parameters Used to Prepare Coatings at Site 1 (S1) and Site 2 (S2)

Parameters	S1 coating†	S2 coating‡
Power (kW)	35	37
Stand-off Distance (cm)	6	10
Arc gas rate (slm)	32 (Ar)	25 (Ar)
Aux gas rate (slm)	8 (H)	21 (He)
Powder carrier gas rate (slm)	5 (Ar)	6 (Ar)
Powder feed rate (rpm)	3	1.5

†S1 coatings were made at Northwestern University and sprayed with an A-3000 Plasma Technik control system (Siegen, Germany) equipped with an F4 gun mounted on an ASEA Brown and Boveri IRB 2000 robot. The injector angle (angle between the powder injector and plasma plume) was 30°C and the offset of the injector was 8 mm.[22] ‡S2 coatings were made at Ames National Laboratory with a Praxair SG-100 gun mounted to a 2-way stage.

element, τ_k, can be further defined as:

$$\tau_k = \frac{\eta_k}{E_k} \qquad (3)$$

where η_k is the viscosity of the dashpot and E_k is the modulus of the spring for each Maxwell element.

The experimental data from the current study did not fit a model that included only a single Maxwell element (i.e., $n = 1$) because at least two relaxation regimes were observed. For this case, a modified model consisting of either 2 or 3 Maxwell elements in parallel was used to fit the data. The 2-element Maxwell model is analogous to mathematically adding two separate stress-relaxation behaviors (fast and slow relaxation) at each time increment, yielding one stress-relaxation curve. For the case where relaxation did not occur within the time frame of the experiment, three Maxwell elements were required to fit the data.

Individual τ_k and w_k were fit to a given set of experimental data using a least-squares curve fitting method by minimizing R as defined in the following equation:

$$R = \sum_{t=0}^{t=t_f} \sqrt{\left| [\sigma_{exp}(t)]^2 - [\sigma_{mod}(t)]^2 \right|} \qquad (4)$$

where $\sigma_{exp}(t)$ and $\sigma_{mod}(t)$ represent stress values obtained experimentally and from numerical modeling, respectively, from the start of the test ($t = 0$) until the end of the test ($t = t_f$). The R value has no physical meaning, but is minimized to provide the best fit of the model to the experimental data. Stress values were squared to give higher stress values greater weight in determining a proper fit. The square root of this difference for each time step is summed over the complete test time, yielding an R value. This value can be minimized by iteratively varying τ_k and w_k, and thereby providing the best fit of the model to the experimental stress-relaxation data. For samples tested at stresses greater than 20 MPa, the least-squares fit approach was applied first to the initial relaxation regime data to determine τ_1, followed by application of this approach to the entire data set. This was necessary to sufficiently weight the rapid relaxation observed during the initial regime where the number of data points recorded was greatly exceeded by those in the slow relaxation regime.

III. Results and Discussion

(1) Physical Properties of YSZ Coatings

The average stand-alone tube thickness, density, and porosity of plasma-sprayed coatings made at S1 and S2 are presented in Table II. S1 and S2 coating thicknesses ranged from an average of 360 to 551 μm, respectively. Coatings taken from the center portion of the rod were generally thicker than those taken from end positions. S2 coatings were more dense (and therefore less porous) than S1 coatings. Sample heights for both coatings were ~17 mm.

(2) Stress-Relaxation Response of Compression-Loaded YSZ Coatings

(A) Effect of Temperature and Density on Stress-Relaxation Response: Representative experimental stress responses

Table II. Average Physical Properties of the As-Sprayed Stand-Alone Coatings Made at Site 1 and Site 2

Property	S1 coating properties	S2 coating properties
Tube thickness (μm)	360 ± 41	551 ± 25
Bulk density (g/cm^3)	5.0 ± 0.3	5.5 ± 0.1
Open porosity (%)	10.3 ± 3.3	2.2 ± 1.3
Closed porosity (%)	7.0 ± 5.2	7.2 ± 2.3
Total porosity (%)	17.3 ± 5.1	9.4 ± 1.9

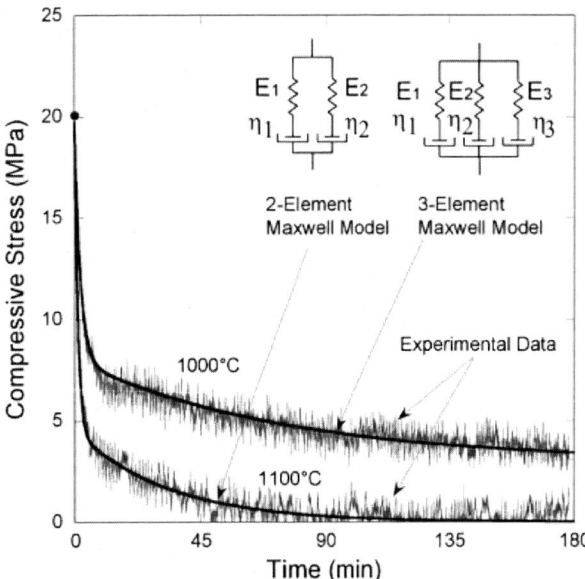

Fig. 1. Plot of experimental data for YSZ S1 coatings relaxed from 20 MPa at 1000°C and 1100°C. A 2-element Maxwell model fit the experimental results for the data obtained at 1100°C while a 3-element Maxell model fit the 1000°C data.

for as-sprayed YSZ S1 coatings heated to 1000° and 1100°C with an initial compressive stress of 20 MPa are presented in Fig. 1. There is some uncertainty in the actual stress measured, as evident by the width of the experimental load data. From our observations and experiences testing these coatings, this error is estimated to be ± 1 MPa. Complete relaxation of the initial stress at 1100°C was observed (i.e., the stress in the coating relaxed to within the measurement error of 1 MPa), while the YSZ coating tested at 1000°C did not relax fully within the 180 min test. The experimental data from each sample tested at elevated temperatures exhibited at least two stress-relaxation regimes. At early times (~0–10 min), a fast stress-relaxation regime was observed, followed by a slower relaxation regime at later times. The curve fit of stress vs. time for a 2-element Maxwell model (using Eq. (1)) is plotted in Fig. 1 along with the experimental data for a YSZ tube under an initial compressive stress of 20 MPa at 1100°C; good agreement was observed between the experimental data and the fit of the 2-element model. When complete relaxation was not observed within the time frame of the test, as shown in Fig. 1 for the experimental data at 1000°C, a 3-element Maxwell model was required to fit the experimental data.

For clarity, Fig. 2 shows only the average modeled stress-relaxation behavior of YSZ S2 tubes at an initial stress of 20 MPa for test temperatures of 25°, 1050°, 1100°, and 1200°C. Each temperature-dependent response represents the averaged modeled stress relaxation behavior for three samples at each temperature. The compressive stress axis was plotted on a log scale to further accentuate the slow and fast relaxation behavior; the lower threshold for the plot was set at 1 MPa, the resolution of the experimental results. In the current work, S2 coatings tested at 25°C relaxed ~4 MPa under the initial 20 MPa stress during the 3 h test. Relaxation of plasma-sprayed YSZ coatings at 25°C was also observed by Rejda *et al.*[18]

As the test temperature was increased, the magnitude of stress relaxed during the fast relaxation regime also increased. For example, it varied from 8.8 ± 0.3 to 13 ± 0.0 MPa at 1050° and 1200°C, respectively. The fact that the magnitude of relaxed stress increased, as compared with the room-temperature sample, suggests that elevated temperatures thermally activate relaxation mechanisms.

Figure 3 compares the stress-relaxation behavior of YSZ S1 and S2 coatings at 1050°C and 1200°C. At 1050°C, the less dense

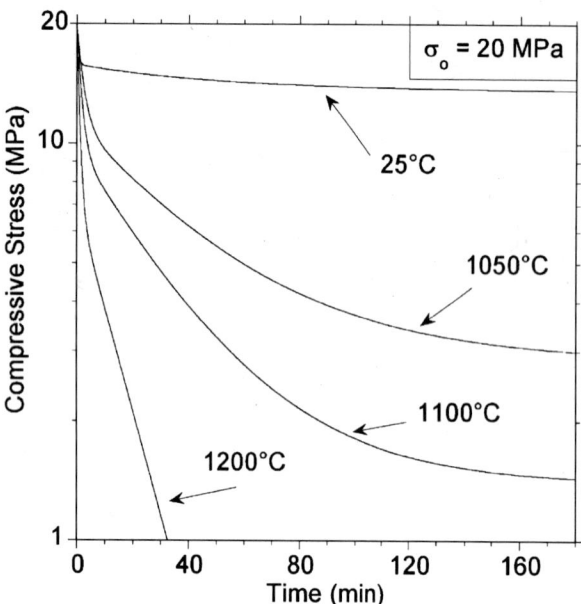

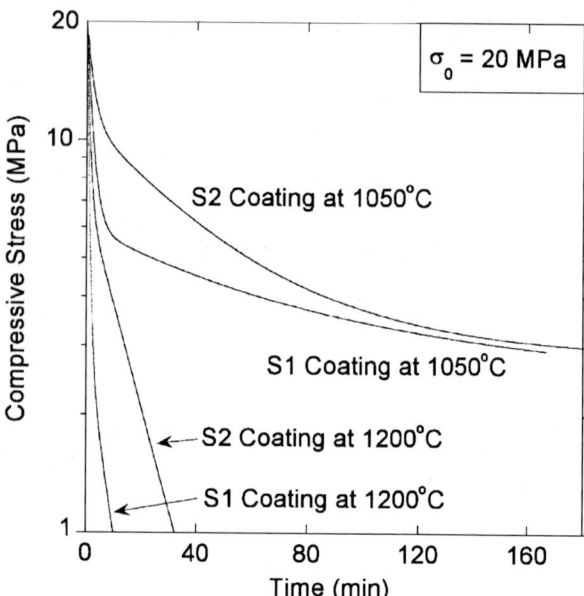

Fig. 2. Plot of the average fit of 2-element or 3-element Maxwell models as a function of test temperature for YSZ S2 coatings. A 2-element Maxwell model was used to fit data at 1200°C; a 3-element Maxwell model was used to fit data at 25°, 1050°, and 1100°C because samples tested at these temperatures did not fully relax within the 180 min test. The initial stress was 20 MPa for all samples represented in this figure.

Fig. 3. Comparison of the modeled stress-relaxation response of YSZ S1 and S2 coatings when relaxed from 20 MPa at 1050° and 1200°C. The bulk densities of YSZ S1 and S2 coatings were 5.0 ± 0.3 and 5.5 ± 0.1 g/cm^3, respectively.

YSZ S1 coating demonstrates a faster initial relaxation, followed by a slower secondary relaxation, as compared with the denser YSZ S2 coating. At 1200°C, the YSZ S1 coating was observed to relax completely in 10 min as compared with the denser YSZ S2 coating, which relaxed in 32 min. Thus, it appears that the density strongly influences the relaxation behavior of plasma-sprayed YSZ coatings.

Table III shows the fitting parameters for both YSZ S1 and S2 coatings relaxed from 20 MPa. For coatings sprayed at either site, the magnitude of the initial fast relaxation stress, given by $w_1\sigma_0$, increased as the test temperature was increased. For example, the S2 coatings relaxed 8.8 ± 0.3 and 13 ± 0 MPa when tested at 1050° and 1200°C, respectively. The magnitude of relaxation observed during slow relaxation, assumed to be $w_2\sigma_0 + w_3\sigma_0$, was less than $w_1\sigma_0$ for the S1 coatings. For YSZ S2 coatings, the magnitude of stress relaxation during slow relaxation was approximately equal to $w_1\sigma_0$ for coatings that did not relax fully during the 3 h test.

The time constants that mathematically describe the initial fast relaxation, given by τ_1, and secondary slower relaxation, given by τ_2 and τ_3, were observed to decrease with increasing temperature for each coating type. That the time constants decrease with increasing temperature shows quantitatively that stress relaxation is accelerated by elevated temperature. At each test temperature, it was also observed for each coating type that fully relaxed within the 3 h test that τ_2 was greater than τ_1 by at least one order of magnitude. At 1200°C, for example, τ_1 and τ_2

were 62 ± 8 and 1000 ± 265 s, respectively, for the YSZ S2 coatings. Thus, the secondary relaxation occurred much more slowly than the initial relaxation event. For the case of coatings that did not fully relax within the 3 h test, it was observed that $\tau_3 > \tau_2 > \tau_1$. τ_3 was observed to be several orders of magnitude greater than τ_2, an indication that full relaxation of the specimen will take an extended amount of time.

A comparison of the fitting parameters in Table III indicates a difference between the stress-relaxation response of the YSZ S1 and S2 coatings that appears to be due to density differences (see Table II). For example, YSZ S2 coatings relaxed 10.3 ± 0.6 MPa at 1100°C as compared with the YSZ S1 coating that relaxed 16 MPa. At all temperatures, the fast relaxation time constants, given by τ_1, were observed to increase for the denser YSZ S2 coatings as compared with YSZ S1 coatings. τ_2 values for the YSZ S2 coatings were of similar magnitude to τ_2 values for the YSZ S1 coatings.

(B) Effect of Initial Stress on Stress-Relaxation Response: The modeled stress-relaxation response as a function of initial stress at 1100°C for YSZ S2 coatings is presented in Fig. 4(a). Samples tested at 20, 60, and 80 MPa initial stress levels did not fully relax at 1100°C within 180 min. The fitting parameters for the data presented in Fig. 4(a) are presented in Table IV. It is noteworthy that τ_1, τ_2, and τ_3 values are essentially identical, independent of the starting stress. Figure 4(b) is an isochronal plot of the data in Fig. 4(a); the material demonstrates linear viscoelastic stress-relaxation behavior for the testing parameters analyzed.

(C) Effect of Stress Relaxation on Modulus: The evolution of the YSZ S1 coating modulus during a stress-relaxation

Table III. Fitting Parameters of As-Sprayed Coatings Relaxed from 20 MPa as a Function of Temperature

Temp. (°C)	Coating	$w_1\sigma_0$ (MPa)	τ_1 (s)	$w_2\sigma_0$ (MPa)	τ_2 (s)	$w_3\sigma_0$ (MPa)	τ_3 (s)
1000	S1	12.0	125	5.0	4500	3.0	1×10^{14}
1050	S1	14.0	123	3.0	3800	3.0	1×10^{5}
1100	S1	16.0	93	4.0	2200	0	0
1200	S1	17.0	50	3.0	549	0	0
1050	S2	8.8 ± 0.3	143 ± 3	8.3 ± 0.6	2600 ± 391	2.8 ± 0.3	$1 \times 10^{6} \pm 0$
1100	S2	10.3 ± 0.6	102 ± 6	8.3 ± 1.2	2000 ± 250	1.3 ± 0	$1 \times 10^{8} \pm 0$
1200	S2	13 ± 0	62 ± 8	7 ± 0	1000 ± 265	0	0

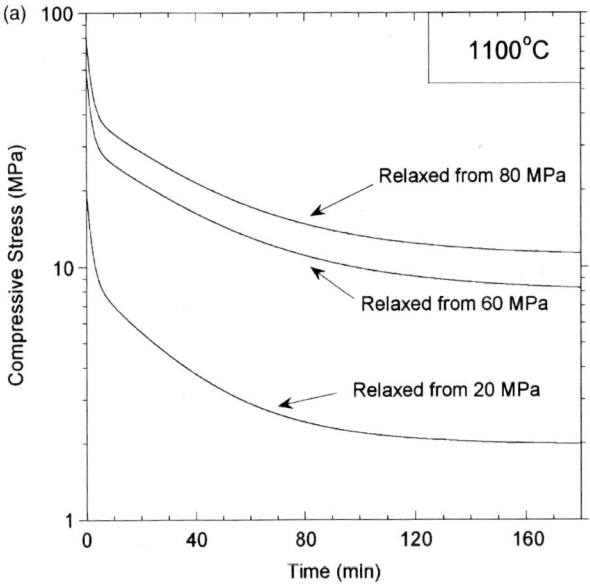

(a)

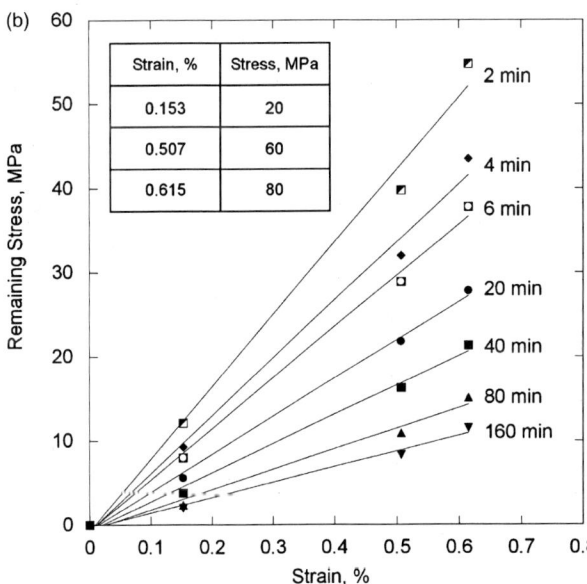

(b)

Fig. 4. (a) Plot of the fit of 3-element Maxwell models as a function of initial stress level at 1100°C for YSZ S2 coatings and (b) corresponding isochronal plot of remaining stress versus constant strain in each of the samples.

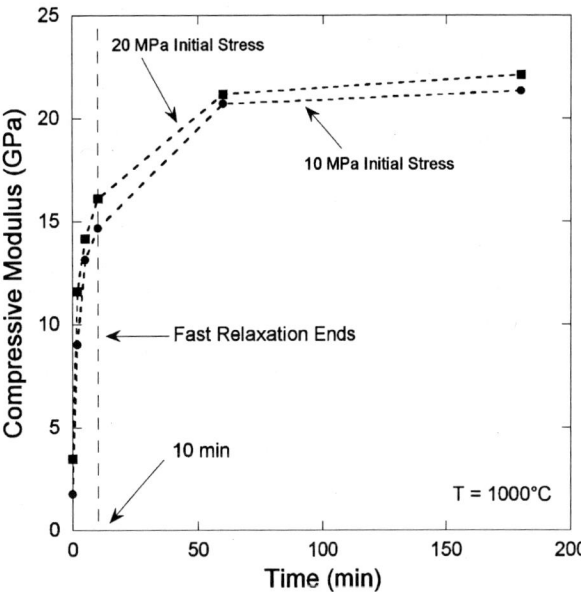

Fig. 5. Plot of modulus as a function of time at 1000°C for three initial stress levels in YSZ S1 coatings. Note that lines joining the data points have been added to guide the eye.

(3) SEM Analysis of YSZ Coatings

The top surfaces of as-sprayed, heat-treated (3 h at 1100°C), and stress-relaxed (from a 60 MPa initial stress, 3 h at 1100°C) standalone YSZ S2 coatings were analyzed using SEM. The as-sprayed coating exhibited extensive intralamellar cracking in all orientations as shown in Fig. 6(a). As indicated in the micrograph, it appears that some of the microcracks are open wider than others (i.e., coarse microcracks), while some of them appear to be much finer. Figure 6(b) shows the top surface of the YSZ coating that was heat treated for 3 h at 1100°C. Intralamellar cracking was observed in all orientations in this sample; however, there is some evidence that the smaller microcracks in all orientations have partially or completely closed, consistent with observations by prior work.[7] An SEM micrograph of a coating stress relaxed from 60 MPa at 1100°C after 3 h is presented in Fig. 6(c); in some areas, it appears that fine microcracks oriented perpendicular to the applied load have completely disappeared. However, it is very difficult to quantify any of the changes in these micrographs using SEM techniques; a technique that would take into account a larger volume (like SANS) would be particularly beneficial to delineate crack closure.

IV. Discussion

(A) Mechanism of Stress Relaxation at 25°C:
The stress-relaxation behavior at 25°C (see Fig. 2) cannot be attributed to any thermally activated mechanism within the YSZ as the temperature is too low. In a prior study, acoustic emission sensors attached to a plasma-sprayed alumina tube continued to detect the formation of microcracks after the peak stress was reached.[19] The orientation of these new cracks with respect to

test at 1000°C for initial stress levels of 10 and 20 MPa is presented in Fig. 5. Modulus values were initially < 5 GPa, but increased to values just above 20 GPa for samples relaxed from initial applied stress levels of 10 and 20 MPa. Most of the increase in modulus was observed during the fast relaxation regime. The average difference in modulus observed between the samples tested at 10 and 20 MPa was ∼1.3 GPa, showing that the change in modulus during stress relaxation was not a strong function of initial stress within the levels tested.

Table IV. Fitting Parameters of As-Sprayed Coatings Tested at 1100°C as a Function of Initial Stress

Stress (MPa)	Coating	$w_1\sigma_0$ (MPa)	τ_1 (s)	$w_2\sigma_0$ (MPa)	τ_2 (s)	$w_3\sigma_0$ (MPa)	τ_3 (s)
20	S2	10.3±0.6	102±6	8.3±1.2	2000±250	1.3±0	$1 \times 10^8 \pm 0$
60	S2	30	98	22	2450	8	1×10^8
80	S2	40	102	29	2350	11	1×10^8

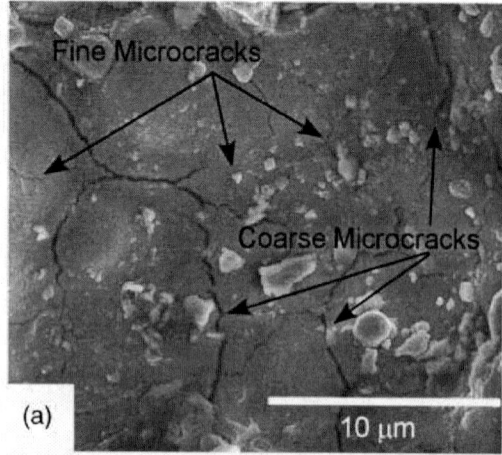

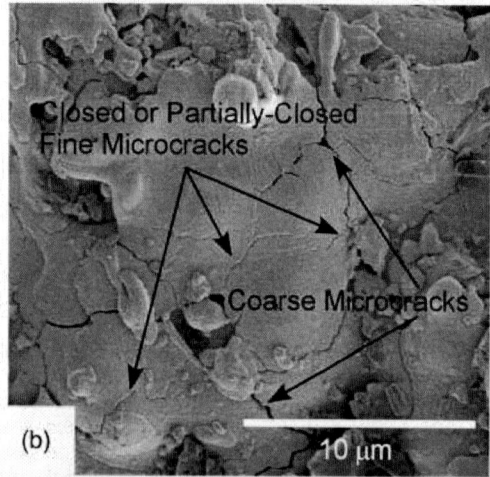

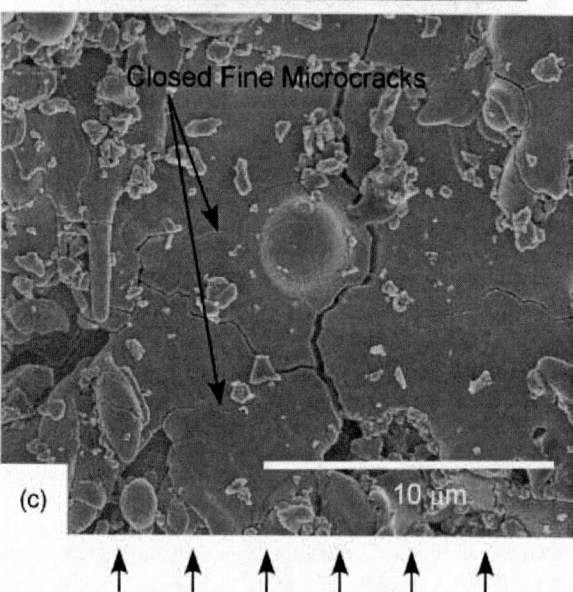

Fig. 6. SEM micrographs of the top surface of YSZ S2 coatings (a) in the as-sprayed condition, (b) after heat treatment for 3 h at 1100°C, and (c) after a 3 h stress-relaxation test (initial stress of 60 MPa) at 1100°C.

the loading direction was directly observed by Levin *et al.* [20] in plasma-sprayed YSZ, where new microcracks were observed to form parallel to the applied stress. Therefore, it is proposed that the mechanism responsible for stress relaxation at 25°C is the formation of microcracks and the concomitant reduction in strain energy stored in the sample.

(B) Mechanism of Stress Relaxation at Elevated Temperatures: The mechanism for stress relaxation at 25°C, i.e., the formation of new microcracks oriented parallel to the applied load, likely contributes at least some of the stress relaxation observed at elevated temperatures. However, the stress relaxation data presented in Figs. 1–4 clearly show that significantly more relaxation is occurring at test temperatures of 1000°C and beyond, by what can reasonably be assumed to be a thermally activated process.

It has been previously shown[19,20] that cracks oriented perpendicular to a uniaxial load in plasma-sprayed coatings will partially close at 25°C; these cracks will open up again upon unloading, as detected in tests on plasma-sprayed alumina coatings by acoustic emission sensors.[19] However, at elevated temperatures, when these cracks are partially closed by a mechanical load, crack tips and adjacent surfaces in contact would be expected to close by sintering.[21] As cracks oriented perpendicular to the applied load are preferentially closed, these would be expected to sinter first due to stress-enhanced sintering. This is also consistent with the modulus measurements presented in Fig. 5, where the increase in modulus after time at 1000°C can be attributed to a larger fraction of the load actually being carried by the YSZ coating. The increased amount of stress relaxation at temperatures equal to or greater than 1000°C can be explained by the closure and sintering of cracks. As the test temperature was increased, more complete relaxation was observed as evidenced by the reduction in τ_1 observed in Table III. It is certainly plausible that the kinetics of the sintering process would increase as temperature increased.

V. Conclusions

Uniaxial compression stress-relaxation testing was performed on stand-alone plasma-sprayed tubes of 7 wt% Y_2O_3–ZrO_2 with different densities at 1000°, 1050°, 1100°, and 1200°C, and stresses ranging from 10 through 80 MPa. The duration of the test was held constant at 3 h. For all temperatures and stresses investigated, time-dependent stress behavior was observed. All YSZ coatings demonstrated an initial fast stress-relaxation regime, typically within the first 10 min of the test, followed by a slower stress-relaxation regime at later times. A Maxwell model, generally used to model the stress-relaxation behavior of polymeric materials, was modified by adding additional Maxwell elements in parallel in order to provide an accurate fit to the experimental stress-relaxation curves. Using scanning electron microscopy and observations from the mechanical data, the mechanism for stress relaxation at all test temperatures, particularly during fast relaxation, was proposed to be the formation of cracks parallel to the applied load. In addition to this mechanism, stress relaxation at elevated temperatures was attributed to partial or complete closure of cracks oriented perpendicular to the applied stress.

References

[1]R. A. Miller, "Current Status of Thermal Barrier Coatings—An Overview," *Surf. Coat. Tech.*, **30**, 1–11 (1987).
[2]R. A. Miller, "Thermal Barrier Coatings for Aircraft Engines–History and Directions," *J. Therm. Spray Tech.*, **6** [1] 35–42 (1997).
[3]M. J. Stiger, N. M. Yanar, M. G. Topping, F. G. Pettit, and G. H. Meier, "Thermal Barrier Coatings for the 21st Century," *Z. Metallkd.*, **90** (1999).
[4]W. Beele, G. Marijnissen, and A. Van Lieshout, "The Evolution of Thermal Barrier Coatings—Status and Upcoming Solutions for Today's Key Issues," *Surf. Coat. Tech.*, **120**, 61–7 (1999).
[5]R. McPherson, "A Review of Microstructure and Properties of Plasma-Sprayed Ceramic Coatings," *Surf. Coat. Tech.*, **39/40**, 173–81 (1989).
[6]R. McPherson, "Relationship Between The Mechanism of Formation, Microstructure and Properties of Plasma-Sprayed Coatings," *Thin Solid Films*, **83** [3] 297–310 (1981).
[7]J. Ilavsky, G. G. Long, A. Allen, and C. Berndt, "Evolution of the Void Structure in Plasma-Sprayed YSZ Deposits During Heating," *Mater. Sci. Eng. A.*, **272**, 215–21 (1999).
[8]K Kokini and Y. R. Takeuchi, "Transient Thermal Fracture of an Interface Crack in the Presence of a Surface Crack," *J. Therm. Stress*, **17**, 63–74 (1994).

[9]K. Kokini, A. Banerjee, and T. A. Taylor, "Thermal Fracture of Interfaces in Precracked Thermal Barrier Coatings," *Mater. Sci. Eng. A*, **323**, 70–82 (2002).

[10]D. Zhu and R. A. Miller, "Determination of Creep Behavior of Thermal Barrier Coatings Under Laser Imposed High Thermal and Stress Gradient Conditions," *J. Mater. Res.*, **14** [1] (1999).

[11]K. R. Kokini, Y. R. Takeuchi, and B. D. Choiles, "Surface Thermal Cracking of Thermal Barrier Coatings Owing to Stress-Relaxation: Zirconia vs. Mullite," *Surf. Coat. Tech.*, **82**, 77–82 (1995).

[12]B. D. Choules, K. Kokini, and T. A. Taylor, "Thermal Fracture of Ceramic Thermal Barrier Coatings under High Heat Flux with Time-Dependent Behavior. Part 1. Experimental results," *Mater. Sci. Eng. A*, 296–304 (2001).

[13]J. B. Wachtman, *Mechanical Properties of Ceramics*, p. 392 Wiley, New York, 1996.

[14]R. M. Christensen, *Theory of Viscoelasticity—An Introduction*. Academic Press, New York, 1971.

[15]Y. C. Fung, *Foundations of Solid Mechanics*. Prentice-Hall, Englewood Cliffs, NJ, 1965.

[16]A. S. Krausz and H. Eyring, *Deformation Kinetics*. John Wiley & Sons, New York, 1975.

[17]N. G. McCrum, C. P. Buckley, and C. B. Bucknall, *Principles of Polymer Engineering*, 2nd edition, Oxford Science Publications, Oxford, 1997.

[18]E. F. Rejda, D. F. Socie, and T. Itoh, "Deformation Behavior of Plasma-Sprayed Thick Thermal Barrier Coatings," *Surf. Coat. Tech.*, **113**, 218–26 (1999).

[19]R. Trice, D. Prine, and K. Faber, "Deformation Mechanisms in Compression-Loaded Stand-Alone Plasma-Sprayed Alumina Coatings," *J. Am. Ceram. Soc.*, **83** [12] 3057–64 (2000).

[20]J. Levin, G. Dickinson, and R. Trice, "In-situ Observation of Crack Behavior in Plasma-Sprayed 7 wt% Yttria-Stabilized Zirconia," *J. Am Ceram. Soc.*, **87** [5] 960–2 (2004).

[21]G. Thurn, G. A. Schneider, and F. Aldinger, "High-Temperature Deformation of Plasma-Sprayed ZrO2 Thermal Barrier Coatings," *Mater. Sci. Eng. A.*, **A233**, 176–82 (1997).

[22]T. F. Bernecki and D. R. Marron, "Small-Particle Plasma-Spray Apparatus, Method and Coated Article," U.S. Patent No. 5 744 777, 1998, and U.S. Patent No. 5 858 470, 1999. □

International Journal of
Applied Ceramic Technology
Ceramic Product Development and Commercialization

Mechanical Properties/Database of Plasma-Sprayed ZrO$_2$-8wt% Y$_2$O$_3$ Thermal Barrier Coatings

Sung R. Choi,* Dongming Zhu, and Robert A. Miller

NASA Glenn Research Center, Cleveland, OH 44135

Mechanical behavior of free-standing, plasma-sprayed ZrO$_2$-8wt% Y$_2$O$_3$ thermal barrier coatings, including strength, fracture toughness, fatigue, constitutive relation, elastic modulus, and directionality, have been determined under various loading-specimen configurations. This paper presents and describes a summary of mechanical properties of the plasma-sprayed coating material to provide them as a design database.

Introduction

Thermal barrier coatings (TBCs) have attracted ever-increasing attention for advanced gas turbine and diesel engine applications because of their ability to provide thermal insulation to engine components.[1-3] The merits of using ceramic TBCs are well recognized and include a potential increase in engine operating temperature with reduced cooling requirements, resulting in significant improvements in thermal efficiency, performance, and reliability. Plasma sprayed zirconia-based ceramics are one of the most important coating materials because of their low thermal conductivity, relatively high thermal expansivity, and unique microstructure as a result of the plasma spraying process. However, the durability of TBCs under severe thermal and mechanical loading conditions encountered in heat engines remains one of the major problems. As a result, the development of TBCs requires better understanding of both mechanical and thermal behavior of the coating materials to ensure life and reliability of the related components.

During the past one to two decades, various plasma-sprayed TBC systems have been developed and characterized to determine their mechanical properties, such as strength, creep, deformation, elastic modulus, fatigue, and interfacial toughness at ambient and/or elevated temperatures.[4-23] These attempts have been made using either free-standing TBCs or TBC/substrate systems. However, few studies have been done to focus on as many mechanical properties of any given TBC as possible in a single place from which the properties can be conveniently used as a design database.

Mechanical and thermal properties of plasma-sprayed ZrO$_2$-8wt% Y$_2$O$_3$ TBCs have been evaluated under various test conditions at the NASA Glenn Research Center. Mechanical properties of free-standing, as-sprayed TBCs are summarized in this paper as simply as possible and presented as a design database. The mechanical properties include strength, fracture toughness, fatigue (slow crack growth), constitutive relation (deformation), and elastic modulus and Poisson's response, determined under various loading-specimen configurations. TBCs exhibit their unique microstructure with microcracking, porosity, and lamellar-like splat morphology, so that they

* Corresponding author: NASA Resident Principal Scientist, Ohio Aerospace Institute; Tel (216) 433-8366, Fax (216) 433-8300; Email address: sung.r.choi@grc.nasa.gov

would be expected to reveal some directionality of mechanical properties. The directionality effect has been quantified through fracture toughness and Knoop hardness measurements and is also presented in this paper. Some of the properties presented here have appeared elsewhere and those desiring additional details should see [24-29] while the data on strength (in part), elastic modulus, Poisson's response, and directionality are appearing here for the first time. Comparison of mechanical properties with other or similar TBC systems was beyond the scope of this paper and thus was not made here, by focusing only on the ZrO_2-8wt% Y_2O_3 TBC system.

Experimental Techniques

Unlike dense ceramics, the stress-strain behavior of as-sprayed TBCs is not linear-elastic. Therefore, applying conventional test methods that are employed for dense materials may not be pertinent, in a rigorous sense, to TBCs. However, since no standardized test methods are currently available for TBCs, most of the experimental techniques used here to determine their mechanical properties were followed in principle in accordance with ASTM test standards that are primarily applied to dense advanced ceramics commonly regarded to be isotropic and homogeneous.

Material

The ZrO_2-8wt% Y_2O_3 powder (Zircoa Inc., Solon, OH), with an average particle size of 60 μm, fabricated by sintering and crushing, was first plasma-sprayed on a graphite substrate measuring 150 by 100 by 6.5 mm to a thickness of about 6 mm, using a Sulzer-Metco ATC-1 plasma coating system (Sulzer Metco, Inc., Westbury, NY) with an industrial robot. The plasma-spray conditions can be found elsewhere.[30] Free-standing, as-sprayed ceramic billets were then obtained by burning away the graphite substrate at 680°C in air for 24 h. The billets were machined into test specimen configurations such as cylindrical rods, flexure beams or disks, depending on the test matrix. Fig. 1 shows a typical layered structure, a polished surface, and a fracture surface, showing the microstructure of as-processed coatings, in which a large number of microcracks and pores are characterized in conjunction with a splat (platelet) structure. It has been found that the coatings primarily consist of tetragonal t′ phase[8] and have a density of 5.147 ± 0.020 g/cm³ with a porosity of about 10%.

↓ SD

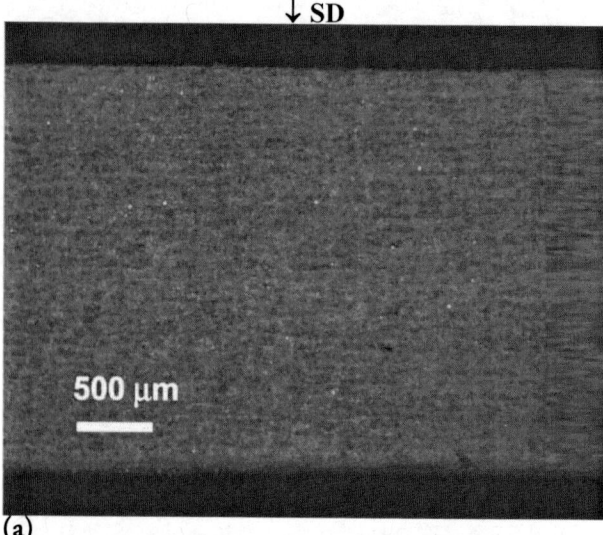

(a)

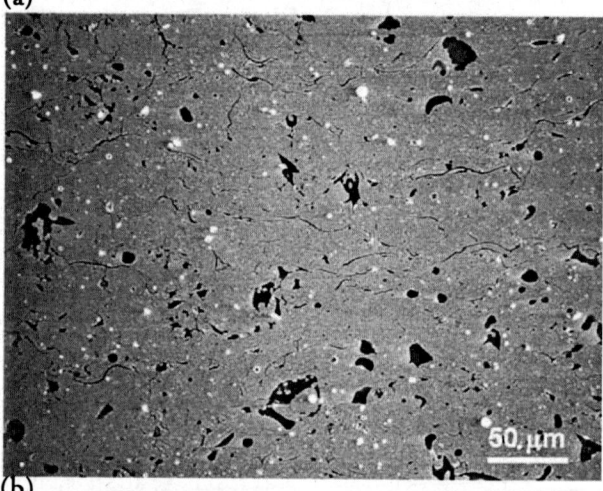

(b)

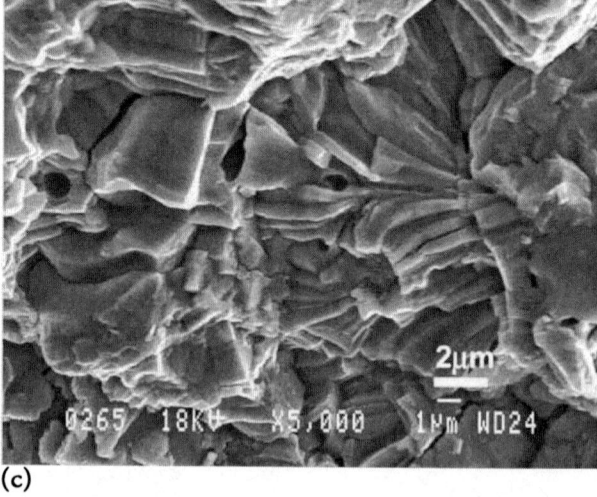

(c)

Fig. 1. Typical microstructures of plasma-sprayed ZrO_2-8wt% Y_2O_3 thermal barrier coatings: (a) side view of flexure specimen showing lamellae (layers); (b) polished top surface; (c) fracture surface. "SD" is plasma-spray direction.

Strength

Five different loading-specimen configurations used in strength testing at ambient temperature (25°C) were uniaxial tension, trans-thickness tension, uniaxial compression, four-point uniaxial flexure, and ring-on-ring biaxial flexure. Schematics of test specimens and direction of applied load with respect to plasma-spray direction are shown in Fig. 2. Dimensions of test specimens and related test methods (ASTM C 1273, C 1468, C 1424, and C 1161)[31-34] are shown in Table I. Note that in certain cases some variations in test specimens' dimensions from the specifications of related ASTM test standards were inevitable, due to limited size of the coating material. Final finish of test specimens was achieved using #500 diamond grinding wheel. For test specimens with round section, a combination of surface and cylindrical grinding was used. Testing was performed at a test rate of 0.5 mm/min using an electromechanical test frame (Model 8562, Instron, Canton, MA). The magnitude of strength was found to be almost independent of whether the top surface was placed in tension or the bottom surface previously in contact with the graphite substrate was placed in tension, as determined with four-point flexure bars.

Fracture Toughness

The method used to determine fracture toughness

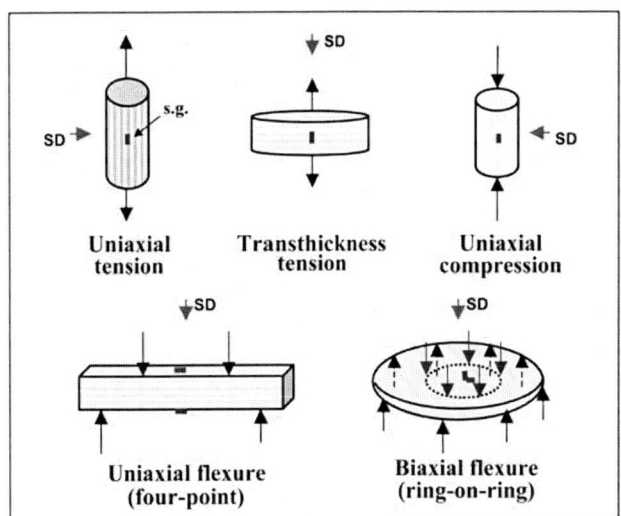

Fig. 2. Five different loading specimen configurations used in strength testing of plasma-sprayed ZrO_2-8wt% Y_2O_3 thermal barrier coatings. The arrows ("SD") indicate plasma-spray direction. Strain gages ("s.g.") attached are also indicated. Note that the biaxial strain gage was placed in the tension side of a biaxial disk.

in modes I and II at 25, 800, and 1316°C was the single-edge V-notched beam (SEVNB) technique.[35] Sharp V notches in which final notch depth and root radius were 2.0 mm and 20 to 50 μm, respectively, were introduced

Table I. Results of Strength Testing in Five Different Loading-Specimen Configurations for Plasma-Sprayed ZrO_2-8wt% Y_2O_3 Thermal Barrier Coatings at Ambient Temperature

Type of Testing	Test Method	Specimen Dimensions[a] (mm)	No. of Test Specimens	Direction of Fracture[b]	Average Strength (MPa)	Weibull Modulus, m
Uniaxial tension	ASTM C 1273	5 × 15 (Dia. × L)	10	P	15(1)[c]	7
Trans-thickness tension	ASTM C 1468	15 × 3 (Dia. × T)	10	N	11(1)	13
Uniaxial compression	ASTM C 1424	5 × 10 (Dia. × L)	10	P/N	300(77)	4
Uniaxial flexure (four-point)	ASTM C 1161	3 × 4 × 25-50 (H×W×L) (20/40 or 10/20 mm spans)	30	P	33(7)	6
Biaxial flexure (ring-on-ring)	-	25 × 3 (Dia. × T) (11/22 mm rings)	10	P	40(4)	12

Notes:
[a] Descriptions for specimen dimensions: Dia: diameter, L: length, T: thickness; H: height; W: width.
[b] "Direction of fracture" indicates the direction of fracture with respect to plasma-spray direction. P: parallel; N: normal
[c] The numbers in parentheses indicate ±1.0 standard deviation.

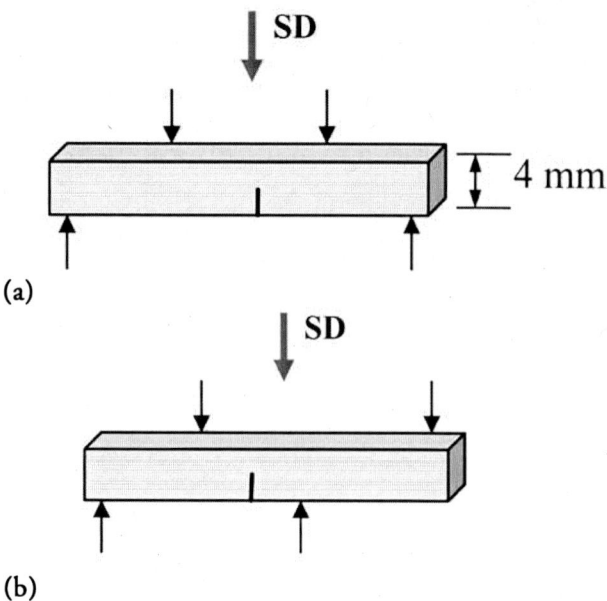

(a)

(b)

Fig. 3. Schematics of test specimens and loading configurations used in fracture toughness testing for plasma-sprayed ZrO_2-8wt% Y_2O_3 thermal barrier coatings: (a) mode I testing (K_{Ic}) in symmetric four-point flexure; (b) mode II testing (K_{IIc}) in asymmetric four-point flexure. "SD" is plasma-spray direction.

in flexure test specimens (3 mm × 4 mm [depth] × 25-50 mm). Through-the-thickness sharp notches thus prepared were aligned parallel with respect to the plasma spraying direction. Other methods used to generate sharp cracks to estimate fracture toughness, such as the single-edge-precracked-beam and the surface crack in flexure techniques (ASTM C 1421)[36] by indentation, were not feasible for the coating material, as indentation response was very poor because of the material's significant porosity and microcracks. Modes I and II fracture toughnesses (K_{Ic} and K_{IIc}) were determined in symmetric and asymmetric four-point flexure, respectively, as shown in Fig. 3. Testing was performed in air at a test rate of 0.5 mm/min using the electromechanical test frame. At elevated-temperature testing, each test specimen was heated at a rate of about 20°C/min and held for 15 min at test temperature for about 15 min prior to testing. The number of test specimens was typically ≥4 for each mode at a given temperature.

Fatigue/Slow Crack Growth

Constant stress rate ("dynamic fatigue") testing in accordance with ASTM C 1368 and C 1465[37,38] was performed in flexure at 25 and 800°C in air to evaluate slow

crack growth behavior of the coating material. The dimensions of test specimens were identical to those used in uniaxial four-point flexure strength or fracture toughness testing. The number of stress rates was three, ranging from 50 MPa/s to 0.0005 MPa/s, and the number of test specimens was five at each stress rate. Cyclic fatigue testing was also conducted in tension at 25°C in air using uniaxial tensile test specimens (see Fig. 2) with a stress ratio of R = 0.1 at a frequency of 10 Hz. Dynamic and cyclic fatigue testing were conducted using electromechanical (Model 8562) and servohydraulic (Model 8502, Instron, Canton, MA) test frames.

Constitutive Relation

Strain gages were used to determine constitutive relations (stress-strain curves or deformation behavior) for different loading-specimen configurations used in the strength testing (Fig. 2). Typically, one test specimen with one strain gage aligned along the principal stress direction was used for each configuration. In ring-on-ring biaxial flexure, a biaxial strain gage was placed at the center of the disk in tension side. Particularly, in four-point flexure loading, two strain gages—one in tension side and another in compression—were employed to determine the constitute relations in both sides of a test specimen.

Elastic Modulus and Poisson's Ratio

Elastic modulus was determined from the slopes of the stress-strain curves obtained from the constitutive relation testing. In addition, the impulse-excitation technique, ASTM C 1259,[39] was employed using flexure test specimens (3 mm × 4 mm × 50 mm) to determine elastic modulus in both directions at ambient temperature. Direction 1 corresponds to the case that excitation in flexure mode is made on the 4-mm side of a test specimen, while direction 2 is the case when excitation is made on the 3-mm side. Five test specimens were used.

Poisson's ratio was estimated with a uniaxial compression test specimen (Fig. 2) by determining both longitudinal and transverse strains via two strain gages that were oriented perpendicular to each other. A test specimen annealed at 1316°C in air for 500 h was also used for comparison.

Directionality

Directionality of as-sprayed TBCs was examined using flexure test specimens by determining both mode I

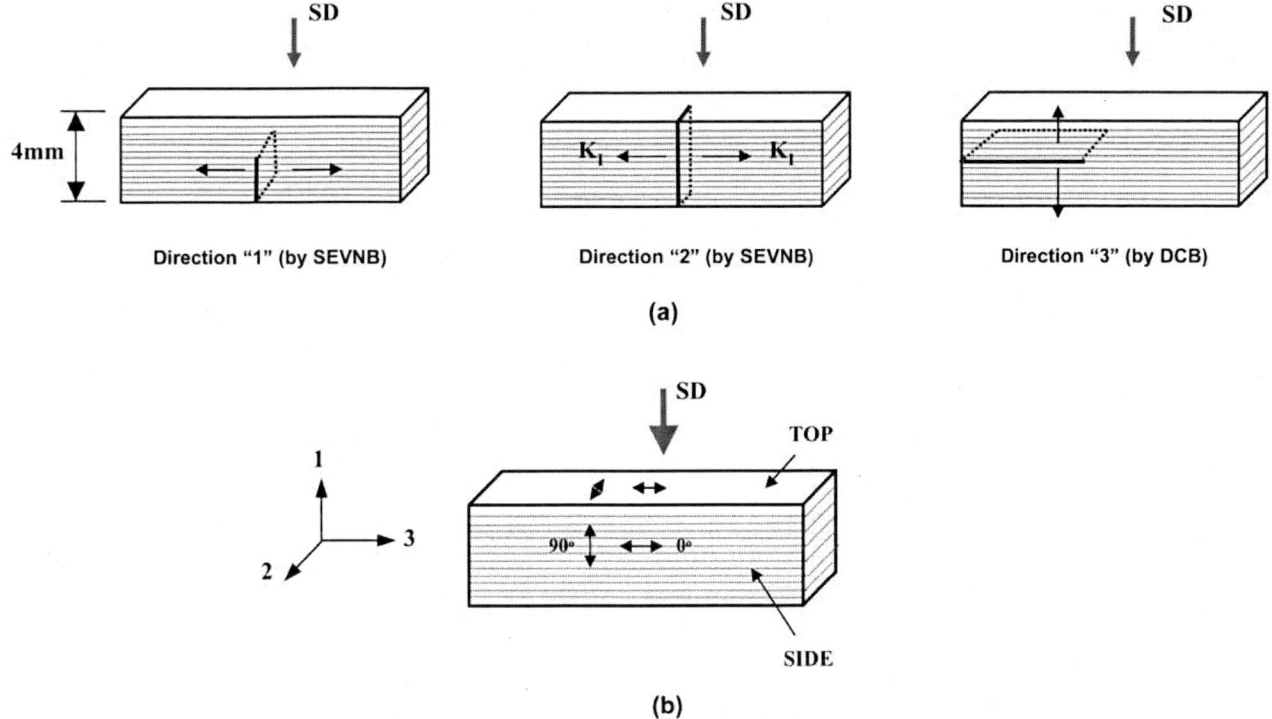

Fig. 4. *Schematics to examine the directionality effect of as-sprayed ZrO₂-8wt% Y₂O₃ thermal barrier coatings: (a) mode I fracture toughness testing in three orientations; (b) Knoop microhardness determined at two different orientations of 0 and 90° at both top and side surfaces. Horizontal lines in test specimens show lamellar-like layers, oriented normal to plasma-spray direction. "SD" is plasma-spray direction.*

fracture toughness and microhardness as a function of material axis, as shown in Fig. 4. Fracture toughness in direction 2 was determined by the SEVNB method, whereas fracture toughness in direction 3 (i.e., the interlamina direction) was determined by the double cantilever beam (DCB) method with a sharp V notch introduced in the midplane of the test specimens. Four to five flexure test specimens were tested for each direction. Fracture toughness in the DCB specimens was calculated based on the formula by Murakami.[40]

Knoop microhardness, in accordance with ASTM C 1326,[41] was determined with a four-point flexure test specimen at both 0 and 90° on its top (perpendicular to plasma-spraying direction) and side (parallel to plasma-spraying direction) surfaces (Fig. 4). An indentation load of 9.8 N was used with a total of five indents at each orientation for a given specimen surface.

Descriptions of Mechanical Properties

Strength

A summary of strength for five different loading-speci-men configurations is shown in Table I. Of the five loading-specimen configurations considered, trans-thickness tensile strength is lowest (11 ± 1 MPa) but somewhat comparable to uniaxial tensile strength (15 ± 1 MPa). Ring-on-ring biaxial flexure strength was highest at 40 ± 4 MPa. Uniaxial flexure strength, determined with a total of 30 test specimens, was 33 ± 7 MPa, which is significantly lower than that (270 MPa) of hot-pressed 10 mol% yttria-stabilized zirconia (YSZ).[42] Uniaxial flexural strength has been observed to be little influenced by vintage, indicating consistency in plasma-spray processing over the years.[24,25,29]

Two-parameter Weibull strength distributions, notwithstanding the insufficient number (about 10, except for four-point flexure) of test specimens used, were made for comparison and are shown in Fig. 5. Weibull modulus of TBCs ranged from m = 6 for uniaxial flexure to m = 13 for trans-thickness tension, similar to a typical range observed for many dense monolithic ceramics. Effect of specimen size on strength is depicted in Fig. 6, in which strength was plotted against (a) effective area and (b) effective volume for various test specimens used. The effec-

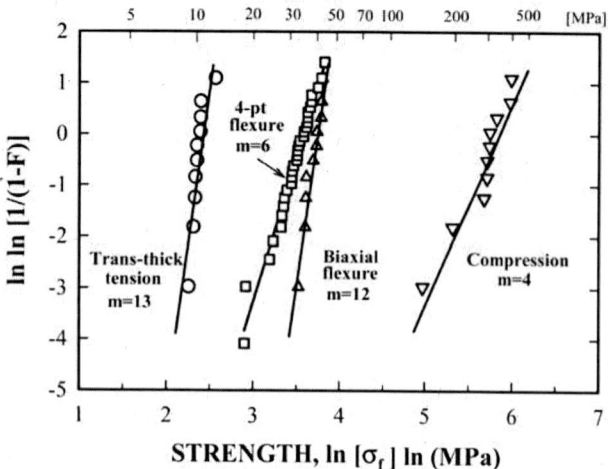

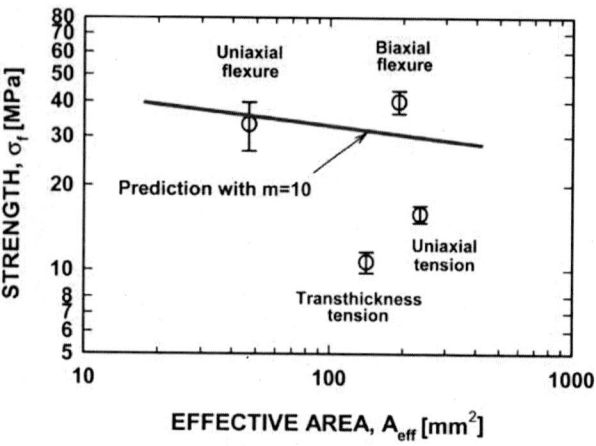

(a)

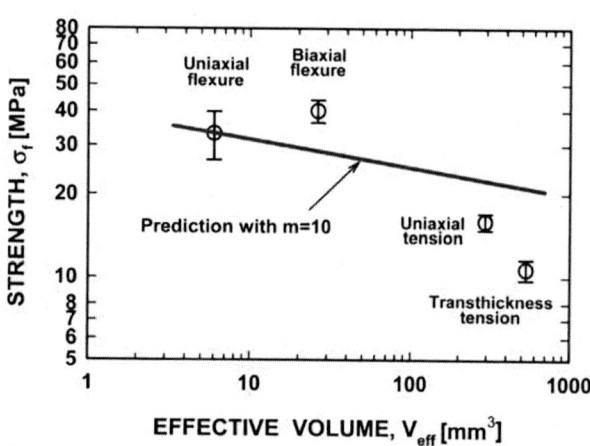

(b)

Fig. 5. Weibull strength distributions determined in trans-thickness tension, uniaxial flexure, biaxial flexure, and compression for plasma-sprayed ZrO_2-8wt% Y_2O_3 thermal barrier coatings. F indicates failure probability.

Fig. 6. Plots of strength against (a) effective surface area and (b) effective volume in uniaxial tension, trans-thickness tension, uniaxial flexure, and biaxial flexure for plasma-sprayed ZrO_2-8wt% Y_2O_3 thermal barrier coatings. Prediction of strength based on the principle of independent action (PIA) with a Weibull modulus of m = 10 for uniaxial flexure was presented for each case.

tive area and effective volume for both uniaxial and bi-axial flexure were calculated based on an average value of Weibull modulus, m = 10. Prediction of strength for each case was made with a reference strength value in uniaxial flexure together with m = 10, based on the principle of independent action (PIA). No reasonable agreement between the prediction and the data was found for the surface-flaw controlled case, as shown in Fig. 6a. However, a generalized trend was observed for the case of volume flaws, as seen in Fig. 6b, implying that a common notion of the size effect that is employed to dense brittle materials may be applicable to TBCs at least for volume flaws. In other words, the difference in strength of TBCs between different specimen geometries would be explained at minimum by the size effect associated with volume flaws. However, the deviation of prediction from the data still indicates that inconsistency in flaw populations was much more predominant in TBCs with complicated structures than in relatively homogeneous dense brittle ceramics.

Typical examples of fracture surfaces and fracture modes of various tested specimens are shown in Fig. 7. Fracture surfaces of TBCs are typified such that exact fracture origins and their nature were hardly identifiable, due to loosely connected open structure that also yielded significantly low strength. Probable fracture origins are indicated by arrows in the figure. For dense brittle materials, fracture origin can be easily identified from a well-developed region of fracture mirror, and its nature can be explored in relative simplicity. Hence, unlike dense brittle

ceramics, TBCs would give rise to an enormous challenge in fractography.

Fracture Toughness

A summary of modes I and II fracture toughness is presented in Fig. 8. The values of K_{Ic} and K_{IIc} represent the average of four and eight measurements, respectively, at 25°C and the average of four measurements for each tested at 1316°C. The values of fracture toughness were K_{Ic} = 1.15 ± 0.07, 1.03 ± 0.07, and 0.98 ± 0.13 MPa√m at 25, 800, and 1316°C, respectively; whereas, K_{IIc} = 0.73 ± 0.10 and 0.65 ± 0.04 MPa√m at 25 and 1316°C, re-

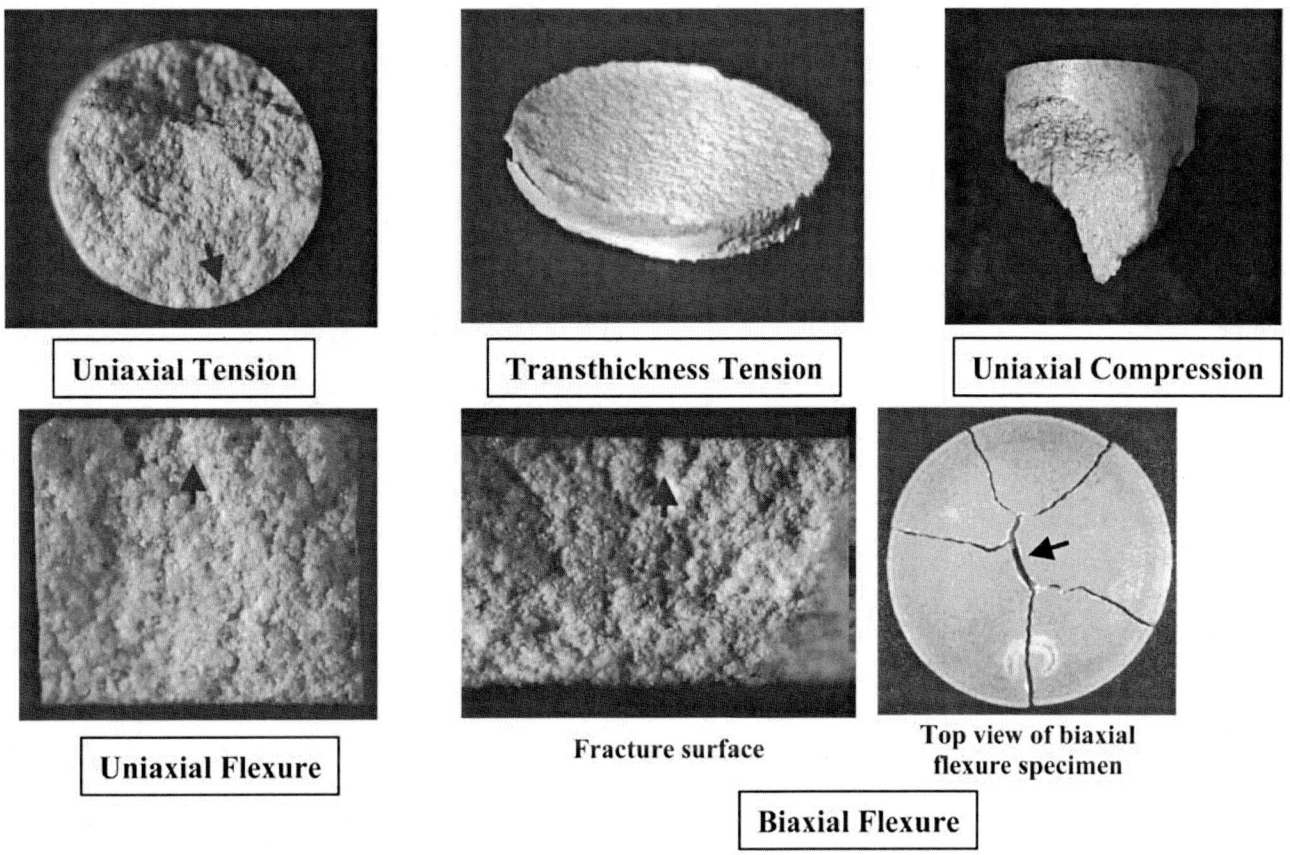

Fig. 7. *Modes of fracture and fracture surfaces of tested specimens subjected to different loadings for plasma-sprayed ZrO₂-8wt% Y₂O₃ thermal barrier coatings at ambient temperature. Arrows indicate probable fracture origins.*

spectively. This indicates that fracture toughness in either mode I or mode II remained almost consistent regardless of temperature up to 1316°C. K_{IIc} was about 35% lower than K_{Ic}. It has been reported that for some dense ceramics K_{IIc} was greater than K_{Ic}, presumably attributed to frictional interaction between the two facing crack planes.[43] However, the previous studies on advanced dense ceramics including silicon nitrides, alumina, and zirconia[44,45] showed a different result that K_{IIc} was almost identical to K_{Ic}, indicative of an insignificant frictional effect on K_{IIc} by either coarse-grained or fine-grained ceramics. The coating material, however, did not exhibit a similar value in both K_{IIc} and K_{Ic}, but rather yielded a lower value in K_{IIc} than in K_{Ic} at both 25 and 1316°C. Hot-pressed 10 mol% YSZ showed a value of K_{Ic} = 1.7 MPa√m at 25°C,[42] consistent with other dense YSZs (with >8 mol% yttria) exhibiting K_{Ic} = 1 2 MPa√m.[46,47]

It must be noted that the effect of sintering on fracture toughness during elevated-temperature testing par-

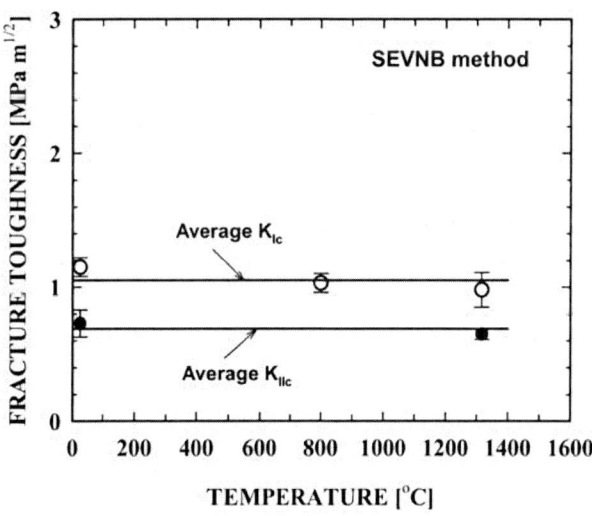

Fig. 8. *Fracture toughnesses of K_{Ic} and K_{IIc} as a function of test temperature for plasma-sprayed ZrO₂-8wt% Y₂O₃ thermal barrier coatings.*

ticularly at 1316°C was found to be negligible based on a sintering study in which fracture toughness increase of the coating material was about 0.12 MPa√m/h for the first 5 h annealing at 1316°C in air.[29] This indicates that fracture toughness increase during testing at 1316°C would be about 3%. However, care must be exercised in interpreting test results when higher test temperature and longer test duration are used. This caution applies to any mechanical testing for plasma-sprayed TBCs at elevated temperatures.

Fatigue/Slow Crack Growth

Fatigue/slow crack growth behavior of TBCs at 25°C in tension and at 800°C in flexure is shown in Fig. 9. The

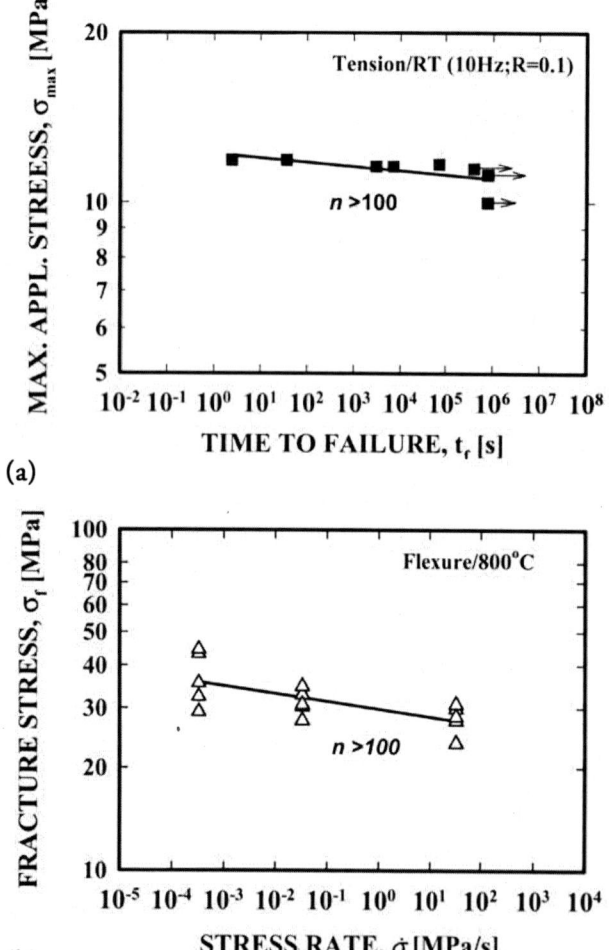

(a)

(b)

Fig. 9. Results of fatigue/slow crack growth testing for plasma-sprayed ZrO₂-8wt% Y₂O₃ thermal barrier coatings: (a) cyclic fatigue testing in tension at 25°C in air; (b) constant stress-rate (dynamic fatigue) testing in flexure at 800°C in air.

coating material exhibited a negligible susceptibility to cyclic fatigue in tension at 25°C with a significant fatigue parameter n > 100, as shown in Fig. 9a. The fatigue parameter n is from the conventional power-law type of crack growth, expressed as follows:

$$V = \alpha \left[K_I / K_{Ic} \right]^n \qquad (1)$$

where V is crack velocity, α is another fatigue parameter, and K_I is mode I stress intensity factor. At 800°C, the coating material did not exhibit any fatigue (i.e., n > 100) but rather strength increase with decreasing applied test rate, attributed to some minor sintering effect at lower test rates, as shown in Fig. 9b. These results indicate that the resistance to fatigue or slow crack growth of as-sprayed TBCs was very significant at both temperatures, implying that slow crack growth under either cyclic tension or monotonic flexure would be hindered by frequent entrapping of a growing crack at the pore- and/or microcrack-rich regions.

Constitutive Relation/Deformation Behavior

Fig. 10 shows the stress-strain curves in tension, compression, biaxial flexure, and uniaxial flexure. TBCs did not exhibit any idealized linear elasticity in either loading or unloading, resulting in appreciable hysteresis in one full loading/unloading sequence. A similar nonlinear elastic behavior was also found previously by the authors (unpublished) in sandstone that possesses a TBC-like, loosely connected, open microstructure. Despite their nonlinearity, as-sprayed TBCs revealed almost elastic behavior with little plastic deformation. The nonlinear elastic behavior was independent of loading rate, number of cycles, and type of loading configurations in either tension, compression, or flexure.[24,26,27] The initial stress-strain curves were altered slightly as the number of loading-and-unloading cycles increased, as seen in Figs. 10a and c. However, once the number of cycles was beyond ten or so, the difference in the shape of stress-strain curves remained almost unchanged. This indicates that any notable damage evolution, such as crack propagation, micro-cracking and/or mechanical-interlocking loosening, etc., would not have occurred under repeated loading-and-unloading sequences. The stress-strain curve of a uniaxial flexure test specimen showed stiffer in compression side than in tension, evidence of the loosely connected, open microstructure of TBCs. Several sequences of loading and unloading with different levels of compressive stress are shown in Fig. 11. This figure shows

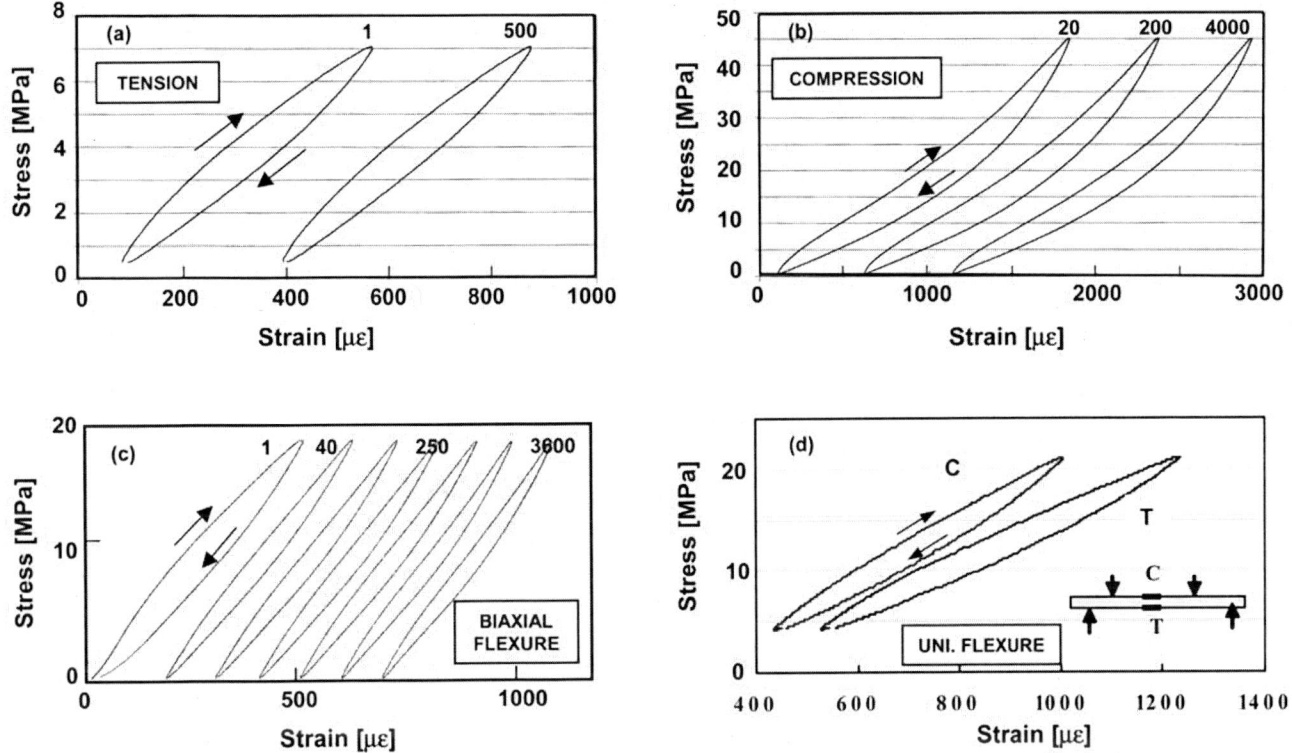

Fig. 10. *Stress vs. strain curves determined for plasma-sprayed ZrO$_2$-8wt% Y$_2$O$_3$ thermal barrier coatings: (a) uniaxial tension; (b) compression; (c) biaxial flexure; (d) uniaxial flexure showing responses of tension and compression sides. The numbers of loading-unloading cycles are indicated at the top of the curves from (a) to (c). T is tension; C is compression in (d).*

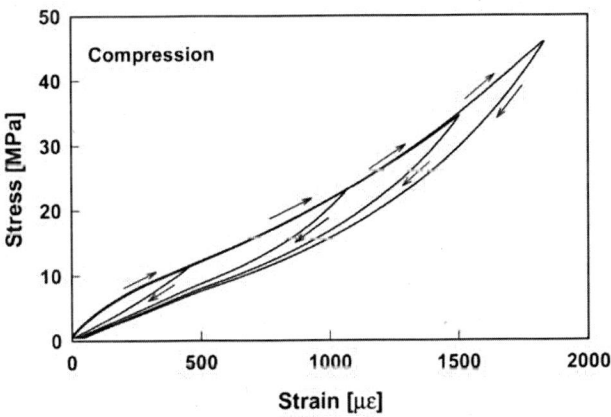

Fig. 11. *Stress vs. strain curve determined in compression using a cylindrical rod test specimen of plasma-sprayed ZrO$_2$-8wt% Y$_2$O$_3$ thermal barrier coatings with different loading-unloading sequences with different peak stresses.*

that regardless of applied stress, each loading curve follows the same loading path; whereas, the unloading curve follows its unique path upon unloading from the respective peak stress, eventually returning back to its original starting point.

Elastic Modulus and Poisson's Ratio

Elastic Modulus

Due to its significant nonlinear behavior of stress-strain curves, a single-valued elastic modulus of the coating material cannot be determined. This results in different values of elastic modulus, depending on the type of loading and on the magnitude of applied stress. The most accurate approach, of course, is to use the *tangent modulus* (E = dσ/dϵ) of elasticity which is the slope of a stress-strain curve at any point for a given loading or unloading sequence.[22,26] With this approach, an assessment of elastic modulus could be made for a given range of applied stress, consistently from specimen to specimen. However, this approach is not practical, particularly in design, analysis, and lifing in which a single value of elastic modulus is predominantly desirable.

Alternatively, if the variation of slopes in the curves is not significant from the lowest stress to the peak, the slope between these two extreme points, i.e., the *secant modulus* of elasticity in a broad sense, could be used as an approximated value of elastic modulus. This alternative approach, applied to tension and compression in each of their loading sequences in Figs. 10a and b, yielded elastic modulus of $E_t \approx 13$ GPa in tension and $E_c \approx 25$ GPa in compression. The variation of elastic modulus with different levels of applied stress, for example in compression, was negligible with $E_c \approx 22\text{-}25$ GPa at applied stresses of 10, 20, 30, and 45 MPa. It is noted that elastic modulus in pure tension, $E_t = 13$ GPa, is comparable to $E_t = 16$ GPa in trans-thickness tension[26] where the latter was evaluated between applied stresses of 0-5 MPa. However, this secant-modulus approach does not account for the hysteresis and would not be appropriate for the unloading sequences. Because of the difference in elastic modulus between tension and compression, a neutral-axis shift should be taken into account in calculating stresses in uniaxial and biaxial flexure. The stresses in both uniaxial and biaxial flexure shown in Figs. 10c and d were calculated based on the linear elastic approach without considering the neutral-axis shift for simplicity.

The situation that elastic modulus varies with applied stress could be changed when TBCs are annealed at elevated temperatures, so that gradually increasing linearity develops with increasing annealing time, as shown in Fig. 12.[29] The stress-strain curves in the figure were obtained in compression at 25°C after annealing cylindrical compression test specimens at 1316°C in air. This change in constitutive relation was attributed to sintering in which the coating material was converted from loosely connected to more closely connected microstructure. The values of elastic modulus were $E \approx 45$, 55, 75, and 77 GPa, respectively, for 5, 20, 100, and 500 h in annealing time; while the as-sprayed value was $E \approx 25$ GPa, as aforementioned. In view of this systematic change in constitutive relation, it would be reasonable to use the two extreme points of a curve as a first order of approximation, in order to estimate an approximated value of elastic modulus of as-sprayed TBCs. The effect of sintering on various mechanical properties of TBCs has been determined in detail from a previous study.[29]

The value of elastic modulus by the impulse-excitation technique was $E = 36 \pm 2$ GPa and 39 ± 2 GPa, respectively, for directions 1 and 2. This insignificant difference in elastic modulus between two directions is also indicative of a negligible directionality effect of TBCs

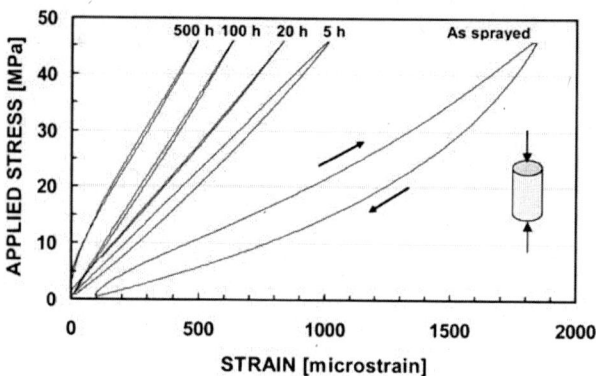

Fig. 12. *Applied stress vs. strain responses for different annealing time for plasma-sprayed ZrO_2-8wt% Y_2O_3 thermal barrier coatings annealed at 1316°C in air, determined at 25°C in compression by strain gaging.*

when it comes to elastic modulus. The value of elastic modulus by the impulse-excitation was greater than that ($E_t \approx 13$ GPa and $E_c \approx 25$ GPa) by strain gaging. The reason for this discrepancy has not been explored yet, but is believed to be due to the fact that elastic modulus estimated by the impulse-excitation technique was involved in flexural mode of vibration, a combination of both tension and compression displacements. This may require a modification of the fundamental flexural-resonance equation in the impulse-excitation method if it is used for TBCs that exhibit a difference in elastic modulus between tension and compression. Note that the impulse-excitation technique is primarily for isotropic, homogeneous materials. Elastic modulus of hot-pressed 10 mol% YSZ was found to be $E = 219$ GPa by the impulse-excitation technique.[42]

Poisson's Ratio

Fig. 13 shows both longitudinal and transverse strains of cylindrical compression test specimens as-sprayed and annealed at 1316°C for 500 h in air, determined by strain gaging. Despite the appreciable nonlinearity in stress-strain curves of the as-sprayed test specimen, Poisson's ratio, which is defined as $\nu = |\varepsilon_2/\varepsilon_1|$ (see the figure) for each pair of loading curves, was very low but relatively consistent with $\nu = 0.04$, regardless of level of applied stress. By contrast, the annealed test specimen exhibited an increased value of Poisson's ratio, $\nu = 0.2$, which is close to Poisson's ratio of many dense brittle materials. This again shows a change in microstructure from a loosely connected to a closely connected network, due to the sintering effect on high-temperature annealing.

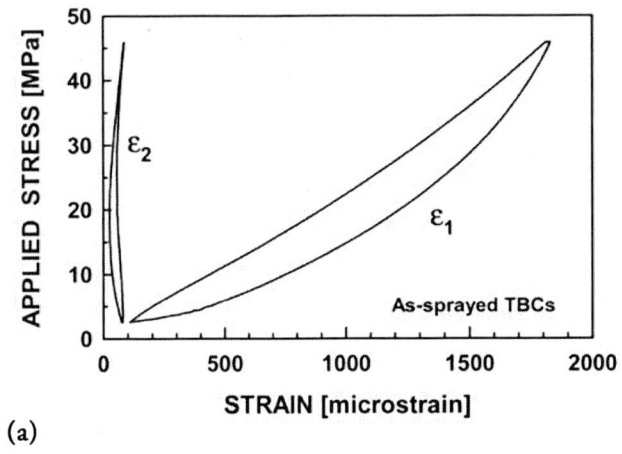

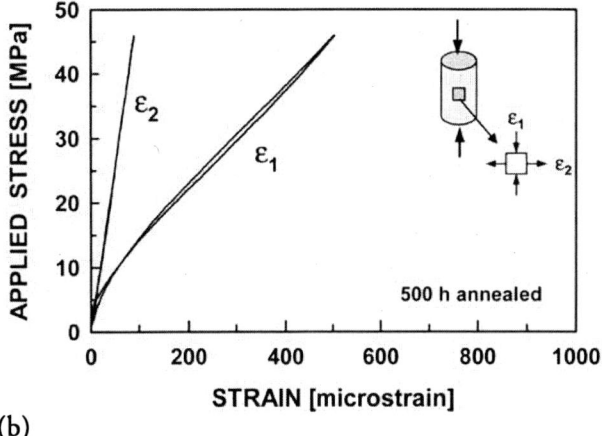

(a)   (b)

Fig. 13. Stress-strain curves in compression for both longitudinal (ε_1) and transverse (ε_2) directions of plasma-sprayed ZrO_2-8wt% Y_2O_3 thermal barrier coatings: (a) as-sprayed condition; (b) annealed at 1316°C for 500 h in air.

Directionality

TBCs were fabricated via plasma-spray process so that anisotropy might be expected due to the prevailing open, splat microstructure, preferably aligned perpendicular with respect to plasma-spray direction. The effect of such anisotropy, as described in Experimental Techniques, was determined for as-sprayed TBCs through mode I fracture toughness and Knoop microhardness testing. The results are summarized in Table II.

The values of fracture toughness additionally determined in directions 2 and 3 were K_{Ic} = 1.04 ± 0.11 and

1.04 ± 0.05 MPa√m, respectively. These values compare well to K_{Ic} = 1.15 ± 0.07 MPavm in direction 1, described in the foregoing Fracture Toughness section. Therefore, fracture toughness yielded an almost consistent value with only a 10% variation in three directions, showing an insignificant directionality in response to fracture toughness. Note that the expected weakest direction (interlaminar, direction 3) yielded a value of fracture toughness similar to the presumably strongest counterpart (direction 1 or 2).

As to Knoop microhardness as a function of material axis, a particular interest is in the side surface: the value

Table II. Summary of Fracture Toughness and Knoop Microhardness with Respect to Material Direction for Plasma-Sprayed ZrO_2-8wt% Y_2O_3 Thermal Barrier Coatings at Ambient Temperature

Type of Testing	Direction or Surface	Method	Number of Tests	Average Values[a]
Fracture toughness	1	SEVNB	5	K_{Ic} = 1.15(0.07) MPa√m
	2	SEVNB	4	K_{Ic} = 1.04(0.11) MPa√m
	3	DCB	4	K_{Ic} = 1.04(0.05) MPa√m
			Overall average	K_{Ic} = 1.08(0.06) MPa√m
Microhardness[b]	0°; side surface	Knoop	5	H = 3.25(0.32) GPa
	90°; side surface	Knoop	5	H = 3.19(0.86) GPa
	0°; top surface	Knoop	5	H = 3.31(0.55) GPa
	90°; top surface	Knoop	5	H = 3.68(1.27) GPa
			Overall average	H = 3.36(0.45) GPa

[a] The numbers in parentheses indicate ±1.0 standard deviation.

[b] An indentation load of 9.8 N was used.

(H = 3.25 ± 0.32 GPa) of microhardness parallel to the layers (at 0° angle) was not significantly different from that (3.19 ± 0.86 GPa) perpendicular to the layers (at 90° angle). Furthermore, Knoop microhardness exhibited little difference between 0 and 90° directions in the top surface, resulting in similar values to those in the side surface. This indicates that as-sprayed TBCs exhibited macroscopically homogeneity in response to microhardness deformation. However, the somewhat significant scatter in hardness, ranging from COV (coefficient of variation) = 10-35%, should not be neglected. Also, it must be stated that the value of microhardness was found to be dependent on applied indentation load because of localized densification of TBCs involved upon indentation: less localized densification would occur with a lower indentation load, resulting in a higher microhardness value, and vice versa.[29] A value of H = 2.7 GPa was reported for as-sprayed ZrO_2-8wt% Y_2O_3 by Knoop indentation with an indent load of 9.8 N.[12]

As already seen, elastic modulus of the coating material was similar both in uniaxial tension (E ≈ 13 GPa) and in trans-thickness (interlaminar) tension (E ≈ 16 GPa) by strain gaging, and was also in reasonable agreement between direction 1 (E = 36 GPa) and direction 2 (E = 39 GPa) by the impulse excitation technique.

Conclusions

Mechanical properties of free-standing as-sprayed ZrO_2-8wt% Y_2O_3 thermal barrier coatings were determined and described in conjunction with strength, fracture toughness, fatigue, constitutive relation, elastic modulus, and directionality. The nature of TBCs exhibiting a loosely connected open microstructure resulted in unique responses to mechanical loading, so many mechanical properties/responses of TBCs were unique and different from those of dense polycrystalline ceramics. From the continuum-mechanics point of view, directionality of as-sprayed TBCs was found to be macroscopically negligible through the determinations of fracture toughness, Knoop microhardness, and elastic modulus with respect to material axis.

Acknowledgments

The authors are grateful to both Ralph Pawlik for experimental work and George Leissler for processing TBC billets. This work was supported by the Ultra-Efficient Engine Technology (UEET) Program, NASA Glenn Research Center, Cleveland, OH.

References

1. R. A. Miller, "Current Status of Thermal Barrier Coatings–An Overview," *Surface and Coating Technology*, 30 1-11 (1987).
2. R. A. Miller, "Thermal Barrier Coatings for Aircraft Engines–History and Direction," pp17-34 in NASA CP-3312, ed. W. J. Brindley, National Aeronautics and Space Administration, Glenn Research Center, Cleveland, OH, 1995.
3. T. M. Yonushonis, "Thick Thermal Barrier Coatings for Diesel Components," NASA CR-187111, National Aeronautics and Space Administration, Glenn Research Center, Cleveland, OH, 1991.
4. R. F. Firestone, W. G. Logan, J. W. Adams, and R. C. Bill, "Creep of Plasma-Sprayed ZrO_2 Thermal Barrier Coatings," *Ceram. Eng. Sci. Proc.*, 3 758-771 (1982).
5. H. E. Eaton and R. C. Novak, "Sintering Studies of Plasma-Sprayed Zirconia," *Surf. Coat. Technol.*, 32 227-236 (1987).
6. R. C. Brink, "Material Property Evaluation of Thick Thermal Barrier Coating Systems," *J. Eng. Gas Turbines & Power*, 111 570-577 (1989).
7. T. A. Cruse, B. P. Johnsen, and A. Nagy, "Mechanical Properties Testing and Results for Thermal Barrier Coatings," *J. Thermal. Spray Technol.*, 6 [1] 57-66 (1997).
8. D. Zhu and R. A. Miller, "Sintering and Creep Behavior of Plasma-Sprayed Zirconia- and Hafnia-Based Thermal Barrier Coatings, *Surf. Coat. Technol.*, 108-109 114-120 (1998).
9. A. Kicuk, C. C. Berndt, U. Senturk, R. S. Lima, and C. R. C. Lima, "Influence of Plasma Spray Parameters on Mechanical Properties of Yttria Stabilized Zirconia Coatings. I: Four Point Bend Test," *Mat. Sci. Eng.*, A284 29-40 (2000).
10 E. F. Rybicki, J. R. Shadley, Y. Xiong, and D. J. Greving, "In Situ Evaluations of Young's Modulus and Poison's Ratio Using a Cantilever Beam Specimen," *Proc. 8th Nat. Thermal Spray Conf.*, ed. by C. C. Berndt and S. Sampath, pp 409-414, ASM International, Materials Park, OH, 1995.
11. C. A. Johnson, J. A. Rudd, A. C. Kaya, and H. G. deLorenzi, "A Method for Measuring Non-Linear Elastic Properties of Thermal Barrier Coatings," *Proc. 8th Nat. Thermal Spray Conf.*, ed. by C. C. Berndt and S. Sampath, pp 425-420, ASM International, Materials Park, OH, 1995.
12. S. A. Leigh, C. W. Lin, and C. C. Berndt, "Elastic Response of Thermal Spray Deposits under Indentation Tests," *J. Am. Ceram. Soc.*, 80 [8] 2093-2099 (1997).
13. E. F. Rejda, D. F. Socie, and T. Itoh, "Deformation Behavior of Plasma-Sprayed Thick Thermal Barrier Coatings," *Surf. Coat. Technol.*, 113 218-226 (1999).
14. R. W. Trice, D. W. Prine, and K. T. Faber, "Deformation Mechanisms in Compression-Loaded, Stand-Alone Plasma-Sprayed Alumina Coatings," *J. Am. Ceram. Soc.*, 83 [12] 3057-64 (2000).
15. T. A. Cruse, S. E. Stewart, and M. Ortiz, "Thermal Barrier Coating Life Prediction Model Development," *J. Eng. Gas Turbines & Power*, 110 610-616 (1988).
16. K. F. Wesling, D. F. Socie, and B. Beardsley, "Fatigue of Thick Thermal Barrier Coatings," *J. Am. Ceram. Soc.*, 77 [7] 1863-1868 (1994).
17. B. P. Johnsen, T. A. Cruse, R. A. Miller, and W. J. Brindley, "Compressive Fatigue of a Plasma-Sprayed ZrO_2-8wt% Y_2O_3 and ZrO_2-10wt% NiCrAlCoY TTBC," *J. Eng. Mater. Technol.*, 117 [7] 305-310 (1995).
18. L. L. Shaw, B. Barber, E. H. Jordan, and M. Gell, "Measurement of the Interfacial Fracture Energy of Thermal Barrier Coatings," *Scrip. Mater.*, 39 [10] 1427-1434 (1998).
19. P. J. Callus, and C. C. Berndt, "Relationship Between the Mode II Fracture Toughness and Microstructure of Thermal Spray Coatings," *Surf. Coat. Technol.*, 114 114–128 (1999).
20. G. Thurn, G. A. Schneider, H. A. Bahr, and F. Aldinger, "Toughness Anisotropy and Behavior of Plasma Sprayed ZrO_2 Thermal Barrier Coatings," *Surf. Coat. Technol.*, 123, 147–158 (2000).
21. Y. C. Zhou, T. Tonomori, A. Yoshida, L. Liu, G. Bignall, and T. Hashida, "Fracture Characteristics of Thermal Barrier Coating after Tensile and Bend Tests," *Surf. Copt. Technol.*, 157 118-127 (2002).
22. A. G. Evans, D. R. Mumm, J. W. Hutchinson, G. H. Meier, and F. S. Pettit, "Mechanisms Controlling the Durability of Thermal Barrier Coatings," *Progress in Mater. Sci.*, 46 503-553 (2001).
23. R. W. Steinbrech, "Thermomechanical Behavior of Plasma Sprayed Thermal Barrier Coatings," *Ceram. Eng. Sci. Proc.*, 23 [4] 397-408 (2002).
24. S. R. Choi, D. Zhu, and R. A. Miller, "High-Temperature Slow Crack Growth,

Fracture Toughness and Room-Temperature Deformation Behavior of Plasma-Sprayed ZrO_2-8wt% Y_2O_3," *Ceram. Eng. Sci. Proc.*, 19 [4] 293-301 (1998).

25. S. R. Choi, D. Zhu, and R. A. Miller, "Flexural and Compressive Strength, and Room-Temperature Creep/Relaxation Properties of Plasma-Sprayed ZrO_2-8wt% Y_2O_3," *ibid*, 20 [3] 365-372 (1999).

26. S. R. Choi, D. Zhu, and R. A. Miller, "Deformation and Strength Behavior of Plasma-Sprayed ZrO_2-8wt% Y_2O_3 Thermal Barrier Coatings in Biaxial Flexure and Trans-Thickness Tension," *ibid*, 21 [4] 653-661 (2000).

27. S. R. Choi, D. Zhu, and R. A. Miller "Deformation and Tensile Cycle Fatigue of Plasma-Sprayed ZrO_2-8wt% Y_2O_3 Thermal Barrier Coatings," *ibid*, 22 [4] 427-434 (2001).

28. S. R. Choi, D. Zhu, and R. A. Miller, "Mode I, Mode II, and Mixed-Mode Fracture of Plasma-Sprayed Thermal Barrier Coatings at Ambient and Elevated Temperatures," presented at the 8th International Symposium on Fracture Mechanics of Ceramics, February 25-28, 2003, Houston, TX; To be published in *Fracture Mechanics of Ceramics*, Vol. 14, Kluwer Academi/Plenum Publisher, New York, 2004; also in NASA/TM-2003-212185, National Aeronautics and Space Administration, Glenn Research Center, Cleveland, OH, 2003.

29. S. R. Choi, D-M Zhu, and R. A. Miller, "Effect of Sintering on Mechanical and Physical Properties of Plasma-Sprayed Thermal Barrier Coatings," NASA/TM—2004-212625, National Aeronautics & Space Administration, Glenn Research Center, Cleveland, Ohio, 2004.

30. D. Zhu and R. A. Miller, "Thermal Conductivity and Elastic Modulus Evolution of Thermal Barrier Coatings under High Heat Flux Conditions," *J. Thermal Spray Technology*, 9 [2] 175-180 (2000).

31. ASTM C 1273, "Test Method for Tensile Strength of Monolithic Advanced Ceramics at Ambient Temperature," *Annual Book of ASTM Standards*, Vol. 15.01, American Society for Testing and Materials, West Conshohocken, PA, 2004.

32. ASTM C 1468, "Test Method for Trans-thickness Tensile Strength of Continuous Fiber-Reinforced Advanced Ceramics at Ambient Temperature," *Annual Book of ASTM Standards*, Vol. 15.01, American Society for Testing and Materials, West Conshohocken, PA, 2004.

33. ASTM C 1424, "Test Method for Monotonic Compressive Strength of Advanced Ceramics at Ambient Temperature," *Annual Book of ASTM Standards*, Vol. 15.01, American Society for Testing and Materials, West Conshohocken, PA, 2004.

34. ASTM C 1161 "Test Method for Flexural Strength of Advanced Ceramics at Ambient Temperature," *Annual Book of ASTM Standards*, Vol. 15.01, American Society for Testing and Materials, West Conshohocken, PA, 2004.

35. J. Kübler, (a) "Fracture Toughness of Ceramics Using the SEVNB Method: Preliminary Results," *Ceram. Eng. Sci. Proc.*, 18 [4] 155-162 (1997); (b) "Fracture Toughness of Ceramics Using the SEVNB Method; Round Robin," VAMAS Report No. 37, EMPA, Swiss Federal Laboratories for Materials Testing & Research, Dübendorf, Switzerland, 1999.

36. ASTM C 1421 "Test Method for Determination of Fracture Toughness of Advanced Ceramics at Ambient Temperature," *Annual Book of ASTM Standards*, Vol. 15.01, American Society for Testing and Materials, West Conshohocken, PA, 2004.

37. ASTM C 1368, "Test Method for Determination of Slow Crack Growth Parameters of Advanced Ceramics by Constant Stress-Rate Flexural Testing at Ambient Temperature," *Annual Book of ASTM Standards*, Vol. 15.01, American Society for Testing and Materials, West Conshohocken, PA, 2004.

38. ASTM C 1465, "Test Method for Determination of Slow Crack Growth Parameters of Advanced Ceramics by Constant Stress-Rate Flexural Testing at Elevated Temperatures," *Annual Book of ASTM Standards*, Vol. 15.01, American Society for Testing and Materials, West Conshohocken, PA, 2004.

39. ASTM C 1259, "Test Method for Dynamic Young's Modulus, Shear Modulus, and Poisson's Ratio for Advanced Ceramics by Impulse Excitation of Vibration," *Annual Book of ASTM Standards*, Vol. 15.01, American Society for Testing and Materials, West Conshohocken, PA, 2004.

40. Y. Murakami, ed., *Stress Intensity Factors Handbook*, Vol. 1, p. 16, Pergamon Press, New York, 1987.

41. ASTM C 1326, "Test Method Knoop Indentation Hardness of Advanced Ceramics," *Annual Book of ASTM Standards*, Vol. 15.01, American Society for Testing and Materials, West Conshohocken, PA, 2004.

42. S. R. Choi and N. P. Bansal, "Strength and Fracture Toughness of Zirconia/Alumina Composites for Solid Oxide Fuel Cells," *Ceram. Eng. Sci. Proc.*, 23 [3] 741-750 (2002); "Processing and Mechanical Properties of Various Zirconia/Alumina Composites for Fuel Cells Applications," NASA/TM-2002-211580, National Aeronautics and Space Administration, Glenn Research Center, Cleveland, OH, 2002.

43. D. K. Shetty, A. R. Rosenfield, and W. H. Duckworth, "Mixed-Mode Fracture of Ceramics in Diametral Compression," *J. Am. Ceram. Soc.*, 69 [6] 437-443 (1986).

44. V. Tikare and S. R. Choi, "Combined Mode I and Mode II Fracture of Monolithic Ceramics," *J. Am. Ceram. Soc.*, 76 [9] 2265-2272 (1993).

45. V. Tikare and S. R. Choi, "Combined Mode I-Mode II Fracture of 12-mol-%-Ceria-Doped Tetragonal Zirconia Polycrystalline Ceramic," *J. Am. Ceram. Soc.*, 80 [6] 1624-1626 (1997).

46. F. F. Lange, "Transformation Toughening; Part 3. Experimental Observation in the ZrO_2-Y_2O_3 System," *J. Mater. Sci.*, 17 240-256 (1982).

47. M. Shimada, K. Matsushita, S. Kuratani, T. Okamoto, K. Tsukuma, and T. Tsukidate, "Temperature Dependence of Young's Modulus and Internal Friction in Alumina, Silicon Nitride, and Partially Stabilized Zirconia Ceramics," *J. Am. Ceram. Soc.*, 67 [2] C-23-C-24 (1984).

48. K. F. Wesling, D. F. Socie, and B. Beardsley, "Fatigue of Thick Thermal Barrier Coatings," *J. Am. Ceram. Soc.*, 77 [7] 1863-1868 (1994).

Thermal Properties

THERMAL AND MECHANICAL PROPERTIES OF ZIRCONIA COATINGS PRODUCED BY ELECTROPHORETIC DEPOSITION

Bernd Baufeld & Omer van der Biest
Metaalkunde en Toegepaste Materiaalkunde
Katholieke Universiteit Leuven
Kasteelpark Arenberg 44
3001 Leuven, Belgium

Hans-Joachim Rätzer-Scheibe
Institute of Materials Research
German Aerospace Center (DLR)
Linder Höhe
51147 Cologne, Germany

ABSTRACT

The topic of this paper is electrophoretic deposition (EPD) as a cheap and fast coating procedure to obtain thermal barrier coatings. In EPD a coating is obtained by deposition of powder from a suspension under the influence of an electric field and a subsequent consolidation by sintering. Crack free, up to 0.15 mm thick coatings with homogenous morphology and high porosity were obtained.

The elastic modulus of an EPD coating was determined to be 22 GPa at room temperature decreasing to 18 GPa at 1000°C. The thermal conductivity has, depending on porosity, values between 0.5 and 0.6 W/(m·K) at room temperature, which decrease slightly with temperature. After annealing in air at 1100°C for 100 h the thermal conductivity has been increased by about 50 %.

The low elastic modulus and the exceptionally low thermal conductivity, both related to the high porosity, make the EPD coatings a promising candidate for thermal barrier coatings.

INTRODUCTION

The thermal barrier coatings (TBC) for gas turbine engine applications usually consist of partially yttria stabilized zirconia due to its low thermal conductivity and its relatively high thermal expansion coefficient. The most common techniques for depositing TBCs are either air plasma spraying (APS) or electron beam physical vapor deposition (EB-PVD), the first mostly applied for energy transformation and the latter for aeronautics. In the as-coated condition the APS coatings have at room temperature thermal conductivity values in the range of 0.6–1.4 W/(m·K)[1-5], while the EB-PVD coatings usually have higher values of 1.5–2.0 W/(m·K)[1, 4-7]. However, these reliable and established coating techniques are cost and time intensive. As a cheaper and faster coating procedure, the use of electrophoretic deposition (EPD) was suggested[8].

EPD is a colloidal deposition process, where in a first step the powder in a suspension is deposited under the influence of an electric field on an electrode and then, in a second step, the coating is consolidated by a heat treatment. A wide range of metallic as well as ceramic powders are already studied for EPD, either in aqueous or organic liquids[9-11].

Until now, not much work has been performed in investigating the material properties of ceramic coatings fabricated by EPD. A few authors report about hardness and elastic modulus of EPD coatings and composites, determined with indenter techniques[12-15]. Preliminary results concerning elastic modulus and damping derived from impulse excitation technique (IET) are published by some of the present authors[16]. Reports about the thermal conductivity of ceramic EPD coatings were not found in the open literature.

Yet, for the application of ceramic coatings prepared by EPD these parameters are essential. In this paper results about thermal conductivity measured by the laser flash method and elastic modulus derived from IET will be presented in dependence on the temperature.

EXPERIMENTAL PROCEDURE
Specimen preparation

Partially yttria stabilized zirconia powder (5 mol% Y_2O_3 stabilized grade Melox 5Y XZO 99.8%) with the addition of 0.75 wt% cobalt oxide nanopowder as a sintering aid (Aldrich Cobalt (II, III) oxide 99.8%) was used. Two different suspensions were investigated, one methyl-ethyl-ketone (MEK) based and one ethanol based. The powder was mixed and ball milled with zirconia balls in MEK or in ethanol with a multidirectional mixer (Turbula type) for one day. For each EPD session fresh suspensions were prepared by adding n-butylamine (BA) and a respective suspension stabilization additive (see Tab. I). The suspensions were first mechanically, then ultrasonically, and finally again mechanically stirred, for 15 minutes each sequence.

The EPD cells consisted of non-conductive containers with the substrate as one electrode. The specimens for the laser flash experiment were placed in a cell with vertical plane electrodes, while the specimen for the IET was installed horizontally in the center of a cylindrical counter-electrode. More details of the latter cell can be found elsewhere[16]. During EPD, the suspension was subjected in both cases to further mechanical stirring. The experimental conditions and the results are presented in Tab. I.

For the laser flash method a special type of sample was used, which has been called quasi-free-standing coating[4], and allows to measure the thermal conductivity of fragile and thin coatings. This sample type consists of a sapphire support (diameter 12.7 mm, thickness 1 mm) on which the ceramic EPD coating is applied (Fig. 1). For the laser flash measurement of semi-transparent zirconia it is necessary to add a thin platinum layer on both sides of the ceramic coating, which were sputter coated before and after the EPD process with a thickness of about 5 μm. At the front surface this Pt layer prevents the laser beam penetration into the interior of the sample and ensures an effective and uniform absorption of the laser pulse. The thin platinum

Tab. I Conditions and results of the EPD (for the porosity a density of 6.05 g/cm^3 of fully dense zirconia[17] was assumed)

Specimen name	EPD1	EPD2	EPD3
Test type	IET	Laser flash	Laser flash
Electrical field strength [V/mm]	17	8.6	2.8
Suspension based on	MEK (20 vol% BA)		Ethanol (3 vol% BA)
Suspension additives	1 wt% nitro-cellulose		1 wt% Dolapix
Powder load	63 g/l		96 g/l
EPD time [s]	50	120	120
EPD coating thickness [mm]	0.15	0.11	0.10
porosity	0.42	0.47	0.38

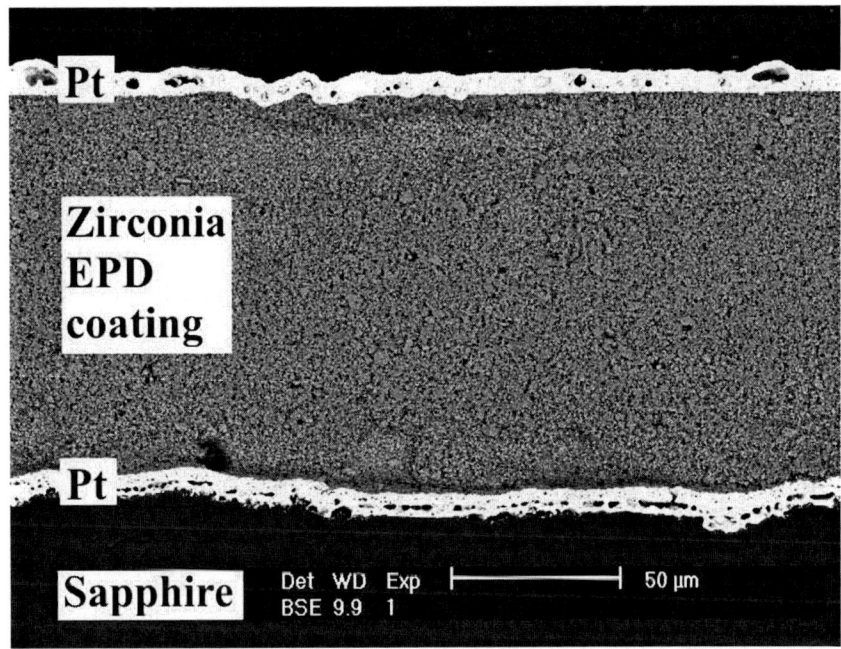

Fig. 1 Cross section of a quasi-free standing EPD coating (specimen EPD2).

layer on the rear EPD surface between sapphire and coating serves as emission layer for measuring the temperature signal by the infrared detector. In addition, this opaque layer prevents the infrared detector from viewing into the sample interior and thus not giving an accurate temperature rise curve for the rear surface. Furthermore, for the EPD process the metal layer on the electrical nonconductive sapphire forms the necessary electrode.

For the coating of an IET specimen, a cylindrical rod of polycrystalline Ni based superalloy IN625 with a length of 65 mm and a diameter of 6 mm was used.

After EPD, the containers were drained and the specimens dried in air. The green specimens were sintered in a conventional resistance furnace in hydrogen atmosphere at 1200°C for 6 h. The density of the EPD coating was determined by geometrical means.

For the IET experiments, the EPD coated specimen was cut to a length of 47 mm and a second, uncoated IN625 specimen (diameter 6 mm) to a length of 54 mm.

Impulse Excitation Technique

IET is a standard test method for the assessment of the dynamic elastic modulus of isotropic homogeneous materials[18]. Basis of IET is the stimulation of the resonance frequency by gently hitting it, and analyzing the resultant vibration signal. The elastic modulus is proportional to the square of the resonance frequency and depends furthermore on mass, geometry and Poisson's ration of the specimen. In the case of coated cylindrical specimens it has been shown by Schrooten et al.[19], that the elastic modulus of the coating E_c can be expressed as a function of the total elastic moduli in the flexural mode of the coated system E_{tot} and of the substrate E_s, of the radius R of the substrate and of the thickness of the coating t:

$$E_C = \frac{E_{tot}(R+t)^4 - E_s R^4}{(R+t)^4 - R^4} \quad (1)$$

Therefore, in order to obtain the elastic modulus of the coating two specimens, one with and one without coating, have to be tested.

The IET tests were performed in a graphite furnace HTVP-1750C in Ar atmosphere up to 1000°C, with a heating and cooling rate of 2°C/min, and analyzed by the RFDA software (both IMCE, Diepenbeek, Belgium) resulting in elastic modulus and damping in dependence of the temperature. More details about this set up can be found elsewhere[20].

Laser flash technique

The thermal diffusivity of the EPD coating was measured using the laser flash method (LFA427, NETZSCH). In this technique, the front side of a plan-parallel disk is heated by a short laser pulse (neodymium doped gallium gadolinium garnet laser with a wavelength of 1.064 μm). The heat diffuses through the sample and the resulting temperature rise of the rear surface is recorded by an infrared detector (2 to 5 μm wavelength range). The thermal diffusivity is determined from the measured temperature rise and analyzed by the NETZSCH LFA Proteus® software. Thermal conductivity of the coating k was then calculated using the relationship $k = c_p \rho \alpha$, where c_p is the temperature dependent specific heat[21], ρ the density, and α the thermal diffusivity. The analysis is based on a one-dimensional heat flow and takes the heat loss from all surfaces of the disk-shaped sample into account. In contrast to former analyses[4], the effect of the radiative heat transfer in the interior of the sample, which is especially noticeable at higher temperatures[22], was considered by this software.

The tests were performed in vacuum or in Ar gas at atmospheric pressure from room temperature up to 1150°C. In order to study sintering effects on the thermal conductivity, which are reported for conventional TBC systems[5], specimen EPD3 was not only investigated in the as-received condition, but also after 100 h annealing in air at 1100°C.

RESULTS

Electrophoretic deposition

Under the chosen conditions up to 0.15 mm thick coatings with smooth surface and without visible cracks were obtained (Tab. I). The coatings have a high porosity with evenly distributed

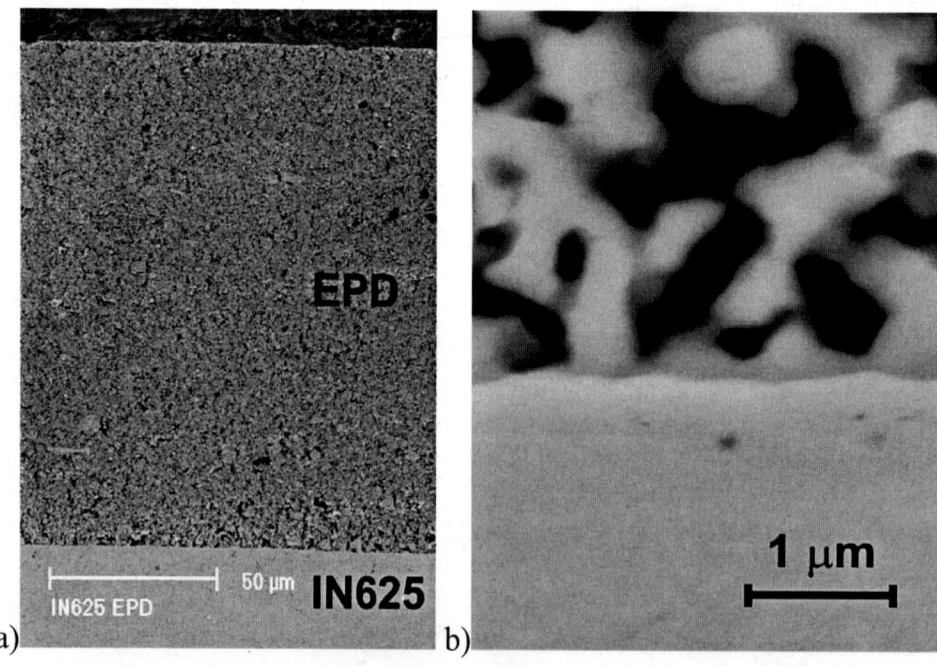

Fig. 2 Cross section of the IET specimen EPD1 showing an overview of the EPD coating (a) and in detail the interface between EPD coating and substrate (b).

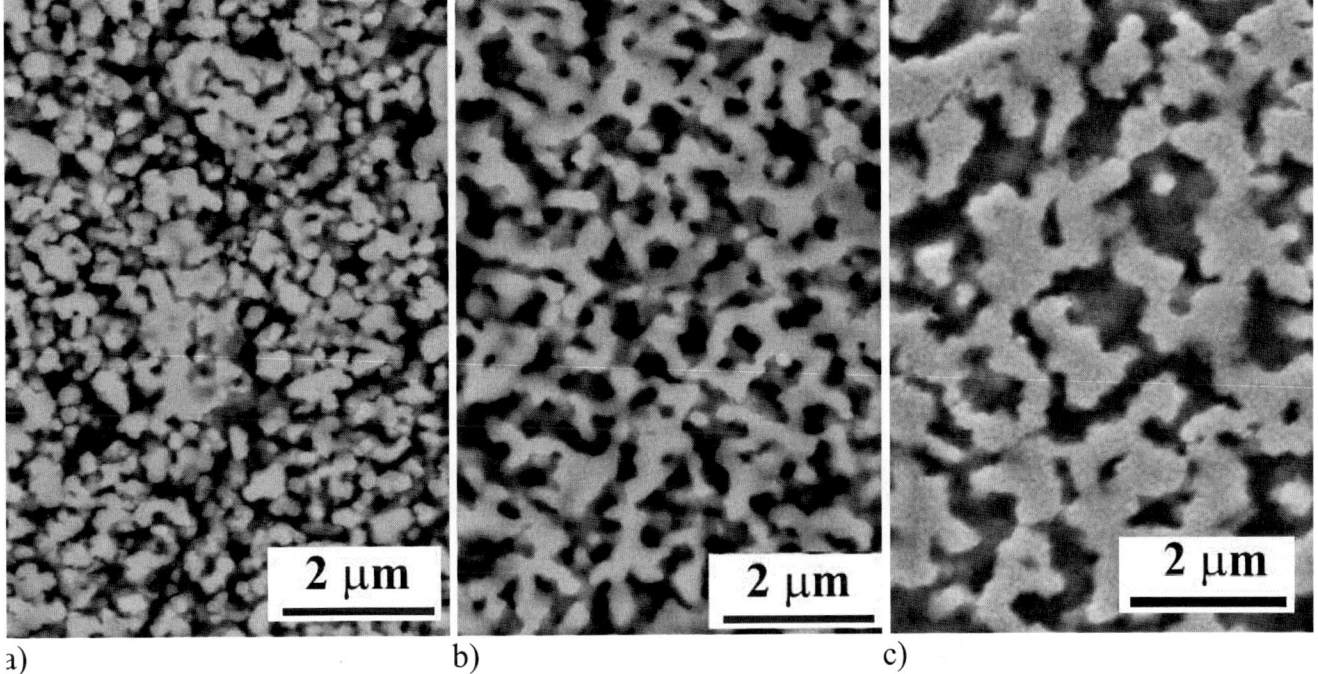

a)   b)   c)

Fig. 3 Microstructure of EPD2 (a) and EPD3 (b) in the as-received state, and EPD3 after 100 h heat treatment at 1100°C in air (c).

pores and some agglomerates (Fig. 1,Fig. 2, Fig. 3a and b). The adherence of the coating is sufficient to allow, for example, mechanical cutting of a specimen. For EPD1 a compressive residual stress of (-91 ± 17) MPa was measured, using XRD. The 100h heat treatment at 1100°C resulted in a significant microstructural change of the coating with increased grain and pore sizes (Fig. 3c).

Impulse excitation technique

The elastic modulus of the coated specimen and of the substrate alone was determined for room temperature to be 186 GPa and 209 GPa, respectively. According to equation (1), the elastic modulus of the EPD coating is calculated to be 22 GPa. With increasing temperature the elastic modulus of IN625 and of the coated system decreases significantly (Fig. 4a). The elastic modulus of the coating, however, decreases only slightly to 18 GPa at 1000°C. While the elastic modulus of the EPD coating is much smaller than the one of dense zirconia, it is in the same order as reported for air plasma sprayed[23] or EB-PVD zirconia coatings[24].

For the coated as well as for the uncoated specimen the damping increases with temperature (Fig. 4b). At temperatures below 850°C, however, the damping of the coated specimen is significantly higher than for the uncoated specimen, supposedly due to energy dissipating processes within the porous ceramic coating. Such an increased damping is a beneficial property, since this may reduce vibration or noise in a turbine. For example, air plasma sprayed zirconia coatings designed for this task were studied by Yu et. al.[25].

Laser flash method

The thermal conductivity of the EPD coatings prepared from different suspensions are at room temperature roughly 0.5 and 0.6 W/(m·K) for the MEK based and the ethanol based EPD coating, respectively, decreasing with temperature (Fig. 5a). Due to the lower porosity the MEK suspension resulted in a coating with slightly lower thermal conductivity than the one of the

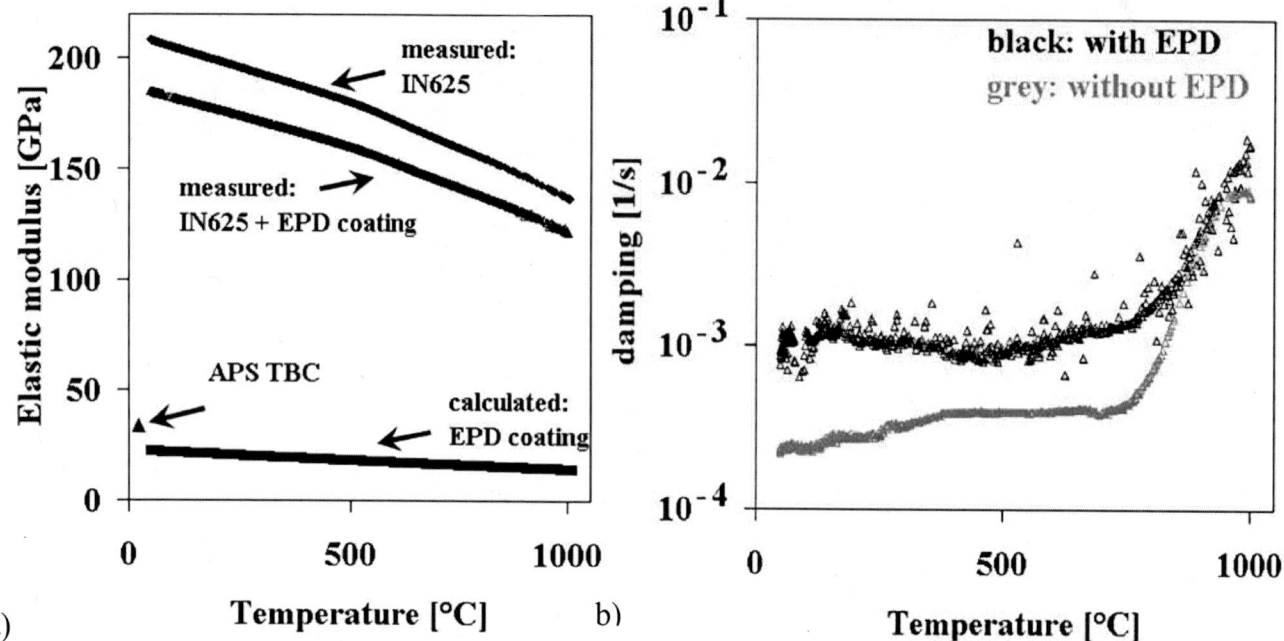

a)   b)

Fig. 4 a: Temperature dependence of the elastic modulus of the coated system, the uncoated substrate, and of the coating. For comparison, the elastic modulus of an APS zirconia coating is given[23]. b: Temperature dependence of the damping of a coated and an uncoated specimen.

ethanol based suspension. The testing atmosphere has only a minor importance with slightly higher thermal conductivity at lower temperatures in Ar atmosphere compared to vacuum (Fig. 5a).

In comparison with typical APS or EB-PVD coatings (Fig. 5b), the as-received EPD coatings have a remarkably low thermal conductivity. However, annealing at 1100°C for 100 h in air substantially increased the thermal conductivity by about 50 %. The thermal conductivity is still lower than for commonly used TBC systems, for which also the thermal conductivity

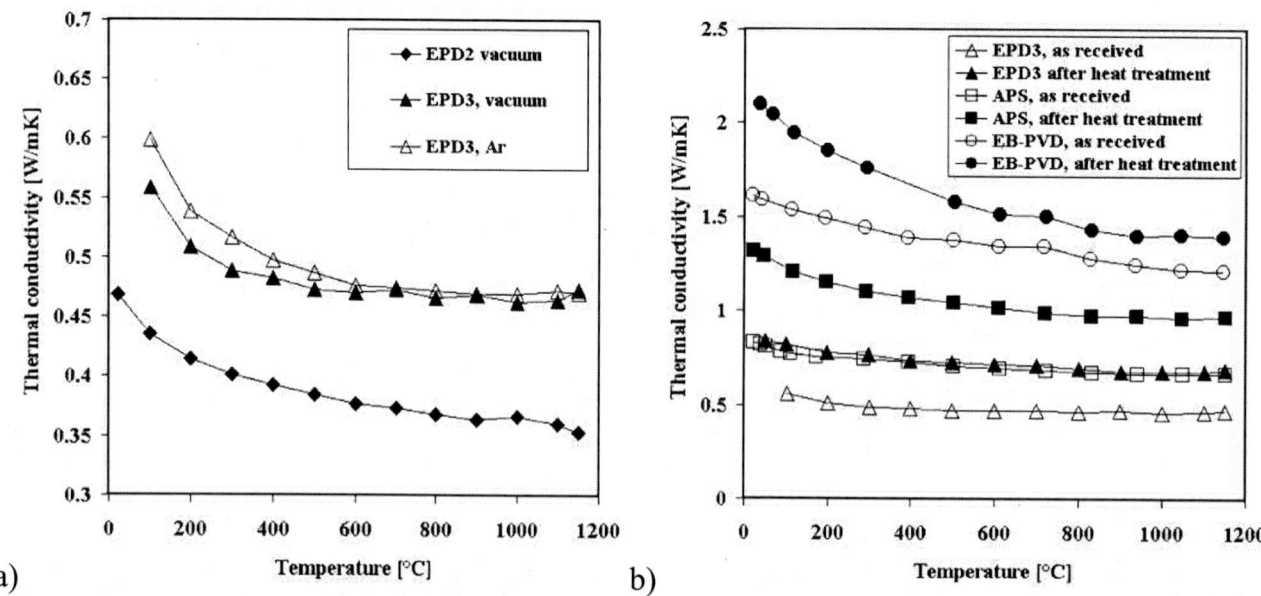

a)   b)

Fig. 5 Thermal conductivity in dependence on temperature. a: comparison between EPD2 and EPD3 in vacuum and in Ar atmosphere. b: comparison of EPD, APS and EB-PVD zirconia coatings in vacuum for as-received specimens and after 100 h annealing in air at 1100°C.

increases significantly (Fig. 5b). Details about the EB-PVD and APS measurements can be found elsewhere[5]. It is worthwhile to note, that in contrast to previously reported results[4], the thermal conductivity of the EB-PVD coating (0.280 mm thick) does not increase at higher temperatures. This is due to the fact, that in the present work the data were corrected by taking the contribution of radiation to the heat conduction[22] into account.

CONCLUSIONS

Crack free, but porous ceramic EPD coatings of more than 0.1 mm thickness have been obtained by MEK, as well as by ethanol based suspensions. The mechanical properties of the coatings were studied with IET resulting in a low elastic modulus of 22 GPa decreasing only slightly with increasing temperature. The EPD coating increased the damping properties of the coated system. The thermal conductivity of the EPD coating proved to be remarkably low. Annealing at 1100°C, however, increased the thermal conductivity by about 50 %. The thermal and mechanical properties of the EPD coatings are very promising for the application as TBC. However, adherence, especially for thicker coatings, is an issue to be studied in future.

ACKNOWLEDGMENTS

B. Baufeld acknowledges an individual Marie-Curie fellowship of the European Commission Nr. MEIF-CT-2005-010277.

REFERENCES

[1]R. B. Dinwiddie, S. C. Beecher, W. D. Porter and B. A. Nagaraj, "The effect of thermal aging on the thermal conductivity of plasma sprayed and EB-PVD thermal barrier coatings", in *Turbo Expo '96* (ASME, Birmingham UK, 1996).

[2]A. J. Slifka, B. J. Filla, J. M. Phelps, G. Bancke and C. C. Berndt, "Thermal conductivity of a zirconia thermal barrier coating", *J. Therm. Spr. Tech.*, **7**, 43-46 (1998).

[3]A. A. Kulkarni, A. Vaidya, A. Goland, S. Sampath and H. Herman, "Processing effects on porosity-property correlations in plasma sprayed yttria-stabilized zirconia coatings", *Mat. Sci. Eng. A*, **359**, 100-11 (2003).

[4]H. J. Rätzer-Scheibe, U. Schulz and T. Krell, "The effect of coating thickness on the thermal conductivity of EB-PVD PYSZ thermal barrier coatings", *Surf. Coat. Tech.*, **200**, 5636-44 (2006).

[5]H. J. Rätzer-Scheibe and U. Schulz, "The effect of heat treatment and gaseous atmosphere on the thermal conductivity of APS and EB-PVD thermal barrier coatings", *Surf. Coat. Tech.*, submitted, (2007).

[6]K. An, K. S. Ravichandran, R. E. Dutton and S. L. Semiatin, "Microstructure, texture, and thermal conductivity of single-layer and multilayer thermal barrier coatings of Y2O3-stabilized ZrO2 and Al2O3 made by physical vapor deposition", *J. Am. Ceram. Soc.*, **82**, 399-406 (1999).

[7]J. R. Nicholls, K. J. Lawson, A. Johnstone and D. S. Rickerby, "Low Thermal Conductivity EB-PVD Thermal Barrier Coatings", *Mat. Sci. For.*, **369-372**, 595-606 (2001).

[8]O. van der Biest, E. Joost, J. Vleugels and B. Baufeld, "Electrophoretic deposition of zirconia layers for thermal barrier coatings", *J. Mat. Sci.*, **41**, 8086-8092 (2006).

[9]O. Van der Biest and L. Vanderperre, "Electrophoretic deposition of materials", *Annu. Rev. Mater. Sci.*, **29**, 327-52 (1999).

[10]A. R. Boccaccini and I. Zhitomirsky, "Application of electrophoretic and electrolytic deposition techniques in ceramics processing", *Cur. Op. Sol. St. Mat. Sci.*, **6**, 251-60 (2002).

[11]L. Besra and M. Liu, "A review on fundamentals and applications of electrophoretic deposition (EPD)", *Prog. Mat. Sci.*, **52**, 1-61 (2007).

[12]X. Wang, P. Xiao, M. Schmidt and L. Li, "Laser processing of yttria stabilised zirconia/alumina coatings on Fecralloy substrates", *Surf. Coat. Tech.*, **187**, 370-76 (2004).

[13]S. Put, J. Vleugels, G. Anne and O. Van der Biest, "Functionally graded ceramic and ceramic-metal composites shaped by electrophoretic deposition", *Coll. Surf. A*, **222**, 223-32 (2003).

[14]X.-J. Lu, X. Wang and P. Xiao, "Nanoindentation and residual stress measurements of yttria-stablized zirconia composite coatings produced by electrophoretic deposition", *Thin Solid Films*, **494**, 223-27 (2006).

[15]P. Hvizdos, J.-M. Calderon Moreno, J. Ocenasek, L. Ceseracciu and G. Anne, "Mechanical Properties of Alumina/Zirconia Functionally Graded Material Prepared by Electrophoretic Deposition", *Key. Eng. Mat.*, **290**, 332-35 (2005).

[16]B. Baufeld and O. van der Biest, "Development of thin ceramic coatings for the protection against temperature and stress induced rumpling of the metal surface of turbine blades", *Key. Eng. Mat.*, **333**, 273-76 (2007).

[17]R. P. Ingel and D. I. Lewis, "Lattice parameters and density for Y_2O_3-stabilized ZrO_2", *J. Am. Ceram. Soc.*, **69**, 325-32 (1986).

[18]ASTM, "Standard Test Method for Dynamic Young's Modulus, Shear Modulus, and Poisson's Ration by Impulse Excitation of Vibration", **E 1876-99**, 1075-84 (1999).

[19]J. Schrooten, G. Roebben and J. A. Helsen, "Young's modulus of bioactive glass coated oral implants: porosity corrected bulk modulus versus resonance frequency analysis", *Scr. Mat.*, **41**, 1047-53 (1999).

[20]G. Roebben, B. Bollen, A. Brebels, J. Van Humbeeck and O. Van der Biest, "Impulse excitation apparatus to measure resonant frequencies, elastic moduli and internal friction at room and high temperature", *Rev. Sci. Instrum.*, **68**, 4511-15 (1997).

[21]T. Krell, "Thermische und thermophysikalische Eigenschaften von elektronenstrahl-gedampften chemisch gradierten Al_2O_3/PYSZ-Wärmedämmschichten", Dissertation RWTH Aachen, (2000).

[22]F. Schmitz, D. Hehn and H. R. Maier, "Evaluation of laser-flash measurements by means of numerical solutions of the heat conduction equation", *High Temp., High Pres.*, **31**, 203-11 (1999).

[23]J. S. Wallace and J. Ilavsky, "Elastic Modulus Measurements in Plasma Sprayed Deposits", *J. Therm. Spr. Tech.*, **7**, 521-26 (1998).

[24]U. Schulz, K. Fritscher, C. Leyens and M. Peters, "Influence of processing on microstructure and performance of EB-PVD thermal barrier coatings", *J. Eng. Gas Turb. Pow.*, **124**, 1-8 (2000).

[25]L. Yu, Y. Ma, C. Zhou and H. Xu, "Damping capacity and dynamic mechanical characteristics of the plasma-sprayed coatings", *Mat. Sci. Eng. A*, **408**, 42-46 (2005).

EFFECT OF AN OPAQUE REFLECTING LAYER ON THE THERMAL BEHAVIOR OF A THERMAL BARRIER COATING

Charles M. Spuckler
NASA Glenn Research Center
21000 Brookpark Rd.
Cleveland, OH 44145

ABSTRACT

A parametric study using a two-flux approximation of the radiative transfer equation was performed to examine the effects of an opaque reflective layer on the thermal behavior of a typical semitransparent thermal barrier coating on an opaque substrate. Some ceramic materials are semitransparent in the wavelength ranges where thermal radiation is important. Even with an opaque layer on each side of the semitransparent thermal barrier coating, scattering and absorption can have an effect on the heat transfer. In this work, a thermal barrier coating that is semitransparent up to a wavelength of 5 micrometers is considered. Above 5 micrometers wavelength, the thermal barrier coating is opaque. The absorption and scattering coefficient of the thermal barrier was varied. The thermal behavior of the thermal barrier coating with an opaque reflective layer is compared to a thermal barrier coating without the reflective layer. For a thicker thermal barrier coating with lower convective loading, which would be typical of a combustor liner, a reflective layer can significantly decrease the temperature in the thermal barrier coating and substrate if the scattering is weak or moderate and for strong scattering if the absorption is large. The layer without the reflective coating can be about as effective as the layer with the reflective coating if the absorption is small and the scattering strong. For low absorption, some temperatures in the thermal barrier coating system can be slightly higher with the reflective layer. For a thin thermal barrier coating with high convective loading, which would be typical of a blade or vane that sees the hot sections of the combustor, the reflective layer is not as effective. The reflective layer reduces the surface temperature of the reflective layer for all conditions considered. For weak and moderate scattering, the temperature of the TBC-substrate interface is reduced but for strong scattering, the temperature of the substrate is increased slightly.

INTRODUCTION

Thermal barrier coatings (TBCs) are being developed for use in gas turbine engines. TBCs can be made more effective by decreasing the heat conducted and/or radiated through them. Some thermal barrier coatings are partially transparent to thermal radiation. For example, for thermal radiation purposes zirconia can be semitransparent up to around 5 μm (refs. 1 and 2). In semitransparent materials, both thermal radiation and heat conduction determine the temperatures and the heat transferred. Scattering, absorption, emission, and the refractive index determine the radiative heat transfer in a semitransparent material. The external and internal reflection of an interface between two semitransparent materials depends on the refractive index of the materials on each side of the interface. If thermal radiation is going from a material with a higher refractive index to one with a lower refractive index, there is a total reflection of the radiation at angles greater than the critical angle. Also, the thermal radiation emitted internally by the material depends on the square of the refractive index. The internal thermal radiation passing through the semitransparent interface is decreased by internal surface reflections, which includes total internal reflection, so the energy emitted by the semitransparent layer can not exceed that of a blackbody. If there is an opaque layer on the semitransparent material the radiation emitted into the material depends on the refractive index squared and the emissivity of the opaque layer. The refractive index can have a considerable effect on the temperature profile in a semitransparent layer.

The scattering and absorption coefficients determine the amount of thermal radiation absorbed, emitted, and scattered by a semitransparent material. These coefficients have units of reciprocal length. The reciprocal of the coefficients can be considered as the mean distance traveled before absorption or scattering occurs (ref. 3 page 424). The smaller the coefficient the larger the distance thermal radiation will travel before being absorbed or scattered. When thermal radiation is absorbed or emitted by a material its temperature changes. Absorption and emission therefore have a direct effect on the temperature of a material. Scattered thermal radiation has no effect on the temperature of a material unless it is absorbed. Scattering in some cases can augment the absorption because it increases the path length of radiation through the material. Here scattering will act as additional absorption in determining the temperature profiles in a material ref. 4.

Putting a highly reflecting layer on the TBC is being considered as a method to improve their performance by reducing the radiative heat flux through the TBC. In reference 5 a gray semitransparent highly reflectance multilayered TBC system was designed and modeled. The results indicate the coating has potential for reducing the metal temperature. Here an opaque reflecting layer is considered. The absorption coefficient, scattering coefficient, and index of refraction still determine the radiative heat transfer through the semitransparent layer even though it is between opaque layers. Because scattering depends on the material structure and the absorption is affected by impurities and temperature, the absorption and scattering coefficients are increased and decreased from the base line values of $a = 0.1346$ cm^{-1} for the absorption coefficient and $\sigma_s = 94.38$ cm^{-1} for the scattering coefficient. These coefficients are a wavelength integrated average of those in ref. 2 for zirconia in wavelengths where it is semitransparent. The thermal behavior of a TBC with a highly reflecting opaque layer is compared to a normal TBC as a function of scattering and absorption.

MODEL

The models used, figure 1, are semi-infinite semitransparent layers on a substrate. One semitransparent layer does not have a reflective coating and the other has an opaque reflective coating. The semitransparent layer is semitransparent for thermal radiation up to 5 μm. For radiation above 5 μm the semitransparent layer is opaque. There is diffuse radiative and convective heat transfer on each side of the layers. The external radiative heating is q_{r1}° and q_{r2}°. The hot side gas and surrounding temperatures, T_{s1} and T_{g1}, cold side temperatures T_{s2} and T_{g2}, heat transfer coefficients, thickness of reflective coating d_{rf}, TBC thickness d_{TBC}, and substrate thickness d_{sub} are given in Table 1. TBCs of the type that would be on a combustor liner and on a blade or vane are considered. For both the combustor and the blade and the emissivity of the back side of the metal substrate, ε_m, is 0.6. The thermal conductivity of TBC and the substrate are 0.8 w/mK and 33 w/mK. These conditions including those in Table 1 except for the reflective layer thickness were used by Siegel (ref. 6) to determine internal radiation effects in a zirconia based TBC. The thermal conductivity of the reflective coating, which is based on a dense zirconia-alumina multilayer coating, is 2.8 w/mK. The refractive index, n, of the semitransparent layer is 2.1. The refractive index of the gas is assumed to be one. The

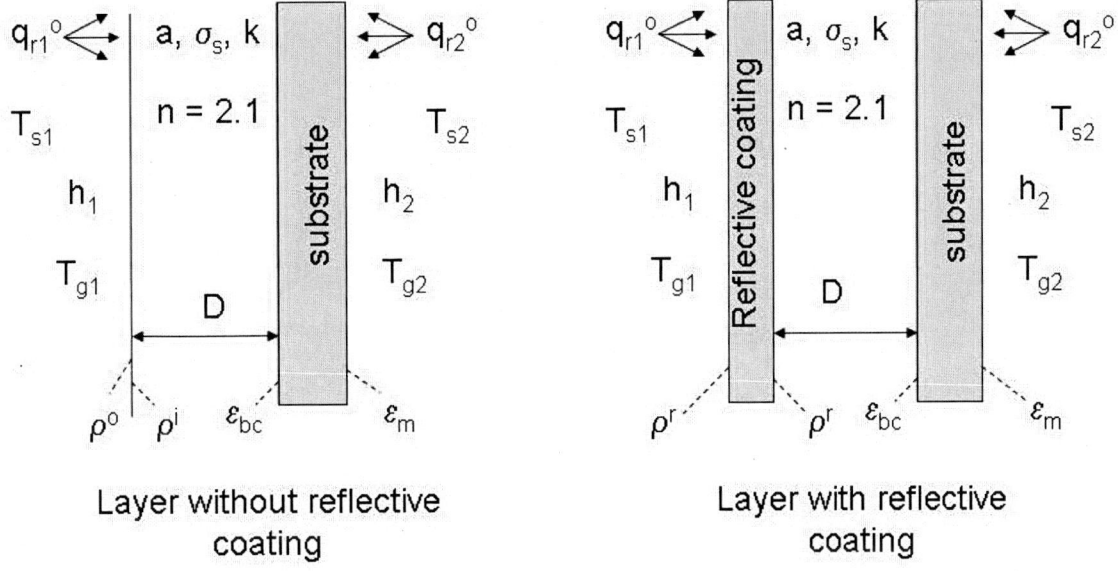

Layer without reflective   Layer with reflective
        coating   coating

Figure 1 Heat transfer model

emissivity of the bond coat ε_{bc} is 0.3. The reflectance on both sides of the opaque highly reflective coating ρ^r is assumed to be 0.9. The external surface reflection for the layer without a reflective coating, ρ^o, was calculated using Fresnel's equation for a non-absorbing layer. This assumption should be good for the absorption coefficients used here (ref. 7 and ref. 3 page 88). The internal surface reflection, ρ^i, for the uncoated TBC was determined from a relationship

Table 1

	T_{s1} K	T_{g1} K	T_{s2} K	T_{g2} K	h_1 w/(m²·K)	h_2 w/(m²·K)	d_{rc} mm	d_{TBC} mm	d_{sub} mm
Combustor	2000	2000	800	800	250	110	0	1.0	0.794
Combustor	2000	2000	800	800	250	110	0.111	0.889	0.794
Blade	2000	2000	1000	1000	3014	3768	0	0.25	0.762
Blade	2000	2000	1000	1000	3014	3768	0.111	0.139	0.762

using the refractive index and external surface reflection in ref. 8. A two flux approximation to the radiative transfer equation was used to calculate the heat flux and the temperature profiles. The boundary conditions for the two flux equations in ref. 9 were modified to account for the opaque substrate and reflective coating.

EFFECT OF ABSORPTION AND SCATTERING ON COMBUSTOR LINER TBC WITH AND WITHOUT AN OPAQUE HIGHLY REFLECTIVE COATING

Temperature profiles

The temperature profiles in the TBC and substrate for a system with and without a highly reflective coating on a combustor liner as a function of absorption are shown in figure 2 for the base line scattering coefficient of 94.38 cm^{-1}. The temperature in the TBC and substrate decrease significantly when a 0.111 mm. opaque layer with a reflectivity of 0.9 is used. Depending on the absorption the decrease in the gas-TBC interface temperature is between about 127 K for 0.0013 cm^{-1} absorption coefficient and 277 K for 67.26 cm^{-1} absorption coefficient.

The TBC-substrate interface temperature decrease is approximately 173 K for 0.0013 cm^{-1} absorption coefficient and 213 K for 13.46 cm^{-1} absorption coefficient. The temperature

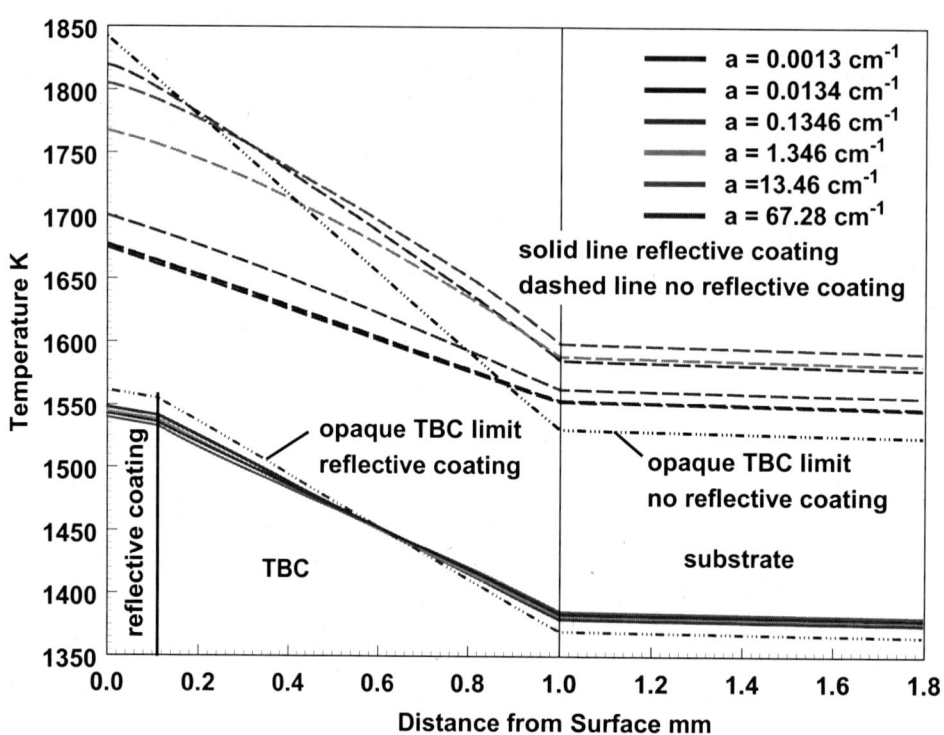

Figure 2. Temperature profiles in reflective coating, TBC, and combustor liner as a function of absorption for $\sigma_s = 94.38$ cm^{-1}.

difference due to a change in absorption also decreases significantly with a reflective coating compared to no reflective coating. Without a reflective coating the temperature change at the gas-TBC interface due to varying the absorption is about 145 K compared to 9 K with a coating. At the TBC-substrate interface, this temperature change is about 46 K without the reflective coating and about 6 K with the coating. Also, for this scattering coefficient, when there is no reflective coating the curvature of the temperature profiles indicate thermal radiation is playing a role. When an opaque reflective coating is present the thermal radiation role is decreased.

Interface temperatures and heat flux

The effect of a highly reflective coating compared to no coating on the gas-TBC interface temperature as a function of absorption and scattering is presented in figure 3. The reflective coating reduces the gas-TBC interface temperatures significantly for higher absorption and all scattering considered. For lower absorption the effects of the reflective coating are reduced as the scattering is increased. For very low absorption and very strong scattering (high scattering coefficients) the gas-TBC interface temperatures are about the same for layers with and without a reflective coating. When a reflective coating is applied to a semitransparent layer the effects of absorption and scattering on the gas-TBC interface temperatures are reduced significantly compared to the uncoated TBC. Only for weaker scattering (smaller scattering coefficients) does absorption have an effect on the gas-TBC interface temperature when an opaque reflective

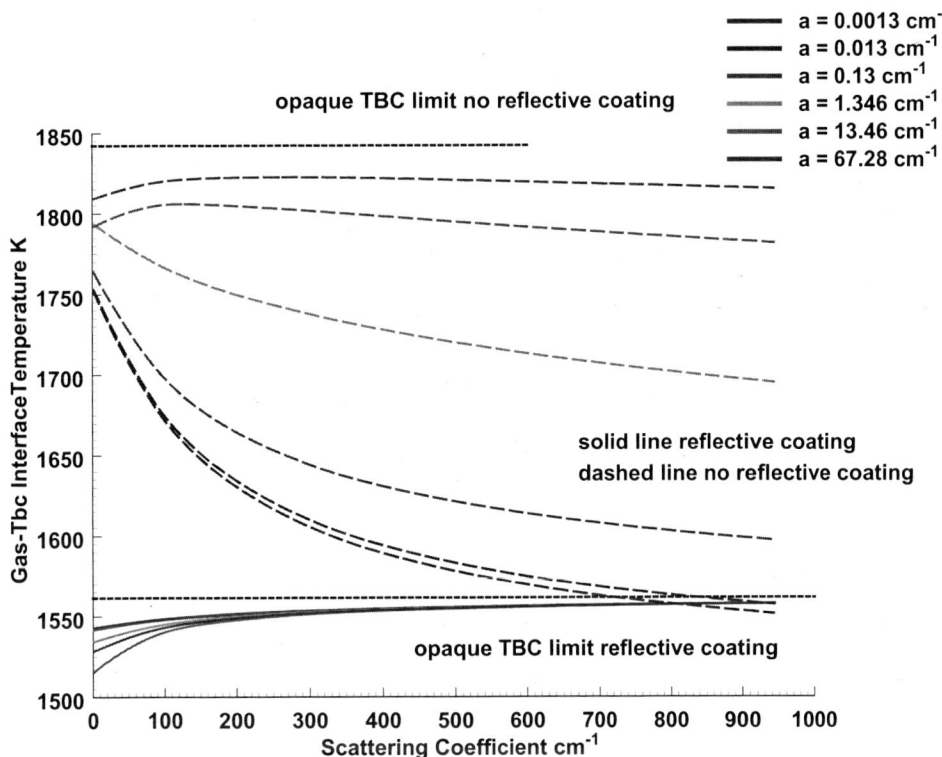

Figure 3. Gas-TBC interface temperature as a function of scattering and absorption for a TBC on a combustor liner

coating is present. The opaque limit shown in the figure is the temperature the TBC-gas interface would have if the TBC was opaque. For both the TBC without the reflective coating and the layer with the reflective coating the gas-TBC interface temperatures are less than the opaque limit, but for the TBC with a reflective coating the temperatures are approaching the opaque limit for strong scattering (higher scattering coefficients).

The temperatures of the TBC-substrate interface for two the layers are shown in figure 4. There is a large decrease in the TBC substrate interface temperature for weak scattering when a reflective coating is used. This temperature difference can be as high as 289 K. For very strong scattering and very low absorption this temperature difference decreases to around 10 K. The temperature difference decreases as scattering increases for all except the highest absorption and low scattering where the temperature difference decreases as absorption increases. When there is a reflective coating the effect of absorption and scattering is decreased substantially compared to a TBC without a reflective coating. For the layer with a reflective coating the effect of absorption decreases with scattering and only with weaker scattering is there an effect. For a TBC without a reflective coating the effect of absorption in general increases as the scattering increases. For a layer with an opaque reflective coating the TBC-substrate temperatures are higher for semitransparent TBC than an opaque TBC. For the TBC without the opaque reflective layer the TBC-substrate temperatures are lower than the opaque limit for some scattering and absorption coefficients. This was shown before in ref 10. The TBC-substrate interface temperature for a semitransparent TBC being higher than the opaque limit indicates an opaque TBC would perform better for these scattering and absorption conditions. The temperature profiles for the back of the substrate are similar to the TBC-Substrate temperatures and not presented here.

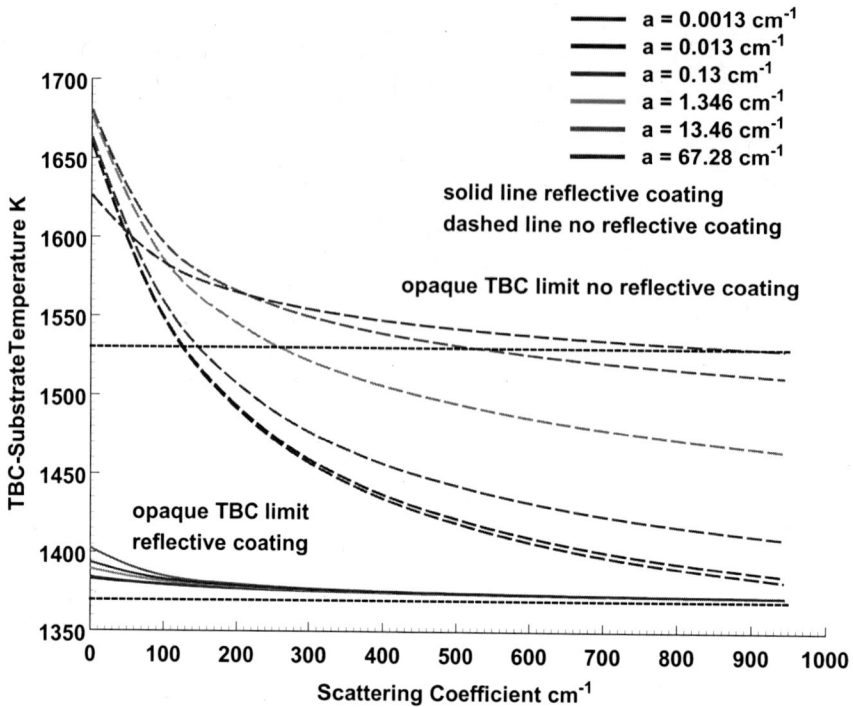

Figure 4 TBC-substrate interface temperatures as a function of scattering and absorption for a TBC on a combustor liner

The heat flux for layers with and without a highly reflective opaque layer as a function of absorption and scattering is shown in figure 5. The profiles and results are similar to the TBC -

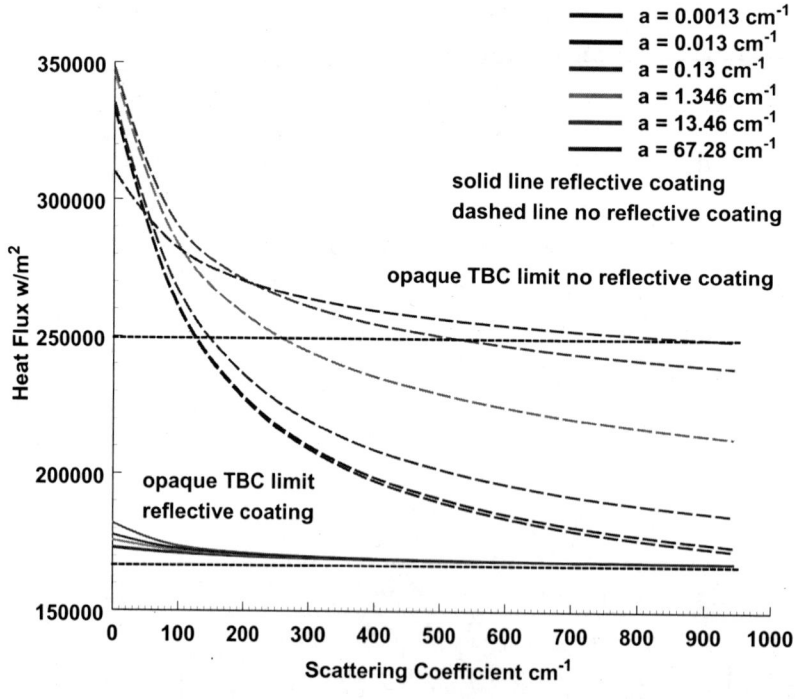

Figure 5 Heat flux as a function of scattering and absorption for TBC on a combustor liner

substrate temperatures. There is a significant decrease in the heat flux through the layer when a highly reflective coating is used especially for smaller scattering coefficients. The difference in heat flux decreases as the scattering coefficient is increased and decreases as the absorption coefficient is decreased except for the highest absorption coefficient and lower scattering coefficients. The effects of scattering and absorption are small and decrease as the scattering coefficient is increased when there is a reflective coating on the TBC. When there is no reflective coating on the TBC the effects of scattering and absorption can be large. With a reflective coating, the heat flux for a semitransparent TBC is larger than an opaque TBC with a reflective coating. This means it would be better to have an opaque TBC as far as the heat flux is concerned. With no reflective coating the heat flux for the semitransparent TBC is higher mainly for lower scattering coefficients. For the conditions where the heat flux is higher than the opaque limit, an opaque TBC would perform better.

EFFECT OF ABSORPTION AND SCATTERING ON BLADE OR VANE TBC WITH AND WITHOUT AN OPAQUE HIGHLY REFLECTIVE COATING

Temperature profiles
         The temperature profiles for a TBC and substrate system for a vane or blade with and without a reflective coating are shown in figure 6. The thickness of the TBC is decreased from

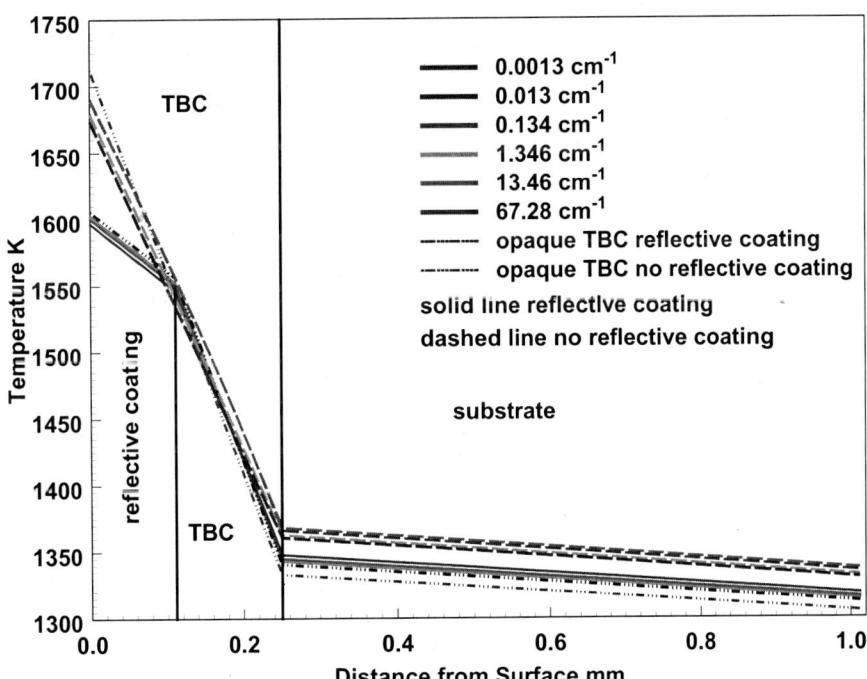

Figure 6 Temperature profiles in reflective coating, TBC, and blade or vane as a function of absorption for σ_s = 94.38 cm^{-1}.

that of a combustor and the convective loads are increased. Here the reflective coating is nearly half of the TBC thickness. The vane and blades are at the front of the turbine so they have a high radiation load for the combustor and soot. When a reflective coating is on the TBC, the surface temperature of the TBC for the base line scattering coefficient of 94.38 cm^{-1} is reduced between

70 and 94 K depending on the absorption. The temperature at the back of the reflective coating is about the same as that of a TBC with out a reflective coating at the same distance 0.111 mm. The TBC-substrate interface temperature decreases between 17 and 24 K depending on the absorption coefficient. The decrease in temperature at the back surface is between 15 and 22 K which is similar to the TBC-substrate interface temperature drop.

Interface temperatures and heat flux

The effect of scattering and absorption on the gas-TBC and gas-reflective coating

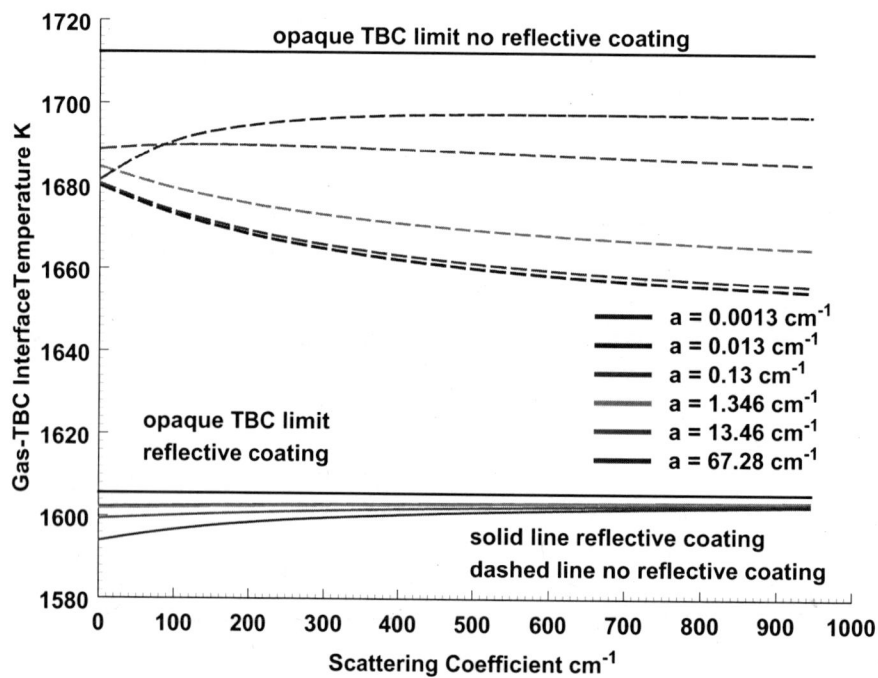

Figure 7. Gas-TBC interface temperature as a function of scattering and absorption for vane or blade

interface temperatures is shown in Figure 7. When there is no reflective coating, the temperature decreases with scattering except for the two highest absorptions coefficients used where the temperatures at first increases with scattering then decreases. The temperature increases with absorption except for the highest absorption used with weak scattering. The effects of absorption are small for absorption coefficients between 0.0013 cm^{-1} and 0.13 cm^{-1}. When there is a reflective coating the effect of absorption on the temperature is small and decreases as the scattering increases. For weak scattering the temperature variation due to absorption is about 12 K. For strong scattering there is essentially no temperature difference. For the TBC without a reflective coating and the TBC with the reflective coating the gas-surface interface temperature is lower than those that would occur if the TBC was opaque. The reduction in the gas TBC interface temperature with a reflective coating can be as high as about 97 K and as low as about 51 K depending on the scattering and absorption.

The temperature of the TBC-substrate interface as a function of scattering and absorption for a layer with and without a reflective coating are shown in Figure 8. The TBC-substrate interface temperature decreases with increased scattering and increases with increased absorption for a TBC with and without a reflective layer. The only exception is for a layer without a reflective coating where the temperature decreases for the highest absorption used when the scattering is moderate to weak. For weak scattering the TBC-substrate interface temperature is at most 40 K lower with a reflective coating. For strong scattering this reverses and the interface temperature is nearly 19 K lower without the reflective coating. For the TBC with the reflective layer the TBC-substrate interface temperature is slightly higher than the temperature that would occur if the TBC was opaque. For strong scattering and low absorption the temperature for the layer without a reflective coating is lower than the opaque limit. This indicates that for some conditions considered an opaque TBC without a reflective coating would perform better than a

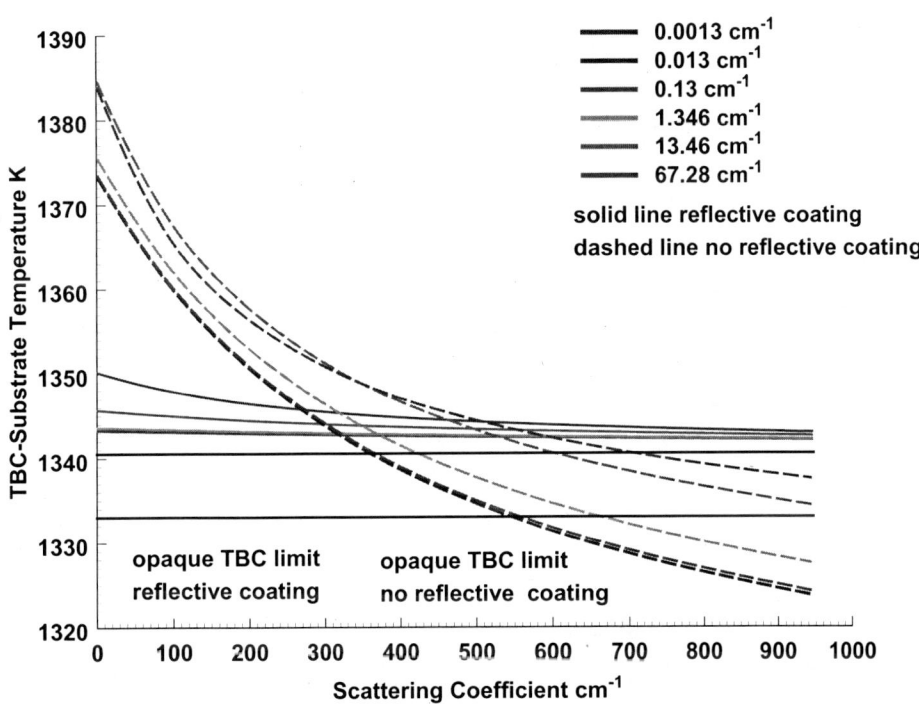

Figure 8 TBC-substrate interface temperatures as a function of scattering and absorption for a TBC on vane or blade

semitransparent TBC with or without a reflective coating if the thermal conductivity remained the same. It is also interesting to note that the temperature for an opaque TBC with a reflective coating is slightly higher that an opaque layer without a reflective coating. This is caused by the higher thermal conductivity coating taking up nearly half of the TBC thickness. Using a simple opaque analysis may give some insight on what thickness of a reflective coating can be tolerated. For the back of the substrate the results are similar to those for the TBC-substrate interface and are not presented.

The heat flux as a function of scattering and absorption is given in figure 9. The profiles and the results are similar to those for the TBC-substrate interface temperatures. For a TBC without a reflective coating the heat flux decreases as the scattering increases. The heat flux also increases as the absorption increases except the largest absorption coefficient when the heat flux decreases when the absorption is increased for weak to moderate scattering. When there is a reflective layer on the TBC the heat flux decreases as the scattering increases and in general increases as the absorption is increased. When there is a reflective coating, the effects of scattering on the heat flux are decreased significantly compared to a TBC without a reflective coating. Also the effects of absorption are reduced for strong scattering when a reflective coating is used. Here as with the TBC-substrate interface temperatures, the heat flux for a TBC with a reflective coating is decreased for weak to moderate scattering. For moderate to strong scattering depending on the absorption, the TBC without a reflective coating has a lower heat

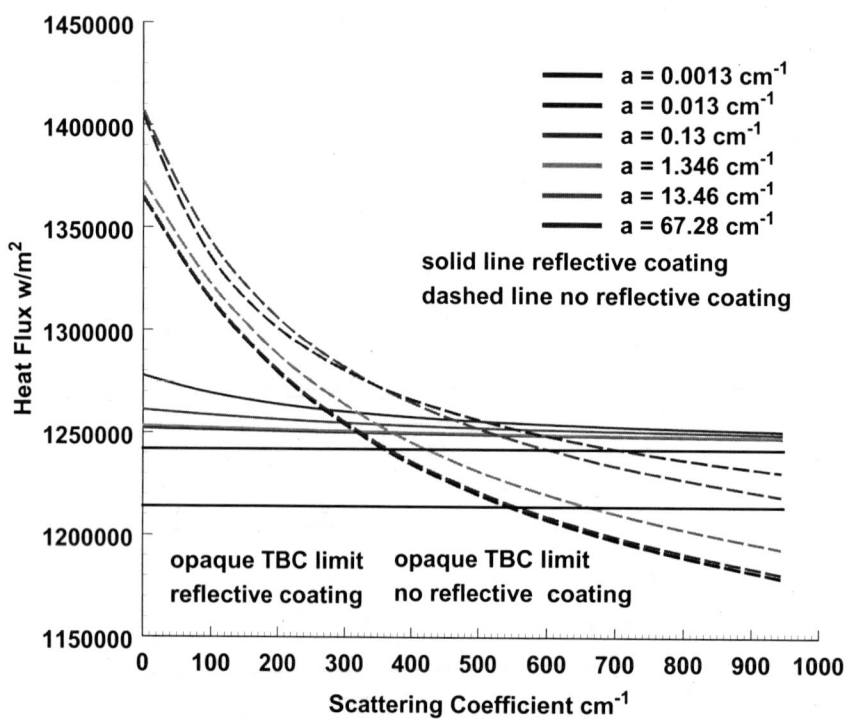

Figure 9. Heat flux as a function of scattering and absorption for TBC on a vane or blade

flux. For a TBC without a reflective coating with strong scattering and low to moderate absorption, the heat flux is lower than that of an opaque TBC. Here like the TBC- substrate temperatures, the heat flux for an opaque TBC with a reflective coating is higher than the heat flux for an uncoated opaque TBC. Also an opaque TBC will perform better than a semitransparent TBC over some of the conditions considered if the thermal conductivity remains the same. It seems like the scattering and absorption condition at which the change in performance occurs is the same for TBC-substrate temperatures and the heat fluxes.

SUMMARY AND CONCLUSIONS

A parametric study using a two-flux approximation to the radiative transfer equation was performed to examine the effects of an opaque reflective layer on the thermal behavior of a typical semitransparent thermal barrier coating on an opaque substrate as a function of scattering and absorption of the TBC. A one dimensional model was used. A TBC 1 mm thick on a 0.794 mm thick substrate with lower convection coefficients was used for the combustor TBC model. A 0.25 mm thick TBC on a 0.762 mm substrate with high convection coefficients was used to model the vane or blade TBC. The highly reflective coating was 0.111mm thick and this reduced the thickness of the underlying TBC to keep the total coating thickness unchanged. The substrate emissivity was 0.3 on the TBC side and 0.6 on the back side. The highly reflective coating has a reflectivity of 0.9 on both sides. The TBC was semitransparent up to 5 μm. Above 5 μm the TBC was opaque. There is radiative and convective heat transfer on each side of the layers. The absorption and scattering coefficients were varied.

For a combustor with a 1.0 mm thick TBC and convection coefficients 250 w/m^{2}K on the hot side and 110 w/m^{2}K on the cold side, the use of a 0.9 reflectivity coating 0.111 mm thick as part of the TBC can be effective in reducing the temperatures and heat flux in a TBC. The reflective coating is very effective at weak or low scattering at all absorption coefficients considered. For these conditions the gas-layer interface temperature decrease can be as high as about 280 K. For moderate to low absorption, the effectiveness of the uncoated semitransparent TBC increases with scattering so that adding a highly reflective coating does not add a significant benefit. For low absorption and strong scattering the gas-layer interface temperature can be higher with a reflective coating than without the reflective coating. At the TBC-substrate interface, the temperature decrease with a highly reflectivity coating can be as high as about 290 K for weak scattering depending on the absorption. The effectiveness of the reflective coating is reduced as the scattering is increased. For the lowest absorption and the highest scattering considered the temperature is 10 K lower when a reflective layer is used. The reduction in heat flux when a highly reflective coating is used is similar to the temperature reductions at the TBC-substrate interface. The largest decreases in heat flux occur for weak scattering. The reduction in heat flux with strong scattering can be quite significant if the absorption is large. It was also noted that when a reflective coating is put on a TBC the effects of the semitransparency, that is varying the scattering and absorption, are reduced drastically.

For a vane or blade with a 0.25 mm thick TBC, convection coefficients 3014 w/m^{2}K on the hot side and 3768 w/m^{2}K on the cold side, and a 0.9 reflectivity coating 0.111mm thick as part of the TBC, the reflective coating has mixed results on the temperatures and heat flux. The gas-TBC interface temperature is reduced by at least 50 K for all scattering and absorption considered. The largest difference in temperature in general occurs for high absorption. The TBC-substrate temperature deceases at least about 40 K for weak scattering. As the scattering increases the positive effects of the reflective coating decrease and become negative when the TBC-substrate temperature for the TBC with the reflective coating becomes greater than the TBC-substrate interface temperature for the uncoated TBC. The same phenomena occur for the heat flux. The decrease in effectiveness is probably due to the higher thermal conductivity reflective coating taking up almost half of the TBC thickness.

The TBC-substrate temperature and the heat flux for the combustor liner TBC without a reflective coating and the TBC–substrate temperatures and heat flux for a blade or vane TBC with and without a reflective layer were higher than that predicted for an opaque TBC material for some scattering and absorption conditions. For these conditions an opaque TBC with the same thermal conductivity would perform better than a semitransparent TBC.

REFERENCES

[1]Wahiduzzaman, S and Morel,T., Effect of Translucence of Engineering Ceramics on Heat Transfer in Diesel Engines, ORNL/Sub/88-22042/2, April 1992

[2]Makino, T., Kunitomo, T., Sakai, I., and Kinoshita, H., Thermal Radiation Properties of Ceramic Materials, *Heat Transfer-Japanese Research,* **13,** [4] 33-50 (1984)

[3]Siegel, R. and Howell, J. R. *Thermal Radiation Heat Transfer,* 4th ed. Taylor & Frances, New York, 2002

[4]Spuckler, C. M. and Siegel, R., "Refractive Index and Scattering Effects on Radiative Behavior of a Semitransparent Layer," *Journal of Thermophysics and Heat Transfer*, 7[2], 302-10 (1993)

5 Wang, D., Huang, X., and Patnaik, P. "Design and Modeling of Multiple Layered TBC System with High Reflectance," *Journal of Material Science,* 41, [19] 6245-55 (2006)

[6]Siegel, R. "Internal Radiation Effects in Zirconia Thermal Barrier Coatings," *Journal of Thermophysics and Heat Transfer*, 10[4], 707-9 (1996)

[7]Cox, R. L., "Fundamentals of Thermal Radiation in Ceramic Materials,"; pp. 83-101 in Symposium on Thermal Radiation of Solids, edited by S. Katzoff, NASA SP-55, 1965

[8]Richmond, J. C., "Relation of Emittance to Other Optical Properties," *Journal of Research of the National Bureau of Standards-C. Engineering and Instrumentation*, 67C [3], 217-26 (1963)

[9]Siegel, R. and Spuckler, C. M., "Approximate Solution Methods for Spectral Radiative Transfer in High Refractive Index Layers," *International Journal of Heat and Mass Transfer*, 37 [Suppl. 1] 403-13 (1994)

[10]Spuckler, C. M., "Effect of Scattering on the Heat Transfer Behavior of a Typical Semitransparent TBC Material on a Substrate," Ceramic Engineering and Science Proceedings, 26[3], 47-54 (2005)

OPTIMIZING OF THE REFLECTIVITY OF AIR PLASMA SPRAYED CERAMIC THERMAL BARRIER COATINGS

A. Stuke, R. Carius, J.-L. Marqués, G. Mauer, M. Schulte, D. Sebold, R. Vaßen, D. Stöver
Institut für Energieforschung (IEF)
Forschungszentrum Jülich GmbH
Jülich, Germany

ABSTRACT

For gas turbine applications, metallic components are coated with a ceramic thermal barrier coating (TBC). The mostly used ceramic material, yttria partially stabilized zirconia (YSZ), absorbs radiation of wavelengths below 5µm only weakly. However, it is in this wavelength range where most of the radiation by walls and gas is emitted within the gas turbine at service temperatures. The aim of this work is to optimize the diffuse reflectivity of the air plasma sprayed (APS) TBC by improving the coating microstructure such that it leads to an increase in the reflectivity of radiation and thus yields a more efficient thermal insulation of the underlying metallic substrate. Powder of different grain size distributions has been air plasma-sprayed under two different spray distances to produce different porosities. The transmission and reflection in the near infrared has been measured in an IR-spectrometer. Additionally, the absorption has been independently measured by means of the photothermal deflection spectroscopy (PDS). The influence on absorption by vacancies in the coating's as-sprayed state has been also investigated. By using the Kubelka-Munk two flux model, the scattering and absorption coefficient of the sprayed TBC corresponding to such model have been determined and correlated to the measured porosimetry. These two coefficients are used to estimate the stationary temperature distribution across the coating by solving numerically a two-flux model containing radiation.

INTRODUCTION

The increase in operating temperature in the gas turbines requires a better thermal insulation of the turbine's metallic components, which can be achieved by depositing on top of them a ceramic thermal barrier coating (TBC), usually made of yttria partially stabilized zirconia (YSZ). In the near infrared (IR) range of wavelengths below 5µm, YSZ is nearly transparent to radiation. Since it is within these wavelengths where most of the radiation in the combustion chamber is emitted, YSZ permits an important part of the energy flow to reach directly the underlying metallic substrate, being absorbed there. Hence a TBC optimization regarding radiation is necessary. Air plasma spraying (APS) as coating process offers the possibility of modifying the microstructure of the deposited TBC (pores and micro-cracks) in order to enhance the backscattering of radiation and thus better shield the metallic substrate.

This work is organized as follows: firstly the different sprayed powders and the used spraying parameters are discussed, together with the resulting TBC microstructures. The optical properties (reflectance, transmittance and absorbance) are investigated by means of two different techniques. The experimental results are then correlated to the Kubelka-Munk model in order to obtain an estimation for the absorption and scattering coefficients, which in the last section will serve to calculate the expected stationary temperature distribution within the coating system using the two flux model similar to the model of R. Siegel and C.M. Spuckler[1].

EXPERIMENTAL: POWDERS, TECHNIQUES AND COATINGS

The feed stocks were two 8wt% YSZ powders with different size distribution and structure: hollow spherical powder manufactured by spray drying (Sulzer Metco) and dense fused and crushed powder (Treibacher). The YSZ coatings were deposited on a steel substrate by atmospheric plasma spraying using a Triplex II gun (Sulzer Metco), with current 500A and a power of 57kW. The stand-off distance was chosen 150mm (in-house standard) and 300mm. The spray parameters and the resulting porosity of the coatings are given in Table I. Coatings from the spraying of hollow spherical powder at a spray distance of 150mm are defined as our standard. The last 3 samples where annealed in a furnace for 1 hour at 600°C.

Table I. TBC samples investigated

sample	injected powder with grain size distribution	spray distance	thickness	total porosity	heat treatment	optical measurement
05_09t	hollow sphere, spray dried $d_{10,50,90}$=7, 40, 77µm	300mm	390µm	22.1%	as-sprayed	IR
06_250t	hollow sphere, spray dried $d_{10,50,90}$=7, 40, 77µm	150mm	400µm	12.7%	as-sprayed	PDS
06_256t	hollow sphere, spray dried $d_{10,50,90}$=7, 40, 77µm	300mm	380µm	20.8%	as-sprayed	IR, PDS
06_508t	hollow sphere, spray dried $d_{10,50,90}$=12, 45, 88µm	150mm	390µm	12.0%	annealed 1h at 600°C	IR, PDS
06_509t	hollow sphere, spray dried $d_{10,50,90}$=12, 45, 88µm	300mm	350µm	19.9%	annealed 1h at 600°C	IR, PDS
06_510t	dense fused & crushed $d_{10,50,90}$=9, 20, 42µm	150mm	390µm	8.7%	annealed 1h at 600°C	IR, PDS

The in-flight velocity and temperature of the injected powder particles were measured during thermal spraying by means of the diagnostics system DPV2000 (Tecnar Automation, Canada), evaluating a total of 5000 particles at a location shortly before the molten particles impinge on the substrate to build the ceramic TBC. The temperature was measured by means of a two-color pyrometer (wavelengths: 787 and 995nm; temperature range: 2000–4000°C); the velocity was obtained using the time-of-flight method. For the hollow spherical powder, the shorter stand-off spray distance (150mm) leads to a larger particle velocity and temperature than for 300mm, since in the last case particles are already decelerating and cooling down. For the dense powder, also smaller in size, the velocity and temperature are significantly larger, due to a better injection and a longer stay within the plasma jet core.

The microstructures of the deposited YSZ-coatings were characterized using optical microscopy and scanning electronic microscopy (SEM). Both the pore distribution and the open porosity of free-standing coatings were determined by means of mercury intrusion porosimetry, with the results represented in Fig. 2; these porosimetry measurements are quite reproducible, as shown for samples sprayed under similar conditions. Polished cross sections with the microstructure of three representative YSZ-coatings (06_508t, 509t & 510t) are shown in Fig. 3. With increased spray distance, the porosity of the coating is clearly increased (both displayed in porosimetry and micrographs): this is a result of the decreased flattening of the impinging particles, since they are slower, as well as of the higher presence of non-molten particles. On the other hand, coating 06_510t shows a more dense structure, characterized by a reduced porosity; this agrees with the particle velocity measurement where a higher velocity for smaller particles leads to a higher flattening degree and a closer contact between deposited particles.

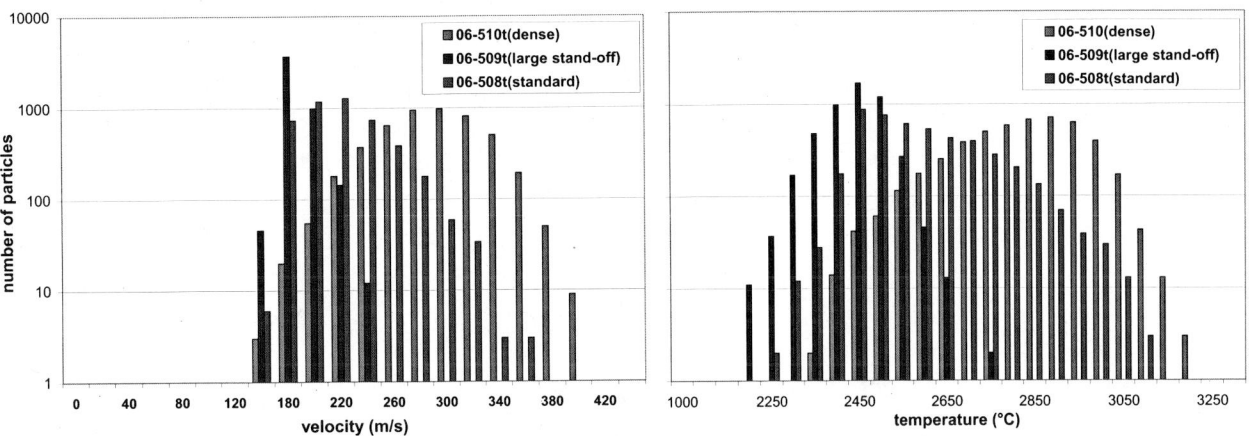

Figure 1: Particle velocity (left) and temperature (right) at impact for the hollow powder with standard (150mm) and large stand-off distance (300mm) as well as for the dense powder.

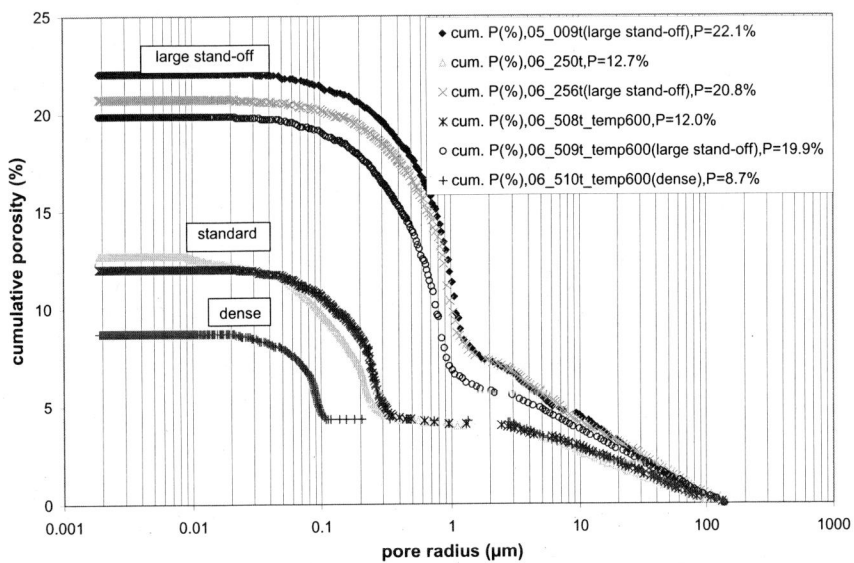

Figure 2. Double logarithmic representation of the cumulated pore distribution for the coatings.

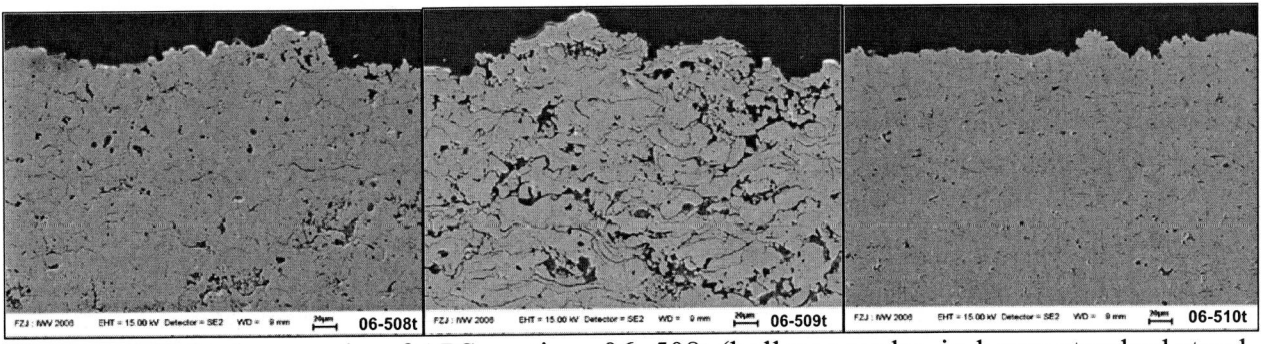

Figure 3: SEM-micrographs of APS coatings 06_508t (hollow powder, in-house standard stand-off), 06_509t (hollow powder, large stand-off) and 06_510t (dense powder, standard stand-off).

Thermal Properties

For the optical characterization of the sprayed coatings (both as-sprayed as well as annealed) two different spectroscopic methods were used. The diffuse reflectance and transmittance measurements were performed by means of a Perkin Elmer UV-VIS-IR spectrometer, Lambda 950 in the wavelength range of 0.25 to 2.5µm. For collecting the scattered light, an integration sphere of 150mm diameter coated with TiO_2 was used.

Additionally, the absorbance of the samples were independently measured (0.5 to 2µm) using the photothermal deflection spectroscopy (PDS), particularly suitable for the determination of low absorbance ($\alpha*d=10^{-5}$).[2,3] Since this is a rather non conventional technique, a short description follows. The external surface of the sample to be measured is immersed in a liquid deflection medium (CCl_4). The sample is periodically excited (for instance by a laser modulated by a chopper). The absorption of the pump beam causes a periodic change of temperature at the sample surface, which for its part produces a gradient in refractive index in the deflection medium in contact to the surface (Fig. 4, left). The resulting periodic deflection of a laser passing along the sample surface is detected with position sensors (measuring both amplitude and phase). The deflection intensity is directly correlated to the optical absorption of the sample.

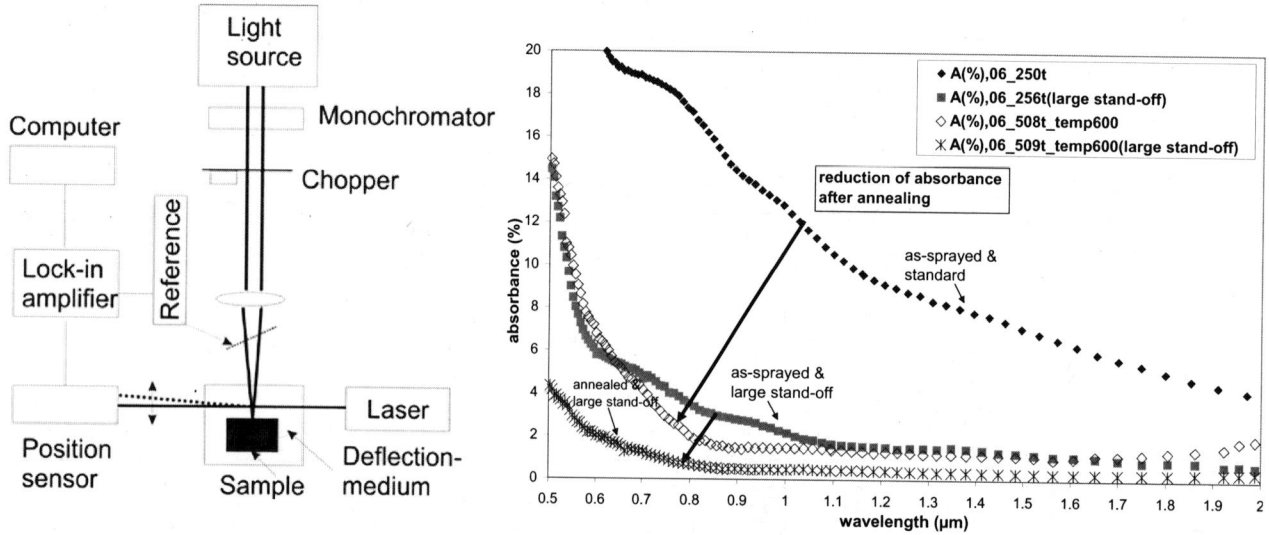

Figure 4: Left, schematic principle of the photothermal deflection spectroscopy (PDS). Right, influence of annealing on the absorbance of coatings sprayed at standard and large stand-off.

OPTICAL PROPERTIES OF THE SPRAYED COATINGS

The reflectance R and transmittance T for as-sprayed as well as annealed TBCs is represented in Fig. 5. Both properties display clearly the effect of the microstructure: with increasing porosity (from 8.7% to 19.9%) the reflectance increases and the transmittance decreases. In order to check the reproducibility of the measurements two samples (05-09t and 06-256t) sprayed with the same powder and stand-off were considered, showing a good agreement of two measurements. Further, similar results by Debout et al[4] has been included as reference. The sharp jump at 0.8µm is due to the change of the detector, and the strong data scattering for wavelengths above 2µm is due to the limited sensitivity of the IR detector. The superimposed periodic fluctuations in particular in the transmittance at 1.4, 1.9, 2.3µm are probably due to vibration modes of water-related species.

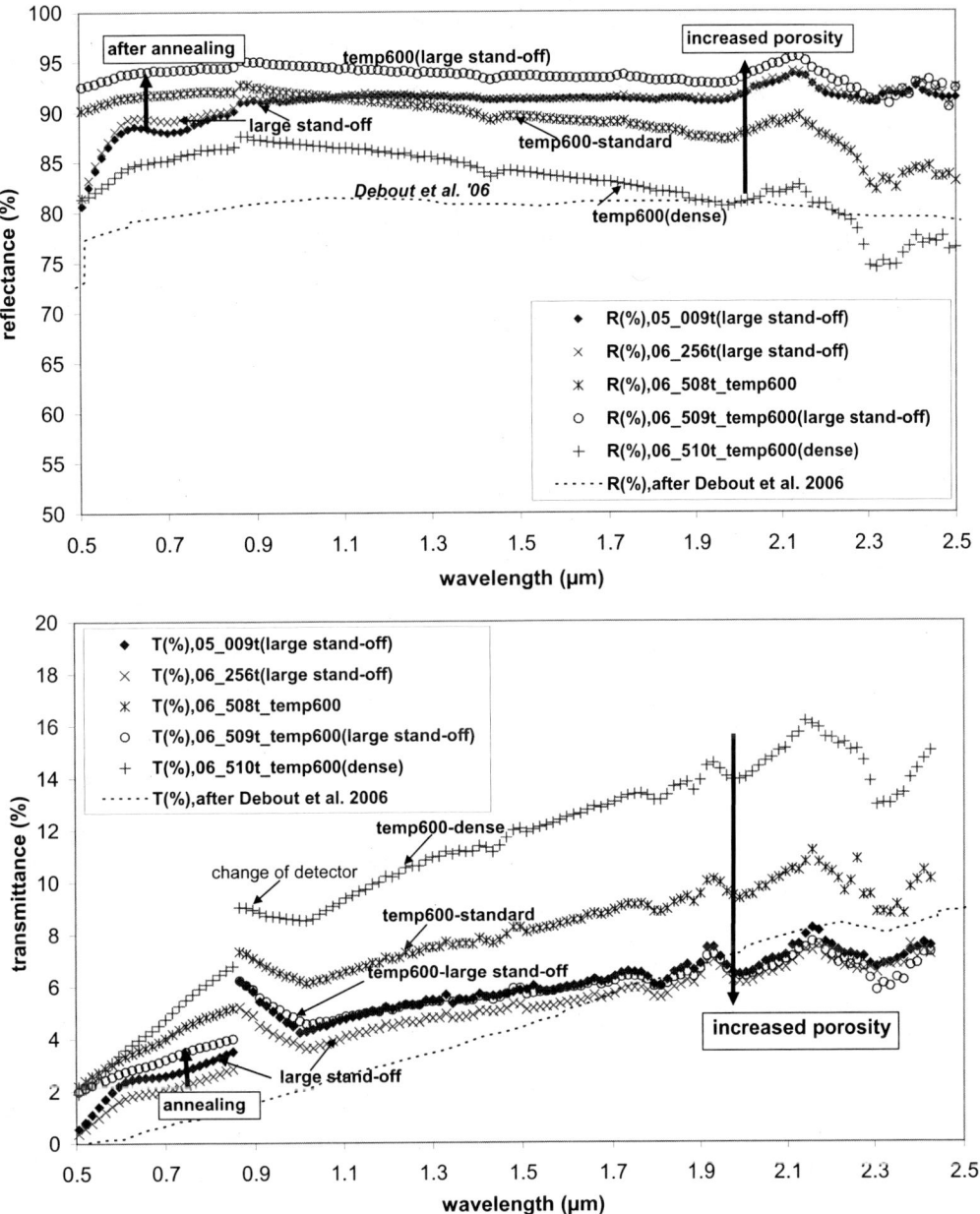

Figure 5. Reflectance (above) and transmittance (below) of as-sprayed and annealed coatings with standard and large spray distances measured with the IR integration sphere.

The coating 06_509t, annealed for 1h at 600°C, shows a higher reflectance than the corresponding as-sprayed coating 06_256t (the similar effect is not so clear in transmittance). This effect can be related to the sub-stoichiometry of oxygen, where a higher number of vacancies increase the absorbance of the coating, reducing thus the reflectance; this has been also discussed in Debout *et al*[4]. This interpretation is further supported when comparing the coating deposited at a shorter spray distance with that at larger stand-off: coating 06_508t, with a shorter stay in the plasma core region where atmospheric oxygen cannot diffuse inwards easily and therefore a higher deficiency in oxygen is expected, display a lower reflectance and a higher absorbance (the low transmittance is not able to point out this tendency clearly).

The previously discussed vacancies effect has been measured mainly in diffuse reflectance, which might not be sensitive enough since the considered coating's reflectance lies above 95%. The PDS method with its high sensibility at low absorption was consulted to confirm this effect. Comparing the IR and PDS spectra both methods provide the same trends. However, in the transition from VIS to IR, the absorbance A calculated from the IR integration sphere data (as $A=100-R-T$) lies above the absorbance directly measured with PDS (Fig. 6). The PDS measurement may lead to a slight under-estimation when applied on highly porous coatings: the open porosity will be filled with the fluid used for the deflection measurement. This reduced the reflectivity and may decrease the light pace resulting in an under-estimation of the absorbance. The effect of the oxygen deficiency is nevertheless confirmed in the PDS spectra (Fig. 4, right), both for standard as well as large stand-off distance.

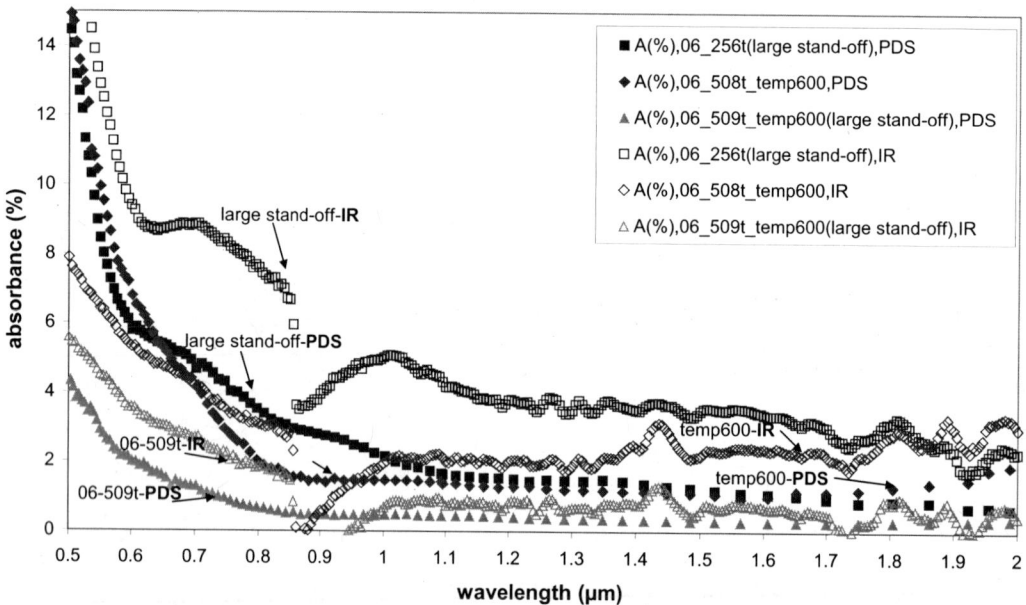

Figure 6. Absorbance of as-sprayed and annealed coatings with standard and large spray distances measured with PDS and compared to the IR integration sphere measurements.

ABSORPTION AND SCATTERING COEFFICIENTS OF THE SPRAYED COATINGS

For the following discussion, the annealed samples 06_508t, 509t and 510t have been selected in order to determine quantitatively the influence of the coating microstructure on the optical properties and on the estimated temperature distribution resulting from such optical properties. Only the annealed coatings correspond to the conditions in a gas turbine. The wavelength dependent absorption and scattering coefficients of the coating, $\kappa_{abs,\lambda}$ and $\kappa_{sca,\lambda}$, are obtained by fitting the IR integration sphere measurements to the Kubelka-Munk model for diffuse reflectivity and transmissivity of a coating. This model describes the 1-dimensional flow of radiation in forward and backward direction within an uniform system diffusely illuminated.[5] The absorption and scattering behavior of the system are effectively described by two parameters K and S which, assuming isotropic scattering, are related to the actual absorption and scattering coefficients by $\kappa_{abs,\lambda} = \frac{1}{2}K_{(\lambda)}$ and $\kappa_{sca,\lambda} = \frac{4}{3}S_{(\lambda)} + \frac{1}{6}K_{(\lambda)}$ (see Appendix B). The steps for obtaining K and S from the IR measurement of reflectance and transmittance are discussed in detail in the Appendix B. The results are represented in Fig. 7: taking as a reference the sample sprayed at the standard distance of 150mm with hollow powder (06_508t), the sample with the higher porosity (06_509t, sprayed at a larger stand-off distance) display as expected a higher scattering coefficient, whereas the dense sample (06_510t, sprayed with dense powder) has the highest absorption and the lowest scattering coefficients.

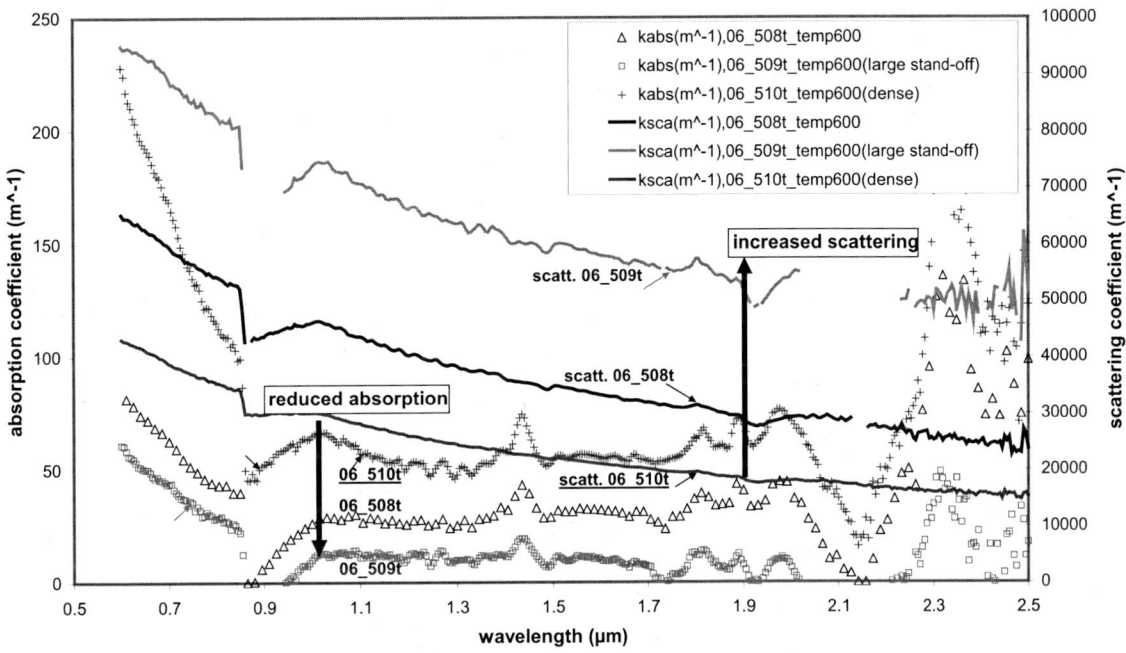

Figure 7. Absorption and scattering coefficients derived from the IR integration sphere measurements adjusted to the Kubelka-Munk model.

STATIONARY TEMPERATURE DISTRIBUTION IN SEMI-TRANSPARENT COATING

After having characterized the optical properties of the sprayed TBC coatings, the resulting stationary temperature distribution across the coating will be calculated in order to determine quantitatively the influence of the microstructure and its scattering ability. The model coupling the radiation transport within the semi-transparent coating to the temperature distribution has been developed by Siegel & Spuckler[1] and is discussed in some detail in Appendix A, together with its numerical solution. The system considered is formed by a ceramic TBC of thickness x_{TBC}=400µm (divided into 700 cells for the numerical solution) attached to a metallic substrate of thickness x_{SUB}=3.15mm.

The TBC is assumed to be semi-transparent for wavelengths $\lambda \leq \lambda_{cut}$=5µm (above this wavelength, the ceramic coating displays nearly black body behavior).[1] In order to smooth out the small jumps in the wavelength dependence of the absorption and scattering coefficients, the data of Fig. 8 between 0.6µm and 2µm are fitted to a power law $\kappa_{abs/sca,\lambda} = A\lambda^{-m}$ fit which is used to extrapolate both coefficients up to λ_{cut}=5µm. This extrapolation is justified by the fact that the slow decrease in the scattering coefficient should be maintained for longer wavelengths (it is the result of the pore distribution) and the absorption coefficient has already achieved almost saturation for wavelengths above 1.5µm.

The other model parameters are[1] (notation in Appendix B)

T_{gas}=2000K	$T_{sur,hot}$=1600K	$T_{sur,cold}$=800K	$T_{gas,cold}$=300K
ε_{gas}=0.22	$\varepsilon_{surr,hot}$=1	$\varepsilon_{surr,cold}$=0.6	$\varepsilon_{gas,cold}$=0
$h_{conv,hot}$=250W/m²K	n_{TBC}=n=2.1	ε_{SUB}=0.3	$h_{conv,cold}$=110W/m²K

with typical thermal conductivities λ_{TBC}=1.1W/mK for TBC and λ_{SUB}=26.2W/mK for the metal; the gas emissivity, based on the emission of CO_2 and H_2O, corresponds to a combustion chamber operating at a pressure of 10atm[6]. Within the TBC coating, the total energy flux is 0.231MW/m² for our standard spray distance (06_508t), 0.211MW/m² for the increased spray distance (06_509t) and 0.255MW/m² for the dense powder coating (06_510t). Compared to the our standard coating (06_508t), the sample with the higher porosity yields a lower temperature at the metallic substrate of about 36°C, whereas the sample dense sprayed has a temperature 7°C higher. In Fig. 8 the temperature distribution inside the TBC is represented, including a transparent TBC as comparison, and showing the better screening of radiation and the resulting lower substrate temperature for the coating of higher porosity. This reduces the thermal load on the metallic substrate and decreases the thermally activated growth of oxide scales at the interface ceramic-metal, leading to a higher life time of the coating system[8].

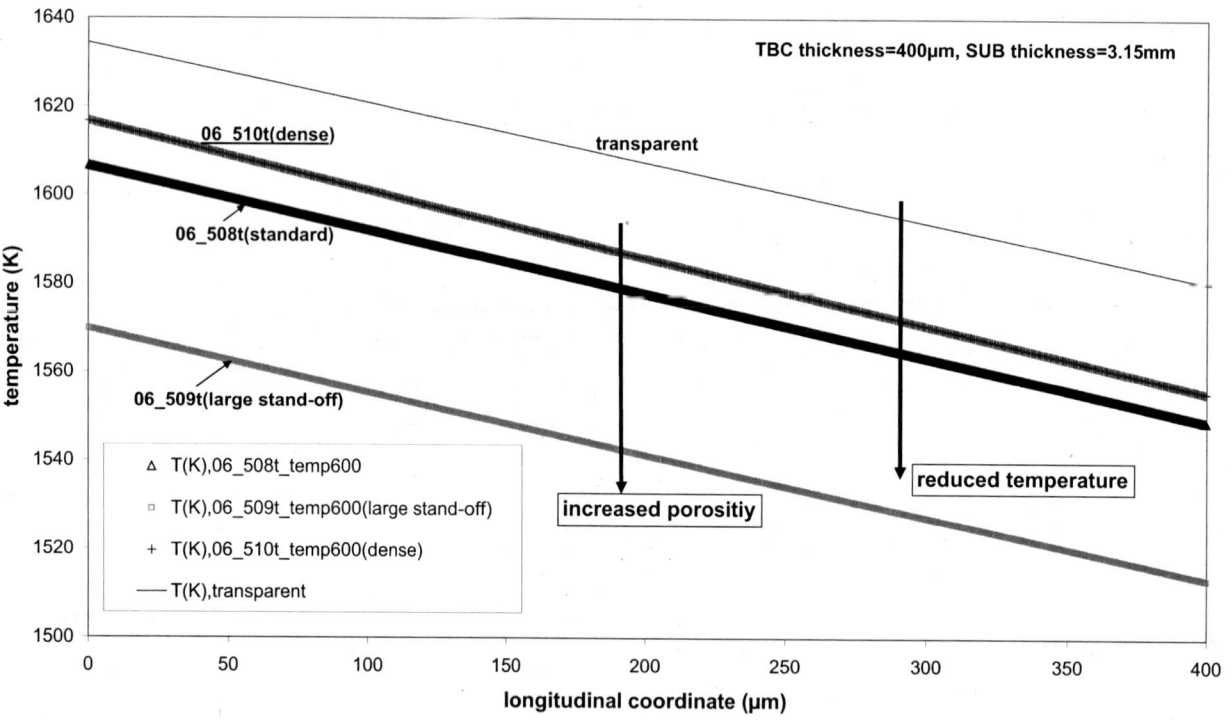

Figure 8. Calculated stationary temperature distribution across semi-transparent TBC (d=400μm) for the absorption and scattering coefficients of Fig. 7.

CONCLUSIONS

The IR optical properties of TBC coatings air plasma sprayed with different powders and stand-off distances have been investigated by means of two different optical techniques. The correlation to the microstructure as well as the effect of annealing was discussed and the absorption and scattering coefficients of the different coatings were determined. Finally these experimental results have been used to estimate numerically the stationary temperature distribution for the different considered TBC microstructures. The sample with the larger spray distance yields promising results for improving the shield of IR radiation by a modified APS TBC.

ACKNOWLEDGMENTS

The authors thank Mr F. Vondahlen and Mr. K.-H. Rauwald for the sample manufacture, Mr. M. Kappertz for the cross-section preparations and Mr. J. Klomfaß for the PDS measurements.

REFERENCES

[1]R. Siegel, C.M. Spuckler, "Analysis of thermal radiation effects on temperatures in turbine engine thermal barrier coatings," *Mater. Sci. Eng.* **A245**, 150-159 (1998).

[2]W.B. Jackson, N.M. Amer, A.C. Boccara, D. Fournier, *Applied Optics* **20** 1333 (1981).

[3]F. Becker, R. Carius, J.-T. Zettler, J. Klomfaß, "Photothermal Deflection Spectroscopy on Amorphous Semiconductor Heterojunctions and Determination of the Interface Defect Densities", *Materials Science Forum*, **173/174**, 177 (1995).

[4]V. Debout, A. Vardelle, P. Abélard, P. Fauchais, "Optical properties of Yttria-Stabilizied Zirconia Plasma-Sprayed Coatings" Proceedings of the *International Spray Conference ITSC2006*, Seattle, Washington (2006).

[5]R. Molenaar, J.J. ten Bosch, J.R. Zijp, "Determination of Kubelka-Munk scattering and absorption coefficients by diffuse illumination," *Applied Optics* **38**, 2068-2077 (1999).

[6]J.G. Knudsen *et al*, "Heat and Mass Transfer," section 5 of *Heat Transmission*, McGraw-Hill, New York (1997).

[7]M. F. Modest, *Radiative Heat Transfer*, Academic Press, Amsterdam (2003).

[8]F. Traeger, M. Ahrens, R. Vaßen, D. Stöver, "A Life Time Model for Ceramic Thermal Barrier Coatings," *Mater. Sci. Eng.* **A358**, 255-265 (2003).

APPENDIX A. TWO-FLUX RADIATION MODEL AND COUPLING TO STATIONARY TEMPERATURE DISTRIBUTION WITHIN SEMI-TRANSPARENT COATINGS

Let $I_\lambda(x;\hat{s})d\lambda$ be the radiation intensity at the location x along direction $\hat{s}$ and for wavelengths between λ and $\lambda+d\lambda$. The stationary change in intensity when moving to a neighbor location $x+dx$ is given by[7]

$$\cos\theta\frac{\partial I_\lambda(x;\hat{s})}{\partial x} = -\left(\kappa_{abs,\lambda}+\kappa_{sca,\lambda}\right)I_\lambda(x;\hat{s})+\kappa_{abs,\lambda}n^2 I_{bb,\lambda}(x)+\kappa_{sca,\lambda}\iint_{\text{whole sphere}}\Phi(\hat{s},\hat{s}')I_\lambda(x;\hat{s}')d\Omega' \quad (A1)$$

with the cosine function for the projection of the intensity flow direction on the 1-dimensional coordinate x. The first term on the right hand side of (A1) describes the loss in intensity due to absorption and scattering, with respective spectral (i.e. wavelength dependent) coefficients $\kappa_{abs,\lambda}$ and $\kappa_{sca,\lambda}$. The second term carries the positive contribution due to the local black body emission $I_{bb,\lambda}(x)$ weighted by the absorption and thus emission coefficient at the considered wavelength

$\kappa_{abs,\lambda}$, with n the refractive index of the material in which the radiation is being transported. This term is considered since the considered coating usually operates at high temperatures and thus the emission of radiation by the system itself cannot be neglected. The last term in (A1) contains the intensity increase as a result from the scattering of intensities in all other directions $\hat{s}'$ back to the considered $\hat{s}$, where $\Phi(\hat{s},\hat{s}')$ represents the probability density for such processes fulfilling $\iint\limits_{\text{whole sphere}} \Phi(\hat{s},\hat{s}')d\Omega' = 1$.

In order to solve (A1) in a closed way, the following two simplifications are assumed: the scattering is assumed isotropic, such that $\Phi(\hat{s},\hat{s}')$ is a constant equal to $1/4\pi$. And the direction dependence in the intensity is reduced to only two main directions, forward and backward

$$I_\lambda(x;\hat{s}) \approx \begin{cases} I_\lambda^{(+)}(x) & \hat{s} \text{ in forward semi-sphere} \\ I_\lambda^{(-)}(x) & \hat{s} \text{ in backward semi-sphere} \end{cases} \tag{A2}$$

which is called the two-flux approximation. Now the number of unknowns at each location and wavelength has been reduced to $I_\lambda^{(+)}(x)$ and $I_\lambda^{(-)}(x)$ (or to their sum and difference), and thus two equations are required. The first one is obtained by integrating (A1) over the whole sphere

$$\pi\frac{\partial}{\partial x}\left(I_\lambda^{(+)} - I_\lambda^{(-)}\right) = -2\pi\left(\kappa_{abs,\lambda} + \kappa_{sca,\lambda}\right)\left(I_\lambda^{(+)} + I_\lambda^{(-)}\right) + 4\pi\kappa_{abs,\lambda}n^2 I_{bb,\lambda} + 2\pi\kappa_{sca,\lambda}\left(I_\lambda^{(+)} + I_\lambda^{(-)}\right)$$
$$= \kappa_{total,\lambda}\left(1 - \omega_\lambda\right)\left[4\pi n^2 I_{bb,\lambda} - 2\pi\left(I_\lambda^{(+)} + I_\lambda^{(-)}\right)\right] \tag{A3}$$

with $\kappa_{total,\lambda} = \kappa_{abs,\lambda} + \kappa_{sca,\lambda}$ the (spectral) extinction coefficient and $\omega_\lambda = \dfrac{\kappa_{sca,\lambda}}{\kappa_{total,\lambda}}$ the scattering albedo. The second equation results from the multiplication of (A1) with $\cos\theta$ and integrating again over the whole sphere

$$\frac{2\pi}{3}\frac{\partial}{\partial x}\left(I_\lambda^{(+)} + I_\lambda^{(-)}\right) = -\pi\kappa_{total,\lambda}\left(I_\lambda^{(+)} - I_\lambda^{(-)}\right) \tag{A4}$$

By applying a further derivative on (A4), and together with (A3), it results

$$-\frac{1}{3\kappa_{total,\lambda}^2}\frac{\partial^2}{\partial x^2}\left[2\pi\left(I_\lambda^{(+)} + I_\lambda^{(-)}\right)\right] = \left(1 - \omega_\lambda\right)\left[4\pi n^2 I_{bb,\lambda} - 2\pi\left(I_\lambda^{(+)} + I_\lambda^{(-)}\right)\right] \tag{A5}$$

which only contains the combination $G_\lambda(x) = 2\pi\left(I_\lambda^{(+)}(x) + I_\lambda^{(-)}(x)\right)$ as unknown. The energy flow due to radiation, equal to the integration of the radiation intensity over all directions, is related to the difference of $I_\lambda^{(+)}(x)$ and $I_\lambda^{(-)}(x)$ within the two-flux approximation

$$\dot{q}_{rad,\lambda}(x) = \iint_{\substack{\text{whole sphere}}} I_\lambda(x;\hat{s})d\Omega \approx \pi\left(I_\lambda^{(+)} - I_\lambda^{(-)}\right) \overset{\text{eq.(A4)}}{=} -\frac{1}{3\kappa_{total,\lambda}}\frac{\partial G_\lambda}{\partial x} \tag{A6}$$

Now let us consider a semi-transparent ceramic coating heated up at the external surface located at $x=0$ (corresponding to the combustion chamber) and attached on the other coating side at $x=x_{TBC}$ to a metallic substrate of thickness x_{SUB} which is cooled down. The external surface is heated due to convection and radiation by a flame at temperature T_{gas}, convection coefficient $h_{conv,hot}$ and emissivity ε_{gas} as well as due to radiation by the hot surrounding walls at temperature T_{sur} and emissivity ε_{sur}. The metallic substrate on the other side of the ceramic coating absorbs all the radiation impinging on it and has an emissivity ε_{sub}; on its free surface it interchanges energy to a cold gas flow at temperature $T_{gas,cold}$ (only convection, no radiation) and to the cold surrounding walls at temperature $T_{surr,cold}$ and emissivity $\varepsilon_{gas,cold}$ (radiation). The solution of the 2nd order differential equation (A5) for such a system requires two boundary conditions, for instance by giving the radiation flow $\dot{q}_{rad,\lambda}$ at the two coating boundaries[1]

$$-\frac{1}{3}\frac{1}{\kappa_{total,\lambda}}\frac{\partial G_\lambda}{\partial x}\bigg|_{x=0} = -\frac{(1-\rho_{int})}{2(1+\rho_{int})}\left[G_\lambda(x=0) - 4\pi\frac{1-\rho_{ext}}{1-\rho_{int}}\left(\varepsilon_{surr}I_{bb,\lambda}(T_{surr}) + \varepsilon_{gas}I_{bb,\lambda}(T_{gas})\right)\right]$$

$$-\frac{1}{3}\frac{1}{\kappa_{total,\lambda}}\frac{\partial G_\lambda}{\partial x}\bigg|_{x=x_{TBC}} = \frac{\varepsilon_{sub}}{2(2-\varepsilon_{sub})}\left[G_\lambda(x=x_{TBC}) - 4\pi n^2 I_{bb,\lambda}(T(x=x_{TBC}))\right] \tag{A7}$$

with the Fresnel diffuse reflectivities at the air-coating interface outwards as well as inwards[7]

$$\rho_{ext} = \frac{1}{2} + \frac{(3n_r+1)(n_r-1)}{6(n_r+1)^2} + \frac{n_r^2(n_r^2-1)^2}{(n_r^2+1)^3}\ln\left(\frac{n_r-1}{n_r+1}\right) - \frac{2n_r^3(n_r^2+2n_r-1)}{(n_r^2+1)(n_r^4-1)} + \frac{8n_r^4(n_r^4+1)}{(n_r^2+1)(n_r^4-1)^2}\ln n_r$$

$$\rho_{int} = 1 - \frac{1}{n_r^2}(1-\rho_{ext}) \tag{A8}$$

and with the relative refractive index at such interface $n_r = n/n_{air} = n$. Within the coating, the complete radiation energy flow integrated over the whole wavelength spectrum can be written as

$$\dot{q}_{rad}(x) = \int_0^\infty \dot{q}_{rad,\lambda}d\lambda = -\frac{1}{3}\frac{\partial}{\partial x}\int_0^\infty\frac{G_\lambda}{\kappa_{total,\lambda}}d\lambda = -\frac{1}{3}\frac{\partial}{\partial x}\int_0^{\lambda_1}\frac{G_\lambda}{\kappa_{total,\lambda}}d\lambda - \frac{1}{3}\frac{\partial}{\partial x}\int_{\lambda_1}^{\lambda_2}\frac{G_\lambda}{\kappa_{total,\lambda}}d\lambda - \ldots - \frac{1}{3}\frac{\partial}{\partial x}\int_{\lambda_{M-1}}^{\lambda_{cut}}\frac{G_\lambda}{\kappa_{total,\lambda}}d\lambda$$

with λ_{cut} the largest wavelength at which the ceramic coating is still semi-transparent; for $\lambda > \lambda_{cut}$ the coating displays black body behavior and absorbs and emits immediately every wavelength perfectly, effect described by an infinite absorption (and thus extinction) coefficient. In the previous equation, the semi-transparent $[0,\lambda_{cut}]$ band has been divided in M windows $\{[0,\lambda_1], [\lambda_1,\lambda_2],\ldots, [\lambda_{M-1},\lambda_M=\lambda_{cut}]\}$ within each one both the spectral extinction coefficient and the scattering albedo can be considered constant.

The total energy flow incoming on the coating consists of a convective part, due to the energy transfer from the hot gas flame, and a radiation part. The latter is divided into two contributions: that arising from the wavelength band where the coating is opaque such that its surface operates as a gray body of emissivity $1-\rho_{ext}$, absorbing the energy emitted by the gas and the walls in this band and radiating itself from the coating's heated surface (coordinate $x=0$); and the second contribution corresponding to the wavelength range where the coating is semi-transparent and therefore the radiation will be able to penetrate and propagate

$$\dot{q}_{total,in} = h_{conv,hot}\left(T_{gas,hot} - T\big|_{x=0}\right) + (1-\rho_{ext})\pi \int\limits_{\lambda_{cut}}^{\infty}\left(\varepsilon_{surr}I_{bb,\lambda}(T_{surr}) + \varepsilon_{gas}I_{bb,\lambda}(T_{gas}) - I_{bb,\lambda}\big|_{x=0}\right)d\lambda + \int\limits_{0}^{\lambda_{cut}}\dot{q}_{rad,\lambda}d\lambda$$

Within the coating, the total energy flow consists of a conductive part as well as the just mentioned radiation able to propagate in the semi-transparent band, this time written as the separated contributions from each of the corresponding wavelength windows where the extinction coefficient is taken as constant

$$\dot{q}_{total,TBC} = \lambda_{TBC}\frac{T\big|_{x=0} - T\big|_{x=x_{TBC}}}{x_{TBC}} + \frac{1}{3}\int\limits_{0}^{\lambda_1}\frac{G_\lambda\big|_{x=0} - G_\lambda\big|_{x=x_{TBC}}}{\kappa_{total,\lambda}}d\lambda + \ldots + \frac{1}{3}\int\limits_{\lambda_{M-1}}^{\lambda_{cut}}\frac{G_\lambda\big|_{x=0} - G_\lambda\big|_{x=x_{TBC}}}{\kappa_{total,\lambda}}d\lambda \quad (A9)$$

with λ_{TBC} the thermal conductivity of the ceramic coating. The metallic substrate absorbs all the incoming radiation and thus the total energy flow is $\dot{q}_{total,SUB} = \lambda_{SUB}\dfrac{T\big|_{x=x_{TBC}} - T_{metal,cold}}{x_{SUB}}$, i.e. only thermal conduction is relevant, with λ_{SUB} its thermal conductivity and $T_{metal,cold}$ the temperature at the externally cooled side. Finally, the total energy flow leaving the coating system through the cooled side consists of a convective part as well as the radiation as a gray body with emissivity ε_{sub} for the energy emitted by the metal surface together with the absorbed radiation emitted by the surrounding cold walls

$\dot{q}_{total,out} = h_{conv,cold}\left(T_{metal,cold} - T_{gas,cold}\right) + \varepsilon_{SUB}\sigma_{SB}\left(T^4_{metal,cold} - \varepsilon_{surr}T^4_{surr,cold}\right)$. In the stationary case, following holds

$$\dot{q}_{total,in} = \dot{q}_{total,TBC} = \dot{q}_{total,SUB} = \dot{q}_{total,out} \quad (A10)$$

which couples thus the radiation energy flow inside the TBC coating to the temperature distribution in the combustion chamber, in the metallic substrate and at the cooled side. This coupled equation system has to be solved numerically, discretizing the differential equation into a system of algebraic equations. For that the TBC coating thickness is divided into N cells of thickness $\Delta x[i]$ ($i=1,\ldots,N$), inside each one the temperature and energy flow is considered constant (although different from those in the neighbors cells). In the case of the radiation field variables, another index $[a]$ ($a=1,\ldots,M$) carries the information about the wavelength window within the semi-transparent band. The variables to be solved are

$$T(x) \to T[i], \qquad \pi \int_{\lambda_{a-1}}^{\lambda_a} I_{bb,\lambda}(T(x))d\lambda \to \dot{q}_{bb}[a][i], \qquad \int_{\lambda_{a-1}}^{\lambda_a} \frac{G_\lambda(x)}{\kappa_{total,\lambda}}d\lambda \to g[a][i] \tag{A11}$$

$$\int_{\lambda_{a-1}}^{\lambda_a} \dot{q}_{rad,\lambda}(x)d\lambda \to \dot{q}_{rad}[a][i] = -\frac{1}{3}\frac{g[a][i+1]-g[a][i-1]}{\Delta x[i+1]/2+\Delta x[i]+\Delta x[i-1]/2}$$

And the discretization of equation (A5), incorporating the boundary conditions (A7), reads

$$\begin{pmatrix} b[1] & c[1] & 0 & \cdots & & \cdots & & \cdots \\ a[2] & b[2] & c[2] & 0 & & \cdots & & \cdots \\ 0 & a[3] & b[3] & c[3] & 0 & & \cdots \\ \vdots & \vdots & \vdots & \vdots & & \vdots & & \vdots \\ \cdots & \cdots & 0 & a[N-1] & b[N-1] & c[N-1] \\ \cdots & \cdots & \cdots & 0 & a[N] & b[N] \end{pmatrix} \begin{pmatrix} g[a][1] \\ g[a][2] \\ g[a][3] \\ \vdots \\ g[a][N-1] \\ g[a][N] \end{pmatrix} = \begin{pmatrix} r[1] \\ r[2] \\ r[3] \\ \vdots \\ r[N-1] \\ r[N] \end{pmatrix} \tag{A12}$$

$$a[i] = \begin{cases} 0 & i=1 \\ -\dfrac{1}{3\Delta x[i]}\dfrac{1}{\Delta x[i]/2+\Delta x[i-1]/2} & i \neq 1 \end{cases} \qquad c[i] = \begin{cases} 0 & i=N \\ -\dfrac{1}{3\Delta x[i]}\dfrac{1}{\Delta x[i+1]/2+\Delta x[i]/2} & i \neq N \end{cases}$$

$$b[i] = -a[i]-c[i]+\kappa_{total,\lambda_a}^2\left(1-\omega_{\lambda_a}\right)+ \begin{cases} \dfrac{\kappa_{total,\lambda_a}}{\Delta x[i]}\dfrac{(1-\rho_{int})}{2(1+\rho_{int})} & i=1 \\[2ex] \dfrac{\kappa_{total,\lambda_a}}{\Delta x[i]}\dfrac{\varepsilon_{sub}}{2(2-\varepsilon_{sub})} & i=N \\[2ex] 0 & i \neq 1,N \end{cases}$$

$$r[i] = \kappa_{total,\lambda_a}\left(1-\omega_{\lambda_a}\right)4n^2 \dot{q}_{bb}[a][i]+ \begin{cases} \dfrac{1}{\Delta x[i]}\dfrac{(1-\rho_{int})}{2(1+\rho_{int})}4\pi\dfrac{1-\rho_{ext}}{1-\rho_{int}}\displaystyle\int_{\lambda_{a-1}}^{\lambda_a}\left(\varepsilon_{surr}I_{bb,\lambda}(T_{surr})+\varepsilon_{gas}I_{bb,\lambda}(T_{gas})\right)d\lambda & i=1 \\[3ex] \dfrac{1}{\Delta x[i]}\dfrac{\varepsilon_{sub}}{2(2-\varepsilon_{sub})}4n^2 \dot{q}_{bb}[a][i] & i=N \\[3ex] 0 & i \neq 1,N \end{cases}$$

Initially, for a guessed temperature distribution, equation system (A12) is solved for each wavelength window [a] in order to obtain the whole (provisional) distribution of g[a][i] through the coating, for which the radiation flow is calculated from (A9). Subsequently, the new temperature distribution is obtained from (A10), which in its discretized form reads

$$T[i] = T[i=1]-\dot{q}_{total}\frac{x[i]}{\lambda_{TBC}}+\frac{1}{3\lambda_{TBC}}\sum_{a=1}^{a=M}\left(g[a][i=1]-g[a][i]\right) \tag{A13}$$

Using this (new) current temperature distribution, the complete process is iteratively repeated until the change in the temperature distribution between two consecutive iterations is below 10^{-4}.

APPENDIX B. DETERMINATION OF THE ABSORPTION AND SCATTERING COEFFICIENTS FOR THE TWO-FLUX KUBELKA-MUNK MODEL

The two-flux Kubelka-Munk model is a simplification of the model in the previous Appendix, equations (A3) and (A4), without considering the radiation by the coating itself, since it is usually applied on systems operating at room temperature. Re-writing (A3) and (A4)

$$\frac{\partial I_\lambda^{(+)}}{\partial x} = -\left(\kappa_{abs,\lambda} + \frac{3}{4}\kappa_{total,\lambda}\right)I_\lambda^{(+)} + \left(\frac{3}{4}\kappa_{total,\lambda} - \kappa_{abs,\lambda}\right)I_\lambda^{(-)}$$

$$\frac{\partial I_\lambda^{(-)}}{\partial x} = +\left(\kappa_{abs,\lambda} + \frac{3}{4}\kappa_{total,\lambda}\right)I_\lambda^{(-)} - \left(\frac{3}{4}\kappa_{total,\lambda} - \kappa_{abs,\lambda}\right)I_\lambda^{(+)}$$

(B1)

which corresponds to the Kubelka-Munk model for the effective parameters K and S defined as $K = 2\kappa_{abs,\lambda}$, $S = \frac{3}{4}\kappa_{total,\lambda} - \kappa_{abs,\lambda} = \frac{3}{4}\kappa_{sca,\lambda} - \frac{1}{4}\kappa_{abs,\lambda}$. Introducing $a = \frac{K+S}{S}$ and $b = \sqrt{a^2-1}$, the reflectivity and transmissivity of a coating of thickness d solving (B1) are given by[7]

$$R_{KM} = \frac{\sinh(bSd)}{a\sinh(bSd) + b\cosh(bSd)}, \quad T_{KM} = \frac{b}{a\sinh(bSd) + b\cosh(bSd)}$$

(B2)

These equations can be considered as functions of the three independent variables d, b, $\zeta = bSd$. For a given thickness d the task now consists in searching for the combination of ζ and b that substituted in (B2) matches the measured reflectivity and transmissivity. For that it is worth noting that R_{KM} is a monotonous decreasing function of b and T_{KM} is monotonous decreasing in ζ. The search for ζ and b is carried out iteratively along the following steps:

1. Initializing the upper and lower limits of ζ: $\zeta_{up}=1\times d\times 10^6$ and $\zeta_{down}=0$.

2. Begin of the iterative search for ζ: for the current $\zeta = \frac{\zeta_{up} + \zeta_{down}}{2}$, initializing the limits of the iterative search for b: $b_{up}=1000$ and $b_{down}=0$.

   a. With the current $b = \frac{b_{up} + b_{down}}{2}$, calculate $R_{KM} = \frac{\sinh\zeta}{\sqrt{b^2+1}\sinh\zeta + b\cosh\zeta}$ and $T_{KM} = \frac{b}{\sqrt{b^2+1}\sinh\zeta + b\cosh\zeta}$. Determine the next limits b_{down} and b_{up}: if R_{KM} is larger than the experimental reflectivity R then $b_{down}=b$ (R_{KM} is a decreasing function of b), otherwise $b_{up}=b$.

   b. Repeat the previous step until b_{down} and b_{up} are near enough to each other such that $2\frac{b_{up} - b_{down}}{b_{up} + b_{down}} < 10^{-4}$ holds.

3. Determine the next limits ζ_{up} and ζ_{down}: if the calculated T_{KM} is larger than the measured transmissivity T then $\zeta_{down}= \zeta$ (T_{KM} is a decreasing function of ζ), otherwise $\zeta_{up}=\zeta$.

4. Repeat steps 2 and 3 until $2\dfrac{\zeta_{up} - \zeta_{down}}{\zeta_{up} + \zeta_{down}} < 10^{-4}$ is fulfilled. Once finished, calculate

$$S = \frac{\zeta}{bd} \text{ and } K = S(a-1) = S\left(\sqrt{b^2 +1} -1\right).$$

THERMAL CONDUCTIVITY OF NANOPOROUS YSZ THERMAL BARRIER COATINGS FABRICATED BY EB-PVD

Byung-Koog Jang and Hideaki Matsubara
Materials Research and Development Laboratory,
Japan Fine Ceramics Center (JFCC)
2-4-1 Mutsuno, Atsuta-ku, Nagoya, 456-8587, Japan

ABSTRACT

ZrO_2-4mol% Y_2O_3 (YSZ) coatings were deposited by EB-PVD. The YSZ coatings consist of porous-columnar structure containing nano pores and intercolumnar gaps between columnar grains. The laser flash method and differential scanning calorimeter were used to measure the thermal diffusivity and specific heat of the coated samples. The thermal conductivity of EB-PVD coatings decreased with increasing measuring temperature as well as porosity. The thermal conductivity of the coating layer alone was also calculated based on thermal diffusion results for the double layers specimens consisting of the combined coatings and substrate. The response function method was employed to obtain an accurate value for the thermal conductivity of the coatings layer. The thermal conductivity of coatings layer showed increasing tendency with increasing the coatings thickness.

INTRODUCTION

Generally, for improvement of the thermal efficiency in gas turbine engine, higher operating temperature is necessary. However, metal component is always exposed at very severe operating temperature, resulting in the limit of metal substrate. In addition, a strong compressor is necessary for cooling which is negative for the thermal efficiency in the gas turbine engine. To overcome these handicaps, thermal barrier coatings (TBCs) have been developed for advanced gas turbine engine components to improve the thermal efficiency by increasing the gas turbine inlet temperature and reducing the amount of cooling air [1][3].

Currently, TBCs manufactured by electron beam-physical vapor deposition (EB-PVD) are being paid a great deal of attention because their columnar microstructure offers the advantage of a superior tolerance against thermal shock [4-6]. For superior TBCs, the development of coatings with the low thermal conductivity is very important. Therefore, thermal conductivity measurement of coatings is also necessary. Generally, it is known that the coatings have non-uniform porous structure so that they are easily damaged and broken due to the poor strength during measurement or handling. Actually, the evaluation of thermal properties of coatings is very critical issue.

The purpose of this work is to investigate the influence of temperature as well as porosity on thermal conductivity of EB-PVD ZrO_2-4mol% Y_2O_3 coatings by laser flash method. In addition, this work describes the thermal conductivity of coatings derived from heat diffusion results of the combined coatings and substrate specimens.

EXPERIMENTAL

ZrO_2-4mol% Y_2O_3 coatings were deposited by EB-PVD onto zirconia disk substrates of 10 mm diameter and 1mm thickness. The coating thickness was about 30~700 μm. The substrates were first preheated at 900~1000°C in a preheating vacuum chamber using graphite heating element. An electron beam evaporation process was conducted in a coating chamber using 30~60 KW of electron beam power. The target material was heated above its evaporation temperature of 3500°C, and the resulting vapors were condensed on substrate. Oxygen flow with 300 ccm/min could be fed into the coating chamber during deposition to control the stoichiometry of the YSZ coatings. Deposition was generally conducted in condition of 0~20 rpm of the substrate rotation for obtaining the different porosity of coatings. The substrate temperature was 950°C.

Free standing YSZ coatings layers were obtained by machining the substrate from the coated specimen prior to thermal conductivity measurements. The laser flash method is used to measure the thermal diffusivity of coated samples. The well-known laser flash method relies on the generation of a thermal pulse on one face of a thin sample and on the observation of the temperature history. The thermal diffusivity is determined from the time required to reach one-half of the peak temperature in resulting temperature rise curve for the rear surface as illustrated in Fig.1 [7].

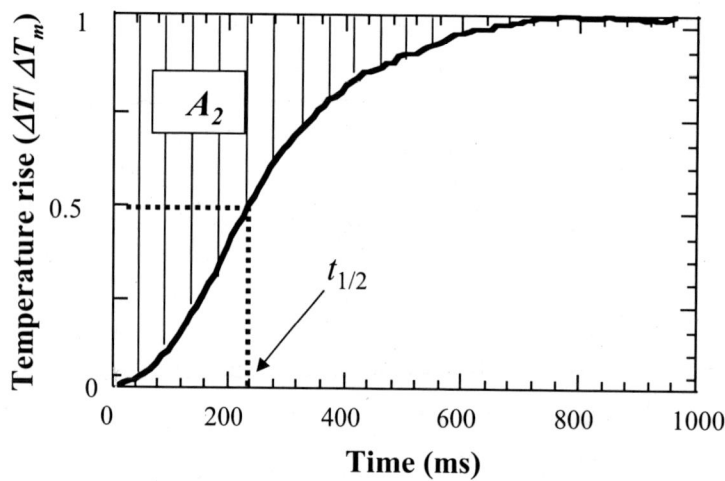

Figure 1. Temperature response behavior as a function of time at the rear surface of EB-PVD coated specimen after laser pulse heating.

All the measurements of the coated samples of 10 mm diameter were carried out between 25°C and 1000°C in 200 degree intervals in a vacuum chamber. Because of the translucency of the specimens to the laser, the specimens were sputter-coated with a thin layer of silver and colloidal graphite spraying to ensure complete and uniform absorption of the laser pulse prior to thermal diffusivity measurement. Specific heat measurements of the coated samples were made with differential scanning calorimeter (DSC) using sapphire as the reference material in an argon gas condition. The thermal conductivity of the coatings was then determined using

$$k = \alpha C \rho \qquad (1)$$

where k is the thermal conductivity, α is the thermal diffusivity, C is the specific heat, and ρ is the density of the coatings, respectively. The density of each coated sample was determined by measuring the mass and the volume of coated samples by a micrometer. The microstructure of the coated samples was observed by SEM. Raman spectroscopy was used to determine the crystal structures of the phases in the coatings.

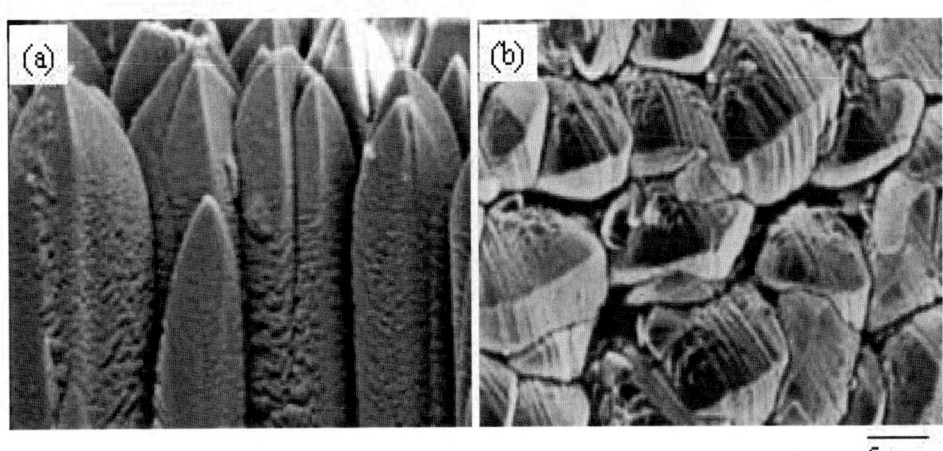

Figure 2. SEM micrographs of microstructure of ZrO_2-4 mol% Y_2O_3 coatings obtained by EB-PVD: (a) side view and (b) surface view.

RESULTS AND DISCUSSION

Typical microstructures in the coatings observed from surface and side regions of EB-PVD coatings are shown in Fig. 2. The top surfaces of the coatings consist of square-pyramidal or cone-like grains. In particular, the morphology of side regions of coatings deposited on substrates have a crystalline columnar texture with all columnar grains oriented in the same direction, namely perpendicular to the substrate, and with of a predominantly open porosity.

The columnar grains increase in size toward the top of column from the substrate, resulting in a tapered columnar structure. This result indicates that an epitaxial growth of YSZ films does not occur when deposited on substrates by EB-PVD. Gaps between columnar grains can also be clearly observed, particularly towards the top of the coatings. These gaps contribute to the porosity of EB-PVD coatings. The feather-like structure on both sides of the columnar grains contains many micro-sized, as well as nano-sized pores [8, 9].

Fig. 3 shows the result of Raman spectrum of the top surfaces of ZrO_2-4mol% Y_2O_3 coatings. This indicates that the observed phase in all coatings materials is the tetragonal phase of zirconia because its spectra were dominated by the relatively sharp tetragonal Raman modes.

Fig. 4 shows the porosity dependence on thermal conductivity for the free standing coatings layers separated from substrate. The porosity was calculated from the difference of density. The different porosity in the coatings was obtained by changing the rotation speed of substrate and multilayer between 300~700 μm.

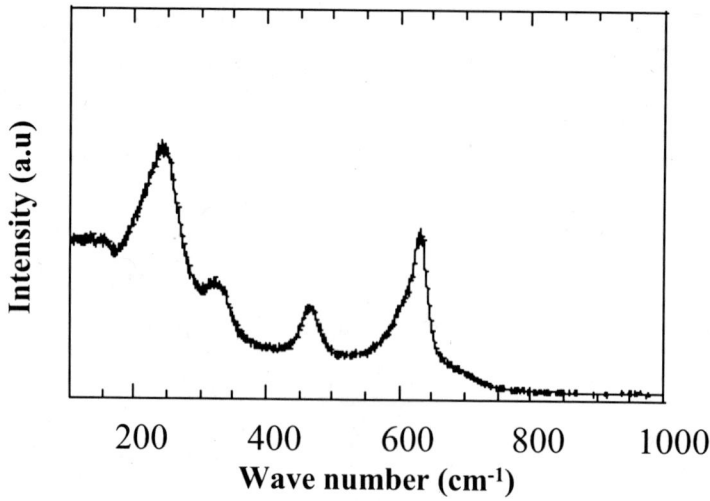

Figure 3. Raman pattern of ZrO_2-4 mol% Y_2O_3 coatings obtained by EB-PVD.

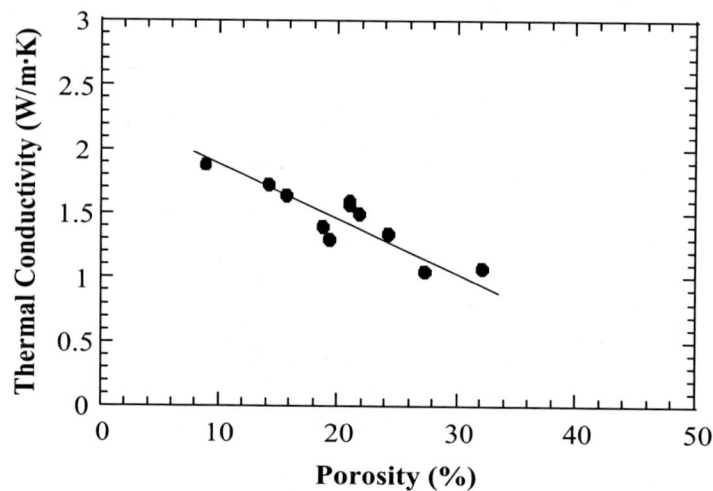

Figure 4. Relationship between thermal conductivity and porosity of
ZrO_2-4mol%Y_2O_3 coatings specimens obtained by EB-PVD.

The porosity showed decreasing tendency with increasing the rotation speed and multilayer. It is seen that a wide variation in thermal conductivity is obtained in close relation with porosity. The thermal conductivity of the free standing coatings layers decreases with increase of porosity. Based on this result, it can be explained that the porosity is the dominant factor in determining thermal conductivity for coatings by EB-PVD. This is consistent with results showing that the porosity provides a major contribution to the reduction of the thermal conductivity of zirconia coatings [10, 11].

Fig. 5 shows the results of thermal conductivity for free standing coated specimens from room temperature to 1000°C. The thermal conductivity decreases slightly with increasing temperature as well as porosity. It is readily apparent that the thermal conductivity of the coatings is well below that of sintered ZrO$_2$-4mol% Y$_2$O$_3$ with full density.

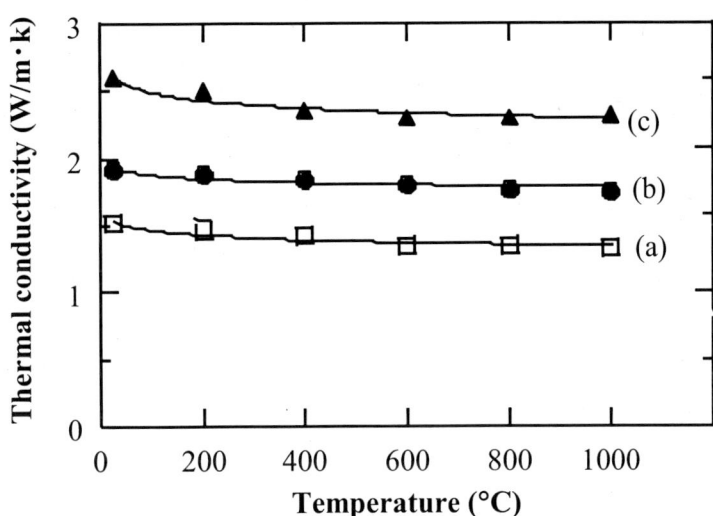

Figure 5. Thermal conductivity vs temperature of ZrO$_2$-4mol%Y$_2$O$_3$ specimens:
(a) coatings with 22% porosity, (b) coatings with 9% porosity obtained
by EB-PVD and (c) sintered bodies with full density.

For free standing specimens of very thinner coatings, the measurement of thermal conductivity is difficult because thinner specimens is easily broken or damaged during measuring or handling. For this reason, some researches on thermal conductivity with the laser flash method in the multi-layers system were reported [12,13].

When calculating the thermal diffusivity in the Fig.1, the method of determining $t_{1/2}$ values assumes that heat diffusion occurs across a uniform, pure and isotropic material. If this is the case, its value is reliable. However, the double layers specimens are non-uniform materials that consist of a porous coating layer and dense substrate similar to many multi-layer materials. Consequently, the method of estimating heat diffusion in the coated specimen to calculate the thermal conductivity of the coating layer must be reconsidered.

We suggested the calculation of thermal conductivity of coatings alone using the double layers specimens (coatings and substrate) using the response function method in the laser flash measurements [14].

Therefore, the thermal conductivity (λ_2) for coatings layer alone can be derived according to Eq. (2) using double layers specimens based on the response function method which the detailed theory was written in the previous work [7,15].

$$\lambda_2 = \frac{d_2^2 C_2 \rho_2 \left(3 d_1 C_1 \rho_1 + d_2 C_2 \rho_2\right)}{6 A_2 \left(d_1 C_1 \rho_1 + d_2 C_2 \rho_2\right) - d_1^2 C_1 \rho_1 \left(d_1 C_1 \rho_1 + 3 d_2 C_2 \rho_2\right)/\lambda_1} \tag{2}$$

where d_1, C_1, ρ_1 and d_2, C_2, ρ_2 correspond to the thickness, specific heat and density of substrate and coatings layer, respectively. λ_1 is the thermal conductivity of substrate. For double layers specimens, the area bounded by the temperature rise curve and the maximum temperature line at the rear face of the coated specimen after the laser pulse heating, designated A_2 in Fig. 1, can be obtained by integration. This area is called the "areal thermal diffusion time".

The areal thermal diffusion time as a function of coatings thickness in the present double layers specimens was given in Table 1.

Table 1. Calculated areal thermal diffusion time values as a function of coatings thickness for coated specimens.

Coating thickness (μm)	Areal thermal diffusion time (A_2)
79	0.208
119	0.223
249	0.279
297	0.298
302	0.305
337	0.324
368	0.333
497	0.415
613	0.489

Fig. 6 shows the correlation between the calculated and measured thermal conductivities for coatings layer alone with coating thickness based on Eq. (2). The calculated values of coatings show the good agreement with experimentally measured values. In the present results, the thermal conductivity of free standing coatings below 300 μm thickness cannot measure because of damage of specimens. The calculated thermal conductivity of coatings tends to increase with increasing of coatings thickness. The present results are consistent with the report of coating thickness on the thermal conductivity studied by Krell et al [16].

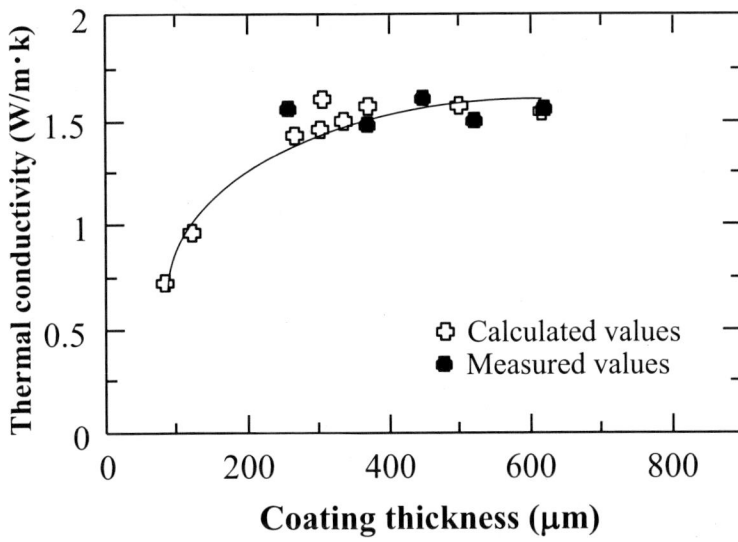

Figure 6. Thermal conductivity as a function of coatings thickness for
ZrO$_2$-4mol%Y$_2$O$_3$ coatings specimens obtained by EB-PVD.

In addition, the thicker coatings layer reveal higher thermal conductivity than that of thinner coated specimens. This reason can be considered that thinner coatings layer have very fine columnar grains and many columns boundaries without grain growth in column. Generally, phonons interact with imperfections such as dislocation, vacancy, pore and grain boundary. Therefore, many columns boundaries in thinner coating layers significantly make to reduce the mean free path by phonon scattering, resulting in the reduction of thermal conductivity [17].

The low thermal conductivity mechanism in the present porous coatings can be considered as following. The thermal conductivity can usually be reduced by decreasing the mean free path due to phonon scattering at pores. Many models exist to describe phonon heat conduction through porous media. The classical Maxwell model is the simplest model of thermal conduction for isolated pores dispersed in a continuous solid phase [18]. This model assumes an even distribution of random spherical pores and neglects pore geometry. Neglecting the heat conduction through the pores, this model can be expressed as:

$$k_{eff} \cong k_s \left[2(1-\phi)/(2+\phi) \right] \qquad (3)$$

where ϕ is porosity, k_{eff} is effective thermal conductivity of porous materials and k_s is thermal conductivity of the solid.

Fig. 7 shows the comparison of experimental thermal conductivity of the present specimens and theoretical thermal conductivity of porous materials based on Eq. (3). The theoretical thermal conductivity of the porous material including random spherical pores decreases with increasing porosity. The experimental thermal conductivity of the present specimens remarkably decreases with increasing porosity. The present specimens show lower thermal conductivity than theoretical values of the porous materials. The reason is that EB-PVD coatings consist of porous columnar grains with a feather-like structure containing evenly dispersed elongated pores as well as intracolumnar pores inside of the columns, resulting in the decrease of thermal conductivity due to the effective disturbance of heat flow.

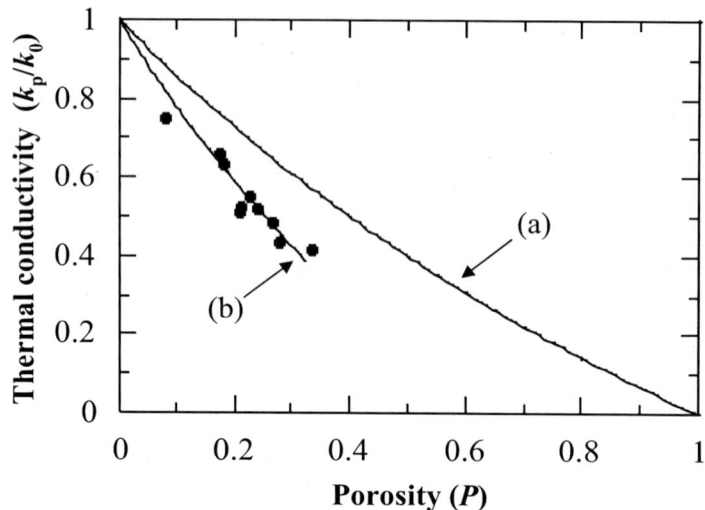

Figure 7. Normalized thermal conductivity as a function of porosity: (a) theoretical thermal conductivity and (b) experimental thermal conductivity of the present specimens.

CONCLUSIONS

ZrO_2-4 mol% Y_2O_3 coatings were deposited by EB-PVD. The coated layers had a columnar microstructure with intercolumnar gaps between columnar grains. Nano sized pores < 50 nm could be observed around feather-like grains as well as inside of columnar grains. The thermal conductivity of the coatings tended to decrease with increasing porosity. The thermal conductivity for coatings decreased slightly with increasing temperature between room temperature to 1000°C. The thermal conductivity of the coatings layer could be successfully calculated using the response function method applied to the combined coatings and substrate specimen. The calculated thermal conductivities of the coatings layer were in good agreement with experimental results from the laser flash method. Thermal conductivity of coatings increased with increasing coatings thickness. Increase of columns boundaries and pores in coatings layer mainly led to reduced thermal conductivity by the reduction of mean free path by phonon scattering.

ACKNOWLEDGMENTS

The authors acknowledge the financial support of the New Energy and Industrial Technology Development Organization (NEDO), Japan.

REFERENCES

[1] U. Schulz, B. Saruhan, K. Fritscher and C. Leyens, "Review on Advanced EB-PVD Ceramic Topcoats for TBC Applications," *Int. J. Appl. Ceram. Technol.*, **1**, 302-15 (2004).

[2] C. G. Levi, " Emerging Materials and Processes for Thermal Barrier Systems," *Current Opinion in Solid State and Mat. Sci.*, **8**, 77-91 (2004).

[3] D. D. Hass, P. A. Parrish and H. N. G. Wadley, "Electron Beam Directed Vapor Deposition

of Thermal Barrier Coatings," *J. Vac. Sci. Technol.*, **16**, 3396-3401 (1998).

[4]J. Singh, D. E. Wolfe, R. A. Miller, J. I. Eldridge and D. M. Zhu, "Tailored Microstructure of Zirconia and Hafnia-based Thermal Barrier Coatings with Low Thermal Conductivity and High Hemispherical Reflectance by EB-PVD,"*J. Mater. Sci.*, **39**, 1975-1985(2004).

[5]O. Unal, T. E. Mitchell and A. H. Heuer, "Microstructure of Y_2O_3-Stabilized ZrO_2 Electron Beam-Physical Vapor Deposition Coatings on Ni-Based Superallys," *J. Am. Ceram. Soc.*, **77**, 984-92 (1994).

[6]T. J. Lu, C. G. Levi, H. N. G. Wadley and A. G. Evans, "Distributed Porosity as a Control Parameter for Oxide Thermal Barriers Made by Physical Vapor Deposition," *J. Am. Ceram. Soc.*, **84**, 2937-2046 (2001).

[7]B. K. Jang, M. Yoshiya, N. Yamaguchi and H. Matsubara, "Evaluation of Thermal Conductivity of Zirconia Coating Layers Deposited by EB-PVD," *J. Mater. Sci.*, **39**, 1823-1825 (2004).

[8]B. K. Jang and H. Matsubara, "Influence of Rotation Speed on Microstructure and Thermal Conductivity of Nano-Porous Zirconia Layers Fabricated by EB-PVD," *Scripta Mater.*, **52**, 553-558 (2005).

[9]B. Saruhan, P. Francois, K. Fritscher and U. Schulz, "EB-PVD Processing of Pyrochlore-Structured $La_2Zr_2O_7$-Based TBCs," *Surf. Coat. Technol.*, **182**, 175-183 (2004).

[10]K. An, K. S. Ravichandran, R. E. Dutton and S. L. Semiatin, "Microstructure, Texture, and Thermal Conductivity of Single-Layer and Multilayer Thermal Barrier Coatings of Y_2O_3-Stabilized ZrO_2 and Al_2O_3 Made by Physical Vapor Deposition," *J. Am. Ceram. Soc.*, **82**, 399-406 (1999).

[11]K. W. Schlichting, N. P. Padture and P. G. Klemens, "Thermal Conductivity of Dense and Porous Yttria-Stabilized Zirconia," *J. Mater. Sci.*, **36**, 3003-3010 (2001).

[12]R. F. Bulmer and R. Taylor, "Measurement by the Flash Method of Thermal Diffusivity in Two-Layer Composite Sample," *High Temp.-High Press.*, **6**, 491-497 (1974).

[13]J. Hartmann, O. Nilsson, J. Fricke, "Thermal Diffusivity Measurements on Two-Layered and Three-Layered Systems with the Laser-Flash Method," *High Temp.-High Press.*, **25**, 403-410 (1993).

[14] T. Baba, N. Taketoshi and A. Ono, "Analysis of Heat Diffusion Across Three Layer Thin Films by the Response Function Method"*21st Jpn. Symp. Thermophysical. Properties*, 229-231 (2000).

[15]B. K. Jang, and H. Matsubara, "Analysis of Thermal Conductivity and Thermal Diffusivity of EB-PVD Coating Materials," *Trans. of Mater. Res. Soc. Jpn.*, **29**, 417-420 (2004).

[16]H. J. R. Scheibe, U. Schulz and T. Krell, "The Effect of Coating Thickness on The Thermal Conductivity of EB-PVD PYSZ Thermal Barrier Coatings," *Surf. Coat. Technol.*, **200**, 5636-5644 (2006).

[17]J. R. Nicholls, K. J. Lawson, A. Johnstone and D. S. Rickerby, "Methods to Reduce the Thermal Conductivity of EB-PVD TBCs," *Surf. Coat. Technol.*, **151-152**, 383-391 (2002).

[18]H. Szelagowski, I. Arvanitidis and S. Seetharaman, "Effective Thermal Conductivity of Porous Strontium Oxide and Strontium Carbonate Samples," *J. Appl. Phys.*, **85**, 193-198 (1985).

COMPARISON OF THE RADIATIVE TWO-FLUX AND DIFFUSION APPROXIMATIONS

Charles M. Spuckler
NASA Glenn Research Center
21000 Brookpark Rd.
Cleveland Ohio 44145

ABSTRACT

Approximate solutions are sometimes used to determine the heat transfer and temperatures in a semitransparent material in which conduction and thermal radiation are acting. A comparison of the Milne-Eddington two-flux approximation and the diffusion approximation for combined conduction and radiation heat transfer in a ceramic material was preformed to determine the accuracy of the diffusion solution. A plane gray semitransparent layer without a substrate and a non-gray semitransparent plane layer on an opaque substrate were considered. For the plane gray layer the material is semitransparent for all wavelengths and the scattering and absorption coefficients do not vary with wavelength. For the non-gray plane layer the material is semitransparent with constant absorption and scattering coefficients up to a specified wavelength. At higher wavelengths the non-gray plane layer is assumed to be opaque. The layers are heated on one side and cooled on the other by diffuse radiation and convection. The scattering and absorption coefficients were varied. The error in the diffusion approximation compared to the Milne-Eddington two flux approximation was obtained as a function of scattering coefficient and absorption coefficient. The percent difference in interface temperatures and heat flux through the layer obtained using the Milne-Eddington two-flux and diffusion approximations are presented as a function of scattering coefficient and absorption coefficient. The largest errors occur for high scattering and low absorption except for the back surface temperature of the plane gray layer where the error is also larger at low scattering and low absorption. It is shown that the accuracy of the diffusion approximation can be improved for some scattering and absorption conditions if a reflectance obtained from a Kubelka-Munk type two flux theory is used instead of a reflection obtained from the Fresnel equation. The Kubelka-Munk reflectance accounts for surface reflection and radiation scattered back by internal scattering sites while the Fresnel reflection only accounts for surface reflections.

INTRODUCTION

Thermal barrier coatings (TBCs) are being developed for high temperature applications in gas turbine engines. Some of the ceramic materials being considered for TBCs are semitransparent in the wavelength range where thermal radiation can be important. For example, zirconia can be transparent up to about 5 μm (refs 1 and 2). In a semitransparent material, combined conduction and radiation determine the temperature inside and the heat transferred through the material. The radiative heat transfer in a semitransparent material is determined by the absorption, emission, scattering, and refractive index. The reflection at an interface is determined by the refractive index of the materials on each side of the interface. For diffuse thermal radiation going from a material with a higher refractive index to a lower refractive index the surface reflection is increased by the total reflection of the incident radiation at angles greater than the critical angle. The thermal radiation emitted internally and by an opaque material into a semitransparent material depends on the refractive index squared. The internal thermal radiation transmitted through the interface of a semitransparent layer is reduced by the internal surface reflection, which may include total internal reflection, so the energy emitted will not exceed that of a blackbody. Therefore, the refractive index of a semitransparent material can have a considerable effect on the temperatures in a material.

The amount of thermal energy absorbed, emitted, and scattered by a material is determined by the absorption and scattering coefficients. The coefficients have units of reciprocal length. The reciprocal of these coefficients can be considered to be the mean distance traveled before absorption or scattering occur if the coefficients don't vary along their path (ref. 3 page 424). Because the temperature changes as a result of absorption or emission, these processes have a direct effect on the temperature. Scattered thermal radiation does not affect the temperature unless it is absorbed. Scattering in some cases can act as additional absorption in determining the temperature profiles in a material ref. 4. Part of the scattered radiation that is not absorbed will be scattered back out of the layer increasing its reflectivity.

The setting up and the solution of the exact spectral radiative transfer equation that include absorption, emission, and scattering is complex. Approximate solutions, such as the two-flux and diffusion methods, which are easer to solve, have been developed. The diffusion approximation is the simplest approximation with radiation treated as a diffusion process and absorption, emission, and reflection of thermal radiation occurring at the surfaces of the material. The two-flux approximation, which includes absorption emission and scattering, is more complicated and requires a computer solution. In the two–flux approximation, it is assumed that there is a radiative flux traveling in the positive and negative x-directions with radiation absorbed emitted and scattered inside the material and reflections occurring at the internal and external surfaces of the layer. The Milne-Eddington two-flux, diffusion, and exact solutions for an absorbing, emitting, and scattering plane layer were compared for a plane layer in ref. 5. The two-flux method was in good agreement with the exact solution for the conditions considered. The diffusion method was found to give good predictions for large optical thicknesses [optical thickness = (absorption coefficient + scattering coefficient) x thickness]. The diffusion approximation and the discrete ordinate method for a plane non-scattering glass layer on an opaque substrate and a two dimensional non-scattering glass in an opaque container were compared in refs. 6 and 7. The results from the diffusion approximation were reasonable for thick glass layers and greatly under predicted the temperature and heat flux for thin layers or layers with small opacity. The discrete ordinate and the diffusion method were used to predict the heat transfer in a cylindrical partially stabilized zirconia piece under going laser assisted machining ref. 8. The temperatures predicted by the diffusion solution are 130 K higher than those predicted by the discrete ordinate method, while the measured quasi-steady temperature approached an intermediate value.

In this paper diffusion and Milne-Eddington two flux solutions are compared for an emitting, absorbing, and scattering plane gray layer and a non-gray plane layer on a substrate. An absorption coefficient of $a = 0.1346$ cm^{-1} and a scattering coefficient of $\sigma_s = 94.38$ cm^{-1} were used as a base line. These coefficients are in the range of those of zirconia in the wavelengths where it is semitransparent ref. 2. To determine how scattering and absorption affect the accuracy of the diffusion solution compared to the Milne-Eddington two flux solution the absorption and scattering coefficients are increased and decreased from the base line. To try to get better agreement between the results of the diffusion and Milne-Eddington two-flux solutions, the reflectance used for the radiative heat input to the layer for the diffusion solution was changed from one that only has surface reflections to one that also included reflections from internal scattering and/or a substrate.

MODEL

The models used are a semi-infinite plane gray semitransparent layer figure 1a and a

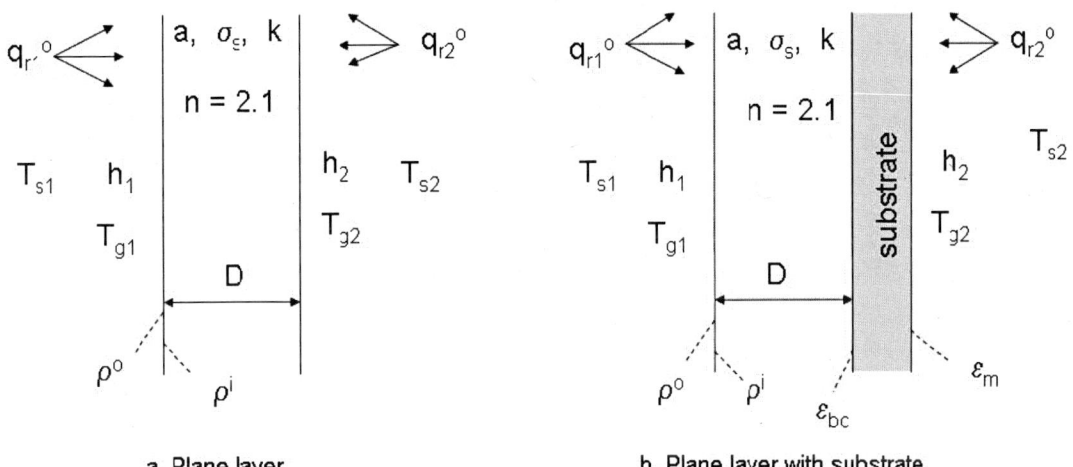

a. Plane layer   b. Plane layer with substrate

Figure 1 Heat transfer models

semi-infinite non-gray plane layer on a substrate figure 1b. The non-gray plane layer is opaque for wavelengths greater than 5 μm and semitransparent for wave lengths less than 5 μm. There is diffuse radiative and convective heat transfer on each side. The external radiative heating is q_{r1}^o and q_{r2}^o. The hot side gas and surrounding temperatures, T_{s1} and T_{g1}, are 2000K and the cold side temperatures T_{s2} and T_{g2} are 800K. The heat transfer coefficients are h_1 = 250 w/m^{2}K on the hot side and h_2 = 110 w/m^{2}K on the cold side. The plane layer and the layer on the substrate are 1 mm thick and have a thermal conductivity k = 0.8 w/mK. The substrate is 0.794 mm thick and has a thermal conductivity of 33 w/mK. The emissivity of the back side of the metal substrate, ε_m, is 0.6. The conditions on the layer with a substrate were used by Siegel (ref. 9) to determine internal radiation effects in a zirconia based TBC on a combustor liner. An infinitely thin bond coat is assumed between the semitransparent layer and the substrate and its emissivity is assumed to be 0.7 or 0.3. The refractive index, n, of the semitransparent layer is assumed to 2.1 which is in the expected range for zirconia ref. 1. The refractive index of the surrounding gas is assumed to be one. The external interface reflection, ρ^o, was obtained using the Fresnel equation for a non-absorbing layer (ref. 3 page 87). The non-absorption assumption should be good for the absorption coefficients used here (refs. 10 and ref. 3 page 88). The internal interface reflection which includes total internal reflections was obtained using eq. 35 in ref. 11. The Milne-Eddington two flux equations used are in ref. 5 and 12 and a method of solution is in reference 5. The Kubelka-Munk type two flux equations are in reference 1 and reference 13 page 193. The equation for the reflectance obtained from solving the Kubelka-Munk two flux equations is in the appendix. The diffusion equations were solved for both Fresnel reflectance and a Kubelka-Munk reflectance. The Fresnel reflectance only takes into account the surface reflection while the Kubelka-Munk reflectance takes into account surface reflection, radiation internally scattered back by the layer and substrate reflection if a substrate is present.

PLANE LAYER WITHOUT A SUBSTRATE

For the plane layer without a substrate, the Milne-Eddington two-flux solution was compared to the solution of the exact radiative transfer equations for an absorbing emitting and scattering layer for scattering coefficients from 0.944 cm^{-1} to 943.81 cm^{-1} and absorption coefficients from 0.0013 cm^{-1} to 13.46 cm^{-1}. The exact radiative transfer equations along with a solution method are in ref. 4. The percent difference in the interface temperatures [100 x ($T_{2\text{-flux}}$−T_{exact})/T_{exact}] was between - 0.075 and 0.28% for front gas-layer interface, between -0.44 and 0.21% for the back gas-layer interface and between 0.14 and 3.6% for the heat flux. Calculations indicate that if the number of spatial increments used in the solution of the exact equations is increased the percent difference in the heat flux would decrease. This shows that for the plane layer the results of the two flux solution are in good agreement with the exact solution. In the remainder of the paper the Milne-Eddington two-flux and diffusion solutions will be compared to determine the accuracy of the diffusion solution.

The percent difference in the temperature between the diffusion and Milne-Eddington two-flux solutions [100 x ($T_{diffusion}$ − $T_{2\text{-flux}}$)/ $T_{2\text{-flux}}$] for the front gas-layer interface of a gray

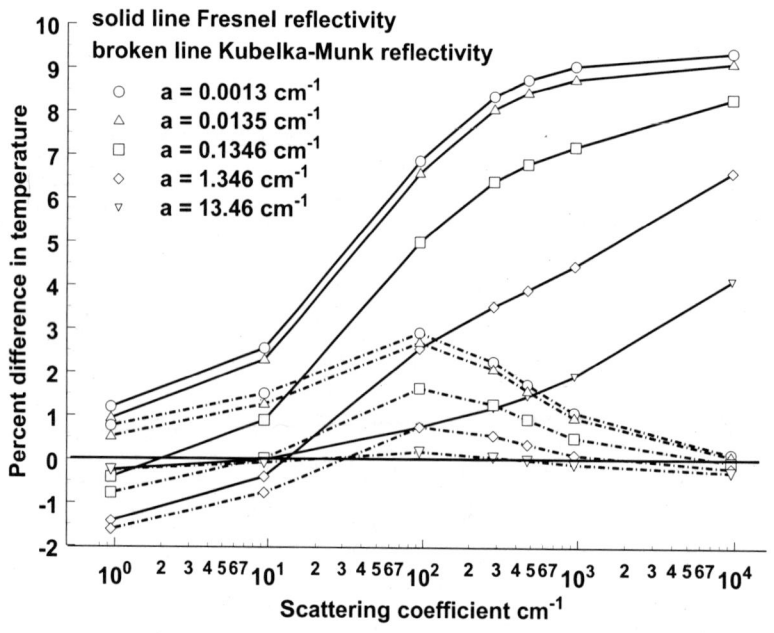

Figure 2 Percent difference between diffusion and 2-flux solutions for front surface of plane layer

plane layer is shown in figure 2. The symbols indicate points for which calculations were made. The percent difference in temperature calculated using Fresnel surface reflectivity and the Kubelka-Munk layer reflectivity for the radiative input in the diffusion solution are shown. The absolute value of the percent difference is less than 10% for all scattering and absorption coefficients considered. Using the Fresnel reflectivity, the percent difference in temperature for moderate to high scattering, increases with increased scattering and decreased absorption. For high scattering and low absorption the percent difference is leveling off at 9 to 10%. When the Kubelka-Munk reflectance was used, the maximum percent difference in temperature was less than 3% and that occurred for a scattering coefficient of 94.38 cm^{-1} and the lowest absorption. For any absorption combined with high scattering the percent difference is near zero using the Kubelka-Munk reflectivity. For high absorption and all scattering considered and for high scattering and all absorption considered the diffusion approximation with a Kubelka-Munk reflectance gives good results, because radiation scattered back is taken into account.

The percent difference in the temperature of the back surface of a plane gray layer for the diffusion and Milne-Eddington two flux solutions is shown in figure 3. The absolute value of the

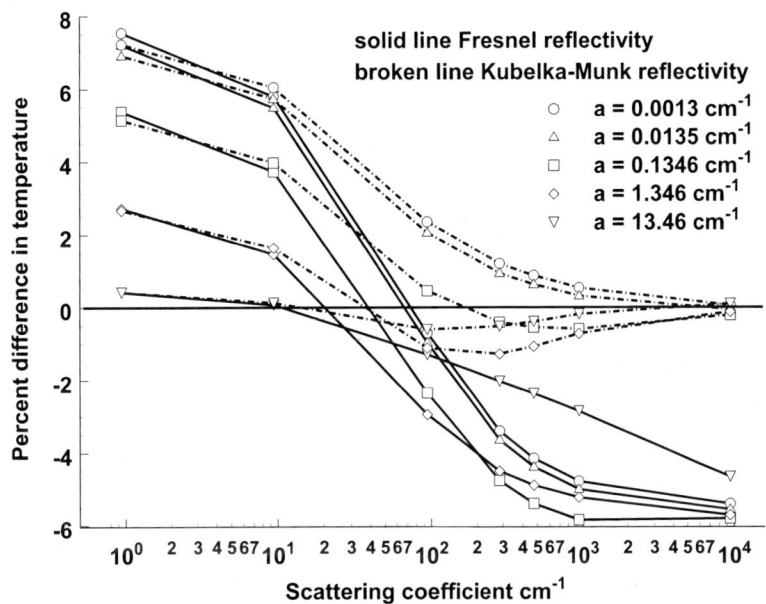

Figure 3 Percent difference between diffusion and 2-flux solutions for back surface of plane layer

percent difference is less than 8%. When the Fresnel reflectivity is used the percent difference goes from a positive value at low scattering to a negative value at high scattering. For low scattering the diffusion solution yields a higher temperature than the Milne-Eddington two-flux solution and the opposite occurs for high scattering. When the Fresnel reflectivity is used, the percent difference in temperature approaches a value of -5 to -6% for high scattering for all except the highest absorption considered where the percent difference is lower. For low scattering the percent difference in temperature is about the same whether the Fresnel or the Kubelka-Munk reflectivity are used. As the scattering is increased, the percent differences in temperature for the two different reflectivities diverge. When the Kubelka-Munk reflectance is used, the percent difference in temperature between the Milne-Eddington two-flux and diffusion solution is small for highest absorption used and all scattering considered. Also, for the Kubelka-Munk reflectance the percent difference approaches zero for all absorption considered and high scattering.

The percent difference in the heat flux calculated using the Milne-Eddington two-flux and the diffusion solutions, is shown in figure 4. The percent difference in heat flux using the Fresnel reflectivity for the diffusion solution is less than -8% for all absorption considered and a scattering coefficient of 94.38 cm^{-1} or less. For a scattering coefficient greater than 94.38 cm^{-1} the percent difference increases rapidly reaching over 200% for low to moderate absorption. The reason for this high a difference is that radiation scattered back is not accounted for when the Fresnel reflectivity is used. When the Kubelka-Munk reflectance is used the percent error is relatively constant for a scattering coefficient less than around 94.38 cm^{-1} and is nearly -47% for the lowest absorption. The percent difference decreases as the absorption increases. Above scattering coefficient 94.38 cm^{-1}, the percent difference decreases with increased scattering reaching less than -10% for high scattering. For the highest absorption coefficient used the percent difference between the two solutions is less than 10%. For low to moderate scattering,

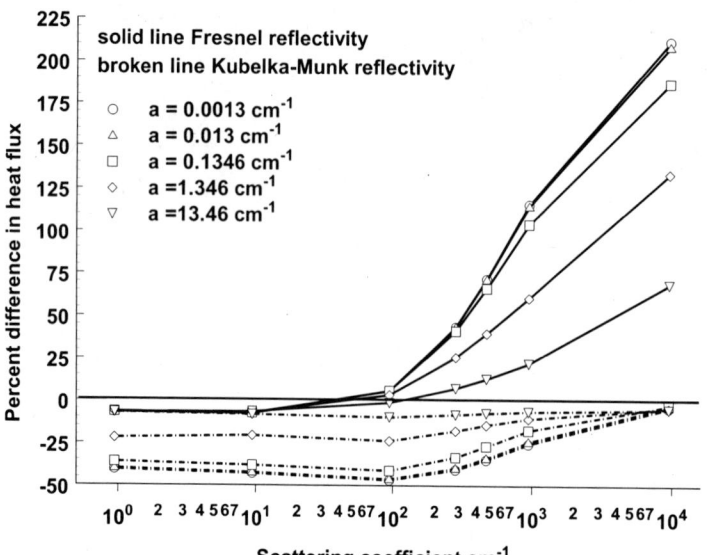

**Figure 4 Percent difference in heat flux between diffusion and
2-flux solutions for plane layer**

using the Fresnel reflectivity gives the best agreement between the diffusion and two flux solutions. When the Kubelka-Munk reflectivity is used the best agreement occurs for highest absorption and all scattering considered and for the highest scattering and all absorption used.

PLANE LAYER WITH SUBSTRATE

For the non-gray plane layer on a substrate, the layer was assumed to be semitransparent up to 5 μm wavelength. Above 5μm the material was considered to be opaque. In the Milne-Eddington two-flux solution, the Fresnel reflectance was used for all wavelengths. In the

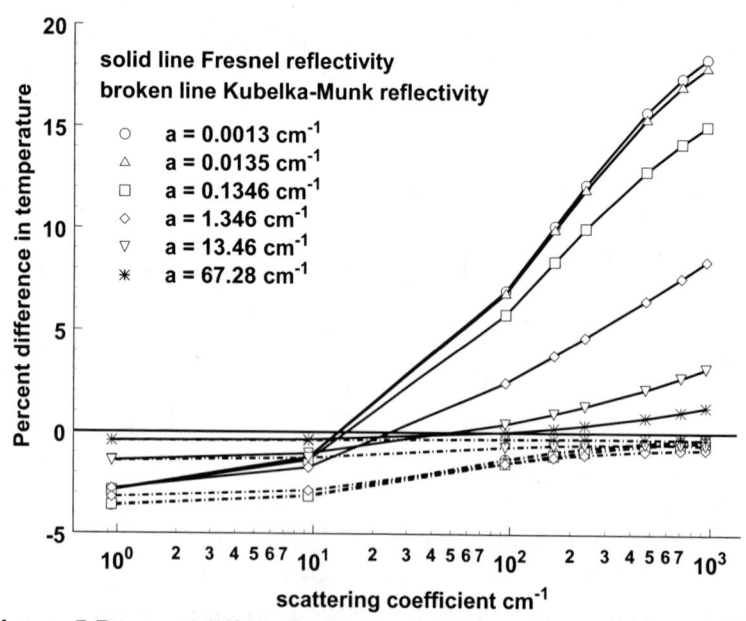

**Figure 5 Percent difference between diffusion and 2-flux solutions
for front surface of layer with substrate reflectivity changes at 5μm**

diffusion solution when the term Fersnel reflectivity is used, means that the Fresnel reflectance

Progress in Thermal Barrier Coatings

was used for wavelengths above and below 5 μm. But, in the wavelength range less than 5 μm the material is semitransparent and internal scattering increases the reflectance of the layer. To try to account for this increase in reflectance the diffusion solution was also solved using the Kubelka-Munk reflectance at wavelengths less than 5 μm and the Fresnel reflectance for wavelengths greater than 5 μm; this is termed the Kubelka-Munk reflectivity.

The percent difference in temperature obtained using the Milne-Eddington two-flux and the diffusion solution for the front gas-layer interface of a layer on a substrate with a bond coat emissivity of 0.7 is in figure 5. Using the Fresnel reflection for all wavelengths, the percent difference in temperature goes from about a -3% at low scattering for all except higher absorptions to almost 20% for high scattering and low absorption. When The Kubelka-Munk reflectance was used for wavelengths less than 5 μm, the absolute value of the percent difference in temperature was less than -4% and decreased with scattering except for absorption coefficients of 13.46 cm^{-1} and higher where there was a slight increase in the absolute percent difference for higher scattering. For high scattering coefficients the percent difference in temperature is less than -1%.

For the interface between the semitransparent layer and the substrate, the percent difference in the temperature using the diffusion and Milne-Eddington two-flux solutions for a bond coat emissivity of 0.7 is shown in figure 6. When the Fresnel reflectivity was used, the

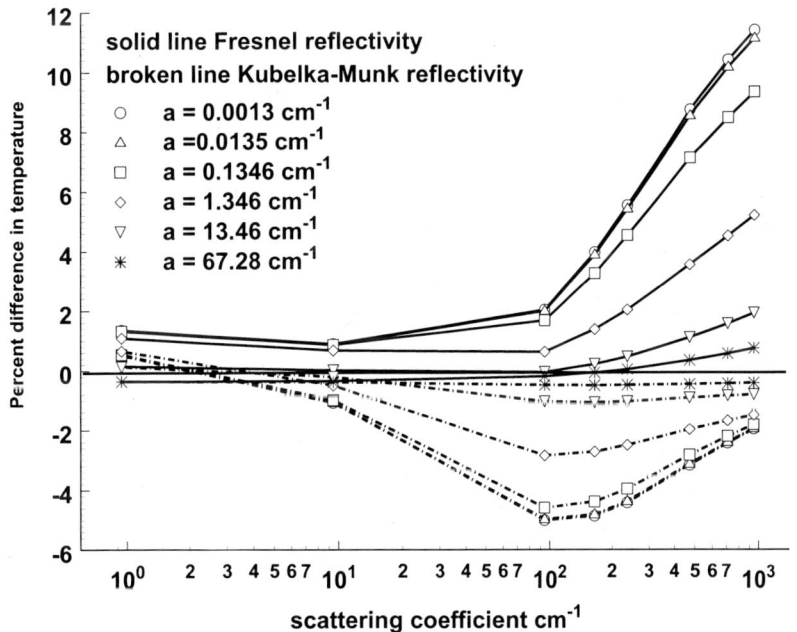

Figure 6 Percent difference in temperature between diffusion and 2-flux solutions for front of substrate reflectivity changes at 5 μm

percent difference in temperature is less than 1.4% for the lowest scattering. The percent difference at first decreases slightly with increasing scattering before increasing. The highest percent error occurs for high scattering and low absorption where the difference reaches about 12%. When the Kubelka-Munk reflectivity is used at the lower wavelengths, the percent difference in temperature is less than 0.7% for the lowest scattering. It decreases with scattering becoming negative and reaching a maximum negative value for a 94.38 cm^{-1} scattering coefficient. The maximum negative percent difference occurs for the lowest absorption. For high absorption and all scattering the diffusion and Milne-Eddington two-flux solution are in good agreement. The percent difference in temperature for the back surface of the substrate is quite similar to the percent difference in temperature for the front surface of the substrate.

The percent difference in the heat flux using the diffusion and Milne-Eddington two-flux solutions for a layer on a substrate with a bond emissivity of 0.7 is shown in figure 7. The

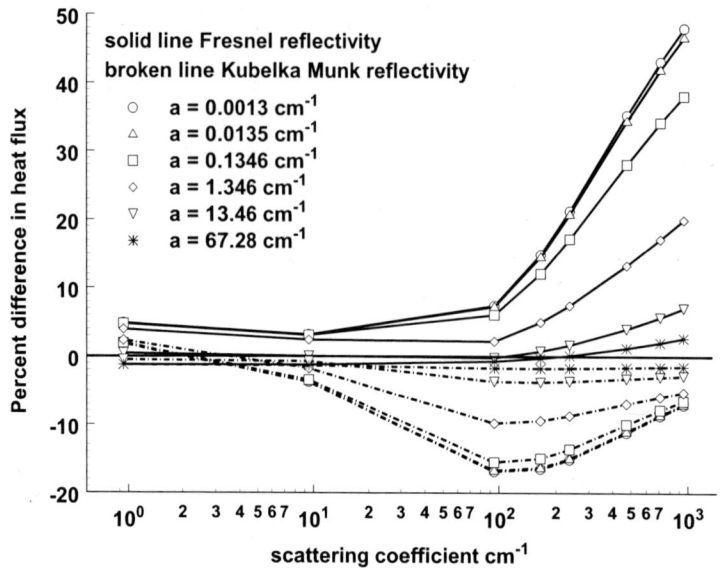

Figure 7 Percent difference in heat flux between diffusion and 2-flux solutions for layer with substrate reflectivity changes at 5μm

shapes of the curves are similar to those in figure 6 for the percent difference in temperature of the substrate, but the percent difference in heat flux is 3 to 4 times higher than the percent difference in temperature. When the Fresnel reflectivity is used for the diffusion solution, the absolute value of the percent difference in heat flux is as high as 6% for the lowest scattering. For a 94.38 cm^{-1} scattering coefficient, the percent difference in heat flux begins to increase rapidly for lower absorption. The percent difference reaches nearly 50% for the lowest absorption used. When the Kubelka-Munk reflectivity is used for the diffusion solution, the absolute value of the percent difference in heat flux is less than about 3% for the lowest scattering. The maximum percent difference which occurs at a 94.38 cm^{-1} scattering coefficient is around -17%. For higher scattering the percent difference decreases with scattering. For absorption coefficients greater than 13.46 cm^{-1} and all scattering the percent difference in heat flux was less than -4%.

When the emissivity of the bond coat was change from 0.7 to 0.3 and the Fresnel reflectivity was used, the change in the absolute value of the percent difference between the diffusion and Milne-Eddington two-flux solution was less than 3% for the surfaces temperatures and was 11.3% for the heat flux. When the Kubelka-Munk reflectivity was used the absolute value of the percent difference between the diffusion and Milne-Eddington two-flux solution was less than 2% for the surfaces temperatures and was 4.0% for the heat flux. This indicates that the Kubelka-Munk reflectivity can handle a change in bond coat emissivity better than the Fresnel reflectivity.

CONCLUSIONS

A study was performed to determine the accuracy of the diffusion solution compared to the Milne-Eddington two-flux solution. A one dimensional model was used and a gray plane layer and a non-gray plane layer on a substrate were considered. The layer was 1 mm thick and the substrate was 0.794 mm thick. There is convective and diffuse radiative heat transfer on each side of the layer. For the diffusion solution Fresnel reflectivity, which only takes surface reflection into account, and Kubelka-Munk type reflectivity which take surface reflection and radiation scattered back into account were used. When the non-gray layer on a substrate was considered, Fresnel reflectivity or Kubelka-Munk type reflectivity was used for the wavelengths where the material was semitransparent and Fresnel reflectivity was used for the opaque regions for the diffusion solution. For the plane layer the Milne-Eddington two-flux solution was compared to the exact solution and found to be in good agreement with a percent difference less than 0.45% for the temperature and less than 3.6% for the heat flux. For the plane gray layer, the largest percent difference between the Milne-Eddington two-flux and the diffusion solutions occurred when Fresnel reflectivity was used and there was high scattering and low absorption, except for the back surface of the plane layer were the difference was higher for low scattering and low absorption. The percent difference in interface temperatures for the diffusion and Milne-Eddington two-flux solution for the plane layer were small for high absorption and all scattering considered, and for high scattering and all absorption considered if the Kubelka-Munk type reflectivity was used. The percent difference in heat fluxes was also small for high scattering if the Kubelka-Munk type reflectivity was used. For the non-gray layer on the substrate, the percent difference in temperature between the Milne-Eddington two-flux and the diffusion solutions were less than 5% for low scattering when either the Fresnel or Kubelka-Munk reflectivities were used. When the Kubelka-Munk reflectivity was used the percent difference in temperature was less than 6% for all scattering and absorption used, while the percent difference was nearly 12% for high scattering and low absorption when the Fresnel reflectivity was used. The percent difference in heat flux was nearly 50% for high scattering and low absorption when the Fresnel reflectivity was used. When the Kubelka-Munk type reflectivity was used the percent difference in heat flux was less than 17%. The diffusion solution can be improved under some conditions if a Kubelka-Munk type reflectivity is used. Also using the Kubelka-Munk type reflectivity seems to account for a change in substrate emissivity better than using the Fresnel reflection.

For the heating conditions of combined external radiative heat load and convection the absorption thickness (absorption coefficient x layer thickness) not the optical thickness [(absorption coefficient + scattering coefficient) x layer thickness] is the crucial factor in determining whether the use of the diffusion approximation is appropriate. The most inaccurate results from the diffusion approximation are generally for high scattering thickness (scattering coefficient x layer thickness) with no absorption or low absorption thickness (absorption coefficient x layer thickness) with only surface reflections accounted for. Using a Kubelka-Munk reflectivity, which includes radiation scattered back, can increase the accuracy of the diffusion solution approximation for large scattering thicknesses.

REFERENCES
[1]Wahiduzzaman, S and Morel,T., Effect of Translucence of Engineering Ceramics on Heat Transfer in Diesel Engines, ORNL/Sub/88-22042/2, April 1992.

[2]Makino, T., Kunitomo, T., Sakai, I., and Kinoshita, H., Thermal Radiation Properties of Ceramic Materials, *Heat Transfer-Japanese Research,* **13**, [4] 33-50 (1984).

[3]Siegel, R. and Howell, J. R. *Thermal Radiation Heat Transfer,* 4th ed. Taylor & Frances, New York, 2002.

[4]Spuckler, C. M. and Siegel, R., "Refractive Index and Scattering Effects on Radiative Behavior of a Semitransparent Layer," *Journal of Thermophysics and Heat Transfer*, **7**[2], 302-10 (1993).

[5]Siegel, R. and Spuckler, C. M., "Approximate Solution Methods for Spectral Radiative Transfer in High Refractive Index Layers," *International Journal of Heat and Mass Transfer*, **37** [Suppl. 1] 403-13 (1994).

[6]Lee, K.H and Viskanta, R. "Comparison of the diffusion approximation and the discrete ordinates method for investigation of heat transfer in glass," *Glastech. Ber. Glass Technol.* 72[8], 254-266 (1999).

[7]Lee, K.H and Viskanta, "Two-dimensional combined conduction and radiation heat transfer: comparison of the discrete ordinates method and the diffusion approximation methods," *Numerical Heat Transfer, Part A*, 39, 205-225 (2001).

[8]Pfefferkorn, F. E., Incropera, F. P., and Shin, Y. C. "Heat Transfer model of semi-transparent ceramics undergoing laser-assisted machining," *International Journal of Heat and Mass Transfer*, 48, 1999-2012 (2005).

[9]Siegel, R. "Internal Radiation Effects in Zirconia Thermal Barrier Coatings," *Journal of Thermophysics and Heat Transfer*, **10**[4], 707-9 (1996).

[10]Cox, R. L., "Fundamentals of Thermal Radiation in Ceramic Materials,"; pp. 83-101 in Symposium on Thermal Radiation of Solids, edited by S. Katzoff, NASA SP-55, 1965.

[11]Richmond, J. C., "Relation of Emittance to Other Optical Properties," *Journal of Research of the National Bureau of Standards-C. Engineering and Instrumentation*, **67C** [3], 217-26 (1963).

[12]Siddall, R. G. "Flux methods for the analysis of radiant heat transfer," *Proceedings of the Fourth Symposium on Flames and Industry*, Paper 16, pp. 169-179, The Institute of Fuel (1972).

[13]Ishimaru A., *Wave Propagation and Scattering in Random Media Vol. 1 Single Scattering and Transport Theory*, Academic Press, New York, 1978.

APPENDIX EQUATIONS

The equation for the reflectance obtained from solving the Kubelka-Munk type two-flux equations is

$$R = \rho^o + \frac{\left(1-\rho^i\right)}{2}\left\{ A\left[\frac{\sqrt{1-\omega^2}}{\left(1-\omega\right)}-1\right] - B\left[\frac{\sqrt{1-\omega^2}}{\left(1-\omega\right)}+1\right]\right\} \tag{1}$$

Where A and B are

$$A = \frac{\dfrac{2.0\left(1-\rho^o\right)\left(1-\omega\right)}{\left(1-\rho^i\right)\sqrt{1-\omega^2}+\left(1+\rho^i\right)\left(1-\omega\right)}}{1.0-\left[\dfrac{\left(1+\rho_{bc}\right)\left(1-\omega\right)-\left(1-\rho_{bc}\right)\sqrt{1-\omega^2}}{\left(1+\rho_{bc}\right)\left(1-\omega\right)+\left(1-\rho_{bc}\right)\sqrt{1-\omega^2}}\right]\cdot\left[\dfrac{\left(1+\rho^i\right)\left(1-\omega\right)-\left(1-\rho^i\right)\sqrt{1-\omega^2}}{\left(1+\rho^i\right)\left(1-\omega\right)+\left(1-\rho^i\right)\sqrt{1-\omega^2}}\right]\cdot e^{-2\sqrt{1-\omega^2}\cdot\tau_L}} \tag{2}$$

$$B = -\frac{\dfrac{2.0\left(1-\rho^o\right)\left(1-\omega\right)}{\left(1-\rho^i\right)\sqrt{1-\omega^2}+\left(1+\rho^i\right)\left(1-\omega\right)}\,e^{-2\sqrt{1-\omega^2}\cdot\tau_L}}{\left[\dfrac{\left(1+\rho_{bc}\right)\left(1-\omega\right)+\left(1-\rho_{bc}\right)\sqrt{1-\omega^2}}{\left(1+\rho_{bc}\right)\left(1-\omega\right)-\left(1-\rho_{bc}\right)\sqrt{1-\omega^2}}\right]-\left[\dfrac{\left(1+\rho^i\right)\left(1-\omega\right)-\left(1-\rho^i\right)\sqrt{1-\omega^2}}{\left(1+\rho^i\right)\left(1-\omega\right)+\left(1-\rho^i\right)\sqrt{1-\omega^2}}\right]\cdot e^{-2\sqrt{1-\omega^2}\cdot\tau_L}} \qquad (3)$$

ρ^o = external interface reflection

ρ_{bc} = reflectivity of the bond coat for a layer with a substrate

ρ_{bc} = ρ^i for a plane layer without a substrate

ρ^i = internal interface reflection

$\tau_L = \left(2a+\sigma_s\right)\cdot D$

$\omega = \dfrac{\sigma_s}{\sigma_s+2a}$

a = absorption coefficient

D = thickness of the semitransparent layer

σ_s = scattering coefficient

Thermal Properties

RELATION OF THERMAL CONDUCTIVITY WITH PROCESS INDUCED ANISOTROPIC VOID SYSTEMS IN EB-PVD PYSZ THERMAL BARRIER COATINGS

A. Flores Renteria, B. Saruhan
Institute of Materials Research,
German Aerospace Center
Linder Hoehe, Porz-Wahnheide
Cologne, NRW 51147, Germany

J. Ilavsky
X-Ray Operations and Research (XOR), Experimental Facilities Division,
Advanced photon Source (APS)
Argonne National Laboratory
9700 S. Cass Avenue, bldg 438E
Argonne, IL 60439, USA

ABSTRACT

Thermal barrier coatings (TBCs) deposited by Electron-beam physical deposition (EB-PVD) protect the turbine blades situated at the high pressure sector of the aircraft and stationary turbines. It is an important task to uphold low thermal conductivity in TBCs during long-term service at elevated temperatures. One of the most promising methods to fulfil this task is to optimize the properties of PYSZ-based TBC by tailoring its microstructure. Thermal conductivity of the EB-PVD produced PYSZ TBCs is influenced mainly by the size, shape, orientation and volume of the various types of porosity present in the coatings. These pores can be classified as open (inter-columnar and between feather arms gaps) and closed (intra-columnar pores). Since such pores are located within the three-dimensionally deposited columns and enclose large differences in their sizes, shapes, distribution and anisotropy, the accessibility for their characterization is very complex and requires the use of sophisticated methods. In this work, three different EB-PVD TBC microstructures were manufactured by varying the process parameters, yielding various characteristics of their pores. The corresponding thermal conductivities in as-coated state and after ageing at 1100C/1h and 100h were measured via Laser Flash Analysis Method (LFA). The pore characteristics and their individual effect on the thermal conductivity are analysed by USAXS which is supported by subsequent modelling and LFA methods, respectively. Evident differences in the thermal conductivity values of each microstructure were found in as-coated and aged conditions. In summary, broader columns introduce higher values in thermal conductivity. In general, thermal conductivity increases after ageing for all three investigated microstructures, although those with initial smaller pore surface area show smaller changes.

INTRODUCTION

Attractive thermo mechanical properties of electron beam - physical vapour phase deposited (EB-PVD) thermal barrier coatings (TBCs) are attributed to their unique microstructure. The primary columns present in these microstructures are separated by inter-columnar gaps oriented perpendicular to the substrate's plane, leading to excellent thermo-shock resistance of these coatings under thermal cyclic conditions. It is noteworthy that the inter-columnar gaps are oriented parallel to the heat flux, and thus, unfavourable for the capability of the TBCs as thermal insulators. The heat flux is principally transported through the solid column material being equilibrated by intra-columnar pores and voids between feather-arms.

During the EB-PVD coating process, primary columns start to grow on the substrate surface following a nucleation stage. The columnar growth occurs in a preferred direction, typically perpendicular to the plane of the substrate. Inter-columnar gaps and feather-arm features are created due to the shadowing effect of the neighbouring column tips which impede the vapour flux to reach the bottom of the valley between the columns[1]. Moreover, the formation and growth of the voids between feather-arms are influenced by the next factors:

(a) The sunshine-sunset shadowing effect of the neighbouring columns tips, resulting in different morphological sequences at the directions parallel and perpendicular to the plane of vapour incidence (PVI);

(b) The substrate temperature, that regulates the diffusion of atoms at the growing surface and

(c) The rotation speed, influencing the solid material growth and yielding the "banana" shape pores due to shadowing after each completed rotation movement.

Since the feather-arm gaps are located at the periphery of the columns, the column-tip shadowing effect will be enhanced during the growing process with the support of the other two mentioned factors (i.e. those given by b and c). Therefore, this phenomenon is significant at the ultimate edge of the columns, forming their conical cross section. For this reason, the feather-arm gaps can be designated as open intra-columnar pores created at the columns periphery due to lower vapour flux available for the complete solid material deposition. In addition, intra-columnar pores are created inside the columns at regions of highest vapour incidence angles (VIA), where the edge of the deposited vapour after each complete rotation overlaps with the initial growth generated by the next rotation's movement. They grow through the deposited material parallel to the feather-arms in an elongated "banana" shape following the altered direction of the vapour incidence angle (VIA) in a sunrise-sunset pattern.

Thermal exposure of the EB-PVD deposited TBCs results in morphological changes which consequently affect the properties such as thermal conductivity and Young's Modulus. These may be due to the occurrence of a series of thermal processes varying from formation of bridges between the columns, formation of sintering necks at contact points between the feather-arms, and eventually changes in pore geometry and sizes of the intra-columnar pores[2]. All these thermally activated processes generate surface area reduction and may follow in a similar way as the sintering processes[3].

Previous studies[4-6] indicate that the microstructural configuration of the EB-PVD PYSZ TBCs significantly contribute to the intrinsic thermal properties of the material. Thus, the quantitative information on the spatial and geometrical characteristics of the pores within these coatings is required to correlate those with their thermal properties. Sophisticated techniques such as USAXS and USANS have shown to be effective in thorough characterization of as-coated EB-PVD TBCs[7-9]. However, no study is known up-to-date which benefits from these techniques to determine the morphological alterations on ageing and to precisely identify the role of each morphological feature and alteration on the in-service stability of the thermal conductivity in EB-PVD manufactured PYSZ TBCs.

In this study, three different microstructures of EB-PVD produced PYSZ TBCs were manufactured by altering coating process parameters. Their morphological characterization was carried out in as-coated and aged (1100°C/100h) conditions via Ultra Small-Angle X-rays Scattering Method (USAXS). The measurements were carried out in two orthogonal directions per specimens due to the anisotropy of the pores. The resulting raw data were fitted employing a computer based model[10] capable to determine the mentioned characterization of the pores; i.e., volume, size, aspect ratio, shape and orientation of the inter-columnar gaps, gaps between feather-arms, and intra-columnar pores. Correspondingly, thermal conductivity of specimens under the same conditions was measured via Laser Flash Analysis Method (LFA).

MATERIALS AND METHODS

Materials and Processing

Partially Yttria Stabilized Zirconia (PYSZ) coatings were manufactured via EB-PVD process by employing "von Ardenne" pilot plant equipment having a maximum EB-power of 150 kW. Evaporation was carried out from a single evaporation source having the ingot dimensions of 62.5 mm diameter and 150 mm length. The chemical composition of the ingot was standard 7-8 wt.%Y_2O_3 stabilized ZrO_2. Deposition of the vapor phase on flat substrates was carried out under conventional rotating mode by mounting the substrates on a holder with its horizontal axis perpendicular to the evaporation source as described in[11] (i.e. perpendicular to PVI). During the coating process, the substrates were rotated at different speeds and heated to different temperatures (Table I). Table I lists the designation of the samples and the applied process parameters to manufacture these three investigated EB-PVD-morphologies.

Table I. EB-PVD coating conditions and designation for the three manufactured microstructures

Morphology	Chamber Pressure (mbar)	Substrate Temperature (°C)	Rotation Speed (rpm)
Feathery	8×10^{-3}	850	30
Intermediate	8×10^{-3}	950	12
Coarse	8×10^{-3}	1000	3

For USAXS characterization, the EB-PVD PYSZ coatings of app. 400 μm thickness were deposited on Ni-basis substrates which were previously coated with a NiCoCrAlY bond coat. USAXS specimens were prepared by cutting and polishing the coatings into 200μm thickness slices. Two orthogonal slices (e.g. perpendicular and parallel to the PVI) per specimen were obtained through this process. For thermal conductivity characterization by Laser-Flash-Analysis (LFA), the coatings were deposited on 12.7 mm diameter discs of FeCrAl-alloys (without bond coat) in the same run. For LFA sample preparation, the FeCrAl-alloy substrates were chemically etched to obtain coatings in free-standing conditions. Finally, these free-standing coatings were additionally coated on both sides with a thin Pt-layer via sputter method to avoid laser penetration during the measurements. Subsequently, corresponding specimens were heat treated in air at 1100°C/100h using a heating rate of 5°C/min.

Methods of Characterization

Microstructural Characterization

The coating microstructures were characterized visually by using a Field-Emission Scanning Electron Microscope (FE-SEM, LEITZ LEO 982).

Ultra-Small Angle X-ray Scattering (USAXS)

The effective pinhole-collimated USAXS instrument which is equipped with a data processing system and available at the UNICAT 33-ID of the Advance Photon Source, ANL, was employed for our measurements[8]. Effective-pinhole USAXS enables the characterization of anisotropic microstructures with nearly the same resolution previously available for isotropic materials. This instrument utilizes the advantages of Bonse-Hart[12] double-crystal diffraction optics to extend its range of the scattering vector Q ($Q = (4\pi/\lambda)(\sin\theta)$, where λ is the wavelength of the incident X-rays and 2θ is the scattering angle) to noticeably lower values ($1{,}2 \times 10^{-4}$ Å^{-1} < $|Q| < 0{,}1$ Å^{-1}) by decoupling the resolution of the instrument from the primary beam size. This effective-pinhole configuration allows the measurement of the scattering vector (Q) in one direction (1D), which is perpendicular to the substrate's plane within the plane of the coating. Therefore, to determine the 2D distribution of the scattering intensities, rotation of the specimens in small increments of an azimuthal angle (α) is required.

In order to obtain the complete characterization of spatial and geometrical anisotropic microstructures such as EB-PVD TBCs, two measuring methods are usually employed:
(1) The scattering intensity, I(Q) is measured as function of the azimuthal angle (α) for a constant Q value (aniso-scans) to determine the principal max. and min. scattering intensities and their respective α values;
(2) The scattering intensity (I(Q)) is measured as a function of Q at the principal α values.
By combining the results obtained by these methods a quantitative map of the microstructural anisotropy as function of the pore sizes can be done[8].

Since the analyzed coatings enclose anisotropic stereometric characteristics at the azimuthal and polar spatial domains, each specimen was measured at two orthogonal directions (perpendicular and parallel to the plane of vapor incidence). Several (i.e. five) aniso-scans at different fixed Q-values were done, which allowed the characterization of the complete size range of the pores within the coatings. Additionally, the scattering intensity at important anisotropic azimuthal orientations was measured.

Scattering form-factor functions have been already derived for different shapes of scattering elements such as spheroids, rods, discs, networks, etc. and reported in previous studies [13-16]. Moreover, it is known that the scattering structure-factor function is dependent of local order, describing spatial correlations that may exist between scattering elements, e.g. monodispersed population of spherically shaped scatterers[17], arrays of parallel cylindrical shaped scatterers[18], or a fractal system[19]. Since real stereometric characteristics of every pore population enclose a certain deviation, the applied model considers Gaussian distribution for the calculation of the orientation and size values. It uses idealized particle shapes for which small-angle scattering can be reasonably well calculated (e.g. as ellipsoidal oblate and prolate, and spheres), and allows the optimization of the parameters by a last square fitting as given in[10]. The model allows the use of five independent pore populations composed of individual spheroid shapes with R_0, R_0 and βR_0 axes. In the employed coordinate system the x axis, and y and z axes lay perpendicular (in other words; parallel to the column axis), and parallel to the substrate's plane, respectively. Furthermore, the orientation of the pores in the space is described by its βR_0 axis with respect to the coordinate system by two independent angles α (azimuthal angle) and ω (polar angle). An anisotropic orientation model was applied in a study[7] which assumes the solution of the differential scattering cross-section as function of the orientation for each scattering pore population. Due to uncertainties in the calibration of the USAXS-specimens thickness, the total volumes of the pores were measured by Archimedes Method. Thus, in this work, the USAXS-model predicts the volume fraction of each pore population.

Laser Flash Analysis Method (LFA)

The thermal diffusivity of the Pt-coated free-standing specimens was measured employing a Netsch-LFA 427 instrument. The calculation of the thermal conductivity was calculated through the formula:

$$\lambda = \alpha \cdot \rho \cdot C_p \qquad (5)$$

Where, λ is the thermal conductivity (W/m·K), in this case ρ represents the bulk density (gr/cm^3) of the free-standing coatings measured by Archimedes Method and Cp is the specific heat (J/g·K) measured by Differential Scanning Calorimeter (DSC).

RESULTS AND DISCUSSION

Microstructural observation of the coatings in the as-coated state displays the typical EB-PVD PYSZ morphology for all three intended microstructures. The main differences are being in the column diameter and in the feather-arm feature configuration. The microstructure "coarse" showed larger column diameters and the microstructure "feathery" more defined feather-arm features (Fig. 1). The resulting volume fractions of each pore population calculated with the USAXS-model were scaled to the corresponding total volume measured with Archimedes Method for each microstructure, which were 27.46% for the "feathery", 27.15% for the "intermediate", and 22.55% for the "coarse", respectively. As SEM micrographs also indicate, USAXS analysis deliver numerical data on the fact that basically all investigated microstructures show variations in their pore distribution and in their column density at the cross-sections, perpendicular and parallel to the plane of vapour incidence (PVI). At all three analysed coating, in the as-coated state, the inter-columnar gaps are slightly broader, and contain evidently higher volumes at the direction perpendicular to the plane of vapour incidence (PE-PVI) compared with that at the parallel direction (PA-PVI). This can also be observed in the micrographs shown in Fig. 1. As a matter of fact, some previous studies has delivered similar qualitative results solely relying on microstructural investigations[1].

For simplicity purposes in this context, we compared the quantitative data obtained on the cross-sections of three investigated microstructures perpendicular to PVI (PE-PVI). Fig. 2 shows the measured and modeled polar distribution of the scattered intensities versus the azimuthal angle (α) at different scattering vector (Q) values for as-coated state of the "coarse" microstructure (top row) and after ageing at 1100°C/100h (bottom row). The intensities drawn with symbols are fittings obtained by modeling. It has to be realized that the scattering intensities corresponding to the different mean opening dimensions (MOD) include scatterings with a specific range of similar sizes. The model uses Gaussian distribution averaging "size distribution" of 40-60%. Scattering intensities oriented at 90° and 270° azimuthal angles correspond to the inter-columnar gaps; while, the scattering intensities at 45-54° and 225-245°

azimuthal angles correspond to those from the feather-arms as well as intra-columnar pores which typically align behind the feather-arm features.

The results of the USAXS-modeling are given in Table II. The obtained quantitative USAXS results confirm clearly the anisotropic character of the porosity at EB-PVD TBCs, especially at the Q ranges between 0.00149 and 0.01285 (corresponding to a main opening dimension of 420nm and 48nm, respectively) (see Table II). These features display somewhat smaller growth-orientation angles (42°) at the microstructures "intermediate" and "feathery". The size distribution of the different pores was in some cases so broad, that these have to be modeled in two separated void populations. These result show that the thickness of the inter-columnar gaps at all investigated microstructures lie in the range of 602-670nm in as-coated conditions, making 4.08% - 8.38% of volume fraction. In the case of feather-arm openings, there are two representative dimensional ranges for the microstructures "intermediate" and "coarse, whereas only one dimension for the microstructure "feathery". Their volume fraction is a factor of four or higher at the microstructures "intermediate" and "coarse" (5.39% and 9.56%, respectively) than that at the microstructure "feathery" (1.40%). The measured coarser opening dimensions for the feather-arm features are given in the following after the schema [diameter/thickness (aspect ratio)]. It is found that these are 1.3/0.12μm (0.09) at the microstructure "intermediate", 1.6/0.15μm (0.09) at the microstructures "coarse" and 1.8/0.09μm (0.05) at the microstructure "feathery". Furthermore, the finer dimensional range for the "intermediate" and "coarse" is represented with 0.7/0.05μm (0.07) (see Table II).

Finally, the intra-columnar pores in the three investigated coatings show also two dimensional ranges. Although these can not be packed into specific dimensional groups being representative for all three coatings, nevertheless, it is distinctive that there is one group of intra-columnar pores yielding prolate ellipsoid shape (aspect ratio = 20) at all coatings. The dimension of those at the microstructure "coarse" (0.4/0.02μm) is approx. twice as large as those in the other two microstructures, counting solely for the smallest volume fraction (5.80%). It is noticeable that a higher volume fraction (19.82%) of the porosity at the microstructure "feathery" corresponds to these prolate ellipsoid shaped intra-columnar pores [diameter/thickness = 0.17/0.008μm]. Moreover, the intra-columnar pores at the microstructure "intermediate" fall in their dimension and volume fraction (0.2/0.01μm and 11.63%) between those of the other two microstructures. These pores are mostly located as aligned rows behind the voids between feather-arms and, thus, oriented at the same angles as the feather-arm voids (see Table II and Fig. 3). The second group of intra-columnar pores display a disc shape with an intermediate aspect ratio value. These are also oriented at similar angles as the feather-arms. The determination of the pore shapes has been carried out by fitting the modelling curves with those of the measured USAXS scattering curves, since each of the modelling shapes (oblate ellipsoid, sphere, and prolate ellipsoid) affects the shape factor P(Q) of the measured scattering intensities I(Q) by modifying the slope of fitted curves[20].

The USAXS results for the coatings in the as-coated state can be summarised as such that the "feathery" microstructure contains the finest dimension and highest volume of intra-columnar pores followed by the "intermediate" and finally by the "coarse" microstructure. Thus, due to their dimensions and aspect ratios, it can be assumed that, on thermal exposure, the intra-columnar pores of the first opening dimension are predestined to break into quasi-spherical pores due to the occurrence of sintering process[21].

By considering the pores as thermal insulators (i.e. insignificant radiation contribution), it can be stated that the spatial and geometrical distribution of all pore types at EB-PVD TBCs will contribute, according to their effectiveness, to the reduction of the thermal conductivity by interruption of phonon flow through the coating. This means that, as well as the finer dimensions, the higher volume of these cylindrical intra-columnar pores contributes significantly in thermal conductivity reduction of EB-PVD TBCs. This hypothesis is supported by the fact that this relationship defined by USAXS-modeling analysis agrees well with those experimentally determined thermal conductivity values of the three as-coated microstructures (see Fig. 4).

In the case of the microstructures "intermediate" and "coarse", larger intra-columnar pore sizes are anticipated and measured due to the applied lower rotation speeds during processing. Moreover, since these microstructures are manufactured at higher substrate temperatures, only a fraction of these pores are able to form at each rotation phase and will only survive until the next array of new pores are created. These both coatings consist of a lower volume fraction of intra-columnar pores which are heterogeneously distributed. Thus, it is plausible that their measured thermal conductivity values are higher.

After heat treatment at 1100°C for 100 hours, the sintering process becomes active, leading to morphological changes at all pore types[3]. Mass transfer occurs through bridging at the contact points between primary columns forming two dimensional groups of inter-columnar pores. This phenomenon is enhanced by the fact that the finer columns, which are cumulated mostly at the bottom zone of the coating and interrupted from growing through the coating thickness, tend to pull together and create finer channels between them. The volume fraction of such fine inter-columnar gaps is higher at the microstructures "feathery" and "intermediate" (15.58% and 14.08%, respectively) than those at the microstructure "coarse" (8.55%) (see Fig. 5).

On ageing, significant changes occur at the finer feather-arm regions and intra-columnar pores. The effect of sintering is obvious, considering the decrease in the USAXS-modeled aspect ratios which is controlled by the surface area reduction of the pores. The dimension group addressing the finer gaps disappears completely after ageing. These break into arrays of quasi-spherical pores at their inner pyramidal ends, leaving isolated openings with lower aspect ratio at the edge of the feather-arms. The secondary columns of feather-arm features appear to sinter into groups, leaving broader but shorter gaps between the feather-arms (see Fig. 5).

Finally, high aspect ratio cylindrical intra-columnar pores (e.g. fine "banana" shaped pores) which are created at each rotation phase and connect the two pore rows break into quasi-spherical pores (see Fig. 6). Moreover, on ageing, the volume fraction of these pores is drastically reduced, especially at the microstructure "feathery" from 19.82% in the as-coated state to 5.41% after heat-treatment. Consequently, a different equilibrium configuration of the pores results from the sintering process altering the effectiveness of the pores in interruption of the heat transfer through the coatings.

Considering the changes at the thermal conductivity of the coatings after heat-treatment at 1100°C/100h (Fig. 4) and in their morphological changes (Figs. 3 and 6), it can be clearly stated that the changes at the intra-columnar cylindrical pores, aligned behind the feather-arm features are responsible for the drastic thermal conductivity increase in the microstructure "feathery". These changes occur not only in their shape (from prolate ellipsoids to quasi-spheres) but also in their volume fraction, resulting in a decrease reaching to nearly a factor four (from 19.82% to 5.41%). Their significant influence can be attributed to the disappearance of heat-flux hindering paths and formation of isotropic features which facilitate heat transfer.

CONCLUSIONS

In this study employed USAXS-analysis supported with modeling was able to quantitatively and rather accurately determine the geometry and location of anisotropic voids within the three EB-PVD TBC microstructures in the as-coated and aged conditions. The variation of the EB-PVD process parameters produces evident differences in the spatial and geometrical characteristics of the porosity within the manufactured TBCs. Also the heat treatment of these at 1100°C/100h induce irreversible thermal activated processes (i.e. sintering), modifying the distribution and geometry of the pores. According to the results of the thermal conductivity measurements and those from the USAXS–modeling, it is discernible that the intra-columnar pores are the principal constructors of the heat-flux (phonons) through the coatings enhanced by their high volume concentration and distribution.

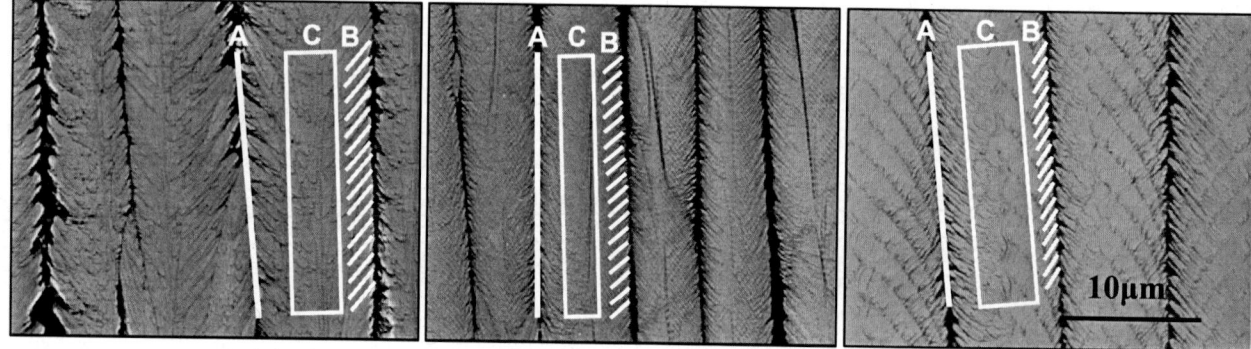

Figure 1: Scanning electron micrographs of EB-PVD TBCs cross sections at the direction perpendicular to the PVI showing the inter-columnar pores (A), pores between feather-arms (B) and intra-columnar pores (C): feathery (left), intermediate (middle), and coarse (right).

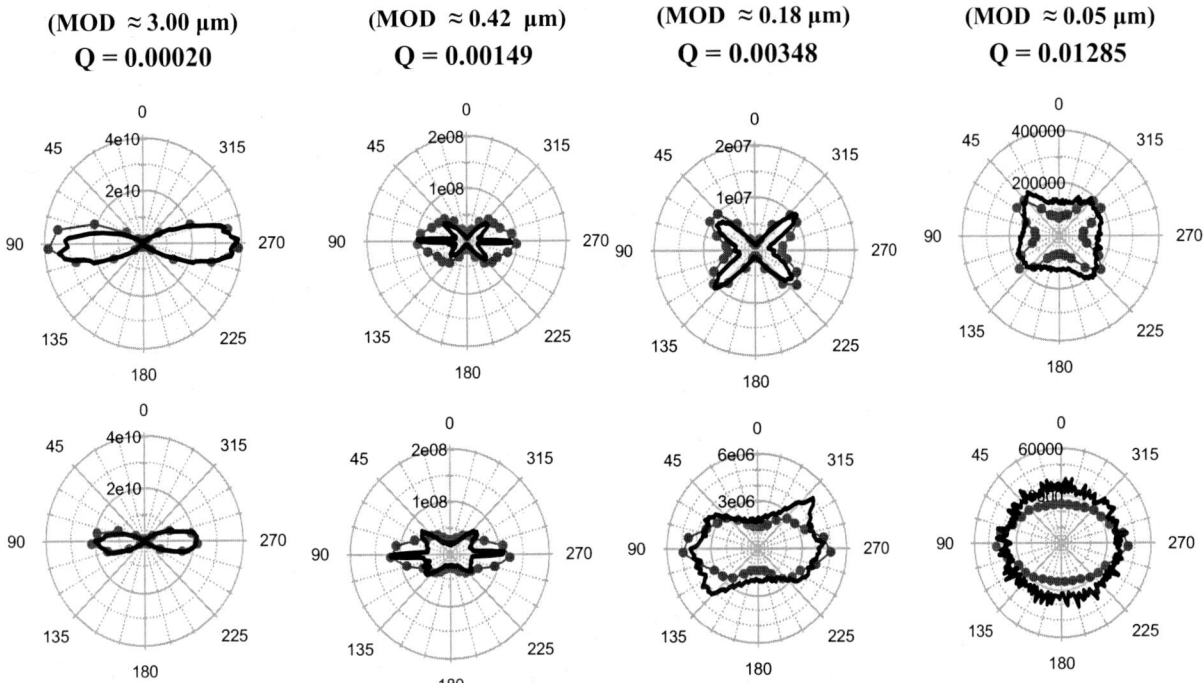

(MOD ≈ 3.00 µm)
Q = 0.00020

(MOD ≈ 0.42 µm)
Q = 0.00149

(MOD ≈ 0.18 µm)
Q = 0.00348

(MOD ≈ 0.05 µm)
Q = 0.01285

Figure 2: Measured (——) and modeled (—●—) polar distribution of the USAXS-scattered intensities versus the azimuthal angle (α) at different Q values and the corresponding main open dimensions (MOD) for the "coarse" microstructure in cross-section perpendicular to PVI: (a) as-coated and (b) after heat treated conditions at 1100°C/100h.

Table II: Calculated values of spatial and geometrical characteristics of the porosity within the three analyzed EB-PVD TBCs in as coated and after ageing conditions via USAXS-modelling.

Pore type as-coated "Intermediate"	Ellipsoidal Shape	Azimuthal angle	Diameter (nm)	Thickness (nm)	Aspect Ratio	Volume fraction
Inter-columnar-1	Oblate	82	-	673.00	0.05	8.38
Between feather-arms-1	Oblate	42	-	118.30	0.091	1.21
Between feather-arms-2	Oblate	42	705.70	45.87	0.065	4.18
Intra-columnar-1	Prolate	42	13.00	260.00	20.00	11.63
Intra-columnar-2	Oblate	42	33.50	15.10	0.45	1.74
Pore type heat-treated "Intermediate"	Shape	Azimuthal angle	Diameter (nm)	Thickness (nm)	Aspect Ratio	Volume fraction
Inter-columnar-1	Oblate	84	-	469.18	0.05	8.65
Inter-columnar-2	Oblate	84	-	124.80	0.26	5.43
Between feather-arms-1	Oblate	45	580.00	174.00	0.30	3.08
Intra-columnar-1	Oblate	45	120.52	84.36	0.70	4.65
Intra-columnar-2	Oblate	45	44.00	35.20	0.80	5,32
Pore type as-coated "coarse"	Shape	Azimuthal angle	Diameter (nm)	Thickness (nm)	Aspect Ratio	Volume fraction
Inter-columnar-1	Oblate	86	-	602	0.05	5.92
Between feather-arms-1	Oblate	54	-	154.11	0.095	2.78
Between feather-arms-2	Oblate	54	718.65	53.907	0.075	6.78
Intra-columnar-1	Oblate	54	210.17	63.05	0.30	1.26
Intra-columnar-2	Prolate	54	22.31	446.33	20.00	5.80
Pore type heat-treated "coarse"	Shape	Azimuthal angle	Diameter (nm)	Thickness (nm)	Aspect Ratio	Volume fraction
Inter-columnar-1	Oblate	92	-	563.00	0.05	7.05
Inter-columnar-2	Oblate	92	-	156.00	0.26	1.50
Between feather-arms-1	Oblate	54	872.75	174.55	0.20	4.95
Intra-columnar-1	Oblate	54	261.30	209.04	0.80	5.27
Intra-columnar-2	Oblate	54	51.43	46.28	0.90	3.78
Pore type as-coated "feathery"	Shape	Azimuthal angle	Diameter (nm)	Thickness (nm)	Aspect Ratio	Volume fraction
Inter-columnar-1	Oblate	84	-	602.45	0.07	4.08
Inter-columnar-2	Oblate	84	-	61.70	0.14	2.14
Between feather-arms-1	Oblate	42	1868.80	93.44	0.05	0.60
Between feather-arms-2	Oblate	42	222.71	31.18	0.14	0.80
Intra-columnar-1	Prolate	42	8.58	171.71	20.00	19.82
Pore type heat-treated "feathery"	Shape	Azimuthal angle	Diameter (nm)	Thickness (nm)	Aspect Ratio	Volume fraction
Inter-columnar-1	Oblate	88	-	591.33	0.05	4.48
Inter-columnar-2	Oblate	88	-	85.00	0.22	11.10
Between feather-arms-1	Oblate	52	1467.36	161.41	0.11	1.14
Intra-columnar-1	Oblate	52	123.45	98.76	0.80	5.31
Intra-columnar-2	Oblate	52	37,36	29,88	0,80	5.41

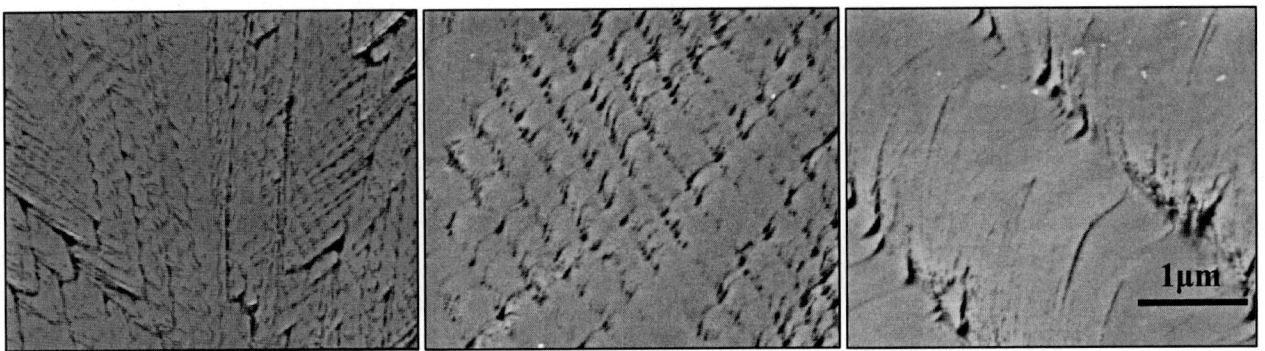

Figure 3: Scanning electron micrographs of EB-PVD TBCs cross sections at the direction perpendicular to the PVI showing the intra-columnar pores: feathery (left), intermediate (middle), and coarse (right).

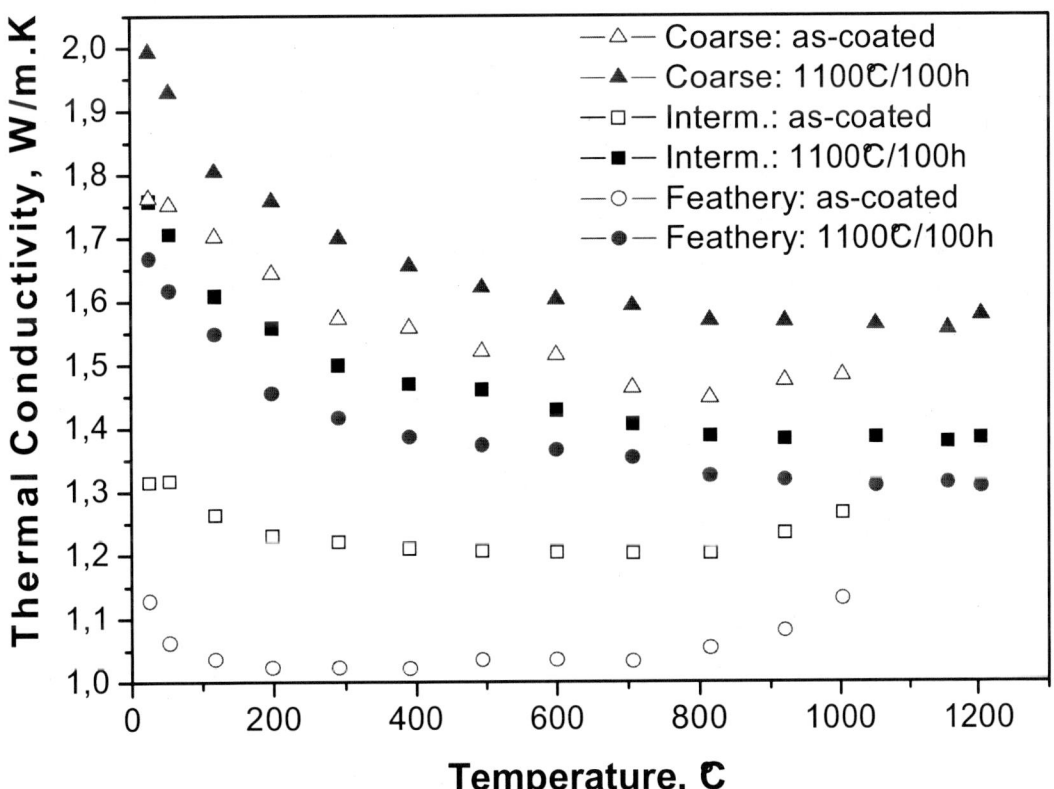

Figure 4: Thermal conductivities vs. measuring temperature values of the three analyzed microstructures: feathery, intermediate and coarse.

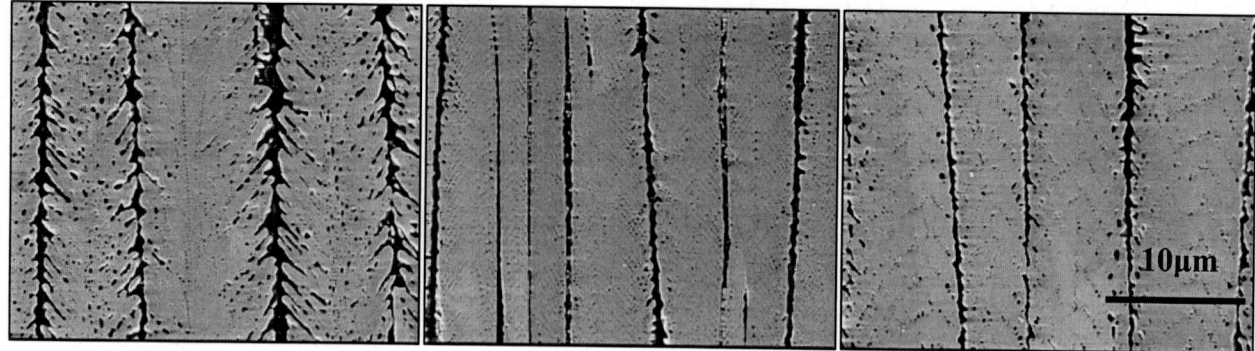

Figure 5: Scanning electron micrographs of heat treated (1100°C/100h) EB-PVD TBCs cross sections after heat treatment at the direction perpendicular to the PVI: feathery (left), intermediate (middle), and coarse (right).

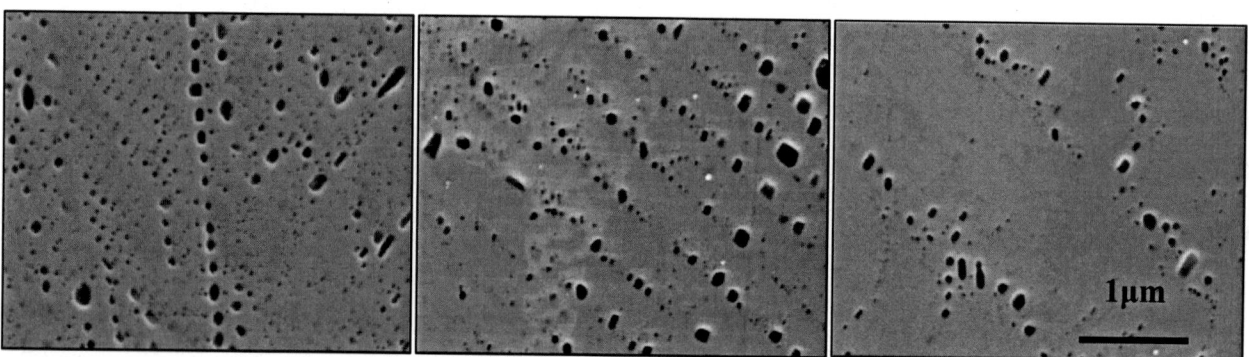

Figure 6: Scanning electron micrographs of heat treated (1100°C/100h) EB-PVD TBCs cross sections at the direction perpendicular to the PVI showing the intra-columnar pores: feathery (left), intermediate (middle), and coarse (right).

REFERENCES

[1] S.G. Terry, Evolution of microstructure during the growth of thermal barrier coatings by Electron-Beam Physical Vapor Deposition, Materials Department, University of California, Santa Barbara, 2001, p. 197.

[2] A.F. Renteria, B. Saruhan, U. Schulz, H.-J. Raetzer-Scheibe, Effect of Morphology on Thermal Conductivity of EB-PVD PYSZ TBCs, Surface and Coatings Technology, accepted for publication, (2006).

[3] A.F. Renteria, B. Saruhan, Effect of ageing on microstructure changes in EB-PVD manufactured standard PYSZ top coat of thermal barrier coatings, Journal of the European Ceramic Society, in Press, Vol.: 26, (2006).

[4] H.-J. Ratzer-Scheibe, U. Schulz, T. Krell, The effect of coating thickness on the thermal conductivity of EB-PVD PYSZ thermal barrier coatings, Surface and Coatings Technology In Press, Corrected Proof (2005).

[5]U. Schulz, C. Leyens, K. Fritscher, M. Peters, B. Saruhan-Brings, O. Lavigne, J.-M. Dorvaux, M. Poulain, R. Mévrel, M. Caliez, Some recent trends in research and technology of advanced thermal barrier coatings, Aerospace Science and Technology, 7, 73-80 (2003).

[6]C.G. Levi, Emerging materials and processes for thermal barrier coatings, Current Opinion in Solid State and Materials Science, 8, 77-91 (2004).

[7]T.A. Dobbins, A.J. Allen, J. Ilavsky, G.G. Long, A. Kulkarni, H. Herman, P.R. Jemian, Recent developments in the characterization of anisotropic void population in thermal barrier coatings using ultra-small angle x-rays scattering, in: Ceram. Eng. and Sci. Proc. (ed. by W.M. Kriven, H.-T. Lin), The American Ceramic Society, pp. 517-524 (2003).

[8]J. Ilavsky, A.J. Allen, G.G. Long, P.R. Gemian, Effective pinhole-collimated ultrasmall-angle x-ray scattering instrument for measuring anisotropic microstructures, Review of scientific instruments, 73, 1-3 (2002).

[9]A.J. Allen, Characterization of ceramics by x-ray and neutron small-angle scattering, J. Am. Ceram. Soc., 88, 1367-1381 (2005).

[10]J. Ilavsky, A.J. Allen, A. Kulkarni, T. Dobbins and H. Herman, Microstructure characterization of thermal barrier coatings deposits - practical models from measurements, in: E. Lugscheider, A. International, M. Park (Eds.), International Thermal Spray Conference, Basel, Switzerland, 2005.

[11]U. Schulz, S.G. Terry, C.G. Levi, Microstructure and texture of EB-PVD TBCs grown under different rotation modes, Materials Science and Engineering A 360, 319-329 (2003).

[12]U. Bonse, M. Hart, An X-ray interferometer with long separated interfering beam paths, Appl. Phys. Lett., 7, 99-100 (1965).

[13]A.J. Allen, J. Ilavsky, G.G. Long, J.S. Wallace, C.C. Berndt, H. Herman, Microstructural characterization of yttria-stabilized zirconia plasma-sprayed deposits using multiple small-angle neutron scattering, Acta Materialia, 49, 1661-1675 (2001).

[14]G. Porod, General theory, in: O. Glatter, O. Kratky (Eds.), Small-angle X-rays scattering, Academic Press, London, (1982).

[15]J.S. Pedersen, Form factors of block copolymer micelles with spherical, ellipsoidal and cylindrical cores, J. Appl. Cryst., 33, 637-640 (2000)

[16]G. Beaucage, Approximations leading to a unified exponential/power law approach to small-angle scattering, J. Appl. Cryst., 28, 717-728 (1995).

[17]A.J. Allen, S. Krueger, G. Skandan, G.G. Long, H. Hahn, H.M. Kerch, J.C. Parker, M.N. Ali, Microstructure evolution during sintering of nanostructured ceramic oxides, J. Am. Ceram. Soc.,9, 1201-1212 (1996).

[18]D. Marchal, B. Demé, Small- angle neutron scattering by porous alumina membranes of aligned cylindrical channels, International Union of Crystallography, pp. 713-717 (2003)

[19]D. Sen, A.K. Patra, S. Mazumder, S. Ramanathan, Pore morphology in sintered ZrO_2-8% mol Y_2O_3 ceramic: a Small-Angle Neutron Scattering investigation, Journal of Alloys and Compounds, 340, 236-241 (2002).

[20]R.-J. Roe, Small-Angle Scattering, in: J.E. Mark (Ed.), Methods of X-ray and neutron scattering in polymer science, Oxford University Press, New York, pp. 155-209 (2000).

[21]J.S. Stoelken, A.M. Glaeser, The morphological evolution of cylindrical rods with anisotropic surface free energy via surface diffusion, Scripta Metallurgica et Materialia, 27, 449-454 (1992).

THERMAL PROPERTIES OF NANOPOROUS YSZ COATINGS FABRICATED BY EB-PVD

Byung-Koog Jang, Norio Yamaguchi and Hideaki Matsubara
Materials Research and Development Laboratory,
Japan Fine Ceramics Center (JFCC)
2-4-1 Mutsuno, Atsuta-ku, Nagoya, 456-8587, Japan

ABSTRACT

This study investigates the specific heat, thermal diffusivity and thermal conductivity of ZrO_2-4mol% Y_2O_3 coating layers fabricated by electron beam physical vapor deposition (EB-PVD) in the temperature range between room temperature to 1000°C. The laser flash method and differential scanning calorimeter are used to measure the thermal diffusivity and specific heat of the coated samples, respectively. EB-PVD coatings reveal the porous columnar microstructure. Coatings are found to contain nano sized pores as well as micron sized pores. The thermal conductivities and thermal diffusivities of EB-PVD coatings decreased with increasing measuring temperature.

INTRODUCTION

Thermal barrier coatings (TBCs) have received a large attention because they increase the thermal efficiency of gas turbine engines by increasing the gas turbine inlet temperature and reducing the amount of cooling air required for the hot section components. Among the various coating processes for producing TBCs, electron beam physical vapor deposition (EB-PVD) is widely used because it has several advantages in comparison with plasma sprayed coatings, including high deposition rate, use of high melting point oxides and excellent thermal shock resistance behavior due to porous columnar microstructure of the coatings [1-3].

The thermophysical property of coatings is one of the most important properties for obtaining superior TBCs. Specially, low thermal conductivity is one of the most important properties for obtaining superior TBCs. It is well known that thermal conductivity of materials closely depends on microstructural properties such as porosity, pore architecture and morphology. In particular, the thermal conductivity of a coatings film is sensitive to the deposition method, microstructural morphology, density and coating composition [4-7]. The purpose of this work is to investigate the influence of temperature on thermal diffusivity and thermal conductivity of nanoporous 4mol%Y_2O_3-ZrO_2 coatings fabricated by EB-PVD.

EXPERIMENTAL

The coatings on zirconia substrates were deposited by EB-PVD (electron beam-physical vapor deposition) using commercially available 4 mol% Y_2O_3 stabilized zirconia targets. The substrates were first preheated at $900 \sim 1000°C$ in a preheating vacuum chamber using graphite heating element. The substrates were then moved to the coating chamber for deposition. An electron beam evaporation process was conducted in a coating chamber under a vacuum level of 1 Pa using 45 KW of electron beam power. The target material was heated above its evaporation temperature of $3500°C$, and the resulting vapors were condensed on rotating substrates. Deposition was generally conducted in condition of rate of 5 μm/min and substrate rotation of 5 rpm. The coating thickness is about 200~300 μm. The substrate temperature was $950°C$.

The thermal diffusivity was determined by the laser flash method using a thermal analyzer (Kyoto Densi, LFA-501). The laser flash method involves heating one side of the sample with a laser pulse of short duration and measuring the temperature rise on the rear surface with an infrared detector. The thermal diffusivity is determined from the time required to reach one-half of the peak temperature in resulting temperature rise curve for the rear surface as illustrated in Fig.1.

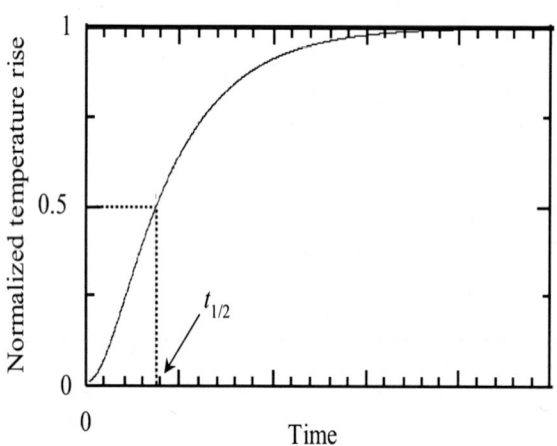

Fig. 1. Temperature response as a function of time at the opposite side of a coated specimen after laser pulse heating.

All the measurements of the coated samples of 10 mm diameter were carried out between $25°C$ and $1000°C$ in 200 degree intervals in a vacuum chamber. Because of the translucency of the specimens to the laser, the specimens were sputter-coated with a thin layer of silver and colloidal graphite spraying to ensure complete and uniform absorption of the laser pulse prior to thermal diffusivity measurement. Specific heat measurements of the coated samples of 5mm diameter and 300 μm thickness were made with differential scanning calorimeter (Netzsch, DSC 404C) using sapphire as the reference material. Measurements were made between room temperature and $1000°C$ in $200°C$ increments in an argon gas condition.

The thermal conductivity of the coatings was then determined using

$$k = \alpha C \rho \tag{1}$$

where k is the thermal conductivity, α is the thermal diffusivity, C is the specific heat, and ρ is the density of the coatings, respectively. Quantitative analysis of the pore size distributions in the coated layers was performed using a mercury porosimeter. The microstructure of the coated samples was observed by SEM.

RESULTS AND DISCUSSION

The typical microstructure of 4mol%Y_2O_3-ZrO_2 coating deposited by EB-PVD is shown in Fig. 2. The top surfaces of EB-PVD coatings consist of square-pyramidal or cone-like grains. The coatings clearly reveal a columnar microstructure with all columnar grains oriented in the same direction, i.e., perpendicular to the substrate. The columnar microstructure characteristic of EB-PVD imparts the low thermal conductivity and excellent strain tolerance to the materials.

Fig.2. SEM micrographs of microstructure of EB-PVD coatings of 4mol%Y_2O_3-ZrO_2: (a) surface view and (b) side view.

X-ray diffraction patterns of top surface of coating layers were obtained as shown in Fig. 3. The spectrum contains (200) and (400) diffraction and no other phase were observed in coated surface. It is clear that the primary growth direction of the coating layers is dominant along the (200) with maximum diffraction intensity. The analysis of XRD patterns indicates that the observed phase in all coatings materials is the tetragonal phase.

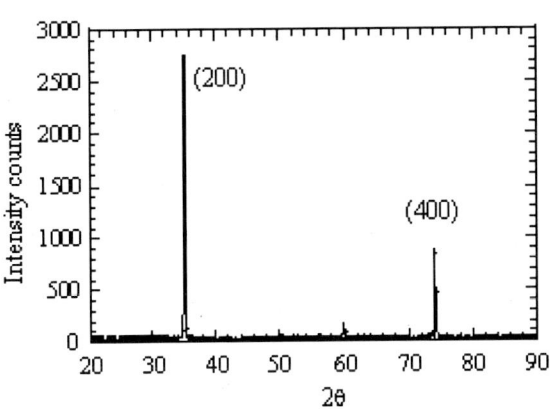

Fig. 3. XRD result for EB-PVD coating layer.

The formation mechanism of the porous column by EB-PVD can be explained as follows. Porous structures in coatings are formed during non-equilibrium condensation of the vapor phase. One of the main mechanisms of porous structure formation is based on the so-called "shadowing" effect. The porous columns can be formed by a flux shadowing mechanism. During nucleation and subsequent growth of various crystallographic faces of the nuclei at different rates, consolidation of grains occurs on the condensation surface.

The faces and micro protrusions, growing at a maximum growth speed, screen the adjacent regions of the surface from the surface of the vapor flow. This results in formation of inner micro-voids in the shadow region. The shadowing effect is also determined by the angle of the vapor flow incidence on the deposited surface. During coating procedure, the vapor flow is a non-uniform, as it is disturbed by the substrate rotation, resulting in various vapor incident angles. The shadowing can be significantly enhanced so that second phase particles form and grow on the deposited surface. Consequently, the shadowing effect contributes to the formation of a columnar coating layer with a highly porous structure.

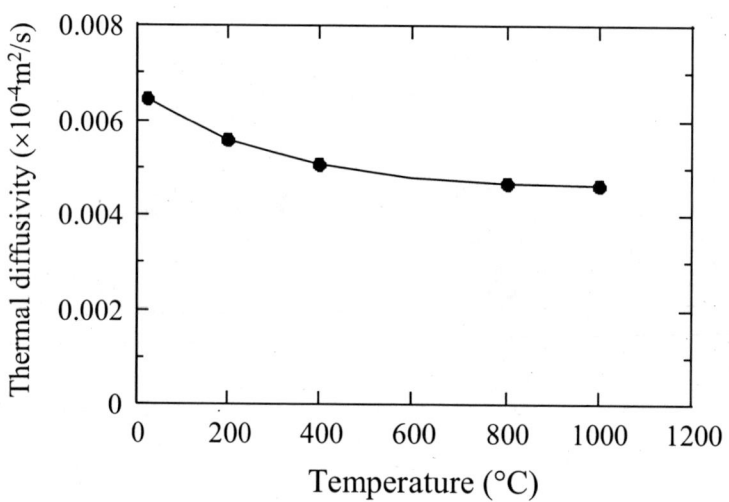

Fig. 4. Thermal diffusivity vs temperature of coatings specimens of $4mol\%Y_2O_3$-ZrO_2.

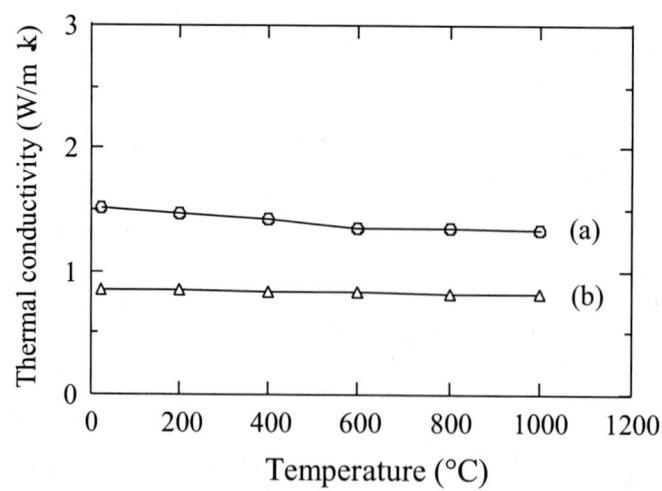

Fig. 5. Thermal conductivity vs temperature of coatings specimens of $4mol\%Y_2O_3$-ZrO_2; (a) EB-PVD coatings and (b) plasma sprayed coatings.

The specific heat increases between 0.45~0.62 kJ/(kg· K) as the temperature increase up to about 600°C and show the almost constant above 600°C. These values are similar to that of plasma sprayed TBCs [8]. The thermal diffusivity of the coated specimens is plotted in Fig. 4 as a function of temperature. It shows that thermal diffusivity of the specimens decrease with increasing the temperature. Fig. 5 shows the thermal conductivity of the coated specimens as a function of temperature. However, thermal conductivity shows a little decrease as the temperature increase.

The reason of smooth decrease of thermal conductivity was attributed to the results of competitive compensation between decrease of thermal diffusivity and increase of specific heat with increasing the temperature. Thermal conductivity of plasma sprayed coatings [9] is significantly lower than that of EB-PVD coatings at given temperature in the Fig. 5. This reason is the difference of pores morphology as well as the microstructure, that is, EB-PVD coatings and plasma sprayed coatings are columnar and splat lamellar microstructure, respectively [10].

We considered the two reasons for the detailed explanation on the decrease of thermal conductivity. As first reason, lower thermal conductivity of plasma sprayed coatings in comparison with EB-PVD coatings is caused by the high porosity of plasma sprayed coatings, resulting in the enhancement of phonon scattering at pores. As second reason, the splat-boundaries between splat lamellar grains for plasma sprayed microstructure are thought to be contribution for increase of phonon scattering because they can intersect the heat flow path. In addition, the flat-pancake shaped pores perpendicular to the temperature gradient in plasma sprayed coatings, lead to the greater decrease in thermal conductivity than that for EB-PVD coatings.

Fig. 6 shows SEM micrographs of columnar gap and the elongated pores of the coated specimen. Gap type-pores between columnar grains are observed which extend from the substrate to the coating surface, as shown in Fig.6 (a). The gap type-pores evolve with the aligned columnar structure.

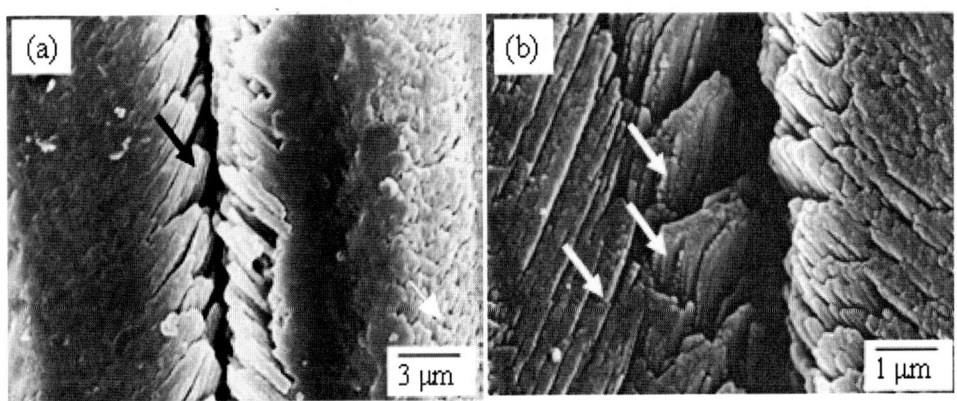

Fig. 6. SEM micrographs of microstructure of EB-PVD coatings of 4mol%Y_2O_3-ZrO_2:
(a) columnar gap and (b) elongated nano pores at feather-like grains.

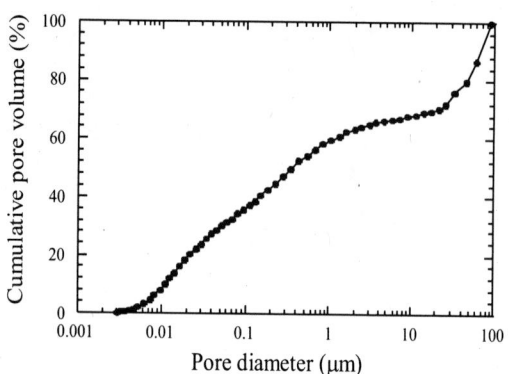

Fig. 7. Pore size distribution for EB-PVD coating layer by mercury porosimeter.

On both sides of the columnar grains, a pronounced dendrite structure, including many micro-pores as well as the elongated nano pores was formed, as shown in Fig.6 (b). The quantitative analysis of pore distribution is necessary to estimate the influence of porosity on thermal conductivity in coating layers. Fig. 7 indicates the result of pore distribution for coating layer with 200 μm coating thickness. According to pore distribution analysis, the volume portion of the nano pores (for consideration of size < 100 nm) on total porosity is approximately 38% and an average pore size is about 300 nm.

It seems that such nano pores exist mainly around feather-like columns as well as intracolumnar pores at inside of columns as shown in Fig. (6). The nano pores can contribute to phonon scattering. The effect of porosity on thermal conductivity can be explained as follows. The thermal conductivity can be usually reduced by decreasing the mean free path due to the phonon scattering. The phenomenon of phonon scattering occurs when phonons interact with lattice defects. Such defects include vacancies, dislocations, pores, grain boundaries, and atoms of different masses. The total mean free path by phonon scattering can therefore be reduced by introduction of pores as lattice imperfections. Thus, the present samples with higher porosity above 20% exhibit lower thermal conductivity. In particular, the elongated nano pores (Fig. 6) at both side of dendrite column are thought to be contribution for increase of phonon scattering because they can intersect the heat flow path due to the inclined arrangement of the elongated nano pores.

This can be explained from the theory of thermal conductivity by phonons where the thermal conductivity is limited by intrinsic phonon scattering. The thermal conductivity (K_p) due to lattice vibration can be described by following expression:

$$K_p = \frac{1}{3} \int_0^{\omega_D} C(\omega)v(\omega)l(\omega)d\omega \qquad (2)$$

where $C(\omega)d\omega$ represents the contribution to the lattice specific heat in the frequency range $d\omega$ at ω, $v(\omega)$ is the sound velocity and $l(\omega)$ is the phonon mean-free path. The integration is performed between zero and ω_D (the latter of which is the Debye frequency). From equation (2), if the phonon mean-free path by increase of phonon scattering at pores can be decrease, it can be concluded that thermal conductivity is decreased.

CONCLUSIONS

Thermal properties of 4 mol% Y_2O_3-stabilized zirconia coatings fabricated by EB-PVD have been investigated in the temperature range between 25°C to 1000°C. The EB-PVD 4mol%Y_2O_3-ZrO_2 coatings had a nanoporous columnar microstructure with gaps between columnar grains. The thermal diffusivity and thermal conductivity of EB-PVD coatings showed the decreasing tendency with increasing temperature. The thermal conductivity of EB-PVD coatings was found to be higher than those of plasma sprayed coatings. It could be concluded that a presence of pores in coating layers mainly leads to reduced thermal conductivity by the reduction of mean free path.

ACKNOWLEDGMENTS

The authors acknowledge the financial support of the New Energy and Industrial Technology Development Organization (NEDO), Japan.

REFERENCES

[1] J. Singh, D.E. Wolfe and J. Singh, "Architecture of thermal barrier coatings produced by electron beam-physical vapor deposition (EB-PVD)," *J.Mater.Sci.*, **37**, 3261-3267 (2002).

[2] C.G.Levi, " Emerging materials and processes for thermal barrier systems," *Current Opinion in Solid State and Mat. Sci.*, **8**, 77-91 (2004).

[3] D.D. HASS, P.A.PARRISH and H.N.G. WADLEY, "Electron Beam Directed Vapor Deposition of Thermal Barrier Coatings," *J.Vac. Sci. Technol.*, **16**, 3396-3401 (1998)

[4] B.K.Jang and H. Matsubara, "Influence of Rotation Speed on Microstructure and Thermal Conductivity of Nano-Porous Zirconia Layers Fabricated by EB-PVD" *Scripta Mater.*, **52**, 553-558 (2005).

[5] J. R. Nicholls, K. J. Lawson, A. Johnstone and D. S. Rickerby, "Methods to reduce the thermal conductivity of EB-PVD TBCs," *Surface and Coatings Technol.*, **151-152**, 383-391 (2002).

[6] U. Schulz, "Phase Transformation in EB-PVD Yttria Partially Stabilized Zirconia Thermal Barrier Coatings during Annealing," *J. Am. Ceram. Soc.*, **83**, 904-910 (2000).

[7] B.K.Jang, M. Yoshiya and H. Matsubara, "Influence of Number of Layers on Thermal Properties of Nano-Pore Dispersed Zirconia Coating Layer Fabricated by EB-PVD" *J. Jpn. Inst. Metals.* **69**, 101-105(2005) .

[8] B. Alzyab, C.H. Perry and R.P.Ingel, "High-Pressure Phase Transitions in Zirconia and Yttria-Doped Zirconia," *J. Am. Ceram. Soc.*, **70**, 760-765 (1987).

[9] R.B. Dinwiddie, S.C. Beecher, W.D. Poter, B.A. Nagaraj, ASME-96-GT-282.

[10] P.S. Anderson, X.Wang, P. Xiao, "Effect of Isothermal Heat Treatment on Plasma-Sprayed Yttria-Stabilized Zirconia Studied by Impedance Spectroscopy," *J. Am. Ceram. Soc.* **88**, 324-330 (2005).

Thermochemical Interaction of Thermal Barrier Coatings with Molten CaO–MgO–Al$_2$O$_3$–SiO$_2$ (CMAS) Deposits

Stephan Krämer,[†] James Yang, and Carlos G. Levi

Materials Department, University of California, Santa Barbara, Santa Barbara, California 93106-5050

Curtis A. Johnson

General Electric Global Research Center, Niskayuna, New York 12309

Thermal barrier coatings (TBCs) are increasingly susceptible to degradation by molten calcium–magnesium alumino silicate (CMAS) deposits in advanced engines that operate at higher temperatures and in environments laden with siliceous debris. This paper investigates the thermochemical aspects of the degradation phenomena using a model CMAS composition and ZrO$_2$–7.6%YO$_{1.5}$ (7YSZ) grown by vapor deposition on alumina substrates. The changes in microstructure and chemistry are characterized after isothermal treatments of 4 h at 1200°–1400°C. It is found that CMAS rapidly penetrates the open structure of the coating as soon as melting occurs, whereupon the original 7YSZ dissolves in the CMAS and reprecipitates with a different morphology and composition that depends on the local melt chemistry. The attack is minimal in the bulk of the coating but severe near the surface and the interface with the substrate, which is also partially dissolved by the melt. The phase evolution is discussed in terms of available thermodynamic information.

I. Introduction

THERMAL barrier coatings (TBC) are now fully recognized as enabling materials for enhancing the performance and durability of gas turbine engines,[1,2] a pre-eminent mode of propulsion technology as well as electricity generation and industrial power. In their coming of age, TBCs have had to overcome a reliability problem that limited their application to extending component life and precluded full exploitation of their capability for increasing gas path temperatures.[3,4] As confidence in their durability and reliability has grown, engine design has become increasingly dependent on TBCs for improving efficiency through higher operating temperatures. The ensuing benefits, however, have been accompanied by the emergence of new degradation and failure mechanisms. Notable among the latter is the attack by calcium–magnesium alumino silicate (CMAS) deposits resulting from the ingestion of siliceous minerals (dust, sand, volcanic ash, runway debris) with the intake air, especially in aircraft engines.[5] At lower temperatures, these contaminants can cause erosive wear or local spallation of the TBC when impacting as solid particles.[6–10] As engine temperatures increase, the siliceous debris adheres to the TBC surfaces and yields glassy melts.[11] The latter can penetrate the micro-

structural features that induce compliance in the coating—microcracks in TBCs deposited by atmospheric plasma spray (APS) and columnar segmentation in those deposited by electron-beam physical vapor deposition (EB-PVD)—leading to a loss of strain tolerance.[5]

The envisaged damage mechanism involves exfoliation of discrete surface layers infiltrated with the molten CMAS as the latter freezes upon cooling and builds up stresses due to the thermal expansion mismatch with the substrate.[5,12] Recent work documented this damage mode[13] but also suggested that thermal shock may play a more important role in this mode of failure than the simple buildup of stress upon cooling. Nevertheless, the mechanism is essentially thermomechanical in origin and relevant to any molten deposit that penetrates the open spaces in the TBC and freezes within them. The extent of penetration is dictated by the interplay of the thermal gradient imposed across the thermal barrier with the fluidity of the melt. If siliceous debris continues to deposit on the surfaces exposed by the exfoliated layers, the process can be repeated until the TBC is removed.

Examination of coatings on engine parts subjected to CMAS reveals that there is also a thermochemical form of damage occurring at high temperature that further degrades the coating and could exacerbate the thermomechanical effects. Of the few studies available in the open literature, the earlier ones were primarily motivated by the concern that siliceous deposits would plug the cooling holes in airfoils and increase their temperature beyond tolerable limits,[14] with potential for hot corrosion of the metallic coating.[15,16] (These components were aluminized but did not have TBCs.) Stott et al. provided the first significant discussion of the corrosive effects of molten silicates on 7 wt% yttria-stabilized zirconia (7YSZ) TBCs.[11,17,18] Laboratory tests were performed by heating self-standing TBCs in contact with different sands for periods of 20–120 h at temperatures in the range 1400°–1500°C. Both APS and EB-PVD coatings were studied, but much of the discussion focused on APS. A salient claim was that YSZ dissolved in the silicate melt, preferentially along grain boundaries, and, depending on the melt chemistry, monoclinic ZrO$_2$ precipitated with a lower Y content.[11]

Studies on actual engine hardware with APS TBCs exposed to siliceous deposits focused on the occurrence of spallation as a result of coating infiltration by the melt and outlined the dominant failure regimes as a function of the TBC surface temperature.[5] A notable observation was that for seemingly disparate sources and overall compositions of mineral intake, the infiltrating melts were of similar composition and contained primarily CaO, MgO, Al$_2$O$_3$, and SiO$_2$. Minor components of the melt included Fe and Ni, originating from the upstream engine metallic components, as well as Zr and Y from the TBC. Incipient melting temperatures for the deposits in this study were of order ~1200°C, whereas others reported temperatures as high as 1275°C[11] or as low as ~1136°C.[16] For reference, the lowest melting eutectics in the ternaries CaO–Al$_2$O$_3$–SiO$_2$, CaO–MgO–

I Smialek—contributing editor

Manuscript No. 21305. Received December 30, 2005; approved March 16, 2006.
Research sponsored by the Office of Naval Research under contracts N00014-99-1-0471 and MURI/N00014-00-1-0438, monitored by Dr. S. G. Fishman. The project benefited from the use of the UCSB-MRL Central Facilities supported by NSF under award No. DMR00-80034.
†Author to whom correspondence should be addressed. e-mail: skraemer@engineering.ucsb.edu

SiO_2, and $MgO-Al_2O_3-SiO_2$ are 1170°, 1320°, and 1355°C, respectively, all involving silica (tridymite), the binary monosilicate (CS or MS), and a ternary silicate (CA$_2$S$_2$, CMS$_2$, or M$_2$A$_4$S$_5$)[‡] (cf. figs. 630, 598, and 712 Levin et al.[19]). The lowest quaternary eutectic is reported at ∼1150°C (fig. 908 Levin et al.[19]) and corresponds to the reaction L→S+MS+CMS$_2$+CA$_2$S$_2$. A cursory analysis of the relevant phase equilibria literature suggests that silicate mixtures over a relatively broad composition range could have incipient melting points of order ∼1200°C (especially when the effect of Fe is considered), comparable with expected surface temperatures for TBCs in state-of-the-art aero- and land-based engines. The CMAS problem is thus expected to become more pervasive as engine temperatures continue to increase in response to the demand for higher performance.

While there is evidence of thermochemical interactions between current TBC materials and molten CMAS, understanding of the ensuing corrosion mechanisms and resulting products is rather limited, especially for EB-PVD coatings. The present study aims to provide fundamental insight into these mechanisms with a focus on EB-PVD TBCs, which are arguably more susceptible to molten deposit penetration because of the open channels in their "segmented" microstructure. Experiments using a model CMAS composition on 7YSZ deposited on alumina substrates were used to explore the evolution of the TBC microstructure under controlled conditions. The results are discussed in light of available information on the relevant phase equilibria, outlining feasible paths leading to the evolution of the observed microstructures.

II. Experimental Procedures

All TBC specimens comprised 7YSZ deposited by EB-PVD on polycrystalline alumina substrates (99.5% purity, CoorsTek, Golden, CO) using an in-house dedicated facility.[20] The use of ceramic substrates allows (i) the system to be heated isothermally above the melting point of the intended CMAS corrodent (>1200°C), which would cause excessive oxidation and microstructural degradation if superalloy substrates were used, and (ii) to assess the potential interactions if molten CMAS reaches the thermally grown aluminum oxide (TGO) in a TBC system. Conversely, the use of alumina substrates does not replicate the residual stresses that arise in the CMAS-infiltrated TBC on a superalloy substrate upon cooling, but such stresses and the related spallation phenomena were not the focus of the present study.

The 7YSZ sources were ceramic ingots, 25 mm diameter × 200 mm long, with ∼38% porosity (Trans-Tech, Adamstown, MD). Major impurities reported (in weight percent) are 1.35HfO$_2$, 0.08TiO$_2$, 0.02SiO$_2$, and ≤0.01 each of CaO, MgO, Al$_2$O$_3$, Fe$_2$O$_3$, Na$_2$O, U, and Th. 200 µm thick coatings were deposited at ∼2 µm/min on alumina substrates (25 mm × 25 mm × 0.6 mm) held at 1000°C and mounted on a tubular ceramic holder rotating over the source at a rate of 8 rpm to achieve the characteristic columnar, strain-tolerant microstructure desired in gas turbine applications.

The model CMAS selected had a chemical composition of 33CaO–9MgO–13AlO$_{1.5}$–45SiO$_2$ or C$_{33}$M$_9$A$_{13}$S$_{45}$ (all compositions henceforth in mole percent of single cation oxide formula units). It was based on the average of deposits on aircraft turboshaft shrouds operated in a desert environment, as reported Borom et al.,[5] excluding the minor components believed to originate mainly from the engine (Fe and Ni). The composition falls near the intersection of the primary crystallization domains for pseudowollastonite (α-CS), pyroxene/diopside (CMS$_2$), melilite (C$_2$MS$_2$), and anorthite (CA$_2$S$_2$) in the quaternary liquidus projection (fig. 2647 Levin et al.[21]), with a crystallization temperature between 1200° and 1300°C. The CMAS was prepared by mixing reagent-grade fine powders of the individual oxides and

milling them in water to form a thick paste, which was subsequently applied to the surface of the TBC specimens to an area density of ∼40 mg/cm^2, well in excess of the amount estimated as needed to infiltrate all the porosity in the coating. After drying, the specimens were heated in a furnace to various temperatures in the range 1200°–1300°C, as well as 1400°C, and held in each case for 4 h. The nominal heating and cooling rate was 6°C/min.

Cross-sections of the as-deposited TBC and the various specimens exposed to CMAS were cut along a plane perpendicular to the rotation axis. These were embedded in epoxy and subsequently polished for examination by scanning electron microscopy (SEM) in both secondary (SE) and back-scattered electron (BSE) imaging modes, as well as by Raman spectroscopy. Transmission electron microscopy (TEM) specimens were cut from selected areas of the polished cross sections using a focused ion beam (FIB). This technique is particularly advantageous in case of the present specimens because it allows precise sampling of locations exhibiting specific microstructural features or reaction products. Chemical analysis was performed in both SEM and TEM using energy-dispersive X-ray spectroscopy (EDS).

III. Results

The first notable observation is the abruptness with which CMAS penetrates and interacts with the TBC as soon as melting occurs. No evidence of melting, penetration, or even adhesion of the "deposit" to the coating surface was detected under 1230°C. Partial melting was noted at ∼1235°C and completed by 1240°C, whereupon the TBC was fully impregnated and noticeably corroded by the CMAS after only 4 h. The excess contaminant on the surface had a glassy appearance with no residual solid and a few bubbles trapped upon cooling. Absent a gradient, the CMAS readily penetrated down to the substrate and visibly attacked it, as further elaborated below, promoting the detachment of parts of the TBC. The fracture surfaces along the rotation axis, where the segmentation in a pristine coating is most pronounced,[20] revealed that essentially all the intercolumnar gaps had been filled by CMAS and the TBC had consequently turned into a monolith upon cooling and lost its strain tolerance. The general features of the microstructure are similar at higher temperatures, with a concomitant increase in the severity of the interaction, as described below. Fewer bubbles appear trapped in the CMAS above 1300°C, presumably because of the concomitant decrease in melt viscosity.

The general appearance of the impregnated TBC is depicted in the polished cross section of Fig. 1, corresponding to the sample heated to 1300°C/4 h. The bulk of the coating appears to retain its columnar structure, albeit impregnated by CMAS, but two distinct interaction layers are evident at the interfaces with the bulk CMAS and the substrate. These are designated as the upper and lower interaction zones, respectively.

The features of the upper interaction zone are illustrated in Figs. 2 and 3. In this region, the columns lose their identity, with their characteristic tips (Fig. 2(a)) replaced by a conglomerate of much smaller globular particles embedded in CMAS (Fig. 2(b)). The relative proportion of phases is more clearly evident in the polished cross section of Fig. 2(c), where both seem to form interpenetrating networks. The upper reaction layer increases in thickness with temperature, from ∼14 µm at 1240°C to ∼21 µm at 1300°C and ∼42 µm at 1400°C. However, its general appearance is relatively independent of temperature and quite similar to the equivalent region in a CMAS-degraded TBC on a service airfoil (Fig. 2(d)) even though the composition of the deposit is significantly different in the latter.[22] This gives confidence that the "model" CMAS replicates reasonably well the phenomena observed in real components, absent the thermal gradient.

TEM analysis (Fig. 3) reveals that the small globular particles in the top layer are fully dense, diverging substantially in appearance from the original columns (cf. Fig. 5). They are also monoclinic in nature and exhibit the twinned structure typical of the tetragonal→monoclinic martensitic transformation upon

---

[‡]The abbreviated formulation adopted here is based on single cation oxide formula units, i.e., A = AlO$_{1.5}$, C = CaO, M = MgO, S = SiO$_2$ and thus differs from that used conventionally in ceramics, based on the full oxide formula unit. Hence, CA$_2$S$_2$ = CaAl$_2$Si$_2$O$_8$ (anorthite), CMS$_2$ = CaMgSi$_2$O$_6$ (diopside), and M$_2$A$_4$S$_5$ = Mg$_2$Al$_4$Si$_5$O$_{18}$ (cordierite).

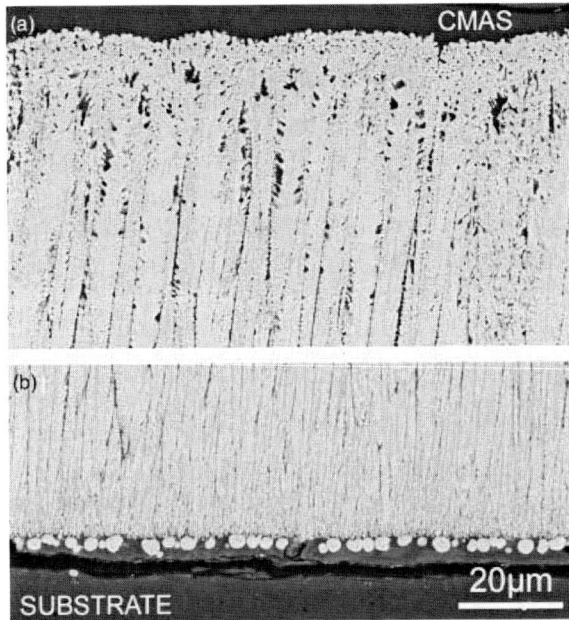

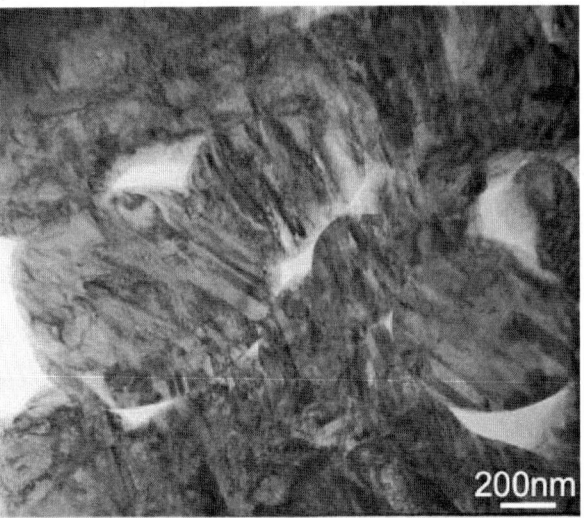

Fig. 1. Cross-section scanning electron microscopy images of a thermal barrier coatings (TBC) after exposure to calcium–magnesium alumino silicate (CMAS) at 1300°C for 4 h, highlighting the regions of severe attack near the outer surface (a) and at the TBC/substrate interface (b). The residual CMAS on top, barely shown, is ~150 μm thick.

Fig. 3. Transmission electron microscopy image of the interaction zone in Fig. 2(b), where the classical martensitic structure produced by the tetragonal→monoclinic transformation is clearly evident. Note that the globules are dense, and the intermediate spaces are filled with calcium–magnesium alumino silicate.

cooling.[23] EDS analysis reveals that they have a lower $YO_{1.5}$ content ($<3\%$) than the original 7YSZ and also incorporate some CaO (~1%) in solid solution. The spaces within the particle network are filled with an amorphous phase, essentially identical in composition to the glassy bulk CMAS on top, both containing minor amounts of Zr and Y in addition to

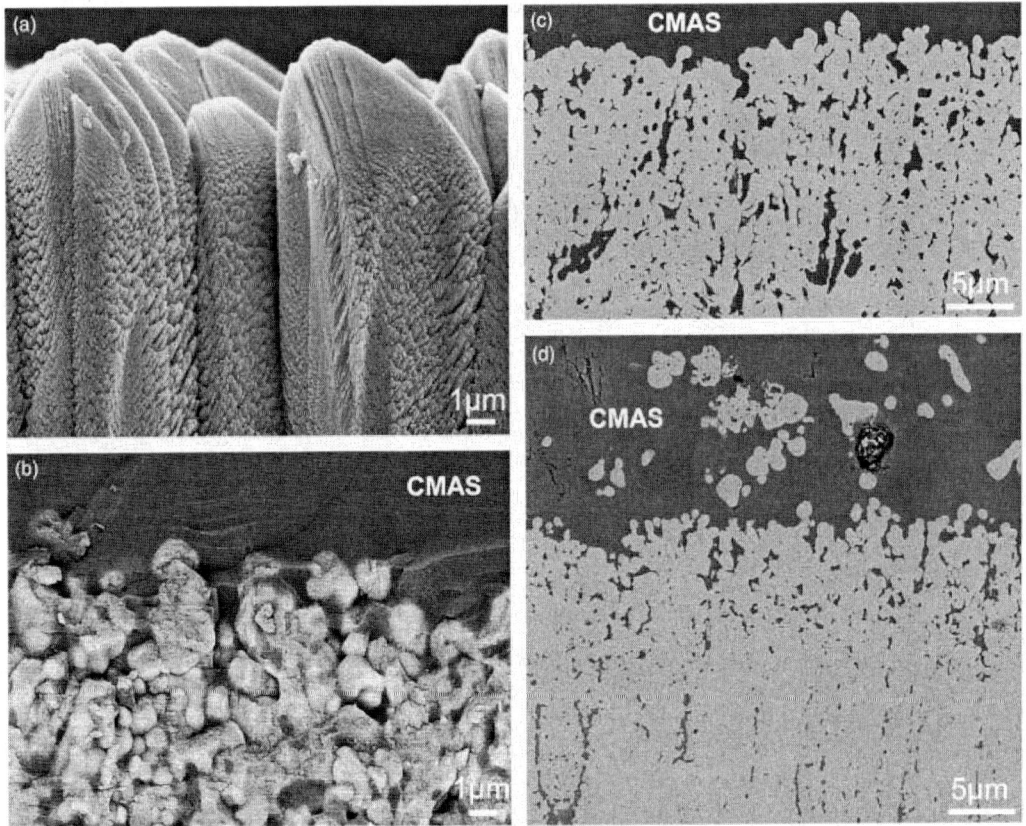

Fig. 2. Comparison of the column tips (a) in the pristine condition and (b, c) after exposure to the model calcium–magnesium alumino silicate (CMAS) at 1300°C for 4 h (BSEI). The globular grains in (b) are about an order of magnitude smaller than the original column diameter and depleted in Y (Table I). They form a network interpenetrated by CMAS (c), and appear quite similar to structures observed near the surface of a thermal barrier coatings attacked by CMAS during actual engine operation (d).

Thermal Properties

Table I. Chemical Compositions of Constituents in the Regions of the Interacting TBC-CMAS System. Values are given in mol%

Region	Constituent	CaO	MgO	AlO$_{1.5}$	SiO$_2$	ZrO$_2$	YO$_{1.5}$
CMAS deposit		33	9	13	45		
Layer above TBC	CMAS	33	7	13	43	2	2
Prior column tips	CMAS	32	7	13	44	2	2
	Zirconia	1				98	2
Layer under corroded tips	CMAS	32	6	14	44	2	3
	Zirconia (column core)					92	8
Column roots	CMAS	32	7	9	46	3	3
	Zirconia					93	7
Reaction zone with substrate	CMAS	34	9	5	45	3	4
	Large globular zirconia	3				85	12
	Small YSZ particles	–				92	8
	Elongated crystals	22		33	43	2	–

TBC, thermal barrier coatings; CMAS, calcium–magnesium alumino silicate; YSZ, yttria-stabilized zirconia.

the original components (Table I). The Y:Zr ratio in the CMAS is ∼1, compared with ∼1:12 in the original TBC. An area EDS analysis in the SEM further revealed that the overall Y:Zr ratio in the upper reaction zone (amorphous+crystalline phases) is also lower than the original 7YSZ, consistent with the observation that significant amounts of Y and Zr migrated out into the large volume of molten CMAS above the coating.

Macroscopically, there is a relatively abrupt transition between the upper corrosion layer and the columnar structure retained through much of the underlying coating, although partially consumed column cores are often seen projecting into the globular structure at its lower boundary—e.g., Fig. 1(a). The transition is most clearly evident at the upper end of the narrower intercolumnar gaps, as in Fig. 4(a) where elongated bright pockets filled with CMAS delineate the boundary between two columns. The structure at the top is low in Y (Table I) and largely monoclinic (Fig. 4(b)). The porous structure and initial composition of the original column are still evident on the right-hand side in Fig. 4(c), but a thin layer of Y-depleted zirconia denoted by arrows has formed at the boundary between the column and the CMAS pocket. It could not be conclusively ascertained that this layer is monoclinic, but its Y content increases and its thickness disappears with increasing distance from the surface. Indeed, the cross-section ∼1 μm above the column roots (Fig. 5(a)) shows little evidence of attack, manifested mostly by smoothing of the column surfaces and the presence of some Y and Zr in the amorphous phase (Table I). Higher magnification of these columnar sections (Fig. 5(b)) reveals that the fine-scale intracolumnar porosity remains largely unaffected and has not been penetrated by CMAS even though the intergranular pores are all impregnated. No monoclinic phase was detected in this region, and both the local and bulk Y:Zr ratios remain essentially at their original value.

The structure of the lower interaction zone in Fig. 6 is markedly different from the layers above. Four different constituents are detected in this figure. The brighter regions are Zr rich and appear in two distinct morphologies, namely large globules and smaller particles comparable in size with the roots of the TBC columns (most evident in Fig. 6(a)). The darker background contains two constituents of different composition, at least one seemingly crystalline with a faceted acicular or tabular morphology (Fig. 6(b)). The overall thickness of this zone increases with temperature, as seen by comparing the 1240° and 1300°C specimens in Figs. 6(a) and (b), respectively. All four constituents are present in both samples; the smaller Zr-rich particles are more abundant at the lower temperature (Fig. 6(a)) but the thinner reaction zone makes it more difficult to distinguish the darker constituents in the background.

TEM examination reveals the larger globular particles in Fig. 6 to be dense single crystals of cubic ZrO$_2$ (Fig. 7(b)) containing on average ∼12% YO$_{1.5}$ and ∼3% CaO (Table I). Conversely, the smaller crystals are t'-YSZ with the original

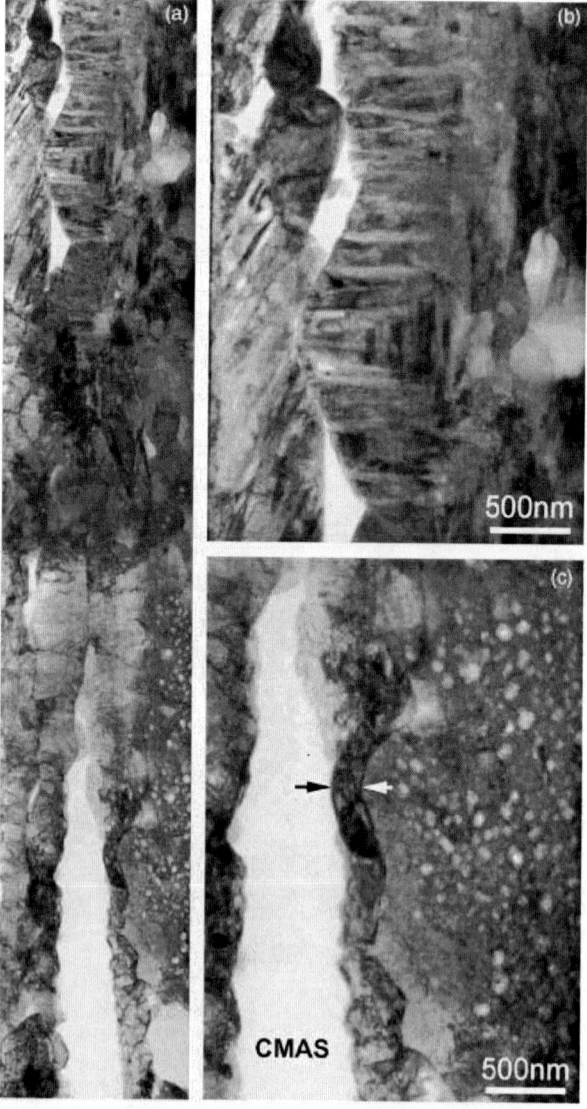

Fig. 4. Transmission electron microscopy images of the transition region at the bottom of the upper interaction layer in the specimen exposed to calcium–magnesium alumino silicate (CMAS) at 1300°C for 4 h. (a) depicts the general view along a former intercolumnar gap, while (b) and (c) represent closer views of the top and bottom areas in (a). The structure in (b) exhibits the same martensitic appearance as Fig. 3. Conversely, the porous structure of the original column is still evident in (c), where a thin layer of Y-depleted yttria-stabilized zirconia, denoted by arrows, has formed at the interface with the CMAS pocket.

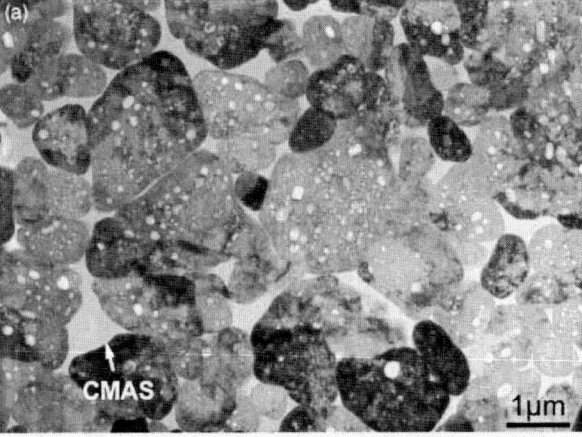

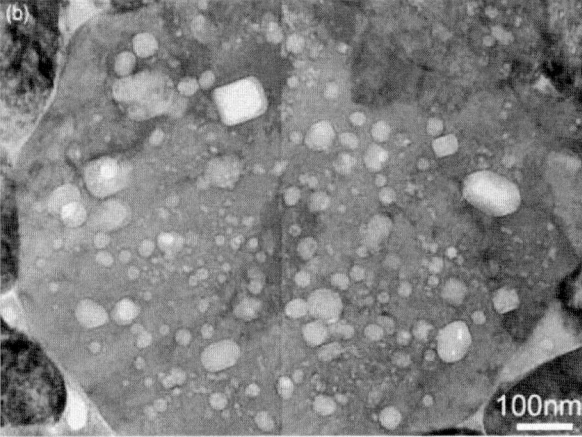

Fig. 5. Transmission electron microscopy views of the cross section of columns above the lower interaction zone, showing complete infiltration of the intercolumnar gaps by calcium–magnesium alumino silicate (a) but no significant attack or Y-depleted layer (b). The columns retain their t' structure and original composition, as well as non-infiltrated intracolumnar porosity as shown in (b).

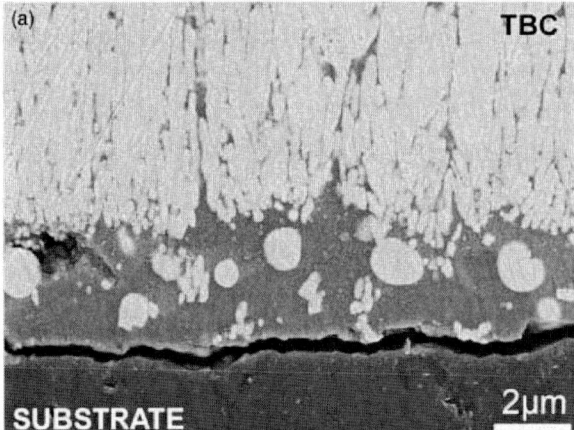

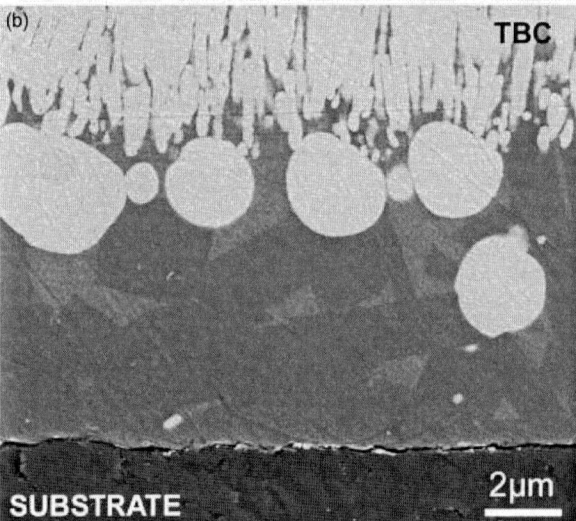

Fig. 6. Comparison of the lower interaction zones in the specimens heat treated at (a) 1240°C and (b) 1300°C, both for 4h (BSEI). Note the difference in scale between the initial column roots and the globules reprecipitated from the melt. The smaller bright particles, more evident in (a), are partially dissolved bits detached from the column roots. The darker faceted phase, more evident in (b), is an aluminosilicate tentatively identified as anorthite.

Y content and no incorporated Ca. No signs of monoclinic or Y-depleted zirconia were found, in contrast to the top part of the sample. It is further noted in Fig. 7(a) that the rounded roots of the t' columns (porous regions) suggest dissolution into the CMAS (glassy matrix) to feed the growth of the neighboring and much larger cubic grains. The composition of the glassy matrix is similar to that of the original CMAS, except for a lower Al content and the incorporation of minor amounts of Zr and Y (Table I). This matrix corresponds to the slightly brighter shade of gray in the background (Fig. 6(b)). The fourth and final constituent in Fig. 7(b) consists of large elongated crystals with Ca, Al, and Si in approximate ratios of 1:2:2, respectively (Table I). The chemistry and diffraction evidence are suggestive of anorthite (CA_2S_2), but the identity of this phase remains to be conclusively ascertained. The area EDS analysis across the lower interaction zone indicates that the average Y:Zr ratio is essentially the same as in the bulk of the columnar structure, in contrast with the upper layer that is significantly depleted in Y.

Raman spectroscopy confirmed that the local TEM observations are representative of the phase constitution in the relevant regions, as shown in Fig. 8. The characteristic signature of monoclinic around 200 cm^{-1} appears down to ~ 20 μm (Fig. 8(b)) mixed with residual tetragonal, but not in the bulk of the coating (Fig. 8(c)) or at the lower interaction zone (Fig. 8(d)). The cubic phase does not have a particularly distinct Raman signature but its generalized presence along the interface is clearly evident in Fig. 1(b).

IV. Discussion

The following salient issues emerge from the above results: (i) CMAS rapidly and extensively infiltrates the TBC structure as soon as melting occurs; (ii) CMAS severely attacks the TBC at lower temperatures (1240°C) and in much shorter times (4 h) than those anticipated from previous reports in the literature[11,17,18]; (iii) CMAS obliterates the columnar morphology and converts the original t'-YSZ into monoclinic globular particles in the upper region of the coating, to a thickness that depends on temperature, but (iv) the attack is largely suppressed in the bulk of the TBC; and (v) it reactivates again at the bottom of the coating, where CMAS dissolves the alumina substrate and converts the original t' into larger cubic YSZ globules. These issues are discussed below.

(1) Infiltration Behavior

Examination of the infiltrated microstructures suggests that molten CMAS readily "wets" 7YSZ, spreading over the column surfaces and penetrating into any small open capillaries, e.g., those represented by the pristine "feathery" structure in Fig. 2(a). If the classical representation of wetting by a "contact

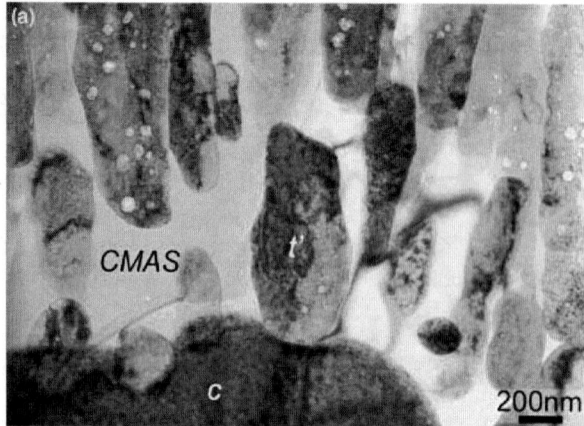

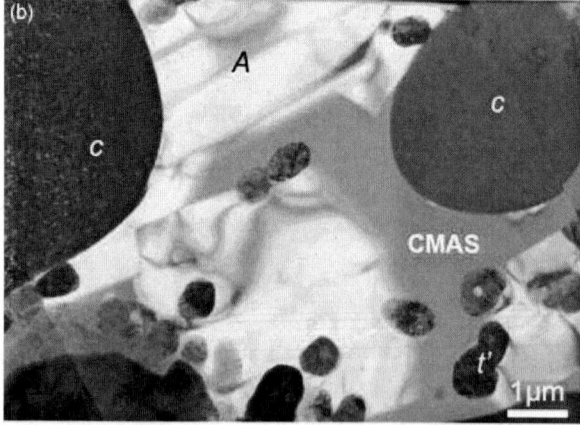

Fig. 7. Transmission electron microscopy images of the lower reaction zone in the specimen heat treated for 4 h at 1300°C showing (a) the roots of the t' columns right above the top of a large c-YSZ particle, and (b) the lower interaction zone comprising the reprecipitated cubic zirconia (large globules, marked "c"), the residual (t') YSZ bits, the crystalline aluminosilicate (A), and the residual glassy calcium–magnesium alumino silicate background. YSZ, yttria-stabilized zirconia.

angle" θ is invoked, the implication is that θ is small and cos(θ → 1.§ Following standard treatments for the infiltration of porous media by fluids, the time needed for a wetting liquid (cosθ → 1) to penetrate to a depth L into the coating due to capillary action alone may be estimated as a first approximation from[25]:

$$t \approx \left[\frac{k_t}{8D_c} \left(\frac{1-\omega}{\omega} \right)^2 L^2 \right] \frac{\eta}{\sigma_{LV}} \qquad (1)$$

where ω is the pore fraction open to flow (essentially the area fraction of intercolumnar gaps, ~0.1), D_c is the capillary diameter (~1 μm), k_t is a tortuosity factor (1–10) reflecting the increased resistance to flow when the capillaries are not straight tubes, η is the viscosity of the fluid, and σ_{LV} is its surface tension. Note that the terms within the square brackets in Eq. (1) are only dependent on the geometry of the system, whereas the terms outside are only dependent on the fluid.

The viscosity and surface tension of silicate melts have been extensively studied in the literature. It is reported that σ_{LV} of silicate glasses at 1400°C may be estimated from the following expression[26]:

$$\sigma_{LV}(mJ/m^2) = 271.2 + 3.32[CaO] + 1.96[MgO]$$
$$+ 3.47[Al_2O_3] + K \qquad (2)$$

---
§One can readily show from an interfacial energy balance that the infiltration of a melt originally deposited on a surface into open capillaries normal to it is always favored as long as θ < π/2, (e.g., Brada and Clarke[24]), and thus near-ideal wetting (θ → 0) is not required.

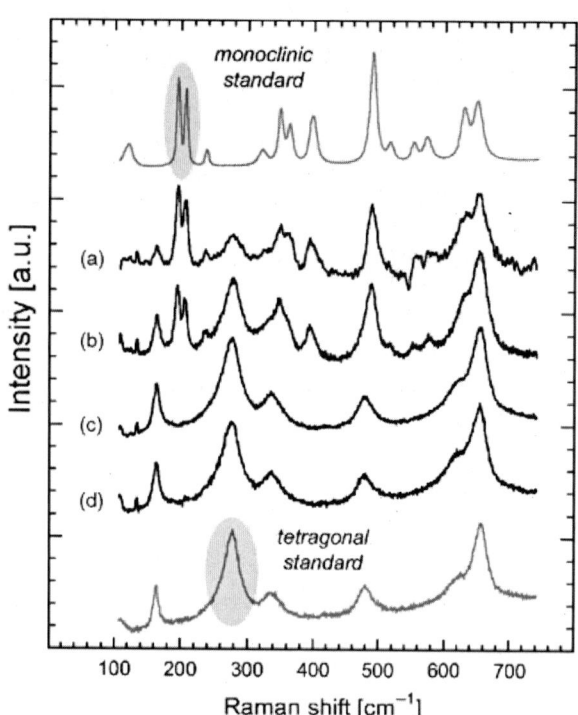

Fig. 8. Raman spectra taken from the polished cross section of the sample exposed to calcium–magnesium alumino silicate (CMAS) at 1300°C/4 h, shown in Fig. 1. The positions correspond to (a) 5 μm and (b) 20 μm below the CMAS/TBC boundary; (c) middle of the TBC; and (d) at the lower interaction zone. The standard monoclinic and tetragonal patterns are given at the top and bottom of the figure. (The cubic pattern is not particularly distinct and not useful in this context.) TBC, thermal barrier coatings

where only the relevant oxides have been included here and all the concentrations are in mole percent. For the model CMAS melt in this study, σ_{LV} is then ~0.4 J/m². (While some dependence on temperature is expected, the order of magnitude is not expected to change significantly within the range investigated.) The viscosities of CMAS can also be estimated from the chemical composition, as described in Turkdogan.[27] The resulting values at 1240°, 1300°, and 1400°C are ~15, ~6, and ~2 N·(s·m²)⁻¹, respectively. On that basis and assuming k_t ~3 for the intercolumnar gaps, one can estimate that the infiltration time for a 200 μm coating would be less than 1 min even for the lowest temperature investigated. Moreover, at the rate of heating used in these experiments (6°C/min), the CMAS is expected to penetrate the coating down to the substrate shortly after melting, before the higher hold temperatures are reached. This explains in part why the attack at the TBC/substrate interface is substantial even at temperatures just above melting (1240°C).

Field experiences suggest that real CMAS could melt at even lower temperatures and, while its chemical composition may vary, the anticipated viscosities upon melting are expected to be similar in magnitude to the values above, and perhaps lower if significant amounts of other modifiers like FeO were incorporated.[25] Conversely, the thermal gradient would preclude penetration down to the TGO, although the infiltration front may not necessarily stop at the isotherm corresponding to the melting point when the melt does not crystallize upon cooling, as in the present case. These issues are currently under investigation.

(2) CMAS Interactions with the TBC Columns
The three common crystallographic forms of YSZ are found present in the coating after exposure to CMAS, each in a different region and with a distinct morphology and Y content. Much of the coating away from the substrate and the exposed surface retains the initial t' structure and chemical composition

(~7.6% YO$_{1.5}$), as well as the columnar morphology with nanoscale intragranular porosity, indicating that the thermochemical attack was minimal in these regions. However, the YSZ near the bulk CMAS deposit is monoclinic, depleted in Y, and that closest to the substrate is cubic, enriched in Y relative to the original composition. Minor amounts of Ca are incorporated into solid solution for both of these forms, but not to any detectable degree in the retained t' form. No other crystalline phases bearing major amounts of Zr or Y were found in the areas of the coating affected by CMAS.

The cumulative evidence suggests that both the Y-enriched and Y-depleted zirconia phases evolve through crystallization from the CMAS melt into which the original t' is concurrently dissolved, with the characteristics of the precipitated phase determined by the local chemistry. The situations for the bulk and upper regions are discussed first, as the lower interaction zone also involves the precipitation of a second crystalline phase based on Al$_2$O$_3$ rather than ZrO$_2$. A complete analysis requires detailed knowledge of the phase equilibria in the system Al$_2$O$_3$–CaO–MgO–SiO$_2$–Y$_2$O$_3$–ZrO$_2$, which is not available. The concept, however, can be illustrated qualitatively with the aid of the CaO–SiO$_2$–ZrO$_2$ liquidus projection in Fig. 9. CMAS is represented by a binary liquid with the same C:S ratio (~0.73) in contact with ZrO$_2$, depicted by the tie line in this figure. At equilibrium, this composition should actually be solid below ~1430°C, consisting of a mixture of pseudowollastonite (α-CS) and tridymite (fig. 10359 Roth[28]). However, the crystallization of these phases is often suppressed kinetically and one may reasonably assume that the composition selected may exist as a supercooled liquid with chemical characteristics similar to the CMAS melt. In that case, the actual boundary of the t-ZrO$_2$ liquidus would not be given by the position of the L+CS+Z twofold saturation line in Fig. 9, but shifted toward the C–S binary at temperatures below 1400°C, as indicated by the arrows. It is then readily apparent from Fig. 9 that the amount of Z needed to saturate the CS melt is quite modest (a few percent), in agreement with the ZrO$_2$ content detected in much of the CMAS within the coating. A corollary is that the presence of

MgO and Al$_2$O$_3$ in the original CMAS does not seem to affect substantially the solubility of ZrO$_2$ in the melt.

The above scenario only explains the dissolution, but not the re-precipitation of YSZ with a different composition. The underlying driver in that case is the absence of equilibrium between the original YSZ composition and the initial melt, whereupon t'-7YSZ would tend to dissolve and ZrO$_2$ to precipitate back with a composition (lower in Y and with some Ca) that reduces the overall free energy of the system. The problem is conceptually analogous to the concurrent dissolution of α and reprecipitation of β in the liquid-phase-assisted sintering of Si$_3$N$_4$ ceramics,[29] with some important differences. First, the composition of the YSZ changes during the process but not its crystal structure, as inferred from the clear evidence that the monoclinic YSZ crystals of the upper interaction zone (Fig. 3(c)) evolved through a martensitic transformation of a tetragonal phase precipitated from the melt at the reaction temperature. This clarifies prior suggestions in the literature that the monoclinic phase forms directly from the melt.[11¶] A second notable difference is that the process is limited to the top layer. Evidently, the saturation limit for Zr is reached first near the dissolution interface, arguably providing a driving force for nucleation of Y-depleted t-YSZ at discrete points on the column surfaces. The differences in chemical potential within the melt in front of the t' parent phase and the new t precipitate continue to drive the dissolution of the former to feed the growth of the latter. (One may argue that precipitation could occur upon cooling but more extensive analysis of the progress of this mechanism, to be presented in a forthcoming publication, reveals that precipitation is actually occurring at temperature.) While the process should be conceptually identical within the intercolumnar spaces away from the surface, the melt volume is much smaller in this case and could reach equilibrium with precipitation of a minimal amount of depleted YSZ (detectable only at the upper end of the surviving columns, as noted in Fig. 4(c)). Conversely, the bulk CMAS on the surface is a large sink for the excess Y and allows the process to continue over the time scale of the experiment, gradually destroying the large columnar crystals and converting them into smaller globules.

The absence of penetration of the fine porosity within the columns in Figs. 4(c) and 5 indicates that the feathery porosity, clearly open near the tip (Fig. 2(a)), breaks down into closed pores before CMAS reaches the surface of the columns in those areas. This phenomenon may start during deposition, wherein the lower ends of the columns age *in situ* as the thickness continues to build up. However, the deposition temperature is rather modest, ~1000°C, and much of the evolution is likely to occur during the relatively slow heating to the CMAS melting temperature. Indeed, a sample exposed to a 1240°C/4 h treatment without CMAS (not shown) revealed substantial evolution of the feathery pores even at the top of the columns. The implication is that the extent of the globular upper zone may depend on the degree of aging of the TBC before contaminants deposit on it. This is currently under investigation.

(3) CMAS Interactions at the TBC/Substrate Interface

In principle, the severity of attack should be comparable at the TBC/substrate interface and within the bulk coating if the relative volume fraction of CMAS remained similarly small. However, two important differences arise because of the attack of the substrate by the melt. The first one is a result of the ensuing local separation of the TBC from the substrate, driven by the combination of substrate dissolution and thermal mismatch stresses,‖ which allows further buildup of molten CMAS to form an

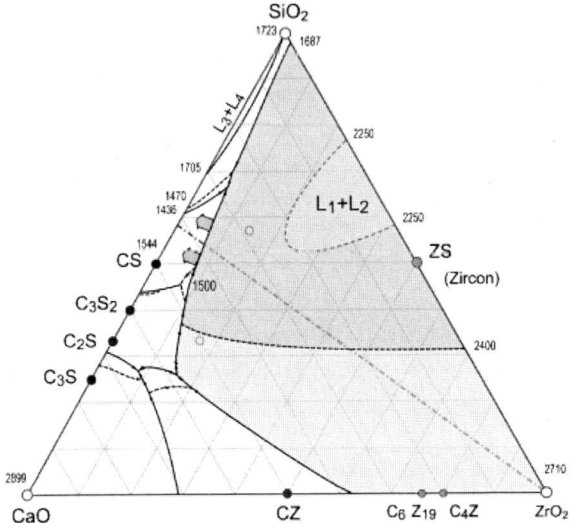

Fig. 9. Schematically modified liquidus projection for the ternary CaO–SiO$_2$–ZrO$_2$. The original diagram (fig. Zr-287 Ondik and McMurdie[36]), as many published around that time, ignores the existence of the cubic ↔ tetragonal transformation at high temperature as well as the existence of a two-phase miscibility gap in the liquid for the SiO$_2$–ZrO$_2$ binary. The schematic presented here incorporates qualitatively more recent evidence from partial isopleths and revised binaries. In general, the location of the relevant L+CS+Z two-phase saturation line on the liquidus is reasonably close to that in the original diagram. The tie line shown has the same C:S ratio as the model calcium–magnesium alumino silicate.

¶It is indeed possible to form monoclinic crystals directly from the melt, as observed in corrosion studies with sulfate–vanadate deposits,[30] but this occurs at lower temperatures (900°C) and the characteristics of the crystals are radically different from those in this study. Conversely, monoclinic can be observed even in the absence of CMAS due to the purely thermal destabilization of t'-YSZ at the more extreme combination of temperatures and times used Stott et al.,[11] as suggested by a variety of phase stability studies on TBCs.[31–34]

‖Local separation of the coating from the substrate does not necessarily lead to a detachment of the attacked region. However, formation of the reaction zone did weaken the interface and resulted in cracking along the substrate interface upon cooling, as illustrated in Figs. 1 and 6. The cracked regions appeared to be sufficiently interlocked with the rest of the sample to avoid spallation.

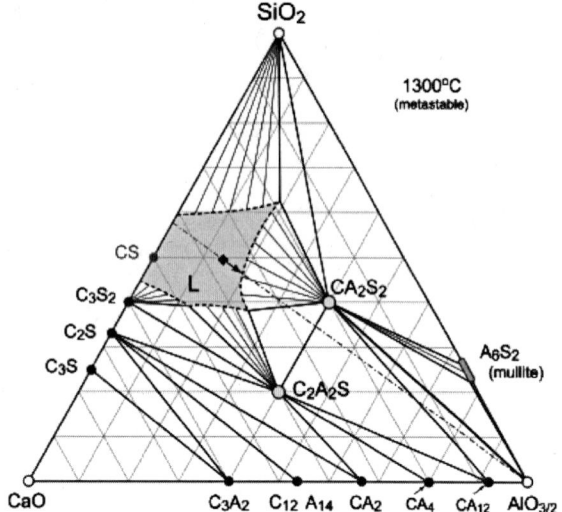

Fig. 10. Metastable isothermal cross section for the AlO$_{1.5}$–CaO–SiO$_2$ system, based on fig. 630 Levin *et al.*[19] The pseudowollastonite liquidus has been suppressed whereupon the equilibrium liquid field has been extended metastably to the CaO–SiO$_2$ binary. The projection of the quaternary calcium–magnesium alumino silicate (CMAS) composition onto this ternary is shown by the diamond symbol. The tie line between this composition and the alumina corner reflects the process that occurs when CMAS dissolves the alumina substrate and, upon saturation, precipitates anorthite (CA$_2$S$_2$) at the lower interaction zone.

interfacial layer several times thicker than the typical intercolumnar gap. This would explain the reactivation of the dissolution–reprecipitation process when the CMAS reaches the bottom of the coating, but not the changes in the phase constitution and chemical composition of the precipitated zirconia. These are arguably linked to the injection of excess alumina into the melt and the subsequent precipitation of an aluminate crystalline phase.

Insight can be gained by examining first the reaction scenario for CMAS with the substrate absent YSZ. In this case, there is reasonable knowledge of the quaternary diagram but visualization is greatly enhanced by focusing on the CAS metastable ternary section in Fig. 10. It is assumed again that crystallization of the primary pseudowollastonite phase is kinetically suppressed, and hence the liquid region at 1300°C is metastably extended up to the C–S binary, as illustrated in Fig. 10. The projection of the quaternary CMAS composition onto the ternary is also marked in this figure, and a tie line representing the interaction with the alumina substrate is drawn to the corresponding corner of the diagram. The driver for dissolution–reprecipitation is immediately evident. Al$_2$O$_3$ dissolves into the melt and shifts its composition toward the boundary of the anorthite (CA$_2$S$_2$)+ liquid field. After the requisite supersaturation is achieved to nucleate CA$_2$S$_2$, the system will tend to establish local equilibrium between anorthite and the melt, continuing to dissolve Al$_2$O$_3$ and precipitate CA$_2$S$_2$ as the overall composition shifts toward the Al$_2$O$_3$ corner. It is further noted that the Al$_2$O$_3$ content of the residual CMAS in the near-substrate region is significantly lower than that throughout the rest of the coating. The inference is that the anorthite crystals continued to grow upon cooling after the reaction hold, gradually depleting the liquid composition beyond the boundary indicated in Fig. 10, as expected from the slope of the liquidus surface in that region.

The physical characteristics of the anorthite and *c*-YSZ crystals suggest that both precipitate at temperature, although their sequence of nucleation is not completely clear. Nevertheless, the change in the local chemistry induced by the dissolution of alumina and reprecipitation of anorthite is arguably responsible for the shift in the redistribution pattern of Y and Zr upon crystallization of YSZ. The area EDS indicates that there is no significant change in the overall Y:Zr ratio relative to the original

coating, so the Y enrichment in *c*-YSZ is presumably compensated by the incorporation of Zr into the anorthite (Table I). It is also evident from the volume of cubic YSZ particles that, while significant, the extent of YSZ dissolution in this region is much smaller than that in the upper reaction layer, consistent with the lower volume of CMAS available. The survival of much of the column root structure (Fig. 7(a)) and small *t'* grains detached from this region into the CMAS melt (Figs. 6(a) and 7(b)) further supports the view that the dissolution rate is moderated by the incorporation of alumina. This suggests possible mitigation strategies that will be explored in subsequent publications.[††]

V. Conclusions

The tendency of molten CMAS deposits to wet 7YSZ is sufficiently strong that columnar TBC structures can be infiltrated just above the onset of melting within times that are negligible compared with typical operation cycles of gas turbine engines. Infiltration is complete in isothermal exposures, but would be limited in a real operation to a depth dictated by the thermal gradient across the coating and the viscosity of the melt.

Thermochemical interactions between CMAS and the TBC occur at lower temperatures (1240°C) and shorter times (<4 h) than suggested by previous studies in the literature. In general, the mechanism involves dissolution of the metastable *t'* phase and re-precipitation with a composition and structure that depends on the local chemistry. Interactions in the bulk of the coating are minimized by the small volume of melt in relation to the amount of TBC material. Where larger volumes of CMAS are available, as in the near-surface region, the reprecipitated YSZ is sufficiently depleted in Y that it transforms to monoclinic upon cooling. The associated volume change could, in principle, contribute to the strains that motivate exfoliation of the coating upon thermal cycling. Near the substrate, the local chemistry is different due to the dissolution of the underlying alumina by the CMAS, inducing precipitation of a crystalline aluminosilicate and globules of a Y-enriched, non-transformable cubic YSZ.

The present study suggests a number of relevant research directions, ranging from understanding of the early stages of interaction, the effect of varying CMAS chemistry in the rate of dissolution and the nature of the crystalline products, the sequence in which these products evolve, and the potential implications for the mitigation of CMAS attack on thermal barrier coatings. These issues are presently under investigation and will be discussed in forthcoming publications.

Acknowledgments

Enlightening discussions with Prof. A. G. Evans and Ms. S. Faulhaber (UCSB), as well as Drs. R. Darolia, B. Nagaraj, M. Gorman, D. Wortman (GE-Aviation), and M. Thompson (GE-Global Research) are gratefully acknowledged.

References

[1]R. Schafrik and R. Sprague, "Saga of Gas Turbine Materials, Part III," *Adv. Mater. Processes*, **162** [5] 29–33 (2004).
[2]C. G. Levi, "Emerging Materials and Processes for Thermal Barrier Systems," *Curr. Opin. Solid State Mater. Sci.*, **8** [1] 77–91 (2004).
[3]P. K. Wright and A. G. Evans, "Mechanisms Governing the Performance of Thermal Barrier Coatings," *Curr. Opin. Solid State Mater. Sci.*, **4**, 255–65 (1999).
[4]A. G. Evans, D. R. Mumm, J. W. Hutchinson, G. H. Meier, and F. S. Pettit, "Mechanisms Controlling the Durability of Thermal Barrier Coatings," *Prog. Mater. Sci.*, **46** [5] 505–53 (2001).
[5]M. P. Borom, C. A. Johnson, and L. A. Peluso, "Role of Environmental Deposits and Operating Surface Temperature in Spallation of Air Plasma Sprayed Thermal Barrier Coatings," *Surf. Coatings Technol.*, **86–87**, 116–26 (1996).
[6]J. R. Nicholls, M. J. Deakin, and D. S. Rickerby, "A Comparison Between the Erosion Behavior of Thermal Spray and Electron-Beam Physical Vapour Deposition Thermal Barrier Coatings," *Wear*, **233–235**, 352–61 (1999).

[††]Approaches to control CMAS infiltration by providing a sacrificial layer of Al$_2$O$_3$ on the TBC surface have been proposed in the patent literature,[35] but the rationale is based primarily on promoting crystallization to reduce penetration into the coating, rather than mitigation of the thermochemical attack.

[7]R. G. Wellman and J. R. Nicholls, "A Mechanism for the Erosion of EB PVD TBCs," *Mater. Sci. Forum*, **369-372** [Part 1] 531–8 (2001).

[8]X. Chen, R. Wang, N. Yao, A. G. Evans, J. W. Hutchinson, and R. W. Bruce, "Foreign Object Damage in a Thermal Barrier System: Mechanisms and Simulations," *Mater. Sci. Eng.*, **A352** [1–2] 221–31 (2003).

[9]X. Chen, M. Y. He, I. T. Spitsberg, N. A. Fleck, J. W. Hutchinson, and A. G. Evans, "Mechanisms Governing the High Temperature Erosion of Thermal Barrier Coatings," *Wear*, **256** [7–8] 735–46 (2004).

[10]A. G. Evans, N. A. Fleck, S. Faulhaber, N. Vermaak, M. Maloney, and R. Darolia, "Scaling Laws Governing the Erosion and Impact Resistance of Thermal Barrier Coatings," *Wear*, **260** [7–8] 886–94 (2006).

[11]F. H. Stott, D. J. de Wet, and R. Taylor, "Degradation of Thermal-Barrier Coatings at Very High Temperatures," *MRS Bull.*, **19** [10] 46–9 (1994).

[12]R. Darolia and B. A. Nagaraj, "Method of Forming a Coating Resistant to Deposits and Coating Formed Thereby"; U.S. Patent 6,720,038, 2004.

[13]C. Mercer, S. Faulhaber, A. G. Evans, and R. Darolia, "A Delamination Mechanism for Thermal Barrier Coatings Subject to Calcium–Magnesium-Alumino-Silicate (CMAS) Infiltration," *Acta Mater.*, **53** [4] 1029–39 (2005).

[14]J. Kim, M. G. Dunn, A. J. Baran, D. P. Wade, and E. L. Tremba, "Deposition of Volcanic Materials in the Hot Sections of Two Gas Turbine Engines," *J. Eng. Gas Turbines Power*, **115** [7] 641–51 (1993).

[15]J. L. Smialek, F. A. Archer, and R. G. Garlick, "The Chemistry of Saudi Arabian Sand: A Deposition Problem on Helicopter Turbine Airfoils"; in *3rd International SAMPE Metals and Metals Processing Conference*, Edited by T.S. Reinhart, M. Rosenow, R.A. Cull, and E. Struckholt. Society for the Advancement of Material and Process Engineering (SAMPE), Toronto, Canada, 1992.

[16]J. L. Smialek, F. A. Archer, and R. G. Garlick, "Turbine Airfoil Degradation in the Persian Gulf War," *J. Min. Metal. Mater. Soc.*, **46** [12] 39–41 (1994).

[17]F. H. Stott, D. J. de Wet, and R. Taylor, "The Effect of Molten Silicate Deposits on the Stability of Thermal Barrier Coatings for Turbine Applications at Very High Temperatures"; in *3rd International SAMPE Metals and Metals Processing Conference*, Edited by T.S. Reinhart, M. Rosenow, R.A. Cull, and E. Struckholt. Society for the Advancement of Material and Process Engineering (SAMPE), Toronto, Canada, 1992.

[18]D. J. de Wet, R. Taylor, and F. H. Stott, "Corrosion Mechanisms of ZrO_2–Y_2O_3 Thermal Barrier Coatings in the Presence of Molten Middle-East Sand," *J. Phys. IV*, **3** [C9] 655–63 (1993).

[19]E. M. Levin, C. R. Robbins, and H. F. McMurdie, eds. *Phase Diagrams for Ceramists*, Vol. I, pp. 210, 219, 246. The American Ceramic Society, Columbus, OH, 1964.

[20]S. G. Terry, "Evolution of Microstructure During the Growth of Thermal Barrier Coatings by Electron-Beam Physical Vapor Deposition"; Doctoral Dissertation in Materials, University of California, Santa Barbara, CA, 2001.

[21]E. M. Levin, C. R. Robbins, and H. F. McMurdie, eds. *Phase Diagrams for Ceramists*, Vol. II, p. 185. The American Ceramic Society, Columbus, OH, 1969.

[22]S. Krämer and S. Faulhaber. Work in progress, to be published.

[23]M. Rühle and A. H. Heuer, "Phase Transformations in ZrO_2-Containing Ceramics: II, the Martensitic Reaction in t-ZrO_2"; pp. 14–32 in *Science and Technology of Zirconia II*, Edited by N. Claussen, M. Rühle, and A. H. Heuer. The American Ceramic Society, Columbus, OH, 1984.

[24]M. P. Brada and D. R. Clarke, "A Thermodynamic Approach to the Wetting and De-Wetting of Grain Boundaries," *Acta Mater.*, **45** [6] 2501–8 (1997).

[25]D. R. Poirier and G. H. Geiger, *Transport Phenomena in Materials Processing*. The Minerals, Metals and Materials Society, Warrendale, PA, 1994.

[26]A. Kucuk, A. G. Clare, and L. Jones, "An Estimation of the Surface Tension of Silicate Glass Melts at 1400°C Using Statistical Analysis," *Glass Technol.*, **40** [5] 149–53 (1999).

[27]E. T. Turkdogan, *Physical Chemistry of High Temperature Technology*. Academic Press, New York, 1980.

[28]R. S. Roth, ed. *Phase Equilibria Diagrams, Volume XIII: Oxides*, p. 91. The American Ceramic Society, Westerville, OH, 2001.

[29]P. Drew and M. H. Lewis, "The Microstructures of Silicon Nitride Ceramics During Hot-Pressing Transformations," *J. Mater. Sci.*, **9**, 261–9 (1974).

[30]F. M. Pitek, S. Krämer, and C. G. Levi. Work in progress, to be published.

[31]R. A. Miller, J. L. Smialek, and R. G. Garlick, "Phase Stability in Plasma-Sprayed, Partially Stabilized Zirconia-Yttria"; pp. 241–53 in *Science and Technology of Zirconia*, Edited by A. H. Heuer and L. W. Hobbs. The American Ceramic Society, Inc., Columbus, OH, 1981.

[32]U. Schulz, "Phase Transformation in EB-PVD Yttria Partially Stabilized Zirconia Thermal Barrier Coatings During Annealing," *J. Am. Ceram. Soc.*, **83** [4] 904–10 (2000).

[33]N. R. Rebollo, "Phase Stability of t′-Zirconia with Trivalent Oxide Additions for Use in Thermal Barrier Coatings"; Doctoral Dissertation in Materials, University of California, Santa Barbara, CA, 2005.

[34]V. Lughi and D. R. Clarke, "High Temperature Aging of YSZ Coatings and Subsequent Transformation at Low Temperature," *Surf. Coatings Technol.*, **200** [5–6] 1287–91 (2005).

[35]W. C. Hasz, C. A. Johnson, and M. P. Borom, "Protection of Thermal Barrier Coating by a Sacrificial Surface Coating"; U.S. Patent 5,660,885, 1997.

[36]H. M. Ondik, and H. F. McMurdie, eds. *Phase Diagrams for Zirconium and Zirconia Systems*, p. 198. The American Ceramic Society, Westerville, OH, 1998.

□